A LIBRARY OF ELEMENTARY FUNCTIONS

BASIC FUNCTIONS

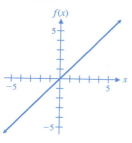

Identity function
$f(x) = x$

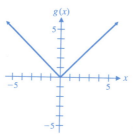

Absolute value function
$g(x) = |x|$

Square function
$h(x) = x^2$

Cube function
$m(x) = x^3$

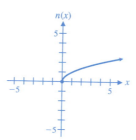

Square root function
$n(x) = \sqrt{x}$

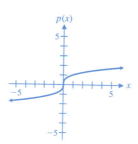

Cube root function
$p(x) = \sqrt[3]{x}$

FIRST-DEGREE POLYNOMIAL FUNCTIONS

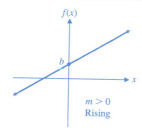

$m > 0$
Rising

Linear function
$f(x) = mx + b$

$m < 0$
Falling

Linear function
$f(x) = mx + b$

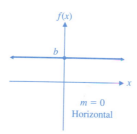

$m = 0$
Horizontal

Constant function
$f(x) = b$

SECOND-DEGREE POLYNOMIAL FUNCTIONS

$a < 0$
Opens downward

$a > 0$
Opens upward

$f(x) = ax^2 + bx + c = a(x - h)^2 + k$

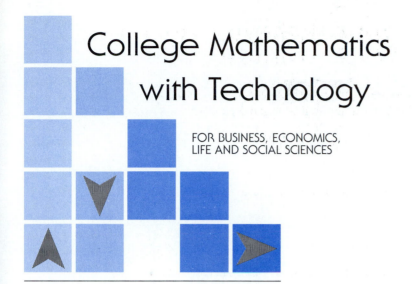

College Mathematics
with Technology

FOR BUSINESS, ECONOMICS,
LIFE AND SOCIAL SCIENCES

College Mathematics with Technology

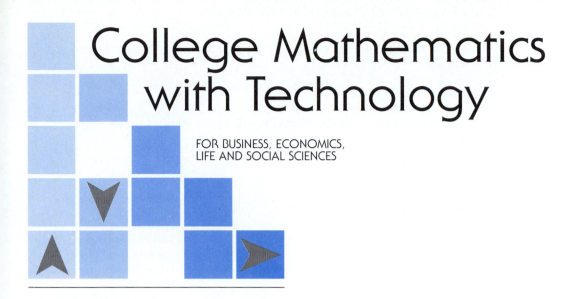

FOR BUSINESS, ECONOMICS,
LIFE AND SOCIAL SCIENCES

RAYMOND A. BARNETT

Merritt College

MICHAEL R. ZIEGLER

Marquette University

with the assistance of
KARL E. BYLEEN
Marquette University

PRENTICE HALL
Upper Saddle River, NJ 07458

LIBRARY OF CONGRESS CATALOGING-IN-PUBLICATION DATA

Barnett, Raymond A.
    College mathematics with technology / Raymond A. Barnett, Michael
R. Ziegler.
      p. cm.
    Includes index.
    ISBN 0-13-570391-3
    1. Mathematics. I. Ziegler, Michael R. II. Byleen, Karl.
III. Title.
QA37.2.B3646   1997   96-45225
510′.78′5—dc21   CIP

Acquisitions Editor:	George Lobell
Editorial Assistant:	Gale Epps
Supplements Editor:	Audra J. Walsh
Editorial Director:	Tim Bozik
Editor-in-Chief:	Jerome Grant
Assistant Vice President of Production and Manufacturing:	David W. Riccardi
Editorial Production/Supervision:	Phyllis Niklas
Managing Editor:	Linda Mihatov Behrens
Executive Managing Editor:	Kathleen Schiaparelli
Manufacturing Buyer:	Alan Fischer
Manufacturing Manager:	Trudy Pisciotti
Director of Marketing:	John Tweeddale
Marketing Assistant:	Diana Penha
Creative Director:	Paula Maylahn
Art Director:	Amy Rosen
Assistant to the Art Director:	Rod Hernandez
Art Manager:	Gus Vibal
Cover Designer:	Joe Sengotta
Interior Design and Layout:	Janet Bollow and Amy Rosen
Interior Illustrations:	Scientific Illustrators
Cover Art:	Melvin Prueitt

 © 1997 Prentice-Hall, Inc.
Simon & Schuster/A Viacom Company
Upper Saddle River, NJ 07458

Printed in the United States of America
10 9 8 7 6 5 4 3 2 1

ISBN 0-13-570391-3

Prentice-Hall International (UK) Limited, *London*
Prentice-Hall of Australia Pty. Limited, *Sydney*
Prentice-Hall Canada, Inc., *Toronto*
Prentice-Hall Hispanoamericano, S. A., *Mexico*
Prentice-Hall of India Private Limited, *New Delhi*
Prentice-Hall of Japan, Inc., *Tokyo*
Simon & Schuster Asia Pte. Ltd., *Singapore*
Editora Prentice-Hall do Brasil, Ltda., *Rio de Janeiro*

CONTENTS

The following material is not included in this book, but is available online at: http://www.prenhall.com/barnett

CHAPTER DEPENDENCIES

PART ONE   A LIBRARY OF ELEMENTARY FUNCTIONS*

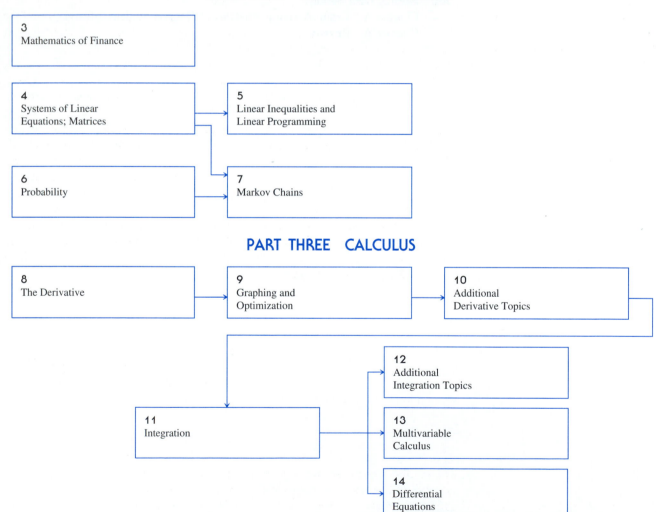

1
A Beginning Library of
Elementary Functions

2
Additional Elementary
Functions

PART TWO   FINITE MATHEMATICS

3
Mathematics of Finance

4
Systems of Linear
Equations; Matrices

5
Linear Inequalities and
Linear Programming

6
Probability

7
Markov Chains

PART THREE   CALCULUS

8
The Derivative

9
Graphing and
Optimization

10
Additional
Derivative Topics

11
Integration

12
Additional
Integration Topics

13
Multivariable
Calculus

14
Differential
Equations

---

* Selected topics from Part One may be referred to as needed in Part Two or reviewed systematically before starting Part Two.

PREFACE

The widespread availability of graphing calculators and computers is revolutionizing mathematics education. Each year, more and more students enter college mathematics courses with a wide variety of technological experiences and an expectation of using and building on these experiences as they continue their mathematical education. Because of this trend, **we have written this book under the assumption that each student has access to an electronic device capable of displaying graphs and performing certain matrix operations.** (We will refer to any such device as a **graphing utility**—see the sections titled ''Technology'' and ''To the Student'' later in this Preface.) However, we do not assume any previous experience with a graphing utility. Instead, we carefully demonstrate through discussions and examples how a graphing utility often gives additional insight into mathematical concepts and processes. (Specific instructions for a number of popular graphing calculators are included in a supplement—see the section titled ''Supplements for the Student'' later in this Preface.) While we assume that the student will attack appropriate problems with a graphing utility at hand and we present problem-solving techniques that will take advantage of this technology, we also include many activities that promote understanding of concepts independent of the use of any technology.

This book is designed for a two-term course in finite mathematics and calculus that emphasizes the topics most useful to students in business, economics, life sciences, and social sciences, and for students who have had $1\frac{1}{2}-2$ years of high school algebra, or the equivalent.

■ EMPHASIS AND STYLE

The text is **written for student comprehension.** Great care has been taken to write a book that is mathematically correct and accessible to students. Emphasis is on computational skills, ideas, and problem solving rather than mathematical theory. General results are usually presented only after particular cases have been discussed.

■ EXAMPLES AND MATCHED PROBLEMS

Over 350 completely worked examples are used to introduce concepts and to demonstrate problem-solving techniques. Many examples have multiple parts, significantly increasing the total number of worked examples. Each example is followed by a similar **matched problem for the student to work** while reading the material. This actively involves the student in the learning process. The answers to these matched problems are included at the end of each section for easy reference.

■ EXPLORATION AND DISCUSSION

Every section contains **Explore–Discuss** boxes interspersed at appropriate places to encourage the student to think about a relationship or process before a result is stated, or to investigate additional consequences of a development in the text. **Verbalization** of mathematical concepts, results, and processes is encouraged in these Explore–Discuss boxes, as well as in some matched problems, and in some problems in almost every exercise set. The Explore–Discuss material also can be used as in-class or out-of-class **group activities.** In addition, at the end of every chapter (before the chapter review), we have included a special **chapter group activity** that involves several of the concepts discussed in the chapter. All these special activities are highlighted with color shading to emphasize their importance.

■ EXERCISE SETS

The book contains over 4,700 problems. Many problems have multiple parts, significantly increasing the total number of problems. Each exercise set is designed so that an average or below-average student will experience success and a very capable student will be challenged. Exercise sets are divided into A (routine, easy mechanics), B (more difficult mechanics), and C (difficult mechanics and some theory) levels. Most exercise sets contain some problems that require a graphing utility for their solution and others that could be solved either with or without a graphing utility. Since a student must learn when to use a graphing utility, as well as how to use one, **the text generally does not identify problems that must be solved with the aid of a graphing utility.**

■ APPLICATIONS

A major objective of this book is to give the student substantial experience in **modeling and solving real-world problems.** Enough applications are included to convince even the most skeptical student that mathematics is really useful (see the Applications Index inside the back cover). Worked examples involving applications are identified by ▶. **Almost every exercise set contains application problems,** usually divided into business and economics, life science, and social science groupings. An instructor with students from all three disciplines can let them choose applications from their own field of interest; if most students are from one of the three areas, then special emphasis can be placed there. Most of the applications are simplified versions of actual real-world problems taken from professional journals and books, and many involve real data. No specialized experience is required to solve any of the applications.

■ TECHNOLOGY

The generic term **graphing utility** is used to refer to any of the various graphing calculators or computer software packages that might be available to a student using this book. **We assume that each student has access to a graphing utility that can perform the following operations:**

- Simultaneously display several graphs in a user-selected viewing window
- Explore graphs using zoom and trace
- Approximate roots
- Approximate intersection points
- Approximate values of the derivative of a function
- Approximate maxima and minima
- Approximate definite integrals
- Plot data sets and find associated regression equations
- Perform basic matrix operations

Most popular graphing calculators support all of these operations, as do *Explorations in Finite Mathematics* and *Visual Calculus,* the software supplements for this text described later in this Preface.

■ GRAPHS

All graphs are computer-generated to ensure mathematical accuracy. Graphing utility screens displayed in the text are actual output from a graphing calculator — in most cases, the Texas Instruments TI-82 graphics calculator.

■ STUDENT AIDS

Annotation of examples and developments, in small color type, is found throughout the text to help students through critical stages (see Sections 1-1 and 4-2). **Think boxes** (dashed boxes) are used to enclose steps that are usually performed mentally (see Sections 1-2 and 4-1). **Boxes** are used to highlight important definitions, theorems, results, and step-by-step processes (see Sections 1-2 and 1-4). **Caution** statements appear throughout the text where student errors often occur (see Sections 4-3 and 4-5). **Functional use of color** improves the clarity of many illustrations, graphs, and developments, and guides students through certain critical steps (see Sections 1-2 and 4-2). **Boldface type** is used to introduce new terms and highlight important comments. **Chapter review** sections include a review of all important terms and symbols and a comprehensive review exercise. **Answers to most review exercises,** keyed to appropriate sections, are included in the back of the book. Answers to all other odd-numbered problems are also in the back of the book.

■ CONTENT

The text begins with an introduction to graphing utilities and the development of a library of elementary functions in Chapters 1 and 2. The development focuses on a discussion of the properties of these functions, related applications, and graphing utility techniques, including **zoom and trace, root approximation, intersection points,** and **regression techniques** for data analysis. Students are encouraged to investigate mathematical ideas and processes **graphically** and **numerically,** as well as **algebraically.** This development substantially enriches the study of mathematics. Depending on the syllabus for the course and the background of the students, some or all of this material can be covered at the beginning of a course, or selected portions can be referred to as needed later in the course.

The material in Part Two (Finite Mathematics) can be thought of as four units: **mathematics of finance** (Chapter 3); linear algebra, including **matrices, linear systems, and linear programming** (Chapters 4 and 5); **probability and statistics** (Chapter 6); and applications of linear algebra and probability to **Markov chains** (Chapter 7). The first three units are independent of each other, while the last chapter is dependent on some of the earlier chapters (see the Chapter Dependencies chart preceding this Preface).

Chapter 3 presents a thorough treatment of simple and compound interest and present and future value of ordinary annuities.

Chapter 4 covers linear systems and matrices with an **emphasis on using Gauss – Jordan elimination** to solve systems. The row operations discussed in Sections 4-2 and 4-3 are required for the simplex method in Chapter 5. Matrix multiplication, matrix inverses, and systems of equations are required for Markov chains in Chapter 7.

Chapter 5 provides **broad and flexible coverage of linear programming.** The first two sections cover two-variable graphing techniques. Instructors who wish to emphasize techniques can cover the basic simplex method in Sections 5-3 and 5-4 and then discuss any or all of the following: the dual method (Section 5-5), the big M method (Section 5-6), or the two-phase simplex method (Chapter 5 Group Activity). Those who want to emphasize modeling can discuss the formation of the mathematical model for any of the application examples in Sections 5-4, 5-5, and 5-6, and either omit the solution or use software to find the solution (see the description of the software that accompanies this text later in this Preface). To facilitate this approach, **all the answers in the back of the book to application problems** in Exercises 5-4, 5-5, 5-6, and the Chapter 5 Review Exercise **contain both the mathematical model and the numeric solution.**

Chapter 6 covers **counting techniques and basic probability,** including Bayes' formula and random variables. Some of the topics discussed in Chapter 6 are required for Chapter 7.

Chapter 7 ties together concepts developed in earlier chapters and applies them to **Markov chains.** This provides an excellent unifying conclusion to the finite mathematics portion of the text.

The material in Part Three (Calculus) consists of **differential calculus** (Chapters 8 – 10), **integral calculus** (Chapters 11 and 12), **multivariable calculus** (Chapter 13), and **differential equations** (Chapter 14). In general, Chapters 8 – 11 must be covered in sequence; however, certain sections can be omitted or given brief treatments, as pointed out in the discussion that follows (see also the Chapter Dependencies chart preceding this Preface).

Chapter 8 introduces the **derivative,** covers the **limit properties** essential to understanding the definition of the derivative, develops the **rules of differentiation** (including the chain rule for power forms), discusses **numerical differentiation** on a graphing utility, and introduces **applications** of derivatives in business and economics. The interplay between graphical, numerical, and algebraic concepts is emphasized here and throughout the text.

Chapter 9 focuses on **graphical analysis** and **optimization.** The first three sections cover **continuity, and first-derivative and second-derivative graph properties** while emphasizing **polynomial functions.** Analysis of graphs of **rational functions** is covered in Section 9-4. In a course that does not emphasize graphical analysis of rational functions, this section can be omitted or given a brief treatment. Optimization is covered in Section 9-5, including examples and problems involving end-point solutions.

The first three sections of Chapter 10 extend the derivative concepts discussed in Chapters 8 and 9 to **exponential and logarithmic functions** (including the general form of the chain rule). This material is required for all the remaining chapters. **Implicit differentiation** is introduced in Section 10-4 and applied to **related rate problems** in Section 10-5. These topics are not referred to elsewhere in the text and can be omitted without loss of continuity.

Chapter 11 introduces **integration.** The first two sections cover **antidifferentiation** techniques essential to the remainder of the text. Section 11-3 discusses some applications involving **differential equations** that can be omitted. Sections 11-4 and 11-5 discuss the **definite integral** in terms of **Riemann sums,** including **approximations** with various types of sums, **simple error estimation,** and **numerical integration** on a graphing utility. As before, the interplay between the graphical, numerical, and algebraic properties is emphasized. These two sections are required for the remaining chapters in the text.

Chapter 12 covers **additional integration topics** and is organized to provide maximum flexibility for the instructor. The first section extends the **area** concepts introduced in Chapter 11 to the area between two curves and related applications. Section 12-2 covers three more **applications** of integration, and Sections 12-3 and 12-4 deal with additional **techniques of integration.** Any or all of the topics in Chapter 12 can be omitted.

The first five sections of Chapter 13 deal with **differential multivariable calculus** and can be covered any time after Section 10-3 has been completed. Section 13-5 applies multivariable optimization techniques to **linear regression,** providing a foundation for the regression techniques introduced in Section 2-4. Section 13-6 requires the **integration** concepts discussed in Chapter 11.

After introducing the basic concepts and terminology used in the study of **differential equations** in Section 14-1, **separable differential equations** (Section 14-2) and **first-order linear differential equations** (Section 14-3) and related applications are covered thoroughly. All the growth laws introduced in Section 11-3 are covered again, this time with more emphasis on student recognition of the relevant growth laws.

■ TO THE STUDENT

A graphing utility is a powerful tool for analyzing graphs, but like any good tool, it must be used properly. The display on most graphing utilities produces only rough approximations to graphs. One of the skills you must develop as you study the material in this book is the ability to visualize the correct appearance of a graph, based on the rough graph produced by a graphing utility and your understanding of the mathematics under discussion. For example, consider Figures 1A and 1B (taken from page 95).

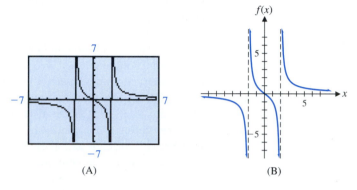

FIGURE 1
$$f(x) = \frac{3x}{x^2 - 4}$$

The graphing utility graph in Figure 1A seems to indicate that the graph is one continuous curve. However, the more accurate graph in Figure 1B clearly shows that the graph consists of three separate curves. When you study this function in the text, you will learn that f is not defined at $x = -2$ and $x = 2$, but has vertical asymptotes at these points. Armed with this knowledge, you will be able to look at the graphing utility graph in Figure 1A, visualize the more accurate graph in Figure 1B, and sketch a good representation of the graph by hand. To assist you in the development of visualization and sketching skills, the text contains hundreds of both graphing utility graphs and more accurate graphs, and also contains many exercises that ask you to sketch graphs by hand.

■ SUPPLEMENTS FOR THE STUDENT

1. A **Student's Solutions Manual** by Garret J. Etgen is available through a book store. The manual includes detailed solutions to all odd-numbered problems and all review exercises.
2. **Computer software** for IBM-compatible computers is available at a nominal cost through a book store. *Explorations in Finite Mathematics* and *Visual Calculus* by David Schneider each contain over twenty routines that provide additional insight into the topics discussed in the text. Although these software packages have much of the computing power of standard mathematical software packages, they are primarily teaching tools that focus on understanding mathematical concepts, rather than on computing. All the routines in these software packages are menu-driven and very easy to use. Included in *Explorations in Finite Mathematics* are routines for Gaussian elimination, matrix inversion, solution of linear programming problems by both the geometric method and the simplex

method, Markov chains, probability and statistics, and mathematics of finance. The matrix routines use and display rational numbers, and matrices may be saved and printed. The *Visual Calculus* routines incorporate graphics whenever possible to illustrate topics such as secant lines; tangent lines; velocity; optimization; the relationship between the graphs of f, f', and f''; and the various approaches to approximating definite integrals. Both software packages are accompanied by manuals with instructions and additional exercises for the student. Hardware requirements are an IBM-compatible computer with at least 384K of memory and a graphics adapter: CGA, EGA, VGA, or Hercules.

3. A **Graphics Calculator Manual** by Carolyn L. Meitler contains examples illustrating the use of a graphics calculator to solve problems similar to those discussed in the text. The manual follows the chapter organization of the text, making it easy to find examples in the manual illustrating appropriate calculator solution methods for problems in the text. The manual includes keystrokes for the TI-81, TI-82, TI-83, and TI-85 calculators. However, the examples and techniques can be used with any graphing utility.

4. A **Supplemental Applications and Topics** manual by Jon E. Baum is available at a nominal cost through a book store. Part I of the manual expands the application exercises in the text and reinforces the important role of the mathematics presented. These exercises provide the student with a richer and more varied experience in solving real-world problems. Part II of the manual presents some applications that are not covered in the text, including transportation problems, assignment problems, sensitivity analysis, and a variety of finance topics. After completing the prerequisite material in the text, students interested in these more specialized topics will realize substantial benefits by studying this portion of the manual.

■ SUPPLEMENTS FOR THE INSTRUCTOR

For a summary of all available supplementary materials and detailed information regarding examination copy requests and orders, see page xx.

1. **PH Custom Test, a menu-driven random test system** for either IBM-compatible or Macintosh computers is available to instructors without cost. The test system has been greatly expanded and now offers **on-line testing.** Carefully constructed algorithms use random-number generators to produce different, yet equivalent, versions of each of these problems. In addition, the system incorporates a unique **editing function** that allows the instructor to create additional problems, or alter any of the existing problems in the test, using a full set of mathematical notation. The test system offers **free-response, multiple-choice, and mixed exams.** An almost unlimited number of quizzes, review exercises, chapter tests, midterms, and final examinations, each different from the other, can be generated quickly and easily. At the same time, the system will produce answer keys, student worksheets, and a gradebook for the instructor, if desired.

2. An **Instructor's Resource Manual** provides over 100 transparency masters and all the answers not included in the text, as well as hard copy of test items available in PH Custom Test. This manual is available to instructors without charge.

3. A **Student's Solutions Manual** by Garret J. Etgen (see ''Supplements for the Student'') is available to instructors without charge from the publisher.

4. **Computer software** and accompanying **manuals** for *Explorations in Finite Mathematics* and *Visual Calculus* by David Schneider (see ''Supplements for the Student'') are available to instructors without charge. These manuals contain complete instructions for using the software, eliminating the need to spend class time discussing these details, and examples and exercises for the student. In addition to providing students with the opportunity to use the computer as an effective tool in the learning process, instructors will find the software very useful for activities such as preparing examples for class, constructing test questions, and classroom demonstrations.

5. A **Graphics Calculator Manual** by Carolyn L. Meitler (see ''Supplements for the Student'') is available to instructors without charge from the publisher. The manual contains all the necessary information for a student with no previous experience with a graphic calculator, eliminating the need for the instructor to prepare materials related to calculator usage. In particular, separate appendixes for the TI-81, TI-82, TI-83, and TI-85 graphic calculators contain detailed instructions, including calculator-specific keystrokes, for performing the various operations required to effectively use each of these calculators to solve problems in the text. Furthermore, the methods illustrated for these calculators are easily adapted to other graphing utilities. The manual is very effective both for a class where all students purchase the same calculator and in a setting where students are using a variety of different calculators—an important consideration as more and more students arrive at college having already purchased a graphic calculator.

6. A **Supplemental Applications and Topics** manual by Jon E. Baum (see ''Supplements for the Student'') is available to instructors without charge from the publisher. Instructors can use Part I of this manual to supplement the exercise sets in the text, providing students with additional experience in solving applications utilizing the mathematics presented in the text. Part II of the manual can be used to provide coverage of applications not covered in the text, such as transportation problems, assignment problems, sensitivity analysis, and a variety of finance topics, either as part of the syllabus for a course or as subjects for independent study.

■ ERROR CHECK

Because of the careful checking and proofing by a number of mathematics instructors (acting independently), the authors and publisher believe this book to be substantially error-free. For any errors remaining, the authors would be grateful if they were sent to: Michael R. Ziegler, 509 W. Dean Court, Fox Point, WI 53217; or, by e-mail, to: michael@mscs.mu.edu

■ ACKNOWLEDGMENTS

In addition to the authors, many others are involved in the successful publication of a book. We wish to thank personally:

Robert D. Adams, University of Kansas
Eileene Marie Connor, St. Joseph's College
Catherine Cron, Fairfield University
Lucinda Gallagher, Florida State University
Debbie Garrison, Valencia Community College
Virginia Harris, Hillsborough Community College
Estela S. Llinás, University of Pittsburgh at Greensburg
Caren McClure, Rancho Santiago Community College
Carol Myers, Georgia State University
Andris Niedra, Robert Morris College
Ronald L. Persky, Christopher Newport University
Jack Porter, University of Kansas
Arthur Rosenthal, Salem State College
Murphy Waggoner, Simpson College

We also wish to thank:

Stephen Merrill, Robert Mullins, and Caroline Woods for providing a careful and thorough check of all the mathematical calculations in the book, the student solutions manual, and the answer manual (a tedious but extremely important job).

Jon Baum, Garret Etgen, Carolyn Meitler, and David Schneider for developing the supplemental manuals that are so important to the success of a text.

Jeanne Wallace for accurately and efficiently producing most of the manuals that supplement the text.

George Morris and his staff at Scientific Illustrators for their effective illustrations and accurate graphs.

Janet Bollow for another outstanding book design.

Karl Byleen for providing major assistance in the preparation of this book.

Phyllis Niklas for guiding the book smoothly through all publication details.

All the people at Prentice Hall who contributed their efforts to the production of this book.

Producing this book with the help of all these extremely competent people has been a most satisfying experience.

R. A. Barnett
M. R. Ziegler

■ ORDERING INFORMATION

When requesting examination copies or placing orders for this text or any of the related supplementary material listed below, please refer to the corresponding ISBN numbers.

TITLE	ISBN NUMBER
College Mathematics with Technology for Business, Economics, Life and Social Sciences	0-13-570391-3
Computer-generated random test system for College Mathematics with Technology	
PH Custom Test IBM Windows	0-13-592841-9
PH Custom Test MAC	0-13-592858-3
Instructor's Resource Manual to accompany College Mathematics with Technology	0-13-592833-8
Visual Calculus (3.5 inch disk and manual)	0-13-232760-0
Student's Solution Manual to accompany College Mathematics with Technology	0-13-592874-5
Graphics Calculator Manual to accompany College Mathematics with Technology	0-13-749517-X
Supplemental Applications and Topics to accompany the Barnett and Ziegler College Mathematics Series	

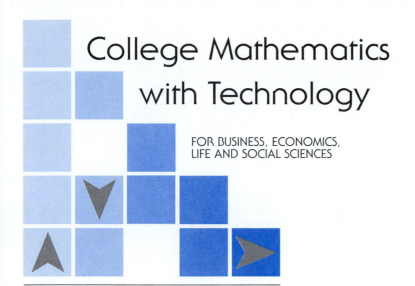

College Mathematics
with Technology

FOR BUSINESS, ECONOMICS,
LIFE AND SOCIAL SCIENCES

A Library of Elementary Functions

CHAPTER 1

A Beginning Library

of Elementary Functions

The function concept is one of the most important ideas in mathematics. The study of mathematics beyond the elementary level requires a firm understanding of a basic list of elementary functions, their properties, and their graphs. See the inside front cover of this book for a list of the functions that form our library of elementary functions. Most functions in the list will be introduced to you by the end of Chapter 2 and should become a part of your mathematical toolbox for use in this and most future courses or activities that involve mathematics. A few more elementary functions may be added to these, but the functions listed inside the front cover are more than sufficient for most of the applications in this text.

This chapter starts with an introduction to graphing and the use of *graphing utilities,* a technological breakthrough that will add considerably to your understanding of functions, their graphs, and their use. As indicated in the Preface, this book assumes that you own or have access to a graphing calculator or graphing software and a computer.

SECTION 1-1 ■ ## Introduction to Graphing Utilities

- CARTESIAN COORDINATE SYSTEM
- GRAPHING: POINT-BY-POINT
- GRAPHING UTILITIES
- SCREEN COORDINATES
- THE TRACE AND ZOOM FEATURES

In this section we will briefly review point-by-point graphing and then move on to graphing and the exploration of graphs using a graphing utility. The use of technology as an aid in drawing and analyzing graphs is revolutionizing mathematics education. Your ability to interpret mathematical concepts and to discover patterns will be greatly increased as you become proficient in the use of an electronic graphing device.

■ CARTESIAN COORDINATE SYSTEM

Recall that a **Cartesian (rectangular) coordinate system** in a plane is formed by taking two mutually perpendicular real number lines **(coordinate axes)**—one horizontal and one vertical—intersecting at their origins, and then assigning unique **ordered pairs** of numbers **(coordinates)** to each point P in the plane (Fig. 1). The first coordinate **(abscissa)** is the distance of P from the vertical axis, and the second coordinate **(ordinate)** is the distance of P from the horizontal axis. In Figure 1, the coordinates of point P are (a, b). By reversing the process, each ordered pair of real numbers can be associated with a unique point in the plane. The coordinate axes divide the plane into four parts **(quadrants),** numbered I to IV in a counterclockwise direction. Out of the above discussion we can conclude:

> **There is a one-to-one correspondence between the points in a plane and the elements in the set of all ordered pairs of real numbers.**

This is often referred to as the **fundamental theorem of analytic geometry.**

FIGURE 1
The Cartesian coordinate system

■ GRAPHING: POINT-BY-POINT

The fundamental theorem of analytic geometry allows us to look at algebraic forms geometrically and to look at geometric forms algebraically. We begin by considering an algebraic form, an equation in two variables:

$$y = 9 - x^2 \tag{1}$$

A **solution** to equation (1) is an ordered pair of real numbers (a, b) such that

$$b = 9 - a^2$$

The **solution set** for equation (1) is the set of all these ordered pairs.

To find a solution to equation (1), we replace x with a number and calculate the value of y. For example, if $x = 2$, then $y = 9 - 2^2 = 5$, and the ordered pair $(2, 5)$ is a solution. Similarly, if $x = -3$, then $y = 9 - (-3)^2 = 0$, and $(-3, 0)$ is a solution. Since any real number substituted for x in equation (1) will produce a solution, the solution set must have an infinite number of elements. We use a rectangular coordinate system to provide a geometric representation of this set.

The **graph of an equation** is the graph of its solution set. To **sketch the graph of an equation,** we plot enough points from its solution set in a rectangular coordinate system so that the total graph is apparent and then connect these points with a smooth curve. This process is called **point-by-point plotting.**

EXAMPLE 1 ➤ Point-by-Point Plotting Sketch a graph of $y = 9 - x^2$.

Solution Make up a table of solutions—that is, ordered pairs of real numbers that satisfy the given equation. For easy mental calculation, choose integer values for x.

x	-4	-3	-2	-1	0	1	2	3	4
y	-7	0	5	8	9	8	5	0	-7

FIGURE 2

After plotting these solutions, if there are any portions of the graph that are unclear, plot additional points until the shape of the graph is apparent. Then join all the plotted points with a smooth curve as shown in Figure 2. Arrowheads are used to indicate that the graph continues beyond the portion shown here with no significant changes in shape.

The curve in Figure 2 is called a *parabola*. Notice that if we fold the paper along the y axis, the right side will match the left side. We say that the graph is *symmetric with respect to the y axis* and call the y axis the *axis of the parabola*. More will be said about parabolas later in this chapter. ◀

MATCHED PROBLEM 1* ➤ Sketch a graph of $y = x^2 - 4$ using point-by-point plotting. ◀

■ GRAPHING UTILITIES

We now turn to the use of electronic graphing devices to graph equations. We will refer to any electronic device capable of displaying graphs as a **graphing utility.** The two most common graphing utilities are hand-held graphing calculators and computers with appropriate software. You should have such a device as you proceed through this book.

---

* Answers to matched problems are found near the end of each section, before the exercise set.

We will discuss graphing utilities only in general terms. Refer to the manual or to the graphing utility supplement accompanying this text for specific details relative to your own graphing utility.

An image on the screen of a graphing utility is made up of darkened rectangles called **pixels** (see Fig. 3). All the pixel rectangles are the same size, and they do not change in size during any application. Graphing utilities use pixel-by-pixel plotting to produce graphs.

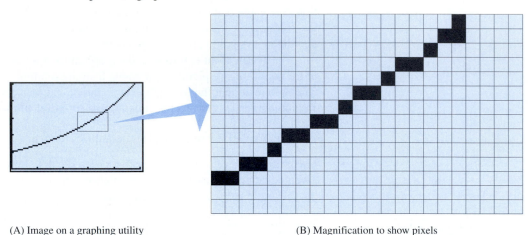

(A) Image on a graphing utility   (B) Magnification to show pixels

FIGURE 3
Pixel-by-pixel plotting on a graphing utilty

The accuracy of the graph depends on the resolution of the graphing utility. Most graphing utilities have screen resolutions of between 50 and 75 pixels per inch, which results in fairly rough but very useful graphs. Some computer systems can print high-quality graphs with resolutions over 1,000 pixels per inch.

Most graphing utility screens are rectangular in shape. The screen on a graphing utility represents a portion of the plane in the rectangular coordinate system. But this representation is an approximation, since pixels are not really points, as is clearly shown in Figure 3. Points are geometric objects without dimensions, whereas a pixel has dimensions. The coordinates of a pixel are usually taken at the center of the pixel and represent all the infinitely many geometric points within the pixel. This will not cause much of a problem, as we will see.

The portion of a rectangular coordinate system displayed on the graphing screen is called a **viewing window.** This is determined by assigning values to six **window variables:** the lower limit, upper limit, and scale for the x axis; the lower limit, upper limit, and scale for the y axis. Figure 4A illustrates the names and values of **standard window variables,** and Figure 4B shows the resulting **standard viewing window.**

The names **Xmin, Xmax, Xscl, Ymin, Ymax,** and **Yscl** will be used for the six window variables. The window variables may be displayed slightly differently for your graphing utility (consult your manual). Even though many graphing utilities do not display viewing window dimensions, we have chosen to do so to make our graphing utility graphs easier to read. These labels always indicate Xmin, Xmax,

(A) Standard window
variable values

(B) Standard viewing window

FIGURE 4
A standard viewing window and its dimensions

Ymin, and Ymax, and are always centered on the viewing window, irrespective of the location of the axes.

We now turn to the use of a graphing utility to graph equations that can be written in the form

$$y = \text{Some expression in } x \tag{2}$$

Consult your manual to determine how equations of this type are entered in your graphing utility. Graphing an equation of type (2) using a graphing utility is a simple three-step process:

Graphing Equations Using a Graphing Utility

Step 1. Enter the equation.

Step 2. Enter values for the window variables.
(A rule-of-thumb for choosing Xscl and Yscl, unless there are reasons to the contrary, is to choose each about one-tenth the corresponding variable range.)

Step 3. Press the graph command.

The following example illustates this procedure for graphing the equation $y = 9 - x^2$, the equation we graphed using point-by-point plotting in Example 1.

EXAMPLE 2 ➤ **Graphing with a Graphing Utility** Use a graphing utility to graph $y = 9 - x^2$ for $-5 \leqslant x \leqslant 5$ and $-10 \leqslant y \leqslant 10$.

Solution Enter the equation, as shown in Figure 5, step 1 (page 8), and the values for the window variables (Figure 5, step 2) in the graphing utility; then press the graph command to obtain the graph in Figure 5, step 3. (The form of the screens in steps 1, 2, and 3 may vary slightly, depending on the graphing utility used.)

Step 1:
Enter equation(s)

Step 2:
Enter window variables

Step 3:
Press the graph command

FIGURE 5
Graphing is a three-step process   ◄

MATCHED PROBLEM 2 ▶ Use a graphing utility to graph $y = x^2 - 4$ for $-5 \le x \le 5$ and $-5 \le y \le 15$. ◄

_____

Remark

For Example 2, we displayed a viewing window for each step in the graphing procedure. In the future, we will show only the final results, as illustrated in Figure 5, step 3.

The next example illustrates how a graphing utility can be used as an aid to sketching the graph of an equation by hand. Copies of graphs of equations (and sometimes tables of values) may be needed for reports, presentations, research papers, and so on. The example illustrates the use of algebraic, numeric, and graphic approaches, which add considerably to the understanding of a problem.

EXAMPLE 3 ▶ Using a Graphing Utility as an Aid to Hand Graphing—Net Cash Flow   The net cash flow, y (in millions of dollars), of a small high-tech company from 1986 to 1994 is given approximately by the following equation:

$$y = 0.4x^3 - 2x + 1 \qquad -4 \le x \le 4 \tag{3}$$

where x represents the number of years before and after 1990, the year when the board of directors appointed a new chief executive officer (CEO).

(A) Construct a table of values for equation (3) for each year from 1986 to 1994, inclusive. Compute y to one decimal place.

(B) Obtain a graph of equation (3) in the viewing window of your graphing utility. Plot the table values from part (A) by hand on graph paper, then join these points with a smooth curve using the graph in the viewing window as an aid.

Solution   (A) Table values can be computed directly in many graphing utilities after an equation has been entered (consult your manual). Using these values, we obtain Table 1, which contains specific information that would be useful in a report.

(B) To create a graph of equation (3) in the viewing window of a graphing utility,

TABLE 1
Net Cash Flow

Year	1986	1987	1988	1989	1990	1991	1992	1993	1994
x	−4	−3	−2	−1	0	1	2	3	4
y (million $)	−16.6	−3.8	1.8	2.6	1	−0.6	0.2	5.8	18.6

we select values for the window variables that cover a little more than the values shown in Table 1:

$$\text{Xmin} = -5, \quad \text{Xmax} = 5, \quad \text{Ymin} = -20, \quad \text{Ymax} = 20$$

We add a grid to the viewing window (consult your manual) to obtain the graphing utility graph shown in Figure 6A. The corresponding hand-drawn sketch is shown in Figure 6B. Notice in the hand sketch, we can add arrowheads, label axes, and add other embellishments as we wish.

(A) Graphing utility graph   (B) Hand sketch

FIGURE 6
Net cash flow

Remark

Table 1 gives us specific details, and the equation with its graph gives us an overview. Each viewpoint has its specific use.

MATCHED PROBLEM 3 ➤ Given the equation $y = 1 + 1.9x - 0.2x^3$, complete a table of values for the integers from −4 to 4, plot these points by hand, and then hand sketch the graph of the equation with the aid of a graphing utility. ◀

The choice of the viewing window has a significant effect on the shape of a graph. Graph $y = x^3 - 4x$ in each of the following viewing windows:

(A) $-1 \leqslant x \leqslant 1, -1 \leqslant y \leqslant 1$
(B) $-10 \leqslant x \leqslant 10, -10 \leqslant y \leqslant 10$
(C) $-100 \leqslant x \leqslant 100, -100 \leqslant y \leqslant 100$

Which window gives the best view of the graph of this equation, and why?

■ SCREEN COORDINATES

We now take a closer look at *screen coordinates* of pixels. Earlier, we indicated that the coordinates of the center point of a pixel are usually used as the **screen coordinates of the pixel,** and these coordinates represent all points within the pixel. As expected, screen coordinates of pixels change as you change values of window variables.

To find screen coordinates of various pixels, move a *cursor* around the viewing window and observe the coordinates displayed on the screen. A **cursor** is a special symbol, such as a plus sign or an arrow, that locates one pixel on the screen at a time. As the cursor is moved around the screen, it moves from pixel to pixel. To illustrate this, set the window variables in your graphing utility so that $-5 \leqslant x \leqslant 5$ and $-5 \leqslant y \leqslant 5$, and activate a grid for the screen. Move the cursor as close as you can to the point (2, 2) and observe what happens. Figure 7 shows the screen coordinates we obtained for the four pixels closest to (2, 2). The coordinates displayed on your screen may vary slightly from these, depending on the graphing utility used.

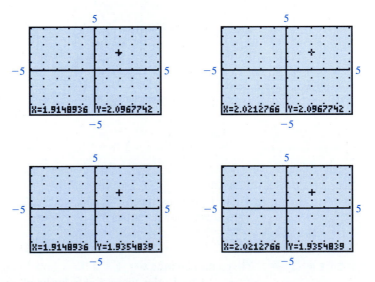

FIGURE 7
Screen coordinates of pixels near (2, 2)

Any of the four pixels in Figure 7 can be used to approximate the point (2, 2), but it is not possible to find a pixel in this viewing window with screen coordinates that are exactly (2, 2)! It is instructive to repeat this exercise with different window variables, say $-7 \leq x \leq 7$ and $-7 \leq y \leq 7$. It turns out that it is possible to carefully select window variables so that screen coordinates are integers (some graphing utilities have this option built in), but this is not necessary for most of our work.

Remark

We think it is important that actual output from existing graphing utilities be used in this book. You may not always be able to produce an exact replica of text figures on your graphing utility, but the differences will be minor and should cause no difficulties.

■ THE TRACE AND ZOOM FEATURES

Analyzing a graph of an equation frequently involves finding coordinates of points on the graph. Using the **trace feature** on a graphing utility is one way to accomplish this. The trace feature places a cursor directly on the graph and only permits movement left and right along the graph. The coordinates displayed during the tracing movement are coordinates of points that satisfy the equation. In most cases, these coordinates are not the same as the pixel screen coordinates displayed using the unrestricted cursor movement discussed earlier. Consult the manual for your graphing utility to learn how to use its trace feature.

explore – discuss 2

Graph the equation $y = x$ in a standard viewing window.

(A) Without selecting the trace feature, move the cursor to a point on the screen that appears to lie on the graph of $y = x$ and is as close to (5, 5) as possible. Record these coordinates. Do these coordinates satisfy the equation $y = x$?

(B) Now select the trace feature and move the cursor along the graph of $y = x$ to a point that has the same x coordinate found in part (A). Is the y coordinate of this point the same as you found in part (A)? Do the coordinates of the point using trace satisfy the equation $y = x$?

(C) Explain the difference in using the trace feature along a curve and trying to use unrestricted movement of a cursor along a curve.

Example 4 illustrates a practical application of the trace feature in approximating coordinates of points on a graph of an equation.

▶ EXAMPLE 4 ➤ **Using Trace Techniques—Manufacturing** Rectangular open boxes are to be manufactured from 11 by 17 inch sheets of cardboard by cutting x by x inch squares out of the corners and folding up the sides, as shown in Figure 8.

FIGURE 8
Template for boxes

(A) Write an equation for the volume y of the resulting box in terms of the size of the square x. Indicate appropriate restrictions on x.

(B) Graph the equation for appropriate values of x. Adjust the window variables for y to include the entire graph of interest.

(C) Use trace to estimate x (to one decimal place) for the smallest square that can be cut out to produce a box of volume 150 cubic inches.

FIGURE 9
Actual box with dimensions added

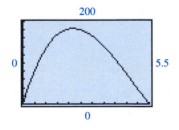

FIGURE 10
Graph showing volume relative to x

Solution   (A) The box is shown in Figure 9 with the sides folded up and dimensions added. From this figure we can write the equation of the volume in terms of x and establish the restrictions on x. No dimension can be negative; that is, $x \geqslant 0$, $11 - 2x \geqslant 0$, and $17 - 2x \geqslant 0$. These three inequalities imply that $0 \leqslant x \leqslant 5.5$. Thus, the volume of the box is given by

$$y = x(17 - 2x)(11 - 2x) \qquad 0 \leqslant x \leqslant 5.5 \qquad (4)$$

(B) Entering this equation in a graphing utility (it does not need to be multiplied out) and evaluating it for several integers between 0 and 5, we find that a good choice for the window dimensions of y appears to be $0 \leqslant y \leqslant 200$. This choice can easily be changed if the viewing window shows too much space above the graph or if part of the graph we are interested in is out of the viewing window. Figure 10 shows the graph of equation (4) in the viewing window selected above.

(C) Select the trace command and move the cursor along the whole curve to observe the changes in the volume y as x moves from 0 to 5.5. Notice that y passes through 150 on the left side of the high point in the curve and again on the right side of the curve. Thus, there are two solutions to the problem, but we are interested only in the smaller value of x. Tracing along the curve from left to right, starting at $x = 0$, we observe the two points with y coordinates closest to 150. These are shown in Figures 11A and 11B:

(A) Lower estimate   (B) Upper estimate

FIGURE 11

If x_1 is the exact value of x that produces the volume 150 cubic inches, then the point $(x_1, 150)$ must lie on the graph somewhere between the two points found in Figures 11A and 11B. Thus, we have

$$1.1702128 < x_1 < 1.2287234$$

which implies that $x = 1.2$, correct to one decimal place. Evaluating y at $x = 1.2$ produces 150.672. The value of y is not exact, because 1.2 is an approximation of the size of the square that will produce a volume of 150 cubic inches. ◄

MATCHED PROBLEM 4 ➤ Use the methods outlined in Example 4 to approximate the size of the largest square that can be cut out to produce a box of volume 150 cubic inches. ◄

The results and methods we used in Example 4 are very significant. If we had attempted to solve the problem algebraically, we would have been confronted with solving

$$150 = x(17 - 2x)(11 - 2x)$$

a third-degree polynomial equation that is very difficult to solve algebraically. In Example 5, we use *zoom techniques* to continue the graphical process to find x in Example 4 to three decimal place accuracy.

▶ EXAMPLE 5 ➤ **Using Zoom Techniques—Manufacturing** Use zoom techniques to find x in Example 4 to three decimal places.

Solution In Figure 11, we found the two points closest to $(x_1, 150)$ to be the points listed in Table 2. We were therefore able to conclude that $x = 1.2$ to one decimal place accuracy. How can we achieve greater accuracy on a graphing utility? We *zoom in*, as shown in Figure 12 (at the top of the next page).

Zooming in can be accomplished by choosing new window variables based on Table 2. Choose Xmin a little smaller than the smaller x value in Table 2 and Xmax

TABLE 2

x	1.1702128	1.2287234
y	148.55345	152.64496

FIGURE 12
Zooming in on a graph

a little larger than the larger x value in Table 2. Choose Ymin and Ymax in the same way. Proceeding in this way, we obtain

Xmin $= 1.17$ Xmax $= 1.23$

Ymin $= 148$ Ymax $= 153$

(We can ignore Xscl and Yscl in the zooming process.)

We have the same number of pixels covering the viewing window as before, but now they are spread over a smaller portion of the curve containing $(x_1, 150)$, thus creating a magnification of the curve. Now we press the graph command and use trace to find the two points closest to $(x_1, 150)$. Figure 13 shows the results.

(A)   (B)

FIGURE 13
First zoom on $(x_1, 150)$

TABLE 3

x	1.1904255	1.1910638
y	149.99912	150.04422

From Figure 13, we form Table 3, and are able to conclude that $x = 1.19$ to two decimal places. We repeat the process one more time, choosing window variables based on Table 3 as follows:

Xmin $= 1.1904$ Xmax $= 1.1911$

Ymin $= 149.999$ Ymax $= 150.045$

Press the graph command, and then use trace to find the two points closest to $(x_1, 150)$. Figure 14 shows the results.

(A)   (B)

FIGURE 14
Second zoom on $(x_1, 150)$

TABLE 4

x	1.1904372	1.1904447
y	149.99995	150.00047

From Figure 14, we form Table 4, and we are able to conclude that $x = 1.190$ to the three decimal places requested. Calculating y for this value, we obtain $y = 149.96904$, which rounded to one decimal place is 150.0. ◀

MATCHED PROBLEM 5 ➤ Proceeding as in Example 5, compute the value of x for the larger square (see Matched Problem 4) to three decimal places. ◀

To facilitate understanding, we have included quite a bit of detail in describing the trace and zoom procedures in Examples 4 and 5. In practice, the process proceeds much more rapidly, with just the new viewing window variables being written down at each zoom stage.

Many graphing utilities have built-in zooming routines (read your manual). Each time a built-in zoom routine is activated, it is by a constant preset factor (usually 4) that can be changed to smaller or larger values, if desired. The method outlined in Example 5 is often faster than using a built-in zoom routine; however, both methods will achieve the desired results, and each method has advantages in certain situations. The main thing to remember, irrespective of which zoom method is used, is that zooming (in or out) is simply a matter of changing viewing window variables.

Answers to Matched Problems

1.

$y = x^2 - 4$

2.

3.

x	-4	-3	-2	-1	0	1	2	3	4
y	6.2	0.7	-1.2	-0.7	1	2.7	3.2	1.3	-4.2

4. $x \approx 3.3$ in.   5. $x \approx 3.319$ in.

EXERCISE   1-1

A   *For each equation in Problems 1–6, make up a table of solutions using x = − 3, − 2, − 1, 0, 1, 2, and 3. Plot these solutions, and then sketch a graph of the equation through these points.*

1. $y = x - 2$ **2.** $y = 2 - x$
3. $y = 4 - x^2$ **4.** $y = x^2 - 4$
5. $y = (x - 1)(x + 1)$ **6.** $y = (x + 2)(x - 2)$

In Problems 7 and 8, use the graph to estimate (to the nearest integer) the missing coordinate of each point. There may be more than one answer for some parts.

7. (A) $(2, ?)$ (B) $(3, ?)$ (C) $(-7, ?)$
       (D) $(0, ?)$ (E) $(?, 0)$ (F) $(?, -4)$

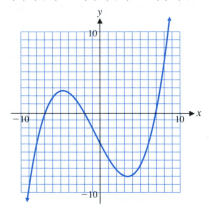

Figure for 7

8. (A) $(3, ?)$ (B) $(-7, ?)$ (C) $(6, ?)$
       (D) $(?, 0)$ (E) $(?, 3)$ (F) $(?, -10)$

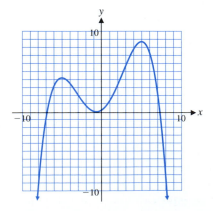

Figure for 8

In Problems 9–14, determine whether the indicated point lies in the viewing window of a graphing utility with the following window variables: Xmin = − 5, Xmax = 10, Ymin = − 5, Ymax = 10.

9. $(0, -5)$ **10.** $(-5, 0)$ **11.** $(-6, -6)$
12. $(7, -7)$ **13.** $(0, 0)$ **14.** $(-4, -4)$

In Problems 15–18, find the smallest viewing window that will contain all the points in the table. State your answers in terms of the window variables Xmin, Xmax, Ymin, and Ymax.

15.

x	4	− 3	0	− 5
y	7	− 2	− 1	3

16.

x	− 6	0	10	− 7
y	12	0	− 13	4

17.

x	− 2	− 7	4	6	− 8	12
y	7	− 2	− 1	3	0	− 5

18.

x	11	− 14	− 15	0	3	10
y	10	− 22	− 18	3	5	15

B   *In Problems 19–24, graph each equation in a standard viewing window.*

19. $y = -x$ **20.** $y = 0.5x$
21. $y = 9 - 0.4x^2$ **22.** $y = 0.3x^2 - 4$
23. $y = 2\sqrt{x + 5}$ **24.** $y = -2\sqrt{x + 5}$

For each equation in Problems 25–30, construct a table of values over the indicated interval, computing y values to the nearest tenth of a unit. Plot these points on graph paper, then with the aid of a graphing utility, complete the hand sketch of the graph.

25. $y = (3 - x)^2, 0 \leq x \leq 6$
       (Use even integers for the table.)

26. $y = (x - 4)^2, 0 \leq x \leq 8$
       (Use even integers for the table.)

27. $y = 0.4x^2 + 0.3x - 4, -5 \leq x \leq 5$
       (Use odd integers for the table.)

28. $y = 4.5 - 0.4x - 0.3x^2$, $-5 \leqslant x \leqslant 5$
(Use odd integers for the table.)

29. $y = 0.5x(4 - x)(x + 2)$, $-3 \leqslant x \leqslant 5$
(Use odd integers for the table.)

30. $y = 0.5x(x + 3.5)(2.8 - x)$, $-4 \leqslant x \leqslant 4$
(Use even integers for the table.)

In Problems 31–34, graph the equation in a standard view-ing window. Use trace and zoom to approximate (to two decimal places) the x coordinate of the points in the window that are on the graph of the equation and have the indicated y coordinate.

31. $y = 2\sqrt{x + 4}$
(A) $(x, 5)$ (B) $(x, 3)$

32. $y = 3\sqrt{5 - x}$
(A) $(x, 4)$ (B) $(x, 8)$

33. $y = x^3 - 3x + 2$
(A) $(x, 5)$ (B) $(x, 1)$, $x > 0$

34. $y = 2 + 3x - x^3$
(A) $(x, 5)$ (B) $(x, 3)$, $x > 0$

C *In Problems 35–38, use trace and zoom to find all real solutions of each equation (to two decimal places).*

35. $2x^3 - 5x^2 + 2 = 0$

36. $3x^3 - 7x^2 + 3 = 0$

37. $0.01x^4 - 0.45x^2 + 2 = 0$

38. $-0.01x^4 + 0.56x^2 - 3 = 0$

39. The point $(\sqrt{5}, 5)$ is on the graph of $y = x^2$. Use trace and zoom to approximate $\sqrt{5}$ to five decimal places. Compare your result with the direct calculator evaluation of $\sqrt{5}$.

40. The point $(\sqrt[3]{2}, 2)$ is on the graph of $y = x^3$. Use trace and zoom to approximate $\sqrt[3]{2}$ to five decimal places. Compare your result with the direct calculator evaluation of $\sqrt[3]{2}$.

41. Briefly describe the difference between the mathematical coordinates of a point in a rectangular coordinate system and the screen coordinates of a pixel.

42. Briefly discuss the difference between the coordinates displayed during unrestricted cursor movement and those displayed during the trace procedure.

▶ **A P P L I C A T I O N S**

Business & Economics

43. *Manufacturing.* Rectangular open-top boxes are to be con-structed out of 8.5 by 11 inch sheets of thin cardboard by cutting x inch squares out of each corner and bending the sides up (as in Figures 8 and 9 in Example 4). What size squares (to two decimal places) should be cut out to produce a box with a volume of 55 cubic inches? Give the dimen-sions (to two decimal places) of all possible boxes with the given volume. Check your answers.

44. *Manufacturing.* Rectangular open-top boxes are to be con-structed out of 9 by 12 inch sheets of thin cardboard by cutting x inch squares out of each corner and bending the sides up (as shown in Figures 8 and 9 in Example 4). What size squares (to two decimal places) should be cut out to produce a box with a volume of 72 cubic inches? Give the dimensions (to two decimal places) of all possible boxes with the given volume. Check your answers.

45. *Manufacturing.* A box with a lid is to be cut out of a 12 by 24 inch sheet of thin cardboard by cutting out six x inch squares and folding as indicated in the figure. What are the dimensions of all possible boxes that will have a volume of 100 cubic inches? Check your answers.

Figure for 45

46. *Manufacturing.* A box with a lid is to be cut out of a 10 by 20 inch sheet of thin cardboard by cutting out six x inch squares and folding. (Refer to the figure for Problem 45.) What are the dimensions of all possible boxes that will have a volume of 75 cubic inches? Check your answers.

47. *Hospital costs.* The figure (page 18) shows a typical hospi-tal cost curve for total costs C (in millions of dollars) for a hospital handling x thousand cases per year. [This problem is based on an article on health care pricing by Yves Nievergelt that appeared in *UMAP (The Journal of Undergraduate Mathematics and Its Applications)* Vol. 9, No. 1, 1988.] Notice that costs rise more rapidly at lower and higher case loads because of inefficient use of facilities. Equations of the type accompanying the graph can be obtained from research data using curve-fitting techniques. Use the equation shown, and trace and zoom on a graphing utility to estimate the annual case load (to the nearest person) that results in an annual cost of $5 million.

48. *Hospital costs.* Using the equation in the figure and trace and zoom in a graphing utility, estimate the annual case load (to the nearest person) that results in an annual cost of $3.5 million.

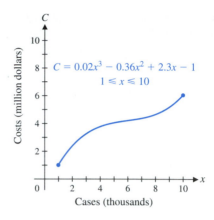

Figure for 47 and 48

S E C T I O N   1-2   ■   ## Functions

■ DEFINITION OF A FUNCTION
■ FUNCTIONS SPECIFIED BY EQUATIONS
■ FUNCTION NOTATION
■ APPLICATIONS

■ DEFINITION OF A FUNCTION

The function concept is one of the most important ideas in mathematics. Central to the concept of function is correspondence. You have already had experiences with correspondences in daily living. For example:

> To each person there corresponds an annual income.
>
> To each item in a supermarket there corresponds a price.
>
> To each student there corresponds a grade-point average.
>
> To each day there corresponds a maximum temperature.
>
> For the manufacture of x items there corresponds a cost.
>
> For the sale of x items there corresponds a revenue.
>
> To each square there corresponds an area.
>
> To each number there corresponds its cube.

One of the most important aspects of any science is the establishment of correspondences among various types of phenomena. Once a correspondence is known, predictions can be made. A cost analyst would like to predict costs for various levels of output in a manufacturing process; a medical researcher would like to know the correspondence between heart disease and obesity; a psychologist would like to predict the level of performance after a subject has repeated a task a given number of times; and so on.

What do all the above examples have in common? Each describes the matching of elements from one set with the elements in a second set. Consider the tables of the cube, square, and square root given in Tables 1–3.

T A B L E 1	
DOMAIN	RANGE
Number	*Cube*

-2	→	-8
-1	→	-1
0	→	0
1	→	1
2	→	8

T A B L E 2	
DOMAIN	RANGE
Number	*Square*

T A B L E 3	
DOMAIN	RANGE
Number	*Square Root*

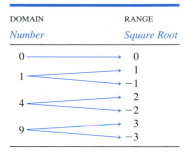

Tables 1 and 2 specify functions, but Table 3 does not. Why not? The definition of the term *function* will explain.

Definition of a Function

A **function** is a rule (process or method) that produces a correspondence between two sets of elements such that to each element in the first set there corresponds one and only one element in the second set.

The first set is called the **domain,** and the set of corresponding elements in the second set is called the **range.**

Tables 1 and 2 specify functions, since to each domain value there corresponds exactly one range value (for example, the cube of -2 is -8 and no other number). On the other hand, Table 3 does not specify a function, since to at least one domain value there corresponds more than one range value (for example, to the domain value 9 there corresponds -3 and 3, both square roots of 9).

Remark
Some graphing utilities use the term *range* to refer to the window variables. In this book, *range* will always refer to the range of a function.

explore – discuss 1

Consider the set of students enrolled in a college and the set of faculty members of that college. Suppose we define a correspondence between the two sets by saying that a student corresponds to a faculty member if the student is currently enrolled in a course taught by that faculty member. Is this correspondence a function? Discuss.

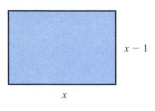

$x - 1$

x

FIGURE 1

■ FUNCTIONS SPECIFIED BY EQUATIONS

Most of the domains and ranges included in this text will be (infinite) sets of real numbers, and the rules associating range values with domain values will be equations in two variables. Consider, for example, the equation for the area of a rectangle with width 1 inch less than its length (Figure 1). If x is the length, then the area y is given by

$$y = x(x - 1) \qquad x \geq 1$$

For each **input** x (length), we obtain an **output** y (area). For example:

If $x = 5$,   then $y = 5(5 - 1) = 5 \cdot 4 = 20.$

If $x = 1$,   then $y = 1(1 - 1) = 1 \cdot 0 = 0.$

If $x = 4.32$,   then $y = 4.32(4.32 - 1) = (4.32)(3.32)$
$$= 14.3424.$$

The input values are domain values, and the output values are range values. The equation (a rule) assigns each domain value x a range value y. The variable x is called an *independent variable* (since values can be "independently" assigned to x from the domain), and y is called a *dependent variable* (since the value of y "depends" on the value assigned to x). In general, any variable used as a placeholder for domain values is called an **independent variable;** any variable that is used as a placeholder for range values is called a **dependent variable.**

When does an equation specify a function?

Functions Defined by Equations

If in an equation in two variables, we get exactly one output (value for the dependent variable) for each input (value for the independent variable), then the equation defines a function.

If we get more than one output for a given input, the equation does not define a function.

EXAMPLE 1 ➤ **Functions and Equations**   Determine which of the following equations specify functions with independent variable x.

(A) $4y - 3x = 8$, x a real number   (B) $y^2 - x^2 = 9$, x a real number

Solution   (A) Solving for the dependent variable y, we have

$$4y - 3x = 8 \tag{1}$$
$$4y = 8 + 3x$$
$$y = 2 + \tfrac{3}{4}x$$

Since each input value x corresponds to exactly one output value ($y = 2 + \tfrac{3}{4}x$), we see that equation (1) specifies a function.

(B) Solving for the dependent variable y, we have

$$y^2 - x^2 = 9 \qquad\qquad (2)$$
$$y^2 = 9 + x^2$$
$$y = \pm\sqrt{9 + x^2}$$

Since $9 + x^2$ is always a positive real number for any real number x and since each positive real number has two square roots, to each input value x there corresponds two output values ($y = -\sqrt{9 + x^2}$ and $y = \sqrt{9 + x^2}$). For example, if $x = 4$, then equation (2) is satisfied for $y = 5$ and for $y = -5$. Thus, equation (2) does not specify a function. ◄

MATCHED PROBLEM 1 ➤ Determine which of the following equations specify functions with independent variable x.

(A) $y^2 - x^4 = 9$, x a real number   (B) $3y - 2x = 3$, x a real number ◄

Since the graph of an equation is the graph of all the ordered pairs that satisfy the equation, it is very easy to determine whether an equation specifies a function by examining its graph. The graphs of the two equations we considered in Example 1 are shown in Figure 2. (To produce Figure 2B on a graphing utility, enter $y_1 = \sqrt{9 + x^2}$ and $y_2 = -\sqrt{9 + x^2}$.)

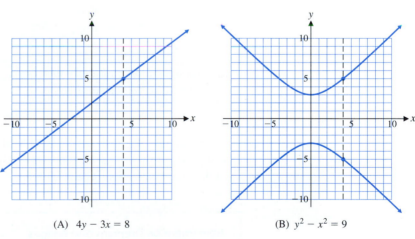

(A) $4y - 3x = 8$ (B) $y^2 - x^2 = 9$

FIGURE 2

In Figure 2A notice that any vertical line will intersect the graph of the equation $4y - 3x = 8$ in exactly one point. This shows that to each x value there corresponds exactly one y value and confirms our conclusion that this equation specifies a function. On the other hand, Figure 2B shows that there exist vertical lines that intersect the graph of $y^2 - x^2 = 9$ in two points. This indicates that there exist x values to which there correspond two different y values and verifies our conclusion that this equation does not specify a function. These observations are generalized in Theorem 1.

theorem 1

Vertical-Line Test for a Function

An equation defines a function if each vertical line in the coordinate system passes through at most one point on the graph of the equation.

If any vertical line passes through two or more points on the graph of an equation, then the equation does not define a function.

explore – discuss 2

(A) Use a graphing utility to graph $y = 4 - x^2$ for $-5 \leqslant x \leqslant 5$ and $-5 \leqslant y \leqslant 5$. Trace along the curve and observe the changes in coordinates as the cursor moves along the curve. Discuss how these coordinates are related to the function defined by the equation.

(B) Graph $y_1 = \sqrt{25 - x^2}$ and $y_2 = -\sqrt{25 - x^2}$ for $-5 \leqslant x \leqslant 5$ and $-5 \leqslant y \leqslant 5$. [Adjust Xmin and Xmax until the graph looks like a circle, or do this directly by ''squaring'' the graph (see your manual).] Using trace, find two points on the circle with the same x coordinate and different y coordinates.

(C) Is it possible to graph a single equation of the form

$$y = \text{Expression in } x$$

on your graphing utility and obtain a graph that is not the graph of a function? Explain your answer.

In Example 1, the domains were explicitly stated along with the given equations. In many cases, this will not be done. Unless stated to the contrary, we shall adhere to the following convention regarding domains and ranges for functions specified by equations:

Agreement on Domains and Ranges

If a function is specified by an equation and the domain is not indicated, then we assume that the domain is the set of all real number replacements of the independent variable (inputs) that produce real values for the dependent variable (outputs). The range is the set of all outputs corresponding to input values.

In many applied problems the domain is determined by practical considerations within the problem (see Example 6).

EXAMPLE 2 ➤ **Finding a Domain** Find the domain of the function specified by the equation $y = \sqrt{4 - x}$, assuming x is the independent variable.

Solution For y to be real, $4 - x$ must be greater than or equal to 0; that is,

$$4 - x \geqslant 0$$
$$-x \geqslant -4$$
$$x \leqslant 4 \qquad \text{\textit{Sense of inequality reverses when both sides are divided by} } -1.$$

Thus,

Domain: $x \leqslant 4$ or $(-\infty, 4]$

Figure 3 confirms this result. [*Note:* Most, but not all, graphing utilities ignore nonreal values when graphing. If this is not the case with yours, consult your manual.] ◄

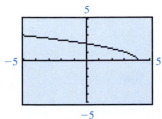

FIGURE 3

$y = \sqrt{4 - x}$

MATCHED PROBLEM 2 ➤ Find the domain of the function specified by the equation $y = \sqrt{x - 2}$, assuming x is the independent variable. Confirm your results with a graphing utility. ◄

■ FUNCTION NOTATION

We have just seen that a function involves two sets, a domain and a range, and a rule of correspondence that enables us to assign to each element in the domain exactly one element in the range. We use different letters to denote names for numbers; in essentially the same way, we will now use different letters to denote names for functions. For example, f and g may be used to name the functions specified by the equations $y = 2x + 1$ and $y = x^2 + 2x - 3$:

$$f: \quad y = 2x + 1$$
$$g: \quad y = x^2 + 2x - 3$$

(3)

If x represents an element in the domain of a function f, then we frequently use the symbol

$$f(x)$$

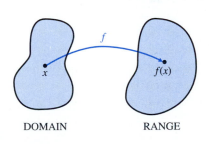

DOMAIN   RANGE

FIGURE 4

in place of y to designate the number in the range of the function f to which x is paired (Fig. 4). This symbol does not represent the product of f and x. The symbol $f(x)$ is read as "f of x," "f at x," or "the value of f at x." Whenever we write $y = f(x)$, we assume that the variable x is an independent variable and that both y and $f(x)$ are dependent variables.

Using function notation, we can now write functions f and g in (3) in the form

$$f(x) = 2x + 1 \qquad \text{and} \qquad g(x) = x^2 + 2x - 3$$

Let us find $f(3)$ and $g(-5)$. To find $f(3)$, we replace x with 3 wherever x occurs in $f(x) = 2x + 1$ and evaluate the right side:

$$f(x) = 2x + 1$$
$$f(\mathbf{3}) = 2 \cdot \mathbf{3} + 1$$
$$= 6 + 1 = 7 \quad \text{\textit{For input 3, the output is 7.}}$$

Thus,

$f(3) = 7$ *The function f assigns the range value 7 to the domain value 3.*

To find $g(-5)$, we replace x by -5 wherever x occurs in $g(x) = x^2 + 2x - 3$ and evaluate the right side:

$$g(x) = x^2 + 2x - 3$$
$$g(-5) = (-5)^2 + 2(-5) - 3$$
$$= 25 - 10 - 3 = 12$$ *For input -5, the output is 12.*

Thus,

$g(-5) = 12$ *The function g assigns the range value 12 to the domain value -5.*

It is very important to understand and remember the definition of $f(x)$:

The Symbol $f(x)$

For any element x in the domain of the function f, the symbol $f(x)$ represents the element in the range of f corresponding to x in the domain of f. If x is an input value, then $f(x)$ is the corresponding output value. If x is an element that is not in the domain of f, then f is *not defined at x* and $f(x)$ *does not exist*.

EXAMPLE 3 ➤ Function Evaluation   If

$$f(x) = \frac{12}{x-2} \qquad g(x) = 1 - x^2 \qquad h(x) = \sqrt{x-1}$$

then:

(A) $f(6) = \dfrac{12}{6-2}^{\,*} = \dfrac{12}{4} = 3$

(B) $g(-2) = 1 - (-2)^2 = 1 - 4 = -3$

(C) $h(-2) = \sqrt{-2-1} = \sqrt{-3}$

Since $\sqrt{-3}$ is not a real number, -2 is not in the domain of h and $h(-2)$ does not exist.

---

* Dashed boxes are used throughout the book to represent steps that are usually performed mentally.

(D) $f(\mathbf{0}) + g(\mathbf{1}) - h(\mathbf{10}) = \dfrac{12}{\mathbf{0} - 2} + (1 - \mathbf{1}^2) - \sqrt{\mathbf{10} - 1}$

$$= \dfrac{12}{-2} + 0 - \sqrt{9}$$

$$= -6 - 3 = -9 \qquad \blacktriangleleft$$

MATCHED PROBLEM 3 ➤ Use the functions in Example 3 to find:

(A) $f(-2)$ (B) $g(-1)$ (C) $h(-8)$ (D) $\dfrac{f(3)}{h(5)}$ ◀

EXAMPLE 4 ➤ **Finding Domains**   Find the domains of functions f, g, and h:

$$f(x) = \dfrac{12}{x - 2} \qquad g(x) = 1 - x^2 \qquad h(x) = \sqrt{x - 1}$$

Solution   *Domain of f:* $12/(x - 2)$ represents a real number for all replacements of x by real numbers except for $x = 2$ (division by 0 is not defined). Thus, $f(2)$ does not exist, and the domain of f is the set of all real numbers except 2. We often indicate this by writing

$$f(x) = \dfrac{12}{x - 2} \qquad x \ne 2$$

Domain of g: The domain is R, the set of all real numbers, since $1 - x^2$ represents a real number for all replacements of x by real numbers.

Domain of h: The domain is the set of all real numbers x such that $\sqrt{x - 1}$ is a real number—that is, such that

$$x - 1 \geqslant 0$$
$$x \geqslant 1 \quad \text{or} \quad [1, \infty) \qquad \blacktriangleleft$$

MATCHED PROBLEM 4 ➤ Find the domains of functions F, G, and H:

$$F(x) = x^2 - 3x + 1 \qquad G(x) = \dfrac{5}{x + 3} \qquad H(x) = \sqrt{2 - x} \qquad ◀$$

In addition to evaluating functions at specific numbers, it is important to be able to evaluate functions at expressions that involve one or more variables. For example, the **difference quotient**

$$\dfrac{f(x + h) - f(x)}{h} \qquad x \text{ and } x + h \text{ in the domain of } f, h \ne 0$$

is studied extensively in calculus.

explore – discuss 3

Let x and h be real numbers.

(A) If $f(x) = 4x + 3$, which of the following is true?

    (1) $f(x + h) = 4x + 3 + h$
    (2) $f(x + h) = 4x + 4h + 3$
    (3) $f(x + h) = 4x + 4h + 6$

(B) If $g(x) = x^2$, which of the following is true?

    (1) $g(x + h) = x^2 + h$
    (2) $g(x + h) = x^2 + h^2$
    (3) $g(x + h) = x^2 + 2hx + h^2$

(C) If $M(x) = x^2 + 4x + 3$, describe the operations that must be performed to evaluate $M(x + h)$.

EXAMPLE 5 ➤   Using Function Notation   For $f(x) = x^2 - 2x + 7$, find:

(A) $f(a)$ (B) $f(a + h)$ (C) $\dfrac{f(a + h) - f(a)}{h}$

Solution   (A) $f(a) = a^2 - 2a + 7$

(B) $f(a + h) = (a + h)^2 - 2(a + h) + 7$
$$= a^2 + 2ah + h^2 - 2a - 2h + 7$$

(C) $\dfrac{f(a + h) - f(a)}{h} = \dfrac{(a^2 + 2ah + h^2 - 2a - 2h + 7) - (a^2 - 2a + 7)}{h}$

$$= \frac{2ah + h^2 - 2h}{h} = \frac{h(2a + h - 2)}{h} = 2a + h - 2 \; \blacktriangleleft$$

MATCHED PROBLEM 5 ➤   Repeat Example 5 for $f(x) = x^2 - 4x + 9$.   ◀

■ APPLICATIONS

We now turn to the important concepts of **break-even** and **profit–loss** analysis, which we will return to a number of times in this text. Any manufacturing company has **costs,** C, and **revenues,** R. The company will have a **loss** if $R < C$, will **break even** if $R = C$, and will have a **profit** if $R > C$. Costs include **fixed costs** such as plant overhead, product design, setup, and promotion; and **variable costs,** which are dependent on the number of items produced at a certain cost per item. In addition, **price–demand** functions, usually established by financial departments using historical data or sampling techniques, play an important part in profit–loss analysis. We will let x, the number of units manufactured and sold, represent the independent variable. Cost functions, revenue functions, profit functions, and price–demand functions are often stated in the following forms, where a, b, m, and n are appropriately selected constants:

Cost Function

$$C = \text{(Fixed costs)} + \text{(Variable costs)}$$
$$= a + bx$$

Price–Demand Function

$$p = m - nx \qquad \text{x is the number of items that can be sold at \$p per item.}$$

Revenue Function

$$R = \text{(Number of items sold)} \times \text{(Price per item)}$$
$$= xp = x(m - nx)$$

Profit Function

$$P = R - C$$
$$= x(m - nx) - (a + bx)$$

Example 6 and Matched Problem 6 explore the relationships among the algebraic definition of a function, the numerical values of the function, and graphical representation of the function. The interplay among algebraic, numeric, and graphic is an important aspect of our treatment of functions and their use. In this example, we also see how a function can be used to describe data from the real world, a process that is often referred to as *mathematical modeling*. The material in Example 6 will be returned to in subsequent sections so that we can analyze it in greater detail and from different points of view.

EXAMPLE 6 ▶ **Price–Demand and Revenue Modeling** A manufacturer of a popular point-and-shoot camera wholesales the camera to retail outlets throughout the United States. Using statistical methods, the financial department in the company produced the price–demand data in Table 4, where p is the wholesale price per camera at which x million cameras are sold. Notice that as the price goes down, the number sold goes up.

Using special analytical techniques (regression analysis), an analyst arrived at the following price–demand function that models the Table 4 data:

$$p(x) = 94.8 - 5x \qquad 1 \leqslant x \leqslant 15 \tag{4}$$

(A) Plot the data in Table 4. Then sketch a graph of the price–demand function in the same coordinate system.

(B) What is the company's revenue function for this camera, and what is the domain of this function?

TABLE 4
Price–Demand

x (million)	p ($)
2	87
5	68
8	53
12	37

(C) Complete Table 5, computing revenues to the nearest million dollars.

TABLE 5
Revenue

x (million)	R(x) (million $)
1	90
3	
6	
9	
12	
15	

(D) Plot the data in Table 5. Then sketch a graph of the revenue function using a graphing utility as an aid.

Solution (A) The plot of the data and the graph of the price–demand function are shown in Figure 5. Notice that the model approximates the actual data in Table 4, and it is assumed that it gives realistic and useful results for all other values of x between 1 million and 15 million.

(B) $R(x) = xp(x) = x(94.8 - 5x)$ million dollars

Domain: $1 \leq x \leq 15$

[Same domain as the price–demand function, equation (4).]

(C) TABLE 5
Revenue

x (million)	R(x) (million $)
1	90
3	239
6	389
9	448
12	418
15	297

p(x)

100

50

Price per camera ($)

5   10   15   *x*

Million cameras

FIGURE 5
Price–demand

(D)

MATCHED PROBLEM 6 ➤

TABLE 6
Cost Data

x (million)	C(x) (million $)
1	175
5	260
8	305
12	395

The financial department in Example 6, using statistical techniques, produced the data in Table 6, where $C(x)$ is the cost in millions of dollars for manufacturing and selling x million cameras.

Using special analytical techniques (regression analysis), an analyst produced the following cost function to model the data:

$$C(x) = 156 + 19.7x \qquad 1 \leq x \leq 15 \qquad (5)$$

(A) Plot the data in Table 6. Then sketch a graph of equation (5) in the same coordinate system.

(B) What is the company's profit function for this camera, and what is its domain?

(C) Complete Table 7, computing profits to the nearest million dollars.

(D) Plot the points from part (C). Then sketch a graph of the profit function through these points using a graphing utility as an aid.

TABLE 7
Profit

x (million)	P(x) (million $)
1	− 86
3	
6	
9	
12	
15	

Answers to Matched Problems

1. (A) Does not specify a function   (B) Specifies a function
2. $x \geq 2$ (inequality notation) or $[2, \infty)$ (interval notation)

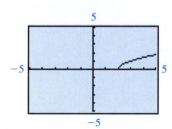

3. (A) − 3   (B) 0   (C) Does not exist   (D) 6
4. Domain of F: R; Domain of G: All real numbers except − 3; Domain of H: $x \leq 2$ (inequality notation) or $(-\infty, 2]$ (interval notation)
5. (A) $a^2 - 4a + 9$ (B) $a^2 + 2ah + h^2 - 4a - 4h + 9$ (C) $2a + h - 4$
6. (A)

(B) $P(x) = R(x) - C(x) = x(94.8 - 5x) - (156 + 19.7x)$; Domain: $1 \leqslant x \leqslant 15$

(C) T A B L E 7
Profit

x *(million)*	P(x) *(million $)*
1	− 86
3	24
6	115
9	115
12	25
15	− 155

(D)

E X E R C I S E   1 - 2

A   *Indicate whether each table specifies a function.*

1. DOMAIN   RANGE

3 ⟶ 0
5 ⟶ 1
7 ⟶ 2

2. DOMAIN   RANGE

−1 ⟶ 5
−2 ⟶ 7
−3 ⟶ 9

3. DOMAIN   RANGE

4. DOMAIN   RANGE

8 ⟶ 0
9 ⟶ 1
     ⟶ 2
10 ⟶ 3

5. DOMAIN   RANGE

6. DOMAIN   RANGE

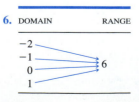

Indicate whether each graph in Problems 7–12 specifies a function

7.

8.

9.

10.

11.

12.

If $f(x) = 3x - 2$ and $g(x) = x - x^2$, find each of the expressions in Problems 13–30.

13. $f(2)$ **14.** $f(1)$ **15.** $f(-1)$
16. $f(-2)$ **17.** $g(3)$ **18.** $g(1)$
19. $f(0)$ **20.** $f(\frac{1}{3})$ **21.** $g(-3)$
22. $g(-2)$ **23.** $f(1) + g(2)$ **24.** $g(1) + f(2)$
25. $g(2) - f(2)$ **26.** $f(3) - g(3)$ **27.** $g(3) \cdot f(0)$

28. $g(0) \cdot f(-2)$ **29.** $\dfrac{g(-2)}{f(-2)}$ **30.** $\dfrac{g(-3)}{f(2)}$

In Problems 31–38, use the following graph of a function f to determine x or y to the nearest integer, as indicated. Some problems may have more than one answer.

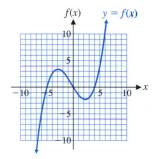

31. $y = f(-5)$ **32.** $y = f(4)$ **33.** $y = f(5)$
34. $y = f(-2)$ **35.** $0 = f(x)$ **36.** $3 = f(x), x < 0$
37. $-4 = f(x)$ **38.** $4 = f(x)$

B *In Problems 39–50, find the domain of each function.*

39. $F(x) = 2x^3 - x^2 + 3$ **40.** $H(x) = 7 - 2x^2 - x^4$

41. $f(x) = \dfrac{x - 2}{x + 4}$ **42.** $g(x) = \dfrac{x + 1}{x - 2}$

43. $F(x) = \dfrac{x + 2}{x^2 + 3x - 4}$ **44.** $G(x) = \dfrac{x - 7}{x^2 + x - 6}$

45. $h(x) = \dfrac{2x - 1}{x^2 + 6x + 9}$ **46.** $F(x) = \dfrac{1 - 3x}{x^2 + 8x + 16}$

47. $g(x) = \sqrt{7 - x}$ **48.** $f(x) = \sqrt{5 + x}$

49. $G(x) = \dfrac{1}{\sqrt{7 - x}}$ **50.** $F(x) = \dfrac{1}{\sqrt{5 + x}}$

51. Two people are discussing the function

$$f(x) = \frac{x^2 - 4}{x^2 - 9}$$

and one says to the other, ''$f(2)$ exists but $f(3)$ does not.'' Explain what they are talking about.

52. Referring to the function in Problem 51, do $f(-2)$ and $f(-3)$ exist? Explain.

The verbal statement "function f multiplies the square of the domain element by 3 and then subtracts 7 from the result" and the algebraic statement "$f(x) = 3x^2 - 7$" define the same function. In Problems 53–56, translate each verbal definition of a function into an algebraic definition.

53. Function g subtracts 5 from twice the cube of the domain element.

54. Function f multiplies the domain element by -3 and adds 4 to the result.

55. Function G multiplies the square root of the domain element by 2 and subtracts the square of the domain element from the result.

56. Function F multiplies the cube of the domain element by -8 and adds 3 times the square root of 3 to the result.

In Problems 57–60, translate each algebraic definition of the function into a verbal definition.

57. $f(x) = 2x - 3$

58. $g(x) = -2x + 7$

59. $F(x) = 3x^3 - 2\sqrt{x}$

60. $G(x) = 4\sqrt{x} - x^2$

Determine which of the equations in Problems 61–70 specify functions with independent variable x. For those that do, find the domain. For those that do not, find a value of x to which there corresponds more than one value of y.

61. $4x - 5y = 20$

62. $3y - 7x = 15$

63. $x^2 - y = 1$

64. $x - y^2 = 1$

65. $x + y^2 = 10$

66. $x^2 + y = 10$

67. $xy - 4y = 1$

68. $xy + y - x = 5$

69. $x^2 + y^2 = 25$

70. $x^2 - y^2 = 16$

71. If $F(t) = 4t + 7$, find:

$$\frac{F(3 + h) - F(3)}{h}$$

72. If $G(r) = 3 - 5r$, find:

$$\frac{G(2 + h) - G(2)}{h}$$

73. If $g(w) = w^2 - 4$, find:

$$\frac{g(1 + h) - g(1)}{h}$$

74. If $f(m) = 2m^2 + 5$, find:

$$\frac{f(4 + h) - f(4)}{h}$$

75. If $Q(x) = x^2 - 5x + 1$, find:

$$\frac{Q(2 + h) - Q(2)}{h}$$

76. If $P(x) = 2x^2 - 3x - 7$, find:

$$\frac{P(3 + h) - P(3)}{h}$$

C *In Problems 77–84, find and simplify:* $\dfrac{f(a + h) - f(a)}{h}$

77. $f(x) = 4x - 3$

78. $f(x) = -3x + 9$

79. $f(x) = 4x^2 - 7x + 6$

80. $f(x) = 3x^2 + 5x - 8$

81. $f(x) = x^3$

82. $f(x) = x^3 - x$

83. $f(x) = \sqrt{x}$

84. $f(x) = \dfrac{1}{x}$

Problems 85–88 refer to the area A and perimeter P of a rectangle with length l and width w (see the figure).

$$A = lw$$
$$P = 2l + 2w$$

85. The area of a rectangle is 25 square inches. Express the perimeter $P(w)$ as a function of the width w, and state the domain of this function.

86. The area of a rectangle is 81 square inches. Express the perimeter $P(l)$ as a function of the length l, and state the domain of this function.

87. The perimeter of a rectangle is 100 meters. Express the area $A(l)$ as a function of the length l, and state the domain of this function.

88. The perimeter of a rectangle is 160 meters. Express the area $A(w)$ as a function of the width w, and state the domain of this function.

APPLICATIONS

Business & Economics

89. *Price–demand.* A company manufactures memory chips for microcomputers. Its marketing research department, using statistical techniques, collected the data shown in Table 8, where p is the wholesale price per chip at which x million chips can be sold. Using special analytical techniques (regression analysis), an analyst produced the following price–demand function to model the data:

$$p(x) = 119 - 6x \qquad 1 \leq x \leq 15$$

TABLE 8
Price–Demand

x (million)	p ($)
1	115
6	80
10	65
15	31

Plot the data points in Table 8, and sketch a graph of the price–demand function in the same coordinate system. What would be the estimated price per chip for a demand of 8 million chips? For a demand of 11 million chips?

90. *Price–demand.* A company manufactures "Notebook" computers. Its marketing research department, using statistical techniques, collected the data shown in Table 9, where p is the wholesale price per computer at which x thousand computers can be sold. Using special analytical techniques (regression analysis), an analyst produced the following price–demand function to model the data:

$$p(x) = 1,190 - 36x \qquad 1 \leq x \leq 25$$

TABLE 9
Price–Demand

x (thousand)	p ($)
2	1,110
5	1,030
10	815
14	695
21	435

Plot the data points in Table 9, and sketch a graph of the price–demand function in the same coordinate system. What would be the estimated price per computer for a demand of 9 thousand computers? For a demand of 18 thousand computers?

91. *Revenue.*
(A) Using the price–demand function $p(x) = 119 - 6x$, $1 \leq x \leq 15$, from Problem 89, write the company's revenue function and indicate its domain.
(B) Complete Table 10, computing revenues to the nearest million dollars.

TABLE 10
Revenue

x (million)	R(x) (million $)
1	113
3	
6	
9	
12	
15	

(C) Plot the points from part (B) and sketch a graph of the revenue function through these points. Choose millions for the units on the horizontal and vertical axes.

92. *Revenue.*
(A) Using the price–demand function $p(x) = 1,190 - 36x$, $1 \leq x \leq 25$, from Problem 90, write the company's revenue function and indicate its domain.
(B) Complete Table 11, computing revenues to the nearest thousand dollars.

TABLE 11
Revenue

x (thousand)	R(x) (thousand $)
2	2,236
5	
10	
15	
20	
25	

(C) Plot the points from part (B) and sketch a graph of the revenue function through these points. Choose thousands for the units on the horizontal and vertical axes.

93. *Profit.* The financial department for the company in Problems 89 and 91 established the following cost function for producing and selling x million memory chips: $C(x) = 234 + 23x$ million dollars.
(A) Write a profit function for producing and selling x million memory chips, and indicate its domain.

(B) Complete Table 12, computing profits to the nearest million dollars.

TABLE 12
Profit

x (million)	P(x) (million $)
1	− 144
3	
6	
9	
12	
15	

(C) Plot the points in part (B) and sketch a graph of the profit function through these points.

94. *Profit.* The financial department for the company in Problems 90 and 92 established the following cost function for producing and selling x thousand "Notebook" computers: $C(x) = 4{,}320 + 146x$ thousand dollars.

(A) Write a profit function for producing and selling x thousand "Notebook" computers, and indicate the domain of this function.

(B) Complete Table 13, computing profits to the nearest thousand dollars.

TABLE 13
Profit

x (thousand)	P(x) (thousand $)
2	− 2,376
5	
10	
15	
20	
25	

(C) Plot the points in part (B) and sketch a graph of the profit function through these points.

95. *Packaging.* A candy box is to be made out of a piece of cardboard that measures 8 by 12 inches. Equal-sized squares x inches on a side will be cut out of each corner, and then the ends and sides will be folded up to form a rectangular box.

(A) Express the volume of the box V(x) in terms of x.

(B) What is the domain of the function V (determined by the physical restrictions)?

(C) Complete Table 14.

TABLE 14
Volume

x	V(x)
1	
2	
3	

(D) Plot the points in part (C) and sketch a graph of the volume function through these points.

96. *Packaging.* A parcel delivery service will only deliver packages with length plus girth (distance around) not exceeding 108 inches. A rectangular shipping box with square ends x inches on a side is to be used.

Length

x

Girth

x

(A) If the full 108 inches is to be used, express the volume of the box V(x) in terms of x.

(B) What is the domain of the function V (determined by the physical restrictions)?

(C) Complete Table 15.

TABLE 15
Volume

x	V(x)
5	
10	
15	
20	
25	

(D) Plot the points in part (C) and sketch a graph of the volume function through these points.

Life Sciences

97. *Muscle contraction.* In a study of the speed of muscle contraction in frogs under various loads, noted British biophysicist and Nobel prize winner A. W. Hill determined that the weight w (in grams) placed on the muscle and the speed of contraction v (in centimeters per second) are approximately related by an equation of the form

$$(w + a)(v + b) = c$$

where a, b, and c are constants. Suppose that for a certain muscle, $a = 15$, $b = 1$, and $c = 90$. Express v as a function of w. Find the speed of contraction if a weight of 16 grams is placed on the muscle.

Social Sciences

98. *Politics.* The percentage s of seats in the House of Representatives won by Democrats and the percentage v of votes cast for Democrats (when expressed as decimal fractions) are related by the equation

$$5v - 2s = 1.4 \qquad 0 < s < 1, \quad 0.28 < v < 0.68$$

(A) Express v as a function of s, and find the percentage of votes required for the Democrats to win 51% of the seats.

(B) Express s as a function of v, and find the percentage of seats won if Democrats receive 51% of the votes.

SECTION 1-3 ■ Elementary Functions: Graphs and Transformations

■ A BEGINNING LIBRARY OF ELEMENTARY FUNCTIONS
■ VERTICAL AND HORIZONTAL SHIFTS
■ REFLECTIONS, EXPANSIONS, AND CONTRACTIONS

The functions

$$g(x) = x^2 - 4 \qquad h(x) = (x - 4)^2 \qquad k(x) = -4x^2$$

all can be expressed in terms of the function $f(x) = x^2$ as follows:

$$g(x) = f(x) - 4 \qquad h(x) = f(x - 4) \qquad k(x) = -4f(x)$$

In this section we will see that the graphs of functions g, h, and k are closely related to the graph of function f. Insight gained by understanding these relationships will help us analyze and interpret the graphs of many different functions.

■ A BEGINNING LIBRARY OF ELEMENTARY FUNCTIONS

As you progress through this book, and most any other mathematics course beyond this one, you will repeatedly encounter a relatively small list of elementary functions. We will identify these functions, study their basic properties, and include them in a library of elementary functions (see the inside front cover). This library will become an important addition to your mathematical toolbox and can be used in any course or activity where mathematics is applied.

Figure 1 (page 36) shows six basic functions that you will encounter frequently. You should know the definition, domain, and range of each, and be able to recognize their graphs. For Figure 1B, recall the definition of *absolute value*:

$$|x| = \begin{cases} x & \text{if } x \geq 0 \\ -x & \text{if } x < 0 \end{cases}$$

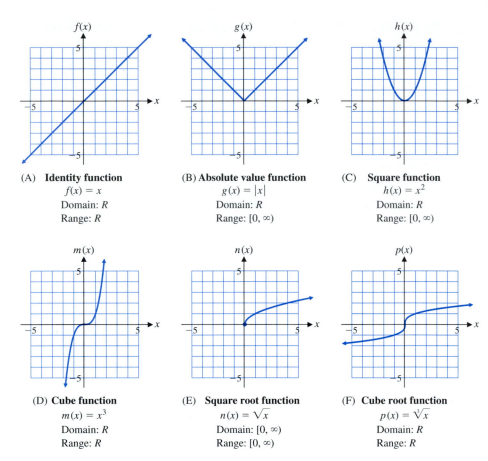

(A) **Identity function**
$$f(x) = x$$
Domain: R
Range: R

(B) **Absolute value function**
$$g(x) = |x|$$
Domain: R
Range: $[0, \infty)$

(C) **Square function**
$$h(x) = x^2$$
Domain: R
Range: $[0, \infty)$

(D) **Cube function**
$$m(x) = x^3$$
Domain: R
Range: R

(E) **Square root function**
$$n(x) = \sqrt{x}$$
Domain: $[0, \infty)$
Range: $[0, \infty)$

(F) **Cube root function**
$$p(x) = \sqrt[3]{x}$$
Domain: R
Range: R

FIGURE 1
Some basic functions and their graphs. *Note:* Letters used to designate these functions may vary from context to context; R is the set of all real numbers.

You should graph each basic function in Figure 1 on your graphing utility. Most graphing utilities allow you to define a number of functions, usually denoted by y_1, y_2, y_3, and so on. You can select certain functions for graphing and suppress others. Also, two or more functions can be graphed simultaneously (consult your manual). Many investigations in this section involve the graphing of two or more functions simultaneously.

■ VERTICAL AND HORIZONTAL SHIFTS

If a new function is formed by performing an operation on a given function, then the graph of the new function is called a **transformation** of the graph of the original function. For example, graphs of both $y = f(x) + k$ and $y = f(x + h)$ are transformations of the graph of $y = f(x)$.

explore – discuss 1

Let $f(x) = x^2$, and use a standard viewing window for the following graphs.

(A) Graph $y = f(x) + k$ for $k = -4$, 0, and 2 simultaneously in the same viewing window. Experiment with some additional values of k, both positive and negative. Describe the relationship between the graph of $y = f(x)$ and the graph of $y = f(x) + k$ for k any real number.

(B) Graph $y = f(x + h)$ for $h = -4$, 0, and 2 simultaneously in the same viewing window. Experiment with some additional values of h, both positive and negative. Describe the relationship between the graph of $y = f(x)$ and the graph of $y = f(x + h)$ for h any real number.

EXAMPLE 1 ➤ **Vertical and Horizontal Shifts**

(A) How are the graphs of $y = |x| + 4$ and $y = |x| - 5$ related to the graph of $y = |x|$? Confirm your answer by graphing all three functions simultaneously in the same viewing window.

(B) How are the graphs of $y = |x + 4|$ and $y = |x - 5|$ related to the graph of $y = |x|$? Confirm your answer by graphing all three functions simultaneously in the same viewing window.

Solution

(A) The graph of $y = |x| + 4$ is the same as the graph of $y = |x|$ shifted upward 4 units, and the graph of $y = |x| - 5$ is the same as the graph of $y = |x|$ shifted downward 5 units. Figure 2 confirms these conclusions. [It appears that the graph of $y = f(x) + k$ is the graph of $y = f(x)$ shifted up if k is positive and down if k is negative.]

(B) The graph of $y = |x + 4|$ is the same as the graph of $y = |x|$ shifted to the left 4 units, and the graph of $y = |x - 5|$ is the same as the graph of $y = |x|$ shifted to the right 5 units. Figure 3 confirms these conclusions. [It appears that the graph of $y = f(x + h)$ is the graph of $y = f(x)$ shifted right if h is negative and left if h is positive — the opposite of what you might expect.]

FIGURE 2
Vertical shifts

FIGURE 3
Horizontal shifts

MATCHED PROBLEM 1 ➤ (A) How are the graphs of $y = \sqrt{x} + 5$ and $y = \sqrt{x} - 4$ related to the graph of $y = \sqrt{x}$? Confirm your answer by graphing all three functions simultaneously in the same viewing window.

(B) How are the graphs of $y = \sqrt{x + 5}$ and $y = \sqrt{x - 4}$ related to the graph of $y = \sqrt{x}$? Confirm your answer by graphing all three functions simultaneously in the same viewing window. ◄

Comparing the graphs of $y = f(x) + k$ with the graph of $y = f(x)$, we see that the graph of $y = f(x) + k$ can be obtained from the graph of $y = f(x)$ by **vertically translating** (shifting) the graph of the latter upward k units if k is positive and downward $|k|$ units is k is negative. Comparing the graphs of $y = f(x + h)$ with the graph of $y = f(x)$, we see that the graph of $y = f(x + h)$ can be obtained from the graph of $y = f(x)$ by **horizontally translating** (shifting) the graph of $y = f(x)$ h units to the left if h is positive and $|h|$ units to the right if h is negative.

EXAMPLE 2 ➤ **Vertical and Horizontal Translations (Shifts)**   The graphs in Figure 4 are either horizontal or vertical shifts of the graph of $f(x) = x^2$. Write appropriate equations for functions H, G, M, and N in terms of f.

   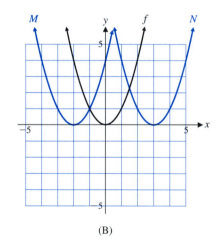

(A)   (B)

FIGURE 4
Vertical and horizontal shifts

Solution   Functions H and G are vertical shifts given by

$$H(x) = x^2 + 2 \qquad G(x) = x^2 - 4$$

Functions M and N are horizontal shifts given by

$$M(x) = (x + 2)^2 \qquad N(x) = (x - 3)^2$$ ◄

MATCHED PROBLEM 2 ➤ The graphs in Figure 5 are either horizontal or vertical shifts of the graph of $f(x) = \sqrt[3]{x}$. Write appropriate equations for functions H, G, M, and N in terms of f.

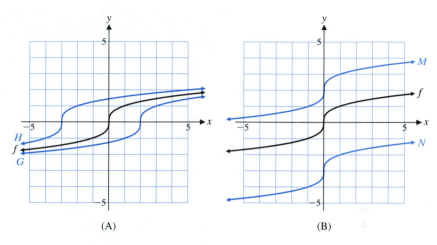

(A)   (B)

FIGURE 5
Vertical and horizontal shifts

■ REFLECTIONS, EXPANSIONS, AND CONTRACTIONS

We now investigate how the graph of $y = Af(x)$ is related to the graph of $y = f(x)$ for different real numbers A.

explore – discuss 2

Use a standard viewing window for the following graphs.

(A) Graph $y = Ax^2$ for $A = 1, 4$, and 0.25 simultaneously in the same viewing window.

(B) Graph $y = Ax^2$ for $A = -1, -4$, and -0.25 simultaneously in the same viewing window.

(C) Experiment with some additional values of A.

(D) Describe the relationship between the graph of $h(x) = x^2$ and the graph of $G(x) = Ax^2$ for A any nonzero real number.

Comparing $y = Af(x)$ to $y = f(x)$, we see that the graph of $y = Af(x)$ can be obtained from the graph of $y = f(x)$ by multiplying each ordinate value of the latter by A. The result is a **vertical expansion** of the graph of $y = f(x)$ if $A > 1$, a **vertical contraction** of the graph of $y = f(x)$ if $0 < A < 1$, and a **reflection in the x axis** if $A = -1$. If A is a negative number other than -1, then the result is a combination of a reflection in the x axis and either a vertical expansion or a vertical contraction.

EXAMPLE 3 ➤ Reflections, Expansions, and Contractions

(A) How are the graphs of $y = 2|x|$ and $y = 0.5|x|$ related to the graph of $y = |x|$? Confirm your answer by graphing all three functions simultaneously in the same viewing window.

(B) How is the graph of $y = -2|x|$ related to the graph of $y = |x|$? Confirm your answer by graphing both functions simultaneously in the same viewing window.

Solution   (A) The graph of $y = 2|x|$ is a vertical expansion of the graph of $y = |x|$ by a factor of 2, and the graph of $y = 0.5|x|$ is a vertical contraction of the graph of $y = |x|$ by a factor of 0.5. Figure 6 confirms this conclusion.

(B) The graph of $y = -2|x|$ is a reflection in the x axis and a vertical expansion of the graph of $y = |x|$. Figure 7 confirms this conclusion.

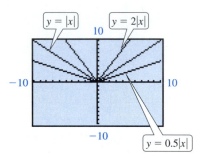

FIGURE 6
Vertical expansion and contraction

FIGURE 7
Reflection and vertical expansion

MATCHED PROBLEM 3   (A) How are the graphs of $y = 2x$ and $y = 0.5x$ related to the graph of $y = x$? Confirm your answer by graphing all three functions simultaneously in the same viewing window.

(B) How is the graph of $y = -0.5x$ related to the graph of $y = x$? Confirm your answer by graphing both functions in the same viewing window.

The various transformations considered above are summarized in the following box for easy reference:

Graph Transformations (Summary)

VERTICAL TRANSLATION:

$y = f(x) + k$ $\begin{cases} k > 0 & \text{Shift graph of } y = f(x) \text{ up } k \text{ units.} \\ k < 0 & \text{Shift graph of } y = f(x) \text{ down } |k| \text{ units.} \end{cases}$

HORIZONTAL TRANSLATION:

$y = f(x + h)$ $\begin{cases} h > 0 & \text{Shift graph of } y = f(x) \text{ left } h \text{ units.} \\ h < 0 & \text{Shift graph of } y = f(x) \text{ right } |h| \text{ units.} \end{cases}$

Graph Transformations (Summary) (Continued)

REFLECTION:

$y = -f(x)$ Reflect the graph of $y = f(x)$ in the x axis.

VERTICAL EXPANSION AND CONTRACTION:

$$y = Af(x) \begin{cases} A > 1 & \text{Vertically expand graph of } y = f(x) \\ & \text{by multiplying each ordinate value by } A. \\ 0 < A < 1 & \text{Vertically contract graph of } y = f(x) \\ & \text{by multiplying each ordinate value by } A. \end{cases}$$

---

explore – discuss 3

Use a graphing utility to explore the graph of $y = A(x + h)^2 + k$ for various values of the constants A, h, and k. Discuss how the graph of $y = A(x + h)^2 + k$ is related to the graph of $y = x^2$.

EXAMPLE 4 ➤ **Combining Graph Transformations** Discuss how the graph of $y = -|x - 3| + 1$ is related to the graph of $y = |x|$. Confirm your answer by graphing both functions simultaneously in the same viewing window.

Solution The graph of $y = -|x - 3| + 1$ is a reflection in the x axis, a horizontal translation of 3 units to the right, and a vertical translation of 1 unit upward of the graph of $y = |x|$. Figure 8 confirms this description.

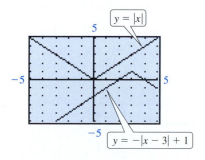

FIGURE 8
Combined transformations

◀

MATCHED PROBLEM 4 ➤ The graph of $y = G(x)$ in Figure 9 (page 42) involves a reflection and a translation of the graph of $y = x^3$. Describe how the graph of function G is related to the graph of $y = x^3$ and find an equation of the function G.

FIGURE 9
Combined transformations

Answers to Matched Problems   1. (A) The graph of $y = \sqrt{x} + 5$ is the same as the graph of $y = \sqrt{x}$ shifted upward 5 units, and the graph of $y = \sqrt{x} - 4$ is the same as the graph of $y = \sqrt{x}$ shifted downward 4 units. The figure confirms these conclusions.

(B) The graph of $y = \sqrt{x + 5}$ is the same as the graph of $y = \sqrt{x}$ shifted to the left 5 units, and the graph of $y = \sqrt{x - 4}$ is the same as the graph of $y = \sqrt{x}$ shifted to the right 4 units. The figure confirms these conclusions.

2. $H(x) = \sqrt[3]{x + 3}$, $G(x) = \sqrt[3]{x - 2}$, $M(x) = \sqrt[3]{x} + 2$, $N(x) = \sqrt[3]{x} - 3$

3. (A) The graph of $y = 2x$ is a vertical expansion of the graph of $y = x$, and the graph of $y = 0.5x$ is a vertical contraction of the graph of $y = x$. The figure at the top of the next page confirms these conclusions.

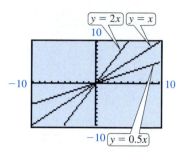

(B) The graph of $y = -0.5x$ is a vertical contraction and a reflection in the x axis of the graph of $y = x$. The figure confirms this conclusion.

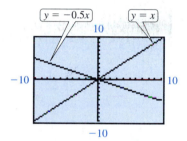

4. The graph of function G is a reflection in the x axis and a horizontal translation of 2 units to the left of the graph of $y = x^3$. An equation for G is $G(x) = -(x + 2)^3$.

EXERCISE 1-3

A *Without looking back in the text, indicate the domain and range of each of the following functions. (Making rough sketches on scratch paper may help.)*

1. $f(x) = 0.4x$
2. $g(x) = 3x$
3. $h(x) = -x^2$
4. $m(x) = -|x|$
5. $g(x) = -2\sqrt{x}$
6. $f(x) = -0.5\sqrt[3]{x}$
7. $F(x) = -0.1x^3$
8. $G(x) = 5x^3$

Graph each of the functions in Problems 9–20 using the graphs of functions f and g below:

9. $y = f(x) + 2$
10. $y = g(x) - 1$
11. $y = f(x + 2)$
12. $y = g(x - 1)$

13. $y = g(x - 3)$ **14.** $y = f(x + 3)$
15. $y = g(x) - 3$ **16.** $y = f(x) + 3$
17. $y = -f(x)$ **18.** $y = -g(x)$
19. $y = 0.5g(x)$ **20.** $y = 2f(x)$

B   *In Problems 21–28, indicate how the graph of each*
function is related to the graph of one of the six basic functions
in Figure 1 (at the beginning of this section). (Check your
description by graphing each function on a graphing utility.)

21. $g(x) = -|x + 3|$ **22.** $h(x) = -|x - 5|$
23. $f(x) = (x - 4)^2 - 3$ **24.** $m(x) = (x + 3)^2 + 4$
25. $f(x) = 7 - \sqrt{x}$ **26.** $g(x) = -6 + \sqrt[3]{x}$
27. $h(x) = -3|x|$ **28.** $m(x) = -0.4x^2$

Each graph in Problems 29–36 is the result of applying a
sequence of transformations to the graph of one of the six basic
functions in Figure 1 (at the beginning of this section). Identify
the basic function and describe the transformation verbally.
Write an equation for the given graph. (Check by graphing the
equation on a graphing utility.)

29.

30.

31.

32.

33.

34.

35.

36.

In Problems 37–42, the graph of the function g is formed by applying the indicated sequence of transformations to the given function f. Find an equation for the function g and graph g on a graphing utility using $-5 \leq x \leq 5$ *and* $-5 \leq y \leq 5.$

37. The graph of $f(x) = \sqrt{x}$ is shifted 2 units to the right and 3 units down.

38. The graph of $f(x) = \sqrt[3]{x}$ is shifted 3 units to the left and 2 units up.

39. The graph of $f(x) = |x|$ is reflected in the x axis and shifted to the left 3 units.

40. The graph of $f(x) = |x|$ is reflected in the x axis and shifted to the right 1 unit.

41. The graph of $f(x) = x^3$ is reflected in the x axis and shifted 2 units to the right and down 1 unit.

42. The graph of $f(x) = x^2$ is reflected in the x axis and shifted to the left 2 units and up 4 units.

C   *Each of the graphs in Problems 43–48 involves a reflection in the x axis and/or a vertical expansion or contraction of one of the basic functions in Figure 1 (at the beginning of this section). Identify the basic function, and describe the transformation verbally. Write an equation for the given graph. (Check by graphing the equation on a graphing utility.)*

43.

44.

45.

46.

47.

48.

APPLICATIONS

Business & Economics

49. *Price–demand.* A retail chain sells CD players. The retail price $p(x)$ (in dollars) and the weekly demand x for a particular model are related by

$$p(x) = 115 - 4\sqrt{x} \qquad 9 \leqslant x \leqslant 289$$

(A) Describe how the graph of function p can be obtained from the graph of one of the basic functions in Figure 1 (at the beginning of this section).

(B) Sketch a graph of function p using part (A) as an aid.

50. *Price–supply.* The manufacturers of the CD players in Problem 49 are willing to supply x players at a price of $p(x)$ as given by the equation

$$p(x) = 4\sqrt{x} \qquad 9 \leqslant x \leqslant 289$$

(A) Describe how the graph of function p can be obtained from the graph of one of the basic functions in Figure 1 (at the beginning of this section).

(B) Sketch a graph of function p using part (A) as an aid.

51. *Hospital costs.* Using statistical methods, the financial department of a hospital arrived at the cost equation

$$C(x) = 0.00048(x - 500)^3 + 60,000 \qquad 100 \leqslant x \leqslant 1,000$$

where $C(x)$ is the cost in dollars for handling x cases per month.

(A) Describe how the graph of function C can be obtained from the graph of one of the basic functions in Figure 1 (at the beginning of this section).

(B) Sketch a graph of function C using part (A) and a graphing utility as aids.

52. *Price–demand.* A company manufactures and sells roller-blade skates. Their financial department has established the price–demand function

$$p(x) = 190 - 0.013(x - 10)^2 \qquad 10 \leqslant x \leqslant 100$$

where $p(x)$ is the price at which x thousand skates can be sold.

(A) Describe how the graph of function p can be obtained from the graph of one of the basic functions in Figure 1 (at the beginning of this section).

(B) Sketch a graph of function p using part (A) and a graphing utility as aids.

Life Sciences

53. *Physiology.* A good approximation of the normal weight of a person 60 inches or taller but not taller than 80 inches is

given by $w(x) = 5.5x - 220$, where x is height in inches and $w(x)$ is weight in pounds.

(A) Describe how the graph of function w can be obtained from the graph of one of the basic functions in Figure 1 (at the beginning of this section).

(B) Sketch a graph of function w using part (A) as an aid.

54. *Physiology.* The average weight of a particular species of snake is given by $w(x) = 463x^3$, $0.2 \leqslant x \leqslant 0.8$, where x is length in meters and $w(x)$ is weight in grams.

(A) Describe how the graph of function w can be obtained from the graph of one of the basic functions in Figure 1 (at the beginning of this section).

(B) Sketch a graph of function w using part (A) as an aid.

Social Sciences

55. *Safety research.* Under ideal conditions, if a person driving a vehicle slams on the brakes and skids to a stop, the speed of the vehicle $v(x)$ (in miles per hour) is given approximately by $v(x) = C\sqrt{x}$ where x is the length of skid marks (in feet) and C is a constant that depends on the road conditions and the weight of the vehicle. For a particular vehicle, $v(x) = 7.08\sqrt{x}$ and $4 \leqslant x \leqslant 144$.

(A) Describe how the graph of function v can be obtained from the graph of one of the basic functions in Figure 1 (at the beginning of this section).

(B) Sketch a graph of function v using part (A) as an aid.

56. *Learning.* A production analyst has found that on the average it takes a new person $T(x)$ minutes to perform a particular assembly operation after x performances of the operation, where $T(x) = 10 - \sqrt[3]{x}$, $0 \leqslant x \leqslant 155$.

(A) Describe how the graph of function T can be obtained from the graph of one of the basic functions in Figure 1 (at the beginning of this section).

(B) Sketch a graph of function T using part (A) as an aid.

S E C T I O N   1-4   ■   ## Linear Functions and Straight Lines

- INTERCEPTS
- LINEAR FUNCTIONS, EQUATIONS, AND INEQUALITIES
- GRAPHS OF $Ax + By = C$
- SLOPE OF A LINE
- EQUATIONS OF LINES—SPECIAL FORMS
- APPLICATIONS

In this section we will add another important class of functions to our basic list of elementary functions. These functions are called *linear functions* and include the identity function $f(x) = x$ as a special case. We will investigate the relationship between linear functions and the solutions to linear equations and inequalities. Finally, we will review the concept of slope and some of the standard equations of straight lines. These new tools will be applied to a variety of significant applications, including cost and price – demand functions.

■ INTERCEPTS

Figure 1 illustrates the graphs of three functions f, g, and h.

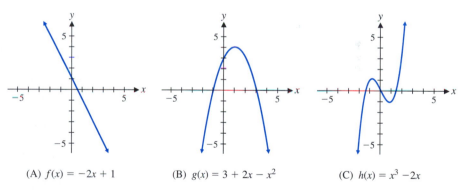

(A) $f(x) = -2x + 1$ (B) $g(x) = 3 + 2x - x^2$ (C) $h(x) = x^3 - 2x$

FIGURE 1
Graphs of several functions

If the graph of a function f crosses the x axis at a point with x coordinate a, then a is called an **x intercept** of f. If the graph of f crosses the y axis at a point with y coordinate b, then b is called the **y intercept.** It is common practice to refer to both the numbers a and b and the points $(a, 0)$ and $(0, b)$ as the x and y intercepts. If the y intercept exists, then 0 must be in the domain of f and the y intercept is simply $f(0)$. Thus, the graph of a function can have at most one y intercept. The x intercepts are all real solutions or roots of $f(x) = 0$, which may vary from none to an unlimited number. In Figure 1, function f has one y intercept and one x intercept; function g has one y intercept and two x intercepts; and function h has one y intercept and three x intercepts.

■ LINEAR FUNCTIONS, EQUATIONS, AND INEQUALITIES

In Figure 1, the graph of $f(x) = -2x + 1$ is a straight line, and because of this, we choose to call this type of function a *linear function*. In general:

Linear and Constant Functions

A function f is a **linear function** if

$$f(x) = mx + b \qquad m \neq 0$$

where m and b are real numbers. The **domain** is the set of all real numbers, and the **range** is the set of all real numbers. If $m = 0$, then f is called a **constant function,**

$$f(x) = b$$

which has the set of all real numbers as its **domain** and the constant b as its **range.**

Since $mx + b$, $m \neq 0$, is a first-degree polynomial, linear functions are also called **first-degree functions.** Figure 2 shows the graphs of two linear functions f and g, and a constant function h.

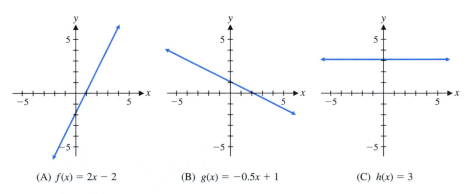

(A) $f(x) = 2x - 2$ (B) $g(x) = -0.5x + 1$ (C) $h(x) = 3$

FIGURE 2
Two linear functions and a constant function

In Section 1-3 we saw that the graph of the identity function $f(x) = x$ is a straight line. The graph of a linear function $g(x) = mx + b$, $m \neq 0$, is the same as the graph of $f(x) = x$ reflected in the x axis if m is negative, vertically expanded or contracted by a factor of $|m|$, and shifted up or down $|b|$ units. In general, it can be shown that:

The graph of a linear function is a straight line that is neither horizontal nor vertical.

The graph of a constant function is a horizontal straight line.

What about vertical lines? Recall from Section 1-2 that the graph of a function cannot contain two points with the same x coordinate and different y coordinates. Since *all* points on a vertical line have the same x coordinate, the graph of a function can never be a vertical line. Later in this section we will discuss equations of vertical lines, but these equations never define functions.

explore – discuss 1

(A) Is it possible for a linear function to have two x intercepts? No x intercepts? If either of your answers is yes, give an example.
(B) Is it possible for a linear function to have two y intercepts? No y intercept? If either of your answers is yes, give an example.
(C) Discuss the possible number of x and y intercepts for a constant function.

EXAMPLE 1 ➤ Intercepts, Equations, and Inequalities

(A) Graph $f(x) = \frac{3}{2}x - 4$ in a standard viewing window.
(B) Find the x and y intercepts (to two decimal places) using trace and zoom or an appropriate built-in routine in your graphing utility.
(C) Find the x and y intercepts algebraically (to two decimal places).
(D) Solve $\frac{3}{2}x - 4 \leq 0$ graphically (to two decimal places).

Solution (A) Graph:

(B) Finding intercepts graphically: Rather than using zoom and trace, as discussed in Section 1-1, we use a built-in routine to find the x intercept, or **root,** of this function (consult the graphing utility manual for this book or the manual for your graphing utility for details). To find the y intercept, we simply evaluate the function at $x = 0$. The graphs are shown at the top of the next page.

x intercept: 2.67

y intercept: −4

(C) Finding intercepts algebraically:

y intercept: $f(0) = \frac{3}{2}(0) - 4 = -4$

x intercept:

$$f(x) = 0$$
$$\tfrac{3}{2}x - 4 = 0$$
$$\tfrac{3}{2}x = 4$$
$$x = \tfrac{8}{3} \approx 2.67$$

(D) Solving $\frac{3}{2}x - 4 \leqslant 0$ graphically: The linear inequality $\frac{3}{2}x - 4 \leqslant 0$ holds for those values of x for which the graph of $f(x) = \frac{3}{2}x - 4$ [see the figure in part (A)] is at or below the x axis. This happens for x less than or equal to the x intercept found in both parts (B) and (C). Thus, the solution set for the linear inequality is $x \leqslant 2.67$ or $(-\infty, 2.67]$. ◄

MATCHED PROBLEM 1 ➤

(A) Graph $f(x) = -\frac{4}{3}x + 5$ in a standard viewing window.

(B) Find the x and y intercepts (to two decimal places) using trace and zoom or an appropriate built-in routine in your graphing utility.

(C) Find the x and y intercepts algebraically (to two decimal places).

(D) Solve $-\frac{4}{3}x + 5 \geqslant 0$ graphically (to two decimal places). ◄

■ GRAPHS OF $Ax + By = C$

We now investigate graphs of linear, or first-degree, equations in two variables:

$$Ax + By = C \tag{1}$$

where A and B are not both 0. Depending on the values of A and B, this equation defines a linear function, a constant function, or no function at all. If $A \neq 0$ and $B \neq 0$, then equation (1) can be written as

$$y = -\frac{A}{B}x + \frac{C}{B} \qquad \text{Linear function (slanted line)} \tag{2}$$

which is in the form $f(x) = mx + b$, $m \neq 0$, and hence, is a linear function. If $A = 0$ and $B \neq 0$, then equation (1) can be written as

$$0x + By = C$$

$$y = \frac{C}{B} \quad \textit{Constant function (horizontal line)} \qquad (3)$$

which is in the form $g(x) = b$, and hence, is a constant function. If $A \neq 0$ and $B = 0$, then equation (1) can be written as

$$Ax + 0y = C$$

$$x = \frac{C}{A} \quad \textit{Not a function (vertical line)} \qquad (4)$$

We can see that the graph of (4) is a vertical line, since the equation is satisfied for any value of y as long as x is the constant C/A. Hence, this form does not define a function.

The following theorem is a generalization of the above discussion:

t h e o r e m 1

Graph of a Linear Equation in Two Variables

The graph of any equation of the form

Standard Form $\quad Ax + By = C$ $\qquad\qquad\qquad$ (5)

where A, B, and C are real constants (A and B not both 0), is a straight line. Every straight line in a Cartesian coordinate system is the graph of an equation of this type.

Vertical and horizontal lines have particularly simple equations, which are special cases of equation (5):

Horizontal line with y intercept $C/B = b$: $y = b$
Vertical line with x intercept $C/A = a$: $x = a$

e x p l o r e – d i s c u s s 2

Graph the following three special cases of $Ax + By = C$ in the same coordinate system:

(A) $3x + 2y = 6$
(B) $0x - 3y = 12$
(C) $2x + 0y = 10$

Which cases define functions? Explain why, or why not.

Graph each case in the same viewing window using a graphing utility. (Check your manual on how to graph vertical lines.)

Sketching the graphs of equations of either form

$$Ax + By = C \qquad \text{or} \qquad y = mx + b$$

is very easy, since the graph of each equation is a straight line. All that is necessary is to plot any two points from the solution set and use a straightedge to draw a line through these two points. The x and y intercepts are usually the easiest to find.

EXAMPLE 2 ➤   Sketching Graphs of Lines

(A) Graph $x = -4$ and $y = 6$ simultaneously in the same rectangular coordinate system. Graph by hand and in a graphing utility.

(B) Write the equations of the vertical and horizontal lines that pass through the point $(7, -5)$.

(C) Graph the equation $2x - 3y = 12$ by hand and in a graphing utility.

Solution   (A) Graphing $x = -4$ and $y = 6$:

By hand

In a graphing utility

(B) Horizontal line through $(7, -5)$: $y = -5$
Vertical line through $(7, -5)$: $x = 7$

(C) Graphing $2x - 3y = 12$:   For the hand-drawn graph, find the intercepts by first letting $x = 0$ and solving for y and then letting $y = 0$ and solving for x. Then draw a line through the intercepts. To graph in a graphing utility, solve the equation for y in terms of x and enter the result.

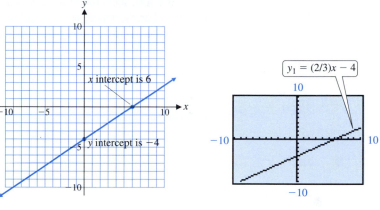

By hand   In a graphing utility

MATCHED PROBLEM 2 ➤
(A) Graph $x = 5$ and $y = -3$ simultaneously in the same rectangular coordinate system. Graph by hand and in a graphing utility.
(B) Write the equations of the vertical and horizontal lines that pass through the point $(-8, 2)$.
(C) Graph the equation $3x + 4y = 12$ by hand and in a graphing utility. ◄

■ SLOPE OF A LINE

If we take two points $P_1(x_1, y_1)$ and $P_2(x_2, y_2)$ on a line, then the ratio of the change in y to the change in x as the point moves from point P_1 to point P_2 is called the **slope** of the line. In a sense, slope provides a measure of the "steepness" of a line. The change in x is often called the **run** and the change in y the **rise.**

Slope of a Line

If a line passes through two distinct points $P_1(x_1, y_1)$ and $P_2(x_2, y_2)$, then its slope is given by the formula

$$m = \frac{y_2 - y_1}{x_2 - x_1} \qquad x_1 \neq x_2$$

$$= \frac{\text{Vertical change (Rise)}}{\text{Horizontal change (Run)}}$$

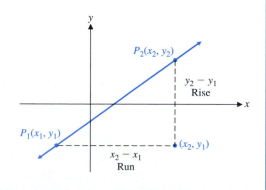

For a horizontal line, y does not change; hence, its slope is 0. For a vertical line, x does not change; hence, $x_1 = x_2$ and its slope is not defined. In general, the slope of a line may be positive, negative, 0, or not defined. Each case is illustrated geometrically in Table 1.

T A B L E 1
Geometric Interpretation of Slope

LINE	SLOPE	EXAMPLE
Rising as x moves from left to right	Positive	
Falling as x moves from left to right	Negative	
Horizontal	0	
Vertical	Not defined	

In using the formula to find the slope of the line through two points, it does not matter which point is labeled P_1 or P_2, since changing the labeling will change the sign in both the numerator and denominator of the slope formula, resulting in equivalent expressions. In addition, it is important to note that the definition of slope does not depend on the two points chosen on the line as long as they are distinct. This follows from the fact that the ratios of corresponding sides of similar triangles are equal.

EXAMPLE 3 ➤ **Finding Slopes** Sketch a line through each pair of points, and find the slope of each line.

(A) $(-3, -2), (3, 4)$ (B) $(-1, 3), (2, -3)$
(C) $(-2, -3), (3, -3)$ (D) $(-2, 4), (-2, -2)$

Solution

(A)

$$m = \frac{4 - (-2)}{3 - (-3)} = \frac{6}{6} = 1$$

(B)

$$m = \frac{-3 - 3}{2 - (-1)} = \frac{-6}{3} = -2$$

(C)

$$m = \frac{-3 - (-3)}{3 - (-2)} = \frac{0}{5} = 0$$

(D)

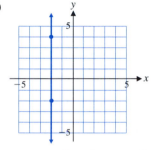

$$m = \frac{-2 - 4}{-2 - (-2)} = \frac{-6}{0}$$

Slope is not defined ◄

MATCHED PROBLEM 3 ► Find the slope of the line through each pair of points.

(A) $(-2, 4), (3, 4)$ (B) $(-2, 4), (0, -4)$

(C) $(-1, 5), (-1, -2)$ (D) $(-1, -2), (2, 1)$ ◄

■ EQUATIONS OF LINES—SPECIAL FORMS

Let us start by investigating why $y = mx + b$ is called the *slope–intercept form* for a line.

explore – discuss 3

(A) Using a graphing utility, graph $y = x + b$ for $b = -5, -3, 0, 3,$ and 5 simultaneously in a standard viewing window. Verbally describe the geometric significance of b.

(B) Using a graphing utility, graph $y = mx - 1$ for $m = -2, -1, 0, 1,$ and 2 simultaneously in a standard viewing window. Verbally describe the geometric significance of m.

As you can see from the above exploration, constants m and b in $y = mx + b$ have special geometric significance, which we now explicitly state.

If we let $x = 0$, then $y = b$, and we observe that the graph of $y = mx + b$ crosses the y axis at $(0, b)$. The constant b is the y *intercept*. For example, the y intercept of the graph of $y = -4x - 1$ is -1.

To determine the geometric significance of m, we proceed as follows: If $y = mx + b$, then by setting $x = 0$ and $x = 1$, we conclude that $(0, b)$ and $(1, m + b)$ lie on its graph (a line). Hence, the slope of this graph (line) is given by:

$$\text{Slope} = \frac{y_2 - y_1}{x_2 - x_1} = \frac{(m + b) - b}{1 - 0} = m$$

Thus, m is the slope of the line given by $y = mx + b$.

Slope–Intercept Form

The equation

$$y = mx + b \qquad m = \text{Slope}, \; b = y \text{ intercept} \tag{6}$$

is called the **slope–intercept form** of an equation of a line.

EXAMPLE 4 ➤   Using the Slope–Intercept Form

(A) Find the slope and y intercept, and graph $y = -\frac{2}{3}x - 3$.

(B) Write the equation of the line with slope $\frac{2}{3}$ and y intercept -2.

Solution

(A) Slope $= m = -\frac{2}{3}$ (B) $m = \frac{2}{3}$ and $b = -2$; thus, $y = \frac{2}{3}x - 2$

   y intercept $= b = -3$

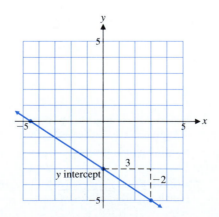

MATCHED PROBLEM 4 ▶ Write the equation of the line with slope $\frac{1}{2}$ and y intercept -1. Graph. ◀

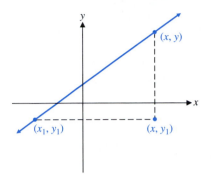

FIGURE 3

Suppose a line has slope m and passes through a fixed point (x_1, y_1). If the point (x, y) is any other point on the line (Fig. 3), then

$$\frac{y - y_1}{x - x_1} = m$$

that is,

$$y - y_1 = m(x - x_1) \tag{7}$$

We now observe that (x_1, y_1) also satisfies equation (7) and conclude that equation (7) is an equation of a line with slope m that passes through (x_1, y_1).

Point–Slope Form

An equation of a line with slope m that passes through (x_1, y_1) is

$$y - y_1 = m(x - x_1) \tag{7}$$

which is called the **point–slope form** of an equation of a line.

The point–slope form is extremely useful, since it enables us to find an equation for a line if we know its slope and the coordinates of a point on the line or if we know the coordinates of two points on the line.

EXAMPLE 5 ▶ Using the Point–Slope Form

(A) Find an equation for the line that has slope $\frac{1}{2}$ and passes through $(-4, 3)$. Write the final answer in the form $Ax + By = C$.

(B) Find an equation for the line that passes through the two points $(-3, 2)$ and $(-4, 5)$. Write the resulting equation in the form $y = mx + b$.

Solution (A) Use $y - y_1 = m(x - x_1)$. Let $m = \frac{1}{2}$ and $(x_1, y_1) = (-4, 3)$. Then

$$y - 3 = \tfrac{1}{2}[x - (-4)]$$
$$y - 3 = \tfrac{1}{2}(x + 4) \qquad \text{Multiply by 2.}$$
$$2y - 6 = x + 4$$
$$-x + 2y = 10 \quad \text{or} \quad x - 2y = -10$$

(B) First, find the slope of the line by using the slope formula:

$$m = \frac{y_2 - y_1}{x_2 - x_1} = \frac{5 - 2}{-4 - (-3)} = \frac{3}{-1} = -3$$

Now use $y - y_1 = m(x - x_1)$ with $m = -3$ and $(x_1, y_1) = (-3, 2)$:

$$y - 2 = -3[x - (-3)]$$
$$y - 2 = -3(x + 3)$$
$$y - 2 = -3x - 9$$
$$y = -3x - 7$$
◄

MATCHED PROBLEM 5 ➤   (A) Find an equation for the line that has slope $\frac{2}{3}$ and passes through $(6, -2)$. Write the resulting equation in the form $Ax + By = C, A > 0$.
(B) Find an equation for the line that passes through $(2, -3)$ and $(4, 3)$. Write the resulting equation in the form $y = mx + b$. ◄

The various forms of the equation of a line that we have discussed are summarized in Table 2 for convenient reference.

TABLE 2
Equations of a Line

Standard form	$Ax + By = C$	A and B not both 0
Slope–intercept form	$y = mx + b$	Slope: m; y intercept: b
Point–slope form	$y - y_1 = m(x - x_1)$	Slope: m; Point: (x_1, y_1)
Horizontal line	$y = b$	Slope: 0
Vertical line	$x = a$	Slope: Undefined

■ APPLICATIONS

We will now see how equations of lines occur in certain applications.

EXAMPLE 6 ➤   **Cost Equation**   The management of a company that manufactures roller skates has fixed costs (costs at 0 output) of $300 per day and total costs of $4,300 per day at an output of 100 pairs of skates per day. Assume that cost C is linearly related to output x.

(A) Find the slope of the line joining the points associated with outputs of 0 and 100; that is, the line passing through $(0, 300)$ and $(100, 4,300)$.
(B) Find an equation of the line relating output to cost. Write the final answer in the form $C = mx + b$.
(C) Graph the cost equation from part (B) for $0 \leq x \leq 200$.

Solution   (A) $m = \dfrac{y_2 - y_1}{x_2 - x_1} = \dfrac{4{,}300 - 300}{100 - 0} = \dfrac{4{,}000}{100} = 40$

(B) We must find an equation of the line that passes through $(0, 300)$ with slope 40. We use the slope–intercept form:

$$C = mx + b$$
$$C = 40x + 300$$

(C)

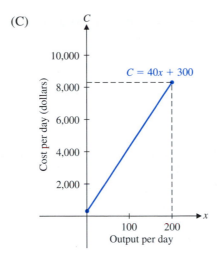

In Example 6, the **fixed cost** of $300 per day covers plant cost, insurance, and so on. This cost is incurred whether or not there is any production. The **variable cost** is $40x$, which depends on the day's output. Note that the slope 40 is the cost of producing one pair of skates; that is, the cost per unit output.

MATCHED PROBLEM 6 ➤ Answer parts (A) and (B) in Example 6 for fixed costs of $250 per day and total costs of $3,450 per day at an output of 80 pairs of skates per day.

➤ EXAMPLE 7 ➤ **Price–Demand**   A company manufactures and sells a specialty watch. The financial research department, using statistical and analytical methods, determined that at a price of $88 each, the demand would be 2 thousand watches, and at $38 each, 12 thousand watches. Assuming a linear relationship between price and demand, find a linear function that models the price–demand relationship in the form $p(x) = mx + b$. What would be the price at a demand of 8 thousand watches? 15 thousand watches?

Solution   Find the equation of the line that passes through $(2, 88)$ and $(12, 38)$. We first find the slope of the line:

$$m = \frac{38 - 88}{12 - 2} = \frac{-50}{10} = -5$$

Use the point–slope form to find the equation of the line:

$$y - y_1 = m(x - x_1)$$
$$y - 88 = -5(x - 2)$$
$$y - 88 = -5x + 10$$
$$y = -5x + 98$$

or

$$p(x) = -5x + 98 \qquad \text{Price–demand equation}$$
$$p(8) = -5(8) + 98 = \$58$$
$$p(15) = -5(15) + 98 = \$23$$

Thus, the price is $58 when the demand is 8,000 and $23 when the demand is 15,000. ◄

MATCHED PROBLEM 7 ➤ The company in Example 7 also manufactures and sells a watch designed for sailboat racing. The financial analyst found that the company could sell 3 hundred watches at a wholesale price of $140 each, and 11 hundred watches at a wholesale price of $92 each. Assuming a linear relationship between price and demand, find a linear function that models the price–demand relationship in the form $p(x) = mx + b$. What would be the price at a demand of 7 hundred watches? 12 hundred watches? ◄

Answers to Matched Problems

1. (A)

(B) x intercept: 3.75; y intercept: 5
(C) x intercept: 3.75; y intercept: 5
(D) $x \leqslant 3.75$ or $(-\infty, 3.75]$

2. (A)

By hand   In a graphing utility

(B) Horizontal line: $y = 2$; Vertical line: $x = -8$

(C)

 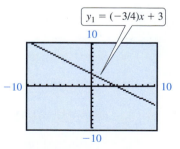

By hand   In a graphing utility

3. (A) 0   (B) -4 (C) Not defined   (D) 1

4. $y = \frac{1}{2}x - 1$

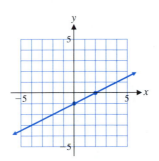

5. (A) $2x - 3y = 18$ (B) $y = 3x - 9$
6. (A) $m = 40$ (B) $C = 40x + 250$
7. $p(x) = -6x + 158$; $p(7) = \$116$; $p(12) = \$86$

EXERCISE 1-4

A   *Problems 1–4 refer to graphs (A)–(D).*

(A)

(B)

(C)

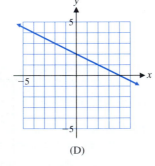

(D)

1. Identify the graph(s) of linear functions with a negative slope.

2. Identify the graph(s) of linear functions with a positive slope.

3. Identify the graph(s) of any constant functions. What is the slope of the graph?

4. Identify any graphs that are not the graphs of functions. What can you say about their slopes?

Sketch a graph of each equation in a rectangular coordinate system. Check by graphing each equation in a graphing utility.

5. $y = 2x - 3$ **6.** $y = \frac{x}{2} + 1$

7. $2x + 3y = 12$ **8.** $8x - 3y = 24$

Find the slope and y intercept of the graph of each equation.

9. $y = 2x - 3$ **10.** $y = \frac{x}{2} + 1$

11. $y = -\frac{2}{3}x + 2$ **12.** $y = \frac{3}{4}x - 2$

Write an equation of the line with the indicated slope and y intercept.

13. Slope $= -2$
     y intercept $= 4$
14. Slope $= -\frac{2}{3}$
     y intercept $= -2$
15. Slope $= -\frac{3}{5}$
     y intercept $= 3$
16. Slope $= 1$
     y intercept $= -2$

B   *Sketch a graph of each equation or pair of equations in a rectangular coordinate system. Check by graphing each equation in a graphing utility.*

17. $y = -\frac{2}{3}x - 2$ **18.** $y = -\frac{3}{2}x + 1$
19. $3x - 2y = 10$ **20.** $5x - 6y = 15$
21. $x = 3$; $y = -2$ **22.** $x = -3$; $y = 2$

In Problems 23–26, find the slope of the graph of each equation. (First write the equation in the form $y = mx + b$.)

23. $3x + y = 5$ **24.** $2x - y = -3$
25. $2x + 3y = 12$ **26.** $3x - 2y = 10$

27. (A) Graph $f(x) = 1.2x - 4.2$ in a standard viewing window.
　　(B) Find the x and y intercepts (to one decimal place) using trace and zoom or an appropriate built-in routine in your graphing utility.
　　(C) Find the x and y intercepts algebraically (to one decimal place).
　　(D) Using the results of parts (A)–(C), find the solution set for the linear inequality

$$1.2x - 4.2 > 0$$

28. (A) Graph $f(x) = -0.8x + 5.2$ in a standard viewing window.
　　(B) Find the x and y intercepts (to one decimal place) using trace and zoom or an appropriate built-in routine in your graphing utility.
　　(C) Find the x and y intercepts algebraically (to one decimal place).
　　(D) Using the results of parts (A)–(C), find the solution set for the linear inequality

$$-0.8x + 5.2 < 0$$

Write the equations of the vertical and horizontal lines through each point.

29. $(3, -5)$ **30.** $(-2, 7)$
31. $(-1, -3)$ **32.** $(96, -4)$

Write the equation of the line through each indicated point with the indicated slope. Write the final answer in the form $y = mx + b$.

33. $m = -3; (4, -1)$ **34.** $m = -2; (-3, 2)$
35. $m = \frac{2}{3}; (-6, -5)$ **36.** $m = \frac{1}{2}; (-4, 3)$
37. $m = 0; (3, -5)$ **38.** $m = 0; (-4, 7)$

Find the slope of the line that passes through the given points.

39. $(1, 3)$ and $(7, 5)$
40. $(2, 1)$ and $(10, 5)$
41. $(-5, -2)$ and $(5, -4)$
42. $(3, 7)$ and $(-6, 4)$
43. $(2, 7)$ and $(2, -3)$
44. $(-2, 3)$ and $(-2, -1)$
45. $(2, 3)$ and $(-5, 3)$
46. $(-3, -3)$ and $(0, -3)$

Write an equation of the line through each indicated pair of points. Write the final answer in the form $Ax + By = C$.

47. $(1, 3)$ and $(7, 5)$
48. $(2, 1)$ and $(10, 5)$
49. $(-5, -2)$ and $(5, -4)$
50. $(3, 7)$ and $(-6, 4)$
51. $(2, 7)$ and $(2, -3)$
52. $(-2, 3)$ and $(-2, -1)$
53. $(2, 3)$ and $(-5, 3)$
54. $(-3, -3)$ and $(0, -3)$

In Problems 55–60, graph the equations obtained from Problems 49–54 and indicate which define a linear function, a constant function, or no function at all.

55. Problem 49   **56.** Problem 50
57. Problem 51   **58.** Problem 52
59. Problem 53   **60.** Problem 54

C

61. (A) Graph the following equations simultaneously in a squared viewing window:

$$3x + 2y = 6 \qquad 3x + 2y = 3$$
$$3x + 2y = -6 \qquad 3x + 2y = -3$$

　　(B) From your observations in part (A), describe the family of lines obtained by varying C in $Ax + By = C$ while holding A and B fixed.

62. (A) Graph the following two equations simultaneously in a squared viewing window.

$$3x + 4y = 12 \qquad 4x - 3y = 12$$

　　(B) Graph the following two equations simultaneously in a squared viewing window.

$$2x + 3y = 12 \qquad 3x - 2y = 12$$

　　(C) From your observations in parts (A) and (B), describe the apparent relationship of the graphs of

$$Ax + By = C \qquad \text{and} \qquad Bx - Ay = C$$

▶ APPLICATIONS

Business & Economics

63. *Simple interest.* If $P (the principal) is invested at an interest rate of r, then the amount A that is due after t years is given by

$$A = Prt + P$$

If $100 is invested at 6% ($r = 0.06$), then $A = 6t + 100$, $t \geqslant 0$.
(A) What will $100 amount to after 5 years? After 20 years?
(B) Sketch a graph of $A = 6t + 100$ for $0 \leqslant t \leqslant 20$.
(C) What is the slope of the graph? (The slope indicates the increase in amount A for each additional year of investment.)

64. *Simple interest.* Use the simple interest formula from Problem 63. If $1,000 is invested at 7.5% ($r = 0.075$), then $A = 75t + 1,000$, $t \geqslant 0$.
(A) What will $1,000 amount to after 5 years? After 20 years?
(B) Sketch a graph of $A = 75t + 1,000$ for $0 \leqslant t \leqslant 1,000$.
(C) What is the slope of the graph? (The slope indicates the increase in amount A for each additional year of investment.)

65. *Cost function.* The management of a company that manufactures surfboards has fixed costs (at 0 output) of $200 per day and total costs of $3,800 per day at a daily output of 20 boards.
(A) Assuming the total cost per day, $C(x)$, is linearly related to the total output per day, x, write an equation for the cost function.
(B) What are the total costs for an output of 12 boards per day?
(C) Graph the cost function for $0 \leqslant x \leqslant 20$.

66. *Cost function.* Repeat Problem 65 if the fixed cost is $300 per day and the total cost per day at an output of 20 boards is $5,100.

67. *Price–demand function.* A manufacturing company is interested in introducing a new power mower. Its market research department gave the management the price–demand forecast listed in Table 3.

TABLE 3
Price–Demand

DEMAND	WHOLESALE PRICE ($)
x	$p(x)$
0	200
2,400	160
4,800	120
7,800	70

(A) Plot these points, letting $p(x)$ represent the price at which x number of mowers can be sold (demand). Label the horizontal axis x.
(B) Note that the points in part (A) lie along a straight line. Find an equation for the price–demand function.
(C) What would be the price for a 3,000 unit demand?

[*Note:* The slope of the line found in part (B) indicates the change in price per unit change in demand.]

68. *Depreciation.* Office equipment was purchased for $20,000 and is assumed to have a scrap value of $2,000 after 10 years. If its value is depreciated linearly (for tax purposes) from $20,000 to $2,000:
(A) Find the linear equation that relates value (V) in dollars to time (t) in years.
(B) What would be the value of the equipment after 6 years?
(C) Graph the equation for $0 \leqslant t \leqslant 10$.

[*Note:* The slope found in part (A) indicates the decrease in value per year.]

Life Sciences

69. *Nutrition.* In a nutrition experiment, a biologist wants to prepare a special diet for the experimental animals. Two food mixes, A and B, are available. If mix A contains 20% protein and mix B contains 10% protein, what combination of each mix will provide exactly 20 grams of protein? Let x be the amount of A used and let y be the amount of B used.

Then write a linear equation relating x, y and 20. Graph this equation for $x \geq 0$ and $y \geq 0$.

70. *Ecology.* As one descends into the ocean, pressure increases linearly. The pressure is 15 pounds per square inch on the surface and 30 pounds per square inch 33 feet below the surface.

(A) If p is the pressure in pounds and d is the depth below the surface in feet, write an equation that expresses p in terms of d. [*Hint:*   Find an equation of the line that passes through (0, 15) and (33, 30).]

(B) What is the pressure at 12,540 feet (the average depth of the ocean)?

(C) Graph the equation for $0 \leq d \leq 12{,}540$.

[*Note:*   The slope found in part (A) indicates the change in pressure for each additional foot of depth.]

Social Sciences

71. *Psychology.* In an experiment on motivation, J. S. Brown trained a group of rats to run down a narrow passage in a cage to obtain food in a goal box. Using a harness, he then connected the rats to an overhead wire that was attached to a spring scale. A rat was placed at different distances d (in centimeters) from the goal box, and the pull p (in grams) of the rat toward the food was measured. Brown found that the relationship between these two variables was very close to being linear and could be approximated by the equation

$$p = -\tfrac{1}{5}d + 70 \qquad 30 \leq d \leq 175$$

(See J. S. Brown, *Journal of Comparative and Physiological Psychology,* 1948, 41:450–465.)

(A) What was the pull when $d = 30$? When $d = 175$?

(B) Graph the equation.

(C) What is the slope of the line?

S E C T I O N   1-5   ■   **Quadratic Functions**

■ QUADRATIC FUNCTIONS, EQUATIONS, AND INEQUALITIES
■ PROPERTIES OF QUADRATIC FUNCTIONS AND THEIR GRAPHS
■ APPLICATIONS

If the degree of a linear function is increased by one, we obtain a *second-degree function,* usually called a *quadratic function,* another basic function that we will need in our library of elementary functions. We will investigate relationships between quadratic functions and the solutions to quadratic equations and inequalities. Other important properties of quadratic functions will also be investigated, including maximum and minimum properties. We will then be in a position to solve important practical problems such as finding production levels that will produce maximum revenue or maximum profit.

■ **QUADRATIC FUNCTIONS, EQUATIONS, AND INEQUALITIES**

A quadratic function is defined as follows:

> ### Quadratic Function
>
> A function f is a **quadratic function** if
>
> $$f(x) = ax^2 + bx + c \qquad a \neq 0$$
>
> where a, b, and c are real numbers. The domain of a quadratic function is the set of all real numbers.

The graphs of three quadratic functions are shown in Figure 1.

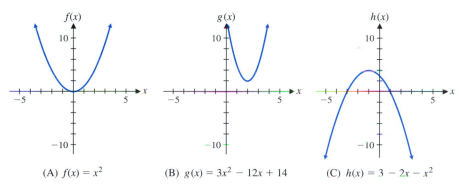

(A) $f(x) = x^2$ (B) $g(x) = 3x^2 - 12x + 14$ (C) $h(x) = 3 - 2x - x^2$

FIGURE 1
Graphs of quadatric functions

The graph of a quadratic function is called a **parabola.** We will discuss this in more detail later in this section.

EXAMPLE 1 ➤   Intercepts, Equations, and Inequalities

(A) Graph $f(x) = -x^2 + 5x + 3$ in a standard viewing window.
(B) Find the x and y intercepts (to two decimal places) using trace and zoom or an appropriate built-in routine in your graphing utility.
(C) Find the x and y intercepts algebraically (to two decimal places).
(D) Solve the quadratic inequality $-x^2 + 5x + 3 \geq 0$ graphically (to two decimal places).

Solution   (A) Graphing in a graphing utility:

(B) Finding intercepts graphically using a graphing utility:

| x intercept: -0.54 | x intercept: 5.54 | y intercept: 3 |

(C) Finding intercepts algebraically:

y intercept: $f(0) = -(0)^2 + 5(0) + 3 = 3$

The **quadratic formula** can be used to find the x intercepts of a quadratic function:

$$f(x) = 0 \qquad \text{Quadratic equation: } ax^2 + bx + c = 0$$

$$-x^2 + 5x + 3 = 0$$

$$x = \frac{-5 \pm \sqrt{5^2 - 4(-1)(3)}}{2(-1)} \qquad \text{Quadratic formula: } x = \frac{-b \pm \sqrt{b^2 - 4ac}}{2a}$$

$$= \frac{-5 \pm \sqrt{37}}{-2} = -0.54 \quad \text{or} \quad 5.54 \quad \text{x intercepts}$$

(D) Solving $-x^2 + 5x + 3 \geqslant 0$ graphically: The quadratic inequality

$$-x^2 + 5x + 3 \geqslant 0$$

holds for those values of x for which the graph of $f(x) = -x^2 + 5x + 3$ in the figure in part (A) is at or above the x axis. This happens for x between the two x intercepts [found in parts (B) and (C)], including the two x intercepts. Thus, the solution set for the quadratic inequality is $-0.54 \leqslant x \leqslant 5.54$ or $[-0.54, 5.54]$. ◄

MATCHED PROBLEM 1 ➤ (A) Graph $g(x) = 2x^2 - 5x - 5$ in a standard viewing window.
(B) Find the x and y intercepts (to two decimal places) using trace and zoom or an appropriate built-in routine in your graphing utility.
(C) Find x and y intercepts algebraically (to two decimal places).
(D) Solve $2x^2 - 5x - 5 \geqslant 0$ graphically (to two decimal places). ◄

explore – discuss 1

How many x intercepts can the graph of a quadratic function have? How many y intercepts? Explain your reasoning.

■ PROPERTIES OF QUADRATIC FUNCTIONS AND THEIR GRAPHS

Many useful properties of the quadratic function can be uncovered by transforming

$$f(x) = ax^2 + bx + c \qquad a \neq 0 \tag{1}$$

into the form

$$f(x) = a(x - h)^2 + k \tag{2}$$

The process of *completing the square* is central to the transformation. We illustrate the process through a specific example and then generalize the results.

Consider the quadratic function given by

$$f(x) = -2x^2 + 16x - 24 \tag{3}$$

We start by transforming equation (3) into the form (2) by completing the square:

$$
\begin{aligned}
f(x) &= -2x^2 + 16x - 24 && \text{Factor the coefficient of } x^2 \text{ out of the} \\
&= -2(x^2 - 8x) - 24 && \text{first two terms.} \\
&= -2(x^2 - 8x + \,?) - 24 && \text{Add 16 to complete the square inside} \\
& && \text{the parentheses. Because of the } -2 \\
& && \text{outside the parentheses, we have} \\
& && \text{actually added } -32, \text{ so we must add} \\
&= -2(x^2 - 8x + 16) - 24 + 32 && \text{32 to the outside.} \\
&= -2(x - 4)^2 + 8 && \text{The transformation is complete and can} \\
& && \text{be checked by multiplying out.}
\end{aligned}
$$

Thus,

$$f(x) = -2(x - 4)^2 + 8 \tag{4}$$

If $x = 4$, then $-2(x - 4)^2 = 0$ and $f(4) = 8$. For any other value of x, the negative number $-2(x - 4)^2$ is added to 8, making it smaller. (Think about this.) Therefore,

$$f(4) = 8$$

is the *maximum value* of $f(x)$ for all x—a very important result! Furthermore, if we choose any two x values that are the same distance from 4, we will obtain the same function value. For example, $x = 3$ and $x = 5$ are each one unit from $x = 4$ and their function values are

$$
\begin{aligned}
f(3) &= -2(3 - 4)^2 + 8 = 6 \\
f(5) &= -2(5 - 4)^2 + 8 = 6
\end{aligned}
$$

Thus, the vertical line $x = 4$ is a line of symmetry. That is, if the graph of equation (3) is drawn on a piece of paper and the paper is folded along the line $x = 4$, then the two sides of the parabola will match exactly. All these results are illustrated by graphing equations (3) and (4) and the line $x = 4$ simultaneously in the same coordinate system (Figure 2).

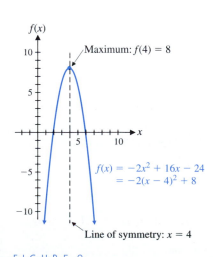

FIGURE 2
Graph of a quadratic function

From the above discussion, we see that as x moves from left to right, $f(x)$ is increasing on $(-\infty, 4]$ and decreasing on $[4, \infty)$, and that $f(x)$ can assume no value greater than 8. Thus,

Range of f: $y \leq 8$ or $(-\infty, 8]$

In general, the graph of a quadratic function is a parabola with line of symmetry parallel to the vertical axis. The lowest or highest point on the parabola, whichever exists, is called the **vertex.** The maximum or minimum value of a quadratic function always occurs at the vertex of the parabola. The line of symmetry through the vertex is called the **axis** of the parabola. In the above example, $x = 4$ is the axis of the parabola and $(4, 8)$ is its vertex.

Applying the graph transformation properties discussed in Section 1-3 to the transformed equation,

$$f(x) = -2x^2 + 16x - 24$$
$$= -2(x - 4)^2 + 8$$

we see that the graph of $f(x) = -2x^2 + 16x - 24$ is the graph of $g(x) = x^2$ vertically expanded by a factor of 2, reflected in the x axis, and shifted to the right 4 units and up 8 units, as shown in Figure 3.

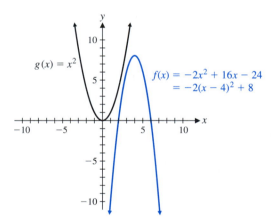

FIGURE 3
Graph of f is the graph of g transformed

Note the important results we have obtained by transforming equation (3) into equation (4):

- The vertex of the parabola
- The axis of the parabola
- The maximum value of $f(x)$
- The range of the function f
- The relationship between the graph of $g(x) = x^2$ and the graph of $f(x) = -2x^2 + 16x - 24$

Now, let us explore the effects of changing the constants a, h, and k on the graph of $f(x) = a(x - h)^2 + k$.

(A) Let $a = 1$ and $h = 5$. Graph $f(x) = a(x - h)^2 + k$ for $k = -4$, 0, and 3 simultaneously in a standard viewing window. Explain the effect of changing k on the graph of f.

(B) Let $a = 1$ and $k = 2$. Graph $f(x) = a(x - h)^2 + k$ for $h = -4$, 0, 5 simultaneously in a standard viewing window. Explain the effect of changing h on the graph of f.

(C) Let $h = 5$ and $k = -2$. Graph $f(x) = a(x - h)^2 + k$ for $a = 0.25$, 1, and 3 simultaneously in a standard viewing window. Graph function f for $a = 1$, -1, and -0.25 simultaneously in a standard viewing window. Explain the effect of changing a on the graph of f.

The above discussion is generalized for all quadratic functions in the following box:

Properties of a Quadratic Function and Its Graph

Given a quadratic function

$$f(x) = ax^2 + bx + c \qquad a \neq 0$$

and the form obtained by completing the square,

$$f(x) = a(x - h)^2 + k$$

we summarize general properties as follows:

1. The graph of f is a parabola:

 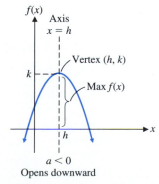

$a > 0$
Opens upward

$a < 0$
Opens downward

> **Properties of a Quadratic Function and Its Graph *(Continued)***
>
> ───────────────────────────────────────
>
> 2. Vertex: (h, k) (Parabola increases on one side of the vertex and decreases on the other)
> 3. Axis (of symmetry): $x = h$ (Parallel to y axis)
> 4. $f(h) = k$ is the minimum if $a > 0$ and the maximum if $a < 0$
> 5. Domain: All real numbers
>   Range: $(-\infty, k]$ if $a < 0$ or $[k, \infty)$ if $a > 0$
> 6. The graph of f is the graph of $g(x) = ax^2$ translated horizontally h units and vertically k units.

EXAMPLE 2 ➤ Analyzing a Quadratic Function   Given the quadratic function

$$f(x) = 0.5x^2 - 6x + 21$$

(A) Graph function f using a suitable viewing window.
(B) Find the vertex and the maximum or minimum (to the nearest integer) graphically using trace and zoom or an appropriate built-in routine. State the range of f.
(C) Find the vertex and the maximum or minimum (to the nearest integer) algebraically by completing the square. State the range of f.
(D) Referring to the completed square form in part (C), describe how the graph of function f can be obtained from the graph of $g(x) = x^2$ using transformations discussed in Section 1-3.

Solution   (A) Graph in a graphing utility:

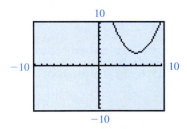

(B) Finding the vertex, minimum, and range graphically using a graphing utility:

Vertex: (6, 3); minimum: $f(6) = 3$; range: $y \geq 3$ or $[3, \infty)$.

(C) Finding the vertex, minimum, and range algebraically: Complete the square:

$$f(x) = 0.5x^2 - 6x + 21$$
$$= 0.5(x^2 - 12x + \textbf{?}) + 21$$
$$= 0.5(x^2 - 12x + 36) + 21 - 18$$
$$= 0.5(x - 6)^2 + 3$$

From the last form, we see that $h = 6$ and $k = 3$. Thus, vertex: $(6, 3)$; minimum: $f(6) = 3$; range: $y \geqslant 3$ or $[3, \infty)$.

(D) The graph of $f(x) = 0.5(x - 6)^2 + 3$ is the same as the graph of $g(x) = x^2$ vertically contracted by a factor of 0.5, and shifted to the right 6 units and up 3 units.   ◀

MATCHED PROBLEM 2 ➤ Given the quadratic function

$$f(x) = -0.25x^2 - 2x + 2$$

(A) Graph function f using a suitable viewing window.
(B) Find the vertex and the maximum or minimum (to the nearest integer) graphically using trace and zoom or an appropriate built-in routine. State the range of f.
(C) Find the vertex and the maximum or minimum (to the nearest integer) algebraically by completing the square. State the range of f.
(D) Referring to the completed square form in part (C), describe how the graph of function f can be obtained from the graph of $g(x) = x^2$ using transformations discussed in Section 1-3.   ◀

■ APPLICATIONS

EXAMPLE 3 ➤ **Maximum Revenue**   This is a continuation of Example 7 in Section 1-2. Recall that the financial department in the company that produces a point-and-shoot camera arrived at the following price–demand function and the corresponding revenue function:

$$p(x) = 94.8 - 5x \qquad \text{Price–demand function}$$
$$R(x) = xp(x) = x(94.8 - 5x) \quad \text{Revenue function}$$

where $p(x)$ is the wholesale price per camera at which x million cameras can be sold and $R(x)$ is the corresponding revenue (in million dollars). Both functions have domain $1 \leqslant x \leqslant 15$.

(A) Graph the revenue function using an appropriate viewing window.
(B) Find the output to the nearest thousand cameras that will produce the maximum revenue. What is the maximum revenue to the nearest thousand dollars? Solve the problem graphically using trace and zoom or an appropriate built-in routine.
(C) Solve part (B) algebraically by completing the square.
(D) What is the wholesale price per camera (to the nearest dollar) that produces the maximum revenue?

Solution   (A) Graph in a graphing utility:

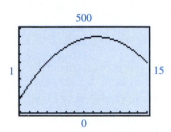

(B) Graphical solution using a graphing utility:

An output of 9.480 million cameras (9,480,000 cameras) will produce a maximum revenue of 449.352 million dollars ($449,352,000).

(C) Algebraic solution:

$$
\begin{aligned}
R(x) &= x(94.8 - 5x) \\
&= -5x^2 + 94.8x \\
&= -5(x^2 - 18.96x + \textbf{?}) \\
&= -5(x^2 - 18.96x + 89.8704) + 449.352 \\
&= -5(x - 9.48)^2 + 449.352
\end{aligned}
$$

The maximum revenue of 449.352 million dollars ($449,352,000) occurs when $x = 9.480$ million cameras (9,480,000 cameras).

(D) Finding the wholesale price per camera: Use the price–demand function for an output of 9.480 million cameras:

$$
\begin{aligned}
p(x) &= 94.8 - 5x \\
p(9.480) &= 94.8 - 5(9.480) \\
&= \$47
\end{aligned}
$$

◀

MATCHED PROBLEM 3 ➤ The financial department in Example 3, using statistical and analytical techniques (see Matched Problem 7 in Section 1-2), arrived at the cost function

$$C(x) = 156 + 19.7x \quad \textit{Cost function}$$

where $C(x)$ is the cost (in million dollars) for manufacturing and selling x million cameras.

(A) Using the revenue function from Example 3 and the cost function above, write an equation for the profit function.
(B) Graph the profit function using an appropriate viewing window.
(C) Find the output to the nearest thousand cameras that will produce the maximum profit. What is the maximum profit to the nearest thousand dollars? Solve the problem graphically using trace and zoom or an appropriate built-in routine.
(D) Solve part (C) algebraically by completing the square.
(E) What is the wholesale price per camera (to the nearest dollar) that produces the maximum profit?   ◀

EXAMPLE 4 ➤ **Break-Even Analysis**   Use the revenue function from Example 3 and the cost function from Matched Problem 3:

$$R(x) = x(94.8 - 5x) \quad \textit{Revenue function}$$
$$C(x) = 156 + 19.7x \quad \textit{Cost function}$$

Both have domain $1 \leqslant x \leqslant 15$.

(A) Plot both functions simultaneously in the same viewing window.
(B) **Break-even points** occur when $R(x) = C(x)$. Find the break-even points graphically and algebraically (to the nearest thousand cameras).
(C) Recall that a loss occurs if $R(x) < C(x)$ and a profit occurs if $R(x) > C(x)$. For what outputs (to the nearest thousand cameras) will a loss occur? A profit?

Solution (A) Graph in a graphing utility:

(B) Graphical solution: Most graphing utilities have a built-in routine that will find intersection points quickly and accurately (see the graphing utility supplement for this book or the manual for your graphing utility for details). The graphs at the top of the next page indicate that the company breaks even at $x = 2.490$ and 12.530 million cameras.

Algebraic solution: Find x such that $R(x) = C(x)$:

$$x(94.8 - 5x) = 156 + 19.7x$$

$$-5x^2 + 75.1x - 156 = 0$$

$$x = \frac{-75.1 \pm \sqrt{75.1^2 - 4(-5)(-156)}}{2(-5)} \qquad \text{Quadratic formula}$$

$$= \frac{-75.1 \pm \sqrt{2520.01}}{-10}$$

$$x = 2.490 \quad \text{or} \quad 12.530$$

The company breaks even at $x = 2.490$ and 12.530 million cameras.

(C) Use the results from parts (A) and (B):

Loss: $1 \leqslant x < 2.490$ or $12.530 < x \leqslant 15$

Profit: $2.490 < x < 12.530$

MATCHED PROBLEM 4 ➤   Use the profit equation from Matched Problem 3:

$$P(x) = R(x) - C(x)$$
$$= -5x^2 + 75.1x - 156 \quad \text{Profit function}$$

Domain: $1 \leqslant x \leqslant 15$

(A) Plot the profit function in an appropriate viewing window.
(B) Break-even points occur when $P(x) = 0$. Find the break-even points graphically and algebraically (to the nearest thousand cameras).
(C) A loss occurs if $P(x) < 0$, and a profit occurs if $P(x) > 0$. For what outputs (to the nearest thousand cameras) will a loss occur? A profit?

Answers to Matched Problems   1. (A)

(B) x intercepts: -0.77, 3.27; y intercept: -5

(C) x intercepts: $-0.77, 3.27$; y intercept: -5

(D) $x \leq -0.77$ or $x \geq 3.27$; or $(-\infty, -0.77]$ or $[3.27, \infty)$

2. (A)

(B) Vertex: $(-4, 6)$; Maximum: $f(-4) = 6$;
Range: $y \leq 6$ or $(-\infty, 6]$

(C) $f(x) = -0.25(x + 4)^2 + 6$; Vertex: $(-4, 6)$; Maximum: $f(-4) = 6$;
Range: $y \leq 6$ or $(-\infty, 6]$

(D) The graph of $f(x) = -0.25(x + 4)^2 + 6$ is the same as the graph of $g(x) = x^2$ vertically contracted by a factor of 0.25, reflected in the x axis, and shifted 4 units to the left and 6 units up.

3. (A) $P(x) = R(x) - C(x) = -5x^2 + 75.1x - 156$

(B)

(C)

An output of 7.510 million cameras will produce a maximum profit of 126.001 million dollars. (Notice that maximum profit does not occur at the same output as maximum revenue.)

(D) $P(x) = R(x) - C(x) = -5(x - 7.51)^2 + 126.0005$; an output of 7.510 million cameras will produce a maximum profit of 126.001 million dollars.

(E) $p(7.510) = \$57$

4. (A)

(B) Graphically: Find x intercepts in the figure for part (A).
Algebraically: Solve the equation $P(x) = 0$.
Both methods produce break-even points $x = 2.490$ or   12.530 million cameras.

(C) Use the results from part (B): Loss: $1 \leq x < 2.490$ or $12.530 < x \leq 15$;
Profit: $2.490 < x < 12.530$

EXERCISE 1-5

A

1. Indicate which equations define a quadratic function with x the independent variable.
    (A) $y = 3 - x^2$ (B) $y^2 + x^2 = 4$
    (C) $y = (2 - 3x)^2$ (D) $2x - y = 3$
    (E) $y = 2x(3 - x)$ (F) $y = -2x^2 + 5x - 1$

2. Indicate which equations define a quadratic function with x the independent variable.
    (A) $x^2 - y^2 = 9$ (B) $y = x(3x - 5)$
    (C) $y = 1 - \sqrt{x}$ (D) $y = 4 - 3x - 2x^2$
    (E) $y = 2(x - 3)^2$ (F) $y = 8.3 - 0.3x^2$

3. Match each equation with a graph of one of the functions f, g, m, or n in the figure.
    (A) $y = -(x + 2)^2 + 1$ (B) $y = (x - 2)^2 - 1$
    (C) $y = (x + 2)^2 - 1$ (D) $y = -(x - 2)^2 + 1$

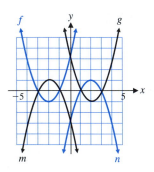

Figure for 3

4. Match each equation with a graph of one of the functions f, g, m, or n in the figure.
    (A) $y = (x - 3)^2 - 4$ (B) $y = -(x + 3)^2 + 4$
    (C) $y = -(x - 3)^2 + 4$ (D) $y = (x + 3)^2 - 4$

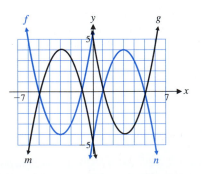

Figure for 4

For the functions indicated in Problems 5–8, find each of the following to the nearest integer by referring to the graphs for Problems 3 and Problem 4.
(A) Intercepts   *(B) Vertex*
(C) Maximum or minimum   *(D) Range*
(E) Increasing interval   *(F) Decreasing interval*

5. Function n in the figure for Problem 3

6. Function m in the figure for Problem 4

7. Function f in the figure for Problem 3

8. Function g in the figure for Problem 4

In Problems 9–12, find each of the following algebraically (to the nearest integer) without referring to any graphs.
(A) Intercepts   *(B) Vertex*
(C) Maximum or minimum   *(D) Range*

9. $f(x) = -(x - 2)^2 + 1$

10. $g(x) = -(x + 3)^2 + 4$

11. $M(x) = (x + 2)^2 - 1$

12. $N(x) = (x - 3)^2 - 4$

B   *In Problems 13–16, write an equation for each graph in the form $y = a(x - h)^2 + k$, where a is either 1 or −1 and h and k are integers.*

13.

14.

15.

16.

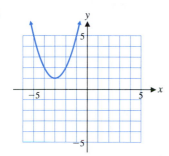

Graph each quadratic function in Problems 17–22 in a graphing utility; then find each of the following (to one decimal place). Use trace and zoom or an appropriate built-in routine for parts (A)–(C).

(A) *Intercepts*   (B) *Vertex*

(C) *Maximum or minimum*   (D) *Range*

17. $f(x) = x^2 - 8x + 13$

18. $g(x) = x^2 + 10x + 20$

19. $M(x) = 1 - 6x - x^2$

20. $N(x) = -10 + 8x - x^2$

21. $G(x) = 0.5x^2 - 4x + 10$

22. $H(x) = -0.5x^2 - 2x - 3$

In Problems 23–28, first write each function in the form $f(x) = a(x - h)^2 + k$; then find each of the following algebraically (to one decimal place) without using any graphs:

(A) *Intercepts*   (B) *Vertex*

(C) *Maximum or minimum*   (D) *Range*

23. $f(x) = x^2 - 8x + 13$

24. $g(x) = x^2 + 10x + 20$

25. $M(x) = 1 - 6x - x^2$

26. $N(x) = -10 + 8x - x^2$

27. $G(x) = 0.5x^2 - 4x + 10$

28. $H(x) = -0.5x^2 - 2x - 3$

Solve Problems 29–34 graphically to two decimal places using a graphing utility. Do not change the form of any equation or inequality.

29. $2 - 5x - x^2 = 0$

30. $7 + 3x - 2x^2 = 0$

31. $1.9x^2 - 1.5x - 5.6 < 0$

32. $3.4 + 2.9x - 1.1x^2 \geq 0$

33. $2.8 + 3.1x - 0.9x^2 \leq 0$

34. $1.8x^2 - 3.1x - 4.9 > 0$

35. Explain under what graphical conditions a quadratic function has exactly one real zero.

36. Explain under what graphical conditions a quadratic function has no real zeros.

C   *Graph each quadratic function in Problems 37–40 in a graphing utility; then find each of the following (to two decimal places). Use trace and zoom or an appropriate built-in routine for parts (A)–(C).*

(A) *Intercepts*   (B) *Vertex*

(C) *Maximum or minimum*   (D) *Range*

37. $g(x) = 0.25x^2 - 1.5x - 7$

38. $m(x) = 0.20x^2 - 1.6x - 1$

39. $f(x) = -0.12x^2 + 0.96x + 1.2$

40. $n(x) = -0.15x^2 - 0.90x + 3.3$

In Problems 41–44, first write each function in the form $f(x) = a(x - h)^2 + k$; then find each of the following algebraically (to two decimal places) without using any graphs:

(A) *Intercepts*   (B) *Vertex*

(C) *Maximum or minimum*   (D) *Range*

41. $g(x) = 0.25x^2 - 1.5x - 7$

42. $m(x) = 0.20x^2 - 1.6x - 1$

43. $f(x) = -0.12x^2 + 0.96x + 1.2$

44. $n(x) = -0.15x^2 - 0.90x + 3.3$

In Problems 45–48, graph functions f and g simultaneously in the same viewing window; then solve each of the following graphically (to two decimal places) using trace and zoom or an appropriate built-in routine:

(A) $f(x) = g(x)$ (B) $f(x) > g(x)$ (C) $f(x) < g(x)$

45. $f(x) = -0.4x(x - 10)$
$g(x) = 0.3x + 5$
$0 \leqslant x \leqslant 10$

46. $f(x) = -0.7x(x - 7)$
$g(x) = 0.5x + 3.5$
$0 \leqslant x \leqslant 7$

47. $f(x) = -0.9x^2 + 7.2x$
$g(x) = 1.2x + 5.5$
$0 \leqslant x \leqslant 8$

48. $f(x) = -0.7x^2 + 6.3x$
$g(x) = 1.1x + 4.8$
$0 \leqslant x \leqslant 9$

49. Give a simple example of a quadratic function that has no real zeros. Explain how its graph is related to the x axis.

50. Give a simple example of a quadratic function that has exactly one real zero. Explain how its graph is related to the x axis.

 APPLICATIONS

Business & Economics

51. *Revenue.* Refer to Problems 89 and 91, Exercise 1-2. We found that the marketing research department for the company that manufactures and sells memory chips for microcomputers established the following price–demand and revenue functions:

$p(x) = 119 - 6x$ Price–demand
$R(x) = xp(x) = x(119 - 6x)$ Revenue

where $p(x)$ is the wholesale price in dollars at which x million chips can be sold, and $R(x)$ is in millions of dollars. Both functions have domain $1 \leqslant x \leqslant 15$.

(A) Graph the revenue function using an appropriate viewing window.

(B) Find the output (to the nearest thousand chips) that will produce the maximum revenue. What is the maximum revenue to the nearest thousand dollars? Solve the problem graphically.

(C) Solve part (B) algebraically by completing the square.

(D) What is the wholesale price per chip (to the nearest dollar) that produces the maximum revenue?

52. *Revenue.* Refer to Problems 90 and 92, Exercise 1-2. We found that the marketing research department for the company that manufactures and sells "Notebook" com-

puters established the following price–demand and revenue functions:

$p(x) = 1,190 - 36x$ Price–demand
$R(x) = xp(x) = x(1,190 - 36x)$ Revenue

where $p(x)$ is the wholesale price in dollars at which x thousand computers can be sold, and $R(x)$ is in thousands of dollars. Both functions have domain $1 \leqslant x \leqslant 25$.

(A) Graph the revenue function using an appropriate viewing window.

(B) Find the output (to the nearest hundred computers) that will produce the maximum revenue. What is the maximum revenue to the nearest thousand dollars? Solve the problem graphically.

(C) Solve part (B) algebraically by completing the square.

(D) What is the wholesale price per computer (to the nearest dollar) that produces the maximum revenue?

53. *Break-even analysis.* Use the revenue function from Problem 51 in this exercise and the cost function from Problem 93, Exercise 1-2:

$R(x) = x(119 - 6x)$ Revenue
$C(x) = 234 + 23x$ Cost

where x is in millions of chips and $R(x)$ and $C(x)$ are in millions of dollars. Both functions have domain $1 \leqslant x \leqslant 15$.

(A) Graph both functions simultaneously in the same viewing window.

(B) Find the break-even points graphically and algebraically (to the nearest thousand chips).

(C) For what outputs will a loss occur? A profit?

54. *Profit–loss analysis.* Use the revenue function from Problem 52 in this exercise and the cost function from Problem 94, Exercise 1-2:

$$R(x) = x(1,190 - 36x) \quad \text{Revenue}$$
$$C(x) = 4,320 + 146x \quad \text{Cost}$$

where x is in thousands of computers and $R(x)$ and $C(x)$ are in thousands of dollars. Both functions have domain $1 \leqslant x \leqslant 25$.

(A) Form a profit function and graph it in an appropriate viewing window.

(B) Find the break-even points, $P(x) = 0$, graphically and algebraically (to the nearest hundred computers).

(C) For what outputs will a loss occur? A profit?

(D) For what output (to the nearest hundred computers) will a maximum profit occur? What is the maximum profit to the nearest thousand dollars? [Note that the maximum profit and maximum revenue (see Problem 52) do not occur at the same outputs!]

Life Sciences

55. *Medicine.* The French physician Poiseuille was the first to discover that blood flows faster near the center of an artery than near the edge. Experimental evidence has shown that the rate of flow v (in centimeters per second) at a point x centimeters from the center of an artery (see the figure) is given by

$$v = f(x) = 1,000(0.04 - x^2) \qquad 0 \leqslant x \leqslant 0.2$$

(A) Find the distance from the center (to two decimal places) where the rate of flow is 20 centimeters per second. Solve graphically by graphing the velocity function f and $v = 20$ simultaneously in the same viewing window and finding the point of intersection of the two graphs.

(B) Solve part (A) algebraically.

Figure for 55 and 56

56. *Medicine.* Refer to Problem 55.

(A) Find the distance from the center (to two decimal places) where the rate of flow is 30 centimeters per second. Solve graphically by graphing the velocity function f and $v = 30$ simultaneously in the same viewing window and finding the point of intersection of the two graphs.

(B) Solve part (A) algebraically.

CHAPTER 1
GROUP ACTIVITY

Mathematical Modeling in Business

A manufacturing company manufactures and sells mountain bikes. The management would like to have price–demand and cost functions for break-even and profit–loss analysis. Price–demand and cost functions may be established by collecting appropriate data at different levels of output, and then finding a model in the form of a basic elementary function (from our library of elementary functions) that "closely fits" the collected data. The finance department, using statistical techniques, arrived at the price–demand and cost data in Tables 1 and 2 (page 80), where p is the wholesale price of a bike for a demand of x hundred bikes, $0 \leqslant x \leqslant 220$, and C is the cost, in hundreds of dollars, of producing and selling x hundred bikes, $0 \leqslant x \leqslant 220$.

(A) *Building a Mathematical Model for Price–Demand.* Plot the data in Table 1 and observe that the relationship between p and x is almost linear. After observing a relationship between variables, analysts often try to

TABLE 1
Price–Demand

x (hundred)	p ($)
0	525
64	370
125	270
185	130

TABLE 2
Cost

x (hundred)	C (hundred $)
0	8,470
30	13,510
120	19,140
180	22,580
220	28,490

model the relationship in terms of a basic function from a portfolio of elementary functions that "best fits" the data.

(1) **Linear regression lines** are frequently used to model linear phenomena. This is a process of fitting a straight line to a set of data that minimizes the sum of the squares of the distances of all the points in the graph of the data to the line using the **method of least squares.** Many graphing utilities have this routine built in. Read your user's manual for your particular graphing utility and discuss among the members of the group how this is done. After obtaining the linear regression line for the data in Table 1, graph the line and the data in the same viewing window.

(2) The linear regression line defines a linear function. Interpret the slope of the line. Discuss its domain and range. Using the mathematical model, determine the price for a demand of 10,000 bikes. For a demand of 20,000 bikes.

(B) *Building a Mathematical Model for Cost*. Plot the data in Table 2 in a rectangular coordinate system. Which type of function appears to best fit the data?

(1) Fit a linear regression line to the data in Table 2. Then plot the data points and the line in the same viewing window.

(2) Interpret the slope of the cost equation function. Discuss its domain and range. Using the mathematical model, determine the cost for an output and sales of 10,000 bikes. For an output and sales of 20,000 bikes.

(C) *Break-Even and Profit–Loss Analysis*. Write an equation for the revenue function, and state its domain. Write the equation for the profit function, and state its domain.

(1) Graph the revenue function and the cost function simultaneously in the same viewing window. Graphically determine at what outputs (to the nearest unit) the company breaks even, and where costs exceed revenues and revenues exceed costs.

(2) Graph the profit function in a graphing utility. Graphically determine at what outputs (to the nearest unit) the company breaks even, and where losses occur and profits occur. At what output and price will a maximum profit occur? Does the maximum revenue and maximum profit occur for the same output? Discuss.

CHAPTER   1   REVIEW   ■   Important Terms and Symbols

1-1 *Introduction to Graphing Utilities.* Cartesian or rectangular coordinate system; coordinate axes; ordered pairs; coordinates; abscissa; ordinate; quadrants; fundamental theorem of analytic geometry; solution and solution set for an equation in two variables; graph of an equation in two variables; point-by-point plotting; graphing utility; pixel; viewing window; window variables; standard viewing window; graphing with a graphing utility; screen coordinates of a pixel; cursor; trace; zoom

1-2 *Functions.* Function; domain; range; functions specified by equations; independent variable; dependent variable; vertical-line test; function notation; cost function; price–demand function; revenue function; profit function

$$f(x); \ C = a + bx; \ p = m - nx; \ R = xp; \ P = R - C$$

1-3 *Elementary Functions: Graphs and Transformations.* Six basic functions: identity, absolute value, square, cube, square root, cube root; transformations; horizontal shift or translation; vertical shift or translation; reflection in the x axis; vertical expansion; vertical contraction

$$f(x) = x; \quad g(x) = |x|; \quad h(x) = x^2;$$
$$m(x) = x^3; \quad n(x) = \sqrt{x}; \quad p(x) = \sqrt[3]{x}$$

See the inside front cover for the graphs of these functions.

1-4 *Linear Functions and Straight Lines.* x intercepts; y intercept; linear function; constant function; first-degree function; graph of a linear function; graph of a constant function; graphs of linear equations in two variables; standard form for the equation of a line; equations of vertical lines and horizontal lines; slope; slope–intercept form; point–slope form

$$f(x) = mx + b, m \neq 0; \quad f(x) = b; \quad x = a;$$
$$Ax + By = C; \quad y = mx + b; \quad y - y_1 = m(x - x_1);$$
$$m = \frac{y_2 - y_1}{x_2 - x_1}, x_1 \neq x_2$$

1-5 *Quadratic Functions.* Quadratic function; parabola; finding intercepts; vertex; axis; maximum or minimum; range; break-even points

$$f(x) = ax^2 + bx + c, a \neq 0;$$
$$f(x) = a(x - h)^2 + k, a \neq 0$$

C H A P T E R   1   ■   Review Exercise

Work through all the problems in this chapter review and check your answers in the back of the book. Answers to all review problems are there along with section numbers in italics to indicate where each type of problem is discussed. Where weaknesses show up, review appropriate sections in the text.

A

1. Use point-by-point plotting to sketch a graph of $y = 5 - x^2$. Use integer values for x from -3 to 3.

2. Graph $y = 5 - x^2$ in a graphing utility using $-5 \leqslant x \leqslant 5$ and $-5 \leqslant y \leqslant 5$.

3. Find the smallest viewing window that will contain all the points in the table. State your answer in terms of the window variables Xmin, Xmax, Ymin, and Ymax.

x	-2	8	0	-7	3	-9	-5
y	5	-1	-10	3	0	2	-5

4. Indicate whether each graph specifies a function:

(A)

(B)

(C)

(D)

$f(x)$

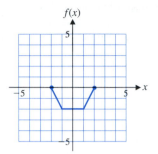

Figure for 7

8. Refer to the figure below.

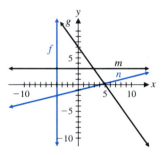

Figure for 8

5. For $f(x) = 2x - 1$ and $g(x) = x^2 - 2x$, find:
   (A) $f(-2) + g(-1)$ (B) $f(0) \cdot g(4)$
   (C) $\dfrac{g(2)}{f(3)}$ (D) $\dfrac{f(3)}{g(2)}$

6. Use the graph of function f in the figure to determine (to the nearest integer) x or y as indicated.

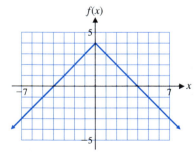

Figure for 6

   (A) $y = f(0)$ (B) $4 = f(x)$
   (C) $y = f(3)$ (D) $3 = f(x)$
   (E) $y = f(-6)$ (F) $-1 = f(x)$

7. Sketch a graph of each of the functions in parts (A)–(D) using the graph of function f in the figure at the top of the next column.
   (A) $y = -f(x)$ (B) $y = f(x) + 4$
   (C) $y = f(x - 2)$ (D) $y = -f(x + 3) - 3$

(A) Identify the graphs of any linear functions with positive slopes.

(B) Identify the graphs of any linear functions with negative slopes.

(C) Identify the graphs of any constant functions. What are their slopes?

(D) Identify any graphs that are not graphs of functions. What can you say about their slopes?

9. Write an equation in the form $y = mx + b$ for a line with slope $-\frac{2}{3}$ and y intercept 6.

10. Write the equations of the vertical line and the horizontal line that pass through $(-6, 5)$.

11. Sketch a graph of $2x - 3y = 18$. What are the intercepts and slope of the line?

12. Indicate which equations define a quadratic function with x the independent variable.
    (A) $y^2 = 4 - x^2$
    (B) $x^2 - 3 = y$
    (C) $y = -2(x - 3)^2 + 1$
    (D) $y = x(x + 7)$
    (E) $y = 3x + 9$
    (F) $y = 0.2x^2 - 5.7$

13. Match each equation with a graph of one of the functions f, g, m, or n in the figure.
(A) $y = (x - 2)^2 - 4$ (B) $y = -(x + 2)^2 + 4$
(C) $y = -(x - 2)^2 + 4$ (D) $y = (x + 2)^2 - 4$

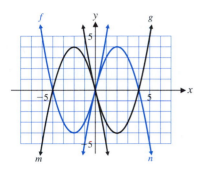

Figure for 13

14. Referring to the graph of function f in the figure for Problem 13 and using known properties of quadratic functions, find each of the following to the nearest integer:
(A) Intercepts (B) Vertex
(C) Maximum or minimum (D) Range
(E) Increasing interval (F) Decreasing interval

B

15. Indicate which of the following equations define a linear function or a constant function:
(A) $2x - 3y = 5$ (B) $x = -2$
(C) $y = 4 - 3x$ (D) $y = -5$
(E) $x = 3y + 5$ (F) $\dfrac{x}{2} - \dfrac{y}{3} = 1$

16. Find the domain of each function:
(A) $f(x) = \dfrac{2x - 5}{x^2 - x - 6}$

(B) $g(x) = \dfrac{3x}{\sqrt{5 - x}}$

17. The function g is defined by $g(x) = 2x - 3\sqrt{x}$. Translate into a verbal definition.

18. Describe the graphs of $x = -3$ and $y = 2$. Graph both simultaneously in the same rectangular coordinate system.

19. Sketch a graph of $y = 0.4x(x + 4)(2 - x)$ on graph paper by first plotting points using odd integer values of x from -3 to 3. Complete the graph using a graphing utility as an aid.

20. Graph $y = 0.3x(x + 4)(4 - x)$ in a standard viewing window. Use trace and zoom to approximate (to two decimal places) the x coordinates of all points on the graph of the form $(x, 6)$.

21. Find $\dfrac{f(2 + h) - f(2)}{h}$ for $f(x) = 3 - 2x$.

22. Find $\dfrac{f(a + h) - f(a)}{h}$ for $f(x) = x^2 - 3x + 1$.

23. Explain how the graph of $m(x) = -|x - 4|$ is related to the graph of $y = |x|$.

24. Explain how the graph of $g(x) = 0.3x^3 + 3$ is related to the graph of $y = x^3$.

25. The following graph is the result of applying a sequence of transformations to the graph of $y = x^2$. Describe the transformations verbally and write an equation for the given graph.

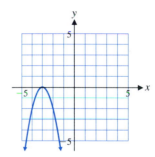

Figure for 25

26. The graph of a function f is formed by vertically expanding the graph of $y = \sqrt{x}$ by a factor of 2, and shifting it to the left 3 units and down 1 unit. Find an equation for function f and graph it on a graphing utility for $-5 \le x \le 5$ and $-5 \le y \le 5$.

27. Write the equation of a line through each indicated point with the indicated slope. Write the final answer in the form $y = mx + b$.
(A) $m = -\frac{2}{3}$; $(-3, 2)$
(B) $m = 0$; $(3, 3)$

28. Write the equation of the line through the two indicated points. Write the final answer in the form $Ax + By = C$.
(A) $(-3, 5), (1, -1)$
(B) $(-1, 5), (4, 5)$
(C) $(-2, 7), (-2, -2)$

29. Write an equation for the graph shown in the form $y = a(x - h)^2 + k$, where a is either -1 or $+1$ and h and k are integers.

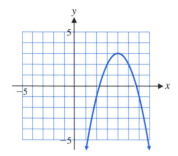

Figure for 29

30. Graph $f(x) = -0.4x^2 + 3.2x + 1.2$ in a graphing utility and find the following (to one decimal place) using trace and zoom or an appropriate graphing utility:
(A) Intercepts   (B) Vertex
(C) Maximum or minimum   (D) Range

31. Given $f(x) = -0.4x^2 + 3.2x + 1.2$, find the following algebraically (to one decimal place) without referring to a graph:
(A) Intercepts   (B) Vertex
(C) Maximum or minimum   (D) Range

32. Solve $2.8 + 3.1x - 0.9x^2 > 0$ graphically (to two decimal places).

C

33. The following graph is the result of applying a sequence of transformations to the graph of $y = \sqrt[3]{x}$. Describe the transformations verbally, and write an equation for the graph.

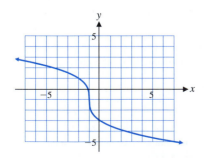

Figure for 33

34. Graph
$$y = mx + b$$
and
$$y = -\frac{1}{m}x + b$$
simultaneously in a squared viewing window for b fixed and several different values of m, $m \neq 0$. Describe the apparent relationship between the graphs of the two equations.

35. Find and simplify
$$\frac{f(x + h) - f(x)}{h}$$
for each of the following functions:
(A) $f(x) = \sqrt{x}$ (B) $f(x) = \dfrac{1}{x}$

36. Graph $G(x) = 0.3x^2 + 1.2x - 6.9$ in a standard viewing window. Then find each of the following (to one decimal place) using trace or zoom or an appropriate built-in routine:
(A) Intercepts
(B) Vertex
(C) Maximum or minimum
(D) Range
(E) Increasing and decreasing intervals

37. Solve Problem 36 algebraically by completing the square.

38. The point $(\sqrt[3]{3}, 3)$ is on the graph of $y = x^3$. Use trace and zoom to approximate $\sqrt[3]{3}$ to four decimal places. (Compare your result with the direct evaluation of $\sqrt[3]{3}$.)

39. Use trace and zoom or an appropriate built-in routine to graphically solve each of the following (to two decimal places):
(A) $3x^3 - 7x^2 - 1 = 0$
(B) $3x^3 - 7x^2 - 1 < 0$

▶ APPLICATIONS

Business & Economics

40. *Linear depreciation.* A computer system was purchased by a small business for $12,000 and, for tax purposes, is assumed to have a salvage value of $2,000 after 8 years. If its value is depreciated linearly from $12,000 to $2,000:
(A) Find the linear equation that relates the value V in dollars to the time t in years. Then graph the equation in a rectangular coordinate system.
(B) What would be the value of the system after 5 years?

41. *Compound interest.* If $1,000 is invested at $100r\%$ compounded annually, at the end of 3 years it will grow to $A = 1,000 (1 + r)^3$.
(A) Graph the equation in a graphing utility for $0 \leq r \leq 0.25$; that is, for money invested at between 0% and 25% compounded annually.
(B) At what rate of interest would money have to be invested to amount to $1,500 in 3 years? Solve for r graphically (to four decimal places) using trace and zoom or an appropriate built-in routine.

42. *Markup.* A sporting goods store sells a tennis racket that cost $130 for $208 and court shoes that cost $50 for $80.

(A) If the markup policy of the store for items that cost over $10 is assumed to be linear and is reflected in the pricing of these two items, write an equation that relates retail price R to cost C.
(B) What would be the retail price of a pair of roller-blade skates that cost $120?
(C) What would be the cost of a pair of cross-country skis that had a retail price of $176?
(D) What is the slope of the graph of the equation found in part (A)? Interpret the slope relative to the problem.

43. *Break-even analysis.* A video production company is planning to produce an instructional videotape. The producer estimates that it will cost $84,000 to shoot the video and $15 per unit to copy and distribute the tape. The wholesale price of the tape is $50 per unit.
(A) Write cost and revenue equations, and graph both simultaneously in the same viewing window.

(B) Graphically determine when $R = C$, $R < C$, and $R > C$.
(C) Algebraically determine when $R = C$, $R < C$, and $R > C$.

44. *Break-even analysis.* The research department in a company that manufactures AM/FM clock radios established the following price–demand, cost, and revenue functions:

$$p(x) = 50 - 1.25x \qquad \text{Price–demand}$$
$$C(x) = 160 + 10x \qquad \text{Cost}$$
$$R(x) = xp(x)$$
$$\quad\quad = x(50 - 1.25x) \quad \text{Revenue}$$

where x is in thousands of units and $C(x)$ and $R(x)$ are in thousands of dollars. All three functions have domain $1 \leq x \leq 40$.
(A) Graph the cost function and the revenue function simultaneously in the same viewing window.
(B) Graphically determine when $R = C$. Then, with the aid of part (A), determine when $R < C$ and $R > C$ to the nearest unit.
(C) Graphically determine the maximum revenue (to the nearest thousand dollars) and the output (to the nearest unit) that produces the maximum revenue. What is the wholesale price of the radio (to the nearest dollar) at this output?

45. *Profit–loss analysis.* Use the cost and revenue functions from Problem 44.
(A) Write a profit function and graph it in a graphing utility.
(B) Graphically determine when $P = 0$, $P < 0$, and $P > 0$ to the nearest unit.
(C) Graphically determine the maximum profit (to the nearest thousand dollars) and the output (to the nearest unit) that produces the maximum profit. What is the wholesale price of the radio (to the nearest dollar) at this output? [Compare with Problem 44(C).]

46. *Construction.* A construction company has 840 feet of chain-link fence that is used to enclose storage areas for equipment and materials at construction sites. The supervisor wants to set up two identical rectangular storage areas sharing a common fence (see the figure on the next page). Assuming all fencing is used:
(A) Express the total area $A(x)$ enclosed by both pens as a function of x.
(B) From physical considerations, what is the domain of the function A?

(C) Graph function A in a graphing utility and graphically determine the dimensions of the storage areas that have the maximum total combined area. What is the maximum area?

Life Sciences

47. *Air pollution.* On an average summer day in a large city, the pollution index at 8:00 AM is 20 parts per million, and it increases linearly by 15 parts per million each hour until 3:00 PM. Let $P(x)$ be the amount of pollutants in the air x hours after 8:00 AM.

(A) Express $P(x)$ as a linear function of x.

(B) What is the air pollution index at 1:00 PM?

(C) Graph the function P for $0 \leqslant x \leqslant 7$.

(D) What is the slope of the graph? (The slope is the amount of increase in pollution for each additional hour of time.)

Social Sciences

48. *Psychology—sensory perception.* One of the oldest studies in psychology concerns the following question: Given a certain level of stimulation (light, sound, weight lifting, electric shock, and so on), how much should the stimulation be increased for a person to notice the difference? In the middle of the nineteenth century, E. H. Weber (a German physiologist) formulated a law that still carries his name: If Δs is the change in stimulus that will just be noticeable at a stimulus level s, then the ratio of Δs to s is a constant:

$$\frac{\Delta s}{s} = k$$

Hence, the amount of change that will be noticed is a linear function of the stimulus level, and we note that the greater the stimulus, the more it takes to notice a difference. In an experiment on weight lifting, the constant k for a given individual was found to be $\frac{1}{30}$.

(A) Find Δs (the difference that is just noticeable) at the 30 pound level. At the 90 pound level.

(B) Graph $\Delta s = s/30$ for $0 \leqslant s \leqslant 120$.

(C) What is the slope of the graph?

CHAPTER 2

Additional Elementary Functions

In this chapter we add the following four general classes of functions to the beginning library of elementary functions started in Chapter 1: polynomial, rational, exponential, and logarithmic. This expanded library of elementary functions should take care of most of your function needs in this and many other courses. The linear and quadratic functions studied in Chapter 1 are special cases of the more general class of functions called *polynomials*. In the last section we will show how a graphing utility can be used to model certain sets of raw data using functions from our library of elementary functions.

SECTION 2-1 ■ **Polynomial and Rational Functions**

- POLYNOMIAL FUNCTIONS
- RATIONAL FUNCTIONS
- APPLICATION

■ POLYNOMIAL FUNCTIONS

In Chapter 1 you were introduced to the basic functions

$$f(x) = b \qquad\qquad\qquad\qquad \text{Constant function}$$
$$f(x) = ax + b \qquad a \neq 0 \qquad \text{Linear function}$$
$$f(x) = ax^2 + bx + c \qquad a \neq 0 \qquad \text{Quadratic function}$$

as well as some special cases of

$$f(x) = ax^3 + bx^2 + cx + d \qquad a \neq 0 \quad \text{Cubic function}$$

Most of the earlier applications we considered, including cost, revenue, profit, loss, and packaging applications, made use of these functions. Notice the evolving pattern

87

going from the constant function to the cubic function—the terms in each equation are of the form ax^n, where n is a nonnegative integer and a is a real number. All these functions are special cases of the general class of functions called *polynomial functions:*

> ### Polynomial Function
>
> A **polynomial function** is a function of the form
>
> $$f(x) = a_n x^n + a_{n-1} x^{n-1} + \cdots + a_1 x + a_0$$
>
> for n a nonnegative integer, called the **degree** of the polynomial. The coefficients $a_0, a_1, \ldots, a_n$ are real numbers with $a_n \neq 0$. The **domain** of a polynomial function is the set of all real numbers.

The shape of the graph of a polynomial function is connected to the degree of the polynomial. The shapes of odd-degree polynomial functions have something in common and the shapes of even-degree polynomial functions have something in common. Figure 1 shows graphs of representative polynomial functions from degrees 1 to 6, and suggests some general properties of graphs of polynomial functions. It is instructive to reproduce these graphs on your own graphing utility.

(A) $f(x) = x - 2$

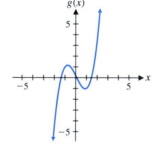

(B) $g(x) = x^3 - 2x$

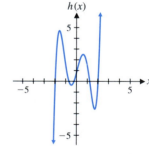

(C) $h(x) = x^5 - 5x^3 + 4x + 1$

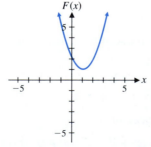

(D) $F(x) = x^2 - 2x + 2$

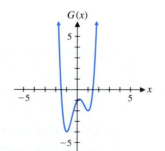

(E) $G(x) = 2x^4 - 4x^2 + x - 1$

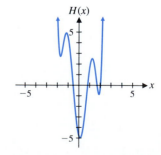

(F) $H(x) = x^6 - 7x^4 + 14x^2 - x - 5$

FIGURE 1
Graphs of polynomial functions

Notice that the odd-degree polynomial graphs start negative, end positive, and cross the x axis at least once. The even-degree polynomial graphs start positive, end positive, and may not cross the x axis at all. In all cases in Figure 1, the coefficient of the highest-degree term was chosen positive. If any leading coefficient had been chosen negative, then we would have a similar graph but flipped over.

The graph of a polynomial function is **continuous,** with no holes or breaks. That is, the graph can be drawn without removing a pen from the paper. Also, the graph of a polynomial has no sharp corners. Figure 2 shows the graphs of two functions, one that is not continuous, and the other that is continuous, but with a sharp corner. Neither function is a polynomial.

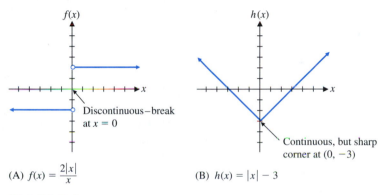

(A) $f(x) = \dfrac{2|x|}{x}$ (B) $h(x) = |x| - 3$

FIGURE 2
Discontinuous and sharp-corner functions

Figure 1 gives examples of polynomial functions with graphs containing the maximum number of *turning points* possible for a polynomial of that degree. A **turning point** on a continuous graph is a point that separates an increasing portion from a decreasing portion or vice versa. In general, it can be shown that:

The graph of a polynomial function of positive degree n can have at most $(n - 1)$ turning points and can cross the x axis at most n times.

explore – discuss 1

(A) What is the least number of turning points an odd-degree polynomial function can have? An even-degree polynomial function?
(B) What is the maximum number of x intercepts the graph of a polynomial function of degree n can have?
(C) What is the maximum number of real solutions an nth-degree polynomial equation can have?
(D) What is the least number of x intercepts the graph of a polynomial function of odd degree can have? Of even degree?
(E) What is the least number of real solutions a polynomial function of odd degree can have? Of even degree?

We now use a graphing utility to compare the graphs of two polynomial functions relative to points close to the origin and then zoom out to compare points distant from the origin. Compare the graphs in Figure 3.

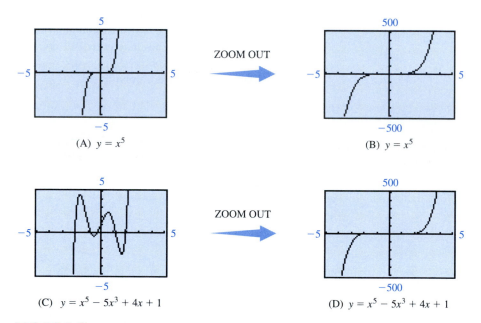

FIGURE 3
Close and distant comparisons

Figure 3 clearly shows that the highest-degree term in the polynomial dominates all other terms combined in the polynomial. As we zoom out, the graph of $y = x^5 - 5x^3 + 4x + 1$ looks more and more like the graph of $y = x^5$. This is a general property of polynomial functions.

explore–discuss 2

Compare the graphs of $y = x^6$ and $y = x^6 - 7x^4 + 14x^2 - x - 5$ in the following two viewing windows:

(A) $-5 \leq x \leq 5, -5 \leq y \leq 5$ (B) $-5 \leq x \leq 5, -500 \leq y \leq 500$

■ RATIONAL FUNCTIONS

Just as rational numbers are defined in terms of quotients of integers, rational functions are defined in terms of quotients of polynomials. The following equations define rational functions:

$$f(x) = \frac{1}{x} \qquad g(x) = \frac{x - 2}{x^2 - x - 6} \qquad h(x) = \frac{x^3 - 8}{x}$$

$$p(x) = 3x^2 - 5x \qquad q(x) = 7 \qquad r(x) = 0$$

Rational Function

A **rational function** is any function of the form

$$f(x) = \frac{n(x)}{d(x)} \qquad d(x) \neq 0$$

where $n(x)$ and $d(x)$ are polynomials. The **domain** is the set of all real numbers such that $d(x) \neq 0$. We assume $n(x)/d(x)$ is reduced to lowest terms.

EXAMPLE 1 ➤ **Domains and Intercepts** Find the domain and intercepts for the rational function

$$f(x) = \frac{x - 2}{x + 1}$$

Solution *Domain:* The denominator is 0 at $x = -1$. Therefore, the domain is the set of all real numbers except -1. The graph of f cannot cross the vertical line $x = -1$.

x intercepts: Find x such that $f(x) = 0$. This happens only if $x - 2 = 0$, that is, at $x = 2$. Thus, 2 is the only x intercept.

y-intercept: The y intercept is

$$f(0) = \frac{0 - 2}{0 + 1} = -2 \qquad\qquad ◀$$

MATCHED PROBLEM 1 ➤ Find the domain and intercepts for the rational function

$$g(x) = \frac{2x}{x - 2} \qquad\qquad ◀$$

In the next example we investigate the graph of $f(x) = (x - 2)/(x + 1)$ near the *point of discontinuity, $x = -1$*, and the behavior of the graph as x increases or decreases without bound. Using this information, we can complete a sketch of the graph of function f with little additional trouble. The investigation uncovers some characteristic features of graphs of rational functions.

EXAMPLE 2 ➤ **Graph of a Rational Function** Given function f from Example 1: $f(x) = \dfrac{x - 2}{x + 1}$

(A) Investigate the graph of f near the point of discontinuity, $x = -1$.
(B) Investigate the graph of f as x increases or decreases without bound.

(C) Sketch a graph of function f.

(D) Confirm the graph in part (C) on a graphing utility.

Solution   (A) *Let x approach − 1 from the left:*

x	-2	-1.1	-1.01	-1.001	-1.0001	-1.00001
$f(x)$	4	31	301	3,001	30,001	300,001

We see that $f(x)$ increases without bound as x approaches − 1 from the left. Symbolically,

$$f(x) \to \infty \qquad \text{as} \qquad x \to -1^-$$

Let x approach − 1 from the right:

x	0	-0.9	-0.99	-0.999	-0.9999	-0.99999
$f(x)$	-2	-29	-299	$-2,999$	$-29,999$	$-299,999$

We see that $f(x)$ decreases without bound as x approaches − 1 from the right. Symbolically,

$$f(x) \to -\infty \qquad \text{as} \qquad x \to -1^+$$

The vertical line $x = -1$ is called a *vertical asymptote*. The graph of f gets closer to this line as x gets closer to − 1. Sketching the vertical asymptote provides an aid to drawing the graph of f near the asymptote (Fig. 4).

(B) Divide each term in the numerator and denominator of $f(x)$ by x, the highest power of x to occur in the numerator and denominator:

$$f(x) = \frac{x - 2}{x + 1} = \frac{1 - \dfrac{2}{x}}{1 + \dfrac{1}{x}}$$

As x increases or decreases without bound, $2/x$ and $1/x$ approach 0 and $f(x)$ gets closer and closer to 1. The horizontal line $y = 1$ is called a *horizontal asymptote*. The graph of $y = f(x)$ gets closer to this line as x decreases or increases without bound. But how does the graph of $y = f(x)$ approach the horizontal line $y = 1$? Does it approach from above? From below? Or from both? To answer these questions we investigate table values as follows:

Let x approach ∞:

x	10	100	1,000	10,000	100,000
$f(x)$	0.72727	0.97030	0.99700	0.99970	0.99997

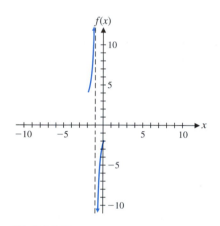

FIGURE 4

Graph near vertical asymptote $x = -1$

The graph of $y = f(x)$ approaches the line $y = 1$ from below as x increases without bound.

Let x approach $-\infty$:

x	-10	-100	$-1,000$	$-10,000$	$-100,000$
$f(x)$	1.33333	1.03030	1.00300	1.00030	1.00003

The graph of $y = f(x)$ approaches the line $y = 1$ from above as x decreases without bound.

Sketching the horizontal asymptote first provides an aid to drawing the graph of f as x moves away from the origin (Fig. 5).

(C) It is now easy to complete the sketch of the graph of f using the intercepts from Example 1 and filling in any points of uncertainty (Fig. 6).

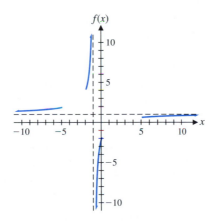

FIGURE 5
Graph of $y = f(x)$ near horizontal asymptote $y = 1$

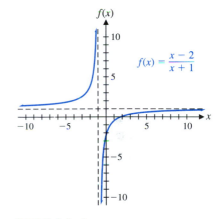

FIGURE 6
Sketch of the rational function f

(D) Graphing utility confirmation:

FIGURE 7
Graph of f (connected mode)

FIGURE 8
Graph of f (dot mode)

Remark

In Figure 7 it appears that the calculator has also drawn the vertical asymptote, but this is not the case. Most graphing calculators, when set in *connected mode,* calculate points on a graph and connect these points with line segments. The last point plotted to the left of the asymptote and the first plotted to the right of the asymptote will usually have very large y coordinates. If these y coordinates have opposite signs, then the calculator will connect the two points with a nearly vertical line segment, which gives the appearance of an asymptote. If you wish, you can set the calculator in *dot mode* to plot the points without the connecting line segments, as illustrated in Figure 8.

MATCHED PROBLEM 2 ▶   Given function g from Matched Problem 1: $g(x) = \dfrac{2x}{x - 2}$

(A) Investigate the graph of g near the point of discontinuity, $x = 2$.
(B) Investigate the graph of g as x increases or decreases without bound.
(C) Sketch a graph of function g.
(D) Confirm the graph in part (C) on a graphing utility.   ◀

Graphing rational functions is considerably aided by locating vertical and horizontal asymptotes first, if they exist. The following general procedures are suggested by Example 2:

Vertical and Horizontal Asymptotes and Rational Functions

Given the rational function

$$f(x) = \frac{n(x)}{d(x)}$$

where $n(x)$ and $d(x)$ are polynomials without common factors:

1. If a is a real number such that $d(a) = 0$, then the line $x = a$ is a **vertical asymptote** of the graph of $y = f(x)$.
2. **Horizontal asymptotes,** if any exist, can be found by dividing each term of the numerator $n(x)$ and denominator $d(x)$ by the highest power of x that appears in the numerator and denominator, and then proceeding as in Example 2.

EXAMPLE 3 ▶   Graphing Rational Functions   Given the rational function: $f(x) = \dfrac{3x}{x^2 - 4}$

(A) Find intercepts and equations for any vertical and horizontal asymptotes.
(B) Using the information from part (A) and a graphing utility as an aid, sketch a graph of f for $-7 \leqslant x \leqslant 7$ and $-7 \leqslant y \leqslant 7$.

Solution   (A) *x intercepts:* $f(x) = 0$ only if $3x = 0$, or $x = 0$. Thus, the only x intercept is 0.

y intercept: $f(0) = \dfrac{3 \cdot 0}{0^2 - 4} = \dfrac{0}{-4} = 0$

Thus, the y intercept is 0.

Vertical asymptotes: $f(x) = \dfrac{3x}{x^2 - 4} = \dfrac{3x}{(x - 2)(x + 2)}$

The denominator is 0 at $x = -2$ and $x = 2$; hence, $x = -2$ and $x = 2$ are vertical asymptotes.

Horizontal asymptotes: Divide each term in the numerator and denominator by x^2, the highest power of x in the numerator and denominator:

$$f(x) = \frac{3x}{x^2 - 4} = \frac{\dfrac{3x}{x^2}}{\dfrac{x^2}{x^2} - \dfrac{4}{x^2}} = \frac{\dfrac{3}{x}}{1 - \dfrac{4}{x^2}}$$

As x increases or decreases without bound, the numerator tends to 0 and the denominator tends to 1, thus, $f(x)$ tends to 0. The line $y = 0$ is a horizontal asymptote.

(B) Using the information from part (A) and the graphing utility graph shown in Figure 9, we complete the graph as shown in Figure 10.

FIGURE 9   FIGURE 10

MATCHED PROBLEM 3   ►   Given the rational function: $g(x) = \dfrac{3x + 3}{x^2 - 9}$

(A) Find all intercepts and equations for any vertical and horizontal asymptotes.
(B) Using the information from part (A) and a graphing utility graph as an aid, sketch a graph of g for $-10 \le x \le 10$ and $-10 \le y \le 10$.   ◄

■ APPLICATION

Rational functions occur naturally in many types of applications. Example 4 provides an illustration.

EXAMPLE 4 ➤   **Employee Training**   A company that manufactures computers has established that, on the average, a new employee can assemble $N(t)$ components per day after t days of on-the-job training, as given by

$$N(t) = \frac{50t}{t + 4} \qquad t \geq 0$$

Sketch a graph of N, $0 \leq t \leq 100$, including any vertical or horizontal asymptotes. What does $N(t)$ approach as t increases without bound?

Solution   *Vertical asymptotes:* None for $t \geq 0$.

Horizontal asymptote:

$$N(t) = \frac{50t}{t + 4} = \frac{50}{1 + \dfrac{4}{t}}$$

$N(t)$ approaches 50 as t increases without bound. Thus, $y = 50$ is a horizontal asymptote.

Sketch of graph:

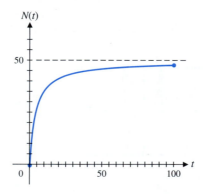

$N(t)$ approaches 50 as t increases without bound. It appears that 50 components per day would be the upper limit that an employee would be expected to assemble.   ◄

MATCHED PROBLEM 4 ➤   Repeat Example 4 for

$$N(t) = \frac{25t + 5}{t + 5} \qquad t \geq 0$$ ◄

Answers to Matched Problems

1. Domain: All real numbers except 2; x intercept: 0; y intercept: 0
2. (A) Vertical asymptote: $x = 2$ (B) Horizontal asymptote: $y = 2$

(C)

(D)

3. (A) x intercept: -1; y intercept: $-\frac{1}{3}$; Vertical asymptotes: $x = -3$ and $x = 3$; Horizontal asymptote: $y = 0$

(B)

4. No vertical asymptotes for $t \geqslant 0$; $y = 25$ is a horizontal asymptote. $N(t)$ approaches 25 as t increases without bound. It appears that 25 components per day would be the upper limit that an employee would be expected to assemble.

E X E R C I S E   2 - 1

A *For each polynomial function in Problems 1–6, find the following:*

(A) *Degree of the polynomial*
(B) *Maximum number of turning points of the graph*
(C) *Maximum number of x intercepts of the graph*
(D) *Minimum number of x intercepts of the graph*
(E) *Maximum number of y intercepts of the graph*
(F) *Minimum number of y intercepts of the graph*

1. $f(x) = ax^2 + bx + c,\ a \neq 0$
2. $f(x) = ax + b,\ a \neq 0$
3. $f(x) = ax^5 + bx^4 + cx^3 + dx^2 + ex + f,\quad a \neq 0$
4. $f(x) = ax^4 + bx^3 + cx^2 + dx + e,\quad a \neq 0$
5. $f(x) = ax^6 + bx^5 + cx^4 + dx^3 + ex^2 + fx + g,\quad a \neq 0$
6. $f(x) = ax^3 + bx^2 + cx + d,\ a \neq 0$

Each graph in Problems 7–14 is the graph of a polynomial function. Answer the following questions for each graph:

(A) *How many turning points are on the graph?*
(B) *What is the minimum degree of a polynomial function that could have the graph?*
(C) *Is the leading coefficient of the polynomial negative or positive?*

7.

8.

9.

10.

11.

12.

13.

14.

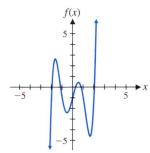

B *For each rational function in Problems 15–20:*

(A) *Find the intercepts for the graph.*
(B) *Determine the domain.*
(C) *Find any vertical or horizontal asymptotes for the graph.*
(D) *Graph $y = f(x)$ in a standard viewing window using a graphing utility.*
(E) *Sketch any asymptotes as dashed lines. Then sketch a graph of $y = f(x)$ for $-10 \leq x \leq 10$ and $-10 \leq y \leq 10$.*

15. $f(x) = \dfrac{x + 2}{x - 2}$ **16.** $f(x) = \dfrac{x - 3}{x + 3}$

17. $f(x) = \dfrac{3x}{x + 2}$ **18.** $f(x) = \dfrac{2x}{x - 3}$

19. $f(x) = \dfrac{4 - 2x}{x - 4}$ **20.** $f(x) = \dfrac{3 - 3x}{x - 2}$

21. How does the graph of $f(x) = 2x^4 - 5x^2 + x + 2$ compare to the graph of $y = 2x^4$ as we zoom out (see Fig. 3)?

22. How does the graph of $f(x) = x^3 - 2x + 2$ compare to the graph of $y = x^3$ as we zoom out?

23. How does the graph of $f(x) = -x^5 + 4x^3 - 4x + 1$ compare to the graph of $y = -x^5$ as we zoom out?

24. How does the graph of $f(x) = -x^5 + 5x^3 + 4x - 1$ compare to the graph of $y = -x^5$ as we zoom out?

25. Compare the graphs of the functions $y = 2x^4$ and $y = 2x^4 - 5x^2 + x + 2$ in the following two viewing windows:
(A) $-5 \leq x \leq 5$, $-5 \leq y \leq 5$
(B) $-5 \leq x \leq 5$, $-500 \leq y \leq 500$

26. Compare the graphs of $y = x^3$ and $y = x^3 - 2x + 2$ in the following two viewing windows:
(A) $-5 \leq x \leq 5$, $-5 \leq y \leq 5$
(B) $-5 \leq x \leq 5$, $-500 \leq y \leq 500$

27. Compare the graphs of the functions $y = -x^5$ and $y = -x^5 + 4x^3 - 4x + 1$ in the following two viewing windows:
(A) $-5 \leq x \leq 5$, $-5 \leq y \leq 5$
(B) $-5 \leq x \leq 5$, $-500 \leq y \leq 500$

28. Compare the graphs of the functions $y = -x^5$ and $y = -x^5 + 5x^3 - 5x + 2$ in the following two viewing windows:
(A) $-5 \leq x \leq 5$, $-5 \leq y \leq 5$
(B) $-5 \leq x \leq 5$, $-500 \leq y \leq 500$

In Problems 29–32:

(A) *State the maximum number of x intercepts and turning points the graph of each polynomial function can have.*
(B) *Graph each polynomial for $-5 \leq x \leq 5$ in a graphing utility, and find all x intercepts and turning points (to two decimal places).*

29. $P(x) = x^3 - 2x + 2$

30. $P(x) = -x^2 + 3x + 2$

31. $P(x) = x^4 - 3x - 1$

32. $P(x) = -x^4 + 3x^2 - x + 1$

C *For each rational function in Problems 33–38:*

(A) *Find any intercepts for the graph.*
(B) *Find any vertical and horizontal asymptotes for the graph.*
(C) *Graph the function in a standard viewing window using a graphing utility.*
(D) *Sketch any asymptotes as dashed lines. Then sketch a graph of f for $-10 \leq x \leq 10$ and $-10 \leq y \leq 10$.*

33. $f(x) = \dfrac{2x^2}{x^2 - x - 6}$ **34.** $f(x) = \dfrac{3x^2}{x^2 + x - 6}$

35. $f(x) = \dfrac{6 - 2x^2}{x^2 - 9}$ **36.** $f(x) = \dfrac{3 - 3x^2}{x^2 - 4}$

37. $f(x) = \dfrac{-4x}{x^2 + x - 6}$ **38.** $f(x) = \dfrac{5x}{x^2 + x - 12}$

39. Write an equation for the lowest-degree polynomial function with the graph and intercepts shown in the figure.

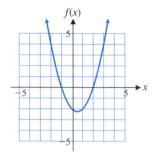

Figure for 39

40. Write an equation for the lowest-degree polynomial function with the graph and intercepts shown in the figure.

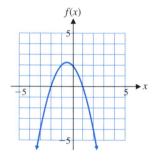

Figure for 40

41. Write an equation for the lowest-degree polynomial function with the graph and intercepts shown in the figure.

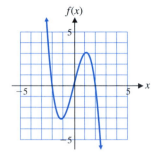

Figure for 41

42. Write an equation for the lowest-degree polynomial function with the graph and intercepts shown in the figure.

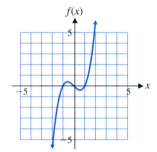

Figure for 42

▶ **APPLICATIONS**

Business & Economics

43. *Average cost.* A company manufacturing snowboards has fixed costs of $200 per day and total costs of $3,800 per day at a daily output of 20 boards.
   (A) Assuming the total cost per day, $C(x)$, is linearly related to the total output per day, x, write an equation for the cost function.
   (B) The average cost per board for an output of x boards is given by $\overline{C}(x) = C(x)/x$. Find the average cost function.
   (C) Sketch a graph of the average cost function, including any asymptotes for $1 \leq x \leq 30$.
   (D) What does the average cost per board tend to as production increases?

44. *Average cost.* A company manufacturing surfboards has fixed costs of $300 per day and total costs of $5,100 per day at a daily output of 20 boards.
   (A) Assuming the total cost per day, $C(x)$, is linearly related to the total output per day, x, write an equation for the cost function.
   (B) The average cost per board for an output of x boards is given by $\overline{C}(x) = C(x)/x$. Find the average cost function.
   (C) Sketch a graph of the average cost function, including any asymptotes for $1 \leq x \leq 30$.
   (D) What does the average cost per board tend to as production increases?

45. *Replacement time.* An office copier has an initial price of $2,500. A service contract costs $200 for the first year and increases $50 per year thereafter. It can be shown that the total cost of the copier after n years is given by

$$C(n) = 2,500 + 175n + 25n^2$$

The average cost per year for n years is $\overline{C}(n) = C(n)/n$.
(A) Find the rational function $\overline{C}$.
(B) Graph the average cost function $\overline{C}$ for $2 \leq n \leq 20$ in a graphing utility, and use trace and zoom or an appropriate built-in routine to find when the average annual cost is minimum. What is the minimum average annual cost? The time when the average cost is minimum is frequently referred to as the **replacement time** for the piece of equipment.

46. *Minimum average cost.* Financial analysts in a company that manufactures audio CD players arrived at the following daily cost equation for manufacturing x CD players per day:

$$C(x) = x^2 + 2x + 2,000$$

The average cost per unit at a production level of x players per day is $\overline{C}(x) = C(x)/x$.
(A) Find the rational function $\overline{C}$.
(B) Graph the average cost function $\overline{C}$ for $5 \leq x \leq 150$ in a graphing utility, and use trace and zoom or an appropriate built-in routine to find the daily production level (to the nearest integer) at which the average cost per unit is minimum. What is the minimum average cost (to the nearest cent)?

47. *Minimum average cost.* A consulting firm, using statistical methods, provided a veterinary clinic with the cost equation

$$C(x) = 0.00048(x - 500)^3 + 60,000 \qquad 100 \leq x \leq 1,000$$

where $C(x)$ is the cost in dollars for handling x cases per month. The average cost per case is given by $\overline{C}(x) = C(x)/x$.
(A) Write the equation for the average cost function $\overline{C}$.
(B) Graph $\overline{C}$ on a graphing utility.
(C) Use trace and zoom or an appropriate built-in routine to find the monthly case load for the minimum average cost per case. What is the minimum average cost per case?

48. *Minimum average cost.* The financial department of a hospital, using statistical methods, arrived at the cost equation

$$C(x) = 20x^3 - 360x^2 + 2,300x - 1,000 \qquad 1 \leq x \leq 12$$

where $C(x)$ is the cost in thousands of dollars for handling x thousand cases per month. The average cost per case is given by $\overline{C}(x) = C(x)/x$.
(A) Write the equation for the average cost function $\overline{C}$.
(B) Graph $\overline{C}$ on a graphing utility.
(C) Use trace and zoom or an appropriate built-in routine to find the monthly case load for the minimum average cost per case. What is the minimum average cost per case to the nearest dollar?

Life Sciences

49. *Physiology.* In a study on the speed of muscle contraction in frogs under various loads, researchers W. O. Fems and J. Marsh found that the speed of contraction decreases with increasing loads. In particular, they found that the relationship between speed of contraction v (in centimeters per second) and load x (in grams) is given approximately by

$$v(x) = \frac{26 + 0.06x}{x} \qquad x \geq 5$$

(A) What does $v(x)$ approach as x increases?
(B) Sketch a graph of function v.

Social Sciences

50. *Learning theory.* In 1917, L. L. Thurstone, a pioneer in quantitative learning theory, proposed the rational function

$$f(x) = \frac{a(x + c)}{(x + c) + b}$$

to model the number of successful acts per unit time that a person could accomplish after x practice sessions. Suppose that for a particular person enrolled in a typing class,

$$f(x) = \frac{55(x + 1)}{(x + 8)} \qquad x \geq 0$$

where $f(x)$ is the number of words per minute the person is able to type after x weeks of lessons.
(A) What does $f(x)$ approach as x increases?
(B) Sketch a graph of function f, including any vertical or horizontal asymptotes.

SECTION   2-2   ■   # Exponential Functions

- ■ EXPONENTIAL FUNCTIONS
- ■ BASE e EXPONENTIAL FUNCTIONS
- ■ GROWTH AND DECAY APPLICATIONS
- ■ COMPOUND INTEREST
- ■ CONTINUOUS COMPOUND INTEREST

This section introduces the important class of functions called *exponential functions*. These functions are used extensively in modeling and solving a wide variety of real-world problems, including growth of money at compound interest; growth of populations of people, animals, and bacteria; radioactive decay; and learning associated with the mastery of such devices as a new computer or an assembly process in a manufacturing plant.

■ EXPONENTIAL FUNCTIONS

We start by noting that

$$f(x) = 2^x \qquad \text{and} \qquad g(x) = x^2$$

are not the same function. Whether a variable appears as an exponent with a constant base or as a base with a constant exponent, makes a big difference. The function g is a quadratic function, which we have already discussed. The function f is a new type of function called an *exponential function*. In general:

> **Exponential Function**
> _____
>
> The equation
>
> $$f(x) = b^x \qquad b > 0, b \neq 1$$
>
> defines an **exponential function** for each different constant b, called the **base.** The **domain** of f is the set of all real numbers, and the **range** of f is the set of all positive real numbers.

We require the base b to be positive to avoid imaginary numbers such as $(-2)^{1/2} = \sqrt{-2} = i\sqrt{2}$. We exclude $b = 1$ as a base, since $f(x) = 1^x = 1$ is a constant function, which we have already considered.

Many students, if asked to hand sketch graphs of equations such as $y = 2^x$ or $y = 2^{-x}$, would not hesitate at all. [*Note:* $2^{-x} = 1/2^x = (1/2)^x$.] They would likely make up tables by assigning integers to x, plot the resulting points, and then join these points with a smooth curve as in Figure 1. The only problem with this process is the definition of 2^x for any real number x. If x is a rational number of the form $x = p/q$, $q \neq 0$, then we define $2^{p/q}$ as

$$2^{p/q} = \sqrt[q]{x^p}$$

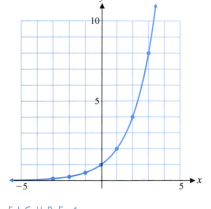

FIGURE 1
$y = 2^x$

But how do we define $2^{\sqrt{2}}$? The question is not easy to answer at this time. In fact, a precise definition of $2^{\sqrt{2}}$ must wait for more advanced courses, where it is shown that

$$2^x$$

names a positive real number for x any real number, and that the graph of $y = 2^x$ is indeed as indicated in Figure 1.

It is useful to compare the graphs of $y = 2^x$ and $y = 2^{-x}$ by plotting both on the same set of coordinate axes, as shown in Figure 2A. The graph of

$$f(x) = b^x \qquad b > 1 \text{ (Fig. 2B)}$$

looks very much like the graph of $y = 2^x$, and the graph of

$$f(x) = b^x \qquad 0 < b < 1 \text{ (Fig. 2B)}$$

looks very much like the graph of $y = 2^{-x}$. Note that in both cases the x axis is a horizontal asymptote for the graphs.

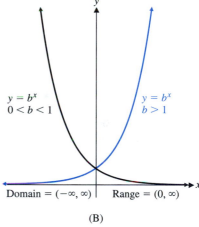

(A)   (B)

FIGURE 2
Exponential functions

The graphs in Figure 2 suggest the following important general properties of exponential functions, which we state without proof:

Basic Properties of the Graph of $f(x) = b^x$, $b > 0$, $b \neq 1$

1. All graphs will pass through the point $(0, 1)$. $b^0 = 1$ for any permissible base b.

2. All graphs are continuous curves, with no holes or jumps.
3. The x axis is a horizontal asymptote.
4. If $b > 1$, then b^x increases as x increases.
5. If $0 < b < 1$, then b^x decreases as x increases.

The use of a graphing utility makes the graphing of exponential functions almost routine.

EXAMPLE 1 ➤ **Graphing Exponential Functions** Sketch a graph of $y = (\frac{1}{2})4^x$, $-2 \le x \le 2$.

Solution

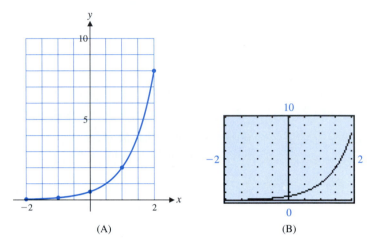

(A)

(B)

FIGURE 3
Graph of $y = (\frac{1}{2})4^x$

◄

MATCHED PROBLEM 1 ➤ **Sketch a graph of $y = (\frac{1}{2})4^{-x}$, $-2 \le x \le 2$.**

◄

explore – discuss 1

Graph the functions $f(x) = 2^x$ and $g(x) = 3^x$ in the same viewing window. At which values of x do the graphs intersect? For which values of x is the graph of f above the graph of g? Below the graph of g? Are the graphs close together as x increases without bound? Are the graphs close together as x decreases without bound? Discuss.

Exponential functions, which allow irrational exponents as well as rational exponents, obey the familiar laws of exponents for rational exponents. We summarize these exponent laws here and add two other important and useful properties.

Exponential Function Properties

For a and b positive, $a \ne 1$, $b \ne 1$, and x and y real:

1. Exponent laws:

$$a^x a^y = a^{x+y} \qquad \frac{a^x}{a^y} = a^{x-y} \qquad \frac{4^{2y}}{4^{5y}} = 4^{2y-5y} = 4^{-3y}$$

$$(a^x)^y = a^{xy} \qquad (ab)^x = a^x b^x \qquad \left(\frac{a}{b}\right)^x = \frac{a^x}{b^x}$$

Exponential Function Properties *(Continued)*

---

2. $a^x = a^y$ if and only if $x = y$ If $7^{5t+1} = 7^{3t-3}$, then

$$5t + 1 = 3t - 3, \text{ and } t = -2.$$

3. For $x \neq 0$,

$$a^x = b^x \qquad \text{if and only if} \qquad a = b \quad \text{If } a^5 = 2^5, \text{ then } a = 2.$$

■ BASE e EXPONENTIAL FUNCTIONS

Of all the possible bases b we can use for the exponential function $y = b^x$, which ones are the most useful? If you look at the keys on a scientific or graphing calculator, you will likely see $\boxed{10^x}$ and $\boxed{e^x}$. It is clear why base 10 would be important, because our number system is a base 10 system. But what is e, and why is it included as a base? It turns out that base e is used more frequently than all other bases combined. The reason for this is that certain formulas and the results of certain processes found in calculus and more advanced mathematics take on their simplest form if this base is used. This is why you will see e used extensively in expressions and formulas that model real-world phenomena. In fact, its use is so prevalent that you will often hear people refer to $y = e^x$ as *the* exponential function.

The base e is an irrational number, and like π, it cannot be represented exactly by any finite decimal fraction. However, e can be approximated as closely as we like by evaluating the expression

$$\left(1 + \frac{1}{x}\right)^x \tag{1}$$

for sufficiently large x. What happens to the value of expression (1) as x increases without bound? Think about this for a moment before proceeding. Maybe you guessed that the value approaches 1, because

$$1 + \frac{1}{x}$$

approaches 1, and 1 raised to any power is 1. Let us see if this reasoning is correct by actually calculating the value of the expression for larger and larger values of x. Table 1 summarizes the results.

TABLE 1

x	$\left(1 + \dfrac{1}{x}\right)^x$
1	2
10	2.593 74. . .
100	2.704 81. . .
1,000	2.716 92. . .
10,000	2.718 14. . .
100,000	2.718 27. . .
1,000,000	2.718 28. . .
⋮	⋮

e x p l o r e – d i s c u s s   2

Graph

$$y = \left(1 + \frac{1}{x}\right)^x \qquad 1 \leq x \leq 100$$

on a graphing utility. Observe the value of y using trace as x varies from 1 to 100. Does it appear that the graph has a horizontal asymptote? If so, what does its equation appear to be? (Approximate value to two decimal places.)

From Table 1 and explore–discuss 2, it is clear that the value of expression (1) is never close to 1, but seems to be approaching a number close to 2.7183. In fact, as x increases without bound, the value of expression (1) approaches an irrational number that we call e. The irrational number e to twelve decimal places is

$$e = 2.718\ 281\ 828\ 459$$

Compare this value of e with the value of e^1 from a calculator. Exactly who discovered the constant e is still being debated. It is named after the great Swiss mathematician Leonhard Euler (1707–1783).

Exponential Function with Base e

Exponential functions with base e and base $1/e$ are respectively defined by

$$y = e^x \quad \text{and} \quad y = e^{-x}$$

Domain: $(-\infty, \infty)$

Range: $(0, \infty)$

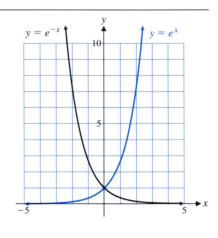

---

explore – discuss 3

Graph the functions $f(x) = e^x$, $g(x) = 2^x$, and $h(x) = 3^x$ in the same viewing window. At which values of x do the graphs intersect? For positive values of x, which of the three graphs lies above the other two? Below the other two? How does your answer change for negative values of x?

■ GROWTH AND DECAY APPLICATIONS

Most exponential growth and decay problems are modeled using base e exponential functions. We present two applications here and many more in Exercise 2-2.

EXAMPLE 2 ➤   Exponential Growth   Cholera, an intestinal disease, is caused by a cholera bacterium that multiplies exponentially by cell division as given approximately by

$$N = N_0 e^{1.386t}$$

where N is the number of bacteria present after t hours and N_0 is the number of bacteria present at the start ($t = 0$). If we start with 25 bacteria, how many bacteria (to the nearest unit) will be present:

(A) In 0.6 hour?   (B) In 3.5 hours?

Solution   Substituting $N_0 = 25$ into the above equation, we obtain

$$N = 25e^{1.386t}$$ The graph is shown in Figure 4, which you should confirm on a graphing utility.

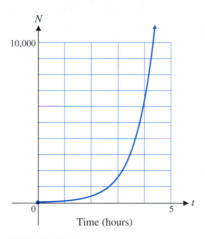

FIGURE 4

(A) Solve for N when $t = 0.6$:
$$N = 25e^{1.386(0.6)}$$ Use a calculator.
$$= 57 \text{ bacteria}$$

(B) Solve for N when $t = 3.5$:
$$N = 25e^{1.386(3.5)}$$ Use a calculator.
$$= 3{,}197 \text{ bacteria}$$

MATCHED PROBLEM 2 ➤   Refer to the exponential growth model for cholera in Example 2. If we start with 55 bacteria, how many bacteria (to the nearest unit) will be present:

(A) In 0.85 hour?   (B) In 7.25 hours?

EXAMPLE 3 ➤   **Exponential Decay**   Cosmic-ray bombardment of the atmosphere produces neutrons, which in turn react with nitrogen to produce radioactive carbon-14 (^{14}C). Radioactive ^{14}C enters all living tissues through carbon dioxide, which is first absorbed by plants. As long as a plant or animal is alive, ^{14}C is maintained in the living organism at a constant level. Once the organism dies, however, ^{14}C decays according to the equation

$$A = A_0 e^{-0.000124t}$$

where A is the amount present after t years and A_0 is the amount present at time $t = 0$. If 500 milligrams of ^{14}C are present in a sample from a skull at the time of death, how many milligrams will be present in the sample in:

(A) 15,000 years?   (B) 45,000 years?

Compute answers to two decimal places.

Solution Substituting $A_0 = 500$ in the decay equation, we have

$$A = 500e^{-0.000124t}$$ *See the graph in Figure 5, which you should confirm on a graphing utility.*

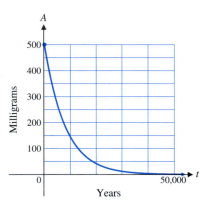

FIGURE 5

(A) Solve for A when $t = 15,000$:

$$A = 500e^{-0.000124(15,000)}$$ *Use a calculator.*
$$= 77.84 \text{ milligrams}$$

(B) Solve for A when $t = 45,000$:

$$A = 500e^{-0.000124(45,000)}$$ *Use a calculator.*
$$= 1.89 \text{ milligrams}$$ ◀

MATCHED PROBLEM 3 ▶ Refer to the exponential decay model in Example 3. How many milligrams of ^{14}C would have to be present at the beginning in order to have 25 milligrams present after 18,000 years? Compute the answer to the nearest milligram.   ◀

explore – discuss 4

(A) In the same viewing window, graph the three decay equations $A = A_0 e^{-0.35t}$, $t \geqslant 0$, for $A_0 = 10$, 20, and 30.
(B) Identify any asymptotes for the three graphs in part (A).
(C) Discuss the long-term behavior for the equations in part (A).

■ COMPOUND INTEREST

We now turn to the growth of money at compound interest. The fee paid to use another's money is called **interest.** It is usually computed as a percent (called **interest rate**) of the principal over a given period of time. If, at the end of a payment period, the interest due is reinvested at the same rate, then the interest earned as well as the principal will earn interest during the next payment period. Interest paid on interest reinvested is called **compound interest,** and may be calculated using the following compound interest formula:

Compound Interest

If a **principal** *P* (**present value**) is invested at an annual **rate** *r* (expressed as a decimal) compounded *m* times a year, then the **amount** *A* (**future value**) in the account at the end of *t* years is given by

$$A = P \left(1 + \frac{r}{m} \right)^{mt}$$

[*Note:* *P* could be replaced by A_0, but convention dictates otherwise.]

EXAMPLE 4 ➤ Compound Growth If $1,000 is invested in an account paying 10% compounded monthly, how much will be in the account at the end of 10 years? Compute the answer to the nearest cent.

Solution We use the compound interest formula as follows:

$$A = P \left(1 + \frac{r}{m} \right)^{mt}$$

$$= 1,000 \left(1 + \frac{0.10}{12} \right)^{(12)(10)} \qquad \text{Use a calculator.}$$

$$= \$2,707.04$$

The graph of

$$A = 1,000 \left(1 + \frac{0.10}{12} \right)^{12t}$$

for $0 \leq t \leq 20$ is shown in Figure 6, which you should confirm on a graphing utility.

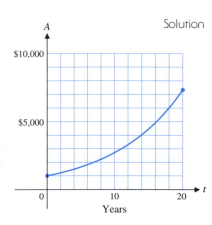

FIGURE 6

MATCHED PROBLEM 4 ➤ If you deposit $5,000 in an account paying 9% compounded daily, how much will you have in the account in 5 years? Compute the answer to the nearest cent. ◀

explore – discuss 5

Suppose that $1,000 is deposited in a savings account at an annual rate of 5%. Guess the amount in the account at the end of 1 year if interest is compounded (1) quarterly, (2) monthly, (3) daily, (4) hourly. Use the compound interest formula to compute the amounts at the end of 1 year to the nearest cent. Discuss the accuracy of your initial guesses.

■ CONTINUOUS COMPOUND INTEREST

Returning to the compound interest formula,

$$A = P\left(1 + \frac{r}{m}\right)^{mt}$$

suppose the principal P, the annual rate r, and the time t are held fixed, and the number of compounding periods per year m is increased without bound. Will the amount A increase without bound, or will it tend to some limiting value?

Starting with $P = \$100$, $r = 0.08$, and $t = 2$ years, we construct Table 2 for several values of m with the aid of a calculator. Notice that the largest gain appears in going from annual to semiannual compounding. Then, the gains slow down as m increases. It appears that A gets closer and closer to $117.35 as m gets larger and larger.

TABLE 2

COMPOUNDING FREQUENCY	m	$A = 100\left(1 + \dfrac{0.08}{m}\right)^{2m}$
Annually	1	$116.6400
Semiannually	2	116.9859
Quarterly	4	117.1659
Weekly	52	117.3367
Daily	365	117.3490
Hourly	8,760	117.3510

It can be shown that

$$P\left(1 + \frac{r}{m}\right)^{mt}$$

gets closer and closer to Pe^{rt} as the number of compounding periods m gets larger and larger. The latter is referred to as the **continuous compound interest formula,** a formula that is widely used in business, banking, and economics.

Continuous Compound Interest Formula

If a principal P is invested at an annual rate r (expressed as a decimal) compounded continuously, then the amount A in the account at the end of t years is
given by

$$A = Pe^{rt}$$

EXAMPLE 5 ▶ **Compounding Daily and Continuously** What amount will an account have after
2 years if $5,000 is invested at an annual rate of 8%:

(A) Compounded daily? (B) Compounded continuously?

Compute answers to the nearest cent.

Solution (A) Use the compound interest formula

$$A = P\left(1 + \frac{r}{m}\right)^{mt}$$

with $P = 5{,}000$, $r = 0.08$, $m = 365$, and $t = 2$:

$$A = 5{,}000\left(1 + \frac{0.08}{365}\right)^{(365)(2)} \quad \textit{Use a calculator.}$$

$$= \$5{,}867.45$$

(B) Use the continuous compound interest formula

$$A = Pe^{rt}$$

with $P = 5{,}000$, $r = 0.08$, and $t = 2$:

$$A = 5{,}000e^{(0.08)(2)} \quad \textit{Use a calculator.}$$

$$= \$5{,}867.55$$

◀

MATCHED PROBLEM 5 ▶ What amount will an account have after 1.5 years if $8,000 is invested at an annual
rate of 9%:

(A) Compounded weekly? (B) Compounded continuously?

Compute answers to the nearest cent. ◀

The formulas for simple interest, compound interest, and continuous compound
interest are summarized in the box for convenient reference.

Interest Formulas

Simple interest	$A = P(1 + rt)$
Compound interest	$A = P\left(1 + \dfrac{r}{m}\right)^{mt}$
Continuous compound interest	$A = Pe^{rt}$

Answers to Matched Problems 1.

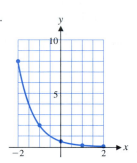

2. (A) 179 bacteria (B) 1,271,659 bacteria
3. 233 mg 4. $7,841.13
5. (A) $9,155.23 (B) $9,156.29

EXERCISE 2-2

A *Graph each function in Problems 1–12 over the indicated interval using a graphing utility as an aid.*

1. $y = 5^x$; $[-2, 2]$ **2.** $y = 3^x$; $[-3, 3]$
3. $y = (\frac{1}{5})^x = 5^{-x}$; $[-2, 2]$ **4.** $y = (\frac{1}{3})^x = 3^{-x}$; $[-3, 3]$
5. $f(x) = -5^x$; $[-2, 2]$ **6.** $g(x) = -3^{-x}$; $[-3, 3]$
7. $y = -e^{-x}$; $[-3, 3]$ **8.** $y = -e^x$; $[-3, 3]$
9. $y = 100e^{0.1x}$; $[-5, 5]$ **10.** $y = 10e^{0.2x}$; $[-10, 10]$
11. $g(t) = 10e^{-0.2t}$; $[-5, 5]$
12. $f(t) = 100e^{-0.1t}$; $[-5, 5]$

Simplify each expression in Problems 13–18.

13. $(4^{3x})^{2y}$ **14.** $10^{3x-1}10^{4-x}$ **15.** $\dfrac{e^{x-3}}{e^{x-4}}$

16. $\dfrac{e^x}{e^{1-x}}$ **17.** $(2e^{1.2t})^3$ **18.** $(3e^{-1.4x})^2$

B *In Problems 19–26, describe the transformations that can be used to obtain the graph of g from the graph of f (see Section 1-3). Check the answers by graphing each pair of functions in the same viewing window of a graphing utility.*

19. $g(x) = -2^x$; $f(x) = 2^x$ **20.** $g(x) = 2^{x-2}$; $f(x) = 2^x$
21. $g(x) = 3^{x+1}$; $f(x) = 3^x$ **22.** $g(x) = -3^x$; $f(x) = 3^x$
23. $g(x) = e^x + 1$; $f(x) = e^x$ **24.** $g(x) = e^x - 2$; $f(x) = e^x$
25. $g(x) = 2e^{-(x+2)}$; $f(x) = e^{-x}$
26. $g(x) = 0.5e^{-(x-1)}$; $f(x) = e^{-x}$

In Problems 27–36, graph each function over the indicated interval using a graphing utility as an aid.

27. $f(t) = 2^{t/10}$; $[-30, 30]$
28. $G(t) = 3^{t/100}$; $[-200, 200]$
29. $y = -3 + e^{1+x}$; $[-4, 2]$ **30.** $y = 2 + e^{x-2}$; $[-1, 5]$
31. $y = e^{|x|}$; $[-3, 3]$ **32.** $y = e^{-|x|}$; $[-3, 3]$
33. $C(x) = \dfrac{e^x + e^{-x}}{2}$; $[-5, 5]$

34. $M(x) = e^{x/2} + e^{-x/2}$; $[-5, 5]$

35. $y = e^{-x^2}$; $[-3, 3]$

36. $y = 2^{-x^2}$; $[-3, 3]$

37. A graphing utility was used to graph the exponential functions $f(x) = 2^x$ and $g(x) = e^x$ in the same viewing window, as shown in the figure. Which curve belongs to which function? Explain.

Figure for 37

38. A graphing utility was used to graph the exponential functions $f(x) = 2^x$ and $g(x) = e^x$ in the same viewing window, as shown in the figure. Which curve belongs to which function? Explain.

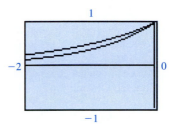

Figure for 38

39. A graphing utility was used to graph the exponential functions $f(x) = 2^{-x}$ and $g(x) = e^{-x}$ in the same viewing window, as shown in the figure. Which curve belongs to which function? Explain.

Figure for 39

40. A graphing utility was used to graph the exponential functions $f(x) = 2^{-x}$ and $g(x) = e^{-x}$ in the same viewing window, as shown in the figure. Which curve belongs to which function? Explain.

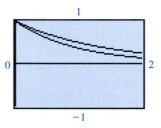

Figure for 40

Algebraically solve each equation in Problems 41–46 for x.

41. $10^{2-3x} = 10^{5x-6}$

42. $5^{3x} = 5^{4x-2}$

43. $4^{5x-x^2} = 4^{-6}$

44. $7^{x^2} = 7^{2x+3}$

45. $5^3 = (x + 2)^3$

46. $(1 - x)^5 = (2x - 1)^5$

C *Algebraically solve each equation in Problems 47–50 for x. (Remember, $e^x \neq 0$ and $e^{-x} \neq 0$.)*

47. $(x - 3)e^x = 0$

48. $2xe^{-x} = 0$

49. $3xe^{-x} + x^2e^{-x} = 0$

50. $x^2e^x - 5xe^x = 0$

Graph each function in Problems 51–54 over the indicated interval using a graphing utility as an aid.

51. $h(x) = x(2^x)$; $[-5, 0]$

52. $m(x) = x(3^{-x})$; $[0, 3]$

53. $N = \dfrac{100}{1 + e^{-t}}$; $[0, 5]$

54. $N = \dfrac{200}{1 + 3e^{-t}}$; $[0, 5]$

▶ A P P L I C A T I O N S

Business & Economics

55. *Finance.* Suppose $2,500 is invested at 7% compounded quarterly. How much money will be in the account in:
(A) $\frac{3}{4}$ year?   (B) 15 years?
Compute answers to the nearest cent.

56. *Finance.* Suppose $4,000 is invested at 11% compounded weekly. How much money will be in the account in:
(A) $\frac{1}{2}$ year?   (B) 10 years?
Compute answers to the nearest cent.

57. *Money growth.* If you invest $7,500 in an account paying 8.35% compounded continuously, how much money will be in the account at the end of:
(A) 5.5 years?   (B) 12 years?

58. *Money growth.* If you invest $5,250 in an account paying 11.38% compounded continuously, how much money will be in the account at the end of:
(A) 6.25 years?   (B) 17 years?

59. *Finance.* A person wishes to have $15,000 cash for a new car 5 years from now. How much should be placed in an account now, if the account pays 9.75% compounded weekly? Compute the answer to the nearest dollar.

60. *Finance.* A couple just had a baby. How much should they invest now at 8.25% compounded daily in order to have $40,000 for the child's education 17 years from now? Compute the answer to the nearest dollar.

61. *Money growth.* *Barron's* (a national business and financial weekly) published the following ''Top Savings Deposit Yields'' for 1 year certificate of deposit accounts:

(A) Alamo Savings, 8.25% compounded quarterly
(B) Lamar Savings, 8.05% compounded continuously

Compute the value of $10,000 invested in each account at the end of 1 year.

62. *Money growth.* Refer to Problem 61. In another issue of *Barron's,* $2\frac{1}{2}$ year certificate of deposit accounts included the following:
(A) Gill Saving, 8.30% compounded continuously
(B) Richardson Savings and Loan, 8.40% compounded quarterly
(C) USA Savings, 8.25% compounded daily

Compute the value of $1,000 invested in each account at the end of $2\frac{1}{2}$ years.

63. *Present value.* A promissory note will pay $50,000 at maturity $5\frac{1}{2}$ years from now. How much should you be willing to pay for the note now if money is worth 10% compounded continuously?

64. *Present value.* A promissory note will pay $30,000 at maturity 10 years from now. How much should you be willing to pay for the note now if money is worth 9% compounded continuously?

65. *Advertising.* A company is trying to introduce a new product to as many people as possible through television advertising in a large metropolitan area with 2 million possible viewers. A model for the number of people N (in millions) who are aware of the product after t days of advertising was found to be

$$N = 2(1 - e^{-0.037t})$$

Graph this function for $0 \leq t \leq 50$. What value does N tend to as t increases without bound?

66. *Learning curve.* People assigned to assemble circuit boards for a computer manufacturing company undergo on-the-job training. From past experience it was found that the learning curve for the average employee is given by

$$N = 40(1 - e^{-0.12t})$$

where N is the number of boards assembled per day after t days of training. Graph this function for $0 \leq t \leq 30$. What is the maximum number of boards an average employee can be expected to produce in 1 day?

Life Sciences

67. *Marine biology.* Marine life is dependent upon the microscopic plant life that exists in the *photic zone,* a zone that goes to a depth where about 1% of the surface light still remains. In some waters with a great deal of sediment, the photic zone may go down only 15–20 feet. In some murky harbors, the intensity of light *d* feet below the surface is given approximately by

$$I = I_0 e^{-0.23d}$$

What percentage of the surface light will reach a depth of:
(A) 10 feet?   (B) 20 feet?

68. *Marine biology.* Refer to Problem 67. Light intensity *I* relative to depth *d* (in feet) for one of the clearest bodies of water in the world, the Sargasso Sea in the West Indies, can be approximated by

$$I = I_0 e^{-0.00942d}$$

where I_0 is the intensity of light at the surface. What percent of the surface light will reach a depth of:
(A) 50 feet?   (B) 100 feet?

69. *AIDS epidemic.* The U.S. Department of Health and Human Services reports that prior to 1993 there were about 40,000 cases of AIDS among intravenous drug users in the United States. It was estimated that the disease was spreading in this group at about 21% compounded continuously. Let 1992 be year 0 and assume this rate does not change.
(A) Write an equation that models the growth of AIDS in this group, starting in 1992.
(B) Based on the model, how many cases (to the nearest thousand) should we expect by the end of the year 2000? The year 2005?
(C) Sketch a graph of the equation found in part (A). Cover the years from 1992 through 2005.

70. *AIDS epidemic.* The U.S. Department of Health and Human Services reports that prior to 1993 there were about 245,000 cases of AIDS among the general population in the United States. It was estimated that the disease was spreading at about 18% compounded continuously. Let 1992 be year 0 and assume this rate does not change.
(A) Write an equation that models the growth of AIDS, starting in 1992.
(B) Based on the model, how many cases (to the nearest hundred thousand) should we expect by the end of the year 2000? The year 2005?
(C) Sketch a graph of the equation found in part (A). Cover the years from 1992 through 2005.

Social Sciences

71. *World population growth.* It took from the dawn of humanity to 1830 for the population to grow to the first billion people, just 100 more years (by 1930) for the second billion, and 3 billion more were added in only 60 more years (by 1990). In 1995, the estimated world population was 5.7 billion. In 1994, the World Bank estimated the world population would be growing at 1.14% compounded continuously until 2030.
(A) Write an equation that models the world population growth, letting 1995 be year 0.
(B) Based on the model, what is the expected world population (to the nearest hundred million) in 2010? In 2030?
(C) Sketch a graph of the equation found in part (A). Cover the years from 1995 through 2030.

72. *Population growth in Ethiopia.* In 1995, the estimated population in Ethiopia was 88 million people. In 1994, the World Bank estimated the population would grow at 1.67% compounded continuously until 2030.
(A) Write an equation that models the population growth in Ethiopia, letting 1995 be year 0.
(B) Based on the model, what is the expected population in Ethiopia (to the nearest million) in 2010? In 2030?
(C) Sketch a graph of the equation found in part (A). Cover the years from 1995 through 2030.

S E C T I O N   2-3   ■   # Logarithmic Functions

- ■ INVERSE FUNCTIONS
- ■ LOGARITHMIC FUNCTIONS
- ■ PROPERTIES OF LOGARITHMIC FUNCTIONS
- ■ CALCULATOR EVALUATION OF LOGARITHMS
- ■ APPLICATION

Find the exponential function keys $\boxed{10^x}$ and $\boxed{e^x}$ on your calculator. Close to these keys you will find $\boxed{\text{LOG}}$ and $\boxed{\text{LN}}$ keys. The latter represent *logarithmic functions,* and each is closely related to the exponential function it is near. In fact, the exponential function and the corresponding logarithmic function are said to be *inverses* of each other. In this section we will develop the concept of inverse functions and use it to define a logarithmic function as the inverse of an exponential function. We will then investigate basic properties of logarithmic functions, use a calculator to evaluate them for particular values of x, and apply them to real-world problems.

Logarithmic functions are used in modeling and solving many types of problems. For example, the decibel scale is a logarithmic scale used to measure sound intensity, and the Richter scale is a logarithmic scale used to measure the strength of the force of an earthquake. An important business application has to do with finding the time it takes money to double if it is invested at a certain rate compounded a given number of times a year or compounded continuously. This requires the solution of an exponential equation, and logarithms play a central role in the process.

■ INVERSE FUNCTIONS

Look at the graphs of $f(x) = \dfrac{x}{2}$ and $g(x) = \dfrac{|x|}{2}$ in Figure 1:

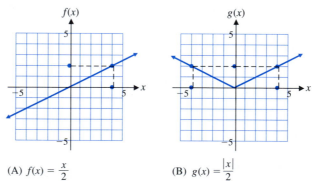

(A) $f(x) = \dfrac{x}{2}$ (B) $g(x) = \dfrac{|x|}{2}$

FIGURE 1

Because both f and g are functions, each domain value corresponds to exactly one range value. For which function does each range value correspond to exactly one domain value? This is the case only for function f. Note that for the range value 2, the corresponding domain value is 4. For function g the range value 2 corresponds to both -4 and 4. Function f is said to be *one-to-one*. In general:

One-to-One Functions

A function f is said to be **one-to-one** if each range value corresponds to exactly one domain value.

It can be shown that any continuous function that is either increasing or decreasing for all domain values is one-to-one. If a continuous function increases for some domain values and decreases for others, it cannot be one-to-one. Figure 1 shows an example of each case.

e x p l o r e – d i s c u s s 1

Graph $f(x) = 2^x$ and $g(x) = x^2$. For a range value of 4, what are the corresponding domain values for each function? Which of the two functions is one-to-one? Explain why.

Starting with a one-to-one function f we can obtain a new function called the *inverse* of f as follows:

Inverse of a Function

If f is a one-to-one function, then the **inverse** of f is the function formed by interchanging the independent and dependent variables for f. Thus, if (a, b) is a point on the graph of f, then (b, a) is a point on the graph of the inverse of f.

[*Note:* If f is not one-to-one, then f **does not have an inverse.**]

A number of important functions in any library of elementary functions are the inverses of other basic functions in the library. In this course, we are interested in the inverses of exponential functions, called *logarithmic functions*.

■ LOGARITHMIC FUNCTIONS

If we start with the exponential function f defined by

$$y = 2^x \tag{1}$$

and interchange the variables, we obtain the inverse of f:

$$x = 2^y \tag{2}$$

We call the inverse the **logarithmic function with base 2,** and write

$$y = \log_2 x \qquad \text{if and only if} \qquad x = 2^y$$

We can graph $y = \log_2 x$ by graphing $x = 2^y$, since they are equivalent. Any ordered pair of numbers on the graph of the exponential function will be on the graph of the logarithmic function if we interchange the order of the components. For example, $(3, 8)$ satisfies equation (1) and $(8, 3)$ satisfies equation (2). The graphs of $y = 2^x$ and $y = \log_2 x$ are shown in Figure 2. Note that if we fold the paper along the dashed line $y = x$ in Figure 2, the two graphs match exactly. The line $y = x$ is a line of symmetry for the two graphs.

EXPONENTIAL FUNCTION		LOGARITHMIC FUNCTION	
x	$y = 2^x$	$x = 2^y$	y
-3	$\frac{1}{8}$	$\frac{1}{8}$	-3
-2	$\frac{1}{4}$	$\frac{1}{4}$	-2
-1	$\frac{1}{2}$	$\frac{1}{2}$	-1
0	1	1	0
1	2	2	1
2	4	4	2
3	8	8	3

Ordered pairs reversed

FIGURE 2

In general, since the graphs of all exponential functions of the form $f(x) = b^x$, $b \neq 1$, $b > 0$, are either increasing or decreasing (see Section 2-2), exponential functions have inverses.

Logarithmic Functions

The inverse of an exponential function is called a **logarithmic function.** For $b > 0$ and $b \neq 1$,

Logarithmic form
$$y = \log_b x \qquad \text{is equivalent to}$$

Exponential form
$$x = b^y$$

The log to the base b of x is the exponent to which b must be raised to obtain x. [*Remember:* A logarithm is an exponent.] The **domain** of the logarithmic function is the set of all positive real numbers, which is also the range of the corresponding exponential function; and the **range** of the logarithmic function is the set of all real numbers, which is also the domain of the corresponding

Logarithmic Functions *(Continued)*

exponential function. Typical graphs of an exponential function and its inverse, a logarithmic function, are shown in the figure, and the domains and ranges are summarized in the table:

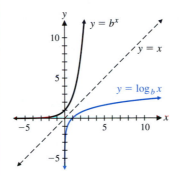

	DOMAIN	RANGE
$y = \log_b x$	$(0, \infty)$	$(-\infty, \infty)$
$y = e^x$	$(-\infty, \infty)$	$(0, \infty)$

Base $b > 1$

The following examples involve converting logarithmic forms to equivalent exponential forms and vice versa.

EXAMPLE 1 ➤ **Logarithmic–Exponential Conversions** Change each logarithmic form to an equivalent exponential form:

(A) $\log_5 25 = 2$ (B) $\log_9 3 = \frac{1}{2}$ (C) $\log_2(\frac{1}{4}) = -2$

Solution   (A) $\log_5 25 = 2$ is equivalent to $25 = 5^2$
(B) $\log_9 3 = \frac{1}{2}$ is equivalent to $3 = 9^{1/2}$
(C) $\log_2(\frac{1}{4}) = -2$ is equivalent to $\frac{1}{4} = 2^{-2}$ ◀

MATCHED PROBLEM 1 ➤ Change each logarithmic form to an equivalent exponential form:

(A) $\log_3 9 = 2$ (B) $\log_4 2 = \frac{1}{2}$ (C) $\log_3(\frac{1}{9}) = -2$ ◀

EXAMPLE 2 ➤ **Exponential–Logarithmic Conversions** Change each exponential form to an equivalent logarithmic form:

(A) $64 = 4^3$ (B) $6 = \sqrt{36}$ (C) $\frac{1}{8} = 2^{-3}$

Solution   (A) $64 = 4^3$ is equivalent to $\log_4 64 = 3$
(B) $6 = \sqrt{36}$ is equivalent to $\log_{36} 6 = \frac{1}{2}$
(C) $\frac{1}{8} = 2^{-3}$ is equivalent to $\log_2(\frac{1}{8}) = -3$ ◀

MATCHED PROBLEM 2 ➤ Change each exponential form to an equivalent logarithmic form:

(A) $49 = 7^2$ (B) $3 = \sqrt{9}$ (C) $\frac{1}{3} = 3^{-1}$ ◀

To gain a little deeper understanding of logarithmic functions and their relationship to the exponential functions, we consider a few problems where we want to find x, b, or y in $y = \log_b x$, given the other two values. All values are chosen so that the problems can be solved exactly without a calculator.

EXAMPLE 3 ➤ **Solutions of the Equation $y = \log_b x$** Find y, b, or x, as indicated.

(A) Find y: $y = \log_4 16$ (B) Find x: $\log_2 x = -3$
(C) Find y: $y = \log_8 4$ (D) Find b: $\log_b 100 = 2$

Solution   (A) $y = \log_4 16$ is equivalent to $16 = 4^y$. Thus,

$$y = 2$$

(B) $\log_2 x = -3$ is equivalent to $x = 2^{-3}$. Thus,

$$x = \frac{1}{2^3} = \frac{1}{8}$$

(C) $y = \log_8 4$ is equivalent to

$$4 = 8^y \quad \text{or} \quad 2^2 = 2^{3y}$$

Thus,

$$3y = 2$$
$$y = \tfrac{2}{3}$$

(D) $\log_b 100 = 2$ is equivalent to $100 = b^2$. Thus,

$$b = 10 \quad \text{Recall that } b \text{ cannot be negative.}$$ ◄

MATCHED PROBLEM 3 ➤ Find y, b, or x, as indicated.

(A) Find y: $y = \log_9 27$ (B) Find x: $\log_3 x = -1$
(C) Find b: $\log_b 1,000 = 3$ ◄

■ PROPERTIES OF LOGARITHMIC FUNCTIONS

Logarithmic functions have many powerful and useful properties. We list eight basic properties in Theorem 1.

theorem 1

Properties of Logarithmic Functions

If b, M, and N are positive real numbers, $b \neq 1$, and p and x are real numbers, then:

1. $\log_b 1 = 0$
2. $\log_b b = 1$
3. $\log_b b^x = x$
4. $b^{\log_b x} = x, \quad x > 0$
5. $\log_b MN = \log_b M + \log_b N$
6. $\log_b \dfrac{M}{N} = \log_b M - \log_b N$
7. $\log_b M^p = p \log_b M$
8. $\log_b M = \log_b N$ if and only if $M = N$

The first four properties in Theorem 1 follow directly from the definition of a logarithmic function. Here we will sketch a proof of property 5. The other properties are established in a similar way. Let

$$u = \log_b M \qquad \text{and} \qquad v = \log_b N$$

Or, in equivalent exponential form,

$$M = b^u \qquad \text{and} \qquad N = b^v$$

Now, see if you can provide reasons for each of the following steps:

$$\log_b MN = \log_b b^u b^v = \log_b b^{u+v} = u + v = \log_b M + \log_b N$$

EXAMPLE 4 ➤ Using Logarithmic Properties

(A) $\log_b \dfrac{wx}{yz} = \log_b wx - \log_b yz$

$\qquad\qquad = \log_b w + \log_b x - (\log_b y + \log_b z)$

$\qquad\qquad = \log_b w + \log_b x - \log_b y - \log_b z$

(B) $\log_b(wx)^{3/5} = \tfrac{3}{5}\log_b wx = \tfrac{3}{5}(\log_b w + \log_b x)$ ◀

MATCHED PROBLEM 4 ➤ Write in simpler logarithmic forms, as in Example 4.

(A) $\log_b \dfrac{R}{ST}$ (B) $\log_b\left(\dfrac{R}{S}\right)^{2/3}$ ◀

The following examples and problems, though somewhat artificial, will give you additional practice in using basic logarithmic properties.

EXAMPLE 5 ➤ Solving Logarithmic Equations   Find x so that:

$$\tfrac{3}{2}\log_b 4 - \tfrac{2}{3}\log_b 8 + \log_b 2 = \log_b x$$

Solution

$\tfrac{3}{2}\log_b 4 - \tfrac{2}{3}\log_b 8 + \log_b 2 = \log_b x$

$\log_b 4^{3/2} - \log_b 8^{2/3} + \log_b 2 = \log_b x$ *Property 7*

$\log_b 8 - \log_b 4 + \log_b 2 = \log_b x$

$\log_b \dfrac{8 \cdot 2}{4} = \log_b x$ *Properties 5 and 6*

$\log_b 4 = \log_b x$

$x = 4$ *Property 8* ◀

MATCHED PROBLEM 5 ➤ Find x so that: $3\log_b 2 + \tfrac{1}{2}\log_b 25 - \log_b 20 = \log_b x$ ◀

EXAMPLE 6 ➤   Solving Logarithmic Equations   Solve: $\log_{10} x + \log_{10}(x + 1) = \log_{10} 6$

Solution

$$\log_{10} x + \log_{10}(x + 1) = \log_{10} 6$$
$$\log_{10}[x(x + 1)] = \log_{10} 6 \quad \text{Property 5}$$
$$x(x + 1) = 6 \quad \text{Property 8}$$
$$x^2 + x - 6 = 0 \quad \text{Solve by factoring.}$$
$$(x + 3)(x - 2) = 0$$
$$x = -3, 2$$

We must exclude $x = -3$, since the domain of the function $\log_{10}(x + 1)$ is $x > -1$ or $(-1, \infty)$; hence, $x = 2$ is the only solution.   ◀

MATCHED PROBLEM 6 ➤   Solve: $\log_3 x + \log_3(x - 3) = \log_3 10$ ◀

explore – discuss 2

Discuss the relationship between each of the following pairs of expressions. If the two expressions are equivalent, explain why. If they are not, give an example.

(A) $\log_b M - \log_b N$; $\dfrac{\log_b M}{\log_b N}$ (B) $\log_b M - \log_b N$; $\log_b \dfrac{M}{N}$

(C) $\log_b M + \log_b N$; $\log_b MN$ (D) $\log_b M + \log_b N$; $\log_b(M + N)$

■ CALCULATOR EVALUATION OF LOGARITHMS

Of all possible logarithmic bases, the base e and the base 10 are used almost exclusively. Before we can use logarithms in certain practical problems, we need to be able to approximate the logarithm of any positive number either to base 10 or to base e. And conversely, if we are given the logarithm of a number to base 10 or base e, we need to be able to approximate the number. Historically, tables were used for this purpose, but now calculators make computations faster and far more accurate.

Common logarithms (also called **Briggsian logarithms**) are logarithms with base 10. **Natural logarithms** (also called **Napierian logarithms**) are logarithms with base e. Most calculators have a key labeled "log" (or "LOG") and a key labeled "ln" (or "LN"). The former represents a common (base 10) logarithm and the latter a natural (base e) logarithm. In fact, "log" and "ln" are both used extensively in mathematical literature, and whenever you see either used in this book without a base indicated they will be interpreted as follows:

Logarithmic Notation

Common logarithm: $\log x = \log_{10} x$

Natural logarithm: $\ln x = \log_e x$

Finding the common or natural logarithm using a calculator is very easy. On some calculators, you simply enter a number from the domain of the function and press $\boxed{\text{LOG}}$ or $\boxed{\text{LN}}$. On other calculators, you press either $\boxed{\text{LOG}}$ or $\boxed{\text{LN}}$, enter a number from the domain, and then press $\boxed{\text{ENTER}}$. Check the user's manual for your calculator.

EXAMPLE 7 ➤ **Calculator Evaluation of Logarithms** Use a calculator to evaluate each to six decimal places:

(A) log 3,184   (B) ln 0.000 349   (C) log(− 3.24)

Solution   (A) log 3,184 = 3.502 973   (B) ln 0.000 349 = − 7.960 439
(C) log(− 3.24) = Error*   *− 3.24 is not in the domain of the log function.* ◀

MATCHED PROBLEM 7 ➤ Use a calculator to evaluate each to six decimal places:

(A) log 0.013 529   (B) ln 28.693 28   (C) ln(− 0.438) ◀

We now turn to the second problem mentioned above: Given the logarithm of a number, find the number. We make direct use of the logarithmic–exponential relationships, which follow from the definition of logarithmic function given at the beginnning of this section.

Logarithmic–Exponential Relationships

$$\log x = y \quad \text{is equivalent to} \quad x = 10^y$$
$$\ln x = y \quad \text{is equivalent to} \quad x = e^y$$

EXAMPLE 8 ➤ **Solving $\log_b x = y$ for x** Find x to four decimal places, given the indicated logarithm:

(A) log x = − 2.315   (B) ln x = 2.386

Solution   (A) log x = − 2.315   *Change to equivalent exponential form.*
$\quad\quad\quad x = 10^{-2.315}$ *Evaluate with a calculator.*
$\quad\quad\quad x = 0.0048$

(B) ln x = 2.386   *Change to equivalent exponential form.*
$\quad\quad\quad x = e^{2.386}$ *Evaluate with a calculator.*
$\quad\quad\quad x = 10.8699$ ◀

---

* Some calculators use a more advanced definition of logarithms involving complex numbers and will display an ordered pair of real numbers as the value of log(− 3.24). You should interpret such a result as an indication that the number entered is not in the domain of the logarithm function as we have defined it.

MATCHED PROBLEM 8 ➤ Find x to four decimal places, given the indicated logarithm:

(A) $\ln x = -5.062$ (B) $\log x = 2.0821$ ◀

EXAMPLE 9 ➤ **Solving Exponential Equations** Solve for x to four decimal places:

(A) $10^x = 2$ (B) $e^x = 3$ (C) $3^x = 4$

Solution   (A) $10^x = 2$ *Take common logarithms of both sides.*

$\log 10^x = \log 2$ *Property 3*

$x = \log 2$ *Use a calculator.*

$x = 0.3010$

(B) $e^x = 3$ *Take natural logarithms of both sides.*

$\ln e^x = \ln 3$ *Property 3*

$x = \ln 3$ *Use a calculator.*

$x = 1.0986$

(C) $3^x = 4$ *Take either natural or common logarithms of both sides. (We choose common logarithms.)*

$\log 3^x = \log 4$ *Property 7*

$x \log 3 = \log 4$ *Solve for x.*

$x = \dfrac{\log 4}{\log 3}$ *Use a calculator.*

$x = 1.2619$ ◀

Exponential equations can also be solved graphically by graphing both sides of an equation and finding the points of intersection. Figure 3 illustrates this approach for the equations in Example 9.

(A) $y_1 = 10^x$
$y_2 = 2$

(B) $y_1 = e^x$
$y_2 = 3$

(C) $y_1 = 3^x$
$y_2 = 4$

FIGURE 3
Graphical solution of exponential equations

MATCHED PROBLEM 9 ➤ Solve for x algebraically and graphically (to four decimal places):

(A) $10^x = 7$ (B) $e^x = 6$ (C) $4^x = 5$ ◀

explore – discuss 3

> Discuss how you could find $y = \log_5 38.25$ using either natural or common logarithms on a calculator. [*Hint:* Start by rewriting the equation in exponential form.]

■ APPLICATION

A convenient and easily understood way of comparing different investments is to use their **doubling times**—the length of time it takes the value of an investment to double. Logarithm properties, as you will see in Example 10, provide us with just the right tool for solving some doubling-time problems.

EXAMPLE 10 ▶ **Doubling Time for an Investment**   How long (to the next whole year) will it take money to double if it is invested at 20% compounded annually? Solve algebraically and graphically.

Solution   We use the compound interest formula discussed in Section 2-2:

$$A = P\left(1 + \frac{r}{m}\right)^{mt} \qquad \textit{Compound interest}$$

The problem is to find t, given $r = 0.20$, $m = 1$, and $A = 2p$; that is,

$$2P = P(1 + 0.2)^t$$
$$2 = 1.2^t$$
$$1.2^t = 2 \qquad \textit{Solve for t by taking the natural or common}$$
$$\ln 1.2^t = \ln 2 \qquad \textit{logarithm of both sides (we choose the}$$
$$t \ln 1.2 = \ln 2 \qquad \textit{natural logarithm).}$$
$$ \qquad \textit{Property 7}$$
$$t = \frac{\ln 2}{\ln 1.2} \qquad \textit{Use a calculator.}$$
$$= 3.8 \text{ years} \qquad [\textit{Note:} \ (\ln 2)/(\ln 1.2) \neq \ln 2 - \ln 1.2]$$
$$\approx 4 \text{ years} \qquad \textit{To the next whole year}$$

Intersection
X=3.801784   Y=2

FIGURE 4
$y_1 = 1.2^x$
$y_2 = 2$

To solve graphically, we graph both sides of the equation $2 = 1.2^t$, and find the intersection point (Fig. 4).

When interest is paid at the end of 3 years, the money will not be doubled; when paid at the end of 4 years, the money will be slightly more than doubled. ◀

MATCHED PROBLEM 10 ▶ How long (to the next whole year) will it take money to triple if it is invested at 13% compounded annually? Solve algebraically and graphically. ◀

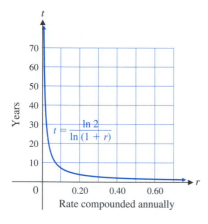

FIGURE 5
Doubling time (in years) at various rates of interest compounded annually

It is interesting and instructive to graph the doubling times for various rates compounded annually. We proceed as follows:

$$A = P(1 + r)^t$$
$$2P = P(1 + r)^t$$
$$2 = (1 + r)^t$$
$$(1 + r)^t = 2$$
$$\ln(1 + r)^t = \ln 2$$
$$t \ln(1 + r) = \ln 2$$
$$t = \frac{\ln 2}{\ln(1 + r)}$$

Figure 5 shows the graph of this equation (doubling time in years) for interest rates compounded annually from 1% to 70% (expressed as decimals). Note the dramatic change in doubling time as rates change from 1% to 20% (from 0.01 to 0.20).

Answers to Matched Problems

1. (A) $9 = 3^2$ (B) $2 = 4^{1/2}$ (C) $\frac{1}{9} = 3^{-2}$
2. (A) $\log_7 49 = 2$ (B) $\log_9 3 = \frac{1}{2}$ (C) $\log_3(\frac{1}{3}) = -1$
3. (A) $y = \frac{3}{2}$ (B) $x = \frac{1}{3}$ (C) $b = 10$
4. (A) $\log_b R - \log_b S - \log_b T$ (B) $\frac{2}{3}(\log_b R - \log_b S)$ 5. $x = 2$
6. $x = 5$ 7. (A) $-1.868\ 734$ (B) $3.356\ 663$ (C) Not defined
8. (A) 0.0063 (B) 120.8092
9. (A) 0.8451 (B) 1.7918 (C) 1.1610 10. 9 yr

EXERCISE 2-3

A   *Rewrite in equivalent exponential form:*

1. $\log_3 27 = 3$
2. $\log_2 32 = 5$
3. $\log_{10} 1 = 0$
4. $\log_e 1 = 0$
5. $\log_4 8 = \frac{3}{2}$
6. $\log_9 27 = \frac{3}{2}$

Rewrite in equivalent logarithmic form:

7. $49 = 7^2$
8. $36 = 6^2$
9. $8 = 4^{3/2}$
10. $9 = 27^{2/3}$
11. $A = b^u$
12. $M = b^x$

Evaluate each of the following without a calculator:

13. $\log_{10} 1$
14. $\log_e 1$
15. $\log_e e$
16. $\log_{10} 10$
17. $\log_{0.2} 0.2$
18. $\log_{13} 13$
19. $\log_{10} 10^3$
20. $\log_{10} 10^{-5}$
21. $\log_2 2^{-3}$
22. $\log_3 3^5$
23. $\log_{10} 1,000$
24. $\log_6 36$

Write in terms of simpler logarithmic forms, as in Example 4.

25. $\log_b \dfrac{P}{Q}$
26. $\log_b FG$
27. $\log_b L^5$

28. $\log_b w^{15}$
29. $\log_b \dfrac{p}{qrs}$
30. $\log_b PQR$

B   *Find x, y, or b without a calculator.*

31. $\log_3 x = 2$
32. $\log_2 x = 2$
33. $\log_7 49 = y$
34. $\log_3 27 = y$
35. $\log_b 10^{-4} = -4$
36. $\log_b e^{-2} = -2$
37. $\log_4 x = \frac{1}{2}$
38. $\log_{25} x = \frac{1}{2}$
39. $\log_{1/3} 9 = y$
40. $\log_{49}(\frac{1}{7}) = y$
41. $\log_b 1,000 = \frac{3}{2}$
42. $\log_b 4 = \frac{2}{3}$

Write in terms of simpler logarithmic forms, going as far as you can with logarithmic properties (see Example 4).

43. $\log_b \dfrac{x^5}{y^3}$
44. $\log_b(x^2 y^3)$

45. $\log_b \sqrt[3]{N}$
46. $\log_b \sqrt[5]{Q}$

47. $\log_b(x^2 \sqrt[3]{y})$
48. $\log_b \sqrt[3]{\dfrac{x^2}{y}}$

49. $\log_b(50 \cdot 2^{-0.2t})$
50. $\log_b(100 \cdot 1.06^t)$
51. $\log_b[P(1 + r)^t]$
52. $\log_e Ae^{-0.3t}$
53. $\log_e 100e^{-0.01t}$
54. $\log_{10}(67 \cdot 10^{-0.12x})$

Find x.

55. $\log_b x = \frac{2}{3} \log_b 8 + \frac{1}{2} \log_b 9 - \log_b 6$

56. $\log_b x = \frac{2}{3} \log_b 27 + 2 \log_b 2 - \log_b 3$

57. $\log_b x = \frac{3}{2} \log_b 4 - \frac{2}{3} \log_b 8 + 2 \log_b 2$

58. $\log_b x = 3 \log_b 2 + \frac{1}{2} \log_b 25 - \log_b 20$

59. $\log_b x + \log_b(x - 4) = \log_b 21$

60. $\log_b(x + 2) + \log_b x = \log_b 24$

61. $\log_{10}(x - 1) - \log_{10}(x + 1) = 1$

62. $\log_{10}(x + 6) - \log_{10}(x - 3) = 1$

Graph Problems 63 and 64 by converting to exponential form first.

63. $y = \log_2(x - 2)$ **64.** $y = \log_3(x + 2)$

65. Explain how the graph of the equation in Problem 63 can be obtained from the graph of $y = \log_2 x$ using a simple transformation (see Section 1-3).

66. Explain how the graph of the equation in Problem 64 can be obtained from the graph of $y = \log_3 x$ using a simple transformation (see Section 1-3).

67. What are the domain and range of the function defined by $y = 1 + \ln(x + 1)$?

68. What are the domain and range of the function defined by $y = \log(x - 1) - 1$?

Evaluate to five decimal places using a calculator.

69. (A) log 3,527.2
   (B) log 0.006 913 2
   (C) ln 277.63
   (D) ln 0.040 883

70. (A) log 72.604
   (B) log 0.033 041
   (C) ln 40,257
   (D) ln 0.005 926 3

Find x to four decimal places.

71. (A) $\log x = 1.1285$
   (B) $\log x = -2.0497$
   (C) $\ln x = 2.7763$
   (D) $\ln x = -1.8879$

72. (A) $\log x = 2.0832$
   (B) $\log x = -1.1577$
   (C) $\ln x = 3.1336$
   (D) $\ln x = -4.3281$

Solve each equation algebraically and graphically to four decimal places.

73. $10^x = 12$ **74.** $10^x = 153$

75. $e^x = 4.304$ **76.** $e^x = 0.3059$

77. $1.03^x = 2.475$ **78.** $1.075^x = 1.837$

79. $1.005^{12t} = 3$ **80.** $1.02^{4t} = 2$

Graph Problems 81–88 using a graphing utility as an aid. Indicate increasing and decreasing intervals.

81. $y = \ln x$ **82.** $y = -\ln x$

83. $y = |\ln x|$ **84.** $y = \ln|x|$

85. $y = 2 \ln(x + 2)$ **86.** $y = 2 \ln x + 2$

87. $y = 4 \ln x - 3$ **88.** $y = 4 \ln(x - 3)$

C

89. Explain why the calculator display shown does not contradict the logarithmic property

$$\log \frac{M}{N} = \log M - \log N$$

Figure for 89

90. Explain why the calculator display shown does not contradict the logarithmic property

$$\log MN = \log M + \log N$$

Figure for 90

91. Explain why the logarithm of 1 for any permissible base is 0.

92. Explain why 1 is not a suitable logarithmic base.

93. Write $\log_{10} y - \log_{10} c = 0.8x$ in an exponential form that is free of logarithms.

94. Write $\log_e x - \log_e 25 = 0.2t$ in an exponential form that is free of logarithms.

95. Let $p(x) = \ln x$, $q(x) = \sqrt{x}$, and $r(x) = x$. Use a graphing utility to draw graphs of all three functions in the same viewing window for $1 \leq x \leq 16$. Discuss what it means for one function to be larger than another on an interval,

and then order the three functions from largest to smallest for $1 < x \leq 16$.

96. Let $p(x) = \log x$, $q(x) = \sqrt[3]{x}$, and $r(x) = x$. Use a graphing utility to draw graphs of all three functions in the same viewing window for $1 \leq x \leq 16$. Discuss what it means for one function to be smaller than another on an interval, and then order the three functions from smallest to largest for $1 < x \leq 16$.

▶ APPLICATIONS

Business & Economics

97. *Doubling time.* How long (to the next whole year) will it take money to double if it is invested at 6% interest compounded annually?

98. *Doubling time.* How long (to the next whole year) will it take money to double if it is invested at 3% interest compounded annually?

99. *Investing.* How many years (to two decimal places) will it take $1,000 to grow to $1,800 if it is invested at 6% compounded quarterly? Compounded continuously?

100. *Investing.* How many years (to two decimal places) will it take $5,000 to grow to $7,500 if it is invested at 8% compounded semiannually? Compounded continuously?

101. *Investment.* A newly married couple wishes to have $20,000 in 8 years for the down payment on a house. At what rate of interest compounded continuously (to three decimal places) must $10,000 be invested now to accomplish this goal?

102. *Investment.* The parents of a newborn child want to have $45,000 for the child's college education 17 years from now. At what rate of interest compounded continuously (to three decimal places) must a grandparent's gift of $10,000 be invested now to achieve this goal?

Life Sciences

103. *Sound intensity—decibels.* Because of the extraordinary range of sensitivity of the human ear (a range of over 1,000 million millions to 1), it is helpful to use a logarithmic

scale, rather than an absolute scale, to measure sound intensity over this range. The unit of measure is called the *decibel*, after the inventor of the telephone, Alexander Graham Bell. If we let N be the number of decibels, I the power of the sound in question (in watts per square centimeter), and I_0 the power of sound just below the threshold of hearing (approximately 10^{-16} watt per square centimeter), then

$$I = I_0 10^{N/10}$$

Show that this formula can be written in the form

$$N = 10 \log \frac{I}{I_0}$$

104. *Sound intensity—decibels.* Use the formula in Problem 103 (with $I_0 = 10^{-16}$ watt/cm^2) to find the decibel ratings of the following sounds:
(A) Whisper: 10^{-13} watt/cm^2
(B) Normal conversation: 3.16×10^{-10} watt/cm^2
(C) Heavy traffic: 10^{-8} watt/cm^2
(D) Jet plane with afterburner: 10^{-1} watt/cm^2

Social Sciences

105. *World population.* If the world population is now 5.8 billion people and if it continues to grow at 1.14% compounded continuously, how long (to the nearest year) will it take before there is only 1 square yard of land per person? (The Earth contains approximately 1.68×10^{14} square yards of land.)

106. *Archaeology—carbon-14 dating.* The radioactive carbon-14 (^{14}C) in an organism at the time of its death decays according to the equation

$$A = A_0 e^{-0.000124t}$$

where t is time in years and A_0 is the amount of ^{14}C present at time $t = 0$. (See Example 3 in Section 2-2.) Estimate the age of a skull uncovered in an archaeological site if 10% of the original amount of ^{14}C is still present. [*Hint:* Find t such that $A = 0.1A_0$.]

SECTION 2-4 ■ ## Modeling Data Using Regression Techniques

- LINEAR REGRESSION
- POLYNOMIAL REGRESSION
- LOGARITHMIC, EXPONENTIAL, AND POWER REGRESSION

In real-world applications it is often necessary to collect data in table form, and then look for an elementary function to model the data. The very powerful mathematical tool, *regression analysis,* may be used for this purpose. Use of *regression techniques* was mentioned a number of times in examples and exercises in Chapter 1, and the group activity at the end of that chapter required the use of *linear regression* to find the ''best fit'' of a straight line to a set of data points. However, regression analysis is not limited to fitting data sets with straight lines. Now that we have a variety of elementary functions at our disposal, we can consider fitting other types of functions to given data sets.

Regression analysis is the process of fitting an elementary function to a set of data points using the **method of least squares.** This method is a process for selecting a particular function such that the sum of the squares of the distances from the points in a data set to the graph of the function is minimal. The method of least squares is developed in detail later in this book. Our interests now are in the mechanics of using regression techniques. Many graphing utilities have built-in routines for this purpose. (We assume you have such a graphing utility.)

Table 1 lists the various types of regression equations that graphing utilities are

TABLE 1
Regression Equations

REGRESSION PROCESS	REGRESSION EQUATION
Linear regression	$y = ax + b$
Quadratic regression	$y = ax^2 + bx + c$
Cubic regression	$y = ax^3 + bx^2 + cx + d$
Quartic regression	$y = ax^4 + bx^3 + cx^2 + dx + e$
Logarithmic regression	$y = a + b \ln(x), x > 0$
Exponential regression	$y = ab^x, y > 0$
Power regression	$y = ax^b, x > 0, y > 0$

capable of producing. The form of an equation and the names associated with each type of equation may vary slightly among the various types of graphing utilities. (Read the user's manual for your particular graphing utility.)

We will give examples of fitting several of the equations to data sets, and once you understand the process you should have little trouble in dealing with the equations we do not discuss.

■ LINEAR REGRESSION

Recall that the graph of a linear function is a straight line with either a positive slope or a negative slope. Example 1 illustrates the key steps in fitting a straight line to a given set of data points.

EXAMPLE 1 ➤ **Linear Regression**   In the Group Activity for Chapter 1, we saw that the finance department of a mountain bike manufacturing company arrived at the price–demand data shown in Table 2, where p is the wholesale price of a bike for a demand of x hundred bikes.

TABLE 2
Price–Demand

x (hundred)	p ($)
0	525
64	370
125	270
185	130

(A) Enter the data set in a graphing utility, find a linear regression equation for the data set, and plot the data set and the linear regression equation in the same viewing window.

(B) Estimate the demand (to the nearest unit) at a price level of $300. Solve graphically in a graphing utility, and round the answer to the nearest unit at the end of the solution process.

Solution   (A) We use the linear regression feature on a graphing utility. (Read your user's manual for the particular details for each step.) Figure 1A shows the data entered in two lists, and Figure 1B shows the results of applying the linear regression routine to the data. Figure 1C shows the linear regression equation transferred to the function editor. Figure 1D shows the plotted data set and the graph of the linear regression equation.

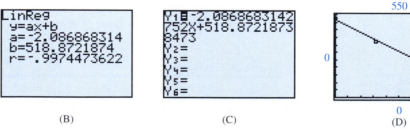

| (A) | (B) | (C) | (D) |

FIGURE 1

[*Note:*   The four steps shown in Figure 1 are included in this first example, however, all four steps will not be repeated in subsequent examples or answers to matched problems or exercise sets.]

(B) We solve the problem graphically by adding the graph of $y_2 = 300$ to the graph of the linear regression equation shown in Figure 1D, and then using the built-in intersect routine. Figure 2 shows the result. Thus, the demand at a price of $300 is 104.88 hundred, or 10,488 bikes.

FIGURE 2

MATCHED PROBLEM 1 ▶

TABLE 3
Cost

x (hundred)	C (hundred $)
0	8,470
30	13,510
120	19,140
180	22,580
220	28,490

Table 3 shows costs at various outputs for the mountain bikes in Example 1.

(A) Enter the data set in a graphing utility, find a linear regression equation for the data set, and plot the data set and the linear regression equation in the same viewing window.
(B) Estimate the cost (to the nearest hundred dollars) for an output of 15,000 bikes. Solve graphically in a graphing utility, and round the answer to the nearest hundred dollars at the end of the solution process. ◀

■ POLYNOMIAL REGRESSION

Many graphing calculators can fit second-, third-, and fourth-degree polynomials to given data sets. Example 2 illustrates fitting a quadratic regression curve (parabola) to a data set.

 EXAMPLE 2 ▶

TABLE 4
Price–Demand

x (thousand)	p ($)
10	200
35	177
50	171
73	135
94	95

Quadratic Regression A company manufactures and sells roller-blade skates. Table 4 is a price–demand table established by their finance department, where p is the price (in dollars) at which x thousand pairs of skates can be sold.

(A) Enter the data set in a graphing utility, find a quadratic regression equation for the data set, and plot the data set and the quadratic regression equation in the same viewing window.
(B) Estimate the price (to the nearest dollar) at a demand level of 65 thousand units. Solve in a graphing utility, and round the answer to the nearest dollar at the end of the solution process.

Solution (A) Use the quadratic regression feature on a graphing utility. Figure 3A shows the data entered in two lists. Figure 3B gives the quadratic regression equation in the form $y = ax^2 + bx + c$. Figure 3C shows the plotted data and the graph of the quadratic regression equation.

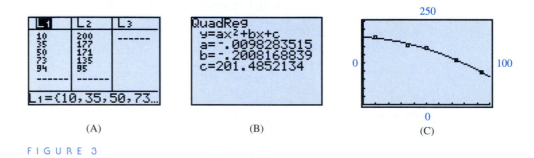

(A)   (B)   (C)

FIGURE 3

(B) See Figure 4. The price (to the nearest dollar) for a demand level of 65 thousand units is $147.

FIGURE 4

MATCHED PROBLEM 2 ▶

Table 5 is a price–supply table established by the finance department of the roller-blade manufacturing company in Example 2, where p is the price (in dollars) at which x thousand pairs of skates will be supplied.

(A) Enter the data set in a graphing utility, find a quadratic regression equation for the data set, and plot the data set and the quadratic regression equation in the same viewing window.

(B) Estimate the price (to the nearest dollar) at a supply level of 55 thousand units. Solve in a graphing utility, and round the answer to the nearest dollar at the end of the solution process. ◀

TABLE 5
Price–Supply

x (thousand)	p ($)
8	72
30	81
45	86
68	121
90	159

explore – discuss 1

Referring to Example 2, estimate the demand x (to the nearest thousand) for a price of $p = \$120$ per pair of roller-blade skates. Discuss how you would solve the problem graphically using a graphing utility, and solve the problem this way.

■ LOGARITHMIC, EXPONENTIAL, AND POWER REGRESSION

We now turn to other types of regression equations. Logarithmic regression will be considered in Example 3 and exponential regression in Matched Problem 3. Power regression occurs in several problems in Exercise 2-4.

▶ E X A M P L E 3 ➤ Logarithmic Regression   The yearly cash flow for each of the first 5 years of a business for a small electronics company is shown in Table 6, where y is the annual cash flow (in millions of dollars) x years after start-up.

T A B L E 6
Cash Flow

x (year)	y (million $)
1	− 2.8
2	− 0.4
3	1.6
4	2.5
5	3.4

(A) Enter the data set in a graphing utility, find a logarithmic regression equation $[y = a + b \ln(x)]$ for the data set, and plot the data set and the logarithmic regression equation in the same viewing window.

(B) Estimate the cash flow (to one-tenth million dollars) 8 years after start-up.* Solve in a graphing utility, and round the answer to the nearest tenth of a million dollars at the end of the solution process.

Solution   (A) Use the logarithmic regression feature on a graphing utility. Figure 5A shows the data entered in two lists. Figure 5B gives the logarithmic regression equation. Figure 5C shows the plotted data and the graph of the logarithmic regression equation.

(A)

(B)

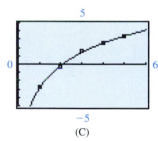

(C)

F I G U R E 5

(B) In Figure 6, we have extended the window dimensions to cover $x = 8$. The estimated cash flow (to one-tenth million dollars) 8 years after start-up is $5.2 million.

F I G U R E 6

---

* The projected cash flow in Example 3B is just that—a projection. It assumes that the cash flow will continue to grow as it did for the first 5 years. If other factors influence the cash flow during the last 3 years, then this projection may not be very accurate.

MATCHED PROBLEM 3 ▶   Table 7 shows the world population for selected years since 1900, where P is the population (in billions) x years after 1900.

TABLE 7
World Population

DATE	x (years after 1900)	P (billion)
1900	0	1.6
1930	30	2.0
1950	50	2.6
1980	80	4.5
1995	95	5.7

(A) Enter the data set in a graphing utility, find an exponential regression equation ($P = ab^x$) for the data set, and plot the data set and the exponential regression equation in the same viewing window.

(B) Estimate the world population (to the nearest one-tenth billion) in the year 2025. Solve in a graphing utility, and round the answer to the nearest tenth of a billion at the end of the solution process.   ◀

Answers to Matched Problems

1. (A)

(B) $2,168,900

2. (A)

(B) $101

3. (A)

(B) 8.2 billion people

EXERCISE 2-4

 APPLICATIONS

In all the problems that follow, unless stated to the contrary, solve as much of the problem as possible within a graphing utility using built-in routines. Round answers to the desired accuracy at the end of the solution process.

Business & Economics

1. *Price–demand.* A manufacturer of a popular point-and-shoot camera wholesales the camera to retail outlets throughout the United States. The finance department in the company produced the price–demand data in Table 8, where p is the wholesale price per camera at which x million cameras are sold.

TABLE 8
Price–Demand

x (million)	p ($)
2	87
5	68
8	53
12	37

(A) Enter the data set in a graphing utility, find a linear regression equation for the data set, and plot the data set and the linear regression equation in the same viewing window.

(B) Estimate the demand (to the nearest million) at a price level of $49.

2. *Costs.* The finance department in the point-and-shoot camera manufacturing company in Problem 1 produced the cost data in Table 9, where C is the cost (in millions of dollars) for manufacturing and selling x million cameras.

TABLE 9
Costs

x (million)	C (million $)
1	175
5	260
8	305
12	395

(A) Enter the data set in a graphing utility, find a linear regression equation for the data set, and plot the data

set and the linear regression equation in the same viewing window.

(B) Estimate the cost (to the nearest million dollars) at a production level of 9 million cameras.

3. *Hospital costs.* The finance department of a hospital produced Table 10, which indicates costs at selected annual case loads, where C is the cost in millions of dollars for handling x thousand cases per year.

TABLE 10
Hospital Costs

x (thousand)	C (million $)
1	1.1
3	2.9
5	4.1
7	4.4
10	6.3

(A) Enter the data set in a graphing utility, find a cubic regression equation for the data set, and plot the data set and the cubic regression equation in the same viewing window.

(B) Estimate the cost (to the nearest tenth of a million dollars) for an annual case load of 9 thousand.

4. *Price–supply.* A company manufactures and wholesales CD players. Their finance department established a price–supply table (Table 11) for selected outputs, where x is the number of CD players the company is willing to supply per day at a price of p dollars each.

TABLE 11
Price–Supply

x	p ($)
10	13
30	23
70	31
110	38
170	55
230	65

(A) Enter the data set in a graphing utility, find a power regression equation ($p = ax^b$) for the data set, and plot the data set and the power regression equation in the same viewing window.

(B) Estimate the price (to the nearest dollar) for a daily supply of 250 CD players.

Problems 5 and 6 involve the following concepts: In a free competitive market, the price of a product is determined by the relationship between supply and demand. If $(\bar{x}, \bar{p})$ is the intersection point of the price–demand equation, $p = D(x)$, and the price–supply equation, $p = S(x)$, then $(\bar{x}, \bar{p})$ is called the **equilibrium point,** *$\bar{p}$ is called the* **equilibrium price,** *and $\bar{x}$ is called the* **equilibrium quantity.** *The equilibrium price is the price at which the supply and the demand are the same.*

5. *Equilibrium point.* A particular CD is sold through a chain of stores in a city. A marketing company established the price–demand and price–supply tables for selected prices given in Tables 12 and 13, where x is the daily number of CD's people are willing to buy and the store is willing to sell at a price of p dollars per CD.

TABLE 12
Price–Demand

x	$p = D(x)$ (\$)
25	19.50
100	14.25
175	10.00
250	8.25

TABLE 13
Price–Supply

x	$p = S(x)$ (\$)
25	2.10
100	3.80
175	8.50
250	15.70

(A) Find a linear regression equation for the data in Table 12 and a quadratic regression equation for the data in Table 13. Transfer each into the function editor.

(B) Plot both equations from part (A) in the same viewing window, and use an appropriate built-in routine to find the point of intersection. Write the equilibrium price to the nearest cent and the equilibrium quantity to the nearest unit.

6. *Equilibrium point.* Repeat Problem 5 with the price–demand and price–supply tables given in Tables 14 and 15, except find a quadratic regression equation for the data in Table 14 and a linear regression equation for the data in Table 15.

TABLE 14
Price–Demand

x	$p = D(x)$ (\$)
0	24
40	23
65	20
115	11

TABLE 15
Price–Supply

x	$p = S(x)$ (\$)
0	5
40	10
65	12
115	16

Life Sciences

7. *Medicine.* Cardiovascular research has shown that above 210 cholesterol level, coronary risk increases with increased cholesterol level. A particular study produced Table 16, which lists associated risk levels R at particular cholesterol levels C.

TABLE 16
Coronary Risk

C	R
210	0.164
233	0.201
264	0.251
280	0.268

(A) Enter the data set in a graphing utility, find a linear regression equation for the data set, and plot the data set and the linear regression equation in the same viewing window.

(B) Estimate the coronary risk (to the three decimal places) for a person with a cholesterol level of 255.

8. *AIDS epidemic.* The *World Almanac* for 1995 gives the information on AIDS (rounded to the nearest thousand) shown in Table 17.

TABLE 17
New AIDS Cases in the United States, 1985–1992*

YEAR	1985	1986†	1987	1988
NEW CASES (THOUSANDS)	8	—	21	31

YEAR	1989	1990	1991	1992
NEW CASES (THOUSANDS)	34	42	44	46

* AIDS case definition was changed in 1993, so we include data through 1992 only.

† 1986 data not available.

(A) Let x represent the number of years after 1985 and N the number of new cases of AIDS. Enter the appropriate data set in a graphing utility, find a quadratic regression equation for the data set, and plot the data set and the quadratic regression equation in the same viewing window.

(B) Estimate the number (to the nearest thousand) of new cases for 1986 (a year that the *World Almanac* omitted from their table).

Social Sciences

9. *Safety research.* In a study relating the length of skid marks and speed when a person slams on the brakes, a safety research team came up with the data in Table 18 for a particular weight car, where x is the length of the skid marks (in feet), and v is the speed of the car (in miles per hour).

TABLE 18
Skid Marks and Velocity

x (feet)	v (miles per hour)
8.2	20
18.1	30
50.7	45
86.3	65

(A) Enter the data set in a graphing utility, find a power regression equation ($y = ax^b$) for the data set, and plot the data set and the power regression equation in the same viewing window.

(B) Estimate the speed of a car (to the nearest mile per hour) leaving skid marks measuring 191.5 feet.

10. *Population growth.* Table 19 shows the California population for selected years since 1930, where P is the population (in millions) x years after 1930.

TABLE 19
California Population

DATE	x (years after 1930)	P (million)
1930	0	5
1950	20	10
1970	40	20
1990	60	30

(A) Enter the data set in a graphing utility, find an exponential regression equation ($y = ab^x$) for the data set, and plot the data set and the exponential regression equation in the same viewing window.

(B) Estimate the population in California (to the nearest million) in the year 2010.

CHAPTER 2
GROUP ACTIVITY

Profit and Loss Analysis

A company manufactures and sells a particular CD featuring a newly discovered acoustical guitar player. Using test marketing and statistical cost analysis, the finance department arrived at the data in Tables 1 and 2, where x is the quantity (in thousands) that can be sold at p dollars per album and C is the cost (in thousands of dollars) to manufacture and sell x thousand albums.

TABLE 1
Price–Demand

x (thousand)	p ($)
0	30
35	28
60	23
120	12

TABLE 2
Cost

x (thousand)	C (thousand $)
0	375
35	750
60	980
120	1620

(A) Find a quadratic regression equation for the data in Table 1, and plot it and the data set in the same viewing window. Discuss the relationship between higher sales and price.

(B) Use the price–demand function from part (A) to define the revenue function. Graph the revenue equation (with appropriate adjustments in window dimensions), and find the maximum revenue graphically by using an appropriate built-in routine. What production output produces the maximum revenue? Discuss why revenues drop at higher sales levels.

(C) Find a linear regression equation for the cost data in Table 2, and plot it and the data set in the same viewing window.

(D) Graph the cost equation from part (C) and the revenue equation from part (B) in the same viewing window. Find the following graphically:
  (1) Break-even points
  (2) Intervals over which the company has a loss
  (3) Intervals over which the company has a profit

(E) Referring to the cost equation in part (C), discuss the variable and fixed costs. What are they and what type of expenses are included?

(F) Find the maximum profit graphically by using an appropriate built-in routine. What sales level produces the maximum profit? Discuss why profits drop at higher sales levels.

(G) Does the maximum revenue and maximum profit occur at the same number of albums sold? Discuss the discrepancy.

CHAPTER 2 REVIEW ■ Important Terms and Symbols

2-1 *Polynomial and Rational Functions.* Polynomial function; degree; continuity; turning point; rational function; points of discontinuity; vertical and horizontal asymptotes

$$f(x) = a_n x^n + a_{n-1} x^{n-1} + \cdots + a_1 x + a_0, a_n \neq 0;$$

$$f(x) = \frac{n(x)}{d(x)}, d(x) \neq 0$$

2-2 *Exponential Functions.* Exponential function; base; basic graphs; horizontal asymptote; basic properties; irrational number e; exponential function with base e; exponential growth; exponential decay; compound interest; continuous compound interest

$$f(x) = b^x, b > 0, b \neq 1; \quad y = e^x; \quad N = N_0 e^{kt};$$

$$A = A_0 e^{-kt}; \quad A = P\left(1 + \frac{r}{m}\right)^{mt}; \quad A = Pe^{rt}$$

2-3 *Logarithmic Functions.* Inverse functions; one-to-one functions; logarithmic function; base; equivalent exponential form; properties; common logarithm; natural logarithm; calculator evaluation

$$\log_b x, b > 0, b \neq 1; \quad \log x; \quad \ln x$$

2-4 *Modeling Data Using Regression Techniques.* Regression analysis; method of least squares; equilibrium point; equilibrium price; equilibrium quantity

See Table 1, page 129, for regression equations.

CHAPTER 2 ■ Review Exercise

Work through all the problems in this chapter review and check your answers in the back of the book. Answers to all review problems are there along with section numbers in italics to indicate where each type of problem is discussed. Where weaknesses show up, review appropriate sections in the text.

A

1. Write in logarithmic form using base e: $u = e^v$
2. Write in logarithmic form using base 10: $x = 10^y$
3. Write in exponential form using base e: $\ln M = N$
4. Write in exponential form using base 10: $\log u = v$

Simplify.

5. $\dfrac{5^{x+4}}{5^{4-x}}$

6. $\left(\dfrac{e^u}{e^{-u}}\right)^u$

Solve for x exactly without the use of a calculator.

7. $\log_3 x = 2$ **8.** $\log_x 36 = 2$ **9.** $\log_2 16 = x$

Solve for x to three decimal places.

10. $10^x = 143.7$ **11.** $e^x = 503{,}000$
12. $\log x = 3.105$ **13.** $\ln x = -1.147$

For each polynomial function in Problems 14 and 15, find the following:

(A) The degree of the polynomial
(B) The maximum number of turning points of the graph
(C) The maximum number of x intercepts of the graph
(D) The minimum number of x intercepts of the graph
(E) The maximum number of y intercepts of the graph
(F) The minimum number of y intercepts of the graph

14. $p(x) = ax^3 + bx^2 + cx + d,\ a \neq 0$
15. $p(x) = ax^4 + bx^3 + cx^2 + dx + e,\ a \neq 0$

Each graph in Problems 16 and 17 is the graph of a polynomial function. Answer the following questions for each graph:

(A) How many turning points are on the graph?
(B) What is the minimum degree of a polynomial function that could have the graph?
(C) Is the leading coefficient of the polynomial negative or positive?

16.

17.

B

For each rational function in Problems 18 and 19:

(A) Find the intercepts for the graph.
(B) Determine the domain.
(C) Find any vertical or horizontal asymptotes for the graph.
(D) Graph $y = f(x)$ in a standard viewing window using a graphing utility.
(E) Sketch any asymptotes as dashed lines. Then sketch a graph of f for $-10 \leqslant x \leqslant 10$ and $-10 \leqslant y \leqslant 10$.

18. $f(x) = \dfrac{x+4}{x-2}$ **19.** $f(x) = \dfrac{3x-4}{2+x}$

Solve for x exactly without the use of a calculator.

20. $\log(x+5) = \log(2x-3)$
21. $2\ln(x-1) = \ln(x^2-5)$
22. $9^{x-1} = 3^{1+x}$ **23.** $e^{2x} = e^{x^2-3}$
24. $2x^2 e^x = 3xe^x$ **25.** $\log_{1/3} 9 = x$
26. $\log_x 8 = -3$ **27.** $\log_9 x = \frac{3}{2}$

Solve Problems 28–37 for x to four decimal places.

28. $x = 3(e^{1.49})$ **29.** $x = 230(10^{-0.161})$
30. $\log x = -2.0144$ **31.** $\ln x = 0.3618$
32. $35 = 7(3^x)$ **33.** $0.01 = e^{-0.05x}$
34. $8{,}000 = 4{,}000(1.08^x)$ **35.** $5^{2x-3} = 7.08$
36. $x = \log_2 7$ **37.** $x = \log_{0.2} 5.321$

38. How does the graph of $f(x) = x^4 - 4x^2 + 1$ compare to the graph of $y = x^4$ as we zoom out?

39. Compare the graphs of $y = x^4$ and $y = x^4 - 4x^2 + 1$ in the following two viewing windows:
(A) $-5 \leqslant x \leqslant 5,\ -5 \leqslant y \leqslant 5$
(B) $-5 \leqslant x \leqslant 5,\ -500 \leqslant y \leqslant 500$

Simplify.

40. $e^x(e^{-x} + 1) - (e^x + 1)(e^{-x} - 1)$
41. $(e^x - e^{-x})^2 - (e^x + e^{-x})(e^x - e^{-x})$

Graph Problems 42–44 over the indicated interval. Indicate increasing and decreasing intervals.

42. $y = 2^{x-1};\ [-2, 4]$ **43.** $f(t) = 10e^{-0.08t};\ t \geqslant 0$
44. $y = \ln(x+1);\ (-1, 10]$

C

45. Noting that $\pi = 3.141\ 592\ 654 \ldots$ and $\sqrt{2} = 1.414\ 213\ 562 \ldots$, explain why the calculator results shown here are obvious. Discuss similar connections between the natural logarithmic function and the exponential function with base e.

Figure for 45

Solve Problems 46–49 exactly without the use of a calculator.

46. $\log x - \log 3 = \log 4 - \log(x + 4)$

47. $\ln(2x - 2) - \ln(x - 1) = \ln x$

48. $\ln(x + 3) - \ln x = 2 \ln 2$

49. $\log 3x^2 = 2 + \log 9x$

50. Write $\ln y = -5t + \ln c$ in an exponential form free of logarithms. Then solve for y in terms of the remaining variables.

51. Explain why 1 cannot be used as a logarithmic base.

▶ APPLICATIONS

Business & Economics

The two formulas below will be of use in some of the problems that follow:

$$A = P\left(1 + \frac{r}{m}\right)^{mt} \qquad \text{Compound interest}$$

$$A = Pe^{rt} \qquad \text{Continuous compound interest}$$

52. *Money growth.* If $5,000 is invested at 12% compounded weekly, how much (to the nearest cent) will be in the account 6 years from now?

53. *Money growth.* If $5,000 is invested at 12% compounded continuously, how much (to the nearest cent) will be in the account 6 years from now?

54. *Finance.* Find the tripling time (to the next whole year) for money invested at 15% compounded annually.

55. *Finance.* Find the doubling time (to two decimal places) for money invested at 10% compounded continuously.

56. *Minimum average cost.* The financial department of a company that manufactures roller-blade skates has fixed costs of $300 per day and total costs of $4,300 per day at an output of 100 pairs of skates per day. Assume the cost $C(x)$ is linearly related to output x.

(A) Find an expression for the cost function $C(x)$ and the average cost function $\overline{C}(x) = C(x)/x$.

(B) Sketch a graph of the average cost function for $5 \leqslant x \leqslant 200$.

(C) Identify any asymptotes.

(D) What does the average cost approach as production increases?

57. *Minimum average cost.* The cost $C(x)$ in thousands of dollars for operating a hospital for a year is given by

$$C(x) = 20x^3 - 360x^2 + 2,300x - 1,000$$

where x is the number of cases per year (in thousands). The average cost function $\overline{C}$ is given by $\overline{C}(x) = C(x)/x$.

(A) Write an equation for the average cost function.

(B) Graph the average cost function for $1 \leqslant x \leqslant 12$.

(C) Use trace and zoom or a built-in routine to find the number of cases per year the hospital should handle to have the minimum average cost. What is the minimum average cost?

58. *Price–demand.* The marketing department of a manufacturer of disposable cameras test markets the camera in selected cities in the United States. Using statistical techniques, the department arrived at the results shown in Table 1, where x is the number of cameras (in millions) that can be sold at p dollars per camera.

(A) Enter the data set in a graphing utility. Then find a linear regression equation for the data set and transfer it into the function editor. Plot the data set and the linear regression equation in the same viewing window. (Leave the linear regression equation in the function editor to be used in Problem 60.)

TABLE 1
Price–Demand

x (million)	p = D(x) ($)
0	17
2	13
4	11
6	8
8	5

(B) Estimate the demand (to the nearest million) at a price of $9.49 using graphical methods in the graphing utility.

59. *Price–supply.* The marketing department for the manufacturer of disposable cameras in Problem 58 also established the results shown in Table 2, where x is the number of cameras (in millions) that the supplier is willing to sell at p dollars per camera.

TABLE 2
Price–Supply

x (million)	p = S(x) ($)
0	5
2	6
4	7
6	8
8	12

(A) Enter the data set in a graphing utility. Then find a quadratic regression equation for the data set and transfer it into the function editor. Plot the data set and the quadratic regression equation in the same viewing window. (Leave the quadratic regression equation in the function editor to be used in Problem 60.)

(B) Estimate the price (to the nearest dollar) at a supply level of 7 million cameras.

60. *Equilibrium price and quantity.* Plot the equations from part (A) in Problems 58 and 59 in the same viewing window, and use an appropriate built-in routine to find the point of intersection. Write the equilibrium price to the nearest cent and the equilibrium quantity to the nearest tenth of a million.

Life Sciences

61. *Medicine.* One leukemic cell injected into a healthy mouse will divide into 2 cells in about $\frac{1}{2}$ day. At the end of the day these 2 cells will divide into 4. This doubling continues until 1 billion cells are formed; then the animal dies with leukemic cells in every part of the body.

(A) Write an equation that will give the number N of leukemic cells at the end of t days.

(B) When, to the nearest day, will the mouse die?

62. *Marine biology.* The intensity of light entering water is reduced according to the exponential equation

$$I = I_0 e^{-kd}$$

where I is the intensity d feet below the surface, I_0 is the intensity at the surface, and k is the coefficient of extinction. Measurements in the Sargasso Sea in the West Indies have indicated that half of the surface light reaches a depth of 73.6 feet. Find k (to five decimal places), and find the depth (to the nearest foot) at which 1% of the surface light remains.

63. *Cholera.* Cholera, an intestinal disease, is caused by a bacterium that multiplies by cell division. A laboratory experiment, starting with 10 bacteria, produced the results shown in Table 3, where N is the number of bacteria present after t hours in the intestine.

TABLE 3
Cholera Bacteria Growth

t (hours)	N
0.0	10
0.5	24
1.0	36
2.0	180
3.0	615

(A) Enter the data set in a graphing calculator, find an exponential regression equation ($y = ab^x$) for the data set, and plot the data set and the exponential regression equation in the same viewing window.

(B) Estimate the number of cholera bacteria that will be present at the end of 4 hours.

Social Sciences

64. *Population growth.* Many countries have a population growth rate of 3% (or more) per year. At this rate, how many years (to the nearest tenth of a year) will it take a population to double? Use the annual compounding growth model $P = P_0(1 + r)^t$.

65. *Population growth.* Repeat Problem 64 using the continuous compounding growth model $P = P_0 e^{rt}$.

Finite Mathematics

CHAPTER 3

Mathematics of Finance

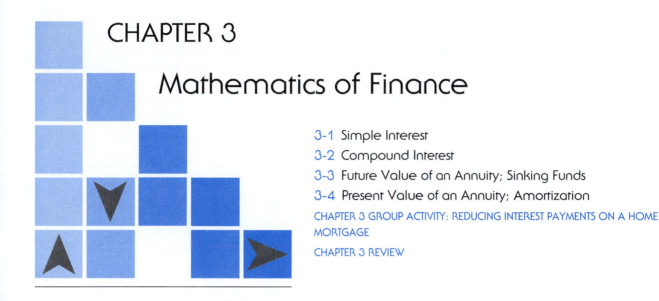

This chapter is independent of the others; you can study it at any time. In particular, we do not assume that you have studied Chapter 2, where a few of the topics in this chapter were briefly discussed as applications of exponential and logarithmic functions.

A graphing utility is an excellent tool for solving problems on compound interest, annuities, amortization, and so on. In addition to using graphs to visualize the growth of an investment or the amortization of a loan, we will also use graphical approximation techniques to find interest rates, growth times, and other related quantities. Most graphing utilities also have special routines called *equation solvers* (illustrated in Figure 6, Section 3-3) that can be used to approximate solutions to equations without first drawing a graph. Equation solvers are used extensively when working with financial equations. Consult the manual for your graphing utility for specific details.

To avoid repeating the statement many times, we now point out:

Throughout the chapter, interest rates are to be converted to decimal form before they are used in a formula.

SECTION 3-1 ■ Simple Interest

Simple interest is generally used only on short-term notes—often of duration less than 1 year. The concept of simple interest, however, forms the basis of much of the rest of the material developed in this chapter, for which time periods may be much longer than a year.

If you deposit a sum of money P in a savings account or if you borrow a sum of money P from a lending agent, then P is referred to as the **principal.** When money is borrowed—whether it is a savings institution borrowing from you when you deposit money in your account or you borrowing from a lending agent—a fee is charged for the money borrowed. This fee is rent paid for the use of another's money, just as rent is paid for the use of another's house. The fee is called **interest.** It is usually computed as a percentage (called the **interest rate**)* of the principal over a given period of time. The interest rate, unless otherwise stated, is an annual rate. **Simple interest** is given by the following formula:

Simple Interest

$$I = Prt \qquad (1)$$

where

P = Principal

r = Annual simple interest rate (written as a decimal)

t = Time in years

For example, the interest on a loan of $100 at 12% for 9 months would be

$I = Prt$

$= (100)(0.12)(0.75)$ *Convert 12% to a decimal (0.12)*

$= \$9$ *and 9 months to years ($\frac{9}{12}$ = 0.75).*

At the end of 9 months, the borrower would repay the principal ($100) plus the interest ($9), or a total of $109.

In general, if a principal P is borrowed at a rate r, then after t years the borrower will owe the lender an amount A that will include the principal P (the **face value** of the note) plus the interest I (the rent paid for the use of the money). Since P is the amount that is borrowed now and A is the amount that must be paid back in the future, P is often referred to as the **present value** and A as the **future value.** The formula relating A and P is as follows:

Amount—Simple Interest

$$A = P + Prt$$
$$ = P(1 + rt) \qquad (2)$$

where

P = Principal, or present value

r = Annual simple interest rate (written as a decimal)

t = Time in years

A = Amount, or future value

---

* If r is the interest rate written as a decimal, then $100r\%$ is the rate using %. For example, if $r = 0.12$, then using the percent symbol, %, we have $100r\% = 100(0.12)\% = 12\%$. The expressions 0.12 and 12% are equivalent.

Given any three of the four variables *A, P, r,* and *t* in (2), we can solve for the fourth. The following examples illustrate several types of common problems that can be solved by using formula (2).

EXAMPLE 1 ➤ Total Amount Due on a Loan Find the total amount due on a loan of $800 at 18% simple interest at the end of 4 months.

Solution To find the amount *A* (future value) due in 4 months, we use formula (2) with $P = 800$, $r = 0.18$, and $t = \frac{4}{12} = \frac{1}{3}$ year. Thus,

$$A = P(1 + rt)$$
$$= 800[1 + 0.18(\tfrac{1}{3})]$$
$$= 800(1.06)$$
$$= \$848$$

MATCHED PROBLEM 1 ➤ Find the total amount due on a loan of $500 at 12% simple interest at the end of 30 months.

explore – discuss 1

(A) Your dear sister has loaned you $1,000 with the understanding that the principal plus 4% simple interest are to be repaid when you are able. How much would you owe her if you repaid the loan after 1 year? After 2 years? After 5 years? After 10 years?
(B) How is the interest after 10 years related to the interest after 1 year? After 2 years? After 5 years?
(C) Explain why your answers are consistent with the fact that for simple interest the graph of future value as a function of time is a straight line (see Fig. 1).

FIGURE 1

EXAMPLE 2 ➤ Present Value of an Investment If you want to earn an annual rate of 10% on your investments, how much (to the nearest cent) should you pay for a note that will be worth $5,000 in 9 months?

Solution   We again use formula (2), but now we are interested in finding the principal P (present value), given $A = \$5,000$, $r = 0.1$, and $t = \frac{9}{12} = 0.75$ year. Thus,

$$A = P(1 + rt)$$
$$5,000 = P[1 + 0.1(0.75)] \quad \text{Replace A, r, and t with the given values,}$$
$$5,000 = (1.075)P \quad \text{and solve for P.}$$
$$P = \$4,651.16 \quad \blacktriangleleft$$

MATCHED PROBLEM 2 ➤   Repeat Example 2 with a time period of 6 months. ◀

▶ EXAMPLE 3 ➤   **Interest Rate Earned on a Note**   If you must pay $960 for a note that will be worth $1,000 in 6 months, what annual simple interest rate will you earn? (Express the answer as a percentage, correct to two decimal places.)

Solution   Again we use formula (2), but this time we are interested in finding r, given $P = \$960$, $A = \$1,000$, and $t = \frac{6}{12} = 0.5$ year. Thus,

$$A = P(1 + rt) \quad \text{Replace P, A, and t with the given values, and}$$
$$1,000 = 960[1 + r(0.5)] \quad \text{solve for r.}$$
$$1,000 = 960 + 960r(0.5)$$
$$40 = 480r$$
$$r = \frac{40}{480} \approx 0.0833 \quad \text{or } 8.33\% \quad \blacktriangleleft$$

MATCHED PROBLEM 3 ➤   Repeat Example 3 assuming you have paid $952 for the note. ◀

▶ EXAMPLE 4 ➤   **Interest Rate Earned on an Investment**   Suppose after buying a new car you decide to sell your old car to a friend. You accept a 270 day note for $3,500 at 10% simple interest as payment. (Both principal and interest will be paid at the end of 270 days.) Sixty days later you find that you need the money and sell the note to a third party for $3,550. What annual interest rate will the third party receive for the investment? (Express the answer as a percentage, correct to three decimal places.)

Solution   Step 1.   Find the amount that will be paid at the end of 270 days to whomever has the note. Some financial institutions use a 365 day year and others a 360 day year. In all the problems in this section involving days, we assume a 360 day year.*

$$A = P(1 + rt)$$
$$= \$3,500\left[1 + (0.1)\left(\frac{270}{360}\right)\right]$$
$$= \$3,762.50$$

---

* In other sections we will use a 365 day year. The choice will always be clearly stated.

Step 2. For the third party we are to find the annual rate of interest r required to make \$3,550 grow to \$3,762.50 in 210 days $(270 - 60)$; that is, we are to find r (which is to be converted to $100r\%$), given $A = \$3,762.50$, $P = \$3,550$, and $t = \frac{210}{360}$.

$$A = P + Prt \quad \text{Solve for } r.$$

$$r = \frac{A - P}{Pt}$$

$$r = \frac{3,762.50 - 3,550}{(3,550)(\frac{210}{360})} = 0.102\ 62 \quad \text{or } 10.262\%$$ ◀

MATCHED PROBLEM 4 ▶ Repeat Example 4 assuming that 90 days after it was initially signed, the note was sold to a third party for \$3,500. ◀

explore – discuss 2

(A) Starting with formula (2), derive each of the following formulas:

$$P = \frac{A}{1 + rt} \qquad r = \frac{A - P}{Pt} \qquad t = \frac{A - P}{Pr}$$

(B) Explain why it is unnecessary to memorize the above formulas for P, r, and t if you know formula (2).

Answers to Matched Problems 1. \$650   2. \$4,761.90   3. 10.08%   4. 15.0%

EXERCISE 3-1

A *In Problems 1–4, make the indicated conversions assuming a 360 day year.*

1. $9.5\% = ?$ (decimal); 60 days = ? year
2. $8.75\% = ?$ (decimal); 3 quarters = ? year
3. $0.18 = ?$ (percentage); 5 months = ? year
4. $0.0525 = ?$ (percentage); 240 days = ? year

Using formula (1) for simple interest, find each of the indicated quantities in Problems 5–8.

5. $P = \$500$; $r = 8\%$; $t = 6$ months; $I = ?$
6. $P = \$900$; $r = 10\%$; $t = 9$ months; $I = ?$
7. $I = \$80$; $P = \$500$; $t = 2$ years; $r = ?$
8. $I = \$40$; $P = \$400$; $t = 4$ years; $r = ?$

B *Use formula (2) in an appropriate form to find the indicated quantities in Problems 9–12.*

9. $P = \$100$; $r = 8\%$; $t = 18$ months; $A = ?$
10. $P = \$6,000$; $r = 6\%$; $t = 8$ months; $A = ?$
11. $A = \$1,000$; $r = 10\%$; $t = 15$ months; $P = ?$
12. $A = \$8,000$; $r = 12\%$; $t = 7$ months; $P = ?$

C *In Problems 13–16, solve each formula for the indicated variable.*

13. $I = Prt$; for r
14. $I = Prt$; for P
15. $A = P + Prt$; for P
16. $A = P + Prt$; for r

17. Discuss the similarities and differences in the graphs of future value A as a function of time t if $1,000 is invested at simple interest at rates of 4%, 8%, and 12%, respectively (see the figure).

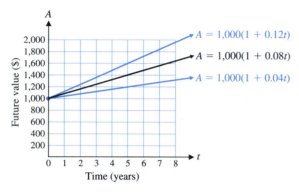

Figure for 17

18. Discuss the similarities and differences in the graphs of future value A as a function of time t for loans of $400, $800, and $1,200, respectively, each at 7.5% simple interest (see the figure).

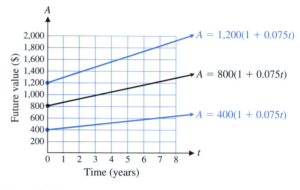

Figure for 18

▶ A P P L I C A T I O N S*

Business & Economics

In all problems involving days, a 360 day year is assumed. When annual rates are requested as an answer, express the rate as a percentage, correct to three decimal places.

19. If $3,000 is loaned for 4 months at a 14% annual rate, how much interest is earned?

20. If $5,000 is loaned for 10 months at a 10% annual rate, how much interest is earned?

21. How much interest will you have to pay for a credit card balance of $554 that is 1 month overdue, if a 20% annual rate is charged?

22. A department store charges an 18% annual rate for overdue accounts. How much interest will be owed on an $835 account that is 2 months overdue?

23. A loan of $7,250 was repaid at the end of 8 months. What size repayment check (principal and interest) was written, if a 9% annual rate of interest was charged?

24. A loan of $10,000 was repaid at the end of 14 months. What amount (principal and interest) was repaid if a 12% annual rate of interest was charged?

25. A loan of $4,000 was repaid at the end of 8 months with a check for $4,270. What annual rate of interest was charged?

26. A check for $3,262.50 was used to retire a 15 month $3,000 loan. What annual rate of interest was charged?

27. If you paid $30 to a loan company for the use of $1,000 for 60 days, what annual rate of interest did they charge?

28. If you paid $120 to a loan company for the use of $2,000 for 90 days, what annual rate of interest did they charge?

29. A radio commercial for a loan company states: "You only pay 50¢ a day for each $500 borrowed." If you borrow $1,500 for 120 days, what amount will you repay, and what annual interest rate is the company actually charging?

30. George finds a company that charges 70¢ per day for each $1,000 borrowed. If he borrows $3,000 for 60 days, what amount will he repay, and what annual interest rate will he be paying the company?

31. You are interested in buying a 13 week T-bill (treasury bill) from the U.S. Treasury Department. If you buy a T-bill with a maturity value of $10,000 for $9,776.94, what annual interest rate will you earn?

---

* The authors wish to thank Professor Roy Luke of Pierce College for his many useful suggestions of applications in this chapter.

32. If you buy a 26 week T-bill with a maturity value of $10,000 for $9,562.56 from the U.S. Treasury Department, what annual interest rate will you earn?

33. If an investor wants to earn an annual interest rate of 12.63% on a 13 week T-bill with a maturity value of $10,000, how much should the investor pay for the bill?

34. If an investor wants to earn an annual interest rate of 10.58% on a 26 week T-bill with a maturity value of $10,000, how much should the investor pay for the bill?

35. For services rendered, an attorney accepts a 90 day note for $5,500 at 12% simple interest from a client. (Both interest and principal will be repaid at the end of 90 days.) Wishing to be able to use her money sooner, the attorney sells the note to a third party for $5,540 after 30 days. What annual interest rate will the third party receive for the investment?

36. To complete the sale of a house, the seller accepts a 180 day note for $10,000 at 10% simple interest. (Both interest and principal will be repaid at the end of 180 days.) Wishing to be able to use the money sooner for the purchase of another house, the seller sells the note to a third party for $10,100 after 60 days. What annual interest rate will the third party receive for the investment?

The following buying and selling commission schedule is from a well-known discount brokerage house:

DOLLAR RANGE PER TRANSACTION	COMMISSION
$0–2,500	$19 + 1.6% of principal amount
$2,501–6,000	$44 + 0.6% of principal amount
$6,001–22,000	$62 + 0.3% of principal amount
$22,001–50,000	$84 + 0.2% of principal amount
$50,001–500,000	$134 + 0.1% of principal amount
$500,000 +	$234 + 0.08% of principal amount

Example: The commission on 500 shares at $15 per share is $84.50.

Taking into consideration the buying and selling commissions in the schedule, find the annual rate of interest earned by each investment in Problems 37–40.

37. An investor purchases 500 shares at $14.20 a share, holds the stock for 39 weeks, and then sells the stock for $16.84 a share.

38. An investor purchases 450 shares at $64.84 a share, holds the stock for 26 weeks, and then sells the stock for $72.08 a share.

39. An investor purchases 2,000 shares at $23.75 a share, holds the stock for 300 days, and then sells the stock for $26.15 a share.

40. An investor purchases 75 shares at $31.50 a share, holds the stock for 150 days, and then sells the stock for $35.40 a share.

SECTION 3-2 ■ Compound Interest

■ COMPOUND INTEREST
■ EFFECTIVE RATE
■ GROWTH AND TIME

■ COMPOUND INTEREST

If at the end of a payment period the interest due is reinvested at the same rate, then the interest as well as the original principal will earn interest during the next payment period. Interest paid on interest reinvested is called **compound interest.**

For example, suppose you deposit $1,000 in a bank that pays 8% compounded quarterly. How much will the bank owe you at the end of a year? *Compounding quarterly* means that earned interest is paid to your account at the end of each 3 month period and that interest as well as the principal earns interest for the next quarter. Using the simple interest formula (2) from the previous section, we compute the amount in the account at the end of the first quarter after interest has been paid:

$$A = P(1 + rt)$$
$$= 1,000[1 + 0.08(\tfrac{1}{4})]$$
$$= 1,000(1.02) = \$1,020$$

Now, $1,020 is your new principal for the second quarter. At the end of the second quarter, after interest is paid, the account will have

$$A = \$1,020[1 + 0.08(\tfrac{1}{4})]$$
$$= \$1,020(1.02) = \$1,040.40$$

Similarly, at the end of the third quarter, you will have

$$A = \$1,040.40[1 + 0.08(\tfrac{1}{4})]$$
$$= \$1,040.40(1.02) = \$1,061.21$$

Finally, at the end of the fourth quarter, the account will have

$$A = \$1,061.21([1 + 0.08(\tfrac{1}{4})]$$
$$= \$1,061.21(1.02) = \$1,082.43$$

How does this compound amount compare with simple interest? The amount with simple interest would be

$$A = P(1 + rt)$$
$$= \$1,000[1 + 0.08(1)]$$
$$= \$1,000(1.08) = \$1,080$$

We see that compounding quarterly yields $2.43 more than simple interest would provide.

Let us look over the above calculations for compound interest to see if we can uncover a pattern that might lead to a general formula for computing compound interest for arbitrary cases:

$$A = 1,000(1.02) \qquad \text{End of first quarter}$$
$$A = [1,000(1.02)](1.02) = 1,000(1.02)^2 \qquad \text{End of second quarter}$$
$$A = [1,000(1.02)^2](1.02) = 1,000(1.02)^3 \qquad \text{End of third quarter}$$
$$A = [1,000(1.02)^3](1.02) = 1,000(1.02)^4 \qquad \text{End of fourth quarter}$$

It appears that at the end of n quarters, we would have

$$A = 1,000(1.02)^n \qquad \text{End of nth quarter}$$

or

$$A = 1,000[1 + 0.08(\tfrac{1}{4})]^n$$
$$= 1,000[1 + \tfrac{0.08}{4}]^n$$

where $\frac{0.08}{4} = 0.02$ is the interest rate per quarter. Since interest rates are generally quoted as *annual nominal rates,* the **rate per compounding period** is found by dividing the annual nominal rate by the number of compounding periods per year.

In general, if P is the principal earning interest compounded m times a year at an annual rate of r, then (by repeated use of the simple interest formula, using $i = r/m$, the rate per period) the amount A at the end of each period is

$$A = P(1 + i) \qquad\qquad \text{End of first period}$$
$$A = [P(1 + i)](1 + i) = P(1 + i)^2 \qquad \text{End of second period}$$
$$A = [P(1 + i)^2](1 + i) = P(1 + i)^3 \qquad \text{End of third period}$$
$$\vdots$$
$$A = [P(1 + i)^{n-1}](1 + i) = P(1 + i)^n \quad \text{End of nth period}$$

We summarize this important result in the following box:

Amount—Compound Interest

$$A = P(1 + i)^n \qquad\qquad\qquad\qquad (1)$$

where $i = r/m$ and

$r = $ Annual nominal rate*

$m = $ Number of compounding periods per year

$i = $ Rate per compounding period

$n = $ Total number of compounding periods

$P = $ Principal (present value)

$A = $ Amount (future value) at the end of n periods

---

* This is often shortened to "annual rate" or just "rate."

Several examples will illustrate different uses of formula (1). If any three of the four variables in (1) are given, we can solve for the fourth using some algebra and a calculator. In particular, if A, P, and i are given, we can solve for the exponent n using properties of the logarithm and a calculator.

E X A M P L E 1 ➤ Comparing Interest for Various Compounding Periods   If $1,000 is invested at 8% compounded

(A) annually   (B) semiannually   (C) quarterly   (D) monthly

what is the amount after 5 years? Write answers to the nearest cent.

Solution   (A) Compounding annually means that there is one interest payment period per year. Thus, $n = 5$ and $i = r = 0.08$.

$$A = P(1 + i)^n$$
$$= 1,000(1 + 0.08)^5 \quad \textit{Use a calculator.}$$
$$= \$1,469.33 \qquad \textit{Interest earned} = A - P = \$469.33$$

(B) Compounding semiannually means that there are two interest payment periods per year. Thus, the number of payment periods in 5 years is $n = 2(5) = 10$, and the interest rate per period is

$$i = \frac{r}{m} = \frac{0.08}{2} = 0.04$$

So,

$$A = P(1 + i)^n$$
$$= 1,000(1 + 0.04)^{10} \quad \textit{Use a calculator.}$$
$$= \$1,480.24 \qquad \textit{Interest earned} = A - P = \$480.24$$

(C) Compounding quarterly means that there are four interest payments per year. Thus, $n = 4(5) = 20$ and $i = \frac{0.08}{4} = 0.02$. So,

$$A = P(1 + i)^n$$
$$= 1,000(1 + 0.02)^{20} \quad \textit{Use a calculator.}$$
$$= \$1,485.95 \qquad \textit{Interest earned} = A - P = \$485.95$$

(D) Compounding monthly means that there are twelve interest payments per year. Thus, $n = 12(5) = 60$ and $i = \frac{0.08}{12} = 0.006\ 66\overline{6}$.* So,

$$A = P(1 + i)^n$$
$$= 1,000\left(1 + \frac{0.08}{12}\right)^{60} \quad \textit{Use a calculator.}$$
$$= \$1,489.85 \qquad \textit{Interest earned} = A - P = \$489.85 \qquad \blacktriangleleft$$

MATCHED PROBLEM 1 ➤ Repeat Example 1 with an annual interest rate of 6% over an 8 year period. ◀

Notice the rather significant increase in interest earned in going from annual compounding to monthly compounding. One might wonder what happens if we compound daily, or every minute, or every second, and so on. Figure 1 shows the effect of increasing the number of compounding periods per year when $1,000 is invested at 8% compound interest for 5 years.

---

*Recall that the bar over the 6 indicates a repeating decimal expansion. Rounding i to a small number of decimal places, such as 0.007 or 0.0067, can result in round-off errors. To avoid this, use as many decimal places for i as your calculator is capable of displaying.

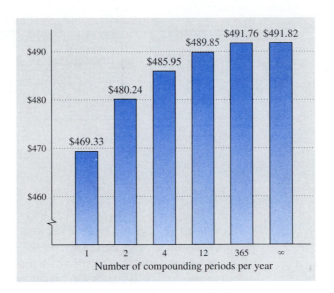

FIGURE 1
Interest on $1,000 for 5 years at 8% with various compounding periods

Note from Figure 1 that the difference in interest earned between annual and semiannual compounding is $480.24 − $469.33 = $10.91, but the difference between semiannual and quarterly compounding is only $5.71. Furthermore, the difference between quarterly and monthly compounding is only $3.90, even though the number of periods is tripled, and the difference between monthly and daily compounding is only $1.91, even though the number of compounding periods has increased dramatically. These facts suggest that the interest earned approaches a limit. The limit is reached at compounding *continuously,* which yields $491.82—only 6¢ more than compounding daily. (If interest is compounded continuously, then $A = Pe^{rt}$, as discussed in Section 2-2.) Compare the results in Figure 1 with simple interest earned over the same time period:

$$I = Prt = 1,000(0.08)5 = \$400$$

EXAMPLE 2 ➤ Visualizing Investments with a Graphing Utility

(A) Use a graphing utility to graph the growth of investments of $1,000 at 8% simple interest and of $1,000 at 8% compounded monthly, both over a period of 5 years.

(B) Use a graphing utility to graph the growth of both investments in part (A) during the fifth year.

Solution (A) To graph the first investment, the formula for simple interest, $A = P(1 + rt)$, is entered as $y_1 = 1000*(1 + .08*x)$, where the variable x represents time in years and y_1 represents the amount after x years. Similarly, to graph the second investment, the formula for compound interest, $A = P(1 + i)^n$, is

(A)   (B)   (C)

FIGURE 2

entered as $y_2 = 1000*(1+.08/12)\hat{\ }(12*x)$. Note that n, the number of periods, is entered as $12*x$ (12 months per year times x years), making the variable x have the same meaning in both equations. This allows us to compare y_1 and y_2 at various values of x using the trace feature of the graphing utility, for example. Figure 2 shows the equations that represent each investment, the window variables, and the resulting graphs. (An alternative approach would be to let the variable x represent months rather than years in both equations, and to use $y_1 = 1000*(1+.08*(x/12))$, $y_2 = 1000*(1+.08/12)\hat{\ }x$, Xmin=0, Xmax=60.)

(B) Only the window variables need to be modified. From Figure 2 we see that the investment at simple interest is worth something more than $1,300 when x = 4. By setting Xmin=4, Xmax=5, Ymin=1300, Ymax=1500 our viewing window will show the growth of both investments during the fifth year. We set Xscl=$\frac{1}{12}$ to obtain a scale for the x axis in months, and set Yscl=20 to obtain a scale for the y axis in increments of $20. Figure 3 shows the equations that represent each investment, the window variables, and the resulting graphs.

(A)   (B)   (C)

FIGURE 3   ◄

MATCHED PROBLEM 2 ➤   Use a graphing utility to graph the growth of an investment of $5,000 at 10% compounded semiannually, and of a second investment of $5,000 at 10% compounded monthly, both over a period of 30 years.   ◄

explore – discuss 1

(A) Which would be the better way to invest $1,000: at 9% simple interest for 10 years, or at 7% compounded monthly for 10 years?

(B) Explain why the graph of future value as a function of time is a straight line for simple interest, but for compound interest the graph curves upward (see Fig 4).

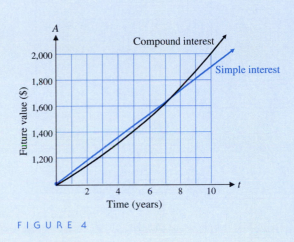

FIGURE 4

Another use of the compound interest formula is in determining how much you should invest now to have a given amount at a future date.

  EXAMPLE 3 ➤ **Finding Present Value**   How much should you invest now at 10% compounded quarterly to have $8,000 toward the purchase of a car in 5 years?

Solution We are given a future value $A = \$8,000$ for a compound interest investment, and we need to find the present value (principal P) given $i = \frac{0.10}{4} = 0.025$ and $n = 4(5) = 20$.

$$A = P(1 + i)^n$$

$$8,000 = P(1 + 0.025)^{20}$$

$$P = \frac{8,000}{(1 + 0.025)^{20}} \qquad \textit{Use a calculator.}$$

$$= \$4,882.17$$

Thus, your initial investment of $4,882.17 will grow to $8,000 in 5 years, as illustrated in Figure 5, at the top of the next page.

	A	B	C
1	Period	Interest	Amount
2	0		$4,882.17
3	1	$122.05	$5,004.22
4	2	$125.11	$5,129.33
5	3	$128.23	$5,257.56
6	4	$131.44	$5,389.00
7	5	$134.73	$5,523.73
8	6	$138.09	$5,661.82
9	7	$141.55	$5,803.37
10	8	$145.08	$5,948.45
11	9	$148.71	$6,097.16
12	10	$152.43	$6,249.59
13	11	$156.24	$6,405.83
14	12	$160.15	$6,565.98
15	13	$164.15	$6,730.13
16	14	$168.25	$6,898.38
17	15	$172.46	$7,070.84
18	16	$176.77	$7,247.61
19	17	$181.19	$7,428.80
20	18	$185.72	$7,614.52
21	19	$190.36	$7,804.88
22	20	$195.12	$8,000.00

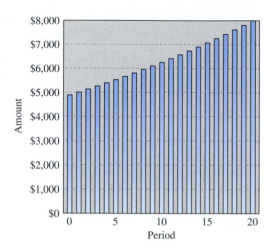

FIGURE 5
Interest on $4,882.17 at 10% compounded quarterly for 5 years   ◄

MATCHED PROBLEM 3   ➤   How much should new parents invest now at 8% compounded semiannually to have $80,000 toward their child's college education in 17 years?   ◄

The table and chart in Figure 5 were created by spreadsheet software on a personal computer. The concepts and formulas discussed in this chapter are used extensively in spreadsheets to produce tables and charts for a wide variety of business applications.

explore – discuss 2

(A) To become a millionaire, you intend to deposit an amount P at rate r compounded quarterly on your 25th birthday, and withdraw 1 million dollars on your 75th birthday. How much would you have to deposit if $r = 4\%$? If $r = 8\%$? If $r = 12\%$?

(B) Suppose you deposit $2,500 on your 25th birthday. What rate r compounded quarterly must your deposit earn in order to grow to 1 million dollars by your 75th birthday?

■ EFFECTIVE RATE

Suppose you read in the newspaper that one investment pays 15% compounded monthly and another pays 15.2% compounded semiannually. Which has the better return? A good way to compare investments is to determine their **effective rates**—

the simple interest rates that would produce the same returns in *1 year* if the same principal had been invested at simple interest without compounding. (Effective rates are also called **annual yields** or **true interest rates.**)

If principal *P* is invested at an annual (nominal) rate *r* compounded *m* times a year, then in 1 year,

$$A = P\left(1 + \frac{r}{m}\right)^m$$

What simple interest rate will produce the same amount *A* in 1 year? We call this simple interest rate the effective rate, and denote it by r_e. To find r_e we proceed as follows:

$$\begin{pmatrix} \text{Amount at} \\ \text{simple interest} \\ \text{after 1 year} \end{pmatrix} = \begin{pmatrix} \text{Amount at} \\ \text{compound interest} \\ \text{after 1 year} \end{pmatrix}$$

$$P(1 + r_e) = P\left(1 + \frac{r}{m}\right)^m \qquad \textit{Divide both sides by P.}$$

$$1 + r_e = \left(1 + \frac{r}{m}\right)^m \qquad \textit{Isolate } r_e \textit{ on the left side.}$$

$$r_e = \left(1 + \frac{r}{m}\right)^m - 1$$

Effective Rate

If principal *P* is invested at the annual (nominal) rate *r* compounded *m* times a year, then the effective rate r_e is given by

$$r_e = \left(1 + \frac{r}{m}\right)^m - 1$$

If interest is compounded continuously at the annual rate *r*, then the effective rate is given in terms of the exponential function with base *e* (introduced in Section 2-2) by $r_e = e^r - 1$.

EXAMPLE 4 ➤ **Computing Effective Rate** A savings and loan pays 8% compounded quarterly. What is the effective rate? (Express the answer as a percentage, correct to three decimal places.)

Solution

$$r_e = \left(1 + \frac{r}{m}\right)^m - 1$$

$$= (1 + \tfrac{0.08}{4})^4 - 1$$

$$= (1.02)^4 - 1 \qquad \textit{Use a calculator.}$$

$$= 0.082\ 432 \quad \text{or } 8.243\%$$

This shows that money invested at 8.243% simple interest earns the same amount of interest in *1 year* as money invested at 8% compounded quarterly. Thus, the effective rate of 8% compounded quarterly is 8.243%.   ◄

MATCHED PROBLEM 4   ➤   What is the effective rate of money invested at 6% compounded quarterly?   ◄

EXAMPLE 5   ➤   **Computing the Annual Nominal Rate Given the Effective Rate**
A savings and loan wants to offer a CD (certificate of deposit) with a monthly compounding rate that has an effective rate of 7.5%. What annual nominal rate compounded monthly should they use?

Solution   The effective rate for an annual nominal rate r compounded monthly is given by

$$r_e = \left(1 + \frac{r}{12}\right)^{12} - 1$$

We graph

$$y_1 = \left(1 + \frac{x}{12}\right)^{12} - 1$$

and

$$y_2 = 0.075$$

on a graphing utility, and locate the point of intersection as indicated in Figure 6. This shows that an annual nominal rate of 7.254% compounded monthly is equivalent to an effective rate of 7.5%.   ◄

0.2

0   0.2

Intersection
X=.07253903 Y=.075

0

FIGURE 6

MATCHED PROBLEM 5   ➤   What is the annual nominal rate compounded quarterly for a bond that has an effective rate of 8.8%?   ◄

EXAMPLE 6   ➤   **Comparing Two Investments**   An investor has an opportunity to purchase two different notes: Note A pays 15% compounded monthly, and note B pays 15.2% compounded semiannually. Which is the better investment, assuming all else is equal?

Solution   Nominal rates with different compounding periods cannot be compared directly. We must first find the effective rate of each nominal rate and then compare the effective rates to determine which investment will yield the larger return.

$$\text{\textit{Effective Rate for Note A:}} \quad r_e = \left(1 + \frac{r}{m}\right)^m - 1$$

$$= (1 + \tfrac{0.15}{12})^{12} - 1$$
$$= (1.0125)^{12} - 1 \qquad \text{Use a calculator.}$$
$$= 0.160\ 755 \quad \text{or } 16.076\%$$

Effective Rate for Note B: $\quad r_e = \left(1 + \dfrac{r}{m}\right)^m - 1$

$$= (1 + \tfrac{0.152}{2})^2 - 1$$
$$= (1.076)^2 - 1$$
$$= 0.157\ 776 \quad \text{or } 15.778\%$$

Since the effective rate for note A is greater than the effective rate for note B, note A is the preferred investment. ◀

MATCHED PROBLEM 6 ➤ Repeat Example 6 if note A pays 9% compounded monthly and note B pays 9.2% compounded semiannually. ◀

■ GROWTH AND TIME

Investments are also compared by computing their **growth time**—the time it takes a given principal to grow to a particular value (the shorter the time, the greater the return on the investment). Example 7 illustrates two methods for making this calculation.

EXAMPLE 7 ➤ Computing Growth Time   How long will it take $10,000 to grow to $12,000 if it is invested at 9% compounded monthly?

Solution   Method 1.   Use logarithms and a calculator.

$$A = P(1 + i)^n$$

$$12,000 = 10,000\left(1 + \frac{0.09}{12}\right)^n$$

$$1.2 = 1.0075^n$$

Now solve for n by taking logarithms of both sides:

$\ln 1.2 = \ln 1.0075^n$ *Logarithms to any base can be used; we choose the natural*
$\ln 1.2 = n \ln 1.0075$ *logarithm (base e) and use the property $\log_b M^p = p \log_b M$.*

$$n = \frac{\ln 1.2}{\ln 1.0075}$$

$$= 24.40 \approx 25 \text{ months} \quad \text{or 2 years and 1 month}$$

[*Note:* 24.40 is rounded up to 25 to guarantee reaching $12,000, since interest is paid at the end of each month.]

Method 2.   Use graphical approximation techniques. On a graphing utility we locate the intersection point of the graphs of $y_1 = 10000*(1 + .09/12)\hat{\ }x$ and $y_2 = 12000$ as indicated in Figure 7, at the top of the next page. Again, we find that $x = n = 24.40$ months, and we conclude that the growth time is 25 months.

(A)

(B)

FIGURE 7   ◄

MATCHED PROBLEM 7 ➤   How long will it take $10,000 to grow to $25,000 if it is invested at 18% compounded quarterly?   ◄

Caution

Each compound interest problem involves two interest rates. Referring to Example 7, $r = 0.09$ or 9% is the annual nominal compounding rate, and $i = r/12 = 0.0075$ or 0.75% is the interest rate per month. Do not confuse these two rates by using r in place of i in the compound interest formula. If interest is compounded annually, then $i = r/1 = r$. In all other cases, r and i are not the same.

Answers to Matched Problems   1. (A) $1,593.85   (B) $1,604.71   (C) $1,610.32   (D) $1,614.14
2.

(A)

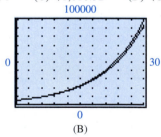

(B)

3. $21,084.17   4. 6.136%   5. 8.524%
6. Note B is better (effective rate of note A is 9.381% and of note B is 9.412%).
7. $20.82 \approx 21$ quarters, or 5 years and 3 months

EXERCISE   3-2

Find all dollar amounts to the nearest cent. When an interest rate is requested as an answer, express the rate as a percentage correct to two decimal places, unless directed otherwise. In all problems involving days, use a 365 day year.

A   *In Problems 1–8, use compound interest formula (1) to find each of the indicated values.*

1. $P = \$100$; $i = 0.01$; $n = 12$; $A = ?$
2. $P = \$1,000$; $i = 0.015$; $n = 20$; $A = ?$

3. $P = \$800$; $i = 0.06$; $n = 25$; $A = ?$
4. $P = \$10,000$; $i = 0.08$; $n = 30$; $A = ?$
5. $A = \$10,000$; $i = 0.03$; $n = 48$; $P = ?$
6. $A = \$1,000$; $i = 0.015$; $n = 60$; $P = ?$
7. $A = \$18,000$; $i = 0.01$; $n = 90$; $P = ?$
8. $A = \$50,000$; $i = 0.005$; $n = 70$; $P = ?$

Given the annual rate and the compounding period in Problems 9–12, find i, the interest rate per compounding period.

9. 9% compounded monthly
10. 15% compounded annually
11. 7% compounded quarterly
12. 11% compounded semiannually

Given the rate per compounding period in Problems 13–16, find r, the annual rate.

13. 0.8% per month
14. 5% per year
15. 4.5% per half-year
16. 2.3% per quarter

B

17. If $100 is invested at 6% compounded
(A) annually   (B) quarterly   (C) monthly
what is the amount after 4 years? How much interest is earned?

18. If $2,000 is invested at 7% compounded
(A) annually   (B) quarterly   (C) monthly
what is the amount after 5 years? How much interest is earned?

19. If $5,000 is invested at 18% compounded monthly, what is the amount after
(A) 2 years?   (B) 4 years?

20. If $20,000 is invested at 6% compounded monthly, what is the amount after
(A) 5 years?   (B) 8 years?

21. Discuss the similarities and the differences in the graphs of future value A as a function of time t if $1,000 is invested for 8 years and interest is compounded monthly at annual rates of 4%, 8%, and 12%, respectively (see the figure).

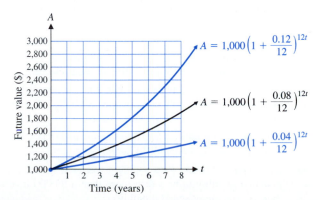

Figure for 21

22. Discuss the similarities and differences in the graphs of future value A as a function of time t for loans of $4,000, $8,000, and $12,000, respectively, each at 7.5% compounded monthly for 8 years (see the figure).

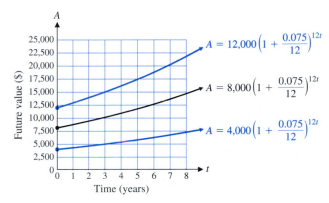

Figure for 22

23. If an investment company pays 8% compounded semiannually, how much should you deposit now to have $10,000
(A) 5 years from now?   (B) 10 years from now?

24. If an investment company pays 10% compounded quarterly, how much should you deposit now to have $6,000
(A) 3 years from now?   (B) 6 years from now?

25. What is the effective rate of interest for money invested at
(A) 10% compounded quarterly?
(B) 12% compounded monthly?

26. What is the effective rate of interest for money invested at
(A) 6% compounded monthly?
(B) 14% compounded semiannually?

27. How long will it take $4,000 to grow to $9,000 if it is invested at 15% compounded monthly?

28. How long will it take $5,000 to grow to $7,000 if it is invested at 8% compounded quarterly?

C *In Problems 29 and 30, use the compound interest formula (1) to find n to the nearest larger integer value.*

29. $A = 2P$; $i = 0.06$; $n = ?$
30. $A = 2P$; $i = 0.05$; $n = ?$

31. How long will it take money to double if it is invested at
(A) 10% compounded quarterly?
(B) 12% compounded quarterly?

32. How long will it take money to double if it is invested at
(A) 14% compounded semiannually?
(B) 10% compounded semiannually?

► APPLICATIONS

Business & Economics

33. A newborn child receives a $5,000 gift toward a college education from her grandparents. How much will the $5,000 be worth in 17 years if it is invested at 9% compounded quarterly?

34. A person with $8,000 is trying to decide whether to purchase a car now, or to invest the money at 12% compounded semiannually and then buy a more expensive car. How much will be available for the purchase of a car at the end of 3 years?

35. What will a $110,000 house cost 10 years from now if the inflation rate over that period averages 6% compounded annually?

36. If the inflation rate averages 8% per year compounded annually for the next 5 years, what will a car costing $10,000 now cost 5 years from now?

37. Rental costs for office space have been going up at 7% per year compounded annually for the past 5 years. If office space rent is now $20 per square foot per month, what were the rental rates 5 years ago?

38. In a suburb of a city, housing costs have been increasing at 8% per year compounded annually for the past 8 years. A house with a $160,000 value now would have had what value 8 years ago?

39. If the population in a particular third-world country is growing at 4% compounded annually, how long will it take the population to double? (Round up to the next higher year if not exact.)

40. If the world population is now about 5 billion people and is growing at 2% compounded annually, how long will it take the population to grow to 8 billion people? (Round up to the next higher year if not exact.)

41. Which is the better investment and why: 9% compounded monthly or 9.3% compounded annually?

42. Which is the better investment and why: 8% compounded quarterly or 8.3% compounded annually?

43. (A) If an investment of $100 were made in the year the Declaration of Independence was signed, and if it earned 3% compounded quarterly, how much would it be worth in 1998?

(B) Discuss the effect of compounding interest monthly, daily, and continuously (rather than quarterly) on the $100 investment.

(C) Graph the growth of the investment of part (A).

44. (A) Starting with formula (1), derive each of the following formulas:

$$P = \frac{A}{(1 + i)^n}$$

$$i = \left(\frac{A}{P}\right)^{1/n} - 1 \qquad n = \frac{\ln A - \ln P}{\ln(1 + i)}$$

(B) Explain why it is unnecessary to memorize the above formulas for P, i, and n if you know formula (1).

45. You have saved $7,000 toward the purchase of a car costing $9,000. How long will the $7,000 have to be invested at 9% compounded monthly to grow to $9,000? (Round up to the next higher month if not exact.)

46. A newly married couple has $15,000 toward the purchase of a house. For the type of house they are interested in buying, they estimate that a $20,000 down payment will be necessary. How long will the money have to be invested at 10% compounded quarterly to grow to $20,000? (Round up to the next higher quarter if not exact.)

47. An Individual Retirement Account (IRA) has $20,000 in it and the owner decides not to add any more money to the account other than interest earned at 8% compounded daily. How much will be in the account 35 years from now when the owner reaches retirement age?

48. If $1 had been placed in a bank account at the birth of Christ and forgotten until now, how much would be in the account at the end of 2010 if money earned 2% interest compounded annually? 2% simple interest? (Now you can see the power of compounding and see why inactive accounts are closed after a relatively short period of time.)

49. How long will it take money to double if it is invested at 14% compounded daily? 15% compounded annually? (Compute answers in years to three decimal places.)

50. How long will it take money to triple if it is invested at 10% compounded daily? 11% compounded annually? (Compute answers in years to three decimal places.)

51. How long will it take for $10,000 to grow to $15,000 if it is invested at 9.6% compounded quarterly?

52. How long will it take for a $4,800 investment at 10% compounded monthly to be worth more than a $5,000 investment at 7% compounded monthly?

53. What is the annual nominal rate compounded daily for a bond that has an effective rate of 7.4%?

54. What is the annual nominal rate compounded monthly for a CD that has an effective rate of 8.2%?

*Problems 55–58 refer to zero coupon bonds. A **zero coupon bond** is a bond that is sold now at a discount and will pay its face value at some time in the future when it matures—no interest payments are made.*

55. Parents wishing to have enough money for their child's college education 17 years from now decide to buy a $30,000 face value zero coupon bond. If money is worth 10% compounded annually, what should they pay for the bond?

56. How much should a $20,000 face value zero coupon bond, maturing in 10 years, be sold for now if its rate of return is to be 8% compounded annually?

57. If the parents in Problem 55 pay $6,844.79 for the $30,000 face value zero coupon bond, what annual compound rate of return will they be receiving?

58. If you pay $5,893.24 for a $12,000 face value zero coupon bond that matures in 7 years, what is your annual compound rate of return?

59. *Barron's* (a national business and financial weekly) published the following "Top Savings Deposit Yields" for money market deposit accounts:
(A) Virginia Beach Federal S&L
     8.28% compounded monthly
(B) Franklin Savings
     8.25% compounded daily
(C) Guaranty Federal Savings
     8.25% compounded monthly
What is the effective yield for each?

60. *Barron's* also published the following "Top Savings Deposit Yields" for 1 year CD accounts:
(A) Spindletop Savings
     9.00% compounded daily
(B) Alamo Savings of Texas
     9.10% compounded quarterly
(C) Atlas Savings and Loan
     9.00% compounded quarterly
What is the effective yield for each?

61. If you just sold a stock for $32,456.32 (net) that cost you $24,766.81 (net) 2 years ago, what annual compound rate of return did you make on your investment?

62. If you just sold a stock for $27,339.79 (net) that cost you $14,664.76 (net) 4 years ago, what annual compound rate of return did you make on your investment?

63. What annual nominal rate compounded monthly has the same effective rate as 8% compounded quarterly?

64. What annual nominal rate compounded daily has the same effective rate as 9.6% compounded monthly?

SECTION   3-3   ■   # Future Value of an Annuity; Sinking Funds

■ FUTURE VALUE OF AN ANNUITY
■ SINKING FUNDS

■ FUTURE VALUE OF AN ANNUITY

An **annuity** is any sequence of equal periodic payments. If payments are made at the end of each time interval, then the annuity is called an **ordinary annuity.** We will consider only ordinary annuities in this book. The amount, or **future value,** of an annuity is the sum of all payments plus all interest earned.

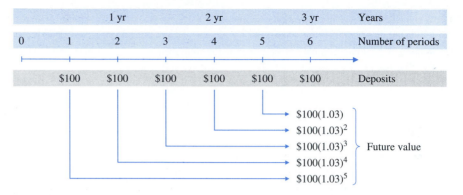

FIGURE 1

Suppose you decide to deposit $100 every 6 months into an account that pays 6% compounded semiannually. If you make six deposits, one at the end of each interest payment period, over 3 years, how much money will be in the account after the last deposit is made? To solve this problem, let us look at it in terms of a time line. Using the compound amount formula $A = P(1 + i)^n$, we can find the value of each deposit after it has earned compound interest up through the sixth deposit, as shown in Figure 1.

We could, of course, evaluate each of the future values in Figure 1 using a calculator and then add the results to find the amount in the account at the time of the sixth deposit—a tedious project at best. Instead, we take another approach that leads directly to a formula that will produce the same result in a few steps (even when the number of deposits is very large). We start by writing the total amount in the account after the sixth deposit in the form

$$S = 100 + 100(1.03) + 100(1.03)^2 + 100(1.03)^3 + 100(1.03)^4 + 100(1.03)^5 \quad (1)$$

We would like a simple way to sum these terms. Let us multiply each side of (1) by 1.03 to obtain

$$1.03S = 100(1.03) + 100(1.03)^2 + 100(1.03)^3 + 100(1.03)^4 + 100(1.03)^5 + 100(1.03)^6 \quad (2)$$

Subtracting (1) from (2), left side from left side and right side from right side, we obtain

$$1.03S - S = 100(1.03)^6 - 100 \qquad \text{Notice how many terms drop out.}$$
$$0.03S = 100[(1.03)^6 - 1]$$
$$S = 100 \frac{(1 + 0.03)^6 - 1}{0.03} \qquad \text{We write } S \text{ in this form to observe a general pattern.} \quad (3)$$

In general, if R is the periodic deposit, i the rate per period, and n the number of periods, then the future value is given by

$$S = R + R(1 + i) + R(1 + i)^2 + \cdots + R(1 + i)^{n-1} \qquad \text{Note how this compares to (1).}$$

and proceeding as in the above example, we obtain the general formula for the future value of an ordinary annuity:

$$S = R \frac{(1 + i)^n - 1}{i} \qquad \textit{Note how this compares to (3).} \qquad (4)$$

Returning to the example above, we use a calculator to complete the problem:

$$S = 100 \frac{(1.03)^6 - 1}{0.03} \qquad \textit{See the figure in the margin.}$$
$$= \$646.84$$

An alternative to the above computation of S is provided by the use of tables of values of certain functions of the mathematics of finance. One such function is the fractional factor of (4), denoted by the symbol $s_{\overline{n}|i}$, read ''s angle n at i'':

$$s_{\overline{n}|i} = \frac{(1 + i)^n - 1}{i}$$

Tables found in books on finance and mathematical handbooks list values of $s_{\overline{n}|i}$ for various values of n and i. To complete the computation of S, R is multiplied by $s_{\overline{n}|i}$ as indicated by (4). (A main advantage of using a calculator rather than tables is that a calculator can handle many more situations than a table, no matter how large the table.)

It is common to use FV (future value) for S and PMT (payment) for R in formula (4). Making these changes, we have the following:

Future Value of an Ordinary Annuity

$$FV = PMT \frac{(1 + i)^n - 1}{i} = PMTs_{\overline{n}|i} \qquad (5)$$

where

PMT = Periodic payment
i = Rate per period
n = Number of payments (periods)
FV = Future value (amount)

[*Note:* Payments are made at the end of each period.]

E X A M P L E   1 **Future Value of an Ordinary Annuity** What is the value of an annuity at the end of 20 years if $2,000 is deposited each year into an account earning 8.5% compounded annually? How much of this value is interest?

Solution   To find the value of the annuity, use formula (5) with $PMT = \$2,000$, $i = r = 0.085$, and $n = 20$.

$$FV = PMT \frac{(1 + i)^n - 1}{i}$$

$$= 2,000 \frac{(1.085)^{20} - 1}{0.085} \quad \text{Use a calculator.}$$

$$= \$96,754.03$$

To find the amount of interest earned, subtract the total amount deposited in the annuity (20 payments of $2,000) from the total value of the annuity after the 20th payment.

$$\text{Deposits} = 20(2,000) = \$40,000$$

$$\begin{aligned} \text{Interest} &= \text{Value} - \text{Deposits} \\ &= 96,754.03 - 40,000 = \$56,754.03 \end{aligned}$$

Figure 2, which was generated using spreadsheet software, illustrates the growth of this account over 20 years.

	A	B	C	D
1	Period	Payment	Interest	Balance
2	1	$2,000.00	$0.00	$2,000.00
3	2	$2,000.00	$170.00	$4,170.00
4	3	$2,000.00	$354.45	$6,524.45
5	4	$2,000.00	$554.58	$9,079.03
6	5	$2,000.00	$771.72	$11,850.75
7	6	$2,000.00	$1,007.31	$14,858.06
8	7	$2,000.00	$1,262.94	$18,120.99
9	8	$2,000.00	$1,540.28	$21,661.28
10	9	$2,000.00	$1,841.21	$25,502.49
11	10	$2,000.00	$2,167.71	$29,670.20
12	11	$2,000.00	$2,521.97	$34,192.17
13	12	$2,000.00	$2,906.33	$39,098.50
14	13	$2,000.00	$3,323.37	$44,421.87
15	14	$2,000.00	$3,775.86	$50,197.73
16	15	$2,000.00	$4,266.81	$56,464.54
17	16	$2,000.00	$4,799.49	$63,264.02
18	17	$2,000.00	$5,377.44	$70,641.47
19	18	$2,000.00	$6,004.52	$78,645.99
20	19	$2,000.00	$6,684.91	$87,330.90
21	20	$2,000.00	$7,423.13	$96,754.03
22	Totals	$40,000.00	$56,754.03	

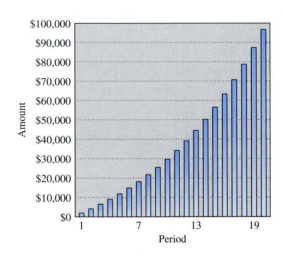

FIGURE 2
Ordinary annuity at 8.5% compounded annually for 20 years   ◀

MATCHED PROBLEM 1 ➤   What is the value of an annuity at the end of 10 years if $1,000 is deposited every 6 months into an account earning 8% compounded semiannually? How much of this value is interest?   ◀

The table in Figure 2 is called a **balance sheet.** Let's take a closer look at the construction of this table. The first line is a special case because the payment is made at the end of the period and no interest is earned. Each subsequent line of the table is computed as follows:

Payment + Interest   + Old balance = New balance
2,000   + 0.085(2,000) + 2,000   = 4,170   *Period 2*
2,000   + 0.085(4,170) + 4,170   = 6,524.45   *Period 3*

And so on. The amounts at the bottom of each column in the balance sheet agree with the results we obtained by using formula (5), as you would expect. Although balance sheets are appropriate for certain situations, we will concentrate on applications of formula (5). There are many important problems that can be solved only by using this formula.

explore – discuss 1

(A) Discuss the similarities and differences in the graphs of future value FV as a function of time t for ordinary annuities in which $100 is deposited each month for 8 years and interest is compounded monthly at annual rates of 4%, 8%, and 12%, respectively (see Fig. 3).

(B) Discuss the connections between the graph of the equation $y = 100t$, where t is time in months, and the graphs of part (A).

FIGURE 3

■ SINKING FUNDS

The formula for the future value of an ordinary annuity has another important application. Suppose the parents of a newborn child decide that on each of the child's birthdays up to the 17th year, they will deposit $PMT in an account that pays 6% compounded annually. The money is to be used for college expenses. What should the annual deposit $PMT be in order for the amount in the account to be $80,000 after the 17th deposit?

We are given FV, i, and n in formula (5), and our problem is to find PMT. Thus,

$$FV = PMT \frac{(1+i)^n - 1}{i}$$

$$80,000 = PMT \frac{(1.06)^{17} - 1}{0.06} \qquad \text{Solve for PMT.}$$

$$PMT = 80,000 \frac{0.06}{(1.06)^{17} - 1} \qquad \text{Use a calculator.}$$

$$= \$2,835.58 \text{ per year}$$

An annuity of 17 annual deposits of $2,835.58 at 6% compounded annually will amount to $80,000 in 17 years.

This is one of many examples of a similar type that are referred to as *sinking fund problems*. In general, any account that is established for accumulating funds to meet future obligations or debts is called a **sinking fund.** If the payments are to be made in the form of an ordinary annuity, then we have only to solve formula (5) for the **sinking fund payment** *PMT*:

$$PMT = FV \frac{i}{(1+i)^n - 1} \qquad\qquad (6)$$

It is important to understand that formula (6), which is convenient to use, is simply a variation of formula (5). You can always find the sinking fund payment by first substituting the appropriate values into formula (5) and then solving for *PMT*, as we did in the college fund example discussed above. Or you can substitute directly into formula (6), as we do in the next example. Use whichever method is easier for you.

EXAMPLE 2 ▶ **Computing the Payment for a Sinking Fund**   A company estimates that it will have to replace a piece of equipment at a cost of $10,000 in 5 years. To have this money available in 5 years, a sinking fund is established by making fixed monthly payments into an account paying 6% compounded monthly. How much should each payment be?

Solution   To find *PMT*, we can use either formula (5) or (6). We choose formula (6) with $FV = \$10,000$, $i = \frac{0.06}{12} = 0.005$, and $n = 5(12) = 60$:

$$PMT = FV \frac{i}{(1+i)^n - 1}$$

$$PMT = 10,000 \frac{0.005}{(1.005)^{60} - 1} \qquad \text{Use a calculator.}$$

$$= \$143.33 \text{ per month}$$ ◀

MATCHED PROBLEM 2 ▶ A bond issue is approved for building a marina in a city. The city is required to make regular payments every 6 months into a sinking fund paying 6% compounded semi-annually. At the end of 10 years, the bond obligation will be retired at a cost of $5,000,000. What should each payment be? ◀

explore – discuss 2

Suppose you intend to establish an ordinary annuity earning 7.5% compounded monthly which will be worth one million dollars on your seventieth birthday. How much would your monthly payments be if you started making payments at age twenty? At age thirty-five? At age fifty?

Using formulas (5) and (6) to find *FV* and *PMT* requires straightforward calculations. What happens if we need to solve for *n* or *i*? Algebraic properties of logarithms can be used to solve for *n*, but there is no algebraic solution for *i*. The next two examples illustrate the use of graphical approximation techniques to find *n* and *i*.

EXAMPLE 3 ➤ **Finding the Growth Time of an Ordinary Annuity** You decide to deposit $200 each month into an account paying 7.2% compounded monthly in order to save money for the purchase of a car. How long will it take for the amount in the account to grow to $10,000?

Solution To begin, we use formula (5) with $PMT = \$200$, $i = 0.072/12 = 0.006$, and $FV = \$10,000$:

$$10,000 = 200 \frac{(1.006)^n - 1}{0.006}$$

To solve for *n*, we graph $y_1 = 200[(1.006)^x - 1]/0.006$ and $y_2 = 10,000$ in a graphing utility (see Fig. 4) and locate the intersection point. (Note that since 50 payments of $200 would produce $10,000 in deposits without taking any interest into account, the value of *n* must be less than 50.) We conclude that $n = x \approx 43.9$, and the amount in the account will grow to $10,000 in 44 months.

FIGURE 4
$y_1 = 200[(1.006)^x - 1]/0.006$

$y_2 = 10,000$

MATCHED PROBLEM 3 ➤ A company plans to deposit $2,500 each quarter into a sinking fund paying 8% compounded monthly. How long will it take for the amount in the account to grow to $100,000?

EXAMPLE 4 ➤ **Finding the Interest Rate for an Ordinary Annuity**   An individual makes annual deposits of $500 into an annuity. After 6 years, the annuity is worth $4,000. What annual nominal compound interest rate has the annuity earned?

Solution   Substituting $PMT = \$500$, $n = 6$, and $FV = \$4,000$ in formula (5) produces the following equation:

$$4,000 = 500\,\frac{(1 + i)^6 - 1}{i}$$

Proceeding as before, we graph $y_1 = 500[(1 + x)^6 - 1]/x$ and $y_2 = 4,000$, and find the intersection point. From Figure 5, we see that $i = x \approx 0.114338$ and conclude that this account has earned 11.434% during this 6 year period.

FIGURE 5
$y_1 = 500[(1 + x)^6 - 1]/x$
$y_2 = 4,000$

MATCHED PROBLEM 4 ➤ An individual makes annual deposits of $1,500 into an annuity. After 4 years, the annuity is worth $7,000. What annual nominal compound interest rate has the annuity earned?

Graphical approximation techniques work on any graphing utility and often give additional insight into a process, such as growth time. However, many graphing utilities also have routines for approximating the solutions of equations without using a graph. These routines are usually referred to as **equation solvers** and can be used to solve any equation, including all the financial equations we have considered. Figure 6A illustrates the solution to Example 4 obtained by using a simple equation solver on a popular graphing calculator. Figure 6B shows a more sophisticated equation solver that allows you to input an equation and, after you specify values for all but one variable, will solve for the remaining variable. Finally, Figure 6C shows a special financial equation solver that is built into this particular graphing calculator. This type of solver is often used in finance courses. As you can see from Figure 6, the details for using one of these equation solvers vary greatly from one graphing utility to another. Consult the manual for your graphing utility to determine whether it has an equation solver and, if so, how to use it.

(A) TI-82

(B) TI-85

(C) TI-83

FIGURE 6
Equation solvers on graphing calculators

Answers to Matched Problems   1. Value: $29,778.08; interest: $9,778.08   2. $186,078.54 every 6 mo for 10 yr
3. 30 quarters   4. 10.375%

EXERCISE 3-3

In Problems 1–26, use formula (5) or (6) and a calculator to solve each problem.

A

1. $n = 20$; $i = 0.03$; $PMT = \$500$; $FV = ?$
2. $n = 25$; $i = 0.04$; $PMT = \$100$; $FV = ?$
3. $n = 40$; $i = 0.02$; $PMT = \$1,000$; $FV = ?$
4. $n = 30$; $i = 0.01$; $PMT = \$50$; $FV = ?$

B

5. $FV = \$3,000$; $n = 20$; $i = 0.02$; $PMT = ?$
6. $FV = \$8,000$; $n = 30$; $i = 0.03$; $PMT = ?$

7. $FV = \$5,000$; $n = 15$; $i = 0.01$; $PMT = ?$
8. $FV = \$2,500$; $n = 10$; $i = 0.08$; $PMT = ?$

C

9. $FV = \$4,000$; $i = 0.02$; $PMT = 200$; $n = ?$
10. $FV = \$8,000$; $i = 0.04$; $PMT = 500$; $n = ?$

▶ APPLICATIONS

Business & Economics

11. What is the value of an ordinary annuity at the end of 10 years if $500 per quarter is deposited into an account earning 8% compounded quarterly? How much of this value is interest?

12. What is the value of an ordinary annuity at the end of 20 years if $1,000 per year is deposited into an account earning 7% compounded annually? How much of this value is interest?

13. In order to accumulate enough money for a down payment on a house, a couple deposits $300 per month into an account paying 6% compounded monthly. If payments are made at the end of each period, how much money will be in the account in 5 years?

14. A self-employed person has a Keogh retirement plan. (This type of plan is free of taxes until money is withdrawn.) If deposits of $7,500 are made each year into an account paying 8% compounded annually, how much will be in the account after 20 years?

15. In 5 years a couple would like to have $25,000 for a down payment on a house. What fixed amount should be deposited each month into an account paying 9% compounded monthly?

16. A person wishes to have $200,000 in an account for retirement 15 years from now. How much should be deposited quarterly in an account paying 8% compounded quarterly?

17. A company estimates it will need $100,000 in 8 years to replace a computer. If it establishes a sinking fund by making fixed monthly payments into an account paying 12% compounded monthly, how much should each payment be?

18. Parents have set up a sinking fund in order to have $120,000 in 15 years for their children's college education. How much should be paid semiannually into an account paying 10% compounded semiannually?

19. Beginning in January, a person plans to deposit $100 at the end of each month into an account earning 9% compounded monthly. Each year taxes must be paid on the interest earned during that year. Find the interest earned during each year for the first 3 years.

20. If $500 is deposited each quarter into an account paying 12% compounded quarterly for 3 years, find the interest earned during each of the 3 years.

21. Why does it make sense to open an Individual Retirement Account (IRA) early in one's life? (For people in certain income brackets, money deposited into an IRA and earnings from an IRA are tax-deferred until withdrawal.) Compare the following:
(A) Jane deposits $2,000 a year into an IRA earning 9% compounded annually. She makes her first deposit on her 24th birthday and her last deposit on her 31st birthday (8 deposits in all). Making no additional deposits, she leaves the accumulated amount from the 8 deposits in the account, earning interest at 9% compounded annually, until her 65th birthday. How much (to the nearest dollar) will be in her account on her 65th birthday?
(B) John procrastinates and does not make his first $2,000 deposit into an IRA until he is 32, but then he continues to deposit $2,000 on every birthday until he is 65 (34 deposits in all). If his account also earns 9% compounded annually, how much (to the nearest dollar) will he have in his account when he makes his last deposit on his 65th birthday?
(*Surprise*—Jane will have more money than John!)

22. Starting on his 24th birthday, and continuing on every birthday up to and including his 65th, a person deposits $2,000 a year into an IRA. How much (to the nearest dollar) will be in the account on the 65th birthday, if the account earns:
(A) 6% compounded annually?
(B) 8% compounded annually?
(C) 10% compounded annually?
(D) 12% compounded annually?

23. You wish to have $10,000 in 4 years to buy a car. How much should you deposit each month into an account paying 8% compounded monthly? How much interest will the account earn in the 4 years?

24. A company establishes a sinking fund to upgrade a plant in 5 years at an estimated cost of $1,500,000. How much should be invested each quarter into an account paying 9.15% compounded quarterly? How much interest will the account earn in the 5 years?

25. You can afford monthly deposits of only $150 into an account that pays 8.5% compounded monthly. How long will it be until you have $7,000 to buy a boat? (Round to the next higher month if not exact.)

26. A company establishes a sinking fund for upgrading office equipment with monthly payments of $1,000 into an account paying 10% compounded monthly. How long will it be before the account has $100,000? (Round up to the next higher month if not exact.)

27. An individual makes annual payments of $1,000 into an ordinary annuity. At the end of 5 years, the amount in the annuity is $6,300. What annual nominal compounding rate has this annuity earned?

28. An individual invests $2,000 annually in an IRA. At the end of 6 years, the amount in the fund is $15,000. What annual nominal compounding rate has this fund earned?

29. At the end of each month, an employee deposits $50 into a Christmas club fund. At the end of the year, the fund contains $620. What annual nominal rate compounded monthly has this fund earned?

30. At the end of each month, an employee deposits $80 into a credit union account. At the end of 2 years, the account contains $2,100. What annual nominal rate compounded monthly has this account earned?

31. Use graphical approximation techniques to determine when an ordinary annuity consisting of quarterly payments of $500 at 6% compounded quarterly would be worth more than a principal of $5,000 invested at 4% simple interest.

32. Use graphical approximation techniques to determine when an ordinary annuity consisting of monthly payments of

$200 at 5% compounded monthly would be worth more than a principal of $10,000 invested at 7.5% compounded monthly.

33. An individual wants to make monthly payments into an IRA in order to accumulate $500,000 for retirement.
   (A) If the account pays 8.7% compounded monthly and the payments are made over 30 years, what is the monthly payment?
   (B) If the account pays 8.7% compounded monthly and the monthly payments are $400, how long (to the nearest month) must the payments continue?
   (C) If the monthly payment is $200 and the payments are made over 30 years, what is the annual nominal compounding rate?

34. An individual wants to make monthly payments of $250 into an IRA.
   (A) If the account pays 9.3% compounded monthly, how much is in the account after 25 years?
   (B) If the account pays 9.3%, how long (to the nearest month) will it take until the account contains $400,000?
   (C) If the amount in the account after 20 years is $200,000, what is the annual nominal compounding rate?

SECTION 3-4 ■ **Present Value of an Annuity; Amortization**

- PRESENT VALUE OF AN ANNUITY
- AMORTIZATION
- AMORTIZATION SCHEDULES
- A GENERAL PROBLEM-SOLVING STRATEGY

■ **PRESENT VALUE OF AN ANNUITY**

How much should you deposit in an account paying 6% compounded semiannually in order to be able to withdraw $1,000 every 6 months for the next 3 years? (After the last payment is made, no money is to be left in the account.)

Actually, we are interested in finding the **present value** of each $1,000 that is paid out during the 3 years. We can do this by solving for P in the compound interest formula:

$$A = P(1 + i)^n$$

$$P = \frac{A}{(1 + i)^n} = A(1 + i)^{-n}$$

The rate per period is $i = \frac{0.06}{2} = 0.03$. The present value P of the first payment is $1,000(1.03)^{-1}$, the present value of the second payment is $1,000(1.03)^{-2}$, and so on. Figure 1 (on the next page) shows this in terms of a time line.

We could evaluate each of the present values in Figure 1 using a calculator and add the results to find the total present values of all the payments (which will be the amount that is needed now to buy the annuity). Since this is generally a tedious process, particularly when the number of payments is large, we will use the same device we used in the last section to produce a formula that will accomplish the same result in a couple of steps. We start by writing the sum of the present values in the form

$$P = 1,000(1.03)^{-1} + 1,000(1.03)^{-2} + \cdots + 1,000(1.03)^{-6} \tag{1}$$

Multiplying both sides of (1) by 1.03, we obtain

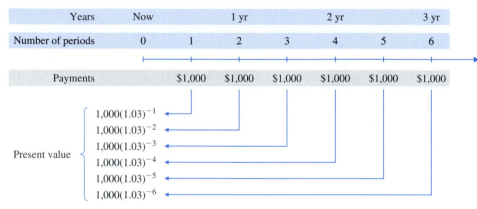

FIGURE 1

$$1.03P = 1,000 + 1,000(1.03)^{-1} + \cdots + 1,000(1.03)^{-5} \qquad (2)$$

Now subtract (1) from (2):

$$1.03P - P = 1,000 - 1,000(1.03)^{-6} \quad \textit{Notice how many terms drop out.}$$

$$0.03P = 1,000[1 - (1 + 0.03)^{-6}]$$

$$P = 1,000 \frac{1 - (1 + 0.03)^{-6}}{0.03} \quad \begin{array}{l}\textit{We write P in this form to} \\ \textit{observe a general pattern.}\end{array} \qquad (3)$$

In general, if R is the periodic payment, i the rate per period, and n the number of periods, then the present value of all payments is given by

$$P = R(1 + i)^{-1} + R(1 + i)^{-2} + \cdots + R(1 + i)^{-n} \quad \textit{Note how this compares to (1).}$$

Proceeding as in the above example, we obtain the general formula for the present value of an ordinary annuity:

$$P = R \frac{1 - (1 + i)^{-n}}{i} \quad \textit{Note how this compares to (3).} \qquad (4)$$

Returning to the example above, we use a calculator to complete the problem:

$$P = 1,000 \frac{1 - (1.03)^{-6}}{0.03}$$

$$= \$5,417.19$$

The fractional factor of (4) may be denoted by the symbol $a_{\overline{n}|i}$, read ''a angle n at i'':

$$a_{\overline{n}|i} = \frac{1 - (1 + i)^{-n}}{i}$$

An alternative approach to the above computation of P is to use a table that gives values of $a_{\overline{n}|i}$ for various values of n and i. The computation of P is then completed by multiplying R by $a_{\overline{n}|i}$ as indicated by (4). (As we noted earlier, a main advantage of using a calculator is that it can handle many more situations than any table.)

It is common to use PV (present value) for P and PMT (payment) for R in formula (4). Making these changes, we have the following:

Present Value of an Ordinary Annuity

$$PV = PMT \frac{1 - (1 + i)^{-n}}{i} = PMTa_{\overline{n}|i} \tag{5}$$

where

$PMT =$ Periodic payment
$\quad i =$ Rate per period
$\quad n =$ Number of periods
$PV =$ Present value of all payments

[*Note:* Payments are made at the end of each period.]

EXAMPLE 1 ➤ **Present Value of an Annuity** What is the present value of an annuity that pays $200 per month for 5 years if money is worth 6% compounded monthly?

Solution To solve this problem, use formula (5) with $PMT = \$200$, $i = \frac{0.06}{12} = 0.005$, and $n = 12(5) = 60$:

$$PV = PMT \frac{1 - (1 + i)^{-n}}{i}$$

$$= 200 \frac{1 - (1.005)^{-60}}{0.005} \quad \text{Use a calculator.}$$

$$= \$10,345.11$$ ◀

MATCHED PROBLEM 1 ➤ How much should you deposit in an account paying 8% compounded quarterly in order to receive quarterly payments of $1,000 for the next 4 years? ◀

■ **AMORTIZATION**

The present value formula for an ordinary annuity (5) has another important use. Suppose you borrow $5,000 from a bank to buy a car and agree to repay the loan in 36 equal monthly payments, including all interest due. If the bank charges 1% per month on the unpaid balance (12% per year compounded monthly), how much should each payment be to retire the total debt including interest in 36 months?

Actually, the bank has bought an annuity from you. The question is: If the bank pays you $5,000 (present value) for an annuity paying them PMT per month for 36 months at 12% interest compounded monthly, what are the monthly payments (PMT)? (Note that the value of the annuity at the end of 36 months is zero.)

To find *PMT,* we have only to use formula (5) with $PV = \$5{,}000$, $i = 0.01$, and $n = 36$:

$$PV = PMT\,\frac{1 - (1 + i)^{-n}}{i}$$

$$5{,}000 = PMT\,\frac{1 - (1.01)^{-36}}{0.01} \qquad \textsf{Solve for PMT and use a calculator.}$$

$$= \$166.07 \text{ per month}$$

At \$166.07 per month, the car will be yours after 36 months. That is, you have *amortized* the debt in 36 equal monthly payments. (*Mort* means "death"; you have "killed" the loan in 36 months.) In general, **amortizing a debt** means that the debt is retired in a given length of time by equal periodic payments that include compound interest. We are usually interested in computing the equal periodic payments. Solving the present value formula (5) for *PMT* in terms of the other variables, we obtain the following **amortization formula:**

$$PMT = PV\,\frac{i}{1 - (1 + i)^{-n}} \tag{6}$$

Formula (6) is simply a variation of formula (5), and either formula can be used to find the periodic payment *PMT.*

EXAMPLE 2 ➤ **Monthly Payment and Total Interest on an Amortized Debt** Assume that you buy a television set for \$800 and agree to pay for it in 18 equal monthly payments at $1\frac{1}{2}\%$ interest per month on the unpaid balance.

(A) How much are your payments? (B) How much interest will you pay?

Solution (A) Use formula (6) with $PV = \$800$, $i = 0.015$, and $n = 18$:

$$PMT = PV\,\frac{i}{1 - (1 + i)^{-n}}$$

$$= 800\,\frac{0.015}{1 - (1.015)^{-18}} \qquad \textsf{Use a calculator.}$$

$$= \$51.04 \text{ per month}$$

(B) Total interest paid = Amount of all payments − Initial loan
$$= 18(\$51.04) - \$800$$
$$= \$118.72$$ ◀

MATCHED PROBLEM 2 ➤ If you sell your car to someone for \$2,400 and agree to finance it at 1% per month on the unpaid balance, how much should you receive each month to amortize the loan in 24 months? How much interest will you receive? ◀

explore – discuss 1

To purchase a home, a family will sign a mortgage of $70,000 at 8% on the unpaid balance. Discuss the advantages and disadvantages of a 20 year mortgage as opposed to a 30 year mortgage. Include a comparison of monthly payments and total interest paid.

■ AMORTIZATION SCHEDULES

What happens if you are amortizing a debt with equal periodic payments and at some point decide to pay off the remainder of the debt in one lump-sum payment? This occurs each time a home with an outstanding mortgage is sold. In order to understand what happens in this situation, we must take a closer look at the amortization process. We begin with an example that is simple enough to allow us to examine the effect each payment has on the debt.

EXAMPLE 3 ➤ **Constructing an Amortization Schedule** If you borrow $500 that you agree to repay in 6 equal monthly payments at 1% interest per month on the unpaid balance, how much of each monthly payment is used for interest and how much is used to reduce the unpaid balance?

Solution First, we compute the required monthly payment using formula (5) or (6). We choose formula (6) with $PV = \$500$, $i = 0.01$, and $n = 6$:

$$PMT = PV \frac{i}{1 - (1 + i)^{-n}}$$

$$= 500 \frac{0.01}{1 - (1.01)^{-6}} \quad \text{Use a calculator.}$$

$$= \$86.27 \text{ per month}$$

At the end of the first month, the interest due is

$$\$500(0.01) = \$5.00$$

The amortization payment is divided into two parts, payment of the interest due and reduction of the unpaid balance (repayment of principal):

Monthly payment	Interest due	Unpaid balance reduction
$86.27 =	$5.00	+ $81.27

The unpaid balance for the next month is

Previous unpaid balance	Unpaid balance reduction	New unpaid balance
$500.00 −	$81.27	= $418.73

At the end of the second month, the interest due on the unpaid balance of $418.73 is

$$\$418.73(0.01) = \$4.19$$

Thus, at the end of the second month, the monthly payment of $86.27 covers interest and unpaid balance reduction as follows:

$$\$86.27 = \$4.19 + \$82.08$$

and the unpaid balance for the third month is

$$\$418.73 - \$82.08 = \$336.65$$

This process continues until all payments have been made and the unpaid balance is reduced to zero. The calculations for each month are listed in Table 1, which is referred to as an **amortization schedule.**

TABLE 1
Amortization Schedule

PAYMENT NUMBER	PAYMENT	INTEREST	UNPAID BALANCE REDUCTION	UNPAID BALANCE
0				$500.00
1	$ 86.27	$ 5.00	$ 81.27	418.73
2	86.27	4.19	82.08	336.65
3	86.27	3.37	82.90	253.75
4	86.27	2.54	83.73	170.02
5	86.27	1.70	84.57	85.45
6	86.30	0.85	85.45	0.00
Totals	$517.65	$17.65	$500.00	

Notice that the last payment had to be increased by $0.03 in order to reduce the unpaid balance to zero. This small discrepancy is due to round-off errors that occur in the computations. In almost all cases, the last payment must be adjusted slightly in order to obtain a final unpaid balance of exactly zero. ◄

MATCHED PROBLEM 3 ➤ Construct the amortization schedule for a $1,000 debt that is to be amortized in 6 equal monthly payments at 1.25% interest per month on the unpaid balance. ◄

EXAMPLE 4 ➤ **Equity in a Home** A family purchased a home 10 years ago for $80,000. The home was financed by paying 20% down and signing a 30 year mortgage at 9% on the unpaid balance. The net market value of the house (amount received after subtracting all costs involved in selling the house) is now $120,000, and the family wishes to sell the house. How much equity (to the nearest dollar) does the family have in the house now after making 120 monthly payments? [**Equity** = (Current net market value) − (Unpaid loan balance).]*

_____

* We use the word ''equity'' in keeping with common usage. If a family wants to sell a house and buy another more expensive house, then the price of a new house that they can afford to buy will often depend on their equity in the first house, where equity is defined by the equation given here. In refinancing a house or taking out an ''equity loan,'' the new mortgage (or second mortgage) often will be based on the equity in the house. Other, more technical definitions of equity do not concern us here.

Solution How can we find the unpaid loan balance after 10 years or 120 monthly payments? One way to proceed would be to construct an amortization schedule, but this would require a table with 120 lines. Fortunately, there is an easier way. The unpaid balance after 120 payments is the amount of the loan that can be paid off with the remaining 240 monthly payments (20 remaining years on the loan). Since the lending institution views a loan as an annuity that they bought from the family, **the unpaid balance of a loan with n remaining payments is the present value of that annuity and can be computed by using formula (5).** Since formula (5) requires knowledge of the monthly payment, we compute PMT first using formula (6).

Step 1. Find the monthly payment:

$$PMT = PV \frac{i}{1 - (1+i)^{-n}}$$

$PV = (0.80)(\$80,000) = \$64,000$
$i = \frac{0.09}{12} = 0.0075$
$n = 12(30) = 360$

$$PMT = 64,000 \frac{0.0075}{1 - (1.0075)^{-360}}$$ Use a calculator.

$$= \$514.96 \text{ per month}$$

Step 2. Find the present value of a \$514.96 per month 20 year annuity:

$$PV = PMT \frac{1 - (1+i)^{-n}}{i}$$

$PMT = \$514.96$
$n = 12(20) = 240$
$i = \frac{0.09}{12} = 0.0075$

$$PV = 514.96 \frac{1 - (1.0075)^{-240}}{0.0075}$$ Use a calculator.

$$= \$57,235$$ Unpaid loan balance

Step 3. Find the equity:

Equity = (Current net market value) − (Unpaid loan balance)
$$= \$120,000 - \$57,235$$
$$= \$62,765$$

Thus, if the family sells the house for \$120,000 net, they will have \$62,765 after paying off the unpaid loan balance of \$57,235. ◀

MATCHED PROBLEM 4 ➤ A couple purchased a home 20 years ago for \$65,000. The home was financed by paying 20% down and signing a 30 year mortgage at 8% on the unpaid balance. The net market value of the house is now \$130,000, and the couple wishes to sell the house. How much equity (to the nearest dollar) does the couple have in the house now after making 240 monthly payments? ◀

The unpaid loan balance in Example 4 may seem a surprisingly large amount to owe after having made payments for 10 years, but long-term amortizations start out with very small reductions in the unpaid balance. For example, the interest due at the end of the very first period of the loan in Example 4 was

$$\$64{,}000(0.0075) = \$480.00$$

The first monthly payment was divided as follows:

		Unpaid
Monthly	Interest	balance
payment	due	reduction

$$\$514.96 - \$480.00 = \$34.96$$

Thus, only $34.96 was applied to the unpaid balance.

explore – discuss 2

(A) A family has an $85,000, 30 year mortgage at 9.6% compounded monthly. Show that the monthly payments are $720.94.

(B) Explain why the equation

$$y = 720.94 \,\frac{1 - (1.008)^{-12(30 - x)}}{0.008}$$

gives the unpaid balance of the loan after x years.

(C) Find the unpaid balance after 5 years, after 10 years, and after 15 years.

(D) When is the unpaid balance exactly half of the original $85,000? (See Fig. 2.)

FIGURE 2

■ A GENERAL PROBLEM-SOLVING STRATEGY

After working the problems in Exercise 3-4, it is very important that you work the problems in the Chapter 3 Review Exercise. This will give you valuable experience in distinguishing between the various types of problems we have considered in this chapter. It is impossible to completely categorize all the problems you will encounter, but you may find the following guidelines helpful in determining which of the four basic formulas is involved in a particular problem. Be aware that some problems may involve more than one of these formulas and others may not involve any of them.

A Strategy for Solving Mathematics of Finance Problems

Step 1. Determine if the problem involves a single payment or a sequence of equal periodic payments. Simple and compound interest problems involve a single present value and a single future value. Ordinary annuities may be concerned with a present value or a future value but always involve a sequence of equal periodic payments.

Step 2. If a single payment is involved, determine if simple or compound interest is used. Simple interest is usually used for durations of a year or less and compound interest for longer periods.

Step 3. If a sequence of periodic payments is involved, determine if the payments are being made into an account that is increasing in value—a future value problem—or if payments are being made out of an account that is decreasing in value—a present value problem. Remember that amortization problems always involve the present value of an ordinary annuity.

Steps 1–3 will help you choose the correct formula for a problem, as indicated in Figure 3. At this point you must determine the values of the quantities in the formula that are given in the problem and those that must be computed, and then solve the problem.

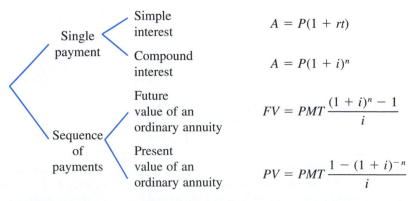

$$A = P(1 + rt)$$

$$A = P(1 + i)^n$$

$$FV = PMT \frac{(1 + i)^n - 1}{i}$$

$$PV = PMT \frac{1 - (1 + i)^{-n}}{i}$$

FIGURE 3
Selecting the correct formula for a problem

Answers to Matched Problems
1. $13,577.71
2. *PMT* = $112.98/mo; Total interest = $311.52
3.

PAYMENT NUMBER	PAYMENT	INTEREST	UNPAID BALANCE REDUCTION	UNPAID BALANCE
0				$1,000.00
1	$ 174.03	$12.50	$ 161.53	838.47
2	174.03	10.48	163.55	674.92
3	174.03	8.44	165.59	509.33
4	174.03	6.37	167.66	341.67
5	174.03	4.27	169.76	171.91
6	174.06	2.15	171.91	0.00
Totals	$1,044.21	$44.21	$1,000.00	

4. $98,551

EXERCISE 3-4

Use formula (5) or (6) and a calculator to solve each problem.

A

1. $n = 30$; $i = 0.04$; *PMT* = $200; *PV* = ?
2. $n = 40$; $i = 0.01$; *PMT* = $400; *PV* = ?
3. $n = 25$; $i = 0.025$; *PMT* = $250; *PV* = ?
4. $n = 60$; $i = 0.0075$; *PMT* = $500; *PV* = ?

B

5. $PV = $6,000$; $n = 36$; $i = 0.01$; *PMT* = ?
6. $PV = $1,200$; $n = 40$; $i = 0.025$; *PMT* = ?
7. $PV = $40,000$; $n = 96$; $i = 0.0075$; *PMT* = ?
8. $PV = $14,000$; $n = 72$; $i = 0.005$; *PMT* = ?
9. $PV = $5,000$; $i = 0.01$; *PMT* = $200; $n = ?$
10. $PV = $20,000$; $i = 0.0175$; *PMT* = $500; $n = ?$

 APPLICATIONS

Business & Economics

11. A relative wills you an annuity paying $4,000 per quarter for the next 10 years. If money is worth 8% compounded quarterly, what is the present value of this annuity?

12. How much should you deposit in an account paying 12% compounded monthly in order to receive $1,000 per month for the next 2 years?

13. Parents of a college student wish to set up an annuity that will pay $350 per month to the student for 4 years. How much should they deposit now at 9% interest compounded monthly to establish this annuity? How much will the student receive in the 4 years?

14. A person pays $120 per month for 48 months for a car, making no down payment. If the loan costs 1.5% interest per month on the unpaid balance, what was the original cost of the car? How much total interest will be paid?

15. (A) If you buy a stereo set for $600 and agree to pay for it in 18 equal installments at 1% interest per month on the unpaid balance, how much are your monthly payments? How much interest will you pay?
   (B) Repeat part (A) for 1.5% interest per month on the unpaid balance.

16. (A) A company buys a large copy machine for $12,000 and finances it at 12% interest compounded monthly. If the loan is to be amortized in 6 years in equal monthly payments, how much is each payment? How much interest will be paid?
   (B) Repeat part (A) with 18% interest compounded monthly.

17. A sailboat costs $16,000. You pay 25% down and amortize the rest with equal monthly payments over a 6 year period. If you must pay 1.5% interest per month on the unpaid balance (18% compounded monthly), what is your monthly payment? How much interest will you pay over the 6 years?

18. A law firm buys a computerized word-processing system costing $10,000. If the firm pays 20% down and amortizes the rest with equal monthly payments over 5 years at 9% compounded monthly, what will be the monthly payment? How much interest will the firm pay?

19. Construct the amortization schedule for a $5,000 debt that is to be amortized in 8 equal quarterly payments at 4.5% interest per quarter on the unpaid balance.

20. Construct the amortization schedule for a $10,000 debt that is to be amortized in 6 equal quarterly payments at 3.5% interest per quarter on the unpaid balance.

21. A woman borrows $6,000 at 12% compounded monthly, which is to be amortized over 3 years in equal monthly payments. For tax purposes, she needs to know the amount of interest paid during each year of the loan. Find the interest paid during the first year, the second year, and the third year of the loan. [*Hint:* Find the unpaid balance after 12 payments and after 24 payments.]

22. A man establishes an annuity for retirement by depositing $50,000 into an account that pays 9% compounded monthly. Equal monthly withdrawals will be made each month for 5 years, at which time the account will have a zero balance. Each year taxes must be paid on the interest earned by the account during that year. How much interest was earned during the first year? [*Hint:* The amount in the account at the end of the first year is the present value of a 4-year annuity.]

23. Some friends tell you that they paid $25,000 down on a new house and are to pay $525 per month for 30 years. If interest is 9.8% compounded monthly, what was the selling price of the house? How much interest will they pay in 30 years?

24. A family is thinking about buying a new house costing $120,000. They must pay 20% down, and the rest is to be amortized over 30 years in equal monthly payments. If money costs 9.6% compounded monthly, what will their monthly payment be? How much total interest will be paid over the 30 years?

25. A student receives a federally backed student loan of $6,000 at 3.5% interest compounded monthly. After finishing col-

lege in 2 years, the student must amortize the loan in the next 4 years by making equal monthly payments. What will the payments be and what total interest will the student pay? [*Hint:* This is a two-part problem. First find the amount of the debt at the end of the first 2 years; then amortize this amount over the next 4 years.]

26. A person establishes a sinking fund for retirement by contributing $7,500 per year at the end of each year for 20 years. For the next 20 years, equal yearly payments are withdrawn, at the end of which time the account will have a zero balance. If money is worth 9% compounded annually, what yearly payments will the person receive for the last 20 years?

27. A family has a $75,000, 30 year mortgage at 13.2% compounded monthly. Find the monthly payment. Also find the unpaid balance after:
(A) 10 years   (B) 20 years   (C) 25 years

28. A family has a $50,000, 20 year mortgage at 10.8% compounded monthly. Find the monthly payment. Also find the unpaid balance after:
(A) 5 years   (B) 10 years   (C) 15 years

29. A family has a $30,000, 20 year mortgage at 15% compounded monthly.
(A) Find the monthly payment and the total interest paid.
(B) Suppose the family decides to add an extra $100 to its mortgage payment each month starting with the very first payment. How long will it take the family to pay off the mortgage? How much interest will the family save?

30. At the time they retire, a couple has $200,000 in an account that pays 8.4% compounded monthly.
(A) If they decide to withdraw equal monthly payments for 10 years, at the end of which time the account will have a zero balance, how much should they withdraw each month?
(B) If they decide to withdraw $3,000 a month until the balance in the account is zero, how many withdrawals can they make?

31. A couple wishes to borrow money using the equity in their home for collateral. A loan company will loan them up to 70% of their equity. They purchased their home 12 years ago for $79,000. The home was financed by paying 20% down and signing a 30 year mortgage at 12% on the unpaid balance. Equal monthly payments were made to amortize the loan over the 30 year period. The net market value of the house is now $100,000. After making their 144th payment, they applied to the loan company for the maximum loan. How much (to the nearest dollar) will they receive?

32. A person purchased a house 10 years ago for $100,000. The house was financed by paying 20% down and signing a 30 year mortgage at 9.6% on the unpaid balance. Equal monthly payments were made to amortize the loan over a 30 year period. The owner now (after the 120th payment) wishes to refinance the house because of a need for additional cash. If the loan company agrees to a new 30 year mortgage of 80% of the new appraised value of the house, which is $136,000, how much cash (to the nearest dollar) will the owner receive after repaying the balance of the original mortgage?

33. Discuss the similarities and differences in the graphs of unpaid balance as a function of time for 30 year mortgages of $50,000, $75,000, and $100,000, respectively, each at 9% compounded monthly (see the figure). Include computations of the monthly payment and total interest paid in each case.

Figure for 33

34. Discuss the similarities and differences in the graphs of unpaid balance as a function of time for 30 year mortgages of $60,000 at rates of 7%, 10%, and 13%, respectively (see the figure). Include computations of the monthly payment and total interest paid in each case.

Figure for 34

35. A discount electronics store offers to let you pay for a $1,000 stereo in 12 equal $90 installments. The store claims that since you repay $1,080 in 1 year, the $80 finance charge represents an 8% annual rate. This would be true if you repaid the loan in a single payment at the end of the year. But since you start repayment after 1 month, this is an amortized loan, and 8% is not the correct rate. What is the annual nominal compounding rate for this loan? [Did you expect the rate to be this high? Loans of this type are called **add-on interest loans** and were very common before Congress enacted the **Truth in Lending Act.** Now credit agreements must fully disclose all credit costs and must express interest rates in terms of the rates used in the amortization process.]

36. A $2,000 computer can be financed by paying $100 per month for 2 years. What is the annual nominal compounding rate for this loan?

37. The owner of a small business has received two offers of purchase. The first prospective buyer offers to pay the owner $100,000 in cash now. The second offers to pay the owner $10,000 now and monthly payments of $1,200 for 10 years. In effect, the second buyer is asking the owner for a $90,000 loan. If the owner accepts the second offer, what annual nominal compounding rate will the owner receive for financing this purchase?

38. At the time they retire, a couple has $200,000 invested in an annuity. They can take the entire amount in a single payment, or they can receive monthly payments of $2,000 for 15 years. If they elect to receive the monthly payments, what annual nominal compounding rate will they earn on the money invested in the annuity?

39. You borrow $100,000 for 30 years at 5% compounded monthly.
(A) If you repay this loan in a single payment at the end of 30 years, how much do you owe?
(B) If the single payment in part (A) is divided by 360 to create monthly payments, what is the monthly payment?
(C) If you make the payment from part (B) each month for 30 years, what is the actual nominal interest rate compounded monthly? Would you agree to such a repayment plan?

CHAPTER 3
GROUP ACTIVITY

Reducing Interest Payments on a Home Mortgage

In appreciation for her parents' financial support of her college education, a recent graduate intends to save her parents some money on the financing of their home. The parents have just made the 180th payment on a 30 year loan of $80,000 at 11% compounded monthly.

(A) How much would be saved in interest payments if the graduate were able to find a lender willing to refinance the unpaid balance over 15 years at 9% compounded monthly? At 8% compounded monthly?

(B) At what interest rate would the refinancing reduce the total interest payments by $10,000?

(C) How much could the graduate save her parents in interest payments if she were to contribute $100 each month to reduce the principal until the loan were paid off? How long would she contribute? How much would she contribute?

(D) How much would the graduate need to contribute each month in order to save her parents a total of $10,000 in interest payments?

Ask a parent, relative, or friend about the terms of their mortgage. Then do the calculations necessary to develop specific strategies for reducing the interest payments on the loan by refinancing and/or early payment of the principal.

CHAPTER 3 REVIEW ■ Important Terms and Symbols

3-1 *Simple Interest.* Principal; interest; interest rate; simple interest; face value; present value; future value

$$I = Prt; \quad A = P(1 + rt)$$

3-2 *Compound Interest.* Compound interest; rate per compounding period; principal (present value); amount (future value); annual nominal rate; effective rate (or annual yield); growth time; zero coupon bond

$$A = P(1 + i)^n; \quad i = \frac{r}{m}; \quad r_e = \left(1 + \frac{r}{m}\right)^m - 1$$

3-3 *Future Value of an Annuity; Sinking Funds.* Annuity; ordinary annuity; future value; balance sheet; sinking fund; sinking fund payment

$$\text{Future value:} \quad FV = PMT\frac{(1 + i)^n - 1}{i}$$

$$\text{Sinking fund:} \quad PMT = FV\frac{i}{(1 + i)^n - 1}$$

3-4 *Present Value of an Annuity; Amortization.* Present value; amortizing a debt; amortization schedule; equity; problem-solving strategy

$$\text{Present value:} \quad PV = PMT\frac{1 - (1 + i)^{-n}}{i}$$

$$\text{Amortization:} \quad PMT = PV\frac{i}{1 - (1 + i)^{-n}}$$

C H A P T E R   3   ■   **Review Exercise**

Work through all the problems in this chapter review and check your answers in the back of the book. Answers to all review problems are there along with section numbers in italics to indicate where each type of problem is discussed. Where weaknesses show up, review appropriate sections in the text.

A *In Problems 1–4, find the indicated quantity, given A = P(1 + rt).*

1. A = ?; P = $100; r = 9%; t = 6 months
2. A = $808; P = ?; r = 12%; t = 1 month
3. A = $212; P = $200; r = 8%; t = ?
4. A = $4,120; P = $4,000; r = ?; t = 6 months

B *In Problems 5 and 6, find the indicated quantity, given A = P(1 + i)ⁿ.*

5. A = ?; P = $1,200; i = 0.005; n = 30
6. A = $5,000; P = ?; i = 0.0075; n = 60

In Problems 7 and 8, find the indicated quantity, given

$$FV = PMT \frac{(1 + i)^n - 1}{i}$$

7. FV = ?; PMT = $1,000; i = 0.005; n = 60
8. FV = $8,000; PMT = ?; i = 0.015; n = 48

In Problems 9 and 10, find the indicated quantity, given

$$PV = PMT \frac{1 - (1 + i)^{-n}}{i}$$

9. PV = ?; PMT = $2,500; i = 0.02; n = 16
10. PV = $8,000; PMT = ?; i = 0.0075; n = 60

C

11. Solve the equation $2,500 = 1,000(1.06)^n$ for n to the nearest integer using:
    (A) Logarithms
    (B) Approximation techniques on a graphing utility

12. Solve the equation

$$5,000 = 100 \frac{(1.01)^n - 1}{0.01}$$

    for n to the nearest integer using:
    (A) Logarithms
    (B) Approximation techniques on a graphing utility

 APPLICATIONS

Business & Economics

Find all dollar amounts correct to the nearest cent. When an interest rate is requested as an answer, express the rate as a percentage, correct to two decimal places.

13. If you borrow $3,000 at 14% simple interest for 10 months, how much will you owe in 10 months? How much interest will you pay?

14. Grandparents deposited $6,000 into a grandchild's account toward a college education. How much money (to the nearest dollar) will be in the account 17 years from now if the account earns 9% compounded monthly?

15. How much should you deposit initially in an account paying 10% compounded semiannually in order to have $25,000 in 10 years?

16. An individual wants to establish an annuity for retirement purposes. He wants to make quarterly deposits for 20 years

so that he can then make quarterly withdrawals of $5,000 for 10 years. The annuity earns 12% interest compounded quarterly.
    (A) How much will have to be in the account at the time he retires?
    (B) How much should be deposited each quarter for 20 years in order to accumulate the required amount?
    (C) What is the total amount of interest earned during the 30 year period?

17. A $10,000 retirement account is left to earn interest at 7% compounded daily. How much money will be in the account 40 years from now when the owner reaches 65? (Use a 365 day year and round answer to the nearest dollar.)

18. Which is the better investment and why: 9% compounded quarterly or 9.25% compounded annually?

19. What is the value of an ordinary annuity at the end of 8 years if $200 per month is deposited into an account earn-

ing 9% compounded monthly? How much of this value is interest?

20. A credit card company charges a 22% annual rate for overdue accounts. How much interest will be owed on a $635 account 1 month overdue?

21. What will an $8,000 car cost (to the nearest dollar) 5 years from now if the inflation rate over that period averages 5% compounded annually?

22. What would the $8,000 car in Problem 21 have cost (to the nearest dollar) 5 years ago if the inflation rate over that period had averaged 5% compounded annually?

23. A loan of $2,500 was repaid at the end of 10 months with a check for $2,812.50. What annual rate of interest was charged?

24. A car salesperson tells you that you can buy the car you are looking at for $3,000 down and $200 a month for 48 months. If interest is 14% compounded monthly, what is the selling price of the car and how much interest will you pay during the 48 months?

25. You have $2,500 toward the purchase of a boat that will cost $3,000. How long will it take the $2,500 to grow to $3,000 if it is invested at 9% compounded quarterly? (Round up to the next higher quarter if not exact.)

26. How long will it take money to double if it is invested at 12% compounded monthly? 18% compounded monthly? (Round up to the next higher month if not exact.)

27. Starting on his 21st birthday, and continuing on every birthday up to and including his 65th, John deposits $2,000 a year into an IRA. How much (to the nearest dollar) will be in the account on the 65th birthday, if the account earns:
(A) 7% compounded annually?
(B) 11% compounded annually?

28. If you just sold a stock for $17,388.17 (net) that cost you $12,903.28 (net) 3 years ago, what annual compound rate of return did you make on your investment?

29. If you paid $100 to a loan company for the use of $1,500 for 120 days, what annual rate of interest did they charge? (Use a 360 day year.)

30. A scholarship committee wishes to establish a scholarship that will pay $1,500 per quarter to a student for 2 years. How much should they deposit now at 8% compounded quarterly in order to make the scholarship payments starting 3 months from now? How much will the student receive in the 2 years?

31. A state-of-art compact disk stereo system costs $3,000. You pay one-third down and amortize the rest with equal monthly payments over a 2 year period. If you are charged 1.5% interest per month on the unpaid balance, what is your monthly payment? How much interest will you pay over the 2 years?

32. A company decides to establish a sinking fund to replace a piece of equipment in 6 years at an estimated cost of $50,000. To accomplish this, they decide to make fixed monthly payments into an account that pays 9% compounded monthly. How much should each payment be?

33. How long will it take money to double if it is invested at 10% compounded daily? 10% compounded annually? (Express answers in years to two decimal places and assume a 365 day year.)

34. A student receives a student loan for $8,000 at 5.5% interest compounded monthly to help her finish the last 1.5 years of college. Starting 1 year after finishing college, the student must amortize the loan in the next 5 years by making equal monthly payments. What will the payments be and what total interest will the student pay?

35. A company makes a payment of $1,200 each month into a sinking fund that earns 6% compounded monthly. When will the fund be worth $100,000?

36. A couple has a $50,000, 20 year mortgage at 9% compounded monthly. When will the unpaid balance drop below $10,000?

37. A loan company advertises in the paper that you will pay only 8¢ a day for each $100 borrowed. What annual rate of interest are they charging? (Use a 360 day year.)

38. Construct the amortizaton schedule for a $1,000 debt that is to be amortized in 4 equal quarterly payments at 2.5% interest per quarter on the unpaid balance.

39. You can afford monthly deposits of only $200 into an account that pays 7.98% compounded monthly. How long will it be until you will have $2,500 to purchase a used car? (Round to the next higher month if not exact.)

40. A company establishes a sinking fund for plant retooling in 6 years at an estimated cost of $850,000. How much should be invested semiannually into an account paying 8.76% compounded semiannually? How much interest will the account earn in the 6 years?

41. Security Savings & Loan pays 9.38% compounded monthly and West Lake Savings & Loan pays 9.35% compounded daily. Which is the better investment? For each investment, express the effective rate as a percentage, correct to three decimal places. Use a 365 day year.

42. If you buy a 13 week T-bill with a maturity value of $5,000 for $4,899.08 from the U.S. Treasury Department, what annual interest rate will you earn?

43. In order to save enough money for the down payment on a condominium, a young couple deposits $200 each month into an account that pays 9% interest compounded monthly. If they need $10,000 for a down payment, how many deposits will they have to make?

44. A business borrows $80,000 at 15% interest compounded monthly for 8 years.
   (A) What is the monthly payment?
   (B) What is the unpaid balance at the end of the first year?
   (C) How much interest was paid during the first year?

45. You unexpectedly inherit $10,000 just after you have made the 72nd monthly payment on a 30-year mortgage of $60,000 at 10.25% compounded monthly. Discuss the relative merits of using the inheritance to reduce the principal of the loan, or to buy a certificate of deposit paying 8.75 compounded monthly.

46. Your parents are considering a $75,000, 30 year mortgage to purchase a new home. The bank at which they have done business for years offers a rate of 12% compounded monthly. A competitor is offering 11.25% compounded monthly. Would it be worthwhile for your parents to switch banks?

47. How much should a $5,000 face value zero coupon bond, maturing in 5 years, be sold for now, if its rate of return is to be 9.5% compounded annually?

48. If you pay $4,476.20 for a $10,000 face value zero coupon bond that matures in 10 years, what is your annual compound rate of return?

49. If an investor wants to earn an annual interest rate of 10.76% on a 26 week T-bill with a maturity value of $5,000, how much should the investor pay for the bill?

50. Two years ago you borrowed $10,000 at 12% interest compounded monthly, which was to be amortized over 5 years. Now you have acquired some additional funds and decide that you want to pay off this loan. What is the unpaid balance after making equal monthly payments for 2 years?

51. A savings and loan company pays 9% compounded monthly. What is the effective rate?

52. (A) A man deposits $2,000 in an IRA on his 21st birthday and on each subsequent birthday up to, and including, his 29th (9 deposits in all). The account earns 8% compounded annually. If he then leaves the money in the account without making any more deposits, how much will he have on his 65th birthday, assuming the account continues to earn the same rate of interest?

   (B) How much would be in the account (to the nearest dollar) on his 65th birthday if he had started the deposits on his 30th birthday and continued making deposits on each birthday until (and including) his 65th birthday?

53. In a new housing development, the houses are selling for $100,000 and require a 20% down payment. The buyer is given a choice of 30 year or 15 year financing, both at 10.75% compounded monthly.
   (A) What is the monthly payment for the 30 year choice? For the 15 year choice?
   (B) What is the unpaid balance after 10 years for the 30 year choice? For the 15 year choice?

54. A loan company will loan up to 60% of the equity in a home. A family purchased their home 8 years ago for $83,000. The home was financed by paying 20% down and signing a 30 year mortgage at 11.25% for the balance. Equal monthly payments were made to amortize the loan over the 30 year period. The market value of the house is now $95,000. After making their 96th payment, the family applied to the loan company for the maximum loan. How much (to the nearest dollar) will they receive?

55. A $600 stereo is financed for 6 months by making monthly payments of $110. What is the annual nominal compounding rate for this loan?

56. An individual deposits $2,000 each year for 25 years into an IRA. When he retires immediately after making the 25th deposit, the IRA is worth $220,000.
   (A) Find the interest rate earned by the IRA over the 25 year period leading up to retirement.
   (B) Assume that the IRA continues to earn the interest rate found in part (A). How long can the retiree withdraw $30,000 per year? How long can he withdraw $24,000 per year?

CHAPTER 4

Systems of Linear

Equations; Matrices

In this chapter we first review how systems of linear equations involving two variables are solved using techniques learned in elementary algebra. (Basic properties of linear equations and lines are reviewed in Section 1-4.) Because these techniques are not suitable for linear systems involving larger numbers of equations and variables, we then turn to a different method of solution involving the concept of an *augmented matrix* and the use of a graphing utility to perform the calculations. We then study *matrices* and *matrix operations* in their own right as a new mathematical form. With these new operations added to our mathematical toolbox, we return to systems of equations from a fresh point of view. Finally, we discuss Wassily Leontief's Nobel prize-winning application of matrices to an important economics problem.

SECTION 4-1 ■ **Review: Systems of Linear Equations in Two Variables**

- SYSTEMS IN TWO VARIABLES
- GRAPHING
- SUBSTITUTION
- ELIMINATION BY ADDITION
- APPLICATIONS

■ SYSTEMS IN TWO VARIABLES

To establish basic concepts, consider the following simple example: If 2 adult tickets and 1 child ticket cost $8, and if 1 adult ticket and 3 child tickets cost $9, what is the price of each?

Let: x = Price of adult ticket

y = Price of child ticket

Then: $2x + \ y = 8$

$x + 3y = 9$

We now have a system of two linear equations in two variables. It is easy to find ordered pairs (x, y) that satisfy one or the other of these equations. For example, the ordered pair $(4, 0)$ satisfies the first equation, but not the second, and the ordered pair $(6, 1)$ satisfies the second, but not the first. To solve this system, we must find all ordered pairs of real numbers that satisfy both equations at the same time. In general, we have the following definition:

Systems of Two Equations in Two Variables

Given the **linear system**

$$ax + by = h$$

$$cx + dy = k$$

where a, b, c, d, h, and k are real constants, a pair of numbers $x = x_0$ and $y = y_0$ [also written as an ordered pair (x_0, y_0)] is a **solution** of this system if each equation is satisfied by the pair. The set of all such ordered pairs is called the **solution set** for the system. To **solve** a system is to find its solution set.

We will consider three methods of solving such systems: *graphing, substitution,* and *elimination by addition.* Each method has certain advantages, depending on the situation.

■ GRAPHING

Recall that the graph of a line is a graph of all the ordered pairs that satisfy the equation of the line. To solve the ticket problem by graphing, we graph both equations in the same coordinate system. The coordinates of any points that the graphs have in common must be solutions to the system, since they must satisfy both equations.

EXAMPLE 1 ➤   Solving a System by Graphing   Solve the ticket problem by graphing:

$$2x + \ y = 8$$

$$x + 3y = 9$$

Solution   Graph each equation in the same coordinate system (both graphs are straight lines). Then estimate the coordinates of any common points on the two lines.

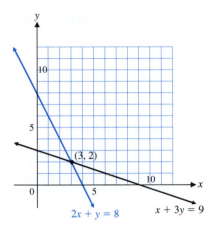

$x = \$3$ Adult ticket

$y = \$2$ Child ticket

Check

$2x + y = 8$	$x + 3y = 9$
$2(3) + 2 \stackrel{?}{=} 8$	$3 + 3(2) \stackrel{?}{=} 9$
$8 \stackrel{\checkmark}{=} 8$	$9 \stackrel{\checkmark}{=} 9$

(3, 2) must satisfy each original equation for a complete check

MATCHED PROBLEM 1 ➤ Solve by graphing and check:

$$2x - y = -3$$
$$x + 2y = -4$$

It is clear that Example 1 has exactly one solution, since the lines have exactly one point of intersection. In general, lines in a rectangular coordinate system are related to each other in one of the three ways illustrated in the next example.

EXAMPLE 2 ➤ Solving Systems by Graphing Solve each of the following systems by graphing:

(A) $x - 2y = 2$ (B) $x + 2y = -4$ (C) $2x + 4y = 8$
    $x + y = 5$ $2x + 4y = 8$ $x + 2y = 4$

Solution

(A)

Intersection at one point only—exactly one solution

(B)

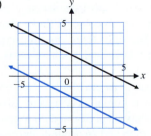

Lines are parallel (each has slope $-\frac{1}{2}$)—no solutions

(C)

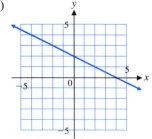

Lines coincide—infinite number of solutions

MATCHED PROBLEM 2 ➤ Solve each of the following systems by graphing:

(A) $x + y = 4$ (B) $6x - 3y = 9$ (C) $2x - y = 4$
     $2x - y = 2$ $2x - y = 3$ $6x - 3y = -18$ ◀

We now define some terms that we can use to describe the different types of solutions to systems of equations that we will encounter.

Systems of Linear Equations: Basic Terms

A system of linear equations is **consistent** if it has one or more solutions and **inconsistent** if no solutions exist. Furthermore, a consistent system is said to be **independent** if it has exactly one solution (often referred to as the **unique solution**) and **dependent** if it has more than one solution.

Referring to the three systems in Example 2, the system in part (A) is a consistent and independent system with the unique solution $x = 4$, $y = 1$. The system in part (B) is inconsistent. And the system in part (C) is a consistent and dependent system with an infinite number of solutions (all the points on the two coinciding lines).

explore – discuss 1

Can a consistent and dependent system have exactly two solutions? Exactly three solutions? Explain.

By geometrically interpreting a system of two linear equations in two variables, we gain useful information about what to expect in the way of solutions to the system. In general, any two lines in a coordinate plane must intersect in exactly one point, be parallel, or coincide (have identical graphs). Thus, the systems in Example 2 illustrate the only three possible types of solutions for systems of two linear equations in two variables. These ideas are summarized in Theorem 1 (at the top of the next page).

In the past, one drawback of the graphical solution method was the inaccuracy of hand-drawn graphs. The recent advent of graphing utilities has changed that. Graphical solutions performed on a graphing utility provide both a useful geometrical interpretation and an accurate approximation of the solution to a system of linear equations in two variables. Example 3 (on the next page) will demonstrate such a solution.

theorem 1

Possible Solutions to a Linear System

The linear system

$$ax + by = h$$
$$cx + dy = k$$

must have:

(A) Exactly one solution   **Consistent and independent**

Or:

(B) No solution   **Inconsistent**

Or:

(C) Infinitely many solutions   **Consistent and dependent**

There are no other possibilities.

EXAMPLE 3 ➤ **Solving a System Using a Graphing Utility**   Solve graphically to two decimal places:

$$5x + 2y = 15$$
$$2x - 3y = 16$$

(A) Equation definitions

(B) Intersection point

FIGURE 1

Solution First, solve each equation for y:

$$5x + 2y = 15 \qquad\qquad 2x - 3y = 16$$
$$2y = -5x + 15 \qquad\qquad -3y = -2x + 16$$
$$y = -2.5x + 7.5 \qquad\qquad y = \tfrac{2}{3}x - \tfrac{16}{3}$$

Next, enter each equation in the graphing utility (Fig. 1A), graph in an appropriate viewing window, and approximate the intersection point (Fig. 1B).

Rounding the values in Figure 1B to two decimal places, we see that the solution is $x = 4.05$ and $y = -2.63$, or $(4.05, -2.63)$.

Check

$$5x + 2y = 15 \qquad\qquad 2x - 3y = 16$$
$$5(4.05) + 2(-2.63) \overset{?}{=} 15 \qquad\qquad 2(4.05) - 3(-2.63) \overset{?}{=} 16$$
$$14.99 \overset{\checkmark}{\approx} 15 \qquad\qquad 15.99 \overset{\checkmark}{\approx} 16$$

The checks are not exact because the values of x and y are approximations. ◀

MATCHED PROBLEM 3 ➤ Solve graphically to two decimal places:

$$2x - 5y = -25$$
$$4x + 3y = 5$$

◀

Graphic methods help us visualize a system and its solutions, frequently reveal relationships that might otherwise be hidden, and, with the assistance of a graphing utility, provide very accurate approximations to solutions.

■ SUBSTITUTION

Now we will review an algebraic method that is easy to use and provides exact solutions to a system of two equations in two variables, provided solutions exist. In this method, we choose one of two equations in a system and solve for one variable in terms of the other. (We make a choice that avoids fractions, if possible.) Then, we **substitute** the result into the other equation and solve the resulting linear equation in one variable. Finally, we substitute this result back into either of the original equations to find the second variable. An example should make the process clear.

EXAMPLE 4 ➤ Solving a System by Substitution   Solve by substitution:

$$5x + y = 4$$
$$2x - 3y = 5$$

Solution   Solve either equation for one variable in terms of the other; then substitute into the remaining equation. In this problem we can avoid fractions by choosing the first equation and solving for y in terms of x:

$$5x + y = 4 \qquad \text{Solve the first equation for } y \text{ in terms of } x.$$
$$y = 4 - 5x \qquad \text{Substitute into second equation.}$$

$$2x - 3y = 5 \qquad \text{Second equation}$$
$$2x - 3(\mathbf{4 - 5x}) = 5 \qquad \text{Solve for } x.$$
$$2x - 12 + 15x = 5$$
$$17x = 17$$
$$\boldsymbol{x = 1}$$

Now, replace x with 1 in $y = 4 - 5x$ to find y:

$$y = 4 - 5x$$
$$y = 4 - 5(\mathbf{1})$$
$$\boldsymbol{y = -1}$$

Thus, the solution is $x = 1$, $y = -1$ or $(1, -1)$.

Check

$$5x + y = 4 \qquad\qquad 2x - 3y = 5$$
$$5(1) + (-1) \overset{?}{=} 4 \qquad\qquad 2(1) - 3(-1) \overset{?}{=} 5$$
$$4 \overset{\checkmark}{=} 4 \qquad\qquad\qquad 5 \overset{\checkmark}{=} 5$$

◀

MATCHED PROBLEM 4 ➤ Solve by substitution:

$$3x + 2y = -2$$
$$2x - y = -6$$

◀

explore – discuss 2

Return to Example 2 and solve each system by substitution. Based on your results, describe how you can recognize a dependent system or an inconsistent system when using substitution.

■ ELIMINATION BY ADDITION

The methods of graphing and substitution both work well for systems involving two variables. However, neither is easily extended to larger systems. Now we turn to **elimination by addition.** This is probably the most important method of solution. It readily generalizes to larger systems and forms the basis for computer-based solution methods.

You are already familiar with operations that can be performed on a single equation without changing its solution set. Similarly, elimination by addition involves performing appropriate operations on a system of equations to produce new and simpler *equivalent* systems with the same solution set. In general, we say that two systems of equations are **equivalent** if they have the same solution set. Theorem 2 lists three useful operations that produce equivalent systems.

theorem 2

Operations That Produce Equivalent Systems

A system of linear equations is transformed into an equivalent system if:

(A) Two equations are interchanged.
(B) An equation is multiplied by a nonzero constant.
(C) A constant multiple of one equation is added to another equation.

Any one of the three operations in Theorem 2 can be used to produce an equivalent system, but the operations in parts (B) and (C) will be of most use to us now. Part (A) becomes useful when we apply the theorem to larger systems. The use of Theorem 2 is best illustrated by examples.

EXAMPLE 5 ➤ **Solving a System Using Elimination by Addition** Solve the following system using elimination by addition:

$$3x - 2y = 8$$
$$2x + 5y = -1$$

Solution We use Theorem 2 to eliminate one of the variables, thus obtaining a system with an obvious solution:

$$3x - 2y = 8$$
$$2x + 5y = -1$$

Multiply the top equation by 5 and the bottom equation by 2 (Theorem 2B).

$$\mathbf{5}(3x - 2y) = \quad \mathbf{5}(8)$$
$$\mathbf{2}(2x + 5y) = \mathbf{2}(-1)$$

$$15x - 10y = \quad 40$$ Add the top equation to the bottom equation
$$\underline{ 4x + 10y = \;-2}$$ (Theorem 2C), eliminating the y terms.
$$19x \quad\quad\;\; = \quad 38$$ Divide both sides by 19, which is the same as
    multiplying the equation by $\frac{1}{19}$ (Theorem 2B).

$$x = 2$$ This equation paired with either of the two
    original equations produces a system
    equivalent to the original system.

Knowing that $x = 2$, we substitute this number back into either of the two original equations (we choose the second) to solve for y:

$$2(\mathbf{2}) + 5y = -1$$
$$5y = -5$$
$$\mathbf{y = -1}$$

Thus, the solution is $x = 2$, $y = -1$ or $(2, -1)$.

Check
$$3x - \quad 2y = 8 \qquad\qquad 2x + \quad 5y = -1$$
$$3(2) - 2(-1) \stackrel{?}{=} 8 \qquad\quad 2(2) + 5(-1) \stackrel{?}{=} -1$$
$$8 \stackrel{\checkmark}{=} 8 \qquad\qquad\qquad\quad -1 \stackrel{\checkmark}{=} -1$$
◄

MATCHED PROBLEM 5 ➤   Solve the following system using elimination by addition:

$$5x - 2y = 12$$
$$2x + 3y = \quad 1$$
◄

Let us see what happens in the elimination process when a system has either no solution or infinitely many solutions. Consider the following system:

$$2x + 6y = -3$$
$$x + 3y = \quad 2$$

Multiplying the second equation by -2 and adding, we obtain

$$2x + 6y = -3$$
$$\underline{-2x - 6y = -4}$$
$$0 = -7 \quad \text{Not possible}$$

We have obtained a contradiction. The assumption that the original system has solutions must be false (otherwise, we have proved that $0 = -7$!). Thus, the system has no solutions, and its solution set is the empty set. The graphs of the equations are parallel and the system is inconsistent.

Now consider the system

$$x - \tfrac{1}{2}y = 4$$
$$-2x + y = -8$$

If we multiply the top equation by 2 and add the result to the bottom equation, we obtain

$$
\begin{aligned}
2x - y &= 8 \\
-2x + y &= -8 \\
\hline
0 &= 0
\end{aligned}
$$

Obtaining $0 = 0$ by addition implies that the equations are equivalent; that is, their graphs coincide and the system is dependent. If we let $x = k$, where k is any real number, and solve either equation for y, we obtain $y = 2k - 8$. Thus, $(k, 2k - 8)$ is a solution for any real number k. The variable k is called a **parameter,** and replacing k with a real number produces a **particular solution** to the system. For example, some particular solutions to this system are

$k = -1$	$k = 2$	$k = 5$	$k = 9.4$
$(-1, -10)$	$(2, -4)$	$(5, 2)$	$(9.4, 10.8)$

■ APPLICATIONS

Many real-world problems are readily solved by applying two-equation–two-variable methods. We shall discuss two applications in detail.

EXAMPLE 6 ➤

Diet   A woman wants to use milk and orange juice to increase the amount of calcium and vitamin A in her daily diet. An ounce of milk contains 37 milligrams of calcium and 57 micrograms* of vitamin A. An ounce of orange juice contains 5 milligrams of calcium and 65 micrograms of vitamin A. How many ounces of milk and orange juice should the woman drink each day to provide exactly 500 milligrams of calcium and 1,200 micrograms of vitamin A?

Solution

First, we define the relevant variables:

$x = $ Number of ounces of milk

$y = $ Number of ounces of orange juice

Next, we summarize the given information in the table in the margin. It is convenient to organize the table so that the quantities represented by the variables correspond to columns in the table (rather than to rows), as shown.

	MILK	ORANGE JUICE	TOTAL NEEDED
CALCIUM	37	5	500
VITAMIN A	57	65	1,200

---

* A microgram (μg) is one-millionth (10^{-6}) of a gram.

Now we use the information in the table to form equations involving x and y:

$$\left(\begin{array}{c}\text{Calcium in } x \text{ oz}\\ \text{of milk}\end{array}\right) + \left(\begin{array}{c}\text{Calcium in } y \text{ oz}\\ \text{of orange juice}\end{array}\right) = \left(\begin{array}{c}\text{Total calcium}\\ \text{needed (mg)}\end{array}\right)$$

$$37x \quad + \quad 5y \quad = \quad 500$$

$$\left(\begin{array}{c}\text{Vitamin A in } x \text{ oz}\\ \text{of milk}\end{array}\right) + \left(\begin{array}{c}\text{Vitamin A in } y \text{ oz}\\ \text{of orange juice}\end{array}\right) = \left(\begin{array}{c}\text{Total Vitamin A}\\ \text{needed } (\mu g)\end{array}\right)$$

$$57x \quad + \quad 65y \quad = \quad 1,200$$

To solve graphically, we solve each equation for y, enter in a graphing utility, and locate the intersection point.

$$y = \frac{500 - 37x}{5} \qquad \textit{Calcium}$$

$$y = \frac{1,200 - 57x}{65} \qquad \textit{Vitamin A}$$

Drinking 12.5 ounces of milk and 7.5 ounces of orange juice each day will provide the required amounts of calcium and vitamin A.

Check

$$37x + 5y = 500 \qquad\qquad 57x + 65y = 1,200$$
$$37(12.5) + 5(7.5) \stackrel{?}{=} 500 \qquad 57(12.5) + 65(7.5) \stackrel{?}{=} 1,200$$
$$500 \stackrel{\checkmark}{=} 500 \qquad\qquad 1,200 \stackrel{\checkmark}{=} 1,200 \qquad \blacktriangleleft$$

MATCHED PROBLEM 6 ➤ A man wants to use cottage cheese and yogurt to increase the amount of protein and calcium in his daily diet. An ounce of cottage cheese contains 3 grams of protein and 15 milligrams of calcium. An ounce of yogurt contains 1 gram of protein and 41 milligrams of calcium. How many ounces of cottage cheese and yogurt should he eat each day to provide exactly 62 grams of protein and 760 milligrams of calcium?

◀

► **EXAMPLE 7** ➤ **Supply and Demand** The quantity of a product that people are willing to buy during some period of time depends on its price. Generally, the higher the price, the less the demand; the lower the price, the greater the demand. Similarly, the quantity of a product that a supplier is willing to sell during some period of time also depends on the price. Generally, a supplier will be willing to supply more of a product at higher prices and less of a product at lower prices. The simplest supply and demand model is a linear model where the graphs of a demand equation and a supply equation are straight lines.

Suppose we are interested in analyzing the sale of cherries each day in a particular city. Using special analytical techniques (regression analysis) and collected data, an analyst arrives at the following price–demand and price–supply equations:

$$p = -0.2q + 4 \qquad \textit{Price–demand equation (consumer)}$$
$$p = 0.07q + 0.76 \qquad \textit{Price–supply equation (supplier)}$$

where q represents the quantity in thousands of pounds and p represents the price in dollars. For example, we see that consumers will purchase 10 thousand pounds ($q = 10$) when the price is $p = -0.2(10) + 4 = \$2$ per pound. On the other hand, suppliers will be willing to supply 17.714 thousand pounds of cherries at \$2 per pound (solve $2 = 0.07q + 0.76$). Thus, at \$2 per pound the suppliers are willing to supply more cherries than consumers are willing to purchase. The supply exceeds the demand at that price and the price will come down. At what price will cherries stabilize for the day? That is, at what price will supply equal demand? This price, if it exists, is called the **equilibrium price,** and the quantity sold at that price is called the **equilibrium quantity.** How do we find these quantities? We solve the linear system

$$p = -0.2q + 4 \qquad \textcolor{blue}{\textit{Demand equation}}$$
$$p = 0.07q + 0.76 \qquad \textcolor{blue}{\textit{Supply equation}}$$

by using substitution (substituting $p = -0.2q + 4$ into the second equation):

$$-0.2q + 4 = 0.07q + 0.76$$
$$-0.27q = -3.24$$
$$\boldsymbol{q = 12 \textbf{ thousand pounds}} \qquad \textcolor{blue}{\textit{Equilibrium quantity}}$$

Now substitute $q = 12$ back into either of the original equations in the system and solve for p (we choose the first equation):

$$p = -0.2(\boldsymbol{12}) + 4$$
$$\boldsymbol{p = \$1.60 \textbf{ per pound}} \qquad \textcolor{blue}{\textit{Equilibrium price}}$$

These results are interpreted geometrically in Figure 2.

FIGURE 2

If the price is above the equilibrium price of \$1.60 per pound, the supply will exceed the demand and the price will come down. If the price is below the equilibrium price of \$1.60 per pound, the demand will exceed the supply and the price will rise. Thus, the price will reach equilibrium at \$1.60. At this price, suppliers will supply 12 thousand pounds of cherries and consumers will purchase 12 thousand pounds. ◀

MATCHED PROBLEM 7 ➤ Repeat Example 7 (including drawing the graph) given:

$$p = -0.1q + 3 \qquad \textit{Demand equation}$$
$$p = 0.08q + 0.66 \quad \textit{Supply equation}$$
◀

Answers to Matched Problems

1. $x = -2, y = -1$

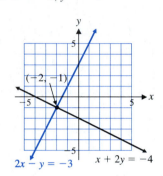

$(-2, -1)$
$2x - y = -3$
$x + 2y = -4$

Check:
$$2x - y = -3$$
$$2(-2) - (-1) \overset{?}{=} -3$$
$$-3 \overset{\checkmark}{=} -3$$

$$x + 2y = -4$$
$$(-2) + 2(-1) \overset{?}{=} -4$$
$$-4 \overset{\checkmark}{=} -4$$

2. (A) $x = 2, y = 2$ (B) Infinitely many solutions   (C) No solution
3. $x = -1.92, y = 4.23$
4. $x = -2, y = 2$
5. $x = 2, y = -1$
6. 16.5 oz of cottage cheese, 12.5 oz of yogurt
7. Equilibrium quantity = 13 thousand pounds; Equilibrium price = \$1.70 per pound

EXERCISE 4-1

A *Match each system in Problems 1–4 with one of the following graphs, and use the graph to solve the system.*

(A)

(B)

(C)

(D)

1. $-4x + 2y = 8$
$2x - y = 0$

2. $x + y = 3$
$2x - y = 0$

3. $-x + 2y = 5$
$2x + 3y = -3$

4. $2x - 4y = -10$
$-x + 2y = 5$

Solve Problems 5–10 by graphing.

5. $x + y = 5$
$x - y = 1$

6. $x - y = 2$
$x + y = 6$

7. $3x - y = 2$
$x + 2y = 10$

8. $3x - 2y = 12$
$7x + 2y = 8$

9. $m + 2n = 4$
$2m + 4n = -8$

10. $3u + 5v = 15$
$6u + 10v = -30$

Solve Problems 11–14 using substitution.

11. $y = 2x - 3$
$x + 2y = 14$

12. $y = x - 4$
$x + 3y = 12$

13. $2x + y = 6$
$x - y = -3$

14. $3x - y = 7$
$2x + 3y = 1$

Solve Problems 15–18 using elimination by addition.

15. $3u - 2v = 12$
$7u + 2v = 8$

16. $2x - 3y = -8$
$5x + 3y = 1$

17. $2m - n = 10$
$m - 2n = -4$

18. $2x + 3y = 1$
$3x - y = 7$

B *Solve Problems 19–34 using substitution or elimination by addition.*

19. $9x - 3y = 24$
$11x + 2y = 1$

20. $4x + 3y = 26$
$3x - 11y = -7$

21. $2x - 3y = -2$
$-4x + 6y = 7$

22. $3x - 6y = -9$
$-2x + 4y = 12$

23. $3x + 8y = 4$
$15x + 10y = -10$

24. $7m + 12n = -1$
$5m - 3n = 7$

25. $-6x + 10y = -30$
$3x - 5y = 15$

26. $2x + 4y = -8$
$x + 2y = 4$

27. $y = 0.07x$
$y = 80 + 0.05x$

28. $y = 0.08x$
$y = 100 + 0.04x$

29. $x + y = 1$
$0.3x - 0.4y = 0$

30. $x + y = 1$
$0.5x - 0.4y = 0$

31. $0.2x - 0.5y = 0.07$
$0.8x - 0.3y = 0.79$

32. $0.3u - 0.6v = 0.18$
$0.5u + 0.2v = 0.54$

33. $\frac{2}{5}x + \frac{3}{2}y = 2$
$\frac{7}{3}x - \frac{5}{4}y = -5$

34. $\frac{7}{2}x - \frac{5}{6}y = 10$
$\frac{2}{5}x + \frac{4}{3}y = 6$

Solve Problems 35–38 by graphing. Round answers to two decimal places.

35. $3x - 2y = 5$
$4x + 3y = 13$

36. $3x - 7y = -20$
$2x + 5y = 8$

37. $-2.4x + 3.5y = 0.1$
$-1.7x + 2.6y = -0.2$

38. $4.2x + 5.4y = -12.9$
$6.4x + 3.7y = -4.5$

C   *In Problems 39–44, graph all three equations in the same coordinate system, and then find the coordinates of any points where two or more lines intersect.*

39. $x - 2y = -6$
$2x + y = 8$
$x + 2y = -2$

40. $x + y = 3$
$x + 3y = 15$
$3x - y = 5$

41. $x + y = 1$
$x - 2y = -8$
$3x + y = -3$

42. $x - y = 6$
$x - 2y = 8$
$x + 4y = -4$

43. $4x - 3y = -24$
$2x + 3y = 12$
$8x - 6y = 24$

44. $2x + 3y = 18$
$2x - 6y = -6$
$4x + 6y = -24$

45.   The coefficients of the three systems given below are very similar. One might guess that the solution sets to the three systems would also be nearly identical. Develop evidence for or against this guess by considering graphs of the system and solutions obtained using substitution or elimination by addition.

(A) $5x + 4y = 4$
$11x + 9y = 4$

(B) $5x + 4y = 4$
$11x + 8y = 4$

(C) $5x + 4y = 4$
$10x + 8y = 4$

46.   Repeat Problem 45 for the following systems:

(A) $6x - 5y = 10$
$-13x + 11y = -20$

(B) $6x - 5y = 10$
$-13x + 10y = -20$

(C) $6x - 5y = 10$
$-12x + 10y = -20$

▶ **A P P L I C A T I O N S**

Business & Economics

47.   *Supply and demand.* Suppose the supply and demand equations for printed T-shirts in a resort town for a particular week are

$p = 0.7q + 3$ *Supply equation*

$p = -1.7q + 15$ *Demand equation*

where p is the price in dollars and q is the quantity in hundreds.
(A) Find the equilibrium price and quantity.
(B) Graph the two equations in the same coordinate system and identify the equilibrium point, supply curve, and demand curve.

48.   *Supply and demand.* Repeat Problem 47 with the following supply and demand equations:

$p = 0.4q + 3.2$ *Supply equation*

$p = -1.9q + 17$ *Demand equation*

49.   *Supply and demand.* At $0.60 per bushel, the daily supply for wheat is 450 bushels, and the daily demand is 570 bushels. When the price is raised to $0.75 per bushel, the daily supply increases to 600 bushels, and the daily demand decreases to 495 bushels. Assume that the supply and demand equations are linear.

(A) Find the supply equation. [*Hint:*   Write the supply equation in the form $p = aq + b$ and solve for a and b.]
(B) Find the demand equation.
(C) Find the equilibrium price and quantity.
(D) Graph the two equations in the same coordinate system and identify the equilibrium point, supply curve, and demand curve.

50.   *Supply and demand.* At $1.40 per bushel, the daily supply for oats is 850 bushels, and the daily demand is 580 bushels. When the price falls to $1.20 per bushel, the daily supply decreases to 350 bushels, and the daily demand increases to 980 bushels. Assume that the supply and demand equations are linear.

(A) Find the supply equation.
(B) Find the demand equation.
(C) Find the equilibrium price and quantity.
(D) Graph the two equations in the same coordinate system and identify the equilibrium point, supply curve, and demand curve.

51.   *Break-even analysis.* A small company manufactures portable home computers. The plant has fixed costs (leases, insurance, and so on) of $48,000 per month and variable costs (labor, materials, and so on) of $1,400 per unit produced. The computers are sold for $1,800 each. Thus, the cost and revenue equations are

$$y = 48,000 + 1,400x \quad \text{Cost equation}$$
$$y = 1,800x \quad \text{Revenue equation}$$

where x is the total number of computers produced and sold each month, and the monthly costs and revenue are in dollars.

(A) How many units must be manufactured and sold each month for the company to break even?
(B) Graph both equations in the same coordinate system and show the break-even point. Interpret the regions between the lines to the left and to the right of the break-even point.

52. *Break-even analysis.* Repeat Problem 51 with the cost and revenue equations

$$y = 65,000 + 1,100x \quad \text{Cost equation}$$
$$y = 1,600x \quad \text{Revenue equation}$$

53. *Break-even analysis.* A mail-order company markets videotapes that sell for $19.95, including shipping and handling. The monthly fixed costs (advertising, rent, and so on) are $24,000, and the variable costs (materials, shipping, and so on) are $7.45 per tape.
(A) How many tapes must be sold each month for the company to break even?
(B) Graph the cost and revenue equations in the same coordinate system and show the break-even point. Interpret the regions between the lines to the left and to the right of the break-even point.

54. *Break-even analysis.* Repeat Problem 53 if the monthly fixed costs increase to $27,200, the variable costs increase to $9.15, and the company raises the selling price of the tapes to $21.95.

Life Sciences

55. *Nutrition.* Animals in an experiment are to be kept under a strict diet. Each animal is to receive, among other things, 20 grams of protein and 6 grams of fat. The laboratory technician is able to purchase two food mixes of the following compositions: Mix A has 10% protein and 6% fat; mix B has 20% protein and 2% fat. How many grams of each mix should be used to obtain the right diet for a single animal?

56. *Nutrition—plants.* A fruit grower can use two types of fertilizer in an orange grove, brand A and brand B. Each bag of brand A contains 8 pounds of nitrogen and 4 pounds of phosphoric acid. Each bag of brand B contains 7 pounds of nitrogen and 6 pounds of phosphoric acid. Tests indicate that the grove needs 720 pounds of nitrogen and 500 pounds of phosphoric acid. How many bags of each brand should be used to provide the required amounts of nitrogen and phosphoric acid?

Social Sciences

57. *Psychology—approach and avoidance.* People often approach certain situations with "mixed emotions." For example, public speaking often brings forth the positive response of recognition and the negative response of failure. Which dominates? J. S. Brown, in an experiment on approach and avoidance, trained rats by feeding them from a goal box. Then the rats received mild electric shocks from the same goal box. This established an approach–avoidance conflict relative to the goal box. Using appropriate apparatus, Brown arrived at the following relationships:

$$p = -\tfrac{1}{5}d + \ 70 \quad \text{Approach equation}$$
$$p = -\tfrac{4}{3}d + 230 \quad \text{Avoidance equation}$$

where $30 \leq d \leq 172.5$. The approach equation gives the pull (in grams) toward the food goal box when the rat is placed d centimeters away from it. The avoidance equation gives the pull (in grams) away from the shock goal box when the rat is placed d centimeters from it.
(A) Graph the approach equation and the avoidance equation in the same coordinate system.
(B) Find the value of d for the point of intersection of these two equations.
(C) What do you think the rat would do when placed the distance d from the box found in part (B)?

(For additional discussion of this phenomenon, see J. S. Brown, "Gradients of Approach and Avoidance Responses and Their Relation to Motivation," *Journal of Comparative and Physiological Psychology,* 1948, 41:450–465.)

S E C T I O N   4-2   ■ **Systems of Linear Equations and Augmented Matrices**

■ MATRICES
■ SOLVING LINEAR SYSTEMS USING AUGMENTED MATRICES
■ SUMMARY

Most linear systems of any consequence involve large numbers of equations and variables. In the past, these systems were solved on large computers, since hand methods were impractical. Now a wide array of readily available hardware and software can be used to solve linear systems. This ranges from graphing calculators, such as the Texas Instrument TI-82, to sophisticated mathematical software for personal computers, such as Matlab, Mathematica, and Maple. In addition, spreadsheet programs such as Excel and Lotus 1-2-3 can solve linear systems. All these various approaches have one thing in common: **The user is expected to be familiar with the solution techniques used to solve large linear systems.**

In the remainder of this chapter, we will develop several methods for solving large systems that are commonly used with computers. It is important to keep in mind that we are not presenting these as more efficient methods for solving linear systems manually—there are none. Instead, we discuss methods that will help you effectively use a variety of software to solve systems. At the same time, you will gain experience in formulating problems in terms of linear systems and in interpreting the results—two activities that computers cannot perform for you.

■ MATRICES

In solving systems of equations using elimination by addition, the coefficients of the variables and the constant terms played a central role. The process can be made more efficient for generalization and computer work by the introduction of a mathematical form called a *matrix*. A **matrix** is a rectangular array of numbers written within brackets. Two examples are

$$A = \begin{bmatrix} 1 & -4 & 5 \\ 7 & 0 & -2 \end{bmatrix} \qquad B = \begin{bmatrix} -4 & 5 & 12 \\ 0 & 1 & 8 \\ -3 & 10 & 9 \\ -6 & 0 & -1 \end{bmatrix} \tag{1}$$

Each number in a matrix is called an **element** of the matrix. Matrix A has 6 elements arranged in 2 rows and 3 columns. Matrix B has 12 elements arranged in 4 rows and 3 columns. If a matrix has m rows and n columns, it is called an **$m \times n$ matrix** (read "m by n matrix"). The expression $m \times n$ is called the **size** of the matrix, and the numbers m and n are called the **dimensions** of the matrix. It is important to note that the number of rows is always given first. Referring to (1) above, A is a 2×3 matrix and B is a 4×3 matrix. A matrix with n rows and n columns is called a **square matrix of order n.** A matrix with only 1 column is called a **column matrix,** and a matrix with only 1 row is called a **row matrix**. These definitions are illustrated by the following:

$$3 \times 3$$
$$\begin{bmatrix} 0.5 & 0.2 & 1.0 \\ 0.0 & 0.3 & 0.5 \\ 0.7 & 0.0 & 0.2 \end{bmatrix}$$
Square matrix of order 3

$$4 \times 1$$
$$\begin{bmatrix} 3 \\ -2 \\ 1 \\ 0 \end{bmatrix}$$
Column matrix

$$1 \times 4$$
$$[2 \quad \tfrac{1}{2} \quad 0 \quad -\tfrac{2}{3}]$$
Row matrix

The **position** of an element in a matrix is given by the row and column containing the element. This is usually denoted using **double subscript notation** a_{ij}, where i is the row and j is the column containing the element a_{ij}, as illustrated below:

$$A = \begin{bmatrix} 1 & -4 & 5 \\ 7 & 0 & -2 \end{bmatrix} \qquad \begin{matrix} a_{11} = 1, & a_{12} = -4, & a_{13} = 5 \\ a_{21} = 7, & a_{22} = 0, & a_{23} = -2 \end{matrix}$$

Note that a_{12} is read "a sub one two" (*not* "a sub twelve"). The elements $a_{11} = 1$ and $a_{22} = 0$ make up the *principal diagonal* of A. In general, the **principal diagonal** of a matrix A consists of the elements $a_{11}, a_{22}, a_{33}, \ldots$.

Remark

Most graphing calculators and computers* are capable of storing and manipulating matrices. (Graphing calculators are actually small computers; and, in addition to graphing, they can deal with many other mathematical forms, including matrices.) Figure 1 shows matrix A displayed in the editing screen of a graphing calculator. The size of the matrix is given at the top of the screen and the position of the currently selected element is given at the bottom. Notice that a comma is used in the notation for the position. This is common practice on graphing calculators and computers, but not in mathematical literature.

FIGURE 1
Matrix notation on a graphing calculator.

The coefficients and constant terms in a system of linear equations can be used to form several matrices of interest. Related to the system

$$\begin{aligned} 2x - 3y &= 5 \\ x + 2y &= -3 \end{aligned} \qquad (2)$$

are the following matrices:

Coefficient matrix
$$\begin{bmatrix} 2 & -3 \\ 1 & 2 \end{bmatrix}$$

Constant matrix
$$\begin{bmatrix} 5 \\ -3 \end{bmatrix}$$

Augmented coefficient matrix
$$\left[\begin{array}{cc|c} 2 & -3 & 5 \\ 1 & 2 & -3 \end{array}\right]$$

---

* See the Preface for a description of the computer software that accompanies this text.

The augmented coefficient matrix contains all the essential parts of the system—both the coefficients and the constants. The vertical bar is included only as a visual aid to help us separate the coefficients from the constant terms. (Matrices entered and displayed on a graphing calculator or computer will not display this line.) Later in this chapter we will make use of the coefficient and constant matrices. For now, we will find the augmented coefficient matrix sufficient for our needs.

For ease of generalization to the larger systems in the following sections, we are now going to change the notation for the variables in (2) to a subscript form. (We would soon run out of letters, but we will not run out of subscripts.) That is, in place of x and y, we use x_1 and x_2, respectively, and (2) is rewritten as

$$2x_1 - 3x_2 = 5$$
$$x_1 + 2x_2 = -3$$

In general, associated with each linear system of the form

$$a_{11}x_1 + a_{12}x_2 = k_1$$
$$a_{21}x_1 + a_{22}x_2 = k_2 \tag{3}$$

where x_1 and x_2 are variables, is the **augmented matrix** of the system:

This matrix contains the essential parts of system (3). Our objective is to learn how to manipulate augmented matrices in order to solve system (3), if a solution exists. The manipulative process is a direct outgrowth of the elimination process discussed in Section 4-1.

Recall that two linear systems are said to be **equivalent** if they have exactly the same solution set. How did we transform linear systems into equivalent linear systems? We used the operations listed below (Theorem 2, Section 4-1):

Operations That Produce Equivalent Systems

A system of linear equations is transformed into an equivalent system if:

(A) Two equations are interchanged.
(B) An equation is multiplied by a nonzero constant.
(C) A constant multiple of one equation is added to another equation.

Paralleling the earlier discussion, we say that two augmented matrices are **row-equivalent,** denoted by the symbol $\sim$ placed between the two matrices, if they are augmented matrices of equivalent systems of equations. (Think about this.) How do we transform augmented matrices into row-equivalent matrices? We use Theorem 1, which is a direct consequence of the operations listed above.

theorem 1

> **Operations That Produce Row-Equivalent Matrices**
>
> An augmented matrix is transformed into a row-equivalent matrix by performing any of the following **row operations:**
>
> (A) Two rows are interchanged ($R_i \leftrightarrow R_j$).
> (B) A row is multiplied by a nonzero constant ($kR_i \rightarrow R_i$).
> (C) A constant multiple of one row is added to another row ($kR_j + R_i \rightarrow R_i$).
>
> [*Note:* The arrow $\rightarrow$ means "replaces."]

■ SOLVING LINEAR SYSTEMS USING AUGMENTED MATRICES

The use of Theorem 1 in solving systems in the form of (3) is best illustrated by examples.

EXAMPLE 1 ➤ **Solving a System Using Augmented Matrix Methods** Solve using augmented matrix methods:

$$3x_1 + 4x_2 = 1$$
$$x_1 - 2x_2 = 7 \tag{4}$$

Solution We start by writing the augmented matrix corresponding to (4):

$$\begin{bmatrix} 3 & 4 & | & 1 \\ 1 & -2 & | & 7 \end{bmatrix} \tag{5}$$

Our objective is to use row operations from Theorem 1 to try to transform (5) into the form

$$\begin{bmatrix} 1 & 0 & | & m \\ 0 & 1 & | & n \end{bmatrix} \tag{6}$$

where m and n are real numbers. The solution to system (4) will then be obvious, since matrix (6) will be the augmented matrix of the following system (a row in an augmented matrix always corresponds to an equation in a linear system):

$$x_1 = m \qquad x_1 + 0x_2 = m$$
$$x_2 = n \qquad 0x_1 + x_2 = n$$

```
[A]
    [[3 4  1]
     [1 -2 7]]
rowSwap([A],1,2)
→[A]
    [[1 -2 7]
     [3 4  1]]
```

```
[A]
    [[1 -2 7]
     [3 4  1]]
*row+(-3,[A],1,2
)→[A]
    [[1 -2 7 ]
     [0 10 -20]]
```

```
[A]
    [[1 -2 7 ]
     [0 10 -20]]
*row(.1,[A],2)→[
A]
    [[1 -2 7 ]
     [0 1 -2]]
```

```
[A]
    [[1 -2 7 ]
     [0 1 -2]]
*row+(2,[A],2,1)

    [[1 0 3 ]
     [0 1 -2]]
```

FIGURE 2
Row operations on a graphing utility

We now proceed to use row operations to transform (5) into form (6). (In this first example, we will also display the output produced by using a graphing utility to perform the row operations. Consult your manual for the details of performing row operations on your graphing utility.)

Step 1. To get a 1 in the upper left corner, we interchange R_1 and R_2 (Theorem 1A):

$$\begin{bmatrix} 3 & 4 & | & 1 \\ 1 & -2 & | & 7 \end{bmatrix} \quad \begin{matrix} R_1 \leftrightarrow R_2 \\ \sim \end{matrix} \quad \begin{bmatrix} 1 & -2 & | & 7 \\ 3 & 4 & | & 1 \end{bmatrix}$$

Step 2. To get a 0 in the lower left corner, we multiply R_1 by (-3) and add to R_2 (Theorem 1C)—this changes R_2 but not R_1. Some people find it useful to write $(-3)R_1$ outside the matrix to help reduce errors in arithmetic, as shown:

$$\begin{bmatrix} 1 & -2 & | & 7 \\ 3 & 4 & | & 1 \end{bmatrix} \quad \begin{matrix} \sim \\ (-3)R_1 + R_2 \to R_2 \end{matrix} \quad \begin{bmatrix} 1 & -2 & | & 7 \\ 0 & 10 & | & -20 \end{bmatrix}$$
$$-3 \quad 6 \qquad -21 \leftarrow - - - - -$$

Step 3. To get a 1 in the second row, second column, we multiply R_2 by $\frac{1}{10}$ (Theorem 1B):

$$\begin{bmatrix} 1 & -2 & | & 7 \\ 0 & 10 & | & -20 \end{bmatrix} \quad \begin{matrix} \sim \\ \frac{1}{10}R_2 \to R_2 \end{matrix} \quad \begin{bmatrix} 1 & -2 & | & 7 \\ 0 & 1 & | & -2 \end{bmatrix}$$

Step 4. To get a 0 in the first row, second column, we multiply R_2 by 2 and add the result to R_1 (Theorem 1C)—this changes R_1 but not R_2:

$$0 \quad 2 \qquad -4 \leftarrow - - - -$$
$$\begin{bmatrix} 1 & -2 & | & 7 \\ 0 & 1 & | & -2 \end{bmatrix} \quad \begin{matrix} 2R_2 + R_1 \to R_1 \\ \sim \end{matrix} \quad \begin{bmatrix} 1 & 0 & | & 3 \\ 0 & 1 & | & -2 \end{bmatrix}$$

We have accomplished our objective! The last matrix is the augmented matrix for the system

$$
\begin{aligned}
x_1 &= 3 & x_1 + 0x_2 &= 3 \\
x &= -2 & 0x_1 + x_2 &= -2
\end{aligned}
\tag{7}
$$

Since system (7) is equivalent to system (4), our starting system, we have solved (4); that is, $x_1 = 3$ and $x_2 = -2$. The check is left to the reader. ◄

The above process may be written more compactly as follows:

Step 3:
Need a 1 here.

$$\sim \begin{bmatrix} 1 & -2 & | & 7 \\ 0 & 10 & | & -20 \end{bmatrix} \quad \tfrac{1}{10}R_2 \longrightarrow R_2$$

$$0 \quad 2 \qquad -4 \longleftarrow$$

Step 4:
Need a 0 here.

$$\sim \begin{bmatrix} 1 & -2 & | & 7 \\ 0 & 1 & | & -2 \end{bmatrix} \quad 2R_2 + R_1 \longrightarrow R_1$$

$$\sim \begin{bmatrix} 1 & 0 & | & 3 \\ 0 & 1 & | & -2 \end{bmatrix}$$

Therefore, $x_1 = 3$ and $x_2 = -2$.

MATCHED PROBLEM 1 ➤ Solve using augmented matrix methods:

$$2x_1 - x_2 = -7$$
$$x_1 + 2x_2 = 4$$

◄

---

explore – discuss 1

The summary following Example 1 shows five augmented coefficient matrices. Write the linear system that each matrix represents, solve each system graphically, and discuss the relationships among these solutions.

EXAMPLE 2 ➤ **Solving a System Using Augmented Matrix Methods** Solve using augmented matrix methods:

$$2x_1 - 3x_2 = 6$$
$$3x_1 + 4x_2 = \tfrac{1}{2}$$

Solution

Step 1:
Need a 1 here.

$$\begin{bmatrix} 2 & -3 & | & 6 \\ 3 & 4 & | & \tfrac{1}{2} \end{bmatrix} \quad \tfrac{1}{2}R_1 \longrightarrow R_1$$

Step 2:
Need a 0 here.

$$\sim \begin{bmatrix} 1 & -\tfrac{3}{2} & | & 3 \\ 3 & 4 & | & \tfrac{1}{2} \end{bmatrix} \quad (-3)R_1 + R_2 \longrightarrow R_2$$

$$-3 \quad \tfrac{9}{2} \qquad -9 \longleftarrow$$

Step 3:
Need a 1 here.

$$\sim \begin{bmatrix} 1 & -\tfrac{3}{2} & | & 3 \\ 0 & \tfrac{17}{2} & | & -\tfrac{17}{2} \end{bmatrix} \quad \tfrac{2}{17}R_2 \longrightarrow R_2$$

$$0 \quad \tfrac{3}{2} \qquad -\tfrac{3}{2} \longleftarrow$$

Step 4:
Need a 0 here.

$$\sim \begin{bmatrix} 1 & -\tfrac{3}{2} & | & 3 \\ 0 & 1 & | & -1 \end{bmatrix} \quad \tfrac{3}{2}R_2 + R_1 \longrightarrow R_1$$

$$\sim \begin{bmatrix} 1 & 0 & | & \tfrac{3}{2} \\ 0 & 1 & | & -1 \end{bmatrix}$$

Thus, $x_1 = \tfrac{3}{2}$ and $x_2 = -1$. The check is left to the reader. ◄

MATCHED PROBLEM 2 ➤ Solve using augmented matrix methods:

$$5x_1 - 2x_2 = 11$$
$$2x_1 + 3x_2 = \tfrac{5}{2}$$

◄

EXAMPLE 3 ➤ **Solving a System Using Augmented Matrix Methods** Solve using augmented matrix methods:

$$
\begin{aligned}
2x_1 - x_2 &= 4 \\
-6x_1 + 3x_2 &= -12
\end{aligned}
\tag{8}
$$

Solution

$$
\begin{bmatrix}
2 & -1 & \bigm| & 4 \\
-6 & 3 & \bigm| & -12
\end{bmatrix}
\quad
\begin{aligned}
&\tfrac{1}{2}R_1 \to R_1 \text{ (to get a 1 in the upper left corner)} \\
&\tfrac{1}{3}R_2 \to R_2 \text{ (this simplifies } R_2)
\end{aligned}
$$

$$
\sim
\begin{bmatrix}
1 & -\tfrac{1}{2} & \bigm| & 2 \\
-2 & 1 & \bigm| & -4
\end{bmatrix}
\quad
2R_1 + R_2 \to R_2 \text{ (to get a 0 in the lower left corner)}
$$

$$
\begin{array}{ccc}
2 & -1 & 4 \\
\end{array}
$$

$$
\sim
\begin{bmatrix}
1 & -\tfrac{1}{2} & \bigm| & 2 \\
0 & 0 & \bigm| & 0
\end{bmatrix}
$$

The last matrix corresponds to the system

$$
\begin{aligned}
x_1 - \tfrac{1}{2}x_2 &= 2 && x_1 - \tfrac{1}{2}x_2 = 2 \\
0 &= 0 && 0x_1 + 0x_2 = 0
\end{aligned}
\tag{9}
$$

This system is equivalent to the original system. Geometrically, the graphs of the two original equations coincide and there are infinitely many solutions. In general, if we end up with a row of zeros in an augmented matrix for a two-equation–two-variable system, the system is dependent and there are infinitely many solutions.

We represent the infinitely many solutions using the same method that was used in Section 4-1; that is, by introducing a parameter. We start by solving $x_1 - \tfrac{1}{2}x_2 = 2$, the first equation in (9), for either variable in terms of the other. We choose to solve for x_1 in terms of x_2 because it is easier:

$$x_1 = \tfrac{1}{2}x_2 + 2 \tag{10}$$

Now we introduce a parameter t (we can use other letters, such as k, s, p, q, and so on, to represent a parameter just as well). If we let $x_2 = t$, then for t any real number,

$$
\begin{aligned}
x_1 &= \tfrac{1}{2}t + 2 \\
x_2 &= t
\end{aligned}
\tag{11}
$$

represents a solution of system (8). Using ordered pair notation, we may also write: For any real number t,

$$(\tfrac{1}{2}t + 2,\ t) \tag{12}$$

is a solution of system (8). And more formally, we may write:

$$\text{Solution set} = \{(\tfrac{1}{2}t + 2, t) | t \in R\} \tag{13}$$

We will generally use the less formal forms (11) and (12) to represent the solution set for problems of this type.

Check The following is a check that (11) provides a solution for system (8) for any real number t:

$$
\begin{array}{ll}
2x_1 - x_2 = 4 & -6x_1 + 3x_2 = -12 \\
2(\tfrac{1}{2}t + 2) - t \overset{?}{=} 4 & -6(\tfrac{1}{2}t + 2) + 3t \overset{?}{=} -12 \\
t + 4 - t \overset{?}{=} 4 & -3t - 12 + 3t \overset{?}{=} -12 \\
4 \overset{\checkmark}{=} 4 & -12 \overset{\checkmark}{=} -12
\end{array}
$$
◄

MATCHED PROBLEM 3 ► Solve using augmented matrix methods:

$$
\begin{array}{rcr}
-2x_1 + 6x_2 = & 6 \\
3x_1 - 9x_2 = & -9
\end{array}
$$
◄

explore – discuss 2

The solution of Example 3 involved three augmented coefficient matrices. Write the linear system that each matrix represents, solve each system graphically, and discuss the relationships among these solutions.

EXAMPLE 4 ► **Solving a System Using Augmented Matrix Methods** Solve using augmented matrix methods:

$$
\begin{array}{rcr}
2x_1 + 6x_2 = & -3 \\
x_1 + 3x_2 = & 2
\end{array}
$$

Solution

$$
\begin{bmatrix} 2 & 6 & | & -3 \\ 1 & 3 & | & 2 \end{bmatrix} \quad R_1 \leftrightarrow R_2
$$

$$
\sim \begin{bmatrix} 1 & 3 & | & 2 \\ 2 & 6 & | & -3 \end{bmatrix} \quad (-2)R_1 + R_2 \rightarrow R_2
$$
$$
\phantom{\sim \begin{bmatrix} 1 & 3 & | \end{bmatrix}} -2 \quad -6 \quad -4
$$

$$
\sim \begin{bmatrix} 1 & 3 & | & 2 \\ 0 & 0 & | & -7 \end{bmatrix} \quad R_2 \text{ implies the contradiction } 0 = -7.
$$

This is the augmented matrix of the system

$$
\begin{array}{rcr}
x_1 + 3x_2 = & 2 \\
0 = & -7
\end{array}
\qquad
\begin{array}{l}
x_1 + 3x_2 = 2 \\
0x_1 + 0x_2 = -7
\end{array}
$$

The second equation is not satisfied by any ordered pair of real numbers. Hence, the original system is inconsistent and has no solution—otherwise, we have proved that $0 = -7$! Thus, if in a row of an augmented matrix we obtain all zeros to the left of the vertical bar and a nonzero number to the right, then the system is inconsistent and there are no solutions. ◄

MATCHED PROBLEM 4 ➤ Solve using augmented matrix methods:

$$2x_1 - x_2 = 3$$
$$4x_1 - 2x_2 = -1$$

◄

■ SUMMARY

> ### Summary
>
> | Form 1 | Form 2 | |
> | --- | --- | --- |
> | **A Unique Solution** (Consistent and Independent) | **Infinitely Many Solutions** (Consistent and Dependent) | **Form 3** **No Solution** (Inconsistent) |
> | $\begin{bmatrix} 1 & 0 & m \\ 0 & 1 & n \end{bmatrix}$ | $\begin{bmatrix} 1 & m & n \\ 0 & 0 & 0 \end{bmatrix}$ | $\begin{bmatrix} 1 & m & n \\ 0 & 0 & p \end{bmatrix}$ |
>
> m, n, p real numbers; $p \neq 0$

The process of solving systems of equations described in this section is referred to as **Gauss–Jordan elimination.** We will formalize this method in the next section so that it will apply to systems of any size, including systems where the number of equations and the number of variables are not the same.

Answers to Matched Problems 1. $x_1 = -2, x_2 = 3$ 2. $x_1 = 2, x_2 = -\frac{1}{2}$
3. The system is dependent. For t any real number, a solution is $x_1 = 3t - 3, x_2 = t$.
4. Inconsistent—no solution

EXERCISE   4-2

A   *Problems 1–10 refer to the following matrices:*

$$A = \begin{bmatrix} 2 & -4 & 0 \\ 6 & 1 & -5 \end{bmatrix} \qquad B = \begin{bmatrix} -1 & 9 & 0 \\ -4 & 8 & 7 \\ 2 & 4 & 0 \end{bmatrix}$$

$$C = \begin{bmatrix} 2 & -3 & 0 \end{bmatrix} \qquad D = \begin{bmatrix} -5 \\ 8 \end{bmatrix}$$

1. What is the size of A? Of C?
2. What is the size of B? Of D?
3. Identify all row matrices.
4. Identify all column matrices.
5. Identify all square matrices.
6. How many additional rows would matrix A have to have to be a square matrix?

7. For matrix A, find a_{12} and a_{23}.

8. For matrix A, find a_{21} and a_{13}.

9. Find the elements on the principal diagonal of matrix B.

10. Find the elements on the principal diagonal of matrix A.

Perform the row operations indicated in Problems 11–22 on the following matrix:

$$\begin{bmatrix} 1 & -3 & \bigm| & 2 \\ 4 & -6 & \bigm| & -8 \end{bmatrix}$$

11. $R_1 \leftrightarrow R_2$ **12.** $\frac{1}{2}R_2 \rightarrow R_2$

13. $-4R_1 \rightarrow R_1$ **14.** $-2R_1 \rightarrow R_1$

15. $2R_2 \rightarrow R_2$ **16.** $-1R_2 \rightarrow R_2$

17. $(-4)R_1 + R_2 \rightarrow R_2$ **18.** $(-\frac{1}{2})R_2 + R_1 \rightarrow R_1$

19. $(-2)R_1 + R_2 \rightarrow R_2$ **20.** $(-3)R_1 + R_2 \rightarrow R_2$

21. $(-1)R_1 + R_2 \rightarrow R_2$ **22.** $R_1 + R_2 \rightarrow R_2$

Solve Problems 23 and 24 using augmented matrix methods. Write the linear system represented by each augmented matrix in your solution, and solve each of these systems graphically. Discuss the relationships among the solutions of these systems.

23. $\begin{aligned} x_1 + x_2 &= 5 \\ x_1 - x_2 &= 1 \end{aligned}$ **24.** $\begin{aligned} x_1 - x_2 &= 2 \\ x_1 + x_2 &= 6 \end{aligned}$

B   *Solve Problems 25–44 using augmented matrix methods.*

25. $\begin{aligned} x_1 - 2x_2 &= 1 \\ 2x_1 - x_2 &= 5 \end{aligned}$ **26.** $\begin{aligned} x_1 + 3x_2 &= 1 \\ 3x_1 - 2x_2 &= 14 \end{aligned}$

27. $\begin{aligned} x_1 - 4x_2 &= -2 \\ -2x_1 + x_2 &= -3 \end{aligned}$ **28.** $\begin{aligned} x_1 - 3x_2 &= -5 \\ -3x_1 - x_2 &= 5 \end{aligned}$

29. $\begin{aligned} 3x_1 - x_2 &= 2 \\ x_1 + 2x_2 &= 10 \end{aligned}$ **30.** $\begin{aligned} 2x_1 + x_2 &= 0 \\ x_1 - 2x_2 &= -5 \end{aligned}$

31. $\begin{aligned} x_1 + 2x_2 &= 4 \\ 2x_1 + 4x_2 &= -8 \end{aligned}$ **32.** $\begin{aligned} 2x_1 - 3x_2 &= -2 \\ -4x_1 + 6x_2 &= 7 \end{aligned}$

33. $\begin{aligned} 2x_1 + x_2 &= 6 \\ x_1 - x_2 &= -3 \end{aligned}$ **34.** $\begin{aligned} 3x_1 - x_2 &= -5 \\ x_1 + 3x_2 &= 5 \end{aligned}$

35. $\begin{aligned} 3x_1 - 6x_2 &= -9 \\ -2x_1 + 4x_2 &= 6 \end{aligned}$ **36.** $\begin{aligned} 2x_1 - 4x_2 &= -2 \\ -3x_1 + 6x_2 &= 3 \end{aligned}$

37. $\begin{aligned} 4x_1 - 2x_2 &= 2 \\ -6x_1 + 3x_2 &= -3 \end{aligned}$ **38.** $\begin{aligned} -6x_1 + 2x_2 &= 4 \\ 3x_1 - x_2 &= -2 \end{aligned}$

39. $\begin{aligned} 2x_1 + x_2 &= 1 \\ 4x_1 - x_2 &= -7 \end{aligned}$ **40.** $\begin{aligned} 2x_1 - x_2 &= -8 \\ 2x_1 + x_2 &= 8 \end{aligned}$

41. $\begin{aligned} 4x_1 - 6x_2 &= 8 \\ -6x_1 + 9x_2 &= -10 \end{aligned}$ **42.** $\begin{aligned} 2x_1 - 4x_2 &= -4 \\ -3x_1 + 6x_2 &= 4 \end{aligned}$

43. $\begin{aligned} -4x_1 + 6x_2 &= -8 \\ 6x_1 - 9x_2 &= 12 \end{aligned}$ **44.** $\begin{aligned} -2x_1 + 4x_2 &= 4 \\ 3x_1 - 6x_2 &= -6 \end{aligned}$

C   *Solve Problems 45–54 using augmented matrix methods.*

45. $\begin{aligned} 3x_1 - x_2 &= 7 \\ 2x_1 + 3x_2 &= 1 \end{aligned}$ **46.** $\begin{aligned} 2x_1 - 3x_2 &= -8 \\ 5x_1 + 3x_2 &= 1 \end{aligned}$

47. $\begin{aligned} 3x_1 + 2x_2 &= 4 \\ 2x_1 - x_2 &= 5 \end{aligned}$ **48.** $\begin{aligned} 4x_1 + 3x_2 &= 26 \\ 3x_1 - 11x_2 &= -7 \end{aligned}$

49. $\begin{aligned} 0.2x_1 - 0.5x_2 &= 0.07 \\ 0.8x_1 - 0.3x_2 &= 0.79 \end{aligned}$ **50.** $\begin{aligned} 0.3x_1 - 0.6x_2 &= 0.18 \\ 0.5x_1 - 0.2x_2 &= 0.54 \end{aligned}$

51. $\begin{aligned} 0.8x_1 + 2.88x_2 &= 4 \\ 1.25x_1 + 4.34x_2 &= 5 \end{aligned}$

52. $\begin{aligned} 2.7x_1 - 15.12x_2 &= 27 \\ 3.25x_1 - 18.52x_2 &= 33 \end{aligned}$

53. $\begin{aligned} 4.8x_1 - 40.32x_2 &= 295.2 \\ -3.75x_1 + 28.7x_2 &= -211.2 \end{aligned}$

54. $\begin{aligned} 5.7x_1 - 8.55x_2 &= -35.91 \\ 4.5x_1 + 5.73x_2 &= 76.17 \end{aligned}$

SECTION 4-3 ■ # Gauss–Jordan Elimination

■ REDUCED MATRICES
■ SOLVING SYSTEMS BY GAUSS–JORDAN ELIMINATION
■ APPLICATION

Now that you have had some experience with row operations on simple augmented matrices, we will consider systems involving more than two variables. In addition, we will not require that a system have the same number of equations as variables. It turns out that the results for two-variable–two-equation linear systems stated in Theorem 1, Section 4-1, actually hold for linear systems of any size.

Possible Solutions to a Linear System

It can be shown that any linear system must have exactly one solution, no solution, or an infinite number of solutions, regardless of the number of equations or number of variables in the system. The terms *unique solution, consistent, inconsistent, dependent,* and *independent* are used to describe these solutions, just as in the two-variable case.

■ REDUCED MATRICES

In the last section we used row operations to transform the augmented coefficient matrix for a system of two equations in two variables,

$$\begin{bmatrix} a_{11} & a_{12} & | & k_1 \\ a_{21} & a_{22} & | & k_2 \end{bmatrix} \qquad \begin{aligned} a_{11}x_1 + a_{12}x_2 &= k_1 \\ a_{21}x_1 + a_{22}x_2 &= k_2 \end{aligned}$$

into one of the following simplified forms:

$$
\begin{array}{ccc}
\text{Form 1} & \text{Form 2} & \text{Form 3} \\[4pt]
\begin{bmatrix} 1 & 0 & | & m \\ 0 & 1 & | & n \end{bmatrix} &
\begin{bmatrix} 1 & m & | & n \\ 0 & 0 & | & 0 \end{bmatrix} &
\begin{bmatrix} 1 & m & | & n \\ 0 & 0 & | & p \end{bmatrix}
\end{array}
\qquad (1)
$$

where m, n, and p are real numbers, $p \neq 0$. Each of these reduced forms represents a system that has a different type of solution set, and no two of these forms are row-equivalent. Thus, we consider each of these to be a different simplified form. Now we want to consider larger systems with more variables and more equations.

---

explore–discuss 1

Forms 1, 2, and 3 in (1) represent systems that have a unique solution, an infinite number of solutions, and no solution, respectively. Discuss the number of solutions for the systems of three equations in three variables represented by the following augmented coefficient matrices:

$$
(A)\ \begin{bmatrix} 1 & 2 & 3 & | & 5 \\ 0 & 0 & 0 & | & 6 \\ 0 & 0 & 0 & | & 0 \end{bmatrix}
\qquad
(B)\ \begin{bmatrix} 1 & 2 & 3 & | & 5 \\ 0 & 0 & 0 & | & 0 \\ 0 & 0 & 0 & | & 0 \end{bmatrix}
\qquad
(C)\ \begin{bmatrix} 1 & 0 & 0 & | & 5 \\ 0 & 1 & 0 & | & 6 \\ 0 & 0 & 1 & | & 7 \end{bmatrix}
$$

Since there is no upper limit on the number of variables or the number of equations in a linear system, it is not feasible to explicitly list all possible "simplified forms" for larger systems, as we did for systems of two equations in two variables.

Instead, we state a general definition of a simplified form called a *reduced matrix* that can be applied to all matrices and systems, regardless of size.

Reduced Matrix

A matrix is in **reduced form** if:

1. Each row consisting entirely of zeros is below any row having at least one nonzero element.
2. The leftmost nonzero element in each row is 1.
3. All other elements in the column containing the leftmost 1 of a given row are zeros.
4. The leftmost 1 in any row is to the right of the leftmost 1 in the row above.

The following matrices are in reduced form. Check each one carefully to convince yourself that the conditions in the definition are met.

$$\left[\begin{array}{cc|c} 1 & 0 & 2 \\ 0 & 1 & -3 \end{array}\right] \qquad \left[\begin{array}{ccc|c} 1 & 0 & 0 & 2 \\ 0 & 1 & 0 & -1 \\ 0 & 0 & 1 & 3 \end{array}\right] \qquad \left[\begin{array}{cc|c} 1 & 0 & 3 \\ 0 & 1 & -1 \\ 0 & 0 & 0 \end{array}\right]$$

$$\left[\begin{array}{cccc|c} 1 & 4 & 0 & 0 & -3 \\ 0 & 0 & 1 & 0 & 2 \\ 0 & 0 & 0 & 1 & 6 \end{array}\right] \qquad \left[\begin{array}{ccc|c} 1 & 0 & 4 & 0 \\ 0 & 1 & 3 & 0 \\ 0 & 0 & 0 & 1 \end{array}\right]$$

EXAMPLE 1 ➤ **Reduced Forms** The matrices below are not in reduced form. Indicate which condition in the definition is violated for each matrix. State the row operation(s) required to transform the matrix into reduced form and find the reduced form.

(A) $\left[\begin{array}{cc|c} 0 & 1 & -2 \\ 1 & 0 & 3 \end{array}\right]$ (B) $\left[\begin{array}{ccc|c} 1 & 2 & -2 & 3 \\ 0 & 0 & 1 & -1 \end{array}\right]$

(C) $\left[\begin{array}{cc|c} 1 & 0 & -3 \\ 0 & 0 & 0 \\ 0 & 1 & -2 \end{array}\right]$ (D) $\left[\begin{array}{ccc|c} 1 & 0 & 0 & -1 \\ 0 & 2 & 0 & 3 \\ 0 & 0 & 1 & -5 \end{array}\right]$

Solution   (A) Condition 4 is violated: The leftmost 1 in Row 2 is not to the right of the leftmost 1 in Row 1. Perform the row operation $R_1 \leftrightarrow R_2$ to obtain

$$\left[\begin{array}{cc|c} 1 & 0 & 3 \\ 0 & 1 & -2 \end{array}\right]$$

(B) Condition 3 is violated: The column containing the leftmost 1 in Row 2 has a nonzero element above the 1. Perform the row operation $2R_2 + R_1 \rightarrow R_1$ to obtain

$$\left[\begin{array}{ccc|c} 1 & 2 & 0 & 1 \\ 0 & 0 & 1 & -1 \end{array}\right]$$

(C) Condition 1 is violated: The second row contains all zeros and it is not below any row having at least one nonzero element. Perform the row operation $R_2 \leftrightarrow R_3$ to obtain

$$\begin{bmatrix} 1 & 0 & | & -3 \\ 0 & 1 & | & -2 \\ 0 & 0 & | & 0 \end{bmatrix}$$

(D) Condition 2 is violated: The leftmost nonzero element in Row 2 is not a 1. Perform the row operation $\frac{1}{2}R_2 \rightarrow R_2$ to obtain

$$\begin{bmatrix} 1 & 0 & 0 & | & -1 \\ 0 & 1 & 0 & | & \frac{3}{2} \\ 0 & 0 & 1 & | & -5 \end{bmatrix}$$

◀

MATCHED PROBLEM 1 ▶ The matrices below are not in reduced form. Indicate which condition in the definition is violated for each matrix. State the row operation(s) required to transform the matrix into reduced form and find the reduced form.

(A) $\begin{bmatrix} 1 & 0 & | & 2 \\ 0 & 3 & | & -6 \end{bmatrix}$
(B) $\begin{bmatrix} 1 & 5 & 4 & | & 3 \\ 0 & 1 & 2 & | & -1 \\ 0 & 0 & 0 & | & 0 \end{bmatrix}$

(C) $\begin{bmatrix} 0 & 1 & 0 & | & -3 \\ 1 & 0 & 0 & | & 0 \\ 0 & 0 & 1 & | & 2 \end{bmatrix}$
(D) $\begin{bmatrix} 1 & 2 & 0 & | & 3 \\ 0 & 0 & 0 & | & 0 \\ 0 & 0 & 1 & | & 4 \end{bmatrix}$

◀

■ SOLVING SYSTEMS BY GAUSS–JORDAN ELIMINATION

We are now ready to outline the Gauss–Jordan method for solving systems of linear equations. The method systematically transforms an augmented matrix into a reduced form. The system corresponding to a reduced augmented coefficient matrix is called a **reduced system.** As we shall see, reduced systems are easy to solve.

The Gauss–Jordan elimination method is named after the German mathematician Carl Friedrich Gauss (1777–1885) and the French mathematician Camille Jordan (1838–1922). Gauss, one of the greatest mathematicians of all time, used a method of solving systems of equations that was later generalized by Jordan to the method we present here.

EXAMPLE 2 ▶ Solving a System Using Gauss–Jordan Elimination   Solve by Gauss–Jordan elimination:

$$\begin{aligned} 2x_1 - 2x_2 + x_3 &= 3 \\ 3x_1 + x_2 - x_3 &= 7 \\ x_1 - 3x_2 + 2x_3 &= 0 \end{aligned}$$

Solution   Write the augmented matrix and follow the steps indicated at the right.

Need a 1 here.

$$\begin{bmatrix} 2 & -2 & 1 & 3 \\ 3 & 1 & -1 & 7 \\ 1 & -3 & 2 & 0 \end{bmatrix} \quad R_1 \leftrightarrow R_3$$

Step 1.   Choose the leftmost nonzero column and get a 1 at the top.

Need 0's here.

$$\sim \begin{bmatrix} 1 & -3 & 2 & 0 \\ 3 & 1 & -1 & 7 \\ 2 & -2 & 1 & 3 \end{bmatrix} \quad \begin{matrix}(-3)R_1 + R_2 \rightarrow R_2 \\ (-2)R_1 + R_3 \rightarrow R_3\end{matrix}$$

Step 2.   Use multiples of the row containing the 1 from step 1 to get zeros in all remaining places in the column containing this 1.

Need a 1 here.

$$\sim \begin{bmatrix} 1 & -3 & 2 & 0 \\ 0 & 10 & -7 & 7 \\ 0 & 4 & -3 & 3 \end{bmatrix} \quad 0.1R_2 \rightarrow R_2$$

Step 3.   Repeat step 1 with the *submatrix* formed by (mentally) deleting the top row.

Need 0's here.

$$\sim \begin{bmatrix} 1 & -3 & 2 & 0 \\ 0 & 1 & -0.7 & 0.7 \\ 0 & 4 & -3 & 3 \end{bmatrix} \quad \begin{matrix}3R_2 + R_1 \rightarrow R_1 \\ \\ (-4)R_2 + R_3 \rightarrow R_3\end{matrix}$$

Step 4.   Repeat step 2 with the *entire matrix*.

Need a 1 here.

$$\sim \begin{bmatrix} 1 & 0 & -0.1 & 2.1 \\ 0 & 1 & -0.7 & 0.7 \\ 0 & 0 & -0.2 & 0.2 \end{bmatrix} \quad (-5)R_3 \rightarrow R_3$$

Step 3.   Repeat step 1 with the *submatrix* formed by (mentally) deleting the top two rows.

Need 0's here.

$$\sim \begin{bmatrix} 1 & 0 & -0.1 & 2.1 \\ 0 & 1 & -0.7 & 0.7 \\ 0 & 0 & 1 & -1 \end{bmatrix} \quad \begin{matrix}0.1R_3 + R_1 \rightarrow R_1 \\ 0.7R_3 + R_2 \rightarrow R_2\end{matrix}$$

Step 4.   Repeat step 2 with the *entire matrix*.

$$\sim \begin{bmatrix} 1 & 0 & 0 & 2 \\ 0 & 1 & 0 & 0 \\ 0 & 0 & 1 & -1 \end{bmatrix}$$

The matrix is now in reduced form, and we can proceed to solve the corresponding reduced system.

$$\begin{aligned} x_1 &= 2 \\ x_2 &= 0 \\ x_3 &= -1 \end{aligned}$$

The solution to this system is $x_1 = 2$, $x_2 = 0$, $x_3 = -1$. You should check this solution in the original system. ◀

Gauss–Jordan Elimination

Step 1. Choose the leftmost nonzero column and use appropriate row operations to get a 1 at the top.

Step 2. Use multiples of the row containing the 1 from step 1 to get zeros in all remaining places in the column containing this 1.

Step 3. Repeat step 1 with the **submatrix** formed by (mentally) deleting the row used in step 2 and all rows above this row.

Step 4. Repeat step 2 with the **entire matrix,** including the mentally deleted rows. Continue this process until it is impossible to go further.

[*Note:* If at any point in this process we obtain a row with all zeros to the left of the vertical line and a nonzero number to the right, we can stop, since we will have a contradiction: $0 = n, n \neq 0$. We can then conclude that the system has no solution.]

Remarks

1. Even though each matrix has a unique reduced form, the sequence of steps (algorithm) presented here for transforming a matrix into a reduced form is not unique. That is, other sequences of steps (using row operations) can produce a reduced matrix. (For example, it is possible to use row operations in such a way that computations involving fractions are minimized.) But we emphasize again that we are not interested in the most efficient hand methods for transforming small matrices into reduced forms. Our main interest is in giving you a little experience with a method that is suitable for solving large-scale systems on a computer or graphing calculator.

2. Most graphing calculators and computers have the ability to find reduced forms, either directly or with some programming. Figure 1 illustrates the solution of Example 2 on a graphing calculator that has a built-in routine for finding reduced forms. Notice that in Row 2 and Column 4 of the reduced form the graphing calculator has displayed the very small number $-3.5E - 13$, instead of the exact value 0. This is a common occurrence on a graphing calculator and causes no problems. Just replace any very small numbers displayed in scientific notation with 0.

FIGURE 1

Gauss–Jordan elimination on a graphing calculator.

MATCHED PROBLEM 2 ▶ Solve by Gauss–Jordan elimination:

$$3x_1 + x_2 - 2x_3 = 2$$
$$x_1 - 2x_2 + x_3 = 3$$
$$2x_1 - x_2 - 3x_3 = 3$$

◀

EXAMPLE 3 ➤ **Solving a System Using Gauss–Jordan Elimination** Solve by Gauss–Jordan elimination:

$$2x_1 - 4x_2 + x_3 = -4$$
$$4x_1 - 8x_2 + 7x_3 = 2$$
$$-2x_1 + 4x_2 - 3x_3 = 5$$

Solution

$$\begin{bmatrix} 2 & -4 & 1 & | & -4 \\ 4 & -8 & 7 & | & 2 \\ -2 & 4 & -3 & | & 5 \end{bmatrix} \quad 0.5R_1 \to R_1$$

$$\sim \begin{bmatrix} 1 & -2 & 0.5 & | & -2 \\ 4 & -8 & 7 & | & 2 \\ -2 & 4 & -3 & | & 5 \end{bmatrix} \quad \begin{matrix} (-4)R_1 + R_2 \to R_2 \\ 2R_1 + R_3 \to R_3 \end{matrix}$$

$$\sim \begin{bmatrix} 1 & -2 & 0.5 & | & -2 \\ 0 & 0 & 5 & | & 10 \\ 0 & 0 & -2 & | & 1 \end{bmatrix} \quad 0.2R_2 \to R_2$$

Note that Column 3 is the left-most nonzero column in this sub-matrix.

$$\sim \begin{bmatrix} 1 & -2 & 0.5 & | & -2 \\ 0 & 0 & 1 & | & 2 \\ 0 & 0 & -2 & | & 1 \end{bmatrix} \quad \begin{matrix} -0.5R_2 + R_1 \to R_1 \\ \\ 2R_2 + R_3 \to R_3 \end{matrix}$$

$$\sim \begin{bmatrix} 1 & -2 & 0 & | & -3 \\ 0 & 0 & 1 & | & 2 \\ 0 & 0 & 0 & | & 5 \end{bmatrix}$$

We stop the Gauss–Jordan elimination, even though the matrix is not in reduced form, since the last row produces a contradiction. ◀

MATCHED PROBLEM 3 ➤ Solve by Gauss–Jordan elimination:

$$2x_1 - 4x_2 - x_3 = -8$$
$$4x_1 - 8x_2 + 3x_3 = 4$$
$$-2x_1 + 4x_2 + x_3 = 11$$

◀

FIGURE 2
Recognizing contradictions on a graphing calculator

Remarks

1. Figure 2 shows the solution to Example 3 on a graphing calculator with a built-in reduced form routine. Notice that the graphing calculator does not stop when a contradiction first occurs, but continues on to find the reduced form. Nevertheless, the last row in the reduced form still produces a contradiction. Do not confuse this type of reduced form with one that represents a consistent system (see Fig. 1).

2. Since not all graphing calculators can find the reduced form of a matrix directly, we will continue to display row operations in matrix reductions. If you are using a graphing calculator or computer that can find the reduced form directly, then you can ignore these intermediate steps and proceed directly from the initial augmented coefficient matrix to the reduced form in a single step, as we did in Figures 1 and 2.

EXAMPLE 4 ▶ **Solving a System Using Gauss–Jordan Elimination** Solve by Gauss–Jordan elimination:

$$3x_1 + 6x_2 - 9x_3 = 15$$
$$2x_1 + 4x_2 - 6x_3 = 10$$
$$-2x_1 - 3x_2 + 4x_3 = -6$$

Solution

$$\begin{bmatrix} 3 & 6 & -9 & | & 15 \\ 2 & 4 & -6 & | & 10 \\ -2 & -3 & 4 & | & -6 \end{bmatrix} \quad \tfrac{1}{3}R_1 \rightarrow R_1$$

$$\sim \begin{bmatrix} 1 & 2 & -3 & | & 5 \\ 2 & 4 & -6 & | & 10 \\ -2 & -3 & 4 & | & -6 \end{bmatrix} \quad \begin{matrix} (-2)R_1 + R_2 \rightarrow R_2 \\ 2R_1 + R_3 \rightarrow R_3 \end{matrix}$$

$$\sim \begin{bmatrix} 1 & 2 & -3 & | & 5 \\ 0 & 0 & 0 & | & 0 \\ 0 & 1 & -2 & | & 4 \end{bmatrix} \quad R_2 \leftrightarrow R_3$$

Note that we must interchange Rows 2 and 3 to obtain a nonzero entry at the top of the second column of this submatrix.

$$\sim \begin{bmatrix} 1 & 2 & -3 & | & 5 \\ 0 & 1 & -2 & | & 4 \\ 0 & 0 & 0 & | & 0 \end{bmatrix} \quad (-2)R_2 + R_1 \rightarrow R_1$$

$$\sim \begin{bmatrix} 1 & 0 & 1 & | & -3 \\ 0 & 1 & -2 & | & 4 \\ 0 & 0 & 0 & | & 0 \end{bmatrix}$$

The matrix is now in reduced form. Write the corresponding reduced system and solve.

$$x_1 \qquad + \quad x_3 = -3$$
$$x_2 - 2x_3 = \quad 4$$

We discard the equation corresponding to the third (all zero) row in the reduced form, since it is satisfied by all values of x_1, x_2, and x_3.

Note that the leftmost variable in each equation appears in one and only one equation. We solve for the leftmost variables x_1 and x_2 in terms of the remaining variable, x_3:

$$x_1 = -x_3 - 3$$
$$x_2 = 2x_3 + 4$$

If we let $x_3 = t$, then for any real number t,

$$x_1 = -t - 3$$
$$x_2 = 2t + 4$$
$$x_3 = t$$

You should check that $(-t - 3, 2t + 4, t)$ is a solution of the original system for any real number t. Some particular solutions are

$t = 0$	$t = -2$	$t = 3.5$
$(-3, 4, 0)$	$(-1, 0, -2)$	$(-6.5, 11, 3.5)$

◀

In general:

If the number of leftmost 1's in a reduced augmented coefficient matrix is less than the number of variables in the system and there are no contradictions, then the system is dependent and has infinitely many solutions.

Describing the solution set to this type of system is not difficult. In a reduced system, the *leftmost variables* correspond to the leftmost 1's in the corresponding reduced augmented matrix. The definition of reduced form for an augmented matrix ensures that each leftmost variable in the corresponding reduced system appears in one and only one equation of the system. Solving for each leftmost variable in terms of the remaining variables and writing a general solution to the system is usually easy. (Example 5 illustrates a slightly more involved case.)

MATCHED PROBLEM 4 ➤ Solve by Gauss–Jordan elimination:

$$2x_1 - 2x_2 - 4x_3 = -2$$
$$3x_1 - 3x_2 - 6x_3 = -3$$
$$-2x_1 + 3x_2 + x_3 = 7$$

◀

explore – discuss 2

Explain why the definition of reduced form ensures that each leftmost variable in a reduced system appears in one and only one equation and no equation contains more than one leftmost variable. Discuss methods for determining whether a consistent system is independent or dependent by examining the reduced form.

EXAMPLE 5 ➤ Solving a System Using Gauss–Jordan Elimination Solve by Gauss–Jordan elimination:

$$x_1 + 2x_2 + 4x_3 + x_4 - x_5 = 1$$
$$2x_1 + 4x_2 + 8x_3 + 3x_4 - 4x_5 = 2$$
$$x_1 + 3x_2 + 7x_3 + 3x_5 = -2$$

Solution

$$\begin{bmatrix} 1 & 2 & 4 & 1 & -1 & | & 1 \\ 2 & 4 & 8 & 3 & -4 & | & 2 \\ 1 & 3 & 7 & 0 & 3 & | & -2 \end{bmatrix} \begin{matrix} \\ (-2)R_1 + R_2 \to R_2 \\ (-1)R_1 + R_3 \to R_3 \end{matrix}$$

$$\sim \begin{bmatrix} 1 & 2 & 4 & 1 & -1 & | & 1 \\ 0 & 0 & 0 & 1 & -2 & | & 0 \\ 0 & 1 & 3 & -1 & 4 & | & -3 \end{bmatrix} \quad R_2 \leftrightarrow R_3$$

$$\sim \begin{bmatrix} 1 & 2 & 4 & 1 & -1 & | & 1 \\ 0 & 1 & 3 & -1 & 4 & | & -3 \\ 0 & 0 & 0 & 1 & -2 & | & 0 \end{bmatrix} \quad (-2)R_2 + R_1 \to R_1$$

$$\sim \begin{bmatrix} 1 & 0 & -2 & 3 & -9 & \vline & 7 \\ 0 & 1 & 3 & -1 & 4 & \vline & -3 \\ 0 & 0 & 0 & 1 & -2 & \vline & 0 \end{bmatrix} \quad \begin{array}{l} (-3)R_3 + R_1 \rightarrow R_1 \\ R_3 + R_2 \rightarrow R_2 \end{array}$$

$$\sim \begin{bmatrix} 1 & 0 & -2 & 0 & -3 & \vline & 7 \\ 0 & 1 & 3 & 0 & 2 & \vline & -3 \\ 0 & 0 & 0 & 1 & -2 & \vline & 0 \end{bmatrix} \quad \text{Matrix is in reduced form.}$$

$$\begin{aligned} x_1 \quad - 2x_3 \quad - 3x_5 &= \quad 7 \\ x_2 + 3x_3 \quad + 2x_5 &= -3 \\ x_4 - 2x_5 &= \quad 0 \end{aligned}$$

Solve for the leftmost variables x_1, x_2, and x_4 in terms of the remaining variables x_3 and x_5:

$$\begin{aligned} x_1 &= \quad 2x_3 + 3x_5 + 7 \\ x_2 &= -3x_3 - 2x_5 - 3 \\ x_4 &= \quad 2x_5 \end{aligned}$$

If we let $x_3 = s$ and $x_5 = t$, then for any real numbers s and t,

$$\begin{aligned} x_1 &= 2s + 3t + 7 \\ x_2 &= -3s - 2t - 3 \\ x_3 &= s \\ x_4 &= 2t \\ x_5 &= t \end{aligned}$$

is a solution. The check is left for you to perform. ◀

MATCHED PROBLEM 5 ➤   Solve by Gauss–Jordan elimination:

$$\begin{aligned} x_1 - \quad x_2 + 2x_3 \qquad\quad - 2x_5 &= \quad 3 \\ -2x_1 + 2x_2 - 4x_3 - x_4 + \quad x_5 &= -5 \\ 3x_1 - 3x_2 + 7x_3 + x_4 - 4x_5 &= \quad 6 \end{aligned}$$

◀

■ APPLICATION

▶   EXAMPLE 6 ➤   **Production Scheduling**   A casting company produces three different bronze sculptures. The casting department has available a maximum of 350 labor-hours per week, and the finishing department has a maximum of 150 labor-hours available per week. Sculpture *A* requires 30 hours for casting and 10 hours for finishing; sculpture *B* requires 10 hours for casting and 10 hours for finishing; and sculpture *C* requires 10 hours for casting and 30 hours for finishing. If the plant is to operate at maximum capacity, how many of each sculpture should be produced each week?

Solution First, we summarize the relevant manufacturing data in a table:

	LABOR-HOURS PER SCULPTURE			MAXIMUM LABOR-HOURS AVAILABLE PER WEEK
	A	B	C	
CASTING DEPARTMENT	30	10	10	350
FINISHING DEPARTMENT	10	10	30	150

Let: x_1 = Number of sculpture A produced per week

x_2 = Number of sculpture B produced per week

x_3 = Number of sculpture C produced per week

Then: $30x_1 + 10x_2 + 10x_3 = 350$ *Casting department*

$10x_1 + 10x_2 + 30x_3 = 150$ *Finishing department*

Now we can form the augmented matrix of the system and solve by using Gauss–Jordan elimination:

$$\begin{bmatrix} 30 & 10 & 10 & \bigm| & 350 \\ 10 & 10 & 30 & \bigm| & 150 \end{bmatrix}$$ $\frac{1}{10}R_1 \to R_1$ *Simplify each row.*
$\frac{1}{10}R_2 \to R_2$

$$\sim \begin{bmatrix} 3 & 1 & 1 & \bigm| & 35 \\ 1 & 1 & 3 & \bigm| & 15 \end{bmatrix}$$ $R_1 \leftrightarrow R_2$

$$\sim \begin{bmatrix} 1 & 1 & 3 & \bigm| & 15 \\ 3 & 1 & 1 & \bigm| & 35 \end{bmatrix}$$ $(-3)R_1 + R_2 \to R_2$

$$\sim \begin{bmatrix} 1 & 1 & 3 & \bigm| & 15 \\ 0 & -2 & -8 & \bigm| & -10 \end{bmatrix}$$ $-\frac{1}{2}R_2 \to R_2$

$$\sim \begin{bmatrix} 1 & 1 & 3 & \bigm| & 15 \\ 0 & 1 & 4 & \bigm| & 5 \end{bmatrix}$$ $(-1)R_2 + R_1 \to R_1$

$$\sim \begin{bmatrix} 1 & 0 & -1 & \bigm| & 10 \\ 0 & 1 & 4 & \bigm| & 5 \end{bmatrix}$$ *Matrix is in reduced form.*

$x_1 \quad - x_3 = 10 \qquad$ or $\qquad x_1 = \quad x_3 + 10$

$\quad x_2 + 4x_3 = 5 \qquad\qquad\qquad x_2 = -4x_3 + 5$

Let $x_3 = t$. Then for t any real number,

$x_1 = t + 10$

$x_2 = -4t + 5$

$x_3 = t$

is a solution—or is it? We cannot produce a negative number of sculptures. If we also assume that we cannot produce a fractional number of sculptures, then t must be a nonnegative whole number. And because of the middle equation ($x_2 = -4t + 5$), t can only assume the values 0 and 1. Thus, for $t = 0$, we have $x_1 = 10$, $x_2 = 5$, $x_3 = 0$; and for $t = 1$, we have $x_1 = 11$, $x_2 = 1$, $x_3 = 1$. These are the only possible production schedules that utilize the full capacity of the plant. ◀

MATCHED PROBLEM 6 ▶ Repeat Example 6 given a casting capacity of 400 labor-hours per week and a finishing capacity of 200 labor-hours per week. ◀

Answers to Matched Problems

1. (A) Condition 2 is violated: The 3 in Row 2 and Column 2 should be a 1. Perform the operation $\frac{1}{3} R_2 \rightarrow R_2$ to obtain

$$\begin{bmatrix} 1 & 0 & | & 2 \\ 0 & 1 & | & -2 \end{bmatrix}$$

(B) Condition 3 is violated: The 5 in Row 1 and Column 2 should be a 0. Perform the operation $(-5)R_2 + R_1 \rightarrow R_1$ to obtain

$$\begin{bmatrix} 1 & 0 & -6 & | & 8 \\ 0 & 1 & 2 & | & -1 \\ 0 & 0 & 0 & | & 0 \end{bmatrix}$$

(C) Condition 4 is violated. The leftmost 1 in the second row is not to the right of the leftmost 1 in the first row. Perform the operation $R_1 \leftrightarrow R_2$ to obtain

$$\begin{bmatrix} 1 & 0 & 0 & | & 0 \\ 0 & 1 & 0 & | & -3 \\ 0 & 0 & 1 & | & 2 \end{bmatrix}$$

(D) Condition 1 is violated: The all-zero second row should be at the bottom. Perform the operation $R_2 \leftrightarrow R_3$ to obtain

$$\begin{bmatrix} 1 & 2 & 0 & | & 3 \\ 0 & 0 & 1 & | & 4 \\ 0 & 0 & 0 & | & 0 \end{bmatrix}$$

2. $x_1 = 1, x_2 = -1, x_3 = 0$ 3. Inconsistent; no solution
4. $x_1 = 5t + 4, x_2 = 3t + 5, x_3 = t, t$ any real number
5. $x_1 = s + 7, x_2 = s, x_3 = t - 2, x_4 = -3t - 1, x_5 = t, s$ and t any real numbers
6. $x_1 = t + 10, x_2 = -4t + 10, x_3 = t,$ where $t = 0, 1, 2$; that is, $(x_1, x_2, x_3) = (10, 10, 0),$ $(11, 6, 1),$ or $(12, 2, 2)$

EXERCISE 4-3

A   *In Problems 1–8, indicate whether each matrix is in reduced form.*

1. $\begin{bmatrix} 1 & 0 & | & 2 \\ 0 & 1 & | & -1 \end{bmatrix}$

2. $\begin{bmatrix} 0 & 1 & | & 2 \\ 1 & 0 & | & -1 \end{bmatrix}$

3. $\begin{bmatrix} 1 & 0 & 2 & | & 3 \\ 0 & 0 & 0 & | & 0 \\ 0 & 1 & -1 & | & 4 \end{bmatrix}$

4. $\begin{bmatrix} 1 & 0 & 0 & | & -2 \\ 0 & 1 & 0 & | & 0 \\ 0 & 0 & 1 & | & 1 \end{bmatrix}$

5. $\begin{bmatrix} 0 & 1 & 0 & | & 2 \\ 0 & 0 & 3 & | & -1 \\ 0 & 0 & 0 & | & 0 \end{bmatrix}$

6. $\begin{bmatrix} 1 & 3 & 0 & | & 0 \\ 0 & 0 & 1 & | & 0 \\ 0 & 0 & 0 & | & 1 \end{bmatrix}$

7. $\begin{bmatrix} 1 & 2 & 0 & 3 & | & 2 \\ 0 & 0 & 1 & -1 & | & 0 \end{bmatrix}$

8. $\begin{bmatrix} 0 & 1 & 2 & | & 1 \\ 1 & 0 & -3 & | & 2 \end{bmatrix}$

Write the linear system corresponding to each reduced augmented matrix in Problems 9–16 and solve.

9. $\begin{bmatrix} 1 & 0 & 0 & | & -2 \\ 0 & 1 & 0 & | & 3 \\ 0 & 0 & 1 & | & 0 \end{bmatrix}$

10. $\begin{bmatrix} 1 & 0 & 0 & 0 & | & -2 \\ 0 & 1 & 0 & 0 & | & 0 \\ 0 & 0 & 1 & 0 & | & 1 \\ 0 & 0 & 0 & 1 & | & 3 \end{bmatrix}$

11. $\begin{bmatrix} 1 & 0 & -2 & | & 3 \\ 0 & 1 & 1 & | & -5 \\ 0 & 0 & 0 & | & 0 \end{bmatrix}$

12. $\begin{bmatrix} 1 & -2 & 0 & | & -3 \\ 0 & 0 & 1 & | & 5 \\ 0 & 0 & 0 & | & 0 \end{bmatrix}$

13. $\begin{bmatrix} 1 & 0 & | & 0 \\ 0 & 1 & | & 0 \\ 0 & 0 & | & 1 \end{bmatrix}$

14. $\begin{bmatrix} 1 & 0 & | & 5 \\ 0 & 1 & | & -3 \\ 0 & 0 & | & 0 \end{bmatrix}$

15. $\begin{bmatrix} 1 & -2 & 0 & -3 & | & -5 \\ 0 & 0 & 1 & 3 & | & 2 \end{bmatrix}$

16. $\begin{bmatrix} 1 & 0 & -2 & 3 & | & 4 \\ 0 & 1 & -1 & 2 & | & -1 \end{bmatrix}$

B *Use row operations to change each matrix in Problems 17–22 to reduced form.*

17. $\begin{bmatrix} 1 & 2 & | & -1 \\ 0 & 1 & | & 3 \end{bmatrix}$ **18.** $\begin{bmatrix} 1 & 3 & | & 1 \\ 0 & 2 & | & -4 \end{bmatrix}$

19. $\begin{bmatrix} 1 & 0 & -3 & | & 1 \\ 0 & 1 & 2 & | & 0 \\ 0 & 0 & 3 & | & -6 \end{bmatrix}$

20. $\begin{bmatrix} 1 & 0 & 4 & | & 0 \\ 0 & 1 & -3 & | & -1 \\ 0 & 0 & -2 & | & 2 \end{bmatrix}$

21. $\begin{bmatrix} 1 & 2 & -2 & | & -1 \\ 0 & 3 & -6 & | & 1 \\ 0 & -1 & 2 & | & -\frac{1}{3} \end{bmatrix}$

22. $\begin{bmatrix} 0 & -2 & 8 & | & 1 \\ 2 & -2 & 6 & | & -4 \\ 0 & -1 & 4 & | & \frac{1}{2} \end{bmatrix}$

Solve Problems 23–42 using Gauss–Jordan elimination.

23.
$$2x_1 + 4x_2 - 10x_3 = -2$$
$$3x_1 + 9x_2 - 21x_3 = 0$$
$$x_1 + 5x_2 - 12x_3 = 1$$

24.
$$3x_1 + 5x_2 - x_3 = -7$$
$$x_1 + x_2 + x_3 = -1$$
$$2x_1 + 11x_3 = 7$$

25.
$$3x_1 + 8x_2 - x_3 = -18$$
$$2x_1 + x_2 + 5x_3 = 8$$
$$2x_1 + 4x_2 + 2x_3 = -4$$

26.
$$2x_1 + 6x_2 + 15x_3 = -12$$
$$4x_1 + 7x_2 + 13x_3 = -10$$
$$3x_1 + 6x_2 + 12x_3 = -9$$

27.
$$2x_1 - x_2 - 3x_3 = 8$$
$$x_1 - 2x_2 = 7$$

28.
$$2x_1 + 4x_2 - 6x_3 = 10$$
$$3x_1 + 3x_2 - 3x_3 = 6$$

29.
$$2x_1 - x_2 = 0$$
$$3x_1 + 2x_2 = 7$$
$$x_1 - x_2 = -1$$

30.
$$2x_1 - x_2 = 0$$
$$3x_1 + 2x_2 = 7$$
$$x_1 - x_2 = -2$$

31.
$$3x_1 - 4x_2 - x_3 = 1$$
$$2x_1 - 3x_2 + x_3 = 1$$
$$x_1 - 2x_2 + 3x_3 = 2$$

32.
$$3x_1 + 7x_2 - x_3 = 11$$
$$x_1 + 2x_2 - x_3 = 3$$
$$2x_1 + 4x_2 - 2x_3 = 10$$

33.
$$3x_1 - 2x_2 + x_3 = -7$$
$$2x_1 + x_2 - 4x_3 = 0$$
$$x_1 + x_2 - 3x_3 = 1$$

34.
$$2x_1 + 3x_2 + 5x_3 = 21$$
$$x_1 - x_2 - 5x_3 = -2$$
$$2x_1 + x_2 - x_3 = 11$$

35.
$$2x_1 + 4x_2 - 2x_3 = 2$$
$$-3x_1 - 6x_2 + 3x_3 = -3$$

36.
$$3x_1 - 9x_2 + 12x_3 = 6$$
$$-2x_1 + 6x_2 - 8x_3 = -4$$

37.
$$4x_1 - x_2 + 2x_3 = 3$$
$$-4x_1 + x_2 - 3x_3 = -10$$
$$8x_1 - 2x_2 + 9x_3 = -1$$

38.
$$4x_1 - 2x_2 + 2x_3 = 5$$
$$-6x_1 + 3x_2 - 3x_3 = -2$$
$$10x_1 - 5x_2 + 9x_3 = 4$$

39.
$$2x_1 - 5x_2 - 3x_3 = 7$$
$$-4x_1 + 10x_2 + 2x_3 = 6$$
$$6x_1 - 15x_2 - x_3 = -19$$

40.
$$-4x_1 + 8x_2 + 10x_3 = -6$$
$$6x_1 - 12x_2 - 15x_3 = 9$$
$$-8x_1 + 14x_2 + 19x_3 = -8$$

41.
$$5x_1 - 3x_2 + 2x_3 = 13$$
$$2x_1 - x_2 - 3x_3 = 1$$
$$4x_1 - 2x_2 + 4x_3 = 12$$

42.
$$4x_1 - 2x_2 + 3x_3 = 3$$
$$3x_1 - x_2 - 2x_3 = -10$$
$$2x_1 + 4x_2 - x_3 = -1$$

43. Consider a consistent system of three linear equations in three variables. Discuss the nature of the system and its solution set if the reduced form of the augmented coefficient matrix has:
(A) One leftmost 1
(B) Two leftmost 1's
(C) Three leftmost 1's
(D) Four leftmost 1's

44. Consider a system of three linear equations in three variables. Give examples of two reduced forms that are not row-equivalent if the system is:
(A) Consistent and dependent
(B) Inconsistent

C *Solve Problems 45–50 using Gauss–Jordan elimination.*

45.
$$x_1 + 2x_2 - 4x_3 - x_4 = 7$$
$$2x_1 + 5x_2 - 9x_3 - 4x_4 = 16$$
$$x_1 + 5x_2 - 7x_3 - 7x_4 = 13$$

46.
$$2x_1 + 4x_2 + 5x_3 + 4x_4 = 8$$
$$x_1 + 2x_2 + 2x_3 + x_4 = 3$$

47.
$$x_1 - x_2 + 3x_3 - 2x_4 = 1$$
$$-2x_1 + 4x_2 - 3x_3 + x_4 = 0.5$$
$$3x_1 - x_2 + 10x_3 - 4x_4 = 2.9$$
$$4x_1 - 3x_2 + 8x_3 - 2x_4 = 0.6$$

48.
$$x_1 + x_2 + 4x_3 + x_4 = 1.3$$
$$-x_1 + x_2 - x_3 = 1.1$$
$$2x_1 + x_3 + 3x_4 = -4.4$$
$$2x_1 + 5x_2 + 11x_3 + 3x_4 = 5.6$$

49.
$$x_1 - 2x_2 + x_3 + x_4 + 2x_5 = 2$$
$$-2x_1 + 4x_2 + 2x_3 + 2x_4 - 2x_5 = 0$$
$$3x_1 - 6x_2 + x_3 + x_4 + 5x_5 = 4$$
$$-x_1 + 2x_2 + 3x_3 + x_4 + x_5 = 3$$

50.
$$x_1 - 3x_2 + x_3 + x_4 + 2x_5 = 2$$
$$-x_1 + 5x_2 + 2x_3 + 2x_4 - 2x_5 = 0$$
$$2x_1 - 6x_2 + 2x_3 + 2x_4 + 4x_5 = 4$$
$$-x_1 + 3x_2 - x_3 - x_5 = -3$$

▶ **APPLICATIONS**

Solve all the following problems using Gauss–Jordan elimination.

Business & Economics

51. *Production scheduling.* A small manufacturing plant makes three types of inflatable boats: one-person, two-person, and four-person models. Each boat requires the services of three departments, as listed in the table. The cutting, assembly, and packaging departments have available a maximum of 380, 330, and 120 labor-hours per week, respectively. How many boats of each type must be produced each week for the plant to operate at full capacity?

	ONE-PERSON BOAT	TWO-PERSON BOAT	FOUR-PERSON BOAT
CUTTING DEPARTMENT	0.5 hr	1.0 hr	1.5 hr
ASSEMBLY DEPARTMENT	0.6 hr	0.9 hr	1.2 hr
PACKAGING DEPARTMENT	0.2 hr	0.3 hr	0.5 hr

52. *Production scheduling.* Repeat Problem 51 assuming the cutting, assembly, and packaging departments have available a maximum of 350, 330, and 115 labor-hours per week, respectively.

53. *Production scheduling.* Work Problem 51 assuming the packaging department is no longer used.

54. *Production scheduling.* Work Problem 52 assuming the packaging department is no longer used.

55. *Production scheduling.* Work Problem 51 assuming the four-person boat is no longer produced.

56. *Production scheduling.* Work Problem 52 assuming the four-person boat is no longer produced.

57. *Purchasing.* A chemical manufacturer wants to purchase a fleet of 24 railroad tank cars with a combined carrying capacity of 250,000 gallons. Tank cars with three different carrying capacities are available: 6,000 gallons, 8,000 gallons, and 18,000 gallons. How many of each type of tank car should be purchased?

58. *Purchasing.* A commuter airline wants to purchase a fleet of 30 airplanes with a combined carrying capacity of 960 passengers. The three available types of planes carry 18, 24, and 42 passengers, respectively. How many of each type of plane should be purchased?

59. *Income tax.* A corporation has a taxable income of $7,650,000. At this income level, the federal income tax rate is 50%, the state tax rate is 20%, and the local tax rate is 10%. If each tax rate is applied to the total taxable income, the resulting tax liability for the corporation would be 80% of taxable income. However, it is customary to deduct taxes paid to one agency before computing taxes for the other agencies. Assume that the federal taxes are based on the income that remains after the state and local taxes are deducted, and that state and local taxes are computed in a similar manner. What is the tax liability of the corporation (as a percentage of taxable income) if these deductions are taken into consideration?

60. *Income tax.* Repeat Problem 59 if local taxes are not allowed as a deduction for federal and state taxes.

Life Sciences

61. *Nutrition.* A dietitian in a hospital is to arrange a special diet composed of three basic foods. The diet is to include exactly 340 units of calcium, 180 units of iron, and 220 units of vitamin A. The number of units per ounce of each special ingredient for each of the foods is indicated in the table. How many ounces of each food must be used to meet the diet requirements?

	UNITS PER OUNCE		
	Food A	*Food B*	*Food C*
CALCIUM	30	10	20
IRON	10	10	20
VITAMIN A	10	30	20

62. *Nutrition.* Repeat Problem 61 if the diet is to include exactly 400 units of calcium, 160 units of iron, and 240 units of vitamin A.

63. *Nutrition.* Solve Problem 61 with the assumption that food C is no longer available.

64. *Nutrition.* Solve Problem 62 with the assumption that food C is no longer available.

65. *Nutrition.* Solve Problem 61 with the assumption that the vitamin A requirement is deleted.

66. *Nutrition.* Solve Problem 62 with the assumption that the vitamin A requirement is deleted.

67. *Nutrition—plants.* A farmer can buy four types of plant food. Each barrel of mix *A* contains 30 pounds of phosphoric acid, 50 pounds of nitrogen, and 30 pounds of potash; each barrel of mix *B* contains 30 pounds of phosphoric acid, 75 pounds of nitrogen, and 20 pounds of potash; each barrel of mix *C* contains 30 pounds of phosphoric acid, 25 pounds of nitrogen, and 20 pounds of potash; and each barrel of mix *D* contains 60 pounds of phosphoric acid, 25 pounds of nitrogen, and 50 pounds of potash. Soil tests indicate that a particular field needs 900 pounds of phosphoric acid, 750 pounds of nitrogen, and 700 pounds of potash. How many barrels of each type of food should the farmer mix together to supply the necessary nutrients for the field?

68. *Nutrition—animals.* In a laboratory experiment, rats are to be fed 5 packets of food containing a total of 80 units of vitamin E. There are four different brands of food packets that can be used. A packet of brand *A* contains 5 units of vitamin E, a packet of brand *B* contains 10 units of vitamin E, a packet of brand *C* contains 15 units of vitamin E, and a packet of brand *D* contains 20 units of vitamin E. How many packets of each brand should be mixed and fed to the rats?

Social Sciences

69. *Sociology.* Two sociologists have grant money to study school busing in a particular city. They wish to conduct an opinion survey using 600 telephone contacts and 400 house contacts. Survey company *A* has personnel to do 30 tele-phone and 10 house contacts per hour; survey company *B* can handle 20 telephone and 20 house contacts per hour. How many hours should be scheduled for each firm to pro-duce exactly the number of contacts needed?

70. *Sociology.* Repeat Problem 69 if 650 telephone contacts and 350 house contacts are needed.

71. *Traffic flow.* The rush-hour traffic flow for a network of four one-way streets in a city is shown in the figure. The numbers next to each street indicate the number of vehicles per hour that enter and leave the network on that street. The variables x_1, x_2, x_3, and x_4 represent the flow of traffic be-tween the four intersections in the network.

Figure for 71

(A) For a smooth traffic flow, the number of vehicles en-tering each intersection should always equal the num-ber leaving. For example, since 1,500 vehicles enter

the intersection of 5th Street and Washington Avenue each hour and $x_1 + x_4$ vehicles leave this intersection, we see that $x_1 + x_4 = 1,500$. Find the equations determined by the traf-fic flow at each of the other three intersections.

(B) Find the solution to the system in part (A).

(C) What is the maximum number of vehicles that can travel from Washington Avenue to Lincoln Avenue on 5th Street? What is the minimum number?

(D) If traffic lights are adjusted so that 1,000 vehicles per hour travel from Washington Avenue to Lincoln Ave-nue on 5th Street, determine the flow around the rest of the network.

72. *Traffic flow.* Refer to Problem 71. Closing Washington Av-enue east of 6th Street for construction changes the traffic flow for the network as indicated in the figure. Repeat parts (A)–(D) of Problem 71 for this traffic flow.

Figure for 72

S E C T I O N   4-4   ■   Matrices: Basic Operations

■ ADDITION AND SUBTRACTION
■ PRODUCT OF A NUMBER *k* AND A MATRIX *M*
■ MATRIX PRODUCT

In the last two sections we introduced the important new idea of matrices. In this and the following sections, we will develop this concept further. Matrices are both a very ancient and a very current mathematical concept. References to matrices and sys-tems of equations can be found in Chinese manuscripts dating back to around 200 B.C. More recently, the advent of computers has made matrices a very useful tool for a wide variety of applications. Most graphing calculators and computers are capable of performing calculations with matrices.

As we will see, matrix addition and multiplication are similar to real number addition and multiplication in many respects, but there are some important differ-ences.

■ ADDITION AND SUBTRACTION

Before we can discuss arithmetic operations for matrices, we have to define equality for matrices. Two matrices are **equal** if they have the same size and their corresponding elements are equal. For example,

$$
\overset{2 \times 3}{\begin{bmatrix} a & b & c \\ d & e & f \end{bmatrix}} = \overset{2 \times 3}{\begin{bmatrix} u & v & w \\ x & y & z \end{bmatrix}} \quad \text{if and only if} \quad \begin{matrix} a = u & b = v & c = w \\ d = x & e = y & f = z \end{matrix}
$$

The **sum of two matrices of the same size** is the matrix with elements that are the sum of the corresponding elements of the two given matrices.

Addition is not defined for matrices of different sizes.

EXAMPLE 1 ➤ Matrix Addition

(A) $\begin{bmatrix} a & b \\ c & d \end{bmatrix} + \begin{bmatrix} w & x \\ y & z \end{bmatrix} = \begin{bmatrix} (a + w) & (b + x) \\ (c + y) & (d + z) \end{bmatrix}$

(B) $\begin{bmatrix} 2 & -3 & 0 \\ 1 & 2 & -5 \end{bmatrix} + \begin{bmatrix} 3 & 1 & 2 \\ -3 & 2 & 5 \end{bmatrix} = \begin{bmatrix} 5 & -2 & 2 \\ -2 & 4 & 0 \end{bmatrix}$

(C) $\begin{bmatrix} 5 & 0 & -2 \\ 1 & -3 & 8 \end{bmatrix} + \begin{bmatrix} -1 & 7 \\ 0 & 6 \\ -2 & 8 \end{bmatrix}$ Not defined

◀

MATCHED PROBLEM 1 ➤ Add: $\begin{bmatrix} 3 & 2 \\ -1 & -1 \\ 0 & 3 \end{bmatrix} + \begin{bmatrix} -2 & 3 \\ 1 & -1 \\ 2 & -2 \end{bmatrix}$

◀

FIGURE 1
Addition on a graphing
calculator

Graphing calculators and computers can be used to solve problems involving matrix operations. Figure 1 illustrates the solution to Example 1B on a graphing calculator.

Because we add two matrices by adding their corresponding elements, it follows from the properties of real numbers that matrices of the same size are commutative and associative relative to addition. That is, if A, B, and C are matrices of the same size, then

Commutative: $\qquad A + B = B + A$

Associative: $\quad (A + B) + C = A + (B + C)$

A matrix with elements that are all zeros is called a **zero matrix.** For example,

$$
\begin{bmatrix} 0 & 0 & 0 \end{bmatrix} \qquad \begin{bmatrix} 0 & 0 \\ 0 & 0 \end{bmatrix} \qquad \begin{bmatrix} 0 \\ 0 \\ 0 \\ 0 \end{bmatrix} \qquad \begin{bmatrix} 0 & 0 & 0 & 0 \\ 0 & 0 & 0 & 0 \\ 0 & 0 & 0 & 0 \end{bmatrix}
$$

are zero matrices of different sizes. [*Note:* The simpler notation "0" is often used to denote the zero matrix of an arbitrary size.] The **negative of a matrix M,** denoted by $-M$, is a matrix with elements that are the negatives of the elements in M. Thus, if

$$M = \begin{bmatrix} a & b \\ c & d \end{bmatrix} \quad \text{then} \quad -M = \begin{bmatrix} -a & -b \\ -c & -d \end{bmatrix}$$

Note that $M + (-M) = 0$ (a zero matrix).

If A and B are matrices of the same size, then we define **subtraction** as follows:

$$A - B = A + (-B)$$

Thus, to subtract matrix B from matrix A, we simply add the negative of B to A.

EXAMPLE 2 ➤ Matrix Subtraction

$$\begin{bmatrix} 3 & -2 \\ 5 & 0 \end{bmatrix} - \begin{bmatrix} -2 & 2 \\ 3 & 4 \end{bmatrix} = \begin{bmatrix} 3 & -2 \\ 5 & 0 \end{bmatrix} + \begin{bmatrix} 2 & -2 \\ -3 & -4 \end{bmatrix} = \begin{bmatrix} 5 & -4 \\ 2 & -4 \end{bmatrix}$$ ◀

MATCHED PROBLEM 2 ➤ Subtract: $[2 \quad -3 \quad 5] - [3 \quad -2 \quad 1]$ ◀

■ PRODUCT OF A NUMBER k AND A MATRIX M

The **product of a number k and a matrix M,** denoted by kM, is a matrix formed by multiplying each element of M by k.

EXAMPLE 3 ➤ Multiplication of a Matrix by a Number

$$-2 \begin{bmatrix} 3 & -1 & 0 \\ -2 & 1 & 3 \\ 0 & -1 & -2 \end{bmatrix} = \begin{bmatrix} -6 & 2 & 0 \\ 4 & -2 & -6 \\ 0 & 2 & 4 \end{bmatrix}$$ ◀

MATCHED PROBLEM 3 ➤ Find: $10 \begin{bmatrix} 1.3 \\ 0.2 \\ 3.5 \end{bmatrix}$ ◀

explore – discuss 1

Multiplication of two numbers can be interpreted as repeated addition if one of the numbers is a positive integer. That is,

$$2a = a + a \qquad 3a = a + a + a \qquad 4a = a + a + a + a$$

and so on. Discuss this interpretation for the product of an integer k and a matrix M. Use specific examples to illustrate your remarks.

The next example illustrates the use of matrix operations in an applied setting.

EXAMPLE 4 ➤ **Sales Commissions** Ms. Smith and Mr. Jones are salespeople in a new-car agency that sells only two models. August was the last month for this year's models, and next year's models were introduced in September. Gross dollar sales for each month are given in the following matrices:

$$
\begin{array}{c}
\text{August sales}\\
\begin{array}{cc}
\quad\ Compact & Luxury
\end{array}\\
\begin{array}{c}
\text{Ms. Smith}\\
\text{Mr. Jones}
\end{array}
\begin{bmatrix}
\$54{,}000 & \$88{,}000\\
\$126{,}000 & 0
\end{bmatrix} = A
\end{array}
\qquad
\begin{array}{c}
\text{September sales}\\
\begin{array}{cc}
Compact & Luxury
\end{array}\\
\begin{bmatrix}
\$228{,}000 & \$368{,}000\\
\$304{,}000 & \$322{,}000
\end{bmatrix} = B
\end{array}
$$

(For example, Ms. Smith had $54,000 in compact sales in August, and Mr. Jones had $322,000 in luxury car sales in September.)

(A) What were the combined dollar sales in August and September for each salesperson and each model?

(B) What was the increase in dollar sales from August to September?

(C) If both salespeople receive 5% commissions on gross dollar sales, compute the commission for each person for each model sold in September.

Solution

(A) $A + B =
\begin{array}{c}
\begin{array}{cc}
\ Compact & Luxury
\end{array}\\
\begin{bmatrix}
\$282{,}000 & \$456{,}000\\
\$430{,}000 & \$322{,}000
\end{bmatrix}
\begin{array}{c}
\text{Ms. Smith}\\
\text{Mr. Jones}
\end{array}
\end{array}$

(B) $B - A =
\begin{array}{c}
\begin{array}{cc}
\ Compact & Luxury
\end{array}\\
\begin{bmatrix}
\$174{,}000 & \$280{,}000\\
\$178{,}000 & \$322{,}000
\end{bmatrix}
\begin{array}{c}
\text{Ms. Smith}\\
\text{Mr. Jones}
\end{array}
\end{array}$

(C) $0.05B =
\begin{bmatrix}
(0.05)(\$228{,}000) & (0.05)(\$368{,}000)\\
(0.05)(\$304{,}000) & (0.05)(\$322{,}000)
\end{bmatrix}$

$\qquad =
\begin{bmatrix}
\$11{,}400 & \$18{,}400\\
\$15{,}200 & \$16{,}100
\end{bmatrix}
\begin{array}{c}
\text{Ms. Smith}\\
\text{Mr. Jones}
\end{array}$

MATCHED PROBLEM 4 ➤ Repeat Example 4 with

$$
A =
\begin{bmatrix}
\$45{,}000 & \$77{,}000\\
\$106{,}000 & \$22{,}000
\end{bmatrix}
\quad \text{and} \quad
B =
\begin{bmatrix}
\$190{,}000 & \$345{,}000\\
\$266{,}000 & \$276{,}000
\end{bmatrix}
$$

Example 4 involved an agency with only 2 salespeople and 2 models. A more realistic problem might involve 20 salespeople and 15 models. Problems of this size are often solved with the aid of an electronic spreadsheet program on a personal computer. Figure 2 illustrates such a computer solution for Example 4.

	1	2	3	4	5	6	7
1		August Sales		September Sales		September Commissions	
2		Compact	Luxury	Compact	Luxury	Compact	Luxury
3	Smith	$54,000	$88,000	$228,000	$368,000	$11,400	$18,400
4	Jones	$126,000	$0	$304,000	$322,000	$15,200	$16,100
5		Combined Sales		Sales Increase			
6	Smith	$282,000	$456,000	$174,000	$280,000		
7	Jones	$430,000	$322,000	$178,000	$322,000		

FIGURE 2

■ MATRIX PRODUCT

Now we are going to introduce a matrix multiplication that will seem rather strange at first. Despite this apparent strangeness, this operation is well-founded in the general theory of matrices and, as we will see, is extremely useful in many practical problems.

Historically, matrix multiplication was introduced by the English mathematician Arthur Cayley (1821–1895) in studies of systems of linear equations and linear transformations. In Section 4-6, you will see how matrix multiplication is central to the process of expressing systems of linear equations as matrix equations and to the process of solving matrix equations. Matrix equations and their solutions provide us with an alternate method of solving linear systems with the same number of variables as equations.

We start by defining the product of two special matrices, a row matrix and a column matrix.

Product of a Row Matrix and a Column Matrix

The **product** of a $1 \times n$ row matrix and an $n \times 1$ column matrix is a 1×1 matrix given by

$$
\underset{1 \times n}{[a_1 \quad a_2 \cdots a_n]}
\underset{n \times 1}{\begin{bmatrix} b_1 \\ b_2 \\ \vdots \\ b_n \end{bmatrix}}
= [a_1 b_1 + a_2 b_2 + \cdots + a_n b_n]
$$

Note that the number of elements in the row matrix and in the column matrix must be the same for the product to be defined.

EXAMPLE 5 ➤ Product of a Row Matrix and a Column Matrix

$$
[2 \quad -3 \quad 0] \begin{bmatrix} -5 \\ 2 \\ -2 \end{bmatrix} = [(2)(-5) + (-3)(2) + (0)(-2)]
$$
$$
= [-10 - 6 + 0] = [-16] \quad ◀
$$

MATCHED PROBLEM 5 ➤

$$
[-1 \quad 0 \quad 3 \quad 2] \begin{bmatrix} 2 \\ 3 \\ 4 \\ -1 \end{bmatrix} = ? \quad ◀
$$

Refer to Example 5. The distinction between the real number -16 and the 1×1 matrix $[-16]$ is a technical one, and it is common to see 1×1 matrices written as real numbers without brackets. In the work that follows, we will frequently refer to 1×1 matrices as real numbers and omit the brackets whenever it is convenient to do so.

EXAMPLE 6 ➤ **Labor Costs** A factory produces a slalom water ski that requires 3 labor-hours in the assembly department and 1 labor-hour in the finishing department. Assembly personnel receive $9 per hour and finishing personnel receive $6 per hour. Total labor cost per ski is given by the product:

$$[3 \quad 1] \begin{bmatrix} 9 \\ 6 \end{bmatrix} = [(3)(9) + (1)(6)] = [27 + 6] = [33] \quad \text{or } \$33 \text{ per ski} \quad ◀$$

MATCHED PROBLEM 6 ➤ If the factory in Example 6 also produces a trick water ski that requires 5 labor-hours in the assembly department and 1.5 labor-hours in the finishing department, write a product between appropriate row and column matrices that will give the total labor cost for this ski. Compute the cost. ◀

We now use the product of a $1 \times n$ row matrix and an $n \times 1$ column matrix to extend the definition of matrix product to more general matrices.

Matrix Product

If A is an $m \times p$ matrix and B is a $p \times n$ matrix, then the **matrix product** of A and B, denoted AB, is an $m \times n$ matrix whose element in the ith row and jth column is the real number obtained from the product of the ith row of A and the jth column of B. If the number of columns in A does not equal the number of rows in B, then the matrix product AB is **not defined.**

It is important to check sizes before starting the multiplication process. If A is an $a \times b$ matrix and B is a $c \times d$ matrix, then if $b = c$, the product AB will exist and will be an $a \times d$ matrix (see Fig. 3). If $b \neq c$, then the product AB does not exist. The definition is not as complicated as it might first seem. An example should help clarify the process.

For

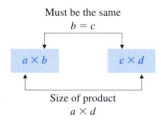

Must be the same
$b = c$

Size of product
$a \times d$

FIGURE 3

$$A = \begin{bmatrix} 2 & 3 & -1 \\ -2 & 1 & 2 \end{bmatrix} \quad \text{and} \quad B = \begin{bmatrix} 1 & 3 \\ 2 & 0 \\ -1 & 2 \end{bmatrix}$$

A is 2×3 and B is 3×2, so AB is 2×2. To find the first row of AB, we take the product of the first row of A with every column of B and write each result as a real number, not as a 1×1 matrix. The second row of AB is computed in the same manner. The four products of row and column matrices used to produce the four elements in AB are shown in the following dashed box. These products are usually calculated mentally or with the aid of a calculator, and need not be written out. The shaded portions highlight the steps involved in computing the element in the first row and second column of AB.

$$
\underset{2 \times 3}{\begin{bmatrix} 2 & 3 & -1 \\ -2 & 1 & 2 \end{bmatrix}} \underset{3 \times 2}{\begin{bmatrix} 1 & 3 \\ 2 & 0 \\ -1 & 2 \end{bmatrix}} = \begin{bmatrix} \begin{bmatrix} 2 & 3 & -1 \end{bmatrix}\begin{bmatrix} 1 \\ 2 \\ -1 \end{bmatrix} & \begin{bmatrix} 2 & 3 & -1 \end{bmatrix}\begin{bmatrix} 3 \\ 0 \\ 2 \end{bmatrix} \\ \begin{bmatrix} -2 & 1 & 2 \end{bmatrix}\begin{bmatrix} 1 \\ 2 \\ -1 \end{bmatrix} & \begin{bmatrix} -2 & 1 & 2 \end{bmatrix}\begin{bmatrix} 3 \\ 0 \\ 2 \end{bmatrix} \end{bmatrix} = \underset{2 \times 2}{\begin{bmatrix} 9 & 4 \\ -2 & -2 \end{bmatrix}}
$$

EXAMPLE 7 ➤ Matrix Multiplication

(A) $\underset{3 \times 2}{\begin{bmatrix} 2 & 1 \\ 1 & 0 \\ -1 & 2 \end{bmatrix}} \underset{2 \times 4}{\begin{bmatrix} 1 & -1 & 0 & 1 \\ 2 & 1 & 2 & 0 \end{bmatrix}} = \underset{3 \times 4}{\begin{bmatrix} 4 & -1 & 2 & 2 \\ 1 & -1 & 0 & 1 \\ 3 & 3 & 4 & -1 \end{bmatrix}}$

(B) $\underset{2 \times 4}{\begin{bmatrix} 1 & -1 & 0 & 1 \\ 2 & 1 & 2 & 0 \end{bmatrix}} \underset{3 \times 2}{\begin{bmatrix} 2 & 1 \\ 1 & 0 \\ -1 & 2 \end{bmatrix}}$

Not defined.

(C) $\begin{bmatrix} 2 & 6 \\ -1 & -3 \end{bmatrix}\begin{bmatrix} 1 & 2 \\ 3 & 6 \end{bmatrix} = \begin{bmatrix} 20 & 40 \\ -10 & -20 \end{bmatrix}$

(D) $\begin{bmatrix} 1 & 2 \\ 3 & 6 \end{bmatrix}\begin{bmatrix} 2 & 6 \\ -1 & -3 \end{bmatrix} = \begin{bmatrix} 0 & 0 \\ 0 & 0 \end{bmatrix}$ (E) $\begin{bmatrix} 2 & -3 & 0 \end{bmatrix}\begin{bmatrix} -5 \\ 2 \\ -2 \end{bmatrix} = \begin{bmatrix} -16 \end{bmatrix}$

(F) $\begin{bmatrix} -5 \\ 2 \\ -2 \end{bmatrix}\begin{bmatrix} 2 & -3 & 0 \end{bmatrix} = \begin{bmatrix} -10 & 15 & 0 \\ 4 & -6 & 0 \\ -4 & 6 & 0 \end{bmatrix}$ ◄

MATCHED PROBLEM 7 ➤ Find each product, if it is defined:

(A) $\begin{bmatrix} -1 & 0 & 3 & -2 \\ 1 & 2 & 2 & 0 \end{bmatrix}\begin{bmatrix} -1 & 1 \\ 2 & 3 \\ 1 & 0 \end{bmatrix}$

(B) $\begin{bmatrix} -1 & 1 \\ 2 & 3 \\ 1 & 0 \end{bmatrix}\begin{bmatrix} -1 & 0 & 3 & -2 \\ 1 & 2 & 2 & 0 \end{bmatrix}$

(C) $\begin{bmatrix} 1 & 2 \\ -1 & -2 \end{bmatrix} \begin{bmatrix} -2 & 4 \\ 1 & -2 \end{bmatrix}$ (D) $\begin{bmatrix} -2 & 4 \\ 1 & -2 \end{bmatrix} \begin{bmatrix} 1 & 2 \\ -1 & -2 \end{bmatrix}$

(E) $\begin{bmatrix} 3 & -2 & 1 \end{bmatrix} \begin{bmatrix} 4 \\ 2 \\ 3 \end{bmatrix}$ (F) $\begin{bmatrix} 4 \\ 2 \\ 3 \end{bmatrix} \begin{bmatrix} 3 & -2 & 1 \end{bmatrix}$ ◄

FIGURE 4
Multiplication on a graphing calculator

Figure 4 illustrates a graphing calculator solution to Example 7A. What would you expect to happen if you tried to solve Example 7B on a graphing calculator?

In the arithmetic of real numbers it does not matter in which order we multiply; for example, $5 \times 7 = 7 \times 5$. In matrix multiplication, however, it does make a difference. That is, AB does not always equal BA, even if both multiplications are defined and both products are the same size (see Examples 7C and 7D). Thus:

Matrix multiplication is not commutative.

Also, AB may be zero with neither A nor B equal to zero (see Example 7D). Thus:

The zero property does not hold for matrix multiplication.

In addition to the commutative and zero properties, there are other significant differences between real number multiplication and matrix multiplication.

(A) In real number multiplication, the only real number whose square is 0 is the real number 0 ($0^2 = 0$). Find at least one 2×2 matrix A with all elements nonzero such that $A^2 = 0$, where 0 is the 2×2 zero matrix.
(B) In real number multiplication, the only nonzero real number that is equal to its square is the real number 1 ($1^2 = 1$). Find at least one 2×2 matrix A with all elements nonzero such that $A^2 = A$.

We will continue our discussion of properties of matrix multiplication later in this chapter. Now we consider an application of matrix multiplication.

EXAMPLE 8 ➤ Labor Costs Let us combine the time requirements for slalom and trick water skis discussed in Example 6 and Matched Problem 6 into one matrix:

Labor-hours per ski

	Assembly department	Finishing department	
Trick ski	5 hr	1.5 hr	$= L$
Slalom ski	3 hr	1 hr	

Now suppose the company has two manufacturing plants, one in California and the other in Wisconsin, and that their hourly rates for each department are given in the following matrix:

Hourly wages

$$
\begin{array}{c}
 & \text{California} \quad \text{Wisconsin} \\
\begin{matrix} \text{Assembly department} \\ \text{Finishing department} \end{matrix}
\begin{bmatrix} \$12 & \$9 \\ \$8 & \$6 \end{bmatrix} = H
\end{array}
$$

Since H and L are both 2×2 matrices, we can take the product of H and L in either order and the result will be a 2×2 matrix:

$$
HL = \begin{bmatrix} 12 & 9 \\ 8 & 6 \end{bmatrix} \begin{bmatrix} 5 & 1.5 \\ 3 & 1 \end{bmatrix} = \begin{bmatrix} 87 & 27 \\ 58 & 18 \end{bmatrix}
$$

$$
LH = \begin{bmatrix} 5 & 1.5 \\ 3 & 1 \end{bmatrix} \begin{bmatrix} 12 & 9 \\ 8 & 6 \end{bmatrix} = \begin{bmatrix} 72 & 54 \\ 44 & 33 \end{bmatrix}
$$

How can we interpret the elements in these products? Let's begin with the product HL. The element 87 in the first row and first column of HL is the product of the first row matrix of H and the first column matrix of L:

$$
\begin{array}{c}
\text{CA} \quad \text{WI} \\
[12 \quad 9\,]
\end{array}
\begin{bmatrix} 5 \\ 3 \end{bmatrix}
\begin{matrix} \text{Trick} \\ \text{Slalom} \end{matrix}
= 12(5) + 9(3) = 60 + 27 = 87
$$

Notice that $60 is the labor cost for assembling a trick ski at the California plant and $27 is the labor cost for assembling a slalom ski at the Wisconsin plant. Although both numbers represent labor costs, it makes no sense to add them together. They do not pertain to the same type of ski or to the same plant. Thus, even though the product HL happens to be defined mathematically, it has no useful interpretation in this problem.

Now let's consider the product LH. The element 72 in the first row and first column of LH is given by the following product:

$$
\begin{array}{c}
\text{Assembly} \quad \text{Finishing} \\
[5 \qquad\qquad 1.5 \qquad\quad]
\end{array}
\begin{bmatrix} 12 \\ 8 \end{bmatrix}
\begin{matrix} \text{Assembly} \\ \text{Finishing} \end{matrix}
= 5(12) + 1.5(8) = 60 + 12 = 72
$$

This time, $60 is the labor cost for assembling a trick ski at the California plant and $12 is the labor cost for finishing a trick ski at the California plant. Thus, the sum is the total labor cost for producing a trick ski at the California plant. The other elements in LH also represent total labor costs, as indicated by the row and column labels shown below:

Labor costs per ski

$$
\begin{array}{c}
 & \text{CA} \quad \text{WI} \\
LH = \begin{bmatrix} \$72 & \$54 \\ \$44 & \$33 \end{bmatrix}
\begin{matrix} \text{Trick} \\ \text{Slalom} \end{matrix}
\end{array}
$$

MATCHED PROBLEM 8 ➤ Refer to Example 8. The company wants to know how many hours to schedule in each department in order to produce 2,000 trick skis and 1,000 slalom skis. These production requirements can be represented by either of the following matrices:

$$
\begin{matrix} \text{Trick} & \text{Slalom} \\ \text{skis} & \text{skis} \end{matrix}
$$
$$
P = [\,2{,}000 \quad 1{,}000\,] \qquad Q = \begin{bmatrix} 2{,}000 \\ 1{,}000 \end{bmatrix} \begin{matrix} \text{Trick skis} \\ \text{Slalom skis} \end{matrix}
$$

Using the labor-hour matrix L from Example 8, find PL or LQ, whichever has a meaningful interpretation for this problem, and label the rows and columns accordingly. ◄

Caution

Example 8 and Matched Problem 8 illustrate an important point about matrix multiplication. Even if you are using a graphing calculator or a computer to perform the calculations in a matrix product, it is still necessary for you to know the definition of matrix multiplication so that you can interpret the results correctly.

Answers to Matched Problems

1. $\begin{bmatrix} 1 & 5 \\ 0 & -2 \\ 2 & 1 \end{bmatrix}$ 2. $[-1 \quad -1 \quad 4]$ 3. $\begin{bmatrix} 13 \\ 2 \\ 35 \end{bmatrix}$

4. (A) $\begin{bmatrix} \$235{,}000 & \$422{,}000 \\ \$372{,}000 & \$298{,}000 \end{bmatrix}$ (B) $\begin{bmatrix} \$145{,}000 & \$268{,}000 \\ \$160{,}000 & \$254{,}000 \end{bmatrix}$

(C) $\begin{bmatrix} \$9{,}500 & \$17{,}250 \\ \$13{,}300 & \$13{,}800 \end{bmatrix}$

5. $[8]$ 6. $[5 \quad 1.5]\begin{bmatrix} 9 \\ 6 \end{bmatrix} = [54]$, or \$54

7. (A) Not defined (B) $\begin{bmatrix} 2 & 2 & -1 & 2 \\ 1 & 6 & 12 & -4 \\ -1 & 0 & 3 & -2 \end{bmatrix}$ (C) $\begin{bmatrix} 0 & 0 \\ 0 & 0 \end{bmatrix}$

(D) $\begin{bmatrix} -6 & -12 \\ 3 & 6 \end{bmatrix}$ (E) $[11]$ (F) $\begin{bmatrix} 12 & -8 & 4 \\ 6 & -4 & 2 \\ 9 & -6 & 3 \end{bmatrix}$

8. $PL = \begin{matrix} \text{Assembly} & \text{Finishing} \\ [13{,}000 & 4{,}000] \end{matrix}$ Labor-hours

EXERCISE 4-4

A *Perform the indicated operations in Problems 1–18, if possible.*

1. $\begin{bmatrix} 2 & -1 \\ 3 & 0 \end{bmatrix} + \begin{bmatrix} -3 & 1 \\ 2 & -3 \end{bmatrix}$

2. $\begin{bmatrix} -1 & 4 \\ 2 & -6 \end{bmatrix} + \begin{bmatrix} 1 & -2 \\ 0 & 5 \end{bmatrix}$

3. $\begin{bmatrix} 4 & -1 & 0 \\ 2 & 1 & 3 \end{bmatrix} + \begin{bmatrix} -2 & 1 & 3 \\ 5 & 6 & -8 \end{bmatrix}$

4. $\begin{bmatrix} -3 & 5 \\ 2 & 0 \\ 1 & 4 \end{bmatrix} + \begin{bmatrix} 2 & 1 \\ -6 & 3 \\ 0 & -5 \end{bmatrix}$

5. $\begin{bmatrix} 4 & -1 & 0 \\ 2 & 1 & 3 \end{bmatrix} + \begin{bmatrix} 2 & 1 \\ -6 & 3 \\ 0 & -5 \end{bmatrix}$

6. $\begin{bmatrix} -3 & 5 \\ 2 & 0 \\ 1 & 4 \end{bmatrix} + \begin{bmatrix} -2 & 1 & 3 \\ 5 & 6 & -8 \end{bmatrix}$

7. $\begin{bmatrix} 4 & -5 \\ 1 & 0 \\ 1 & -3 \end{bmatrix} - \begin{bmatrix} -1 & 2 \\ 6 & -2 \\ 1 & -7 \end{bmatrix}$

8. $\begin{bmatrix} 6 & 2 & -3 \\ 0 & -4 & 5 \end{bmatrix} - \begin{bmatrix} 4 & -1 & 2 \\ -5 & 1 & -2 \end{bmatrix}$

9. $5\begin{bmatrix} 1 & -2 & 0 & 4 \\ -3 & 2 & -1 & 6 \end{bmatrix}$

10. $10\begin{bmatrix} 2 & -1 & 3 \\ 0 & -4 & 5 \end{bmatrix}$

11. $[2 \quad 4]\begin{bmatrix} 3 \\ 1 \end{bmatrix}$

12. $[1 \quad 5]\begin{bmatrix} 6 \\ 2 \end{bmatrix}$

13. $\begin{bmatrix} 3 & 4 \\ -1 & -2 \end{bmatrix}\begin{bmatrix} -1 \\ 2 \end{bmatrix}$

14. $\begin{bmatrix} -1 & 1 \\ 2 & -3 \end{bmatrix}\begin{bmatrix} 4 \\ -2 \end{bmatrix}$

15. $\begin{bmatrix} 2 & -3 \\ 1 & 2 \end{bmatrix}\begin{bmatrix} 1 & -1 \\ 0 & -2 \end{bmatrix}$

16. $\begin{bmatrix} -3 & 2 \\ 4 & -1 \end{bmatrix}\begin{bmatrix} -2 & 5 \\ -1 & 3 \end{bmatrix}$

17. $\begin{bmatrix} 1 & -1 \\ 0 & -2 \end{bmatrix}\begin{bmatrix} 2 & -3 \\ 1 & 2 \end{bmatrix}$

18. $\begin{bmatrix} -2 & 5 \\ -1 & 3 \end{bmatrix}\begin{bmatrix} -3 & 2 \\ 4 & -1 \end{bmatrix}$

B *Find the products in Problems 19–26.*

19. $[5 \quad -2]\begin{bmatrix} -3 \\ -4 \end{bmatrix}$

20. $[-4 \quad 3]\begin{bmatrix} -2 \\ 1 \end{bmatrix}$

21. $\begin{bmatrix} -3 \\ -4 \end{bmatrix}[5 \quad -2]$

22. $\begin{bmatrix} -2 \\ 1 \end{bmatrix}[-4 \quad 3]$

23. $[3 \quad -2 \quad -4]\begin{bmatrix} 1 \\ 2 \\ -3 \end{bmatrix}$

24. $[1 \quad -2 \quad 2]\begin{bmatrix} 2 \\ -1 \\ 1 \end{bmatrix}$

25. $\begin{bmatrix} 1 \\ 2 \\ -3 \end{bmatrix}[3 \quad -2 \quad -4]$

26. $\begin{bmatrix} 2 \\ -1 \\ 1 \end{bmatrix}[1 \quad -2 \quad 2]$

Problems 27–44 refer to the following matrices:

$$A = \begin{bmatrix} 2 & -1 & 3 \\ 0 & 4 & -2 \end{bmatrix} \qquad B = \begin{bmatrix} -3 & 1 \\ 2 & 5 \end{bmatrix}$$

$$C = \begin{bmatrix} -1 & 0 & 2 \\ 4 & -3 & 1 \\ -2 & 3 & 5 \end{bmatrix} \qquad D = \begin{bmatrix} 3 & -2 \\ 0 & -1 \\ 1 & 2 \end{bmatrix}$$

Perform the indicated operations, if possible.

27. AC **28.** CA **29.** AB

30. BA **31.** $B^2 = BB$ **32.** $C^2 = CC$

33. $B + AD$ **34.** $C + DA$ **35.** $(0.1)DB$

36. $(0.2)CD$ **37.** $(3)BA + (4)AC$

38. $(2)DB + (5)CD$ **39.** $(-2)BA + (6)CD$

40. $(-1)AC + (3)DB$ **41.** ACD

42. CDA **43.** DBA

44. BAD

45. Find a, b, c, and d so that

$$\begin{bmatrix} a & b \\ c & d \end{bmatrix} + \begin{bmatrix} 2 & -3 \\ 0 & 1 \end{bmatrix} = \begin{bmatrix} 1 & -2 \\ 3 & -4 \end{bmatrix}$$

46. Find w, x, y, and z so that

$$\begin{bmatrix} 4 & -2 \\ -3 & 0 \end{bmatrix} + \begin{bmatrix} w & x \\ y & z \end{bmatrix} = \begin{bmatrix} 2 & -3 \\ 0 & 5 \end{bmatrix}$$

47. Find x and y so that

$$\begin{bmatrix} 2x & 4 \\ -3 & 5x \end{bmatrix} + \begin{bmatrix} 3y & -2 \\ -2 & -y \end{bmatrix} = \begin{bmatrix} -5 & 2 \\ -5 & 13 \end{bmatrix}$$

48. Find x and y so that

$$\begin{bmatrix} 5 & 3x \\ 2x & -4 \end{bmatrix} + \begin{bmatrix} 1 & -4y \\ 7y & 4 \end{bmatrix} = \begin{bmatrix} 6 & -7 \\ 5 & 0 \end{bmatrix}$$

C

49. Find x and y so that

$$\begin{bmatrix} x & -1 \\ 1 & 0 \end{bmatrix}\begin{bmatrix} 2 & 1 \\ 4 & 1 \end{bmatrix} = \begin{bmatrix} y & y \\ 2 & 1 \end{bmatrix}$$

50. Find x and y so that

$$\begin{bmatrix} 1 & 3 \\ -2 & -2 \end{bmatrix}\begin{bmatrix} x & 1 \\ 3 & 2 \end{bmatrix} = \begin{bmatrix} y & 7 \\ y & -6 \end{bmatrix}$$

51. Find a, b, c, and d so that

$$\begin{bmatrix} 1 & -2 \\ 2 & -3 \end{bmatrix}\begin{bmatrix} a & b \\ c & d \end{bmatrix} = \begin{bmatrix} 1 & 0 \\ 3 & 2 \end{bmatrix}$$

52. Find *a*, *b*, *c*, and *d* so that

$$\begin{bmatrix} 1 & 3 \\ 1 & 4 \end{bmatrix} \begin{bmatrix} a & b \\ c & d \end{bmatrix} = \begin{bmatrix} 6 & -5 \\ 7 & -7 \end{bmatrix}$$

53. A square matrix is a **diagonal matrix** if all elements not on the principal diagonal are zero. Thus, a 2 × 2 diagonal matrix has the form

$$A = \begin{bmatrix} a & 0 \\ 0 & d \end{bmatrix}$$

where *a* and *d* are any real numbers. Discuss the validity of each of the following statements. If the statement is always true, explain why. If not, give examples.
(A) If *A* and *B* are 2 × 2 diagonal matrices, then *A* + *B* is a 2 × 2 diagonal matrix.
(B) If *A* and *B* are 2 × 2 diagonal matrices, then *A* + *B* = *B* + *A*.
(C) If *A* and *B* are 2 × 2 diagonal matrices, then *AB* is a 2 × 2 diagonal matrix.

(D) If *A* and *B* are 2 × 2 diagonal matrices, then *AB* = *BA*.

54. A square matrix is an **upper triangular matrix** if all elements below the principal diagonal are zero. Thus, a 2 × 2 upper triangular matrix has the form

$$A = \begin{bmatrix} a & b \\ 0 & d \end{bmatrix}$$

where *a*, *b*, and *d* are any real numbers. Discuss the validity of each of the following statements. If the statement is always true, explain why. If not, give examples.
(A) If *A* and *B* are 2 × 2 upper triangular matrices, then *A* + *B* is a 2 × 2 upper triangular matrix.
(B) If *A* and *B* are 2 × 2 upper triangular matrices, then *A* + *B* = *B* + *A*.
(C) If *A* and *B* are 2 × 2 upper triangular matrices, then *AB* is a 2 × 2 upper triangular matrix.
(D) If *A* and *B* are 2 × 2 upper triangular matrices, then *AB* = *BA*.

▶ **APPLICATIONS**

Business & Economics

55. *Cost analysis.* A company with two different plants manufactures guitars and banjos. Its production costs for each instrument are given in the following matrices:

	Plant X		Plant Y	
	Guitar	Banjo	Guitar	Banjo
Materials	$30	$25	$36	$27
Labor	$60	$80	$54	$74

$$\begin{bmatrix} \$30 & \$25 \\ \$60 & \$80 \end{bmatrix} = A \qquad \begin{bmatrix} \$36 & \$27 \\ \$54 & \$74 \end{bmatrix} = B$$

Find $\frac{1}{2}(A + B)$, the average cost of production for the two plants.

56. *Cost analysis.* If both labor and materials at plant *X* in Problem 55 are increased by 20%, find $\frac{1}{2}(1.2A + B)$, the new average cost of production for the two plants.

57. *Markup.* An import car dealer sells three models of a car. The retail prices and the current dealer invoice prices (costs) for the basic models and indicated options are given in the following two matrices (where "Air" means airconditioning):

	Retail price			
	Basic car	Air	AM/FM radio	Cruise control
Model A	$10,900	$683	$253	$195
Model B	$13,000	$738	$382	$206
Model C	$16,300	$867	$537	$225

$$= M$$

	Dealer invoice price			
	Basic car	Air	AM/FM radio	Cruise control
Model A	$9,400	$582	$195	$160
Model B	$11,500	$621	$295	$171
Model C	$14,100	$737	$420	$184

$$= N$$

We define the markup matrix to be *M* − *N* (**markup** is the difference between the retail price and the dealer invoice price). Suppose the value of the dollar has had a sharp decline and the dealer invoice price is to have an across-the-board 15% increase next year. In order to stay competitive with domestic cars, the dealer increases the retail prices only 10%. Calculate a markup matrix for next year's models and the indicated options. (Compute results to the nearest dollar.)

58. *Markup.* Referring to Problem 57, what is the markup matrix resulting from a 20% increase in dealer invoice prices and an increase in retail prices of 15%? (Compute results to the nearest dollar.)

59. *Labor costs.* A company with manufacturing plants located in different parts of the country has labor-hour and wage requirements for the manufacturing of three types of inflatable boats as given in the following two matrices:

Labor-hours per boat

	Cutting department	Assembly department	Packaging department	
$M =$	0.6 hr	0.6 hr	0.2 hr	One-person boat
	1.0 hr	0.9 hr	0.3 hr	Two-person boat
	1.5 hr	1.2 hr	0.4 hr	Four-person boat

Hourly wages

	Plant I	Plant II	
$N =$	$8	$9	Cutting department
	$10	$12	Assembly department
	$5	$6	Packaging department

(A) Find the labor costs for a one-person boat manufactured at plant I.

(B) Find the labor costs for a four-person boat manufactured at plant II.

(C) Discuss possible interpretations of the elements in the matrix products MN and NM.

(D) If either of the products MN or NM has a meaningful interpretation, find the product and label its rows and columns.

60. *Inventory value.* A personal computer retail company sells five different computer models through three stores located in a large metropolitan area. The inventory of each model

on hand in each store is summarized in matrix M. Wholesale (W) and retail (R) values of each model computer are summarized in matrix N.

Model

	A	B	C	D	E	
$M =$	4	2	3	7	1	Store 1
	2	3	5	0	6	Store 2
	10	4	3	4	3	Store 3

	W	R		
$N =$	$700	$840	A	
	$1,400	$1,800	B	
	$1,800	$2,400	C	Model
	$2,700	$3,300	D	
	$3,500	$4,900	E	

(A) What is the retail value of the inventory at store 2?

(B) What is the wholesale value of the inventory at store 3?

(C) Discuss possible interpretations of the elements in the matrix products MN and NM.

(D) If either of the products MN or NM has a meaningful interpretation, find the product and label its rows and columns.

(E) Discuss methods of matrix multiplication that can be used to find the total inventory of each model on hand at all three stores. State the matrices that can be used and perform the necessary operations.

(F) Discuss methods of matrix multiplication that can be used to find the total inventory of all five models at each store. State the matrices that can be used and perform the necessary operations.

61. *Air freight.* A nationwide air freight service has connecting flights between five cities as illustrated in the diagram. To represent this schedule in matrix form, we construct a 5 × 5 **incidence matrix** A, where the rows represent the origin of each flight and the columns represent the destinations. We place a 1 in the ith row and jth column of this matrix if there is a connecting flight from the ith city to the jth city; otherwise, insert a 0. We also place 0's on the principal diagonal, because a connecting flight with the same origin and destination makes no sense. With the schedule represented in the mathematical form of a matrix, we can perform operations on this matrix to obtain information about the schedule.

Atlanta   Baltimore

①———————→②

③ Chicago

④←—————⑤

Denver   El Paso

Destination

$$
\begin{array}{c}
 & \begin{array}{ccccc} 1 & 2 & 3 & 4 & 5 \end{array} \\
\text{Origin} \begin{array}{c} 1 \\ 2 \\ 3 \\ 4 \\ 5 \end{array} & \left[\begin{array}{ccccc} 0 & 1 & 0 & 1 & 0 \\ 0 & 0 & 1 & 0 & 0 \\ 1 & 0 & 0 & 0 & 1 \\ 0 & 0 & 1 & 0 & 0 \\ 0 & 0 & 0 & 1 & 0 \end{array}\right] = A
\end{array}
$$

(A) Find A^2. What does the 1 in Row 2 and Column 1 of A^2 indicate about the schedule? What does the 2 in Row 1 and Column 3 indicate about the schedule? In general, how would you interpret each element not on the principal diagonal of A^2? [*Hint:* Examine the diagram for possible connections between the ith city and the jth city.]

(B) Find A^3. What does the 1 in Row 4 and Column 2 of A^3 indicate about the schedule? What does the 2 in Row 1 and Column 5 indicate about the schedule? In general, how would you interpret each element not on the principal diagonal of A^3?

(C) Compute A, $A + A^2$, $A + A^2 + A^3$, . . . , until you obtain a matrix with no 0 elements (except possibly on the principal diagonal), and interpret.

62. *Air freight.* Find the incidence matrix A for the flight schedule illustrated in the diagram. Compute A, $A + A^2$, $A + A^2 + A^3$, . . . , until you obtain a matrix with no 0 elements (except possibly on the principal diagonal), and interpret.

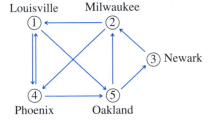

Louisville   Milwaukee

① ←——— ②

③ Newark

④ ——→ ⑤

Phoenix   Oakland

Life Sciences

63. *Nutrition.* A nutritionist for a cereal company blends two cereals in three different mixes. The amounts of protein, carbohydrate, and fat (in grams per ounce) in each cereal are given by matrix M. The amounts of each cereal used in the three mixes are given by matrix N.

$$
M = \begin{array}{c} \\ \\ \\ \end{array}
\overset{\text{Cereal A \quad Cereal B}}{\left[\begin{array}{cc} 4\text{ g/oz} & 2\text{ g/oz} \\ 20\text{ g/oz} & 16\text{ g/oz} \\ 3\text{ g/oz} & 1\text{ g/oz} \end{array}\right]}
\begin{array}{c} \text{Protein} \\ \text{Carbohydrate} \\ \text{Fat} \end{array}
$$

$$
N = \overset{\text{Mix X \quad Mix Y \quad Mix Z}}{\left[\begin{array}{ccc} 15\text{ oz} & 10\text{ oz} & 5\text{ oz} \\ 5\text{ oz} & 10\text{ oz} & 15\text{ oz} \end{array}\right]}
\begin{array}{c} \text{Cereal A} \\ \text{Cereal B} \end{array}
$$

(A) Find the amount of protein in mix X.

(B) Find the amount of fat in mix Z.

(C) Discuss possible interpretations of the elements in the matrix products MN and NM.

(D) If either of the products MN or NM has a meaningful interpretation, find the product and label its rows and columns.

64. *Heredity.* Gregor Mendel (1822–1884), an Austrian monk and botanist, made discoveries that revolutionized the science of genetics. In one experiment, he crossed dihybrid yellow round peas (yellow and round are dominant characteristics; the peas also contained genes for the recessive characteristics green and wrinkled) and obtained 560 peas of the types indicated in the matrix:

$$
\begin{array}{c} \text{Yellow} \\ \text{Green} \end{array}
\overset{\text{Round \quad Wrinkled}}{\left[\begin{array}{cc} 319 & 101 \\ 108 & 32 \end{array}\right]} = M
$$

Suppose he carried out a second experiment of the same type and obtained 640 peas of the types indicated in this matrix:

$$
\begin{array}{c} \text{Yellow} \\ \text{Green} \end{array}
\overset{\text{Round \quad Wrinkled}}{\left[\begin{array}{cc} 370 & 124 \\ 110 & 36 \end{array}\right]} = N
$$

If the results of the two experiments are combined, write the resulting matrix $M + N$. Compute the percentage of the total number of peas (1,200) in each category of the combined results.

Social Sciences

65. *Politics.* In a local California election, a group hired a public relations firm to promote its candidate in three ways: telephone calls, house calls, and letters. The cost per contact is given in matrix M, and the number of contacts of each type made in two adjacent cities is given in matrix N (see the next page).

$$M = \begin{bmatrix} \$0.40 \\ \$1.00 \\ \$0.35 \end{bmatrix} \begin{matrix} \text{Telephone call} \\ \text{House call} \\ \text{Letter} \end{matrix}$$

Cost per contact

$$N = \begin{bmatrix} 1{,}000 & 500 & 5{,}000 \\ 2{,}000 & 800 & 8{,}000 \end{bmatrix} \begin{matrix} \text{Berkeley} \\ \text{Oakland} \end{matrix}$$

Telephone call   House call   Letter

(A) Find the total amount spent in Berkeley.

(B) Find the total amount spent in Oakland.

(C) Discuss possible interpretations of the elements in the matrix products MN and NM.

(D) If either of the products MN or NM has a meaningful interpretation, find the product and label its rows and columns.

(E) Discuss methods of matrix multiplication that can be used to find the total number of telephone calls, house calls, and letters. State the matrices that can be used and perform the necessary operations.

(F) Discuss methods of matrix multiplication that can be used to find the total number of contacts in Berkeley and in Oakland. State the matrices that can be used and perform the necessary operations.

66. *Averaging tests.* A teacher has given four tests to a class of five students and stored the results in the following matrix:

	1	2	3	4
Ann	78	84	81	86
Bob	91	65	84	92
Carol	95	90	92	91
Dan	75	82	87	91
Eric	83	88	81	76

Tests $= M$

Discuss methods of matrix multiplication that the teacher can use to obtain the information indicated below. In each case, state the matrices to be used and then perform the necessary operations.

(A) The average on all four tests for each student, assuming that all four tests are given equal weight

(B) The average on all four tests for each student, assuming that the first three tests are given equal weight and the fourth is given twice this weight

(C) The class average on each of the four tests

SECTION   4-5   ■   **Inverse of a Square Matrix**

■ IDENTITY MATRIX FOR MULTIPLICATION
■ INVERSE OF A SQUARE MATRIX
■ APPLICATION: CRYPTOLOGY

■ **IDENTITY MATRIX FOR MULTIPLICATION**

We know that

$$1a = a1 = a \qquad \text{for all real numbers } a$$

The number 1 is called the **identity** for real number multiplication. Does the set of all matrices of a given size have an identity element for multiplication? That is, if M is an arbitrary $m \times n$ matrix, does M have an identity element I such that $IM = MI = M$? The answer, in general, is no. However, the set of all **square matrices of order n** (matrices with n rows and n columns) does have an identity, and it is given as follows:

> The **identity element for multiplication** for the set of all square matrices of order n is the square matrix of order n, denoted by I, with 1's along the principal diagonal (from the upper left corner to the lower right) and 0's elsewhere.

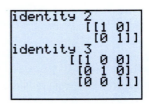

FIGURE 1
Identity matrices

For example,

$$\begin{bmatrix} 1 & 0 \\ 0 & 1 \end{bmatrix} \quad \text{and} \quad \begin{bmatrix} 1 & 0 & 0 \\ 0 & 1 & 0 \\ 0 & 0 & 1 \end{bmatrix}$$

are the identity matrices for all square matrices of order 2 and 3, respectively.

Most graphing calculators have a built-in command for generating the identity matrix of a given order (see Fig. 1).

EXAMPLE 1 ➤ Identity Matrix Multiplication

(A) $\begin{bmatrix} 1 & 0 & 0 \\ 0 & 1 & 0 \\ 0 & 0 & 1 \end{bmatrix} \begin{bmatrix} a & b & c \\ d & e & f \\ g & h & i \end{bmatrix} = \begin{bmatrix} a & b & c \\ d & e & f \\ g & h & i \end{bmatrix}$

(B) $\begin{bmatrix} a & b & c \\ d & e & f \\ g & h & i \end{bmatrix} \begin{bmatrix} 1 & 0 & 0 \\ 0 & 1 & 0 \\ 0 & 0 & 1 \end{bmatrix} = \begin{bmatrix} a & b & c \\ d & e & f \\ g & h & i \end{bmatrix}$

(C) $\begin{bmatrix} 1 & 0 \\ 0 & 1 \end{bmatrix} \begin{bmatrix} a & b & c \\ d & e & f \end{bmatrix} = \begin{bmatrix} a & b & c \\ d & e & f \end{bmatrix}$

(D) $\begin{bmatrix} a & b & c \\ d & e & f \end{bmatrix} \begin{bmatrix} 1 & 0 & 0 \\ 0 & 1 & 0 \\ 0 & 0 & 1 \end{bmatrix} = \begin{bmatrix} a & b & c \\ d & e & f \end{bmatrix}$ ◀

MATCHED PROBLEM 1 ➤ Multiply:

(A) $\begin{bmatrix} 1 & 0 \\ 0 & 1 \end{bmatrix} \begin{bmatrix} 2 & -3 \\ 5 & 7 \end{bmatrix} \quad \text{and} \quad \begin{bmatrix} 2 & -3 \\ 5 & 7 \end{bmatrix} \begin{bmatrix} 1 & 0 \\ 0 & 1 \end{bmatrix}$

(B) $\begin{bmatrix} 1 & 0 & 0 \\ 0 & 1 & 0 \\ 0 & 0 & 1 \end{bmatrix} \begin{bmatrix} 4 & 2 \\ 3 & -5 \\ 6 & 8 \end{bmatrix} \quad \text{and} \quad \begin{bmatrix} 4 & 2 \\ 3 & -5 \\ 6 & 8 \end{bmatrix} \begin{bmatrix} 1 & 0 \\ 0 & 1 \end{bmatrix}$ ◀

In general, we can show that if M is a square matrix of order n and I is the identity matrix of order n, then

$IM = MI = M$

If M is an $m \times n$ matrix that is not square ($m \neq n$), then it is still possible to multiply M on the left and on the right by an identity matrix, but not with the same size identity matrix (see Examples 1C and 1D). In order to avoid the complications involved with associating two different identity matrices with each nonsquare matrix, we restrict our attention in this section to square matrices.

The only real number solutions to the equation $x^2 = 1$ are $x = 1$ and $x = -1$.

(A) Show that $A = \begin{bmatrix} 0 & 1 \\ 1 & 0 \end{bmatrix}$ satisfies $A^2 = I$, where I is the 2×2 identity.

(B) Show that $B = \begin{bmatrix} 0 & -1 \\ -1 & 0 \end{bmatrix}$ satisfies $B^2 = I$.

(C) Find a 2×2 matrix with all elements nonzero whose square is the 2×2 identity matrix.

■ INVERSE OF A SQUARE MATRIX

In the set of real numbers, we know that for each real number a (except 0) there exists a real number a^{-1} such that

$$a^{-1}a = 1$$

The number a^{-1} is called the **inverse** of the number a relative to multiplication, or the **multiplicative inverse** of a. For example, 2^{-1} is the multiplicative inverse of 2, since $2^{-1} \cdot 2 = 1$. We use this idea to define the inverse of a square matrix.

Inverse of a Square Matrix

Let M be a square matrix of order n and I be the identity matrix of order n. If there exists a matrix M^{-1} (read ''M inverse'') such that

$$M^{-1}M = MM^{-1} = I$$

then M^{-1} is called the **multiplicative inverse of M** or, more simply, the **inverse of M.**

The multiplicative inverse of a nonzero real number a also can be written as $1/a$. This notation is not used for matrix inverses.

Let us use the above definition to find M^{-1} for

$$M = \begin{bmatrix} 2 & 3 \\ 1 & 2 \end{bmatrix}$$

We are looking for

$$M^{-1} = \begin{bmatrix} a & c \\ b & d \end{bmatrix}$$

such that

$$MM^{-1} = M^{-1}M = I$$

Thus, we write

$$
\overset{M}{\begin{bmatrix} 2 & 3 \\ 1 & 2 \end{bmatrix}}
\overset{M^{-1}}{\begin{bmatrix} a & c \\ b & d \end{bmatrix}}
=
\overset{I}{\begin{bmatrix} 1 & 0 \\ 0 & 1 \end{bmatrix}}
$$

and try to find a, b, c, and d so that the product of M and M^{-1} is the identity matrix I. Multiplying M and M^{-1} on the left side, we obtain

$$
\begin{bmatrix} (2a + 3b) & (2c + 3d) \\ (a + 2b) & (c + 2d) \end{bmatrix}
=
\begin{bmatrix} 1 & 0 \\ 0 & 1 \end{bmatrix}
$$

which is true only if

$$2a + 3b = 1 \qquad 2c + 3d = 0$$
$$a + 2b = 0 \qquad c + 2d = 1$$

Solving these two systems, we find that $a = 2$, $b = -1$, $c = -3$, and $d = 2$. Thus,

$$
M^{-1} = \begin{bmatrix} 2 & -3 \\ -1 & 2 \end{bmatrix}
$$

as is easily checked:

$$
\overset{M}{\begin{bmatrix} 2 & 3 \\ 1 & 2 \end{bmatrix}}
\overset{M^{-1}}{\begin{bmatrix} 2 & -3 \\ -1 & 2 \end{bmatrix}}
=
\overset{I}{\begin{bmatrix} 1 & 0 \\ 0 & 1 \end{bmatrix}}
=
\overset{M^{-1}}{\begin{bmatrix} 2 & -3 \\ -1 & 2 \end{bmatrix}}
\overset{M}{\begin{bmatrix} 2 & 3 \\ 1 & 2 \end{bmatrix}}
$$

Unlike nonzero real numbers, inverses do not always exist for square matrices. For example, if

$$
N = \begin{bmatrix} 2 & 1 \\ 4 & 2 \end{bmatrix}
$$

then, proceeding as above, we are led to the systems

$$2a + b = 1 \qquad 2c + d = 0$$
$$4a + 2b = 0 \qquad 4c + 2d = 1$$

These are both inconsistent and have no solution. Hence, N^{-1} does not exist.

Being able to find inverses, when they exist, leads to direct and simple solutions to many practical problems. In the next section, we shall show how inverses can be used to solve systems of linear equations.

The method outlined above for finding M^{-1}, if it exists, gets very involved for matrices of order larger than 2. Now that we know what we are looking for, we can introduce the idea of the augmented matrix (considered in Sections 4-2 and 4-3) to make the process more efficient.

EXAMPLE 2 ➤ **Finding the Inverse of a Matrix**   Find the inverse, if it exists, of the matrix

$$M = \begin{bmatrix} 1 & -1 & 1 \\ 0 & 2 & -1 \\ 2 & 3 & 0 \end{bmatrix}$$

Solution   We start as before and write

$$\underset{M}{\begin{bmatrix} 1 & -1 & 1 \\ 0 & 2 & -1 \\ 2 & 3 & 0 \end{bmatrix}} \underset{M^{-1}}{\begin{bmatrix} a & d & g \\ b & e & h \\ c & f & i \end{bmatrix}} = \underset{I}{\begin{bmatrix} 1 & 0 & 0 \\ 0 & 1 & 0 \\ 0 & 0 & 1 \end{bmatrix}}$$

which is true only if

$$
\begin{array}{lll}
a - b + c = 1 & d - e + f = 0 & g - h + i = 0 \\
2b - c = 0 & 2e - f = 1 & 2h - i = 0 \\
2a + 3b \quad = 0 & 2d + 3e \quad = 0 & 2g + 3h \quad = 1
\end{array}
$$

Now we write augmented matrices for each of the three systems:

First

$$\begin{bmatrix} 1 & -1 & 1 & | & 1 \\ 0 & 2 & -1 & | & 0 \\ 2 & 3 & 0 & | & 0 \end{bmatrix}$$

Second

$$\begin{bmatrix} 1 & -1 & 1 & | & 0 \\ 0 & 2 & -1 & | & 1 \\ 2 & 3 & 0 & | & 0 \end{bmatrix}$$

Third

$$\begin{bmatrix} 1 & -1 & 1 & | & 0 \\ 0 & 2 & -1 & | & 0 \\ 2 & 3 & 0 & | & 1 \end{bmatrix}$$

Since each matrix to the left of the vertical bar is the same, exactly the same row operations can be used on each augmented matrix to transform it into a reduced form. We can speed up the process substantially by combining all three augmented matrices into the single augmented matrix form below:

$$\begin{bmatrix} 1 & -1 & 1 & | & 1 & 0 & 0 \\ 0 & 2 & -1 & | & 0 & 1 & 0 \\ 2 & 3 & 0 & | & 0 & 0 & 1 \end{bmatrix} = [M \,|\, I] \tag{1}$$

We now try to perform row operations on matrix (1) until we obtain a row-equivalent matrix that looks like matrix (2):

$$\begin{bmatrix} 1 & 0 & 0 & | & a & d & g \\ 0 & 1 & 0 & | & b & e & h \\ 0 & 0 & 1 & | & c & f & i \end{bmatrix} = [I \,|\, B] \tag{2}$$

If this can be done, then the new matrix B to the right of the vertical bar will be M^{-1}! Now let us try to transform (1) into a form like (2). We will follow the same sequence of steps as we did in the solution of linear systems by Gauss–Jordan elimination (see Section 4-3).

$$\begin{array}{cc} & M & \\ \end{array}$$

$$\begin{bmatrix} 1 & -1 & 1 & | & 1 & 0 & 0 \\ 0 & 2 & -1 & | & 0 & 1 & 0 \\ 2 & 3 & 0 & | & 0 & 0 & 1 \end{bmatrix} \quad (-2)R_1 + R_3 \rightarrow R_3$$

$$\sim \begin{bmatrix} 1 & -1 & 1 & | & 1 & 0 & 0 \\ 0 & 2 & -1 & | & 0 & 1 & 0 \\ 0 & 5 & -2 & | & -2 & 0 & 1 \end{bmatrix} \quad \tfrac{1}{2}R_2 \rightarrow R_2$$

$$\sim \begin{bmatrix} 1 & -1 & 1 & | & 1 & 0 & 0 \\ 0 & 1 & -\tfrac{1}{2} & | & 0 & \tfrac{1}{2} & 0 \\ 0 & 5 & -2 & | & -2 & 0 & 1 \end{bmatrix} \quad \begin{array}{l} R_2 + R_1 \rightarrow R_1 \\ \\ (-5)R_2 + R_3 \rightarrow R_3 \end{array}$$

$$\sim \begin{bmatrix} 1 & 0 & \tfrac{1}{2} & | & 1 & \tfrac{1}{2} & 0 \\ 0 & 1 & -\tfrac{1}{2} & | & 0 & \tfrac{1}{2} & 0 \\ 0 & 0 & \tfrac{1}{2} & | & -2 & -\tfrac{5}{2} & 1 \end{bmatrix} \quad 2R_3 \rightarrow R_3$$

$$\sim \begin{bmatrix} 1 & 0 & \tfrac{1}{2} & | & 1 & \tfrac{1}{2} & 0 \\ 0 & 1 & -\tfrac{1}{2} & | & 0 & \tfrac{1}{2} & 0 \\ 0 & 0 & 1 & | & -4 & -5 & 2 \end{bmatrix} \quad \begin{array}{l} (-\tfrac{1}{2})R_3 + R_1 \rightarrow R_1 \\ \tfrac{1}{2}R_3 + R_2 \rightarrow R_2 \end{array}$$

$$\sim \begin{bmatrix} 1 & 0 & 0 & | & 3 & 3 & -1 \\ 0 & 1 & 0 & | & -2 & -2 & 1 \\ 0 & 0 & 1 & | & -4 & -5 & 2 \end{bmatrix} \quad = [I \,|\, B]$$

Converting back to systems of equations equivalent to our three original systems (we will not have to do this step in practice), we have

$$\begin{array}{lll} a = 3 & d = 3 & g = -1 \\ b = -2 & e = -2 & h = 1 \\ c = -4 & f = -5 & i = 2 \end{array}$$

And these are just the elements of M^{-1} that we are looking for! Hence,

$$M^{-1} = \begin{bmatrix} 3 & 3 & -1 \\ -2 & -2 & 1 \\ -4 & -5 & 2 \end{bmatrix}$$

Note that this is the matrix to the right of the vertical line in the last augmented matrix. That is, $M^{-1} = B$.

Since the definition of matrix inverse requires that

$$M^{-1}M = I \quad \text{and} \quad MM^{-1} = I \tag{3}$$

it appears that we must compute both $M^{-1}M$ and MM^{-1} to check our work. However, it can be shown that if one of the equations in (3) is satisfied, then the other is also satisfied. Thus, for checking purposes, it is sufficient to compute either $M^{-1}M$ or MM^{-1}; we do not need to do both.

Check
$$M^{-1}M = \begin{bmatrix} 3 & 3 & -1 \\ -2 & -2 & 1 \\ -4 & -5 & 2 \end{bmatrix} \begin{bmatrix} 1 & -1 & 1 \\ 0 & 2 & -1 \\ 2 & 3 & 0 \end{bmatrix} = \begin{bmatrix} 1 & 0 & 0 \\ 0 & 1 & 0 \\ 0 & 0 & 1 \end{bmatrix} = I$$
◄

MATCHED PROBLEM 2 ► Let: $M = \begin{bmatrix} 3 & -1 & 1 \\ -1 & 1 & 0 \\ 1 & 0 & 1 \end{bmatrix}$

(A) Form the augmented matrix $[M|I]$.
(B) Use row operations to transform $[M|I]$ into $[I|B]$.
(C) Verify by multiplication that $B = M^{-1}$ (that is, show that $BM = I$). ◄

The procedure shown in Example 2 can be used to find the inverse of any square matrix, if the inverse exists, and will also indicate when the inverse does not exist. These ideas are summarized in Theorem 1.

theorem 1

Inverse of a Square Matrix M

If $[M|I]$ is transformed by row operations into $[I|B]$, then the resulting matrix B is M^{-1}. However, if we obtain 0's in one or more rows to the left of the vertical line, then M^{-1} does not exist.

explore – discuss 2

(A) Suppose that the square matrix M has a row of all zeros. Explain why M has no inverse.
(B) Suppose that the square matrix M has a column of all zeros. Explain why M has no inverse.

EXAMPLE 3 ► **Finding a Matrix Inverse** Find M^{-1}, given: $M = \begin{bmatrix} 4 & -1 \\ -6 & 2 \end{bmatrix}$

Solution
$$\begin{bmatrix} 4 & -1 & | & 1 & 0 \\ -6 & 2 & | & 0 & 1 \end{bmatrix} \quad \tfrac{1}{4}R_1 \to R_1$$

$$\sim \begin{bmatrix} 1 & -\tfrac{1}{4} & | & \tfrac{1}{4} & 0 \\ -6 & 2 & | & 0 & 1 \end{bmatrix} \quad 6R_1 + R_2 \to R_2$$

$$\sim \begin{bmatrix} 1 & -\tfrac{1}{4} & | & \tfrac{1}{4} & 0 \\ 0 & \tfrac{1}{2} & | & \tfrac{3}{2} & 1 \end{bmatrix} \quad 2R_2 \to R_2$$

$$\sim \begin{bmatrix} 1 & -\tfrac{1}{4} & | & \tfrac{1}{4} & 0 \\ 0 & 1 & | & 3 & 2 \end{bmatrix} \quad \tfrac{1}{4}R_2 + R_1 \to R_1$$

$$\sim \begin{bmatrix} 1 & 0 & | & 1 & \tfrac{1}{2} \\ 0 & 1 & | & 3 & 2 \end{bmatrix}$$

Thus,

$$M^{-1} = \begin{bmatrix} 1 & \frac{1}{2} \\ 3 & 2 \end{bmatrix}$$

Check by showing $M^{-1}M = I$. ◀

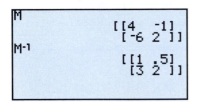

FIGURE 2

Finding a matrix inverse on a graphing calculator

When finding the inverse of a matrix on a graphing calculator or computer, we could perform each row operation required to find the reduced form of $[M|I]$. Or, if there is a built-in row reduction routine available, we could use that routine to find the reduced form of $[M|I]$. Fortunately, most graphing calculators and computers have built-in routines that will compute the inverse directly. Figure 2 illustrates this process on a graphing calculator. For those who are performing individual row operations, we will continue to display the row operations required to find the inverse.

MATCHED PROBLEM 3 ➤ Find M^{-1}, given: $M = \begin{bmatrix} 2 & -6 \\ 1 & -2 \end{bmatrix}$ ◀

EXAMPLE 4 ➤ Finding a Matrix Inverse   Find M^{-1}, given:

$$M = \begin{bmatrix} 2 & -4 \\ -3 & 6 \end{bmatrix}$$

Solution

$$\begin{bmatrix} 2 & -4 & | & 1 & 0 \\ -3 & 6 & | & 0 & 1 \end{bmatrix} \quad \tfrac{1}{2}R_1 \rightarrow R_1$$

$$\sim \begin{bmatrix} 1 & -2 & | & \frac{1}{2} & 0 \\ -3 & 6 & | & 0 & 1 \end{bmatrix} \quad 3R_1 + R_2 \rightarrow R_2$$

$$\sim \begin{bmatrix} 1 & -2 & | & \frac{1}{2} & 0 \\ 0 & 0 & | & \frac{3}{2} & 1 \end{bmatrix}$$

We have all 0's in the second row to the left of the vertical bar; therefore, the inverse does not exist. ◀

MATCHED PROBLEM 4 ➤ Find N^{-1}, given: $N = \begin{bmatrix} 3 & 1 \\ 6 & 2 \end{bmatrix}$ ◀

Caution

Matrices that do not have inverses are called **singular matrices.** Graphing calculators and computers usually recognize singular matrices and display an appropriate error message. However, in some cases, roundoff error may cause a graphing calculator or computer to display an incorrect inverse when the matrix is singular or to display an error message when the inverse does, in fact, exist. Figure 3 (at the top of the next page) illustrates the results of solving Example 4 and Matched Problem 4 on a particular graphing calculator.

(A) Solution to Example 4   (B) Solution to Matched Problem 4

FIGURE 3

Notice that the calculator gives the correct answer to Example 4, but displays an inverse for the singular matrix N in Matched Problem 4. The extremely large numbers in N^{-1} suggest that an error has occurred, and a quick check on the graphing calculator shows that NN^{-1} is not the identity. Whenever the results of finding an inverse on a graphing calculator or computer are in doubt, use Theorem 1 to find the inverse.

■ APPLICATION: CRYPTOLOGY

Matrix inverses can provide a simple and effective procedure for encoding and decoding messages. To begin, assign the numbers 1–26 to the letters in the alphabet, as shown below. Also assign the number 27 to a blank to provide for space between words. (A more sophisticated code could include both capital and lowercase letters and punctuation symbols.)

A	B	C	D	E	F	G	H	I	J	K	L	M	N	O	P	Q	R	S	T	U	V	W	X	Y	Z	Blank
1	2	3	4	5	6	7	8	9	10	11	12	13	14	15	16	17	18	19	20	21	22	23	24	25	26	27

Thus, the message "SECRET CODE" corresponds to the sequence

19 5 3 18 5 20 27 3 15 4 5

Any matrix whose elements are positive integers and whose inverse exists can be used as an **encoding matrix.** For example, to use the 2×2 matrix

$$A = \begin{bmatrix} 4 & 3 \\ 1 & 1 \end{bmatrix}$$

to encode the above message, first divide the numbers in the sequence into groups of two and use these groups as the columns of a matrix with 2 rows. (Notice that we added an extra blank at the end of the message to make the columns come out even.) Then multiply this matrix on the left by A:

$$\begin{bmatrix} 4 & 3 \\ 1 & 1 \end{bmatrix} \begin{bmatrix} 19 & 3 & 5 & 27 & 15 & 5 \\ 5 & 18 & 20 & 3 & 4 & 27 \end{bmatrix} = \begin{bmatrix} 91 & 66 & 80 & 117 & 72 & 101 \\ 24 & 21 & 25 & 30 & 19 & 32 \end{bmatrix}$$

The coded message is

$$91 \ 24 \ 66 \ 21 \ 80 \ 25 \ 117 \ 30 \ 72 \ 19 \ 101 \ 32$$

This message can be decoded simply by putting it back into matrix form and multiplying on the left by the **decoding matrix** A^{-1}. Since A^{-1} is easily determined if A is known, the encoding matrix A is the only key needed to decode messages encoded in this manner. Although simple in concept, codes of this type can be very difficult to crack.

Answers to Matched Problems

1. (A) $\begin{bmatrix} 2 & -3 \\ 5 & 7 \end{bmatrix}$ (B) $\begin{bmatrix} 4 & 2 \\ 3 & -5 \\ 6 & 8 \end{bmatrix}$

2. (A) $\left[\begin{array}{rrr|rrr} 3 & -1 & 1 & 1 & 0 & 0 \\ -1 & 1 & 0 & 0 & 1 & 0 \\ 1 & 0 & 1 & 0 & 0 & 1 \end{array}\right]$

  (B) $\left[\begin{array}{rrr|rrr} 1 & 0 & 0 & 1 & 1 & -1 \\ 0 & 1 & 0 & 1 & 2 & -1 \\ 0 & 0 & 1 & -1 & -1 & 2 \end{array}\right]$

  (C) $\begin{bmatrix} 1 & 1 & -1 \\ 1 & 2 & -1 \\ -1 & -1 & 2 \end{bmatrix} \begin{bmatrix} 3 & -1 & 1 \\ -1 & 1 & 0 \\ 1 & 0 & 1 \end{bmatrix} = \begin{bmatrix} 1 & 0 & 0 \\ 0 & 1 & 0 \\ 0 & 0 & 1 \end{bmatrix}$

3. $\begin{bmatrix} -1 & 3 \\ -\frac{1}{2} & 1 \end{bmatrix}$ 4. Does not exist

---

EXERCISE 4-5

A *Perform the indicated operations in Problems 1–8.*

1. $\begin{bmatrix} 1 & 0 \\ 0 & 1 \end{bmatrix} \begin{bmatrix} 2 & -3 \\ 4 & 5 \end{bmatrix}$ 2. $\begin{bmatrix} 1 & 0 \\ 0 & 1 \end{bmatrix} \begin{bmatrix} -1 & 6 \\ 0 & 2 \end{bmatrix}$

3. $\begin{bmatrix} 2 & -3 \\ 4 & 5 \end{bmatrix} \begin{bmatrix} 1 & 0 \\ 0 & 1 \end{bmatrix}$ 4. $\begin{bmatrix} -1 & 6 \\ 0 & 2 \end{bmatrix} \begin{bmatrix} 1 & 0 \\ 0 & 1 \end{bmatrix}$

5. $\begin{bmatrix} 1 & 0 & 0 \\ 0 & 1 & 0 \\ 0 & 0 & 1 \end{bmatrix} \begin{bmatrix} -2 & 1 & 3 \\ 2 & 4 & -2 \\ 5 & 1 & 0 \end{bmatrix}$

6. $\begin{bmatrix} 1 & 0 & 0 \\ 0 & 1 & 0 \\ 0 & 0 & 1 \end{bmatrix} \begin{bmatrix} 3 & -4 & 0 \\ 1 & 2 & -5 \\ 6 & -3 & -1 \end{bmatrix}$

7. $\begin{bmatrix} -2 & 1 & 3 \\ 2 & 4 & -2 \\ 5 & 1 & 0 \end{bmatrix} \begin{bmatrix} 1 & 0 & 0 \\ 0 & 1 & 0 \\ 0 & 0 & 1 \end{bmatrix}$

8. $\begin{bmatrix} 3 & -4 & 0 \\ 1 & 2 & -5 \\ 6 & -3 & -1 \end{bmatrix} \begin{bmatrix} 1 & 0 & 0 \\ 0 & 1 & 0 \\ 0 & 0 & 1 \end{bmatrix}$

In Problems 9–14, show that the two matrices are inverses of each other by showing that their product is the identity matrix I.

9. $\begin{bmatrix} 3 & -4 \\ -2 & 3 \end{bmatrix}$; $\begin{bmatrix} 3 & 4 \\ 2 & 3 \end{bmatrix}$

10. $\begin{bmatrix} 5 & -7 \\ -2 & 3 \end{bmatrix}$; $\begin{bmatrix} 3 & 7 \\ 2 & 5 \end{bmatrix}$

11. $\begin{bmatrix} -5 & 2 \\ -8 & 3 \end{bmatrix}$; $\begin{bmatrix} 3 & -2 \\ 8 & -5 \end{bmatrix}$

12. $\begin{bmatrix} 7 & 4 \\ -5 & -3 \end{bmatrix}$; $\begin{bmatrix} 3 & 4 \\ -5 & -7 \end{bmatrix}$

13. $\begin{bmatrix} 1 & -1 & 1 \\ 0 & 2 & -1 \\ 2 & 3 & 0 \end{bmatrix}$; $\begin{bmatrix} 3 & 3 & -1 \\ -2 & -2 & 1 \\ -4 & -5 & 2 \end{bmatrix}$

14. $\begin{bmatrix} 3 & 3 & -1 \\ -2 & -2 & 1 \\ -4 & -5 & 2 \end{bmatrix}$; $\begin{bmatrix} 1 & -1 & 1 \\ 0 & 2 & -1 \\ 2 & 3 & 0 \end{bmatrix}$

B *Given M in Problems 15–24, find M^{-1} and show that $M^{-1}M = I$.*

15. $\begin{bmatrix} -1 & 0 \\ -3 & 1 \end{bmatrix}$

16. $\begin{bmatrix} 1 & -5 \\ 0 & -1 \end{bmatrix}$

17. $\begin{bmatrix} 1 & 2 \\ 1 & 3 \end{bmatrix}$

18. $\begin{bmatrix} 2 & 1 \\ 5 & 3 \end{bmatrix}$

19. $\begin{bmatrix} 1 & 3 \\ 2 & 7 \end{bmatrix}$

20. $\begin{bmatrix} 2 & 1 \\ 1 & 1 \end{bmatrix}$

21. $\begin{bmatrix} 1 & -3 & 0 \\ 0 & 1 & 1 \\ 2 & -1 & 4 \end{bmatrix}$

22. $\begin{bmatrix} 2 & 3 & 0 \\ 1 & 2 & 3 \\ 0 & -1 & -5 \end{bmatrix}$

23. $\begin{bmatrix} 1 & 1 & 0 \\ 2 & 3 & -1 \\ 1 & 0 & 2 \end{bmatrix}$

24. $\begin{bmatrix} 1 & 0 & -1 \\ 2 & -1 & 0 \\ 1 & 1 & -2 \end{bmatrix}$

Find the inverse of each matrix in Problems 25–30, if it exists.

25. $\begin{bmatrix} 4 & 3 \\ -3 & -2 \end{bmatrix}$

26. $\begin{bmatrix} -4 & 3 \\ -5 & 4 \end{bmatrix}$

27. $\begin{bmatrix} 2 & 6 \\ 3 & 9 \end{bmatrix}$

28. $\begin{bmatrix} 2 & -4 \\ -3 & 6 \end{bmatrix}$

29. $\begin{bmatrix} 2 & 1 \\ 4 & 3 \end{bmatrix}$

30. $\begin{bmatrix} -5 & 3 \\ 2 & -2 \end{bmatrix}$

C *Find the inverse of each matrix in Problems 31–38, if it exists.*

31. $\begin{bmatrix} -5 & -2 & -2 \\ 2 & 1 & 0 \\ 1 & 0 & 1 \end{bmatrix}$

32. $\begin{bmatrix} 2 & -2 & 4 \\ 1 & 1 & 1 \\ 1 & 0 & 1 \end{bmatrix}$

33. $\begin{bmatrix} 2 & 1 & 1 \\ 1 & 1 & 0 \\ -1 & -1 & 0 \end{bmatrix}$

34. $\begin{bmatrix} 1 & -1 & 0 \\ 2 & -1 & 1 \\ 0 & 1 & 1 \end{bmatrix}$

35. $\begin{bmatrix} -1 & -2 & 2 \\ 4 & 3 & 0 \\ 4 & 0 & 4 \end{bmatrix}$

36. $\begin{bmatrix} 4 & 2 & 2 \\ 4 & 2 & 0 \\ 5 & 0 & 5 \end{bmatrix}$

37. $\begin{bmatrix} 2 & -1 & -2 \\ -4 & 2 & 8 \\ 6 & -2 & -1 \end{bmatrix}$

38. $\begin{bmatrix} -1 & -1 & 4 \\ 3 & 3 & -22 \\ -2 & -1 & 19 \end{bmatrix}$

39. Show that $(A^{-1})^{-1} = A$ for:
$$A = \begin{bmatrix} 4 & 3 \\ 3 & 2 \end{bmatrix}$$

40. Show that $(AB)^{-1} = B^{-1}A^{-1}$ for:
$$A = \begin{bmatrix} 4 & 3 \\ 3 & 2 \end{bmatrix} \quad \text{and} \quad B = \begin{bmatrix} 2 & 5 \\ 3 & 7 \end{bmatrix}$$

41. Discuss the existence of M^{-1} for 2×2 diagonal matrices of the form
$$M = \begin{bmatrix} a & 0 \\ 0 & d \end{bmatrix}$$
Generalize your conclusions to $n \times n$ diagonal matrices.

42. Discuss the existence of M^{-1} for 2×2 upper triangular matrices of the form
$$M = \begin{bmatrix} a & b \\ 0 & d \end{bmatrix}$$
Generalize your conclusions to $n \times n$ upper triangular matrices.

Find the inverse of each matrix in Problems 43–46, if it exists.

43. $\begin{bmatrix} 6 & 2 & 0 & 4 \\ 5 & 3 & 2 & 1 \\ 0 & -1 & 1 & -2 \\ 2 & -3 & 1 & 0 \end{bmatrix}$

44. $\begin{bmatrix} -2 & 4 & 0 & -1 \\ 2 & -1 & 2 & 5 \\ 0 & 2 & -1 & 7 \\ 2 & -3 & 0 & 5 \end{bmatrix}$

45. $\begin{bmatrix} 3 & 2 & 3 & 4 & 4 \\ 5 & 4 & 3 & 2 & 1 \\ -1 & -1 & 2 & -2 & 3 \\ 3 & -3 & 1 & 0 & 1 \\ 1 & 1 & 2 & 0 & 2 \end{bmatrix}$

46. $\begin{bmatrix} 1 & 2 & 3 & 4 & 5 \\ 2 & 6 & 4 & 5 & 6 \\ -1 & -2 & -1 & 2 & 3 \\ 1 & 6 & 1 & 6 & 4 \\ 1 & -4 & 3 & -7 & -4 \end{bmatrix}$

A P P L I C A T I O N S

Social Sciences

Problems 47–50 refer to the encoding matrix:

$$A = \begin{bmatrix} 1 & 2 \\ 1 & 3 \end{bmatrix}$$

47. *Cryptography.* Encode the message "THE SUN ALSO RISES" using matrix A.

48. *Cryptography.* Encode the message "THE GRAPES OF WRATH" using matrix A.

49. *Cryptography.* The following message was encoded with matrix A. Decode this message:

37 52 24 29 73 96 49 69 62 89 36 44 59 86 41 50 22 26

50. *Cryptography.* The following message was encoded with matrix A. Decode this message:

9 13 40 49 29 34 29 30 22 26 33 36 43 57 29 34 54 74

To use the 5 × 5 encoding matrix B given below, form a matrix with 5 rows and as many columns as necessary to accommodate each message.

$$B = \begin{bmatrix} 1 & 0 & 1 & 0 & 1 \\ 0 & 1 & 1 & 0 & 3 \\ 2 & 1 & 1 & 1 & 1 \\ 0 & 0 & 1 & 0 & 2 \\ 1 & 1 & 1 & 2 & 1 \end{bmatrix}$$

51. *Cryptography.* Encode the message "THE BEST YEARS OF OUR LIVES" using matrix B.

52. *Cryptography.* Encode the message "THE BRIDGE ON THE RIVER KWAI" using matrix B.

53. *Cryptography.* The following message was encoded with matrix B. Decode this message:

32 34 87 19 94 24 21 67 11 69 54 71 112 43 112 56 92 109 55 109 48 66 98 41 89 62 135 124 81 143

54. *Cryptography.* The following message was encoded with matrix B. Decode this message:

28 22 83 11 90 57 108 102 69 105 57 78 91 57 89 66 111 112 79 125

S E C T I O N   4-6   ■   **Matrix Equations and Systems of Linear Equations**

■ MATRIX EQUATIONS
■ MATRIX EQUATIONS AND SYSTEMS OF LINEAR EQUATIONS
■ APPLICATION

The identity matrix and inverse matrix discussed in the last section can be put to immediate use in the solution of certain simple matrix equations. Being able to solve a matrix equation gives us another important method of solving systems of equations, provided the system is independent and has the same number of variables as equations. If the system is dependent or if it has either fewer or more variables than equations, we must return to the Gauss–Jordan method of elimination.

■ MATRIX EQUATIONS

Before we discuss the solution of matrix equations, you will probably find it helpful to briefly review the basic procedures for solving real number equations.

Let a, b, and c be real numbers, with $a \neq 0$. Solve each equation for x.

(A) $ax = b$ (B) $ax + b = c$

Solving simple matrix equations follows very much the same procedures used in solving real number equations. We have, however, less freedom with matrix equations, because matrix multiplication is not commutative. In solving matrix equations, we will be guided by the properties of matrices summarized in Theorem 1.

t h e o r e m   1

Basic Properties of Matrices

Assuming all products and sums are defined for the indicated matrices A, B, C, I, and 0, then:

ADDITION PROPERTIES

Associative:	$(A + B) + C = A + (B + C)$
Commutative:	$A + B = B + A$
Additive Identity:	$A + 0 = 0 + A = A$
Additive Inverse:	$A + (-A) = (-A) + A = 0$

MULTIPLICATION PROPERTIES

Associative Property:	$A(BC) = (AB)C$
Multiplicative Identity:	$AI = IA = A$
Multiplicative Inverse:	If A is a square matrix and A^{-1} exists, then $AA^{-1} = A^{-1}A = I$.

COMBINED PROPERTIES

Left Distributive:	$A(B + C) = AB + AC$
Right Distributive:	$(B + C)A = BA + CA$

EQUALITY

Addition:	If $A = B$, then $A + C = B + C$.
Left Multiplication:	If $A = B$, then $CA = CB$.
Right Multiplication:	If $A = B$, then $AC = BC$.

The use of these properties in the solution of matrix equations is best illustrated by an example.

E X A M P L E   1 ➤ **Solving a Matrix Equation** Given an $n \times n$ matrix A and $n \times 1$ column matrices B and X, solve $AX = B$ for X. Assume all necessary inverses exist.

Solution   We are interested in finding a column matrix X that satisfies the matrix equation $AX = B$. To solve this equation, we multiply both sides on the left by A^{-1}, assuming it exists, to isolate X on the left side.

$$AX = B \qquad \text{\textit{Use the left multiplication property.}}$$
$$A^{-1}(AX) = A^{-1}B \qquad \text{\textit{Use the associative property.}}$$
$$(A^{-1}A)X = A^{-1}B \qquad \text{\textit{$A^{-1}A = I$}}$$
$$IX = A^{-1}B \qquad \text{\textit{$IX = X$}}$$
$$X = A^{-1}B \qquad \blacktriangleleft$$

MATCHED PROBLEM 1 ➤   Given an $n \times n$ matrix A and $n \times 1$ column matrices B, C, and X, solve $AX + C = B$ for X. Assume all necessary inverses exist.   ◀

Caution

Do not mix the left multiplication property and the right multiplication property. If $AX = B$, then

$$A^{-1}(AX) \neq BA^{-1}$$

▓ MATRIX EQUATIONS AND SYSTEMS OF LINEAR EQUATIONS

We now show how independent systems of linear equations with the same number of variables as equations can be solved. First, convert the system into a matrix equation of the form $AX = B$, and then use $X = A^{-1}B$ as obtained in Example 1.

EXAMPLE 2 ➤   **Using Inverses to Solve Systems of Equations**   Use matrix inverse methods to solve the system:

$$\begin{aligned} x_1 - x_2 + x_3 &= 1 \\ 2x_2 - x_3 &= 1 \\ 2x_1 + 3x_2 \quad\;\; &= 1 \end{aligned} \qquad (1)$$

Solution   The inverse of the coefficient matrix

$$A = \begin{bmatrix} 1 & -1 & 1 \\ 0 & 2 & -1 \\ 2 & 3 & 0 \end{bmatrix}$$

provides an efficient method for solving this system. To see how, we convert system (1) into a matrix equation:

$$\begin{array}{ccc} A & X & B \end{array}$$
$$\begin{bmatrix} 1 & -1 & 1 \\ 0 & 2 & -1 \\ 2 & 3 & 0 \end{bmatrix} \begin{bmatrix} x_1 \\ x_2 \\ x_3 \end{bmatrix} = \begin{bmatrix} 1 \\ 1 \\ 1 \end{bmatrix} \qquad (2)$$

FIGURE 1
Using inverse methods on a graphing calculator

Check that matrix equation (2) is equivalent to system (1) by finding the product of the left side and then equating corresponding elements on the left with those on the right. Now you see another important reason for defining matrix multiplication as we did.

We are interested in finding a column matrix X that satisfies the matrix equation $AX = B$. In Example 1 we found that if $AX = B$ and if A^{-1} exists, then

$$X = A^{-1}B$$

The inverse of A was found in Example 2, Section 4-5, to be

$$A^{-1} = \begin{bmatrix} 3 & 3 & -1 \\ -2 & -2 & 1 \\ -4 & -5 & 2 \end{bmatrix}$$

Thus,

$$\begin{array}{ccc} X & A^{-1} & B \end{array}$$
$$\begin{bmatrix} x_1 \\ x_2 \\ x_3 \end{bmatrix} = \begin{bmatrix} 3 & 3 & -1 \\ -2 & -2 & 1 \\ -4 & -5 & 2 \end{bmatrix}\begin{bmatrix} 1 \\ 1 \\ 1 \end{bmatrix} = \begin{bmatrix} 5 \\ -3 \\ -7 \end{bmatrix}$$

and we can conclude that $x_1 = 5$, $x_2 = -3$, and $x_3 = -7$. Check this result in system (1). See Figure 1 for a graphing calculator solution. ◀

MATCHED PROBLEM 2 ➤ Use matrix inverse methods to solve the system:

$$\begin{array}{rcl} 3x_1 - x_2 + x_3 &=& 1 \\ -x_1 + x_2 &=& 3 \\ x_1 \quad\quad + x_3 &=& 2 \end{array}$$

[*Note*: The inverse of the coefficient matrix was found in Matched Problem 2, Section 4-5.] ◀

At first glance, using matrix inverse methods seems to require the same amount of effort as using Gauss–Jordan elimination. In either case, row operations must be applied to an augmented matrix involving the coefficients of the system. The advantage of the inverse matrix method becomes readily apparent when solving a number of systems with a common coefficient matrix and different constant terms.

EXAMPLE 3 ➤ **Using Inverses to Solve Systems of Equations**   Use matrix inverse methods to solve each of the following systems:

(A) $\quad x_1 - x_2 + x_3 = 3$
$\quad\quad\quad 2x_2 - x_3 = 1$
$\quad 2x_1 + 3x_2 \quad = 4$

(B) $\quad x_1 - x_2 + x_3 = -5$
$\quad\quad\quad 2x_2 - x_3 = 2$
$\quad 2x_1 + 3x_2 \quad = -3$

Solution   Notice that both systems have the same coefficient matrix A as system (1) in Example 2. Only the constant terms have changed. Thus, we can use A^{-1} to solve these systems just as we did in Example 2.

(A) X A^{-1} B

$$\begin{bmatrix} x_1 \\ x_2 \\ x_3 \end{bmatrix} = \begin{bmatrix} 3 & 3 & -1 \\ -2 & -2 & 1 \\ -4 & -5 & 2 \end{bmatrix} \begin{bmatrix} 3 \\ 1 \\ 4 \end{bmatrix} = \begin{bmatrix} 8 \\ -4 \\ -9 \end{bmatrix}$$

Thus, $x_1 = 8$, $x_2 = -4$, and $x_3 = -9$.

(B) X A^{-1} B

$$\begin{bmatrix} x_1 \\ x_2 \\ x_3 \end{bmatrix} = \begin{bmatrix} 3 & 3 & -1 \\ -2 & -2 & 1 \\ -4 & -5 & 2 \end{bmatrix} \begin{bmatrix} -5 \\ 2 \\ -3 \end{bmatrix} = \begin{bmatrix} -6 \\ 3 \\ 4 \end{bmatrix}$$

Thus, $x_1 = -6$, $x_2 = 3$, and $x_3 = 4$.   ◀

MATCHED PROBLEM 3 ➤   Use matrix inverse methods to solve each of the following systems (see Matched Problem 2):

(A) $3x_1 - x_2 + x_3 = 3$ (B) $3x_1 - x_2 + x_3 = -5$

      $-x_1 + x_2 \qquad = -3$ $-x_1 + x_2 \qquad = 1$

      $x_1 \qquad + x_3 = 2$ $x_1 \qquad + x_3 = -4$ ◀

    As Examples 2 and 3 illustrate, inverse methods are very convenient for hand calculations because once the inverse is found, it can be used to solve any new system formed by changing only the constant terms. Since most graphing calculators and computers can compute the inverse of a matrix, this method also adapts readily to graphing calculator and computer solutions. However, if your graphing calculator or computer also has a built-in procedure for finding the reduced form of an augmented coefficient matrix, then it is just as convenient to use Gauss–Jordan elimination. Furthermore, Gauss–Jordan elimination can be used in all cases and, as noted below, matrix inverse methods cannot always be used.

Using Inverse Methods to Solve Systems of Equations

If the number of equations in a system equals the number of variables and the coefficient matrix has an inverse, then the system will always have a unique solution that can be found by using the inverse of the coefficient matrix to solve the corresponding matrix equation.

    Matrix Equation   Solution

         $AX = B$ $X = A^{-1}B$

Remark

What happens if the coefficient matrix does not have an inverse? In this case, it can be shown that the system does not have a unique solution and is either dependent or inconsistent. Gauss–Jordan elimination must be used to determine which is the case. Also, as we mentioned earlier, Gauss–Jordan elimination always must be used if the number of variables is not the same as the number of equations.

■ APPLICATION

The following application will illustrate the usefulness of the inverse matrix method for solving systems of equations.

EXAMPLE 4 ➤   Investment Analysis   An investment advisor currently has two types of investments available for clients: a conservative investment A that pays 10% per year and an investment B of higher risk that pays 20% per year. Clients may divide their investments between the two to achieve any total return desired between 10% and 20%. However, the higher the desired return, the higher the risk. How should each client listed in the table invest to achieve the indicated return?

| | CLIENT | | | |
	1	*2*	*3*	*k*
TOTAL INVESTMENT	$20,000	$50,000	$10,000	k_1
ANNUAL RETURN DESIRED	$ 2,400	$ 7,500	$ 1,300	k_2
	(12%)	(15%)	(13%)	

Solution   We will solve the problem for an arbitrary client k by finding an inverse matrix. Then we will apply the result to the three specific clients.

Let: x_1 = Amount invested in A

  x_2 = Amount invested in B

Then: $x_1 +$ $x_2 = k_1$ Total invested

    $0.1x_1 + 0.2x_2 = k_2$ Total annual return desired

Write as a matrix equation:

$$\begin{matrix} A & X & B \end{matrix}$$
$$\begin{bmatrix} 1 & 1 \\ 0.1 & 0.2 \end{bmatrix} \begin{bmatrix} x_1 \\ x_2 \end{bmatrix} = \begin{bmatrix} k_1 \\ k_2 \end{bmatrix}$$

If A^{-1} exists, then

  $x = A^{-1}B$

We now find A^{-1} by starting with the augmented matrix $[A\,|\,I]$ and proceeding as discussed in Section 4-5:

$$\begin{bmatrix} 1 & 1 & \Big| & 1 & 0 \\ 0.1 & 0.2 & \Big| & 0 & 1 \end{bmatrix} \quad 10R_2 \rightarrow R_2$$

$$\sim \begin{bmatrix} 1 & 1 & \Big| & 1 & 0 \\ 1 & 2 & \Big| & 0 & 10 \end{bmatrix} \quad (-1)R_1 + R_2 \rightarrow R_2$$

$$\sim \begin{bmatrix} 1 & 1 & \Big| & 1 & 0 \\ 0 & 1 & \Big| & -1 & 10 \end{bmatrix} \quad (-1)R_2 + R_1 \rightarrow R_1$$

$$\sim \begin{bmatrix} 1 & 0 & \Big| & 2 & -10 \\ 0 & 1 & \Big| & -1 & 10 \end{bmatrix}$$

Thus

$$A^{-1} = \begin{bmatrix} 2 & -10 \\ -1 & 10 \end{bmatrix} \quad Check: \quad \overset{A^{-1}}{\begin{bmatrix} 2 & -10 \\ -1 & 10 \end{bmatrix}} \overset{A}{\begin{bmatrix} 1 & 1 \\ 0.1 & 0.2 \end{bmatrix}} = \overset{I}{\begin{bmatrix} 1 & 0 \\ 0 & 1 \end{bmatrix}}$$

and

$$\overset{X}{\begin{bmatrix} x_1 \\ x_2 \end{bmatrix}} = \overset{A^{-1}}{\begin{bmatrix} 2 & -10 \\ -1 & 10 \end{bmatrix}} \overset{B}{\begin{bmatrix} k_1 \\ k_2 \end{bmatrix}}$$

To solve each client's investment problem, we replace k_1 and k_2 with appropriate values from the table and multiply by A^{-1}:

Client 1

$$\begin{bmatrix} x_1 \\ x_2 \end{bmatrix} = \begin{bmatrix} 2 & -10 \\ -1 & 10 \end{bmatrix} \begin{bmatrix} 20,000 \\ 2,400 \end{bmatrix} = \begin{bmatrix} 16,000 \\ 4,000 \end{bmatrix}$$

Solution: $x_1 = \$16,000$ in investment A, $x_2 = \$4,000$ in investment B

Client 2

$$\begin{bmatrix} x_1 \\ x_2 \end{bmatrix} = \begin{bmatrix} 2 & -10 \\ -1 & 10 \end{bmatrix} \begin{bmatrix} 50,000 \\ 7,500 \end{bmatrix} = \begin{bmatrix} 25,000 \\ 25,000 \end{bmatrix}$$

Solution: $x_1 = \$25,000$ in investment A, $x_2 = \$25,000$ in investment B

Client 3

$$\begin{bmatrix} x_1 \\ x_2 \end{bmatrix} = \begin{bmatrix} 2 & -10 \\ -1 & 10 \end{bmatrix} \begin{bmatrix} 10,000 \\ 1,300 \end{bmatrix} = \begin{bmatrix} 7,000 \\ 3,000 \end{bmatrix}$$

Solution: $x_1 = \$7,000$ in investment A, $x_2 = \$3,000$ in investment B

To solve this problem on a graphing calculator or computer (with appropriate built-in routines), simply enter the matrix A (Fig. 2); then, for each client, enter B and compute $A^{-1}B$ (see Fig. 3, at the top of the next page).

FIGURE 2

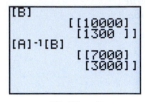

(A) Client 1   (B) Client 2   (C) Client 3

FIGURE 3

MATCHED PROBLEM 4   ➤   Repeat Example 4 with investment A paying 8% and investment B paying 24%.

Answers to Matched Problems

1. $AX + C = B$

$(AX + C) - C = B - C$
$AX + (C - C) = B - C$
$\quad\quad AX + 0 = B - C$

$AX = B - C$

$A^{-1}(AX) = A^{-1}(B - C)$
$(A^{-1}A)X = A^{-1}(B - C)$
$\quad\quad IX = A^{-1}(B - C)$

$X = A^{-1}(B - C)$

2. $x_1 = 2, x_2 = 5, x_3 = 0$

3. (A) $x_1 = -2, x_2 = -5, x_3 = 4$ (B) $x_1 = 0, x_2 = 1, x_3 = -4$

4. $A^{-1} = \begin{bmatrix} 1.5 & -6.25 \\ -0.5 & 6.25 \end{bmatrix}$; Client 1; \$15,000 in A and \$5,000 in B; Client 2: \$28,125 in A and \$21,875 in B; Client 3: \$6,875 in A and \$3,125 in B

EXERCISE   4-6

A   *Write Problems 1–4 as systems of linear equations without matrices.*

1. $\begin{bmatrix} 3 & 1 \\ 2 & -1 \end{bmatrix} \begin{bmatrix} x_1 \\ x_2 \end{bmatrix} = \begin{bmatrix} 5 \\ -4 \end{bmatrix}$

2. $\begin{bmatrix} -2 & 1 \\ -3 & 4 \end{bmatrix} \begin{bmatrix} x_1 \\ x_2 \end{bmatrix} = \begin{bmatrix} -5 \\ 7 \end{bmatrix}$

3. $\begin{bmatrix} -3 & 1 & 0 \\ 2 & 0 & 1 \\ -1 & 3 & -2 \end{bmatrix} \begin{bmatrix} x_1 \\ x_2 \\ x_3 \end{bmatrix} = \begin{bmatrix} 3 \\ -4 \\ 2 \end{bmatrix}$

4. $\begin{bmatrix} 2 & -1 & 0 \\ -2 & 3 & -1 \\ 4 & 0 & 3 \end{bmatrix} \begin{bmatrix} x_1 \\ x_2 \\ x_3 \end{bmatrix} = \begin{bmatrix} 6 \\ -4 \\ 7 \end{bmatrix}$

Write each system in Problems 5–8 as a matrix equation of the form AX = B.

5. $3x_1 - 4x_2 = 1$
$2x_1 + x_2 = 5$

6. $2x_1 + x_2 = 8$
$-5x_1 + 3x_2 = -4$

7. $x_1 - 3x_2 + 2x_3 = -3$
$-2x_1 + 3x_2 = 1$
$x_1 + x_2 + 4x_3 = -2$

8. $3x_1 + 2x_3 = 9$
$-x_1 + 4x_2 + x_3 = -7$
$-2x_1 + 3x_2 = 6$

Find x_1 and x_2 in Problems 9–12.

9. $\begin{bmatrix} x_1 \\ x_2 \end{bmatrix} = \begin{bmatrix} 3 & -2 \\ 1 & 4 \end{bmatrix} \begin{bmatrix} -2 \\ 1 \end{bmatrix}$

10. $\begin{bmatrix} x_1 \\ x_2 \end{bmatrix} = \begin{bmatrix} -2 & 1 \\ -1 & 2 \end{bmatrix} \begin{bmatrix} 3 \\ -2 \end{bmatrix}$

11. $\begin{bmatrix} x_1 \\ x_2 \end{bmatrix} = \begin{bmatrix} -2 & 3 \\ 2 & -1 \end{bmatrix} \begin{bmatrix} 3 \\ 2 \end{bmatrix}$

12. $\begin{bmatrix} x_1 \\ x_2 \end{bmatrix} = \begin{bmatrix} 3 & -1 \\ 0 & 2 \end{bmatrix} \begin{bmatrix} -2 \\ 1 \end{bmatrix}$

B *In Problems 13–20, write each system as a matrix equation and solve by using inverses.*

13. $x_1 + 2x_2 = k_1$
$x_1 + 3x_2 = k_2$
(A) $k_1 = 1, k_2 = 3$
(B) $k_1 = 3, k_2 = 5$
(C) $k_1 = -2, k_2 = 1$

14. $2x_1 + x_2 = k_1$
$5x_1 + 3x_2 = k_2$
(A) $k_1 = 2, k_2 = 13$
(B) $k_1 = 2, k_2 = 4$
(C) $k_1 = 1, k_2 = -3$

15. $x_1 + 3x_2 = k_1$
$2x_1 + 7x_2 = k_2$
(A) $k_1 = 2, k_2 = -1$
(B) $k_1 = 1, k_2 = 0$
(C) $k_1 = 3, k_2 = -1$

16. $2x_1 + x_2 = k_1$
$x_1 + x_2 = k_2$
(A) $k_1 = -1, k_2 = -2$
(B) $k_1 = 2, k_2 = 3$
(C) $k_1 = 2, k_2 = 0$

17. $x_1 - 3x_2 = k_1$
$x_2 + x_3 = k_2$
$2x_1 - x_2 + 4x_3 = k_3$
(A) $k_1 = 1, k_2 = 0, k_3 = 2$
(B) $k_1 = -1, k_2 = 1, k_3 = 0$
(C) $k_1 = 2, k_2 = -2, k_3 = 1$

18. $2x_1 + 3x_2 = k_1$
$x_1 + 2x_2 + 3x_3 = k_2$
$-x_2 - 5x_3 = k_3$
(A) $k_1 = 0, k_2 = 2, k_3 = 1$
(B) $k_1 = -2, k_2 = 0, k_3 = 1$
(C) $k_1 = 3, k_2 = 1, k_3 = 0$

19. $x_1 + x_2 = k_1$
$2x_1 + 3x_2 - x_3 = k_2$
$x_1 + 2x_3 = k_3$
(A) $k_1 = 2, k_2 = 0, k_3 = 4$
(B) $k_1 = 0, k_2 = 4, k_3 = -2$
(C) $k_1 = 4, k_2 = 2, k_3 = 0$

20. $x_1 - x_3 = k_1$
$2x_1 - x_2 = k_2$
$x_1 + x_2 - 2x_3 = k_3$
(A) $k_1 = 4, k_2 = 8, k_3 = 0$
(B) $k_1 = 4, k_2 = 0, k_3 = -4$
(C) $k_1 = 0, k_2 = 8, k_3 = -8$

C *For n × n matrices A and B, and n × 1 column matrices C, D, and X, solve each matrix equation in Problems 21–26 for X. Assume all necessary inverses exist.*

21. $AX - BX = C$
22. $AX + BX = C$
23. $AX + X = C$
24. $AX - X = C$
25. $AX - C = D - BX$
26. $AX + C = BX + D$

27. Use matrix inverse methods to solve the following system for the indicated values of k_1 and k_2.

$$x_1 + 2.001x_2 = k_1$$
$$x_1 + 2x_2 = k_2$$

(A) $k_1 = 1, k_2 = 1$
(B) $k_1 = 1, k_2 = 0$
(C) $k_1 = 0, k_2 = 1$

Discuss the effect of small changes in the constant terms on the solution set of this system.

28. Repeat Problem 27 for the following system:

$$x_1 - 3.001x_2 = k_1$$
$$x_1 - 3x_2 = k_2$$

In Problems 29–32, write each system as a matrix equation and solve by using the inverse coefficient matrix.

29.
$$x_1 + 8x_2 + 7x_3 = 135$$
$$6x_1 + 6x_2 + 8x_3 = 155$$
$$3x_1 + 4x_2 + 6x_3 = 75$$

30.
$$5x_1 + 3x_2 - 2x_3 = 112$$
$$7x_1 + 5x_2 \qquad = 70$$
$$3x_1 + x_2 - 9x_3 = 96$$

31.
$$6x_1 + 9x_2 + 7x_3 + 5x_4 = 250$$
$$6x_1 + 4x_2 + 7x_3 + 3x_4 = 195$$
$$4x_1 + 5x_2 + 3x_3 + 2x_4 = 145$$
$$4x_1 + 3x_2 + 8x_3 + 2x_4 = 125$$

32.
$$3x_1 + 3x_2 + 6x_3 + 5x_4 = 10$$
$$4x_1 + 5x_2 + 8x_3 + 2x_4 = 15$$
$$3x_1 + 6x_2 + 7x_3 + 4x_4 = 30$$
$$4x_1 + x_2 + 6x_3 + 3x_4 = 25$$

 A P P L I C A T I O N S

Solve Problems 33–38 using systems of equations and matrix inverses.

Business & Economics

33. *Resource allocation.* A concert hall has 10,000 seats. If tickets are $4 and $8, how many of each type of ticket should be sold (assuming all seats can be sold) to bring in each of the returns indicated in the table?

| | CONCERT | | |
	1	2	3
TICKETS SOLD	10,000	10,000	10,000
RETURN REQUIRED	$56,000	$60,000	$68,000

34. *Parking receipts.* Parking fees at a municipal zoo are $5.00 for local residents and $7.50 for all others. At the end of each day, the total number of vehicles parked that day and the gross receipts for the day are recorded, but the number of vehicles in each category is not. The following table contains the relevant information for a recent 4 day period. How many vehicles in each category used the zoo's parking facilities each day?

| | DAY | | | |
	1	2	3	4
VEHICLES PARKED	1,200	1,550	1,740	1,400
GROSS RECEIPTS	$7,125	$9,825	$11,100	$8,650

35. *Production scheduling.* A supplier for the automobile industry manufactures car and truck frames at two different plants. The production rates in frames per hour for each plant are given in the table:

PLANT	CAR FRAMES	TRUCK FRAMES
A	10	5
B	8	8

How many hours should each plant be scheduled to operate to exactly fill each of the orders in the following table?

| | ORDERS | | |
	1	2	3
CAR FRAMES	3,000	2,800	2,600
TRUCK FRAMES	1,600	2,000	2,200

36. *Production scheduling.* Labor and material costs for manufacturing two guitar models are given in the table:

GUITAR MODEL	LABOR COST	MATERIAL COST
A	$30	$20
B	$40	$30

If a total of $3,000 a week is allowed for labor and material, how many of each model should be produced each week to use exactly each of the allocations of the $3,000 indicated in the table at the top of the next page?

	WEEKLY ALLOCATION		
	1	*2*	*3*
LABOR	$1,800	$1,750	$1,720
MATERIAL	$1,200	$1,250	$1,280

37. *Incentive plan.* A small company provides an incentive plan for its top executives. Each executive receives as a bonus a percentage of the portion of the annual profit that remains after the bonuses for the other executives have been deducted (see the table). If the company has an annual profit of $2 million, find the bonus for each executive. Round each bonus to the nearest hundred dollars.

OFFICER	BONUS
President	3%
Executive vice president	2.5%
Associative vice president	2%
Assistant vice president	1.5%

38. *Incentive plan.* Repeat Problem 37 if the company decides to include a 1% bonus for the sales manager in the incentive plan.

39. *Taxable income.* As a result of several mergers and acquisitions, stock in four companies has been distributed among the companies. Each row of the following table gives the percentage of stock in the four companies that a particular company owns and the annual net income of each company (in millions of dollars):

COMPANY	PERCENTAGE OF STOCK OWNED IN COMPANY				ANNUAL NET INCOME
	A	*B*	*C*	*D*	Million $
A	71	8	3	7	3.2
B	12	81	11	13	2.6
C	11	9	72	8	3.8
D	6	2	14	72	4.4

Thus, company *A* holds 71% of its own stock, 8% of the stock in company *B*, 3% of the stock in company *C*, and so on. For the purpose of assessing a state tax on corporate income, the taxable income of each company is defined to

be its share of its own annual net income plus its share of the taxable income of each of the other companies, as determined by the percentages in the table. What is the taxable income of each company to the nearest thousand dollars?

40. *Taxable income.* Repeat Problem 39 if tax law is changed so that the taxable income of a company is defined to be all of its own annual net income plus its share of the taxable income of each of the other companies.

41. *Operating costs.* A durable-goods manufacturer has two production departments, freezers and refrigerators, and two service departments, accounting and maintenance. The monthly direct cost of operating each department is given in the table, along with the percentage of service time each department receives from each service department:

DEPARTMENT	DIRECT COSTS *Monthly*	PERCENTAGE OF SERVICE TIME RECEIVED FROM	
		Accounting	*Maintenance*
Freezers	$260,000	38	42
Refrigerators	$190,000	34	28
Accounting	$55,000	16	13
Maintenance	$95,000	12	17

The total cost of operating each department consists of its direct costs plus the cost of services received, which are determined by the percentages in the table. For example,

$$
\begin{pmatrix} \text{Total costs for} \\ \text{freezer department} \end{pmatrix}
= \begin{pmatrix} \text{Direct costs for} \\ \text{freezer department} \end{pmatrix}
$$

$$
+ 0.38 \begin{pmatrix} \text{Total costs for} \\ \text{accounting department} \end{pmatrix}
$$

$$
+ 0.42 \begin{pmatrix} \text{Total costs for} \\ \text{maintenance department} \end{pmatrix}
$$

(A) Find the total cost of operating each department. Round each amount to the nearest thousand dollars.

(B) Compare the sum of the direct costs of all four departments with the sum of the total costs of the two production departments determined in part (A), and interpret.

42. *Operating costs.* The manufacturer in Problem 41 adds another department that produces stoves. The relevant information is given in the table.

DEPARTMENT	DIRECT COSTS Monthly	PERCENTAGE OF SERVICE TIME RECEIVED FROM Accounting	Maintenance
Freezers	$260,000	24	28
Refrigerators	$190,000	26	23
Stoves	$210,000	22	19
Accounting	$75,000	16	13
Maintenance	$115,000	12	17

(A) Find the total cost of operating each department. Round each amount to the nearest thousand dollars.
(B) Compare the sum of the direct costs of all five departments with the sum of the total costs of the three production departments determined in part (A), and interpret.

Life Sciences

43. *Diets.* A biologist has available two commercial food mixes containing the percentages of protein and fat given in the table:

MIX	PROTEIN (%)	FAT (%)
A	20	2
B	10	6

How many ounces of each mix should be used to prepare each of the diets listed in the following table?

	DIET 1	2	3
PROTEIN	20 oz	10 oz	10 oz
FAT	6 oz	4 oz	6 oz

Social Sciences

44. *Education—resource allocation.* A state university system is planning to hire new faculty at the rank of lecturer or instructor for several of its 2 year community colleges. The number of sections taught and the annual salary (in thousands of dollars) for each rank are given in the table:

	RANK Lecturer	Instructor
SECTIONS TAUGHT	3	4
ANNUAL SALARY *Thousand $*	20	25

The number of sections that must be taught by the new faculty and the amount budgeted for salaries (in thousands of dollars) at each of the colleges are given in the following table. How many faculty of each rank should be hired at each college to exactly meet the demand for sections and completely exhaust the salary budget?

	COMMUNITY COLLEGE 1	2	3
DEMAND FOR SECTIONS	30	33	35
SALARY BUDGET *Thousand $*	200	210	220

S E C T I O N   4-7   ■   # Leontief Input–Output Analysis

- ■ INTRODUCTION
- ■ TWO-INDUSTRY MODEL
- ■ THREE-INDUSTRY MODEL

■ INTRODUCTION

A very important application of matrices and their inverses is found in the branch of applied mathematics called **input–output analysis.** Wassily Leontief, the primary force behind these new developments, was awarded the Nobel prize in economics in 1973 because of the significant impact his work had on economic planning for industrialized countries. Among other things, he conducted a comprehensive study of how 500 sectors of the American economy interacted with each other. Of course, large-scale computers played an important role in this analysis.

Our investigation will be more modest. In fact, we will start with an economy comprised of only two industries. From these humble beginnings, ideas and definitions will evolve that can be readily generalized for more realistic economics. Input–output analysis attempts to establish equilibrium conditions under which industries in an economy have just enough output to satisfy each other's demands in addition to final (outside) demands. Given the internal demands within the industries for each other's output, the problem is to determine output levels that will meet various levels of final (outside) demands.

■ TWO-INDUSTRY MODEL

To make the problem concrete, let us start with a hypothetical economy with only two industries—electric company E and water company W. Output for both companies is measured in dollars. The electric company uses both electricity and water (input) in the production of electricity (output), and the water company uses both electricity and water (input) in the production of water (output). Suppose that the production of each dollar's worth of electricity requires \$0.30 worth of electricity and \$0.10 worth of water, and the production of each dollar's worth of water requires \$0.20 worth of electricity and \$0.40 worth of water. If the final demand from the outside sector of the economy (the demand from all other users of electricity and water) is

$d_1 = \$12$ million for electricity

$d_2 = \$8$ million for water

how much electricity and water should be produced to meet this final demand?

To begin, suppose the electricity company produces \$12 million worth of electricity and the water company produces \$8 million worth of water (the final demand).

Then the production processes of the companies would require

Electricity	Electricity
required to	required to
produce	produce
electricity	water

$$0.3(12) \ + \ 0.2(8) \ = \$5.2 \text{ million of electricity}$$

and

Water	Water
required to	required to
produce	produce
electricity	water

$$0.1(12) \ + \ 0.4(8) \ = \$4.4 \text{ million of water}$$

leaving only $6.8 million of electricity and $3.6 million of water to satisfy the final demand of the outside sector. Thus, in order to meet the internal demands of both companies and to end up with enough electricity for the final outside demand, both companies must produce more than just the amount demanded by the outside sector. In fact, they must produce exactly enough to meet their own internal demands plus that demanded by the outside sector.

explore – discuss 1

Suppose the electric company doubles its production to $24 million of electricity and the water company doubles its production to $16 million of water. How much electricity and water is consumed in the production process? How much is left for the final demand of the outside sector? Try other values to see if you can determine by trial and error the production levels that will result in a final demand of $12 million for electricity and $8 million for water.

We now state the main problem of input–output analysis.

Basic Input–Output Problem

Given the internal demands for each industry's output, determine output levels for the various industries that will meet a given final (outside) level of demand as well as the internal demand.

If

$x_1 =$ Total output from electric company

$x_2 =$ Total output from water company

then reasoning as above, the internal demands are

$$0.3x_1 + 0.2x_2 \quad \text{Internal demand for electricity}$$
$$0.1x_1 + 0.4x_2 \quad \text{Internal demand for water}$$

Combining the internal demand with the final demand produces the following system of equations:

Total   Internal   Final
output   demand   demand

$$
\begin{aligned}
x_1 &= 0.3x_1 + 0.2x_2 + d_1 \\
x_2 &= 0.1x_1 + 0.4x_2 + d_2
\end{aligned}
\tag{1}
$$

or, in matrix form,

$$
\begin{bmatrix} x_1 \\ x_2 \end{bmatrix} = \begin{bmatrix} 0.3 & 0.2 \\ 0.1 & 0.4 \end{bmatrix} \begin{bmatrix} x_1 \\ x_2 \end{bmatrix} + \begin{bmatrix} d_1 \\ d_2 \end{bmatrix}
$$

or

$$X = MX + D \tag{2}$$

where

$$D = \begin{bmatrix} d_1 \\ d_2 \end{bmatrix} \qquad \text{Final demand matrix}$$

$$X = \begin{bmatrix} x_1 \\ x_2 \end{bmatrix} \qquad \text{Output matrix}$$

$$
\begin{array}{cc}
 & \begin{array}{cc} E & W \end{array} \\
M = \begin{array}{c} E \\ W \end{array} & \begin{bmatrix} 0.3 & 0.2 \\ 0.1 & 0.4 \end{bmatrix}
\end{array} \quad \text{Technology matrix}
$$

The **technology matrix** is the heart of input–output analysis. The elements in the technology matrix are determined as follows (read from left to right and then up):

$$
\begin{array}{c}
\text{Output} \\
\begin{array}{cc} E & W \end{array} \\
\uparrow \\
\text{Input} \quad \begin{array}{c} E \longrightarrow \\ \\ W \end{array}
\begin{bmatrix}
\left(\begin{array}{c} \text{Input from } E \\ \text{to produce \$1} \\ \text{of electricity} \end{array} \right) &
\left(\begin{array}{c} \text{Input from } E \\ \text{to produce \$1} \\ \text{of water} \end{array} \right) \\
\left(\begin{array}{c} \text{Input from } W \\ \text{to produce \$1} \\ \text{of electricity} \end{array} \right) &
\left(\begin{array}{c} \text{Input from } W \\ \text{to produce \$1} \\ \text{of water} \end{array} \right)
\end{bmatrix} = M
\end{array}
$$

Now our problem is to solve equation (2) for X. We proceed as in Section 4-6:

$$X = MX + D$$
$$X - MX = D$$
$$IX - MX = D \qquad I = \begin{bmatrix} 1 & 0 \\ 0 & 1 \end{bmatrix}$$
$$(I - M)X = D$$
$$X = (I - M)^{-1}D \qquad \text{Assuming } I - M \text{ has an inverse} \qquad (3)$$

We find

$$I - M = \begin{bmatrix} 0.7 & -0.2 \\ -0.1 & 0.6 \end{bmatrix} \quad \text{and} \quad (I - M)^{-1} = \begin{bmatrix} 1.5 & 0.5 \\ 0.25 & 1.75 \end{bmatrix}$$

Then we have

$$\begin{bmatrix} x_1 \\ x_2 \end{bmatrix} = \begin{bmatrix} 1.5 & 0.5 \\ 0.25 & 1.75 \end{bmatrix} \begin{bmatrix} d_1 \\ d_2 \end{bmatrix} \qquad (4)$$
$$= \begin{bmatrix} 1.5 & 0.5 \\ 0.25 & 1.75 \end{bmatrix} \begin{bmatrix} 12 \\ 8 \end{bmatrix} = \begin{bmatrix} 22 \\ 17 \end{bmatrix}$$

Therefore, the electric company must have an output of \$22 million and the water company must have an output of \$17 million so that each company can meet both internal and final demands.

Check We use equation (2) to check our work:

$$X = MX + D$$
$$\begin{bmatrix} 22 \\ 17 \end{bmatrix} \overset{?}{=} \begin{bmatrix} 0.3 & 0.2 \\ 0.1 & 0.4 \end{bmatrix} \begin{bmatrix} 22 \\ 17 \end{bmatrix} + \begin{bmatrix} 12 \\ 8 \end{bmatrix}$$
$$\begin{bmatrix} 22 \\ 17 \end{bmatrix} \overset{?}{=} \begin{bmatrix} 10 \\ 9 \end{bmatrix} + \begin{bmatrix} 12 \\ 8 \end{bmatrix}$$
$$\begin{bmatrix} 22 \\ 17 \end{bmatrix} \overset{\checkmark}{=} \begin{bmatrix} 22 \\ 17 \end{bmatrix}$$

Actually, (4) solves the original problem for arbitrary final demands d_1 and d_2. This is very useful, since (4) gives a quick solution not only for the final demands stated in the original problem, but also for various other projected final demands. If we had solved system (1) by Gauss–Jordan elimination, then we would have to start over for each new set of final demands.

Suppose in the original problem that the projected final demands 5 years from now are $d_1 = 24$ and $d_2 = 16$. To determine each company's output for this projection, we simply substitute these values into (4) and multiply:

$$\begin{bmatrix} x_1 \\ x_2 \end{bmatrix} = \begin{bmatrix} 1.5 & 0.5 \\ 0.25 & 1.75 \end{bmatrix} \begin{bmatrix} 24 \\ 16 \end{bmatrix} = \begin{bmatrix} 44 \\ 34 \end{bmatrix}$$

We summarize these results for convenient reference.

Solution to a Two-Industry Input–Output Problem

Given two industries, C_1 and C_2, with

$$
M = \begin{array}{c} C_1 \\ C_2 \end{array} \begin{bmatrix} a_{11} & a_{12} \\ a_{21} & a_{22} \end{bmatrix} \qquad X = \begin{bmatrix} x_1 \\ x_2 \end{bmatrix} \qquad D = \begin{bmatrix} d_1 \\ d_2 \end{bmatrix}
$$

Technology matrix C_1 C_2 — Output matrix — Final demand matrix

where a_{ij} is the input required from C_i to produce a dollar's worth of output for C_j, the solution to the input–output matrix equation

Total output — Internal demand — Final demand

$$
X = MX + D \tag{2}
$$

is

$$
X = (I - M)^{-1}D \tag{3}
$$

assuming $I - M$ has an inverse.

To solve an input–output problem on a graphing calculator or computer, simply store matrices M, D, and I in memory; then use equation (3) to find X and equation (2) to check your results. Figure 1 illustrates this process on a graphing calculator.

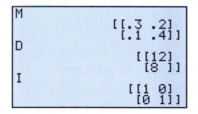

```
M
             [[.3 .2]
              [.1 .4]]
D
             [[12]
              [8 ]]
I
             [[1 0]
              [0 1]]
```

```
(I-M)-1*D→X
             [[22]
              [17]]
M*X+D
             [[22]
              [17]]
```

(A) Store M, D, and I in the graphing calculator's memory   (B) Compute X and check in equation (2)

FIGURE 1

■ THREE-INDUSTRY MODEL

Equations (2) and (3) in the solution to a two-industry input–output problem are the same for a three-industry economy, a four-industry economy, or an economy with n industries (where n is any natural number). The steps we took going from (2) to (3) hold for arbitrary matrices as long as the matrices have the correct sizes and $(I - M)^{-1}$ exists.

If equations (2) and (3) are valid for an economy with n industries, discuss the size of all the matrices in each equation.

The next example illustrates the application of equations (2) and (3) to a three-industry economy.

EXAMPLE 1 ➤ **Input–Output Analysis** An economy is based on three sectors, agriculture (A), energy (E), and manufacturing (M). Production of a dollar's worth of agriculture requires an input of \$0.20 from the agriculture sector and \$0.40 from the energy sector. Production of a dollar's worth of energy requires an input of \$0.20 from the energy sector and \$0.40 from the manufacturing sector. Production of a dollar's worth of manufacturing requires an input of \$0.10 from the agriculture sector, \$0.10 from the energy sector, and \$0.30 from the manufacturing sector. Find the output from each sector that is needed to satisfy a final demand of \$20 billion for agriculture, \$10 billion for energy, and \$30 billion for manufacturing.

Solution Since this is a three-industry problem, the technology matrix will be a 3×3 matrix, and the output and final demand matrices will be 3×1 column matrices. Using the information given in the problem, we can write

$$
\begin{array}{ccc}
\text{Technology matrix} & \text{Final demand} & \text{Output} \\
\begin{array}{c} \ \ A\ \ E\ \ \ M \end{array} & \text{matrix} & \text{matrix}
\end{array}
$$

$$
M = \begin{array}{c} A \\ E \\ M \end{array}\begin{bmatrix} 0.2 & 0 & 0.1 \\ 0.4 & 0.2 & 0.1 \\ 0 & 0.4 & 0.3 \end{bmatrix} \qquad D = \begin{bmatrix} 20 \\ 10 \\ 30 \end{bmatrix} \qquad X = \begin{bmatrix} x_1 \\ x_2 \\ x_3 \end{bmatrix}
$$

where M, X, and D satisfy the input–output equation $X = MX + D$. We store M, D, and I in the memory of a graphing utility (Fig. 2A), calculate $X = (I - M)^{-1}D$, and check by calculating $MX + D$ (Fig. 2B).

(A)

(B)

FIGURE 2

An output of $33 billion for agriculture, $37 billion for energy, and $64 billion for manufacturing will meet the given final demands. ◄

MATCHED PROBLEM 1 ► An economy is based on three sectors, coal, oil, and transportation. Production of a dollar's worth of coal requires an input of $0.20 from the coal sector and $0.40 from the transportation sector. Production of a dollar's worth of oil requires an input of $0.10 from the oil sector and $0.20 from the transportation sector. Production of a dollar's worth of transportation requires an input of $0.40 from the coal sector, $0.20 from the oil sector, and $0.20 from the transportation sector.

(A) Find the technology matrix M.
(B) Find $(I - M)^{-1}$.
(C) Find the output from each sector that is needed to satisfy a final demand of $30 billion for coal, $10 billion for oil, and $20 billion for transportation. ◄

Answers to Matched Problem 1. (A) $\begin{bmatrix} 0.2 & 0 & 0.4 \\ 0 & 0.1 & 0.2 \\ 0.4 & 0.2 & 0.2 \end{bmatrix}$ (B) $\begin{bmatrix} 1.7 & 0.2 & 0.9 \\ 0.2 & 1.2 & 0.4 \\ 0.9 & 0.4 & 1.8 \end{bmatrix}$

(C) $71 billion for coal, $26 billion for oil, and $67 billion for transportation

---

EXERCISE 4-7

A *Problems 1–4 pertain to the following input–output model: Assume an economy is based on two industrial sectors, agriculture (A) and energy (E). The technology matrix M and final demand matrices are (in billions of dollars)*

$$\begin{array}{cc} & A \quad E \\ \begin{matrix} A \\ E \end{matrix} & \begin{bmatrix} 0.4 & 0.2 \\ 0.2 & 0.1 \end{bmatrix} = M \end{array}$$

$$D_1 = \begin{bmatrix} 6 \\ 4 \end{bmatrix}$$

$$D_2 = \begin{bmatrix} 8 \\ 5 \end{bmatrix}$$

1. How much input from A and E are required to produce a dollar's worth of output for A?

2. How much input from A and E are required to produce a dollar's worth of output for E?

3. Find the output for each sector that is needed to satisfy the final demand D_1.

4. Repeat Problem 3 for D_2.

B *Problems 5–8 pertain to the following input–output model: Assume an economy is based on three industrial sectors, agriculture (A), building (B), and energy (E). The technology matrix M and final demand matrices (in billions of dollars) are*

$$\begin{array}{cccc} & A & B & E \\ \begin{matrix} A \\ B \\ E \end{matrix} & \begin{bmatrix} 0.3 & 0.2 & 0.2 \\ 0.1 & 0.1 & 0.1 \\ 0.2 & 0.1 & 0.1 \end{bmatrix} & = M \end{array}$$

$$D_1 = \begin{bmatrix} 5 \\ 10 \\ 15 \end{bmatrix}$$

$$D_2 = \begin{bmatrix} 20 \\ 15 \\ 10 \end{bmatrix}$$

5. How much input from A, B, and E are required to produce a dollar's worth of output for B?

6. How much of each of B's output dollars is required as input for each of the three sectors?

7. Find the output for each sector that is needed to satisfy the final demand D_1.

8. Repeat Problem 7 for D_2.

In Problems 9 and 10, find X.

9. $M = \begin{bmatrix} 0.2 & 0.2 \\ 0.3 & 0.3 \end{bmatrix}$; $D = \begin{bmatrix} 10 \\ 25 \end{bmatrix}$

10. $M = \begin{bmatrix} 0.4 & 0.1 \\ 0.2 & 0.3 \end{bmatrix}$; $D = \begin{bmatrix} 15 \\ 20 \end{bmatrix}$

C

In Problems 11 and 12, find X.

11. $M = \begin{bmatrix} 0.3 & 0.1 & 0.3 \\ 0.2 & 0.1 & 0.2 \\ 0.1 & 0.1 & 0.1 \end{bmatrix}$; $D = \begin{bmatrix} 20 \\ 5 \\ 10 \end{bmatrix}$

12. $M = \begin{bmatrix} 0.3 & 0.2 & 0.3 \\ 0.1 & 0.1 & 0.1 \\ 0.1 & 0.2 & 0.1 \end{bmatrix}$; $D = \begin{bmatrix} 10 \\ 25 \\ 15 \end{bmatrix}$

13. The technology matrix for an economy based on agriculture (A) and manufacturing (M) is

$$M = \begin{array}{cc} & \begin{array}{cc} A & M \end{array} \\ \begin{array}{c} A \\ M \end{array} & \begin{bmatrix} 0.3 & 0.25 \\ 0.1 & 0.25 \end{bmatrix} \end{array}$$

(A) Find the output for each sector that is needed to satisfy a final demand of $40 million for agriculture and $40 million for manufacturing.

(B) Discuss the effect on the final demand if the agriculture output in part (A) is increased by $20 million and the manufacturing output remains unchanged.

14. The technology matrix for an economy based on energy (E) and transportation (T) is

$$M = \begin{array}{cc} & \begin{array}{cc} E & T \end{array} \\ \begin{array}{c} E \\ T \end{array} & \begin{bmatrix} 0.25 & 0.25 \\ 0.4 & 0.2 \end{bmatrix} \end{array}$$

(A) Find the output for each sector that is needed to satisfy a final demand of $50 million for energy and $50 million for transportation.

(B) Discuss the effect on the final demand if the transportation output in part (A) is increased by $40 million and the energy output remains unchanged.

15. The technology matrix for an economy based on energy (E) and mining (M) is

$$M = \begin{array}{cc} & \begin{array}{cc} E & M \end{array} \\ \begin{array}{c} E \\ M \end{array} & \begin{bmatrix} 0.2 & 0.3 \\ 0.4 & 0.3 \end{bmatrix} \end{array}$$

The management of these two sectors would like to set the total output level so that the final demand is always 40% of the total output. Discuss methods that could be used to accomplish this objective.

16. The technology matrix for an economy based on automobiles (A) and construction (C) is

$$M = \begin{array}{cc} & \begin{array}{cc} A & C \end{array} \\ \begin{array}{c} A \\ C \end{array} & \begin{bmatrix} 0.1 & 0.4 \\ 0.1 & 0.1 \end{bmatrix} \end{array}$$

The management of these two sectors would like to set the total output level so that the final demand is always 70% of the total output. Discuss methods that could be used to accomplish this objective.

17. All the technology matrices in the text have elements between 0 and 1. Why is this the case? Would you ever expect to find an element in a technology matrix that is negative? That is equal to 0? That is equal to 1? That is greater than 1?

18. The sum of the elements in a column of any of the technology matrices in the text is less than 1. Why is this the case? Would you ever expect to find a column with a sum equal to 1? Greater than 1? How would you describe an economic system where the sum of the elements in every column of the technology matrix is 1?

APPLICATIONS

Business & Economics

19. An economy is based on two industrial sectors, coal and steel. Production of a dollar's worth of coal requires an input of $0.10 from the coal sector and $0.20 from the steel sector. Production of a dollar's worth of steel requires an input of $0.20 from the coal sector and $0.40 from the steel sector. Find the output for each sector that is needed to satisfy a final demand of $20 billion for coal and $10 billion for steel.

20. An economy is based on two sectors, transportation and manufacturing. Production of a dollar's worth of transportation requires $0.10 of input from each sector and production of a dollar's worth of manufacturing requires an input of $0.40 from each sector. Find the output for each sector that is needed to satisfy a final demand of $5 billion for transportation and $20 billion for manufacturing.

21. The economy of a small island nation is based on two sectors, agriculture and tourism. Production of a dollar's worth of agriculture requires an input of $0.20 from agriculture and $0.15 from tourism. Production of a dollar's worth of tourism requires an input of $0.40 from agriculture and $0.30 from tourism. Find the output from each sector that is needed to satisfy a final demand of $60 million for agriculture and $80 million for tourism.

22. The economy of a country is based on two sectors, agriculture and oil. Production of a dollar's worth of agriculture requires an input of $0.40 from agriculture and $0.35 from oil. Production of a dollar's worth of oil requires an input of $0.20 from agriculture and $0.05 from oil. Find the output from each sector that is needed to satisfy a final demand of $40 million for agriculture and $250 million for oil.

23. An economy is based on three sectors, agriculture, manufacturing, and energy. Production of a dollar's worth of agriculture requires inputs of $0.20 from agriculture, $0.20 from manufacturing, and $0.20 from energy. Production of a dollar's worth of manufacturing requires inputs of $0.40 from agriculture, $0.10 from manufacturing, and $0.10 from energy. Production of a dollar's worth of energy requires inputs of $0.30 from agriculture, $0.10 from manufacturing, and $0.10 from energy. Find the output for each sector that is needed to satisfy a final demand of $10 billion for agriculture, $15 billion for manufacturing, and $20 billion for energy.

24. A large energy company produces electricity, natural gas, and oil. The production of a dollar's worth of electricity requires inputs of $0.30 from electricity, $0.10 from natural gas, and $0.20 from oil. Production of a dollar's worth of natural gas requires inputs of $0.30 from electricity, $0.10 from natural gas, and $0.20 from oil. Production of a dollar's worth of oil requires inputs of $0.10 from each sector. Find the output for each sector that is needed to satisfy a final demand of $25 billion for electricity, $15 billion for natural gas, and $20 billion for oil.

25. An economy is based on four sectors, agriculture (A), energy (E), labor (L), and manufacturing (M). The table gives the input requirements for a dollar's worth of output for each sector, along with the projected final demand (in billions of dollars) for a 3 year period. Find the output for each sector that is needed to satisfy each of these final demands. Round answers to the nearest billion dollars.

| | | OUTPUT | | | FINAL DEMAND | | |
		A	E	L	M	1	2	3
	A	0.05	0.17	0.23	0.09	23	32	55
	E	0.07	0.12	0.15	0.19	41	48	62
INPUT	L	0.25	0.08	0.03	0.32	18	21	25
	M	0.11	0.19	0.28	0.16	31	33	35

26. Repeat Problem 25 with the following table:

| | | OUTPUT | | | FINAL DEMAND | | |
		A	E	L	M	1	2	3
	A	0.07	0.09	0.27	0.12	18	22	37
	E	0.14	0.07	0.21	0.24	26	31	42
INPUT	L	0.17	0.06	0.02	0.21	12	19	28
	M	0.15	0.13	0.31	0.19	41	45	49

Using Matrices to Find Cost, Revenue, and Profit

A toy distributor purchases model train components from various suppliers and packages these components in three different ready-to-run train sets, the Limited, the Empire, and the Comet. The components used in each set are listed in Table 1. For convenience, the total labor time required to prepare a set for shipping is included as a component.

TABLE 1
Product Components

	TRAIN SET		
COMPONENT	Limited	Empire	Comet
Locomotive	1	1	2
Car	5	6	8
Track piece	20	24	32
Track switch	1	2	4
Power pack	1	1	1
Labor (minutes)	15	18	24

The current costs of the components are given in Table 2, and the distributor's selling prices for the sets are given in Table 3.

TABLE 2
Component Costs

COMPONENT	COST PER UNIT ($)
Locomotive	12.52
Car	1.43
Track piece	0.25
Track switch	2.29
Power pack	12.54
Labor (per minute)	0.15

TABLE 3
Selling Prices

SET	PRICE ($)
Limited	54.60
Empire	62.28
Comet	81.15

The distributor has just received the order shown in Table 4 from a retail toy store.

TABLE 4
Customer Order

SET	QUANTITY
Limited	48
Empire	24
Comet	12

The distributor wants to store the information in each table in a matrix and use matrix operations to find the following information:

1. The inventory (parts and labor) required to fill the order
2. The cost (parts and labor) of filling the order
3. The revenue (sales) received from the customer
4. The profit realized on the order

(A) Use a single letter to designate the matrix representing each table and write matrix expressions in terms of these letters that will provide the required information. Discuss the size of the matrix you must use to represent each table so that all the pertinent matrix operations are defined.

(B) Evaluate the matrix expressions in part (A).

Shortly after filling the order in Table 4, a supplier informs the distributor that the cars and locomotives used in these train sets are no longer available. The distributor currently has 30 locomotives and 134 cars in stock.

(C) How many train sets of each type can the distributor produce using all the available locomotives and cars? Assume that the distributor has unlimited quantities of the other components used in these sets.

(D) How much profit will the distributor make if all these sets are sold? If there is more than one way to use all the available locomotives and cars, which one will produce the largest profit?

C H A P T E R   4   R E V I E W   ■   **Important Terms and Symbols**

4-1 *Review: Systems of Linear Equations in Two Variables.* Linear system; solution of a system; solution set; graphing method; consistent; inconsistent; independent; unique solution; dependent; substitution method; elimination by addition; equivalent systems; parameter; particular solution; equilibrium price; equilibrium quantity

4-2 *Systems of Linear Equations and Augmented Matrices.* Matrix; element; size; $m \times n$ matrix; dimensions; square matrix of order n; column matrix; row matrix; position of an element; double subscript notation, a_{ij}; principal diagonal; augmented matrix; equivalent systems; row-equivalent matrices; row operations; Gauss–Jordan elimination

$$R_i \leftrightarrow R_j; \quad kR_i \to R_i; \quad kR_j + R_i \to R_i$$

4-3 *Gauss–Jordan Elimination.* Reduced form; reduced system; Gauss–Jordan elimination; submatrix

4-4 *Matrices: Basic Operations.* Equal matrices; sum of two matrices; zero matrix; negative of a matrix; subtraction of matrices; product of a number and a matrix; product of a row matrix and a column matrix; product of two matrices; diagonal matrix; upper triangular matrix; markup; incidence matrix

4-5 *Inverse of a Square Matrix.* Identity element for multiplication; multiplicative inverse; singular matrix; encoding matrix; decoding matrix

4-6 *Matrix Equations and Systems of Linear Equations.* Matrix equation; basic properties of matrices

4-7 *Leontief Input–Output Analysis.* Input–output analysis; technology matrix; final demand matrix; output matrix

C H A P T E R   4   ■   **Review Exercise**

Work through all the problems in this chapter review and check your answers in the back of the book. Answers to all review problems are there along with section numbers in italics to indicate where each type of problem is discussed. Where weaknesses show up, review appropriate sections in the text.

A

1. Solve the following system by graphing:

$$2x - y = 4$$
$$x - 2y = -4$$

2. Solve the system in Problem 1 by substitition.

In Problems 3–11, perform the operations that are defined, given the following matrices:

$$A = \begin{bmatrix} 1 & 2 \\ 3 & 1 \end{bmatrix} \qquad B = \begin{bmatrix} 2 & 1 \\ 1 & 1 \end{bmatrix}$$

$$C = [2 \quad 3] \qquad D = \begin{bmatrix} 1 \\ 2 \end{bmatrix}$$

3. $A + B$ 4. $B + D$ 5. $A - 2B$
6. AB 7. AC 8. AD
9. DC 10. CD 11. $C + D$

12. Find the inverse of the matrix A given below by appropriate row operations on $[A \mid I]$. Show that $A^{-1}A = I$.

$$A = \begin{bmatrix} 4 & 3 \\ 3 & 2 \end{bmatrix}$$

13. Solve the following system using elimination by addition:

$$4x_1 + 3x_2 = 3$$
$$3x_1 + 2x_2 = 5$$

14. Solve the system in Problem 13 by performing appropriate row operations on the augmented matrix of the system.

15. Solve the system in Problem 13 by writing the system as a matrix equation and using the inverse of the coefficient matrix. Also, solve the system if the constants 3 and 5 are replaced by 7 and 10, respectively. By 4 and 2, respectively.

B
In Problems 16–21, perform the operations that are defined, given the following matrices:

$$A = \begin{bmatrix} 2 & -2 \\ 1 & 0 \\ 3 & 2 \end{bmatrix} \qquad B = \begin{bmatrix} -1 \\ 2 \\ 3 \end{bmatrix} \qquad C = [2 \quad 1 \quad 3]$$

$$D = \begin{bmatrix} 3 & -2 & 1 \\ -1 & 1 & 2 \end{bmatrix} \qquad E = \begin{bmatrix} 3 & -4 \\ -1 & 0 \end{bmatrix}$$

16. $A + D$ 17. $E + DA$ 18. $DA - 3E$
19. BC 20. CB 21. $AD - BC$

22. Find the inverse of the matrix A given below by appropriate row operations on $[A \mid I]$. Show that $A^{-1}A = I$.

$$A = \begin{bmatrix} 1 & 2 & 3 \\ 2 & 3 & 4 \\ 1 & 2 & 1 \end{bmatrix}$$

23. Solve by Gauss–Jordan elimination:

(A) $x_1 + 2x_2 + 3x_3 = 1$ (B) $x_1 + 2x_2 - x_3 = 2$
     $2x_1 + 3x_2 + 4x_3 = 3$ $2x_1 + 3x_2 + x_3 = -3$
     $x_1 + 2x_2 + x_3 = 3$ $3x_1 + 5x_2 = -1$

24. Solve the system in Problem 23A by writing the system as a matrix equation and using the inverse of the coefficient matrix. Also, solve the system if the constants 1, 3, and 3 are replaced by 0, 0, and -2, respectively. By -3, -4, and 1, respectively.

25. Given the technology matrix M and the final demand matrix D (in billions of dollars), find the output matrix X:

$$M = \begin{bmatrix} 0.2 & 0.15 \\ 0.4 & 0.3 \end{bmatrix} \qquad D = \begin{bmatrix} 30 \\ 20 \end{bmatrix}$$

26. Solve graphically to two decimal places:

$$x - 5y = -5$$
$$2x + 3y = 12$$

C

27. Find the inverse of the matrix A given below. Show that $A^{-1}A = I$.

$$A = \begin{bmatrix} 4 & 5 & 6 \\ 4 & 5 & -4 \\ 1 & 1 & 1 \end{bmatrix}$$

28. Solve the system

$$0.04x_1 + 0.05x_2 + 0.06x_3 = 360$$
$$0.04x_1 + 0.05x_2 - 0.04x_3 = 120$$
$$x_1 + x_2 + x_3 = 7,000$$

by writing it as a matrix equation and using the inverse of the coefficient matrix. (Before starting, multiply the first two equations by 100 to eliminate decimals.)

29. Solve Problem 28 by Gauss–Jordan elimination.

30. Given the technology matrix M and the final demand matrix D (in billions of dollars), find the output matrix X:

$$M = \begin{bmatrix} 0.2 & 0 & 0.4 \\ 0.1 & 0.3 & 0.1 \\ 0 & 0.4 & 0.2 \end{bmatrix} \qquad D = \begin{bmatrix} 40 \\ 20 \\ 30 \end{bmatrix}$$

31. Discuss the number of solutions for a system of n equations in n variables if the coefficient matrix:
(A) Has an inverse
(B) Does not have an inverse

32. Discuss the number of solutions for the system corresponding to the reduced form shown below if:
(A) $m \neq 0$ (B) $m = 0$ and $n \neq 0$
(C) $m = 0$ and $n = 0$

$$\begin{bmatrix} 1 & 0 & -2 & | & 5 \\ 0 & 1 & 3 & | & 3 \\ 0 & 0 & m & | & n \end{bmatrix}$$

33. One solution to the input–output equation $X = MX + D$ is given by $X = (I - M)^{-1}D$. Discuss the validity of each step in the following solutions of this equation. (Assume all necessary inverses exist.) Are both solutions correct?

(A)
$$X = MX + D$$
$$X - MX = D$$
$$X(I - M) = D$$
$$X = D(I - M)^{-1}$$

(B)
$$X = MX + D$$
$$-D = MX - X$$
$$-D = (M - I)X$$
$$X = (M - I)^{-1}(-D)$$

 APPLICATIONS

Business & Economics

34. *Break-even analysis.* A cookware manufacturer is preparing to market a new pasta machine. The company's fixed costs for research, development, tooling, etc., are $243,000 and the variable costs are $22.45 per machine. The company sells the pasta machine for $59.95.

(A) Find the cost and revenue equations.
(B) Find the break-even point.
(C) Graph both equations in the same coordinate system and show the break-even point. Use the graph to determine the production levels that will result in a profit and in a loss.

35. *Resource allocation.* A mining company has two mines with ore composition as given in the table. How many tons of each ore should be used to obtain exactly 4.5 tons of nickel and 10 tons of copper? Set up a system of equations and solve using Gauss–Jordan elimination.

ORE	NICKEL (%)	COPPER (%)
A	1	2
B	2	5

36. (A) Set up Problem 35 as a matrix equation and solve using the inverse of the coefficient matrix.
(B) Solve Problem 35 as in part (A) if 2.3 tons of nickel and 5 tons of copper are needed.

37. *Purchasing.* A soft-drink distributor has budgeted $300,000 for the purchase of 12 new delivery trucks. If a model A truck costs $18,000, a model B truck costs $22,000, and a model C truck costs $30,000, how many trucks of each model should the distributor purchase to use exactly all the budgeted funds?

38. *Material costs.* A manufacturer wishes to make two different bronze alloys in a metal foundry. The quantities of copper, tin, and zinc needed are indicated in matrix M. The costs for these materials (in dollars per pound) from two suppliers are summarized in matrix N. The company must choose one supplier or the other.

$$M = \begin{bmatrix} 4{,}800 \text{ lb} & 600 \text{ lb} & 300 \text{ lb} \\ 6{,}000 \text{ lb} & 1{,}400 \text{ lb} & 700 \text{ lb} \end{bmatrix} \begin{array}{l} \text{Alloy 1} \\ \text{Alloy 2} \end{array}$$

with columns labeled *Copper Tin Zinc*

$$N = \begin{bmatrix} \$0.75 & \$0.70 \\ \$6.50 & \$6.70 \\ \$0.40 & \$0.50 \end{bmatrix} \begin{array}{l} \text{Copper} \\ \text{Tin} \\ \text{Zinc} \end{array}$$

with columns labeled *Supplier A Supplier B*

(A) Discuss possible interpretations of the elements in the matrix products MN and NM.
(B) If either of the products MN or NM has a meaningful interpretation, find the product and label its rows and columns.

(C) Discuss methods of matrix multiplication that can be used to determine the supplier that will provide the necessary materials at the lowest cost.

39. *Labor costs.* A company with manufacturing plants in California and Texas has labor-hour and wage requirements for the manufacture of two inexpensive calculators as given in matrices M and N below:

Labor-hours per calculator

	Fabricating department	Assembly department	Packaging department	
$M =$	0.15 hr	0.10 hr	0.05 hr	Model A
	0.25 hr	0.20 hr	0.05 hr	Model B

Hourly wages

	California plant	Texas plant	
$N =$	$15	$12	Fabricating department
	$12	$10	Assembly department
	$ 4	$ 4	Packaging department

(A) Find the labor cost for producing one model B calculator at the California plant.

(B) Discuss possible interpretations of the elements in the matrix products MN and NM.

(C) If either of the products MN or NM has a meaningful interpretation, find the product and label its rows and columns.

40. *Investment analysis.* A person has $5,000 to invest, part at 5% and the rest at 10%. How much should be invested at each rate to yield $400 per year? Set up a system of equations and solve using augmented matrix methods.

41. *Investment analysis.* Solve Problem 40 by using a matrix equation and the inverse of the coefficient matrix.

42. *Resource allocation.* An outdoor amphitheater has 25,000 seats. Ticket prices are $8, $12, and $20, and the number of tickets priced at $8 must equal the number priced at $20. How many tickets of each type should be sold (assuming all seats can be sold) to bring in each of the returns indicated in the table? Set up a system of equations and solve using the inverse of the coefficient matrix.

	CONCERT		
	1	*2*	*3*
TICKETS SOLD	25,000	25,000	25,000
RETURN REQUIRED	$320,000	$330,000	$340,000

43. *Input–output analysis.* An economy is based on two industrial sectors, agriculture and fabrication. Production of a dollar's worth of agriculture requires an input of $0.30 from the agriculture sector and $0.20 from the fabricating sector. Production of a dollar's worth of fabrication requires $0.10 from the agriculture sector and $0.40 from the fabrication sector.

(A) Find the output for each sector that is needed to satisfy a final demand of $50 billion for agriculture and $20 billion for fabrication.

(B) Find the output for each sector that is needed to satisfy a final demand of $80 billion for agriculture and $60 billion for fabrication.

Social Sciences

44. *Cryptology.* The following message was encoded with the matrix B shown below. Decode the message:

25 8 26 24 25 33 21 41 48 41 30 50 21 32 41 52 52 79

$$B = \begin{bmatrix} 1 & 1 & 0 \\ 1 & 0 & 1 \\ 1 & 1 & 1 \end{bmatrix}$$

45. *Traffic flow.* The rush-hour traffic flow (in vehicles per hour) for a network of four one-way streets is shown in the figure.

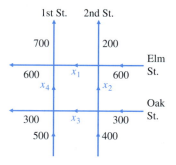

Figure for 45

(A) Write the system of equations determined by the flow of traffic through the four intersections.

(B) Find the solution of the system in part (A).

(C) What is the maximum number of vehicles per hour that can travel from Oak Street to Elm Street on 1st Street? What is the minimum number?

(D) If traffic lights are adjusted so that 500 vehicles per hour travel from Oak Street to Elm Street on 1st Street, determine the flow around the rest of the network.

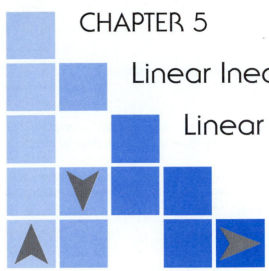

CHAPTER 5

Linear Inequalities and

Linear Programming

In this chapter we will discuss linear inequalities in two and more variables. In addition, we will introduce a relatively new and powerful mathematical tool called *linear programming,* which will be used to solve a variety of interesting practical problems. The row operations on matrices introduced in Chapter 4 will be particularly useful in Sections 5-4, 5-5, and 5-6.

S E C T I O N   5-1   ■ **Systems of Linear Inequalities in Two Variables**

- GRAPHING LINEAR INEQUALITIES IN TWO VARIABLES
- SOLVING SYSTEMS OF LINEAR INEQUALITIES GRAPHICALLY
- APPLICATIONS

Many applications of mathematics involve systems of inequalities rather than systems of equations. A graph is often the most convenient way to represent the solutions of a system of linear inequalities in two variables. In this section we discuss techniques for graphing both a single linear inequality in two variables and a system of linear inequalities in two variables.

■ **GRAPHING LINEAR INEQUALITIES IN TWO VARIABLES**

We know how to graph first-degree equations such as

$$y = 2x - 3 \qquad \text{and} \qquad 2x - 3y = 5$$

281

but how do we graph first-degree inequalities such as the following?

$$y \leq 2x - 3 \qquad \text{and} \qquad 2x - 3y > 5$$

We will find that graphing these inequalities is almost as easy as graphing the equations, but first, we must discuss some important subsets of a plane in a rectangular coordinate system.

A line divides the plane into two halves called **half-planes.** A vertical line divides it into **left** and **right half-planes;** a nonvertical line divides it into **upper** and **lower half-planes** (Fig. 1).

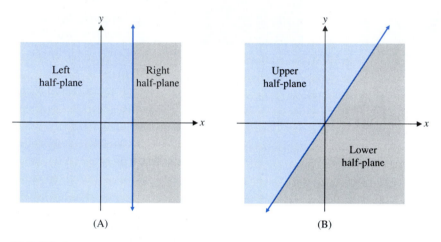

(A)   (B)

FIGURE 1

e x p l o r e – d i s c u s s   1

Consider the following linear equation and related linear inequalities:

(1) $3x - 4y = 24$ (2) $3x - 4y < 24$ (3) $3x - 4y > 24$

(A) Graph the line with equation (1).
(B) Find the point on this line with x coordinate 4 and draw a vertical line through this point. Discuss the relationship between the y coordinates of the points on this line and statements (1), (2), and (3).
(C) Repeat part (B) for $x = -4$. For $x = 12$.
(D) Based on your observations in parts (B) and (C), write a verbal description of all the points in the plane that satisfy equation (1), those that satisfy inequality (2), and those that satisfy inequality (3).

To investigate the half-planes determined by a linear equation such as $y - x = -2$, we rewrite the equation as $y = x - 2$. For any given value of x, there is exactly one value for y such that (x, y) lies on the line. For example, for $x = 4$, we have $y = 4 - 2 = 2$. For the same x and smaller values of y, the point (x, y) will lie below the line, since $y < x - 2$. Thus, the lower half-plane corresponds to the solution of the inequality $y < x - 2$. Similarly, the upper half-plane corresponds to $y > x - 2$, as shown in Figure 2.

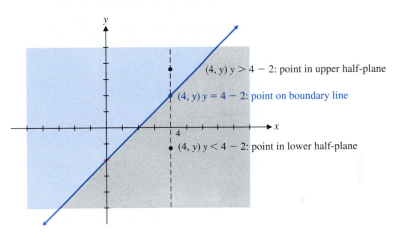

(4, y) $y > 4 - 2$: point in upper half-plane

(4, y) $y = 4 - 2$: point on boundary line

(4, y) $y < 4 - 2$: point in lower half-plane

FIGURE 2

The four inequalities formed from $y = x - 2$ by replacing the $=$ sign by $>$, $\geq$, $<$, and $\leq$, respectively, are

$$y > x - 2 \qquad y \geq x - 2 \qquad y < x - 2 \qquad y \leq x - 2$$

The graph of each is a half-plane, excluding the boundary line for $<$ and $>$, and including the boundary line for $\leq$ and $\geq$. In Figure 3, the half-planes are indicated with small arrows on the graph of $y = x - 2$ and then graphed as shaded regions. Excluded boundary lines are shown as dashed lines, and included boundary lines are shown as solid lines.

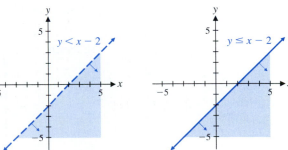

FIGURE 3

The preceding discussion suggests the following theorem, which is stated without proof:

theorem 1

> **Graphs of Linear Inequalities**
>
> The graph of the linear inequality
>
> $$Ax + By < C \qquad \text{or} \qquad Ax + By > C$$
>
> with $B \neq 0$, is either the upper half-plane or the lower half-plane (but not both) determined by the line $Ax + By = C$.
>
> If $B = 0$, the graph of
>
> $$Ax < C \qquad \text{or} \qquad Ax > C$$
>
> is either the left half-plane or the right half-plane (but not both) determined by the line $Ax = C$.

As a consequence of this theorem, we state a simple and fast mechanical procedure for graphing linear inequalities.

> **Procedure for Graphing Linear Inequalities**
>
> Step 1.   First graph $Ax + By = C$ as a dashed line if equality is not included in the original statement or as a solid line if equality is included.
>
> Step 2.   Choose a test point anywhere in the plane not on the line [the origin $(0, 0)$ usually requires the least computation] and substitute the coordinates into the inequality.
>
> Step 3.   The graph of the original inequality includes the half-plane containing the test point if the inequality is satisfied by that point or the half-plane not containing the test point if the inequality is not satisfied by that point.

EXAMPLE 1 ➤   **Graphing a Linear Inequality**   Graph: $2x - 3y \leq 6$

Solution   Step 1.   Graph $2x - 3y = 6$ as a solid line, since equality is included in the original statement:

Step 2. Pick a convenient test point above or below the line. The origin $(0, 0)$ requires the least computation, so substituting $(0, 0)$ into the inequality, we get

$$2x - 3y \leq 6$$
$$2(0) - 3(0) \ = 0 \leq 6$$

This is a true statement; therefore, the point $(0, 0)$ is in the solution set.

Step 3. The line $2x - 3y = 6$ and the half-plane containing the origin form the graph of $2x - 3y \leq 6$, as shown in the figure:

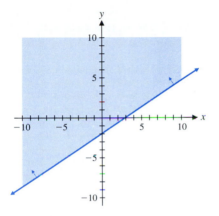

MATCHED PROBLEM 1 ▶ Graph: $6x - 3y > 18$

EXAMPLE 2 ▶ **Graphing Inequalities** Graph: (A) $y > -3$ (B) $2x \leq 5$ (C) $x \leq 3y$

Solution   (A)

(B)

(C)

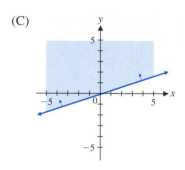

MATCHED PROBLEM 2 ➤   Graph: (A) $y < 4$ (B) $4x \geqslant -9$ (C) $3x \geqslant 2y$

■ SOLVING SYSTEMS OF LINEAR INEQUALITIES GRAPHICALLY

We now consider systems of linear inequalities such as

$$x + y \geqslant 6 \qquad \text{and} \qquad 2x + y \leqslant 22$$
$$2x - y \geqslant 0 \qquad\qquad\qquad x + y \leqslant 13$$
$$2x + 5y \leqslant 50$$
$$x \geqslant 0$$
$$y \geqslant 0$$

We wish to **solve** such systems **graphically;** that is, to find the graph of all ordered pairs of real numbers (x, y) that simultaneously satisfy all the inequalities in the system. The graph is called the **solution region** for the system. (In many applications, the solution region is also called the **feasible region.**) To find the solution region, we graph each inequality in the system and then take the intersection of all the graphs. To simplify the discussion that follows, **we will consider only systems of linear inequalities where equality is included in each statement in the system.**

EXAMPLE 3 ➤   Solving a System of Linear Inequalities Graphically   Solve the following system of linear inequalities graphically:

$$x + y \geqslant 6$$
$$2x - y \geqslant 0$$

Solution   Graph the line $x + y = 6$ and shade the region that satisfies the linear inequality $x + y \geqslant 6$. This region is shaded with gray lines in figure (A). Next, graph the line $2x - y = 0$ and shade the region that satisfies the inequality $2x - y \geqslant 0$. This region is shaded in blue in figure (A). The solution region for the system of inequalities is the intersection of these two regions. This is the region shaded in both gray and blue in figure (A) and redrawn in figure (B) with only the solution region shaded. The coordinates of any point in the shaded region of figure (B) specify a solution to the system. For example, the points (2, 4), (6, 3), and (7.43, 8.56) are three of infinitely many solutions, as can be easily checked. The intersection point (2, 4) is obtained by solving the equations $x + y = 6$ and $2x - y = 0$ simultaneously using any of the techniques discussed in Chapter 4.

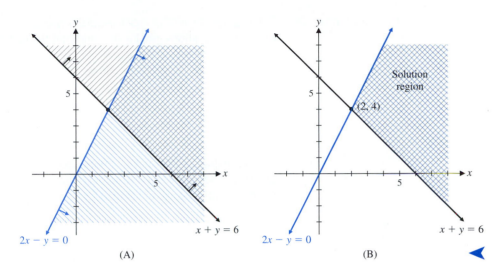

(A)   (B)

MATCHED PROBLEM 3 ▶ Solve the following system of linear inequalities graphically:

$$3x + y \leq 21$$
$$x - 2y \leq 0$$

explore – discuss 2

Refer to Example 3. Graph each boundary line and shade the regions obtained by reversing each inequality. That is, shade the region of the plane that corresponds to the inequality $x + y < 6$ and then shade the region that corresponds to the inequality $2x - y < 0$. What portion of the plane is left unshaded? Compare this method with the one used in the solution to Example 3.

FIGURE 4

As we saw in Section 4-1, a graphing utility is a useful tool for graphing lines and finding intersection points. Most graphing utilities also have some limited capabilities for shading regions that satisfy inequalities. Figure 4 shows a graphing utility solution for Example 3. If a solution region involves more than two boundary lines, it is best to make a hand sketch of the region, including the shading.

The points of intersection of the lines that form the boundary of a solution region will play a fundamental role in the solution of linear programming problems, which are discussed in the next section.

Corner Point

A **corner point** of a solution region is a point in the solution region that is the intersection of two boundary lines.

For example, the point (2, 4) is the only corner point of the solution region in Example 3.

EXAMPLE 4 ▶ **Solving a System of Linear Inequalities Graphically** Solve the following system of linear inequalities graphically, and find the corner points:

$$2x + y \leq 22$$
$$x + y \leq 13$$
$$2x + 5y \leq 50$$
$$x \geq 0$$
$$y \geq 0$$

Solution The inequalities $x \geq 0$ and $y \geq 0$ indicate that the solution region will lie in the first quadrant.* Thus, we can restrict our attention to that portion of the plane. First, using a graphing utility as an aid (Fig. 5), we graph the line associated with each inequality (Fig. 6).

$$2x + y = 22 \quad \text{Enter } y_1 = 22 - 2x.$$
$$x + y = 13 \quad \text{Enter } y_2 = 13 - x.$$
$$2x + 5y = 50 \quad \text{Enter } y_3 = 10 - 0.4x.$$

Next, choosing $(0, 0)$ as a test point, we see that the graph of each of the first three inequalities in the system consists of its corresponding line and the half-plane lying below the line, as indicated by the small arrows in Figure 6. Thus, the solution region of the system consists of the points in the first quadrant that simultaneously lie on or below all three of these lines (see the shaded region in Figure 6).

FIGURE 5

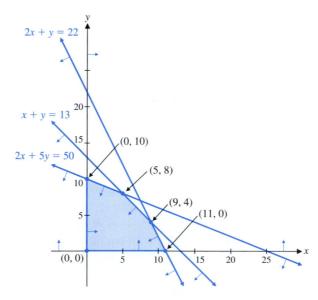

FIGURE 6

---

* The inequalities $x \geq 0$ and $y \geq 0$ occur frequently in applications involving systems of inequalities, since x and y often represent quantities that cannot be negative (number of units produced, number of hours worked, etc.).

The corner points $(0, 0)$, $(0, 10)$, and $(11, 0)$ are easily read from the graph. Using an intersection routine (details omitted), the other two corner points are

$(9, 4)$ Intersection of y_1 and y_2

$(5, 8)$ Intersection of y_2 and y_3

Note that the lines y_1 and y_3 also intersect, but the intersection point is not part of the solution region, and hence, is not a corner point.   ◄

MATCHED PROBLEM 4 ➤ Solve the following system of linear inequalities graphically, and find the corner points:

$$5x + y \geqslant 20$$
$$x + y \geqslant 12$$
$$x + 3y \geqslant 18$$
$$x \geqslant 0$$
$$y \geqslant 0$$

◄

If we compare the solution regions of Example 3 and Example 4, we see that there is a fundamental difference between these two regions. We can draw a circle around the solution region in Example 4; however, it is impossible to include all the points in the solution region in Example 3 in any circle, no matter how large we draw it. This leads to the following definition:

Bounded and Unbounded Solution Regions

A solution region of a system of linear inequalities is **bounded** if it can be enclosed within a circle. If it cannot be enclosed within a circle, then it is **unbounded.**

Thus, the solution region for Example 4 is bounded and the solution region for Example 3 is unbounded. This definition will be important in the next section.

■ APPLICATIONS

EXAMPLE 5 ➤ Medicine   A patient in a hospital is required to have at least 84 units of drug A and 120 units of drug B each day (assume that an overdosage of either drug is harmless). Each gram of substance M contains 10 units of drug A and 8 units of drug B, and each gram of substance N contains 2 units of drug A and 4 units of drug B. How many grams of substances M and N can be mixed to meet the minimum daily requirements?

Solution   To clarify relationships, we summarize the information in the following table:

| | AMOUNT OF DRUG PER GRAM | | MINIMUM DAILY REQUIREMENT |
	Substance M	*Substance N*	
DRUG A	10 units	2 units	84 units
DRUG B	8 units	4 units	120 units

Let: $x =$ Number of grams of substance M used

$y =$ Number of grams of substance N used

Then: $10x =$ Number of units of drug A in x grams of substance M

$2y =$ Number of units of drug A in y grams of substance N

$8x =$ Number of units of drug B in x grams of substance M

$4y =$ Number of units of drug B in y grams of substance N

The following conditions must be satisfied to meet daily requirements:

$$\begin{pmatrix} \text{Number of units of} \\ \text{drug } A \\ \text{in } x \text{ grams of substance } M \end{pmatrix} + \begin{pmatrix} \text{Number of units of} \\ \text{drug } A \\ \text{in } y \text{ grams of substance } N \end{pmatrix} \geqslant 84$$

$$\begin{pmatrix} \text{Number of units of} \\ \text{drug } B \\ \text{in } x \text{ grams of substance } M \end{pmatrix} + \begin{pmatrix} \text{Number of units of} \\ \text{drug } B \\ \text{in } y \text{ grams of substance } N \end{pmatrix} \geqslant 120$$

$$(\text{Number of grams of substance } M \text{ used}) \geqslant 0$$

$$(\text{Number of grams of substance } N \text{ used}) \geqslant 0$$

Converting these verbal statements into symbolic statements by using the variables x and y introduced above, we obtain the following system of linear inequalities:

$$10x + 2y \geqslant 84 \qquad \textit{Drug A restriction}$$
$$8x + 4y \geqslant 120 \qquad \textit{Drug B restriction}$$
$$x \geqslant 0 \qquad \textit{Cannot use a negative amount of M}$$
$$y \geqslant 0 \qquad \textit{Cannot use a negative amount of N}$$

Graphing this system of linear inequalities, we obtain the set of feasible solutions, or the feasible region (solution region), as shown in Figure 7. Thus, any point in the shaded area (including the straight-line boundaries) will meet the daily requirements; any point outside the shaded area will not. For example, 4 units of drug M and 23 units of drug N will meet the daily requirements, but 4 units of drug M and 21 units of drug N will not. (Note that the feasible region is unbounded.)

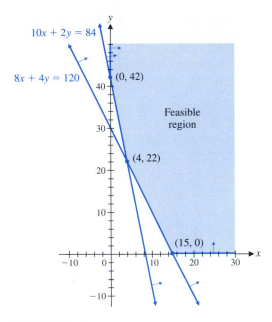

FIGURE 7

MATCHED PROBLEM 5 ➤   **Resource Allocation**   A manufacturing plant makes two types of inflatable boats, a two-person boat and a four-person boat. Each two-person boat requires 0.9 labor-hour in the cutting department and 0.8 labor-hour in the assembly department. Each four-person boat requires 1.8 labor-hours in the cutting department and 1.2 labor-hours in the assembly department. The maximum labor-hours available each month in the cutting and assembly departments are 864 and 672, respectively.

(A) Summarize this information in a table.
(B) If x two-person boats and y four-person boats are manufactured each month, write a system of linear inequalities that reflect the conditions indicated. Find the set of feasible solutions graphically. ◄

Remark

Refer to Example 5 and Matched Problem 5. In Example 5, it is certainly reasonable to consider noninteger values for x and y. The situation is not as clear in Matched Problem 5. How do we interpret a production schedule of 214.5 two-person boats and 347.75 four-person boats? It is not possible to manufacture a fraction of a boat. But it is possible to *average* 214.5 two-person and 347.75 four-person boats per month. In general, we will assume that all points in the feasible region represent acceptable solutions, even though noninteger solutions might require special interpretation.

Answers to Matched Problems

1. Graph $6x - 3y = 18$ as a dashed line (since equality is not included). Choosing the origin $(0, 0)$ as a test point, we see that $6(0) - 3(0) > 18$ is a false statement; thus, the lower half-plane determined by $6x - 3y = 18$ is the graph of $6x - 3y > 18$.

2. (A)

(B)

(C)

3.

4.

5. (A)

| | LABOR-HOURS REQUIRED | | MAXIMUM LABOR- |
	Two-Person Boat	*Four-Person Boat*	HOURS AVAILABLE PER MONTH
CUTTING DEPARTMENT	0.9	1.8	864
ASSEMBLY DEPARTMENT	0.8	1.2	672

(B) $0.9x + 1.8y \leq 864$
$0.8x + 1.2y \leq 672$
$x \geq 0$
$y \geq 0$

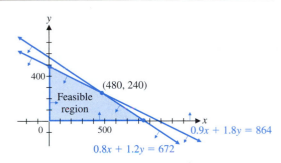

EXERCISE 5-1

A *Graph each inequality in Problems 1–12.*

1. $y \leq x - 1$
2. $y > x + 1$
3. $3x - 2y > 6$
4. $2x - 5y \leq 10$
5. $x \geq -4$
6. $y < 5$
7. $-4 \leq y < 4$
8. $0 \leq x < 6$
9. $6x + 4y \geq 24$
10. $4x + 8y \geq 32$
11. $5x \leq -2y$
12. $6x \geq 4y$

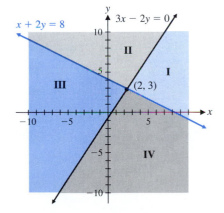

Figure for 13–16

In Problems 13–16, match the solution region of each system of linear inequalities with one of the four regions shown in the figure.

13. $x + 2y \leq 8$
    $3x - 2y \geq 0$

14. $x + 2y \geq 8$
    $3x - 2y \leq 0$

15. $x + 2y \geq 8$
    $3x - 2y \geq 0$

16. $x + 2y \leq 8$
    $3x - 2y \leq 0$

In Problems 17–20, solve each system of linear inequalities graphically.

17. $3x + y \geq 6$
    $x \leq 4$

18. $3x + 4y \leq 12$
    $y \geq -3$

19. $x - 2y \leq 12$
    $2x + y \geq 4$

20. $2x + 5y \leq 20$
    $x - 5y \geq -5$

B *In Problems 21–24, match the solution region of each system of linear inequalities with one of the four regions shown in the figure. Identify the corner points of each solution region.*

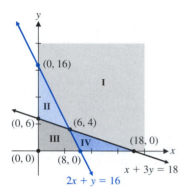

Figure for 21–24

21. $x + 3y \leq 18$
$2x + y \geq 16$
$x \geq 0$
$y \geq 0$

22. $x + 3y \leq 18$
$2x + y \leq 16$
$x \geq 0$
$y \geq 0$

23. $x + 3y \geq 18$
$2x + y \geq 16$
$x \geq 0$
$y \geq 0$

24. $x + 3y \geq 18$
$2x + y \leq 16$
$x \geq 0$
$y \geq 0$

Solve the systems in Problems 25–34 graphically, and indicate whether each solution region is bounded or unbounded. Find the coordinates of each corner point.

25. $2x + 3y \leq 12$
$x \geq 0$
$y \geq 0$

26. $3x + 4y \leq 24$
$x \geq 0$
$y \geq 0$

27. $2x + y \leq 10$
$x + 2y \leq 8$
$x \geq 0$
$y \geq 0$

28. $6x + 3y \leq 24$
$3x + 6y \leq 30$
$x \geq 0$
$y \geq 0$

29. $2x + y \geq 10$
$x + 2y \geq 8$
$x \geq 0$
$y \geq 0$

30. $4x + 3y \geq 24$
$3x + 4y \geq 8$
$x \geq 0$
$y \geq 0$

31. $2x + y \leq 10$
$x + y \leq 7$
$x + 2y \leq 12$
$x \geq 0$
$y \geq 0$

32. $3x + y \leq 21$
$x + y \leq 9$
$x + 3y \leq 21$
$x \geq 0$
$y \geq 0$

33. $2x + y \geq 16$
$x + y \geq 12$
$x + 2y \geq 14$
$x \geq 0$
$y \geq 0$

34. $3x + y \geq 24$
$x + y \geq 16$
$x + 3y \geq 30$
$x \geq 0$
$y \geq 0$

C *Solve the systems in Problems 35–44 graphically, and indicate whether each solution region is bounded or unbounded. Find the coordinates of each corner point.*

35. $x + 4y \leq 32$
$3x + y \leq 30$
$4x + 5y \geq 51$

36. $x + y \leq 11$
$x + 5y \geq 15$
$2x + y \geq 12$

37. $4x + 3y \leq 48$
$2x + y \geq 24$
$x \leq 9$

38. $2x + 3y \geq 24$
$x + 3y \leq 15$
$y \geq 4$

39. $x - y \leq 0$
$2x - y \leq 4$
$0 \leq x \leq 8$

40. $2x + 3y \geq 12$
$-x + 3y \leq 3$
$0 \leq y \leq 5$

41. $-x + 3y \geq 1$
$5x - y \geq 9$
$x + y \leq 9$
$x \leq 5$

42. $x + y \leq 10$
$5x + 3y \geq 15$
$-2x + 3y \leq 15$
$2x - 5y \leq 6$

43. $16x + 13y \leq 119$
$12x + 16y \geq 101$
$-4x + 3y \leq 11$

44. $8x + 4y \leq 41$
$-15x + 5y \leq 19$
$2x + 6y \geq 37$

Problems 45 and 46 introduce an algebraic process for finding the corner points of a solution region without drawing a graph. We will have a great deal more to say about this process later in this chapter.

45. Consider the following system of inequalities and corresponding boundary lines:

$3x + 4y \leq 36$ $\quad$ $3x + 4y = 36$
$3x + 2y \leq 30$ $\quad$ $3x + 2y = 30$
$x \geq 0$ $\quad\quad\quad$ $x = 0$
$y \geq 0$ $\quad\quad\quad$ $y = 0$

(A) Use algebraic methods to find the intersection points (if any exist) for each possible pair of boundary lines. (There are six different possible pairs.)
(B) Test each intersection point in all four inequalities to determine which are corner points.

46. Repeat Problem 45 for

$2x + y \leq 16$ $\quad$ $2x + y = 16$
$2x + 3y \leq 36$ $\quad$ $2x + 3y = 36$
$x \geq 0$ $\quad\quad\quad$ $x = 0$
$y \geq 0$ $\quad\quad\quad$ $y = 0$

APPLICATIONS

Business & Economics

47. *Manufacturing—resource allocation.* A manufacturing company makes two types of water skis, a trick ski and a slalom ski. The trick ski requires 6 labor-hours for fabricating and 1 labor-hour for finishing. The slalom ski requires 4 labor-hours for fabricating and 1 labor-hour for finishing. The maximum labor-hours available per day for fabricating and finishing are 108 and 24, respectively. If x is the number of trick skis and y is the number of slalom skis produced per day, write a system of linear inequalities that indicates appropriate restraints on x and y. Find the set of feasible solutions graphically for the number of each type of ski that can be produced.

48. *Manufacturing—resource allocation.* A furniture manufacturing company manufactures dining room tables and chairs. A table requires 8 labor-hours for assembling and 2 labor-hours for finishing. A chair requires 2 labor-hours for assembling and 1 labor-hour for finishing. The maximum labor-hours available per day for assembly and finishing are 400 and 120, respectively. If x is the number of tables and y is the number of chairs produced per day, write a system of linear inequalities that indicates appropriate restraints on x and y. Find the set of feasible solutions graphically for the number of tables and chairs that can be produced.

49. *Manufacturing—resource allocation.* Refer to Problem 47. The company makes a profit of $50 on each trick ski and a profit of $60 on each slalom ski.
  (A) If the company makes 10 trick skis and 10 slalom skis per day, the daily profit will be $1,100. Are there other production schedules that will result in a daily profit of $1,100? How are these schedules related to the graph of the line $50x + 60y = 1,100$?

  (B) Find a production schedule that will produce a daily profit greater than $1,100 and repeat part (A) for this schedule.

  (C) Discuss methods for using lines like those in parts (B) and (C) to find the largest possible daily profit.

50. *Manufacturing—resource allocation.* Refer to Problem 48. The company makes a profit of $50 on each table and a profit of $15 on each chair.
  (A) If the company makes 20 tables and 20 chairs per day, the daily profit will be $1,300. Are there other production schedules that will result in a daily profit of $1,300? How are these schedules related to the graph of the line $50x + 15y = 1,300$?

  (B) Find a production schedule that will produce a daily profit greater than $1,300 and repeat part (A) for this schedule.

  (C) Discuss methods for using lines like those in parts (B) and (C) to find the largest possible daily profit.

Life Sciences

51. *Nutrition—plants.* A farmer can buy two types of plant food, mix A and mix B. Each cubic yard of mix A contains 20 pounds of phosphoric acid, 30 pounds of nitrogen, and 5 pounds of potash. Each cubic yard of mix B contains 10 pounds of phosphoric acid, 30 pounds of nitrogen, and 10 pounds of potash. The minimum monthly requirements are 460 pounds of phosphoric acid, 960 pounds of nitrogen, and 220 pounds of potash. If x is the number of cubic yards of mix A used and y is the number of cubic yards of mix B used, write a system of linear inequalities that indicates appropriate restraints on x and y. Find the set of feasible solutions graphically for the amounts of mix A and mix B that can be used.

52. *Nutrition—people.* A dietitian in a hospital is to arrange a special diet using two foods. Each ounce of food M contains 30 units of calcium, 10 units of iron, and 10 units of vitamin A. Each ounce of food N contains 10 units of calcium, 10 units of iron, and 30 units of vitamin A. The minimum requirements in the diet are 360 units of calcium, 160 units of iron, and 240 units of vitamin A. If x is the number of ounces of food M used and y is the number of ounces of food N used, write a system of linear inequalities that reflects the conditions indicated. Find the set of feasible solutions graphically for the amount of each kind of food that can be used.

Social Sciences

53. *Psychology.* In an experiment on conditioning, a psychologist uses two types of Skinner (conditioning) boxes with mice and rats. Each mouse spends 10 minutes per day in box A and 20 minutes per day in box B. Each rat spends 20 minutes per day in box A and 10 minutes per day in box B. The total maximum time available per day is 800 minutes for box A and 640 minutes for box B. We are interested in the various numbers of mice and rats that can be used in the experiment under the conditions stated. If we let x be the number of mice used and y the number of rats used, write a system of linear inequalities that indicates appropriate restrictions on x and y. Find the set of feasible solutions graphically.

SECTION 5-2 ■ **Linear Programming in Two Dimensions— A Geometric Approach**

- A LINEAR PROGRAMMING PROBLEM
- LINEAR PROGRAMMING—A GENERAL DESCRIPTION
- GEOMETRIC SOLUTION OF LINEAR PROGRAMMING PROBLEMS
- APPLICATIONS

Several problems discussed in the last section are related to a more general type of problem called a *linear programming problem.* Linear programming is a mathematical process that has been developed to help management in decision-making, and it has become one of the most widely used and best-known tools of management science. We will introduce this topic by considering an example in detail, using an intuitive geometric approach. Insight gained from this approach will prove invaluable when we later consider an algebraic approach that is less intuitive but necessary in solving most real-world problems.

Notation Change

For ease of generalization to the larger problems in later sections, we will now change variable notation from letters such as x and y to subscript forms such as x_1 and x_2.

■ A LINEAR PROGRAMMING PROBLEM

We begin our discussion with a concrete example. The geometric method of solution will suggest two important theorems and a simple general geometric procedure for solving linear programming problems in two variables.

▶ EXAMPLE 1 ➤ **Production Scheduling** A manufacturer of lightweight mountain tents makes a standard model and an expedition model for national distribution. Each standard tent requires 1 labor-hour from the cutting department and 3 labor-hours from the assembly department. Each expedition tent requires 2 labor-hours from the cutting department and 4 labor-hours from the assembly department. The maximum labor-hours available per day in the cutting department and the assembly department are 32 and 84, respectively. If the company makes a profit of $50 on each standard tent and $80 on each expedition tent, how many tents of each type should be manufactured each day to maximize the total daily profit (assuming all tents can be sold)?

Solution

This is an example of a linear programming problem. To see relationships more clearly, we summarize the manufacturing requirements, objectives, and restrictions in Table 1.

T A B L E 1

| | LABOR-HOURS PER TENT | | MAXIMUM |
	Standard Model	Expedition Model	LABOR-HOURS AVAILABLE PER DAY
CUTTING DEPARTMENT	1	2	32
ASSEMBLY DEPARTMENT	3	4	84
PROFIT PER TENT	$50	$80	

We now proceed to formulate a mathematical model for the problem and then to solve it by using geometric methods.

Objective Function

The *objective* of management is to *decide* how many of each tent model should be produced each day in order to maximize profit.

Let: x_1 = Number of standard tents produced per day } Decision
       x_2 = Number of expedition tents produced per day } variables

The following equation gives the total profit for x_1 standard tents and x_2 expedition tents manufactured each day, assuming all tents manufactured are sold:

$P = 50x_1 + 80x_2$ Objective function

Mathematically, the manufacturer needs to decide on values for the **decision variables** (x_1, x_2) that achieve its objective—that is, maximize the **objective function** (profit) $P = 50x_1 + 80x_2$. (Note that P is a function of two independent variables.) It appears that the profit can be made as large as we like by manufacturing more and more tents—or can it?

Constraints

Any manufacturing company, no matter how large or small, has manufacturing limits imposed by available resources, plant capacity, demand, and so forth. These limits are referred to as **problem constraints.**

Cutting Department Constraint

$$\begin{pmatrix} \text{Daily cutting} \\ \text{time for } x_1 \\ \text{standard tents} \end{pmatrix} + \begin{pmatrix} \text{Daily cutting} \\ \text{time for } x_2 \\ \text{expedition tents} \end{pmatrix} \leq \begin{pmatrix} \text{Maximum labor-} \\ \text{hours available} \\ \text{per day} \end{pmatrix}$$

$$1x_1 + 2x_2 \leq 32$$

Assembly Department Constraint

$$\begin{pmatrix} \text{Daily assembly} \\ \text{time for } x_1 \\ \text{standard tents} \end{pmatrix} + \begin{pmatrix} \text{Daily assembly} \\ \text{time for } x_2 \\ \text{expedition tents} \end{pmatrix} \leq \begin{pmatrix} \text{Maximum labor-} \\ \text{hours available} \\ \text{per day} \end{pmatrix}$$

$$3x_1 \quad + \quad 4x_2 \quad \leq \quad 84$$

Nonnegative Constraints

It is not possible to manufacture a negative number of tents; thus, we have the **nonnegative constraints**

$$x_1 \geq 0$$
$$x_2 \geq 0$$

which we usually write in the form

$$x_1, x_2 \geq 0$$

Mathematical Model

We now have a **mathematical model** for the problem under consideration:

Maximize $\quad P = 50x_1 + 80x_2 \quad$ *Objective function*

Subject to $\quad \left.\begin{matrix} x_1 + 2x_2 \leq 32 \\ 3x_1 + 4x_2 \leq 84 \end{matrix}\right\} \quad$ *Problem constraints*

$\qquad\qquad x_1, x_2 \geq 0 \quad$ *Nonnegative constraints*

Graphical Solution

Solving the set of linear inequality constraints **graphically** (see the last section), we obtain the feasible region for production schedules (Fig. 1).

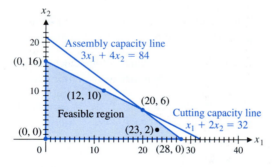

FIGURE 1

By choosing a production schedule (x_1, x_2) from the feasible region, a profit can be determined using the objective function

$$P = 50x_1 + 80x_2$$

For example, if $x_1 = 12$ and $x_2 = 10$, then the profit for the day would be

$$P = 50(12) + 80(10)$$
$$= \$1,400$$

Or if $x_1 = 23$ and $x_2 = 2$, then the profit for the day would be

$$P = 50(23) + 80(2)$$
$$= \$1,310$$

The question is, out of all possible production schedules (x_1, x_2) from the feasible region, which schedule(s) produces the maximum profit? Thus, we have a **maximization problem.** Since point-by-point checking is impossible (there are infinitely many points to check), we must find another way.

By assigning P in $P = 50x_1 + 80x_2$ a particular value and plotting the resulting equation in the coordinate system shown in Figure 1, we obtain a **constant-profit line (isoprofit line).** Every point in the feasible region on this line represents a production schedule that will produce the same profit. By doing this for a number of values for P, we obtain a family of constant-profit lines (Fig. 2) that are parallel to each other, since they all have the same slope. To see this, we write $P = 50x_1 + 80x_2$ in the slope–intercept form

$$x_2 = -\frac{5}{8} x_1 + \frac{P}{80}$$

and note that for any profit P, the constant-profit line has slope $-\frac{5}{8}$. We also observe that as the profit P increases, the x_2 intercept ($P/80$) increases, and the line moves away from the origin.

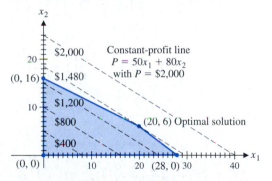

FIGURE 2
Constant-profit lines

Thus, the maximum profit occurs at a point where a constant-profit line is the farthest from the origin but still in contact with the feasible region. In this example, this occurs at $(20, 6)$, as is seen in Figure 2. Thus, if the manufacturer makes 20 standard tents and 6 expedition tents per day, the profit will be maximized at

$$P = 50(20) + 80(6)$$
$$= \$1,480$$

The point $(20, 6)$ is called an **optimal solution** to the problem, because it maximizes the objective (profit) function and is in the feasible region. In general, it appears that a maximum profit occurs at one of the corner points. We also note that the minimum profit ($P = 0$) occurs at the corner point $(0, 0)$. ◄

MATCHED PROBLEM 1 ➤ A manufacturing plant makes two types of inflatable boats, a two-person boat and a four-person boat. Each two-person boat requires 0.9 labor-hour from the cutting department and 0.8 labor-hour from the assembly department. Each four-person boat requires 1.8 labor-hours from the cutting department and 1.2 labor-hours from the assembly department. The maximum labor-hours available per month in the cutting department and the assembly department are 864 and 672, respectively. The company makes a profit of $25 on each two-person boat and $40 on each four-person boat.

(A) Summarize the relevant material in a table similar to Table 1 in Example 1.
(B) Identify the decision variables.
(C) Write the objective function P.
(D) Write the problem constraints and the nonnegative constraints.
(E) Graph the feasible region. Include graphs of the objective function for $P = $5,000$, $P = $10,000$, $P = $15,000$, and $P = $21,600$.
(F) From the graph and constant-profit lines, determine how many boats should be manufactured each month to maximize the profit. What is the maximum profit? ◀

explore – discuss 1

Refer to the feasible region S shown in the figure.

(A) Let $P = x_1 + x_2$. Graph the isoprofit lines through the points $(5, 5)$ and $(10, 10)$. Place a straightedge along the line with the smaller profit and slide it in the direction of increasing profit, without changing its slope. What is the maximum value of P? Where does this maximum value occur?
(B) Repeat part (A) for $P = x_1 + 10x_2$.
(C) Repeat part (A) for $P = 10x_1 + x_2$.

■ LINEAR PROGRAMMING—A GENERAL DESCRIPTION

In Example 1 and Matched Problem 1, the optimal solution occurs at a corner point of the feasible region. Is this always the case? The answer is a qualified yes, as will be seen in Theorem 1. First, we give a few general definitions.

A **linear programming problem** is one that is concerned with finding the **optimal value** (maximum or minimum value) of a linear **objective function** of the form

$$z = c_1x_1 + c_2x_2 + \cdots + c_nx_n$$

where the **decision variables** $x_1, x_2, \ldots x_n$ are subject to **problem constraints** in the form of linear inequalities and equations. In addition, the decision variables must satisfy the **nonnegative constraints** $x_i \geq 0$, $i = 1, 2, \ldots, n$. The set of points satisfying both the problem constraints and the nonnegative constraints is called the **feasible region** for the problem. Any point in the feasible region that produces the optimal value of the objective function over the feasible region is called an **optimal solution.**

theorem 1

> ### Fundamental Theorem of Linear Programming—Version 1
>
> If the optimal value of the objective function in a linear programming problem exists, then that value must occur at one (or more) of the corner points of the feasible region.

Theorem 1 provides a simple procedure for solving a linear programming problem, *provided the problem has a solution—not all do.* In order to use Theorem 1, we must know that the problem under consideration has a solution. Theorem 2 provides some conditions that will ensure that a linear programming problem has a solution.

theorem 2

> ### Existence of Solutions
>
> (A) If the feasible region for a linear programming problem is bounded, then both the maximum value and the minimum value of the objective function always exist.
>
> (B) If the feasible region is unbounded and the coefficients of the objective function are positive, then the minimum value of the objective function exists, but the maximum value does not.
>
> (C) If the feasible region is empty (that is, there are no points that satisfy all the constraints), then both the maximum value and the minimum value of the objective function do not exist.

Theorem 2 does not cover all possibilities. For example, what happens if the feasible region is unbounded and one (or both) of the coefficients of the objective function are negative? Problems of this type must be solved by carefully examining the graph of the objective function for various feasible solutions, as we did in Example 1. Since virtually all applied problems satisfy one of the conditions listed in Theorem 2, we will consider only problems of this type.

■ GEOMETRIC SOLUTION OF LINEAR PROGRAMING PROBLEMS

The discussion above leads to the following procedure for the geometric solution of linear programming problems with two decision variables:

Geometric Solution of a Linear Programming Problem with Two Decision Variables

Step 1.   For an applied problem, summarize relevant material in table form (see Table 1 in Example 1).

Step 2.   Form a mathematical model for the problem:

(A) Introduce decision variables, and write a linear objective function.
(B) Write problem constraints using linear inequalities and/or equations.
(C) Write nonnegative constraints.

Step 3.   Graph the feasible region. Then, if according to Theorem 2 an optimal solution exists, find the coordinates of each corner point.

Step 4.   Make a table listing the value of the objective function at each corner point.

Step 5.   Determine the optimal solution(s) from the table in step 4.

Step 6.   For an applied problem, interpret the optimal solution(s) in terms of the original problem.

Before we consider additional applications, let us use this procedure to solve some linear programming problems where the model has already been determined.

EXAMPLE 2 ➤   Solving a Linear Programming Problem

(A) Minimize and maximize

$$z = 3x_1 + x_2$$

Subject to

$$2x_1 + x_2 \leqslant 20$$
$$10x_1 + x_2 \geqslant 36$$
$$2x_1 + 5x_2 \geqslant 36$$
$$x_1, x_2 \geqslant 0$$

(B) Minimize and maximize

$$z = 10x_1 + 20x_2$$

Subject to

$$6x_1 + 2x_2 \geqslant 36$$
$$2x_1 + 4x_2 \geqslant 32$$
$$x_2 \leqslant 20$$
$$x_1, x_2 \geqslant 0$$

Solution   (A) We begin with step 3, since step 1 does not apply and step 2 has already been done for us.

Step 3.   Graph the feasible region S. Then, after checking Theorem 2 to determine that an optimal solution exists, find the coordinates of each corner point. Since S is bounded, z will have both a maximum and a minimum on S (Theorem 2A) and these will both occur at corner points (Theorem 1).

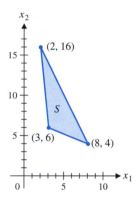

CORNER POINT

(x_1, x_2)	$z = 3x_1 + x_2$
(3, 6)	15
(2, 16)	22
(8, 4)	28

Step 4. Evaluate the objective function at each corner point, as shown in the table in the margin.

Step 5. Determine the optimal solutions from step 4. Examining the values in the table, we see that the minimum value of z is 15 at (3, 6) and the maximum value of z is 28 at (8, 4).

(B) Again, we can begin with step 3.

Step 3. Graph the feasible region S. Then, after checking Theorem 2 to determine that an optimal solution exists, find the coordinates of each corner point. Since S is unbounded and the coefficients of the objective function are positive, z has a minimum value on S but no maximum value (Theorem 2B).

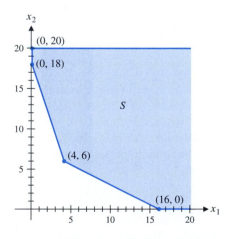

CORNER POINT

(x_1, x_2)	$z = 10x_1 + 20x_2$
(0, 20)	400
(0, 18)	360
(4, 6)	160
(16, 0)	160

Step 4. Evaluate the objective function at each corner point, as shown in the table in the margin.

Step 5. Determine the optimal solution from step 4. The minimum value of z is 160 at (4, 6) and at (16, 0). ◀

This is a **multiple optimal solution.**

In general, if two corner points are both optimal solutions to a linear programming problem, then any point on the line segment joining them is also an optimal solution.

This is the only time that optimal solutions also occur at noncorner points.

MATCHED PROBLEM 2 ▶ (A) Maximize and minimize $z = 4x_1 + 2x_2$ subject to the constraints given in Example 2A.
(B) Maximize and minimize $z = 20x_1 + 5x_2$ subject to the constraints given in Example 2B. ◀

explore – discuss 2

In Example 2B we saw that there was no optimal solution for the problem of maximizing the objective function P over the feasible region S. We want to add an additional constraint to modify the feasible region so that an optimal solution for the maximization problem does exist. Which of the following constraints will accomplish this objective?

(A) $x_1 \le 20$ (B) $x_2 \ge 4$ (C) $x_1 \le x_2$ (D) $x_2 \le x_1$

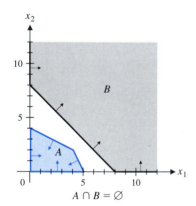

$A \cap B = \varnothing$

For an illustration of Theorem 2C, consider the following:

Maximize $P = 2x_1 + 3x_2$
Subject to $x_1 + x_2 \ge 8$
$x_1 + 2x_2 \le 8$
$2x_1 + x_2 \le 10$
$x_1, x_2 \ge 0$

The intersection of the graphs of the constraint inequalities is the empty set; hence, the feasible region is empty. If this happens, then the problem should be reexamined to see if it has been formulated properly. If it has, then the management may have to reconsider items such as labor-hours, overtime, budget, and supplies allocated to the project in order to obtain a nonempty feasible region and a solution to the original problem.

■ APPLICATIONS

▶ EXAMPLE 3 ▶ **Medicine**   We now convert Example 5 from the preceding section into a linear programing problem. A patient in a hospital is required to have at least 84 units of drug D_1 and 120 units of drug D_2 each day (assume that an overdosage of either drug is harmless). Each gram of substance M contains 10 units of drug D_1 and 8 units of drug D_2, and each gram of substance N contains 2 units of drug D_1 and 4 units of drug D_2. Now, suppose both M and N contain an undesirable drug D_3, 3 units per gram in M and 1 unit per gram in N. How many grams of each of substances M and N should be mixed to meet the minimum daily requirements and at the same time minimize the intake of drug D_3? How many units of the undesirable drug D_3 will be in this mixture?

Solution   **Step 1.** Summarize relevant material in table form:

	AMOUNT OF DRUG PER GRAM		MINIMUM DAILY
	Substance M	*Substance N*	REQUIREMENT
DRUG D_1	10 units	2 units	84 units
DRUG D_2	8 units	4 units	120 units
DRUG D_3	3 units	1 unit	

Step 2. Form a mathematical model for the problem.

Let: $x_1 =$ Number of grams of substance M used } Decision
　　　$x_2 =$ Number of grams of substance N used } variables

We form the linear objective function

$$C = 3x_1 + x_2$$

which gives the amount of the undesirable drug D_3 in x_1 grams of M and x_2 grams of N. Proceeding as before, we formulate the following mathematical model for the problem:

Minimize $C = 3x_1 + x_2$ Objective function
Subject to $10x_1 + 2x_2 \geq 84$ Drug D_1 constraint
　　　　　　$8x_1 + 4x_2 \geq 120$ Drug D_2 constraint
　　　　　　$x_1, x_2 \geq 0$ Nonnegative constraints

Step 3. Graph the feasible region. Then, after checking Theorem 2 to determine that an optimal solution exists, find the coordinates of each corner point. Solving the system of constraint inequalities graphically, we obtain the feasible region shown in Figure 3. Since the feasible region is unbounded and the coefficients of the objective function are positive, this minimization problem has a solution.

FIGURE 3

CORNER POINT (x_1, x_2)	$C = 3x_1 + x_2$
(0, 42)	42
(4, 22)	34
(15, 0)	45

Step 4.   Evaluate the objective function at each corner point, as shown in the table.

Step 5.   Determine the optimal solution from step 4. The optimal solution is $C = 34$ at the corner point (4, 22).

Step 6.   Interpret the optimal solution in terms of the original problem. If we use 4 grams of substance M and 22 grams of substance N, we will supply the minimum daily requirements for drugs D_1 and D_2 and minimize the intake of the undesirable drug D_3 at 34 units. (Any other combination of M and N from the feasible region will result in a larger amount of the undesirable drug D_3.)   ◄

MATCHED PROBLEM 3 ►

Agriculture   A chicken farmer can buy a special food mix A at 20¢ per pound and a special food mix B at 40¢ per pound. Each pound of mix A contains 3,000 units of nutrient N_1 and 1,000 units of nutrient N_2; each pound of mix B contains 4,000 units of nutrient N_1 and 4,000 units of nutrient N_2. If the minimum daily requirements for the chickens collectively are 36,000 units of nutrient N_1 and 20,000 units of nutrient N_2, how many pounds of each food mix should be used each day to minimize daily food costs while meeting (or exceeding) the minimum daily nutrient requirements? What is the minimum daily cost?   ◄

Answers to Matched Problems

1. (A)

	LABOR-HOURS REQUIRED		MAXIMUM LABOR-
	Two-Person Boat	*Four-Person Boat*	HOURS AVAILABLE PER MONTH
CUTTING DEPARTMENT	0.9	1.8	864
ASSEMBLY DEPARTMENT	0.8	1.2	672
PROFIT PER BOAT	$25	$40	

(B) x_1 = Number of two-person boats produced each month
     x_2 = Number of four-person boats produced each month

(C) $P = 25x_1 + 40x_2$

(D) $0.9x_1 + 1.8x_2 \le 864$
     $0.8x_1 + 1.2x_2 \le 672$
     $x_1, x_2 \ge 0$

(E)

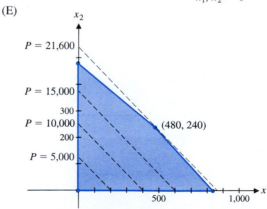

(F) 480 two-person boats, 240 four-person boats; Max $P = \$21,600$ per month

2. (A) Min $z = 24$ at $(3, 6)$; Max $z = 40$ at $(2, 16)$ and $(8, 4)$ (multiple optimal solution)

   (B) Min $z = 90$ at $(0, 18)$; no maximum value

3. 8 lb of mix A, 3 lb of mix B; Min $C = \$2.80$ per day

EXERCISE   5-2

A   *Find the maximum value of each objective function in Problems 1–4 over the feasible region S shown in the figure.*

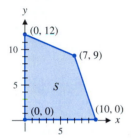

1. $z = x + y$ 2. $z = 4x + y$
3. $z = 3x + 7y$ 4. $z = 9x + 3y$

Find the minimum value of each objective function in Problems 5–8 over the feasible region T shown in the figure.

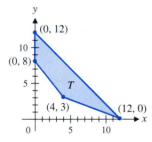

5. $z = 7x + 4y$ 6. $z = 7x + 9y$
7. $z = 3x + 8y$ 8. $z = 5x + 4y$

B   *Solve the linear programming problems stated in Problems 9–26.*

9. Maximize $P = 5x_1 + 5x_2$
    Subject to $2x_1 + x_2 \le 10$
    $\qquad\qquad\; x_1 + 2x_2 \le 8$
    $\qquad\qquad\;\; x_1, x_2 \ge 0$

10. Maximize $P = 3x_1 + 2x_2$
    Subject to $6x_1 + 3x_2 \le 24$
    $\qquad\qquad\; 3x_1 + 6x_2 \le 30$
    $\qquad\qquad\;\; x_1, x_2 \ge 0$

11. Minimize and maximize

    $z = 2x_1 + 3x_2$

    Subject to $2x_1 + x_2 \ge 10$
    $\qquad\qquad\; x_1 + 2x_2 \ge 8$
    $\qquad\qquad\;\; x_1, x_2 \ge 0$

12. Minimize and maximize

    $z = 8x_1 + 7x_2$

    Subject to $4x_1 + 3x_2 \ge 24$
    $\qquad\qquad\; 3x_1 + 4x_2 \ge 8$
    $\qquad\qquad\;\; x_1, x_2 \ge 0$

13. Maximize $P = 30x_1 + 40x_2$
    Subject to $2x_1 + x_2 \le 10$
    $\qquad\qquad\; x_1 + x_2 \le 7$
    $\qquad\qquad\; x_1 + 2x_2 \le 12$
    $\qquad\qquad\;\; x_1, x_2 \ge 0$

14. Maximize $P = 20x_1 + 10x_2$
    Subject to $3x_1 + x_2 \le 21$
    $\qquad\qquad\; x_1 + x_2 \le 9$
    $\qquad\qquad\; x_1 + 3x_2 \le 21$
    $\qquad\qquad\;\; x_1, x_2 \ge 0$

15. Minimize and maximize

    $z = 10x_1 + 30x_2$

    Subject to $2x_1 + x_2 \ge 16$
    $\qquad\qquad\; x_1 + x_2 \ge 12$
    $\qquad\qquad\; x_1 + 2x_2 \ge 14$
    $\qquad\qquad\;\; x_1, x_2 \ge 0$

16. Minimize and maximize

    $z = 400x_1 + 100x_2$

    Subject to $3x_1 + x_2 \ge 24$
    $\qquad\qquad\; x_1 + x_2 \ge 16$
    $\qquad\qquad\; x_1 + 3x_2 \ge 30$
    $\qquad\qquad\;\; x_1, x_2 \ge 0$

17. Minimize and maximize

$$P = 30x_1 + 10x_2$$

Subject to
$$2x_1 + 2x_2 \geqslant 4$$
$$6x_1 + 4x_2 \leqslant 36$$
$$2x_1 + x_2 \leqslant 10$$
$$x_1, x_2 \geqslant 0$$

18. Minimize and maximize

$$P = 2x_1 + x_2$$

Subject to
$$x_1 + x_2 \geqslant 2$$
$$6x_1 + 4x_2 \leqslant 36$$
$$4x_1 + 2x_2 \leqslant 20$$
$$x_1, x_2 \geqslant 0$$

19. Minimize and maximize

$$P = 3x_1 + 5x_2$$

Subject to
$$x_1 + 2x_2 \leqslant 6$$
$$x_1 + x_2 \leqslant 4$$
$$2x_1 + 3x_2 \geqslant 12$$
$$x_1, x_2 \geqslant 0$$

20. Minimize and maximize

$$P = -x_1 + 3x_2$$

Subject to
$$2x_1 - x_2 \geqslant 4$$
$$-x_1 + 2x_2 \leqslant 4$$
$$x_2 \leqslant 6$$
$$x_1, x_2 \geqslant 0$$

21. Minimize and maximize

$$P = 20x_1 + 10x_2$$

Subject to
$$2x_1 + 3x_2 \geqslant 30$$
$$2x_1 + x_2 \leqslant 26$$
$$-2x_1 + 5x_2 \leqslant 34$$
$$x_1, x_2 \geqslant 0$$

22. Minimize and maximize

$$P = 12x_1 + 14x_2$$

Subject to
$$-2x_1 + x_2 \geqslant 6$$
$$x_1 + x_2 \leqslant 15$$
$$3x_1 - x_2 \geqslant 0$$
$$x_1, x_2 \geqslant 0$$

23. Maximize $P = 20x_1 + 30x_2$

Subject to
$$0.6x_1 + 1.2x_2 \leqslant 960$$
$$0.03x_1 + 0.04x_2 \leqslant 36$$
$$0.3x_1 + 0.2x_2 \leqslant 270$$
$$x_1, x_2 \geqslant 0$$

24. Minimize $C = 30x_1 + 10x_2$

Subject to
$$1.8x_1 + 0.9x_2 \geqslant 270$$
$$0.3x_1 + 0.2x_2 \geqslant 54$$
$$0.01x_1 + 0.03x_2 \geqslant 3.9$$
$$x_1, x_2 \geqslant 0$$

25. Maximize $P = 525x_1 + 478x_2$

Subject to
$$275x_1 + 322x_2 \leqslant 3,381$$
$$350x_1 + 340x_2 \leqslant 3,762$$
$$425x_1 + 306x_2 \leqslant 4,114$$
$$x_1, x_2 \geqslant 0$$

26. Maximize $P = 300x_1 + 460x_2$

Subject to
$$245x_1 + 452x_2 \leqslant 4,181$$
$$290x_1 + 379x_2 \leqslant 3,888$$
$$390x_1 + 299x_2 \leqslant 4,407$$
$$x_1, x_2 \geqslant 0$$

C

27. The corner points for the bounded feasible region determined by the system of linear inequalities

$$x_1 + 2x_2 \leqslant 10$$
$$3x_1 + x_2 \leqslant 15$$
$$x_1, x_2 \geqslant 0$$

are $O = (0, 0)$, $A = (0, 5)$, $B = (4, 3)$, and $C = (5, 0)$. If $P = ax_1 + bx_2$ and $a, b > 0$, determine conditions on a and b that will ensure that the maximum value of P occurs:

(A) Only at A (B) Only at B
(C) Only at C (D) At both A and B
(E) At both B and C

28. The corner points for the feasible region determined by the system of linear inequalities

$$x_1 + 4x_2 \geqslant 30$$
$$3x_1 + x_2 \geqslant 24$$
$$x_1, x_2 \geqslant 0$$

are $A = (0, 24)$, $B = (6, 6)$, and $D = (30, 0)$. If $C = ax_1 + bx_2$ and $a, b > 0$, determine conditions on a and b that will ensure that the minimum value of C occurs:

(A) Only at A (B) Only at B
(C) Only at D (D) At both A and B
(E) At both B and D

A P P L I C A T I O N S

Business & Economics

29. *Manufacturing—resource allocation.* A manufacturing company makes two types of water skis, a trick ski and a slalom ski. The relevant manufacturing data are given in the table.

| | LABOR-HOURS PER SKI | | MAXIMUM LABOR-HOURS AVAILABLE PER DAY |
	Trick Ski	*Slalom Ski*	
FABRICATING DEPARTMENT	6	4	108
FINISHING DEPARTMENT	1	1	24

(A) If the profit on a trick ski is $40 and the profit on a slalom ski is $30, how many of each type of ski should be manufactured each day to realize a maximum profit? What is the maximum profit?

(B) Discuss the effect on the production schedule and the maximum profit if the profit on a slalom ski decreases to $25.

(C) Discuss the effect on the production schedule and the maximum profit if the profit on a slalom ski increases to $45.

30. *Manufacturing—resource allocation.* A furniture manufacturing company manufactures dining room tables and chairs. The relevant manufacturing data are given in the table.

| | LABOR-HOURS PER UNIT | | MAXIMUM LABOR-HOURS AVAILABLE PER DAY |
	Table	*Chair*	
ASSEMBLY DEPARTMENT	8	2	400
FINISHING DEPARTMENT	2	1	120
PROFIT PER UNIT	$90	$25	

(A) How many tables and chairs should be manufactured each day to realize a maximum profit? What is the maximum profit?

(B) Discuss the effect on the production schedule and the maximum profit if the marketing department of the company decides that the number of chairs produced should be at least four times the number of tables produced.

31. *Manufacturing—production scheduling.* A furniture company has two plants that produce the lumber used in manufacturing tables and chairs. In 1 day of operation, plant *A* can produce the lumber required to manufacture 20 tables and 60 chairs, and plant *B* can produce the lumber required to manufacture 25 tables and 50 chairs. The company needs enough lumber to manufacture at least 200 tables and 500 chairs.

(A) If it costs $1,000 to operate plant *A* for 1 day and $900 to operate plant *B* for 1 day, how many days should each plant be operated in order to produce a sufficient amount of lumber at a minimum cost? What is the minimum cost?

(B) Discuss the effect on the operating schedule and the minimum cost if the daily cost of operating plant *A* is reduced to $600 and all other data in part (A) remain the same.

(C) Discuss the effect on the operating schedule and the minimum cost if the daily cost of operating plant *B* is reduced to $800 and all other data in part (A) remain the same.

32. *Manufacturing—resource allocation.* An electronics firm manufactures two types of personal computers, a standard model and a portable model. The production of a standard computer requires a capital expenditure of $400 and 40 hours of labor. The production of a portable computer requires a capital expenditure of $250 and 30 hours of labor. The firm has $20,000 capital and 2,160 labor-hours available for production of standard and portable computers.

(A) What is the maximum number of computers the company is capable of producing?

(B) If each standard computer contributes a profit of $320 and each portable model contributes a profit of $220, how much profit will the company make by producing the maximum number of computers determined in part (A)? Is this the maximum profit? If not, what is the maximum profit?

33. *Transportation.* The officers of a high school senior class are planning to rent buses and vans for a class trip. Each bus can transport 40 students, requires 3 chaperones, and costs $1,200 to rent. Each van can transport 8 students, requires 1 chaperone, and costs $100 to rent. Since there are 400 students in the senior class that may be eligible to go on the trip, the officers must plan to accommodate at least 400 students. Since only 36 parents have volunteered to serve as chaperones, the officers must plan to use at most 36 chaperones. How many vehicles of each type should the officers rent in order to minimize the transportation costs? What are the minimal transportation costs?

34. *Investment.* An investor has $24,000 to invest in bonds of AAA and B qualities. The AAA bonds yield on the average 6% and the B bonds yield 10%. The investor's policy requires that she invest at least three times as much money in AAA bonds as in B bonds. How much should she invest in each type of bond to maximize her return? What is the maximum return?

35. *Pollution control.* Because of new federal regulations on pollution, a chemical plant introduced a new, more expensive process to supplement or replace an older process used in the production of a particular chemical. The older process emitted 20 grams of sulfur dioxide and 40 grams of particulate matter into the atmosphere for each gallon of chemical produced. The new process emits 5 grams of sulfur dioxide and 20 grams of particulate matter for each gallon produced. The company makes a profit of 60¢ per gallon and 20¢ per gallon on the old and new processes, respectively.

(A) If the government allows the plant to emit no more than 16,000 grams of sulfur dioxide and 30,000 grams of particulate matter daily, how many gallons of the chemical should be produced by each process to maximize daily profit? What is the maximum daily profit?

(B) Discuss the effect on the production schedule and the maximum profit if the government decides to restrict emissions of sulfur dioxide to 11,500 grams daily and all other data remain unchanged.

(C) Discuss the effect on the production schedule and the maximum profit if the government decides to restrict emissions of sulfur dioxide to 7,200 grams daily and all other data remain unchanged.

36. *Capital expansion.* A fast-food chain plans to expand by opening several new restaurants. The chain operates two types of restaurants, drive-through and full-service. A drive-through restaurant cost $100,000 to construct, requires 5 employees, and has an expected annual revenue of $200,000. A full-service restaurant costs $150,000 to construct, requires 15 employees, and has an expected annual revenue of $500,000. The chain has $2,400,000 in capital available for expansion. Labor contracts require that they hire no more than 210 employees, and licensing restrictions require that they open no more than 20 new restaurants. How many restaurants of each type should the chain open in order to maximize the expected revenue? What is the maximum expected revenue? How much of their capital will they use and how many employees will they hire?

Life Sciences

37. *Nutrition—plants.* A fruit grower can use two types of fertilizer in his orange grove, brand A and brand B. The amounts (in pounds) of nitrogen, phosphoric acid, and chlorine in a bag of each brand are given in the table. Tests indicate that the grove needs at least 1,000 pounds of phosphoric acid and at most 400 pounds of chlorine.

	POUNDS PER BAG	
	Brand A	Brand B
NITROGEN	8	3
PHOSPHORIC ACID	4	4
CHLORINE	2	1

(A) If the grower wants to maximize the amount of nitrogen added to the grove, how many bags of each mix should be used? How much nitrogen will be added?

(B) If the grower wants to minimize the amount of nitrogen added to the grove, how many bags of each mix should be used? How much nitrogen will be added?

38. *Nutrition–people.* A dietitian in a hospital is to arrange a special diet composed of two foods, M and N. Each ounce of food M contains 30 units of calcium, 10 units of iron, 10 units of vitamin A, and 8 units of cholesterol. Each ounce of food N contains 10 units of calcium, 10 units of iron, 30 units of vitamin A, and 4 units of cholesterol. If the minimum daily requirements are 360 units of calcium, 160 units of iron, and 240 units of vitamin A, how many ounces of each food should be used to meet the minimum requirements and at the same time minimize the cholesterol intake? What is the minimum cholesterol intake?

39. *Nutrition—plants.* A farmer can buy two types of plant food, mix A and mix B. Each cubic yard of mix A contains 20 pounds of phosphoric acid, 30 pounds of nitrogen, and 5 pounds of potash. Each cubic yard of mix B contains 10 pounds of phosphoric acid, 30 pounds of nitrogen, and 10 pounds of potash. The minimum monthly requirements are 460 pounds of phosphoric acid, 960 pounds of nitrogen, and 220 pounds of potash. If mix A costs $30 per cubic yard and

mix *B* costs $35 per cubic yard, how many cubic yards of each mix should the farmer blend to meet the minimum monthly requirements at a minimal cost? What is this cost?

40. *Nutrition—animals.* A laboratory technician in a medical research center is asked to formulate a diet from two commercially packaged foods, food *A* and food *B,* for a group of animals. Each ounce of food *A* contains 8 units of fat, 16 units of carbohydrate, and 2 units of protein. Each ounce of food *B* contains 4 units of fat, 32 units of carbohydrate, and 8 units of protein. The minimum daily requirements are 176 units of fat, 1,024 units of carbohydrate, and 384 units of protein. If food *A* costs 5¢ per ounce and food *B* costs 5¢ per ounce, how many ounces of each food should be used to meet the minimum daily requirements at the least cost? What is the cost for this amount of food?

Social Sciences

41. *Psychology.* In an experiment on conditioning, a psychologist uses two types of Skinner boxes with mice and rats. The amount of time (in minutes) each mouse and each rat spends in each box per day is given in the table. What is the maximum number of mice and rats that can be used in this experiment? How many mice and how many rats produce this maximum?

42. *Sociology.* A city council voted to conduct a study on inner-city community problems. A nearby university was contacted to provide sociologists and research assistants. Allocation of time and costs per week are given in the table. How many sociologists and how many research assistants should be hired to minimize the cost and meet the weekly labor-hour requirements? What is the minimum weekly cost?

	LABOR-HOURS		MINIMUM LABOR-HOURS NEEDED PER WEEK
	Sociologist	*Research Assistant*	
FIELDWORK	10	30	180
RESEARCH CENTER	30	10	140
COSTS PER WEEK	$500	$300	

	TIME		MAXIMUM TIME AVAILABLE PER DAY
	Mice	*Rats*	
SKINNER BOX *A*	10 min	20 min	800 min
SKINNER BOX *B*	20 min	10 min	640 min

SECTION 5-3 ■ **A Geometric Introduction to the Simplex Method**

- ■ STANDARD MAXIMIZATION PROBLEMS IN STANDARD FORM
- ■ SLACK VARIABLES
- ■ BASIC AND NONBASIC VARIABLES; BASIC SOLUTIONS AND BASIC FEASIBLE SOLUTIONS
- ■ BASIC FEASIBLE SOLUTIONS AND THE SIMPLEX METHOD

The geometric method of solving linear programming problems provides us with an overview of the subject and some useful terminology. But, practically speaking, the method is useful only for problems involving two decision variables and relatively few problem constraints. What happens when we need more decision variables and more problem constraints? We use an algebraic method called the *simplex method,* which was developed by George B. Dantzig in 1947 while on assignment to the U.S. Department of the Air Force. Ideally suited to computer use, the method is used routinely on applied problems involving hundreds and even thousands of variables and problem constraints.

The algebraic procedures utilized in the simplex method require the problem constraints to be written as equations rather than inequalities. This new form of the linear programming problem also prompts the use of some new terminology. We introduce this new form and associated terminology through a simple example and an appropriate geometric interpretation. From this example we can illustrate what the simplex method does geometrically before we immerse ourselves in the algebraic details of the process.

■ STANDARD MAXIMIZATION PROBLEMS IN STANDARD FORM

We now return to the tent production problem in Example 1 from the last section. Recall the mathematical model for the problem:

$$\begin{aligned}
\text{Maximize} \quad & P = 50x_1 + 80x_2 && \textit{Objective function} \\
\text{Subject to} \quad & x_1 + 2x_2 \le 32 && \textit{Cutting department constraint} \\
& 3x_1 + 4x_2 \le 84 && \textit{Assembly department constraint} \\
& x_1, x_2 \ge 0 && \textit{Nonnegative constraints}
\end{aligned} \tag{1}$$

where the decision variables x_1 and x_2 are the number of standard and expedition tents, respectively, produced each day.

Notice that the problem constraints involve $\le$ inequalities with positive constants to the right of the inequality. Maximization problems that satisfy this condition are called *standard maximization problems*. In this and the next section we will restrict our attention to standard maximization problems.

A Standard Maximization Problem in Standard Form

A linear programming problem is said to be a **standard maximization problem in standard form** if its mathematical model is of the following form:

Maximize the objective function

$$P = c_1x_1 + c_2x_2 + \cdots + c_nx_n$$

Subject to problem constraints of the form

$$a_1x_1 + a_2x_2 + \cdots + a_nx_n \le b \qquad b \ge 0$$

With nonnegative constraints

$$x_1, x_2, \ldots, x_n \ge 0$$

[*Note:* Mathematical model (1) above is a standard maximization problem in standard form. Also note that the coefficients of the objective function can be any real numbers.]

explore – discuss 1

Find an example of a standard maximization problem in standard form involving two variables and one problem constraint such that:

(A) The feasible region is bounded.
(B) The feasible region is unbounded.

Is it possible for a standard maximization problem to have no solution? Explain.

■ SLACK VARIABLES

To adapt a linear programming problem to the matrix methods used in the simplex process (as discussed in the next section), we convert the problem constraint inequalities into a system of linear equations by using a simple device called a *slack variable*. In particular, to convert the system of problem constraint inequalities from (1),

$$x_1 + 2x_2 \leq 32 \quad \textit{Cutting department constraint}$$
$$3x_1 + 4x_2 \leq 84 \quad \textit{Assembly department constraint}$$

(2)

into a system of equations, we add variables s_1 and s_2 to the left sides of (2) to obtain

$$x_1 + 2x_2 + s_1 \quad\ = 32$$
$$3x_1 + 4x_2 \quad\ + s_2 = 84$$

(3)

The variables s_1 and s_2 are called **slack variables** because each makes up the difference (takes up the slack) between the left and right sides of the inequalities in (2). For example, if we produced 20 standard tents ($x_1 = 20$) and 5 expedition tents ($x_2 = 5$), then the number of labor-hours used in the cutting department would be $20 + 2(5) = 30$, leaving a slack of 2 unused labor-hours out of the 32 available. Thus, s_1 would have the value of 2.

Notice that if the decision variables x_1 and x_2 satisfy the system of constraint inequalities (2), then the slack variables s_1 and s_2 are nonnegative. We will have more to say about this later in this discussion.

■ BASIC AND NONBASIC VARIABLES; BASIC SOLUTIONS AND BASIC FEASIBLE SOLUTIONS

Observe that system (3) has infinitely many solutions—just solve for s_1 and s_2 in terms of x_1 and x_2, and then assign x_1 and x_2 arbitrary values. Certain solutions of system (3), called *basic solutions*, are related to the intersection points of the (extended) boundary lines of the feasible region in Figure 1 (on the next page).

How are basic solutions to system (3) determined? System (3) involves four variables and two equations. We divide the four variables into two groups, called *basic variables* and *nonbasic variables,* as follows: Basic variables are selected arbitrarily with the restriction that there be as many basic variables as there are equations. The remaining variables are nonbasic variables.

Since system (3) has two equations, we can select any two of the four variables as basic variables. The remaining two variables are then nonbasic variables. A solution found by setting the two nonbasic variables equal to 0 and solving for the two basic variables is a basic solution. [Note that setting two variables equal to 0 in system (3) results in a system of two equations with two variables, which has (from Chapter 4) exactly one solution, infinitely many solutions, or no solution.]

EXAMPLE 1 ➤ Basic Solutions and the Feasible Region

(A) Find two basic solutions for system (3) by first selecting s_1 and s_2 as basic variables, and then by selecting x_2 and s_1 as basic variables.

(B) Associate each basic solution found in part (A) with an intersection point of the (extended) boundary lines of the feasible region in Figure 1, and indicate which boundary lines produce each intersection point.

(C) Indicate which of the intersection points found in part (B) are in the feasible region.

Solution (A) With s_1 and s_2 selected as basic variables, x_1 and x_2 are nonbasic variables. A basic solution is found by setting the nonbasic variables equal to 0 and solving for the basic variables. If $x_1 = 0$ and $x_2 = 0$, then system (3) becomes

$$
\begin{aligned}
s_1 = 32 \quad & x_1 + 2x_2 + s_1 \qquad = 32 \\
s_2 = 84 \quad & 3x_1 + 4x_2 \qquad + s_2 = 84
\end{aligned}
$$

and the basic solution is

$$x_1 = 0, \quad x_2 = 0, \quad s_1 = 32, \quad s_2 = 84 \tag{4}$$

If we select x_2 and s_1 as basic variables, then x_1 and s_2 are nonbasic variables. Setting the nonbasic variables equal to 0, system (3) becomes

$$2x_2 + s_1 = 32 \qquad \boxed{x_1} + 2x_2 + s_1 \quad \boxed{} = 32$$
$$4x_2 = 84 \qquad \boxed{3x_1} + 4x_2 \quad + \boxed{s_2} = 84$$

Solving, we see that $x_2 = 21$ and $s_1 = -10$, and the basic solution is

$$x_1 = 0, \quad x_2 = 21, \quad s_1 = -10, \quad s_2 = 0 \tag{5}$$

(B) Basic solution (4)—since $x_1 = 0$ and $x_2 = 0$—corresponds to the origin $O(0, 0)$ in Figure 1, which is the intersection of the boundary lines $x_1 = 0$ and $x_2 = 0$. Basic solution (5)—since $x_1 = 0$ and $x_2 = 21$—corresponds to the intersection point $D(0, 21)$, which is the intersection of the boundary lines $x_1 = 0$ and $3x_1 + 4x_2 = 84$.

(C) The intersection point $O(0, 0)$ is in the feasible region; hence, it is natural to call the corresponding basic solution a *basic feasible solution*. The intersection point $D(0, 21)$ is not in the feasible region; hence, the corresponding basic solution is not feasible. ◀

MATCHED PROBLEM 1 ➤ (A) Find two basic solutions for system (3) by first selecting x_1 and s_1 as basic variables, and then by selecting x_1 and s_2 as basic variables.

(B) Associate each basic solution found in part (A) with an intersection point of the (extended) boundary lines of the feasible region in Figure 1, and indicate which boundary lines produce each intersection point.

(C) Indicate which of the intersection points found in part (B) are in the feasible region. ◀

Proceeding systematically as in Example 1 and Matched Problem 1, we can obtain all basic solutions to system (3). The results are summarized in Table 1, which also includes geometric interpretations of the basic solutions relative to Figure 1. Figure 2 (on the next page) summarizes these interpretations. A careful study of Table 1 and Figure 2 is very worthwhile. (Note that to be sure we have listed all

TABLE 1
Basic Solutions

x_1	BASIC SOLUTIONS x_2	s_1	s_2	INTERSECTION POINT	INTERSECTING BOUNDARY LINES	FEASIBLE
0	0	32	84	$O(0, 0)$	$x_1 = 0$ $x_2 = 0$	Yes
0	16	0	20	$A(0, 16)$	$x_1 = 0$ $x_1 + 2x_2 = 32$	Yes
0	21	-10	0	$D(0, 21)$	$x_1 = 0$ $3x_1 + 4x_2 = 84$	No
32	0	0	-12	$E(32, 0)$	$x_2 = 0$ $x_1 + 2x_2 = 32$	No
28	0	4	0	$C(28, 0)$	$x_2 = 0$ $3x_1 + 4x_2 = 84$	Yes
20	6	0	0	$B(20, 6)$	$x_1 + 2x_2 = 32$ $3x_1 + 4x_2 = 84$	Yes

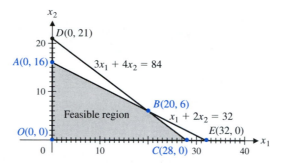

FIGURE 2
Basic feasible solutions: O, A, B, C
Basic solutions that are not feasible: D, E

possible basic solutions in Table 1, it is convenient to organize the table in terms of the zero values of the nonbasic variables.)

Observations from Table 1 and Figure 2 Important to the Development of the Simplex Method

1. In Table 1, observe that a basic solution that is not feasible includes at least one negative value and that a basic feasible solution does not include any negative values. That is, we can determine the feasibility of a basic solution simply by examining the signs of all the variables in the solution.
2. In Table 1 and Figure 2, observe that basic feasible solutions are associated with the corner points of the feasible region, which include the optimal solution to the original linear programming problem.

Before proceeding further, let us formalize the definitions alluded to in the discussion above so that they apply to standard maximization problems in general, without any reference to geometric forms.

Basic Variables and Nonbasic Variables; Basic Solutions and Basic Feasible Solutions

Given a system of linear equations associated with a linear programming problem (such a system will always have more variables than equations):

The variables are divided into two (mutually exclusive) groups, as follows: **Basic variables** are selected arbitrarily with the one restriction that there be as many basic variables as there are equations. The remaining variables are called **nonbasic variables.**

A solution found by setting the nonbasic variables equal to 0 and solving for the basic variables is called a **basic solution.** If a basic solution has no negative values, it is a **basic feasible solution.**

EXAMPLE 2 ➤ **Basic Variables and Basic Solutions** Suppose there is a system of three problem constraint equations with eight (slack and decision) variables associated with a standard maximization problem.

(A) How many basic variables and how many nonbasic variables are associated with the system?

(B) Setting the nonbasic variables equal to 0 will result in a system of how many linear equations with how many variables?

(C) If a basic solution has all nonnegative elements, is it feasible or not feasible?

Solution (A) Since there are three equations in the system, there should be three basic variables and five nonbasic variables.

(B) Three equations with three variables

(C) Feasible ◄

MATCHED PROBLEM 2 ➤ Suppose there is a system of five problem constraint equations with eleven (slack and decision) variables associated with a standard maximization problem.

(A) How many basic variables and how many nonbasic variables are associated with the system?

(B) Setting the nonbasic variables equal to 0 will result in a system of how many linear equations with how many variables?

(C) If a basic solution has one or more negative elements, is it feasible or not feasible? ◄

■ **BASIC FEASIBLE SOLUTIONS AND THE SIMPLEX METHOD**

The following important theorem, which is equivalent to the fundamental theorem (Theorem 1 in the preceding section), is stated without proof:

theorem 1

Fundamental Theorem of Linear Programming—Version 2

If the optimal value of the objective function in a linear programming problem exists, then that value must occur at one (or more) of the basic feasible solutions.

Now you can understand why the concepts of basic and nonbasic variables and basic solutions and basic feasible solutions are so important—the concepts are central to the process of finding optimal solutions to linear programming problems.

explore–discuss 2

If we know that a standard maximization problem has a solution, how could we use a table of basic solutions like Table 1 to find the optimal solution?

We have taken the first step toward finding a general method of solving linear programming problems involving any number of variables and problem constraints. That is, **we have found a method of identifying all the corner points (basic feasible solutions) of a feasible region without drawing its graph.** This is a critical step if we want to consider problems with more than two decision variables. Unfortunately, the number of corner points increases dramatically as the number of variables and constraints increases. In real-world problems, it is not practical to find all the corner points in order to find the optimal solution. Thus, the next step is to find a method of locating the optimal solution without finding every corner point. The procedure for doing this is the simplex method mentioned at the beginning of this section.

The simplex method, using a special matrix and row operations, automatically moves from one basic feasible solution to another — that is, from one corner point of the feasible region to another — each time getting closer to an optimal solution (if one exists), until an optimal solution is reached. Then the process stops. A remarkable property of the simplex method is that in large linear programming problems it usually arrives at an optimal solution (if one exists) by testing relatively few of the large number of basic feasible solutions (corner points) available.

With this background, we are now ready to discuss the algebraic details of the simplex method in the next section.

Answers to Matched Problems

1. (A) Basic solution corresponding to basic variables x_1 and s_1: $x_1 = 28$, $x_2 = 0$, $s_1 = 4$, $s_2 = 0$. Basic solution corresponding to basic variables x_1 and s_2: $x_1 = 32$, $x_2 = 0$, $s_1 = 0$, $s_2 = -12$.

   (B) The first basic solution corresponds to $C(28, 0)$, which is the intersection of the boundary lines $x_2 = 0$ and $3x_1 + 4x_2 = 84$. The second basic solution corresponds to $E(32, 0)$, which is the intersection of the boundary lines $x_2 = 0$ and $x_1 + 2x_2 = 32$.

   (C) $C(28, 0)$ is in the feasible region (hence, the corresponding basic solution is a basic feasible solution); $E(32, 0)$ is not in the feasible region (hence, the corresponding basic solution is not feasible).

2. (A) Five basic variables and six nonbasic variables

   (B) Five equations and five variables   (C) Not feasible

EXERCISE 5-3

A

1. Discuss the relationship between a standard maximization problem with two problem constraints and three decision variables, and the associated system of problem constraint equations. In particular, find each of the following quantities and explain how each was determined:

(A) The number of slack variables that must be introduced to form the system of problem constraint equations

(B) The number of basic variables and the number of nonbasic variables associated with the system

(C) The number of linear equations and the number of variables in the system formed by setting the nonbasic variables equal to 0

2. Repeat Problem 1 for a standard maximization problem with three problem constraints and four decision variables.

3. Discuss the relationship between a standard maximization problem and the associated system of problem constraint equations, if the system of problem constraint equations has nine variables, including five slack variables. In particular, find each of the following quantities and explain how each was determined:

(A) The number of constraint equations in the system

(B) The number of decision variables in the system

(C) The number of basic variables and the number of nonbasic variables associated with the system

(D) The number of linear equations and the number of variables in the system formed by setting the nonbasic variables equal to 0

4. Repeat Problem 3 if the system of problem constraint equations has ten variables, including four slack variables.

5. Listed in the table below are all the basic solutions for the system

$$2x_1 + 3x_2 + s_1 \qquad = 24$$
$$4x_1 + 3x_2 \qquad + s_2 = 36$$

For each basic solution, identify the nonbasic variables and the basic variables. Then classify each basic solution as feasible or not feasible.

	x_1	x_2	s_1	s_2
(A)	0	0	24	36
(B)	0	8	0	12
(C)	0	12	-12	0
(D)	12	0	0	-12
(E)	9	0	6	0
(F)	6	4	0	0

6. Repeat Problem 5 for the system

$$2x_1 + x_2 + s_1 \qquad = 30$$
$$x_1 + 5x_2 \qquad + s_2 = 60$$

whose basic solutions are given in the following table:

	x_1	x_2	s_1	s_2
(A)	0	0	30	60
(B)	0	30	0	-90
(C)	0	12	18	0
(D)	15	0	0	45
(E)	60	0	-90	0
(F)	10	10	0	0

7. Listed in the table below are all the possible choices of nonbasic variables for the system

$$2x_1 + x_2 + s_1 \qquad = 50$$
$$x_1 + 2x_2 \qquad + s_2 = 40$$

In each case, find the values of the basic variables and determine whether the basic solution is feasible.

	x_1	x_2	s_1	s_2
(A)	0	0	?	?
(B)	0	?	0	?
(C)	0	?	?	0
(D)	?	0	0	?
(E)	?	0	?	0
(F)	?	?	0	0

8. Repeat Problem 7 for the system

$$x_1 + 2x_2 + s_1 \qquad = 12$$
$$3x_1 + 2x_2 \qquad + s_2 = 24$$

B *Graph the systems of inequalities in Problems 9–12. Introduce slack variables to convert each system of inequalities to a system of equations, and find all the basic solutions of the system. Construct a table (similar to Table 1) listing each basic solution, the corresponding point on the graph, and whether the basic solution is feasible. (You do not need to list the intersecting lines.)*

9.
$$x_1 + x_2 \leq 16$$
$$2x_1 + x_2 \leq 20$$
$$x_1, x_2 \geq 0$$

10.
$$5x_1 + x_2 \leq 35$$
$$4x_1 + x_2 \leq 32$$
$$x_1, x_2 \geq 0$$

11.
$$2x_1 + x_2 \leq 22$$
$$x_1 + x_2 \leq 12$$
$$x_1 + 2x_2 \leq 20$$
$$x_1, x_2 \geq 0$$

12.
$$4x_1 + x_2 \leq 28$$
$$2x_1 + x_2 \leq 16$$
$$x_1 + x_2 \leq 13$$
$$x_1, x_2 \geq 0$$

SECTION 5-4 ■ **The Simplex Method: Maximization with Problem Constraints of the Form ≤**

- INITIAL SYSTEM
- THE SIMPLEX TABLEAU
- THE PIVOT OPERATION
- INTERPRETING THE SIMPLEX PROCESS GEOMETRICALLY
- THE SIMPLEX METHOD SUMMARIZED
- APPLICATION

We are now ready to develop the simplex method for a standard maximization problem. Specific details in the presentation of the method generally vary from one person to another or from one book to another, even though the underlying process is the same. The presentation developed here emphasizes concept development and understanding.

As was pointed out in the last section, the simplex method is most efficient when used with computers. Consequently, it is not intended that you become expert in manually solving linear programming problems using the simplex method. It is important that you become proficient in setting up linear programming problems so that they can be solved using computers, and it is also important that you develop skill in interpreting the results. The best way to gain this proficiency and interpretive skill is to set up and manually solve a number of fairly simple linear programming problems using the simplex method, and this is the purpose of the remainder of this chapter. (The computer supplement for this book contains a computer program for solving linear programming problems — see the Preface. Furthermore, some graphing calculators can be programmed to solve linear programming problems of the size discussed in this book. However, most real-world linear programming problems are far too complex to be solved on a graphing calculator.)

■ **INITIAL SYSTEM**

We will introduce the concepts and procedures involved in the simplex method through an example — the tent production example we have discussed earlier. We restate the problem here in standard form for convenient reference:

$$\text{Maximize} \quad P = 50x_1 + 80x_2 \quad \textit{Objective function}$$

$$\text{Subject to} \quad \left.\begin{array}{l} x_1 + 2x_2 \leq 32 \\ 3x_1 + 4x_2 \leq 84 \end{array}\right\} \quad \textit{Problem constraints} \tag{1}$$

$$x_1, x_2 \geq 0 \quad \textit{Nonnegative constraints}$$

Introducing slack variables s_1 and s_2, we convert the problem constraint inequalities in (1) into the following system of problem constraint equations:

$$\begin{array}{rl} x_1 + 2x_2 + s_1 = 32 \\ 3x_1 + 4x_2 + s_2 = 84 \\ x_1, x_2, s_1, s_2 \geq 0 \end{array} \tag{2}$$

Since a basic solution of (2) is not feasible if it contains any negative values, we have also included the nonnegative constraints for both the decision variables x_1 and x_2 and the slack variables s_1 and s_2. From our discussion in the last section, we know that out of the infinitely many solutions to system (2), an optimal solution is among the basic feasible solutions, which correspond to the corner points of the feasible region.

As part of the simplex method we add the objective function equation $P = 50x_1 + 80x_2$ in the form $-50x_1 - 80x_2 + P = 0$ to system (2) to create what is called the **initial system:**

$$\begin{aligned}
x_1 + 2x_2 + s_1 &= 32 \\
3x_1 + 4x_2 + s_2 &= 84 \\
-50x_1 - 80x_2 + P &= 0 \\
x_1, x_2, s_1, s_2 &\geq 0
\end{aligned} \tag{3}$$

When we add the objective function equation to system (2), we must slightly modify the earlier definitions of basic solution and basic feasible solution so that they apply to the initial system (3).

Basic Solutions and Basic Feasible Solutions for Initial Systems

1. The objective function variable P is always selected as a basic variable and is never selected as a nonbasic variable.
2. Note that a basic solution of system (3) is also a basic solution of system (2) after P is deleted.
3. If a basic solution of system (3) is a basic feasible solution of system (2) after deleting P, then the basic solution of (3) is called a **basic feasible solution** of (3).
4. A basic feasible solution of system (3) can contain a negative number, but only if it is the value of P, the objective function variable.

These changes lead to a small change in the second version of the fundamental theorem (see Theorem 1 in Section 5-3).

t h e o r e m 1

Fundamental Theorem of Linear Programming—Version 3

If the optimal value of the objective function in a linear programming problem exists, then that value must occur at one (or more) of the basic feasible solutions of the initial system.

With these adjustments understood, we start the simplex process with a basic feasible solution of the initial system (3), which we will refer to as an **initial basic**

feasible solution. An initial basic feasible solution that is easy to find is the one associated with the origin.

Since system (3) has three equations and five variables, it has three basic variables and two nonbasic variables. Looking at the system, we see that x_1 and x_2 appear in all equations and s_1, s_2, and P each appears only once and each in a different equation. A basic solution can be found by inspection by selecting s_1, s_2, and P as the basic variables (remember, P is always selected as a basic variable) and x_1 and x_2 as the nonbasic variables to be set equal to 0. Setting x_1 and x_2 equal to 0 and solving for the basic variables, we obtain the basic solution:

$$x_1 = 0, \quad x_2 = 0, \quad s_1 = 32, \quad s_2 = 84, \quad P = 0$$

This basic solution is feasible since none of the variables (excluding P) are negative. Thus, this is the initial basic feasible solution we seek.

Now you can see why we wanted to add the objective function equation to system (2): A basic feasible solution of (3) not only includes a basic feasible solution of (2), but, in addition, it includes the value of P for that basic feasible solution of (2).

The initial basic feasible solution we just found is associated with the origin. Of course, if we do not produce any tents, we do not expect a profit, so $P = \$0$. Starting with this easily obtained initial basic feasible solution, the simplex process moves through each iteration (each repetition) to another basic feasible solution, each time improving the profit, and the process continues until the maximum profit is reached. Then the process stops.

■ THE SIMPLEX TABLEAU

To facilitate the search for the optimal solution, we now turn to matrix methods discussed in Chapter 4. Our first step is to write the augmented matrix for the initial system (3). This matrix is called the **initial simplex tableau,** and it is simply a tabulation of the coefficients in system (3).

$$
\begin{array}{c}
 \\
s_1 \\
s_2 \\
P
\end{array}
\begin{array}{c}
\begin{array}{ccccc}
x_1 & x_2 & s_1 & s_2 & P
\end{array} \\
\left[
\begin{array}{ccccc|c}
1 & 2 & 1 & 0 & 0 & 32 \\
3 & 4 & 0 & 1 & 0 & 84 \\
\hline
-50 & -80 & 0 & 0 & 1 & 0
\end{array}
\right]
\end{array}
\qquad \text{Initial simplex tableau} \qquad (4)
$$

In tableau (4), the row below the dashed line always corresponds to the objective function. Each of the basic variables we selected above, s_1, s_2, and P, is also placed on the left of the tableau so that the intersection element in its row and column is not 0. For example, we place the basic variable s_1 on the left so that the intersection element of the s_1 row and the s_1 column is 1 and not 0. The basic variable s_2 is similarly placed. The objective function variable P is always placed at the bottom. The reason for writing the basic variables on the left in this way, is that this placement makes it possible to read certain basic feasible solutions directly from the tableau. If $x_1 = 0$ and $x_2 = 0$, then the basic variables on the left of tableau (4) are lined up with their corresponding values, 32, 84, and 0, to the right of the vertical line.

Looking at tableau (4) relative to the choice of s_1, s_2, and P as basic variables, we see that each basic variable is above a column that has all 0 elements except for a single 1 and that no two such columns contain 1's in the same row. These observations lead to a formalization of the process of selecting basic and nonbasic variables that is an important part of the simplex method:

Selecting Basic and Nonbasic Variables for the Simplex Method

Given a simplex tableau:

Step 1. Determine the number of basic variables and the number of nonbasic variables. These numbers do not change during the simplex process.

Step 2. *Selecting basic variables:* A variable can be selected as a basic variable only if it corresponds to a column in the tableau that has exactly one nonzero element (usually 1) and the nonzero element in the column is not in the same row as the nonzero element in the column of another basic variable. (This procedure always selects P as a basic variable, since the P column never changes during the simplex process.)

Step 3. *Selecting nonbasic variables:* After the basic variables are selected in step 2, the remaining variables are selected as the nonbasic variables. (The tableau columns under the nonbasic variables usually contain more than one nonzero element.)

The earlier selection of s_1, s_2, and P as basic variables and x_1 and x_2 as nonbasic variables conforms to this prescribed convention of selecting basic and nonbasic variables for the simplex process.

■ THE PIVOT OPERATION

The simplex method will now switch one of the nonbasic variables, x_1 or x_2, for one of the basic variables, s_1 or s_2 (but not P), as a step toward improving the profit. For a nonbasic variable to be classified as a basic variable we need to perform appropriate row operations on the tableau so that the newly selected basic variable will end up with exactly one nonzero element in its column. In this process, the old basic variable will usually gain additional nonzero elements in its column as it becomes nonbasic.

Which nonbasic variable should we select to become basic? It makes sense to select the nonbasic variable that will increase the profit the most per unit change in that variable. Looking at the objective function

$$P = 50x_1 + 80x_2$$

we see that if x_1 stays a nonbasic variable (set equal to 0) and if x_2 becomes a new basic variable, then

$$P = 50(0) + 80x_2 = 80x_2$$

and for each unit increase in x_2, P will increase \$80. If x_2 stays a nonbasic variable and x_1 becomes a new basic variable, then (reasoning in the same way) for each unit increase in x_1, P will increase only \$50. So, we select the nonbasic variable x_2 to enter the set of basic variables, and call it the **entering variable.** (The basic variable leaving the set of basic variables to become a nonbasic variable is called the **exiting variable.** Exiting variables will be discussed shortly.)

We call the column corresponding to the entering variable the **pivot column.** Looking at the bottom row in tableau (4)—the objective function row below the dashed line—we see that the pivot column is associated with the column to the left of the P column that has the most negative bottom element. In general, the most negative element in the bottom row to the left of the P column *indicates* the variable above it that will produce the greatest increase in P for a unit increase in that variable. For this reason, we call the elements in the bottom row of the tableau to the left of the P column **indicators.**

We illustrate the indicators, the pivot column, the entering variable, and the initial basic feasible solution below:

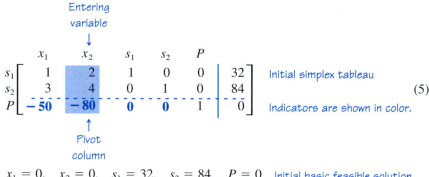

$$x_1 = 0, \quad x_2 = 0, \quad s_1 = 32, \quad s_2 = 84, \quad P = 0 \quad \text{Initial basic feasible solution}$$

Now that we have chosen the nonbasic variable x_2 as the entering variable (the nonbasic variable to become basic), which of the two basic variables, s_1 or s_2, should we choose as the exiting variable (the basic variable to become nonbasic)? We saw above that for $x_1 = 0$, each unit increase in the entering variable x_2 results in an increase of \$80 for P. Can we increase x_2 without limit? No! A limit is imposed by the nonnegative requirements for s_1 and s_2. (Remember that if any of the basic variables except P become negative, we no longer have a feasible solution.) So we rephrase the question and ask: How much can x_2 be increased when $x_1 = 0$ without causing s_1 or s_2 to become negative? To see how much x_2 can be increased, we refer to tableau (5) or system (3) and write the two problem constraint equations with $x_1 = 0$:

$$2x_2 + s_1 = 32$$
$$4x_2 + s_2 = 84$$

Solving for s_1 and s_2, we have

$$s_1 = 32 - 2x_2$$
$$s_2 = 84 - 4x_2$$

For s_1 and s_2 to be nonnegative, x_2 must be chosen so that both $32 - 2x_2$ and $84 - 4x_2$ are nonnegative. That is, so that

$$32 - 2x_2 \geqslant 0 \qquad \text{and} \qquad 84 - 4x_2 \geqslant 0$$
$$-2x_2 \geqslant -32 \qquad\qquad\qquad -4x_2 \geqslant -84$$
$$x_2 \leqslant \tfrac{32}{2} = 16 \qquad\qquad\qquad x_2 \leqslant \tfrac{84}{4} = 21$$

For both inequalities to be satisfied, x_2 must be less than or equal to the smaller of the values, which is 16. Thus, x_2 can increase to 16 without either s_1 or s_2 becoming negative. Now, observe how each value (16 and 21) can be obtained directly from tableau (6) below:

From (6) we can determine the amount the entering variable can increase by choosing the smallest of the quotients obtained by dividing each element in the last column above the dashed line by the corresponding *positive* element in the pivot column. The row with the smallest quotient is called the **pivot row,** and the variable to the left of the pivot row is the exiting variable. In this case, s_1 will be the exiting variable, and the roles of x_2 and s_1 will be interchanged. The element at the intersection of the pivot column and the pivot row is called the **pivot element,** and we circle this element for ease of recognition. Since a negative or 0 element in the pivot column places no restriction on the amount an entering variable can increase, it is not necessary to compute the quotient for negative or 0 values in the pivot column.

A negative or 0 element is never selected for the pivot element.

Tableau (7) illustrates this process, which is summarized in the next box.

Selecting the Pivot Element

Step 1.   Locate the most negative indicator in the bottom row of the tableau to the left of the P column (the negative number with the largest absolute value). The column containing this element is the *pivot column*. If there is a tie for the most negative, choose either.

Step 2.   Divide each *positive* element in the pivot column above the dashed line into the corresponding element in the last column. The *pivot row* is the row corresponding to the smallest quotient obtained. If there is a tie for the smallest quotient, choose either. If the pivot column above the dashed line has no positive elements, then there is no solution, and we stop.

Step 3.   The *pivot* (or *pivot element*) is the element in the intersection of the pivot column and pivot row. [*Note:*   The pivot element is always positive and is never in the bottom row.]

[*Remember:*   The entering variable is at the top of the pivot column, and the exiting variable is at the left of the pivot row.]

In order for x_2 to be classified as a basic variable, we perform row operations on tableau (7) so that the pivot element is transformed into 1 and all other elements in the column into 0's. This procedure for transforming a nonbasic variable into a basic variable is called a *pivot operation,* or *pivoting,* and is summarized in the box.

Performing a Pivot Operation

A **pivot operation,** or **pivoting,** consists of performing row operations as follows:

Step 1.   Multiply the pivot row by the reciprocal of the pivot element to transform the pivot element into a 1. (If the pivot element is already a 1, omit this step.)

Step 2.   Add multiples of the pivot row to other rows in the tableau to transform all other nonzero elements in the pivot column into 0's.

[*Note:*   Rows are not to be interchanged while performing a pivot operation. The only way the (positive) pivot element can be transformed into 1 (if it is not a 1 already) is for the pivot row to be multiplied by the reciprocal of the pivot element.]

Performing a pivot operation has the following effects:

1. The (entering) nonbasic variable becomes a basic variable.
2. The (exiting) basic variable becomes a nonbasic variable.
3. The value of the objective function is increased, or, in some cases, remains the same.

We now carry out the pivot operation on tableau (7). (To facilitate the process, we do not repeat the variables after the first tableau, and we use "Enter" and "Exit" for "Entering variable" and "Exiting variable," respectively.)

$$
\begin{array}{c}
\text{Enter} \\
\downarrow
\end{array}
$$

$$
\begin{array}{c}
\\
\text{Exit} \rightarrow s_1 \\
s_2 \\
P
\end{array}
\begin{array}{c}
x_1 \quad x_2 \quad s_1 \quad s_2 \quad P \\
\left[
\begin{array}{ccccc|c}
1 & ② & 1 & 0 & 0 & 32 \\
3 & 4 & 0 & 1 & 0 & 84 \\
\hline
-50 & -80 & 0 & 0 & 1 & 0
\end{array}
\right]
\begin{array}{l}
\tfrac{1}{2}R_1 \rightarrow R_1
\end{array}
\end{array}
$$

$$
\sim
\left[
\begin{array}{ccccc|c}
\tfrac{1}{2} & ① & \tfrac{1}{2} & 0 & 0 & 16 \\
3 & 4 & 0 & 1 & 0 & 84 \\
\hline
-50 & -80 & 0 & 0 & 1 & 0
\end{array}
\right]
\begin{array}{l}
(-4)R_1 + R_2 \rightarrow R_2 \\
80R_1 + R_3 \rightarrow R_3
\end{array}
$$

$$
\sim
\left[
\begin{array}{ccccc|c}
\tfrac{1}{2} & 1 & \tfrac{1}{2} & 0 & 0 & 16 \\
1 & 0 & -2 & 1 & 0 & 20 \\
\hline
-10 & 0 & 40 & 0 & 1 & 1,280
\end{array}
\right]
$$

We have completed the pivot operation, and now we must insert appropriate variables for this new tableau. Since x_2 replaced s_1, the basic variables are now x_2, s_2, and P, as indicated by the labels on the left side of the new tableau. Note that this selection of basic variables agrees with the procedure outlined on page 323 for selecting basic variables. We write the new basic feasible solution by setting the nonbasic variables x_1 and s_1 equal to 0 and solving for the basic variables by inspection. (Remember, the values of the basic variables listed on the left are the corresponding numbers to the right of the vertical line. To see this, substitute $x_1 = 0$ and $s_1 = 0$ in the corresponding system shown next to the simplex tableau.)

$$
\begin{array}{c}
\\
x_2 \\
s_2 \\
P
\end{array}
\begin{array}{c}
x_1 \quad x_2 \quad s_1 \quad s_2 \quad P \\
\left[
\begin{array}{ccccc|c}
\tfrac{1}{2} & 1 & \tfrac{1}{2} & 0 & 0 & 16 \\
1 & 0 & -2 & 1 & 0 & 20 \\
\hline
-10 & 0 & 40 & 0 & 1 & 1,280
\end{array}
\right]
\end{array}
\begin{array}{l}
\tfrac{1}{2}x_1 + x_2 + \tfrac{1}{2}s_1 \qquad\qquad = 16 \\
x_1 \qquad\quad - 2s_1 + s_2 \qquad = 20 \\
-10x_1 + \qquad 40s_1 \qquad + P = 1,280
\end{array}
$$

$$
x_1 = 0, \quad x_2 = 16, \quad s_1 = 0, \quad s_2 = 20, \quad P = \$1,280
$$

A profit of \$1,280 is a marked improvement over the \$0 profit produced by the initial basic feasible solution. But we can improve P still further, since a negative indicator still remains in the bottom row. To see why, we write out the objective function:

$$
-10x_1 + 40s_1 + P = 1,280
$$

or

$$
P = 10x_1 - 40s_1 + 1,280
$$

If s_1 stays a nonbasic variable (set equal to 0) and x_1 becomes a new basic variable, then

$$P = 10x_1 - 40(0) + 1{,}280 = 10x_1 + 1{,}280$$

and for each unit increase in x_1, P will increase \$10.

We now go through another iteration of the simplex process (that is, we repeat the above sequence of steps) using another pivot element. The pivot element and the entering and exiting variables are shown in the following tableau:

Enter
↓

	x_1	x_2	s_1	s_2	P	
x_2	$\frac{1}{2}$	1	$\frac{1}{2}$	0	0	16
Exit → s_2	①	0	-2	1	0	20
P	-10	0	40	0	1	1,280

$\dfrac{16}{1/2} = 32$

$\dfrac{20}{1} = 20$

We now pivot on (the circled) 1. That is, we perform a pivot operation using this 1 as the pivot element. Since the pivot element is 1, we do not need to perform the first step in the pivot operaton, so we proceed to the second step to get 0's above and below the pivot element 1. As before, to facilitate the process, we omit writing the variables, except for the first tableau.

Enter
↓

	x_1	x_2	s_1	s_2	P		
x_2	$\frac{1}{2}$	1	$\frac{1}{2}$	0	0	16	$(-\frac{1}{2})R_2 + R_1 \rightarrow R_1$
Exit → s_2	①	0	-2	1	0	20	
P	-10	0	40	0	1	1,280	$10R_2 + R_3 \rightarrow R_3$

$\sim$

0	1	$\frac{3}{2}$	$-\frac{1}{2}$	0	6
1	0	-2	1	0	20
0	0	20	10	1	1,480

Since there are no more negative indicators in the bottom row, we are through. Let us insert the appropriate variables for this last tableau and write the corresponding basic feasible solution. The basic variables are now x_1, x_2, and P, so to get the corresponding basic feasible solution, we set the nonbasic variables s_1 and s_2 equal to 0 and solve for the basic variables by inspection.

	x_1	x_2	s_1	s_2	P	
x_2	0	1	$\frac{3}{2}$	$-\frac{1}{2}$	0	6
x_1	1	0	-2	1	0	20
P	0	0	20	10	1	1,480

$x_1 = 20,\quad x_2 = 6,\quad s_1 = 0,\quad s_2 = 0,\quad P = 1{,}480$

To see why this is the maximum, we rewrite the objective function from the bottom row:

$$20s_1 + 10s_2 + P = 1{,}480$$
$$P = 1{,}480 - 20s_1 - 10s_2$$

Since s_1 and s_2 cannot be negative, any increase of either from 0 will make the profit smaller.

Finally, returning to our original problem, we conclude that a production schedule of 20 standard tents and 6 expedition tents will produce a maximum profit of \$1,480 per day, which is the same as the geometric solution obtained in Section 5-2. The fact that the slack variables are both 0 means that for this production schedule, the plant will operate at full capacity—there is no slack in either the cutting department or the assembly department.

■ INTERPRETING THE SIMPLEX PROCESS GEOMETRICALLY

We can now interpret the simplex process just completed geometrically in terms of the feasible region graphed in the preceding section. Table 1 lists the three basic feasible solutions we just found using the simplex method (in the order they were found). The table also includes the corresponding corner points of the feasible region illustrated in Figure 1.

TABLE 1
Basic Feasible Solutions (Obtained Above)

x_1	x_2	s_1	s_2	P	CORNER POINT
0	0	32	84	\$ 0	$O(0, 0)$
0	16	0	20	\$1,280	$A(0, 16)$
20	6	0	0	\$1,480	$B(20, 6)$

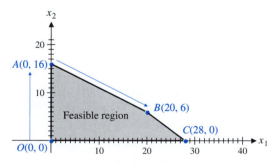

FIGURE 1

Looking at Table 1 and Figure 1, we see that the simplex process started at the origin, moved to the adjacent corner point $A(0, 16)$, and then to the optimal solution $B(20, 6)$ at the next adjacent corner point. This is typical of the simplex process.

■ THE SIMPLEX METHOD SUMMARIZED

Before commencing with further examples, we summarize the important parts of the simplex method schematically in Figure 2.

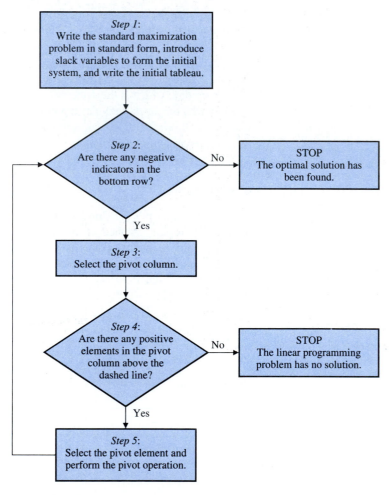

FIGURE 2
Simplex algorithm for standard maximization problems
(Problem constraints are of the ≤ form with nonnegative constants on the right. The coefficients of the objective function can be any real numbers.)

EXAMPLE 1 ➤ **Using the Simplex Method** Solve the following linear programming problem using the simplex method:

$$\text{Maximize} \quad P = 10x_1 + 5x_2$$
$$\text{Subject to} \quad 4x_1 + x_2 \leq 28$$
$$2x_1 + 3x_2 \leq 24$$
$$x_1, x_2 \geq 0$$

Solution Introduce slack variables s_1 and s_2 and write the initial system:

$$
\begin{aligned}
4x_1 + x_2 + s_1 \quad\quad\quad &= 28 \\
2x_1 + 3x_2 \quad + s_2 \quad &= 24 \\
-10x_1 - 5x_2 \quad\quad\quad + P &= 0 \\
x_1, x_2, s_1, s_2 &\geq 0
\end{aligned}
$$

Write the simplex tableau, and identify the first pivot element and the entering and exiting variables:

Enter
↓

	x_1	x_2	s_1	s_2	P		
Exit → s_1	④	1	1	0	0	28	$\frac{28}{4} = 7$
s_2	2	3	0	1	0	24	$\frac{24}{2} = 12$
P	−10	−5	0	0	1	0	

Perform the pivot operation:

Enter
↓

	x_1	x_2	s_1	s_2	P		
Exit → s_1	④	1	1	0	0	28	$\frac{1}{4}R_1 \to R_1$
s_2	2	3	0	1	0	24	
P	−10	−5	0	0	1	0	

	x_1	x_2	s_1	s_2	P		
~	①	0.25	0.25	0	0	7	$(-2)R_1 + R_2 \to R_2$
	2	3	0	1	0	24	$10R_1 + R_3 \to R_3$
	−10	−5	0	0	1	0	

	x_1	x_2	s_1	s_2	P	
x_1	1	0.25	0.25	0	0	7
~ s_2	0	2.5	−0.5	1	0	10
P	0	−2.5	2.5	0	1	70

Since there is still a negative indicator in the last row, we repeat the process by finding a new pivot element:

Enter
↓

	x_1	x_2	s_1	s_2	P		
x_1	1	0.25	0.25	0	0	7	$\frac{7}{0.25} = 28$
Exit → s_2	0	②.⑤	−0.5	1	0	10	$\frac{10}{2.5} = 4$
P	0	−2.5	2.5	0	1	70	

Performing the pivot operation, we obtain

Enter
↓

$$
\begin{array}{c}
\\
x_1 \\
\text{Exit} \rightarrow s_2 \\
P
\end{array}
\begin{array}{ccccc}
x_1 & x_2 & s_1 & s_2 & P \\
\end{array}
\left[
\begin{array}{ccccc|c}
1 & 0.25 & 0.25 & 0 & 0 & 7 \\
0 & \boxed{2.5} & -0.5 & 1 & 0 & 10 \\
\hline
0 & -2.5 & 2.5 & 0 & 1 & 70
\end{array}
\right]
\begin{array}{l}
\\
\frac{1}{2.5}R_2 \rightarrow R_2 \\
\end{array}
$$

$$
\sim
\left[
\begin{array}{ccccc|c}
1 & 0.25 & 0.25 & 0 & 0 & 7 \\
0 & \boxed{1} & -0.2 & 0.4 & 0 & 4 \\
\hline
0 & -2.5 & 2.5 & 0 & 1 & 70
\end{array}
\right]
\begin{array}{l}
(-0.25)R_2 + R_1 \rightarrow R_1 \\
\\
2.5R_2 + R_3 \rightarrow R_3
\end{array}
$$

$$
\begin{array}{c}
x_1 \\
\sim \; x_2 \\
P
\end{array}
\left[
\begin{array}{ccccc|c}
1 & 0 & 0.3 & -0.1 & 0 & 6 \\
0 & 1 & -0.2 & 0.4 & 0 & 4 \\
\hline
0 & 0 & 2 & 1 & 1 & 80
\end{array}
\right]
$$

Since all the indicators in the last row are nonnegative, we stop and read the optimal solution:

$$\text{Max } P = 80 \quad \text{at} \quad x_1 = 6, \quad x_2 = 4, \quad s_1 = 0, \quad s_2 = 0$$

(To see why this makes sense, write the objective function corresponding to the last row to see what happens to P when you try to increase s_1 or s_2.) ◄

MATCHED PROBLEM 1 ➤ Solve the following linear programming problem using the simplex method:

$$\text{Maximize} \quad P = 2x_1 + x_2$$
$$\text{Subject to} \quad 5x_1 + x_2 \leq 9$$
$$x_1 + x_2 \leq 5$$
$$x_1, x_2 \geq 0$$

◄

explore – discuss 1

Graph the feasible region for the linear programming problem in Example 1 and trace the path to the optimal solution determined by the simplex method.

EXAMPLE 2 ➤ **Using the Simplex Method** Solve using the simplex method:

$$\text{Maximize} \quad P = 6x_1 + 3x_2$$
$$\text{Subject to} \quad -2x_1 + 3x_2 \leq 9$$
$$-x_1 + 3x_2 \leq 12$$
$$x_1, x_2 \geq 0$$

Solution   Write the initial system using the slack variables s_1 and s_2:

$$-2x_1 + 3x_2 + s_1 \qquad\qquad = 9$$
$$-x_1 + 3x_2 \qquad + s_2 \qquad = 12$$
$$-6x_1 - 3x_2 \qquad\qquad + P = 0$$

Write the simplex tableau and identify the first pivot element:

$$
\begin{array}{c}
\\
s_1\\
s_2\\
P
\end{array}
\begin{array}{c}
x_1 \quad x_2 \quad s_1 \quad s_2 \quad P \\
\left[\begin{array}{ccccc|c}
-2 & 3 & 1 & 0 & 0 & 9 \\
-1 & 3 & 0 & 1 & 0 & 12 \\
\hline
-6 & -3 & 0 & 0 & 1 & 0
\end{array}\right]
\end{array}
$$

↑
Pivot column

Since both elements in the pivot column above the dashed line are negative, we are unable to select a pivot row. We stop and conclude that there is no solution. ◄

MATCHED PROBLEM 2 ➤   Solve using the simplex method:

$$\text{Maximize} \quad P = 2x_1 + 3x_2$$
$$\text{Subject to} \quad -3x_1 + 4x_2 \leqslant 12$$
$$x_2 \leqslant 2$$
$$x_1, x_2 \geqslant 0$$

◄

explore – discuss 2

In Example 2 we encountered a tableau with a pivot column and no pivot row, indicating a problem with no solution.

(A) What happens if you violate the rule for selecting the pivot row, select the negative element -1 for the pivot, and then continue with the simplex method?

(B) There was another negative indicator in the tableau in Example 2. What happens if you violate the rule for selecting the pivot column, select the second column for a pivot column, and then continue with the simplex method?

Refer to the solutions of Examples 1 and 2. In Example 1 we concluded that we had found the optimal solution because we could not select a pivot column. In Example 2 we concluded that the problem had no solution because we selected a pivot column and then could not select a pivot row. Notice that we do not try to continue with the simplex method by selecting a negative pivot element or using a different column for the pivot column. Remember:

If it is not possible to select a pivot column, the simplex method stops and we conclude that the optimal solution has been found. If the pivot column has been selected and it is not possible to select a pivot row, the simplex method stops and we conclude that there is no solution.

■ APPLICATION

EXAMPLE 3 ➤ Agriculture   A farmer owns a 100 acre farm and plans to plant at most three crops. The seed for crops A, B, and C costs $40, $20, and $30 per acre, respectively. A maximum of $3,200 can be spent on seed. Crops A, B, and C require 1, 2, and 1 workdays per acre, respectively, and there are a maximum of 160 workdays available. If the farmer can make a profit of $100 per acre on crop A, $300 per acre on crop B, and $200 per acre on crop C, how many acres of each crop should be planted to maximize profit?

Solution   Let: x_1 = Number of acres of crop A

x_2 = Number of acres of crop B

x_3 = Number of acres of crop C

P = Total profit

Then we have the following linear programming problem:

Maximize $P = 100x_1 + 300x_2 + 200x_3$ *Objective function*

Subject to
$$\left.\begin{array}{r} x_1 + x_2 + x_3 \leq 100 \\ 40x_1 + 20x_2 + 30x_3 \leq 3{,}200 \\ x_1 + 2x_2 + x_3 \leq 160 \end{array}\right\}\ \text{\textit{Problem constraints}}$$

$x_1, x_2, x_3 \geq 0$ *Nonnegative constraints*

Next, we introduce slack variables and form the initial system:

$$
\begin{array}{rcrcrcrcrcrcr}
x_1 &+& x_2 &+& x_3 &+& s_1 & & & & & &= 100 \\
40x_1 &+& 20x_2 &+& 30x_3 & & &+& s_2 & & & &= 3{,}200 \\
x_1 &+& 2x_2 &+& x_3 & & & & &+& s_3 & &= 160 \\
-100x_1 &-& 300x_2 &-& 200x_3 & & & & & & &+ P &= 0
\end{array}
$$

$x_1, x_2, x_3, s_1, s_2, s_3 \geq 0$

Notice that the initial system has $7 - 4 = 3$ nonbasic variables and 4 basic variables. Now we form the simplex tableau and solve by the simplex method:

Enter
↓

	x_1	x_2	x_3	s_1	s_2	s_3	P	
s_1	1	1	1	1	0	0	0	100
s_2	40	20	30	0	1	0	0	3,200
Exit → s_3	1	②	1	0	0	1	0	160
P	-100	-300	-200	0	0	0	1	0

$0.5R_3 \rightarrow R_3$

$\sim$

	x_1	x_2	x_3	s_1	s_2	s_3	P	
	1	1	1	1	0	0	0	100
	40	20	30	0	1	0	0	3,200
	0.5	①	0.5	0	0	0.5	0	80
	-100	-300	-200	0	0	0	1	0

$(-1)R_3 + R_1 \rightarrow R_1$
$(-20)R_3 + R_2 \rightarrow R_2$

$300R_3 + R_4 \rightarrow R_4$

Enter
↓

	x_1	x_2	x_3	s_1	s_2	s_3	P	
Exit → s_1	0.5	0	⑤.5	1	0	-0.5	0	20
s_2	30	0	20	0	1	-10	0	1,600
x_2	0.5	1	0.5	0	0	0.5	0	80
P	50	0	-50	0	0	150	1	24,000

$2R_1 \rightarrow R_1$

$\sim$

	x_1	x_2	x_3	s_1	s_2	s_3	P	
	1	0	①	2	0	-1	0	40
	30	0	20	0	1	-10	0	1,600
	0.5	1	0.5	0	0	0.5	0	80
	50	0	-50	0	0	150	1	24,000

$(-20)R_1 + R_2 \rightarrow R_2$
$(-0.5)R_1 + R_3 \rightarrow R_3$
$50R_1 + R_4 \rightarrow R_4$

$\sim$

	x_1	x_2	x_3	s_1	s_2	s_3	P	
x_3	1	0	1	2	0	-1	0	40
s_2	10	0	0	-40	1	10	0	800
x_2	0	1	0	-1	0	1	0	60
P	100	0	0	100	0	100	1	26,000

All indicators in the bottom row are nonnegative, and we can now read the optimal solution:

$$x_1 = 0, \quad x_2 = 60, \quad x_3 = 40, \quad s_1 = 0, \quad s_2 = 800, \quad s_3 = 0, \quad P = \$26,000$$

Thus, if the farmer plants 60 acres in crop B, 40 acres in crop C, and no crop A, the maximum profit of \$26,000 will be realized. The fact that $s_2 = 800$ tells us (look at the second row in the equations at the start) that this maximum profit is reached by using only \$2,400 of the \$3,200 available for seed; that is, we have a slack of \$800 that can be used for some other purpose. ◀

MATCHED PROBLEM 3 ➤ Repeat Example 3 modified as follows:

	INVESTMENT PER ACRE			MAXIMUM
	Crop A	*Crop B*	*Crop C*	AVAILABLE
SEED COST	$24	$40	$30	$3,600
WORKDAYS	1	2	2	160
PROFIT	$140	$200	$160	

◄

Remarks

1. It can be shown that the feasible region for the linear programming problem in Example 3 has eight corner points, yet the simplex method found the solution in only two steps. Now you begin to see the power of the simplex method. In larger problems, the difference between the total number of corner points and the number of steps required by the simplex method is even more dramatic. A feasible region may have hundreds or even thousands of corner points, yet the simplex method will often find the optimal solution in 10 or 15 steps.

2. Refer to the second problem constraint in the model for Example 3:

$$40x_1 + 20x_2 + 30x_3 \le 3,200$$

Multiplying both sides of this inequality by $\frac{1}{10}$ before introducing a slack variable simplifies subsequent calculations. However, performing this operation has a side effect—it changes the units of the slack variable from dollars to tens of dollars. To see why this happens, compare the equations

$$40x_1 + 20x_2 + 30x_3 + s_2 = 3,200 \quad s_2 \text{ represents dollars}$$

and

$$4x_1 + 2x_2 + 3x_3 + s_2' = 320 \quad s_2' \text{ represents tens of dollars}$$

In general, if you multiply a problem constraint by a number, remember to take this into account when you interpret the value of the slack variable for that constraint.

3. It is important to realize that in order to keep this introduction as simple as possible, we have purposely avoided certain degenerate cases that lead to difficulties. Discussion and resolution of these problems is left to a more advanced treatment of the subject.

Answers to Matched Problems

1. Max $P = 6$ when $x_1 = 1$ and $x_2 = 4$ 2. No solution
3. 40 acres of crop A, 60 acres of crop B, no crop C; Max $P = \$17,600$ (since $s_2 = 240$, $240 out of the $3,600 will not be spent)

EXERCISE 5-4

A   *For the simplex tableaux in Problems 1–4:*
(A) Identify the basic and nonbasic variables.
(B) Find the corresponding basic feasible solution.
(C) Determine whether the optimal solution has been found, an additional pivot is required, or the problem has no solution.

1.

	x_1	x_2	s_1	s_2	P	
	2	1	0	3	0	12
	3	0	1	-2	0	15
	-4	0	0	4	1	20

2.

	x_1	x_2	s_1	s_2	P	
	1	4	-2	0	0	10
	0	2	3	1	0	25
	0	5	6	0	1	35

3.

	x_1	x_2	x_3	s_1	s_2	s_3	P	
	-2	0	1	3	1	0	0	5
	0	1	0	-2	0	0	0	15
	-1	0	0	4	1	1	0	12
	-4	0	0	2	4	0	1	45

4.

	x_1	x_2	x_3	s_1	s_2	s_3	P	
	0	2	-1	1	4	0	0	5
	0	1	2	0	-2	1	0	2
	1	3	0	0	5	0	0	11
	0	-5	4	0	-3	0	1	27

In Problems 5–8, find the pivot element, identify the entering and exiting variables, and perform one pivot operation.

5.

	x_1	x_2	s_1	s_2	P	
	1	4	1	0	0	4
	3	5	0	1	0	24
	-8	-5	0	0	1	0

6.

	x_1	x_2	s_1	s_2	P	
	1	6	1	0	0	36
	3	1	0	1	0	5
	-1	-2	0	0	1	0

7.

	x_1	x_2	s_1	s_2	s_3	P	
	2	1	1	0	0	0	4
	3	0	1	1	0	0	8
	0	0	2	0	1	0	2
	-4	0	-3	0	0	1	5

8.

	x_1	x_2	s_1	s_2	s_3	P	
	0	0	2	1	1	0	2
	1	0	-4	0	1	0	3
	0	1	5	0	2	0	11
	0	0	-6	0	-5	1	18

In Problems 9–12:
(A) Using slack variables, write the initial system for each linear programming problem.
(B) Write the simplex tableau, circle the first pivot, and identify the entering and exiting variables.
(C) Use the simplex method to solve the problem.

9.   Maximize $P = 15x_1 + 10x_2$
Subject to $2x_1 + x_2 \le 10$
$x_1 + 3x_2 \le 10$
$x_1, x_2 \ge 0$

10.   Maximize $P = 3x_1 + 2x_2$
Subject to $5x_1 + 2x_2 \le 20$
$3x_1 + 2x_2 \le 16$
$x_1, x_2 \ge 0$

11.   Repeat Problem 9 with the objective function changed to $P = 30x_1 + x_2$.

12.   Repeat Problem 10 with the objective function changed to $P = x_1 + 3x_2$.

B   *Solve the linear programming problems in Problems 13–28 using the simplex method.*

13.   Maximize $P = 30x_1 + 40x_2$
Subject to $2x_1 + x_2 \le 10$
$x_1 + x_2 \le 7$
$x_1 + 2x_2 \le 12$
$x_1, x_2 \ge 0$

14.   Maximize $P = 15x_1 + 20x_2$
Subject to $2x_1 + x_2 \le 9$
$x_1 + x_2 \le 6$
$x_1 + 2x_2 \le 10$
$x_1, x_2 \ge 0$

15.   Maximize $P = 2x_1 + 3x_2$
Subject to $-2x_1 + x_2 \le 2$
$-x_1 + x_2 \le 5$
$x_2 \le 6$
$x_1, x_2 \ge 0$

16.   Repeat Problem 15 with $P = -x_1 + 3x_2$.

17.   Maximize $P = -x_1 + 2x_2$
   Subject to $-x_1 + x_2 \leq 2$
   $-x_1 + 3x_2 \leq 12$
   $x_1 - 4x_2 \leq 4$
   $x_1, x_2 \geq 0$

18.   Repeat Problem 17 with $P = x_1 + 2x_2$.

19.   Maximize $P = 5x_1 + 2x_2 - x_3$
   Subject to $x_1 + x_2 - x_3 \leq 10$
   $2x_1 + 4x_2 + 3x_3 \leq 30$
   $x_1, x_2, x_3 \geq 0$

20.   Maximize $P = 4x_1 - 3x_2 + 2x_3$
   Subject to $x_1 + 2x_2 - x_3 \leq 5$
   $3x_1 + 2x_2 + 2x_3 \leq 22$
   $x_1, x_2, x_3 \geq 0$

21.   Maximize $P = 2x_1 + 3x_2 + 4x_3$
   Subject to $x_1 + x_3 \leq 4$
   $x_2 + x_3 \leq 3$
   $x_1, x_2, x_3 \geq 0$

22.   Maximize $P = x_1 + x_2 + 2x_3$
   Subject to $x_1 - 2x_2 + x_3 \leq 9$
   $2x_1 + x_2 + 2x_3 \leq 28$
   $x_1, x_2, x_3 \geq 0$

23.   Maximize $P = 4x_1 + 3x_2 + 2x_3$
   Subject to $3x_1 + 2x_2 + 5x_3 \leq 23$
   $2x_1 + x_2 + x_3 \leq 8$
   $x_1 + x_2 + 2x_3 \leq 7$
   $x_1, x_2, x_3 \geq 0$

24.   Maximize $P = 4x_1 + 2x_2 + 3x_3$
   Subject to $x_1 + x_2 + x_3 \leq 11$
   $2x_1 + 3x_2 + x_3 \leq 20$
   $x_1 + 3x_2 + 2x_3 \leq 20$
   $x_1, x_2, x_3 \geq 0$

C

25.   Maximize $P = 20x_1 + 30x_2$
   Subject to $0.6x_1 + 1.2x_2 \leq 960$
   $0.03x_1 + 0.04x_2 \leq 36$
   $0.3x_1 + 0.2x_2 \leq 270$
   $x_1, x_2 \geq 0$

26.   Repeat Problem 25 with $P = 20x_1 + 20x_2$.

27.   Maximize $P = x_1 + 2x_2 + 3x_3$
   Subject to $2x_1 + 2x_2 + 8x_3 \leq 600$
   $x_1 + 3x_2 + 2x_3 \leq 600$
   $3x_1 + 2x_2 + x_3 \leq 400$
   $x_1, x_2, x_3 \geq 0$

28.   Maximize $P = 10x_1 + 50x_2 + 10x_3$
   Subject to $3x_1 + 3x_2 + 3x_3 \leq 66$
   $6x_1 - 2x_2 + 4x_3 \leq 48$
   $3x_1 + 6x_2 + 9x_3 \leq 108$
   $x_1, x_2, x_3 \geq 0$

In Problems 29 and 30, first solve the linear programming problem by the simplex method, keeping track of the basic feasible solutions at each step. Then graph the feasible region and illustrate the path to the optimal solution determined by the simplex method.

29.   Maximize $P = 2x_1 + 5x_2$
   Subject to $x_1 + 2x_2 \leq 40$
   $x_1 + 3x_2 \leq 48$
   $x_1 + 4x_2 \leq 60$
   $x_2 \leq 14$
   $x_1, x_2 \geq 0$

30.   Maximize $P = 5x_1 + 3x_2$
   Subject to $5x_1 + 4x_2 \leq 100$
   $2x_1 + x_2 \leq 28$
   $4x_1 + x_2 \leq 42$
   $x_1 \leq 10$
   $x_1, x_2 \geq 0$

In Problems 31–34, there is a tie for the choice of the first pivot column. Use the simplex method to solve each problem two different ways: first by choosing column 1 as the first pivot column, and then by choosing column 2 as the first pivot column. Discuss the relationship between these two solutions.

31.   Maximize $P = x_1 + x_2$
   Subject to $2x_1 + x_2 \leq 16$
   $x_1 \leq 6$
   $x_2 \leq 10$
   $x_1, x_2 \geq 0$

32.   Maximize $P = x_1 + x_2$
   Subject to $x_1 + 2x_2 \leq 10$
   $x_1 \leq 6$
   $x_2 \leq 4$
   $x_1, x_2 \geq 0$

33. Maximize $P = 3x_1 + 3x_2 + 2x_3$
    Subject to $x_1 + x_2 + 2x_3 \leqslant 20$
    $2x_1 + x_2 + 4x_3 \leqslant 32$
    $x_1, x_2, x_3 \geqslant 0$

34. Maximize $P = 2x_1 + 2x_2 + x_3$
    Subject to $x_1 + x_2 + 3x_3 \leqslant 10$
    $2x_1 + 4x_2 + 5x_3 \leqslant 24$
    $x_1, x_2, x_3 \geqslant 0$

▶ **APPLICATIONS**

Formulate each of the following as a linear programming problem. (The answers in the back of the book for these application problems include the formulation.) Then solve the problem using the simplex method. Include an interpretation of any nonzero slack variables in the optimal solution.

Business & Economics

35. *Manufacturing—resource allocation.* A small company manufactures three different electronic components for computers. Component A requires 2 hours of fabrication and 1 hour of assembly; component B requires 3 hours of fabrication and 1 hour of assembly; and component C requires 2 hours of fabrication and 2 hours of assembly. The company has up to 1,000 labor-hours of fabrication time and 800 labor-hours of assembly time available per week. The profit on each component, A, B, and C, is $7, $8, and $10, respectively. How many components of each type should the company manufacture each week in order to maximize its profit (assuming all components that it manufactures can be sold)? What is the maximum profit?

36. *Manufacturing—resource allocation.* Solve Problem 35 with the additional restriction that the combined total number of components produced each week cannot exceed 420. Discuss the effect of this restriction on the solution to Problem 35.

37. *Investment.* An investor has at most $100,000 to invest in government bonds, mutual funds, and money market funds. The average yields for government bonds, mutual funds, and money market funds are 8%, 13%, and 15%, respectively. The investor's policy requires that the total amount invested in mutual and money market funds not exceed the amount invested in government bonds. How much should be invested in each type of investment in order to maximize the return? What is the maximum return?

38. *Investment.* Repeat Problem 37 under the additional assumption that no more than $30,000 can be invested in money market funds.

39. *Advertising.* A department store chain has up to $20,000 to spend on television advertising for a sale. All ads will be placed with one television station, where a 30 second ad costs $1,000 on daytime TV and is viewed by 14,000 potential customers, $2,000 on prime-time TV and is viewed by 24,000 potential customers, and $1,500 on late-night TV and is viewed by 18,000 potential customers. The television station will not accept a total of more than 15 ads in all three time periods. How many ads should be placed in each time period in order to maximize the number of potential customers who will see the ads? How many potential customers will see the ads? (Ignore repeated viewings of the ad by the same potential customer.)

40. *Advertising.* Repeat Problem 39 if the department store increases its budget to $24,000 and requires that at least half of the ads be placed in prime-time shows.

41. *Construction—resource allocation.* A contractor is planning to build a new housing development consisting of colonial, split-level, and ranch-style houses. A colonial house requires $\frac{1}{2}$ acre of land, $60,000 capital, and 4,000 labor-hours to construct, and returns a profit of $20,000. A split-level house requires $\frac{1}{2}$ acre of land, $60,000 capital, and 3,000 labor-hours to construct, and returns a profit of $18,000. A ranch house requires 1 acre of land, $80,000 capital, and 4,000 labor-hours to construct, and returns a profit of $24,000. The contractor has available 30 acres of land, $3,200,000 capital, and 180,000 labor-hours.

(A) How many houses of each type should be constructed to maximize the contractor's profit? What is the maximum profit?

(B) A decrease in demand for colonial houses causes the profit on a colonial house to drop from $20,000 to $17,000. Discuss the effect of this change on the number of houses built and on the maximum profit.

(C) An increase in demand for colonial houses causes the profit on a colonial house to rise from $20,000 to $25,000. Discuss the effect of this change on the number of houses built and on the maximum profit.

42. *Manufacturing—resource allocation.* A company manufactures three-speed, five-speed, and ten-speed bicycles.

Each bicycle passes through three departments, fabrication, painting & plating, and final assembly. The relevant manufacturing data are given in the table.

| | LABOR-HOURS PER BICYCLE | | | MAXIMUM LABOR-HOURS AVAILABLE PER DAY |
	Three-Speed	*Five-Speed*	*Ten-Speed*	
FABRICATION	3	4	5	120
PAINTING & PLATING	5	3	5	130
FINAL ASSEMBLY	4	3	5	120
PROFIT PER BICYCLE	$80	$70	$100	

(A) How many bicycles of each type should the company manufacture per day in order to maximize its profit? What is the maximum profit?

(B) Discuss the effect on the solution to part (A) if the profit on a ten-speed bicycle increases to $110 and all other data in part (A) remain the same.

(C) Discuss the effect on the solution to part (A) if the profit on a five-speed bicycle increases to $110 and all other data in part (A) remain the same.

43. *Packaging—product mix.* A candy company makes three types of candy, solid-center, fruit-filled, and cream-filled, and packages these candies in three different assortments. A box of assortment I contains 4 solid-center, 4 fruit-filled, and 12 cream-filled candies, and sells for $9.40. A box of assortment II contains 12 solid-center, 4 fruit-filled, and 4 cream-filled candies, and sells for $7.60. A box of assortment III contains 8 solid-center, 8 fruit-filled, and 8 cream-filled candies, and sells for $11.00. The manufacturing costs per piece of candy are $0.20 for solid-center, $0.25 for fruit-filled, and $0.30 for cream-filled. The company can manufacture 4,800 solid-center, 4,000 fruit-filled, and 5,600 cream-filled candies weekly.

(A) How many boxes of each type should the company produce each week in order to maximize their profit? What is the maximum profit?

(B) Discuss the effect on the solution to part (A) if the number of fruit-filled candies manufactured weekly is increased to 5,000 and all other data in part (A) remain the same.

(C) Discuss the effect on the solution to part (A) if the number of fruit-filled candies and the number of solid-center candies manufactured weekly are each increased to 6,000 and all other data in part (A) remain the same.

44. *Scheduling—resource allocation.* A small accounting firm prepares tax returns for three types of customers: individual, commercial, and industrial. The tax preparation process begins with a 1 hour interview with the customer. The data collected during this interview are entered into a time-sharing computer system, which produces the customer's tax return. It takes 1 hour to enter the data for an individual customer, 2 hours for a commercial customer, and $1\frac{1}{2}$ hours for an industrial customer. It takes 10 minutes of computer time to process an individual return, 25 minutes to process a commercial return, and 20 minutes to process an industrial return. The firm has one employee who conducts the initial interview and two who enter the data into the computer. The interviewer can work a maximum of 50 hours a week, and each of the data-entry employees can work a maximum of 40 hours a week. The computer is available for a maximum of 1,025 minutes a week. The firm makes a profit of $50 on each individual customer, $65 on each commercial customer, and $60 on each industrial customer.

(A) How many customers of each type should the firm schedule each week in order to maximize its profit? What is the maximum profit?

(B) Discuss the effect on the solution to part (A) if the maximum number of hours the interviewer works is decreased to 35 hours per week and all other data in part (A) remain the same.

(C) Discuss the effect on the solution to part (A) if the maximum number of minutes of available computer time is decreased to 450 minutes per week and all other data in part (A) remain the same.

Life Sciences

45. *Nutrition—animals.* The natural diet of a certain animal consists of three foods, *A, B,* and *C.* The number of units of calcium, iron, and protein in 1 gram of each food and the average daily intake are given in the table. A scientist wants to investigate the effect of increasing the protein in the animal's diet while not allowing the units of calcium and iron to exceed their average daily intakes. How many grams of each food should be used to maximize the amount of protein in the diet? What is the maximum amount of protein?

| | UNITS PER GRAM | | | AVERAGE DAILY INTAKE (UNITS) |
	Food A	Food B	Food C	
CALCIUM	1	3	2	30
IRON	2	1	2	24
PROTEIN	3	4	5	60

46. *Nutrition—animals.* Repeat Problem 45 if the scientist wants to maximize the daily calcium intake while not allowing the intake of iron or protein to exceed the average daily intake.

Social Sciences

47. *Opinion survey.* A political scientist has received a grant to fund a research project involving voting trends. The budget of the grant includes $3,200 for conducting door-to-door interviews the day before an election. Undergraduate students, graduate students, and faculty members will be hired to conduct the interviews. Each undergraduate student will conduct 18 interviews and be paid $100. Each graduate student will conduct 25 interviews and be paid $150. Each faculty member will conduct 30 interviews and be paid $200. Due to limited transportation facilities, no more than 20 interviewers can be hired. How many undergraduate students, graduate students, and faculty members should be hired in order to maximize the number of interviews that will be conducted? What is the maximum number of interviews?

48. *Opinion survey.* Repeat Problem 47 if one of the requirements of the grant is that at least 50% of the interviewers be undergraduate students.

S E C T I O N   5-5   ■   # The Dual; Minimization with Problem Constraints of the Form ⩾

■ FORMATION OF THE DUAL PROBLEM
■ SOLUTION OF MINIMIZATION PROBLEMS
■ APPLICATION: A TRANSPORTATION PROBLEM
■ SUMMARY OF PROBLEM TYPES AND SOLUTION METHODS

In the last section we restricted ourselves to standard maximization problems (problem constraints of the form ⩽, with nonnegative constants on the right and objective function coefficients any real numbers). Now we will consider minimization problems with ⩾ problem constraints. These two types of problems turn out to be very closely related.

■ ## FORMATION OF THE DUAL PROBLEM

Associated with each minimization problem with ⩾ constraints is a maximization problem called the **dual problem.** To illustrate the procedure for forming the dual problem, consider the following minimization problem:

$$\text{Minimize} \quad C = 16x_1 + 45x_2$$
$$\text{Subject to} \quad 2x_1 + 5x_2 \geqslant 50$$
$$x_1 + 3x_2 \geqslant 27 \tag{1}$$
$$x_1, x_2 \geqslant 0$$

The first step in forming the dual problem is to construct a matrix by using the problem constraints and the objective function written in the following form:

$$
\begin{aligned}
2x_1 + 5x_2 &\geqslant 50 \\
x_1 + 3x_2 &\geqslant 27 \\
16x_1 + 45x_2 &= C
\end{aligned}
\qquad
A = \left[\begin{array}{cc|c} 2 & 5 & 50 \\ 1 & 3 & 27 \\ \hline 16 & 45 & 1 \end{array}\right]
$$

Caution

Do not confuse matrix A with the simplex tableau. We use a solid horizontal line in matrix A to help distinguish the dual matrix from the simplex tableau. No slack variables are involved in matrix A, and the coefficient of C is in the same column as the constants from the problem constraints.

Now we will form a second matrix called the *transpose of A*. In general, the **transpose** of a given matrix A is the matrix A^T formed by interchanging the rows and corresponding columns of A (first row with first column, second row with second column, and so on).

$$
A = \left[\begin{array}{cc|c} 2 & 5 & 50 \\ 1 & 3 & 27 \\ \hline 16 & 45 & 1 \end{array}\right]
\begin{array}{l}
R_1 \text{ in } A = C_1 \text{ in } A^T \\
R_2 \text{ in } A = C_2 \text{ in } A^T \\
R_3 \text{ in } A = C_3 \text{ in } A^T
\end{array}
$$

A^T is the transpose of A
$$
A^T = \left[\begin{array}{cc|c} 2 & 1 & 16 \\ 5 & 3 & 45 \\ \hline 50 & 27 & 1 \end{array}\right]
$$

Most computers and graphing utilities can find the transpose of a matrix (see Fig. 1). Consult your manual for details.

FIGURE 1
Finding A^T on a graphing calculator

Now, we use the rows of A^T to define a new linear programming problem. This new problem will always be a maximization problem with ⩽ problem constraints. To avoid confusion, we shall use different variables in this new problem:

$$
\begin{array}{ll}
2y_1 + \ y_2 \leqslant 16 \\
5y_1 + \ 3y_2 \leqslant 45 \qquad & A^T = \begin{bmatrix} \overset{y_1}{2} & \overset{y_2}{1} & 16 \\ 5 & 3 & 45 \\ 50 & 27 & 1 \end{bmatrix} \\
50y_1 + 27y_2 = P
\end{array}
$$

The dual of the minimization problem (1) is the following maximization problem:

$$
\begin{array}{ll}
\text{Maximize} & P = 50y_1 + 27y_2 \\
\text{Subject to} & 2y_1 + \ y_2 \leqslant 16 \\
& 5y_1 + 3y_2 \leqslant 45 \\
& y_1, y_2 \geqslant \ 0
\end{array}
\qquad (2)
$$

explore – discuss 1

Excluding the nonnegative constraints, the components of a linear programming problem can be divided into three categories: the coefficients of the objective function, the coefficients of the problem constraints, and the constants on the right side of the problem constraints. Write a verbal description of the relationship between the components of the original minimization problem (1) and the dual maximization problem (2).

The procedure for forming the dual problem is summarized in the box below:

Formation of the Dual Problem

Given a minimization problem with ⩾ problem constraints:

Step 1. Use the coefficients and constants in the problem constraints and the objective function to form a matrix A with the coefficients of the objective function in the last row.

Step 2. Interchange the rows and columns of matrix A to form the matrix A^T, the transpose of A.

Step 3. Use the rows of A^T to form a maximization problem with ⩽ problem constraints.

EXAMPLE 1 ➤ Forming the Dual Problem   Form the dual problem:

$$\text{Minimize} \quad C = 40x_1 + 12x_2 + 40x_3$$
$$\text{Subject to} \quad 2x_1 + x_2 + 5x_3 \geqslant 20$$
$$4x_1 + x_2 + x_3 \geqslant 30$$
$$x_1, x_2, x_3 \geqslant 0$$

Solution   Step 1.   Form the matrix A:

$$A = \begin{bmatrix} 2 & 1 & 5 & | & 20 \\ 4 & 1 & 1 & | & 30 \\ 40 & 12 & 40 & | & 1 \end{bmatrix}$$

Step 2.   Form the matrix A^T, the transpose of A:

$$A^T = \begin{bmatrix} 2 & 4 & | & 40 \\ 1 & 1 & | & 12 \\ 5 & 1 & | & 40 \\ 20 & 30 & | & 1 \end{bmatrix}$$

Step 3.   State the dual problem:

$$\text{Maximize} \quad P = 20y_1 + 30y_2$$
$$\text{Subject to} \quad 2y_1 + 4y_2 \leqslant 40$$
$$y_1 + y_2 \leqslant 12$$
$$5y_1 + y_2 \leqslant 40$$
$$y_1, y_2 \geqslant 0$$ ◀

MATCHED PROBLEM 1 ➤ Form the dual problem:

$$\text{Minimize} \quad C = 16x_1 + 9x_2 + 21x_3$$
$$\text{Subject to} \quad x_1 + x_2 + 3x_3 \geqslant 12$$
$$2x_1 + x_2 + x_3 \geqslant 16$$
$$x_1, x_2, x_3 \geqslant 0$$ ◀

■ SOLUTION OF MINIMIZATION PROBLEMS

The following theorem establishes the relationship between the solution of a minimization problem and the solution of its dual:

theorem 1

The Fundamental Principle of Duality

A minimization problem has a solution if and only if its dual problem has a solution. If a solution exists, then the optimal value of the minimization problem is the same as the optimal value of the dual problem.

The proof of Theorem 1 is beyond the scope of this text. However, we can illustrate Theorem 1 by solving minimization problem (1) and its dual maximization problem (2) geometrically. (Note that Theorem 2 in Section 5-2 guarantees that both problems have solutions.)

Original Problem (1)

Minimize $\quad C = 16x_1 + 45x_2$

Subject to $\quad 2x_1 + 5x_2 \geqslant 50$

$\qquad\qquad x_1 + 3x_2 \geqslant 27$

$\qquad\qquad x_1, x_2 \geqslant 0$

Dual Problem (2)

Maximize $\quad P = 50y_1 + 27y_2$

Subject to $\quad 2y_1 + \quad y_2 \leqslant 16$

$\qquad\qquad 5y_1 + 3y_2 \leqslant 45$

$\qquad\qquad y_1, y_2 \geqslant 0$

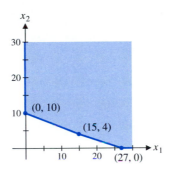

CORNER POINT	
(x_1, x_2)	$C = 16x_1 + 45x_2$
(0, 10)	450
(15, 4)	420
(27, 0)	432

Min $C = 420$ at (15, 4)

CORNER POINT	
(y_1, y_2)	$P = 50y_1 + 27y_2$
(0, 0)	0
(0, 15)	405
(3, 10)	420
(8, 0)	400

Max $P = 420$ at (3, 10)

Thus, the minimum value of C in problem (1) is the same as the maximum value of P in problem (2). Notice that the optimal solutions that produce this optimal value are different: (15, 4) is the optimal solution for problem (1) and (3, 10) is the optimal solution for problem (2). Theorem 1 only guarantees that the optimal values of a minimization problem and its dual are equal, not that the optimal solutions are the same. In general, it is not possible to determine an optimal solution for a minimization problem by examining the feasible set for the dual problem. However, it is possible to apply the simplex method to the dual problem and find both the optimal value and an optimal solution to the original minimization problem. To see how this is done, we will now solve problem (2) by the simplex method.

For reasons that will become clear later, we will use the variables x_1 and x_2 from the original problem as the slack variables in the dual problem:

$$2y_1 + y_2 + x_1 \qquad\qquad = 16 \qquad \text{Initial system for the dual problem}$$
$$5y_1 + 3y_2 \qquad + x_2 \qquad = 45$$
$$-50y_1 - 27y_2 \qquad\qquad + P = 0$$

$$
x_1 \\
x_2 \\
P
\begin{bmatrix}
\begin{array}{ccccc|c}
y_1 & y_2 & x_1 & x_2 & P & \\
② & 1 & 1 & 0 & 0 & 16 \\
5 & 3 & 0 & 1 & 0 & 45 \\
\hdashline
-50 & -27 & 0 & 0 & 1 & 0
\end{array}
\end{bmatrix}
\begin{array}{l}
0.5R_1 \to R_1
\end{array}
$$

$$
\sim
\begin{bmatrix}
\begin{array}{ccccc|c}
① & 0.5 & 0.5 & 0 & 0 & 8 \\
5 & 3 & 0 & 1 & 0 & 45 \\
\hdashline
-50 & -27 & 0 & 0 & 1 & 0
\end{array}
\end{bmatrix}
\begin{array}{l}
(-5)R_1 + R_2 \to R_2 \\
50R_1 + R_3 \to R_3
\end{array}
$$

$$
\begin{array}{c}
y_1 \\
\sim x_2 \\
P
\end{array}
\begin{bmatrix}
\begin{array}{ccccc|c}
1 & 0.5 & 0.5 & 0 & 0 & 8 \\
0 & ⓪.5 & -2.5 & 1 & 0 & 5 \\
\hdashline
0 & -2 & 25 & 0 & 1 & 400
\end{array}
\end{bmatrix}
\begin{array}{l}
2R_2 \to R_2
\end{array}
$$

$$
\sim
\begin{bmatrix}
\begin{array}{ccccc|c}
1 & 0.5 & 0.5 & 0 & 0 & 8 \\
0 & ① & -5 & 2 & 0 & 10 \\
\hdashline
0 & -2 & 25 & 0 & 1 & 400
\end{array}
\end{bmatrix}
\begin{array}{l}
(-0.5)R_2 + R_1 \to R_1 \\
2R_2 + R_3 \to R_3
\end{array}
$$

$$
\begin{array}{c}
y_1 \\
\sim y_2 \\
P
\end{array}
\begin{bmatrix}
\begin{array}{ccccc|c}
1 & 0 & 3 & -1 & 0 & 3 \\
0 & 1 & -5 & 2 & 0 & 10 \\
\hdashline
0 & 0 & 15 & 4 & 1 & 420
\end{array}
\end{bmatrix}
$$

Since all indicators in the bottom row are nonnegative, the solution to the dual problem is

$$y_1 = 3, \quad y_2 = 10, \quad x_1 = 0, \quad x_2 = 0, \quad P = 420$$

which agrees with our earlier geometric solution. Furthermore, examining the bottom row of the final simplex tableau, we see the same optimal solution to the minimization problem that we obtained directly by the geometric method:

$$\text{Min } C = 420 \quad \text{at} \quad x_1 = 15, \quad x_2 = 4$$

This is no accident.

An optimal solution to a minimization problem always can be obtained from the bottom row of the final simplex tableau for the dual problem.

Now we can see that using x_1 and x_2 as slack variables in the dual problem makes it easy to identify the solution of the original problem.

explore – discuss 2

The simplex method can be used to solve any standard maximization problem. Which of the following minimization problems have dual problems that are standard maximization problems (do not solve the problems)?

(A) Minimize $C = 2x_1 + 3x_2$

Subject to $2x_1 - 5x_2 \geqslant 4$

$x_1 - 3x_2 \geqslant -6$

$x_1, x_2 \geqslant 0$

(B) Minimize $C = 2x_1 - 3x_2$

Subject to $-2x_1 + 5x_2 \geqslant 4$

$-x_1 + 3x_2 \geqslant 6$

$x_1, x_2 \geqslant 0$

In general, what conditions must a minimization problem satisfy so that its dual problem is a standard maximization problem?

The procedure for solving a minimization problem by applying the simplex method to its dual problem is summarized in the following box:

Solution of a Minimization Problem

Given a minimization problem with nonnegative coefficients in the objective function:

Step 1. Write all problem constraints as $\geqslant$ inequalities. (This may introduce negative numbers on the right side of some problem constraints.)

Step 2. Form the dual problem.

Step 3. Write the initial system of the dual problem, using the variables from the minimization problem as slack variables.

Step 4. Use the simplex method to solve the dual problem.

Step 5. Read the solution of the minimization problem from the bottom row of the final simplex tableau in step 4. [*Note:* If the dual problem has no solution, then the minimization problem has no solution.]

EXAMPLE 2 ➤   **Solving a Minimization Problem**   Solve the following minimization problem by maximizing the dual:

Minimize $C = 40x_1 + 12x_2 + 40x_3$

Subject to $2x_1 + x_2 + 5x_3 \geqslant 20$

$4x_1 + x_2 + x_3 \geqslant 30$

$x_1, x_2, x_3 \geqslant 0$

Solution   From Example 1 the dual is

$$\text{Maximize} \quad P = 20y_1 + 30y_2$$
$$\text{Subject to} \quad 2y_1 + 4y_2 \le 40$$
$$y_1 + y_2 \le 12$$
$$5y_1 + y_2 \le 40$$
$$y_1, y_2 \ge 0$$

Using x_1, x_2, and x_3 for slack variables, we obtain the initial system for the dual:

$$2y_1 + 4y_2 + x_1 \qquad\qquad\qquad = 40$$
$$y_1 + y_2 \qquad + x_2 \qquad\qquad = 12$$
$$5y_1 + y_2 \qquad\qquad + x_3 \qquad = 40$$
$$-20y_1 - 30y_2 \qquad\qquad\qquad + P = 0$$

Now we form the simplex tableau and solve the dual problem:

$$
\begin{array}{c}
\begin{array}{c} x_1 \\ x_2 \\ x_3 \\ P \end{array}
\left[
\begin{array}{cccccc|c}
y_1 & y_2 & x_1 & x_2 & x_3 & P & \\
2 & ④ & 1 & 0 & 0 & 0 & 40 \\
1 & 1 & 0 & 1 & 0 & 0 & 12 \\
5 & 1 & 0 & 0 & 1 & 0 & 40 \\
\hdashline
-20 & -30 & 0 & 0 & 0 & 1 & 0
\end{array}
\right]
\begin{array}{l} \tfrac{1}{4}R_1 \to R_1 \\ \\ \\ \\ \end{array}
\end{array}
$$

$$
\sim
\left[
\begin{array}{cccccc|c}
\tfrac{1}{2} & ① & \tfrac{1}{4} & 0 & 0 & 0 & 10 \\
1 & 1 & 0 & 1 & 0 & 0 & 12 \\
5 & 1 & 0 & 0 & 1 & 0 & 40 \\
\hdashline
-20 & -30 & 0 & 0 & 0 & 1 & 0
\end{array}
\right]
\begin{array}{l} \\ (-1)R_1 + R_2 \to R_2 \\ (-1)R_1 + R_3 \to R_3 \\ 30R_1 + R_4 \to R_4 \end{array}
$$

$$
\begin{array}{c}
\begin{array}{c} y_2 \\ x_2 \\ x_3 \\ P \end{array}
\sim
\left[
\begin{array}{cccccc|c}
\tfrac{1}{2} & 1 & \tfrac{1}{4} & 0 & 0 & 0 & 10 \\
① & 0 & -\tfrac{1}{4} & 1 & 0 & 0 & 2 \\
\tfrac{9}{2} & 0 & -\tfrac{1}{4} & 0 & 1 & 0 & 30 \\
\hdashline
-5 & 0 & \tfrac{15}{2} & 0 & 0 & 1 & 300
\end{array}
\right]
\begin{array}{l} \\ 2R_2 \to R_2 \\ \\ \\ \end{array}
\end{array}
$$

$$
\sim
\left[
\begin{array}{cccccc|c}
\tfrac{1}{2} & 1 & \tfrac{1}{4} & 0 & 0 & 0 & 10 \\
① & 0 & -\tfrac{1}{2} & 2 & 0 & 0 & 4 \\
\tfrac{9}{2} & 0 & -\tfrac{1}{4} & 0 & 1 & 0 & 30 \\
\hdashline
-5 & 0 & \tfrac{15}{2} & 0 & 0 & 1 & 300
\end{array}
\right]
\begin{array}{l} (-\tfrac{1}{2})R_2 + R_1 \to R_1 \\ \\ (-\tfrac{9}{2})R_2 + R_3 \to R_3 \\ 5R_2 + R_4 \to R_4 \end{array}
$$

$$
\begin{array}{c}
\begin{array}{c} y_2 \\ y_1 \\ x_3 \\ P \end{array}
\sim
\left[
\begin{array}{cccccc|c}
0 & 1 & \tfrac{1}{2} & -1 & 0 & 0 & 8 \\
1 & 0 & -\tfrac{1}{2} & 2 & 0 & 0 & 4 \\
0 & 0 & 2 & -9 & 1 & 0 & 12 \\
\hdashline
0 & 0 & 5 & 10 & 0 & 1 & 320
\end{array}
\right]
\end{array}
$$

From the bottom row of this tableau, we see that

$$\text{Min } C = 320 \quad \text{at} \quad x_1 = 5, \quad x_2 = 10, \quad x_3 = 0$$

◀

MATCHED PROBLEM 2 ➤ Solve the following minimization problem by maximizing the dual (see Matched Problem 1):

$$\text{Minimize} \quad C = 16x_1 + 9x_2 + 21x_3$$
$$\text{Subject to} \quad x_1 + x_2 + 3x_3 \geqslant 12$$
$$2x_1 + x_2 + x_3 \geqslant 16$$
$$x_1, x_2, x_3 \geqslant 0$$

◀

Caution

In the preceding section, we noted that multiplying a problem constraint by a number (usually in order to simplify calculations) changes the units of the slack variable. This requires some special interpretation of the value of the slack variable in the optimal solution, but causes no serious problems. However, when using the dual method, multiplying a problem constraint in the dual problem by a number can have some very serious consequences—the bottom row of the final simplex tableau may no longer give the correct solution to the minimization problem. To see this, refer to the first problem constraint of the dual problem in Example 2:

$$2y_1 + 4y_2 \leqslant 40$$

If we multiply this constraint by $\frac{1}{2}$ and then solve, the final tableau is (verify this):

$$
\begin{array}{ccccccc}
y_1 & y_2 & x_1 & x_2 & x_3 & P & \\
\begin{bmatrix}
0 & 1 & 1 & -1 & 0 & 0 & 8 \\
1 & 0 & -1 & 2 & 0 & 0 & 4 \\
0 & 0 & 4 & -9 & 1 & 0 & 12 \\
\hdashline
0 & 0 & 10 & 10 & 0 & 1 & 320
\end{bmatrix}
\end{array}
$$

The bottom row of this tableau indicates that the optimal solution to the minimization problem is $C = 320$ at $x_1 = 10$ and $x_2 = 10$. This is not the correct answer ($x_1 = 5$ is the correct answer). Thus, **you should never multiply a problem constraint in a maximization problem by a number if that maximization problem is being used to solve a minimization problem.** You may still simplify problem constraints in a minimization problem before forming the dual problem.

EXAMPLE 3 ➤ **Solving a Minimization Problem** Solve the following minimization problem by maximizing the dual:

$$\text{Minimize} \quad C = 5x_1 + 10x_2$$
$$\text{Subject to} \quad x_1 - x_2 \geqslant 1$$
$$-x_1 + x_2 \geqslant 2$$
$$x_1, x_2 \geqslant 0$$

Solution

$$A = \left[\begin{array}{cc|c} 1 & -1 & 1 \\ -1 & 1 & 2 \\ \hline 5 & 10 & 1 \end{array}\right] \qquad A^{\mathrm{T}} = \left[\begin{array}{cc|c} 1 & -1 & 5 \\ -1 & 1 & 10 \\ \hline 1 & 2 & 1 \end{array}\right]$$

The dual problem is

$$\text{Maximize} \quad P = y_1 + 2y_2$$
$$\text{Subject to} \quad y_1 - y_2 \leqslant 5$$
$$-y_1 + y_2 \leqslant 10$$
$$y_1, y_2 \geqslant 0$$

Introduce slack variables x_1 and x_2 and form the initial system for the dual:

$$y_1 - y_2 + x_1 \qquad\qquad = 5$$
$$-y_1 + y_2 \qquad + x_2 \qquad = 10$$
$$-y_1 - 2y_2 \qquad\qquad + P = 0$$

Form the simplex tableau and solve:

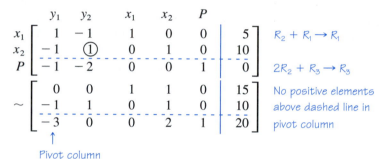

$$
\begin{array}{c}

\end{array}
\begin{array}{c|ccccc|c}
 & y_1 & y_2 & x_1 & x_2 & P & \\
\hline
x_1 & 1 & -1 & 1 & 0 & 0 & 5 \\
x_2 & -1 & \textcircled{1} & 0 & 1 & 0 & 10 \\
\hline
P & -1 & -2 & 0 & 0 & 1 & 0
\end{array}
\quad
\begin{array}{l}
R_2 + R_1 \rightarrow R_1 \\
\\
2R_2 + R_3 \rightarrow R_3
\end{array}
$$

$$
\sim
\begin{array}{c|ccccc|c}
 & 0 & 0 & 1 & 1 & 0 & 15 \\
 & -1 & 1 & 0 & 1 & 0 & 10 \\
\hline
 & -3 & 0 & 0 & 2 & 1 & 20
\end{array}
\quad
\begin{array}{l}
\text{No positive elements} \\
\text{above dashed line in} \\
\text{pivot column}
\end{array}
$$

↑
Pivot column

The -3 in the bottom row indicates that Column 1 is the pivot column. Since no positive elements appear in the pivot column above the dashed line, we are unable to select a pivot row. We stop the pivot operation and conclude that this maximization problem has no solution (see Fig. 2 in Section 5-4). Theorem 1 now implies that the original minimization problem has no solution. The graph of the inequalities in the minimization problem (Fig. 2) shows that the feasible region is empty; thus, it is not surprising that an optimal solution does not exist.

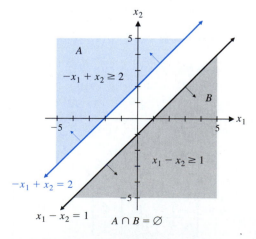

FIGURE 2

MATCHED PROBLEM 3 ➤   Solve the following minimization problem by maximizing the dual:

$$\text{Minimize} \quad C = 2x_1 + 3x_2$$
$$\text{Subject to} \quad x_1 - 2x_2 \geq 2$$
$$-x_1 + x_2 \geq 1$$
$$x_1, x_2 \geq 0$$

◀

■ APPLICATION: A TRANSPORTATION PROBLEM

One of the first applications of linear programming was to the problem of minimizing the cost of transporting materials. Problems of this type are referred to as **transportation problems.**

▶   EXAMPLE 4 ➤   **A Transportation Problem**   A computer manufacturing company has two assembly plants, plant A and plant B, and two distribution outlets, outlet I and outlet II. Plant A can assemble at most 700 computers a month, and plant B can assemble at most 900 computers a month. Outlet I must have at least 500 computers a month, and outlet II must have at least 1,000 computers a month. Transportation costs for shipping one computer from each plant to each outlet are as follows: $6 from plant A to outlet I; $5 from plant A to outlet II; $4 from plant B to outlet I; $8 from plant B to outlet II. Find a shipping schedule that will minimize the total cost of shipping the computers from the assembly plants to the distribution outlets. What is this minimum cost?

Solution   First, we summarize the relevant data in the table in the margin. In order to find a shipping schedule, we must determine the number of computers that should be shipped from each plant to each outlet. This will require the use of four decision variables:

	DISTRIBUTION OUTLET		ASSEMBLY CAPACITY
	I	II	
PLANT A	$6	$5	700
PLANT B	$4	$8	900
MINIMUM REQUIRED	500	1,000	

$x_1 = $ Number of computers shipped from plant A to outlet I

$x_2 = $ Number of computers shipped from plant A to outlet II

$x_3 = $ Number of computers shipped from plant B to outlet I

$x_4 = $ Number of computers shipped from plant B to outlet II

The total number of computers shipped from plant A is $x_1 + x_2$. Since this cannot exceed the assembly capacity at A, we have

$$x_1 + x_2 \leq 700 \quad \text{Number shipped from plant A}$$

Similarly, the total number shipped from plant B must satisfy

$$x_3 + x_4 \leq 900 \quad \text{Number shipped from plant B}$$

The total number shipped to each outlet must satisfy

$$x_1 + x_3 \geq 500 \quad \text{Number shipped to outlet I}$$

and

$$x_2 + x_4 \geq 1,000 \quad \text{Number shipped to outlet II}$$

Using the shipping charges in the table, the total shipping charges are

$$C = 6x_1 + 5x_2 + 4x_3 + 8x_4$$

Thus, we must solve the following linear programming problem:

Minimize $C = 6x_1 + 5x_2 + 4x_3 + 8x_4$

Subject to $x_1 + x_2 \qquad\qquad\leq 700$ Available from A

$\qquad\qquad\qquad x_3 + x_4 \leq 900$ Available from B

$\qquad x_1 \quad + x_3 \qquad\geq 500$ Required at I

$\qquad\qquad x_2 \qquad + x_4 \geq 1{,}000$ Required at II

$\qquad x_1, x_2, x_3, x_4 \geq 0$

Before we can solve this problem, we must multiply the first two constraints by -1 so that all the problem constraints are of the $\geq$ type. This will introduce negative constants into the minimization problem but not into the dual. Since the coefficients of C are nonnegative, the constants in the dual problem will be nonnegative and the dual will be a standard maximization problem. The problem can now be stated as

Minimize $C = 6x_1 + 5x_2 + 4x_3 + 8x_4$

Subject to $-x_1 - x_2 \qquad\qquad \geq -700$

$\qquad\qquad\qquad -x_3 - x_4 \geq -900$

$\qquad x \qquad + x_3 \qquad\geq 500$

$\qquad\qquad x_2 \qquad + x_4 \geq 1{,}000$

$\qquad x_1, x_2, x_3, x_4 \geq 0$

$$A = \begin{bmatrix} -1 & -1 & 0 & 0 & -700 \\ 0 & 0 & -1 & -1 & -900 \\ 1 & 0 & 1 & 0 & 500 \\ 0 & 1 & 0 & 1 & 1{,}000 \\ \hline 6 & 5 & 4 & 8 & 1 \end{bmatrix}$$

$$A^{\mathrm{T}} = \begin{bmatrix} -1 & 0 & 1 & 0 & 6 \\ -1 & 0 & 0 & 1 & 5 \\ 0 & -1 & 1 & 0 & 4 \\ 0 & -1 & 0 & 1 & 8 \\ \hline -700 & -900 & 500 & 1{,}000 & 1 \end{bmatrix}$$

The dual problem is

Maximize $P = -700y_1 - 900y_2 + 500y_3 + 1{,}000y_4$

Subject to $-y_1 \qquad + y_3 \qquad\leq 6$

$\qquad -y_1 \qquad\qquad + y_4 \leq 5$

$\qquad\qquad -y_2 + y_3 \qquad\leq 4$

$\qquad\qquad -y_2 \qquad + y_4 \leq 8$

$\qquad y_1, y_2, y_3, y_4 \geq 0$

Introduce slack variables x_1, x_2, x_3, and x_4 and form the initial system for the dual:

$$
\begin{array}{rcrcrcrcrcr}
-y_1 & & +\ y_3 & & +\ x_1 & & & & & = 6 \\
-y_1 & & & +\ y_4 & +\ x_2 & & & & & = 5 \\
& -y_2\ + & y_3 & & & +\ x_3 & & & & = 4 \\
& -y_2 & & +\ y_4 & & & +\ x_4 & & & = 8 \\
700y_1 & +\ 900y_2 & -\ 500y_3 & -\ 1{,}000y_4 & & & & +\ P & & = 0
\end{array}
$$

Form the simplex tableau and solve:

	y_1	y_2	y_3	y_4	x_1	x_2	x_3	x_4	P		
x_1	-1	0	1	0	1	0	0	0	0	6	
x_2	-1	0	0	①	0	1	0	0	0	5	
x_3	0	-1	1	0	0	0	1	0	0	4	
x_4	0	-1	0	1	0	0	0	1	0	8	$(-1)R_2 + R_4 \rightarrow R_4$
P	700	900	-500	$-1{,}000$	0	0	0	0	1	0	$1{,}000R_2 + R_5 \rightarrow R_5$

	y_1	y_2	y_3	y_4	x_1	x_2	x_3	x_4	P		
x_1	-1	0	1	0	1	0	0	0	0	6	$(-1)R_3 + R_1 \rightarrow R_1$
y_4	-1	0	0	1	0	1	0	0	0	5	
$\sim x_3$	0	-1	①	0	0	0	1	0	0	4	
x_4	1	-1	0	0	0	-1	0	1	0	3	
P	-300	900	-500	0	0	$1{,}000$	0	0	1	$5{,}000$	$500R_3 + R_5 \rightarrow R_5$

	y_1	y_2	y_3	y_4	x_1	x_2	x_3	x_4	P		
x_1	-1	1	0	0	1	0	-1	0	0	2	$R_4 + R_1 \rightarrow R_1$
y_4	-1	0	0	1	0	1	0	0	0	5	$R_4 + R_2 \rightarrow R_2$
$\sim y_3$	0	-1	1	0	0	0	1	0	0	4	
x_4	①	-1	0	0	0	-1	0	1	0	3	
P	-300	400	0	0	0	$1{,}000$	500	0	1	$7{,}000$	$300R_4 + R_5 \rightarrow R_5$

	y_1	y_2	y_3	y_4	x_1	x_2	x_3	x_4	P	
x_1	0	0	0	0	1	-1	-1	1	0	5
y_4	0	-1	0	1	0	0	0	1	0	8
$\sim y_3$	0	-1	1	0	0	0	1	0	0	4
y_1	1	-1	0	0	0	-1	0	1	0	3
P	0	100	0	0	0	700	500	300	1	$7{,}900$

From the bottom row of this tableau, we have

$$\text{Min } C = 7{,}900 \quad \text{at} \quad x_1 = 0, \quad x_2 = 700, \quad x_3 = 500, \quad x_4 = 300$$

The shipping schedule that minimizes the shipping charges is 700 from plant A to outlet II, 500 from plant B to outlet I, and 300 from plant B to outlet II. The total shipping cost is $7,900. ◀

MATCHED PROBLEM 4 ➤ Repeat Example 4 if the shipping charge from plant A to outlet I is increased to $7 and the shipping charge from plant B to outlet II is decreased to $3. ◀

■ SUMMARY OF PROBLEM TYPES AND SOLUTION METHODS

In this and the preceding sections, we have solved both maximization and minimization problems, but with certain restrictions on the problem constraints, constants on the right, and/or objective function coefficients. Table 1 summarizes the types of problems and methods of solution we have considered so far.

TABLE 1 Summary of Problem Types and Simplex Solution Methods

PROBLEM TYPE	PROBLEM CONSTRAINTS	RIGHT-SIDE CONSTANTS	COEFFICIENTS OF OBJECTIVE FUNCTION	METHOD OF SOLUTION
1. Maximization	$\leq$	Nonnegative	Any real numbers	Simplex method with slack variables
2. Minimization	$\geq$	Any real numbers	Nonnegative	Form dual and solve by simplex method with slack variables

The next section develops a generalized version of the simplex method that can handle both maximization and minimization problems with any combination of $\leq$, $\geq$, and $=$ problem constraints.

Answers to Matched Problems

1. Maximize $P = 12y_1 + 16y_2$
   Subject to $\quad y_1 + 2y_2 \leq 16$
   $\qquad\qquad\quad y_1 + y_2 \leq 9$
   $\qquad\qquad 3y_1 + y_2 \leq 21$
   $\qquad\qquad\qquad\quad y_1, y_2 \geq 0$

2. Min $C = 136$ at $x_1 = 4$, $x_2 = 8$, $x_3 = 0$

3. Dual problem:
   Maximize $P = 2y_1 + y_2$
   Subject to $\quad y_1 - y_2 \leq 2$
   $\qquad\qquad -2y_1 + y_2 \leq 3$
   $\qquad\qquad\qquad y_1, y_2 \geq 0$

   No solution

4. 600 from plant A to outlet II, 500 from plant B to outlet I, 400 from plant B to outlet II; total shipping cost is $6,200

EXERCISE 5-5

A *In Problems 1 and 2:*

(A) Form the dual problem.
(B) Write the initial system for the dual problem.
(C) Write the initial simplex tableau for the dual problem and label the columns of the tableau.

1. Minimize $C = 8x_1 + 9x_2$
    Subject to $\quad x_1 + 3x_2 \geq 4$
    $\qquad\qquad 2x_1 + x_2 \geq 5$
    $\qquad\qquad\qquad x_1, x_2 \geq 0$

2. Minimize $C = 12x_1 + 5x_2$
    Subject to $\quad 2x_1 + x_2 \geq 7$
    $\qquad\qquad 3x_1 + x_2 \geq 9$
    $\qquad\qquad\quad x_1, x_2 \geq 0$

In Problems 3 and 4, a minimization problem, the corresponding dual problem, and the final simplex tableau in the solution of the dual problem are given.

(A) Find the optimal solution of the dual problem.
(B) Find the optimal solution of the minimization problem.

3. Minimize $C = 21x_1 + 50x_2$
    Subject to $\quad 2x_1 + 5x_2 \geq 12$
    $\qquad\qquad 3x_1 + 7x_2 \geq 17$
    $\qquad\qquad\qquad x_1, x_2 \geq 0$

    Maximize $P = 12y_1 + 17y_2$
    Subject to $\quad 2y_1 + 3y_2 \leq 21$
    $\qquad\qquad 5y_1 + 7y_2 \leq 50$
    $\qquad\qquad\qquad y_1, y_2 \geq 0$

Tableau for Problem 3:

$$\begin{bmatrix} y_1 & y_2 & x_1 & x_2 & P & \\ 0 & 1 & 5 & -2 & 0 & 5 \\ 1 & 0 & -7 & 3 & 0 & 3 \\ \hline 0 & 0 & 1 & 2 & 1 & 121 \end{bmatrix}$$

4. Minimize $C = 16x_1 + 25x_2$
   Subject to $3x_1 + 5x_2 \geqslant 30$
   $\qquad\qquad 2x_1 + 3x_2 \geqslant 19$
   $\qquad\qquad x_1, x_2 \geqslant 0$

   Maximize $P = 30y_1 + 19y_2$
   Subject to $3y_1 + 2y_2 \leqslant 16$
   $\qquad\qquad 5y_1 + 3y_2 \leqslant 25$
   $\qquad\qquad y_1, y_2 \geqslant 0$

Tableau for Problem 4:

$$\begin{bmatrix} y_1 & y_2 & x_1 & x_2 & P & \\ 0 & 1 & 5 & -3 & 0 & 5 \\ 1 & 0 & -3 & 2 & 0 & 2 \\ \hline 0 & 0 & 5 & 3 & 1 & 155 \end{bmatrix}$$

In Problems 5–12:

(A) Form the dual problem.
(B) Find the solution to the original problem by applying the
simplex method to the dual problem.

5. Minimize $C = 9x_1 + 2x_2$
   Subject to $4x_1 + x_2 \geqslant 13$
   $\qquad\qquad 3x_1 + x_2 \geqslant 12$
   $\qquad\qquad x_1, x_2 \geqslant 0$

6. Minimize $C = x_1 + 4x_2$
   Subject to $x_1 + 2x_2 \geqslant 5$
   $\qquad\qquad x_1 + 3x_2 \geqslant 6$
   $\qquad\qquad x_1, x_2 \geqslant 0$

7. Minimize $C = 7x_1 + 12x_2$
   Subject to $2x_1 + 3x_2 \geqslant 15$
   $\qquad\qquad x_1 + 2x_2 \geqslant 8$
   $\qquad\qquad x_1, x_2 \geqslant 0$

8. Minimize $C = 3x_1 + 5x_2$
   Subject to $2x_1 + 3x_2 \geqslant 7$
   $\qquad\qquad x_1 + 2x_2 \geqslant 4$
   $\qquad\qquad x_1, x_2 \geqslant 0$

9. Minimize $C = 11x_1 + 4x_2$
   Subject to $\qquad 2x_1 + x_2 \geqslant 8$
   $\qquad\qquad -2x_1 + 3x_2 \geqslant 4$
   $\qquad\qquad x_1, x_2 \geqslant 0$

10. Minimize $C = 40x_1 + 10x_2$
   Subject to $2x_1 + x_2 \geqslant 12$
   $\qquad\qquad 3x_1 - x_2 \geqslant 3$
   $\qquad\qquad x_1, x_2 \geqslant 0$

11. Minimize $C = 7x_1 + 9x_2$
   Subject to $-3x_1 + x_2 \geqslant 6$
   $\qquad\qquad x_1 - 2x_2 \geqslant 4$
   $\qquad\qquad x_1, x_2 \geqslant 0$

12. Minimize $C = 10x_1 + 15x_2$
   Subject to $-4x_1 + x_2 \geqslant 12$
   $\qquad\qquad 12x_1 - 3x_2 \geqslant 10$
   $\qquad\qquad x_1, x_2 \geqslant 0$

B *Solve the linear programming problems in Problems 13–*
24 by applying the simplex method to the dual problem.

13. Minimize $C = 3x_1 + 9x_2$
   Subject to $2x_1 + x_2 \geqslant 8$
   $\qquad\qquad x_1 + 2x_2 \geqslant 8$
   $\qquad\qquad x_1, x_2 \geqslant 0$

14. Minimize $C = 2x_1 + x_2$
   Subject to $x_1 + x_2 \geqslant 8$
   $\qquad\qquad x_1 + 2x_2 \geqslant 4$
   $\qquad\qquad x_1, x_2 \geqslant 0$

15. Minimize $C = 7x_1 + 5x_2$
   Subject to $\qquad x_1 + x_2 \geqslant 4$
   $\qquad\qquad x_1 - 2x_2 \geqslant -8$
   $\qquad\qquad -2x_1 + x_2 \geqslant -8$
   $\qquad\qquad x_1, x_2 \geqslant 0$

16. Minimize $C = 10x_1 + 4x_2$
   Subject to $\qquad 2x_1 + x_2 \geqslant 6$
   $\qquad\qquad x_1 - 4x_2 \geqslant -24$
   $\qquad\qquad -8x_1 + 5x_2 \geqslant -24$
   $\qquad\qquad x_1, x_2 \geqslant 0$

17. Minimize $C = 10x_1 + 30x_2$
   Subject to $2x_1 + x_2 \geqslant 16$
   $\qquad\qquad x_1 + x_2 \geqslant 12$
   $\qquad\qquad x_1 + 2x_2 \geqslant 14$
   $\qquad\qquad x_1, x_2 \geqslant 0$

18. Minimize $C = 40x_1 + 10x_2$
   Subject to $3x_1 + x_2 \geqslant 24$
   $\qquad\qquad x_1 + x_2 \geqslant 16$
   $\qquad\qquad x_1 + 4x_2 \geqslant 30$
   $\qquad\qquad x_1, x_2 \geqslant 0$

19. Minimize $C = 5x_1 + 7x_2$
Subject to $x_1 \qquad\ \geqslant 4$
$x_1 + \ x_2 \geqslant 8$
$x_1 + 2x_2 \geqslant 10$
$x_1, x_2 \geqslant 0$

20. Minimize $C = 4x_1 + 5x_2$
Subject to $2x_1 + x_2 \geqslant 12$
$x_1 + x_2 \geqslant 9$
$x_2 \geqslant 4$
$x_1, x_2 \geqslant 0$

21. Minimize $C = 10x_1 + 7x_2 + 12x_3$
Subject to $x_1 + x_2 + 2x_3 \geqslant 7$
$2x_1 + x_2 + \ x_3 \geqslant 4$
$x_1, x_2, x_3 \geqslant 0$

22. Minimize $C = 14x_1 + 8x_2 + 20x_3$
Subject to $x_1 + x_2 + 3x_3 \geqslant 6$
$2x_1 + x_2 + \ x_3 \geqslant 9$
$x_1, x_2, x_3 \geqslant 0$

23. Minimize $C = 5x_1 + 2x_2 + 2x_3$
Subject to $x_1 - 4x_2 + \ x_3 \geqslant 6$
$-x_1 + \ x_2 - 2x_3 \geqslant 4$
$x_1, x_2, x_3 \geqslant 0$

24. Minimize $C = 6x_1 + 8x_2 + 3x_3$
Subject to $-3x_1 - 2x_2 + x_3 \geqslant 4$
$x_1 + \ x_2 - x_3 \geqslant 2$
$x_1, x_2, x_3 \geqslant 0$

25. A minimization problem has 4 variables and 2 problem constraints. How many variables and problem constraints are in the dual problem?

26. A minimization problem has 3 variables and 5 problem constraints. How many variables and problem constraints are in the dual problem?

27. If you want to solve a minimization problem by applying the geometric method to the dual problem, how many variables and problem constraints must be in the original problem?

28. If you want to solve a minimization problem by applying the geometric method to the original problem, how many variables and problem constraints must be in the original problem?

In Problems 29–32, determine whether a minimization problem with the indicated condition can be solved by applying the simplex method to the dual problem. If your answer is yes, describe any necessary modifications that must be made before forming the dual problem. If your answer is no, explain why.

29. A coefficient of the objective function is negative.

30. A coefficient of a problem constraint is negative.

31. A problem constraint is of the $\leqslant$ form.

32. A problem constraint has a negative constant on the right side.

C   *Solve the linear programming problems in Problems 33–36 by applying the simplex method to the dual problem.*

33. Minimize $C = 16x_1 + 8x_2 + 4x_3$
Subject to $3x_1 + 2x_2 + 2x_3 \geqslant 16$
$4x_1 + 3x_2 + \ x_3 \geqslant 14$
$5x_1 + 3x_2 + \ x_3 \geqslant 12$
$x_1, x_2, x_3 \geqslant 0$

34. Minimize $C = 6x_1 + 8x_2 + 12x_3$
Subject to $x_1 + 3x_2 + 3x_3 \geqslant 6$
$x_1 + 5x_2 + 5x_3 \geqslant 4$
$2x_1 + 2x_2 + 3x_3 \geqslant 8$
$x_1, x_2, x_3 \geqslant 0$

35. Minimize $C = 5x_1 + 4x_2 + 5x_3 + 6x_4$
Subject to $x_1 + x_2 \qquad\qquad \leqslant 12$
$x_3 + x_4 \leqslant 25$
$x_1 \qquad + x_3 \qquad \geqslant 20$
$x_2 \qquad + x_4 \geqslant 15$
$x_1, x_2, x_3, x_4 \geqslant 0$

36. Repeat Problem 35 with $C = 4x_1 + 7x_2 + 5x_3 + 6x_4$.

▶ APPLICATIONS

Formulate each of the following as a linear programming problem. (The answers in the back of the book for these application problems include the formulation.) Then solve the problem by applying the simplex method to the dual problem.

Business & Economics

37. *Manufacturing—production scheduling.* A food processing company produces regular and deluxe ice cream at three plants. Per hour of operation, the plant in Cedarburg produces 20 gallons of regular ice cream and 10 gallons of deluxe ice cream, the Grafton plant produces 10 gallons of regular and 20 gallons of deluxe, and the West Bend plant produces 20 gallons of regular and 20 gallons of deluxe. It costs $70 per hour to operate the Cedarburg plant, $75 per hour to operate the Grafton plant, and $90 per hour to operate the West Bend plant.

(A) The company needs at least 300 gallons of regular ice cream and at least 200 gallons of deluxe ice cream each day. How many hours per day should each plant be scheduled to operate in order to produce the required amounts of ice cream and minimize the cost of production? What is the minimum production cost?

(B) Discuss the effect on the production schedule and the minimum production cost if the demand for deluxe ice cream increases to 300 gallons per day and all other data in part (A) remain the same.

(C) Repeat part (B) if the demand for deluxe ice cream increases to 400 gallons per day.

38. *Mining—production scheduling.* A mining company operates two mines, each of which produces three grades of ore. The West Summit mine can produce 2 tons of low-grade ore, 3 tons of medium-grade ore, and 1 ton of high-grade ore per hour of operation. The North Ridge mine can produce 2 tons of low-grade ore, 1 ton of medium-grade ore, and 2 tons of high-grade ore per hour of operation. To satisfy existing orders, the company needs at least 100 tons of low-grade ore, 60 tons of medium-grade ore, and 80 tons of high-grade ore. The cost of operating each mine varies, depending on the conditions encountered while extracting the ore.

(A) If it costs $400 per hour to operate the West Summit mine and $600 per hour to operate the North Ridge mine, how many hours should each mine be operated to supply the required amounts of ore and minimize the cost of production? What is the minimum production cost?

(B) Discuss the effect on the production schedule and the minimum production cost if it costs $300 per hour to operate the West Summit mine, $700 per hour to operate the North Ridge mine, and all other data in part (A) remain the same.

(C) Repeat part (B) if it costs $800 per hour to operate the West Summit mine and $200 per hour to operate the North Ridge mine.

39. *Purchasing.* Acme Micros markets computers with single-sided and double-sided disk drives. The disk drives are supplied by two other companies, Associated Electronics and Digital Drives. Associated Electronics charges $250 for a single-sided disk drive and $350 for a double-sided disk drive. Digital Drives charges $290 for a single-sided disk drive and $320 for a double-sided disk drive. Each month, Associated Electronics can supply at most 1,000 disk drives in any combination of single-sided and double-sided drives. The combined monthly total supplied by Digital Drives cannot exceed 2,000 disk drives. Acme Micros needs at least 1,200 single-sided drives and at least 1,600 double-sided drives each month. How many disk drives of each type should Acme Micros order from each supplier in order to meet its monthly demand and minimize the purchase cost? What is the minimum purchase cost?

40. *Transportation.* A feed company stores grain in elevators located in Ames, Iowa, and Bedford, Indiana. Each month the grain is shipped to processing plants in Columbia, Missouri, and Danville, Illinois. The monthly supply (in tons) of grain at each elevator, the monthly demand (in tons) at each processing plant, and the cost per ton for transporting the grain are given in the table. Determine a shipping schedule that will minimize the cost of transporting the grain. What is the minimum cost?

| | SHIPPING COST (PER TON) | | SUPPLY |
	Columbia	*Danville*	(IN TONS)
AMES	$22	$38	700
BEDFORD	$46	$24	500
DEMAND (IN TONS)	400	600	

Life Sciences

41. *Nutrition—people.* A dietitian in a hospital is to arrange a special diet using three foods, L, M, and N. Each ounce of food L contains 20 units of calcium, 10 units of iron, 10 units of vitamin A, and 20 units of cholesterol. Each ounce of food M contains 10 units of calcium, 10 units of iron, 15 units of vitamin A, and 24 units of cholesterol. Each ounce of food N contains 10 units of calcium, 10 units of iron, 10 units of vitamin A, and 18 units of cholesterol. If the minimum daily requirements are 300 units of calcium, 200 units of iron, and 240 units of vitamin A, how many ounces of each food should be used to meet the minimum requirements and at the same time minimize the cholesterol intake? What is the minimum cholesterol intake?

42. *Nutrition—plants.* A farmer can buy three types of plant food, mix *A*, mix *B*, and mix *C*. Each cubic yard of mix *A* contains 20 pounds of phosphoric acid, 10 pounds of nitrogen, and 10 pounds of potash. Each cubic yard of mix *B* contains 10 pounds of phosphoric acid, 10 pounds of nitrogen, and 15 pounds of potash. Each cubic yard of mix *C* contains 20 pounds of phosphoric acid, 20 pounds of nitrogen, and 5 pounds of potash. The minimum monthly requirements are 480 pounds of phosphoric acid, 320 pounds of nitrogen, and 225 pounds of potash. If mix *A* costs $30 per cubic yard, mix *B* costs $36 per cubic yard, and mix *C* costs $39 per cubic yard, how many cubic yards of each mix should the farmer blend to meet the minimum monthly requirements at a minimal cost? What is the minimum cost?

Social Sciences

43. *Education—resource allocation.* A metropolitan school district has two high schools that are overcrowded and two that are underenrolled. In order to balance the enrollment, the school board has decided to bus students from the overcrowded schools to the underenrolled schools. North Divi-

sion High School has 300 more students than it should have, and South Division High School has 500 more students than it should have. Central High School can accommodate 400 additional students and Washington High School can accommodate 500 additional students. The weekly cost of busing a student from North Division to Central is $5, from North Division to Washington is $2, from South Division to Central is $3, and from South Division to Washington is $4. Determine the number of students that should be bused from each of the overcrowded schools to each of the underenrolled schools in order to balance the enrollment and minimize the cost of busing the students. What is the minimum cost?

44. *Education—resource allocation.* Repeat Problem 43 if the weekly cost of busing a student from North Division to Washington is $7 and all the other information remains the same.

SECTION 5-6 ■ **Maximization and Minimization with Mixed Problem Constraints**

- ■ AN INTRODUCTION TO THE BIG *M* METHOD
- ■ THE BIG *M* METHOD
- ■ MINIMIZATION BY THE BIG *M* METHOD
- ■ SUMMARY OF METHODS OF SOLUTION
- ■ LARGER PROBLEMS—A REFINERY APPLICATION

In the preceding two sections, we have seen how to solve both maximization and minimization problems, but with rather severe restrictions on problem constraints, right-side constants, and/or objective function coefficients (see the summary in Table 1 of the last section). In this section we will present a generalized version of the simplex method that will solve both maximization and minimization problems with any combination of $\leq$, $\geq$, and $=$ problem constraints. The only requirement is that each problem constraint have a nonnegative constant on the right side. (This restriction is easily accommodated, as you will see.)

■ AN INTRODUCTION TO THE BIG *M* METHOD

We introduce the big *M* method through a simple maximization problem with mixed problem constraints. The key parts of the method will then be summarized and applied to more complex problems.

Consider the following problem:

Maximize $P = 2x_1 + x_2$

Subject to $x_1 + x_2 \leqslant 10$

$\qquad -x_1 + x_2 \geqslant 2$

$\qquad x_1, x_2 \geqslant 0$ (1)

To form an equation out of the first inequality, we introduce a slack variable s_1, as before, and write

$$x_1 + x_2 + s_1 = 10$$

How can we form an equation out of the second inequality? We introduce a second variable s_2 and subtract it from the left side so that we can write

$$-x_1 + x_2 - s_2 = 2$$

The variable s_2 is called a **surplus variable,** because it is the amount (surplus) by which the left side of the inequality exceeds the right side.

We now express the linear programming problem (1) as a system of equations:

$$\begin{aligned}
x_1 + x_2 + s_1 \qquad\qquad &= 10 \\
-x_1 + x_2 \qquad - s_2 \qquad &= 2 \\
-2x_1 - x_2 \qquad\qquad + P &= 0 \\
x_1, x_2, s_1, s_2 &\geqslant 0
\end{aligned}$$ (2)

It can be shown that a basic solution of (2) is not feasible if any of the variables (excluding P) are negative. Thus, **a surplus variable is required to satisfy the nonnegative constraint.**

The basic solution found by setting the nonbasic variables x_1 and x_2 equal to 0 is

$$x_1 = 0, \quad x_2 = 0, \quad s_1 = 10, \quad s_2 = -2, \quad P = 0$$

But this basic solution is not feasible, since the surplus variable s_2 is negative (which is a violation of the nonnegative requirements of all variables except P). The simplex method works only when the basic solution for a tableau is feasible, so we cannot solve this problem simply by writing the tableau for (2) and starting pivot operations.

e x p l o r e – d i s c u s s   1

To see that the simplex method will not work when the basic solution is not feasible, write the tableau for (2) and perform a pivot operation. Is the new basic solution feasible? Is it possible to select another pivot element?

In order to use the simplex method on problems with mixed constraints, we turn to an ingenious device called an *artificial variable*. This variable has no physical meaning in the original problem (which explains the use of the word ''artificial'') and is introduced solely for the purpose of obtaining a basic feasible solution so that we can apply the simplex method. An **artificial variable** is a variable introduced into each equation that has a surplus variable. As before, to ensure that we consider only feasible basic solutions, **an artificial variable is required to satisfy the nonnegative constraint.** (As we shall see later, artificial variables are also used to augment equality problem constraints when they are present.)

Returning to the problem at hand, we introduce an artificial variable a_1 into the equation involving the surplus variable s_2:

$$-x_1 + x_2 - s_2 + a_1 = 2$$

To prevent an artificial variable from becoming part of an optimal solution to the original problem, a very large "penalty" is introduced into the objective function. This penalty is created by choosing a positive constant M so large that the artificial variable is forced to be 0 in any final optimal solution of the original problem. (Since the constant M can be made as large as we wish, in computer solutions, M is often selected as the largest number the computer can hold!) We then add the term $-Ma_1$ to the objective function:

$$P = 2x_1 + x_2 - Ma_1$$

We now have a new problem, which we call the **modified problem:**

$$
\begin{aligned}
\text{Maximize} \quad & P = 2x_1 + x_2 - Ma_1 \\
\text{Subject to} \quad & x_1 + x_2 + s_1 \qquad\qquad = 10 \\
& -x_1 + x_2 \qquad - s_2 + a_1 = 2 \\
& x_1, x_2, s_1, s_2, a_1 \geqslant 0
\end{aligned}
\tag{3}
$$

The initial system for the modified problem (3) is

$$
\begin{aligned}
& x_1 + x_2 + s_1 \qquad\qquad\qquad = 10 \\
& -x_1 + x_2 \qquad - s_2 + \ a_1 \qquad\quad = 2 \\
& -2x_1 - x_2 \qquad\qquad\quad + Ma_1 + P = 0 \\
& x_1, x_2, s_1, s_2, a_1 \geqslant 0
\end{aligned}
\tag{4}
$$

We next write the augmented coefficient matrix for (4), which we call the **preliminary simplex tableau** for the modified problem. (The reason we call it the "preliminary" simplex tableau instead of the "initial" simplex tableau will be made clear shortly.)

$$
\begin{array}{ccccccc}
x_1 & x_2 & s_1 & s_2 & a_1 & P & \\
\left[\begin{array}{cccccc|c}
1 & 1 & 1 & 0 & 0 & 0 & 10 \\
-1 & 1 & 0 & -1 & 1 & 0 & 2 \\
\hline
-2 & -1 & 0 & 0 & M & 1 & 0
\end{array}\right] &
\end{array}
\tag{5}
$$

To start the simplex process, including any necessary pivot operations, the preliminary simplex tableau should either meet the two requirements given in the box below or be transformed by row operations into a tableau that meets these two requirements.

Initial Simplex Tableau Requirements

For a system tableau to be considered an **initial simplex tableau,** it must satisfy the following two requirements:

1. The requisite number of basic variables must be selectable by the process described in Section 5-4. That is, a variable can be selected as a basic variable only if it corresponds to a column in the tableau that has exactly one nonzero element and the nonzero element in the column is not in the same row as the nonzero element in the column of another basic variable. The remaining variables are then selected as nonbasic variables to be set equal to 0 in determining a basic solution.
2. The basic solution found by setting the nonbasic variables equal to 0 is feasible.

Tableau (5) satisfies the first initial simplex tableau requirement, since s_1, s_2, and P can be selected as basic variables according to the criterion stated. (Not all preliminary simplex tableaux satisfy the first requirement; see Example 2.) However, tableau (5) does not satisfy the second initial simplex tableau requirement, since the basic solution is not feasible ($s_2 = -2$). To use the simplex method, we must first use row operations to transform (5) into an equivalent matrix that satisfies both initial simplex tableau requirements. **Note that this transformation is not a pivot operation.**

To get an idea of how to proceed, notice in tableau (5) that -1 in the s_2 column is in the same row as 1 in the a_1 column. This is not an accident! The artificial variable a_1 was introduced so that this would happen. If we eliminate M from the bottom of the a_1 column, then the nonbasic variable a_1 will become a basic variable and the troublesome basic variable s_2 will become a nonbasic variable. Thus, we proceed to eliminate M from the a_1 column using row operations:

$$
\begin{array}{ccccccc}
x_1 & x_2 & s_1 & s_2 & a_1 & P & \\
\end{array}
$$

$$
\left[\begin{array}{cccccc|c}
1 & 1 & 1 & 0 & 0 & 0 & 10 \\
-1 & 1 & 0 & -1 & 1 & 0 & 2 \\
-2 & -1 & 0 & 0 & M & 1 & 0
\end{array}\right] \quad (-M)R_2 + R_3 \rightarrow R_3
$$

$$
\sim \left[\begin{array}{cccccc|c}
1 & 1 & 1 & 0 & 0 & 0 & 10 \\
-1 & 1 & 0 & -1 & 1 & 0 & 2 \\
M-2 & -M-1 & 0 & M & 0 & 1 & -2M
\end{array}\right]
$$

From this last matrix we see that the basic variables are s_1, a_1, and P. The basic solution found by setting the nonbasic variables x_1, x_2, and s_2 equal to 0 is

$$
x_1 = 0, \quad x_2 = 0, \quad s_1 = 10, \quad s_2 = 0, \quad a_1 = 2, \quad P = -2M
$$

The basic solution is feasible (remember, P can be negative), and both requirements for an initial simplex tableau are met. We can now commence with the simplex process using pivot operations.

The pivot column is determined by the most negative indicator in the bottom row of the tableau. Since M is a positive number, $-M - 1$ is certainly a negative indicator. What about the indicator $M - 2$? Remember that M is a very large positive number. We will assume that M is so large that any expression of the form $M - k$ is positive. Thus, the only negative indicator in the bottom row is $-M - 1$.

$$
\begin{array}{c}
 x_1 x_2 s_1 s_2 a_1 P \\
\text{Pivot row} \rightarrow
\left[
\begin{array}{cccccc|c}
1 & 1 & 1 & 0 & 0 & 0 & 10 \\
-1 & \textcircled{1} & 0 & -1 & 1 & 0 & 2 \\
M-2 & -M-1 & 0 & M & 0 & 1 & -2M
\end{array}
\right]
\begin{array}{l}
\tfrac{10}{1} = 10 \\
\tfrac{2}{1} = 2
\end{array}
\end{array}
$$

$\uparrow$
Pivot column

Having identified the pivot element, we now begin pivoting:

$$
\begin{array}{c}
 x_1 x_2 s_1 s_2 a_1 P \\
\begin{array}{c} s_1 \\ a_1 \\ P \end{array}
\left[
\begin{array}{cccccc|c}
1 & 1 & 1 & 0 & 0 & 0 & 10 \\
-1 & \textcircled{1} & 0 & -1 & 1 & 0 & 2 \\
M-2 & -M-1 & 0 & M & 0 & 1 & -2M
\end{array}
\right]
\begin{array}{l}
(-1)R_2 + R_1 \rightarrow R_1 \\
\\
(M+1)R_2 + R_3 \rightarrow R_3
\end{array}
\end{array}
$$

$$
\begin{array}{c}
\begin{array}{c} s_1 \\ \sim x_2 \\ P \end{array}
\left[
\begin{array}{cccccc|c}
\textcircled{2} & 0 & 1 & 1 & -1 & 0 & 8 \\
-1 & 1 & 0 & -1 & 1 & 0 & 2 \\
-3 & 0 & 0 & -1 & M+1 & 1 & 2
\end{array}
\right]
\begin{array}{l}
\tfrac{1}{2}R_1 \rightarrow R_1 \\
\\
\\
\end{array}
\end{array}
$$

$$
\begin{array}{c}
\sim
\left[
\begin{array}{cccccc|c}
\textcircled{1} & 0 & \tfrac{1}{2} & \tfrac{1}{2} & -\tfrac{1}{2} & 0 & 4 \\
-1 & 1 & 0 & -1 & 1 & 0 & 2 \\
-3 & 0 & 0 & -1 & M+1 & 1 & 2
\end{array}
\right]
\begin{array}{l}
\\
R_1 + R_2 \rightarrow R_2 \\
3R_1 + R_3 \rightarrow R_3
\end{array}
\end{array}
$$

$$
\begin{array}{c}
\begin{array}{c} x_1 \\ \sim x_2 \\ P \end{array}
\left[
\begin{array}{cccccc|c}
1 & 0 & \tfrac{1}{2} & \tfrac{1}{2} & -\tfrac{1}{2} & 0 & 4 \\
0 & 1 & \tfrac{1}{2} & -\tfrac{1}{2} & \tfrac{1}{2} & 0 & 6 \\
0 & 0 & \tfrac{3}{2} & \tfrac{1}{2} & M-\tfrac{1}{2} & 1 & 14
\end{array}
\right]
\end{array}
$$

Since all the indicators in the last row are nonnegative ($M - \tfrac{1}{2}$ is nonnegative because M is a very large positive number), we can stop and write the optimal solution:

$$\text{Max } P = 14 \quad \text{at} \quad x_1 = 4, \quad x_2 = 6, \quad s_1 = 0, \quad s_2 = 0, \quad a_1 = 0$$

This is an optimal solution to the modified problem (3). How is it related to the original problem (2)? Since $a_1 = 0$ in this solution,

$$x_1 = 4, \quad x_2 = 6, \quad s_1 = 0, \quad s_2 = 0, \quad P = 14 \tag{6}$$

is certainly a feasible solution for (2). [You can verify this by direct substitution into (2).] Surprisingly, it turns out that (6) is an optimal solution to the original problem. To see that this is true, suppose we were able to find feasible values of $x_1, x_2, s_1,$ and s_2 that satisfy the original system (2) and produce a value of $P > 14$. Then by using these same values in (3) along with $a_1 = 0$, we have found a feasible solution of (3) with $P > 14$. This contradicts the fact that $P = 14$ is the maximum value of P for the modified problem. Thus, (6) is an optimal solution for the original problem.

As this example illustrates, if $a_1 = 0$ is an optimal solution for the modified problem, then deleting a_1 produces an optimal solution for the original problem. What happens if $a_1 \neq 0$ in the optimal solution for the modified problem? In this case, it can be shown that the original problem has no solution because its feasible set is empty.

explore – discuss 2

Graph the feasible region for problem (1), plot the (x_1, x_2) coordinates of the basic solution for each simplex tableau in the solution of this problem, and illustrate the path to the optimal solution.

In larger problems, each $\geqslant$ problem constraint will require the introduction of a surplus variable and an artificial variable. If one of the problem constraints is an equation rather than an inequality, there is no need to introduce a slack or surplus variable. However, each $=$ problem constraint will require the introduction of another artificial variable to prevent the initial basic solution from violating the equality constraint—the decision variables are often 0 in the initial basic solution (see Example 2). Finally, each artificial variable also must be included in the objective function for the modified problem. The same constant M can be used for each artificial variable. Because of the role that the constant M plays in this approach, this method is often called the **big M method.**

■ THE BIG M METHOD

We now summarize the key steps of the big M method and use them to solve several problems.

The Big M Method—Introducing Slack, Surplus, and Artificial Variables to Form the Modified Problem

Step 1. If any problem constraints have negative constants on the right side, multiply both sides by -1 to obtain a constraint with a nonnegative constant. (If the constraint is an inequality, this will reverse the direction of the inequality.)

Step 2. Introduce a slack variable in each $\leqslant$ constraint.

Step 3. Introduce a surplus variable and an artificial variable in each $\geqslant$ constraint.

Step 4. Introduce an artificial variable in each $=$ constraint.

Step 5. For each artificial variable a_i, add $-Ma_i$ to the objective function. Use the same constant M for all artificial variables.

EXAMPLE 1 ➤ **Finding the Modified Problem**   Find the modified problem for the following linear programming problem. (Do not attempt to solve the problem.)

$$\text{Maximize}\quad P = 2x_1 + 5x_2 + 3x_3$$
$$\text{Subject to}\quad x_1 + 2x_2 - x_3 \leq 7$$
$$-x_1 + x_2 - 2x_3 \leq -5$$
$$x_1 + 4x_2 + 3x_3 \geq 1$$
$$2x_1 - x_2 + 4x_3 = 6$$
$$x_1, x_2, x_3 \geq 0$$

Solution   First, we multiply the second constraint by -1 to change -5 to 5:

$$(-1)(-x_1 + x_2 - 2x_3) \geq (-1)(-5)$$
$$x_1 - x_2 + 2x_3 \geq 5$$

Next, we introduce the slack, surplus, and artificial variables according to the rules stated in the box:

$$x_1 + 2x_2 - x_3 + s_1 \qquad\qquad\qquad = 7$$
$$x_1 - x_2 + 2x_3 \qquad - s_2 + a_1 \qquad\qquad = 5$$
$$x_1 + 4x_2 + 3x_3 \qquad\qquad - s_3 + a_2 \quad = 1$$
$$2x_1 - x_2 + 4x_3 \qquad\qquad\qquad + a_3 = 6$$

Finally, we add $-Ma_1$, $-Ma_2$, and $-Ma_3$ to the objective function:

$$P = 2x_1 + 5x_2 + 3x_3 - Ma_1 - Ma_2 - Ma_3$$

The modified problem is

$$\text{Maximize}\quad P = 2x_1 + 5x_2 + 3x_3 - Ma_1 - Ma_2 - Ma_3$$
$$\text{Subject to}\quad x_1 + 2x_2 - x_3 + s_1 \qquad\qquad\qquad = 7$$
$$x_1 - x_2 + 2x_3 \qquad - s_2 + a_1 \qquad\qquad = 5$$
$$x_1 + 4x_2 + 3x_3 \qquad\qquad - s_3 + a_2 \quad = 1$$
$$2x_1 - x_2 + 4x_3 \qquad\qquad\qquad + a_3 = 6$$
$$x_1, x_2, x_3, s_1, s_2, s_3, a_1, a_2, a_3 \geq 0$$

◄

MATCHED PROBLEM 1 ➤ Repeat Example 1 for:

$$\text{Maximize}\quad P = 3x_1 - 2x_2 + x_3$$
$$\text{Subject to}\quad x_1 - 2x_2 + x_3 \geq 5$$
$$-x_1 - 3x_2 + 4x_3 \leq -10$$
$$2x_1 + 4x_2 + 5x_3 \leq 20$$
$$3x_1 - x_2 - x_3 = -15$$
$$x_1, x_2, x_3 \geq 0$$

◄

We now list the key steps for solving a problem using the big M method. The various steps and remarks are based on a number of important theorems which we assume without proof. In particular, step 2 is based on the fact that (except for some degenerate cases not considered here) if the modified linear programming problem has a solution, then the preliminary simplex tableau will be transformed into an initial simplex tableau by eliminating the M's from the columns corresponding to the artificial variables in the preliminary simplex tableau. Having obtained an initial simplex tableau, we can then commence with pivot operations.

The Big M Method—Solving the Problem

Step 1. Form the preliminary simplex tableau for the modified problem.

Step 2. Use row operations to eliminate the M's in the bottom row of the preliminary simplex tableau in the columns corresponding to the artificial variables. The resulting tableau is the initial simplex tableau.

Step 3. Solve the modified problem by applying the simplex method to the initial simplex tableau found in step 2.

Step 4. Relate the solution of the modified problem to the original problem.

(A) If the modified problem has no solution, then the original problem has no solution.
(B) If all artificial variables are 0 in the solution to the modified problem, then delete the artificial variables to find a solution to the original problem.
(C) If any artificial variables are nonzero in the solution to the modified problem, then the original problem has no solution.

EXAMPLE 2 ➤ **Using the Big M Method** Solve the following linear programming problem using the big M method:

$$\text{Maximize} \quad P = x_1 - x_2 + 3x_3$$
$$\text{Subject to} \quad x_1 + x_2 \quad\quad \leqslant 20$$
$$x_1 \quad\quad + x_3 = 5$$
$$x_2 + x_3 \geqslant 10$$
$$x_1, x_2, x_3 \geqslant 0$$

Solution State the modified problem:

$$\text{Maximize} \quad P = x_1 - x_2 + 3x_3 - Ma_1 - Ma_2$$
$$\text{Subject to} \quad x_1 + x_2 \quad + s_1 \quad\quad\quad = 20$$
$$x_1 \quad\quad + x_3 \quad\quad + a_1 \quad\quad = 5$$
$$x_2 + x_3 \quad\quad - s_2 + a_2 = 10$$
$$x_1, x_2, x_3, s_1, s_2, a_1, a_2 \geqslant 0$$

Write the preliminary simplex tableau for the modified problem, and find the initial simplex tableau by eliminating the M's from the artificial variable columns:

$$
\begin{array}{cccccccc|c}
x_1 & x_2 & x_3 & s_1 & a_1 & s_2 & a_2 & P & \\
1 & 1 & 0 & 1 & 0 & 0 & 0 & 0 & 20 \\
1 & 0 & 1 & 0 & 1 & 0 & 0 & 0 & 5 \\
0 & 1 & 1 & 0 & 0 & -1 & 1 & 0 & 10 \\
-1 & 1 & -3 & 0 & M & 0 & M & 1 & 0
\end{array}
$$

Eliminate M from the a_1 column.

$(-M)R_2 + R_4 \rightarrow R_4$

$$
\sim
\begin{array}{cccccccc|c}
1 & 1 & 0 & 1 & 0 & 0 & 0 & 0 & 20 \\
1 & 0 & 1 & 0 & 1 & 0 & 0 & 0 & 5 \\
0 & 1 & 1 & 0 & 0 & -1 & 1 & 0 & 10 \\
-M-1 & 1 & -M-3 & 0 & 0 & 0 & M & 1 & -5M
\end{array}
$$

Eliminate M from the a_2 column.

$(-M)R_3 + R_4 \rightarrow R_4$

$$
\sim
\begin{array}{cccccccc|c}
1 & 1 & 0 & 1 & 0 & 0 & 0 & 0 & 20 \\
1 & 0 & 1 & 0 & 1 & 0 & 0 & 0 & 5 \\
0 & 1 & 1 & 0 & 0 & -1 & 1 & 0 & 10 \\
-M-1 & -M+1 & -2M-3 & 0 & 0 & M & 0 & 1 & -15M
\end{array}
$$

From this last matrix we see that the basic variables are s_1, a_1, a_2, and P. The basic solution found by setting the nonbasic variables x_1, x_2, x_3, and s_2 equal to 0 is

$$x_1 = 0, \quad x_2 = 0, \quad x_3 = 0, \quad s_1 = 20, \quad a_1 = 5, \quad s_2 = 0, \quad a_2 = 10, \quad P = -15M$$

The basic solution is feasible, and both requirements for an initial simplex tableau are met. We can now commence with pivot operations to find the optimal solution.

$$
\begin{array}{c}
\\ s_1 \\ a_1 \\ a_2 \\ P
\end{array}
\begin{array}{cccccccc|c}
x_1 & x_2 & x_3 & s_1 & a_1 & s_2 & a_2 & P & \\
1 & 1 & 0 & 1 & 0 & 0 & 0 & 0 & 20 \\
1 & 0 & ① & 0 & 1 & 0 & 0 & 0 & 5 \\
0 & 1 & 1 & 0 & 0 & -1 & 1 & 0 & 10 \\
-M-1 & -M+1 & -2M-3 & 0 & 0 & M & 0 & 1 & -15M
\end{array}
$$

$(-1)R_2 + R_3 \rightarrow R_3$
$(2M+3)R_2 + R_4 \rightarrow R_4$

$$
\sim
\begin{array}{c}
\\ s_1 \\ x_3 \\ a_2 \\ P
\end{array}
\begin{array}{cccccccc|c}
1 & 1 & 0 & 1 & 0 & 0 & 0 & 0 & 20 \\
1 & 0 & 1 & 0 & 1 & 0 & 0 & 0 & 5 \\
-1 & ① & 0 & 0 & -1 & -1 & 1 & 0 & 5 \\
M+2 & -M+1 & 0 & 0 & 2M+3 & M & 0 & 1 & -5M+15
\end{array}
$$

$(-1)R_3 + R_1 \rightarrow R_1$

$(M-1)R_3 + R_4 \rightarrow R_4$

$$
\sim
\begin{array}{c}
\\ s_1 \\ x_3 \\ x_2 \\ P
\end{array}
\begin{array}{cccccccc|c}
2 & 0 & 0 & 1 & 1 & 1 & -1 & 0 & 15 \\
1 & 0 & 1 & 0 & 1 & 0 & 0 & 0 & 5 \\
-1 & 1 & 0 & 0 & -1 & -1 & 1 & 0 & 5 \\
3 & 0 & 0 & 0 & M+4 & 1 & M-1 & 1 & 10
\end{array}
$$

Since the bottom row has no negative indicators, we can stop and write the optimal solution to the modified problem:

$$x_1 = 0, \quad x_2 = 5, \quad x_3 = 5, \quad s_1 = 15, \quad a_1 = 0, \quad s_2 = 0, \quad a_2 = 0, \quad P = 10$$

Since $a_1 = 0$ and $a_2 = 0$, the solution to the original problem is

$$\text{Max } P = 10 \quad \text{at} \quad x_1 = 0, \quad x_2 = 5, \quad x_3 = 5 \quad \blacktriangleleft$$

MATCHED PROBLEM 2 ➤ Solve the following linear programming problem using the big M method:

$$\text{Maximize} \quad P = x_1 + 4x_2 + 2x_3$$
$$\text{Subject to} \quad x_2 + x_3 \leq 4$$
$$x_1 \qquad - x_3 = 6$$
$$x_1 - x_2 - x_3 \geq 1$$
$$x_1, x_2, x_3 \geq 0 \qquad \blacktriangleleft$$

EXAMPLE 3 ➤ **Using the Big M Method** Solve the following linear programming problem using the big M method:

$$\text{Maximize} \quad P = 3x_1 + 5x_2$$
$$\text{Subject to} \quad 2x_1 + x_2 \leq 4$$
$$x_1 + 2x_2 \geq 10$$
$$x_1, x_2 \geq 0$$

Solution Introducing slack, surplus, and artificial variables, we obtain the modified problem:

$$2x_1 + x_2 + s_1 \qquad\qquad\qquad = 4$$
$$x_1 + 2x_2 \qquad - s_2 + a_1 \qquad = 10 \quad \text{Modified problem}$$
$$-3x_1 - 5x_2 \qquad\qquad + Ma_1 + P = 0$$

	x_1	x_2	s_1	s_2	a_1	P		
	2	1	1	0	0	0	4	Preliminary simplex tableau
	1	2	0	-1	1	0	10	Eliminate M in the a_1 column.
	-3	-5	0	0	M	1	0	$(-M)R_2 + R_3 \rightarrow R_3$

	x_1	x_2	s_1	s_2	a_1	P		
s_1	2	①	1	0	0	0	4	Initial simplex tableau
$\sim a_1$	1	2	0	-1	1	0	10	Begin pivot operations.
P	$-M-3$	$-2M-5$	0	M	0	1	$-10M$	$(-2)R_1 + R_2 \rightarrow R_2$
								$(2M+5)R_1 + R_3 \rightarrow R_3$

	x_1	x_2	s_1	s_2	a_1	P	
x_2	2	1	1	0	0	0	4
$\sim a_1$	-3	0	-2	-1	1	0	2
P	$3M+7$	0	$2M+5$	M	0	1	$-2M+20$

The optimal solution of the modified problem is

$$x_1 = 0, \quad x_2 = 4, \quad s_1 = 0, \quad s_2 = 0, \quad a_1 = 2, \quad P = -2M + 20$$

Since $a_1 \neq 0$, the original problem has no solution. Figure 1 (on the next page) shows that the feasible region for the original problem is empty.

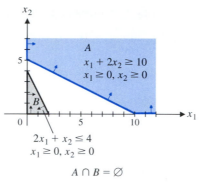

$A \cap B = \varnothing$

FIGURE 1

MATCHED PROBLEM 3 ▶ Solve the following linear programming problem using the big M method:

$$\text{Maximize} \quad P = 3x_1 + 2x_2$$
$$\text{Subject to} \quad x_1 + 5x_2 \le 5$$
$$2x_1 + x_2 \ge 12$$
$$x_1, x_2 \ge 0$$

■ MINIMIZATION BY THE BIG M METHOD

In addition to solving any maximization problem, the big M method can be used to solve minimization problems. To minimize an objective function, we have only to maximize its negative. Figure 2 illustrates the fact that the minimum value of a function f occurs at the same point as the maximum value of the function $-f$. Furthermore, if m is the minimum value of f, then $-m$ is the maximum value of $-f$, and conversely. Thus, we can find the minimum value of a function f by finding the maximum value of $-f$ and then changing the sign of the maximum value.

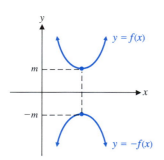

FIGURE 2

EXAMPLE 4 ▶ **Production Scheduling—A Minimization Problem** A small jewelry manufacturing company employs a person who is a highly skilled gem cutter, and it wishes to use this person at least 6 hours per day for this purpose. On the other hand, the polishing facilities can be used in any amounts up to 10 hours per day. The company specializes in three kinds of semiprecious gemstones, J, K, and L. Relevant cutting, polishing, and cost requirements are listed in the table. How many gemstones of each type should be processed each day to minimize the cost of the finished stones? What is the minimum cost?

	J	K	L
CUTTING	1 hr	1 hr	1 hr
POLISHING	2 hr	1 hr	2 hr
COST PER STONE	$30	$30	$10

Solution If we let x_1, x_2, and x_3 represent the number of type J, K, and L stones finished per day, respectively, then we have the following linear programming problem to solve, where C is the cost of the stones:

$$\text{Minimize} \quad C = 30x_1 + 30x_2 + 10x_3 \qquad \textit{Objective function}$$

$$\text{Subject to} \quad \left. \begin{array}{l} x_1 + x_2 + x_3 \ge 6 \\ 2x_1 + x_2 + 2x_3 \le 10 \end{array} \right\} \qquad \textit{Problem constraints}$$

$$x_1, x_2, x_3 \ge 0 \qquad \textit{Nonnegative constraints}$$

We convert this to a maximization problem by letting

$$P = -C = -30x_1 - 30x_2 - 10x_3$$

Thus, we get,

Maximize $P = -30x_1 - 30x_2 - 10x_3$

Subject to $x_1 + x_2 + x_3 \geqslant 6$

$2x_1 + x_2 + 2x_3 \leqslant 10$

$x_1, x_2, x_3 \geqslant 0$

and Min $C = -$ Max P. To solve, we first state the modified problem:

$$x_1 + x_2 + x_3 - s_1 + a_1 = 6$$
$$2x_1 + x_2 + 2x_3 + s_2 = 10$$
$$30x_1 + 30x_2 + 10x_3 + Ma_1 + P = 0$$
$$x_1, x_2, x_3, s_1, s_2, a_1 \geqslant 0$$

	x_1	x_2	x_3	s_1	a_1	s_2	P		
	1	1	1	-1	1	0	0	6	Eliminate M in the a_1
	2	1	2	0	0	1	0	10	column.
	30	30	10	0	M	0	1	0	$(-M)R_1 + R_3 \rightarrow R_3$

Begin pivot operations. Assume M is so large
that $-M + 30$ and $-M + 10$ are negative.

	x_1	x_2	x_3	s_1	a_1	s_2	P		
a_1	1	1	1	-1	1	0	0	6	
$\sim s_2$	2	1	②	0	0	1	0	10	$0.5R_2 \rightarrow R_2$
P	$-M + 30$	$-M + 30$	$-M + 10$	M	0	0	1	$-6M$	

	1	1	1	-1	1	0	0	6	$(-1)R_2 + R_1 \rightarrow R_1$
$\sim$	1	0.5	①	0	0	0.5	0	5	
	$-M + 30$	$-M + 30$	$-M + 10$	M	0	0	1	$-6M$	$(M - 10)R_2 + R_3 \rightarrow R_3$

a_1	0	⓪.5	0	-1	1	-0.5	0	1	$2R_1 \rightarrow R_1$
$\sim x_3$	1	0.5	1	0	0	0.5	0	5	
P	20	$-0.5M + 25$	0	M	0	$0.5M - 5$	1	$-M - 50$	

	0	①	0	-2	2	-1	0	2	
$\sim$	1	0.5	1	0	0	0.5	0	5	$(-0.5)R_1 + R_2 \rightarrow R_2$
	20	$-0.5M + 25$	0	M	0	$0.5M - 5$	1	$-M - 50$	$(0.5M - 25)R_1 + R_3 \rightarrow R_3$

x_2	0	1	0	-2	2	-1	0	2
$\sim x_3$	1	0	1	1	-1	1	0	4
P	20	0	0	50	$M - 50$	20	1	-100

The bottom row has no negative indicators, so the optimal solution for the modified problem is

$$x_1 = 0, \quad x_2 = 2, \quad x_3 = 4, \quad s_1 = 0, \quad a_1 = 0, \quad s_2 = 0, \quad P = -100$$

Since $a_1 = 0$, deleting a_1 produces the solution to the original maximization problem and also to the minimization problem. Thus,

$$\text{Min } C = -\text{Max } P = -(-100) = 100 \quad \text{at} \quad x_1 = 0, \quad x_2 = 2, \quad x_3 = 4$$

That is, a minimum cost of \$100 for gemstones will be realized if no type J, 2 type K, and 4 type L stones are processed each day. ◀

MATCHED PROBLEM 4 ▶ Repeat Example 4 if the gem cutter is to be used for at least 8 hours a day and all other data remain the same. ◀

■ SUMMARY OF METHODS OF SOLUTION

The big M method can be used to solve any minimization problem, including those that can be solved by the dual method. (Note that Example 4 could have been solved by the dual method.) Both methods of solving minimization problems are important. You will be instructed to solve most minimization problems in Exercise 5-6 by the big M method in order to gain more experience with that method. If the method of solution is not specified, then the dual method is usually easier.

Table 1 should help you select the proper method of solution for any linear programming problem.

TABLE 1
Summary of Problem Types and Simplex Solution Methods

PROBLEM TYPE	PROBLEM CONSTRAINTS	RIGHT-SIDE CONSTANTS	COEFFICIENTS OF OBJECTIVE FUNCTION	METHOD OF SOLUTION
1. Maximization	$\leqslant$	Nonnegative	Any real numbers	Simplex method with slack variables
2. Minimization	$\geqslant$	Any real numbers	Nonnegative	Form dual and solve by method 1
3. Maximization	Mixed ($\leqslant, \geqslant, =$)	Nonnegative	Any real numbers	Form modified problem with slack, surplus, and artificial variables and solve by the big M method
4. Minimization	Mixed ($\leqslant, \geqslant, =$)	Nonnegative	Any real numbers	Maximize negative of objective function by method 3

■ LARGER PROBLEMS—A REFINERY APPLICATION

Up to this point, all the problems we have considered could be solved by hand. However, the real value of the simplex method lies in its ability to solve problems with a large number of variables and constraints, where a computer is generally used to perform the actual pivot operations. As a final application, we will consider a problem that would require the use of a computer to complete the solution.

▶ EXAMPLE 5 ▶ Petroleum Blending A refinery produces two grades of gasoline, regular and premium, by blending together two components, A and B. Component A has an octane rating of 90 and costs \$28 a barrel. Component B has an octane rating of 110 and costs \$32 a barrel. The octane rating for regular gasoline must be at least 95, and the octane rating for premium must be at least 105. Regular gasoline sells for $34 a barrel and premium sells for $40 a barrel. Currently, the company has 30,000 barrels of component A and 20,000 barrels of component B. It also has orders for 20,000 barrels of regular and 10,000 barrels of premium that must be filled. Assuming that all the gasoline produced can be sold, determine the maximum possible profit.

Solution   First we organize the information given in the problem in Table 2.

TABLE 2

COMPONENT	OCTANE RATING	COST	AVAILABLE SUPPLY
A	90	$28	30,000 barrels
B	110	$32	20,000 barrels

GRADE	MINIMUM OCTANE RATING	SELLING PRICE	EXISTING ORDERS
Regular	95	$34	20,000 barrels
Premium	105	$40	10,000 barrels

Let: x_1 = Number of barrels of component A used in regular gasoline
x_2 = Number of barrels of component A used in premium gasoline
x_3 = Number of barrels of component B used in regular gasoline
x_4 = Number of barrels of component B used in premium gasoline

The total amount of component A used is $x_1 + x_2$. This cannot exceed the available supply. Thus, one constraint is

$$x_1 + x_2 \le 30,000$$

The corresponding inequality for component B is

$$x_3 + x_4 \le 20,000$$

The amounts of regular and premium gasoline produced must be sufficient to meet the existing orders:

$$x_1 + x_3 \ge 20,000 \quad \text{Regular}$$
$$x_2 + x_4 \ge 10,000 \quad \text{Premium}$$

Now consider the octane ratings. The octane rating of a blend is simply the proportional average of the octane ratings of the components. Thus, the octane rating for regular gasoline is

$$90\,\frac{x_1}{x_1 + x_3} + 110\,\frac{x_3}{x_1 + x_3}$$

where $x_1/(x_1 + x_3)$ is the percentage of component A used in regular gasoline and $x_3/(x_1 + x_3)$ is the percentage of component B. The final octane rating of regular gasoline must be at least 95; thus,

$$90 \, \frac{x_1}{x_1 + x_3} + 110 \, \frac{x_3}{x_1 + x_3} \geqslant 95 \qquad \textcolor{blue}{\text{Multiply by } x_1 + x_3.}$$

$$90x_1 + 110x_3 \geqslant 95(x_1 + x_3) \qquad \textcolor{blue}{\text{Collect like terms on the left side.}}$$

$$-5x_1 + 15x_3 \geqslant 0 \qquad \textcolor{blue}{\text{Octane rating for regular}}$$

The corresponding inequality for premium gasoline is

$$90 \, \frac{x_2}{x_2 + x_4} + 110 \, \frac{x_4}{x_2 + x_4} \geqslant 105$$

$$90x_2 + 110x_4 \geqslant 105(x_2 + x_4)$$

$$-15x_2 + 5x_4 \geqslant 0 \qquad \textcolor{blue}{\text{Octane rating for premium}}$$

The cost of the components used is

$$C = 28(x_1 + x_2) + 32(x_3 + x_4)$$

The revenue from selling all the gasoline is

$$R = 34(x_1 + x_3) + 40(x_2 + x_4)$$

and the profit is

$$
\begin{aligned}
P &= R - C \\
&= 34(x_1 + x_3) + 40(x_2 + x_4) - 28(x_1 + x_2) - 32(x_3 + x_4) \\
&= (34 - 28)x_1 + (40 - 28)x_2 + (34 - 32)x_3 + (40 - 32)x_4 \\
&= 6x_1 + 12x_2 + 2x_3 + 8x_4
\end{aligned}
$$

To find the maximum profit, we must solve the following linear programming problem:

$$
\begin{array}{lll}
\text{Maximize} & P = 6x_1 + 12x_2 + 2x_3 + 8x_4 & \textcolor{blue}{\text{Profit}} \\
\text{Subject to} & x_1 + x_2 \leqslant 30{,}000 & \textcolor{blue}{\text{Available A}} \\
& x_3 + x_4 \leqslant 20{,}000 & \textcolor{blue}{\text{Available B}} \\
& x_1 + x_3 \geqslant 20{,}000 & \textcolor{blue}{\text{Required regular}} \\
& x_2 + x_4 \geqslant 10{,}000 & \textcolor{blue}{\text{Required premium}} \\
& -5x_1 + 15x_3 \geqslant 0 & \textcolor{blue}{\text{Octane for regular}} \\
& -15x_2 + 5x_4 \geqslant 0 & \textcolor{blue}{\text{Octane for premium}} \\
& x_1, x_2, x_3, x_4 \geqslant 0 &
\end{array}
$$

The tableau for this problem would have seven rows and sixteen columns. Solving this problem by hand is possible but would require considerable effort. In actual practice, computers are used to solve most linear programming problems. Table 3 displays typical input and output for a computer solution of this problem. (The computer software that accompanies this book contains a program that can be used to solve this problem.)

TABLE 3

INPUT TO PROGRAM	OUTPUT FROM PROGRAM

```
    * - LINEAR PROGRAMMING - *              * - LINEAR PROGRAMMING - *
         PROBLEM DISPLAY                          SOLUTION DISPLAY

MAXIMIZE                              DECISION VARIABLES    SLACK VARIABLES
P = 6X1 + 12X2 + 2X3 + 8X4           X1 = 26250            S1 = 0
SUBJECT TO:                          X2 = 3750             S2 = 0
  X1 +    X2              <= 30000   X3 = 8750             SURPLUS VARIABLES
              X3 +   X4 <= 20000     X4 = 11250            S3 = 15000
  X1            +   X3      >= 20000                        S4 = 5000
         X2           +  X4 >= 10000                        S5 = 0
 -5X1         + 15X3        >=    0                         S6 = 0
      -  15X2        + 5X4 >=     0   MAXIMUM VALUE OF OBJECTIVE FUNCTION
XI >= 0, I = 1 TO 4                   310000
```

According to the output in Table 3, the refinery should blend 26,250 barrels of component A and 8,750 barrels of component B to produce 35,000 barrels of regular. They should blend 3,750 barrels of component A and 11,250 barrels of component B to produce 15,000 barrels of premium. This will result in a maximum profit of $310,000. ◄

explore–discuss 3

Interpret the values of the slack and surplus variables in the computer solution in Table 3.

MATCHED PROBLEM 5 ➤ Suppose the refinery in Example 5 has 35,000 barrels of component A, which costs $25 a barrel, and 15,000 barrels of component B, which costs $35 a barrel. If all the other information is unchanged, formulate a linear programming problem whose solution is the maximum profit. Do not attempt to solve the problem (unless you have access to a computer and a program that solves linear programming problems). ◄

Answers to Matched Problems

1. Maximize $P = 3x_1 - 2x_2 + x_3 - Ma_1 - Ma_2 - Ma_3$

   Subject to
   $$x_1 - 2x_2 + x_3 - s_1 + a_1 = 5$$
   $$x_1 + 3x_2 - 4x_3 - s_2 + a_2 = 10$$
   $$2x_1 + 4x_2 + 5x_3 + s_3 = 20$$
   $$-3x_1 + x_2 + x_3 + a_3 = 15$$
   $$x_1, x_2, x_3, s_1, a_1, s_2, a_2, s_3, a_3 \geq 0$$

2. Max $P = 22$ at $x_1 = 6$, $x_2 = 4$, $x_3 = 0$ 3. No solution
4. A minimum cost of $200 is realized when no type J, 6 type K, and 2 type L stones are processed each day.

5. Maximize $\quad P = 9x_1 + 15x_2 - x_3 + 5x_4$
   Subject to
   $$
   \begin{aligned}
   x_1 + x_2 &\leqslant 35{,}000 \\
   x_3 + x_4 &\leqslant 15{,}000 \\
   x_1 \phantom{{}+{}} + x_3 \phantom{{}+{}} &\geqslant 20{,}000 \\
   x_2 \phantom{{}+{}} + x_4 &\geqslant 10{,}000 \\
   -5x_1 \phantom{{}+{}} + 15x_3 \phantom{{}+{}} &\geqslant 0 \\
   -15x_2 \phantom{{}+{}} + 5x_4 &\geqslant 0 \\
   x_1, x_2, x_3, x_4 &\geqslant 0
   \end{aligned}
   $$

EXERCISE 5-6

A   *In Problems 1–8:*

(A) *Introduce slack, surplus, and artificial variables and form the modified problem.*
(B) *Write the preliminary simplex tableau for the modified problem and find the initial simplex tableau.*
(C) *Find the optimal solution of the modified problem by applying the simplex method to the initial simplex tableau.*
(D) *Find the optimal solution of the original problem, if it exists.*

1. Maximize $\quad P = 5x_1 + 2x_2$
   Subject to $\quad x_1 + 2x_2 \leqslant 12$
   $$x_1 + x_2 \geqslant 4$$
   $$x_1, x_2 \geqslant 0$$

2. Maximize $\quad P = 3x_1 + 7x_2$
   Subject to $\quad 2x_1 + x_2 \leqslant 16$
   $$x_1 + x_2 \geqslant 6$$
   $$x_1, x_2 \geqslant 0$$

3. Maximize $\quad P = 3x_1 + 5x_2$
   Subject to $\quad 2x_1 + x_2 \leqslant 8$
   $$x_1 + x_2 = 6$$
   $$x_1, x_2 \geqslant 0$$

4. Maximize $\quad P = 4x_1 + 3x_2$
   Subject to $\quad x_1 + 3x_2 \leqslant 24$
   $$x_1 + x_2 = 12$$
   $$x_1, x_2 \geqslant 0$$

5. Maximize $\quad P = 4x_1 + 3x_2$
   Subject to $\quad -x_1 + 2x_2 \leqslant 2$
   $$x_1 + x_2 \geqslant 4$$
   $$x_1, x_2 \geqslant 0$$

6. Maximize $\quad P = 3x_1 + 4x_2$
   Subject to $\quad x_1 - 2x_2 \leqslant 2$
   $$x_1 + x_2 \geqslant 5$$
   $$x_1, x_2 \geqslant 0$$

7. Maximize $\quad P = 5x_1 + 10x_2$
   Subject to $\quad x_1 + x_2 \leqslant 3$
   $$2x_1 + 3x_2 \geqslant 12$$
   $$x_1, x_2 \geqslant 0$$

8. Maximize $\quad P = 4x_1 + 6x_2$
   Subject to $\quad x_1 + x_2 \leqslant 2$
   $$3x_1 + 5x_2 \geqslant 15$$
   $$x_1, x_2 \geqslant 0$$

B   *Use the big M method to solve Problems 9–22.*

9. Minimize and maximize $\quad P = 2x_1 - x_2$
   Subject to $\quad x_1 + x_2 \leqslant 8$
   $$5x_1 + 3x_2 \geqslant 30$$
   $$x_1, x_2 \geqslant 0$$

10. Minimize and maximize $\quad P = -4x_1 + 16x_2$
   Subject to $\quad 3x_1 + x_2 \leqslant 28$
   $$x_1 + 2x_2 \geqslant 16$$
   $$x_1, x_2 \geqslant 0$$

11. Maximize $\quad P = 2x_1 + 5x_2$
   Subject to $\quad x_1 + 2x_2 \leqslant 18$
   $$2x_1 + x_2 \leqslant 21$$
   $$x_1 + x_2 \geqslant 10$$
   $$x_1, x_2 \geqslant 0$$

12. Maximize $\quad P = 6x_1 + 2x_2$
   Subject to $\quad x_1 + 2x_2 \leqslant 20$
   $$2x_1 + x_2 \leqslant 16$$
   $$x_1 + x_2 \geqslant 9$$
   $$x_1, x_2 \geqslant 0$$

13. Maximize $\quad P = 10x_1 + 12x_2 + 20x_3$
   Subject to $\quad 3x_1 + x_2 + 2x_3 \geqslant 12$
   $$x_1 - x_2 + 2x_3 = 6$$
   $$x_1, x_2, x_3 \geqslant 0$$

14. Maximize $P = 5x_1 + 7x_2 + 9x_3$
  Subject to $x_1 - x_2 + x_3 \geq 20$
    $2x_1 + x_2 + 5x_3 = 35$
    $x_1, x_2, x_3 \geq 0$

15. Minimize $C = -5x_1 - 12x_2 + 16x_3$
  Subject to $x_1 + 2x_2 + x_3 \leq 10$
    $2x_1 + 3x_2 + x_3 \geq 6$
    $2x_1 + x_2 - x_3 = 1$
    $x_1, x_2, x_3 \geq 0$

16. Minimize $C = -3x_1 + 15x_2 - 4x_3$
  Subject to $2x_1 + x_2 + 3x_3 \leq 24$
    $x_1 + 2x_2 + x_3 \geq 6$
    $x_1 - 3x_2 + x_3 = 2$
    $x_1, x_2, x_3 \geq 0$

17. Maximize $P = 3x_1 + 5x_2 + 6x_3$
  Subject to $2x_1 + x_2 + 2x_3 \leq 8$
    $2x_1 + x_2 - 2x_3 = 0$
    $x_1, x_2, x_3 \geq 0$

18. Maximize $P = 3x_1 + 6x_2 + 2x_3$
  Subject to $2x_1 + 2x_2 + 3x_3 \leq 12$
    $2x_1 - 2x_2 + x_3 = 0$
    $x_1, x_2, x_3 \geq 0$

19. Maximize $P = 2x_1 + 3x_2 + 4x_3$
  Subject to $x_1 + 2x_2 + x_3 \leq 25$
    $2x_1 + x_2 + 2x_3 \leq 60$
    $x_1 + 2x_2 - x_3 \geq 10$
    $x_1, x_2, x_3 \geq 0$

20. Maximize $P = 5x_1 + 2x_2 + 9x_3$
  Subject to $2x_1 + 4x_2 + x_3 \leq 150$
    $3x_1 + 3x_2 + x_3 \leq 90$
    $-x_1 + 5x_2 + x_3 \geq 120$
    $x_1, x_2, x_3 \geq 0$

21. Maximize $P = x_1 + 2x_2 + 5x_3$
  Subject to $x_1 + 3x_2 + 2x_3 \leq 60$
    $2x_1 + 5x_2 + 2x_3 \geq 50$
    $x_1 - 2x_2 + x_3 \geq 40$
    $x_1, x_2, x_3 \geq 0$

22. Maximize $P = 2x_1 + 4x_2 + x_3$
  Subject to $2x_1 + 3x_2 + 5x_3 \leq 280$
    $2x_1 + 2x_2 + x_3 \geq 140$
    $2x_1 + x_2 \geq 150$
    $x_1, x_2, x_3 \geq 0$

23. Solve Problems 5 and 7 by the geometric method. Compare the conditions in the big M method that indicate no optimal solution exists with the conditions stated in Theorem 2 in Section 5-2.

24. Repeat Problem 23 with Problems 6 and 8.

C *Problems 25–32 are mixed. Some can be solved by the methods presented in Sections 5-4 and 5-5, while others must be solved by the big M method.*

25. Minimize $C = 10x_1 - 40x_2 - 5x_3$
  Subject to $x_1 + 3x_2 \leq 6$
    $4x_2 + x_3 \leq 3$
    $x_1, x_2, x_3 \geq 0$

26. Maximize $P = 7x_1 - 5x_2 + 2x_3$
  Subject to $x_1 - 2x_2 + x_3 \geq -8$
    $x_1 - x_2 + x_3 \leq 10$
    $x_1, x_2, x_3 \geq 0$

27. Maximize $P = -5x_1 + 10x_2 + 15x_3$
  Subject to $2x_1 + 3x_2 + x_3 \leq 24$
    $x_1 - 2x_2 - 2x_3 \geq 1$
    $x_1, x_2, x_3 \geq 0$

28. Minimize $C = -5x_1 + 10x_2 + 15x_3$
  Subject to $2x_1 + 3x_2 + x_3 \leq 24$
    $x_1 - 2x_2 - 2x_3 \geq 1$
    $x_1, x_2, x_3, \geq 0$

29. Minimize $C = 10x_1 + 40x_2 + 5x_3$
  Subject to $x_1 + 3x_2 \geq 6$
    $4x_2 + x_3 \geq 3$
    $x_1, x_2, x_3 \geq 0$

30. Maximize $P = 8x_1 + 2x_2 - 10x_3$
  Subject to $x_1 + x_2 - 3x_3 \leq 6$
    $4x_1 - x_2 + 2x_3 \leq -7$
    $x_1, x_2, x_3 \geq 0$

31. Maximize $P = 12x_1 + 9x_2 + 5x_3$
  Subject to $x_1 + 3x_2 + x_3 \leq 40$
    $2x_1 + x_2 + 3x_3 \leq 60$
    $x_1, x_2, x_3 \geq 0$

32. Minimize $C = 10x_1 + 12x_2 + 28x_3$
  Subject to $4x_1 + 2x_2 + 3x_3 \geq 20$
    $3x_1 - x_2 - 4x_3 \leq 10$
    $x_1, x_2, x_3 \geq 0$

APPLICATIONS

In Problems 33–40, formulate each problem as a linear programming problem. (The answers in the back of the book for these application problems include the formulation.) Then solve the problem using the big M method.

Business & Economics

33. *Manufacturing—resource allocation.* An electronics company manufactures two types of add-on memory modules for microcomputers, a 16k module and a 64k module. Each 16k module requires 10 minutes for assembly and 2 minutes for testing. Each 64k module requires 15 minutes for assembly and 4 minutes for testing. The company makes a profit of $18 on each 16k module and $30 on each 64k module. The assembly department can work a maximum of 2,200 minutes per day, and the testing department can work a maximum of 500 minutes a day. In order to satisfy current orders, the company must produce at least 50 of the 16k modules per day. How many units of each module should the company manufacture each day in order to maximize the daily profit? What is the maximum profit?

34. *Manufacturing—resource allocation.* Discuss the effect on the solution to Problem 33 if the assembly department can work only 2,100 minutes daily.

35. *Advertising.* A company planning an advertising campaign to attract new customers wants to place a total of at most 10 ads in 3 newspapers. Each ad in the *Sentinel* costs $200 and will be read by 2,000 people. Each ad in the *Journal* costs $200 and will be read by 500 people. Each ad in the *Tribune* costs $100 and will be read by 1,500 people. The company wants at least 16,000 people to read its ads. How many ads should it place in each paper in order to minimize the advertising costs? What is the minimum cost?

36. *Advertising.* Discuss the effect on the solution to Problem 35 if the *Tribune* will not accept more than 4 ads from the company.

Life Sciences

37. *Nutrition—people.* An individual on a high-protein, low-carbohydrate diet requires at least 100 units of protein and at most 24 units of carbohydrates daily. The diet will consist entirely of three special liquid diet foods, *A*, *B*, and *C*. The contents and costs of the diet foods are given in the table. How many bottles of each brand of diet food should be consumed daily in order to meet the protein and carbohy-

drate requirements at minimal cost? What is the minimum cost?

	UNITS PER BOTTLE		
	A	*B*	*C*
PROTEIN	10	10	20
CARBOHYDRATES	2	3	4
COST PER BOTTLE	$0.60	$0.40	$0.90

38. *Nutrition—people.* Discuss the effect on the solution to Problem 37 if the cost of brand *C* liquid diet food increases to $1.50 per bottle.

39. *Nutrition—plants.* A farmer can use three types of plant food, mix *A*, mix *B*, and mix *C*. The amounts (in pounds) of nitrogen, phosphoric acid, and potash in a cubic yard of each mix are given in the table. Tests performed on the soil in a large field indicate that the field needs at least 800 pounds of potash. The tests also indicate that no more than 700 pounds of phosphoric acid should be added to the field. The farmer plans to plant a crop that requires a great deal of nitrogen. How many cubic yards of each mix should be added to the field in order to satisfy the potash and phosphoric acid requirements and maximize the amount of nitrogen added? What is the maximum amount of nitrogen?

	POUNDS PER CUBIC YARD		
	A	*B*	*C*
NITROGEN	12	16	8
PHOSPHORIC ACID	12	8	16
POTASH	16	8	16

40. *Nutrition—plants.* Discuss the effect on the solution to Problem 39 if the limit on phosphoric acid is increased to 1,000 pounds.

In Problems 41–47, formulate each problem as a linear programming problem. Do not solve.

Business & Economics

41. *Manufacturing—production scheduling.* A company manufactures car and truck frames at plants in Milwaukee and Racine. The Milwaukee plant has a daily operating budget of $50,000 and can produce at most 300 frames daily in any combination. It costs $150 to manufacture a car frame and $200 to manufacture a truck frame at the Milwaukee plant. The Racine plant has a daily operating budget of $35,000, can produce a maximum combined total of 200 frames daily, and produces a car frame at a cost of $135 and a truck frame at a cost of $180. Based on past demand, the company wants to limit production to a maximum of 250 car frames and 350 truck frames per day. If the company realizes a profit of $50 on each car frame and $70 on each truck frame, how many frames of each type should be produced at each plant to maximize the daily profit?

42. *Finances—loan distributions.* A savings and loan company has $3 million to lend. The types of loans and annual returns offered by the company are given in the table. State laws require that at least 50% of the money loaned for mortgages must be for first mortgages and that at least 30% of the total amount loaned must be for either first or second mortgages. Company policy requires that the amount of signature and automobile loans cannot exceed 25% of the total amount loaned and that signature loans cannot exceed 15% of the total amount loaned. How much money should be allocated to each type of loan in order to maximize the company's return?

TYPE OF LOAN	ANNUAL RETURN
Signature	18%
First mortgage	12%
Second mortgage	14%
Automobile	16%

43. *Blending—petroleum.* A refinery produces two grades of gasoline, regular and premium, by blending together three components, A, B, and C. Component A has an octane rating of 90 and costs $28 a barrel, component B has an octane rating of 100 and costs $30 a barrel, and component C has an octane rating of 110 and costs $34 a barrel. The octane rating for regular must be at least 95 and the octane rating for premium must be at least 105. Regular gasoline sells for $38 a barrel and premium sells for $46 a barrel. The company has 40,000 barrels of component A, 25,000 barrels of component B, and 15,000 barrels of component C, and must produce at least 30,000 barrels of regular and 25,000 barrels of premium. How should the components be blended in order to maximize profit?

44. *Blending—food processing.* A company produces two brands of trail mix, regular and deluxe, by mixing dried fruits, nuts, and cereal. The recipes for the mixes are given in the table. The company has 1,200 pounds of dried fruits, 750 pounds of nuts, and 1,500 pounds of cereal to be used in producing the mixes. The company makes a profit of $0.40 on each pound of regular mix and $0.60 on each pound of deluxe mix. How many pounds of each ingredient should be used in each mix in order to maximize the company's profit?

TYPE OF MIX	INGREDIENTS
Regular	At least 20% nuts
	At most 40% cereal
Deluxe	At least 30% nuts
	At most 25% cereal

Life Sciences

45. *Nutrition—people.* A dietitian in a hospital is to arrange a special diet using the foods L, M, and N. The table gives the nutritional contents and the cost of 1 ounce of each food. The daily requirements for the diet are at least 400 units of calcium, at least 200 units of iron, at least 300 units of vitamin A, at most 150 units of cholesterol, and at most 900 calories. How many ounces of each food should be used in order to meet the requirements of the diet at minimal cost?

	UNITS PER BOTTLE		
	L	M	N
CALCIUM	30	10	30
IRON	10	10	10
VITAMIN A	10	30	20
CHOLESTEROL	8	4	6
CALORIES	60	40	50
COST PER OUNCE	$0.40	$0.60	$0.80

46. *Nutrition—feed mixtures.* A farmer grows three crops, corn, oats, and soybeans, which he mixes together to feed his cows and pigs. At least 40% of the feed mix for the cows must be corn. The feed mix for the pigs must contain at least twice as much soybeans as corn. He has harvested 1,000 bushels of corn, 500 bushels of oats, and 1,000 bushels of soybeans. He needs 1,000 bushels of each feed mix for his livestock. The unused corn, oats, and soybeans can be sold for $4, $3.50, and $3.25 a bushel, respectively (thus, these amounts also represent the cost of the crops used to feed the livestock). How many bushels of each crop should be used in each feed mix in order to produce sufficient food for the livestock at minimal cost?

Social Sciences

47. *Education—resource allocation.* Three towns are forming a consolidated school district with two high schools. Each high school has a maximum capacity of 2,000 students. Town *A* has 500 high school students, town *B* has 1,200, and town *C* has 1,800. The weekly costs of transporting a student from each town to each school are given in the table. In order to keep the enrollment balanced, the school board has decided that each high school must enroll at least 40% of the total student population. Furthermore, no more than 60% of the students in any town should be sent to the same high school. How many students from each town should be enrolled in each school in order to meet these requirements and minimize the cost of transporting the students?

	WEEKLY TRANSPORTATION COST PER STUDENT	
	School I	*School II*
TOWN *A*	$4	$8
TOWN *B*	6	4
TOWN *C*	3	9

CHAPTER 5
GROUP ACTIVITY

The Two-Phase Method (An Alternative to the Big M Method)

In the big *M* method, we added artificial variables to $\geq$ and $=$ constraints in order to obtain a basic feasible solution. Then we added terms of the form $-Ma_i$ to the objective function and assumed that *M* was an arbitrarily large number, which forced the values of the artificial variables to be zero in any optimal solution. There is another commonly used method, called the *two-phase method,* that treats the artificial variables in a different manner. To illustrate this method, consider the following example:

$$\begin{aligned}
\text{Maximize} \quad & P = 2x_1 + x_2 \\
\text{Subject to} \quad & x_1 + x_2 \leq 10 \\
& x_1 \geq 3 \\
& x_2 \geq 4 \\
& x_1, x_2 \geq 0
\end{aligned}$$

(A) Graph the feasible region and solve this problem by the geometric method for comparison with the new method that follows.

If we add slack and surplus variables, we obtain the following system:

$$\begin{aligned}
x_1 + x_2 + s_1 \quad\quad\quad\quad & = 10 \\
x_1 \quad\quad\quad - s_2 \quad\quad\quad & = 3 \\
x_2 \quad\quad\quad - s_3 \quad\quad & = 4 \\
-2x_1 - x_2 \quad\quad\quad\quad\quad + P & = 0
\end{aligned} \tag{1}$$

The basic solution for this system is

$$x_1 = 0, \quad x_2 = 0, \quad s_1 = 10, \quad s_2 = -3, \quad s_3 = -4, \quad P = 0$$

This solution is not feasible, and we cannot start the simplex process. As before, we add artificial variables to obtain a basic feasible solution, but this time we do not modify the objective function:

$$
\begin{aligned}
x_1 + x_2 + s_1 \quad\quad\quad\quad\quad\quad &= 10 \\
x_1 \quad\quad\quad - s_2 + a_1 \quad\quad\quad &= 3 \\
x_2 \quad\quad\quad - s_3 + a_2 \quad &= 4 \\
-2x_1 - x_2 \quad\quad\quad\quad\quad\quad + P &= 0
\end{aligned}
\tag{2}
$$

The basic solution for this system is

$$x_1 = 0, \quad x_2 = 0, \quad s_1 = 10, \quad s_2 = 0, \quad a_1 = 3, \quad s_3 = 0, \quad a_2 = 4, \quad P = 0$$

This solution is feasible, and we can start the simplex process.

(B) Write the simplex tableau for system (2) and apply the simplex procedure. Compare your result with the geometric solution from part (A).

In general, applying the simplex method to system (2) will produce an optimal solution of (2) with one or more of the artificial variables having nonzero values. In most cases, simply deleting the artificial variables will not produce a basic feasible solution to (1). However, if we can find a basic feasible solution of (2) with the artificial variables all equal to zero, then we can delete the artificial variables and find a basic feasible solution to (1). In the big M method, we accomplished this by adding terms to the objective function. In this new method, we consider another objective function:

$$C = a_1 + a_2$$

If we minimize C and the minimum value turns out to be zero, then, since a_1 and a_2 are nonnegative, they will both be zero. To minimize C, we maximize

$$T = -C = -a_1 - a_2$$

We cannot add T to system (2) in this form. To preserve the necessary form for a simplex tableau, we must express T in terms of the nonbasic variables. Solving the second and third equations in (2) for a_1 and a_2, respectively, and adding, we obtain

$$T = x_1 + x_2 - s_2 - s_3 - 7$$

Transposing all the variables to the left side, we add this equation to (2) to form system (3):

$$
\begin{aligned}
x_1 + x_2 + s_1 &= 10 \\
x_1 \quad\quad - s_2 + a_1 &= 3 \\
x_2 \quad\quad - s_3 + a_2 &= 4 \\
-2x_1 - x_2 \quad\quad + P &= 0 \\
-x_1 - x_2 + \quad s_2 \quad + s_3 \quad\quad + T &= -7
\end{aligned}
\tag{3}
$$

(C) Write the simplex tableau for system (3) and apply the simplex method, with the following restriction: *Do not choose any elements in the objective function row as pivot elements*. When you find the optimal solution, write the corresponding system, which is shown below.

$$
\begin{aligned}
s_1 + \ s_2 - \ a_1 + s_3 - a_2 &= 3 \\
x_1 \quad - \ s_2 + \ a_1 &= 3 \\
x_2 \quad\quad - s_3 + a_2 &= 4 \\
- 2s_2 + 2a_1 - s_3 + a_2 + P &= 10 \\
a_1 \quad + a_2 \quad + T &= 0
\end{aligned}
\tag{4}
$$

The basic feasible solution for (4) is

$$
x_1 = 3, \quad x_2 = 4, \quad s_1 = 3, \quad s_2 = 0, \quad a_1 = 0,
$$
$$
s_3 = 0, \quad a_2 = 0, \quad P = 10, \quad T = 0
$$

Note that the maximum value of T is 0 and that it occurs for $a_1 = a_2 = 0$. Thus, deleting the equation for T and the columns of the artificial variables gives us system (5), which is equivalent to system (1) but has a basic solution that is feasible:

$$
\begin{aligned}
s_1 + \ s_2 + s_3 \quad &= 3 \\
x_1 \quad - \ s_2 \quad &= 3 \\
x_2 \quad\quad - s_3 \quad &= 4 \\
- 2s_2 - s_3 + P &= 10
\end{aligned}
\tag{5}
$$

(D) Write the simplex tableau for (5) and apply the simplex procedure to find the optimal solution. Compare the results with the solution in part (A).

This method is called the *two-phase method* because it involves solving two problems:

Phase 1.   Minimize the sum of the artificial variables to obtain a basic feasible solution with all artificial variables equal to zero.

Phase 2.   Delete the artificial variables and complete the solution of the original problem.

(E) Write a step-by-step procedure for using the two-phase method to solve linear programming problems. Discuss situations where this method will not produce an optimal solution. (Considering some simple examples might be helpful.) Use your procedure to solve Problems 13, 15, 17, and 19 in Exercise 5-6.

CHAPTER 5 REVIEW ■ Important Terms and Symbols

5-1 *Systems of Linear Inequalities in Two Variables.* Graph of a linear inequality in two variables; left half-plane; right half-plane; upper half-plane; lower half-plane; graphical solution of systems of linear inequalities; solution region; feasible region; corner point; bounded regions; unbounded regions

5-2 *Linear Programming in Two Dimensions—A Geometric Approach.* Linear programming problem; decision variables; objective function; problem constraints; nonnegative constraints; mathematical model; graphical solution; maximization problem; constant-profit line; optimal solution; feasible region; fundamental theorem of linear programming; multiple optimal solution; empty feasible region; unbounded objective function

5-3 *A Geometric Introduction to the Simplex Method.* Standard maximization problem in standard form; slack variables; basic variables; nonbasic variables; basic solution; basic feasible solution

5-4 *The Simplex Method: Maximization with Problem Constraints of the Form* $\leq$. Initial system; basic solutions and basic feasible solutions for initial systems; initial basic feasible solution; simplex tableau; initial simplex tableau; selecting basic and nonbasic variables for the simplex method; pivot operation; entering variable; exiting variable; pivot column; indicators; pivot row; pivot element; pivoting

5-5 *The Dual; Minimization with Problem Constraints of the Form* $\geq$. Dual problem; transpose; A^T; solution of a minimization problem; transportation problem

5-6 *Maximization and Minimization with Mixed Problem Constraints.* Surplus variable; artificial variable; modified problem; preliminary simplex tableau; initial simplex tableau; big M method

Summary of Problem Types and Simplex Solution Methods

PROBLEM TYPE	PROBLEM CONSTRAINTS	RIGHT-SIDE CONSTANTS	COEFFICIENTS OF OBJECTIVE FUNCTION	METHOD OF SOLUTION
1. Maximization	$\leq$	Nonnegative	Any real numbers	Simplex method with slack variables
2. Minimization	$\geq$	Any real numbers	Nonnegative	Form dual and solve by method 1
3. Maximization	Mixed $(\leq, \geq, =)$	Nonnegative	Any real numbers	Form modified problem with slack, surplus, and artificial variables and solve by the big M method
4. Minimization	Mixed $(\leq, \geq, =)$	Nonnegative	Any real numbers	Maximize negative of objective function by method 3

CHAPTER 5 ■ Review Exercise

Work through all the problems in this chapter review and check your answers in the back of the book. Answers to all review problems are there along with section numbers in italics to indicate where each type of problem is discussed. Where weaknesses show up, review appropriate sections in the text.

A *Solve the systems in Problems 1 and 2 graphically, and indicate whether each solution region is bounded or unbounded. Find the coordinates of each corner point.*

1. $2x_1 + x_2 \leq 8$
$3x_1 + 9x_2 \leq 27$
$x_1, x_2 \geq 0$

2. $3x_1 + x_2 \geq 9$
$2x_1 + 4x_2 \geq 16$
$x_1, x_2 \geq 0$

3. Solve the linear programming problem geometrically:

Maximize $P = 6x_1 + 2x_2$

Subject to $2x_1 + x_2 \leq 8$
$x_1 + 2x_2 \leq 10$
$x_1, x_2 \geq 0$

4. Convert the problem constraints in Problem 3 into a system of equations using slack variables.

5. How many basic variables and how many nonbasic variables are associated with the system in Problem 4?

6. Find all basic solutions for the system in Problem 4, and determine which basic solutions are feasible.

7. Write the simplex tableau for Problem 3, and circle the pivot element. Indicate the entering and exiting variables.

8. Solve Problem 3 using the simplex method.

9. For the simplex tableau below, identify the basic and nonbasic variables. Find the pivot element, the entering and exiting variables, and perform one pivot operation.

$$
\begin{array}{ccccccc|c}
x_1 & x_2 & x_3 & s_1 & s_2 & s_3 & P & \\
2 & 1 & 3 & -1 & 0 & 0 & 0 & 20 \\
3 & 0 & 4 & 1 & 1 & 0 & 0 & 30 \\
2 & 0 & 5 & 2 & 0 & 1 & 0 & 10 \\
-8 & 0 & -5 & 3 & 0 & 0 & 1 & 50
\end{array}
$$

10. Find the basic solution for each tableau. Determine whether the optimal solution has been reached, additional pivoting is required, or the problem has no solution.

(A)
$$
\begin{array}{ccccc|c}
x_1 & x_2 & s_1 & s_2 & P & \\
4 & 1 & 0 & 0 & 0 & 2 \\
2 & 0 & 1 & 1 & 0 & 5 \\
-2 & 0 & 3 & 0 & 1 & 12
\end{array}
$$

(B)
$$
\begin{array}{ccccc|c}
-1 & 3 & 0 & 1 & 0 & 7 \\
0 & 2 & 1 & 0 & 0 & 0 \\
-2 & 1 & 0 & 0 & 1 & 22
\end{array}
$$

(C)
$$
\begin{array}{ccccc|c}
1 & -2 & 0 & 4 & 0 & 6 \\
0 & 2 & 1 & 6 & 0 & 15 \\
0 & 3 & 0 & 2 & 1 & 10
\end{array}
$$

11. Solve the linear programming problem geometrically:

Minimize $C = 5x_1 + 2x_2$
Subject to $x_1 + 3x_2 \geqslant 15$
$2x_1 + x_2 \geqslant 20$
$x_1, x_2 \geqslant 0$

12. Form the dual of Problem 11.

13. Write the initial system for the dual in Problem 12.

14. Write the first simplex tableau for the dual in Problem 12 and label the columns.

15. Use the simplex method to find the optimal solution of the dual in Problem 12.

16. Use the final simplex tableau from Problem 15 to find the optimal solution of Problem 11.

B

17. Solve the linear programming problem geometrically:

Maximize $P = 3x_1 + 4x_2$
Subject to $2x_1 + 4x_2 \leqslant 24$
$3x_1 + 3x_2 \leqslant 21$
$4x_1 + 2x_2 \leqslant 20$
$x_1, x_2 \geqslant 0$

18. Solve Problem 17 using the simplex method.

19. Solve the linear programming problem geometrically:

Minimize $C = 3x_1 + 8x_2$
Subject to $x_1 + x_2 \geqslant 10$
$x_1 + 2x_2 \geqslant 15$
$x_2 \geqslant 3$
$x_1, x_2 \geqslant 0$

20. Form the dual of Problem 19.

21. Solve Problem 19 by applying the simplex method to the dual in Problem 20.

Solve the linear programming Problems 22 and 23.

22. Maximize $P = 5x_1 + 3x_2 - 3x_3$
Subject to $x_1 - x_2 - 2x_3 \leqslant 3$
$2x_1 + 2x_2 - 5x_3 \leqslant 10$
$x_1, x_2, x_3 \geqslant 0$

23. Maximize $P = 5x_1 + 3x_2 - 3x_3$
Subject to $x_1 - x_2 - 2x_3 \leqslant 3$
$x_1 + x_2 \leqslant 5$
$x_1, x_2, x_3 \geqslant 0$

In Problems 24 and 25:

(A) Introduce slack, surplus, and artificial variables and form the modified problem.
(B) Write the preliminary simplex tableau for the modified problem and find the initial simplex tableau.
(C) Find the optimal solution of the modified problem by applying the simplex method to the initial simplex tableau.
(D) Find the optimal solution of the original problem, if it exists.

24. Maximize $P = x_1 + 3x_2$
Subject to $x_1 + x_2 \geqslant 6$
$x_1 + 2x_2 \leqslant 8$
$x_1, x_2 \geqslant 0$

25. Maximize $P = x_1 + x_2$
Subject to $x_1 + x_2 \geqslant 5$
$x_1 + 2x_2 \leqslant 4$
$x_1, x_2 \geqslant 0$

26. Find the modified problem for the following linear programming problem. (Do not solve.)

$$\text{Maximize} \quad P = 2x_1 + 3x_2 + x_3$$
$$\text{Subject to} \quad x_1 - 3x_2 + x_3 \leq 7$$
$$-x_1 - x_2 + 2x_3 \leq -2$$
$$3x_1 + 2x_2 - x_3 = 4$$
$$x_1, x_2, x_3 \geq 0$$

Write a brief verbal description of the type of linear programming problem that can be solved by the method indicated in Problems 27–30. Include the type of optimization, the number of variables, the type of constraints, and any restrictions on the coefficients and constants.

27. Geometric method

28. Basic simplex method with slack variables

29. Dual method

30. Big *M* method

31. Solve the following linear programming problem by the simplex method, keeping track of the obvious basic solution at each step. Then graph the feasible region and illus-

trate the path to the optimal solution determined by the simplex method.

$$\text{Maximize} \quad P = 2x_1 + 3x_2$$
$$\text{Subject to} \quad x_1 + 2x_2 \leq 22$$
$$3x_1 + x_2 \leq 26$$
$$x_1 \leq 8$$
$$x_2 \leq 10$$
$$x_1, x_2 \geq 0$$

32. Solve by the dual method:

$$\text{Minimize} \quad C = 3x_1 + 2x_2$$
$$\text{Subject to} \quad 2x_1 + x_2 \leq 20$$
$$2x_1 + x_2 \geq 9$$
$$x_1 + x_2 \geq 6$$
$$x_1, x_2 \geq 0$$

33. Solve Problem 32 by the big *M* method.

34. Solve by the dual method:

$$\text{Minimize} \quad C = 15x_1 + 12x_2 + 15x_3 + 18x_4$$
$$\text{Subject to} \quad x_1 + x_2 \qquad\qquad \leq 240$$
$$x_3 + x_4 \leq 500$$
$$x_1 \qquad + x_3 \qquad \geq 400$$
$$x_2 \qquad + x_4 \geq 300$$
$$x_1, x_2, x_3, x_4 \geq 0$$

▶ APPLICATIONS

Business & Economics

Formulate Problems 35–39 as linear programming problems. (The answers in the back of the book for these application problems include the formulation.) Then solve the problem by the indicated method.

35. *Manufacturing—resource allocation.* South Shore Sail Loft manufactures regular and competition sails. Each regular sail takes 2 hours to cut and 4 hours to sew. Each competition sail takes 3 hours to cut and 10 hours to sew. There are 150 hours available in the cutting department and 380 hours available in the sewing department. Use the geometric method.

(A) If the loft makes a profit of $100 on each regular sail and $200 on each competition sail, how many sails of each type should the company manufacture to maximize their profit? What is the maximum profit?

(B) An increase in the demand for competition sails causes the profit on a competition sail to rise to $260. Discuss the effect of this change on the number of sails manufactured and on the maximum profit.

(C) A decrease in the demand for competition sails causes the profit on a competition sail to drop to $140. Discuss the effect of this change on the number of sails manufactured and on the maximum profit.

36. *Investment.* An investor has $150,000 to invest in oil stock, steel stock, and government bonds. The bonds are guaranteed to yield 5%, but the yield for each stock can vary. To protect against major losses, the investor decides that the amount invested in oil stock should not exceed $50,000 and that the total amount invested in stock cannot exceed the amount invested in bonds by more than $25,000. Use the simplex method to answer the questions at the top of the next page.

(A) If the oil stock yields 12% and the steel stock yields 9%, how much money should be invested in each alternative in order to maximize the return? What is the maximum return?

(B) Repeat part (A) if the oil stock yields 9% and the steel stock yields 12%.

37. *Transportation—shipping schedule.* A company produces motors for washing machines at factory *A* and factory *B*. The motors are then shipped to either plant *X* or plant *Y*, where the washing machines are assembled. The maximum number of motors that can be produced at each factory monthly, the minimum number required monthly for each plant to meet anticipated demand, and the shipping charges for one motor are given in the table. Determine a shipping schedule that will minimize the cost of transporting the motors from the factories to the assembly plants. Use the dual method.

	PLANT *X*	PLANT *Y*	MAXIMUM PRODUCTION
FACTORY *A*	$5	$8	1,500
FACTORY *B*	$9	$7	1,000
MINIMUM REQUIREMENT	900	1,200	

38. *Blending—food processing.* A company blends long-grain rice and wild rice to produce two brands of rice mixes, brand *A*, which is marketed under the company's name, and brand *B*, which is marketed as a generic brand. Brand *A* must contain at least 10% wild rice, and brand *B* must contain at least 5% wild rice. Long-grain rice costs the company $0.70 per pound, and wild rice costs $3.40 per pound. The company sells brand *A* for $1.50 a pound and brand *B* for $1.20 a pound. The company has 8,000 pounds of long-grain rice and 500 pounds of wild rice on hand. How should the company use the available rice to maximize their profit on these two brands of rice? What is the maximum profit? Use the simplex method.

Life Sciences

39. *Nutrition—animals.* A special diet for laboratory animals is to contain at least 850 units of vitamins, 800 units of minerals, and 1,150 calories. There are two feed mixes available, mix *A* and mix *B*. A gram of mix *A* contains 2 units of vitamins, 2 units of minerals, and 4 calories. A gram of mix *B* contains 5 units of vitamins, 4 units of minerals, and 5 calories. Use the geometric method.

(A) If mix *A* costs $0.04 per gram and mix *B* costs $0.09 per gram, how many grams of each mix should be used to satisfy the requirements of the diet at minimal cost? What is the minimum cost?

(B) If the price of mix *B* decreases to $0.06 per gram, discuss the effect of this change on the solution in part (A).

(C) If the price of mix *B* increases to $0.12 per gram, discuss the effect of this change on the solution in part (A).

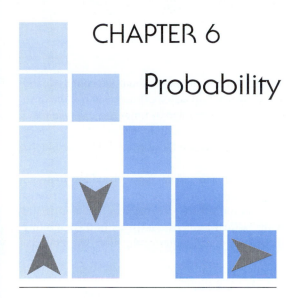

CHAPTER 6

Probability

Probability, like many branches of mathematics, evolved out of practical considerations. Girolamo Cardano (1501–1576), a gambler and physician, produced some of the best mathematics of his time, including a systematic analysis of gambling problems. In 1654, another gambler, the Chevalier de Méré, plagued with bad luck, approached the well-known French philosopher and mathematician Blaise Pascal (1623–1662) regarding certain dice problems. Pascal became interested in these problems, studied them, and discussed them with Pierre de Fermat (1601–1665), another French mathematician. Thus, out of the gaming rooms of western Europe the study of probability was born.

In spite of this lowly birth, probability has matured into a highly respected and immensely useful branch of mathematics. It is used in practically every field. Probability theory can be thought of as the science of uncertainty. If, for example, a card is drawn from a deck of 52 playing cards, it is uncertain which card will be drawn. But suppose a card is drawn and replaced in the deck and a card is again drawn and replaced, and this action is repeated a large number of times. A particular card, say the ace of spades, will be drawn over the long run with a relative frequency that is approximately predictable. Probability theory is concerned with determining the long-run frequency of the occurrence of a given event.

How do we assign probabilities to events? There are two basic approaches to this problem, one theoretical and the other empirical. An example will illustrate the difference between the two approaches.

Suppose you were asked, "What is the probability of obtaining a 2 on a single throw of a die?" Using a *theoretical approach*, we would reason as follows: Since there are 6 *equally likely* ways the die can turn up (assuming the die is fair) and there is only 1 way a 2 can turn up, then the probability of obtaining a 2 is $\frac{1}{6}$. Here, we have arrived at a probability assignment without even rolling a die once; we have used certain assumptions and a reasoning process.

What does the result have to do with reality? We would expect that in the long run (after rolling a die many times), the 2 would appear approximately $\frac{1}{6}$ of the time.

With the *empirical approach,* we make no assumption about the equally likely ways in which the die can turn up. We simply set up an experiment and roll the die a large number of times. Then we compute the percentage of times the 2 appears and use this number as an estimate of the probability of obtaining a 2 on a single roll of the die. Each approach has advantages and drawbacks; these will be discussed in the following sections.

We will first consider the theoretical approach and develop procedures that will lead to the solution of a large variety of interesting problems. These procedures require counting the number of ways certain events can happen, and this is not always easy. However, powerful mathematical tools can assist us in this counting task. The development of these tools is the subject matter of the next two sections.

S E C T I O N   6-1   ■   # Basic Counting Principles

- ADDITION PRINCIPLE
- VENN DIAGRAMS
- MULTIPLICATION PRINCIPLE

■ ADDITION PRINCIPLE

If the enrollment in a college chemistry class consists of 13 males and 15 females, then it is easy to see that there are a total of 28 students enrolled in the class. This is a simple example of a **counting technique**—a method for determining the number of elements in a set without actually enumerating the elements one by one. Set operations play a fundamental role in many counting techniques. For example, if M is the set of male students in the chemistry class and F is the set of female students, then the *union* of sets M and F, denoted $M \cup F$, is the set of all students in the class. Since these sets have no elements in common, the *intersection* of sets M and F, denoted $M \cap F$, is the *empty set* $\varnothing$; we then say that M and F are *disjoint sets.* The total number of students enrolled in the class is the **number of elements** in $M \cup F$, denoted by $n(M \cup F)$ and given by

$$n(M \cup F) = n(M) + n(F) \qquad (1)$$
$$= 13 + 15 = 28$$

Thus, we see that in this example the number of elements in the union of sets M and F is the sum of the number of elements in M and in F. However, this does not work for all pairs of sets. To see why, consider another example. Suppose that the enrollment in a mathematics class consists of 22 math majors and 16 physics majors, and that 7 of these students have majors in both subjects. If M represents the set of math majors and P represents the set of physics majors, then $M \cap P$ represents the set of double majors. It is tempting to proceed as before and conclude that there are $22 + 16 = 38$ students in the class, but this is incorrect. We have counted the double majors twice, once as math majors and again as physics majors. To correct for this double counting, we subtract the number of double majors from this sum. Thus, the total number of students enrolled in this class is given by

$$n(M \cup P) = n(M) + n(P) - n(M \cap P) \tag{2}$$
$$= 22 + 16 - 7 = 31$$

Equation (1) for the chemistry class and equation (2) for the math class are illustrations of the *addition principle* for counting the elements in the union of two sets.

Addition Principle (for Counting)

For any two sets A and B,

$$n(A \cup B) = n(A) + n(B) - n(A \cap B) \tag{3}$$

If A and B are disjoint, then

$$n(A \cup B) = n(A) + n(B) \tag{4}$$

EXAMPLE 1 ➤ **Employee Benefits** According to a survey of business firms in a certain city, 750 firms offer their employees health insurance, 640 offer dental insurance, and 280 offer health insurance and dental insurance. How many firms offer their employees health insurance or dental insurance?

Solution   If H is the set of firms that offer their employees health insurance and D is the set that offer dental insurance, then

$$H \cap D = \text{Set of firms that offer health insurance \textbf{and} dental insurance}$$

$$H \cup D = \text{Set of firms that offer health insurance \textbf{or} dental insurance}$$

Thus,

$$n(H) = 750 \qquad n(D) = 640 \qquad n(H \cap D) = 280$$

and

$$n(H \cup D) = n(H) + n(D) - n(H \cap D)$$
$$= 750 + 640 - 280 = 1,110$$

Thus, 1,110 firms offer their employees health insurance or dental insurance. ◄

MATCHED PROBLEM 1 ➤ The survey in Example 1 also indicated that 345 firms offer their employees group life insurance, 285 offer long-term disability insurance, and 115 offer group life insurance and long-term disability insurance. How many firms offer their employees group life insurance or long-term disability insurance? ◄

■ VENN DIAGRAMS

The next example introduces some additional set concepts that provide useful counting techniques.

EXAMPLE 2 ➤   Market Research   A city has two daily newspapers, the *Sentinel* and the *Journal*. The following information was obtained from a survey of 100 residents of the city: 35 people subscribe to the *Sentinel*, 60 subscribe to the *Journal*, and 20 subscribe to both newspapers.

(A) How many people in the survey subscribe to the *Sentinel* but not to the *Journal*?
(B) How many subscribe to the *Journal* but not to the *Sentinel*?
(C) How many do not subscribe to either paper?
(D) Organize this information in a table.

Solution   Let U be the group of people surveyed, let S be the set of people who subscribe to the *Sentinel*, and let J be the set of people who subscribe to the *Journal*. Since U contains all the elements under consideration, it is the *universal set* for this problem. The *complement* of S, denoted S', is the set of people in the survey group U who do not subscribe to the *Sentinel*. Similarly, J' is the set of people in the group who do not subscribe to the *Journal*. Using the sets S and J, their complements, and set intersection, we can divide the set U into the four disjoint subsets defined below and illustrated in the *Venn diagram* in Figure 1.

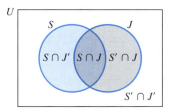

FIGURE 1
Venn diagram for the newspaper survey

$S \cap J$ = Set of people who subscribe to both papers
$S \cap J'$ = Set of people who subscribe to the *Sentinel* but not the *Journal*
$S' \cap J$ = Set of people who subscribe to the *Journal* but not the *Sentinel*
$S' \cap J'$ = Set of people who do not subscribe to either paper

The given survey information can be expressed in the terms of set notation as

$$n(U) = 100 \qquad n(S) = 35 \qquad n(J) = 60 \qquad n(S \cap J) = 20$$

We can use this information and a Venn diagram to answer parts (A)–(C). To begin, we place 20 in $S \cap J$ in the diagram (see Fig. 2). As we proceed through parts (A)–(C), we add each answer to the diagram.

FIGURE 2
Newspaper survey results

(A) Since 35 people subscribe to the *Sentinel* and 20 subscribe to both papers, the number of people who subscribe to the *Sentinel* but not to the *Journal* is

$$n(S \cap J') = 35 - 20 = 15$$

(B) In a similar manner, the number of people who subscribe to the *Journal* but not to the *Sentinel* is

$$n(S' \cap J) = 60 - 20 = 40$$

(C) The total number of newspaper subscribers is $20 + 15 + 40 = 75$. Thus, the number of people who do not subscribe to either paper is

$$n(S' \cap J') = 100 - 75 = 25$$

(D) Venn diagrams are useful tools for determining the number of elements in the various sets in a survey, but often the results must be presented in the form of a table, rather than a diagram. The following table contains the same information as Figure 2, but organized in a different format:

		JOURNAL		
		Subscriber, J	*Nonsubscriber, J'*	*Totals*
SENTINEL	*Subscriber, S*	20	15	35
	Nonsubscriber, S'	40	25	65
	Totals	60	40	100

MATCHED PROBLEM 2 ➤ A small town has two radio stations, an AM station and an FM station. A survey of 100 residents of the town produced the following results: In the last 30 days, 65 people have listened to the AM station, 45 have listened to the FM station, and 30 have listened to both stations.

(A) How many people in the survey have listened to the AM station, but not to the FM station, in this 30 day period?
(B) How many have listened to the FM station, but not to the AM station?
(C) How many have not listened to either station?
(D) Organize this information in a table.

explore – discuss 1

Let A, B, and C be three sets. Use a Venn diagram to explain the following equation:

$$n(A \cup B \cup C) = n(A) + n(B) + n(C) - n(A \cap B) - n(A \cap C)$$
$$- n(B \cap C) + n(A \cap B \cap C)$$

■ MULTIPLICATION PRINCIPLE

As we have just seen, if the elements of a set are determined by the union operation, then addition and subtraction are used to count the number of elements in the set. Now we want to consider sets whose elements are determined by a sequence of operations. We will see that multiplication is used to count the number of elements in sets formed in this manner. The best way to see how this works is to start with an example.

EXAMPLE 3 ➤ Product Mix   A retail store stocks windbreaker jackets in small, medium, large, and extra large, and all are available in blue or red. What are the combined choices and how many combined choices are there?

Solution   To solve the problem we use a tree diagram:

Size Choices (Outcomes)	Color Choices (Outcomes)	Combined Choices (Outcomes)
S	B	(S, B)
S	R	(S, R)
M	B	(M, B)
M	R	(M, R)
L	B	(L, B)
L	R	(L, R)
XL	B	(XL, B)
XL	R	(XL, R)

Start

Thus, there are 8 possible combined choices (outcomes). There are 4 ways a size can be chosen and 2 ways a color can be chosen. The first element in the ordered pair represents a size choice, and the second element represents a color choice. ◀

MATCHED PROBLEM 3 ➤ A company offers its employees health plans from three different companies, R, S, and T. Each company offers two levels of coverage, A and B, with one level requiring additional employee contributions. What are the combined choices and how many choices are there? Solve using a tree diagram. ◀

Now suppose you asked, "From the 26 letters in the alphabet, how many ways can 3 letters appear in a row on a license plate if no letter is repeated?" To try to count the possibilities using a tree diagram would be extremely tedious, to say the least. The following **multiplication principle** will enable us to solve this problem easily; in addition, it forms the basis for several other counting devices that are developed in the next section.

> **Multiplication Principle (for Counting)**
>
> 1. If two operations O_1 and O_2 are performed in order, with N_1 possible outcomes for the first operation and N_2 possible outcomes for the second operation, then there are
>
> $$N_1 \cdot N_2$$
>
> possible combined outcomes of the first operation followed by the second.
> 2. In general, if n operations $O_1, O_2, \ldots, O_n$ are performed in order, with possible number of outcomes $N_1, N_2, \ldots, N_n$, respectively, then there are
>
> $$N_1 \cdot N_2 \cdot \cdots \cdot N_n$$
>
> possible combined outcomes of the operations performed in the given order.

In Example 3, we see that there are 4 possible outcomes in choosing a size (the first operation) and 2 possible outcomes in choosing a color (the second operation); hence, by the multiplication principle, there are $4 \cdot 2 = 8$ possible combined outcomes. Use the multiplication principle to solve Matched Problem 3.
[*Answer:* $3 \cdot 2 = 6$]

To answer the license plate question: There are 26 ways the first letter can be chosen; after a first letter is chosen, there are 25 ways a second letter can be chosen; and after 2 letters are chosen, there are 24 ways a third letter can be chosen. Hence, using the multiplication principle, there are $26 \cdot 25 \cdot 24 = 15{,}600$ possible ways 3 letters can be chosen from the alphabet without repeats.

explore – discuss 2

A state is about to adopt a 6-character format for its license plates, consisting of a block of letters followed by a block of numbers. The format must accommodate up to 20 million vehicles. What size would you recommend for each block?

EXAMPLE 4 ▶ **Computer-Assisted Testing** Many colleges and universities are now using computer-assisted testing procedures. Suppose a screening test is to consist of 5 questions, and a computer stores 5 comparable questions for the first test question, 8 for the second, 6 for the third, 5 for the fourth, and 10 for the fifth. How many different 5-question tests can the computer select? (Two tests are considered different if they differ in one or more questions.)

Solution O_1:   Selecting the first question N_1:   5 ways
                     O_2:   Selecting the second question N_2:   8 ways
                     O_3:   Selecting the third question N_3:   6 ways
                     O_4:   Selecting the fourth question N_4:   5 ways
                     O_5:   Selecting the fifth question N_5:   10 ways

Thus, the computer can generate

$$5 \cdot 8 \cdot 6 \cdot 5 \cdot 10 = 12{,}000 \text{ different tests}$$ ◀

MATCHED PROBLEM 4 ▶ Each question on a multiple-choice test has 5 choices. If there are 5 such questions on a test, how many different response sheets are possible if only 1 choice is marked for each question? ◀

EXAMPLE 5 ▶ **Code Words** How many 3-letter code words are possible using the first 8 letters of the alphabet if:

(A) No letter can be repeated?   (B) Letters can be repeated?
(C) Adjacent letters cannot be alike?

Solution   To form 3-letter code words from the 8 letters available, we select a letter for the first position, one for the second position, and one for the third position. Altogether, there are three operations.

(A) No letter can be repeated:

     O_1:   Selecting the first letter N_1:   8 ways
     O_2:   Selecting the second letter N_2:   7 ways   Since 1 letter has been used

     O_3:   Selecting the third letter N_3:   6 ways   Since 2 letters have been used

Thus, there are

$$8 \cdot 7 \cdot 6 = 336 \text{ possible code words}$$ Possible combined operations

(B) Letters can be repeated:

     O_1:   Selecting the first letter N_1:   8 ways
     O_2:   Selecting the second letter N_2:   8 ways   Repeats allowed
     O_3:   Selecting the third letter N_3:   8 ways   Repeats allowed

Thus, there are

$$8 \cdot 8 \cdot 8 = 8^3 = 512 \text{ possible code words}$$

(C) Adjacent letters cannot be alike:

     O_1:   Selecting the first letter N_1:   8 ways
     O_2:   Selecting the second letter N_2:   7 ways   Cannot be the same as the first

     O_3:   Selecting the third letter N_3:   7 ways   Cannot be the same as the second, but can be the same as the first

Thus, there are

$$8 \cdot 7 \cdot 7 = 392 \text{ possible code words}$$ ◄

MATCHED PROBLEM 5 ➤ How many 4-letter code words are possible using the first 10 letters of the alphabet under the three different conditions stated in Example 5? ◄

Answers to Matched Problems

1. 515
2. (A) 35   (B) 15   (C) 20
   (D)

		FM		
		Listener	Nonlistener	Totals
AM	Listener	30	35	65
	Nonlistener	15	20	35
	Totals	45	55	100

3. There are 6 combined choices:

Company Choices (Outcomes)	Coverage Choices (Outcomes)	Combined Choices (Outcomes)
R	A B	(R, A) (R, B)
Start — S	A B	(S, A) (S, B)
T	A B	(T, A) (T, B)

4. 5^5, or 3,125
5. (A) $10 \cdot 9 \cdot 8 \cdot 7 = 5{,}040$ (B) $10 \cdot 10 \cdot 10 \cdot 10 = 10{,}000$
   (C) $10 \cdot 9 \cdot 9 \cdot 9 = 7{,}290$

EXERCISE 6-1

A *Problems 1–12 refer to the following Venn diagram. Find the number of elements in each of the indicated sets.*

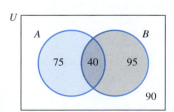

1. A
2. B
3. U
4. A'
5. B'
6. $A \cap B$
7. $A \cup B$
8. $A \cap B'$
9. $A' \cap B$
10. $A' \cap B'$
11. $(A \cap B)'$
12. $(A \cup B)'$

Solve Problems 13–16 two ways: using (A) a tree diagram and (B) the multiplication principle.

13. How many ways can 2 coins turn up—heads, H, or tails, T—if the combined outcome (H, T) is to be distinguished from the outcome (T, H)?

14. How many 2-letter code words can be formed from the first 3 letters of the alphabet if no letter can be used more than once?

15. A coin is tossed with possible outcomes heads, H or tails, T. Then a single die is tossed with possible outcomes 1, 2, 3, 4, 5, or 6. How many combined outcomes are there?

16. In how many ways can 3 coins turn up—heads, H, or tails, T—if combined outcomes such as (H, T, H), (H, H, T), and (T, H, H) are to be considered different?

B *In Problems 17–20, use the given information to determine the number of elements in each of the four disjoint subsets in the following Venn diagram.*

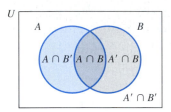

17. $n(A) = 80$, $n(B) = 50$, $n(A \cap B) = 20$, $n(U) = 200$

18. $n(A) = 45$, $n(B) = 35$, $n(A \cap B) = 15$, $n(U) = 100$

19. $n(A) = 25$, $n(B) = 55$, $n(A \cup B) = 60$, $n(U) = 100$

20. $n(A) = 70$, $n(B) = 90$, $n(A \cup B) = 120$, $n(U) = 200$

In Problems 21–24, use the given information to complete the following table.

	A	A'	Totals
B	?	?	?
B'	?	?	?
Totals	?	?	?

21. $n(A) = 70$, $n(B) = 90$, $n(A \cap B) = 30$, $n(U) = 200$

22. $n(A) = 55$, $n(B) = 65$, $n(A \cap B) = 35$, $n(U) = 100$

23. $n(A) = 45$, $n(B) = 55$, $n(A \cup B) = 80$, $n(U) = 100$

24. $n(A) = 80$, $n(B) = 70$, $n(A \cup B) = 110$, $n(U) = 200$

25. A particular new car model is available with 5 choices of color, 3 choices of transmission, 4 types of interior, and 2 types of engine. How many different variations of this model car are possible?

26. A delicatessen serves meat sandwiches with the following options: 3 kinds of bread, 5 kinds of meat, and lettuce or sprouts. How many different sandwiches are possible, assuming one item is used out of each category?

27. How many 4-letter code words are possible from the first 6 letters of the alphabet if no letter is repeated? If letters are repeated? If adjacent letters must be different?

28. How many 5-letter code words are possible from the first 7 letters of the alphabet if no letter is repeated? If letters are repeated? If adjacent letters must be different?

29. A combination lock has 5 wheels, each labeled with the 10 digits from 0 to 9. How many 5-digit opening combinations are possible if no digit is repeated? If digits are repeated? If successive digits must be different?

30. A small combination lock on a suitcase has 3 wheels, each labeled with the 10 digits from 0 to 9. How many 3-digit combinations are possible if no digit is repeated? If digits are repeated? If successive digits must be different?

31. How many different license plates are possible if each contains 3 letters (out of the 26 letters of the alphabet) followed by 3 digits (from 0 to 9)? How many of these license plates contain no repeated letters and no repeated digits?

32. How many 5-digit ZIP code numbers are possible? How many of these numbers contain no repeated digits?

33. In Example 3 does it make any difference in which order the selection operations are performed? That is, if we select a jacket color first and then select a size, are there as many combined choices available as selecting a size first and then a color? Justify your answer using tree diagrams and the multiplication principle.

34. Explain how three sets, A, B, and C, can be related to each other in order for the following equation to hold (Venn diagrams may be helpful):

$$n(A \cup B \cup C) = n(A) + n(B) + n(C) - n(A \cap C) - n(B \cap C)$$

C

35. A group of 75 people includes 32 who play tennis, 37 who play golf, and 8 who play both tennis and golf. How many people in the group play neither sport?

36. A class of 30 music students includes 13 who play the piano, 16 who play the guitar, and 5 who play both the piano and the guitar. How many students in the class play neither instrument?

37. A group of 100 people touring Europe includes 42 people who speak French, 55 who speak German, and 17 who speak neither language. How many people in the group speak both French and German?

38. A high school football team with 40 players includes 16 players who played offense last year, 17 who played defense, and 12 who were not on last year's team. How many players from last year played both offense and defense?

▶ APPLICATIONS

Business & Economics

39. *Management selection.* A management selection service classifies its applicants (using tests and interviews) as high-IQ, middle-IQ, or low-IQ and as aggressive or passive. How many combined classifications are possible?
(A) Solve by using a tree diagram.
(B) Solve by using the multiplication principle.

40. *Management selection.* A corporation plans to fill 2 different positions for vice president, V_1 and V_2, from administrative officers in 2 of its manufacturing plants. Plant A has 6 officers and plant B has 8. How many ways can these 2 positions be filled if the V_1 position is to be filled from plant A and the V_2 position from plant B? How many ways can the 2 positions be filled if the selection is made without regard to plant?

41. *Transportation.* A sales representative who lives in city A wishes to start from home and fly to 3 different cities, B, C, and D. If there are 2 choices of local transportation (drive her own car to and from the airport or use a taxi for both trips), and if all cities are interconnected by airlines, how many travel plans can be constructed to visit each city exactly once and return home?

42. *Transportation.* A manufacturing company in city A wishes to truck its product to 4 different cities, B, C, D, and E. If the cities are all interconnected by roads, how many different route plans can be constructed so that a single truck, starting from A, will visit each city exactly once, then return home?

43. *Market research.* A survey of 1,200 people in a certain city indicates that 850 own microwave ovens, 740 own VCR's, and 580 own microwave ovens and VCR's.
(A) How many people in the survey own either a microwave oven or a VCR?
(B) How many own neither a microwave oven nor a VCR?
(C) How many own a microwave oven and do not own a VCR?

44. *Market research.* A survey of 800 small businesses indicates that 250 own photocopiers, 420 own fax machines, and 180 own photocopiers and fax machines.
(A) How many businesses in the survey own either a photocopier or a fax machine?
(B) How many own neither a photocopier nor a fax machine?
(C) How many own a fax machine and do not own a photocopier?

45. *Communications.* A cable television company has 8,000 subscribers in a suburban community. The company offers two premium channels, HBO and Showtime. If 2,450 subscribers receive HBO, 1,940 receive Showtime, and 5,180 do not receive any premium channel, how many subscribers receive both HBO and Showtime?

46. *Communications.* A local telephone company offers its 10,000 customers two special services: call forwarding and call waiting. If 3,770 customers use call forwarding, 3,250 use call waiting, and 4,530 do not use either of these services, how many customers use both call forwarding and call waiting?

47. *Minimum wage.* The table on the next page gives the number of male and female workers (in thousands) earning at or below the minimum wage for several age categories.

| | WORKERS PER AGE GROUP (IN THOUSANDS) | | | |
	16–19	20–24	25+	Totals
MALES AT MINIMUM WAGE	343	154	237	734
MALES BELOW MINIMUM WAGE	118	102	159	379
FEMALES AT MINIMUM WAGE	367	186	503	1,056
FEMALES BELOW MINIMUM WAGE	251	202	540	993
Totals	1,079	644	1,439	3,162

(A) How many males are age 20–24 and below minimum wage?

(B) How many females are age 20 or older and at minimum wage?

(C) How many workers are age 16–19 or males at minimum wage?

(D) How many workers are below minimum wage?

48. *Minimum wage.* Refer to the table in Problem 47.

(A) How many females are age 16–19 and at minimum wage?

(B) How many males are age 16–24 and below minimum wage?

(C) How many workers are age 20–24 or females below minimum wage?

(D) How many workers are at minimum wage?

Life Sciences

49. *Medicine.* A medical researcher classifies subjects according to male or female; smoker or nonsmoker; and underweight, average weight, or overweight. How many combined classifications are possible?

(A) Solve using a tree diagram.

(B) Solve using the multiplication principle.

50. *Family planning.* A couple is planning to have 3 children. How many boy–girl combinations are possible? Distinguish between combined outcomes such as (B, B, G), (B, G, B), and (G, B, B).

(A) Solve by using a tree diagram.

(B) Solve by using the multiplication principle.

Social Sciences

51. *Politics.* A politician running for a third term in office is planning to contact all contributors to her first two campaigns. If 1,475 individuals contributed to the first campaign, 2,350 individuals contributed to the second campaign, and 920 individuals contributed to the first and second campaigns, how many individuals have contributed to the first or second campaign?

52. *Politics.* If 12,457 people voted for a politician in his first election, 15,322 voted for him in his second election, and 9,345 voted for him in the first and second elections, how many people voted for this politician in the first or second election?

SECTION 6-2 ■ **Permutations and Combinations**

- FACTORIAL
- PERMUTATIONS
- COMBINATIONS
- APPLICATIONS

The multiplication principle discussed in the preceding section can be used to develop two additional devices for counting that are extremely useful in more complicated counting problems. Both of these devices use a function called a *factorial function,* which we introduce first.

■ FACTORIAL

When using the multiplication principle, we encountered expressions of the form

$$26 \cdot 25 \cdot 24 \qquad 8 \cdot 7 \cdot 6$$

where each natural number factor is decreased by 1 as we move from left to right. The factors in the following product continue to decrease by 1 until a factor of 1 is reached:

$$5 \cdot 4 \cdot 3 \cdot 2 \cdot 1$$

Products of this type are encountered so frequently in counting problems that it is useful to be able to express them in a concise notation. The product of the first n natural numbers is called **n factorial** and is denoted by **$n!$**. Also, we define **zero factorial, 0!,** to be 1. Symbolically:

Factorial

For n a natural number,

$$n! = n(n-1)(n-2) \cdot \cdots \cdot 2 \cdot 1 \qquad 4! = 4 \cdot 3 \cdot 2 \cdot 1$$
$$0! = 1$$
$$n! = n \cdot (n-1)!$$

[*Note:* Many calculators have an $\boxed{n!}$ key or its equivalent.]

EXAMPLE 1 ➤ Computing Factorials

(A) $5! = 5 \cdot 4 \cdot 3 \cdot 2 \cdot 1 = 120$

(B) $\dfrac{7!}{6!} = \dfrac{7 \cdot 6!}{6!} = 7$

(C) $\dfrac{8!}{5!} = \dfrac{8 \cdot 7 \cdot 6 \cdot 5!}{5!} = 8 \cdot 7 \cdot 6 = 336$

(D) $\dfrac{52!}{5!47!} = \dfrac{52 \cdot 51 \cdot 50 \cdot 49 \cdot 48 \cdot 47!}{5 \cdot 4 \cdot 3 \cdot 2 \cdot 1 \cdot 47!} = 2{,}598{,}960$ ◀

MATCHED PROBLEM 1 ➤ Find: (A) $6!$ (B) $\dfrac{10!}{9!}$ (C) $\dfrac{10!}{7!}$ (D) $\dfrac{5!}{0!3!}$ (E) $\dfrac{20!}{3!17!}$ ◀

It is interesting and useful to note that $n!$ grows very rapidly. Compare the following:

$$5! = 120 \qquad 10! = 3{,}628{,}800 \qquad 15! = 1{,}307{,}674{,}368{,}000$$

Try 69!, 70!, and 71! on your calculator.

■ PERMUTATIONS

A particular (horizontal) arrangement of a set of paintings on a wall is called a *permutation* of the set of paintings. In general:

> ### A Permutation of a Set of Objects
>
> A **permutation** of a set of distinct objects is an arrangement of the objects in a specific order without repetition.

Suppose 4 pictures are to be arranged from left to right on one wall of an art gallery. How many permutations (ordered arrangements) are possible? Using the multiplication principle, there are 4 ways of selecting the first picture; after the first picture is selected, there are 3 ways of selecting the second picture; after the first 2 pictures are selected, there are 2 ways of selecting the third picture; and after the first 3 pictures are selected, there is only 1 way to select the fourth. Thus, the number of permutations (ordered arrangements) of the set of 4 pictures is

$$4 \cdot 3 \cdot 2 \cdot 1 = 4! = 24$$

In general, how many permutations of a set of n distinct objects are possible? Reasoning as above, there are n ways in which the first object can be chosen, there are $n - 1$ ways in which the second object can be chosen, and so on. Using the multiplication principle, we have the following:

> ### The Number of Permutations of n Objects
>
> The number of permutations of n distinct objects without repetition, denoted by $P_{n,n}$, is
>
> $$P_{n,n} = n(n - 1) \cdot \ \cdots \ \cdot 2 \cdot 1 = n! \quad \text{n factors}$$
>
> Example: The number of permutations of 7 objects is
>
> $$P_{7,7} = 7 \cdot 6 \cdot 5 \cdot 4 \cdot 3 \cdot 2 \cdot 1 = 7! \quad \text{7 factors}$$

Now suppose the museum director decides to use only 2 of the 4 available paintings, and they will be arranged on the wall from left to right. We are now talking about a particular arrangement of 2 paintings out of the 4, which is called a *permutation of 4 objects taken 2 at a time.* In general:

> ### A Permutation of n Objects Taken r at a Time
>
> A permutation of a set of n distinct objects taken r at a time without repetition is an arrangement of the r objects in a specific order.

How many ordered arrangements of 2 pictures can be formed from the 4? That is, how many permutations of 4 objects taken 2 at a time are there? There are 4 ways the first picture can be selected; after selecting the first picture, there are 3 ways the second picture can be selected. Thus, the number of permutations of a set of 4 objects taken 2 at a time, which is denoted by $P_{4,2}$, is given by

$$P_{4,2} = 4 \cdot 3$$

In terms of factorials, multiplying $4 \cdot 3$ by 1 in the form $2!/2!$, we have

$$P_{4,2} = 4 \cdot 3 = \frac{4 \cdot 3 \cdot \mathbf{2!}}{\mathbf{2!}} = \frac{4!}{2!}$$

Reasoning in the same way as in the example, we find that the number of permutations of n distinct objects taken r at a time without repetition ($0 \leq r \leq n$) is given by

$$P_{n,r} = n(n-1)(n-2) \cdot \cdots \cdot (n-r+1) \quad r \text{ factors}$$
$$P_{9,6} = 9(9-1)(9-2) \cdot \cdots \cdot (9-6+1) \quad 6 \text{ factors}$$
$$= 9 \cdot 8 \cdot 7 \cdot 6 \cdot 5 \cdot 4$$

Multiplying the right side by 1 in the form $(n-r)!/(n-r)!$, we obtain a factorial form for $P_{n,r}$:

$$P_{n,r} = n(n-1)(n-2) \cdot \cdots \cdot (n-r+1) \frac{(n-r)!}{(n-r)!}$$

But, since

$$n(n-1)(n-2) \cdot \cdots \cdot (n-r+1)(n-r)! = n!$$

The above expression simplifies to

$$P_{n,r} = \frac{n!}{(n-r)!}$$

We summarize these results in the following box:

Number of Permutations of n Objects Taken r at a Time

The number of permutations of n distinct objects taken r at a time without repetition is given by*

$$P_{n,r} = n(n-1)(n-2) \cdot \cdots \cdot (n-r+1) \quad r \text{ factors}$$
$$P_{5,2} = 5 \cdot 4 \qquad 2 \text{ factors}$$

---

* In place of the symbol $P_{n,r}$, the symbols P_r^n, $_nP_r$, and $P(n, r)$ are often used.

(Continued)

> **Number of Permutations of n Objects Taken r at a Time**   *(Continued)*
>
> or
>
> $$P_{n,r} = \frac{n!}{(n-r)!} \qquad 0 \leq r \leq n \qquad P_{5,2} = \frac{5!}{(5-2)!} = \frac{5!}{3!}$$
>
> [*Note:* $P_{n,n} = \dfrac{n!}{(n-n)!} = \dfrac{n!}{0!} = n!$ permutations of n objects taken n at a time. Remember, by definition, $0! = 1$.]

EXAMPLE 2 ➤ **Permutations**   Given the set $\{A, B, C\}$, how many permutations are there of this set of 3 objects taken 2 at a time? Answer the question:

(A) Using a tree diagram   (B) Using the multiplication principle
(C) Using the two formulas for $P_{n,r}$

Solution   (A) Using a tree diagram:

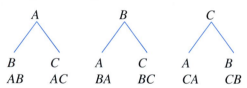

    AB   AC   BA   BC   CA   CB

There are 6 permutations of 3 objects taken 2 at a time.

(B) Using the multiplication principle:

O_1: Fill the first position N_1: 3 ways
O_2: Fill the second position N_2: 2 ways

Thus, there are

$$3 \cdot 2 = 6 \text{ permutations of 3 objects taken 2 at a time}$$

(C) Using the two formulas for $P_{n,r}$:

2 factors
↓

$$P_{3,2} = 3 \cdot 2 = 6 \qquad \text{or} \qquad P_{3,2} = \frac{3!}{(3-2)!} = \frac{3 \cdot 2 \cdot 1}{1} = 6$$

Thus, there are 6 permutations of 3 objects taken 2 at a time.

Of course, all three methods produce the same answer.   ◀

MATCHED PROBLEM 2 ➤ Given the set $\{A, B, C, D\}$, how many permutations are there of this set of 4 objects taken 2 at a time? Answer the question:

(A) Using a tree diagram   (B) Using the multiplication principle
(C) Using the two formulas for $P_{n,r}$ ◀

In Example 2 you probably found the multiplication principle the easiest to use and wonder about the usefulness of the factorial formula for $P_{n,r}$. When the numbers get large you will find that the factorial formula is very useful, since most calculators have an $\boxed{n!}$ key or its equivalent. (In fact, many calculators even have a $\boxed{P_{n,r}}$ key that can evaluate $P_{n,r}$ directly.) Check your calculator and user's manual. The use of a calculator will significantly reduce the burden of computing in this and the following sections. To appreciate this fact, try solving Example 3 without a calculator.

EXAMPLE 3 ➤ **Permutations**   Find the number of permutations of 13 objects taken 8 at a time. Compute the answer using a calculator.

Solution   We use the factorial formula for $P_{n,r}$:

$$P_{13,8} = \frac{13!}{(13-8)!} = \frac{13!}{5!} = 51,891,840$$

Using a tree diagram to solve this problem would involve a monumental effort. Using the multiplication principle would involve multiplying $13 \cdot 12 \cdot 11 \cdot 10 \cdot 9 \cdot 8 \cdot 7 \cdot 6$ (8 factors), which is not too bad. A calculator can provide instant results. ◄

MATCHED PROBLEM 3 ➤ Find the number of permutations of 30 objects taken 4 at a time. Compute the answer exactly using a calculator. ◄

■ COMBINATIONS

Now suppose that an art museum owns 8 paintings by a given artist and another art museum wishes to borrow 3 of these paintings for a special show. In selecting 3 of the 8 paintings for shipment, the order would not matter, and we would simply be selecting a 3 element subset from the set of 8 paintings. That is, we would be selecting what is called *a combination of 8 objects taken 3 at a time*. In general:

A Combination of *n* Objects Taken *r* at a Time

A **combination** of a set of *n* distinct objects taken *r* at a time without repetition is an *r* element subset of the set of *n* objects. The arrangement of the elements in the subset does not matter.

How many ways can the 3 paintings be selected for shipment out of the 8 available? That is, what is the number of combinations of 8 objects taken 3 at a time? To answer this question, and to get a better insight into the general problem, we return to Example 2.

In Example 2 we were given the set $\{A, B, C\}$ and found the number of permutations of 3 objects taken 2 at a time using a tree diagram. From this tree diagram we also can determine the number of combinations of 3 objects taken 2 at a time (the number of 2 element subsets from a 3 element set), and compare it with the number of permutations (see Fig. 1).

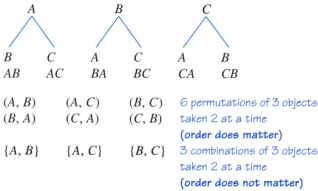

(A, B)	(A, C)	(B, C)
(B, A)	(C, A)	(C, B)
{A, B}	{A, C}	{B, C}

6 permutations of 3 objects taken 2 at a time
(order does matter)

3 combinations of 3 objects taken 2 at a time
(order does not matter)

FIGURE 1

There are fewer combinations than permutations, as we would expect. To each subset (combination) there corresponds two ordered pairs (permutations). We denote the number of combinations in Figure 1 by

$$C_{3,2} \quad \text{or} \quad \binom{3}{2}$$

Our final goal is to find a factorial formula for $C_{n,r}$, the number of combinations of n objects taken r at a time. But first, we will develop a formula for $C_{3,2}$, and then we will generalize from this experience.

We know the number of permutations of 3 objects taken 2 at a time is given by $P_{3,2}$, and we have a formula for computing this number. Now, suppose we think of $P_{3,2}$ in terms of two operations:

O_1:   Selecting a subset of 2 elements N_1: $C_{3,2}$ ways
O_2:   Arranging the subset in a given order N_2:   2! ways

The combined operation, O_1 followed by O_2, produces a permutation of 3 objects taken 2 at a time. Thus,

$$P_{3,2} = C_{3,2} \cdot 2! \quad \text{or} \quad C_{3,2} = \frac{P_{3,2}}{2!}$$

To find $C_{3,2}$, the number of combinations of 3 objects taken 2 at a time, we substitute

$$P_{3,2} = \frac{3!}{(3-2)!}$$

and solve for $C_{3,2}$:

$$C_{3,2} = \frac{\mathbf{3!}}{2!\mathbf{(3-2)!}} = \frac{3 \cdot 2 \cdot 1}{(2 \cdot 1)(1)} = 3$$

This result agrees with the result we got using a tree diagram. Note that the number of combinations of 3 objects taken 2 at a time is the same as the number of permutations of 3 objects taken 2 at a time divided by the number of permutations of the elements in a 2 element subset. This observation also can be made in Figure 1.

Reasoning the same way as in the example, the number of combinations of n objects taken r at a time $(0 \leqslant r \leqslant n)$ is given by

$$C_{n,r} = \frac{P_{n,r}}{r!}$$

$$= \frac{n!}{r!(n-r)!} \quad \text{Since } P_{n,r} = \frac{n!}{(n-r)!}$$

In summary:

Number of Combinations of n Objects Taken r at a Time

The number of combinations of n distinct objects taken r at a time without repetition is given by*

$$C_{n,r} = \binom{n}{r} = \frac{P_{n,r}}{r!} = \frac{n!}{r!(n-r)!} \qquad 0 \leqslant r \leqslant n \qquad C_{52,5} = \binom{52}{5}$$

$$= \frac{P_{52,5}}{5!} = \frac{52!}{5!(52-5)!}$$

---

* In place of the symbols $C_{n,r}$ and $\binom{n}{r}$, the symbols C_r^n, $_nC_r$, and $C(n, r)$ are often used.

Now we can answer the question posed earlier in the museum example. There are

$$C_{8,3} = \frac{8!}{3!(8-3)!} = \frac{8!}{3!5!} = \frac{8 \cdot 7 \cdot 6 \cdot 5!}{3 \cdot 2 \cdot 1 \cdot 5!} = 56$$

ways the 3 paintings can be selected for shipment. That is, there are 56 combinations of 8 objects taken 3 at a time.

EXAMPLE 4 ➤ **Permutations and Combinations** From a committee of 10 people:

(A) In how many ways can we choose a chairperson, a vice chairperson, and a secretary, assuming that one person cannot hold more than one position?

(B) In how many ways can we choose a subcommittee of 3 people?

Solution Note how parts (A) and (B) differ. In part (A), order of choice makes a difference in the selection of the officers. In part (B), the ordering does not matter in choosing a 3 person subcommittee. Thus, in part (A), we are interested in the number of *permutations* of 10 objects taken 3 at a time; and in part (B), we are interested in the number of *combinations* of 10 objects taken 3 at a time. These quantities are computed as follows (and since the numbers are not large, we do not need to use a calculator):

(A) $P_{10,3} = \dfrac{10!}{(10-3)!} = \dfrac{10!}{7!} = \dfrac{10 \cdot 9 \cdot 8 \cdot \cancel{7!}}{\cancel{7!}} = 720$ ways

(B) $C_{10,3} = \dfrac{10!}{3!(10-3)!} = \dfrac{10!}{3!7!} = \dfrac{10 \cdot 9 \cdot 8 \cdot \cancel{7!}}{3 \cdot 2 \cdot 1 \cdot \cancel{7!}} = 120$ ways ◄

MATCHED PROBLEM 4 ➤ From a committee of 12 people:

(A) In how many ways can we choose a chairperson, a vice chairperson, a secretary, and a treasurer, assuming that one person cannot hold more than one position?
(B) In how many ways can we choose a subcommittee of 4 people? ◄

If n and r are other than small numbers (as in Example 4), a calculator will simplify the computation. (Many calculators also have a $\boxed{C_{n,r}}$ key that will simplify the computation even further.)

EXAMPLE 5 ➤ **Combinations** Find the number of combinations of 13 objects taken 8 at a time. Compute the answer exactly, using a calculator.

Solution $\quad C_{13,8} = \dbinom{13}{8} = \dfrac{13!}{8!(13-8)!} = \dfrac{13!}{8!5!} = 1{,}287$ ◄

Compare the result in Example 5 with that obtained in Example 3, and note that $C_{13,8}$ is substantially smaller than $P_{13,8}$.

MATCHED PROBLEM 5 ➤ Find the number of combinations of 30 objects taken 4 at a time. Compute the answer exactly using a calculator. ◄

Remember

In a permutation, order matters.
In a combination, order does not matter.

To determine whether permutations or combinations are involved in a problem, see if rearranging the collection or listing produces a different object. If so, use permutations; if not, use combinations.

---

explore – discuss 1

(A) List, alphabetically by the first letter, all 3-letter license plate codes consisting of 3 different letters chosen from M, A, T, H. Discuss how this list relates to $P_{n,r}$.
(B) Reorganize the list from part (A) so that now all codes without M come first, then all codes without A, then all codes without T, and finally all codes without H. Discuss how this list illustrates the formula $P_{n,r} = r!C_{n,r}$.

■ APPLICATIONS

We now consider some applications of the concepts discussed above. Several applications in this and the following sections involve a standard 52-card deck of playing cards, which is described below:

Standard 52-Card Deck of Playing Cards

A standard deck of 52 cards (see Fig. 2) has four 13-card suits: diamonds, hearts, clubs, and spades. The diamonds and hearts are red, and the clubs and spades are black. Each 13-card suit contains cards numbered from 2 to 10, a jack, a queen, a king, and an ace. The jack, queen, and king are called *face cards*. Depending on the game, the ace may be counted as the lowest and/or the highest card in the suit.

FIGURE 2

EXAMPLE 6 ➤   **Counting Techniques**   How many 5-card hands will have 3 aces and 2 kings?

Solution   The solution involves both the multiplication principle and combinations. Think of selecting the 5-card hand in terms of the following two operations:

O_1:   Choosing 3 aces out of 4 possible N_1: $C_{4,3}$
       (order is not important)

O_2:   Choosing 2 kings out of 4 possible N_2: $C_{4,2}$
       (order is not important)

Using the multiplication principle, we have

$$\text{Number of hands} = C_{4,3} \cdot C_{4,2}$$
$$= \frac{4!}{3!(4-3)!} \cdot \frac{4!}{2!(4-2)!}$$
$$= 4 \cdot 6 = 24$$

◄

MATCHED PROBLEM 6 ➤ How many 5-card hands will have 3 hearts and 2 spades? ◀

EXAMPLE 7 ➤ **Counting Techniques** Serial numbers for a product are to be made using 2 letters followed by 3 numbers. If the letters are to be taken from the first 8 letters of the alphabet with no repeats and the numbers are to be taken from the 10 digits (0–9) with no repeats, how many serial numbers are possible?

Solution The solution involves both the multiplication principle and permutations. Think of selecting a serial number in terms of the following two operations:

O_1: Choosing 2 letters out of 8 available (order is important) $\qquad$ N_1: $P_{8,2}$

O_2: Choosing 3 numbers out of 10 available (order is important) $\qquad$ N_2: $P_{10,3}$

Using the multiplication principle, we have

$$\text{Number of serial numbers} = P_{8,2} \cdot P_{10,3}$$
$$= \frac{8!}{(8-2)!} \cdot \frac{10!}{(10-3)!}$$
$$= 56 \cdot 720 = 40{,}320$$ ◀

MATCHED PROBLEM 7 ➤ Repeat Example 7 under the same conditions, except the serial numbers are now to have 3 letters followed by 2 digits (no repeats). ◀

EXAMPLE 8 ➤ **Counting Techniques** A company has 7 senior and 5 junior officers. An ad hoc legislative committee is to be formed. In how many ways can a 4-officer committee be formed so that it is composed of:

(A) Any 4 officers? $\qquad\qquad$ (B) 4 senior officers?
(C) 3 senior officers and 1 junior officer? $\qquad$ (D) 2 senior and 2 junior officers?
(E) At least 2 senior officers?

Solution (A) Since there are a total of 12 officers in the company, the number of different 4-member committees is

$$C_{12,4} = \frac{12!}{4!(12-4)!} = \frac{12!}{4!8!} = 495$$

(B) If only senior officers can be on the committee, the number of different committees is

$$C_{7,4} = \frac{7!}{4!(7-4)!} = \frac{7!}{4!3!} = 35$$

(C) The 3 senior officers can be selected in $C_{7,3}$ ways, and the 1 junior officer can be selected in $C_{5,1}$ ways. Applying the multiplication principle, the number of ways that 3 senior officers and 1 junior officer can be selected is

$$C_{7,3} \cdot C_{5,1} = \frac{7!}{3!(7-3)!} \cdot \frac{5!}{1!(5-1)!} = \frac{7!5!}{3!4!1!4!} = 175$$

(D) $$C_{7,2} \cdot C_{5,2} = \frac{7!}{2!(7-2)!} \cdot \frac{5!}{2!(5-2)!} = \frac{7!5!}{2!5!2!3!} = 210$$

(E) The committees with *at least* 2 senior officers can be divided into three disjoint collections:

1. Committees with 4 senior officers and 0 junior officers
2. Committees with 3 senior officers and 1 junior officer
3. Committees with 2 senior officers and 2 junior officers

The number of committees of types 1, 2, and 3 were computed in parts (B), (C), and (D), respectively. The total number of committees of all three types is the sum of these quantities:

Type 1   Type 2   Type 3
$$C_{7,4} + C_{7,3} \cdot C_{5,1} + C_{7,2} \cdot C_{5,2} = 35 + 175 + 210 = 420$$ ◄

MATCHED PROBLEM 8 ➤ Given the information in Example 8, answer the following questions:

(A) How many 4-officer committees with 1 senior officer and 3 junior officers can be formed?
(B) How many 4-officer committees with 4 junior officers can be formed?
(C) How many 4-officer committees with at least 2 junior officers can be formed?

◄

explore – discuss 2

(A) Compute the numbers $C_{n,0}, C_{n,1}, C_{n,2}, \ldots, C_{n,n}$ for $n = 4, 5, 6$. Discuss the patterns you observe.
(B) Compute the sum

$$C_{n,0} + C_{n,1} + C_{n,2} + \cdots + C_{n,n}$$

and the alternating sum

$$C_{n,0} - C_{n,1} + C_{n,2} - \cdots + (-1)^n C_{n,n}$$

for $n = 4, 5, 6$. Guess the sum and alternating sum for $n = 7$, and verify your guess.

Answers to Matched Problems

1. (A) 720   (B) 10   (C) 720   (D) 20   (E) 1,140
2. (A)

12 permutations of 4 objects taken 2 at a time

(B) O_1:   Fill first position N_1:   4 ways
        O_2:   Fill second position N_2:   3 ways
        $4 \cdot 3 = 12$

(C) $P_{4,2} = 4 \cdot 3 = 12$; $P_{4,2} = \dfrac{4!}{(4-2)!} = 12$

3. $P_{30,4} = \dfrac{30!}{(30-4)!} = 657{,}720$

4. (A) $P_{12,4} = \dfrac{12!}{(12-4)!} = 11{,}880$ ways   (B) $C_{12,4} = \dfrac{12!}{4!(12-4)!} = 495$ ways

5. $C_{30,4} = \dfrac{30!}{4!(30-4)!} = 27{,}405$ 6. $C_{13,3} \cdot C_{13,2} = 22{,}308$ 7. $P_{8,3} \cdot P_{10,2} = 30{,}240$

8. (A) $C_{7,1} \cdot C_{5,3} = 70$ (B) $C_{5,4} = 5$ (C) $C_{7,2} \cdot C_{5,2} + C_{7,1} \cdot C_{5,3} + C_{5,4} = 285$

EXERCISE   6-2

A

Evaluate the expressions in Problems 1–20.

1. $4!$

2. $6!$

3. $\dfrac{9!}{8!}$

4. $\dfrac{14!}{13!}$

5. $\dfrac{11!}{8!}$

6. $\dfrac{14!}{12!}$

7. $\dfrac{5!}{2!3!}$

8. $\dfrac{6!}{4!2!}$

9. $\dfrac{7!}{4!(7-4)!}$

10. $\dfrac{8!}{3!(8-3)!}$

11. $\dfrac{7!}{7!(7-7)!}$

12. $\dfrac{8!}{0!(8-0)!}$

13. $P_{5,3}$

14. $P_{4,2}$

15. $P_{52,4}$

16. $P_{52,2}$

17. $C_{5,3}$

18. $C_{4,2}$

19. $C_{52,4}$

20. $C_{52,2}$

21. In a horse race, how many different finishes among the first 3 places are possible if 10 horses are running? (Exclude ties.)

22. In a long-distance foot race, how many different finishes among the first 5 places are possible if 50 people are running? (Exclude ties.)

23. How many ways can a 3-person subcommittee be selected from a committee of 7 people? How many ways can a president, vice president, and secretary be chosen from a committee of 7 people?

24. Nine cards are numbered with the digits from 1 to 9. A 3-card hand is dealt, 1 card at a time. How many hands are possible where:
    (A) Order is taken into consideration?
    (B) Order is not taken into consideration?

B

25. Discuss the relative growth rates of $x!$, 3^x, and x^3.

26. Discuss the relative growth rates of $x!$, 2^x, and x^2.

27. From a standard 52-card deck, how many 5-card hands will have all hearts?

28. From a standard 52-card deck, how many 5-card hands will have all face cards? All face cards, but no kings?

29. From a standard 52-card deck, how many 7-card hands have exactly 5 spades and 2 hearts?

30. From a standard 52-card deck, how many 5-card hands will have 2 clubs and 3 hearts?

31. A catering service offers 8 appetizers, 10 main courses, and 7 desserts. A banquet committee is to select 3 appetizers, 4 main courses, and 2 desserts. How many ways can this be done?

32. Three departments have 12, 15, and 18 members, respectively. If each department is to select a delegate and an alternate to represent the department at a conference, how many ways can this be done?

C

33. Eight distinct points are selected on the circumference of a circle.
    (A) How many chords can be drawn by joining the points in all possible ways?

(B) How many triangles can be drawn using these 8 points as vertices?

(C) How many quadrilaterals can be drawn using these 8 points as vertices?

34. Five distinct points are selected on the circumference of a circle.

(A) How many chords can be drawn by joining the points in all possible ways?

(B) How many triangles can be drawn using these 5 points as vertices?

35. How many ways can 2 people be seated in a row of 5 chairs? 3 people? 4 people? 5 people?

36. Each of 2 countries sends 5 delegates to a negotiating conference. A rectangular table is used with 5 chairs on each long side. If each country is assigned a long side of the table (operation 1), how many seating arrangements are possible?

37. A basketball team has 5 distinct positions. Out of 8 players, how many starting teams are possible if:

(A) The distinct positions are taken into consideration?

(B) The distinct positions are not taken into consideration?

(C) The distinct positions are not taken into consideration, but either Mike or Ken (but not both) must start?

38. How many 4-person committees are possible from a group of 9 people if:

(A) There are no restrictions?

(B) Both Jim and Mary must be on the committee?

(C) Either Jim or Mary (but not both) must be on the committee?

39. Find the largest integer k such that your calculator can compute $k!$ without an overflow error.

40. Find the largest integer k such that your calculator can compute $C_{2k,k}$ without an overflow error.

 A P P L I C A T I O N S

Business & Economics

41. *Quality control.* A computer store receives a shipment of 24 laser printers, including 5 that are defective. Three of these printers are selected to be displayed in the store.

(A) How many selections can be made?

(B) How many of these selections will contain no defective printers?

42. *Quality control.* An electronics store receives a shipment of 30 graphing calculators, including 6 that are defective. Four of these calculators are selected to be sent to a local high school.

(A) How many selections can be made?

(B) How many of these selections will contain no defective calculators?

43. *Business closings.* A jewelry store chain with 8 stores in Georgia, 12 in Florida, and 10 in Alabama is planning to close 10 of these stores.

(A) How many ways can this be done?

(B) The company decides to close 2 stores in Georgia, 5 in Florida, and 3 in Alabama. How many ways can this be done?

44. *Employee layoffs.* A real estate company with 14 employees in their central office, 8 in their north office, and 6 in their south office is planning to lay off 12 employees.

(A) How many ways can this be done?

(B) The company decides to lay off 5 employees from the central office, 4 from the north office, and 3 from the south office. How many ways can this be done?

45. *Personnel selection.* Suppose 6 female and 5 male applicants have been successfully screened for 5 positions. In how many ways can the following compositions be selected?

(A) 3 females and 2 males

(B) 4 females and 1 male

(C) 5 females

(D) 5 people regardless of sex

(E) At least 4 females

46. *Committee selection.* A 4-person grievance committee is to be selected out of 2 departments, A and B, with 15 and 20 people, respectively. In how many ways can the following committees be selected?

(A) 3 from A and 1 from B

(B) 2 from A and 2 from B

(C) All from A

(D) 4 people regardless of department

(E) At least 3 from department A

Life Sciences

47. *Medicine.* There are 8 standard classifications of blood type. An examination for prospective laboratory technicians consists of having each candidate determine the type for 3 blood samples. How many different examinations can be given if no 2 of the samples provided for the candidate have the same type? If 2 or more samples can have the same type?

48. *Medical research.* Because of limited funds, 5 research centers are to be chosen out of 8 suitable ones for a study on heart disease. How many choices are possible?

Social Sciences

49. *Politics.* A nominating convention is to select a president and vice president from among 4 candidates. Campaign buttons, listing a president and a vice president, are to be designed for each possible outcome before the convention. How many different kinds of buttons should be designed?

50. *Politics.* In how many different ways can 6 candidates for an office be listed on a ballot?

S E C T I O N   6-3   ■   # Sample Spaces, Events, and Probability

- ■ EXPERIMENTS
- ■ SAMPLE SPACES AND EVENTS
- ■ PROBABILITY OF AN EVENT
- ■ EQUALLY LIKELY ASSUMPTION

This section provides a relatively brief and informal introduction to probability. More detailed and formal treatments can be found in books and courses devoted entirely to the subject. Probability studies involve many subtle notions, and care must be taken at the beginning to understand the fundamental concepts.

■ EXPERIMENTS

Our first step in constructing a mathematical model for probability studies is to describe the type of experiments on which probability studies are based. Some experiments do not yield the same results each time they are performed no matter how carefully they are repeated under the same conditions. These experiments are called **random experiments.** Familiar examples of random experiments are flipping coins, rolling dice, observing the frequency of defective items from an assembly line, or observing the frequency of deaths in a certain age group.

Probability theory is a branch of mathematics that has been developed to deal with outcomes of random experiments, both real and conceptual. In the work that follows, we will simply use the word **experiment** to mean a random experiment.

■ SAMPLE SPACES AND EVENTS

Associated with outcomes of experiments are *sample spaces* and *events*. Our second step in constructing a mathematical model for probability studies is to define these two terms. Set concepts are useful in this regard.

Consider the experiment, ''A wheel with 18 numerals on the perimeter (Fig. 1) is spun and allowed to come to rest so that a pointer points within a numbered sector.''

What outcomes might we observe? When the wheel stops, we might be interested in which number is next to the pointer, or whether that number is an odd number, or whether that number is divisible by 5, or whether that number is prime, or whether the pointer is in a shaded or white sector, and so on. The list of possible outcomes appears endless. In general, there is no unique method of analyzing all possible outcomes of an experiment. Therefore, before conducting an experiment, it is important to decide just what outcomes are of interest.

Suppose we limit our interest to the set of numbers on the wheel and to various subsets of these numbers, such as the set of prime numbers on the wheel or the set of odd numbers. Having decided what to observe, we make a list of outcomes of the experiment, called *simple outcomes* or *simple events,* such that in each trial of the experiment (each spin of the wheel), one and only one of the outcomes on the list will occur. For our stated interests, we choose each number on the wheel as a simple event and form the set

$$S = \{1, 2, 3, \ldots, 17, 18\}$$

The set of simple events S for the experiment is called a *sample space* for the experiment.

Now consider the outcome, ''When the wheel comes to rest, the number next to the pointer is divisible by 4.'' This outcome is not a simple outcome (or simple event), since it is not associated with one and only one element in the sample space S. The outcome will occur whenever any one of the simple events 4, 8, 12, or 16 occurs; that is, whenever an element in the subset

$$E = \{4, 8, 12, 16\}$$

occurs. Subset E is called a *compound event* (and the outcome, a *compound outcome*).

In general:

Sample Spaces and Events

If we formulate a set S of outcomes (events) of an experiment in such a way that in each trial of the experiment one and only one of the outcomes (events) in the set will occur, then we call the set S a **sample space** for the experiment. Each element in S is called a **simple outcome,** or **simple event.**

An **event E** is defined to be any subset of S (including the empty set $\varnothing$ and the sample space S). Event E is a **simple event** if it contains only one element and a **compound event** if it contains more than one element. We say that **an event E occurs** if any of the simple events in E occurs.

We use the terms *event* and *outcome of an experiment* interchangeably. Technically, an event is the mathematical counterpart of an outcome of an experiment (as outlined at the top of the next page), but we will not insist on a strict adherence to this distinction in our development of probability.

Real World	Mathematical Model
Experiment (real or conceptual)	Sample space (set S)
Outcome (simple or compound)	Event (subset of S; simple or compound)

EXAMPLE 1 ➤ **Simple and Compound Events**   Relative to the number wheel experiment (see Fig. 1) and the sample space

$$S = \{1, 2, 3, \ldots , 17, 18\}$$

what is the event E (subset of the sample space S) that corresponds to each of the following outcomes? Indicate whether the event is a simple event or a compound event.

(A) The outcome is a prime number.   (B) The outcome is the square of 4.

Solution   (A) The outcome is a prime number if any of the simple events 2, 3, 5, 7, 11, 13, or 17 occurs.* Thus, to say the event "A prime number occurs" is the same as saying the experiment has an outcome in the set

$$E = \{2, 3, 5, 7, 11, 13, 17\}$$

Since event E has more than one element, it is a compound event.

(B) The outcome is the square of 4 if 16 occurs. Thus, to say the event "The square of 4 occurs" is the same as saying the experiment has an outcome in the set

$$E = \{16\}$$

Since E has only one element, it is a simple event.   ◄

MATCHED PROBLEM 1 ➤ Repeat Example 1 for:

(A) The outcome is a number divisible by 12.
(B) The outcome is an even number greater than 15.   ◄

EXAMPLE 2 ➤ **Sample Spaces**   A nickel and a dime are tossed. How shall we identify a sample space for this experiment? There are a number of possibilities, depending on our interest. We shall consider three.

(A) If we are interested in whether each coin falls heads (H) or tails (T), then, using a tree diagram, we can easily determine an appropriate sample space for the experiment:

Nickel Outcomes	Dime Outcomes	Combined Outcomes
H	H	HH
	T	HT
T	H	TH
	T	TT

---

* Technically, we should write {2}, {3}, {5}, {7}, {11}, {13}, and {17} for the simple events, since there is a logical distinction between an element of a set and a subset consisting of only that element. But we will just keep this in mind and drop the braces for simple events to simplify the notation.

Thus,

$$S_1 = \{HH, HT, TH, TT\}$$

and there are 4 simple events in the sample space.

(B) If we are interested only in the number of heads that appear on a single toss of the two coins, then we can let

$$S_2 = \{0, 1, 2\}$$

and there are 3 simple events in the sample space.

(C) If we are interested in whether the coins match (M) or do not match (D), then we can let

$$S_3 = \{M, D\}$$

and there are only 2 simple events in the sample space.   ◀

In Example 2, which sample space would be appropriate for all three interests? Sample space S_1 contains more information than either S_2 or S_3. If we know which outcome has occurred in S_1, then we know which outcome has occurred in S_2 and S_3. However, the reverse is not true. (Note that the simple events in S_2 and S_3 are compound events in S_1.) In this sense, we say that S_1 is a more **fundamental sample space** than either S_2 or S_3. Thus, we would choose S_1 as an appropriate sample space for all three expressed interests.

Choosing Sample Spaces

There is no one correct sample space for a given experiment. When specifying a sample space for an experiment, we include as much detail as is necessary to answer *all* questions of interest regarding the outcomes of the experiment. When in doubt, choose a sample space with more elements rather than fewer.

MATCHED PROBLEM 2 ➤ An experiment consists of recording the boy–girl composition of 2-child families.

(A) What is an appropriate sample space if we are interested in the sex of each child in the order of their births? Draw a tree diagram.
(B) What is an appropriate sample space if we are interested only in the number of girls in a family?
(C) What is an appropriate sample space if we are interested only in whether the sexes are alike (A) or different (D)?
(D) What is an appropriate sample space for all three interests expressed in parts (A)–(C)?   ◀

EXAMPLE 3 ➤ **Sample Spaces and Events**   Consider an experiment of rolling two dice. A convenient sample space that will enable us to answer many questions about interesting events is shown in Figure 2 (on the next page). Let S be the set of all ordered pairs in the figure. The simple event (3, 2) is to be distinguished from the simple event (2, 3). The former indicates that a 3 turned up on the first die and a 2 on the second, while the latter indicates that a 2 turned up on the first die and a 3 on the second.

SECOND DIE

	(1, 1)	(1, 2)	(1, 3)	(1, 4)	(1, 5)	(1, 6)
(2, 1)	(2, 2)	(2, 3)	(2, 4)	(2, 5)	(2, 6)
(3, 1)	(3, 2)	(3, 3)	(3, 4)	(3, 5)	(3, 6)
(4, 1)	(4, 2)	(4, 3)	(4, 4)	(4, 5)	(4, 6)
(5, 1)	(5, 2)	(5, 3)	(5, 4)	(5, 5)	(5, 6)
(6, 1)	(6, 2)	(6, 3)	(6, 4)	(6, 5)	(6, 6)

FIGURE 2

What is the event (subset of the sample space S) that corresponds to each of the following outcomes?

(A) A sum of 7 turns up.   (B) A sum of 11 turns up.
(C) A sum less than 4 turns up.   (D) A sum of 12 turns up.

Solution (A) By "A sum of 7 turns up," we mean that the sum of all dots on both turned-up faces is 7. This outcome corresponds to the event

$$\{(6, 1), (5, 2), (4, 3), (3, 4), (2, 5), (1, 6)\}$$

(B) "A sum of 11 turns up" corresponds to the event

$$\{(6, 5), (5, 6)\}$$

(C) "A sum less than 4 turns up" corresponds to the event

$$\{(1, 1), (2, 1), (1, 2)\}$$

(D) "A sum of 12 turns up" corresponds to the event

$$\{(6, 6)\}$$ ◄

MATCHED PROBLEM 3 ► Refer to the sample space shown in Figure 2. What is the event that corresponds to each of the following outcomes?

(A) A sum of 5 turns up.
(B) A sum that is a prime number greater than 7 turns up.   ◄

As indicated earlier, we often use the terms *event* and *outcome of an experiment* interchangeably. Thus, in Example 3 we might say "the event, 'A sum of 11 turns up'" in place of "the outcome, 'A sum of 11 turns up,'" or even write

$$E = \text{A sum of 11 turns up} = \{(6, 5), (5, 6)\}$$

■ PROBABILITY OF AN EVENT

The next step in developing our mathematical model for probability studies is the introduction of a *probability function*. This is a function that assigns to an arbitrary event associated with a sample space a real number between 0 and 1, inclusive. We start by discussing ways in which probabilities are assigned to simple events in the sample space S.

Probabilities for Simple Events

Given a sample space

$$S = \{e_1, e_2, \ldots, e_n\}$$

with n simple events, to each simple event e_i we assign a real number, denoted by $P(e_i)$, called the **probability of the event e_i.** These numbers can be assigned in an arbitrary manner as long as the following two conditions are satisfied:

1. The probability of a simple event is a number between 0 and 1, inclusive. That is,

   $$0 \leqslant P(e_i) \leqslant 1$$

2. The sum of the probabilities of all simple events in the sample space is 1. That is,

   $$P(e_1) + P(e_2) + \cdots + P(e_n) = 1$$

Any probability assignment that meets conditions 1 and 2 is said to be an **acceptable probability assignment.**

Our mathematical theory does not explain how acceptable probabilities are assigned to simple events. These assignments are generally based on the expected or actual percentage of times a simple event occurs when an experiment is repeated a large number of times. Assignments based on this principle are called **reasonable.**

Let an experiment be the flipping of a single coin, and let us choose a sample space S to be

$$S = \{H, T\}$$

If a coin appears to be fair, we are inclined to assign probabilities to the simple events in S as follows:

$$P(H) = \tfrac{1}{2} \quad \text{and} \quad P(T) = \tfrac{1}{2}$$

These assignments are based on reasoning that, since there are 2 ways a coin can land, in the long run, a head will turn up half the time and a tail will turn up half the time. These probability assignments are acceptable, since both conditions for acceptable probability assignments stated in the box are satisfied:

1. $0 \leqslant P(\text{H}) \leqslant 1, \qquad 0 \leqslant P(\text{T}) \leqslant 1$
2. $P(\text{H}) + P(\text{T}) = \frac{1}{2} + \frac{1}{2} = 1$

But there are other acceptable assignments. Maybe after flipping a coin 1,000 times, we find that the head turns up 376 times and the tails turns up 624 times. With this result, we might suspect that the coin is not fair and assign the simple events in the sample space S the following probabilities, based on our experimental results:

$$P(\text{H}) = .376 \qquad \text{and} \qquad P(\text{T}) = .624$$

This is also an acceptable assignment. However, the probability assignment

$$P(\text{H}) = 1 \qquad \text{and} \qquad P(\text{T}) = 0$$

though acceptable, is not reasonable (unless the coin has 2 heads). And the assignment

$$P(\text{H}) = .6 \qquad \text{and} \qquad P(\text{T}) = .8$$

is not acceptable, since $.6 + .8 = 1.4$, which violates condition 2 in the box. [*Note:* In probability studies, the 0 to the left of the decimal is usually omitted. Thus, we write .6 and .8 instead of 0.6 and 0.8.]

It is important to keep in mind that out of the infinitely many possible acceptable probability assignments to simple events in a sample space, we are generally inclined to choose one assignment over another based on reasoning or experimental results.

Given an acceptable probability assignment for simple events in a sample space S, how do we define the probability of an arbitrary event E associated with S?

Probability of an Event E

Given an acceptable probability assignment for the simple events in a sample space S, we define the **probability of an arbitrary event E,** denoted by $P(E)$, as follows:

(A) If E is the empty set, then $P(E) = 0$.
(B) If E is a simple event, then $P(E)$ has already been assigned.
(C) If E is a compound event, then $P(E)$ is the sum of the probabilities of all the simple events in E.
(D) If E is the sample space S, then $P(E) = P(S) = 1$ [this is a special case of part (C)].

EXAMPLE 4 ➤ **Probabilities of Events** Let us return to Example 2, the tossing of a nickel and a dime, and the sample space

$$S = \{\text{HH, HT, TH, TT}\}$$

Since there are 4 simple outcomes and the coins are assumed to be fair, it would appear that each outcome would occur 25% of the time, in the long run. Let us assign the same probability of $\frac{1}{4}$ to each simple event in S:

SIMPLE EVENT e_i	HH	HT	TH	TT
$P(e_i)$	$\frac{1}{4}$	$\frac{1}{4}$	$\frac{1}{4}$	$\frac{1}{4}$

This is an acceptable assignment according to conditions 1 and 2, and it is a reasonable assignment for ideal (perfectly balanced) coins or coins close to ideal.

(A) What is the probability of getting 1 head (and 1 tail)?
(B) What is the probability of getting at least 1 head?
(C) What is the probability of getting at least 1 head or at least 1 tail?
(D) What is the probability of getting 3 heads?

Solution (A) E_1 = Getting 1 head = {HT, TH}

Since E_1 is a compound event, we use part (C) in the box and find $P(E_1)$ by adding the probabilities of the simple events in E_1:

$$P(E_1) = P(\text{HT}) + P(\text{TH}) = \tfrac{1}{4} + \tfrac{1}{4} = \tfrac{1}{2}$$

(B) E_2 = Getting at least 1 head = {HH, HT, TH}

$$P(E_2) = P(\text{HH}) + P(\text{HT}) + P(\text{TH}) = \tfrac{1}{4} + \tfrac{1}{4} + \tfrac{1}{4} = \tfrac{3}{4}$$

(C) E_3 = {HH, HT, TH, TT} = S

$$P(E_3) = P(S) = 1 \qquad\qquad \tfrac{1}{4} + \tfrac{1}{4} + \tfrac{1}{4} + \tfrac{1}{4} = 1$$

(D) E_4 = Getting 3 heads = $\varnothing$ Empty set

$$P(\varnothing) = 0$$

◄

Steps for Finding the Probability of an Event E

Step 1. Set up an appropriate sample space S for the experiment.

Step 2. Assign acceptable probabilities to the simple events in S.

Step 3. To obtain the probability of an arbitrary event E, add the probabilities of the simple events in E.

The function P defined in steps 2 and 3 is a **probability function** whose domain is all possible events (subsets) in the sample space S and whose range is a set of real numbers between 0 and 1, inclusive.

MATCHED PROBLEM 4 ➤ Suppose in Example 4 that, after flipping the nickel and dime 1,000 times, we find that HH turns up 273 times, HT turns up 206 times, TH turns up 312 times, and TT turns up 209 times. On the basis of this evidence, we assign probabilities to simple events in S as follows:

SIMPLE EVENT e_i	HH	HT	TH	TT
$P(e_i)$	.273	.206	.312	.209

This is an acceptable and reasonable probability assignment for the simple events in S. What are the probabilities of the following events?

(A) E_1 = Getting at least 1 tail (B) E_2 = Getting 2 tails
(C) E_3 = Getting at least 1 head or at least 1 tail ◄

Example 4 and Matched Problem 4 illustrate two important ways in which acceptable and reasonable probability assignments are made for simple events in a sample space S. Each approach has its advantage in certain situations:

1. *Theoretical.* We use assumptions and a deductive reasoning process to assign probabilities to simple events. No experiments are actually conducted. This is what we did in Example 4.
2. *Empirical.* We assign probabilities to simple events based on the results of actual experiments. This is what we did in Matched Problem 4.

Empirical probability concepts are stated more precisely as follows: If we conduct an experiment n times and event E occurs with **frequency $f(E)$,** then the ratio $f(E)/n$ is called the **relative frequency** of the occurrence of event E in n trials. We define the **empirical probability** of E, denoted by $P(E)$, by the number (if it exists) that the relative frequency $f(E)/n$ approaches as n gets larger and larger. For any particular n, the relative frequency $f(E)/n$ is also called the **approximate empirical probability** of event E.

Empirical Probability Approximation

$$P(E) \approx \frac{\text{Frequency of occurrence of } E}{\text{Total number of trials}} = \frac{f(E)}{n}$$

(The larger n is, the better the approximation.)

For most of this section we will emphasize the theoretical approach. In the next section we will return to the empirical approach.

■ EQUALLY LIKELY ASSUMPTION

In tossing a nickel and a dime (Example 4), we assigned the same probability, $\frac{1}{4}$, to each simple event in the sample space $S = \{HH, HT, TH, TT\}$. By assigning the same probability to each simple event in S, we are actually making the assumption that each simple event is as likely to occur as any other. We refer to this as an **equally likely assumption.** In general, we have the following:

Probability of a Simple Event Under an Equally Likely Assumption

If, in a sample space

$$S = \{e_1, e_2, \ldots , e_n\}$$

with n elements, we assume each simple event e_i is as likely to occur as any other, then we assign the probability $1/n$ to each. That is,

$$P(e_i) = \frac{1}{n}$$

Under an equally likely assumption, we can develop a very useful formula for finding probabilities of arbitrary events associated with a sample space S. Consider the following example.

If a single die is rolled and we assume each face is as likely to come up as any other, then for the sample space

$$S = \{1, 2, 3, 4, 5, 6\}$$

we assign a probability of $\frac{1}{6}$ to each simple event, since there are 6 simple events. The probability of

$$E = \text{Rolling a prime number} = \{2, 3, 5\}$$

is

Number of elements in E
↓

$$P(E) = P(2) + P(3) + P(5) = \tfrac{1}{6} + \tfrac{1}{6} + \tfrac{1}{6} = \tfrac{3}{6} = \tfrac{1}{2}$$

↑
Number of elements in S

Thus, under the assumption that each simple event is as likely to occur as any other, the computation of the probability of the occurrence of any event E in a sample space S is the number of elements in E divided by the number of elements in S.

theorem 1

Probability of an Arbitrary Event Under an Equally Likely Assumption

If we assume each simple event in sample space S is as likely to occur as any other, then the probability of an arbitrary event E in S is given by

$$P(E) = \frac{\text{Number of elements in } E}{\text{Number of elements in } S} = \frac{n(E)}{n(S)}$$

EXAMPLE 5 ➤ **Probabilities and Equally Likely Assumptions** Let us again consider rolling two dice, and assume each simple event in the sample space shown in Figure 2 (page 414) is as likely as any other. Find the probabilities of the following events:

(A) $E_1 = $ A sum of 7 turns up (B) $E_2 = $ A sum of 11 turns up
(C) $E_3 = $ A sum less than 4 turns up (D) $E_4 = $ A sum of 12 turns up

Solution Referring to Figure 2 (page 414) and the results found in Example 3, we find:

(A) $P(E_1) = \dfrac{n(E_1)}{n(S)} = \dfrac{6}{36} = \dfrac{1}{6}$ $E_1 = \{(6, 1), (5, 2), (4, 3), (3, 4), (2, 5), (1, 6)\}$

(B) $P(E_2) = \dfrac{n(E_2)}{n(S)} = \dfrac{2}{36} = \dfrac{1}{18}$ $E_2 = \{(6, 5), (5, 6)\}$

(C) $P(E_3) = \dfrac{n(E_3)}{n(S)} = \dfrac{3}{36} = \dfrac{1}{12}$ $E_3 = \{(1, 1), (2, 1), (1, 2)\}$

(D) $P(E_4) = \dfrac{n(E_4)}{n(S)} = \dfrac{1}{36}$ $E_4 = \{(6, 6)\}$ ◀

MATCHED PROBLEM 5 ➤ Under the conditions in Example 5, find the probabilities of the following events (each event refers to the sum of the dots facing up on both dice):

(A) $E_5 = $ A sum of 5 turns up
(B) $E_6 = $ A sum that is a prime number greater than 7 turns up   ◀

explore – discuss 1

A shipment box contains 12 graphing calculators, out of which 2 are defective. A calculator is drawn at random from the box and then, without replacement, a second calculator is drawn. Discuss whether the equally likely assumption would be appropriate for the sample space $S = \{GG, GD, DG, DD\}$, where G is a good calculator and D is a defective one.

We now turn to some examples that make use of the counting techniques developed in the preceding sections.

EXAMPLE 6 ➤   **Probability and Equally Likely Assumption** In drawing 5 cards from a 52-card deck without replacement, what is the probability of getting 5 spades?

Solution   Let the sample space S be the set of all 5-card hands from a 52-card deck. Since the order in a hand does not matter, $n(S) = C_{52,5}$. Let event E be the set of all 5-card hands from 13 spades. Again, the order does not matter and $n(E) = C_{13,5}$. Thus, assuming each 5-card hand is as likely as any other,

$$P(E) = \frac{n(E)}{n(S)} = \frac{C_{13,5}}{C_{52,5}} = \frac{1,287}{2,598,960} \approx .0005$$

MATCHED PROBLEM 6 ➤   In drawing 7 cards from a 52-card deck without replacement, what is the probability of getting 7 hearts?

EXAMPLE 7 ➤   **Probability and Equally Likely Assumption** The board of regents of a university is made up of 12 men and 16 women. If a committee of 6 is chosen at random, what is the probability that it will contain 3 men and 3 women?

Solution   Let S be the set of all 6-person committees out of 28 people. Then

$$n(S) = C_{28,6}$$

Let E be the set of all 6-person committees with 3 men and 3 women. To find $n(E)$, we use the multiplication principle and the following two operations:

O_1: Select 3 men out of the 12 available N_1: $C_{12,3}$
O_2: Select 3 women out of the 16 available N_2: $C_{16,3}$

Thus,

$$n(E) = N_1 \cdot N_2 = C_{12,3} \cdot C_{16,3}$$

and

$$P(E) = \frac{n(E)}{n(S)} = \frac{C_{12,3} \cdot C_{16,3}}{C_{28,6}} \approx .327$$

MATCHED PROBLEM 7 ➤   What is the probability that the committee in Example 7 will have 4 men and 2 women?

It needs to be pointed out that there are many counting problems for which it is not possible to produce a simple formula that will yield the number of possible cases. In situations of this type, we often revert back to tree diagrams and counting branches.

explore – discuss 2

Five football teams play each other in the Central Division. At the beginning of the season we are interested in discussing what will be the relative standing of the five teams at the end of the season, excluding ties. Describe an appropriate sample space for this discussion. How many simple events does it contain? If each possible final standing of the five teams is as likely as any other, what would be the probability of the occurrence of a particular final standing? Discuss the reasons that a sports writer would or would not use the equally likely assumption in predicting the final standing of the five teams.

Answers to Matched Problems

1. (A) $E = \{12\}$; simple event (B) $E = \{16, 18\}$; compound event

2. (A) $S_1 = \{BB, BG, GB, GG\}$;
   (B) $S_2 = \{0, 1, 2\}$
   (C) $S_3 = \{A, D\}$
   (D) S_1

Sex of First Child	Sex of Second Child	Combined Outcomes
B	B	BB
	G	BG
G	B	GB
	G	GG

3. (A) $\{(4, 1), (3, 2), (2, 3), (1, 4)\}$ (B) $\{(6, 5), (5, 6)\}$
4. (A) .727 (B) .209 (C) 1 5. (A) $P(E_5) = \frac{1}{9}$ (B) $P(E_6) = \frac{1}{18}$
6. $C_{13,7}/C_{52,7} \approx 1.3 \times 10^{-5}$ 7. $C_{12,4}C_{16,2}/C_{28,6} \approx .158$

EXERCISE 6-3

A

1. How would you interpret $P(E) = 1$?

2. How would you interpret $P(E) = 0$?

3. In a family with 2 children, excluding multiple births, what is the probability of having 2 children of the opposite sex? Assume a girl is as likely as a boy at each birth.

4. In a family with 2 children, excluding multiple births, what is the probability of having 2 girls? Assume a girl is as likely as a boy at each birth.

5. A store carries four brands of CD players, J, G, P, and S. From past records, the manager found that the relative frequency of brand choice among customers varied. Which of the following probability assignments for a particular customer choosing a particular brand of CD player would have to be rejected? Why?

(A) $P(J) = .15$, $P(G) = -.35$, $P(P) = .50$, $P(S) = .70$
(B) $P(J) = .32$, $P(G) = .28$, $P(P) = .24$, $P(S) = .30$
(C) $P(J) = .26$, $P(G) = .14$, $P(P) = .30$, $P(S) = .30$

6. Using the probability assignments in Problem 5(C), what is the probability that a random customer will not choose brand S?

7. Using the probability assignments in Problem 5(C), what is the probability that a random customer will choose brand J or brand P?

8. Using the probability assignments in Problem 5(C), what is the probability that a random customer will not choose brand J or brand P?

B

9. In a family with 3 children, excluding multiple births, what is the probability of having 2 boys and 1 girl, in that order? Assume a boy is as likely as a girl at each birth.

10. In a family with 3 children, excluding multiple births, what is the probability of having 2 boys and 1 girl, in any order? Assume a boy is as likely as a girl at each birth.

11. A small combination lock on a suitcase has 3 wheels, each labeled with the 10 digits from 0 to 9. If an opening combination is a particular sequence of 3 digits with no repeats, what is the probability of a person guessing the right combination?

12. A combination lock has 5 wheels, each labeled with the 10 digits from 0 to 9. If an opening combination is a particular sequence of 5 digits with no repeats, what is the probability of a person guessing the right combination?

Refer to the description of a standard deck of 52 cards and Figure 2 on page 405. An experiment consists of dealing 5 cards from a standard 52-card deck. In Problems 13–16, what is the probability of being dealt:

13. 5 black cards?   **14.** 5 hearts?

15. 5 face cards?   **16.** 5 nonface cards?

17. Twenty thousand students are enrolled at a state university. A student is selected at random and his or her birthday (month and day, not year) is recorded. Describe an appropriate sample space for this experiment, and assign acceptable probabilities to the simple events. What are your assumptions in making this assignment?

18. In a hotly contested three-way race for the U.S. Senate, polls indicate the two leading candidates are running neck-and-neck while the third candidate is receiving half the support of either of the others. Registered voters are chosen at random and are asked which of the three will get their vote. Describe an appropriate sample space for this random survey experiment and assign acceptable probabilities to the simple events.

19. Suppose 5 thank-you notes are written and 5 envelopes are addressed. Accidentally, the notes are randomly inserted into the envelopes and mailed without checking the addresses. What is the probability that all the notes will be inserted into the correct envelopes?

20. Suppose 6 people check their coats in a checkroom. If all claim checks are lost and the 6 coats are randomly returned, what is the probability that all the people will get their own coats back?

An experiment consists of rolling two fair dice and adding the dots on the two sides facing up. Using the sample space shown in Figure 2 (page 414) and assuming each simple event is as likely as any other, find the probability of the sum of the dots indicated in Problems 21–36:

21. Sum is 2.   **22.** Sum is 10.

23. Sum is 6.   **24.** Sum is 8.

25. Sum is less than 5.   **26.** Sum is greater than 8.

27. Sum is not 7 or 11.   **28.** Sum is not 2, 4, or 6.

29. Sum is 1.   **30.** Sum is 13.

31. Sum is divisible by 3.   **32.** Sum is divisible by 4.

33. Sum is 7 or 11 (a "natural").

34. Sum is 2, 3, or 12 ("craps").

35. Sum is divisible by 2 or 3.

36. Sum is divisible by 2 and 3.

An experiment consists of tossing three fair (not weighted) coins, except one of the three coins has a head on both sides. Compute the probability of obtaining the indicated results in Problems 37–42.

37. 1 head   **38.** 2 heads

39. 3 heads   **40.** 0 heads

41. More than 1 head   **42.** More than 1 tail

C *An experiment consists of rolling two fair (not weighted) dice and adding the dots on the two sides facing up. Each die has the number 1 on two opposite faces, the number 2 on two opposite faces, and the number 3 on two opposite faces. Compute the probability of obtaining the indicated sums in Problems 43–50.*

43. 2   **44.** 3   **45.** 4

46. 5   **47.** 6   **48.** 7

49. An odd sum   **50.** An even sum

Refer to the description of a standard deck of 52 cards on page 405. An experiment consists of dealing 5 cards from a standard 52-card deck. In Problems 51–58, what is the probability of being dealt:

51. 5 face cards or aces?

52. 5 numbered cards (2 through 10)?

53. 4 aces?

54. Four of a kind (4 queens, 4 kings, and so on)?

55. A 10, jack, queen, king, and ace, all in the same suit?

56. A 2, 3, 4, 5, and 6, all in the same suit?

57. 2 aces and 3 queens?

58. 2 kings and 3 aces?

Business & Economics

59. *Consumer testing.* Twelve popular brands of beer are to be used in a blind taste study for consumer recognition.
- (A) If 4 distinct brands are chosen at random from the 12 and if a consumer is not allowed to repeat any answers, what is the probability that all 4 brands could be identified by just guessing?
- (B) If repeats are allowed in the 4 brands chosen at random from the 12 and if the consumer is allowed to repeat answers, what is the probability of correct identification of all 4 by just guessing?

60. *Consumer testing.* Six popular brands of cola are to be used in a blind taste study for consumer recognition.
- (A) If 3 distinct brands are chosen at random from the 6 and if a consumer is not allowed to repeat any answers, what is the probability that all 3 brands could be identified by just guessing?
- (B) If repeats are allowed in the 3 brands chosen at random from the 6 and if the consumer is allowed to repeat answers, what is the probability of correct identification of all 3 by just guessing?

61. *Personnel selection.* Suppose 6 female and 5 male applicants have been successfully screened for 5 positions. If the 5 positions are selected at random from the 11 finalists, what is the probability of selecting:
- (A) 3 females and 2 males?
- (B) 4 females and 1 male?
- (C) 5 females?
- (D) At least 4 females?

62. *Committee selection.* A 4-person grievance committee is to be composed of employees in 2 departments, A and B, with 15 and 20 employees, respectively. If the 4 people are se-lected at random from the 35 employees, what is the probability of selecting:
- (A) 3 from A and 1 from B?
- (B) 2 from A and 2 from B?
- (C) All from A?
- (D) At least 3 from A?

Life Sciences

63. *Medicine.* A prospective laboratory technician is to be tested on identifying blood types from 8 standard classifications.
- (A) If 3 distinct samples are chosen at random from the 8 types and if the examinee is not allowed to repeat any answers, what is the probability that all 3 could be correctly identified by just guessing?
- (B) If repeats are allowed in the 3 blood types chosen at random from the 8 and if the examinee is allowed to repeat answers, what is the probability of correct identification of all 3 by just guessing?

64. *Medical research.* Because of limited funds, 5 research centers are to be chosen out of 8 suitable ones for a study on heart disease. If the selection is made at random, what is the probability that 5 particular research centers will be chosen?

Social Sciences

65. *Membership selection.* A town council has 11 members, 6 Democrats and 5 Republicans.
- (A) If the president and vice president are selected at random, what is the probability that they are both Democrats?
- (B) If a 3-person committee is selected at random, what is the probability that Republicans make up the majority?

S E C T I O N   6-4   ■   # Union, Intersection, and Complement of Events; Odds

- ■ UNION AND INTERSECTION
- ■ COMPLEMENT OF AN EVENT
- ■ ODDS
- ■ APPLICATIONS TO EMPIRICAL PROBABILITY

Recall that in Section 6-3 we said that given a sample space

$$S = \{e_1, e_2, \ldots, e_n\}$$

any function P defined on S such that

$$0 \leqslant P(e_i) \leqslant 1 \qquad i = 1, 2, \ldots, n$$

and

$$P(e_1) + P(e_2) + \cdots + P(e_n) = 1$$

is called a *probability function*. In addition, we said that any subset of S is called an *event E*, and we defined the probability of E to be the sum of the probabilities of the simple events in E.

■ UNION AND INTERSECTION

Since events are subsets of a sample space, the union and intersection of events are simply the union and intersection of sets as defined in the following box. In this section we will concentrate on the union of events and consider only simple cases of intersection. The latter will be investigated in more detail in the next section.

Union and Intersection of Events

If A and B are two events in a sample space S, then the **union** of A and B, denoted by $A \cup B$, and the **intersection** of A and B, denoted by $A \cap B$, are defined as follows:

$$A \cup B = \{e \in S | e \in A \text{ or } e \in B\} \qquad A \cap B = \{e \in S | e \in A \text{ and } e \in B\}$$

$A \cup B$

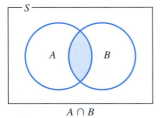
$A \cap B$

Furthermore, we define:

The **event A or B** to be $A \cup B$
The **event A and B** to be $A \cap B$

EXAMPLE 1 ➤ **Probability Involving Union and Intersection** Consider the sample space of equally likely events for the rolling of a single fair die:

$$S = \{1, 2, 3, 4, 5, 6\}$$

(A) What is the probability of rolling a number that is odd **and** exactly divisible by 3?

(B) What is the probability of rolling a number that is odd **or** exactly divisible by 3?

Solution

(A) Let A be the event of rolling an odd number, B the event of rolling a number divisible by 3, and F the event of rolling a number that is odd **and** divisible by 3. Then

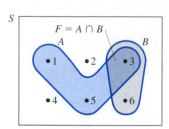

$$A = \{1, 3, 5\} \qquad B = \{3, 6\} \qquad F = A \cap B = \{3\}$$

Thus, the probability of rolling a number that is odd **and** exactly divisible by 3 is

$$P(F) = P(A \cap B) = \frac{n(A \cap B)}{n(S)} = \frac{1}{6}$$

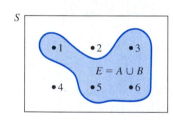

(B) Let A and B be the same events as in part (A), and let E be the event of rolling a number that is odd **or** divisible by 3. Then

$$A = \{1, 3, 5\} \qquad B = \{3, 6\} \qquad E = A \cup B = \{1, 3, 5, 6\}$$

Thus, the probability of rolling a number that is odd **or** exactly divisible by 3 is

$$P(E) = P(A \cup B) = \frac{n(A \cup B)}{n(E)} = \frac{4}{6} = \frac{2}{3} \qquad ◀$$

MATCHED PROBLEM 1 ➤ Use the sample space in Example 1 to answer the following:

(A) What is the probability of rolling an odd number **and** a prime number?

(B) What is the probability of rolling an odd number **or** a prime number? ◀

Suppose

$$E = A \cup B$$

Can we find $P(E)$ in terms of A and B? The answer is almost yes, but we must be careful. It would be nice if

$$P(A \cup B) = P(A) + P(B) \tag{1}$$

This turns out to be true if events A and B are **mutually exclusive (disjoint);** that is, if $A \cap B = \emptyset$ (Fig. 1). In this case, $P(A \cup B)$ is the sum of all the probabilities of simple events in A added to the sum of all the probabilities of simple events in B. But what happens if events A and B are not mutually exclusive; that is, if $A \cap B \neq \emptyset$ (see Fig. 2)? If we simply add the probabilities of the elements in A to the probabili-

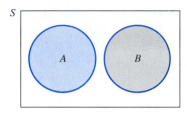

FIGURE 1
Mutually exclusive: $A \cap B = \varnothing$

FIGURE 2
Not mutually exclusive: $A \cap B \neq \varnothing$

ties of the elements in B, we are including some of the probabilities twice, namely those for elements that are in both A and B. To compensate for this double counting, we subtract $P(A \cap B)$ from equation (1) to obtain

$$P(A \cup B) = P(A) + P(B) - P(A \cap B) \tag{2}$$

We notice that (2) holds for both cases, $A \cap B \neq \varnothing$ and $A \cap B = \varnothing$, since (2) reduces to (1) for the latter case $[P(A \cap B) = P(\varnothing) = 0]$. It is better to use (2) if there is any doubt that A and B are mutually exclusive. We summarize this discussion in the box for convenient reference.

Probability of a Union of Two Events

For any events A and B,

$$P(A \cup B) = P(A) + P(B) - P(A \cap B) \tag{2}$$

If A and B are mutually exclusive, then

$$P(A \cup B) = P(A) + P(B) \tag{1}$$

EXAMPLE 2 ➤ Probability Involving Union and Intersection   Suppose two fair dice are rolled:

(A) What is the probability that a sum of 7 or 11 turns up?
(B) What is the probability that both dice turn up the same or that a sum less than 5 turns up?

Solution   (A) If A is the event that a sum of 7 turns up and B is the event that a sum of 11 turns up, then (see Fig. 3, page 428) the event that a sum of 7 or 11 turns up is $A \cup B$, where

$$A = \{(1, 6), (2, 5), (3, 4), (4, 3), (5, 2), (6, 1)\}$$
$$B = \{(5, 6), (6, 5)\}$$

SECOND DIE

	(1, 1)	(1, 2)	(1, 3)	(1, 4)	(1, 5)	(1, 6)
	(2, 1)	(2, 2)	(2, 3)	(2, 4)	(2, 5)	(2, 6)
FIRST DIE	(3, 1)	(3, 2)	(3, 3)	(3, 4)	(3, 5)	(3, 6)
	(4, 1)	(4, 2)	(4, 3)	(4, 4)	(4, 5)	(4, 6)
	(5, 1)	(5, 2)	(5, 3)	(5, 4)	(5, 5)	(5, 6)
	(6, 1)	(6, 2)	(6, 3)	(6, 4)	(6, 5)	(6, 6)

A

B

FIGURE 3

Since events A and B are mutually exclusive, we can use equation (1) to calculate $P(A \cup B)$:

$$P(A \cup B) = P(A) + P(B)$$
$$= \tfrac{6}{36} + \tfrac{2}{36}$$
$$= \tfrac{8}{36} = \tfrac{2}{9}$$

In this equally likely sample space, $n(A) = 6$, $n(B) = 2$, and $n(S) = 36$.

(B) If A is the event that both dice turn up the same and B is the event that the sum is less than 5, then (see Fig. 4) the event that both dice turn up the same or the sum is less than 5 is $A \cup B$, where

$$A = \{(1, 1), (2, 2), (3, 3), (4, 4), (5, 5), (6, 6)\}$$
$$B = \{(1, 1), (1, 2), (1, 3), (2, 1), (2, 2), (3, 1)\}$$

SECOND DIE

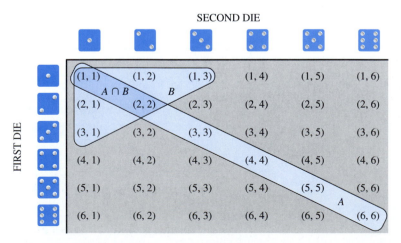

FIGURE 4

Since $A \cap B = \{(1, 1), (2, 2)\}$, A and B are not mutually exclusive and we use equation (2) to calculate $P(A \cup B)$:

$$P(A \cup B) = P(A) + P(B) - P(A \cap B)$$
$$= \tfrac{6}{36} + \tfrac{6}{36} - \tfrac{2}{36}$$
$$= \tfrac{10}{36} = \tfrac{5}{18}$$

◀

MATCHED PROBLEM 2 ➤ Use the sample space in Example 2 to answer the following:

(A) What is the probability that a sum of 2 or 3 turns up?
(B) What is the probability that both dice turn up the same or that a sum greater than 8 turns up?

◀

You no doubt noticed in Example 2 that we actually did not have to use either formula (1) or (2). We could have proceeded as in Example 1 and simply counted sample points in $A \cup B$. The following example illustrates the use of formula (2) in a situation where visual representation of sample points is not practical.

EXAMPLE 3 ➤ **Probability Involving Union and Intersection** What is the probability that a number selected at random from the first 500 positive integers is (exactly) divisible by 3 or 4?

Solution Let A be the event that a drawn integer is divisible by 3 and B the event that a drawn integer is divisible by 4. Note that events A and B are not mutually exclusive, because multiples of 12 are divisible by both 3 and 4. Since each of the positive integers from 1 to 500 is as likely to be drawn as any other, we can use $n(A)$, $n(B)$, and $n(A \cap B)$ to determine $P(A \cup B)$, where (think about this)

$$n(A) = \text{The largest integer less than or equal to } \tfrac{500}{3} = 166$$
$$n(B) = \text{The largest integer less than or equal to } \tfrac{500}{4} = 125$$
$$n(A \cap B) = \text{The largest integer less than or equal to } \tfrac{500}{12} = 41$$

Now we can compute $P(A \cup B)$:

$$P(A \cup B) = P(A) + P(B) - P(A \cap B)$$
$$= \frac{n(A)}{n(S)} + \frac{n(B)}{n(S)} - \frac{n(A \cap B)}{n(S)}$$
$$= \frac{166}{500} + \frac{125}{500} - \frac{41}{500} = \frac{250}{500} = .5$$

◀

MATCHED PROBLEM 3 ➤ What is the probability that a number selected at random from the first 140 positive integers is (exactly) divisible by 4 or 6?

◀

■ COMPLEMENT OF AN EVENT

Suppose we divide a finite sample space

$$S = \{e_1, \ldots, e_n\}$$

into two subsets E and E' such that

$$E \cap E' = \emptyset$$

that is, E and E' are mutually exclusive, and

$$E \cup E' = S$$

Then E' is called the **complement of E** relative to S. Thus, E' contains all the elements of S that are not in E (Fig. 5). Furthermore,

$$P(S) = P(E \cup E')$$
$$= P(E) + P(E') = 1$$

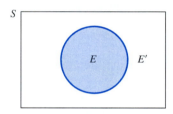

FIGURE 5

Hence, we have:

Complements

$$P(E) = 1 - P(E') \qquad P(E') = 1 - P(E) \tag{3}$$

If the probability of rain is .67, then the probability of no rain is $1 - .67 = .33$; if the probability of striking oil is .01, then the probability of not striking oil is .99. If the probability of having at least 1 boy in a 2-child family is .75, what is the probability of having 2 girls? [*Answer:* .25.]

e x p l o r e – d i s c u s s 1

(A) Suppose that E and F are complementary events. Are E and F necessarily mutually exclusive? Explain why or why not.

(B) Suppose that E and F are mutually exclusive events. Are E and F necessarily complementary? Explain why or why not.

In looking for $P(E)$, there are situations in which it is easier to find $P(E')$ first, and then use equations (3) to find $P(E)$. The next two examples illustrate two such situations.

EXAMPLE 4 ➤   **Quality Control**   A shipment of 45 precision parts, including 9 that are defective, is sent to an assembly plant. The quality control division selects 10 at random for testing and rejects the whole shipment if 1 or more in the sample are found defective. What is the probability that the shipment will be rejected?

Solution   If E is the event that 1 or more parts in a random sample of 10 are defective, then E', the complement of E, is the event that no parts in a random sample of 10 are defective. It is easier to compute $P(E')$ than to compute $P(E)$ directly (try the latter to see why). Once $P(E')$ is found, we will use $P(E) = 1 - P(E')$ to find $P(E)$.

The sample space S for this experiment is the set of all subsets of 10 elements from the set of 45 parts shipped. Thus, since there are $45 - 9 = 36$ nondefective parts,

$$P(E') = \frac{n(E')}{n(S)} = \frac{C_{36,10}}{C_{45,10}} \approx .08$$

and

$$P(E) = 1 - P(E') \approx 1 - .08 = .92 \qquad \blacktriangleleft$$

MATCHED PROBLEM 4 ➤   A shipment of 40 precision parts, including 8 that are defective, is sent to an assembly plant. The quality control division selects 10 at random for testing and rejects the whole shipment if 1 or more in the sample are found defective. What is the probability that the shipment will be rejected?   ◀

EXAMPLE 5 ➤   **Birthday Problem**   In a group of n people, what is the probability that at least 2 people have the same birthday (the same month and day, excluding February 29)? (Make a guess for a class of 40 people, and check your guess with the conclusion of this example.)

Solution   If we form a list of the birthdays of all the people in the group, then we have a simple event in the sample space

$$S = \text{Set of all lists of } n \text{ birthdays}$$

For any person in the group, we will assume that any birthday is as likely as any other, so that the simple events in S are equally likely. How many simple events are in the set S? Since any person could have any one of 365 birthdays (excluding

February 29), the multiplication principle implies that the number of simple events in S is

$$
\begin{array}{ccccc}
\text{1st} & \text{2nd} & \text{3rd} & & \text{nth} \\
\text{person} & \text{person} & \text{person} & & \text{person}
\end{array}
$$

$$
\begin{aligned}
n(S) &= 365 \cdot 365 \cdot 365 \cdots \cdots 365 \\
&= 365^n
\end{aligned}
$$

Now, let E be the event that at least 2 people in the group have the same birthday. Then E' is the event that no 2 people have the same birthday. The multiplication principle also can be used to determine the number of simple events in E':

$$
\begin{array}{ccccc}
\text{1st} & \text{2nd} & \text{3rd} & & \text{nth} \\
\text{person} & \text{person} & \text{person} & & \text{person}
\end{array}
$$

$$
\begin{aligned}
n(E') &= 365 \cdot 364 \cdot 363 \cdots \cdots (366 - n) \\
&= \frac{[365 \cdot 364 \cdot 363 \cdots \cdots (366 - n)](365 - n)!}{(365 - n)!} \\
&= \frac{365!}{(365 - n)!}
\end{aligned}
$$

Multiply numerator and denominator by $(365 - n)!$

Since we have assumed that S is an equally likely sample space,

$$
P(E') = \frac{n(E')}{n(S)} = \frac{\dfrac{365!}{(365 - n)!}}{365^n} = \frac{365!}{365^n(365 - n)!}
$$

Thus,

$$
\begin{aligned}
P(E) &= 1 - P(E') \\
&= 1 - \frac{365!}{365^n(365 - n)!}
\end{aligned}
\tag{4}
$$

Equation (4) is valid for any n satisfying $1 \leq n \leq 365$. [What is $P(E)$ if $n > 365$?] For example, in a group of 6 people,

$$
\begin{aligned}
P(E) &= 1 - \frac{365!}{(365)^6 359!} \\
&= 1 - \frac{365 \cdot 364 \cdot 363 \cdot 362 \cdot 361 \cdot 360 \cdot 359!}{365 \cdot 365 \cdot 365 \cdot 365 \cdot 365 \cdot 365 \cdot 359!} \\
&= .04
\end{aligned}
$$

It is interesting to note that as the size of the group increases, $P(E)$ increases more rapidly than you might expect. Table 1 gives the value of $P(E)$ for selected values of n. Notice that for a group of only 23 people, the probability that 2 or more have the same birthday is greater than $\frac{1}{2}$. ◄

TABLE 1
The Birthday Problem

NUMBER OF PEOPLE IN GROUP n	PROBABILITY THAT 2 OR MORE HAVE SAME BIRTHDAY $P(E)$
5	.027
10	.117
15	.253
20	.411
23	**.507**
30	.706
40	.891
50	.970
60	.994
70	.999

MATCHED PROBLEM 5 ► Use equation (4) to evaluate $P(E)$ for $n = 4$. ◄

e x p l o r e – d i s c u s s   2

> Determine the smallest number n such that in a group of n people the probability that 2 or more have a birthday in the same month is greater than .5. Discuss the assumptions underlying your computation.

■ ODDS

When the probability of an event E is known, it is often customary (particularly in gaming situations) to speak of *odds* for (or against) the event E, rather than the *probability* of the event E. The relationship between these two designations is outlined below:

From Probability to Odds

If $P(E)$ is the probability of the event E, then we define:

(A) Odds for $E = \dfrac{P(E)}{1 - P(E)} = \dfrac{P(E)}{P(E')}$ $P(E) \neq 1$

(B) Odds against $E = \dfrac{P(E')}{P(E)}$ $P(E) \neq 0$

[*Note:* When possible, odds are expressed as ratios of whole numbers.]

E X A M P L E   6 ➤ **Probability and Odds**   If you roll a fair die once, the probability of rolling a 4 is $\frac{1}{6}$, whereas the odds in favor of rolling a 4 are

$$\frac{P(E)}{P(E')} = \frac{\frac{1}{6}}{\frac{5}{6}} = \frac{1}{5} \quad \text{Read as "1 to 5" and also written as "1:5"}$$

and

$$\text{Odds against rolling a 4} = \frac{5}{1}$$

In terms of a fair game, if you bet $1 on a 4 turning up, you would lose $1 to the house if any number other than 4 turns up and you would be paid $5 by the house (and in addition your bet of $1 returned) if a 4 turns up. (An experienced gambler would say that the house pays 5 to 1 on a 4 turning up on a single roll of a die.) ◄

In general, if the odds for an event E are $a:b$, then **a game is fair** if your bet of $a is lost if event E does not happen, but you win $b (and in addition your bet of $a is returned) if event E does happen. (Gamblers would say that the house pays b to a on event E happening.)

MATCHED PROBLEM 6 ➤   (A) What are the odds for rolling a sum of 8 in a single roll of two fair dice?
(B) If you bet $5 that a sum of 8 will turn up, what should the house pay (plus returning your $5 bet) if a sum of 8 does turn up for the game to be fair? ◄

Now we will go in the other direction: If we are given the odds for an event, what is the probability of the event? (The verification of the following formula is left to Problem 53 in Exercise 6-4.)

From Odds to Probability

If the odds for event E are a/b, then the probability of E is

$$P(E) = \frac{a}{a + b}$$

EXAMPLE 7 ➤   **Odds and Probability**   If in repeated rolls of two fair dice the odds for rolling a 5 before rolling a 7 are 2 to 3, then the probability of rolling a 5 before rolling a 7 is

$$P(E) = \frac{a}{a + b} = \frac{2}{2 + 3} = \frac{2}{5}$$ ◄

MATCHED PROBLEM 7 ➤   If in repeated rolls of two fair dice the odds against rolling a 6 before rolling a 7 are 6 to 5, then what is the probability of rolling a 6 before rolling a 7? (Be careful! Read the problem again.) ◄

■ APPLICATIONS TO EMPIRICAL PROBABILITY

In the following discussions, the term *empirical probability* will mean the probability of an event determined by a sample that is used to approximate the probability of the corresponding event in the total population. How does the approximate empirical probability of an event determined from a sample relate to the actual probability of an event relative to the total population? In mathematical statistics an important theorem called the **law of large numbers** (or the **law of averages**) is proved. Informally, it states that the approximate empirical probability can be made as close to the actual probability as we please by making the sample sufficiently large.

EXAMPLE 8 ➤   **Market Research**   From a survey involving 1,000 people in a certain city, it was found that 500 people had tried a certain brand of diet cola, 600 had tried a certain brand of regular cola, and 200 had tried both types of cola. If a resident of the city is selected at random, what is the (empirical) probability that:

(A) The resident has tried the diet or the regular cola? What are the (empirical) odds for this event?
(B) The resident has tried one of the colas but not both? What are the (empirical) odds against this event?

Solution

Let D be the event that a person has tried the diet cola and R the event that a person has tried the regular cola. The events D and R can be used to partition the residents of the city into four mutually exclusive subsets (a collection of subsets is mutually exclusive if the intersection of any two of them is the empty set):

$$D \cap R = \text{Set of people who have tried both colas}$$
$$D \cap R' = \text{Set of people who have tried the diet cola but not the regular cola}$$
$$D' \cap R = \text{Set of people who have tried the regular cola but not the diet cola}$$
$$D' \cap R' = \text{Set of people who have not tried either cola}$$

These sets are displayed in the Venn diagram in Figure 6.
    The sample population of 1,000 residents is also partitioned into four mutually exclusive sets, with $n(D) = 500$, $n(R) = 600$, and $n(D \cap R) = 200$. By using a Venn diagram (Fig. 7), we can determine the number of sample points in the sets $D \cap R'$, $D' \cap R$, and $D' \cap R'$ (see Example 2 in Section 6-1).

FIGURE 6
Total population

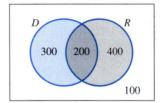

FIGURE 7
Sample population

These frequencies can be conveniently displayed in a table:

		REGULAR R	NO REGULAR R'	Totals
DIET	D	200	300	500
NO DIET	D'	400	100	500
	Totals	600	400	1,000

Assuming that each sample point is equally likely, we form a probability table by dividing each entry in this table by 1,000, the total number surveyed. These are theoretical probabilities for the sample population, which we can use as empirical probabilities to approximate the corresponding probabilities for the total population.

		REGULAR R	NO REGULAR R'	Totals
DIET	D	.2	.3	.5
NO DIET	D'	.4	.1	.5
Totals		.6	.4	1.0

Now we are ready to compute the required probabilities.

(A) The event that a person has tried the diet or the regular cola is $E = D \cup R$. We compute $P(E)$ two ways.

Method 1.   Directly:

$$P(E) = P(D \cup R)$$
$$= P(D) + P(R) - P(D \cap R)$$
$$= .5 + .6 - .2 = .9$$

Method 2.   Using the complement of E:

$$P(E) = 1 - P(E')$$
$$= 1 - P(D' \cap R') \quad E' = (D \cup R)' = D' \cap R' \text{ (see Fig. 6)}$$
$$= 1 - .1 = .9$$

In either case,

$$\text{Odds for } E = \frac{P(E)}{P(E')} = \frac{.9}{.1} = \frac{9}{1} \quad \text{or } 9:1$$

(B) The event that a person has tried one cola but not both is the event that the person has tried diet and not regular cola or has tried regular and not diet cola. In terms of sets, this is event $E = (D \cap R') \cup (D' \cap R)$. Since $D \cap R'$ and $D' \cap R$ are mutually exclusive (look at the Venn diagram in Fig. 6),

$$P(E) = P[(D \cap R') \cup (D' \cap R)]$$
$$= P(D \cap R') + P(D' \cap R)$$
$$= .3 + .4 = .7$$

$$\text{Odds against } E = \frac{P(E')}{P(E)} = \frac{.3}{.7} = \frac{3}{7} \quad \text{or } 3:7$$

◀

MATCHED PROBLEM 8 ▶   If a resident from the city in Example 8 is selected at random, what is the (empirical) probability that:

(A) The resident has not tried either cola? What are the (empirical) odds for this event?

(B) The resident has tried the diet cola or has not tried the regular cola? What are the (empirical) odds against this event? ◀

Answers to Matched Problems   1. (A) $\frac{1}{3}$ (B) $\frac{2}{3}$ 2. (A) $\frac{1}{12}$ (B) $\frac{7}{18}$ 3. $\frac{47}{140} \approx .336$ 4. .92
5. .016   6. (A) 5:31   (B) $31   7. $\frac{5}{11} \approx .455$
8. (A) $P(D' \cap R') = .1$; odds for $D' \cap R' = \frac{1}{9}$ or 1:9
(B) $P(D \cup R') = .6$; odds against $D \cup R' = \frac{2}{3}$ or 2:3

EXERCISE 6-4

A

1. If a manufactured item has the probability of .003 of failing within 90 days, what is the probability that the item will not fail in that time period?

2. In a particular cross of two plants the probability that the flowers will be red is .25. What is the probability that they will not be red?

A spinner is numbered from 1 through 10, and each number is as likely to come up as any other. In Problems 3–6, use equation (1) or (2), indicating which is used, to compute the probability that in a single spin the spinner will stop at:

3. A number less than 3 or larger than 7

4. A 2 or a number larger than 6

5. An even number or a number divisible by 3

6. An odd number or a number divisible by 3

Problems 7–18 refer to the Venn diagram for events A and B in an equally likely sample space S shown below. Find each of the indicated probabilities.

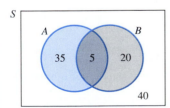

7. $P(A)$
8. $P(A')$
9. $P(B)$
10. $P(B')$
11. $P(A \cap B)$
12. $P(A \cap B')$
13. $P(A' \cap B)$
14. $P(A' \cap B')$
15. $P(A \cup B)$
16. $P(A \cup B')$
17. $P(A' \cup B)$
18. $P(A' \cup B')$

In Problems 19–22, use the equally likely sample space in Example 2 and equation (1) or (2), indicating which is used, to compute the probability of the following events:

19. A sum of 5 or 6   20. A sum of 9 or 10

21. The number on the first die is a 1 or the number on the second die is a 1.

22. The number on the first die is a 1 or the number on the second die is less than 3.

23. Given the following probabilities for an event E, find the odds for and against E:
(A) $\frac{3}{8}$ (B) $\frac{1}{4}$ (C) .4   (D) .55

24. Given the following probabilities for an event E, find the odds for and against E:
(A) $\frac{3}{5}$ (B) $\frac{1}{7}$ (C) .6   (D) .35

25. Compute the probability of event E if the odds in favor of E are:
(A) $\frac{3}{8}$ (B) $\frac{11}{7}$ (C) $\frac{4}{1}$ (D) $\frac{49}{51}$

26. Compute the probability of event E if the odds in favor of E are:
(A) $\frac{5}{9}$ (B) $\frac{4}{3}$ (C) $\frac{3}{7}$ (D) $\frac{23}{77}$

B   *In Problems 27–30, compute the odds in favor of obtaining:*

27. A head in a single toss of a coin

28. A number divisible by 3 in a single roll of a die

29. At least 1 head when a single coin is tossed 3 times

30. 1 head when a single coin is tossed twice

In Problems 31–34, compute the odds against obtaining:

31. A number greater than 4 in a single roll of a die

32. 2 heads when a single coin is tossed twice

33. A 3 or an even number in a single roll of a die

34. An odd number or a number divisible by 3 in a single roll of a die

35. (A) What are the odds for rolling a sum of 5 in a single roll of two fair dice?

(B) If you bet $1 that a sum of 5 will turn up, what should the house pay (plus returning your $1 bet) if a sum of 5 turns up for the game to be fair?

36. (A) What are the odds for rolling a sum of 10 in a single roll of two fair dice?

(B) If you bet $1 that a sum of 10 will turn up, what should the house pay (plus returning your $1 bet) if a sum of 10 turns up for the game to be fair?

A pair of dice are rolled 1,000 times with the following frequencies of outcomes:

SUM	2	3	4	5	6	7	8	9	10	11	12
FREQUENCY	10	30	50	70	110	150	170	140	120	80	70

Use these frequencies to calculate the approximate empirical probabilities and odds for the events in Problems 37 and 38.

37. (A) The sum is less than 4 or greater than 9.

(B) The sum is even or exactly divisible by 5.

38. (A) The sum is a prime number or is exactly divisible by 4.

(B) The sum is an odd number or exactly divisible by 3.

In Problems 39–42, a single card is drawn from a standard 52-card deck. Calculate the probability of and odds for each event.

39. A face card or a club is drawn.

40. A king or a heart is drawn.

41. A black card or an ace is drawn.

42. A heart or a number less than 7 (count an ace as 1) is drawn.

43. What is the probability of getting at least 1 diamond in a 5-card hand dealt from a standard 52-card deck?

44. What is the probability of getting at least 1 black card in a 7-card hand dealt from a standard 52-card deck?

45. What is the probability that a number selected at random from the first 1,000 positive integers is (exactly) divisible by 6 or 8?

46. What is the probability that a number selected at random from the first 600 positive integers is (exactly) divisible by 6 or 9?

47. Explain how the three events A, B, and C from a sample space S are related to each other in order for the following equation to hold:

$$P(A \cup B \cup C) = P(A) + P(B) + P(C) - P(A \cap B)$$

48. Explain how the three events A, B, and C from a sample space S are related to each other in order for the following equation to hold:

$$P(A \cup B \cup C) = P(A) + P(B) + P(C)$$

49. Show that the solution to the birthday problem in Example 5 can be written in the form

$$P(E) = 1 - \frac{P_{365,n}}{365^n}$$

For a calculator that has a $\boxed{P_{n,r}}$ key, explain why this form may be better for direct evaluation than the other form used in the solution to Example 5. Try direct evaluation of both forms on a calculator for $n = 25$.

50. Many (but not all) calculators experience an overflow error when computing $P_{365,n}$ for $n > 39$ and when computing 365^n. Explain how you would evaluate $P(E)$ for any $n > 39$ on such a calculator.

C

51. In a group of n people ($n \leq 12$), what is the probability that at least 2 of them have the same birth month? (Assume any birth month is as likely as any other.)

52. In a group of n people ($n \leq 100$), each person is asked to select a number between 1 and 100, write the number on a slip of paper, and place the slip in a hat. What is the probability that at least 2 of the slips in the hat have the same number written on them?

53. If the odds in favor of an event E occurring are a to b, show that

$$P(E) = \frac{a}{a + b}$$

[*Hint:* Solve the equation $P(E)/P(E') = a/b$ for $P(E)$.]

54. If $P(E) = c/d$, show that the odds in favor of E occurring are c to $d - c$.

APPLICATIONS

Business & Economics

55. *Market research.* From a survey involving 1,000 students at a large university, a market research company found that 750 students owned stereos, 450 owned cars, and 350 owned cars and stereos. If a student at the university is selected at random, what is the (empirical) probability that:
(A) The student owns either a car or a stereo?
(B) The student owns neither a car nor a stereo?

56. *Market research.* If a student at the university in Problem 55 is selected at random, what is the (empirical) probability that:
(A) The student does not own a car?
(B) The student owns a car but not a stereo?

57. *Insurance.* By examining the past driving records of drivers in a certain city, an insurance company has determined the (empirical) probabilities in the table:

		MILES DRIVEN PER YEAR			
		Less than 10,000, M_1	*10,000– 15,000 Inclusive,* M_2	*More than 15,000,* M_3	*Totals*
ACCIDENT	A	.05	.1	.15	.3
NO ACCIDENT	A'	.15	.2	.35	.7
	Totals	.2	.3	.5	1.0

If a driver in this city is selected at random, what is the probability that:
(A) He or she drives less than 10,000 miles per year or has an accident?
(B) He or she drives 10,000 or more miles per year and has no accidents?

58. *Insurance.* Use the (empirical) probabilities in Problem 57 to find the probability that a driver in the city selected at random:
(A) Drives more than 15,000 miles per year or has an accident
(B) Drives 15,000 or fewer miles per year and has an accident

59. *Manufacturing.* Manufacturers of a portable computer provide a 90 day limited warranty covering only the keyboard and the disk drive. Their records indicate that during the warranty period, 6% of their computers are returned because they have defective keyboards, 5% are returned because they have defective disk drives, and 1% are returned because both the keyboard and the disk drive are defective. What is the (empirical) probability that a computer will not be returned during the warranty period?

60. *Product testing.* In order to test a new car, an automobile manufacturer wants to select 4 employees to test drive the car for 1 year. If 12 management and 8 union employees volunteer to be test drivers and the selection is made at random, what is the probability that at least 1 union employee is selected?

61. *Quality control.* A shipment of 60 inexpensive digital watches, including 9 that are defective, is sent to a department store. The receiving department selects 10 at random for testing and rejects the whole shipment if 1 or more in the sample are found defective. What is the probability that the shipment will be rejected?

62. *Quality control.* An automated manufacturing process produces 40 computer circuit boards, including 7 that are defective. The quality control department selects 10 at random (from the 40 produced) for testing and will shut down the plant for trouble shooting if 1 or more in the sample are found defective. What is the probability that the plant will be shut down?

Life Sciences

63. *Medicine.* In order to test a new drug for adverse reactions, the drug was administered to 1,000 test subjects with the following results: 60 subjects reported that their only adverse reaction was a loss of appetite, 90 subjects reported that their only adverse reaction was a loss of sleep, and 800 subjects reported no adverse reactions at all. If this drug is released for general use, what is the (empirical) probability that a person using the drug will suffer both a loss of appetite and a loss of sleep?

64. *Medicine.* Thirty animals are to be used in a medical experiment on diet deficiency: 3 male and 7 female rhesus monkeys, 6 male and 4 female chimpanzees, and 2 male and 8 female dogs. If one animal is selected at random, what is the probability of getting:
(A) A chimpanzee or a dog?
(B) A chimpanzee or a male?
(C) An animal other than a female monkey?

Social Sciences

Problems 65 and 66 refer to the data in the table below, obtained from a random survey of 1,000 residents of a state. The participants were asked their political affiliations and their preferences in an upcoming gubernatorial election.

		DEMO-CRAT D	REPUB-LICAN R	UNAFFIL-IATED U	Totals
CANDIDATE A	A	200	100	85	385
CANDIDATE B	B	250	230	50	530
NO PREFERENCE	N	50	20	15	85
Totals		500	350	150	1,000

65. *Politics.* If a resident of the state is selected at random, what is the (empirical) probability that the resident is:
(A) Not affiliated with a political party or has no preference? What are the odds for this event?
(B) Affiliated with a political party and prefers candidate *A*? What are the odds against this event?

66. *Politics.* If a resident of the state is selected at random, what is the (empirical) probability that the resident is:
(A) A Democrat or prefers candidate *B*? What are the odds for this event?
(B) Not a Democrat and has no preference? What are the odds against this event?

67. *Sociology.* A group of 5 Blacks, 5 Asians, 5 Latinos, and 5 Whites was used in a study on racial influence in small group dynamics. If 3 people are chosen at random, what is the probability that at least 1 is Black?

S E C T I O N   6-5   ■   # Conditional Probability, Intersection, and Independence

- ■ CONDITIONAL PROBABILITY
- ■ INTERSECTION OF EVENTS—PRODUCT RULE
- ■ PROBABILITY TREES
- ■ INDEPENDENT EVENTS
- ■ SUMMARY

In the previous section we asked if the probability of the union of two events could be expressed in terms of the probabilities of the individual events, and we found the answer to be a qualified yes (see page 426). Now we ask the same question for the intersection of two events; that is, can the probability of the intersection of two events be represented in terms of the probabilities of the individual events? The answer again is a qualified yes. But before we find out how and under what conditions, we must investigate a related concept called *conditional probability*.

■ CONDITIONAL PROBABILITY

The probability of an event may change if we are told of the occurrence of another event. For example, if an adult (a person 21 years or older) is selected at random from all adults in the United States, the probability of that person having lung cancer would not be too high. However, if we are told that the person is also a heavy smoker, then we would certainly want to revise the probability upward.

In general, the probability of the occurrence of an event *A*, given the occurrence of another event *B*, is called a **conditional probability** and is denoted by $P(A|B)$.

In the above illustration, events A and B would be

A = Adult has lung cancer

B = Adult is a heavy smoker

And $P(A|B)$ would represent the probability of an adult having lung cancer, given that he or she is a heavy smoker.

Our objective now is to try to formulate a precise definition of $P(A|B)$. It is helpful to start with a relatively simple problem, solve it intuitively, and then generalize from this experience.

What is the probability of rolling a prime number (2, 3, or 5) in a single roll of a fair die? Let

$$S = \{1, 2, 3, 4, 5, 6\}$$

Then the event of rolling a prime number is (see Fig. 1)

$$A = \{2, 3, 5\}$$

Thus, since we assume each simple event in the sample space is equally likely,

$$P(A) = \frac{n(A)}{n(S)} = \frac{3}{6} = \frac{1}{2}$$

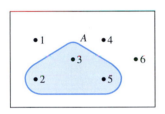

Now suppose you are asked, "In a single roll of a fair die, what is the probability that a prime number has turned up if we are given the additional information that an odd number has turned up?" The additional knowledge that another event has occurred, namely,

B = Odd number turns up

puts the problem in a new light. We are now interested only in the part of event A (rolling a prime number) that is in event B (rolling an odd number). Event B, since we know it has occurred, becomes the new sample space. The Venn diagrams in Figure 2 illustrate the various relationships. Thus, the probability of A given B is the number of A elements in B divided by the total number of elements in B. Symbolically,

$$P(A|B) = \frac{n(A \cap B)}{n(B)} = \frac{2}{3}$$

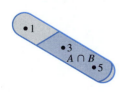

FIGURE 2
B is the new sample space

Dividing the numerator and denominator of $n(A \cap B)/n(B)$ by $n(S)$, the number of elements in the original sample space, we can express $P(A|B)$ in terms of $P(A \cap B)$ and $P(B)$:*

$$P(A|B) = \frac{n(A \cap B)}{n(B)} = \frac{\dfrac{n(A \cap B)}{n(S)}}{\dfrac{n(B)}{n(S)}} = \frac{P(A \cap B)}{P(B)}$$

---

* Note that $P(A|B)$ is a probability based on the new sample space B, while $P(A \cap B)$ and $P(B)$ are both probabilities based on the original sample space S.

Using the right side to compute $P(A|B)$ for the example above, we obtain the same result (as we should):

$$P(A|B) = \frac{P(A \cap B)}{P(B)} = \frac{\frac{2}{6}}{\frac{3}{6}} = \frac{2}{3}$$

We use the above formula to motivate the following definition of *conditional probability*, which applies to any sample space, including those having simple events that are not equally likely (see Example 1).

Conditional Probability

For events A and B in an arbitrary sample space S, we define the conditional probability of A given B by

$$P(A|B) = \frac{P(A \cap B)}{P(B)} \qquad P(B) \neq 0 \qquad (1)$$

EXAMPLE 1 ➤ **Conditional Probability** A pointer is spun once on the circular spinner shown in Figure 3. The probability assigned to the pointer landing on a given integer (from 1 to 6) is the ratio of the area of the corresponding circular sector to the area of the whole circle, as given in the table:

e_i	1	2	3	4	5	6
$P(e_i)$	.1	.2	.1	.1	.3	.2

$S = \{1, 2, 3, 4, 5, 6\}$

(A) What is the probability of the pointer landing on a prime number?
(B) What is the probability of the pointer landing on a prime number, given that it landed on an odd number?

FIGURE 3

Solution   Let the events E and F be defined as follows:

$$E = \text{Pointer lands on a prime number} = \{2, 3, 5\}$$
$$F = \text{Pointer lands on an odd number} = \{1, 3, 5\}$$

(A) $P(E) = P(2) + P(3) + P(5)$
    $= .2 + .1 + .3 = .6$

(B) First note that $E \cap F = \{3, 5\}$.

$$P(E|F) = \frac{P(E \cap F)}{P(F)} = \frac{P(3) + P(5)}{P(1) + P(3) + P(5)}$$

$$= \frac{.1 + .3}{.1 + .1 + .3} = \frac{.4}{.5} = .8$$

◀

MATCHED PROBLEM 1 ➤ Refer to the spinner and table in Example 1.

(A) What is the probability of the pointer landing on a number greater than 4?

(B) What is the probability of the pointer landing on a number greater than 4, given that it landed on an even number? ◄

---

explore – discuss 1

Compare the spinner problem in Example 1 to the die problem at the beginning of the section. Discuss their similarities and differences.

▶ EXAMPLE 2 ➤ **Safety Research** Suppose past records in a large city produced the following probability data on a driver being in an accident on the last day of a Memorial Day weekend:

		ACCIDENT A	NO ACCIDENT A'	Totals
RAIN	R	.025	.335	.360
NO RAIN	R'	.015	.625	.640
	Totals	.040	.960	1.000

$S = \{RA, RA', R'A, R'A'\}$

(A) Find the probability of an accident, rain or no rain.

(B) Find the probability of rain, accident or no accident.

(C) Find the probability of an accident and rain.

(D) Find the probability of an accident, given rain.

Solution (A) Let $A = \{RA, R'A\}$ Event: "accident"

$$P(A) = P(RA) + P(R'A) = .025 + .015 = .040$$

(B) Let $R = \{RA, RA'\}$ Event: "rain"

$$P(R) = P(RA) + P(RA') = .025 + .335 = .360$$

(C) $A \cap R = \{RA\}$ Event: "accident and rain"

$$P(A \cap R) = P(RA) = .025$$

(D) $P(A|R) = \dfrac{P(A \cap R)}{P(R)} = \dfrac{.025}{.360} = .069$ Event: "accident, given rain"

Compare this result with that in part (A). Note that $P(A|R) \neq P(A)$, and the probability of an accident, given rain, is higher than the probability of an accident without the knowledge of rain. ◄

MATCHED PROBLEM 2 ➤ Referring to the table in Example 2 determine the following:

(A) Probability of no rain

(B) Probability of an accident and no rain

(C) Probability of an accident, given no rain [Use formula (1) and the results of parts (A) and (B)]. ◄

■ INTERSECTION OF EVENTS—PRODUCT RULE

We now return to the original problem of this section; that is, representing the probability of an intersection of two events in terms of the probabilities of the individual events. If $P(A) \neq 0$ and $P(B) \neq 0$, then using formula (1) we can write

$$P(A|B) = \frac{P(A \cap B)}{P(B)} \qquad \text{and} \qquad P(B|A) = \frac{P(B \cap A)}{P(A)}$$

Solving the first equation for $P(A \cap B)$ and the second equation for $P(B \cap A)$, we have

$$P(A \cap B) = P(B)P(A|B) \qquad \text{and} \qquad P(B \cap A) = P(A)P(B|A)$$

Since $A \cap B = B \cap A$ for any sets A and B, it follows that

$$P(A \cap B) = P(B)P(A|B) = P(A)P(B|A)$$

and we have the **product rule:**

Product Rule

For events A and B with nonzero probabilities in a sample space S,

$$P(A \cap B) = P(A)P(B|A) = P(B)P(A|B) \qquad (2)$$

and we can use either $P(A)P(B|A)$ or $P(B)P(A|B)$ to compute $P(A \cap B)$.

EXAMPLE 3 ▶ Consumer Survey   If 60% of a department store's customers are female and 75% of the female customers have charge accounts at the store, what is the probability that a customer selected at random is a female and has a charge account?

Solution   Let: S = All store customers

F = Female customers

C = Customers with a charge account

If 60% of the customers are female, then the probability that a customer selected at random is a female is

$$P(F) = .60$$

Since 75% of the female customers have charge accounts, the probability that a customer has a charge account, given that the customer is a female, is

$$P(C|F) = .75$$

Using equation (2), the probability that a customer is a female and has a charge account is

$$P(F \cap C) = P(F)P(C|F) = (.60)(.75) = .45 \qquad \blacktriangleleft$$

MATCHED PROBLEM 3 ➤   If 80% of the male customers of the department store in Example 3 have charge accounts, what is the probability that a customer selected at random is a male and has a charge account?   ◄

■ PROBABILITY TREES

We used tree diagrams in Section 6-1 to help us count the number of combined outcomes in a sequence of experiments. In much the same way we will now use probability trees to help us compute the probabilities of combined outcomes in a sequence of experiments. An example will help make the process of forming and using probability trees clear.

EXAMPLE 4 ➤   Probability Tree   Two balls are drawn in succession, without replacement, from a box containing 3 blue and 2 white balls (Fig. 4). What is the probability of drawing a white ball on the second draw?

FIGURE 4

Solution   We start with a tree diagram (Figure 5) showing the combined outcomes of the two experiments (first draw and second draw):

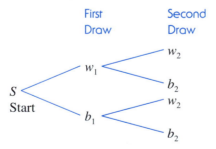

FIGURE 5

We now assign a probability to each branch on the tree (Fig. 6, page 446). For example, we assign the probability $\frac{2}{5}$ to the branch Sw_1, since this is the probability of drawing a white ball on the first draw (there are 2 white balls and 3 blue balls in the box). What probability should be assigned to the branch w_1w_2? This is the conditional probability $P(w_2|w_1)$; that is, the probability of drawing a white ball on the second draw given that a white ball was drawn on the first draw and not replaced. Since the box now contains 1 white ball and 3 blue balls, the probability is $\frac{1}{4}$. Continuing in the same way, we assign probabilities to the other branches of the tree and obtain the following probability tree:

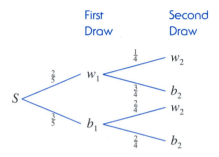

First
Draw

Second
Draw

FIGURE 6

What is the probability of the combined outcome $w_1 \cap w_2$; that is, of drawing a white ball on the first draw and a white ball on the second draw?* Using the product rule (2), we have

$$P(w_1 \cap w_2) = P(w_1)P(w_2|w_1)$$
$$= (\tfrac{2}{5})(\tfrac{1}{4}) = \tfrac{1}{10}$$

The combined outcome $w_1 \cap w_2$ corresponds to the unique path Sw_1w_2 in the tree diagram, and we see that the probability of reaching w_2 along this path is just the product of the probabilities assigned to the branches on the path. Reasoning in the same way, we obtain the probability of each remaining combined outcome by multiplying the probabilities assigned to the branches on the path corresponding to the given combined outcomes. These probabilities are often written at the ends of the paths to which they correspond (Fig. 7).

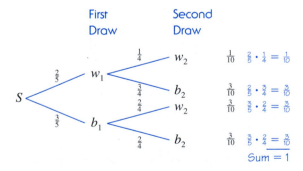

FIGURE 7

_____

* The sample space for the combined outcomes is $S = \{w_1w_2, w_1b_2, b_1w_2, b_1b_2\}$. If we let $w_1 = \{w_1w_2, w_1b_2\}$ and $w_2 = \{w_1w_2, b_1w_2\}$, then $w_1 \cap w_2 = \{w_1w_2\}$.

Now it is an easy matter to complete the problem. A white ball drawn on the second draw corresponds to either the combined outcome $w_1 \cap w_2$ or $b_1 \cap w_2$ occurring. Thus, since these combined outcomes are mutually exclusive:

$$P(w_2) = P(w_1 \cap w_2) + P(b_1 \cap w_2)$$
$$= \tfrac{1}{10} + \tfrac{3}{10} = \tfrac{4}{10} = \tfrac{2}{5}$$

which is just the sum of the probabilities listed at the ends of the two paths terminating in w_2. ◄

MATCHED PROBLEM 4 ➤ Two balls are drawn in succession without replacement from a box containing 4 red and 2 white balls. What is the probability of drawing a red ball on the second draw? ◄

The sequence of two experiments in Example 4 is an example of a *stochastic process*. In general, a **stochastic process** involves a sequence of experiments where the outcome of each experiment is not certain. Our interest is in making predictions about the process as a whole. The analysis in Example 4 generalizes to stochastic processes involving any finite sequence of experiments. We summarize the procedures used in Example 4 for general application:

Constructing Probability Trees

Step 1. Draw a tree diagram corresponding to all combined outcomes of the sequence of experiments.

Step 2. Assign a probability to each tree branch. (This is the probability of the occurrence of the event on the right end of the branch subject to the occurrence of all events on the path leading to the event on the right end of the branch. The probability of the occurrence of a combined outcome that corresponds to a path through the tree is the product of all branch probabilities on the path.*)

Step 3. Use the results in steps 1 and 2 to answer various questions related to the sequence of experiments as a whole.

---

* If we form a sample space S such that each simple event S corresponds to one path through the tree, and if the probability assigned to each simple event in S is the product of the branch probabilities on the corresponding path, then it can be shown that this is not only an acceptable assignment (all probabilities for the simple events in S are nonnegative and their sum is 1), but it is the only assignment consistent with the method used to assign branch probabilities within the tree.

explore – discuss 2

Refer to the table on rain and accidents in Example 2 and use formula (1), where appropriate, to complete the following probability tree:

Discuss the difference between $P(R \cap A)$ and $P(A|R)$.

EXAMPLE 5 ➤ **Product Defects**   A large computer company A subcontracts the manufacturing of its circuit boards to two companies, 40% to company B and 60% to company C. Company B in turn subcontracts 70% of the orders it receives from company A to company D and the remaining 30% to company E, both subsidiaries of company B. When the boards are completed by companies D, E and C, they are shipped to company A to be used in various computer models. It has been found that 1.5%, 1%, and .5% of the boards from D, E, and C, respectively, prove defective during the 90 day warranty period after a computer is first sold. What is the probability that a given board in a computer will be defective during the 90 day warranty period?

Solution   Draw a tree diagram and assign probabilities to each branch (Fig. 8):

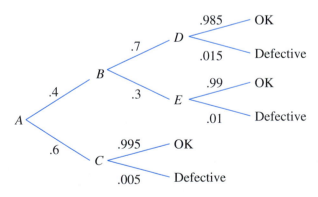

FIGURE 8

There are three paths leading to defective (the board will be defective within the 90 day warranty period). We multiply the branch probabilities on each path and add the three products:

$$P(\text{Defective}) = (.4)(.7)(.015) + (.4)(.3)(.01) + (.6)(.005)$$
$$= .0084$$

MATCHED PROBLEM 5 ➤ In Example 5, what is the probability that a circuit board in a completed computer came from company E or C? ◀

■ INDEPENDENT EVENTS

We return to Example 4, which involved drawing two balls in succession without replacement from a box of 3 blue and 2 white balls (Fig. 4). What difference does "without replacement" and "with replacement" make? Figure 9 shows probability trees corresponding to each case. Go over the probability assignments for the branches in part (B) to convince yourself of their correctness.

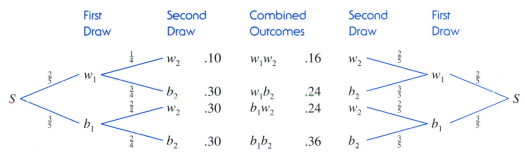

(A) Without replacement   (B) With replacement

FIGURE 9
$S = \{w_1w_2, w_1b_2, b_1w_2, b_1b_2\}$

Let: A = White ball on second draw = $\{w_1w_2, b_1w_2\}$
B = White ball on first draw = $\{w_1w_2, w_1b_2\}$

We now compute $P(A|B)$ and $P(A)$ for each case in Figure 9.

Case 1. *Without Replacement:*

$$P(A|B) = \frac{P(A \cap B)}{P(B)} = \frac{P\{w_1w_2\}}{P\{w_1w_2, w_1b_2\}} = \frac{.10}{.10 + .30} = .25$$

(This is the assignment to branch w_1w_2 we made by looking in the box and counting.)

$$P(A) = P\{w_1w_2, b_1w_2\} = .10 + .30 = .40$$

Note that $P(A|B) \neq P(A)$, and we conclude that the probability of A is affected by the occurrence of B.

Case 2. *With Replacement:*

$$P(A|B) = \frac{P(A \cap B)}{P(B)} = \frac{P\{w_1w_2\}}{P\{w_1w_2, w_1b_2\}} = \frac{.16}{.16 + .24} = .40$$

(This is the assignment to branch w_1w_2 we made by looking in the box and counting.)

$$P(A) = P\{w_1w_2, b_1w_2\} = .16 + .24 = .40$$

Note that $P(A|B) = P(A)$, and we conclude that the probability of A is not affected by the occurrence of B.

Intuitively, if $P(A|B) = P(A)$, then it appears that event A is "independent" of B. Let us pursue this further. If events A and B are such that

$$P(A|B) = P(A)$$

then replacing the left side by its equivalent from equation (1), we obtain

$$\frac{P(A \cap B)}{P(B)} = P(A)$$

After multiplying both sides by $P(B)$, the last equation becomes

$$P(A \cap B) = P(A)P(B)$$

This result motivates the following definition of *independence:*

Independence

If A and B are any events in a sample space S, we say that **A and B are independent** if and only if

$$P(A \cap B) = P(A)P(B) \tag{3}$$

Otherwise, A and B are said to be **dependent.**

From the definition of independence one can prove (see Problems 41 and 42 in Exercise 6-5) the following theorem:

t h e o r e m   1

On Independence

If A and B are independent events with nonzero probabilities in a sample space S, then

$$P(A|B) = P(A) \qquad \text{and} \qquad P(B|A) = P(B) \tag{4}$$

If either equation in (4) holds, then A and B are independent.

In practice, one often has correct intuitive feelings about independence. For example, if you toss a coin twice, the second toss is independent of the first (a coin has no memory); if a card is drawn from a deck twice, with replacement, the second draw is independent of the first (the deck has no memory); if you roll a pair of dice twice, the second roll is independent of the first (dice have no memory); and so on. However, there are pairs of events that are not obviously independent or dependent, so we need equations (3) or (4) to decide these cases. Example 6 considers two events that are obviously independent, and Example 7 considers events where independence or dependence is not obvious.

EXAMPLE 6 ▶ **Testing for Independence**   In two tosses of a single fair coin show that the events "A head on the first toss" and "A head on the second toss" are independent.

Solution   Consider the sample space of equally likely outcomes for the tossing of a fair coin twice,

$$S = (\text{HH, HT, TH, TT})$$

and the two events,

$$A = \text{A head on the first toss} = (\text{HH, HT})$$
$$B = \text{A head on the second toss} = (\text{HH, TH})$$

Then

$$P(A) = \tfrac{2}{4} = \tfrac{1}{2} \qquad P(B) = \tfrac{2}{4} = \tfrac{1}{2} \qquad P(A \cap B) = \tfrac{1}{4}$$

Thus,

$$P(A \cap B) = \tfrac{1}{4} = \tfrac{1}{2} \cdot \tfrac{1}{2} = P(A)P(B)$$

and the two events are independent. (The theory agrees with our intuition—a coin has no memory.)   ◀

MATCHED PROBLEM 6 ▶ In Example 6, compute $P(B|A)$ and compare with $P(B)$.   ◀

EXAMPLE 7 ▶ **Testing for Independence**   A single card is drawn from a standard 52-card deck. Test the following events for independence (try guessing the answer to each part before looking at the solution):

(A) E = The drawn card is a spade.

   F = The drawn card is a face card.

(B) G = The drawn card is a club.

   H = The drawn card is a heart.

Solution   (A) To test E and F for independence, we compute $P(E \cap F)$ and $P(E)P(F)$. If they are equal, then events E and F are independent; if they are not equal, then events E and F are dependent.

$$P(E \cap F) = \tfrac{3}{52} \qquad P(E)P(F) = (\tfrac{13}{52})(\tfrac{12}{52}) = \tfrac{3}{52}$$

Events E and F are independent. (Did you guess this?)

(B) Proceeding as in part (A), we see that

$$P(G \cap H) = P(\varnothing) = 0 \qquad P(G)P(H) = (\tfrac{13}{52})(\tfrac{13}{52}) = \tfrac{1}{16}$$

Events G and H are dependent. (Did you guess this?)   ◀

Caution

Students often confuse *mutually exclusive (disjoint) events* with *independent events*. One does not necessarily imply the other. In fact, it is not difficult to show (see Problem 45, Exercise 6-5) that any two mutually exclusive events A and B, with nonzero probabilities, are always dependent!

MATCHED PROBLEM 7 ➤ A single card is drawn from a standard 52-card deck. Test the following events for independence:

(A) E = The drawn card is a red card.

   F = The drawn card's number is divisible by 5 (face cards are not assigned values).

(B) G = The drawn card is a king.

   H = The drawn card is a queen.   ◄

explore–discuss 3

In college basketball, would it be reasonable to assume that the events

   A = The Golden Eagles win in the first round of the NCAA tournament

   B = The Golden Eagles win in the second round of the NCAA tournament

are independent? Explain why or why not.

The notion of independence can be extended to more than two events:

Independent Set of Events

A set of events is said to be **independent** if for each finite subset $\{E_1, E_2, \ldots, E_k\}$

$$P(E_1 \cap E_2 \cap \cdots \cap E_k) = P(E_1)P(E_2) \cdots \cdot P(E_k) \tag{5}$$

The next example makes direct use of this definition.

▶ EXAMPLE 8 ➤ **Computer Control Systems**   A space shuttle has four independent computer control systems. If the probability of failure (during flight) of any one system is .001, what is the probability of failure of all four systems?

Solution   Let: E_1 = Failure of system 1 E_3 = Failure of system 3

            E_2 = Failure of system 2 E_4 = Failure of system 4

Then, since events E_1, E_2, E_3, and E_4 are given to be independent,

$$P(E_1 \cap E_2 \cap E_3 \cap E_4) = P(E_1)P(E_2)P(E_3)P(E_4)$$
$$= (.001)^4$$
$$= .000\ 000\ 000\ 001$$

M A T C H E D   P R O B L E M   8 ➤   A single die is rolled 6 times. What is the probability of getting the sequence, 1, 2, 3, 4, 5, 6?

■ SUMMARY

The key results in this section are summarized in the following box for an overview and ease of reference:

Summary of Key Concepts

CONDITIONAL PROBABILITY

$$P(A|B) = \frac{P(A \cap B)}{P(B)} \qquad P(B|A) = \frac{P(B \cap A)}{P(A)}$$

[*Note:* $P(A|B)$ is a probability based on the new sample space B, while $P(A \cap B)$ and $P(B)$ are probabilities based on the original sample space S.]

PRODUCT RULE

$$P(A \cap B) = P(A)P(B|A) = P(B)P(A|B)$$

INDEPENDENT EVENTS

■ *A* and *B* are independent if and only if

$$P(A \cap B) = P(A)P(B)$$

■ If *A* and *B* are independent events with nonzero probabilities, then

$$P(A|B) = P(A) \qquad \text{and} \qquad P(B|A) = P(B)$$

■ If *A* and *B* are events with nonzero probabilities and either $P(A|B) = P(A)$ or $P(B|A) = P(B)$, then *A* and *B* are independent.

■ If E_1, E_2, . . . , E_n are independent, then

$$P(E_1 \cap E_2 \cap \cdots \cap E_n) = P(E_1)P(E_2) \cdot \cdots \cdot P(E_n)$$

Answers to Matched Problems   1. (A) .5   (B) .4

2. (A) $P(R') = .640$ (B) $P(A \cap R') = .015$ (C) $P(A|R') = \dfrac{P(A \cap R')}{P(R')} = .023$

3. $P(M \cap C) = P(M)P(C|M) = .32$ 4. $\frac{2}{3}$ 5. .72

6. $P(B|A) = \dfrac{P(A \cap B)}{P(A)} = \dfrac{\frac{1}{4}}{\frac{1}{2}} = \dfrac{1}{2} = P(B)$

7. (A) E and F are independent   (B) G and H are dependent

8. $(\frac{1}{6})^6 \approx .000\ 021\ 4$

EXERCISE 6-5

A   *Given the probabilities in the table below for events in a sample space S, solve Problems 1–16 relative to these probabilities.*

	A	B	C	Totals
D	.10	.04	.06	.20
E	.40	.26	.14	.80
Totals	.50	.30	.20	1.00

In Problems 1–8, find each probability directly from the table.

1. $P(A)$
2. $P(C)$
3. $P(D)$
4. $P(E)$
5. $P(A \cap D)$
6. $P(A \cap E)$
7. $P(C \cap D)$
8. $P(C \cap E)$

In Problems 9–12 compute each probability using equation (1) and appropriate table values:

9. $P(A|D)$
10. $P(A|E)$
11. $P(C|D)$
12. $P(C|E)$

In Problems 13–16, test each pair of events for independence:

13. A and D
14. A and E
15. C and D
16. C and E

17. A fair coin is tossed 8 times.
    (A) What is the probability of tossing a head on the 8th toss, given that the preceding 7 tosses were heads?
    (B) What is the probability of getting 8 heads or 8 tails?

18. A fair die is rolled 5 times.
    (A) What is the probability of getting a 6 on the 5th roll, given that a 6 turned up on the preceding 4 rolls?
    (B) What is the probability that the same number turns up every time?

19. A pointer is spun once on the circular spinner shown below. The probability assigned to the pointer landing on a given integer (from 1 to 5) is the ratio of the area of the corresponding circular sector to the area of the whole circle, as given in the table:

e_i	1	2	3	4	5
$P(e_i)$	.3	.1	.2	.3	.1

Given the events:

E = Pointer lands on an even number

F = Pointer lands on a number less than 4

(A) Find $P(F|E)$.
(B) Test events E and F for independence.

20. Repeat Problem 19 with the following events:

E = Pointer lands on an odd number

F = Pointer lands on a prime number

Compute the indicated probabilities in Problems 21 and 22 by referring to the following probability tree:

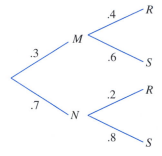

21. (A) $P(M \cap S)$ (B) $P(R)$

22. (A) $P(N \cap R)$ (B) $P(S)$

B

23. A fair coin is tossed twice. Let $S = \{HH, HT, TH, TT\}$ be the sample space of equally likely simple events. We are interested in the following events:

$E_1 =$ A head on the first toss

$E_2 =$ A tail on the first toss

$E_3 =$ A tail on the second toss

$E_4 =$ A head on the second toss

For each pair of events, discuss whether they are independent and whether they are mutually exclusive.
(A) E_1 and E_4 (B) E_1 and E_2

24. For each pair of events (see Problem 23), discuss whether they are independent and whether they are mutually exclusive.
(A) E_1 and E_3 (B) E_3 and E_4

25. In 2 throws of a fair die, what is the probability that you will get an even number on each throw? An even number on the first or second throw?

26. In 2 throws of a fair die, what is the probability that you will get at least 5 on each throw? At least 5 on the first or second throw?

27. Two cards are drawn in succession from a standard 52-card deck. What is the probability that the first card is a club and the second card is a heart:
(A) If the cards are drawn without replacement?
(B) If the cards are drawn with replacement?

28. Two cards are drawn in succession from a standard 52-card deck. What is the probability that both cards are red:
(A) If the cards are drawn without replacement?
(B) If the cards are drawn with replacement?

29. A card is drawn at random from a standard 52-card deck. Events G and H are:

$G =$ The drawn card is black.

$H =$ The drawn card is divisible by 3
    (face cards are not valued).

(A) Find $P(H|G)$.   (B) Test H and G for independence.

30. A card is drawn at random from a standard 52-card deck. Events M and N are:

$M =$ The drawn card is a diamond.

$N =$ The drawn card is even
    (face cards are not valued).

(A) Find $P(N|M)$.   (B) Test M and N for independence.

31. Let A be the event that all of a family's children are the same sex, and let B be the event that the family has at most 1 boy. Assuming the probability of having a girl is the same as the probability of having a boy (both .5), test events A and B for independence if:
(A) The family has 2 children.
(B) The family has 3 children.

32. An experiment consists of tossing n coins. Let A be the event that at least 2 heads turn up, and let B be the event that all the coins turn up the same. Test A and B for independence if:
(A) 2 coins are tossed.   (B) 3 coins are tossed.

Problems 33–36 refer to the following experiment: 2 balls are drawn in succession out of a box containing 2 red and 5 white balls. Let R_i be the event that the ith ball is red, and let W_i be the event that the ith ball is white.

33. Construct a probability tree for this experiment and find the probability of each of the events $R_1 \cap R_2$, $R_1 \cap W_2$, $W_1 \cap R_2$, $W_1 \cap W_2$, given that the first ball drawn was:
(A) Replaced before the second draw
(B) Not replaced before the second draw

34. Find the probability that the second ball was red, given that the first ball was:
(A) Replaced before the second draw
(B) Not replaced before the second draw

35. Find the probability that at least 1 ball was red, given that the first ball was:
(A) Replaced before the second draw
(B) Not replaced before the second draw

36. Find the probability that both balls were the same color, given that the first ball was:
(A) Replaced before the second draw
(B) Not replaced before the second draw

C

37. A box contains 2 red, 3 white, and 4 green balls. Two balls are drawn out of the box in succession without replacement. What is the probability that both balls are the same color?

38. For the experiment in Problem 37, what is the probability that no white balls are drawn?

39. An urn contains 2 one-dollar bills, 1 five-dollar bill, and 1 ten-dollar bill. A player draws bills one at a time without replacement from the urn until a ten-dollar bill is drawn. Then the game stops. All bills are kept by the player.
(A) What is the probability of winning $16?
(B) What is the probability of winning all bills in the urn?
(C) What is the probability of the game stopping at the second draw?

40. Ann and Barbara are playing a tennis match. The first player to win 2 sets wins the match. For any given set, the probability that Ann wins that set is $\frac{2}{3}$. Find the probability that:
(A) Ann wins the match   (B) 3 sets are played.
(C) The player who wins the first set goes on to win the match.

41. Show that if A and B are independent events with nonzero probabilities in a sample space S, then

$$P(A|B) = P(A) \quad \text{and} \quad P(B|A) = P(B)$$

42. Show that if A and B are events with nonzero probabilities in a sample space S, and either $P(A|B) = P(A)$ or $P(B|A) = P(B)$, then events A and B are independent.

43. Show that $P(A|A) = 1$, when $P(A) \neq 0$.

44. Show that $P(A|B) + P(A'|B) = 1$.

45. Show that A and B are dependent if A and B are mutually exclusive and $P(A) \neq 0$, $P(B) \neq 0$.

46. Show that $P(A|B) = 1$ if B is a subset of A and $P(B) \neq 0$.

 APPLICATIONS

Business & Economics

47. *Labor relations.* In a study to determine employee voting patterns in a recent strike election, 1,000 employees were selected at random and the following tabulation was made:

		SALARY CLASSIFICATION			
		Hourly (H)	Salary (S)	Salary + Bonus (B)	Totals
TO STRIKE	Yes (Y)	400	180	20	600
	No (N)	150	120	130	400
	Totals	550	300	150	1,000

(A) Convert this table to a probability table by dividing each entry by 1,000.
(B) What is the probability of an employee voting to strike, given the person is paid hourly?
(C) What is the probability of an employee voting to strike, given the person receives a salary plus bonus?

(D) What is the probability of an employee being on straight salary (S)? Of being on straight salary given he or she voted in favor of striking?
(E) What is the probability of an employee being paid hourly? Of being paid hourly given he or she voted in favor of striking?
(F) What is the probability of an employee being in a salary plus bonus position and voting against striking?
(G) Are events S and Y independent?
(H) Are events H and Y independent?
(I) Are events B and N independent?

48. *Quality control.* An automobile manufacturer produces 37% of its cars at plant A. If 5% of the cars manufactured at plant A have defective emission control devices, what is the probability that one of this manufacturer's cars was manufactured at plant A and has a defective emission control device?

49. *Bonus incentives.* If a salesperson has gross sales of over $600,000 in a year, he or she is eligible to play the company's bonus game: A black box contains 1 twenty-dollar bill, 2 five-dollar bills, and 1 one-dollar bill. Bills are drawn out of the box one at a time without replacement until a twenty-dollar bill is drawn. Then the game stops. The salesperson's bonus is 1,000 times the value of the bills drawn.
  (A) What is the probability of winning a $26,000 bonus?
  (B) What is the probability of winning the maximum bonus, $31,000, by drawing out all bills in the box?
  (C) What is the probability of the game stopping at the third draw?

50. *Personnel selection.* To transfer into a particular technical department, a company requires an employee to pass a screening test. A maximum of 3 attempts are allowed at 6 month intervals between trials. From past records it is found that 40% pass on the first trial; of those that fail the first trial and take the test a second time, 60% pass; and of those that fail on the second trial and take the test a third time, 20% pass. For an employee wishing to transfer:
  (A) What is the probability of passing the test on the first or second try?
  (B) What is the probability of failing on the first 2 trials and passing on the third?
  (C) What is the probability of failing on all 3 attempts?

Life Sciences

51. *Food and Drug Administration.* An ice cream company wishes to use a red dye to enhance the color in its strawberry ice cream. The Food and Drug Administration (FDA) requires the dye to be tested for cancer-producing potential in a laboratory using laboratory rats. The results of one test on 1,000 rats are summarized in the following table:

		DEVELOPED CANCER C	NO CANCER C'	Totals
ATE RED DYE	R	60	440	500
DID NOT EAT RED DYE	R'	20	480	500
	Totals	80	920	1,000

  (A) Convert the table into a probability table by dividing each entry by 1,000.
  (B) Are "developing cancer" and "eating red dye" independent events?

(C) Should the FDA approve or ban the use of the dye? Explain why or why not using $P(C|R)$ and $P(C)$.

(D) Suppose the number of rats that ate red dye and developed cancer was 20, but the number that developed cancer was still 80 and the number that ate red dye was still 500. What should the FDA do, based on these results? Explain why.

52. *Genetics.* In a study to determine frequency and dependency of color-blindness relative to females and males, 1,000 people were chosen at random and the following results were recorded:

		FEMALE F	MALE F'	Totals
COLOR-BLIND	C	2	24	26
NORMAL	C'	518	456	974
	Totals	520	480	1,000

  (A) Convert this table to a probability table by dividing each entry by 1,000.
  (B) What is the probability that a person is a woman, given that the person is color-blind?
  (C) What is the probability that a person is color-blind, given that the person is a male?
  (D) Are the events color-blindness and male independent?
  (E) Are the events color-blindness and female independent?

Social Sciences

53. *Psychology.* In a study to determine the frequency and dependency of IQ ranges relative to males and females, 1,000 people were chosen at random and the following results were recorded:

		IQ			
		Below 90 (A)	90–120 (B)	Above 120 (C)	Totals
FEMALE	F	130	286	104	520
MALE	F'	120	264	96	480
	Totals	250	550	200	1,000

(A) Convert this table to a probability table by dividing each entry by 1,000.

(B) What is the probability of a person having an IQ below 90, given that the person is a female? A male?

(C) What is the probability of a person having an IQ above 120, given that the person is a female? A male?

(D) What is the probability of a person having an IQ below 90?

(E) What is the probability of a person having an IQ between 90 and 120? Of a person having an IQ between 90 and 120, given that the person is a male?

(F) What is the probability of a person being female and having an IQ above 120?

(G) Are any of the events *A, B,* or *C* dependent relative to *F* or *F'*?

54. *Voting patterns.* A survey of the residents of a precinct in a large city revealed that 55% of the residents were members of the Democratic party and that 60% of the Democratic party members voted in the last election. What is the probability that a person selected at random from the residents of this precinct is a member of the Democratic party and voted in the last election?

SECTION 6-6 ■ **Bayes' Formula**

In the preceding section we discussed the conditional probability of the occurrence of an event, given the occurrence of an earlier event. Now we are going to reverse the problem and try to find the probability of an earlier event conditioned on the occurrence of a later event. As you will see before the discussion is over, a number of practical problems are of this form. First, let us consider a relatively simple problem that will provide the basis for a generalization.

EXAMPLE 1 ➤ **Probability of an Earlier Event Given a Later Event** One urn has 3 blue and 2 white balls; a second urn has 1 blue and 3 white balls (Fig. 1). A single fair die is rolled and if 1 or 2 comes up, a ball is drawn out of the first urn; otherwise, a ball is drawn out of the second urn. If the drawn ball is blue, what is the probability that it came out of the first urn? Out of the second urn?

Urn U_1 Urn U_2

FIGURE 1

Solution   We form a probability tree, letting U_1 represent urn 1, U_2 urn 2, B a blue ball, and W a white ball. Then, on the various outcome branches we assign appropriate probabilities. For example, $P(U_1) = \frac{1}{3}$, $P(B|U_1) = \frac{3}{5}$, and so on:

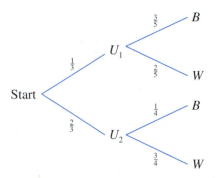

Now we are interested in finding $P(U_1|B)$; that is, the probability that the ball came out of urn 1, given the drawn ball is blue. Using equation (1) from Section 6-5, we can write

$$P(U_1|B) = \frac{P(U_1 \cap B)}{P(B)} \qquad (1)$$

If we look at the tree diagram, we can see that B is at the end of two different branches; thus,

$$P(B) = P(U_1 \cap B) + P(U_2 \cap B) \qquad (2)$$

After substituting equation (2) into equation (1), we get

$$P(U_1|B) = \frac{P(U_1 \cap B)}{P(U_1 \cap B) + P(U_2 \cap B)} \qquad \color{blue}{P(A \cap B) = P(A)P(B|A)}$$

$$= \frac{P(U_1)P(B|U_1)}{P(U_1)P(B|U_1) + P(U_2)P(B|U_2)}$$

$$= \frac{P(B|U_1)P(U_1)}{P(B|U_1)P(U_1) + P(B|U_2)P(U_2)} \qquad (3)$$

Formula (3) is really a lot simpler to use than it looks. You do not need to memorize it; you simply need to understand its form relative to the probability tree above. Referring to the probability tree, we see that

$P(B|U_1)P(U_1) =$ Product of branch probabilities leading to B through U_1
$\qquad = (\frac{3}{5})(\frac{1}{3})$ *We usually start at B and work back through U_1.*

$P(B|U_2)P(U_2) =$ Product of branch probabilities leading to B through U_2
$\qquad = (\frac{1}{4})(\frac{2}{3})$ *We usually start at B and work back through U_2.*

Equation (3) now can be interpreted in terms of the probability tree as follows:

$$P(U_1|B) = \frac{\text{Product of branch probabilities leading to } B \text{ through } U_1}{\text{Sum of all branch products leading to } B}$$

$$= \frac{(\frac{3}{5})(\frac{1}{3})}{(\frac{3}{5})(\frac{1}{3}) + (\frac{1}{4})(\frac{2}{3})} = \frac{6}{11} \approx .55$$

Similarly,

$$P(U_2|B) = \frac{\text{Product of branch probabilities leading to } B \text{ through } U_2}{\text{Sum of all branch products leading to } B}$$

$$= \frac{(\frac{1}{4})(\frac{2}{3})}{(\frac{3}{5})(\frac{1}{3}) + (\frac{1}{4})(\frac{2}{3})} = \frac{5}{11} \approx .45$$

[*Note:* We also could have obtained $P(U_2|B)$ by subtracting $P(U_1|B)$ from 1. Why?] ◄

MATCHED PROBLEM 1 ►   Repeat Example 1, but find $P(U_1|W)$ and $P(U_2|W)$. ◄

explore – discuss 1

Given the probability tree:

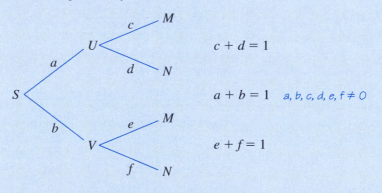

$c + d = 1$

$a + b = 1$ *a, b, c, d, e, f ≠ 0*

$e + f = 1$

(A) Discuss the difference between $P(M|U)$ and $P(U|M)$ and between $P(N|V)$ and $P(V|N)$ in terms of $a, b, c, d, e,$ and f.
(B) Show that $ac + ad + be + bf = 1$. What is the significance of this result?

In generalizing the results in Example 1, it is helpful to look at its structure in terms of the Venn diagram shown in Figure 2. We note that U_1 and U_2 are mutually exclusive (disjoint) and their union forms S. The following two equations can now be interpreted in terms of this diagram:

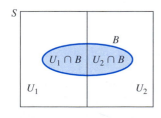

FIGURE 2

$$P(U_1|B) = \frac{P(U_1 \cap B)}{P(B)} = \frac{P(U_1 \cap B)}{P(U_1 \cap B) + P(U_2 \cap B)}$$

$$P(U_2|B) = \frac{P(U_2 \cap B)}{P(B)} = \frac{P(U_2 \cap B)}{P(U_1 \cap B) + P(U_2 \cap B)}$$

Look over the equations and the diagram carefully.

Of course, there is no reason to stop here. Suppose U_1, U_2, and U_3 are three mutually exclusive events whose union is the whole sample space S. Then, for an arbitrary event E in S, with $P(E) \neq 0$, the corresponding Venn diagram looks like Figure 3, and

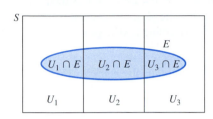

$$P(U_1|E) = \frac{P(U_1 \cap E)}{P(E)} = \frac{P(U_1 \cap E)}{P(U_1 \cap E) + P(U_2 \cap E) + P(U_3 \cap E)}$$

Similar results hold for U_2 and U_3.

Reasoning in the same way, we arrive at the following famous theorem, which was first stated by the Presbyterian minister Thomas Bayes (1702–1763):

F I G U R E 3

t h e o r e m 1

> ## Bayes' Formula
>
> Let U_1, U_2, . . . , U_n be n mutually exclusive events whose union is the sample space S. Let E be an arbitrary event in S such that $P(E) \neq 0$. Then
>
> $$P(U_1|E) = \frac{P(U_1 \cap E)}{P(E)}$$
>
> $$= \frac{P(U_1 \cap E)}{P(U_1 \cap E) + P(U_2 \cap E) + \cdots + P(U_n \cap E)}$$
>
> $$= \frac{P(E|U_1)P(U_1)}{P(E|U_1)P(U_1) + P(E|U_2)P(U_2) + \cdots + P(E|U_n)P(U_n)}$$
>
> Similar results hold for U_2, U_3, . . . , U_n.

You do not need to memorize Bayes' formula. In practice, it is often easier to draw a probability tree and use the following:

> ### Bayes' Formula and Probability Trees
>
> $$P(U_1|E) = \frac{\text{Product of branch probabilities leading to } E \text{ through } U_1}{\text{Sum of all branch products leading to } E}$$
>
> Similar results hold for U_2, U_3, . . . , U_n.

E X A M P L E 2 ➤ Tuberculosis Screening A new, inexpensive skin test is devised for detecting tuberculosis. To evaluate the test before it is put into use, a medical researcher randomly selects 1,000 people. Using precise but more expensive methods already available, it is found that 8% of the 1,000 people tested have tuberculosis. Now each of the 1,000 subjects is given the new skin test and the following results are recorded: The test indicates tuberculosis in 96% of those who have it and in 2% of those who do not. Based on these results, what is the probability of a randomly chosen person having tuberculosis given that the skin test indicates the disease? What is the probability of a person not having tuberculosis given that the skin test indicates the disease? (That is, what is the probability of the skin test giving a *false positive result?*)

Solution   Now we will see the power of Bayes' formula in an important application. To start, we form a tree diagram and place appropriate probabilities on each branch:

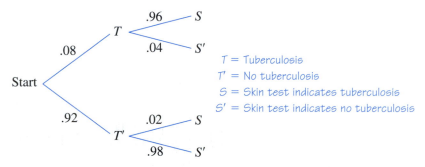

T = Tuberculosis
T' = No tuberculosis
S = Skin test indicates tuberculosis
S' = Skin test indicates no tuberculosis

We are interested in finding $P(T|S)$; that is, the probability of a person having tuberculosis given that the skin test indicates the disease. Bayes' formula for this case is

$$P(T|S) = \frac{\text{Product of branch probabilities leading to } S \text{ through } T}{\text{Sum of all branch products leading to } S}$$

Substituting appropriate values from the probability tree, we obtain

$$P(T|S) = \frac{(.08)(.96)}{(.08)(.96) + (.92)(.02)} = .81$$

The probability of a person not having tuberculosis given that the skin test indicates the disease, denoted by $P(T'|S)$, is

$$P(T'|S) = 1 - P(T|S) = 1 - .81 = .19 \quad P(T|S) + P(T'|S) = 1 \quad \blacktriangleleft$$

Other important questions that need to be answered are indicated in Matched Problem 2.

MATCHED PROBLEM 2 ➤   What is the probability that a person has tuberculosis given that the test indicates no tuberculosis is present? (That is, what is the probability of the skin test giving a *false negative result?*) What is the probability that a person does not have tuberculosis given that the test indicates no tuberculosis is present?   ◀

EXAMPLE 3 ➤   **Product Defects**   A company produces 1,000 refrigerators a week at three plants. Plant *A* produces 350 refrigerators a week, plant *B* produces 250 refrigerators a week, and plant *C* produces 400 refrigerators a week. Production records indicate that 5% of the refrigerators produced at plant *A* will be defective, 3% of those produced at plant *B* will be defective, and 7% of those produced at plant *C* will be defective. All the refrigerators are shipped to a central warehouse. If a refrigerator at the warehouse is found to be defective, what is the probability that it was produced at plant *A*?

Solution We begin by constructing a tree diagram:

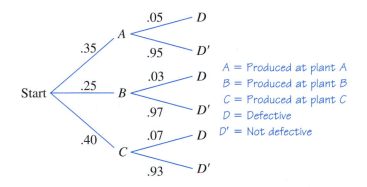

A = Produced at plant A
B = Produced at plant B
C = Produced at plant C
D = Defective
D' = Not defective

The probability that a defective refrigerator was produced at plant A is $P(A|D)$. Bayes' formula for this case is

$$P(A|D) = \frac{\text{Product of branch probabilities leading to } D \text{ through } A}{\text{Sum of all branch products leading to } D}$$

Using the values from the probability tree, we have

$$P(A|D) = \frac{(.35)(.05)}{(.35)(.05) + (.25)(.03) + (.40)(.07)}$$
$$\approx .33$$

◄

MATCHED PROBLEM 3 ➤ In Example 3, what is the probability that a defective refrigerator in the warehouse was produced at plant B? At plant C?

◄

e x p l o r e – d i s c u s s 2

Given the probability tree:

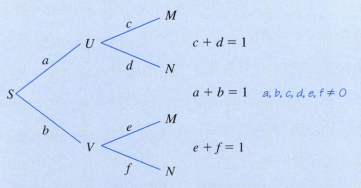

$c + d = 1$

$a + b = 1$ $\quad a, b, c, d, e, f \neq 0$

$e + f = 1$

Suppose U and M are independent events. Discuss the restrictions that their independence forces on the probabilities $a, b, c, d, e,$ and f.

Answers to Matched Problems   1. $P(U_1|W) = \frac{4}{19} \approx .21$; $P(U_2|W) = \frac{15}{19} \approx .79$

2. $P(T|S') = .004$; $P(T'|S') = .996$ 3. $P(B|D) \approx .14$; $P(C|D) \approx .53$

EXERCISE 6-6

A *Find the probabilities in Problems 1–6 by referring to the tree diagram below.*

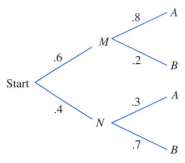

1. $P(M \cap A) = P(M)P(A|M)$
2. $P(N \cap B) = P(N)P(B|N)$
3. $P(A) = P(M \cap A) + P(N \cap A)$
4. $P(B) = P(M \cap B) + P(N \cap B)$
5. $P(M|A) = \dfrac{P(M \cap A)}{P(M \cap A) + P(N \cap A)}$
6. $P(N|B) = \dfrac{P(N \cap B)}{P(N \cap B) + P(M \cap B)}$

Find the probabilities in Problems 7–10 by referring to the following Venn diagram and using Bayes' formula (assume that the simple events in S are equally likely):

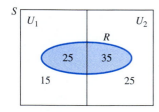

7. $P(U_1|R)$
8. $P(U_2|R)$
9. $P(U_1|R')$
10. $P(U_2|R')$

B *Find the probabilities in Problems 11–16 by referring to the following tree diagram and using Bayes' formula:*

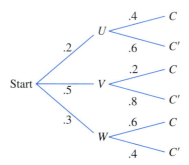

11. $P(U|C)$
12. $P(V|C')$
13. $P(W|C)$
14. $P(U|C')$
15. $P(V|C)$
16. $P(W|C')$

Find the probabilities in Problems 17–22 by referring to the following Venn diagram and using Bayes' formula (assume that the simple events in S are equally likely):

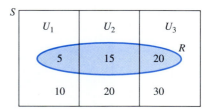

17. $P(U_1|R)$
18. $P(U_2|R')$
19. $P(U_3|R)$
20. $P(U_1|R')$
21. $P(U_2|R)$
22. $P(U_3|R')$

In Problems 23 and 24, use the probabilities on the tree diagram on the left to find the probability of each branch of the tree diagram on the right.

23.

24.

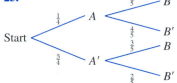

In Problems 25–28, one of two urns is chosen at random with one as likely to be chosen as the other. Then a ball is withdrawn from the chosen urn. Urn 1 contains 1 white and 4 red balls, and urn 2 has 3 white and 2 red balls.

25. If a white ball is drawn, what is the probability that it came from urn 1?

26. If a white ball is drawn, what is the probability that it came from urn 2?

27. If a red ball is drawn, what is the probability that it came from urn 2?

28. If a red ball is drawn, what is the probability that it came from urn 1?

In Problems 29 and 30, an urn contains 4 red and 5 white balls. Two balls are drawn in succession without replacement.

29. If the second ball is white, what is the probability that the first ball was white?

30. If the second ball is red, what is the probability that the first ball was red?

In Problems 31 and 32, urn 1 contains 7 red and 3 white balls. Urn 2 contains 4 red and 5 white balls. A ball is drawn from urn 1 and placed in urn 2. Then a ball is drawn from urn 2.

31. If the ball drawn from urn 2 is red, what is the probability that the ball drawn from urn 1 was red?

32. If the ball drawn from urn 2 is white, what is the probability that the ball drawn from urn 1 was white?

In Problems 33 and 34, refer to the following probability tree:

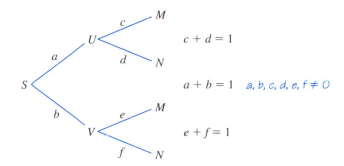

$$c + d = 1$$

$$a + b = 1 \quad a, b, c, d, e, f \neq 0$$

$$e + f = 1$$

33. Suppose that $c = e$. Discuss the dependence or independence of events U and M.

34. Suppose that $c = d = e = f$. Discuss the dependence or independence of events M and N.

C

35. If 2 cards are drawn in succession from a standard 52-card deck without replacement and the second card is a heart, what is the probability that the first card is a heart?

36. A box contains 10 balls numbered 1 through 10. Two balls are drawn in succession without replacement. If the second ball drawn has the number 4 on it, what is the probability that the first ball had a smaller number on it? An even number on it?

37. Show that $P(U_1|R) + P(U_1'|R) = 1$.

38. If U_1 and U_2 are two mutually exclusive events whose union is the equally likely sample space S and if E is an arbitrary event in S such that $P(E) \neq 0$, show that

$$P(U_1|E) = \frac{n(U_1 \cap E)}{n(U_1 \cap E) + n(U_2 \cap E)}$$

APPLICATIONS

In the following applications, the word "probability" is often understood to mean "approximate empirical probability."

Business & Economics

39. *Employee screening.* The management of a company finds that 30% of the secretaries hired are unsatisfactory. The personnel director is instructed to devise a test that will improve the situation. One hundred employed secretaries are chosen at random and are given a newly constructed test. Out of these, 90% of the successful secretaries pass the test and 20% of the unsuccessful secretaries pass. Based on these results, if a person applies for a secretarial job, takes the test, and passes it, what is the probability that he or she is a good secretary? If the applicant fails the test, what is the probability that he or she is a good secretary?

40. *Employee rating.* A company has rated 75% of its employees as satisfactory and 25% as unsatisfactory. Personnel records indicate that 80% of the satisfactory workers had previous work experience, while only 40% of the unsatisfactory workers had any previous work experience. If a person with previous work experience is hired, what is the probability that this person will be a satisfactory employee? If a person with no previous work experience is hired, what is the probability that this person will be a satisfactory employee?

41. *Product defects.* A manufacturer obtains clock–radios from three different subcontractors: 20% from A, 40% from B, and 40% from C. The defective rates for these subcontractors are 1%, 3%, and 2%, respectively. If a defective clock–radio is returned by a customer, what is the probability that it came from subcontractor A? From B? From C?

42. *Product defects.* A computer store sells three types of microcomputers, brand A, brand B, and brand C. Of the computers they sell, 60% are brand A, 25% are brand B, and 15% are brand C. They have found that 20% of the brand A computers, 15% of the brand B computers, and 5% of the brand C computers are returned for service during the warranty period. If a computer is returned for service during the warranty period, what is the probability that it is a brand A computer? A brand B computer? A brand C computer?

Life Sciences

43. *Cancer screening.* A new, simple test has been developed to detect a particular type of cancer. The test must be evaluated before it is put into use. A medical researcher selects a ran-

dom sample of 1,000 adults and finds (by other means) that 2% have this type of cancer. Each of the 1,000 adults is given the test, and it is found that the test indicates cancer in 98% of those who have it and in 1% of those who do not. Based on these results, what is the probability of a randomly chosen person having cancer given that the test indicates cancer? Of a person having cancer given that the test does not indicate cancer?

44. *Pregnancy testing.* In a random sample of 200 women who suspect that they are pregnant, 100 turn out to be pregnant. A new pregnancy test given to these women indicated pregnancy in 92 of the 100 pregnant women and in 12 of the 100 nonpregnant women. If a woman suspects she is pregnant and this test indicates that she is pregnant, what is the probability that she is pregnant? If the test indicates that she is not pregnant, what is the probability that she is not pregnant?

45. *Medical research.* In a random sample of 1,000 people, it is found that 7% have a liver ailment. Of those who have a liver ailment, 40% are heavy drinkers, 50% are moderate drinkers, and 10% are nondrinkers. Of those who do not have a liver ailment, 10% are heavy drinkers, 70% are moderate drinkers, and 20% are nondrinkers. If a person is chosen at random and it is found that he or she is a heavy drinker, what is the probability of that person having a liver ailment? What is the probability for a nondrinker?

46. *Tuberculosis screening.* A test for tuberculosis was given to 1,000 subjects, 8% of whom were known to have tuberculosis. For the subjects who had tuberculosis, the test indicated tuberculosis in 90% of the subjects, was inconclusive for 7%, and indicated no tuberculosis in 3%. For the subjects who did not have tuberculosis, the test indicated tuberculosis in 5% of the subjects, was inconclusive for 10%, and indicated no tuberculosis in the remaining 85%. What is the probability of a randomly selected person having tuberculosis given that the test indicates tuberculosis? Of not having tuberculosis given that the test was inconclusive?

Social Sciences

47. *Police science.* A new lie-detector test has been devised and must be tested before it is put into use. One hundred people are selected at random, and each person draws and keeps a card from a box of 100 cards. Half the cards instruct the person to lie and the others instruct the person to tell the truth. The test indicates lying in 80% of those who lied and in 5% of those who did not. What is the probability that a randomly chosen subject will have lied given that the test indicates lying? That the subject will not have lied given that the test indicates lying?

48. *Politics.* In a given county, records show that of the registered voters, 45% are Democrats, 35% are Republicans, and 20% are independents. In an election, 70% of the Democrats, 40% of the Republicans, and 80% of the independents voted in favor of a parks and recreation bond proposal. If a registered voter chosen at random is found to have voted in favor of the bond, what is the probability that the voter is a Republican? An independent? A Democrat?

SECTION 6-7 ■ Random Variable, Probability Distribution, and Expectation

- RANDOM VARIABLE; PROBABILITY DISTRIBUTION
- EXPECTED VALUE OF A RANDOM VARIABLE
- DECISION-MAKING AND EXPECTED VALUE

■ RANDOM VARIABLE; PROBABILITY DISTRIBUTION

When performing a random experiment, a sample space S is selected in such a way that all probability problems of interest relative to the experiment can be solved. In many situations we may not be interested in each simple event in the sample space S but in some numerical value associated with the event. For example, if 3 coins are tossed, we may be interested in the number of heads that turn up rather than in the particular pattern that turns up. Or, in selecting a random sample of students, we may be interested in the proportion that are women rather than which particular students are women. In the same way, a "craps" player is usually interested in the sum of the dots on the showing faces of the dice rather than the pattern of dots on each face.

In each of these examples, we have a rule that assigns to each simple event in S a single real number. Mathematically speaking, we are dealing with a function (see Section 1-2). Historically, this particular type of function has been called a "random variable."

> **Random Variable**
>
> A **random variable** is a function that assigns a numerical value to each simple event in a sample space S.

The term *random variable* is an unfortunate choice, since it is neither random nor a variable—it is a function with a numerical value, and it is defined on a sample space. But the terminology has stuck and is now standard, so we shall have to live with it. Capital letters, such as X, are used to represent random variables.

Let us return to the experiment of tossing 3 coins. A sample space S of equally likely simple events is indicated in Table 1. Suppose we are interested in the number of heads (0, 1, 2, or 3) appearing on each toss of the 3 coins and the probability of each of these events. We introduce a random variable X (a function) that indicates the

TABLE 1
Number of Heads in the Toss of 3 Coins

SAMPLE SPACE S		NUMBER OF HEADS $X(e_j)$
e_1:	TTT	0
e_2:	TTH	1
e_3:	THT	1
e_4:	HTT	1
e_5:	THH	2
e_6:	HTH	2
e_7:	HHT	2
e_8:	HHH	3

number of heads for each simple event in S (see the second column in Table 1). For example, $X(e_1) = 0$, $X(e_2) = 1$, and so on. The random variable X assigns a numerical value to each simple event in the sample space S.

We are interested in the probability of the occurrence of each image or range value of X; that is, in the probability of the occurrence of 0 heads, 1 head, 2 heads, or 3 heads in the single toss of 3 coins. We indicate this probability by

$$p(x) \qquad \text{where} \quad x \in \{0, 1, 2, 3\}$$

The function p is called the **probability function* of the random variable X.**

What is $p(2)$, the probability of getting exactly 2 heads on the single toss of 3 coins? "Exactly 2 heads occur" is the event

$$E = \{THH, HTH, HHT\}$$

Thus,

$$p(2) = \frac{n(E)}{n(S)} = \frac{3}{8}$$

Proceeding similarly for $p(0)$, $p(1)$, and $p(3)$, we obtain the results in Table 2. This table is called a **probability distribution for the random variable X.** Probability distributions are also represented graphically, as shown in Figure 1. The graph of a probability distribution is often called a **histogram.**

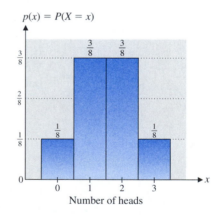

TABLE 2
Probability Distribution

NUMBER OF HEADS x	0	1	2	3
PROBABILITY $p(x)$	$\dfrac{1}{8}$	$\dfrac{3}{8}$	$\dfrac{3}{8}$	$\dfrac{1}{8}$

FIGURE 1
Histogram for a probability distribution

Note from Table 2 or Figure 1 that

1. $0 \leqslant p(x) \leqslant 1, \quad x \in \{0, 1, 2, 3\}$
2. $p(0) + p(1) + p(2) + p(3) = \frac{1}{8} + \frac{3}{8} + \frac{3}{8} + \frac{1}{8} = 1$

These are general properties that any probability distribution of a random variable X associated with a finite sample space must have.

---

* The probability function p of the random variable X is defined by $p(x) = P(\{e_i \in S | X(e_i) = x\})$, which, because of its cumbersome nature, is usually simplified to $p(x) = P(X = x)$ or, simply, $p(x)$. We will use the simplified notation.

> ### Probability Distribution of a Random Variable X
>
> A probability function $P(X = x) = p(x)$ is a **probability distribution of the random variable X** if
>
> 1. $0 \leq p(x) \leq 1, \quad x \in \{x_1, x_2, \ldots, x_n\}$
> 2. $p(x_1) + p(x_2) + \cdots + p(x_n) = 1$
>
> where $\{x_1, x_2, \ldots, x_n\}$ are the (range) values of X (see Fig. 2).

Figure 2 illustrates the process of forming a probability distribution of a random variable.

FIGURE 2
Probability distribution of a random variable for a finite sample space

■ EXPECTED VALUE OF A RANDOM VARIABLE

Suppose the experiment of tossing 3 coins was repeated a large number of times. What would be the average number of heads per toss (the total number of heads in all tosses divided by the total number of tosses)? Consulting the probability distribution in Table 2 or Figure 1, we see that we would expect to toss 0 heads $\frac{1}{8}$ of the time, 1 head $\frac{3}{8}$ of the time, 2 heads $\frac{3}{8}$ of the time, and 3 heads $\frac{1}{8}$ of the time. Thus, in the long run, we would expect the average number of heads per toss of the 3 coins, or the *expected value* $E(X)$, to be given by

$$E(X) = 0(\tfrac{1}{8}) + 1(\tfrac{3}{8}) + 2(\tfrac{3}{8}) + 3(\tfrac{1}{8}) = \tfrac{12}{8} = 1.5$$

It is important to note that the expected value is not a value that will necessarily occur in a single experiment (1.5 heads cannot occur in the toss of 3 coins), but it is an average of what occurs over a large number of experiments. Sometimes we will toss more than 1.5 heads and sometimes less, but if the experiment is repeated many times, the average number of heads per experiment should be close to 1.5.

We now make the above discussion more precise through the following definition of expected value:

Expected Value of a Random Variable X

Given the probability distribution for the random variable X,

x_i	x_1	$x_2 \cdots x_n$
p_i	p_1	$p_2 \cdots p_n$

where $p_i = p(x_i)$, we define the **expected value of X,** denoted $E(X)$, by the formula

$$E(X) = x_1 p_1 + x_2 p_2 + \cdots + x_n p_n$$

We again emphasize that the expected value is not to be expected to occur in a single experiment; it is a long-run average of repeated experiments—it is the weighted average of the possible outcomes, each weighted by its probability.

Steps for Computing the Expected Value of a Random Variable X

Step 1.   Form the probability distribution of the random variable X.

Step 2.   Multiply each image value of X, x_i, by its corresponding probability of occurrence p_i; then add the results.

EXAMPLE 1 ➤ **Expected Value**   What is the expected value (long-run average) of the number of dots facing up for the roll of a single die?

Solution   If we choose

$$S = \{1, 2, 3, 4, 5, 6\}$$

as our sample space, then each simple event is a numerical outcome reflecting our interest, and each is equally likely. The random variable X in this case is just the identity function (each number is associated with itself). Thus, the probability distribution for X is

x_i	1	2	3	4	5	6
p_i	$\frac{1}{6}$	$\frac{1}{6}$	$\frac{1}{6}$	$\frac{1}{6}$	$\frac{1}{6}$	$\frac{1}{6}$

Hence,

$$E(X) = 1(\tfrac{1}{6}) + 2(\tfrac{1}{6}) + 3(\tfrac{1}{6}) + 4(\tfrac{1}{6}) + 5(\tfrac{1}{6}) + 6(\tfrac{1}{6})$$
$$= \tfrac{21}{6} = 3.5$$

◀

MATCHED PROBLEM 1 ➤ Suppose the die in Example 1 is not fair and we obtain (empirically) the following probability distribution for X:

x_i	1	2	3	4	5	6	
p_i	.14	.13	.18	.20	.11	.24	[*Note:* Sum = 1.]

What is the expected value of X?   ◀

explore – discuss 1

In a class of 10 students, 1 student scored 95 on the first exam, 3 scored 85, 1 scored 82, 2 scored 75, 2 scored 73, and 1 scored 65. The probability distribution of an exam score for a student selected at random from the class is as follows:

x_i	65	73	75	82	85	95
p_i	.1	.2	.2	.1	.3	.1

Discuss the relationship between the expected value of the probability distribution and the class average for the exam.

EXAMPLE 2 ➤ **Expected Value** A carton of 20 calculator batteries contains 2 dead ones. A random sample of 3 is selected from the 20 and tested. Let X be the random variable associated with the number of dead batteries found in a sample.

(A) Find the probability distribution of X.
(B) Find the expected number of dead batteries in a sample.

Solution (A) The number of ways of selecting a sample of 3 from 20 (order is not important) is $C_{20,3}$. This is the number of simple events in the experiment, each as likely as the other. A sample will have either 0, 1, or 2 dead batteries. These are the values of the random variable in which we are interested. The probability distribution is computed as follows:

$$p(0) = \frac{C_{18,3}}{C_{20,3}} \approx .716 \qquad p(1) = \frac{C_{2,1}C_{18,2}}{C_{20,3}} \approx .268 \qquad p(2) = \frac{C_{2,2}C_{18,1}}{C_{20,3}} \approx .016$$

We summarize the above results in a convenient table:

x_i	0	1	2	
p_i	.716	.268	.016	[*Note:* .716 + .268 + .016 = 1.]

(B) The expected number of dead batteries in a sample is readily computed as follows:

$$E(X) = (0)(.716) + (1)(.268) + (2)(.016) = .3$$

The expected value is not one of the random variable values; rather, it is a number that the average number of dead batteries in a sample would approach as the experiment is repeated without end. ◄

MATCHED PROBLEM 2 ▶ Repeat Example 2 using a random sample of 4. ◄

EXAMPLE 3 ▶ **Expected Value of a Game** A spinner device is numbered from 0 to 5, and each of the 6 numbers is as likely to come up as any other. A player who bets $1 on any given number wins $4 (and gets the bet back) if the pointer comes to rest on the chosen number; otherwise, the $1 bet is lost. What is the expected value of the game (long-run average gain or loss per game)?

Solution The sample space of equally likely events is

$$S = \{0, 1, 2, 3, 4, 5\}$$

Each sample point occurs with a probability of $\frac{1}{6}$. The random variable X assigns $4 to the winning number and $-$1$ to each of the remaining numbers. Thus, the probability distribution for X, called a **payoff table,** is as shown in the margin. The probability of winning $4 is $\frac{1}{6}$ and of losing $1 is $\frac{5}{6}$. We can now compute the expected value of the game:

$$E(X) = \$4(\tfrac{1}{6}) + (-\$1)(\tfrac{5}{6}) = -\$\tfrac{1}{6} \approx -\$0.1667 \approx -17¢ \text{ per game}$$

Thus, in the long run the player will lose an average of about 17¢ per game. ◄

Using the definition of a fair game in Section 6-4, it can be shown that **a game is fair if and only if $E(X) = 0$.** The game in Example 3 is not fair.

Payoff Table (Probability Distribution for X)

x_i	$4	$-$1
p_i	$\frac{1}{6}$	$\frac{5}{6}$

MATCHED PROBLEM 3 ▶ Repeat Example 3 with the player winning $5 instead of $4 if the chosen number turns up. The loss is still $1 if any other number turns up. Is this now a fair game? ◄

e x p l o r e – d i s c u s s 2

A coin is tossed twice and the number of heads recorded. You pay $1 to play. You keep your dollar if no heads turn up and lose your dollar if exactly one head turns up. What should the house pay if two heads turns up in order for the game to be fair? Discuss the process and reasoning used to arrive at your answer.

▶ EXAMPLE 4 ▶ **Expected Value and Insurance** Suppose you are interested in insuring a car stereo system for $500 against theft. An insurance company charges a premium of $60 for coverage for 1 year, claiming an empirically determined probability of .1 that the stereo will be stolen some time during the year. What is your expected return from the insurance company if you take out this insurance?

Payoff Table

x_i	$440	−$60
p_i	.1	.9

Solution   This is actually a game of chance in which your stake is $60. You have a .1 chance of receiving $440 from the insurance company ($500 minus your stake of $60) and a .9 chance of losing your stake of $60. What is the expected value of this "game"? We form a payoff table (the probability distribution for X) as shown in the margin. Then we compute the expected value as follows:

$$E(X) = (\$440)(.1) + (-\$60)(.9) = -\$10$$

This means that if you insure with this company over many years and the circumstances remain the same, you would have an average net loss to the insurance company of $10 per year. ◄

MATCHED PROBLEM 4 ➤   Find the expected value in Example 4 from the insurance company's point of view.
◄

■ DECISION-MAKING AND EXPECTED VALUE

We conclude this section with an example in decision-making.

EXAMPLE 5 ➤   **Decision Analysis**   An outdoor concert featuring a very popular musical group is scheduled for a Sunday afternoon in a large open stadium. The promoter, worrying about being rained out, contacts a long-range weather forecaster who predicts the chance of rain on that Sunday to be .24. If it does not rain, the promoter is certain to net $100,000; if it does rain, the promoter estimates that the net will be only $10,000. An insurance company agrees to insure the concert for $100,000 against rain at a premium of $20,000. Should the promoter buy the insurance?

Solution   The promoter has a choice between two courses of action; A_1: Insure and A_2: Do not insure. As an aid in making a decision, the expected value is computed for each course of action. Probability distributions are indicated in the following payoff table (read vertically):

Payoff Table

p_i	A_1: INSURE x_i	A_2: DO NOT INSURE x_i
.24 (rain)	$90,000	$ 10,000
.76 (no rain)	$80,000	$100,000

Note that the $90,000 entry comes from the insurance company's payoff ($100,000) minus the premium ($20,000) plus gate receipts ($10,000). The reasons for the other entries should be obvious. The expected value for each course of action is computed as follows:

A_1: Insure

$$E(X) = x_1 p_1 + x_2 p_2$$
$$= (\$90,000)(.24) + (\$80,000)(.76)$$
$$= \$82,400$$

A_2: Do Not Insure

$$E(X) = (\$10,000)(.24) + (\$100,000)(.76)$$
$$= \$78,400$$

It appears that the promoter's best course of action is to buy the insurance at $20,000. The promoter is using a long-run average to make a decision about a single event—a common practice in making decisions in areas of uncertainty. ◀

MATCHED PROBLEM 5 ▶ In Example 5, what is the insurance company's expected value if it writes the policy? ◀

Answers to Matched Problems   1. $E(X) = 3.73$ 2. (A)

x_i	0	1	2
p_i	.632	.337	.032*

Note: Due to roundoff error, sum = 1.001 ≈ 1.

(B) .4

3. $E(X) = \$0$; the game is fair.
4. $E(X) = (-\$440)(.1) + (\$60)(.9) = \$10$ (This amount, of course, is necessary to cover expenses and profit.)
5. $E(X) = (-\$80,000)(.24) + (\$20,000)(.76) = -\$4,000$ (This means the insurance company had other information regarding the weather than the promoter; otherwise, the company would not have written this policy.)

EXERCISE 6-7

Where possible, construct a probability distribution or payoff table for a suitable random variable X; then complete the problem.

A

1. If the probability distribution for the random variable X is given in the table, what is the expected value of X?

x_i	-3	0	4
p_i	.3	.5	.2

2. If the probability distribution for the random variable X is given in the table, what is the expected value of X?

x_i	-2	-1	0	1	2
p_i	.1	.2	.4	.2	.1

3. In tossing 2 fair coins, what is the expected number of heads?

4. In a family with 2 children, excluding multiple births and assuming a boy is as likely as a girl at each birth, what is the expected number of boys?

5. A fair coin is flipped. If a head turns up, you win $1. If a tail turns up, you lose $1. What is the expected value of the game? Is the game fair?

6. Repeat Problem 5, assuming an unfair coin with the probability of a head being .55 and a tail being .45.

B

7. After paying $4 to play, a single fair die is rolled and you are paid back the number of dollars corresponding to the number of dots facing up. For example, if a 5 turns up, $5 is returned to you for a net gain, or payoff, of $1; if a 1 turns up, $1 is returned for a net gain of $-$3$; and so on. What is the expected value of the game? Is the game fair?

8. Repeat Problem 7 with the same game costing $3.50 for each play.

9. Two coins are flipped. You win $2 if either 2 heads or 2 tails turn up; you lose $3 if a head and a tail turn up. What is the expected value of the game?

10. In Problem 9, for the game to be fair, how much *should* you lose if a head and a tail turn up?

11. A friend offers the following game: She wins $1 from you if on 4 rolls of a single die, a 6 turns up at least once; other-

wise, you win $1 from her. What is the expected value of the game to you? To her?

12. On 3 rolls of a single die, you will lose $10 if a 5 turns up at least once, and you will win $7 otherwise. What is the expected value of the game?

13. A single die is rolled once. You win $5 if a 1 or 2 turns up and $10 if a 3, 4, or 5 turns up. How much should you lose if a 6 turns up for the game to be fair? Describe the steps you took to arrive at your answer.

14. A single die is rolled once. You lose $12 if a number divisible by 3 turns up. How much should you win if a number not divisible by 3 turns up for the game to be fair? Describe the process and reasoning used to arrive at your answer.

15. A pair of dice is rolled once. Suppose you lose $10 if a 7 turns up and win $11 if an 11 or 12 turns up. How much should you win or lose if any other number turns up in order for the game to be fair?

16. A coin is tossed 3 times. Suppose you lose $3 if 3 heads appear, lose $2 if 2 heads appear, and win $3 if 0 heads appear. How much should you win or lose if 1 head appears in order for the game to be fair?

17. The payoff table for two courses of action, A_1 or A_2, is given below. Which of the two actions will produce the largest expected value? What is it?

p_i	A_1 x_i	A_2 x_i
.1	− $200	− $100
.2	$100	$200
.4	$400	$300
.3	$100	$200

18. The payoff table for three possible courses of action is given below. Which of the three actions will produce the largest expected value? What is it?

p_i	A_1 x_i	A_2 x_i	A_3 x_i
.2	$ 500	$ 400	$ 300
.4	1,200	1,100	1,000
.3	1,200	1,800	1,700
.1	1,200	1,800	2,400

19. Roulette wheels in Nevada generally have 38 equally spaced slots numbered 00, 0, 1, 2, . . . , 36. A player who bets $1 on any given number wins $35 (and gets the bet back) if the ball comes to rest on the chosen number; otherwise, the $1 bet is lost. What is the expected value of this game?

20. In roulette (see Problem 19) the numbers from 1 to 36 are evenly divided between red and black. A player who bets $1 on black wins $1 (and gets the bet back) if the ball comes to rest on black; otherwise (if the ball lands on red, 0 or 00), the $1 bet is lost. What is the expected value of the game?

21. A game has an expected value to you of $100. It costs $100 to play, but if you win you receive $100,000 (including your $100 bet), for a net gain of $99,900. What is the probability of winning? Would you play this game? Discuss the factors that would influence your decision.

22. A game has an expected value to you of − $0.50. It costs $2 to play, but if you win you receive $20 (including your $2 bet), for a net gain of $18. What is the probability of winning? Would you play this game? Discuss the factors that would influence your decision.

C

23. Five thousand tickets are sold at $1 each for a charity raffle. Tickets are to be drawn at random and monetary prizes awarded as follows: 1 prize of $500, 3 prizes of $100, 5 prizes of $20, and 20 prizes of $5. What is the expected value of this raffle if you buy 1 ticket?

24. Ten thousand raffle tickets are sold at $2 each for a local library benefit. Prizes are awarded as follows: 2 prizes of $1,000, 4 prizes of $500, and 10 prizes of $100. What is the expected value of this raffle if you purchase 1 ticket?

25. A box of 10 flashbulbs contains 3 defective bulbs. A random sample of 2 is selected and tested. Let X be the random variable associated with the number of defective bulbs in the sample.
(A) Find the probability distribution of X.
(B) Find the expected number of defective bulbs in a sample.

26. A box of 8 flashbulbs contains 3 defective bulbs. A random sample of 2 is selected and tested. Let X be the random variable associated with the number of defective bulbs in a sample.
(A) Find the probability distribution of X.
(B) Find the expected number of defective bulbs in a sample.

27. One thousand raffle tickets are sold at $1 each. Three tickets will be drawn at random (without replacement) and each will pay $200. Suppose you buy 5 tickets.
(A) Create a payoff table for 0, 1, 2, and 3 winning tickets among the 5 tickets you purchased. (If you do not have any winning tickets, you lose $5; if you have 1 winning

ticket, you net $195, since your initial $5 will not be returned to you; and so on.)
(B) What is the expected value of the raffle to you?

28. Repeat Problem 27 with the purchase of 10 tickets.

 APPLICATIONS

Business & Economics

29. *Insurance.* The annual premium for a $5,000 insurance policy against the theft of a painting is $150. If the (empirical) probability that the painting will be stolen during the year is .01, what is your expected return from the insurance company if you take out this insurance?

30. *Insurance.* Repeat Problem 29 from the point of view of the insurance company.

31. *Decision analysis.* An oil company, after careful testing and analysis, is considering drilling in two different sites. It is estimated that site A will net $30 million if successful (probability .2) and lose $3 million if not (probability .8); site B will net $70 million if successful (probability .1) and lose $4 million if not (probability .9). Which site should the company choose according to the expected return from each site?

32. *Decision analysis.* Repeat Problem 31, assuming additional analysis caused the estimated probability of success in field B to be changed from .1 to .11.

Life Sciences

33. *Genetics.* Suppose that, at each birth, having a girl is not as likely as having a boy. The probability assignments for the number of boys in a 3-child family are approximated empirically from past records and are given in the table. What is the expected number of boys in a 3-child family?

NUMBER OF BOYS x_i	p_i
0	.12
1	.36
2	.38
3	.14

34. *Genetics.* A pink-flowering plant is of genotype RW. If two such plants are crossed, we obtain a red plant (RR) with probability .25, a pink plant (RW or WR) with probability .50, and a white plant (WW) with probability .25, as shown in the table. What is the expected number of W genes present in a crossing of this type?

NUMBER OF W GENES PRESENT x_i	p_i
0	.25
1	.50
2	.25

Social Sciences

35. *Politics.* A money drive is organized by a campaign committee for a candidate running for public office. Two approaches are considered:

> A_1: A general mailing with a followup mailing
>
> A_2: Door-to-door solicitation with followup telephone calls

From campaign records of previous committees, average donations and their corresponding probabilities are estimated to be:

A_1		A_2	
x_i *(Return per Person)*	p_i	x_i *(Return per Person)*	p_i
$10	.3	$15	.3
5	.2	3	.1
0	.5	0	.6
	1.0		1.0

What are the expected returns? Which course of action should be taken according to the expected returns?

CHAPTER 6
GROUP ACTIVITY

Car and Crying Towels

The host on a television game show gives you (the contestant) a choice of three closed doors labeled *A*, *B*, and *C*. Behind one of the doors is a car, and behind each of the other two, a crying towel. After you have selected a door, which is not opened, the host opens one of the two doors not selected by you. The host, knowing which door hides the car, always chooses a crying towel door. The host then gives you a chance to stick to your original choice or to switch to the other unopened door. Which of the following two strategies would you choose and why?

Strategy 1.   **Stick** to your original choice.

Strategy 2.   **Switch** to the other unopened door.

(A) *Solve by ''Common Sense'' Reasoning.* Let each person in your group indicate their feeling about which strategy would be best or if it does not matter. Present the problem to friends and ask which strategy they would prefer and why. Summarize the results of these discussions.

(B) *Solve by Simulation.* To simulate this problem use a spinner with three equally likely outcomes, and suppose the car is hidden behind door *C*. Your original door choice is made by spinning the spinner once and making the choice indicated by the spinner. [*Note:*   A single die can be used in place of the spinner by choosing door *A* if 1 or 2 dots turn up, door *B* for 3 or 4 dots, and door *C* for 5 or 6 dots.]

We now outline a simulation of the Stick strategy, and leave it to you to do the same for the Switch strategy.

Car prize
behind C

Stick Strategy Simulation

1. Suppose the spinner lands on door *A*. The host opens door *B*, knowing the car is behind door *C*, and reveals a crying towel. You stick to your original choice *A* and lose, getting a crying towel when door *A* is opened.

2. Suppose the spinner lands on door *B*. The host opens door *A*, knowing the car is behind *C*, and reveals a crying towel. You stick to *B* and lose, getting a crying towel when door *B* is opened.
3. Suppose the spinner lands on door *C*. The host opens either door *A* or *B*, revealing a crying towel. You stick to door *C* and win, getting a car when door *C* is opened.

Switch Strategy Simulation   Modify the simulation for the Stick strategy to apply to the Switch strategy.

Each student in your group should run at least 100 trials for each strategy simulation and record the wins and losses. Pool the results of all students in your group. Use the relative frequency of wins for each strategy to write approximate empirical probabilities for a win for that strategy. Now, which strategy do you think you would use and why?

(C) *Solve Theoretically Using Probability Trees*. The car is assumed to be behind door *C*.

Stick Strategy   Complete the following probability tree and compute the probability of a win for the Stick strategy.

Switch Strategy   Construct a probability tree for the Switch strategy, and compute the probability of a win for this strategy.

Compare the results of parts (A), (B), and (C). Which strategy would you now choose and why? Present the problem to a friend and try to convince your friend that the strategy you have chosen is the best strategy.

Now consider the **three cards problem:** Three cards are in a box. Both sides of one card are white, both sides of a second card are black, and one side of a third card is white and the other side is black. A card is drawn from the box at random and placed on the table. A white side is showing. What is the probability that the other side is black? Describe how you would solve this problem by using a simulation approach and by using a theoretical approach. Use each approach to solve the problem.

CHAPTER 6 REVIEW ■ Important Terms and Symbols

6-1 *Basic Counting Principles.* Counting technique; number of elements in a set; addition principle; tree diagram; multiplication principle

$$n(A); \quad n(A \cup B) = n(A) + n(B) - n(A \cap B);$$
$$n(A \cup B) = n(A) + n(B), \text{ if } A \cap B = \varnothing$$

6-2 *Permutations and Combinations.* n factorial; zero factorial; permutation; permutation of n objects; permutation of n objects taken r at a time; combination; combination of n objects taken r at a time

$$n! = n(n-1)(n-2) \cdot \cdots \cdot 2 \cdot 1; \quad 0! = 1;$$

$$P_{n,n} = n!; \quad P_{n,r} = \frac{n!}{(n-r)!};$$

$$C_{n,r} = \binom{n}{r} = \frac{P_{n,r}}{r!} = \frac{n!}{r!(n-r)!}, \, 0 \leqslant r \leqslant n$$

6-3 *Sample Spaces, Events, and Probability.* Random experiment; experiment; sample space; event; simple outcome; simple event; compound event; fundamental sample space; probability of an event; acceptable probability assignment; reasonable probability assignment; probability function; frequency; relative frequency; empirical probability; approximate empirical probability; equally likely assumption

$$P(E); \quad P(E) = \frac{f(E)}{n}; \quad n(E)$$

6-4 *Union, Intersection, and Complement of Events; Odds.* Union; intersection; event A **or** event B; event A **and** event B; mutually exclusive; disjoint; complement of an event; odds; fair game; law of large numbers

$$A \cup B; \quad A \cap B;$$
$$P(A \cup B) = P(A) + P(B) - P(A \cap B);$$
$$P(A \cup B) = P(A) + P(B), \text{ if } A \cap B = \varnothing;$$

$$A'; \quad P(A') = 1 - P(A); \quad \frac{P(E)}{P(E')} \text{ (odds for } E)$$

6-5 *Conditional Probability, Intersection, and Independence.* Conditional probability of A given B; product rule; stochastic process; independent events; dependent events

$$P(A|B) = \frac{P(A \cap B)}{P(B)}, \text{ when } P(B) \neq 0;$$

$$P(A \cap B) = P(A)P(B|A);$$
$$P(A \cap B) = P(A)P(B), \text{ if and only if } A \text{ and } B \text{ are independent;}$$
$$P(A|B) = P(A) \text{ and } P(B|A) = P(B), \text{ if } A \text{ and } B \text{ are independent and } P(A) \neq 0, P(B) \neq 0;$$
$$P(E_1 \cap E_2 \cap \cdots \cap E_n) = P(E_1)P(E_2) \cdots \cdot P(E_n),$$
if $E_1, E_2, \ldots, E_n$ are independent

6-6 *Bayes' Formula.*

$$P(U_1|E)$$

$$= \frac{P(E|U_1)P(U_1)}{P(E|U_1)P(U_1) + P(E|U_2)P(U_2) + \cdots + P(E|U_n)P(U_n)}$$

$$= \frac{\left(\begin{array}{c}\text{Product of branch probabilities leading} \\ \text{to } E \text{ through } U_1\end{array}\right)}{\text{Sum of all branch products leading to } E}$$

6-7 *Random Variable, Probability Distribution, and Expectation.* Random variable; probability function of a random variable X; probability distribution for a random variable; histogram; expected value of a random variable; payoff table; fair game

$$E(X) = x_1p_1 + x_2p_2 + \cdots + x_np_n$$

CHAPTER 6 ■ Review Exercise

Work through all the problems in this chapter review and check your answers in the back of the book. Answers to all review problems are there along with section numbers in italics to indicate where each type of problem is discussed. Where weaknesses show up, review appropriate sections in the text.

A

1. A single die is rolled and a coin is flipped. How many combined outcomes are possible? Solve:
   (A) By using a tree diagram
   (B) By using the multiplication principle

2. Use the Venn diagram to find the number of elements in each of the following sets:

(A) A (B) B
(C) $A \cap B$ (D) $A \cup B$
(E) U (F) A'
(G) $(A \cap B)'$ (H) $(A \cup B)'$

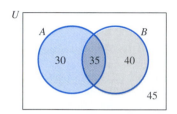

53. Evaluate $C_{6,2}$ and $P_{6,2}$.

4. How many seating arrangements are possible with 6 people and 6 chairs in a row? Solve by using the multiplication principle.

5. Solve Problem 4 using permutations or combinations, whichever is applicable.

6. In a single deal of 5 cards from a standard 52-card deck, what is the probability of being dealt 5 clubs?

7. Betty and Bill are members of a 15-person ski club. If the president and treasurer are selected by lottery, what is the probability that Betty will be president and Bill will be treasurer? (A person cannot hold more than one office.)

8. Each of the first 10 letters of the alphabet is printed on a separate card. What is the probability of drawing 3 cards and getting the code word *dig* by drawing *d* on the first draw, *i* on the second draw, and *g* on the third draw? What is the probability of being dealt a 3-card hand containing the letters *d*, *i*, and *g* in any order?

9. A drug has side effects for 50 out of 1,000 people in a test. What is the approximate empirical probability that a person using the drug will have side effects?

10. A spinning device has 5 numbers, 1, 2, 3, 4, and 5, each as likely to turn up as the other. A person pays $3 and then receives back the dollar amount corresponding to the number turning up on a single spin. What is the expected value of the game? Is the game fair?

11. If A and B are events in a sample space S and $P(A) = .3$, $P(B) = .4$, and $P(A \cap B) = .1$, find:
(A) $P(A')$
(B) $P(A \cup B)$

12. A spinner lands on R with probability .3, on G with probability .5, and on B with probability .2. Find the probability and odds for the spinner landing on either R or G.

13. If in repeated rolls of two fair dice the odds for rolling a sum of 8 before rolling a sum of 7 are 5 to 6, then what is the probability of rolling a sum of 8 before rolling a sum of 7?

Answer Problems 14–22 using the table of probabilities shown below.

	X	Y	Z	$Totals$
S	.10	.25	.15	.50
T	.05	.20	.02	.27
R	.05	.15	.03	.23
$Totals$	.20	.60	.20	1.00

14. Find $P(T)$.
15. Find $P(Z)$.
16. Find $P(T \cap Z)$.
17. Find $P(R \cap Z)$.
18. Find $P(R|Z)$.
19. Find $P(Z|R)$.
20. Find $P(T|Z)$.
21. Are T and Z independent?
22. Are S and X independent?

Answer Problems 23–30 using the following probability tree:

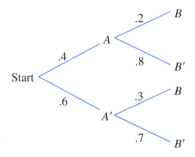

23. $P(A)$ **24.** $P(B|A)$ **25.** $P(B|A')$
26. $P(A \cap B)$ **27.** $P(A' \cap B)$ **28.** $P(B)$
29. $P(A|B)$ **30.** $P(A|B')$

31. (A) If 10 out of 32 students in a class were born in June, July, or August, what is the approximate empirical probability of any student being born in June, July, or August?

(B) If one is as likely to be born in any of the 12 months of a year as any other, what is the theoretical probability of being born in either June, July or August?

(C) Discuss the discrepancy between the answers to parts (A) and (B).

B

32. A player tosses 2 coins and receives $5 if 2 heads turn up, loses $4 if 1 head turns up, and wins $2 if 0 heads turn up. (Would you play this game?) Compute the expected value of the game. Is the game fair?

33. A spinning device has 3 numbers, 1, 2, and 3, each as likely to turn up as the other. If the device is spun twice, what is the probability that:
(A) The same number turns up both times?
(B) The sum of the numbers turning up is 5?

34. In a single draw from a standard 52-card deck, what are the probability and odds for drawing:
(A) A jack or a queen?
(B) A jack or a spade?
(C) A card other than an ace?

35. (A) What are the odds for rolling a sum of 5 on the single roll of two fair dice?
(B) If you bet $1 that a sum of 5 will turn up, what should the house pay (plus return your $1 bet) for the game to be fair?

36. Two coins are flipped 1,000 times with the following frequencies:

2 heads   210
1 head   480
0 heads   310

(A) Compute the empirical probability for each outcome.
(B) Compute the theoretical probability for each outcome.
(C) Using the theoretical probabilities computed in part (B), compute the expected frequency of each outcome, assuming fair coins.

37. A man has 5 children. Each of those children has 3 children, who in turn each have 2 children. Discuss the number of descendants that the man has.

38. A fair coin is tossed 10 times. On each of the first 9 tosses the outcome is heads. Discuss the probability of a head on the 10th toss.

39. An experiment consists of rolling a pair of fair dice. Let X be the random variable associated with the sum of the values that turn up.
(A) Find the probability distribution for X.
(B) Find the expected value of X.

40. Two dice are rolled. The sample space is chosen as the set of all ordered pairs of integers taken from {1, 2, 3, 4, 5, 6}. What is the event A that corresponds to the sum being divisible by 4? What is the event B that corresponds to the sum being divisible by 6? What are $P(A)$, $P(B)$, $P(A \cap B)$, and $P(A \cup B)$?

41. A person tells you that the following approximate empirical probabilities apply to the sample space $\{e_1, e_2, e_3, e_4\}$: $P(e_1) \approx .1$, $P(e_2) \approx -.2$, $P(e_3) \approx .6$, $P(e_4) \approx 2$. There are three reasons why P cannot be a probability function. Name them.

42. Use the following information to complete the table below:

$$n(A) = 50, \quad n(B) = 45,$$
$$n(A \cup B) = 80, \quad n(U) = 100$$

	A	A'	Totals
B			
B'			
Totals			

43. A pointer is spun on a circular spinner. The probabilities of the pointer landing on the integers from 1 to 5 are given in the table below.

e_i	1	2	3	4	5
p_i	.2	.1	.3	.3	.1

(A) What is the probability of the pointer landing on an odd number?
(B) What is the probability of the pointer landing on a number less than 4 given that it landed on an odd number?

44. A card is drawn at random from a standard 52-card deck. If E is the event "The drawn card is red" and F is the event "The drawn card is an ace," then:
(A) Find $P(F|E)$.   (B) Test E and F for independence.

45. How many 3-letter code words are possible using the first 8 letters of the alphabet if no letter can be repeated? If letters can be repeated? If adjacent letters cannot be alike?

46. Solve the following problems using $P_{n,r}$ or $C_{n,r}$:

(A) How many 3-digit opening combinations are possible on a combination lock with 6 digits if the digits cannot be repeated?

(B) Five tennis players have made the finals. If each of the 5 players is to play every other player exactly once, how many games must be scheduled?

In Problems 47–51, urn U_1 contains 2 white balls and 3 red balls; urn U_2 contains 2 white balls and 1 red ball.

47. Two balls are drawn out of urn U_1 in succession. What is the probability of drawing a white ball followed by a red ball if the first ball is:

(A) Replaced? (B) Not replaced?

48. Which of the two parts in Problem 47 involve dependent events?

49. In Problem 47, what is the expected number of red balls if the first ball is:

(A) Replaced? (B) Not replaced?

50. An urn is selected at random by flipping a fair coin; then a ball is drawn from the urn. Compute:

(A) $P(R|U_1)$ (B) $P(R|U_2)$ (C) $P(R)$
(D) $P(U_1|R)$ (E) $P(U_2|W)$ (F) $P(U_1 \cap R)$

51. In Problem 50, are the events "Selecting urn U_1" and "Drawing a red ball" independent?

52. From a standard deck of 52 cards, what is the probability of obtaining a 5-card hand:

(A) Of all diamonds?

(B) Of 3 diamonds and 2 spades?

Write answers in terms of $C_{n,r}$ or $P_{n,r}$; do not evaluate.

53. A group of 10 people includes one married couple. If 4 people are selected at random, what is the probability that the married couple is selected?

54. If 3 operations O_1, O_2, O_3 are performed in order, with possible number of outcomes N_1, N_2, N_3, respectively, determine the number of branches in the corresponding tree diagram.

55. A 5-card hand is drawn from a standard deck. Discuss how you can tell that the following two events are dependent without any computation.

S = Hand consists entirely of spades
H = Hand consists entirely of hearts

C

56. Three fair coins are tossed 1,000 times with the following frequencies of outcomes:

NUMBER OF HEADS	0	1	2	3
FREQUENCY	120	360	350	170

(A) What is the approximate empirical probability of obtaining 2 heads?

(B) What is the theoretical probability of obtaining 2 heads?

(C) What is the expected frequency of obtaining 2 heads?

57. You bet a friend $1 that you will get 1 or more double 6's on 24 rolls of a pair of fair dice. What is your expectation for this game? What is your friend's expectation? Is the game fair?

58. A software development department consists of 6 women and 4 men.

(A) How many ways can they select a chief programmer, a backup programmer, and a programming librarian?

(B) If the positions in part (A) are selected by lottery, what is the probability that women are selected for all 3 positions?

(C) How many ways can they select a team of 3 programmers to work on a particular project?

(D) If the selections in part (C) are made by lottery, what is the probability that a majority of the team members will be women?

59. A group of 150 people includes 52 who play chess, 93 who play checkers, and 28 who play both chess and checkers. How many people in the group play neither game?

60. If 3 people are selected from a group of 7 men and 3 women, what is the probability that at least 1 woman is selected?

Two cards are drawn in succession without replacement from a standard 52-card deck. In Problems 61 and 62, compute the indicated probabilities.

61. The second card is a heart given the first card is a heart.

62. The first card is a heart given the second card is a heart.

63. Two fair (not weighted) dice are each numbered with a 3 on one side, a 2 on two sides, and a 1 on three sides. The dice

are rolled and the numbers on the two up faces are added. If X is the random variable associated with the sample space $S = \{2, 3, 4, 5, 6\}$:
(A) Find the probability distribution of X.
(B) Find the expected value of X.

64. If you pay $3.50 to play the game in Problem 63 (the dice are rolled once) and you are returned the dollar amount corresponding to the sum on the faces, what is the expected value of the game? Is the game fair? If it is not fair, how much should you pay to make the game fair?

65. How many different 5-child families are possible where the sex of each child in the order of their birth is taken into consideration [that is, birth sequences such as (B, G, G, B, B) and (G, B, G, B, B) produce different families]? How many families are possible if the order pattern is not taken into account?

66. Suppose 3 white balls and 1 black ball are placed in a box. Balls are drawn in succession without replacement until a black ball is drawn, and then the game is over. You win if the black ball is drawn on the fourth draw.
(A) What are the probability and odds for winning?
(B) If you bet $1, what should the house pay you for winning (plus return your $1 bet) if the game is to be fair?

67. If each of 5 people is asked to identify his or her favorite book from a list of 10 best-sellers, what is the probability that at least 2 of them identify the same book?

 APPLICATIONS

Business & Economics

68. *Transportation.* A distribution center A wishes to distribute its products to five different retail stores, $B, C, D, E,$ and F, in a city. How many different route plans can be constructed so that a single truck, starting from A, will deliver to each store exactly once, and then return to the center?

69. *Market research.* A survey of 1,000 people indicates that 340 have invested in stocks, 480 have invested in bonds, and 210 have invested in stocks and bonds.
(A) How many people in the survey have invested in stocks or bonds?
(B) How many have invested in neither stocks nor bonds?
(C) How many have invested in bonds and not in stocks?

70. *Market research.* From a survey of 100 residents of a city, it was found that 40 read the daily morning paper, 70 read the daily evening paper, and 30 read both papers. What is the (empirical) probability that a resident selected at random:
(A) Reads a daily paper?
(B) Does not read a daily paper?
(C) Reads exactly one daily paper?

71. *Market research.* A market research firm has determined that 40% of the people in a certain area have seen the advertising for a new product and that 85% of those who have seen the advertising have purchased the product. What is the probability that a person in this area has seen the advertising and purchased the product?

72. *Market analysis.* A clothing company selected 1,000 persons at random and surveyed them to determine a relationship between age of purchaser and annual purchases of jeans. The results are given in the table.

AGE	JEANS PURCHASED ANNUALLY				Totals
	0	*1*	*2*	*Above 2*	
Under 12	60	70	30	10	170
12–18	30	100	100	60	290
19–25	70	110	120	30	330
Over 25	100	50	40	20	210
Totals	260	330	290	120	1,000

Given the events:

   A = Person buys 2 pairs of jeans

   B = Person is between 12 and 18 years old

   C = Person does not buy more than 2 pairs of jeans

   D = Person buys more than 2 pairs of jeans

(A) Find $P(A), P(B), P(A \cap B), P(A|B), P(B|A)$.

(B) Are events A and B independent? Explain.

(C) Find $P(C), P(D), P(C \cap D), P(C|D), P(D|C)$.

(D) Are events C and D mutually exclusive? Independent? Explain.

73. *Decision analysis.* A company sales manager, after careful analysis, presents two sales plans. It is estimated that plan *A* will net $10 million if successful (probability .8) and lose $2 million if not (probability .2); plan *B* will net $12 million if successful (probability .7) and lose $2 million if not (probability .3). What is the expected return for each plan? Which plan should be chosen based on the expected return?

74. *Insurance.* A $300 bicycle is insured against theft for an annual premium of $30. If the probability that the bicycle will be stolen during the year is .08 (empirically determined), what is the expected value of the policy?

75. *Quality control.* Twelve precision parts, including 2 that are substandard, are sent to an assembly plant. The plant will select 4 at random and will return the whole shipment if 1 or more of the sample are found to be substandard. What is the probability that the shipment will be returned?

76. *Quality control.* A dozen computer circuit boards, including 2 that are defective, are sent to a computer service center. A random sample of 3 is selected and tested. Let *X* be the random variable associated with the number of circuit boards in a sample that is defective.
(A) Find the probability distribution of *X*.
(B) Find the expected number of defective boards in a sample.

Life Sciences

77. *Medicine—cardiogram test.* By testing a large number of individuals, it has been determined that 82% of the population have normal hearts, 11% have some minor heart problems, and 7% have severe heart problems. Ninety-five percent of the persons with normal hearts, 30% of those with minor problems, and 5% of those with severe problems will pass a cardiogram test. What is the probability that a person who passes the cardiogram test has a normal heart?

78. *Genetics.* Six men in 100 and 1 woman in 100 are color-blind. A person is selected at random and is found to be color-blind. What is the probability that this person is a man? (Assume the total population contains the same number of women as men.)

Social Sciences

79. *Voter preference.* In a straw poll, the 30 students in a mathematics class are asked to indicate their preference for president of the student government. Approximate empirical probabilities are assigned on the basis of the poll: candidate *A* should receive 53% of the vote, candidate *B* should receive 37%, and candidate *C* should receive 10%. One week later, candidate *B* wins the election. Discuss the factors that may account for the discrepancy between the poll and the election results.

CHAPTER 7

Markov Chains

In this chapter we consider a mathematical model that combines probability and matrices to analyze a *stochastic process,* which consists of a sequence of trials that satisfy certain conditions. The sequence of trials is called a *Markov chain* after the Russian mathematician Andrei Markov (1856–1922), who is credited with supplying much of the foundation work in stochastic processes. Many of the early applications of Markov chains were in the physical sciences. More recent applications of Markov chains involve a wide variety of topics, including finance, market research, genetics, medicine, demographics, psychology, and political science.

In the first section we introduce the basic properties of Markov chains. In the remaining sections we discuss the long-term behavior of two different types of Markov chains.

S E C T I O N   7-1   ■   **Properties of Markov Chains**

■ INTRODUCTION
■ TRANSITION AND STATE MATRICES
■ POWERS OF TRANSITION MATRICES
■ APPLICATION

■ **INTRODUCTION**

In this section we are going to be interested in physical *systems* and their possible *states*. To understand what this means, consider the following examples:

1. A stock listed on the New York Stock Exchange either increases, decreases, or does not change in price each day the exchange is open. The stock can be thought of as a physical system with three possible states: increase, decrease, or no change.

2. A commuter, relative to a rapid transit system, can be thought of as a physical system with two states, a user or a nonuser.
3. A voting precinct casts a simple majority vote during each congressional election for a Republican candidate, a Democratic candidate, or a third-party candidate. The precinct, relative to all congressional elections past, present, and future, constitutes a physical system that can be thought of as being in one (and only one) of three states after each election: Republican, Democratic, or other.

If a system evolves from one state to another in such a way that chance elements are involved in progressing from one state to the next, then the progression of the system through a sequence of states is called a **stochastic process** (*stochos* is the Greek word for "guess"). We will now consider a simple example of a stochastic process in detail, and out of it will arise further definitions and methodology.

A toothpaste company markets a product (brand *A*) that currently has 10% of the toothpaste market. The company hires a market research firm to estimate the percentage of the market it might acquire in the future if it launches an aggressive sales campaign. The research firm uses test marketing and extensive surveys to predict the effect of the campaign. They find that if a person is using brand *A*, then the probability is .8 that this person will buy it again when he or she runs out of toothpaste. On the other hand, a person using another brand will switch to brand *A* with a probability of .6 when he or she runs out of toothpaste. Thus, each toothpaste consumer can be considered to be in one of two possible states:

$$A = \text{Uses brand } A \qquad \text{and} \qquad A' = \text{Uses another brand}$$

The probabilities determined by the market research firm can be represented graphically in a **transition diagram** as shown in Figure 1.

We can also represent the information from the market research firm numerically in a **transition probability matrix:**

FIGURE 1
Transition diagram

$$\text{Current state} \quad \begin{array}{c} \\ A \\ A' \end{array} \overset{\begin{array}{cc} A & A' \end{array}}{\begin{bmatrix} .8 & .2 \\ .6 & .4 \end{bmatrix}} = P$$

Next state

explore – discuss 1

(A) Refer to the transition diagram in Figure 1. What is the probability that a person using brand *A* will switch to another brand when he or she runs out of toothpaste?
(B) Refer to transition probability matrix *P*. What is the probability that a person who is not using brand *A* will not switch to brand *A* when he or she runs out of toothpaste?
(C) In Figure 1, the sum of the probabilities on the arrows leaving each state is 1. Will this be true for any transition diagram? Explain your answer.
(D) In transition probability matrix *P*, the sum of the probabilities in each row is 1. Will this be true for any transition probability matrix? Explain your answer.

The toothpaste company's 10% share of the market at the beginning of the sales campaign can be represented as an **initial-state distribution matrix:**

$$\begin{array}{cc} A & A' \\ \end{array}$$
$$S_0 = [.1 \quad .9]$$

If a person is chosen at random, then the probability that this person uses brand A (state A) is .1, and the probability that this person does not use brand A (state A') is .9. Thus, S_0 also can be interpreted as an **initial-state probability matrix.**

What are the probabilities of a person being in state A or A' on the first purchase after the start of the sales campaign? Let us look at the probability tree given below. [*Note:* A_0 represents state A at the beginning of the campaign, A_1' represents state A' on the first purchase after the campaign, and so on.]

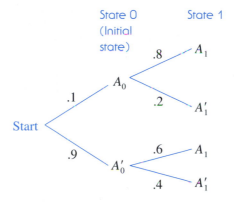

Proceeding as in Chapter 6, we can read the required probabilities directly from the tree:

$$P(A_1) = P(A_0 \cap A_1) + P(A_0' \cap A_1)$$
$$= (.1)(.8) + (.9)(.6) = .62$$

$$P(A_1') = P(A_0 \cap A_1') + P(A_0' \cap A_1')$$
$$= (.1)(.2) + (.9)(.4) = .38$$

[*Note:* $P(A_1) + P(A_1') = 1$, as expected.]

Thus, the **first-state matrix** is

$$\begin{array}{cc} A & A' \\ \end{array}$$
$$S_1 = [.62 \quad .38]$$

This matrix gives us the probabilities of a randomly chosen person being in state A or A' on the first purchase after the start of the campaign. We see that brand A's market share has increased from 10% to 62%.

Now, if you were asked to find the probabilities of a person being in state A or state A' on the tenth purchase after the start of the campaign, you might start to draw additional branches on the probability tree, but you would soon become discouraged —and for good reason, because the number of branches doubles for each successive purchase. By the tenth purchase, there would be $2^{11} = 2{,}048$ branches! Fortunately, we can convert the summing of branch products to matrix multiplication. In particular, if we multiply the initial-state matrix S_0 by the transition matrix P, we obtain the first-state matrix S_1:

$$S_0 P = [.1 \quad .9] \begin{bmatrix} .8 & .2 \\ .6 & .4 \end{bmatrix} = [(.1)(.8) + (.9)(.6) \quad\quad (.1)(.2) + (.9)(.4)] = [.62 \quad .38] = S_1$$

Initial state

Transition matrix

Compare with the tree computations above

First state

As you might guess, we can get the second-state matrix S_2 (for the second purchase) by multiplying the first-state matrix by the transition matrix:

$$S_1 P = [.62 \quad .38] \begin{bmatrix} .8 & .2 \\ .6 & .4 \end{bmatrix} = [.724 \quad .276] = S_2$$

First state

Second state

The third-state matrix S_3 is computed in a similar manner:

$$S_2 P = [.724 \quad .276] \begin{bmatrix} .8 & .2 \\ .6 & .4 \end{bmatrix} = [.7448 \quad .2552] = S_3$$

Second state

Third state

Examining the values in the first three state matrices, we see that brand A's market share increases after each toothpaste purchase. Will the market share for brand A continue to increase until it approaches 100%, or will it level off at some value less than 100%? These questions will be answered in the next section when we develop techniques for determining the long-run behavior of state matrices.

■ TRANSITION AND STATE MATRICES

The sequence of trials (toothpaste purchases) with the constant transition matrix P described above is a special kind of stochastic process called a *Markov chain*. In general, a **Markov chain,** or **process,** is a sequence of experiments, trials, or observations such that the transition probability matrix from one state to the next is constant. A Markov process has no memory. The various matrices associated with a Markov chain are defined in the next box.

Markov Chains

Given a Markov chain with n states, a **kth-state matrix** is a matrix of the form

$$S_k = [s_{k1} \quad s_{k2} \quad \cdots \quad s_{kn}]$$

Each entry s_{ki} is the proportion of the population that is in state i after the kth trial, or, equivalently, the probability of a randomly selected element of the population being in state i after the kth trial. The sum of all the entries in the kth state matrix S_k must be 1.

A **transition matrix** is a constant square matrix P of order n such that the entry in the ith row and jth column indicates the probability of the system moving from the ith state to the jth state on the next observation or trial. The sum of the entries in each row must be 1.

Remarks

1. Since the entries in a kth-state matrix or a transition matrix are probabilities, they must be real numbers between 0 and 1, inclusive.

2. Rearranging the various states and corresponding transition probabilities in a transition matrix will produce a different, but equivalent, transition matrix. For example, both of the following matrices are transition matrices for the toothpaste manufacturer discussed earlier:

$$P = \begin{array}{c} \\ A \\ A' \end{array} \begin{array}{cc} A & A' \\ \begin{bmatrix} .8 & .2 \\ .6 & .4 \end{bmatrix} \end{array}$$

$$P' = \begin{array}{c} \\ A' \\ A \end{array} \begin{array}{cc} A' & A \\ \begin{bmatrix} .4 & .6 \\ .2 & .8 \end{bmatrix} \end{array}$$

Such rearrangements will affect the form of the matrices used in the solution of a problem, but will not affect any of the information obtained from these matrices. In Section 7-3 we will encounter situations where it will be helpful to select a transition matrix that has a special form. For now, you can choose any order for the states in a transition matrix.

As we indicated in the preceding discussion, matrix multiplication can be used to compute the various state matrices of a Markov chain, described in the box at the top of the next page.

Computing State Matrices for a Markov Chain

If S_0 is the initial-state matrix and P is the transition matrix for a Markov chain, then the subsequent state matrices are given by:

$$S_1 = S_0 P \qquad \text{First-state matrix}$$
$$S_2 = S_1 P \qquad \text{Second-state matrix}$$
$$S_3 = S_2 P \qquad \text{Third-state matrix}$$
$$\vdots$$
$$S_k = S_{k-1} P \qquad \text{kth-state matrix}$$

EXAMPLE 1 ▶ **Insurance** An insurance company found that on the average, over a period of 10 years, 23% of the drivers in a particular community who were involved in an accident one year were also involved in an accident the following year. They also found that only 11% of the drivers who were not involved in an accident one year were involved in an accident the following year. Use these percentages as approximate empirical probabilities for the following:

(A) Draw a transition diagram.
(B) Find the transition matrix P.
(C) If 5% of the drivers in the community are involved in an accident this year, what is the probability that a driver chosen at random from the community will be involved in an accident next year? Year after next?

Solution (A)

.77

.23 ⟲ A ⇄ A' ⟲ .89

.11

Next year

A = Accident
A' = No accident

(B)
$$\begin{array}{c} \text{This } A \\ \text{year } A' \end{array} \begin{array}{cc} A & A' \\ \end{array} \begin{bmatrix} .23 & .77 \\ .11 & .89 \end{bmatrix} = P \quad \text{Transition matrix}$$

(C) The initial-state matrix S_0 is

$$S_0 = \begin{array}{cc} A & A' \\ [.05 & .95] \end{array} \quad \text{Initial-state matrix}$$

Thus,

$$S_0 P = \begin{array}{cc} A & A' \\ [.05 & .95] \end{array} \begin{array}{cc} A & A' \\ \begin{bmatrix} .23 & .77 \\ .11 & .89 \end{bmatrix} \end{array} = \begin{array}{cc} A & A' \\ [.116 & .884] \end{array} = S_1$$

This year
(initial state)

Next year
(first state)

$$\overset{}{S_1P} = \underset{}{[.116 \quad .884]} \begin{bmatrix} \overset{A}{.23} & \overset{A'}{.77} \\ .11 & .89 \end{bmatrix} = \underset{}{[.12392} \quad \overset{A'}{.87608]} = S_2$$

Next year   Year after next
(first state)   (second state)

The probability of a driver chosen at random from the community having an accident next year is .116 and having an accident year after next is .12392. That is, it is expected that 11.6% of the drivers in the community will have an accident next year and 12.392% the year after. ◀

MATCHED PROBLEM 1 ➤ An insurance company classifies drivers as low-risk if they are accident-free for 1 year. Past records indicate that 98% of the drivers in low-risk category (L) one year will remain in that category the next year, and 78% of the drivers that are not in the low-risk category (L') one year will be in the low-risk category the next year.

(A) Draw a transition diagram.
(B) Find the transition matrix P.
(C) If 90% of the drivers in the community are in the low-risk category this year, what is the probability that a driver chosen at random from the community will be in the low-risk category next year? Year after next? ◀

■ POWERS OF TRANSITION MATRICES

We will now investigate the effective use of the powers of a transition matrix.

explore – discuss 2

Given the transition matrix P and initial-state matrix S_0, where

$$P = \begin{matrix} A \\ A' \end{matrix} \begin{bmatrix} \overset{A}{.9} & \overset{A'}{.1} \\ .7 & .3 \end{bmatrix} \quad \text{and} \quad S_0 = [\overset{A}{.5} \quad \overset{A'}{.5}]$$

(A) Find S_2, S_3, and S_4.
(B) Find P^2, P^3, and P^4.
(C) Find S_0P^2, S_0P^3, and S_0P^4.
(D) Compare the results of parts (A) and (C). What interpretation of the entries in P^2, P^3, and P^4 does this suggest?

The state matrices for a Markov chain are defined **recursively;** that is, each state matrix is defined in terms of the preceding state matrix. For example, to find the fourth-state matrix S_4, it is necessary to compute the preceding three state matrices:

$$S_1 = S_0P \qquad S_2 = S_1P \qquad S_3 = S_2P \qquad S_4 = S_3P$$

Is there any way to compute a given state matrix directly without first computing all the preceding state matrices? If we substitute the equation for S_1 in the equation for S_2, substitute this new equation for S_2 in the equation for S_3, and so on, a definite pattern emerges:

$$S_1 = S_0P$$
$$S_2 = S_1P = (S_0P)P = S_0P^2$$
$$S_3 = S_2P = (S_0P^2)P = S_0P^3$$
$$S_4 = S_3P = (S_0P^3)P = S_0P^4$$
$$\vdots$$

In general, it can be shown that the kth-state matrix is given by $S_k = S_0P^k$. We summarize this important result in Theorem 1.

theorem 1

> **Powers of a Transition Matrix**
>
> If P is the transition matrix and S_0 is an initial-state matrix for a Markov chain, then the kth-state matrix is given by
>
> $$S_k = S_0P^k$$
>
> The entry in the ith row and jth column of P^k indicates the probability of the system moving from the ith state to the jth state in k observations or trials. The sum of the entries in each row of P^k is 1.

EXAMPLE 2 ➤ **Using P^k to Compute S_k** Use the formula $S_k = S_0P^k$ to find S_2, S_4, and S_8 for

$$P = \begin{matrix} & \begin{matrix} A & B & C \end{matrix} \\ \begin{matrix} A \\ B \\ C \end{matrix} & \begin{bmatrix} .5 & 0 & .5 \\ .2 & .6 & .2 \\ 1 & 0 & 0 \end{bmatrix} \end{matrix} \quad \text{and} \quad S_0 = \begin{matrix} \begin{matrix} A & B & C \end{matrix} \\ [.5 \quad .2 \quad .3] \end{matrix}$$

Round all values in the state matrices to six decimal places.

Solution   First, we store the matrices P and S_0 in the memory of a graphing utility (Fig. 2A). Then we use the formula $S_k = S_0P^k$ to compute the required state matrices on the graphing utility (Fig. 2B).

(A)   (B)

FIGURE 2

The state matrices (with values rounded to six decimal places) are

$$S_2 = S_0 P^2 = [.609 \quad .072 \quad .319]$$
$$S_4 = S_0 P^4 = [.64649 \quad .02592 \quad .32759]$$
$$S_8 = S_0 P^8 = [.664141 \quad .003359 \quad .3325]$$

◀

MATCHED PROBLEM 2 ➤ Use the formula $S_k = S_0 P^k$ to find S_2, S_4, and S_8 for

$$P = \begin{array}{c} \\ A \\ B \\ C \end{array} \begin{array}{ccc} A & B & C \\ \begin{bmatrix} .9 & .1 & 0 \\ 0 & 1 & 0 \\ 0 & .2 & .8 \end{bmatrix} \end{array} \quad \text{and} \quad S_0 = \begin{array}{ccc} A & B & C \\ [.4 & .2 & .4] \end{array}$$

Round all values in the state matrices to six decimal places. ◀

■ APPLICATION

The next example illustrates the use of Theorem 1 in an applied problem.

▶ EXAMPLE 3 ➤ University Enrollment Part-time students admitted to an MBA program in a university are considered to be first-year students until they successfully complete 15 credits. Then they are classified as second-year students and may begin to take more advanced courses and to work on the thesis required for graduation. Past records indicate that at the end of each year 10% of the first-year students (F) drop out of the program (D) and 30% become second-year students (S). Also, 10% of the second-year students drop out of the program and 40% graduate (G) each year. Students that graduate or drop out never return to the program.

(A) Draw a transition diagram.
(B) Find the transition matrix P.
(C) What is the probability that a first-year student graduates within 4 years? Drops out within 4 years?

Solution (A) If 10% of the first-year students drop out and 30% become second-year students, the remaining 60% must continue as first-year students for another year (see the diagram). Similarly, 50% of the second-year students must continue as second-year students for another year. Since students who drop out never return, all the students in state D in one year will continue in that state the next year. We indicate this by placing a 1 on the arrow from D back to D. State G is labeled in the same manner.

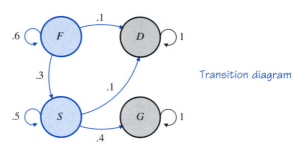

Transition diagram

$$
\text{(B)} \quad P = \begin{array}{c} \\ F \\ D \\ S \\ G \end{array} \begin{array}{cccc} F & D & S & G \\ \left[\begin{array}{cccc} .6 & .1 & .3 & 0 \\ 0 & 1 & 0 & 0 \\ 0 & .1 & .5 & .4 \\ 0 & 0 & 0 & 1 \end{array}\right] \end{array} \quad \text{Transition matrix}
$$

(C) The probability that a first-year student moves from state F to state G within 4 years is the entry in Row 1 and Column 4 of P^4 (Theorem 1). Hand computation of P^4 requires two multiplications:

$$
P^2 = \begin{bmatrix} .6 & .1 & .3 & 0 \\ 0 & 1 & 0 & 0 \\ 0 & .1 & .5 & .4 \\ 0 & 0 & 0 & 1 \end{bmatrix} \begin{bmatrix} .6 & .1 & .3 & 0 \\ 0 & 1 & 0 & 0 \\ 0 & .1 & .5 & .4 \\ 0 & 0 & 0 & 1 \end{bmatrix} = \begin{bmatrix} .36 & .19 & .33 & .12 \\ 0 & 1 & 0 & 0 \\ 0 & .15 & .25 & .6 \\ 0 & 0 & 0 & 1 \end{bmatrix}
$$

$$
P^4 = P^2P^2 = \begin{bmatrix} .36 & .19 & .33 & .12 \\ 0 & 1 & 0 & 0 \\ 0 & .15 & .25 & .6 \\ 0 & 0 & 0 & 1 \end{bmatrix} \begin{bmatrix} .36 & .19 & .33 & .12 \\ 0 & 1 & 0 & 0 \\ 0 & .15 & .25 & .6 \\ 0 & 0 & 0 & 1 \end{bmatrix}
$$

$$
= \begin{bmatrix} .1296 & .3079 & .2013 & .3612 \\ 0 & 1 & 0 & 0 \\ 0 & .1875 & .0625 & .75 \\ 0 & 0 & 0 & 1 \end{bmatrix}
$$

Thus, the probability that a first-year student has graduated within 4 years is .3612. Similarly, the probability that a first-year student has dropped out within 4 years is .3079 (the entry in Row 1 and Column 2 of P^4). ◀

MATCHED PROBLEM 3 ▶ Refer to Example 3. At the end of each year the faculty examines the progress each second-year student has made in writing the required thesis. Past records indicate that 30% of the second-year students (S) have their theses approved (A) and 10% of the students are dropped from the program for insufficient progress (D), never to return. The remaining students continue to work on their theses.

(A) Draw a transition diagram.
(B) Find the transition matrix P.
(C) What is the probability that a second-year student completes the thesis requirement within 4 years? Is dropped from the program for insufficient progress within 4 years? ◀

explore–discuss 3

Refer to Example 3. States D and G are referred to as *absorbing states*, because a student who enters either one of these states never leaves it. Absorbing states are discussed in detail in Section 7-3.

(A) How can absorbing states be recognized from a transition diagram? Draw a transition diagram with two states, one that is absorbing and one that is not, to illustrate.

(B) How can absorbing states be recognized from a transition matrix? Write the transition matrix for the diagram you drew in part (A) to illustrate.

Answers to Matched Problems

1. (A)

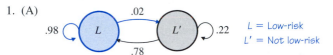

L = Low-risk
L' = Not low-risk

(B)
$$\begin{array}{cc} & \text{Next year} \\ & \begin{array}{cc} L & L' \end{array} \\ \text{This } \begin{array}{c} L \\ \text{year } L' \end{array} & \begin{bmatrix} .98 & .02 \\ .78 & .22 \end{bmatrix} = P \end{array}$$

(C) Next year: .96; year after next: .972

2. $S_2 = [.324 \quad .42 \quad .256]$; $S_4 = [.26244 \quad .57372 \quad .16384]$;
$S_8 = [.172187 \quad .760704 \quad .067109]$

3. (A)

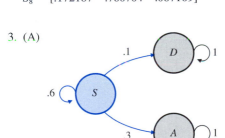

(B)
$$P = \begin{array}{c} \\ S \\ A \\ D \end{array} \begin{array}{c} \begin{array}{ccc} S & A & D \end{array} \\ \begin{bmatrix} .6 & .3 & .1 \\ 0 & 1 & 0 \\ 0 & 0 & 1 \end{bmatrix} \end{array}$$

(C) .6528; .2176

EXERCISE 7-1

A *Problems 1–8 refer to the following transition matrix:*

$$P = \begin{array}{c} \\ A \\ B \end{array} \begin{array}{c} \begin{array}{cc} A & B \end{array} \\ \begin{bmatrix} .8 & .2 \\ .4 & .6 \end{bmatrix} \end{array}$$

In Problems 1–4, find S_1 for the indicated initial-state matrix S_0, and interpret with the aid of a tree diagram.

1. $S_0 = [1 \quad 0]$

2. $S_0 = [0 \quad 1]$

3. $S_0 = [.5 \quad .5]$

4. $S_0 = [.3 \quad .7]$

In Problems 5–8, find S_2 for the indicated initial-state matrix S_0, and explain what it represents.

5. $S_0 = [1 \quad 0]$

6. $S_0 = [0 \quad 1]$

7. $S_0 = [.5 \quad .5]$

8. $S_0 = [.3 \quad .7]$

In Problem 9–12, fill in the missing probabilities in the transition diagram and write the corresponding transition matrix.

9.

10.

11.

12.

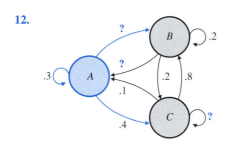

In Problems 13–16, find the values of a, b, and c that will make P a transition matrix, and draw the corresponding transition diagram.

13. $P = \begin{array}{c} A \\ B \\ C \end{array} \begin{bmatrix} \overset{A}{0} & \overset{B}{.5} & \overset{C}{a} \\ b & 0 & .4 \\ .2 & c & .1 \end{bmatrix}$

14. $P = \begin{array}{c} A \\ B \\ C \end{array} \begin{bmatrix} \overset{A}{a} & \overset{B}{0} & \overset{C}{.9} \\ .2 & .3 & b \\ .6 & c & 0 \end{bmatrix}$

15. $P = \begin{array}{c} A \\ B \\ C \end{array} \begin{bmatrix} \overset{A}{0} & \overset{B}{a} & \overset{C}{.3} \\ 0 & b & 0 \\ c & .8 & 0 \end{bmatrix}$

16. $P = \begin{array}{c} A \\ B \\ C \end{array} \begin{bmatrix} \overset{A}{0} & \overset{B}{1} & \overset{C}{a} \\ 0 & 0 & b \\ c & .5 & 0 \end{bmatrix}$

B In Problems 17–20, use the given information to draw the transition diagram and to find the transition matrix.

17. A Markov process has two states, A and B. The probability of going from state A to state B in one trial is .7, and the probability of going from state B to state A in one trial is .9.

18. A Markov process has two states, A and B. The probability of going from state A to state A in one trial is .6, and the probability of going from state B to state B in one trial is .2.

19. A Markov chain has three states, A, B, and C. The probability of going from state A to state B in one trial is .1, the probability of going from state A to state C in one trial is .3, the probability of going from state B to state A in one trial is .2, the probability of going from state B to state C in one trial is .5, and the probability of going from state C to state C in one trial is 1.

20. A Markov chain has three states, A, B, and C. The probability of going from state A to state B in one trial is 1, the probability of going from state B to state A in one trial is .5, the probability of going from state B to state C in one trial is .5, and the probability of going from state C to state A in one trial is 1.

Problems 21–28 refer to the transition matrix P and the powers of P given below:

$$P = \begin{array}{c} A \\ B \\ C \end{array} \begin{bmatrix} \overset{A}{.6} & \overset{B}{.3} & \overset{C}{.1} \\ .2 & .5 & .3 \\ .1 & .2 & .7 \end{bmatrix} \quad P^2 = \begin{array}{c} A \\ B \\ C \end{array} \begin{bmatrix} \overset{A}{.43} & \overset{B}{.35} & \overset{C}{.22} \\ .25 & .37 & .38 \\ .17 & .27 & .56 \end{bmatrix}$$

$$P^3 = \begin{array}{c} A \\ B \\ C \end{array} \begin{bmatrix} \overset{A}{.35} & \overset{B}{.348} & \overset{C}{.302} \\ .262 & .336 & .402 \\ .212 & .298 & .49 \end{bmatrix}$$

21. Find the probability of going from state A to state B in two trials.

22. Find the probability of going from state B to state C in two trials.

23. Find the probability of going from state C to state A in three trials.

24. Find the probability of going from state B to state B in three trials.

25. Find S_2 for $S_0 = [1 \quad 0 \quad 0]$, and explain what it represents.

26. Find S_2 for $S_0 = [0 \quad 1 \quad 0]$, and explain what it represents.

27. Find S_3 for $S_0 = [0 \ \ 0 \ \ 1]$, and explain what it represents.

28. Find S_3 for $S_0 = [1 \ \ 0 \ \ 0]$, and explain what it represents.

In Problems 29–32, given the transition matrix P and initial-state matrix S_0, use P^4 to find S_4.

29. $P = \begin{matrix} A \\ B \end{matrix} \begin{bmatrix} .1 & .9 \\ .6 & .4 \end{bmatrix}$, $S_0 = [.8 \ \ .2]$

where the column headers are $A \quad B$.

30. $P = \begin{matrix} A \\ B \end{matrix} \begin{bmatrix} .8 & .2 \\ .3 & .7 \end{bmatrix}$, $S_0 = [.4 \ \ .6]$

where the column headers are $A \quad B$.

31. $P = \begin{matrix} A \\ B \\ C \end{matrix} \begin{bmatrix} 0 & .4 & .6 \\ 0 & 0 & 1 \\ 1 & 0 & 0 \end{bmatrix}$, $S_0 = [.2 \ \ .3 \ \ .5]$

where the column headers are $A \quad B \quad C$.

32. $P = \begin{matrix} A \\ B \\ C \end{matrix} \begin{bmatrix} 0 & 1 & 0 \\ .8 & 0 & .2 \\ 1 & 0 & 0 \end{bmatrix}$, $S_0 = [.4 \ \ .2 \ \ .4]$

where the column headers are $A \quad B \quad C$.

33. A Markov process with two states has transition matrix P. If the initial-state matrix is $S_0 = [1 \ \ 0]$, discuss the relationship between the entries in the kth-state matrix and the entries in the kth power of P.

34. Repeat Problem 33 if the initial-state matrix is $S_0 = [0 \ \ 1]$.

C

35. Given the transition matrix

$$P = \begin{matrix} A \\ B \\ C \\ D \end{matrix} \begin{bmatrix} .2 & .2 & .3 & .3 \\ 0 & 1 & 0 & 0 \\ .2 & .2 & .1 & .5 \\ 0 & 0 & 0 & 1 \end{bmatrix}$$

with column headers $A \quad B \quad C \quad D$.

(A) Find P^4.
(B) Find the probability of going from state A to state D in four trials.
(C) Find the probability of going from state C to state B in four trials.
(D) Find the probability of going from state B to state A in four trials.

36. Repeat Problem 35 for the transition matrix

$$P = \begin{matrix} A \\ B \\ C \\ D \end{matrix} \begin{bmatrix} .5 & .3 & .1 & .1 \\ 0 & 1 & 0 & 0 \\ 0 & 0 & 1 & 0 \\ .1 & .2 & .3 & .4 \end{bmatrix}$$

with column headers $A \quad B \quad C \quad D$.

*A matrix is called a **probability matrix** if all its entries are real numbers between 0 and 1, inclusive, and the sum of the entries in each row is 1. Thus, transition matrices are square probability matrices and state matrices are probability matrices with one row.*

37. If

$$P = \begin{bmatrix} a & 1 - a \\ 1 - b & b \end{bmatrix}$$

is a probability matrix, show that P^2 is a probability matrix.

38. If

$$P = \begin{bmatrix} a & 1 - a \\ 1 - b & b \end{bmatrix} \quad \text{and} \quad S = [c \ \ 1 - c]$$

are probability matrices, show that SP is a probability matrix.

Use the formula $S_k = S_0 P^k$ (Theorem 1) to compute the required state matrices in Problems 39–42.

39. The transition matrix for a Markov chain is

$$P = \begin{bmatrix} .4 & .6 \\ .2 & .8 \end{bmatrix}$$

(A) If $S_0 = [0 \ \ 1]$, find $S_2, S_4, S_8, \ldots$. Can you identify a state matrix S that the matrices S_k seem to be approaching?
(B) Repeat part (A) for $S_0 = [1 \ \ 0]$.
(C) Repeat part (A) for $S_0 = [.5 \ \ .5]$.
(D) Find SP for any matrix S you identified in parts (A) − (C).
(E) Write a brief verbal description of the long-term behavior of the state matrices of this Markov chain based on your observations in parts (A) − (D).

40. Repeat Problem 39 for

$$P = \begin{bmatrix} .9 & .1 \\ .4 & .6 \end{bmatrix}$$

41. Refer to Problem 39. Find P^k for $k = 2, 4, 8, \ldots$. Can you identify a matrix Q that the matrices P^k are approaching? If so, how is Q related to the results you discovered in Problem 39?

42. Refer to Problem 40. Find P^k for $k = 2, 4, 8, \ldots$. Can you identify a matrix Q that the matrices P^k are approaching? If so, how is Q related to the results you discovered in Problem 40?

 APPLICATIONS

Business & Economics

43. *Scheduling.* An outdoor restaurant in a summer resort closes only on days that it rains. From past records it is found that from May through September, when it rains one day, the probability of rain for the next day is .4; when it does not rain one day, the probability of rain the next day is .06.
(A) Draw a transition diagram.
(B) Write the transition matrix.
(C) If it rains on Thursday, what is the probability that the restaurant will be closed on Saturday? On Sunday?

44. *Scheduling.* Repeat Problem 43 if the probability of rain following a rainy day is .6 and the probability of rain following a nonrainy day is .1.

45. *Advertising.* A vigorous television advertising campaign is conducted during the football season to promote a well-known brand X shaving cream. For each of several weeks, a survey is made, and it is found that each week 80% of those using brand X continue to use it and 20% switch to another brand. It is also found that of those not using brand X, 20% switch to brand X while the other 80% continue using another brand.
(A) Draw a transition diagram.
(B) Write the transition matrix.
(C) If 20% of the people are using brand X at the start of the advertising campaign, what percentage will be using it 1 week later? 2 weeks later?

46. *Car rental.* A car rental agency has rental and return facilities at both Kennedy and LaGuardia airports, two of the principal airports in the New York City area. Assume a car rented at either airport must be returned to one or the other airport. If a car is rented at LaGuardia, the probability that it will be returned there is .8; if a car is rented at Kennedy, the probability that it will be returned there is .7. Assume the company rents all its 100 cars each day and each car is rented (and returned) only once a day. If we start with 50 cars at each airport:
(A) What is the expected distribution the next day?
(B) What is the expected distribution 2 days later?

47. *Homeowners insurance.* The market for homeowners insurance in a particular city is dominated by two companies, National Property and United Family. Currently, National Property insures 50% of the homes in the city, United Family insures 30%, and the remainder are insured by a collection of smaller companies. United Family decides to offer rebates to increase its market share. This has the following effects on insurance purchases for the next several years: each year 25% of National Property's customers switch to United Family and 10% switch to other companies; 10% of United Family's customers switch to National Property and 5% switch to other companies; and 15% of the customers of other companies switch to National Property and 35% switch to United Family.
(A) Draw a transition diagram.
(B) Write the transition matrix.
(C) What percentage of the homes will be insured by National Property next year? The year after next?
(D) What percentage of the homes will be insured by United Family next year? The year after next?

48. *Service contracts.* A small community has two heating services that offer annual service contracts for home heating systems, Alpine Heating and Badger Furnaces. Currently, 25% of the homeowners have service contracts with Alpine, 30% have service contracts with Badger, and the remainder do not have service contracts. Both companies launch aggressive advertising campaigns to attract new customers, with the following effects on service contract purchases for the next several years: each year 35% of the homeowners with no current service contract decide to purchase a contract from Alpine and 40% decide to purchase one from Badger. In addition, 10% of the previous customers at each company decide to switch to the other company, and 5% decide they do not want a service contract.
(A) Draw a transition diagram.
(B) Write the transition matrix.
(C) What percentage of the homes will have service contracts with Alpine next year? The year after next?

(D) What percentage of the homes will have service contracts with Badger next year? The year after next?

49. *Employee training.* A nationwide chain of travel agencies maintains a training program for new travel agents. Initially, all new employees are classified as beginning agents requiring extensive supervision. Every 6 months the performance of each agent is reviewed. Past records indicate that after each semiannual review, 40% of the beginning agents are promoted to intermediate agents requiring only minimal supervision, 10% are terminated for unsatisfactory performance, and the remainder continue as beginning agents. Furthermore, 30% of the intermediate agents are promoted to qualified travel agents requiring no supervision, 10% are terminated for unsatisfactory performance, and the remainder continue as intermediate agents.
(A) Draw a transition diagram.
(B) Write the transition matrix.
(C) What is the probability that a beginning agent is promoted to qualified agent within 1 year? Within 2 years?

50. *Employee training.* All welders in a factory begin as apprentices. Every year the performance of each apprentice is reviewed. Past records indicate that after each review, 10% of the apprentices are promoted to professional welder, 20% are terminated for unsatisfactory performance, and the remainder continue as apprentices.
(A) Draw a transition diagram.
(B) Write the transition matrix.
(C) What is the probability that an apprentice is promoted to professional welder within 2 years? Within 4 years?

Life Sciences

51. *Health insurance.* A midwestern university offers its employees three choices for health care: a clinic-based health maintenance organization (HMO), a preferred provider organization (PPO), and a traditional fee-for-service program (FFS). Each year the university designates an open enrollment period during which employees may change from one health plan to another. Prior to the last open enrollment period, 20% of the employees were enrolled in the HMO, 25% in the PPO, and the remainder in the FFS. During the open enrollment period, 15% of the employees in the HMO switched to the PPO and 5% switched to the FFS; 20% of the employees in the PPO switched to the HMO and 10% to the FFS; and 25% of the employees in the FFS switched to the HMO and 30% switched to the PPO.
(A) Write the transition matrix.
(B) What percentage of employees were enrolled in each plan after the last open enrollment period?

(C) If this trend continues, what percentage of employees will be enrolled in each program after the next open enrollment period?

52. *Dental insurance.* Refer to Problem 51. During the open enrollment period, employees at the university can switch between the two available dental care programs, the low-option plan (LOP) and the high-option plan (HOP). Prior to the last open enrollment period, 40% of the employees were enrolled in the LOP and 60% in the HOP. During the open enrollment program, 30% of the employees in the LOP switched to the HOP and 10% of the employees in the HOP switched to the LOP.
(A) Write the transition matrix.
(B) What percentage of employees were enrolled in each plan after the last open enrollment period?
(C) If this trend continues, what percentage of employees will be enrolled in each program after the next open enrollment period?

Social Sciences

53. *Housing trends.* Home ownership in the United States declined during the 1980's, ending a 40 year period of growth. The 1980 census reported that 45.1% of the households in a particular city were homeowners and the remainder were renters. During the next decade, 12% of the homeowners became renters and the rest continued to be homeowners. Likewise, 5% of the renters became homeowners and the rest continued to rent.
(A) Write the appropriate transition matrix.
(B) According to this transition matrix, what percentage of households were homeowners in 1990?
(C) If the transition matrix remains the same, what percentage of the households will be homeowners in 2000?

54. *Housing trends.* The 1980 census reported that 73.3% of the households in a suburban county were homeowners. During the next decade, 5% of the homeowners became renters and the rest continued to be homeowners. Likewise, 3% of the renters became homeowners and the rest continued to rent.
(A) Write the appropriate transition matrix.
(B) According to this transition matrix, what percentage of households were homeowners in 1990?
(C) If the transition matrix remains the same, what percentage of the households will be homeowners in 2000?

SECTION   7-2   ■   **Regular Markov Chains**

- STATIONARY MATRICES
- REGULAR MARKOV CHAINS
- APPLICATIONS
- GRAPHING UTILITY APPROXIMATIONS

Given a Markov chain with transition matrix P and initial-state matrix S_0, the entries in the state matrix S_k are the probabilities of being in the corresponding states after k trials. What happens to these probabilities as the number of trials k increases? In this section we establish conditions on the transition matrix P that enable us to determine the long-run behavior of both the state matrices S_k and the powers of the transition matrix P^k.

■ **STATIONARY MATRICES**

We begin by considering a concrete example — the toothpaste company discussed in the preceding section. Recall that the transition matrix was given by

$$P = \begin{array}{c} \\ A \\ A' \end{array} \begin{array}{cc} A & A' \\ \begin{bmatrix} .8 & .2 \\ .6 & .4 \end{bmatrix} \end{array} \quad \begin{array}{l} A = \text{Uses brand A toothpaste} \\ A' = \text{Uses another brand} \end{array}$$

Initially, this company had a 10% share of the toothpaste market. If the probabilities in the transition matrix P remain valid over a long period of time, what will happen to the company's market share? Examining the first several state matrices will give us some insight into this situation (matrix multiplication details are omitted):

$$S_0 = [.1 \quad .9]$$
$$S_1 = S_0P = [.62 \quad .38]$$
$$S_2 = S_1P = [.724 \quad .276]$$
$$S_3 = S_2P = [.7448 \quad .2552]$$
$$S_4 = S_3P = [.74896 \quad .25104]$$
$$S_5 = S_4P = [.749792 \quad .250208]$$
$$S_6 = S_5P = [.7499584 \quad .2500416]$$

It appears that the state matrices are getting closer and closer to $S = [.75 \quad .25]$ as we proceed to higher states. Let us multiply the matrix S (the matrix the other state matrices appear to be approaching) by the transition matrix:

$$SP = [.75 \quad .25] \begin{bmatrix} .8 & .2 \\ .6 & .4 \end{bmatrix} = [.75 \quad .25] = S$$

No change occurs! The matrix $[.75 \quad .25]$ is called a **stationary matrix.** If we reach this state or are very close to it, the system is said to be at steady-state — that is, later states either will not change or will not change very much. In terms of this example, this means that in the long run a person will purchase brand A with a probability of

.75; that is, the company can expect to capture 75% of the market assuming the transition matrix does not change.

The general definition of a stationary matrix is given in the next box.

Stationary Matrix for a Markov Chain

The state matrix $S = [s_1 \quad s_2 \quad \ldots \quad s_n]$ is a **stationary matrix** for a Markov chain with transition matrix P if

$$SP = S$$

where $s_i \geqslant 0$, $i = 1, \ldots, n$, and $s_1 + s_2 + \cdots + s_n = 1$.

---

explore – discuss 1

(A) Suppose the toothpaste company started with only 5% of the market instead of 10%. Write the initial-state matrix, find the next six state matrices, and discuss the behavior of these state matrices as you proceed to higher states.

(B) Repeat part (A) if the company started with 90% of the toothpaste market.

■ REGULAR MARKOV CHAINS

Does every Markov chain have a unique stationary matrix? And if a Markov chain has a unique stationary matrix, will the successive state matrices always approach this stationary matrix? Unfortunately, the answer to both these questions is no. (See Problems 27–30 in Exercise 7-2.) However, there is one important type of Markov chain for which both questions always can be answered in the affirmative. These are called *regular Markov chains*.

Regular Markov Chains

A transition matrix P is **regular** if some power of P has only positive entries. A Markov chain is a **regular Markov chain** if its transition matrix is regular.

EXAMPLE 1 ➤ **Recognizing Regular Matrices** Which of the following matrices are regular?

$$(A) \ P = \begin{bmatrix} .8 & .2 \\ .6 & .4 \end{bmatrix} \qquad (B) \ P = \begin{bmatrix} 0 & 1 \\ 1 & 0 \end{bmatrix} \qquad (C) \ P = \begin{bmatrix} .5 & .5 & 0 \\ 0 & .5 & .5 \\ 1 & 0 & 0 \end{bmatrix}$$

Solution (A) This is the transition matrix for the toothpaste company. Since all the entries in P are positive, we can immediately conclude that P is regular.

(B) P has two 0 entries, so we must examine higher powers of P:

$$P^2 = \begin{bmatrix} 1 & 0 \\ 0 & 1 \end{bmatrix} \qquad P^3 = \begin{bmatrix} 0 & 1 \\ 1 & 0 \end{bmatrix} \qquad P^4 = \begin{bmatrix} 1 & 0 \\ 0 & 1 \end{bmatrix} \qquad P^5 = \begin{bmatrix} 0 & 1 \\ 1 & 0 \end{bmatrix}$$

Since the powers of P oscillate between P and I, the 2×2 identity, all powers of P will contain 0 entries. Hence, P is not regular.

(C) Again, we examine higher powers of P:

$$P^2 = \begin{bmatrix} .25 & .5 & .25 \\ .5 & .25 & .25 \\ .5 & .5 & 0 \end{bmatrix} \qquad P^3 = \begin{bmatrix} .375 & .375 & .25 \\ .5 & .375 & .125 \\ .25 & .5 & .25 \end{bmatrix}$$

Since all the entries in P^3 are positive, P is regular.   ◄

MATCHED PROBLEM 1 ► Which of the following matrices are regular?

(A) $P = \begin{bmatrix} .3 & .7 \\ 1 & 0 \end{bmatrix}$ (B) $P = \begin{bmatrix} 1 & 0 \\ 1 & 0 \end{bmatrix}$ (C) $P = \begin{bmatrix} 0 & 1 & 0 \\ .5 & 0 & .5 \\ .5 & 0 & .5 \end{bmatrix}$ ◄

explore–discuss 2

Consider the toothpaste company (discussed at the beginning of this section) with regular transition matrix P and stationary matrix S, where

$$P = \begin{bmatrix} .8 & .2 \\ .6 & .4 \end{bmatrix} \qquad \text{and} \qquad S = [.75 \quad .25]$$

Compute P^2, P^4, and P^8. Discuss any apparent relationships between P^k and S.

The relationships among successive state matrices, powers of the transition matrix, and the stationary matrix for a regular Markov chain are given in Theorem 1. The proof of this theorem is left to more advanced courses.

theorem 1

Properties of Regular Markov Chains

Let P be the transition matrix for a regular Markov chain.

(A) There is a unique stationary matrix S that can be found by solving the equation

$$SP = S$$

> **Properties of Regular Markov Chains** *(Continued)*
>
> (B) Given any initial-state matrix S_0, the state matrices S_k approach the stationary matrix S.
>
> (C) The matrices P^k approach a **limiting matrix** $\overline{P}$, where each row of $\overline{P}$ is equal to the stationary matrix S.

EXAMPLE 2 ➤ Finding the Stationary Matrix   The transition matrix for a Markov chain is

$$P = \begin{bmatrix} .7 & .3 \\ .2 & .8 \end{bmatrix}$$

(A) Find the stationary matrix S.

(B) Discuss the long-run behavior of S_k and P^k.

Solution   (A) Since P is regular, the stationary matrix S must exist. To find it, we must solve the equation $SP = S$. Let

$$S = [s_1 \quad s_2]$$

and write

$$[s_1 \quad s_2] \begin{bmatrix} .7 & .3 \\ .2 & .8 \end{bmatrix} = [s_1 \quad s_2]$$

After multiplying the left side, we obtain

$$[(.7s_1 + .2s_2) \quad (.3s_1 + .8s_2)] = [s_1 \quad s_2]$$

which is equivalent to the system

$$
\begin{array}{llll}
.7s_1 + .2s_2 = s_1 & \text{or} & -.3s_1 + .2s_2 = 0 \\
.3s_1 + .8s_2 = s_2 & \text{or} & .3s_1 - .2s_2 = 0
\end{array}
\tag{1}
$$

System (1) is dependent and has an infinite number of solutions. However, we are looking for a solution that is also a state matrix. This gives us another equation that we can add to (1) to obtain a system with a unique solution.

$$
\begin{array}{l}
-.3s_1 + .2s_2 = 0 \\
.3s_1 - .2s_2 = 0 \\
s_1 + \quad s_2 = 1
\end{array}
\tag{2}
$$

System (2) can be solved using matrix methods or elimination to obtain

$$s_1 = .4 \quad \text{and} \quad s_2 = .6$$

Thus,

$$S = [.4 \quad .6]$$

is the stationary matrix.

Check $$SP = [.4 \quad .6] \begin{bmatrix} .7 & .3 \\ .2 & .8 \end{bmatrix} = [.4 \quad .6] = S$$

(B) Given any initial-state matrix S_0, Theorem 1 guarantees that the state matrices S_k will approach the stationary matrix S. Furthermore,

$$P^k = \begin{bmatrix} .7 & .3 \\ .2 & .8 \end{bmatrix}^k \quad \text{approaches the limiting matrix} \quad \overline{P} = \begin{bmatrix} .4 & .6 \\ .4 & .6 \end{bmatrix} \blacktriangleleft$$

MATCHED PROBLEM 2 ➤ The transition matrix for a Markov chain is

$$P = \begin{bmatrix} .6 & .4 \\ .1 & .9 \end{bmatrix}$$

Find the stationary matrix S and the limiting matrix $\overline{P}$.   ◀

■ APPLICATIONS

▶

EXAMPLE 3 ➤ **Insurance** Refer to Example 1 in Section 7-1, where we found the following transition matrix for an insurance company:

$$\begin{array}{cc} & A \quad A' \\ P = \begin{array}{c} A \\ A' \end{array} & \begin{bmatrix} .23 & .77 \\ .11 & .89 \end{bmatrix} \end{array} \quad \begin{array}{l} A = \text{Accident} \\ A' = \text{No Accident} \end{array}$$

If these probabilities remain valid over a long period of time, what percentage of drivers can the company expect to have an accident during any given year?

Solution To determine what happens in the long run, we find the stationary matrix by solving the following system:

$$[s_1 \quad s_2] \begin{bmatrix} .23 & .77 \\ .11 & .89 \end{bmatrix} = [s_1 \quad s_2] \quad \text{and} \quad s_1 + s_2 = 1$$

which is equivalent to

$$\begin{array}{lll} .23s_1 + .11s_2 = s_1 & \text{or} & -.77s_1 + .11s_2 = 0 \\ .77s_1 + .89s_2 = s_2 & & .77s_1 - .11s_2 = 0 \\ s_1 + \quad s_2 = 1 & & s_1 + \quad s_2 = 1 \end{array}$$

Solving this system, we obtain

$$s_1 = .125 \quad \text{and} \quad s_2 = .875$$

The stationary matrix is $[.125 \quad .875]$, which means, in the long run, assuming the transition matrix does not change, about 12.5% of the drivers in the community will have an accident during any given year.   ◀

MATCHED PROBLEM 3 ▶ Refer to Matched Problem 1 in Section 7-1, where we found the following transition matrix for an insurance company:

$$P = \begin{array}{c} \\ L \\ L' \end{array} \begin{array}{cc} L \quad\quad L' \\ \begin{bmatrix} .98 & .02 \\ .78 & .22 \end{bmatrix} \end{array} \quad \begin{array}{l} L = \text{Low-risk} \\ L' = \text{Not low-risk} \end{array}$$

If these probabilities remain valid over a long period of time, what percentage of drivers can the company expect to be in the low-risk category during any given year?

◀

EXAMPLE 4 ▶ **Employee Evaluation**   A company rates every employee as below-average, average, or above-average. Past performance indicates that each year 10% of the below-average employees will raise their rating to average and 25% of the average employees will raise their rating to above-average. On the other hand, 15% of the average employees will lower their rating to below-average, and 15% of the above-average employees will lower their rating to average. Company policy prohibits rating changes from below-average to above-average, or conversely, in a single year. Over the long run, what percentage of employees will receive below-average ratings? Average ratings? Above-average ratings?.

Solution   First, we find the transition matrix:

$$\begin{array}{c} \\ \\ \text{This} \\ \text{year} \end{array} \begin{array}{c} \\ \quad A^- \\ A^- \\ A \\ A^+ \end{array} \begin{array}{c} \text{Next year} \\ A^- \quad A \quad A^+ \\ \begin{bmatrix} .9 & .1 & 0 \\ .15 & .6 & .25 \\ 0 & .15 & .85 \end{bmatrix} \end{array} \begin{array}{l} A^- = \text{Below-average} \\ A \;\; = \text{Average} \\ A^+ = \text{Above-average} \end{array}$$

To determine what happens over the long run, we find the stationary matrix by solving the following system:

$$[s_1 \quad s_2 \quad s_3] \begin{bmatrix} .9 & .1 & 0 \\ .15 & .6 & .25 \\ 0 & .15 & .85 \end{bmatrix} = [s_1 \quad s_2 \quad s_3] \quad\quad \text{and} \quad\quad s_1 + s_2 + s_3 = 1$$

which is equivalent to

$$\begin{array}{rll} .9s_1 + .15s_2 & = s_1 \\ .1s_1 + .6s_2 + .15s_3 & = s_2 \\ .25s_2 + .85s_3 & = s_3 \\ s_1 + s_2 + s_3 & = 1 \end{array} \quad\quad \text{or} \quad\quad \begin{array}{rl} -.1s_1 + .15s_2 & = 0 \\ .1s_1 - .4s_2 + .15s_3 & = 0 \\ .25s_2 - .15s_3 & = 0 \\ s_1 + s_2 + s_3 & = 1 \end{array}$$

Using Gauss–Jordan elimination to solve this system of four equations with three variables, we obtain

$$s_1 = .36 \quad\quad s_2 = .24 \quad\quad s_3 = .4$$

Thus, in the long run, 36% of the employees will be rated as below-average, 24% as average, and 40% as above-average.   ◀

MATCHED PROBLEM 4 ➤   A mail order company classifies its customers as preferred, standard, or infrequent, depending on the number of orders placed in a year. Past records indicate that each year 5% of the preferred customers are reclassified as standard and 12% as infrequent; 5% of the standard customers are reclassified as preferred and 5% as infrequent; and 9% of the infrequent customers are reclassified as preferred and 10% as standard. Assuming these percentages remain valid, what percentage of customers can the company expect to have in each category in the long run?   ◀

■ GRAPHING UTILITY APPROXIMATIONS

If P is the transition matrix for a regular Markov chain, then the powers of P approach the limiting matrix $\overline{P}$, where each row of $\overline{P}$ is equal to the stationary matrix S (Theorem 1C). We can use this result to approximate S by computing P^k for sufficiently large values of k. The next example illustrates this approach on a graphing utility.

EXAMPLE 5 ➤   **Approximating the Stationary Matrix**   Compute powers of the transition matrix P to approximate $\overline{P}$ and S to four decimal places. Check the approximation in the equation $SP = S$.

$$P = \begin{bmatrix} .5 & .2 & .3 \\ .7 & .1 & .2 \\ .4 & .1 & .5 \end{bmatrix}$$

Solution   To approximate $\overline{P}$ to four decimal places we enter P in a graphing utility (Figure 1A), set the decimal display to four places, and compute powers of P until all three rows of P^k are identical. Examining the output in Figure 1B, we conclude that

$$\overline{P} = \begin{bmatrix} .4943 & .1494 & .3563 \\ .4943 & .1494 & .3563 \\ .4943 & .1494 & .3563 \end{bmatrix} \quad \text{and} \quad S = [.4943 \quad .1494 \quad .3563]$$

Entering S in the graphing utility and computing SP shows that these matrices are correct to four decimal places (see Fig. 1C).

(A) P

(B) P^9

(C) Check: $SP = S$

FIGURE 1   ◀

MATCHED PROBLEM 5 ➤ Repeat Example 5 for: $P = \begin{bmatrix} .3 & .6 & .1 \\ .2 & .3 & .5 \\ .1 & .2 & .7 \end{bmatrix}$ ◄

—————

Remarks

1. We used a relatively small value of k to approximate $\bar{P}$ in Example 5. Many graphing utilities will compute P^k for large values of k almost as rapidly as for small values. However, round-off errors can occur in these calculations. A safe procedure is to start with a relatively small value of k, such as $k = 8$, and then keep doubling k until the rows of P^k are identical to the specified number of decimal places.

2. If any of the entries of P^k are approaching 0, the graphing utility may use scientific notation to display these entries as very small numbers. Figure 2 shows the 100th power of a transition matrix P. The entry in Row 2 and Column 2 of P^{100} is approaching 0, but the graphing utility displays it as 5.1538×10^{-53}. If this occurs, simply change this value to 0 in the corresponding entry in $\bar{P}$. Thus, from the output in Figure 2 we conclude that

$$P^k = \begin{bmatrix} 1 & 0 \\ .7 & .3 \end{bmatrix}^k \quad \text{approaches} \quad \bar{P} = \begin{bmatrix} 1 & 0 \\ 1 & 0 \end{bmatrix}$$

FIGURE 2

Answers to Matched Problems
1. (A) Regular   (B) Not regular   (C) Regular

2. $S = [.2 \quad .8]$, $\bar{P} = \begin{bmatrix} .2 & .8 \\ .2 & .8 \end{bmatrix}$ 3. 97.5%

4. 28% preferred, 43% standard, 29% infrequent

5. $\bar{P} = \begin{bmatrix} .1618 & .2941 & .5441 \\ .1618 & .2941 & .5441 \\ .1618 & .2941 & .5441 \end{bmatrix}$, $S = [.1618 \quad .2941 \quad .5441]$

EXERCISE 7-2

A *In Problems 1–12, determine whether each transition matrix is regular.*

1. $P = \begin{bmatrix} .1 & .9 \\ .7 & .3 \end{bmatrix}$

2. $P = \begin{bmatrix} .8 & .2 \\ .5 & .5 \end{bmatrix}$

3. $P = \begin{bmatrix} .2 & .8 \\ 1 & 0 \end{bmatrix}$

4. $P = \begin{bmatrix} .9 & .1 \\ 0 & 1 \end{bmatrix}$

5. $P = \begin{bmatrix} 1 & 0 \\ .6 & .4 \end{bmatrix}$

6. $P = \begin{bmatrix} 0 & 1 \\ .3 & .7 \end{bmatrix}$

7. $P = \begin{bmatrix} .3 & .4 & .3 \\ 0 & 0 & 1 \\ .4 & .2 & .4 \end{bmatrix}$

8. $P = \begin{bmatrix} 0 & 0 & 1 \\ .1 & .1 & .8 \\ .5 & .3 & .2 \end{bmatrix}$

9. $P = \begin{bmatrix} 0 & 1 & 0 \\ .4 & 0 & .6 \\ 0 & 1 & 0 \end{bmatrix}$

10. $P = \begin{bmatrix} .1 & .3 & .6 \\ 0 & 0 & 1 \\ 1 & 0 & 0 \end{bmatrix}$

11. $P = \begin{bmatrix} 0 & 0 & 1 \\ .8 & .1 & .1 \\ 0 & 1 & 0 \end{bmatrix}$

12. $P = \begin{bmatrix} 0 & .5 & .5 \\ 1 & 0 & 0 \\ 1 & 0 & 0 \end{bmatrix}$

B *For each transition matrix P in Problems 13–20, solve the equation SP = S to find the stationary matrix S and the limiting matrix P̄.*

13. $P = \begin{bmatrix} .1 & .9 \\ .6 & .4 \end{bmatrix}$ **14.** $P = \begin{bmatrix} .8 & .2 \\ .3 & .7 \end{bmatrix}$

15. $P = \begin{bmatrix} .5 & .5 \\ .3 & .7 \end{bmatrix}$ **16.** $P = \begin{bmatrix} .9 & .1 \\ .7 & .3 \end{bmatrix}$

17. $P = \begin{bmatrix} .5 & .1 & .4 \\ .3 & .7 & 0 \\ 0 & .6 & .4 \end{bmatrix}$ **18.** $P = \begin{bmatrix} .4 & .1 & .5 \\ .2 & .8 & 0 \\ 0 & .5 & .5 \end{bmatrix}$

19. $P = \begin{bmatrix} .8 & .2 & 0 \\ .5 & .1 & .4 \\ 0 & .6 & .4 \end{bmatrix}$ **20.** $P = \begin{bmatrix} .2 & .8 & 0 \\ .6 & .1 & .3 \\ 0 & .9 & .1 \end{bmatrix}$

In Problems 21–24, approximate the stationary matrix S for each transition matrix P by computing powers of the transition matrix P. Round matrix entries to four decimal places.

21. $P = \begin{bmatrix} .51 & .49 \\ .27 & .73 \end{bmatrix}$ **22.** $P = \begin{bmatrix} .68 & .32 \\ .19 & .81 \end{bmatrix}$

23. $P = \begin{bmatrix} .5 & .5 & 0 \\ 0 & .5 & .5 \\ .8 & .1 & .1 \end{bmatrix}$ **24.** $P = \begin{bmatrix} .2 & .2 & .6 \\ .5 & 0 & .5 \\ .5 & 0 & .5 \end{bmatrix}$

C

25. A red urn contains 2 red marbles and 3 blue marbles, and a blue urn contains 1 red marble and 4 blue marbles. A marble is selected from an urn, the color is noted, and the marble is returned to the urn from which it was drawn. The next marble is drawn from the urn whose color is the same as the marble just drawn. Thus, this is a Markov process with two states: draw from the red urn or draw from the blue urn.
(A) Draw a transition diagram for this process.
(B) Write the transition matrix.
(C) Find the stationary matrix and describe the long-run behavior of this process.

26. Repeat Problem 25 if the red urn contains 5 red and 3 blue marbles, and the blue urn contains 1 red and 3 blue marbles.

27. Given the transition matrix:
$$P = \begin{bmatrix} 0 & 1 \\ 1 & 0 \end{bmatrix}$$
(A) Discuss the behavior of the state matrices S_1, S_2, S_3, . . . for the initial-state matrix $S_0 = [.2 \quad .8]$.
(B) Repeat part (A) for $S_0 = [.5 \quad .5]$.
(C) Discuss the behavior of P^k, $k = 2, 3, 4, \ldots$
(D) Which of the conclusions of Theorem 1 are not valid for this matrix? Why is this not a contradiction?

28. Given the transition matrix:
$$P = \begin{bmatrix} 0 & 1 & 0 \\ 0 & 0 & 1 \\ 1 & 0 & 0 \end{bmatrix}$$
(A) Discuss the behavior of the state matrices S_1, S_2, S_3, . . . for the initial-state matrix $S_0 = [.2 \quad .3 \quad .5]$.
(B) Repeat part (A) for $S_0 = [\frac{1}{3} \quad \frac{1}{3} \quad \frac{1}{3}]$.
(C) Discuss the behavior of P^k, $k = 2, 3, 4, \ldots$
(D) Which of the conclusions of Theorem 1 are not valid for this matrix? Why is this not a contradiction?

29. The transition matrix for a Markov chain is
$$P = \begin{bmatrix} 1 & 0 & 0 \\ .2 & .2 & .6 \\ 0 & 0 & 1 \end{bmatrix}$$
(A) Show that $R = [1 \quad 0 \quad 0]$ and $S = [0 \quad 0 \quad 1]$ are both stationary matrices for P. Explain why this does not contradict Theorem 1A.
(B) Find another stationary matrix for P. [*Hint:* Consider $T = aR + (1 - a)S$, where $0 < a < 1$.]
(C) How many different stationary matrices does P have?

30. The transition matrix for a Markov chain is
$$P = \begin{bmatrix} .7 & 0 & .3 \\ 0 & 1 & 0 \\ .2 & 0 & .8 \end{bmatrix}$$
(A) Show that $R = [.4 \quad 0 \quad .6]$ and $S = [0 \quad 1 \quad 0]$ are both stationary matrices for P. Explain why this does not contradict Theorem 1A.
(B) Find another stationary matrix for P. [*Hint:* Consider $T = aR + (1 - a)S$, where $0 < a < 1$.]
(C) How many different stationary matrices does P have?

31. Refer to the transition matrix P in Problem 29. What matrix $\overline{P}$ do the powers of P appear to be approaching? Are the rows of $\overline{P}$ stationary matrices for P?

32. Refer to the transition matrix P in Problem 30. What matrix $\overline{P}$ do the powers of P appear to be approaching? Are the rows of $\overline{P}$ stationary matrices for P?

 A P P L I C A T I O N S

Business & Economics

33. *Transportation.* Most railroad cars are owned by individual railroad companies. When a car leaves its home railroad's trackage, it becomes part of a national pool of cars and can be used by other railroads. The rules governing the use of these pooled cars are designed to eventually return the car to the home trackage. A particular railroad found that each month 11% of its box cars on the home trackage left to join the national pool and 29% of its box cars in the national pool were returned to the home trackage. If these percentages remain valid for a long period of time, what percentage of its box cars can this railroad expect to have on its home trackage in the long run?

34. *Transportation.* The railroad in Problem 33 also has a fleet of tank cars. If 14% of the tank cars on the home trackage enter the national pool each month and 26% of the tank cars in the national pool are returned to the home trackage each month, what percentage of its tank cars can the railroad expect to have on its home trackage in the long run?

35. *Labor force.* Table 1 gives the percentage of the female population of the United States who were members of the civilian labor force in the indicated years:

TABLE 1

YEAR	PERCENT
1970	43.3
1980	51.5
1990	57.5

Source: U.S. Bureau of Labor Statistics

The following transition matrix P is proposed as a model for the data, where L represents females who are in the labor force and L' represents females who are not in the labor force:

$$
\begin{array}{cc}
 & \text{Next decade} \\
 & \begin{array}{cc} L & L' \end{array} \\
\begin{array}{c} \text{Current} \ L \\ \text{decade} \ \ L' \end{array} & \begin{bmatrix} .93 & .07 \\ .2 & .8 \end{bmatrix} = P
\end{array}
$$

(A) Let $S_0 = [.433 \quad .567]$, and find S_1 and S_2. (Compute both matrices exactly and then round entries to three decimal places.)

(B) Construct a new table comparing the results from part (A) with the data in Table 1.

(C) According to this transition matrix, what percentage of the female population will be in the labor force in the long run?

36. *Market research.* Table 2 gives the percentage of households that owned a dog over an 8 year period:

TABLE 2

YEAR	PERCENT
1983	42.5
1987	38.2
1991	36.5

Source: American Veterinary Medical Association

The following transition matrix P is proposed as a model for the data, where D represents the households that own a dog:

$$
\begin{array}{cc}
 & \text{Next 4 year period} \\
 & \begin{array}{cc} D & D' \end{array} \\
\begin{array}{c} \text{Current} \ \ \ D \\ \text{4 year period} \ D' \end{array} & \begin{bmatrix} .61 & .39 \\ .21 & .79 \end{bmatrix} = P
\end{array}
$$

(A) Let $S_0 = [.425 \quad .575]$, and find S_1 and S_2. (Compute both matrices exactly and then round entries to three decimal places.)

(B) Construct a new table comparing the results from part (A) with the data in Table 2.

(C) According to this transition matrix, what percentage of the households will own a dog in the long run?

37. *Market share.* Consumers in a certain state can choose between three long-distance telephone services: GTT, NCJ, and Dash. Aggressive marketing by all three companies results in a continual shift of customers among the three services. Each year, GTT loses 5% of its customers to NCJ and 20% to Dash, NCJ loses 15% of its customers to GTT and 10% to Dash, and Dash loses 5% of its customers to GTT and 10% to NCJ. Assuming these percentages remain valid over a long period of time, what is each company's expected market share in the long run?

38. *Market share.* Consumers in a certain area can choose between three package delivery services: APS, GX, and WWP. Each week, APS loses 10% of its customers to GX and 20% to WWP, GX loses 15% of its customers to APS and 10% to WWP, and WWP loses 5% of its customers to APS and 5% to GX. Assuming these percentages remain valid over a long period of time, what is each company's expected market share in the long run?

39. *Insurance.* An auto insurance company classifies its customers in three categories: poor, satisfactory, and preferred. Each year, 40% of those in the poor category are moved to satisfactory, and 20% of those in the satisfactory category are moved to preferred. Also, 20% in the preferred category are moved to the satisfactory category, and 20% in the satisfactory category are moved to the poor category. Customers are never moved from poor to preferred, or conversely, in a single year. Assuming these percentages remain valid over a long period of time, how many customers can the company expect to have in each category in the long run?

40. *Insurance.* Repeat Problem 39 if 40% of the preferred customers are moved to the satisfactory category each year and all other information remains the same.

41. *Market share.* Acme Soap Company markets one brand of soap, called Standard Acme (*SA*), and Best Soap Company markets two brands, Standard Best (*SB*) and Deluxe Best

(*DB*). Currently, Acme has 40% of the market, and the remainder is equally divided between the two Best brands. Acme is considering the introduction of a second brand to get a larger share of the market. A proposed new brand, called brand *X*, was test-marketed in several large cities, producing the following transition matrix for the consumers' weekly buying habits:

$$
P = \begin{array}{c} \\ SB \\ DB \\ SA \\ X \end{array}
\begin{array}{cccc} SB & DB & SA & X \\ \left[\begin{array}{cccc} .4 & .1 & .3 & .2 \\ .3 & .2 & .2 & .3 \\ .1 & .2 & .2 & .5 \\ .3 & .3 & .1 & .3 \end{array}\right] \end{array}
$$

Assuming that P represents the consumers' buying habits over a long period of time, use this transition matrix and the initial-state matrix $S_0 = [.3 \quad .3 \quad .4 \quad 0]$ to compute successive state matrices in order to approximate the elements in the stationary matrix correct to two decimal places. If Acme decides to market this new soap, what is the long-run expected total market share for their two soaps?

42. *Market share.* Refer to Problem 41. The chemists at Acme Soap Company have developed a second new soap, called brand *Y*. Test-marketing this soap against the three established brands produces the following transition matrix:

$$
P = \begin{array}{c} \\ SB \\ DB \\ SA \\ Y \end{array}
\begin{array}{cccc} SB & DB & SA & Y \\ \left[\begin{array}{cccc} .3 & .2 & .2 & .3 \\ .2 & .2 & .2 & .4 \\ .2 & .2 & .4 & .2 \\ .1 & .2 & .3 & .4 \end{array}\right] \end{array}
$$

Proceed as in Problem 41 to approximate the elements in the stationary matrix correct to two decimal places. If Acme decides to market brand *Y*, what is the long-run expected total market share for Standard Acme and brand *Y*? Should Acme market brand *X* or brand *Y*?

Life Sciences

43. *Genetics.* A given plant species has red, pink, or white flowers according to the genotypes RR, RW, and WW, respectively. If each of these genotypes is crossed with a pink-flowering plant (genotype RW), then the transition matrix is

		Next generation		
		Red	Pink	White
This	Red	.5	.5	0
generation	Pink	.25	.5	.25
	White	0	.5	.5

Assuming the plants of each generation are crossed only with pink plants to produce the next generation, show that regardless of the makeup of the first generation, the genotype composition will eventually stabilize at 25% red, 50% pink, and 25% white. (Find the stationary matrix.)

44. *Gene mutation.* Suppose a gene in a chromosome is of type *A* or type *B*. Assume the probability that a gene of type *A* will mutate to type *B* in one generation is 10^{-4} and that a gene of type *B* will mutate to type *A* is 10^{-6}.
   (A) What is the transition matrix?
   (B) After many generations, what is the probability that the gene will be of type *A*? Of type *B*? (Find the stationary matrix.)

Social Sciences

45. *Rapid transit.* A new rapid transit system has just started operating. In the first month of operation, it is found that 25% of the commuters are using the system, while 75% still travel by automobile. The following transition matrix was determined from records of other rapid transit systems:

$$
\begin{array}{cc}
 & \text{Next month} \\
 & \begin{array}{cc} \text{Rapid} \\ \text{transit} & \text{Automobile} \end{array}
\end{array}
$$

$$
\begin{array}{c}
\text{Current} \quad \text{Rapid transit} \\
\text{month} \quad \text{Automobile}
\end{array}
\begin{bmatrix} .8 & .2 \\ .3 & .7 \end{bmatrix}
$$

   (A) What is the initial-state matrix?
   (B) What percentage of the commuters will be using the new system after 1 month? After 2 months?
   (C) Find the percentage of commuters using each type of transportation after it has been in service for a long time.

46. *Politics—filibuster.* The Senate is in the middle of a floor debate and a filibuster is threatened. Senator Hanks, who is still vacillating, has a probability of .1 of changing his mind during the next 5 minutes. If this pattern continues for each 5 minutes that the debate continues, and if a 24 hour filibuster takes place before a vote is taken, what is the probability that Senator Hanks will cast a yes vote? A no vote?
   (A) First, complete the following transition matrix:

$$
\begin{array}{cc}
 & \text{Next 5 minutes} \\
 & \begin{array}{cc} \text{Yes} & \text{No} \end{array}
\end{array}
$$

$$
\begin{array}{c}
\text{Current} \quad \text{Yes} \\
\text{5 minutes} \quad \text{No}
\end{array}
\begin{bmatrix} .9 & .1 \\ & \end{bmatrix}
$$

   (B) Find the stationary matrix and answer the two questions.
   (C) What is the stationary matrix if the probability of Senator Hanks changing his mind (.1) is replaced with an arbitrary probability *p*?

The center of population of the 48 contiguous states of the United States is the point where a flat, rigid map of the contiguous states would balance if the location of each person was represented on the map by a weight of equal measure. In 1790 the population center was 23 miles east of Baltimore, Maryland. By 1990, the center had shifted about 800 miles west and 100 miles south to a point in southeast Missouri. To study this shifting population, the U.S. Bureau of the Census divides the states into the four regions shown in Figure 3. Problems 47 and 48 deal with population shifts among some of these regions.

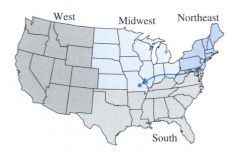

F I G U R E 3
Regions of the United States and the center of population

47. *Population shifts.* Table 3 (at the top of the next page) gives the percentage of the U.S. population living in the south region of the United States in the indicated years:

TABLE 3

YEAR	PERCENT
1970	30.9
1980	33.3
1990	34.4

Source: U.S. Bureau of the Census

The following transition matrix P is proposed as a model for the data, where S represents the population that lives in the south region:

$$\begin{array}{cc} & \text{Next decade} \\ & \begin{array}{cc} S & S' \end{array} \end{array}$$

$$\begin{array}{cc} \text{Current} & S \\ \text{decade} & S' \end{array} \begin{bmatrix} .61 & .39 \\ .21 & .79 \end{bmatrix} = P$$

(A) Let $S_0 = [.309 \quad .691]$, and find S_1 and S_2. (Compute both matrices exactly and then round entries to three decimal places.)

(B) Construct a new table comparing the results from part (A) with the data in Table 3.

(C) According to this transition matrix, what percentage of the population will live in the south region in the long run?

48. *Population shifts.* Table 4 gives the percentage of the U.S. population living in the northeast region of the United States in the indicated years:

TABLE 4

YEAR	PERCENT
1970	24.1
1980	21.7
1990	20.4

Source: U.S. Bureau of the Census

The following transition matrix P is proposed as a model for the data, where N represents the population that lives in the northeast region:

$$\begin{array}{cc} & \text{Next decade} \\ & \begin{array}{cc} N & N' \end{array} \end{array}$$

$$\begin{array}{cc} \text{Current} & N \\ \text{decade} & N' \end{array} \begin{bmatrix} .61 & .39 \\ .09 & .91 \end{bmatrix} = P$$

(A) Let $S_0 = [.241 \quad .759]$, and find S_1 and S_2. (Compute both matrices exactly and then round entries to three decimal places.)

(B) Construct a new table comparing the results from part (A) with the data in Table 4.

(C) According to this transition matrix, what percentage of the population will live in the northeast region in the long run?

SECTION   7-3   ■   # Absorbing Markov Chains

- ■ ABSORBING STATES AND ABSORBING CHAINS
- ■ STANDARD FORM
- ■ LIMITING MATRIX
- ■ GRAPHING UTILITY APPROXIMATIONS

In Section 7-2, we saw that the powers of a regular transition matrix always approach a limiting matrix. Not all transition matrices have this property. In this section we discuss another type of Markov chain called an *absorbing Markov chain*. Although regular and absorbing Markov chains have some differences, they have one important similarity: the powers of the transition matrix for an absorbing Markov chain also approach a limiting matrix. After introducing basic concepts, we will develop methods for finding the limiting matrix and discuss the relationship between the states in the Markov chain and the entries in the limiting matrix.

■ ABSORBING STATES AND ABSORBING CHAINS

A state in a Markov chain is called an **absorbing state** if once the state is entered, it is impossible to leave.

EXAMPLE 1 ➤ **Recognizing Absorbing States** Identify any absorbing states for the following transition matrices:

$$
\text{(A)} \quad P = \begin{array}{c} \\ A \\ B \\ C \end{array}\begin{array}{c} A \quad B \quad C \\ \begin{bmatrix} 1 & 0 & 0 \\ .5 & .5 & 0 \\ 0 & .5 & .5 \end{bmatrix} \end{array}
\qquad
\text{(B)} \quad P = \begin{array}{c} \\ A \\ B \\ C \end{array}\begin{array}{c} A \quad B \quad C \\ \begin{bmatrix} 0 & 0 & 1 \\ 0 & 1 & 0 \\ 1 & 0 & 0 \end{bmatrix} \end{array}
$$

Solution (A) The probability of going from state A to state A is 1, and the probability of going from state A to either state B or state C is 0. Thus, once state A is entered, it is impossible to leave; hence, A is an absorbing state. Since the probability of going from state B to state A is nonzero, it is possible to leave B and B is not an absorbing state. Similarly, the probability of going from state C to state B is nonzero, so C is not an absorbing state.

(B) Reasoning as before, the 1 in Row 2 and Column 2 indicates that state B is an absorbing state. The probability of going from state A to state C and the probability of going from state C to state A are both nonzero. Hence, A and C are not absorbing states. ◀

MATCHED PROBLEM 1 ➤ Identify any absorbing states for the following transition matrices:

$$
\text{(A)} \quad P = \begin{array}{c} \\ A \\ B \\ C \end{array}\begin{array}{c} A \quad B \quad C \\ \begin{bmatrix} .5 & 0 & .5 \\ 0 & 1 & 0 \\ 0 & .5 & .5 \end{bmatrix} \end{array}
\qquad
\text{(B)} \quad P = \begin{array}{c} \\ A \\ B \\ C \end{array}\begin{array}{c} A \quad B \quad C \\ \begin{bmatrix} 0 & 1 & 0 \\ 1 & 0 & 0 \\ 0 & 0 & 1 \end{bmatrix} \end{array}
\qquad ◀
$$

The reasoning used in Example 1 to identify absorbing states is generalized in Theorem 1.

theorem 1

> **Absorbing States and Transition Matrices**
>
> A state in a Markov chain is **absorbing** if and only if the row of the transition matrix corresponding to the state has a 1 on the main diagonal and 0's elsewhere.

The presence of an absorbing state in a transition matrix does not guarantee that the powers of the matrix approach a limiting matrix nor that the state matrices in the corresponding Markov chain approach a stationary matrix. For example, if we square the matrix P from Example 1B, we obtain

$$
P^2 = \begin{bmatrix} 0 & 0 & 1 \\ 0 & 1 & 0 \\ 1 & 0 & 0 \end{bmatrix}\begin{bmatrix} 0 & 0 & 1 \\ 0 & 1 & 0 \\ 1 & 0 & 0 \end{bmatrix} = \begin{bmatrix} 1 & 0 & 0 \\ 0 & 1 & 0 \\ 0 & 0 & 1 \end{bmatrix} = I
$$

Since $P^2 = I$, the 3×3 identity matrix, it follows that

$$P^3 = PP^2 = PI = P \quad \text{Since } P^2 = I$$
$$P^4 = PP^3 = PP = I \quad \text{Since } P^3 = P \text{ and } PP = P^2 = I$$

In general, the powers of this transition matrix P oscillate between P and I and do not approach a limiting matrix.

explore – discuss 1

(A) For the initial-state matrix $S_0 = [a \quad b \quad c]$, find the first four state matrices, S_1, S_2, S_3, and S_4, in the Markov chain with transition matrix

$$P = \begin{bmatrix} 0 & 0 & 1 \\ 0 & 1 & 0 \\ 1 & 0 & 0 \end{bmatrix}$$

(B) Do the state matrices appear to be approaching a stationary matrix? Discuss.

To ensure that transition matrices for Markov chains with one or more absorbing states have limiting matrices, it is necessary to require the chain to satisfy one additional condition, as stated in the following definition.

Absorbing Markov Chains

A Markov chain is an **absorbing chain** if:

1. There is at least one absorbing state.
2. It is possible to go from each nonabsorbing state to at least one absorbing state in a finite number of steps.

As we saw earlier, absorbing states are easily identified by examining the rows of a transition matrix. It is also possible to use a transition matrix to determine whether a Markov chain is an absorbing chain, but this can be a difficult task, especially if the matrix is large. A transition diagram is often a more appropriate tool for determining whether a Markov chain is absorbing. The next example illustrates this approach for the two matrices discussed in Example 1.

EXAMPLE 2 ➤ **Recognizing Absorbing Markov Chains** Use a transition diagram to determine whether P is the transition matrix for an absorbing Markov chain.

$$\text{(A)} \quad P = \begin{array}{c} \\ A \\ B \\ C \end{array} \begin{array}{ccc} A & B & C \\ \begin{bmatrix} 1 & 0 & 0 \\ .5 & .5 & 0 \\ 0 & .5 & .5 \end{bmatrix} \end{array} \qquad \text{(B)} \quad P = \begin{array}{c} \\ A \\ B \\ C \end{array} \begin{array}{ccc} A & B & C \\ \begin{bmatrix} 0 & 0 & 1 \\ 0 & 1 & 0 \\ 1 & 0 & 0 \end{bmatrix} \end{array}$$

Solution (A) From Example 1A, we know that A is the only absorbing state. The second condition in the definition of an absorbing Markov chain is satisfied if we can show that it is possible to go from the nonabsorbing states B and C to the absorbing state A in a finite number of steps. This is easily determined by drawing a transition diagram. Examining the diagram in the margin, we see that it is possible to go from state B to the absorbing state A in one step and from state C to the absorbing state A in two steps. Thus, P is the transition matrix for an absorbing Markov chain.

(B) Again, we draw the transition diagram for P, as shown. From this diagram it is clear that it is impossible to go from either state A or state C to the absorbing state B. Hence, P is not the transition matrix for an absorbing Markov chain.

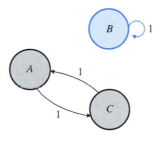

MATCHED PROBLEM 2 ▶ Use a transition diagram to determine whether P is the transition matrix for an absorbing Markov chain.

(A) $\quad P = \begin{array}{c} \\ A \\ B \\ C \end{array} \begin{array}{ccc} A & B & C \\ \begin{bmatrix} .5 & 0 & .5 \\ 0 & 1 & 0 \\ 0 & .5 & .5 \end{bmatrix} \end{array}$ (B) $\quad P = \begin{array}{c} \\ A \\ B \\ C \end{array} \begin{array}{ccc} A & B & C \\ \begin{bmatrix} 0 & 1 & 0 \\ 1 & 0 & 0 \\ 0 & 0 & 1 \end{bmatrix} \end{array}$ ◀

explore–discuss 2

Determine whether each statement is true or false. Use examples and verbal arguments to support your conclusions.

(A) A Markov chain with two states, one nonabsorbing and one absorbing, is always an absorbing chain.

(B) A Markov chain with two states, both of which are absorbing, is always an absorbing chain.

(C) A Markov chain with three states, one nonabsorbing and two absorbing, is always an absorbing chain.

■ STANDARD FORM

The transition matrix for a Markov chain is not unique. Consider the transition diagram in Figure 1. Since there are 4! = 24 different ways to arrange the four states in this diagram, there are 24 different ways to write a transition matrix. (Some of

FIGURE 1

these matrices may have identical entries, but all are different when the row and column labels are taken into account.) For example, matrices M, N, and P shown below are three different transition matrices for this diagram.

$$
M = \begin{array}{c} \\ A \\ B \\ C \\ D \end{array}
\begin{array}{cccc}
A & B & C & D \\
\end{array}
\left[\begin{array}{cccc}
.1 & .6 & .3 & 0 \\
0 & 1 & 0 & 0 \\
.5 & 0 & .2 & .3 \\
0 & 0 & 0 & 1
\end{array}\right]
\qquad
N = \begin{array}{c} \\ D \\ B \\ C \\ A \end{array}
\begin{array}{cccc}
D & B & C & A \\
\end{array}
\left[\begin{array}{cccc}
1 & 0 & 0 & 0 \\
0 & 1 & 0 & 0 \\
.3 & 0 & .2 & .5 \\
0 & .6 & .3 & .1
\end{array}\right]
\qquad
P = \begin{array}{c} \\ B \\ D \\ A \\ C \end{array}
\begin{array}{cccc}
B & D & A & C \\
\end{array}
\left[\begin{array}{cccc}
1 & 0 & 0 & 0 \\
0 & 1 & 0 & 0 \\
.6 & 0 & .1 & .3 \\
0 & .3 & .5 & .2
\end{array}\right]
\quad (1)
$$

In matrices N and P, notice that all the absorbing states precede all the nonabsorbing states. A transition matrix written in this form is said to be a *standard form*. We will find standard forms very useful in determining limiting matrices for absorbing Markov chains. The general definition of standard form is given in the next box.

Standard Forms for Absorbing Markov Chains

A transition matrix for an absorbing Markov chain is a **standard form** if the rows and columns are labeled so that all the absorbing states precede all the nonabsorbing states. (There may be more than one standard form.) Any standard form can always be partitioned into four submatrices:

$$
\begin{array}{c} \\ A \\ N \end{array}
\begin{array}{cc}
A & N \\
\end{array}
\left[\begin{array}{c:c}
I & 0 \\ \hdashline
R & Q
\end{array}\right]
\qquad
\begin{array}{l}
A = \text{All absorbing states} \\
N = \text{All nonabsorbing states}
\end{array}
$$

where I is an identity matrix and 0 is a zero matrix.

Referring to the matrix P in (1), we see that the submatrices in this standard form are

$$
I = \begin{bmatrix} 1 & 0 \\ 0 & 1 \end{bmatrix}
\qquad
0 = \begin{bmatrix} 0 & 0 \\ 0 & 0 \end{bmatrix}
$$

$$
R = \begin{bmatrix} .6 & 0 \\ 0 & .3 \end{bmatrix}
\qquad
Q = \begin{bmatrix} .1 & .3 \\ .5 & .2 \end{bmatrix}
\qquad
P = \begin{array}{c} \\ B \\ D \\ A \\ C \end{array}
\begin{array}{cccc}
B & D & A & C \\
\end{array}
\left[\begin{array}{cc:cc}
1 & 0 & 0 & 0 \\
0 & 1 & 0 & 0 \\ \hdashline
.6 & 0 & .1 & .3 \\
0 & .3 & .5 & .2
\end{array}\right]
$$

e x p l o r e – d i s c u s s 3

We used the diagram in Figure 1 to find the transition matrices M, N, and P in (1). Devise a procedure for transforming matrix M into standard forms N and P by interchanging rows and columns in M without reference to the transition diagram. Discuss the relative merits of using transition diagrams versus using row and column rearrangements to find standard forms.

■ LIMITING MATRIX

We are now ready to discuss the long-run behavior of absorbing Markov chains. We begin with an application.

EXAMPLE 3 ➤ **Real Estate Development**   Two competing real estate companies are trying to buy all the farms in a particular area for future housing development. Each year, 20% of the farmers decide to sell to company A, 30% decide to sell to company B, and the rest continue to farm their land. Neither company ever sells any of the farms they purchase.

(A) Draw a transition diagram for this Markov process and determine whether the associated Markov chain is absorbing.

(B) Write a transition matrix that is in standard form.

(C) If neither company owns any farms at the beginning of this competitive buying process, estimate the percentage of farms that each company will purchase in the long run.

(D) If company A buys 50% of the farms before company B enters this competitive buying process, estimate the percentage of farms that each company will purchase in the long run.

Solution   (A)

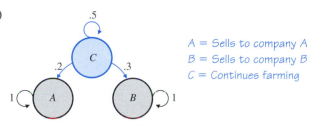

A = Sells to company A
B = Sells to company B
C = Continues farming

The associated Markov chain is absorbing, since there are two absorbing states, A and B, and it is possible to go from the nonabsorbing state C to either A or B in one step.

(B) We use the transition diagram to write a transition matrix that is in standard form:

$$P = \begin{matrix} & \begin{matrix} A & B & C \end{matrix} \\ \begin{matrix} A \\ B \\ C \end{matrix} & \begin{bmatrix} 1 & 0 & 0 \\ 0 & 1 & 0 \\ .2 & .3 & .5 \end{bmatrix} \end{matrix} \quad \text{Standard form}$$

(C) At the beginning of the competitive buying process all the farmers are in state C (own a farm). Thus, $S_0 = [0 \quad 0 \quad 1]$. The successive state matrices are (multiplication details omitted):

$$S_1 = S_0 P = [.2 \quad .3 \quad .5]$$
$$S_2 = S_1 P = [.3 \quad .45 \quad .25]$$
$$S_3 = S_2 P = [.35 \quad .525 \quad .125]$$
$$S_4 = S_3 P = [.375 \quad .5625 \quad .0625]$$

$$S_5 = S_4P = [.3875 \quad .58125 \quad .03125]$$
$$S_6 = S_5P = [.39375 \quad .590625 \quad .015625]$$
$$S_7 = S_6P = [.396875 \quad .5953125 \quad .0078125]$$
$$S_8 = S_7P = [.3984375 \quad .59765625 \quad .00390625]$$
$$S_9 = S_8P = [.39921875 \quad .598828125 \quad .001953125]$$

It appears that these state matrices are approaching the matrix

$$
\begin{array}{ccc}
A & B & C
\end{array}
$$
$$S = [.4 \quad .6 \quad 0]$$

This indicates that in the long run, company A will acquire approximately 40% of the farms and company B will acquire the remaining 60%.

(D) This time, at the beginning of the competitive buying process 50% of the farmers are already in state A and the rest are in state C. Thus, $S_0 = [.5 \quad 0 \quad .5]$. The successive state matrices are (multiplication details omitted):

$$S_1 = S_0P = [.6 \quad .15 \quad .25]$$
$$S_2 = S_1P = [.65 \quad .225 \quad .125]$$
$$S_3 = S_2P = [.675 \quad .2625 \quad .0625]$$
$$S_4 = S_3P = [.6875 \quad .28125 \quad .03125]$$
$$S_5 = S_4P = [.69375 \quad .290625 \quad .015625]$$
$$S_6 = S_5P = [.696875 \quad .2953125 \quad .0078125]$$
$$S_7 = S_6P = [.6984375 \quad .29765625 \quad .00390625]$$
$$S_8 = S_7P = [.69921875 \quad .298828125 \quad .001953125]$$

These state matrices approach a matrix different from the one in part (A):

$$
\begin{array}{ccc}
A & B & C
\end{array}
$$
$$S' = [.7 \quad .3 \quad 0]$$

Because of its head start, company A will now acquire approximately 70% of the farms and company B will acquire the remaining 30%. ◄

MATCHED PROBLEM 3 ► Repeat Example 3 if 10% of the farmers sell to company A each year, 40% sell to company B, and the remainder continue farming. ◄

Recall from Theorem 1 in Section 7-2 that the successive state matrices of a regular Markov chain always approach a stationary matrix. Furthermore, this stationary matrix is unique. That is, changing the initial-state matrix does not change the stationary matrix. The successive state matrices for an absorbing Markov chain also approach a stationary matrix, but this matrix is not unique. To confirm this, consider the transition matrix P and the state matrices S and S' from Example 3:

$$
P = \begin{array}{c} \\ A \\ B \\ C \end{array} \begin{array}{ccc} A & B & C \\ \left[\begin{array}{ccc} 1 & 0 & 0 \\ 0 & 1 & 0 \\ .2 & .3 & .5 \end{array} \right] \end{array} \qquad
\begin{array}{ccc} A & B & C \end{array} \\ S = [.4 \quad .6 \quad 0] \qquad
\begin{array}{ccc} A & B & C \end{array} \\ S' = [.7 \quad .3 \quad 0]
$$

It turns out that S and S' are both stationary matrices, as the following multiplications verify:

$$
SP = [.4 \quad .6 \quad 0] \left[\begin{array}{ccc} 1 & 0 & 0 \\ 0 & 1 & 0 \\ .2 & .3 & .5 \end{array} \right] = [.4 \quad .6 \quad 0] = S
$$

$$
S'P = [.7 \quad .3 \quad 0] \left[\begin{array}{ccc} 1 & 0 & 0 \\ 0 & 1 & 0 \\ .2 & .3 & .5 \end{array} \right] = [.7 \quad .3 \quad 0] = S'
$$

In fact, this absorbing Markov chain has an infinite number of stationary matrices (see Problems 37 and 38 in Exercise 7-3).

Thus, changing the initial-state matrix for an absorbing Markov chain can cause the successive state matrices to approach a different stationary matrix.

In Section 7-2 we used the unique stationary matrix for a regular Markov chain to find the limiting matrix $\overline{P}$. Since an absorbing Markov chain can have many different stationary matrices, we cannot expect this approach to work for absorbing chains. However, it turns out that transition matrices for absorbing chains do have limiting matrices, and they are not very difficult to find. Theorem 2 gives us the necessary tools. The proof of this theorem is left for more advanced courses.

theorem 2

Limiting Matrices for Absorbing Markov Chains

If a standard form P for an absorbing Markov chain is partitioned as

$$
P = \left[\begin{array}{c|c} I & 0 \\ \hline R & Q \end{array} \right]
$$

then P^k approaches a limiting matrix $\overline{P}$ as k increases, where

$$
\overline{P} = \left[\begin{array}{c|c} I & 0 \\ \hline FR & 0 \end{array} \right]
$$

The matrix F is given by $F = (I - Q)^{-1}$ and is called the **fundamental matrix** for P.

The identity matrix used to form the fundamental matrix F must be the same size as the matrix Q.

EXAMPLE 4 ➤ Finding the Limiting Matrix

(A) Find the limiting matrix $\overline{P}$ for the standard form P found in Example 3.
(B) Use $\overline{P}$ to find the limit of the successive state matrices for $S_0 = [0 \quad 0 \quad 1]$.
(C) Use $\overline{P}$ to find the limit of the successive state matrices for $S_0 = [.5 \quad 0 \quad .5]$.

Solution (A) From Example 3, we have

$$P = \left[\begin{array}{cc:c} 1 & 0 & 0 \\ 0 & 1 & 0 \\ \hdashline .2 & .3 & .5 \end{array}\right] \quad \left[\begin{array}{c:c} I & O \\ \hdashline R & Q \end{array}\right]$$

where

$$I = \begin{bmatrix} 1 & 0 \\ 0 & 1 \end{bmatrix} \qquad 0 = \begin{bmatrix} 0 \\ 0 \end{bmatrix} \qquad R = [.2 \quad .3] \qquad Q = [.5]$$

If $I = [1]$ is the 1×1 identity matrix, then $I - Q$ is also a 1×1 matrix, and $F = (I - Q)^{-1}$ is simply the multiplicative inverse of the single entry in $I - Q$. Thus,

$$F = ([1] - [.5])^{-1} = [.5]^{-1} = [2]$$
$$FR = [2][.2 \quad .3] = [.4 \quad .6]$$

and the limiting matrix is

$$\begin{array}{cc} & \begin{array}{ccc} A & B & C \end{array} \\ \overline{P} = \begin{array}{c} A \\ B \\ C \end{array} & \begin{bmatrix} 1 & 0 & 0 \\ 0 & 1 & 0 \\ .4 & .6 & 0 \end{bmatrix} \end{array} \quad \left[\begin{array}{c:c} I & O \\ \hdashline FR & O \end{array}\right]$$

(B) Since the successive state matrices are given by $S_k = S_0 P^k$ (Theorem 1 in Section 7-1) and P^k approaches $\overline{P}$, it follows that S_k approaches

$$S_0 \overline{P} = [0 \quad 0 \quad 1]\begin{bmatrix} 1 & 0 & 0 \\ 0 & 1 & 0 \\ .4 & .6 & 0 \end{bmatrix} = [.4 \quad .6 \quad 0]$$

which agrees with the results in part (C) of Example 3.

(C) This time, the successive state matrices approach

$$S_0 \overline{P} = [.5 \quad 0 \quad .5]\begin{bmatrix} 1 & 0 & 0 \\ 0 & 1 & 0 \\ .4 & .6 & 0 \end{bmatrix} = [.7 \quad .3 \quad 0]$$

which agrees with the results in part (D) of Example 3. ◄

MATCHED PROBLEM 4 ➤ Repeat Example 4 for the standard form P found in Matched Problem 3. ◄

Recall that the limiting matrix for a regular Markov chain contains the long-run probabilities of going from any state to any other state. This is also true for the limiting matrix of an absorbing Markov chain. Let's compare the transition matrix P and its limiting matrix $\bar{P}$ from Example 4:

$$
P = \begin{array}{c} \\ A \\ B \\ C \end{array}\begin{array}{ccc} A & B & C \\ \left[\begin{array}{ccc} 1 & 0 & 0 \\ 0 & 1 & 0 \\ .2 & .3 & .5 \end{array}\right] \end{array} \qquad \text{approaches} \qquad \bar{P} = \begin{array}{c} \\ A \\ B \\ C \end{array}\begin{array}{ccc} A & B & C \\ \left[\begin{array}{ccc} 1 & 0 & 0 \\ 0 & 1 & 0 \\ .4 & .6 & 0 \end{array}\right] \end{array}
$$

The rows of P and $\bar{P}$ corresponding to the absorbing states A and B are identical. That is, if the probability of going from state A to state A is 1 at the beginning of the chain, this probability will remain 1 for all trials in the chain and for the limiting matrix. The entries in the third row of $\bar{P}$ give the long-run probabilities of going from the nonabsorbing state C to states A, B, or C.

The fundamental matrix F provides some additional information about an absorbing chain. Recall from Example 4 that $F = [2]$. It can be shown that the entries in F determine the average number of trials it takes to go from a given nonabsorbing state to an absorbing state. In the case of Example 4, the single entry 2 in F indicates that it will take an average of 2 years for a farmer to go from state C (owns a farm) to one of the absorbing states (sells the farm). Some will reach an absorbing state in 1 year, some will take more than 2 years, but the average will be 2 years. These observations are summarized in Theorem 3, which we state below without proof.

theorem 3

Properties of the Limiting Matrix $\bar{P}$

If P is a transition matrix in standard form for an absorbing Markov chain, F is the fundamental matrix, and $\bar{P}$ is the limiting matrix, then:

(A) The entry in Row i and Column j of $\bar{P}$ is the long-run probability of going from state i to state j. For the nonabsorbing states, these probabilities are also the entries in the matrix FR used to form $\bar{P}$.

(B) The sum of the entries in each row of the fundamental matrix F is the average number of trials it will take to go from each nonabsorbing state to some absorbing state.

[Note that the rows of both F and FR correspond to the nonabsorbing states in the order given in the standard form P.]

Remarks

1. The zero matrix in the lower right corner of the limiting matrix $\bar{P}$ in Theorem 2 indicates that the long-run probability of going from any nonabsorbing state to any other nonabsorbing state is always 0. That is, in the long run, all elements in an absorbing Markov chain end up in one of the absorbing states.

2. If the transition matrix for an absorbing Markov chain is not a standard form, it is still possible to find a limiting matrix (see Problems 33 and 34 in Exercise 7-3). However, it is customary to use a standard form when investigating the limiting behavior of an absorbing chain.

Now that we have developed the necessary tools for analyzing the long-run behavior of an absorbing Markov chain, we apply these tools to an application we considered earlier (see Example 3 in Section 7-1).

▶ EXAMPLE 5 ▶ University Enrollment   The transition diagram for part-time students enrolled in an MBA program in a university is shown below:

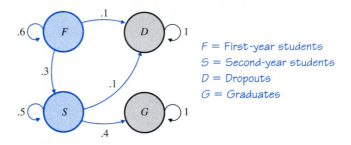

F = First-year students
S = Second-year students
D = Dropouts
G = Graduates

(A) In the long run, what percentage of first-year students will graduate? What percentage of second-year students will not graduate?
(B) What is the average number of years a first-year student will remain in this program? A second-year student?

Solution   (A) First, notice that this is an absorbing Markov chain with two absorbing states, state D and state G. A standard form for this absorbing chain is

$$P = \begin{array}{c} \\ D \\ G \\ F \\ S \end{array} \begin{array}{cc} \begin{array}{cccc} D & G & F & S \end{array} \\ \left[\begin{array}{cc|cc} 1 & 0 & 0 & 0 \\ 0 & 1 & 0 & 0 \\ \hline .1 & 0 & .6 & .3 \\ .1 & .4 & 0 & .5 \end{array} \right] \end{array} \quad \begin{bmatrix} I & \vdots & O \\ \hdashline R & \vdots & Q \end{bmatrix}$$

The submatrices in this partition are

$$I = \begin{bmatrix} 1 & 0 \\ 0 & 1 \end{bmatrix} \qquad O = \begin{bmatrix} 0 & 0 \\ 0 & 0 \end{bmatrix} \qquad R = \begin{bmatrix} .1 & 0 \\ .1 & .4 \end{bmatrix} \qquad Q = \begin{bmatrix} .6 & .3 \\ 0 & .5 \end{bmatrix}$$

Thus,

$$F = (I - Q)^{-1} = \left(\begin{bmatrix} 1 & 0 \\ 0 & 1 \end{bmatrix} - \begin{bmatrix} .6 & .3 \\ 0 & .5 \end{bmatrix} \right)^{-1}$$

$$= \begin{bmatrix} .4 & -.3 \\ 0 & .5 \end{bmatrix}^{-1}$$

$$= \begin{bmatrix} 2.5 & 1.5 \\ 0 & 2 \end{bmatrix}$$

and

$$FR = \begin{bmatrix} 2.5 & 1.5 \\ 0 & 2 \end{bmatrix} \begin{bmatrix} .1 & 0 \\ .1 & .4 \end{bmatrix} = \begin{bmatrix} .4 & .6 \\ .2 & .8 \end{bmatrix}$$

The limiting matrix is

$$\overline{P} = \begin{array}{c} \\ D \\ G \\ F \\ S \end{array} \begin{array}{cccc} D & G & F & S \\ \begin{bmatrix} 1 & 0 & 0 & 0 \\ 0 & 1 & 0 & 0 \\ .4 & .6 & 0 & 0 \\ .2 & .8 & 0 & 0 \end{bmatrix} \end{array} \quad \begin{bmatrix} I & | & 0 \\ \hline FR & | & 0 \end{bmatrix}$$

From this limiting form, we see that in the long run 60% of the first-year students will graduate and 20% of the second-year students will not graduate.

(B) The sum of the entries in the first row of the fundamental matrix F is $2.5 + 1.5 = 4$. According to Theorem 3, this indicates that a first-year student will spend an average of 4 years in the transient states F and S before reaching one of the absorbing states, D or G. The sum of the entries in the second row of F is $0 + 2 = 2$. Thus, a second-year student spends an average of 2 years in the program before either graduating or dropping out. ◀

MATCHED PROBLEM 5 ▶ Repeat Example 5 for the following transition diagram:

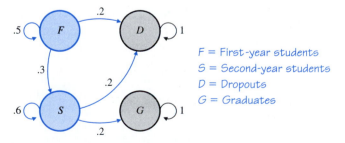

F = First-year students
S = Second-year students
D = Dropouts
G = Graduates

◀

■ GRAPHING UTILITY APPROXIMATIONS

Just as was the case for regular Markov chains, the limiting matrix $\overline{P}$ for an absorbing Markov chain with transition matrix P can be approximated by computing P^k on a graphing utility for sufficiently large values of k. For example, computing P^{50} for the standard form P in Example 5 produces the following results:

$$P^{50} = \begin{bmatrix} 1 & 0 & 0 & 0 \\ 0 & 1 & 0 & 0 \\ .1 & 0 & .6 & .3 \\ .1 & .4 & 0 & .5 \end{bmatrix}^{50} = \begin{bmatrix} 1 & 0 & 0 & 0 \\ 0 & 1 & 0 & 0 \\ .4 & .6 & 0 & 0 \\ .2 & .8 & 0 & 0 \end{bmatrix} = \overline{P}$$

where once again we have replaced very small numbers displayed in scientific notation with 0 (see page 507 in Section 7-2).

_____

Caution

Before you use P^k to approximate $\overline{P}$, be certain to determine that $\overline{P}$ does in fact exist. If you attempt to approximate a limiting matrix when none exists, the results can be misleading. For example, consider the transition matrix

$$P = \begin{bmatrix} 1 & 0 & 0 & 0 & 0 \\ .2 & .2 & 0 & .3 & .3 \\ 0 & 0 & 0 & .5 & .5 \\ 0 & 0 & 1 & 0 & 0 \\ 0 & 0 & 1 & 0 & 0 \end{bmatrix}$$

Computing P^{50} on a graphing utility produces the following matrix:

$$P^{50} = \begin{bmatrix} 1 & 0 & 0 & 0 & 0 \\ .25 & 0 & .625 & .0625 & .0625 \\ 0 & 0 & 1 & 0 & 0 \\ 0 & 0 & 0 & .5 & .5 \\ 0 & 0 & 0 & .5 & .5 \end{bmatrix} \tag{2}$$

It is tempting to stop at this point and conclude that the matrix in (2) must be a good approximation for $\overline{P}$. But to do so would be incorrect! If P^{50} approximates a limiting matrix $\overline{P}$, then P^{51} should also approximate the same matrix. However, computing P^{51} produces quite a different matrix:

$$P^{51} = \begin{bmatrix} 1 & 0 & 0 & 0 & 0 \\ .25 & 0 & .125 & .3125 & .3125 \\ 0 & 0 & 0 & .5 & .5 \\ 0 & 0 & 1 & 0 & 0 \\ 0 & 0 & 1 & 0 & 0 \end{bmatrix} \tag{3}$$

Computing additional powers of P shows that the even powers of P approach the matrix in (2) while the odd powers approach the matrix in (3). Thus, the transition matrix P does not have a limiting matrix and, consequently, is not the transition matrix for an absorbing Markov chain. Examining the transition diagram in Figure 2 (at the top of the next page), we see that it is not possible to go from states C, D, or E to state A, the only absorbing state.

$$P = \begin{array}{c} \\ A \\ B \\ C \\ D \\ E \end{array} \begin{array}{ccccc} A & B & C & D & E \\ \left[\begin{array}{ccccc} 1 & 0 & 0 & 0 & 0 \\ .2 & .2 & 0 & .3 & .3 \\ 0 & 0 & 0 & .5 & .5 \\ 0 & 0 & 1 & 0 & 0 \\ 0 & 0 & 1 & 0 & 0 \end{array}\right] \end{array}$$

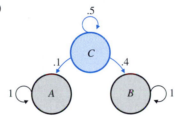

FIGURE 2

Answers to Matched Problems
1. (A) State B is absorbing.   (B) State C is absorbing.
2. (A) Absorbing Markov chain   (B) Not an absorbing Markov chain
3. (A)

$$\text{(B)} \ P = \begin{array}{c} \\ A \\ B \\ C \end{array} \begin{array}{ccc} A & B & C \\ \left[\begin{array}{ccc} 1 & 0 & 0 \\ 0 & 1 & 0 \\ .1 & .4 & .5 \end{array}\right] \end{array}$$

(C) Company A will purchase 20% of the farms and company B will purchase 80%.
(D) Company A will purchase 60% of the farms and company B will purchase 40%.

$$\text{4. (A)} \ \overline{P} = \begin{array}{c} \\ A \\ B \\ C \end{array} \begin{array}{ccc} A & B & C \\ \left[\begin{array}{ccc} 1 & 0 & 0 \\ 0 & 1 & 0 \\ .2 & .8 & 0 \end{array}\right] \end{array} \qquad \text{(B)} \ [.2 \ \ .8 \ \ 0] \qquad \text{(C)} \ [.6 \ \ .4 \ \ 0]$$

5. (A) 30% of the first-year students will graduate; 50% of the second-year students will not graduate.
   (B) A first-year student will spend an average of 3.5 years in the program; a second-year student will spend an average of 2.5 years in the program.

EXERCISE   7-3

A   *In Problems 1–4, identify the absorbing states in the indicated transition matrix.*

1. $P = \begin{array}{c} \\ A \\ B \\ C \end{array} \begin{array}{ccc} A & B & C \\ \left[\begin{array}{ccc} .6 & .3 & .1 \\ 0 & 1 & 0 \\ 0 & 0 & 1 \end{array}\right] \end{array}$

2. $P = \begin{array}{c} \\ A \\ B \\ C \end{array} \begin{array}{ccc} A & B & C \\ \left[\begin{array}{ccc} 0 & 1 & 0 \\ .3 & .2 & .5 \\ 0 & 0 & 1 \end{array}\right] \end{array}$

3. $P = \begin{array}{c} \\ A \\ B \\ C \\ D \end{array} \begin{array}{cccc} A & B & C & D \\ \left[\begin{array}{cccc} 1 & 0 & 0 & 0 \\ 0 & 0 & 1 & 0 \\ .1 & .1 & .5 & .3 \\ 0 & 0 & 0 & 1 \end{array}\right] \end{array}$

4. $P = \begin{array}{c} \\ A \\ B \\ C \\ D \end{array} \begin{array}{cccc} A & B & C & D \\ \left[\begin{array}{cccc} 0 & 1 & 0 & 0 \\ 1 & 0 & 0 & 0 \\ .1 & .2 & .3 & .4 \\ .7 & .1 & .1 & .1 \end{array}\right] \end{array}$

In Problems 5–8, identify the absorbing states for each transition diagram, and determine whether the diagram represents an absorbing Markov chain.

5.

6.

7.

8.

B *In Problems 9–12, find a standard form for the absorbing Markov chain with the indicated transition diagram.*

9.

10.

11.

12.

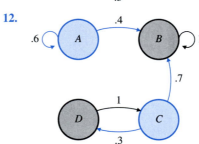

In Problems 13–16, find a standard form for the absorbing Markov chain with the indicated transition matrix.

13. $P = \begin{array}{c} \\ A \\ B \\ C \end{array} \begin{array}{ccc} A & B & C \\ \left[\begin{array}{ccc} .2 & .3 & .5 \\ 1 & 0 & 0 \\ 0 & 0 & 1 \end{array}\right] \end{array}$

14. $P = \begin{array}{c} \\ A \\ B \\ C \end{array} \begin{array}{ccc} A & B & C \\ \left[\begin{array}{ccc} 0 & 0 & 1 \\ 0 & 1 & 0 \\ .7 & .2 & .1 \end{array}\right] \end{array}$

15. $P = \begin{array}{c} \\ A \\ B \\ C \\ D \end{array} \begin{array}{cccc} A & B & C & D \\ \left[\begin{array}{cccc} .1 & .2 & .3 & .4 \\ 0 & 1 & 0 & 0 \\ .5 & .2 & .2 & .1 \\ 0 & 0 & 0 & 1 \end{array}\right] \end{array}$

16. $P = \begin{array}{c} \\ A \\ B \\ C \\ D \end{array} \begin{array}{cccc} A & B & C & D \\ \left[\begin{array}{cccc} 0 & .3 & .3 & .4 \\ 0 & 1 & 0 & 0 \\ 0 & 0 & 1 & 0 \\ .8 & .1 & .1 & 0 \end{array}\right] \end{array}$

In Problems 17–22, find the limiting matrix for the indicated standard form. Find the long-run probability of going from each nonabsorbing state to each absorbing state and the average number of trials needed to go from each nonabsorbing state to an absorbing state.

17. $P = \begin{array}{c} \\ A \\ B \\ C \end{array} \begin{array}{ccc} A & B & C \\ \left[\begin{array}{ccc} 1 & 0 & 0 \\ 0 & 1 & 0 \\ .1 & .4 & .5 \end{array}\right] \end{array}$

18. $P = \begin{array}{c} \\ A \\ B \\ C \end{array} \begin{array}{ccc} A & B & C \\ \left[\begin{array}{ccc} 1 & 0 & 0 \\ 0 & 1 & 0 \\ .3 & .2 & .5 \end{array}\right] \end{array}$

19. $P = \begin{array}{c} \\ A \\ B \\ C \end{array} \begin{array}{ccc} A & B & C \\ \left[\begin{array}{ccc} 1 & 0 & 0 \\ .2 & .6 & .2 \\ .4 & .2 & .4 \end{array}\right] \end{array}$

20. $P = \begin{array}{c} \\ A \\ B \\ C \end{array} \begin{array}{ccc} A & B & C \\ \left[\begin{array}{ccc} 1 & 0 & 0 \\ .1 & .6 & .3 \\ .2 & .2 & .6 \end{array}\right] \end{array}$

21. $P = \begin{array}{c} \\ A \\ B \\ C \\ D \end{array} \begin{array}{cccc} A & B & C & D \\ \left[\begin{array}{cccc} 1 & 0 & 0 & 0 \\ 0 & 1 & 0 & 0 \\ .1 & .2 & .6 & .1 \\ .2 & .2 & .3 & .3 \end{array}\right] \end{array}$

22. $P = \begin{array}{c} \\ A \\ B \\ C \\ D \end{array} \begin{array}{cccc} A & B & C & D \\ \left[\begin{array}{cccc} 1 & 0 & 0 & 0 \\ 0 & 1 & 0 & 0 \\ .1 & .1 & .7 & .1 \\ .3 & .1 & .4 & .2 \end{array}\right] \end{array}$

Problems 23–28 refer to the matrices in Problems 17–22, as indicated. Use the limiting matrix $\overline{P}$ found for each transition matrix P in Problems 17–22 to determine the long-run behavior of the successive state matrices for the indicated initial-state matrices.

23. For matrix *P* from Problem 17 with:
(A) $S_0 = [0 \ \ 0 \ \ 1]$ (B) $S_0 = [.2 \ \ .5 \ \ .3]$

24. For matrix *P* from Problem 18 with:
(A) $S_0 = [0 \ \ 0 \ \ 1]$ (B) $S_0 = [.2 \ \ .5 \ \ .3]$

25. For matrix *P* from Problem 19 with:
(A) $S_0 = [0 \ \ 0 \ \ 1]$ (B) $S_0 = [.2 \ \ .5 \ \ .3]$

26. For matrix *P* from Problem 20 with:
(A) $S_0 = [0 \ \ 0 \ \ 1]$ (B) $S_0 = [.2 \ \ .5 \ \ .3]$

27. For matrix *P* from Problem 21 with:
(A) $S_0 = [0 \ \ 0 \ \ 0 \ \ 1]$ (B) $S_0 = [0 \ \ 0 \ \ 1 \ \ 0]$
(C) $S_0 = [0 \ \ 0 \ \ .4 \ \ .6]$ (D) $S_0 = [.1 \ \ .2 \ \ .3 \ \ .4]$

28. For matrix *P* from Problem 22 with:
(A) $S_0 = [0 \ \ 0 \ \ 0 \ \ 1]$ (B) $S_0 = [0 \ \ 0 \ \ 1 \ \ 0]$
(C) $S_0 = [0 \ \ 0 \ \ .4 \ \ .6]$ (D) $S_0 = [.1 \ \ .2 \ \ .3 \ \ .4]$

In Problems 29–32, use a graphing utility to approximate the limiting matrix for the indicated standard form.

29. $P = \begin{array}{c} \\ A \\ B \\ C \\ D \end{array} \begin{array}{cccc} A & B & C & D \\ \left[\begin{array}{cccc} 1 & 0 & 0 & 0 \\ 0 & 1 & 0 & 0 \\ .5 & .3 & .1 & .1 \\ .6 & .2 & .1 & .1 \end{array}\right] \end{array}$

30. $P = \begin{array}{c} \\ A \\ B \\ C \\ D \end{array} \begin{array}{cccc} A & B & C & D \\ \left[\begin{array}{cccc} 1 & 0 & 0 & 0 \\ 0 & 1 & 0 & 0 \\ .1 & .1 & .5 & .3 \\ 0 & .2 & .3 & .5 \end{array}\right] \end{array}$

31. $P = \begin{array}{c} \\ A \\ B \\ C \\ D \\ E \end{array} \begin{array}{ccccc} A & B & C & D & E \\ \left[\begin{array}{ccccc} 1 & 0 & 0 & 0 & 0 \\ 0 & 1 & 0 & 0 & 0 \\ 0 & .4 & .5 & 0 & .1 \\ 0 & .4 & 0 & .3 & .3 \\ .4 & .4 & 0 & .2 & 0 \end{array}\right] \end{array}$

32. $P = \begin{array}{c} \\ A \\ B \\ C \\ D \\ E \end{array} \begin{array}{ccccc} A & B & C & D & E \\ \left[\begin{array}{ccccc} 1 & 0 & 0 & 0 & 0 \\ 0 & 1 & 0 & 0 & 0 \\ .5 & 0 & 0 & 0 & .5 \\ 0 & .4 & 0 & .2 & .4 \\ 0 & 0 & .1 & .7 & .2 \end{array}\right] \end{array}$

C

33. The following matrix *P* is a nonstandard transition matrix for an absorbing Markov chain:

$$P = \begin{array}{c} \\ A \\ B \\ C \\ D \end{array} \begin{array}{cccc} A & B & C & D \\ \left[\begin{array}{cccc} .2 & .2 & .6 & 0 \\ 0 & 1 & 0 & 0 \\ .5 & .1 & 0 & .4 \\ 0 & 0 & 0 & 1 \end{array}\right] \end{array}$$

To find a limiting matrix for *P*, follow the steps outlined below:

Step 1.   Using a transition diagram as an aid, rearrange the columns and rows of P to produce a standard form for this chain.

Step 2. Find the limiting matrix for this standard form.

Step 3. Using a transition diagram as an aid, reverse the process used in step 1 to produce a limiting matrix for the original matrix P.

34. Repeat Problem 33 for

$$P = \begin{array}{c} \\ A \\ B \\ C \\ D \end{array} \begin{array}{cccc} A & B & C & D \\ \begin{bmatrix} 1 & 0 & 0 & 0 \\ .3 & .6 & 0 & .1 \\ .2 & .3 & .5 & 0 \\ 0 & 0 & 0 & 1 \end{bmatrix} \end{array}$$

35. Verify the results in Problem 33 by computing P^k on a graphing utility for large values of k.

36. Verify the results in Problem 34 by computing P^k on a graphing utility for large values of k.

37. Show that $S = [x \quad 1 - x \quad 0], 0 \leqslant x \leqslant 1$, is a stationary matrix for the transition matrix

$$P = \begin{array}{c} \\ A \\ B \\ C \end{array} \begin{array}{ccc} A & B & C \\ \begin{bmatrix} 1 & 0 & 0 \\ 0 & 1 & 0 \\ .1 & .5 & .4 \end{bmatrix} \end{array}$$

Discuss the generalization of this result to any absorbing Markov chain with two absorbing states and one nonabsorbing state.

38. Show that $S = [x \quad 1 - x \quad 0 \quad 0], 0 \leqslant x \leqslant 1$, is a stationary matrix for the transition matrix

$$P = \begin{array}{c} \\ A \\ B \\ C \\ D \end{array} \begin{array}{cccc} A & B & C & D \\ \begin{bmatrix} 1 & 0 & 0 & 0 \\ 0 & 1 & 0 & 0 \\ .1 & .2 & .3 & .4 \\ .6 & .2 & .1 & .1 \end{bmatrix} \end{array}$$

Discuss the generalization of this result to any absorbing Markov chain with two absorbing states and two nonabsorbing states.

▶ **APPLICATIONS**

Business & Economics

39. *Loans.* A credit union classifies automobile loans into one of four categories: the loan has been paid in full (F), the account is in good standing (G) with all payments up to date, the account is in arrears (A) with one or more missing payments, or the account has been classified as a bad debt (B) and sold to a collection agency. Past records indicate that each month 10% of the accounts in good standing pay the loan in full, 80% remain in good standing, and 10% become in arrears. Furthermore, 10% of the accounts in arrears are paid in full, 40% become accounts in good standing, 40% remain in arrears, and 10% are classified as bad debts.
(A) In the long run, what percentage of the accounts in arrears will pay their loan in full?
(B) In the long run, what percentage of the accounts in good standing will become bad debts?
(C) What is the average number of months an account in arrears will remain in this system before it is either paid in full or classified as a bad debt?

40. *Employee training.* A national chain of automobile muffler and brake repair shops maintains a training program for its mechanics. All new mechanics begin training in muffler repairs. Every 3 months the performance of each mechanic is reviewed. Past records indicate that after each quarterly review, 30% of the muffler repair trainees are rated as qualified to repair mufflers and begin training in brake repairs, 20% are terminated for unsatisfactory performance, and the remainder continue as muffler repair trainees. Also, 30% of the brake repair trainees are rated as fully qualified mechanics requiring no further training, 10% are terminated for unsatisfactory performance, and the remainder continue as brake repair trainees.
(A) In the long run, what percentage of the muffler repair trainees will become fully qualified mechanics?
(B) In the long run, what percentage of the brake repair trainees will be terminated?
(C) What is the average number of quarters a muffler repair trainee will remain in the training program before being either terminated or promoted to fully qualified mechanic?

41. *Marketing.* Three electronics firms are aggressively marketing their graphing calculators to high school and college mathematics departments by offering volume discounts,

complimentary display equipment, and assistance with curriculum development. Due to the amount of equipment involved and the necessary curriculum changes, once a department decides to use a particular calculator in their courses, they never switch to another brand or stop using calculators. Each year, 6% of the departments decide to use calculators from company A, 3% decide to use calculators from company B, 11% decide to use calculators from company C, and the remainder decide not to use any calculators in their courses.
(A) In the long run, what is the market share of each company?
(B) On the average, how many years will it take a department to decide to use calculators from one of these companies in their courses?

42. *Pensions.* Once a year employees at a company are given the opportunity to join one of three pension plans, A, B, or C. Once an employee decides to join one of these plans, the employee cannot drop the plan or switch to another plan. Past records indicate that each year 4% of the employees elect to join plan A, 14% elect to join plan B, 7% elect to join plan C, and the remainder do not join any plan.
(A) In the long run, what percentage of the employees will elect to join plan A? Plan B? Plan C?
(B) On the average, how many years will it take an employee to decide to join a plan?

Life Sciences

43. *Medicine.* After bypass surgery, patients are placed in an intensive care unit (ICU) until their condition stabilizes. Then they are transferred to a cardiac care ward (CCW) where they remain until they are released from the hospital. In a particular metropolitan area, a study of hospital records produced the following data: each day 2% of the patients in the ICU died, 52% were transferred to the CCW, and the remainder stayed in the ICU. Furthermore, each day 4% of the patients in the CCW developed complications and were returned to the ICU, 1% died while in the CCW, 22% were released from the hospital, and the remainder stayed in the CCW.
(A) In the long run, what percentage of the patients in the ICU are released from the hospital?
(B) In the long run, what percentage of the patients in the CCW die without ever being released from the hospital?
(C) What is the average number of days a patient in the ICU will stay in the hospital?

44. *Medicine.* The study discussed in Problem 43 also produced the following data for patients who underwent aortic valve replacements: each day 2% of the patients in the ICU died, 60% were transferred to the CCW, and the remainder stayed in the ICU. Furthermore, each day 5% of the patients in the CCW developed complications and were returned to the ICU, 1% died while in the CCW, 19% were released from the hospital, and the remainder stayed in the CCW.
(A) In the long run, what percentage of the patients in the CCW are released from the hospital?
(B) In the long run, what percentage of the patients in the ICU die without ever being released from the hospital?
(C) What is the average number of days a patient in the CCW will stay in the hospital?

Social Sciences

45. *Psychology.* A rat is placed in room F or room B of the maze shown in the figure. The rat wanders from room to room until it enters one of the rooms containing food, L or R. Assume that the rat chooses an exit from a room at random and that once it enters a room with food it never leaves.
(A) What is the long-run probability that a rat placed in room B ends up in room R?
(B) What is the average number of exits a rat placed in room B will choose until it finds food?

Figure for 45–46

46. *Psychology.* Repeat Problem 45 if the exit from room B to room R is blocked.

Using Markov Chains to Model Games

Gambler's Ruin   Games of chance between two players that continue until one player goes broke are often referred to as **gambler's ruin problems.** For example, suppose two people who have a total of $4 between them are playing a simple coin-tossing game. A fair coin is tossed. If the coin comes up heads, player A pays player B $1. If the coin comes up tails, player B pays player A $1. The game continues until one player has all the money and the other player is broke, or ruined, hence the name *gambler's ruin*.

This game can be analyzed as an absorbing Markov chain, where the states are the amount of money player A has at each stage of the game. Thus, the possible states are $0, $1, $2, $3, and $4. The game ends when player A has $0 or $4, so these two states are absorbing. The transition diagram and a standard form for the transition matrix are shown below.

$$P = \begin{array}{c} \\ \$0 \\ \$4 \\ \$1 \\ \$2 \\ \$3 \end{array} \begin{array}{ccccc} \$0 & \$4 & \$1 & \$2 & \$3 \\ \left[\begin{array}{ccccc} 1 & 0 & 0 & 0 & 0 \\ 0 & 1 & 0 & 0 & 0 \\ .5 & 0 & 0 & .5 & 0 \\ 0 & 0 & .5 & 0 & .5 \\ 0 & .5 & 0 & .5 & 0 \end{array}\right] \end{array}$$

(A) Find the fundamental matrix F and the limiting matrix $\overline{P}$. What is the probability that player A wins all the money if player A starts with $1? With $2? With $3? What is the average number of coin tosses until the game ends if player A starts with $1? With $2? With $3?

(B) Repeat part (A), but assume the players start with a total of $5 between them. Draw the corresponding transition diagram, and find a standard form for the transition matrix. Be certain to consider all the different amounts of money player A could have at the beginning of the game.

(C) Roulette wheels in Nevada generally have 38 equally spaced slots numbered 00, 0, 1, 2, . . . , 36. The numbers from 1 to 36 are evenly divided between red and black. A player who bets $1 on black wins $1 (and gets the bet back) if the ball comes to rest on black; otherwise (if the ball lands on red, 0, or 00), the $1 bet is lost. A player at a roulette wheel starts with $3 and plans to continue to make $1 wagers on black until he either doubles his money or goes broke. Find the probability that the player goes broke. Find the average number of wagers the player will make before either going broke or doubling his money. Solve using Markov chain techniques and use a graphing utility to perform the necessary matrix calculations.

Best Two-Out-of-Three Series Many sporting competitions consist of playing a series of games until one player (or team) has won two games. This can be viewed as a Markov chain with the states consisting of the number of games each player has won. Denote the states as $A0-B0$, $A0-B1$, $A0-B2$, and so on, where $An-Bm$ indicates that player A has won n games and player B has won m games. Let p be the probability that player A wins any particular game and assume that this probability is the same for each game in the series.

(A) List all the possible states, and indicate which are absorbing.

(B) Draw a transition diagram and find a standard form for the transition matrix.

(C) If $p = .5$, find the fundamental matrix F and the limiting matrix $\overline{P}$. What is the probability that player A wins the series? What is the average number of games played in the series?

(D) Repeat part (C) if $p = .6$.

C H A P T E R 7 R E V I E W ▪ **Important Terms and Symbols**

7-1 *Properties of Markov Chains.* Stochastic process; transition diagram; transition probability matrix; initial-state distribution matrix; initial-state probability matrix; first-state matrix; Markov chain, or process; kth-state matrix; recursive definition of state matrices; powers of a transition matrix

$$S_k = S_{k-1}P = S_0P^k$$

7-2 *Regular Markov Chains.* Stationary matrix; regular transition matrix; regular Markov chain; limiting matrix

$$SP = S; \quad P^k \text{ approaches } \overline{P}$$

7-3 *Absorbing Markov Chains.* Absorbing state; absorbing Markov chain; standard form; limiting matrix; fundamental matrix

$$P = \left[\begin{array}{c|c} I & 0 \\ \hline R & Q \end{array}\right]; \quad \overline{P} = \left[\begin{array}{c|c} I & 0 \\ \hline FR & 0 \end{array}\right]; \quad F = (I - Q)^{-1}$$

C H A P T E R 7 ▪ **Review Exercise**

Work through all the problems in this chapter review and check your answers in the back of the book. Answers to all review problems are there along with section numbers in italics to indicate where each type of problem is discussed. Where weaknesses show up, review appropriate sections in the text.

A

1. Given the transition matrix P and initial-state matrix S_0 shown below, find S_1 and S_2 and explain what each represents:

$$P = \begin{array}{c} \\ A \\ B \end{array}\begin{array}{cc} A & B \\ \left[\begin{array}{cc} .6 & .4 \\ .2 & .8 \end{array}\right] \end{array} \qquad S_0 = [.3 \quad .7]$$

In Problems 2–4, P is a transition matrix for a Markov chain. Identify any absorbing states and classify the chain as regular, absorbing, or neither.

2. $P = \begin{array}{c} \\ A \\ B \end{array} \begin{array}{cc} A & B \\ \left[\begin{array}{cc} 1 & 0 \\ .7 & .3 \end{array} \right] \end{array}$

3. $P = \begin{array}{c} \\ A \\ B \end{array} \begin{array}{cc} A & B \\ \left[\begin{array}{cc} 0 & 1 \\ .7 & .3 \end{array} \right] \end{array}$

4. $P = \begin{array}{c} \\ A \\ B \end{array} \begin{array}{cc} A & B \\ \left[\begin{array}{cc} 0 & 1 \\ 1 & 0 \end{array} \right] \end{array}$

In Problems 5–8, write a transition matrix for the indicated transition diagram, identify any absorbing states, and classify each Markov chain as regular, absorbing, or neither.

5.

6.

7.

8.

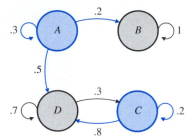

B

9. A Markov chain has three states, A, B, and C. The probability of going from state A to state B in one trial is .2, the probability of going from state A to state C in one trial is .5, the probability of going from state B to state A in one trial is .8, the probability of going from state B to state C in one trial is .2, the probability of going from state C to state A in one trial is .1, and the probability of going from state C to state B in one trial is .3. Draw a transition diagram and write a transition matrix for this chain.

10. Given the transition matrix

$$P = \begin{array}{c} \\ A \\ B \end{array} \begin{array}{cc} A & B \\ \left[\begin{array}{cc} .4 & .6 \\ .9 & .1 \end{array} \right] \end{array}$$

    find the probability of:
    (A) Going from state A to state B in two trials.
    (B) Going from state B to state A in three trials.

In Problems 11 and 12, solve the equation SP = S to find the stationary matrix S and the limiting matrix $\overline{P}$.

11. $P = \begin{array}{c} \\ A \\ B \end{array} \begin{array}{cc} A & B \\ \left[\begin{array}{cc} .4 & .6 \\ .2 & .8 \end{array} \right] \end{array}$

12. $P = \begin{array}{c} \\ A \\ B \\ C \end{array} \begin{array}{ccc} A & B & C \\ \left[\begin{array}{ccc} .4 & .6 & 0 \\ .5 & .3 & .2 \\ 0 & .8 & .2 \end{array} \right] \end{array}$

In Problems 13 and 14, find the limiting matrix for the indicated standard form. Find the long-run probability of going from each nonabsorbing state to each absorbing state and the average number of trials needed to go from each nonabsorbing state to an absorbing state.

13. $P = \begin{array}{c} \\ A \\ B \\ C \end{array} \begin{array}{ccc} A & B & C \\ \left[\begin{array}{ccc} 1 & 0 & 0 \\ 0 & 1 & 0 \\ .3 & .1 & .6 \end{array} \right] \end{array}$

14.

$$P = \begin{array}{c} \\ A \\ B \\ C \\ D \end{array} \begin{array}{cccc} A & B & C & D \\ \left[\begin{array}{cccc} 1 & 0 & 0 & 0 \\ 0 & 1 & 0 & 0 \\ .1 & .5 & .2 & .2 \\ .1 & .1 & .4 & .4 \end{array} \right] \end{array}$$

In Problems 15–18, use a graphing utility to approximate the limiting matrix for the indicated transition matrix.

15. Matrix P from Problem 11

16. Matrix P from Problem 12

17. Matrix P from Problem 13

18. Matrix P from Problem 14

19. Find a standard form for the absorbing Markov chain with transition matrix

$$P = \begin{array}{c} \\ A \\ B \\ C \\ D \end{array} \begin{array}{cccc} A & B & C & D \\ \left[\begin{array}{cccc} .6 & .1 & .2 & .1 \\ 0 & 1 & 0 & 0 \\ .3 & .2 & .3 & .2 \\ 0 & 0 & 0 & 1 \end{array} \right] \end{array}$$

In Problems 20 and 21, determine the long-run behavior of the successive state matrices for the indicated transition matrix and initial-state matrices.

20.

$$P = \begin{array}{c} \\ A \\ B \\ C \end{array} \begin{array}{ccc} A & B & C \\ \left[\begin{array}{ccc} 0 & 1 & 0 \\ 0 & 0 & 1 \\ .2 & .6 & .2 \end{array} \right] \end{array}$$

(A) $S_0 = [0 \ \ 0 \ \ 1]$ (B) $S_0 = [.5 \ \ .3 \ \ .2]$

21.

$$P = \begin{array}{c} \\ A \\ B \\ C \end{array} \begin{array}{ccc} A & B & C \\ \left[\begin{array}{ccc} 1 & 0 & 0 \\ 0 & 1 & 0 \\ .2 & .6 & .2 \end{array} \right] \end{array}$$

(A) $S_0 = [0 \ \ 0 \ \ 1]$ (B) $S_0 = [.5 \ \ .3 \ \ .2]$

C

22. A red urn contains 2 red marbles, 1 blue marble, and 1 green marble. A blue urn contains 1 red marble, 3 blue marbles, and 1 green marble. A green urn contains 6 red marbles, 3 blue marbles, and 1 green marble. A marble is selected from an urn, the color is noted, and the marble is returned to the urn from which it was drawn. The next marble is drawn from the urn whose color is the same as the marble just drawn. Thus, this is a Markov process with three states: draw from the red urn, draw from the blue urn, or draw from the green urn.

(A) Draw a transition diagram for this process.

(B) Write the transition matrix P.

(C) Determine whether this chain is regular, absorbing, or neither.

(D) Find the limiting matrix $\bar{P}$, if it exists, and describe the long-run behavior of this process.

23. Repeat Problem 22 if the blue and green marbles are removed from the red urn.

24. Show that $S = [x \ \ y \ \ z \ \ 0]$, where $x + y + z = 1$, $0 \le x \le 1, 0 \le y \le 1, 0 \le z \le 1$, is a stationary matrix for the transition matrix

$$P = \begin{array}{c} \\ A \\ B \\ C \\ D \end{array} \begin{array}{cccc} A & B & C & D \\ \left[\begin{array}{cccc} 1 & 0 & 0 & 0 \\ 0 & 1 & 0 & 0 \\ 0 & 0 & 1 & 0 \\ .1 & .3 & .4 & .2 \end{array} \right] \end{array}$$

Discuss the generalization of this result to any absorbing chain with three absorbing states and one nonabsorbing state.

In Problems 25–31, either give an example of a Markov chain with the indicated properties or explain why no such chain can exist.

25. A regular Markov chain with an absorbing state.

26. An absorbing Markov chain that is regular.

27. A regular Markov chain with two different stationary matrices.

28. An absorbing Markov chain with two different stationary matrices.

29. A Markov chain with no limiting matrix.

30. A regular Markov chain with no limiting matrix.

31. An absorbing Markov chain with no limiting matrix.

APPLICATIONS

Business & Economics

32. *Product switching.* A company's brand (X) has 20% of the market. A market research firm finds that if a person uses brand X, the probability is .7 that he or she will buy it next time. On the other hand, if a person does not use brand X (represented by X'), the probability is .5 that he or she will switch to brand X the next time.
(A) Draw a transition diagram.
(B) Write a transition matrix.
(C) Write the initial-state matrix.
(D) Find the first-state matrix and explain what it represents.
(E) Find the stationary matrix.
(F) What percentage of the market will brand X have in the long run if the transition matrix does not change?

33. *Marketing.* Recent technological advances have led to the development of three new milling machines, brand A, brand B, and brand C. Due to the extensive retooling and startup costs, once a company converts its machine shop to one of these new machines, it never switches to another brand. Each year 6% of the machine shops convert to brand A machines, 8% convert to brand B machines, 11% convert to brand C machines, and the remainder continue to use their old machines.
(A) In the long run, what is the market share of each brand?
(B) What is the average number of years a company waits before converting to one of the new milling machines?

34. *Market research.* Table 1 gives the percentage of households that owned a VCR in the indicated years:

T A B L E 1

YEAR	PERCENT
1984	10.6
1988	58.0
1992	72.5

Source: Television Bureau of Advertising, Inc.

The following transition matrix P is proposed as a model for the data, where V represents the households that own a VCR:

$$\text{Current year} \quad \begin{matrix} \\ V \\ V' \end{matrix} \overset{\displaystyle \overset{\text{Four years later}}{\begin{matrix} V & \quad V' \end{matrix}}}{\begin{bmatrix} .853 & .147 \\ .553 & .447 \end{bmatrix}} = P$$

(A) Let $S_0 = [.106 \quad .894]$, and find S_1 and S_2. (Compute both matrices exactly and then round entries to three decimal places.)
(B) Construct a new table comparing the results from part (A) with the data in Table 1.
(C) According to the transition matrix, what percentage of the households will own a VCR in the long run?

35. *Employee training.* In order to become a Fellow of the Society of Actuaries, an individual must pass a series of ten examinations given by the Society. Passage of the first two preliminary exams is a prerequisite for employment as a trainee in the actuarial department of a large insurance company. Each year 15% of the trainees complete the next three exams in the program and become associates of the Society of Actuaries, 5% leave the company, never to return, and the remainder continue as trainees. Furthermore, each year 17% of the associates complete the remaining five exams and become fellows of the Society of Actuaries, 3% leave the company, never to return, and the remainder continue as associates.
(A) In the long run, what percentage of the trainees will become fellows?
(B) In the long run, what percentage of the associates will leave the company?
(C) What is the average number of years a trainee remains in this program before either becoming a fellow or being discharged?

Life Sciences

36. *Genetics.* A given plant species has red, pink, or white flowers according to the genotypes RR, RW, and WW, respectively. If each of these genotypes is crossed with a red-flowering plant, then the transition matrix is

$$\text{This generation} \quad \begin{matrix} \text{Red} \\ \text{Pink} \\ \text{White} \end{matrix} \overset{\displaystyle \overset{\text{Next generation}}{\begin{matrix} \text{Red} & \text{Pink} & \text{White} \end{matrix}}}{\begin{bmatrix} 1 & 0 & 0 \\ .5 & .5 & 0 \\ 0 & 1 & 0 \end{bmatrix}}$$

If each generation of the plant is crossed only with red plants to produce the next generation, show that eventually all the flowers produced by the plants will be red. (Find the limiting matrix.)

Social Sciences

37. *Farm population.* Table 2 gives the percentage of the U.S. population living on farms in the indicated years:

TABLE 2

YEAR	PERCENT
1980	2.8
1985	2.3
1990	1.9

Source: U.S. Bureau of the Census

The following transition matrix P is proposed as a model for the data, where F represents the population that lives on a farm:

Five years later

$$\text{Current year} \begin{array}{c} F \\ F' \end{array} \begin{bmatrix} .751 & .249 \\ .001 & .999 \end{bmatrix} = P$$

(A) Let $S_0 = [.028 \quad .972]$, and find S_1 and S_2. (Compute both matrices exactly and then round entries to three decimal places.)

(B) Construct a new table comparing the results from part (A) with the data in Table 2.

(C) According to this transition matrix, what percentage of the population will live on a farm in the long run?

PART THREE
Calculus

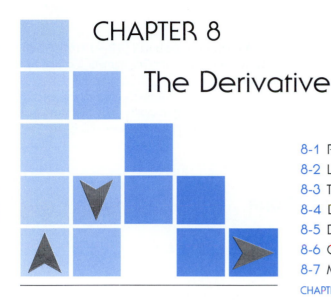

CHAPTER 8

The Derivative

How do algebra and calculus differ? The two words *static* and *dynamic* probably come as close as any in expressing the difference between the two disciplines. In algebra, we solve equations for a particular value of a variable—a static notion. In calculus, we are interested in how a change in one variable affects another variable—a dynamic notion.

Parts (A)–(C) of the figure illustrate three basic problems in calculus. It may surprise you to learn that all three problems—as different as they appear—are mathematically related. The solutions to these problems and the discovery of their relationship required the creation of a new kind of mathematics. Isaac Newton

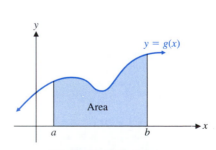

(A) Find the equation of the tangent line at (x_1, y_1) given $y = f(x)$

(B) Find the instantaneous velocity of a falling object

(C) Find the indicated area bounded by $y = g(x)$, $x = a$, $x = b$, and the x axis

539

(1642–1727) of England and Gottfried Wilhelm von Leibniz (1646–1716) of Germany simultaneously and independently developed this new mathematics, called **the calculus**—it was an idea whose time had come.

In addition to solving the problems described in the figure, calculus will enable us to solve many other important problems. Until fairly recently, calculus was used primarily in the physical sciences, but now people in many other disciplines are finding it a useful tool.

S E C T I O N   8-1   ■   # Rate of Change and Slope

■ RATE OF CHANGE
■ SLOPE
■ SUMMARY

To provide a feeling and overview for the subject of calculus, a number of key concepts are introduced informally in this section. These concepts will be considered again in subsequent sections.

■ RATE OF CHANGE

Let us start by considering a simple example.

▶   EXAMPLE 1   ➤   **Revenue Analysis**   The revenue (in dollars) from the sale of x plastic planter boxes is given by

$$R(x) = 20x - 0.02x^2 \qquad 0 \leq x \leq 1,000$$

which is graphed in Figure 1.

(A) What is the change in revenue if production is changed from 100 planters to 400 planters?
(B) What is the average change in revenue for this change in production?

Solution   (A) The change in revenue is given by

$$R(400) - R(100) = 20(400) - 0.02(400)^2 - [20(100) - 0.02(100)^2]$$
$$= 4,800 - 1,800 = \$3,000$$

Thus, increasing production from 100 planters to 400 planters will increase revenue by $3,000.
(B) To find the average change in revenue, we divide the change in revenue by the change in production:

$$\frac{R(400) - R(100)}{400 - 100} = \frac{3,000}{300} = \$10$$

Thus, the average change in revenue is $10 per planter when production is increased from 100 to 400 planters.

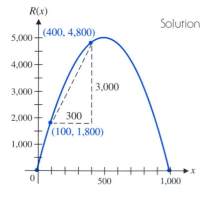

FIGURE 1
$R(x) = 20x - 0.02x^2$

MATCHED PROBLEM 1 Refer to the revenue function in Example 1.

(A) What is the change in revenue if production is changed from 600 planters to 800 planters?

(B) What is the average change in revenue for this change in production? ◀

In general, if we are given a function $y = f(x)$ and if x is changed from a to $a + h$, then y will change from $f(a)$ to $f(a + h)$. The *average rate of change is the ratio of the change in y to the change in x.*

Average Rate of Change

For $y = f(x)$, the **average rate of change from $x = a$ to $x = a + h$** is

$$\frac{f(a + h) - f(a)}{(a + h) - a} = \frac{f(a + h) - f(a)}{h} \qquad h \neq 0 \qquad (1)$$

The mathematical expression in (1) is also referred to as the **difference quotient.** The preceding discussion shows that the difference quotient can be interpreted as an average rate of change. The next example illustrates another interpretation of this quotient: velocity of a moving object.

EXAMPLE 2 ➤ **Average Velocity**   A small steel ball dropped from a tower will fall a distance of y feet in x seconds, as given approximately by the formula (from physics)

$$y = f(x) = 16x^2$$

Figure 2 shows the position of the ball on a coordinate line (positive direction down) at the end of 0, 1, 2, and 3 seconds. Find the average velocity from $x = 2$ seconds to $x = 3$ seconds.

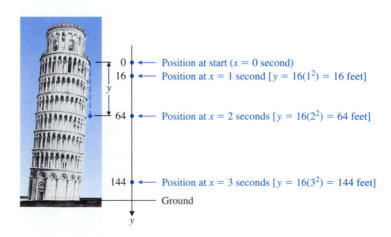

FIGURE 2
Note: Positive y direction is down.

Solution   Recall the formula $d = rt$, which can be written in the equivalent form

$$r = \frac{d}{t} = \frac{\text{Distance covered}}{\text{Elapsed time}} = \text{Average velocity (rate)}$$

For example, if a person drives from San Francisco to Los Angeles (a distance of about 420 miles) in 7 hours, then the average velocity is

$$r = \frac{d}{t} = \frac{420}{7} = 60 \text{ miles per hour}$$

Sometimes the person will be traveling faster and sometimes slower, but the average velocity is 60 miles per hour. In our present problem, the average velocity of the steel ball from $x = 2$ seconds to $x = 3$ seconds is

$$\begin{aligned} \text{Average velocity} &= \frac{\text{Distance covered}}{\text{Elapsed time}} \\ &= \frac{f(3) - f(2)}{3 - 2} \\ &= \frac{16(3)^2 - 16(2)^2}{1} = 80 \text{ feet per second} \end{aligned}$$

Thus, we see that if $y = f(x)$ is the position of the falling ball, then the average velocity is simply the average rate of change of $f(x)$ with respect to time x. And we have another interpretation of the difference quotient (1). ◄

MATCHED PROBLEM 2 ►   For the falling steel ball in Example 2, find the average velocity from $x = 1$ second to $x = 2$ seconds. ◄

In Figure 2, note that the distance the ball falls during each 1 second interval increases with time. Thus, the velocity of the ball is increasing as it falls. Rather than considering average velocity, suppose we are interested in determining the velocity of the ball at the exact instant 2 seconds after it was released. One way to do this is to consider the difference quotient

$$\frac{f(2 + h) - f(2)}{h} \qquad \frac{\text{Distance covered}}{\text{Elapsed time}}$$

which is the average velocity for the change in time from 2 seconds to $2 + h$ seconds if $h > 0$ (or from $2 + h$ seconds to 2 seconds if $h < 0$). We can use these average velocities to approximate the velocity at $x = 2$ seconds. And we would expect that the smaller the choice of h, the better the approximation of this velocity.

EXAMPLE 3 ►   Velocity   For the steel ball in Example 2, find the average velocity for $x = 2$ and $h = \pm 0.1, \pm 0.01, \pm 0.001$. What do these average velocities indicate about the velocity at 2 seconds?

Solution The average velocities are shown in Table 1:

TABLE 1
Average Velocities

h	-0.1	-0.01	$-0.001 \rightarrow 0 \leftarrow 0.001$		0.01	0.1
$\dfrac{f(2 + h) - f(2)}{h}$	62.4	63.84	$63.984 \rightarrow 64 \leftarrow 64.016$		64.16	65.6

Examining these values, it seems reasonable to conclude that the average velocity from 2 seconds to $2 + h$ seconds is approaching 64 feet per second as h approaches 0. We describe this type of behavior verbally by saying that 64 is *the limit of the average velocity as h approaches 0,* and we express it symbolically by writing

$$\frac{f(2 + h) - f(2)}{h} \rightarrow 64 \quad \text{as} \quad h \rightarrow 0$$

or

$$\lim_{h \rightarrow 0} \frac{f(2 + h) - f(2)}{h} = 64$$

The value of this limit is called the **instantaneous velocity** of the ball at the end of 2 seconds. ◀

MATCHED PROBLEM 3 ➤ For the steel ball in Example 2, find the average velocity for $x = 1$ and $h = \pm 0.1$, ± 0.01, ± 0.001. What do these average velocities indicate about the velocity at the end of 1 second? ◀

Remark
The average velocities in Table 1 can be computed with an ordinary calculator. However, utilizing the function evaluation capabilities of a graphing utility will speed up the process considerably. To do this, we store the difference quotient in the function list, using x in place of h (Fig. 3A), and evaluate repeatedly. Figure 3B shows the first three average velocities in Table 1. (To avoid typing each line, we recall the previous entry and edit the value of x.) Some graphing utilities have built-in table routines that make this process even simpler (see Fig. 3C).

(A) Function list (B) Home screen

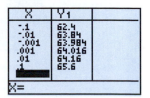

(C) Table routine

FIGURE 3

explore – discuss 1

Recall the revenue function discussed in Example 1: $R(x) = 20x - 0.02x^2$. Approximate

$$\lim_{h \to 0} \frac{R(100 + h) - R(100)}{h}$$

using $h = \pm 10, \pm 1, \pm 0.1, \pm 0.01$. Discuss possible interpretations of this limit.

The ideas introduced in Example 3 are not confined to just average velocity, but can be applied to the average rate of change of any function.

Instantaneous Rate of Change

For $y = f(x)$, the **instantaneous rate of change at $x = a$** is

$$\lim_{h \to 0} \frac{f(a + h) - f(a)}{h} \qquad (2)$$

if the limit exists.

The adjective "instantaneous" is often omitted, with the understanding that the phrase **rate of change** always refers to the instantaneous rate of change and not the average rate of change. In subsequent sections we will develop procedures for finding the exact value of limit (2). For now, we will simply approximate these limits numerically.

EXAMPLE 4 ➤ Agriculture   Using data from the U.S. Census Bureau, an analyst constructed the following function to model the number of farms (in thousands) in the United States:

$$f(t) = 2,300 + 4t - 2t^2$$

where t is time in years and $t = 0$ corresponds to 1974.

(A) Find the number of farms in 1984.
(B) Approximate the instantaneous rate of change of the number of farms with respect to time in 1984 by constructing a table similar to that used in Example 3.
(C) Interpret the results in parts (A) and (B).

Solution   (A) Since $t = 10$ corresponds to 1984, $f(10)$ is the number of farms in 1984. From Figure 4, we see that $f(10) = 2,140$, or 2.14 million farms.

FIGURE 4
$y_1 = 2,300 + 4x - 2x^2$

(B) We construct a table of values for $t = 10$ and h small:

h	-0.1	-0.01	$-0.001 \rightarrow$	$0 \leftarrow$	0.001	0.01	0.1
$\dfrac{f(10 + h) - f(10)}{h}$	-35.8	-35.98	$-35.998 \rightarrow$	$-36 \leftarrow -36.002$		-36.02	-36.2

From the table we conclude that the instantaneous rate of change is approximately -36 or $-36,000$ farms per year.

(C) In 1984 the total number of farms was 2.14 million and was decreasing at the rate of 36,000 farms per year. ◄

MATCHED PROBLEM 4 ➤ Although the total number of farms in the United States has been declining, the number of small farms (under 10 acres) has been growing. The following function gives the number of small farms (in thousands):

$$f(t) = 125 + 10t - 0.4t^2$$

where t is time in years and $t = 0$ corresponds to 1974.

(A) Find the number of small farms in 1984.
(B) Approximate the instantaneous rate of change of the number of small farms with respect to time in 1984 by constructing an appropriate table.
(C) Interpret the results in parts (A) and (B). ◄

One of the major applications of calculus is the calculation and interpretation of instantaneous rates of change. Notice that the brief discussion given in part (C) of Example 4 includes the value of the independent variable (1984), the value of the dependent variable (2.14 million farms), and how this variable is changing (decreasing at the rate of 36,000 farms per year). This information should be included in most interpretations involving rates of change.

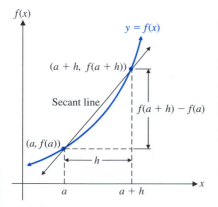

FIGURE 5
Secant line

■ **SLOPE**

So far our interpretations of the difference quotient have been numerical in nature. Now we want to consider a geometric interpretation. A line through two points on the graph of a function is called a **secant line.** If $(a, f(a))$ and $(a + h, f(a + h))$ are two points on the graph of $y = f(x)$, then we can use the slope formula from Section 1-4 to find the slope of the secant line through these points (see Fig. 5).

$$\textbf{Slope of secant line} = \frac{f(a + h) - f(a)}{(a + h) - a}$$

$$= \frac{f(a + h) - f(a)}{h} \quad \textcolor{teal}{\text{Difference quotient}}$$

Thus, the difference quotient can be interpreted as both the average rate of change and the slope of the secant line.

EXAMPLE 5 ➤ Slope of Secant Line   Given $f(x) = 0.5x^2$, find the slopes of the secant lines for $a = 1$ and $h = 2$ and 1, respectively. Graph $y = f(x)$ and the two secant lines.

Solution   For $a = 1$ and $h = 2$, the secant line goes through $(1, f(1)) = (1, 0.5)$ and $(3, f(3)) = (3, 4.5)$, and its slope is

$$\frac{f(1 + 2) - f(1)}{2} = \frac{0.5(3)^2 - 0.5(1)^2}{2} = 2$$

For $a = 1$ and $h = 1$, the secant line goes through $(1, f(1)) = (1, 0.5)$ and $(2, f(2)) = (2, 2)$, and its slope is

$$\frac{f(1 + 1) - f(1)}{1} = \frac{0.5(2)^2 - 0.5(1)^2}{1} = 1.5$$

The graphs of $y = f(x)$ and the two secant lines are shown in Figure 6.

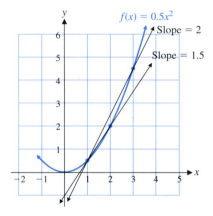

FIGURE 6
Secant lines

MATCHED PROBLEM 5 ➤ Refer to Example 5. Find the slopes of the secant lines for $a = 1$ and $h = -1$ and -0.5, respectively.

Referring to Figure 6 in Example 5, what happens to the slope of the secant line as the point $(1 + h, f(1 + h))$ approaches $(1, f(1))$? A graphing utility can be used to shed some light on this problem. Figure 7A shows the graph of $y = f(x) = 0.5x^2$ and the point $(1, f(1)) = (1, 0.5)$. We zoom in on the point $(1, 0.5)$ as shown in Figure 7B, and then use trace to find the coordinates of another point on the graph of $y = 0.5x^2$ that is very close to the point $(1, 0.5)$. We now compute the slope of the secant line through these two points:

$$\frac{f(1 + h) - f(1)}{(1 + h) - 1} = \frac{0.50021279 - 0.5}{1.0002128 - 1} = 0.99995301$$

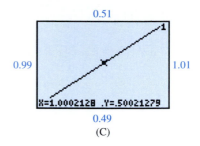

(A) (B) (C)

FIGURE 7
$y_1 = 0.5x^2$

Repeating this process with h getting closer and closer to 0, that is, with the point $(1 + h, f(1 + h))$ getting closer and closer to $(1, 0.5)$, suggests that

$$\lim_{h \to 0} \frac{f(1 + h) - f(1)}{h} = 1$$

The value of this limit is called the *slope of the graph* of $y = f(x)$ at the point $(1, f(1))$. The line with this slope through the point $(1, f(1))$ is called the *tangent line* (see Fig. 8).

FIGURE 8
Tangent line

e x p l o r e – d i s c u s s 2

The equation of the line tangent to the graph of $f(x) = 0.5x^2$ at $x = 1$ is $y = x - 0.5$. Graph $y_1 = 0.5x^2$ and $y_2 = x - 0.5$, and zoom in on the point $(1, 0.5)$ repeatedly. Use trace to discuss the relationship between these graphs near the point $(1, 0.5)$.

The ideas introduced in the preceding discussion are summarized below.

Slope of a Graph

Given $y = f(x)$, the **slope of the graph** at the point $(a, f(a))$ is given by

$$\lim_{h \to 0} \frac{f(a + h) - f(a)}{h} \tag{3}$$

provided the limit exists. The slope of the graph is also the **slope of the tangent line** at the point $(a, f(a))$.

From plane geometry, we know that a line tangent to a circle is a line that passes through one and only one point of the circle (see Fig. 9A). Although this definition cannot be extended to graphs of functions in general, the visual relationship between graphs of functions and their tangent lines is similar to the circle case (see Fig. 9B). Limit (3) provides both a mathematically sound definition for the concept of tangent line and a method for approximating the slope of the tangent line.

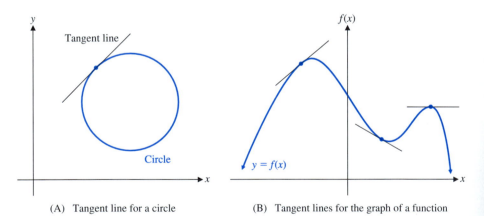

(A) Tangent line for a circle   (B) Tangent lines for the graph of a function

FIGURE 9

If a function is defined by an equation, then the slope of the tangent line at a point on the graph can be estimated by computing slopes of secant lines, as we did earlier (see Fig. 7). It is also important to be able to estimate the slope of a tangent line by examining the graph of the function.

EXAMPLE 6 ➤ **Estimating the Slope of a Graph** Use the graph of $y = f(x)$ in Figure 10 to estimate the slope of the graph at $x = -3$ and $x = 1$.

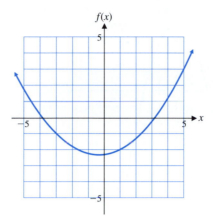

FIGURE 10

Solution Sketch tangent lines at the points $(-3, -1)$ and $(1, -2)$, and estimate the slope of each tangent line by taking the ratio of the rise to the run (see Fig. 11).

$$\text{Slope at } (-3, -1) = \frac{\text{Rise}}{\text{Run}} \approx \frac{-1}{1} = -1$$

$$\text{Slope at } (1, -2) = \frac{\text{Rise}}{\text{Run}} \approx \frac{1}{2}$$

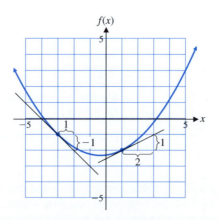

FIGURE 11

MATCHED PROBLEM 6 ➤ Use the graph of $y = f(x)$ in Figure 12 to estimate the slope of the graph at $x = -2$ and $x = 1$.

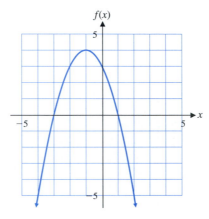

FIGURE 12   ◄

explore – discuss 3

(A) Sketch the graph of a function that has slope 1 at $(-2, 2)$ and slope -3 at $(3, 1)$.

(B) Sketch the graph of a function that has slope $-\frac{1}{2}$ at $(-3, -1)$ and slope 1 at $(3, 1)$.

■ SUMMARY

We have seen that the difference quotient for a function occurs naturally in a number of different situations (see Table 2).

TABLE 2
The Difference Quotient and Its Interpretations

FORMULA OR PROCESS	NUMERIC INTERPRETATION	GEOMETRIC INTERPRETATION
$\dfrac{f(a + h) - f(a)}{h}$	Average rate of change, or average velocity	Slope of secant line
$\lim\limits_{h \to 0} \dfrac{f(a + h) - f(a)}{h}$	Instantaneous rate of change, or velocity	Slope of graph and slope of tangent line

We will encounter many more situations involving the interplay between formulas for difference quotients, numeric values, and geometric visualizations. In this introduction our approach has been intuitive and informal. Before we can make the concepts of instantaneous rate of change and slope of a graph more precise, we must discuss the concept of limit in more detail. This is the subject of the next section.

Answers to Matched Problems   1. (A) $-\$1,600$ (B) $-\$8$ per planter   2. 48 ft/sec

3.

h	-0.1	-0.01	$-0.001 \rightarrow 0 \leftarrow 0.001$	0.01	0.1
$\dfrac{f(1+h) - f(1)}{h}$	30.4	31.84	$31.984 \rightarrow 32 \leftarrow 32.016$	32.16	33.6

The instantaneous velocity after 1 sec is approx. 32 ft/sec.

4. (A) 185,000 farms   (B) 2,000 farms per year
   (C) In 1984 the total number of farms was 185,000 and was increasing at the rate of 2,000 farms per year.

5. For $h = -1$, slope of secant line is 0.5; for $h = -0.5$, slope of secant line is 0.75.

6. Slope at $(-2, 3) \approx 2$; slope at $(1, 0) \approx -4$

EXERCISE 8-1

A   *In Problems 1–6, find the indicated quantities for the function $y = f(x) = 3x^2$.*

1. The change in y if x changes from 1 to 4
2. The change in y if x changes from 2 to 5
3. The average rate of change if x changes from 1 to 4
4. The average rate of change if x changes from 2 to 5
5. The slope of the secant line through the points $(1, f(1))$ and $(4, f(4))$ on the graph of $y = f(x)$
6. The slope of the secant line through the points $(2, f(2))$ and $(5, f(5))$ on the graph of $y = f(x)$

In Problems 7 and 8, complete the following table for $y = f(x) = 3x^2$ and the indicated value of a:

h	-0.1	-0.01	$-0.001 \rightarrow 0 \leftarrow 0.001$	0.01	0.1
$\dfrac{f(a+h) - f(a)}{h}$	?	?	$? \rightarrow ? \leftarrow ?$	?	?

7. $a = 1$ 8. $a = 2$

In Problems 9 and 10, approximate (to the nearest integer) the instantaneous rate of change of $y = f(x) = 3x^2$ with respect to x for the indicated value of x.

9. $x = 1$ 10. $x = 2$

In Problems 11 and 12, the position of an object moving along the y axis is given by $y = f(x) = 3x^2$, where y is distance in feet and x is time in seconds. Approximate (to the nearest integer) the instantaneous velocity for the indicated value of x.

11. $x = 1$ 12. $x = 2$

In Problems 13 and 14, approximate (to the nearest integer) the slope of the graph of $y = f(x) = 3x^2$ at the indicated point.

13. $(1, f(1))$ 14. $(2, f(2))$

B   *Problems 15–18 refer to the following situation: An automobile starts from rest and travels down a straight section of road. The distance y (in feet) of the car from the starting position after x seconds is given by $y = f(x) = 10x^2$.*

15. Find the average velocity of the car from $x = 4$ seconds to $x = 6$ seconds.

16. Find the average velocity of the car from $x = 5$ seconds to $x = 7$ seconds.

17. Use a table of average velocities to approximate (to the nearest integer) the instantaneous velocity at $x = 4$ seconds.

18. Use a table of average velocities to approximate (to the nearest integer) the instantaneous velocity at $x = 5$ seconds.

19. Given $y = f(x) = x^2 - 2x - 4$:
    (A) Find the slope of the secant line through the points $(2, f(2))$ and $(4, f(4))$.
    (B) Find the slope of the secant line through the points $(2, f(2))$ and $(3, f(3))$.
    (C) Use a table of secant line slopes to approximate (to the nearest integer) the slope of the graph at $x = 2$.
    (D) Graph $y = f(x)$, the secant lines from parts (A) and (B), and the tangent line at $x = 2$.

20. Given $y = f(x) = 5 - x^2$:
    (A) Find the slope of the secant line through the points $(1, f(1))$ and $(3, f(3))$.
    (B) Find the slope of the secant line through the points $(1, f(1))$ and $(2, f(2))$.
    (C) Use a table of secant line slopes to approximate (to the nearest integer) the slope of the graph at $x = 1$.
    (D) Graph $y = f(x)$, the secant lines from parts (A) and (B), and the tangent line at $x = 1$.

In Problems 21–24, use the given graph of y = f(x) to approximate (to the nearest integer) the slope of the graph at the points with the indicated x coordinates.

21. x = −1, 3

22. x = −2, 2

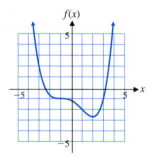

23. x = −3, −1, 1, 3

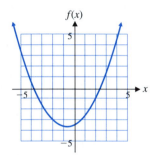

24. x = −2, 0, 2, 4

The tables in Problems 25–28 list points on the graph of a function y = f(x) and the slope of the graph at these points. Sketch a possible graph of y = f(x).

25.

POINT	SLOPE
(−3, 2)	2
(−2, 3)	0
(0, −1)	−4

26.

POINT	SLOPE
(−1, 1)	−4
(1, −3)	0
(2, −2)	2

27.

POINT	SLOPE
(−8, 1)	3
(−5, 5)	0
(−2, 3)	−1
(1, 1)	0
(4, 5)	3

28.

POINT	SLOPE
(−4, 6)	−3
(−1, 2)	0
(2, 4)	1
(5, 6)	0
(8, 2)	−3

In Problems 29–32, use a graphing utility to approximate (to two decimal places) the slope of the graph of y = f(x) at the point (a, f(a)) as follows:

1. *Zoom in on the graph of y = f(x) near (a, f(a)) until the graph appears to be a straight line.*
2. *Use trace to find a second point on the graph of y = f(x) near the point (a, f(a)).*
3. *Find the slope of the secant line through these two points.*

29. $f(x) = 2x^2$; (2, 8)   **30.** $f(x) = 2x^2$; (−1, 2)
31. $f(x) = \sqrt{x}$; (4, 2)   **32.** $f(x) = \sqrt{x}$; (1, 1)

C   *In Problems 33 and 34, use the given graph of y = f(x) to approximate (to the nearest integer) the slope of the graph of y = f(x) at x = −3, −2, −1, 0, 1, 2, and 3. Find a function whose values agree with these slopes.*

33.

34.

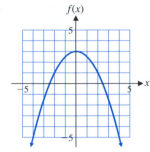

35. Discuss the relationship between the slope of the line $y = mx + b$ and the slope of the graph of the linear function $f(x) = mx + b$, $m \neq 0$.

36. Discuss the relationship between the slope of the horizontal line $y = b$ and the slope of the graph of the constant function $f(x) = b$.

In Problems 37 and 38, find and simplify the difference quotient for $y = f(x) = 3x^2$ at the indicated value of a. Discuss the relationship between the value of the simplified difference quotient at $h = 0$ and the slope of the graph at $x = a$.

37. $a = 1$ **38.** $a = 2$

In Problems 39 and 40, construct a table of secant line slopes at $a = 0$ and discuss the slope of the graph at the point (0, 0).

39. $f(x) = |x|$ **40.** $f(x) = |2x|$

 APPLICATIONS

Business & Economics

41. *Labor.* The average weekly hours and hourly earnings for production workers in the United States from 1970 to 1990 are given in Table 3.

TABLE 3

YEAR	1970	1975	1980	1985	1990
WEEKLY HOURS	37.1	36.1	35.3	34.9	34.5
HOURLY EARNINGS	$3.23	4.53	6.66	8.57	10.01

(A) Find the average rate of change of weekly hours from 1975 to 1985.
(B) Find the average rate of change of hourly earnings from 1975 to 1990.

42. *Labor.* Refer to Table 3.
(A) Find the average rate of change of weekly hours from 1970 to 1980.
(B) Find the average rate of change of hourly earnings from 1970 to 1985.

43. *Labor.* Refer to Table 3.
(A) Let x represent time (in years), with $x = 0$ corresponding to 1970, and let y represent the corresponding weekly hours listed in Table 3. Enter the appropriate data set in a graphing utility and find a quadratic regression equation for the data.
(B) Use the quadratic regression equation to approximate (to two decimal places) the instantaneous rate of change of weekly hours with respect to time in 1980.

44. *Labor.* Refer to Table 3.
(A) Let x represent time (in years), with $x = 0$ corresponding to 1970, and let y represent the corresponding hourly earnings listed in Table 3. Enter the appropriate data set in a graphing utility and find a quadratic regression equation for the data.
(B) Use the quadratic regression equation to approximate (to the nearest cent) the instantaneous rate of change of hourly earnings with respect to time in 1980.

45. *Revenue.* The revenue (in dollars) from the sale of x car seats for infants is given by

$$R(x) = 60x - 0.025x^2 \qquad 0 \leq x \leq 2,400$$

(A) Find the revenue, and approximate (to the nearest integer) the instantaneous rate of change of revenue at a production level of 1,000 car seats. Write a brief verbal interpretation of these results.
(B) Repeat part (A) for a production level of 1,300 car seats.

46. *Profit.* The profit (in dollars) from the sale of x car seats for infants is given by

$$P(x) = 45x - 0.025x^2 - 5,000 \qquad 0 \leq x \leq 2,400$$

(A) Find the profit, and approximate (to the nearest integer) the instantaneous rate of change of profit at a production level of 800 car seats. Write a brief verbal interpretation of these results.
(B) Repeat part (A) for a production level of 1,100 car seats.

47. *Mineral production.* The annual U.S. production of sulfur (in thousands of metric tons) is given approximately by

$$f(t) = -150t^2 + 770t + 10,400$$

where t is time in years and $t = 0$ corresponds to 1987. Find the annual production in 1990, and approximate (to the nearest integer) the instantaneous rate of change of production in 1990. Write a brief verbal interpretation of these results.

48. *Mineral production.* The annual U.S. production of talc (in thousands of metric tons) is given approximately by

$$f(t) = -30t^2 + 100t + 1,170$$

where t is time in years and $t = 0$ corresponds to 1987. Find the annual production in 1991, and approximate (to the nearest integer) the instantaneous rate of change of production in 1991. Write a brief verbal interpretation of these results.

Life Sciences

49. *Health care expenditures.* Table 4 lists U.S. health care expenditures (in billions of dollars) for services and supplies and for research and construction from 1987 to 1991.

TABLE 4

YEAR	1987	1988	1989	1990	1991
SERVICES AND SUPPLIES	476.9	526.2	583.6	652.4	728.6
RESEARCH AND CONSTRUCTION	17.3	19.8	20.7	22.7	23.1

(A) Find the average rate of change of expenditures for services and supplies from 1988 to 1991.

(B) Find the average rate of change of expenditures for research and construction from 1987 to 1989.

50. *Health care expenditures.* Refer to Table 4.

(A) Find the average rate of change of expenditures for services and supplies from 1989 to 1991.

(B) Find the average rate of change of expenditures for research and construction from 1987 to 1990.

51. *Health care expenditures.* Refer to Table 4.

(A) Let x represent time (in years), with $x = 0$ corresponding to 1987, and let y represent the corresponding expenditures for services and supplies listed in Table 4. Enter the appropriate data set in a graphing utility and find a quadratic regression equation for the data.

(B) Use the quadratic regression equation to approximate (to two decimal places) the instantaneous rate of change of expenditures for services and supplies with respect to time in 1989.

52. *Health care expenditures.* Refer to Table 4.

(A) Let x represent time (in years), with $x = 0$ corresponding to 1987, and let y represent the corresponding expenditures for research and construction listed in Table 4. Enter the appropriate data set in a graphing utility and find a quadratic regression equation for the data.

(B) Use the quadratic regression equation to approximate (to two decimal places) the instantaneous rate of change of expenditures for research and construction with respect to time in 1989.

Social Sciences

53. *Infant mortality.* The number of male infant deaths per 100,000 births in the United States is given approximately by

$$f(t) = 0.008t^2 - 0.9t + 29.6$$

where t is time in years and $t = 0$ corresponds to 1960. Find the number of male infant deaths in 1990, and approximate (to the nearest integer) the instantaneous rate of change of the number of male infant deaths in 1990. Write a brief verbal interpretation of these results.

54. *Infant mortality.* The number of female infant deaths per 100,000 births in the United States is given approximately by

$$f(t) = 0.005t^2 - 0.65t + 22.8$$

where t is time in years and $t = 0$ corresponds to 1960. Find the number of female infant deaths in 1990, and approximate (to the nearest integer) the instantaneous rate of change of the number of female infant deaths in 1990. Write a brief verbal interpretation of these results.

SECTION   8-2   ■   ## Limits

- FUNCTIONS AND GRAPHS—A BRIEF REVIEW
- LIMITS
- LIMIT EVALUATION
- LIMITS OF DIFFERENCE QUOTIENTS

In the preceding section we saw that determining the instantaneous rate of change of a function or the slope of a graph involves a limit of a difference quotient:

$$\lim_{h \to 0} \frac{f(a + h) - f(a)}{h} \tag{1}$$

Up to now we have approximated these limits numerically. Before we can proceed further in the study of calculus, we need to discuss the limit concept in more detail. As we will see, limit evaluation is a basic calculus tool that can be used in many different situations. In this section we will develop a combined numerical, graphical, and algebraic approach that will enable us to evaluate a variety of limit forms, including the very important form (1).

■ FUNCTIONS AND GRAPHS—A BRIEF REVIEW

The graph of the function $y = f(x) = x + 2$ is the graph of the set of all ordered pairs $(x, f(x))$. For example, if $x = 2$, then $f(2) = 4$ and $(2, f(2)) = (2, 4)$ is a point on the graph of f. Figure 1 shows $(-1, f(-1))$, $(1, f(1))$, and $(2, f(2))$ plotted on the graph of f. Notice that the domain values -1, 1, and 2 are associated with the x axis, and the range values $f(-1) = 1$, $f(1) = 3$, and $f(2) = 4$ are associated with the y axis.

The trace feature on a graphing utility displays the coordinates of the point on the graph of f corresponding to the cursor location (Fig. 2), providing an instant connection between the graph and the numeric values of the function.

FIGURE 1

FIGURE 2
$y_1 = x + 2$

explore – discuss 1

Reproduce the graph in Figure 2 on your graphing utility.

(A) Trace along the graph from $x = 2$ to $x = 5$, and discuss the relationship between the x and y coordinates displayed on the screen.

(B) There are an infinite number of points on the graph of $y = f(x)$, but a graphing utility can display only a finite number of these points. Use the trace feature to find the closest point displayed by your graphing utility on each side of the point $(2, 4)$.

If we are given the graph of a function, but not its equation, then it is sometimes useful to be able to read the values of the function directly from the graph. Example 1 reviews this process.

EXAMPLE 1 ➤ Finding Values of a Function from Its Graph   Complete the table below using the given graph of the function g.

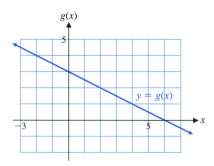

x	$g(x)$
-2	
1	
3	
4	

Solution   To determine $g(x)$, proceed vertically from the x value on the x axis to the graph of g, then horizontally to the corresponding y value, $g(x)$, on the y axis (as indicated by the dashed lines):

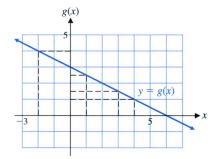

x	$g(x)$
-2	4.0
1	2.5
3	1.5
4	1.0

◄

MATCHED PROBLEM 1 ➤ Complete the table below using the given graph of the function h.

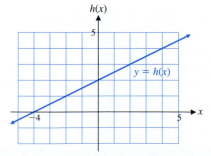

x	$h(x)$
-2	
-1	
0	
1	
2	
3	
4	

◄

■ LIMITS

We introduce the important notion of *limit* through two examples, after which the limit concept will be defined.

EXAMPLE 2 ➤ **Analyzing a Limit** Let $f(x) = x + 2$. Discuss the behavior of $f(x)$ numerically, graphically, and algebraically when x is chosen closer and closer to 2, but not equal to 2.

Solution   To investigate the behavior numerically and graphically for x near 2, we return to the graph in Figure 2 on page 555. Note that the center of the viewing window in Figure 2 is (2, 4). We zoom in on the point (2, 4) by a factor of 5 and find the points on the graph closest to $(2, f(2)) = (2, 4)$ displayed by the graphing utility on the left and on the right (see Fig. 3). These coordinates (rounded to four decimal places) are listed in the first line of Table 1. We repeat this process, always keeping the point (2, 4) in the center of the screen and always zooming by a factor of 5. The coordinates obtained after each zoom are shown in Table 1. (Your graphing utility may give slightly different results, but the pattern will be the same.)

FIGURE 3
Coordinates of points near (2, 4)

TABLE 1

LEFT SIDE		RIGHT SIDE	
x	$y = f(x)$	x	$y = f(x)$
1.9787	3.9787	2.0213	4.0213
1.9957	3.9957	2.0043	4.0043
1.9991	3.9991	2.0009	4.0009
1.9998	3.9998	2.0002	4.0002
↓	↓	↓	↓
2	4	2	4

Thus, we see that $f(x)$ approaches 4 as x approaches 2 from either side of 2. To confirm this algebraically, we note that if x approaches 2, then $x + 2$ must approach $2 + 2 = 4$. Collecting all this evidence, we conclude that

$$\lim_{x \to 2} (x + 2) = 4 \quad \text{or} \quad x + 2 \to 4 \quad \text{as} \quad x \to 2$$

Also, note that $f(2) = 4$. Thus, the value of the function at 2 and the limit of the function at 2 are the same. That is,

$$\lim_{x \to 2} (x + 2) = f(2)$$

Graphically, this means there is no break or hole in the graph of f at $x = 2$.   ◄

MATCHED PROBLEM 2 ➤ Let $f(x) = x + 1$.

(A) Discuss the behavior of $f(x)$ numerically, graphically, and algebraically when x is chosen closer and closer to 1, but not equal to 1.

(B) Use a similar approach to find $\lim_{x \to 0} (x + 1)$ and $\lim_{x \to 3} (x + 1)$.   ◄

Remark

In Example 2, we referred to the points closest to (2, 4) *displayed by the graphing utility.* Remember, a graphing utility screen displays a finite number of the infinite number of points on a graph. There is no point ''closest'' to a given point on the graph of a function, but on any screen there is a closest point that the graphing utility is capable of displaying.

The results found in Example 2 and Matched Problem 2 were fairly obvious. The next example is a little less obvious.

EXAMPLE 3 ➤ Analyzing a Limit   Let

$$g(x) = \frac{x^2 - 4}{x - 2} \qquad x \neq 2$$

Even though the function is not defined when $x = 2$ (both the numerator and denominator are 0), we can still ask how $g(x)$ behaves when x is near 2, but not equal to 2. Can you guess what happens to $g(x)$ as x approaches 2? The numerator tending to 0 is a force pushing the fraction toward 0. The denominator tending to 0 is another force pushing the fraction toward larger values. How do these two forces balance out?

Solution   We begin by graphing $y = g(x)$ in Figure 4. Notice that there is a hole in the graph at $x = 2$ and no y coordinate is displayed.* Proceeding as before, we repeatedly zoom in on the graph of $y = g(x)$ and record the coordinates of the pixels adjacent to (2, 4) in Table 2.

TABLE 2

LEFT SIDE		RIGHT SIDE	
x	$y = g(x)$	x	$y = g(x)$
1.9787	3.9787	2.0213	4.0213
1.9957	3.9957	2.0043	4.0043
1.9991	3.9991	2.0009	4.0009
1.9998	3.9998	2.0002	4.0002
↓	↓	↓	↓
2	4	2	4

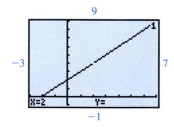

FIGURE 4

$$y_1 = g(x) = \frac{x^2 - 4}{x - 2}$$

---

* To facilitate graphic explorations, most graphing utilities do not display an error message when a point not in the domain of the function is encountered. However, if you try to evaluate $g(2)$ directly, you will probably receive an error message.

Notice that the values in Table 2 agree with the values of $f(x) = x + 2$ in Table 1. This suggests that there is an algebraic relationship between f and g:

$$g(x) = \frac{x^2 - 4}{x - 2} = \frac{(x - 2)(x + 2)}{x - 2} = x + 2 \qquad x \neq 2$$

Thus, we see that $g(x) = f(x)$ for all x, except $x = 2$. This implies that the graph of g is the same as the graph of f (Fig. 1), except that the graph of g has a hole at the point with coordinates (2, 4), as shown in Figure 5.

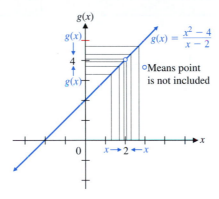

FIGURE 5

Since the behavior of $(x^2 - 4)/(x - 2)$ for x near 2, but not equal to 2, is the same as the behavior of $x + 2$ for x near 2, but not equal to 2, we have

$$\lim_{x \to 2} \frac{x^2 - 4}{x - 2} = \lim_{x \to 2} (x + 2) = 4$$

And we see that the limit of the function g at 2 exists even though the function is not defined there (the graph has a hole at $x = 2$). ◀

MATCHED PROBLEM 3 ➤ Proceed as in Example 3 to find: $\displaystyle \lim_{x \to 1} \frac{x^2 - 1}{x - 1}$ ◀

_____

Remark

When using a graphing utility to investigate the graph of $g(x) = (x^2 - 4)/(x - 2)$, you may not always see a hole in the graph at $x = 2$ (see Fig. 6, p. 560). This is due to the difference between the coordinates of a point in the plane (an infinite set) and the coordinates of a pixel on the screen (a finite set). To make the hole visible in Example 3, we simply choose the window parameters so that $x = 2$ is always the midpoint of the graphing interval (see Fig. 7). See Problems 87 and 88 in Exercise 8-2 for another way to choose window parameters so that holes in graphs will be visible.

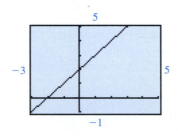

FIGURE 6
$g(x) = \dfrac{x^2 - 4}{x - 2}$ for $-3 \leqslant x \leqslant 5$

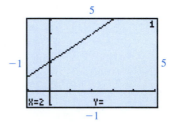

FIGURE 7
$g(x) = \dfrac{x^2 - 4}{x - 2}$ for $-1 \leqslant x \leqslant 5$

We now present an informal definition of the important concept of limit. A precise definition is not needed for our discussion, but one is given in the footnote.*

Limit

We write

$$\lim_{x \to c} f(x) = L \qquad \text{or} \qquad f(x) \to L \quad \text{as} \quad x \to c$$

if the functional value $f(x)$ is close to the single real number L whenever x is close to, but not equal to, c (on either side of c).

[*Note:* The existence of a limit at c has nothing to do with the value of the function at c. In fact, c may not even be in the domain of f (see Example 3). However, the function must be defined on both sides of c.]

The next example involves the **absolute value function:**

$$f(x) = |x| = \begin{cases} -x & \text{if } x < 0 \\ x & \text{if } x \geqslant 0 \end{cases} \qquad \begin{aligned} f(3) &= |3| = 3 \\ f(-2) &= |-2| = -(-2) = 2 \end{aligned}$$

---

* To make the informal definition of limit precise, the use of the word *close* must be made more precise. This is done as follows: We write $\lim_{x \to c} f(x) = L$ if for each $e > 0$, there exists a $d > 0$ such that $|f(x) - L| < e$ whenever $0 < |x - c| < d$. This definition is used to establish particular limits and to prove many useful properties of limits that will be helpful to us in finding particular limits. [Even though intuitive notions of limit existed for a long time, it was not until the nineteenth century that a precise definition was given by the German mathematician, Karl Weierstrass (1815–1897).]

The graph of the absolute value function is shown in Figure 8. Most graphing utilities use abs(x) to represent the absolute value function, but some use |x| (see the manual for your graphing utility).

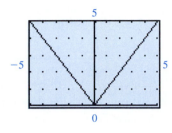

FIGURE 8
$y_1 = |x|$

EXAMPLE 4 ➤ Analyzing a Limit   Let $h(x) = |x|/x$. Explore the behavior of $h(x)$ for x near 0, but not equal to 0, using a table and a graph. Find $\lim_{x \to 0} h(x)$, if it exists.

Solution   The function h is defined for all real numbers except 0. For example,

$$h(-2) = \frac{|-2|}{-2} = \frac{2}{-2} = -1$$

$$h(0) = \frac{|0|}{0} = \frac{0}{0} \qquad \text{Not defined}$$

$$h(2) = \frac{|2|}{2} = \frac{2}{2} = 1$$

In general, $h(x)$ is -1 for all negative x and 1 for all positive x. Figure 9 and Table 3 illustrate the behavior of $h(x)$ for x near 0.

(A) $h(0)$ is not defined   (B) $h(x) = 1$ if $x > 0$ (C) $h(x) = -1$ if $x < 0$

FIGURE 9
$y_1 = h(x) = |x|/x$

TABLE 3

LEFT SIDE		RIGHT SIDE	
x	$y = h(x)$	x	$y = h(x)$
-1	-1	1	1
-0.1	-1	0.1	1
-0.01	-1	0.01	1
-0.001	-1	0.001	1
$\downarrow$	$\downarrow$	$\downarrow$	$\downarrow$
0	-1	0	1

When x is near 0 (on either side of 0), is $h(x)$ near one specific number? The answer is "No," because $h(x)$ is -1 for $x < 0$ and 1 for $x > 0$. Consequently, we say that

$$\lim_{x \to 0} \frac{|x|}{x} \quad \text{does not exist}$$

Thus, neither $h(x)$ nor the limit of $h(x)$ exist at $x = 0$. However, the limit from the left and the limit from the right both exist at 0, but they are not equal. (We will discuss this further below.) ◀

MATCHED PROBLEM 4 ▶   Graph

$$h(x) = \frac{x - 2}{|x - 2|}$$

and find $\lim_{x \to 2} h(x)$, if it exists. ◀

In Example 2, we found it helpful to examine the values of the function $f(x)$ as x approached 2 from the left and then from the right. In Example 4, we saw that the values of the function $h(x)$ approached two different numbers, depending on the direction of approach, and it was natural to refer to these values as "the limit from the left" and "the limit from the right." These experiences suggest that the notion of **one-sided limits** will be very useful when discussing basic limit concepts.

We write

$$\lim_{x \to c^-} f(c) = K$$ $x \to c^-$ is read "x approaches c from the left" and means $x \to c$ and $x < c$.

and call K the **limit from the left** (or **left-hand limit**) if $f(x)$ is close to K whenever x is close to c, but to the left of c on the real number line. We write

$$\lim_{x \to c^+} f(c) = L$$ $x \to c^+$ is read "x approaches c from the right" and means $x \to c$ and $x > c$.

and call L the **limit from the right** (or **right-hand limit**) if $f(x)$ is close to L whenever x is close to c, but to the right of c on the real number line.

We now make the following important observation:

On the Existence of a Limit

In order for a limit to exist, the limit from the left and the limit from the right must exist and be equal.

In Example 4,

$$\lim_{x \to 0^-} \frac{|x|}{x} = -1 \qquad \text{and} \qquad \lim_{x \to 0^+} \frac{|x|}{x} = 1$$

Since the left- and right-hand limits are not the same,

$$\lim_{x \to 0} \frac{|x|}{x} \quad \text{does not exist}$$

EXAMPLE 5 ➤ Analyzing Limits Graphically   Given the graph of the function f shown in Figure 10, we discuss the behavior of $f(x)$ for x near $-1, 1, 2,$ and 3:

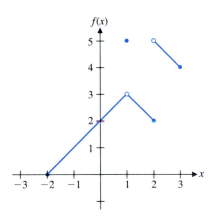

FIGURE 10

(A) Behavior of $f(x)$ for x near -1:

$$\lim_{x \to -1^-} f(x) = 1$$

$$\lim_{x \to -1^+} f(x) = 1$$

$$\lim_{x \to -1} f(x) = 1$$

$$f(-1) = 1$$

(B) Behavior of $f(x)$ for x near 1:

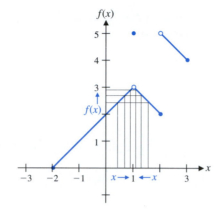

$$\lim_{x \to 1^-} f(x) = 3$$

$$\lim_{x \to 1^+} f(x) = 3$$

$$\lim_{x \to 1} f(x) = 3$$

$$f(1) = 5$$

(C) Behavior of $f(x)$ for x near 2:

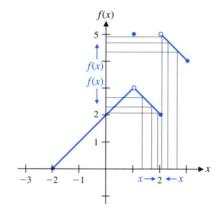

$$\lim_{x \to 2^-} f(x) = 2$$

$$\lim_{x \to 2^+} f(x) = 5$$

$$\lim_{x \to 2} f(x) \quad \text{does not exist}$$

$$f(2) = 2$$

(D) Behavior of $f(x)$ for x near 3:

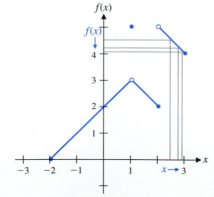

$$\lim_{x \to 3^-} f(x) = 4$$

$$\lim_{x \to 3^+} f(x) \quad \text{does not exist}$$

f is not defined for x > 3

$$\lim_{x \to 3} f(x) \quad \text{does not exist}$$

$$f(3) = 4$$

MATCHED PROBLEM 5 ▶

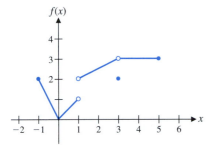

FIGURE 11

Given the graph of the function f shown in Figure 11, discuss the following, as we did in Example 5:

(A) Behavior of $f(x)$ for x near -1
(B) Behavior of $f(x)$ for x near 0
(C) Behavior of $f(x)$ for x near 1
(D) Behavior of $f(x)$ for x near 3

◀

■ LIMIT EVALUATION

Tables and graphs are very useful tools for investigating limits, especially if something unusual happens at the point in question. However, many of the limits encountered in calculus are routine and can be evaluated quickly, using a little algebraic simplification, some intuition, and basic properties of limits. The following list of properties of limits forms the basis for this approach.

theorem 1

Properties of Limits

Let f and g be two functions, and assume that

$$\lim_{x \to c} f(x) = L \qquad \lim_{x \to c} g(x) = M$$

where L and M are real numbers (both limits exist). Then:

1. $\lim_{x \to c} [f(x) + g(x)] = \lim_{x \to c} f(x) + \lim_{x \to c} g(x) = L + M$

2. $\lim_{x \to c} [f(x) - g(x)] = \lim_{x \to c} f(x) - \lim_{x \to c} g(x) = L - M$

3. $\lim_{x \to c} kf(x) = k \lim_{x \to c} f(x) = kL$ for any constant k

4. $\lim_{x \to c} [f(x) \cdot g(x)] = \left[\lim_{x \to c} f(x) \right]\left[\lim_{x \to c} g(x) \right] = LM$

5. $\lim_{x \to c} \dfrac{f(x)}{g(x)} = \dfrac{\lim_{x \to c} f(x)}{\lim_{x \to c} g(x)} = \dfrac{L}{M}$ if $M \neq 0$

6. $\lim_{x \to c} \sqrt[n]{f(x)} = \sqrt[n]{\lim_{x \to c} f(x)} = \sqrt[n]{L}$ $L > 0$ for n even

explore–discuss 2

The properties listed in Theorem 1 can be paraphrased in brief verbal statements. For example, property 1 simply states that *the limit of a sum is equal to the sum of the limits.* Write brief verbal statements for the remaining properties in Theorem 1.

EXAMPLE 6 ▶   Using Limit Properties   Find: $\lim\limits_{x \to 3} (x^2 - 4x)$

Solution   First, note the following obvious limit:

$$\lim_{x \to 3} x = 3$$

(All this says is that if x approaches 3, then x approaches 3.) Now we make use of this limit and the limit properties:

$$\lim_{x \to 3} (x^2 - 4x) = \lim_{x \to 3} x^2 - \lim_{x \to 3} 4x \qquad \text{Property 2}$$

$$= \left(\lim_{x \to 3} x\right) \cdot \left(\lim_{x \to 3} x\right) - 4 \lim_{x \to 3} x \qquad \text{Properties 3 and 4}$$

$$= 3 \cdot 3 - 4 \cdot 3 = -3$$

With a little practice you will soon be able to omit the steps in the dashed boxes and simply write

$$\lim_{x \to 3} (x^2 - 4x) = 3 \cdot 3 - 4 \cdot 3 = -3 \qquad ◀$$

MATCHED PROBLEM 6 ▶   Find: $\lim\limits_{x \to -2} (x^2 + 5x)$ ◀

What happens if we try to evaluate a limit like the one in Example 6, but with x approaching an unspecified number, such as c? Proceeding as we did in Example 6, we have

$$\lim_{x \to c} (x^2 - 4x) = c \cdot c - 4 \cdot c = c^2 - 4c$$

If we let $f(x) = x^2 - 4x$, then we have

$$\lim_{x \to c} f(x) = \lim_{x \to c} (x^2 - 4x) = c^2 - 4c = f(c)$$

That is, this limit can be evaluated simply by evaluating the function f at c. It would certainly simplify the process of evaluating limits if we could identify the functions for which

$$\lim_{x \to c} f(x) = f(c) \tag{2}$$

since we could use this fact to evaluate the limit. It turns out that there are many functions that satisfy (2). We will postpone a detailed discussion of these functions until the next chapter. For now, we note that if

$$f(x) = a_n x^n + a_{n-1} x^{n-1} + \cdots + a_0$$

is a polynomial function, then the properties in Theorem 1 imply that

$$\lim_{x \to c} f(x) = \lim_{x \to c} (a_n x^n + a_{n-1} x^{n-1} + \cdots + a_0)$$

$$= a_n c^n + a_{n-1} c^{n-1} + \cdots + a_0 = f(c)$$

We state this useful result in Theorem 2.

theorem 2

Limit of a Polynomial Function

If $f(x)$ is a polynomial function and c is any real number, then

$$\lim_{x \to c} f(x) = f(c)$$

Now let's see how we can use Theorems 1 and 2 together to simply and quickly evaluate limits.

EXAMPLE 7 ➤ **Evaluating Limits** Find each limit:

(A) $\displaystyle\lim_{x \to 2} (x^3 - 5x - 1)$ (B) $\displaystyle\lim_{x \to -1} \sqrt{2x^2 + 3}$ (C) $\displaystyle\lim_{x \to 4} \frac{2x}{3x + 1}$

Solution (A) $\displaystyle\lim_{x \to 2} (x^3 - 5x - 1) = 2^3 - 5 \cdot 2 - 1 = -3$ Theorem 2

(B) $\displaystyle\lim_{x \to -1} \sqrt{2x^2 + 3} = \sqrt{\lim_{x \to -1} (2x^2 + 3)}$ Property 6

$$= \sqrt{2(-1)^2 + 3}$$ Theorem 2

$$= \sqrt{5}$$

(C) $\displaystyle\lim_{x \to 4} \frac{2x}{3x + 1} = \frac{\displaystyle\lim_{x \to 4} 2x}{\displaystyle\lim_{x \to 4} (3x + 1)}$ Property 5

$$= \frac{2 \cdot 4}{3 \cdot 4 + 1}$$ Theorem 2

$$= \frac{8}{13}$$ ◄

MATCHED PROBLEM 7 ➤ Find each limit:

(A) $\displaystyle\lim_{x \to -1} (x^4 - 2x + 3)$ (B) $\displaystyle\lim_{x \to 2} \sqrt{3x^2 - 6}$ (C) $\displaystyle\lim_{x \to -2} \frac{x^2}{x^2 + 1}$ ◄

It is important to note that there are restrictions on some of the limit properties. In particular, if

$$\lim_{x \to c} f(x) = 0 \qquad \text{and} \qquad \lim_{x \to c} g(x) = 0$$

then finding

$$\lim_{x \to c} \frac{f(x)}{g(x)} \tag{3}$$

may present some difficulties, since limit property 5 (the limit of a quotient) does not apply when $\lim_{x \to c} g(x) = 0$. We often have to use algebraic manipulation or other devices to determine the outcome. Recall from Examples 3 and 4 that

$$\lim_{x \to 2} \frac{x^2 - 4}{x - 2} = \lim_{x \to 2} \frac{(x - 2)(x + 2)}{x - 2} = \lim_{x \to 2} (x + 2) = 4$$

and

$$\lim_{x \to 0} \frac{|x|}{x} \quad \text{does not exist}$$

From these two examples, it is clear that knowing only that $\lim_{x \to c} f(x) = 0$ and $\lim_{x \to c} g(x) = 0$ is not enough to determine limit (3). Depending on the choice of functions f and g, the limit (3) may or may not exist. Consequently, if we are given (3) and $\lim_{x \to c} f(x) = 0$ and $\lim_{x \to c} g(x) = 0$, then (3) is said to be **indeterminate,** or, more specifically, a **0/0 indeterminate form.**

Caution

The expression 0/0 does not represent a real number and should never be used as the value of a limit. If a limit is a 0/0 indeterminate form, then further investigation is always required to determine whether the limit exists and to find its value, if it does exist.

---

explore – discuss 3

Use algebraic, numerical, and/or graphical techniques to analyze each of the following indeterminate forms:

(A) $\displaystyle\lim_{x \to 1} \frac{x - 1}{x^2 - 1}$ (B) $\displaystyle\lim_{x \to 1} \frac{(x - 1)^2}{x^2 - 1}$ (C) $\displaystyle\lim_{x \to 1} \frac{x^2 - 1}{(x - 1)^2}$

■ LIMITS OF DIFFERENCE QUOTIENTS

Now that we have developed some experience working with limits, we are ready to apply these techniques to the limits of difference quotients.

EXAMPLE 8 ➤   Limit of a Difference Quotient   Find the following limit for $f(x) = 4x - 5$.

$$\lim_{h \to 0} \frac{f(3 + h) - f(3)}{h}$$

Solution

$$\lim_{h \to 0} \frac{f(3 + h) - f(3)}{h} = \lim_{h \to 0} \frac{[4(3 + h) - 5] - [4(3) - 5]}{h}$$

Since this is a 0/0 indeterminate form and property 5 in Theorem 1 does not apply, we proceed with algebraic simplification.

$$= \lim_{h \to 0} \frac{12 + 4h - 5 - 12 + 5}{h}$$

$$= \lim_{h \to 0} \frac{4h}{h} = \lim_{h \to 0} 4 = 4$$ ◀

MATCHED PROBLEM 8 ➤ Find the following limit for $f(x) = 7 - 2x$.

$$\lim_{h \to 0} \frac{f(4 + h) - f(4)}{h}$$ ◀

explore – discuss 4

The following is an incorrect solution to Example 8 with the invalid statements indicated by ≠. Explain why each ≠ is used.

$$\lim_{h \to 0} \frac{f(3 + h) - f(3)}{h} \neq \lim_{h \to 0} \frac{4(3 + h) - 5 - 4(3) - 5}{h}$$

$$= \lim_{h \to 0} \frac{-10 + 4h}{h}$$

$$\neq \lim_{h \to 0} \frac{-10 + 4}{1} = -6$$

EXAMPLE 9 ➤ **Limit of a Difference Quotient**   Find the following limit for $f(x) = |x + 5|$.

$$\lim_{h \to 0} \frac{f(-5 + h) - f(-5)}{h}$$

Solution

$$\lim_{h \to 0} \frac{f(-5 + h) - f(-5)}{h} = \lim_{h \to 0} \frac{|(-5 + h) + 5| - |-5 + 5|}{h}$$

Since this is a 0/0 indeterminate form and property 5 in Theorem 1 does not apply, we proceed with algebraic simplification.

$$= \lim_{h \to 0} \frac{|h|}{h} \quad \text{does not exist}$$ *See Example 4.*   ◀

MATCHED PROBLEM 9 ➤ Find the following limit for $f(x) = |x - 1|$.

$$\lim_{h \to 0} \frac{f(1 + h) - f(1)}{h}$$ ◀

E X A M P L E   1 0   ▶   **Limit of a Difference Quotient**   Find the following limit for $f(x) = \sqrt{x}$.

$$\lim_{h \to 0} \frac{f(2 + h) - f(2)}{h}$$

Solution

$$\lim_{h \to 0} \frac{f(2 + h) - f(2)}{h} = \lim_{h \to 0} \frac{\sqrt{2 + h} - \sqrt{2}}{h}$$

This is a 0/0 indeterminate form, so property 5 in Theorem 1 does not apply. Rationalizing the numerator will be of help.

$$= \lim_{h \to 0} \frac{\sqrt{2 + h} - \sqrt{2}}{h} \cdot \frac{\sqrt{2 + h} + \sqrt{2}}{\sqrt{2 + h} + \sqrt{2}}$$

$(A - B)(A + B) = A^2 - B^2$

$$= \lim_{h \to 0} \frac{2 + h - 2}{h(\sqrt{2 + h} + \sqrt{2})}$$

$$= \lim_{h \to 0} \frac{1}{\sqrt{2 + h} + \sqrt{2}}$$

$$= \frac{1}{\sqrt{2} + \sqrt{2}} = \frac{1}{2\sqrt{2}}$$ ◀

M A T C H E D   P R O B L E M   1 0   ▶   Find the following limit for $f(x) = \sqrt{x}$.

$$\lim_{h \to 0} \frac{f(3 + h) - f(3)}{h}$$ ◀

E X A M P L E   1 1   ▶   **Slope of a Graph**   Find the slope of the graph of $y = f(x) = 2x - x^2$ at the point $(2, 0)$ and illustrate with a graph.

Solution   Recall from Section 8-1 that

$$\text{Slope of the graph} = \lim_{h \to 0} \frac{f(2 + h) - f(2)}{h}$$

So we have

$$\lim_{h \to 0} \frac{f(2 + h) - f(2)}{h} = \lim_{h \to 0} \frac{[2(2 + h) - (2 + h)^2] - [2(2) - 2^2]}{h}$$

$$= \lim_{h \to 0} \frac{4 + 2h - 4 - 4h - h^2 - 4 + 4}{h}$$

$$= \lim_{h \to 0} \frac{-2h - h^2}{h}$$

$$= \lim_{h \to 0} \frac{h(-2 - h)}{h}$$

$$= \lim_{h \to 0} (-2 - h)$$

$$= -2$$

Thus, the slope of the graph at the point (2, 0) is -2. The graph of f and the tangent line [the line with slope -2 through the point (2, 0)] are shown in Figure 12. Figure 12A shows a graph of f and the tangent line at (2, 0) produced on a graphing utility. Figure 12B shows the annotations that must be included to clearly illustrate the tangent line graphically.

FIGURE 12

MATCHED PROBLEM 11 ▶ Find the slope of the graph of $y = f(x) = 3 + x - x^2$ at the point $(-1, 1)$ and illustrate with a graph. ◀

Remark

Most graphing utilities have a command of the form Tangent(y1, a), which will graph the line tangent to the graph of y_1 at $x = a$. We used this command to produce Figure 12A. Check the manual for your graphing utility or the graphing utility supplement for this text for details. Of course, you can sketch the tangent line in Figure 12B simply by using your knowledge of slope.

Answers to Matched Problems

1.

x	-2	-1	0	1	2	3	4
$h(x)$	1.0	1.5	2.0	2.5	3.0	3.5	4.0

2. (A)

x	0.9	0.99	$0.999 \rightarrow 1 \leftarrow 1.001$	1.01	1.1
$f(x)$	1.9	1.99	$1.999 \rightarrow 2 \leftarrow 2.001$	2.01	2.1

2. (A) *Continued*

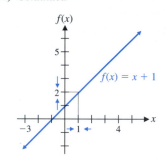

$$\lim_{x \to 1} (x + 1) = 2$$

(B) $\lim_{x \to 0} (x + 1) = 1$;

$$\lim_{x \to 3} (x + 1) = 4$$

3. 2

4.

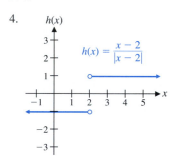

$$\lim_{x \to 2} \frac{x - 2}{|x - 2|} \text{ does not exist}$$

5.   (A) $\lim_{x \to -1^-} f(x)$ does not exist   (B) $\lim_{x \to 0^-} f(x) = 0$

   $\lim_{x \to -1^+} f(x) = 2$ $\lim_{x \to 0^+} f(x) = 0$

   $\lim_{x \to -1} f(x)$ does not exist $\lim_{x \to 0} f(x) = 0$

   $f(-1) = 2$ $f(0) = 0$

   (C) $\lim_{x \to 1^-} f(x) = 1$ (D) $\lim_{x \to 3^-} f(x) = 3$

   $\lim_{x \to 1^+} f(x) = 2$ $\lim_{x \to 3^+} f(x) = 3$

   $\lim_{x \to 1} f(x)$ does not exist $\lim_{x \to 3} f(x) = 3$

   $f(1)$ not defined $f(3) = 2$

6. -6 7. (A) 6   (B) $\sqrt{6}$ (C) $\frac{4}{5}$ 8. -2 9. Does not exist

10. $1/(2\sqrt{3})$ 11. Slope $= 3$

EXERCISE 8-2

A *In Problems 1–6, sketch a possible graph of a function that satisfies the given conditions.*

1. $f(0) = 1$; $\lim\limits_{x \to 0^-} f(x) = 3$; $\lim\limits_{x \to 0^+} f(x) = 1$

2. $f(1) = -2$; $\lim\limits_{x \to 1^-} f(x) = 2$; $\lim\limits_{x \to 1^+} f(x) = -2$

3. $f(2) = 0$; $\lim\limits_{x \to 2^-} f(x) = -3$; $\lim\limits_{x \to 2^+} f(x) = 3$

4. $f(-1) = -1$; $\lim\limits_{x \to -1^-} f(x) = 2$; $\lim\limits_{x \to -1^+} f(x) = -3$

5. $f(-2) = 2$; $\lim\limits_{x \to -2^-} f(x) = 1$; $\lim\limits_{x \to -2^+} f(x) = 1$

6. $f(0) = -1$; $\lim\limits_{x \to 0^-} f(x) = 2$; $\lim\limits_{x \to 0^+} f(x) = 2$

In Problems 7–10, use the graph of the function f shown below to estimate the indicated limits and function values.

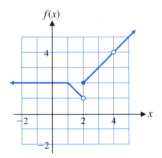

Figure for 7–10

7. (A) $\lim\limits_{x \to 0^-} f(x)$ (B) $\lim\limits_{x \to 0^+} f(x)$

     (C) $\lim\limits_{x \to 0} f(x)$ (D) $f(0)$

8. (A) $\lim\limits_{x \to 1^-} f(x)$ (B) $\lim\limits_{x \to 1^+} f(x)$

     (C) $\lim\limits_{x \to 1} f(x)$ (D) $f(1)$

9. (A) $\lim\limits_{x \to 2^-} f(x)$ (B) $\lim\limits_{x \to 2^+} f(x)$

     (C) $\lim\limits_{x \to 2} f(x)$ (D) $f(2)$

10. (A) $\lim\limits_{x \to 4^-} f(x)$ (B) $\lim\limits_{x \to 4^+} f(x)$

     (C) $\lim\limits_{x \to 4} f(x)$ (D) $f(4)$

In Problems 11–14, use the graph of the function g shown below to estimate the indicated limits and function values.

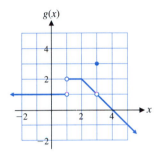

Figure for 11–14

11. (A) $\lim\limits_{x \to 1^-} g(x)$ (B) $\lim\limits_{x \to 1^+} g(x)$

     (C) $\lim\limits_{x \to 1} g(x)$ (D) $g(1)$

12. (A) $\lim\limits_{x \to 2^-} g(x)$ (B) $\lim\limits_{x \to 2^+} g(x)$

     (C) $\lim\limits_{x \to 2} g(x)$ (D) $g(2)$

13. (A) $\lim\limits_{x \to 3^-} g(x)$ (B) $\lim\limits_{x \to 3^+} g(x)$

     (C) $\lim\limits_{x \to 3} g(x)$ (D) $g(3)$

14. (A) $\lim\limits_{x \to 4^-} g(x)$ (B) $\lim\limits_{x \to 4^+} g(x)$

     (C) $\lim\limits_{x \to 4} g(x)$ (D) $g(4)$

Given $\lim\limits_{x \to 3} f(x) = 5$ and $\lim\limits_{x \to 3} g(x) = 9$, find the indicated limits in Problems 15–24.

15. $\lim\limits_{x \to 3} [f(x) - g(x)]$ **16.** $\lim\limits_{x \to 3} [f(x) + g(x)]$

17. $\lim\limits_{x \to 3} 4g(x)$ **18.** $\lim\limits_{x \to 3} (-2)f(x)$

19. $\lim\limits_{x \to 3} \dfrac{f(x)}{g(x)}$ **20.** $\lim\limits_{x \to 3} [f(x) \cdot g(x)]$

21. $\lim\limits_{x \to 3} \sqrt{f(x)}$ **22.** $\lim\limits_{x \to 3} \sqrt{g(x)}$

23. $\lim\limits_{x \to 3} \dfrac{f(x) + g(x)}{2f(x)}$ **24.** $\lim\limits_{x \to 3} \dfrac{g(x) - f(x)}{3g(x)}$

B *In Problems 25–52, use limit properties, algebraic simplification, tables of values, and/or graphs to find each limit, if it exists.*

25. $\lim\limits_{x \to 5} (2x^2 - 3)$

26. $\lim\limits_{x \to 2} (x^2 - 8x + 2)$

27. $\lim\limits_{x \to 2} \dfrac{5x}{2 + x^2}$

28. $\lim\limits_{x \to 10} \dfrac{2x + 5}{3x - 5}$

29. $\lim\limits_{x \to 2} (x + 1)^3 (2x - 1)^2$

30. $\lim\limits_{x \to 3} (x + 2)^2 (2x - 4)$

31. $\lim\limits_{x \to -1} \sqrt{5 - 4x}$

32. $\lim\limits_{x \to 4} \sqrt{25 - x^2}$

33. $\lim\limits_{x \to -3} \dfrac{x^2 - 9}{x + 3}$

34. $\lim\limits_{x \to -5} \dfrac{x^2 - 25}{x + 5}$

35. $\lim\limits_{x \to 1^+} \dfrac{|x - 1|}{x - 1}$

36. $\lim\limits_{x \to 3^-} \dfrac{x - 3}{|x - 3|}$

37. $\lim\limits_{x \to 1^-} \dfrac{|x - 1|}{x - 1}$

38. $\lim\limits_{x \to 3^+} \dfrac{x - 3}{|x - 3|}$

39. $\lim\limits_{x \to 1} \dfrac{|x - 1|}{x - 1}$

40. $\lim\limits_{x \to 3} \dfrac{x - 3}{|x - 3|}$

41. $\lim\limits_{x \to 1} \dfrac{x - 2}{x^2 - 2x}$

42. $\lim\limits_{x \to 1} \dfrac{x + 3}{x^2 + 3x}$

43. $\lim\limits_{x \to 2} \dfrac{x - 2}{x^2 - 2x}$

44. $\lim\limits_{x \to -3} \dfrac{x + 3}{x^2 + 3x}$

45. $\lim\limits_{x \to 2} \dfrac{x^2 - x - 6}{x + 2}$

46. $\lim\limits_{x \to 3} \dfrac{x^2 + x - 6}{x + 3}$

47. $\lim\limits_{x \to -2} \dfrac{x^2 - x - 6}{x + 2}$

48. $\lim\limits_{x \to -3} \dfrac{x^2 + x - 6}{x + 3}$

49. $\lim\limits_{x \to 3} \left(\dfrac{x}{x + 3} + \dfrac{x - 3}{x^2 - 9} \right)$

50. $\lim\limits_{x \to 2} \left(\dfrac{1}{x + 2} + \dfrac{x - 2}{x^2 - 4} \right)$

51. $\lim\limits_{x \to 0} \left(\sqrt{x^2 + 9} - \dfrac{x^2 + 3x}{x} \right)$

52. $\lim\limits_{x \to 1} \left(\dfrac{x^2 - 1}{x - 1} + \sqrt{x^2 + 3} \right)$

Compute the following limit for each function in Problems 53–62:

$$\lim\limits_{h \to 0} \frac{f(2 + h) - f(2)}{h}$$

53. $f(x) = 3x + 1$

54. $f(x) = 5x - 1$

55. $f(x) = x^2 + 1$

56. $f(x) = x^2 - 2$

57. $f(x) = 5$

58. $f(x) = -2$

59. $f(x) = \sqrt{x} - 2$

60. $f(x) = 1 + \sqrt{x}$

61. $f(x) = |x - 2| - 3$

62. $f(x) = 2 + |x - 2|$

In Problems 63–66, find the slope of the graph of $y = f(x)$ at the indicated point. Graph f and the tangent line at this point.

63. $f(x) = x^2 - 3$; $(2, 1)$

64. $f(x) = 5 - x^2$; $(-1, 4)$

65. $f(x) = \sqrt{x}$; $(4, 2)$

66. $f(x) = \sqrt{x}$; $(1, 1)$

In Problems 67 and 68, an automobile starts from rest and travels down a straight section of road. The distance y (in feet) of the car from the starting position after x seconds is given by $y = f(x) = 10x^2$.

67. Find the instantaneous velocity at $x = 4$ seconds.

68. Find the instantaneous velocity at $x = 5$ seconds.

In Problems 69 and 70, use the given graph of f to estimate

$$\lim\limits_{h \to 0} \frac{f(a + h) - f(a)}{h}$$

to the nearest integer for the indicated values of a.

69. $a = -3, -1, 3$

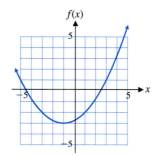

70. $a = -3, 1, 3$

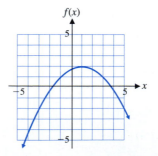

71. Use a graph and a table of values to investigate the behavior of the following limits:

(A) $\lim_{x \to 0^-} \dfrac{1}{x}$ (B) $\lim_{x \to 0^+} \dfrac{1}{x}$

Write a brief verbal description of each limit.

72. Use a graph and a table of values to investigate the behavior of the following limits:

(A) $\lim_{x \to 0^-} \dfrac{1}{x^2}$ (B) $\lim_{x \to 0^+} \dfrac{1}{x^2}$

Write a brief verbal description of each limit.

C *In Problems 73 and 74, put your graphing utility in the dot mode and use it to investigate the indicated limits.*

73. Let $f(x) = \dfrac{(x + m)(x - 1)}{|x - 1|}$, where m is a constant.

(A) Find: $\lim_{x \to 1^-} f(x)$ and $\lim_{x \to 1^+} f(x)$ for $m = -2$

(B) Find: $\lim_{x \to 1^-} f(x)$ and $\lim_{x \to 1^+} f(x)$ for $m = 2$

(C) Find m so that: $\lim_{x \to 1^-} f(x) = \lim_{x \to 1^+} f(x)$

(D) Write a brief verbal description of each graph. How does the graph in part (C) differ from the graphs in parts (A) and (B)?

74. Let $f(x) = \dfrac{(x + 2 - m)(x - 2)}{|x - 2|}$, where m is a constant.

(A) Find: $\lim_{x \to 2^-} f(x)$ and $\lim_{x \to 2^+} f(x)$ for $m = 2$

(B) Find: $\lim_{x \to 2^-} f(x)$ and $\lim_{x \to 2^+} f(x)$ for $m = 5$

(C) Find m so that: $\lim_{x \to 2^-} f(x) = \lim_{x \to 2^+} f(x)$

(D) Write a brief verbal description of each graph. How does the graph in part (C) differ from the graphs in parts (A) and (B)?

Find each limit in Problems 75–78, where a is a real constant.

75. $\lim_{h \to 0} \dfrac{(a + h)^2 - a^2}{h}$

76. $\lim_{h \to 0} \dfrac{[3(a + h) - 2] - (3a - 2)}{h}$

77. $\lim_{h \to 0} \dfrac{\sqrt{a + h} - \sqrt{a}}{h}, a > 0$

78. $\lim_{h \to 0} \dfrac{\dfrac{1}{a + h} - \dfrac{1}{a}}{h}, a \neq 0$

79. Let $f(x) = x^2 - 3x + 1$.

(A) Find the slope of the graph of f at $a = 1$, $a = 2$, and $a = 3$.

(B) Find a formula for the slope of the graph of f at any point $(a, f(a))$.

(C) Use the formula from part (B) to find the slope of the graph at $a = 1$, $a = 2$, and $a = 3$. Write a brief verbal comparison of these two methods for finding the slope of a graph.

80. Let $f(x) = 5x - 2x^2$.

(A) Find the slope of the graph of f at $a = -1$, $a = 0$, and $a = 1$.

(B) Find a formula for the slope of the graph of f at any point $(a, f(a))$.

(C) Use the formula from part (B) to find the slope of the graph at $a = -1$, $a = 0$, and $a = 1$. Write a brief verbal comparison of these two methods for finding the slope of a graph.

In Problems 81–86, use a table of values to estimate each limit to three decimal places.

81. $\lim_{x \to 1} \dfrac{x^{10} - 1}{x - 1}$

82. $\lim_{x \to 1} \dfrac{x^{15} - 1}{x - 1}$

83. $\lim_{x \to 0} \dfrac{2^x - 1}{x}$

84. $\lim_{x \to 0} \dfrac{3^x - 1}{x}$

85. $\lim_{x \to 0} (1 + x)^{1/x}$

86. $\lim_{x \to 0} (1 + 2x)^{1/x}$

When investigating limits with a graphing utility, it is helpful to choose Xmin and Xmax so that coordinates displayed on the screen have finite decimal expansions. Problems 87 and 88 describe a simple way to accomplish this.

87. Let Xmin $= 0$, and use trial and error to find an integer n so that if Xmax $= n$, then the x coordinates of points displayed on the screen are integers.

88. Given any value for Xmin, let Xmax $=$ Xmin $+ h \cdot n$, where n is the integer discovered in Problem 87. Discuss the nature of the x coordinates of points displayed on the screen if $h = 0.5$. If $h = 0.1$.

In Problems 89–94, graph each function and use zoom and trace to investigate the left- and right-hand limits at the indicated value(s) of c. Use the ideas discussed in Problems 87 and 88 to choose Xmin and Xmax.

89. $f(x) = \dfrac{x^4 - 10x^2 + 24}{4 - x^2}; c = -2, c = 2$

90. $f(x) = \dfrac{x^4 - 12x^2 + 27}{x^2 - 9}; c = -3, c = 3$

91. $f(x) = \dfrac{x^2 - 4}{|x - 2|}; c = 2$

92. $f(x) = \dfrac{1 - x^2}{|x + 1|}; c = -1$

93. $f(x) = \dfrac{x^3 - 9x}{|x^2 - 9|}; c = -3, c = 3$

94. $f(x) = \dfrac{4x - x^3}{|x^2 - 4|}; c = -2, c = 2$

▶ A P P L I C A T I O N S

Business & Economics

95. *Revenue.* The revenue (in dollars) from the sale of x variable-speed jigsaws is given by

$$R(x) = 200x - 0.1x^2 \qquad 0 \le x \le 2,000$$

(A) Find the revenue and the instantaneous rate of change of revenue at a production level of 900 jigsaws. Write a brief verbal interpretation of these results and illustrate graphically.

(B) Repeat part (A) for a production level of 1,200 jigsaws.

96. *Profit.* The profit (in dollars) from the sale of x variable-speed jigsaws is given by

$$P(x) = 150x - 0.1x^2 - 5,000 \qquad 0 \le x \le 2,000$$

(A) Find the profit and the instantaneous rate of change of profit at a production level of 700 jigsaws. Write a brief verbal interpretation of these results and illustrate graphically.

(B) Repeat part (A) for a production level of 900 jigsaws.

97. *Consumer debt.* Revolving-credit debt (in billions of dollars) in the United States can be described approximately by

$$f(t) = 0.62t^2 - t + 5.1$$

where t is time in years and $t = 0$ corresponds to 1970. Find the debt and the instantaneous rate of change of the debt in 1990. Write a brief verbal interpretation of these results and illustrate graphically.

98. *Consumer debt.* Credit union debt (in billions of dollars) in the United States can be described approximately by

$$f(t) = 0.5t^2 + 5.6t + 46.6$$

where t is time in years and $t = 0$ corresponds to 1970. Find the debt and the instantaneous rate of change of the debt in 1990. Write a brief verbal interpretation of these results and illustrate graphically.

Life Sciences

99. *Human physiology.* Many people experience headaches, fatigue, and shortness of breath at high altitudes, due to an inadequate supply of oxygen in the bloodstream. The reduced air pressure at high altitudes results in less oxygen being forced through the membranes of the lungs into the blood. The *alveolar pressure* measures the amount of oxygen that can enter the bloodstream at various altitudes. If y is the ratio of the pressure to the pressure at sea level (expressed as a percentage) at an altitude of x thousand feet for an average size person, then y is given approximately by

$$y = 0.04x^2 - 3.66x + 100$$

Notice that $y = 100\%$ at $x = 0$ (sea level). Find y and find the instantaneous rate of change of y with respect to x at an altitude of 9,000 feet. (This is the altitude at which most people will begin to experience discomfort.) Write a brief verbal interpretation of these results and illustrate graphically.

100. *Plant physiology.* High altitudes and the corresponding shorter growing seasons have an effect on the size of plants. For a particular species, the relationship between the height y (in inches) of a full-grown plant and the altitude x (in thousands of feet) is given approximately by

$$y = 0.16x^2 - 5.5x + 48$$

Find y and find the instantaneous rate of change of y with respect to x at an altitude of 5,000 feet. Write a brief verbal interpretation of these results and illustrate graphically.

Social Sciences

101. *Education.* The annual enrollment in secondary schools in the United States (in millions) is given approximately by

$$y = f(t) = 0.05t^2 - 0.44t + 16.6$$

where t is time (in years) and $t = 0$ corresponds to 1985. Find the annual enrollment and the instantaneous rate of change of this population in 1995. Write a brief verbal interpretation of these results and illustrate with a graph.

102. *Education.* The number of students that graduate from high school each year in the United States (in millions) is given approximately by

$$y = f(t) = 0.005t^2 - 0.05t + 2.6$$

where t is time (in years) and $t = 0$ corresponds to 1985. Find the number of high school graduates and the instantaneous rate of change of this population in 1995. Write a brief verbal interpretation of these results and illustrate with a graph.

SECTION 8-3 ■ The Derivative

- THE DERIVATIVE
- NONEXISTENCE OF THE DERIVATIVE
- SUMMARY

In this section we apply the limit techniques developed in Section 8-2 to the difference quotients discussed in Section 8-1. This will lead to the definition of the *derivative,* a fundamental mathematical concept that forms the basis for much of the subject matter of calculus.

■ THE DERIVATIVE

We begin with an example that will review some of the ideas discussed in the preceding two sections.

EXAMPLE 1 ➤ Slope of a Tangent Line Given $f(x) = x^2$:

(A) Find the slope of the tangent line at $x = 1$.
(B) Find the equation of the tangent line at $x = 1$; that is, through $(1, f(1))$.
(C) Sketch the graph of f, the tangent line at $(1, f(1))$, and the secant line passing through $(1, f(1))$ and $(2, f(2))$.

Solution (A) Slope of the tangent line:

Step 1.   Write the difference quotient and simplify:

$$\frac{f(1 + h) - f(1)}{h} = \frac{(1 + h)^2 - 1^2}{h}$$

This is the slope of a secant line passing through $(1, f(1))$ and $(1 + h, f(1 + h))$.

$$= \frac{1 + 2h + h^2 - 1}{h}$$

$$= \frac{2h + h^2}{h} = \frac{h(2 + h)}{h} = 2 + h \qquad h \neq 0$$

Step 2.   Find the limit of the difference quotient:

$$\text{Slope of tangent line} = \lim_{h \to 0} \frac{f(1 + h) - f(1)}{h}$$

$$= \lim_{h \to 0} (2 + h) = 2$$

This is also the slope of the graph of $f(x) = x^2$ at $(1, f(1))$.

(B) Equation of the tangent line: The tangent line passes through $(1, f(1)) = (1, 1)$ with slope $m = 2$ [from part (A)]. The point–slope formula from Section 1-4 gives its equation:

$$y - y_1 = m(x - x_1) \qquad \text{Point–slope formula}$$
$$y - 1 = 2(x - 1)$$
$$y = 2x - 1 \qquad \text{Tangent line equation}$$

(C)

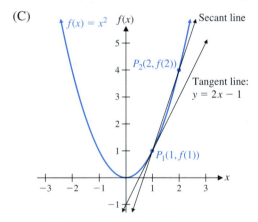

MATCHED PROBLEM 1 ➤ Find the slope of the tangent line for the graph of $f(x) = x^2$ at $x = 2$, and write the equation of the tangent line in the form $y = mx + b$. ◀

Refer to Example 1 and Matched Problem 1. What happens if we want to find the slope of the line tangent to the graph of $y = f(x) = x^2$ at some additional points on the graph of f? Each time we consider a different point, we must evaluate another limit. It would be much more efficient if we could evaluate a single limit that would give us the slope of the tangent line at any point on the graph of f. To do this, we apply the two-step process illustrated in Example 1 at the point $(a, f(a)) = (a, a^2)$, where a is an unspecified, but fixed, real number.

Step 1. Write the difference quotient and simplify:

$$\frac{f(a + h) - f(a)}{h} = \frac{(a + h)^2 - a^2}{h} \qquad \text{\textit{Slope of the secant line}}$$

$$= \frac{a + 2ah + h^2 - a^2}{h}$$

$$= \frac{2ah + h^2}{h} = \frac{h(2a + h)}{h}$$

$$= 2a + h \qquad h \neq 0$$

Step 2. Find the limit of the difference quotient:

$$\text{Slope of tangent line} = \lim_{h \to 0} \frac{f(a + h) - f(a)}{h}$$

$$= \lim_{h \to 0} (2a + h)$$

$$= 2a$$

Thus, we see that if $(a, f(a)) = (a, a^2)$ is any point on the graph of f, then the slope of the tangent line at this point is $2a$. Note that the slope is a function of a, the first coordinate of the point of tangency. Also note that when $a = 1$, the slope is 2, which agrees with the result in Example 1A.

Generalizing the process of finding slopes of tangent lines is not simply a matter of increasing efficiency. Rather, the relationship between a function and the slope of the tangent line at any point on the graph of the function is one of the most fundamental concepts of calculus.

explore – discuss 1

(A) Use the graph of the function f in Figure 1 to estimate (to one decimal place) the slopes of the tangent lines at the points whose x coordinates are given in Table 1.

TABLE 1

x	SLOPE OF TANGENT LINE AT $(x, f(x))$
-4	
-3	
-2	
-1	
0	
1	
2	
3	
4	

FIGURE 1

(B) Find a function whose values agree with the values listed in Table 1.

(C) If $f(x) = 0.25x^2 - 2$ is the function whose graph is shown in Figure 1, find the slope of the tangent line at any x by evaluating

$$\lim_{h \to 0} \frac{f(x + h) - f(x)}{h}$$

Discuss the relationship between the function you found in part (B) and this limit.

We are now ready to define the *derivative of a function*. To follow customary practice, we use x in place of a and think of the difference quotient

$$\frac{f(x + h) - f(x)}{h}$$

as a function of h, with x held fixed as h tends to 0.

The Derivative

For $y = f(x)$, we define the **derivative of f at x,** denoted by $f'(x)$, to be

$$f'(x) = \lim_{h \to 0} \frac{f(x + h) - f(x)}{h} \qquad \text{if the limit exists}$$

If $f'(x)$ exists for each x in the open interval (a, b), then f is said to be **differentiable** over (a, b).

(Differentiability from the left or from the right is defined using $h \to 0^-$ or $h \to 0^+$, respectively, in place of $h \to 0$ in the above definition.)

The process of finding the derivative of a function is called **differentiation.** That is, the derivative of a function is obtained by **differentiating** the function.

Interpretations of the Derivative

The derivative of a function f is a new function f'. The domain of f' is a subset of the domain of f. The derivative has various applications and interpretations, including:

1. *Slope of the tangent line.* For each x in the domain of f', $f'(x)$ is the slope of the line tangent to the graph of f at the point $(x, f(x))$.
2. *Instantaneous rate of change.* For each x in the domain of f', $f'(x)$ is the instantaneous rate of change of $y = f(x)$ with respect to x.

For example, if $f(x) = x^2$, then the derivative is

$$f'(x) = \lim_{h \to 0} \frac{f(x + h) - f(x)}{h} = 2x$$

(See page 579 for the two-step process involved in evaluating this limit.) We can interpret this derivative graphically as follows: The slope of the tangent line at any point (x, x^2) is $2x$. On the other hand, if $y = f(x)$ is the position (measured in feet) at x seconds of an object moving along the y axis, then we can also intepret the derivative as a rate of change: The instantaneous velocity of the object at x seconds is $2x$ feet per second.

EXAMPLE 2 ➤ **Finding a Derivative** Find $f'(x)$, the derivative of f at x, for $f(x) = 4x - x^2$.

Solution To find $f'(x)$, we use the two-step process:

Step 1. Form the difference quotient and simplify:

$$\frac{f(x + h) - f(x)}{h} = \frac{[4(x + h) - (x + h)^2] - (4x - x^2)}{h}$$

$$= \frac{4x + 4h - x^2 - 2xh - h^2 - 4x + x^2}{h}$$

$$= \frac{4h - 2xh - h^2}{h}$$

$$= \frac{h(4 - 2x - h)}{h}$$

$$= 4 - 2x - h \qquad h \neq 0$$

Step 2. Find the limit of the difference quotient:

$$f'(x) = \lim_{h \to 0} \frac{f(x + h) - f(x)}{h}$$

$$= \lim_{h \to 0} (4 - 2x - h) = 4 - 2x$$

Thus, if $f(x) = 4x - x^2$, then $f'(x) = 4 - 2x$. The derivative f' is a new function derived from the function f. ◀

MATCHED PROBLEM 2 ➤ Find $f'(x)$, the derivative of f at x, for $f(x) = 8x - 2x^2$. ◀

EXAMPLE 3 ➤ **Finding Tangent Line Slopes** In Example 2, we started with the function specified by $f(x) = 4x - x^2$ and found the derivative of f at x to be $f'(x) = 4 - 2x$. Thus, the slope of a tangent line to the graph of f at any point $(x, f(x))$ on the graph is

$$m = f'(x) = 4 - 2x$$

Find the slope of the graph of f at $x = 0$, 2, and 3, and illustrate graphically.

Solution   Using $f'(x) = 4 - 2x$, we have

$$f'(0) = 4 - 2(0) = 4 \qquad \textit{Slope at } x = 0$$
$$f'(2) = 4 - 2(2) = 0 \qquad \textit{Slope at } x = 2$$
$$f'(3) = 4 - 2(3) = -2 \quad \textit{Slope at } x = 3$$

Figure 2A shows the graph of f and the three tangent lines, and Figure 2B includes appropriate annotations.

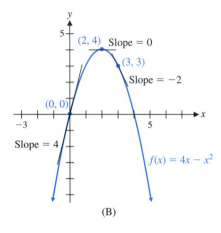

(A)   (B)

FIGURE 2

MATCHED PROBLEM 3 ▶   In Matched Problem 2, we started with the function specified by $f(x) = 8x - 2x^2$. Using the derivative found there, find the slope of the graph of f at $x = 1, 2,$ and 4, and illustrate graphically.

explore – discuss 2

In Example 2 we found that the derivative of $f(x) = 4x - x^2$ is $f'(x) = 4 - 2x$, and in Example 3 we graphed $f(x)$ and several tangent lines.

(A) Graph f and f' on the same set of axes.
(B) The graph of f' is a straight line. Is it a tangent line for the graph of f? Explain.
(C) Find the x intercept for the graph of f'. What is the slope of the line tangent to the graph of f for this value of x? Write a verbal description of the relationship between the slopes of the tangent lines of a function and the x intercepts of the derivative of the function.

Remark: Derivatives and Technology

Refer to Figure 2A. In order to graph the three tangent lines, the graphing utility must be able to determine the slopes of these tangent lines. How does it do this — numerically or algebraically? The answer depends on the particular graphing utility. Some graphing utilities, including most graphing calculators, have a built-in **numerical differentiation** routine that will approximate numerically the value of $f'(x)$ for any given value of x. Figure 3A illustrates this approach on the TI-82 graphing calculator. Other graphing utilities use **symbolic differentiation** to find an algebraic formula for the derivative, and then evaluate this formula at the indicated values of x. Figure 3B illustrates this approach in Maple V, a very powerful mathematical software package.

```
nDeriv(Y1,X,0)
                4
nDeriv(Y1,X,2)
                0
nDeriv(Y1,X,3)
               -2
```

```
• f:=4*x-x^2;d:=diff(f,x);
                2
       f := 4 x - x
       d := 4 - 2 x
• x:=0;d;x:=2;d;x:=3;d;
       x := 0
          4
       x := 2
          0
       x := 3
         -2
```

(A) Numerical differentiation:
   y1=4x-x^2

(B) Symbolic differentiation:
   f:=4*x-x^2

FIGURE 3

Differentiation is as basic to calculus as addition is to algebra, and it is important that you learn to perform this operation for yourself, without using technology. However, you certainly can use numerical differentiation on a graphing utility to check your work. And, in some cases, you will be able to solve entire problems involving derivatives on a graphing utility.

EXAMPLE 4 ➤ Finding a Derivative   Find $f'(x)$, the derivative of f at x, for $f(x) = \sqrt{x} + 2$.

Solution   To find $f'(x)$, we find the following limit using the two-step process.

$$\lim_{h \to 0} \frac{f(x + h) - f(x)}{h}$$

Step 1.   Form the difference quotient and simplify:

$$\frac{f(x + h) - f(x)}{h} = \frac{(\sqrt{x + h} + 2) - (\sqrt{x} + 2)}{h}$$

$$= \frac{\sqrt{x + h} - \sqrt{x}}{h}$$

Since this is a 0/0 indeterminate form, we change the form by rationalizing the numerator:

$$\frac{\sqrt{x + h} - \sqrt{x}}{h} \cdot \frac{\sqrt{x + h} + \sqrt{x}}{\sqrt{x + h} + \sqrt{x}} = \frac{x + h - x}{h(\sqrt{x + h} + \sqrt{x})}$$

$$= \frac{h}{h(\sqrt{x + h} + \sqrt{x})}$$

$$= \frac{1}{\sqrt{x + h} + \sqrt{x}} \qquad h \neq 0$$

Step 2.   Find the limit of the difference quotient:

$$f'(x) = \lim_{h \to 0} \frac{f(x + h) - f(x)}{h}$$

$$= \lim_{h \to 0} \frac{1}{\sqrt{x + h} + \sqrt{x}}$$

$$= \frac{1}{\sqrt{x} + \sqrt{x}} = \frac{1}{2\sqrt{x}} \qquad x > 0$$

Thus, the derivative of $f(x) = \sqrt{x} + 2$ is $f'(x) = 1/(2\sqrt{x})$, a new function. The domain of f is $[0, \infty)$. Since $f'(0)$ is not defined, the domain of f' is $(0, \infty)$, a subset of the domain of f.   ◄

MATCHED PROBLEM 4   ➤   Find $f'(x)$ for $f(x) = \sqrt{x + 4}$.   ◄

▶   EXAMPLE 5   ➤   **Sales Analysis**   The total sales of a company (in millions of dollars) t months from now are given by $S(t) = \sqrt{t} + 2$. Find $S(25)$ and $S'(25)$, and interpret. Use these results to estimate the total sales after 26 months and after 27 months. Use a graphing utility to check your work.

Solution   The total sales function S has the same form as the function f in Example 4—only the letters used to represent the function and the independent variable have been changed. It follows that S' and f' also have the same form:

$$S(t) = \sqrt{t} + 2 \qquad f(x) = \sqrt{x} + 2$$

$$S'(t) = \frac{1}{2\sqrt{t}} \qquad f'(x) = \frac{1}{2\sqrt{x}}$$

Evaluating S and S' at $t = 25$, we have

$$S(25) = \sqrt{25} + 2 = 7 \qquad S'(25) = \frac{1}{2\sqrt{25}} = 0.1$$

Figure 4 shows a check of these results on a graphing utility. Thus, 25 months from now the total sales are $7 million and are increasing at the rate of $0.1 million ($100,000) per month. If this instantaneous rate of change of sales remained constant, then the sales would grow to $7.1 million after 26 months, $7.2 million after 27 months, and so on. Even though $S'(t)$ is not a constant function in this case, these values provide useful estimates of the total sales.   ◄

```
25→X
                  25
Y1
                   7
nDeriv(Y1,X,25)
          .1000000001
```

FIGURE 4
$y_1 = \sqrt{x} + 2$

MATCHED PROBLEM 5   ➤   The total sales of a company (in millions of dollars) t months from now are given by $S(t) = \sqrt{t} + 4$. Find $S(12)$ and $S'(12)$, and interpret. Use these results to estimate the total sales after 13 months and after 14 months. (Use the derivative found in Matched Problem 4.)   ◄

Refer to Example 5. It is instructive to compare the estimates of total sales obtained by using the derivative with the corresponding exact values of $S(t)$:

Exact Values   Estimated Values

$S(26) = \sqrt{26} + 2 = 7.099. . . \approx 7.1$

$S(27) = \sqrt{27} + 2 = 7.196. . . \approx 7.2$

For this function, the estimated values provide very good approximations to the exact values of $S(t)$. For other functions, the approximation might not be as accurate.

Using the instantaneous rate of change of a function at a point to estimate values of the function at nearby points is a simple, but important application of the derivative.

■ NONEXISTENCE OF THE DERIVATIVE

The existence of a derivative at $x = a$ depends on the existence of a limit at $x = a$; that is, on the existence of

$$f'(a) = \lim_{h \to 0} \frac{f(a + h) - f(a)}{h} \tag{1}$$

If the limit does not exist at $x = a$, we say that the function f is **nondifferentiable at** $x = a$, or $f'(a)$ **does not exist.**

explore – discuss 3

Let $f(x) = |x - 1|$.

(A) Graph f.   (B) Complete the following table:

h	-0.1	-0.01	$-0.001 \to 0 \leftarrow 0.001$	0.01	0.1
$\dfrac{f(1 + h) - f(1)}{h}$	?	?	? $\to$? $\leftarrow$?	?	?

(C) Find the following limit, if it exists.

$$\lim_{h \to 0} \frac{f(1 + h) - f(1)}{h}$$

(D) Use the results of parts (A)–(C) to discuss the existence of $f'(1)$.

Repeat parts (A)–(D) for $g(x) = \sqrt[3]{x - 1}$.

How can we recognize the points on the graph of f where $f'(a)$ does not exist? It is impossible to describe all the ways that the limit in (1) can fail to exist. However, we can illustrate some common situations where $f'(a)$ does fail to exist:

1. If the graph of f has a hole or a break at $x = a$, then $f'(a)$ does not exist* (Fig. 5A).
2. If the graph of f has a sharp corner at $x = a$, then $f'(a)$ does not exist and the graph has no tangent line at $x = a$ (Fig. 5B). (In Fig. 5B, the left- and right-hand derivatives exist but are not equal.)
3. If the graph of f has a vertical tangent line at $x = a$, then $f'(a)$ does not exist (Fig. 5C and D).

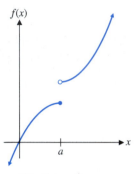

(A) Not continuous at
     $x = a$

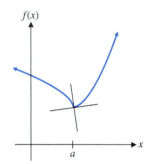

(B) Graph has sharp
     corner at $x = a$

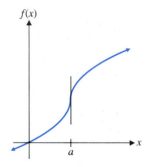

(C) Vertical tangent
     at $x = a$

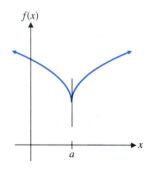

(D) Vertical tangent
     at $x = a$

FIGURE 5
The function f is nondifferentiable at $x = a$.

explore – discuss 4

Refer to Explore–Discuss 3. Graph $f(x) = |x - 1|$ and $g(x) = \sqrt[3]{x - 1}$ on a graphing utility. Use the graphs to explain why $f'(1)$ and $g'(1)$ do not exist. What happens if you use the numerical differentiation routine on the graphing utility to approximate $f'(1)$ or $g'(1)$?

As you should have discovered in Explore–Discuss 4, attempting to approximate the value of a derivative at a point where it does not exist can produce misleading results. Therefore, we issue the following caution:

Caution

Before using a numerical differentiation routine to approximate $f'(a)$, always check that $f'(a)$ does, in fact, exist.

---

* Informally, if the graph of a function has no holes or breaks, then the function is said to be **continuous** (see Section 2-1). We will discuss the formal definition of continuity in Section 9-1. For now, this intuitive formulation will suffice.

■ SUMMARY

Given a differentiable function f and a fixed number x, we have seen three different ways to find or approximate $f'(x)$:

1. Numerically, by computing

$$\frac{f(x + h) - f(x)}{h}$$

for small values of h.

2. Graphically, by estimating the slope of the tangent line at the point $(x, f(x))$.

3. Algebraically, by using the two-step limiting process to evaluate

$$\lim_{h \to 0} \frac{f(x + h) - f(x)}{h}$$

Each of these approaches has its uses. In fact, in order to effectively use derivative concepts in the many and varied applications we will consider, it is necessary to understand all three approaches. The first two methods produce approximate values of the derivative at a single fixed value of x. These methods are especially useful for functions defined by tables or graphs. However, for functions defined by algebraic expressions, the third method has the distinct advantage of producing an algebraic expression for f'. This expression can be used to find the exact value of $f'(x)$ at any number x. For example, once we know that the derivative of $f(x) = x^2$ is $f'(x) = 2x$, then we know the exact value of $f'(x)$ for any real number x.

At this point in our development, finding the algebraic expression for f' requires using the two-step limiting process for each new function we encounter. In the next three sections we will develop formulas and general properties of derivatives that will enable us to find the derivatives of many functions without having to go through the two-step limiting process each time.

Answers to Matched Problems
1. $y = 4x - 4$ 2. $f'(x) = 8 - 4x$
3. $f'(1) = 4, f'(2) = 0, f'(4) = -8$

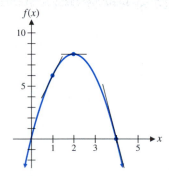

4. $f'(x) = 1/(2\sqrt{x + 4})$
5. $S(12) = 4, S'(12) = 0.125$; 12 months from now the total sales are \$4 million and are increasing at the rate of \$0.125 million (\$125,000) per month. The estimated total sales are \$4.125 million after 13 months and \$4.25 million after 14 months.

EXERCISE 8-3

A *In Problems 1 and 2, find the indicated quantity for y =*
f(x) = x² − 1, and interpret it in terms of the graph of f shown
below.

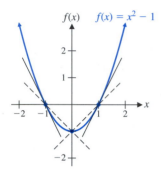

Figure for 1 and 2

1. (A) $\dfrac{f(1) - f(0)}{1 - 0}$ (B) $\dfrac{f(1 + h) - f(1)}{h}$

(C) $\lim\limits_{h \to 0} \dfrac{f(1 + h) - f(1)}{h}$

2. (A) $\dfrac{f(-1) - f(0)}{-1 - 0}$ (B) $\dfrac{f(-1 + h) - f(-1)}{h}$

(C) $\lim\limits_{h \to 0} \dfrac{f(-1 + h) - f(-1)}{h}$

In Problems 3–6, use the given expression for f(x + h) − f(x)
to find f′(x).

3. $f(x + h) - f(x) = 4hx - 3h + 2h^2$
4. $f(x + h) - f(x) = 6hx - 5h + 3h^2$
5. $f(x + h) - f(x) = 3hx^2 - 2hx + 3h^2x - h^2 + h^3$
6. $f(x + h) - f(x) = -6hx^2 + 10hx - 6h^2x + 5h^2 - 2h^3$

In Problems 7–10, find f′(x) using the two-step process:

Step 1. Simplify: $\dfrac{f(x + h) - f(x)}{h}$

Step 2. Evaluate: $\lim\limits_{h \to 0} \dfrac{f(x + h) - f(x)}{h}$

Then find f′(1), f′(2), and f′(3).

7. $f(x) = 2x - 3$ **8.** $f(x) = 4x + 3$
9. $f(x) = 2 - x^2$ **10.** $f(x) = 2x^2 + 5$

In Problems 11–14, use the given graph of y = f(x) to estimate
f′(x) at the indicated values of x and to estimate the value(s)
of x for which f′(x) = 0. Round all estimates to the nearest
integer.

11. $x = -3, 3$

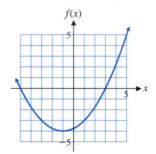

12. $x = -3, 3$

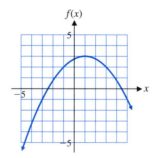

13. $x = -5, 1, 7$

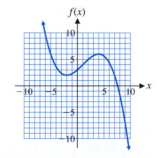

14. $x = -8, -2, 4$

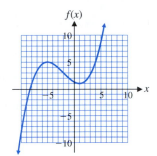

B *Problems 15 and 16 refer to the graph of $y = f(x) = x^2 + x$ shown.*

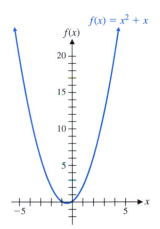

Figure for 15 and 16

15. (A) Find the slope of the secant line joining $(1, f(1))$ and $(3, f(3))$.
   (B) Find the slope of the secant line joining $(1, f(1))$ and $(1 + h, f(1 + h))$.
   (C) Find the slope of the tangent line at $(1, f(1))$.
   (D) Find the equation of the tangent line at $(1, f(1))$.

16. (A) Find the slope of the secant line joining $(2, f(2))$ and $(4, f(4))$.
   (B) Find the slope of the secant line joining $(2, f(2))$ and $(2 + h, f(2 + h))$.
   (C) Find the slope of the tangent line at $(2, f(2))$.
   (D) Find the equation of the tangent line at $(2, f(2))$.

In Problems 17 and 18, suppose an object moves along the y axis so that its location is $y = f(x) = x^2 + x$ at time x (y is in meters and x is in seconds). Find:

17. (A) The average velocity (the average rate of change of y with respect to x) for x changing from 1 to 3 seconds
   (B) The average velocity for x changing from 1 to $1 + h$ seconds
   (C) The instantaneous velocity at $x = 1$ second

18. (A) The average velocity (the average rate of change of y with respect to x) for x changing from 2 to 4 seconds
   (B) The average velocity for x changing from 2 to $2 + h$ seconds
   (C) The instantaneous velocity at $x = 2$ seconds

In Problems 19–24, find $f'(x)$ using the two-step limiting process. Then find $f'(1)$, $f'(2)$, and $f'(3)$, and check with numerical differentiation on a graphing utility.

19. $f(x) = 6x - x^2$ **20.** $f(x) = 2x - 3x^2$
21. $f(x) = \sqrt{x} - 3$ **22.** $f(x) = 2 - \sqrt{x}$

23. $f(x) = \dfrac{-1}{x}$ **24.** $f(x) = \dfrac{1}{x + 1}$

Problems 25–32 refer to the function F in the graph shown. Use the graph to determine whether $F'(x)$ exists at each indicated value of x.

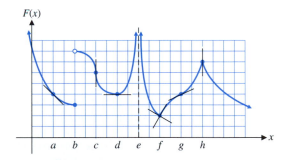

Figure for 25–32

25. $x = a$ **26.** $x = b$ **27.** $x = c$
28. $x = d$ **29.** $x = e$ **30.** $x = f$
31. $x = g$ **32.** $x = h$

33. Given $f(x) = x^2 - 4x$:
   (A) Find $f'(x)$.
   (B) Find the slopes of the tangent lines to the graph of f at $x = 0, 2$, and 4, and illustrate graphically.

34. Given $f(x) = x^2 + 2x$:
(A) Find $f'(x)$.
(B) Find the slopes of the tangent lines to the graph of f at $x = -2, -1$, and 1, and illustrate graphically.

35. If an object moves along a line so that it is at $y = f(x) = 4x^2 - 2x$ at time x (in seconds), find the instantaneous velocity function $v = f'(x)$, and find the velocity at times $x = 1, 3$, and 5 seconds (y is measured in feet).

36. Repeat Problem 35 with $f(x) = 8x^2 - 4x$.

In Problems 37–40, use the information in each table to sketch a possible graph for the function $y = f(x)$.

37.

x	$f(x)$	$f'(x)$
-4	-3	0.5
0	-1	0.5
2	0	0.5

38.

x	$f(x)$	$f'(x)$
-1	1	-2
0	-1	-2
1	-3	-2

39.

x	$f(x)$	$f'(x)$
-3	0	-3
-1	-2	0
3	0	0.75

40.

x	$f(x)$	$f'(x)$
1	-3	-0.75
3	-4	0
4	-3	3

In Problems 41–44, use numerical differentiation to approximate $f'(0)$ to two decimal places.

41. $f(x) = 2^x$
42. $f(x) = 3^x$
43. $f(x) = \sqrt{2 + 2x - x^2}$
44. $f(x) = \sqrt{3 - 2x - x^2}$

45. Let $f(x) = x^2$, $g(x) = x^2 - 3$, and $h(x) = x^2 + 1$.

(A) How are the graphs of these functions related? How would you expect the derivatives of these functions to be related?
(B) Use the two-step process to find the derivative of $m(x) = x^2 + C$, where C is any real number constant.

46. Let $f(x) = 2x$, $g(x) = 2x - 1$, and $h(x) = 2x + 2$.

(A) How are the graphs of these functions related? How would you expect the derivatives of these functions to be related?
(B) Use the two-step process to find the derivative of $m(x) = 2x + C$, where C is any real number constant.

47. (A) Give a geometric explanation of the following statement: If $f(x) = C$ is a constant function, then $f'(x) = 0$.
(B) Use the two-step process to verify the statement in part (A).

48. (A) Give a geometric explanation of the following statement: If $f(x) = mx + b$ is a linear function, then $f'(x) = m$.
(B) Use the two-step process to verify the statement in part (A).

C *In Problems 49 and 50, sketch the graph of f and determine where f is nondifferentiable.*

49. $f(x) = \begin{cases} 2x & \text{if } x < 1 \\ 2 & \text{if } x \geq 1 \end{cases}$

50. $f(x) = \begin{cases} 2x & \text{if } x < 2 \\ 6 - x & \text{if } x \geq 2 \end{cases}$

In Problems 51–54, determine whether f is differentiable at $x = 0$ by considering

$$\lim_{h \to 0} \frac{f(0 + h) - f(0)}{h}$$

51. $f(x) = |x|$
52. $f(x) = 1 - |x|$
53. $f(x) = x^{1/3}$
54. $f(x) = x^{2/3}$

Most graphing utilities perform numerical differentiation by considering the following average of difference quotients:

$$\frac{1}{2}\left[\frac{f(x + h) - f(x)}{h} + \frac{f(x - h) - f(x)}{-h}\right] = \frac{f(x + h) - f(x - h)}{2h}$$

In Problems 55 and 56, what happens if you use this formula to approximate $f'(0)$? What happens if you use numerical differentiation on your graphing utility to approximate $f'(0)$? Compare these results with Problems 51 and 52.

55. $f(x) = |x|$
56. $f(x) = 1 - |x|$

57. Show that $f(x) = 2x - x^2$ is differentiable over the closed interval $[0, 2]$ by showing that each of the following limits exists:

(A) $\lim_{h \to 0} \dfrac{f(x + h) - f(x)}{h}$, $0 < x < 2$

(B) $\lim\limits_{h \to 0^+} \dfrac{f(0 + h) - f(0)}{h}, \qquad x = 0$

(C) $\lim\limits_{h \to 0^-} \dfrac{f(2 + h) - f(2)}{h}, \qquad x = 2$

58. Show that $f(x) = \sqrt{x}$ is differentiable over the interval $(0, \infty)$ but not over the half-closed interval $[0, \infty)$ by considering

$$\lim\limits_{h \to 0} \dfrac{f(x + h) - f(x)}{h} \qquad 0 < x < \infty$$

and

$$\lim\limits_{h \to 0^+} \dfrac{f(0 + h) - f(0)}{h} \qquad x = 0$$

▶ APPLICATIONS

Business & Economics

59. *Sales analysis.* The total sales of a company (in millions of dollars) t months from now are given by

$$S(t) = 2\sqrt{t} + 10$$

(A) Use the two-step process to find $S'(t)$.

(B) Find $S(15)$ and $S'(15)$. Write a brief verbal interpretation of these results.

(C) Use the results in part (B) to estimate the total sales after 16 months and after 17 months.

60. *Sales analysis.* The total sales of a company (in millions of dollars) t months from now are given by

$$S(t) = 2\sqrt{t} + 6$$

(A) Use the two-step process to find $S'(t)$.

(B) Find $S(10)$ and $S'(10)$. Write a brief verbal interpretation of these results.

(C) Use the results in part (B) to estimate the total sales after 11 months and after 12 months.

61. *Compound interest.* If \$100 is invested in an account that earns 6% compounded annually, then the amount in the account after t years is given by

$$A(t) = 100(1.06)^t$$

(A) Find $A(5)$ and use numerical differentiation to approximate $A'(5)$ to the nearest cent.

(B) Write a brief verbal interpretation of the results in part (A) and illustrate graphically.

62. *Compound interest.* If \$500 is invested in an account that earns 8% compounded annually, then the amount in the account after t years is given by

$$A(t) = 500(1.08)^t$$

(A) Find $A(7)$ and use numerical differentiation to approximate $A'(7)$ to the nearest cent.

(B) Write a brief verbal interpretation of the results in part (A) and illustrate graphically.

63. *Financial analysis.* The price of a stock during a 12 month period is graphed in the figure.

Figure for 63 and 64

(A) Use this graph to estimate (to the nearest \$5) the price of the stock and the rate of change of the price in March. Write a brief verbal interpretation of these results.

(B) When did the stock reach its highest price, and what is the rate of change of the price at this point in time?

4. *Financial analysis.* Refer to the figure.

(A) Use this graph to estimate (to the nearest \$5) the price of the stock and the rate of change of the price in June. Write a brief verbal interpretation of these results.

(B) When did the stock reach its lowest price, and what is the rate of change of the price at this point in time?

65. *Residential energy consumption.* A survey by the Energy Information Administration produced the results in Table 2.

(A) Let x represent time (in years), with $x = 0$ corresponding to 1980, and let y represent the corresponding percentage of households with black-and-white television sets. Enter the appropriate data set in a graphing utility and find a quadratic regression equation for the data.

(B) If $y = B(x)$ denotes the regression equation found in part (A), find $B(15)$ and $B'(15)$, and write a brief verbal interpretation of these results. Express answers as percentages correct to one decimal place.

T A B L E 2
Households with Television Sets

YEAR	BLACK-AND-WHITE (%)	COLOR (%)
1980	51	82
1981	48	82
1982	47	85
1984	43	88
1987	36	93
1990	31	96
1993	20	98

66. *Residential energy consumption.* Refer to the survey by the Energy Information Administration in Table 2.

(A) Let x represent time (in years), with $x = 0$ corresponding to 1980, and let y represent the corresponding percentage of households with color television sets. Enter the appropriate data set in a graphing utility and find a quadratic regression equation for the data.

(B) If $y = C(x)$ denotes the regression equation found in part (A), find $C(15)$ and $C'(15)$, and write a brief verbal interpretation of these results. Express answers as percentages correct to one decimal place.

Life Sciences

67. *Air pollution.* The ozone level (in parts per billion) on a summer day in a metropolitan area is given by

$$P(t) = 80 + 12t - t^2$$

where t is time in hours and $t = 0$ corresponds to 9 AM.

(A) Use the two-step process to find $P'(t)$.

(B) Find $P(3)$ and $P'(3)$. Write a brief verbal interpretation of these results and illustrate graphically.

68. *Medicine.* The body temperature (in degrees Fahrenheit) of a patient t hours after being given a fever-reducing drug is given by

$$F(t) = 98 + \frac{4}{t + 1}$$

(A) Use the two-step process to find $F'(t)$.

(B) Find $F(3)$ and $F'(3)$. Write a brief verbal interpretation of these results and illustrate graphically.

Social Sciences

69. *Population.* The number of persons in the United States 65 years old and older can be described by the function

$$P(t) = 26(1.02)^t$$

where P is population in millions and t is time in years since 1980.

(A) Find $P(30)$ and use numerical differentiation to approximate $P'(30)$. Round both quantities to one decimal place.

(B) Write a brief verbal interpretation of the results in part (A).

(C) Use the results in part (B) to estimate the 65 or over population in 2011 and in 2012.

70. *Population.* The number of households in the United States can be described by the function

$$H(t) = 80(1.014)^t$$

where H is the number of households in millions and t is time in years since 1980.

(A) Find $H(25)$ and use numerical differentiation to approximate $H'(25)$. Round both quantities to one decimal place.

(B) Write a brief verbal interpretation of the results in part (A).

(C) Use the results in part (B) to estimate the number of households in 2006 and in 2007.

SECTION 8-4 ■ **Derivatives of Constants, Power Forms, and Sums**

- DERIVATIVE OF A CONSTANT FUNCTION
- POWER RULE
- DERIVATIVE OF A CONSTANT TIMES A FUNCTION
- DERIVATIVES OF SUMS AND DIFFERENCES
- APPLICATIONS

In the preceding section, we defined the derivative of f at x as

$$f'(x) = \lim_{h \to 0} \frac{f(x + h) - f(x)}{h}$$

if the limit exists, and we used this definition and a two-step process to find the derivatives of several functions. In this and the next two sections, we will develop some rules based on this definition that will enable us to determine the derivatives of a rather large class of functions without having to go through the two-step process each time.

Before starting on these rules, we list some symbols that are widely used to represent derivatives:

Derivative Notation

Given $y = f(x)$, then

$$f'(x) \qquad y' \qquad \frac{dy}{dx}$$

all represent the derivative of f at x.

Each of these symbols for the derivative has its particular advantage in certain situations. All of them will become familiar to you after a little experience.

■ **DERIVATIVE OF A CONSTANT FUNCTION**

Suppose

$$f(x) = C \qquad C \text{ a constant} \qquad \textit{A constant function}$$

Geometrically, the graph of $f(x) = C$ is a horizontal straight line with slope 0 (see Fig. 1); hence, we would expect $f'(x) = 0$. We will show that this is actually the case using the definition of the derivative and the two-step process introduced earlier. We want to find

$$f'(x) = \lim_{h \to 0} \frac{f(x + h) - f(x)}{h} \qquad \textit{Definition of } f'(x)$$

Slope = 0 $f(x) = C$

FIGURE 1

Step 1. $\dfrac{f(x + h) - f(x)}{h} = \dfrac{C - C}{h} = \dfrac{0}{h} = 0$ $h \neq 0$

Step 2. $\lim\limits_{h \to 0} 0 = 0$

Thus,

$f'(x) = 0$

We conclude that:

The derivative of any constant function is 0.

Derivative of a Constant Function Rule

If $y = f(x) = C$, then

$f'(x) = 0$

Also, $y' = 0$ and $dy/dx = 0$.

[*Note:* When we write $C' = 0$ or $\dfrac{d}{dx} C = 0$, we mean $y' = \dfrac{dy}{dx} = 0$ where $y = C$.]

EXAMPLE 1 ➤ Differentiating Constant Functions

(A) If $f(x) = 3$, then $f'(x) = 0$.   (B) If $y = -1.4$, then $y' = 0$.

(C) If $y = \pi$, then $\dfrac{dy}{dx} = 0$.   (D) $\dfrac{d}{dx} 23 = 0$

◄

MATCHED PROBLEM 1 ➤ Find:

(A) $f'(x)$ for $f(x) = -24$ (B) y' for $y = 12$

(C) $\dfrac{dy}{dx}$ for $y = -\sqrt{7}$ (D) $\dfrac{d}{dx}(-\pi)$

◄

■ POWER RULE

A function of the form $f(x) = x^k$, where k is a real number, is called a **power function.** The elementary functions (see the inside front cover) listed below are examples of power functions:

$f(x) = x$ $h(x) = x^2$ $m(x) = x^3$ (1)
$n(x) = \sqrt{x}$ $p(x) = \sqrt[3]{x}$

e x p l o r e – d i s c u s s   1

(A) It is clear that the functions f, h, and m in (1) are power functions. Explain why the functions n and p are also power functions.

(B) The domain of a power function depends on the power. Discuss the domain of each of the following power functions:

$$r(x) = x^4 \qquad s(x) = x^{-4} \qquad t(x) = x^{1/4}$$

$$u(x) = x^{-1/4} \qquad v(x) = x^{1/5} \qquad w(x) = x^{-1/5}$$

The definition of the derivative and the two-step process introduced in the preceding section can be used to find the derivatives of many power functions. For example, it can be shown that:

$$\text{If} \quad f(x) = x^2, \quad \text{then} \quad f'(x) = 2x.$$
$$\text{If} \quad f(x) = x^3, \quad \text{then} \quad f'(x) = 3x^2.$$
$$\text{If} \quad f(x) = x^4, \quad \text{then} \quad f'(x) = 4x^3.$$
$$\text{If} \quad f(x) = x^5, \quad \text{then} \quad f'(x) = 5x^4.$$

Notice the pattern in these derivatives. In each case, the power in f becomes the coefficient in f' and the power in f' is 1 less than the power in f. In general, for any positive integer n:

$$\text{If} \quad f(x) = x^n, \quad \text{then} \quad f'(x) = nx^{n-1}. \tag{2}$$

In fact, more advanced techniques can be used to show that (2) holds for *any* real number n. We will assume this general result for the remainder of this book.

Power Rule

If $y = f(x) = x^n$, where n is a real number, then

$$f'(x) = nx^{n-1}$$

Also, $y' = nx^{n-1}$ and $dy/dx = nx^{n-1}$.

---

e x p l o r e – d i s c u s s   2

(A) Write a verbal description of the power rule.

(B) If $f(x) = x$, what is $f'(x)$? Discuss how this derivative can be obtained from the power rule.

EXAMPLE 2 ➤ Differentiating Power Functions

(A) If $f(x) = x^5$, then $f'(x) = 5x^{5-1} = 5x^4$.

(B) If $y = x^{25}$, then $y' = 25x^{25-1} = 25x^{24}$.

(C) If $y = x^{-3}$, then $\dfrac{dy}{dx} = -3x^{-3-1} = -3x^{-4} = -\dfrac{3}{x^4}$

(D) $\dfrac{d}{dx} x^{5/3} = \tfrac{5}{3} x^{(5/3)-1} = \tfrac{5}{3} x^{2/3}$ ◀

MATCHED PROBLEM 2 ➤ Find:

(A) $f'(x)$ for $f(x) = x^6$ (B) y' for $y = x^{30}$

(C) $\dfrac{dy}{dx}$ for $y = x^{-2}$ (D) $\dfrac{d}{dx} x^{3/2}$ ◀

In some cases, properties of exponents must be used to rewrite an expression before the power rule is applied.

EXAMPLE 3 ➤ **Differentiating Power Functions**

(A) If $f(x) = 1/x^4$, then we can write $f(x) = x^{-4}$ and

$$f'(x) = -4x^{-4-1} = -4x^{-5} \quad \text{or} \quad \dfrac{-4}{x^5}$$

(B) If $y = \sqrt{x}$, then we can write $y = x^{1/2}$ and

$$y' = \dfrac{1}{2} x^{(1/2)-1} = \dfrac{1}{2} x^{-1/2} \quad \text{or} \quad \dfrac{1}{2\sqrt{x}}$$

(C) $\dfrac{d}{dx} \dfrac{1}{\sqrt[3]{x}} = \dfrac{d}{dx} x^{-1/3} = -\dfrac{1}{3} x^{(-1/3)-1} = -\dfrac{1}{3} x^{-4/3} \quad \text{or} \quad \dfrac{-1}{3\sqrt[3]{x^4}}$ ◀

MATCHED PROBLEM 3 ➤ Find:

(A) $f'(x)$ for $f(x) = \dfrac{1}{x}$ (B) y' for $y = \sqrt[3]{x^2}$ (C) $\dfrac{d}{dx} \dfrac{1}{\sqrt{x}}$ ◀

■ DERIVATIVE OF A CONSTANT TIMES A FUNCTION

Let $f(x) = ku(x)$; where k is a constant and u is differentiable at x. Then, using the two-step process, we have the following:

Step 1. $\dfrac{f(x+h) - f(x)}{h} = \dfrac{ku(x+h) - ku(x)}{h} = k\left[\dfrac{u(x+h) - u(x)}{h}\right]$

Step 2. $\displaystyle\lim_{h \to 0} \dfrac{f(x+h) - f(x)}{h} = \lim_{h \to 0} k\left[\dfrac{u(x+h) - u(x)}{h}\right]$ $\displaystyle\lim_{x \to c} kg(x) = k \lim_{x \to c} g(x)$

$\qquad\qquad = k \displaystyle\lim_{h \to 0} \left[\dfrac{u(x+h) - u(x)}{h}\right]$ Definition of $u'(x)$

$\qquad\qquad = ku'(x)$

Thus:

> **The derivative of a constant times a differentiable function is the constant times the derivative of the function.**

Constant Times a Function Rule

If $y = f(x) = ku(x)$, then

$$f'(x) = ku'(x)$$

Also,

$$y' = ku' \qquad \frac{dy}{dx} = k\frac{du}{dx}$$

EXAMPLE 4 ➤ Differentiating a Constant Times a Function

(A) If $f(x) = 3x^2$, then $f'(x) = 3 \cdot 2x^{2-1} = 6x$.

(B) If $y = \dfrac{x^3}{6} = \dfrac{1}{6}x^3$, then $\dfrac{dy}{dx} = \dfrac{1}{6} \cdot 3x^{3-1} = \dfrac{1}{2}x^2$.

(C) If $y = \dfrac{1}{2x^4} = \dfrac{1}{2}x^{-4}$, then $y' = \dfrac{1}{2}(-4x^{-4-1}) = -2x^{-5}$ or $\dfrac{-2}{x^5}$.

(D) $\dfrac{d}{dx}\dfrac{4}{\sqrt{x^3}} = \dfrac{d}{dx}\dfrac{4}{x^{3/2}} = \dfrac{d}{dx}4x^{-3/2} = 4\left[-\dfrac{3}{2}x^{(-3/2)-1}\right]$

$$= -6x^{-5/2} \quad \text{or} \quad -\dfrac{6}{\sqrt{x^5}} \qquad \blacktriangleleft$$

MATCHED PROBLEM 4 ➤ Find:

(A) $f'(x)$ for $f(x) = 4x^5$ (B) $\dfrac{dy}{dx}$ for $y = \dfrac{x^4}{12}$

(C) y' for $y = \dfrac{1}{3x^3}$ (D) $\dfrac{d}{dx}\dfrac{9}{\sqrt[3]{x}}$ $\blacktriangleleft$

■ **DERIVATIVES OF SUMS AND DIFFERENCES**

Let $f(x) = u(x) + v(x)$, where $u'(x)$ and $v'(x)$ exist. Then, using the two-step process, we have the following:

Step 1. $\dfrac{f(x + h) - f(x)}{h} = \dfrac{[u(x + h) + v(x + h)] - [u(x) + v(x)]}{h}$

$$= \dfrac{u(x + h) + v(x + h) - u(x) - v(x)}{h}$$

$$= \dfrac{u(x + h) - u(x)}{h} + \dfrac{v(x + h) - v(x)}{h}$$

Step 2. $\displaystyle\lim_{h \to 0} \dfrac{f(x + h) - f(x)}{h} = \lim_{h \to 0} \left[\dfrac{u(x + h) - u(x)}{h} + \dfrac{v(x + h) - v(x)}{h} \right]$

$$\lim_{x \to c} [g(x) + h(x)] = \lim_{x \to c} g(x) + \lim_{x \to c} h(x)$$

$$= \lim_{h \to 0} \dfrac{u(x + h) - u(x)}{h} + \lim_{h \to 0} \dfrac{v(x + h) - v(x)}{h}$$

$$= u'(x) + v'(x)$$

Thus:

The derivative of the sum of two differentiable functions is the sum of the derivatives.

Similarly, we can show that:

The derivative of the difference of two differentiable functions is the difference of the derivatives.

Together, we then have the **sum and difference rule** for differentiation:

Sum and Difference Rule

If $y = f(x) = u(x) \pm v(x)$, then

$$f'(x) = u'(x) \pm v'(x)$$

Also,

$$y' = u' \pm v' \qquad\qquad \dfrac{dy}{dx} = \dfrac{du}{dx} \pm \dfrac{dv}{dx}$$

[*Note:* This rule generalizes to the sum and difference of any given number of functions.]

With this and the other rules stated previously, we will be able to compute the derivatives of all polynomials and a variety of other functions.

EXAMPLE 5 ➤ Differentiating Sums and Differences

(A) If $f(x) = 3x^2 + 2x$, then

$$f'(x) = (3x^2)' + (2x)' = 3(2x) + 2(1) = 6x + 2$$

(B) If $y = 4 + 2x^3 - 3x^{-1}$, then

$$y' = (4)' + (2x^3)' - (3x^{-1})' = 0 + 2(3x^2) - 3(-1)x^{-2} = 6x^2 + 3x^{-2}$$

(C) If $y = \sqrt[3]{x} - 3x$, then

$$\frac{dy}{dx} = \frac{d}{dx}x^{1/3} - \frac{d}{dx}3x = \frac{1}{3}x^{-2/3} - 3 = \frac{1}{3x^{2/3}} - 3$$

(D) $\dfrac{d}{dx}\left(\dfrac{5}{3x^2} - \dfrac{2}{x^4} + \dfrac{x^3}{9}\right) = \dfrac{d}{dx}\dfrac{5}{3}x^{-2} - \dfrac{d}{dx}2x^{-4} + \dfrac{d}{dx}\dfrac{1}{9}x^3$

$$= \frac{5}{3}(-2)x^{-3} - 2(-4)x^{-5} + \frac{1}{9}\cdot 3x^2$$

$$= -\frac{10}{3x^3} + \frac{8}{x^5} + \frac{1}{3}x^2 \qquad \blacktriangleleft$$

MATCHED PROBLEM 5 ➤ Find:

(A) $f'(x)$ for $f(x) = 3x^4 - 2x^3 + x^2 - 5x + 7$

(B) y' for $y = 3 - 7x^{-2}$ (C) $\dfrac{dy}{dx}$ for $y = 5x^3 - \sqrt[4]{x}$

(D) $\dfrac{d}{dx}\left(-\dfrac{3}{4x} + \dfrac{4}{x^3} - \dfrac{x^4}{8}\right)$ $\blacktriangleleft$

■ APPLICATIONS

EXAMPLE 6 ➤ Instantaneous Velocity   An object moves along the y axis (marked in feet) so that its position at time x (in seconds) is

$$f(x) = x^3 - 6x^2 + 9x$$

(A) Find the instantaneous velocity function v.
(B) Find the velocity at $x = 2$ and $x = 5$ seconds.
(C) Find the time(s) when the velocity is 0.

Solution   (A) $v = f'(x) = (x^3)' - (6x^2)' + (9x)' = 3x^2 - 12x + 9$

(B) $f'(2) = 3(2)^2 - 12(2) + 9 = -3$ feet per second
$\quad\ f'(5) = 3(5)^2 - 12(5) + 9 = 24$ feet per second

(C) $v = f'(x) = 3x^2 - 12x + 9 = 0$
$$3(x^2 - 4x + 3) = 0$$
$$3(x - 1)(x - 3) = 0$$
$$x = 1, 3$$
Thus, $v = 0$ at $x = 1$ and $x = 3$ seconds.

MATCHED PROBLEM 6 ➤ Repeat Example 6 for $f(x) = x^3 - 15x^2 + 72x$.

EXAMPLE 7 ➤ **Tangents** Let $g(x) = x^4 - 5x^2 + 3x + 4$.

(A) Find $g'(x)$.
(B) Find the equation of the tangent line at $x = 1$ and check with a graphing utility.
(C) Approximate (to three decimal places) the values of x where the tangent line is horizontal.
(D) Graph $y = g(x)$ and the horizontal tangent lines found in part (C) on a graphing utility.

Solution
(A) $g'(x) = (x^4)' - (5x^2)' + (3x)' + (4)'$

$$= 4x^3 - 10x + 3$$

(B) $y - y_1 = m(x - x_1)$ $y_1 = g(x_1) = g(1) = (1)^4 - 5(1)^2 + 3(1) + 4 = 3$
$y - 3 = -3(x - 1)$ $m = g'(x_1) = g'(1) = 4(1)^3 - 10(1) + 3 = -3$
$y = -3x + 6$ Tangent line at $x = 1$

Figure 2 shows the graph of g and the tangent line at $x = 1$.

(C) The values of x where the tangent line is horizontal are the solutions of the equation

$$g'(x) = 4x^3 - 10x + 3 = 0$$

Since most cubic polynomials cannot be factored, we use a root approximation routine to approximate the solutions of this equation, as illustrated in Figure 3:

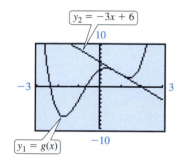

$y_2 = -3x + 6$

$y_1 = g(x)$

FIGURE 2

Root
X=-1.713939 Y=0

Root
X=1.4017712 Y=0

FIGURE 3
$y_1 = g'(x) = 4x^3 - 10x + 3$

Recall from Section 2-1 that a third-degree polynomial has at most three roots. Thus, Figure 3 shows that we have found all the solutions to the equation

$g'(x) = 0$. Rounding the values displayed in Figure 3, we conclude that the graph of $y = g(x)$ has horizontal tangent lines at $x \approx -1.714$, 0.312, and 1.402.

(D)

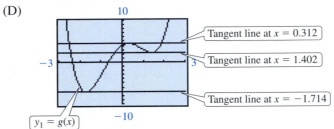

MATCHED PROBLEM 7 ▶ Repeat Example 7 for $f(x) = x^4 + 3x^3 - 2x + 2$. ◀

Remark

In Example 6, we solved the quadratic equation $f'(x) = 0$ by factoring, while in Example 7 we used a graphing utility to approximate the solutions to the cubic equation $g'(x) = 0$. Which method should you use? Since approximation always works and factoring works only in very special cases, it makes sense to make approximation the method of choice for solving equations. Thus, unless the equation is very simple, or instructions specifically state that factoring must be used, you should use root approximation techniques on a graphing utility to solve equations.

In business and economics, the rates at which certain quantities are changing often provide useful insight into various economic systems. A manufacturer, for example, is interested not only in the total cost $C(x)$ at certain production levels, but is also interested in the rate of change of costs at various production levels.

In economics, the word **marginal** refers to a rate of change; that is, to a derivative. Thus, if

$C(x)$ = Total cost of producing x items

then

$C'(x)$ = **Marginal cost**
    = Instantaneous rate of change of total cost $C(x)$ with respect to the number of items produced at a production level of x items

EXAMPLE 8 ▶ **Marginal Cost** Suppose the total cost $C(x)$ (in thousands of dollars) for manufacturing x sailboats per year is given by the function

$$C(x) = 575 + 25x - 0.25x^2 \qquad 0 \leq x \leq 50$$

(see Fig. 4, at the top of the next page).

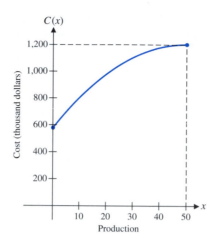

FIGURE 4

(A) Find the marginal cost at a production level of x boats per year.
(B) Find the marginal cost at a production level of 40 boats per year, and interpret the results.

Solution (A) The marginal cost at a production level of x boats is

$$C'(x) = (575)' + (25x)' - (0.25x^2)' = 25 - 0.5x$$

(B) The marginal cost at a production level of 40 boats is

$$C'(40) = 25 - 0.5(40) = 5 \quad \text{or} \quad \$5,000 \text{ per boat}$$

At a production level of 40 boats per year, the total cost is increasing at the rate of $5,000 per boat. ◀

MATCHED PROBLEM 8 ➤ Suppose the total cost $C(x)$ (in thousands of dollars) for manufacturing x sailboats per year is given by the function

$$C(x) = 500 + 24x - 0.2x^2 \qquad 0 \leqslant x \leqslant 50$$

(A) Find the marginal cost at a production level of x boats per year.
(B) Find the marginal cost at a production level of 35 boats per year, and interpret the results. ◀

The interpretation of marginal cost given in Example 8B is the usual way we use the derivative to describe the manner in which a quantity is changing. However, there is a special way to interpret the rate of change of a total cost function that is used extensively in economics. Referring to the results in Example 8B, if the total cost is increasing at the rate of $5,000 per boat when 40 boats are produced per year, and if production is increased to 41 boats per year, then the total cost will increase by approximately $5,000. But the increase in total cost is just the cost of producing the next boat. Thus, the cost of producing the 41st boat must be approximately $5,000, the marginal cost at the production level of 40 boats per year. These ideas are summarized in the following box:

Marginal Cost Function
___

If $C(x)$ is the total cost of producing x items, then the marginal cost function $C'(x)$ approximates the cost of producing one more item at a production level of x items.

▶ EXAMPLE 9 ➤ **Marginal Cost** Refer to the total cost function given in Example 8:

$$C(x) = 575 + 25x - 0.25x^2 \qquad 0 \leqslant x \leqslant 50$$

(A) Use the marginal cost function to approximate the cost of producing the 31st boat.

(B) Use the total cost function to find the exact cost of producing the 31st boat.

Solution (A) From Example 8, $C'(x) = 25 - 0.5x$. Thus,

$$C'(30) = 25 - 0.5(30) = 10 \quad \text{or} \quad \$10,000$$

The cost of producing the 31st boat is approximately $10,000.

(B) The exact cost of producing the 31st boat is

$$\left(\begin{array}{c} \text{Total cost of} \\ \text{producing} \\ \text{31 boats} \end{array} \right) - \left(\begin{array}{c} \text{Total cost of} \\ \text{producing} \\ \text{30 boats} \end{array} \right)$$

$$= C(31) \qquad - \qquad C(30)$$

$$= 1,109.75 - 1,100 = 9.750 \quad \text{or} \quad \$9,750$$

Notice that the marginal cost of $10,000 per boat is a close approximation to this exact cost. ◀

MATCHED PROBLEM 9 ➤ Refer to the cost function given in Matched Problem 8:

$$C(x) = 500 + 24x - 0.2x^2 \qquad 0 \leqslant x \leqslant 50$$

(A) Use the marginal cost function to approximate the cost of producing the 41st boat.

(B) Use the total cost function to find the exact cost of producing the 41st boat. ◀

Remark

A derivative can always be interpreted as an instantaneous rate of change, as we did in Example 8B. The interpretation of the marginal cost function as the approximate cost of producing the next item, as in Example 9A, is a special case that applies to total cost functions. This interpretation also applies to total revenue functions and total profit functions, but not to most of the other economic functions we will consider.

Answers to Matched Problems

1. All are 0.　　2. (A) $6x^5$　　(B) $30x^{29}$　　(C) $-2x^{-3} = -2/x^3$　　(D) $\frac{3}{2}x^{1/2}$
3. (A) $-x^{-2}$, or $-1/x^2$　　(B) $\frac{2}{3}x^{-1/3}$, or $2/(3\sqrt[3]{x})$　　(C) $-\frac{1}{2}x^{-3/2}$, or $-1/(2\sqrt{x^3})$
4. (A) $20x^4$　　(B) $x^3/3$　　(C) $-x^{-4}$, or $-1/x^4$　　(D) $-3x^{-4/3}$, or $-3/\sqrt[3]{x^4}$
5. (A) $12x^3 - 6x^2 + 2x - 5$　　(B) $14x^{-3}$, or $14/x^3$
　　(C) $15x^2 - \frac{1}{4}x^{-3/4}$, or $15x^2 - 1/(4x^{3/4})$　　(D) $3/(4x^2) - (12/x^4) - (x^3/2)$
6. (A) $v = 3x^2 - 30x + 72$　　(B) $f'(2) = 24$ ft/sec; $f'(5) = -3$ ft/sec
　　(C) $x = 4$ and $x = 6$ sec
7. (A) $g'(x) = 4x^3 + 9x^2 - 2$　　(B) $y = 11x - 7$　　(C) $-2.141, -0.541, 0.432$
　　(D)

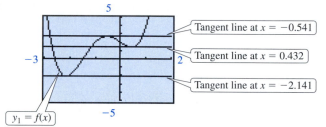

8. (A) $C'(x) = 24 - 0.4x$
　　(B) $C'(35) = 10$, or \$10,000 per boat; at a production level of 35 boats, total cost is increasing at the rate of \$10,000 per boat.
9. (A) $C'(40) = 8$ or \$8,000; the cost of producing the 41st boat is approximately \$8,000.
　　(B) $C(41) - C(40) = 7.8$ or \$7,800

EXERCISE 8-4

A *Find the indicated derivatives in Problems 1–20.*

1. $f'(x)$ for $f(x) = 12$

2. $\dfrac{dy}{dx}$ for $y = -\sqrt{3}$

3. $\dfrac{d}{dx} 23$

4. y' for $y = \pi$

5. $\dfrac{dy}{dx}$ for $y = x^{12}$

6. $\dfrac{d}{dx} x^5$

7. $f'(x)$ for $f(x) = x$

8. y' for $y = x^7$

9. y' for $y = x^{-7}$

10. $f'(x)$ for $f(x) = x^{-11}$

11. $\dfrac{dy}{dx}$ for $y = x^{5/2}$

12. $\dfrac{d}{dx} x^{7/3}$

13. $\dfrac{d}{dx} \dfrac{1}{x^5}$

14. $f'(x)$ for $f(x) = \dfrac{1}{x^9}$

15. $f'(x)$ for $f(x) = 2x^4$

16. $\dfrac{dy}{dx}$ for $y = -3x$

17. $\dfrac{d}{dx}(\frac{1}{3}x^6)$

18. y' for $y = \frac{1}{2}x^4$

19. $\dfrac{dy}{dx}$ for $y = \dfrac{x^5}{15}$

20. $f'(x)$ for $f(x) = \dfrac{x^6}{24}$

Problems 21–26 refer to functions f and g that satisfy $f'(2) = 3$ and $g'(2) = -1$. In each problem, find $h'(2)$ for the indicated function h.

21. $h(x) = 4f(x)$

22. $h(x) = 5g(x)$

23. $h(x) = f(x) + g(x)$

24. $h(x) = g(x) - f(x)$

25. $h(x) = 2f(x) - 3g(x) + 7$

26. $h(x) = -4f(x) + 5g(x) - 9$

B *Find the indicated derivatives in Problems 27–50.*

27. $\dfrac{d}{dx}(2x^{-5})$

28. y' for $y = -4x^{-1}$

29. $f'(x)$ for $f(x) = \dfrac{4}{x^4}$

30. $\dfrac{dy}{dx}$ for $y = \dfrac{-3}{x^6}$

31. $\dfrac{d}{dx}\dfrac{-1}{2x^2}$

32. y' for $y = \dfrac{1}{6x^3}$

33. $f'(x)$ for $f(x) = -3x^{1/3}$

34. $\dfrac{dy}{dx}$ for $y = -8x^{1/4}$

35. $\dfrac{d}{dx}(2x^2 - 3x + 4)$

36. y' for $y = 3x^2 + 4x - 7$

37. $\dfrac{dy}{dx}$ for $y = 3x^5 - 2x^3 + 5$

38. $f'(x)$ for $f(x) = 2x^3 - 6x + 5$

39. $\dfrac{d}{dx}(3x^{-4} + 2x^{-2})$

40. y' for $y = 2x^{-3} - 4x^{-1}$

41. $\dfrac{dy}{dx}$ for $y = \dfrac{1}{2x} - \dfrac{2}{3x^3}$

42. $f'(x)$ for $f(x) = \dfrac{3}{4x^3} + \dfrac{1}{2x^5}$

43. $\dfrac{d}{dx}(3x^{2/3} - 5x^{1/3})$

44. $\dfrac{d}{dx}(8x^{3/4} + 4x^{-1/4})$

45. $\dfrac{d}{dx}\left(\dfrac{3}{x^{3/5}} - \dfrac{6}{x^{1/2}}\right)$

46. $\dfrac{d}{dx}\left(\dfrac{5}{x^{1/5}} - \dfrac{8}{x^{3/2}}\right)$

47. $\dfrac{d}{dx}\dfrac{1}{\sqrt[3]{x}}$

48. y' for $y = \dfrac{10}{\sqrt[5]{x}}$

49. $\dfrac{dy}{dx}$ for $y = \dfrac{12}{\sqrt{x}} - 3x^{-2} + x$

50. $f'(x)$ for $f(x) = 2x^{-3} - \dfrac{6}{\sqrt[3]{x^2}} + 7$

For Problems 51–54, find:
(A) $f'(x)$
(B) *The slope of the graph of f at x = 2 and x = 4*
(C) *The equations of the tangent lines at x = 2 and x = 4*
(D) *The value(s) of x where the tangent line is horizontal*

51. $f(x) = 6x - x^2$

52. $f(x) = 2x^2 + 8x$

53. $f(x) = 3x^4 - 6x^2 - 7$

54. $f(x) = x^4 - 32x^2 + 10$

If an object moves along the y axis (marked in feet) so that its position at time x (in seconds) is given by the indicated function in Problems 55–58, find:
(A) *Instantaneous velocity function $v = f'(x)$*
(B) *The velocity when x = 0 and x = 3 seconds*
(C) *The time(s) when v = 0*

55. $f(x) = 176x - 16x^2$

56. $f(x) = 80x - 10x^2$

57. $f(x) = x^3 - 9x^2 + 15x$

58. $f(x) = x^3 - 9x^2 + 24x$

In Problems 59–68, find the value(s) of x where the graph of f has a horizontal tangent line. Round answers to two decimal places.

59. $f(x) = 3x - 10\sqrt{x} + 5$
60. $f(x) = 10\sqrt{x} - 2x^2$
61. $f(x) = x^3 - x^2 - 6x + 2$
62. $f(x) = x^3 - 6x^2 + 5x + 4$
63. $f(x) = x^4 - x^3 - 5x^2 + 5x + 3$
64. $f(x) = x^4 + 2x^3 - 6x^2 - 8x + 5$
65. $f(x) = 0.5x^4 - 4x^3 + 9x^2 - 8$
66. $f(x) = 1.5x^4 + 8x^3 + 12x^2 - 6$
67. $f(x) = x^5 - 8x^3 + x^2 + 15x$
68. $f(x) = x^5 - 7x^3 - x^2 + 10x$

69. Let $f(x) = ax^2 + bx + c$, $a \neq 0$. Recall that the graph of $y = f(x)$ is a parabola. Use the derivative $f'(x)$ to derive a formula for the x coordinate of the vertex of this parabola.

70. Now that you know how to find derivatives, explain why it is no longer necessary for you to memorize the formula for the x coordinate of the vertex of a parabola.

71. Give an example of a cubic polynomial function that has:
(A) No horizontal tangents
(B) One horizontal tangent
(C) Two horizontal tangents

72. Can a cubic polynomial function have more than two horizontal tangents? Explain.

C *In Problems 73–76, find each derivative.*

73. $f'(x)$ for $f(x) = \dfrac{10x + 20}{x}$

74. $\dfrac{dy}{dx}$ for $y = \dfrac{x^2 + 25}{x^2}$

75. $\dfrac{d}{dx}\dfrac{x^4 - 3x^3 + 5}{x^2}$

76. y' for $y = \dfrac{2x^5 - 4x^3 + 2x}{x^3}$

In Problems 77 and 78, use the definition of derivative and the two-step process to verify each statement.

77. $\dfrac{d}{dx} x^3 = 3x^2$ **78.** $\dfrac{d}{dx} x^4 = 4x^3$

79. The domain of the power function $f(x) = x^{1/3}$ is the set of all real numbers. Find the domain of the derivative $f'(x)$.

Discuss the nature of the graph of $y = f(x)$ for any x values excluded from the domain of $f'(x)$.

80. The domain of the power function $f(x) = x^{2/3}$ is the set of all real numbers. Find the domain of the derivative $f'(x)$. Discuss the nature of the graph of $y = f(x)$ for any x values excluded from the domain of $f'(x)$.

▶ APPLICATIONS

Business & Economics

81. *Marginal cost.* The total cost (in dollars) of producing x tennis rackets per day is

$$C(x) = 800 + 60x - 0.25x^2 \qquad 0 \leqslant x \leqslant 120$$

(A) Find the marginal cost at a production level of x rackets.

(B) Find the marginal cost at a production level of 60 rackets, and interpret the result verbally and graphically.

(C) Find the actual cost of producing the 61st racket, and compare this cost with the result found in part (B).

(D) Find $C'(80)$, and interpret the result verbally and graphically.

82. *Marginal cost.* The total cost (in dollars) of producing x portable radios per day is

$$C(x) = 1{,}000 + 100x - 0.5x^2 \qquad 0 \leqslant x \leqslant 100$$

(A) Find the marginal cost at a production level of x radios.

(B) Find the marginal cost at a production level of 80 radios, and interpret the result verbally and graphically.

(C) Find the actual cost of producing the 81st radio, and compare this cost with the result found in part (B).

(D) Find $C'(50)$, and interpret the result verbally and graphically.

83. *Marginal cost.* The total cost (in dollars) of producing x microwave ovens per week is shown in the figure. Which is greater, the approximate cost of producing the 101st oven or the approximate cost of producing the 401st oven? Does this graph represent a manufacturing process

that is becoming more efficient or less efficient as production levels increase? Explain.

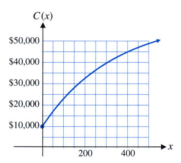

Figure for 83

84. *Marginal cost.* The total cost (in dollars) of producing x electric stoves per week is shown in the figure. Which is greater, the approximate cost of producing the 101st stove or the approximate cost of producing the 401st stove? Does this graph represent a manufacturing process that is becoming more efficient or less efficient as production levels increase? Explain.

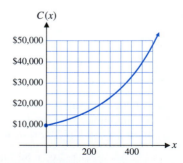

Figure for 84

85. *Advertising.* Using past records, it is estimated that a marine manufacturer will sell $N(x)$ power boats after spending x thousand on advertising, as given by

$$N(x) = 1,000 - \frac{3,780}{x} \qquad 5 \leqslant x \leqslant 30$$

See the figure.

Figure for 85

(A) Find $N'(x)$.

(B) Find $N'(10)$ and $N'(20)$. Write a brief verbal interpretation of these results.

86. *Price–demand equation.* Suppose that in a given gourmet food store, people are willing to buy x pounds of chocolate candy per day at p per quarter pound, as given by the price–demand equation

$$x = 10 + \frac{180}{p} \qquad 2 \leqslant p \leqslant 10$$

Figure for 86

This function is graphed in the figure. Find the demand and the instantaneous rate of change of demand with re-

spect to price when the price is $5. Write a brief verbal interpretation of these results.

Life Sciences

87. *Medicine.* A person x inches tall has a pulse rate of y beats per minute, as given approximately by

$$y = 590x^{-1/2} \qquad 30 \leqslant x \leqslant 75$$

What is the instantaneous rate of change of pulse rate at the:
(A) 36 inch level?   (B) 64 inch level?

88. *Ecology.* A coal-burning electrical generating plant emits sulfur dioxide into the surrounding air. The concentration $C(x)$, in parts per million, is given approximately by

$$C(x) = \frac{0.1}{x^2}$$

where x is the distance from the plant in miles. Find the instantaneous rate of change of concentration at:
(A) $x = 1$ mile   (B) $x = 2$ miles

Social Sciences

89. *Learning.* Suppose a person learns y items in x hours, as given by

$$y = 50\sqrt{x} \qquad 0 \leqslant x \leqslant 9$$

(see the figure). Find the rate of learning at the end of:
(A) 1 hour   (B) 9 hours

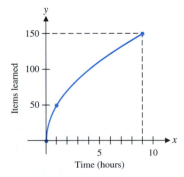

Figure for 89

90. *Learning.* If a person learns y items in x hours, as given by

$$y = 21\sqrt[3]{x^2} \qquad 0 \leqslant x \leqslant 8$$

find the rate of learning at the end of:
(A) 1 hour   (B) 8 hours

S E C T I O N .   8-5   ■   # Derivatives of Products and Quotients

■ DERIVATIVES OF PRODUCTS
■ DERIVATIVES OF QUOTIENTS

The derivative rules discussed in the preceding section added substantially to our ability to compute and apply derivatives to many practical problems. In this and the next section, we will add a few more rules that will increase this ability even further.

■ DERIVATIVES OF PRODUCTS

In Section 8-4 we found that the derivative of a sum is the sum of the derivatives. Is the derivative of a product the product of the derivatives?

e x p l o r e – d i s c u s s   1

Let $F(x) = x^2$, $S(x) = x^3$, and $f(x) = F(x)S(x) = x^5$. Which of the following is $f'(x)$?

(A) $F'(x)S'(x)$ (B) $F(x)S'(x)$
(C) $F'(x)S(x)$ (D) $F(x)S'(x) + F'(x)S(x)$

Comparing the various expressions computed in Explore–Discuss 1, we see that the derivative of a product is not the product of the derivatives, but appears to involve a slightly more complicated form.

Using the definition of the derivative and the two-step process, it can be shown that:

The derivative of the product of two functions is the first function times the derivative of the second function plus the second function times the derivative of the first function.

That is:

Product Rule

If

$$y = f(x) = F(x)S(x)$$

and if $F'(x)$ and $S'(x)$ exist, then

$$f'(x) = F(x)S'(x) + S(x)F'(x)$$

Also,

$$y' = FS' + SF' \qquad \frac{dy}{dx} = F\frac{dS}{dx} + S\frac{dF}{dx}$$

EXAMPLE 1 ➤ **Differentiating a Product** Use two different methods to find $f'(x)$ for $f(x) = 2x^2(3x^4 - 2)$.

Solution **Method 1.** Use the product rule:

$$
\begin{aligned}
f'(x) &= 2x^2(3x^4 - 2)' + (3x^4 - 2)(2x^2)' && \textit{First times derivative of second plus} \\
&= 2x^2(12x^3) + (3x^4 - 2)(4x) && \textit{second times derivative of first} \\
&= 24x^5 + 12x^5 - 8x \\
&= 36x^5 - 8x
\end{aligned}
$$

Method 2. Multiply first; then take derivatives:

$$
\begin{aligned}
f(x) &= 2x^2(3x^4 - 2) = 6x^6 - 4x^2 \\
f'(x) &= 36x^5 - 8x
\end{aligned}
$$
◄

MATCHED PROBLEM 1 ➤ Use two different methods to find $f'(x)$ for $f(x) = 3x^3(2x^2 - 3x + 1)$. ◄

At this point, all the products we will encounter can be differentiated by either of the methods illustrated in Example 1. In the next and later sections, we will see that there are situations where the product rule must be used. Unless instructed otherwise, you should use the product rule to differentiate all products in this section to gain experience with the use of this important differentiation rule.

EXAMPLE 2 ➤ **Tangent Lines** Let $f(x) = (2x - 9)(x^2 + 6)$.

(A) Find the equation of the line tangent to the graph of $f(x)$ at $x = 3$.
(B) Find the value(s) of x where the tangent line is horizontal.

Solution (A) First, find $f'(x)$:

$$
\begin{aligned}
f'(x) &= (2x - 9)(x^2 + 6)' + (x^2 + 6)(2x - 9)' \\
&= (2x - 9)(2x) + (x^2 + 6)(2)
\end{aligned}
$$

Now, find the equation of the tangent line at $x = 3$:

$$
\begin{aligned}
y - y_1 &= m(x - x_1) && y_1 = f(x_1) = f(3) = -45 \\
y - (-45) &= 12(x - 3) && m = f'(x_1) = f'(3) = 12 \\
y &= 12x - 81 && \textit{Tangent line at } x = 3
\end{aligned}
$$

(B) The tangent line is horizontal at any value of x such that $f'(x) = 0$. Since $f'(x)$ is a quadratic polynomial, let's see if we can solve this equation by factoring:

$$
\begin{aligned}
f'(x) = (2x - 9)2x + (x^2 + 6)2 &= 0 \\
6x^2 - 18x + 12 &= 0 \\
x^2 - 3x + 2 &= 0 \\
(x - 1)(x - 2) &= 0 \\
x &= 1, 2
\end{aligned}
$$

The tangent line is horizontal at $x = 1$ and at $x = 2$. ◄

MATCHED PROBLEM 2 ➤ Repeat Example 2 for $f(x) = (2x + 9)(x^2 - 12)$. ◄

As Example 2 illustrates, the way we write $f'(x)$ depends on what we want to do with it. If we are interested only in evaluating $f'(x)$ at specified values of x, the form in part (A) is sufficient. However, if we want to solve $f'(x) = 0$ by factoring, we must multiply and collect like terms, as we did in part (B).

■ DERIVATIVES OF QUOTIENTS

As is the case with a product, the derivative of a quotient of two functions is not the quotient of the derivatives of the two functions.

explore – discuss 2

Let $T(x) = x^5$, $B(x) = x^2$, and

$$f(x) = \frac{T(x)}{B(x)} = \frac{x^5}{x^2} = x^3$$

Which of the following is $f'(x)$?

(A) $\dfrac{T'(x)}{B'(x)}$ (B) $\dfrac{T'(x)B(x)}{[B(x)]^2}$ (C) $\dfrac{T(x)B'(x)}{[B(x)]^2}$

(D) $\dfrac{T'(x)B(x)}{[B(x)]^2} - \dfrac{T(x)B'(x)}{[B(x)]^2} = \dfrac{T'(x)B(x) - T(x)B'(x)}{[B(x)]^2}$

The expressions in Explore–Discuss 2 suggest that the derivative of a quotient leads to a more complicated quotient than you might expect.

In general, if $T(x)$ and $B(x)$ are any two differentiable functions and

$$f(x) = \frac{T(x)}{B(x)}$$

then it can be shown that

$$f'(x) = \frac{B(x)T'(x) - T(x)B'(x)}{[B(x)]^2}$$

Thus:

The derivative of the quotient of two functions is the bottom function times the derivative of the top function minus the top function times the derivative of the bottom function, all over the bottom function squared.

> **Quotient Rule**
>
> If
>
> $$y = f(x) = \frac{T(x)}{B(x)}$$
>
> and if $T'(x)$ and $B'(x)$ exist, then
>
> $$f'(x) = \frac{B(x)T'(x) - T(x)B'(x)}{[B(x)]^2}$$
>
> Also,
>
> $$y' = \frac{BT' - TB'}{B^2} \qquad \frac{dy}{dx} = \frac{B\dfrac{dT}{dx} - T\dfrac{dB}{dx}}{B^2}$$

EXAMPLE 3 ➤ Differentiating Quotients

(A) If $f(x) = \dfrac{x^2}{2x - 1}$, find $f'(x)$. (B) If $y = \dfrac{x^2 - x}{x^3 + 1}$, find y'.

(C) Find $\dfrac{d}{dx} \dfrac{x^2 - 3}{x^2}$ by using the quotient rule and also by splitting the fraction into two fractions.

Solution (A) $f'(x) = \dfrac{(2x - 1)(x^2)' - x^2(2x - 1)'}{(2x - 1)^2}$ *The bottom times the derivative of the top minus the top times the derivative of the bottom, all over the square of the bottom*

$$= \frac{(2x - 1)(2x) - x^2(2)}{(2x - 1)^2}$$

$$= \frac{4x^2 - 2x - 2x^2}{(2x - 1)^2}$$

$$= \frac{2x^2 - 2x}{(2x - 1)^2}$$

(B) $y' = \dfrac{(x^3 + 1)(x^2 - x)' - (x^2 - x)(x^3 + 1)'}{(x^3 + 1)^2}$

$$= \frac{(x^3 + 1)(2x - 1) - (x^2 - x)(3x^2)}{(x^3 + 1)^2}$$

$$= \frac{2x^4 - x^3 + 2x - 1 - 3x^4 + 3x^3}{(x^3 + 1)^2} = \frac{-x^4 + 2x^3 + 2x - 1}{(x^3 + 1)^2}$$

(C) Method 1.   Use the quotient rule:

$$\frac{d}{dx}\frac{x^2 - 3}{x^2} = \frac{x^2 \dfrac{d}{dx}(x^2 - 3) - (x^2 - 3)\dfrac{d}{dx}x^2}{(x^2)^2}$$

$$= \frac{x^2(2x) - (x^2 - 3)2x}{x^4}$$

$$= \frac{2x^3 - 2x^3 + 6x}{x^4} = \frac{6x}{x^4} = \frac{6}{x^3}$$

Method 2.   Split into two fractions:

$$\frac{x^2 - 3}{x^2} = \frac{x^2}{x^2} - \frac{3}{x^2} = 1 - 3x^{-2}$$

$$\frac{d}{dx}(1 - 3x^{-2}) = 0 - 3(-2)x^{-3} = \frac{6}{x^3}$$

Comparing methods 1 and 2, we see that it often pays to change an expression algebraically before blindly using a differentiation formula.   ◄

MATCHED PROBLEM 3 ►   Find:

(A) $f'(x)$ for $f(x) = \dfrac{2x}{x^2 + 3}$ (B) y' for $y = \dfrac{x^3 - 3x}{x^2 - 4}$

(C) $\dfrac{d}{dx}\dfrac{2 + x^3}{x^3}$ two ways   ◄

explore – discuss 3

Explain why ≠ is used below, and then find the correct derivative.

$$\frac{d}{dx}\frac{x^3}{x^2 + 3x + 4} \neq \frac{3x^2}{2x + 3}$$

► EXAMPLE 4 ►   **Sales Analysis**   The total sales S (in thousands of games) for a home video game t months after the game is introduced are given by

$$S(t) = \frac{125t^2}{t^2 + 100}$$

(A) Find $S'(t)$.

(B) Find $S(10)$ and $S'(10)$. Write a brief verbal interpretation of these results and illustrate graphically.

(C) Use the results from part (B) to estimate the total sales after 11 months.

Solution (A) $S'(t) = \dfrac{(t^2 + 100)(125t^2)' - 125t^2(t^2 + 100)'}{(t^2 + 100)^2}$

$= \dfrac{(t^2 + 100)(250t) - 125t^2(2t)}{(t^2 + 100)^2}$

$= \dfrac{250t^3 + 25,000t - 250t^3}{(t^2 + 100)^2}$

$= \dfrac{25,000t}{(t^2 + 100)^2}$

(B) $S(10) = \dfrac{125(10)^2}{10^2 + 100} = 62.5$ and $S'(10) = \dfrac{25,000(10)}{(10^2 + 100)^2} = 6.25$

The total sales after 10 months are 62,500 games, and sales are increasing at the rate of 6,250 games per month. The graph of $S(t)$ and the tangent line at $t = 10$ are shown in Figure 1A, and the corresponding illustration with annotations is shown in Figure 1B.

(A)

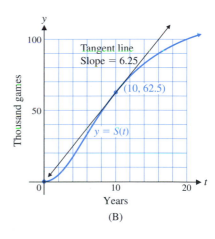

(B)

FIGURE 1

(C) The total sales will increase by approximately 6,250 games during the next month. Thus, the estimated total sales after 11 months are $62,500 + 6,250 = 68,750$ games. ◀

MATCHED PROBLEM 4 ▶ Refer to Example 4. Suppose the total sales S (in thousands of games) t months after the game is introduced are given by

$$S(t) = \frac{150t}{t + 3}$$

(A) Find $S'(t)$.

(B) Find $S(12)$ and $S'(12)$. Write a brief verbal interpretation of these results and illustrate graphically.

(C) Use the results from part (B) to estimate the total sales after 13 months. ◀

```
10→X
            10
Y₁
            62.5
nDeriv(Y₁,X,10)
            6.25
```

FIGURE 2

$$y_1 = \frac{125x^2}{x^2 + 100}$$

Remark: Derivatives and Technology

If we had omitted part (A) in Example 4, then we could have solved parts (B) and (C) on a graphing utility without first finding an algebraic expression for $S'(t)$. Figure 2 shows the necessary calculations. In later chapters, we will encounter problems where it will be necessary to find an algebraic expression for the derivative. Thus, to give you additional practice with the rules of differentiation, you will be instructed to find an algebraic expression for the derivative in most, but not all, problems in this chapter. When this instruction is omitted, you can use a graphing utility to solve the problem numerically.

Answers to Matched Problems

1. $30x^4 - 36x^3 + 9x^2$ 2. (A) $y = 84x - 297$ (B) $x = -4, x = 1$

3. (A) $\dfrac{(x^2 + 3)2 - (2x)(2x)}{(x^2 + 3)^2} = \dfrac{6 - 2x^2}{(x^2 + 3)^2}$

   (B) $\dfrac{(x^2 - 4)(3x^2 - 3) - (x^3 - 3x)(2x)}{(x^2 - 4)^2} = \dfrac{x^4 - 9x^2 + 12}{(x^2 - 4)^2}$ (C) $-\dfrac{6}{x^4}$

4. (A) $S'(t) = \dfrac{450}{(t + 3)^2}$

   (B) $S(12) = 120$; $S'(12) = 2$. After 12 months, the total sales are 120,000 games, and sales are increasing at the rate of 2,000 games per month.

   (C) 122,000 games

EXERCISE 8-5

The answers to most of the problems in this exercise set contain both an unsimplified form and a simplified form of a derivative. When checking your work, first check that you applied the rules correctly, and then check that you performed the algebraic simplification correctly. Unless instructed otherwise, when differentiating a product, use the product rule rather than performing the multiplication first.

A *In Problems 1–16, find $f'(x)$ and simplify.*

1. $f(x) = 2x^3(x^2 - 2)$ 2. $f(x) = 5x^2(x^3 + 2)$
3. $f(x) = (x - 3)(2x - 1)$
4. $f(x) = (3x + 2)(4x - 5)$

5. $f(x) = \dfrac{x}{x - 3}$ 6. $f(x) = \dfrac{3x}{2x + 1}$

7. $f(x) = \dfrac{2x + 3}{x - 2}$ **8.** $f(x) = \dfrac{3x - 4}{2x + 3}$

9. $f(x) = (x^2 + 1)(2x - 3)$
10. $f(x) = (3x + 5)(x^2 - 3)$

11. $f(x) = \dfrac{x^2 + 1}{2x - 3}$ **12.** $f(x) = \dfrac{3x + 5}{x^2 - 3}$

13. $f(x) = (x^2 + 2)(x^2 - 3)$
14. $f(x) = (x^2 - 4)(x^2 + 5)$

15. $f(x) = \dfrac{x^2 + 2}{x^2 - 3}$ **16.** $f(x) = \dfrac{x^2 - 4}{x^2 + 5}$

Problems 17–22 refer to functions f and g that satisfy $f(1) = 4$, $f'(1) = -2$, $g(1) = 2$, and $g'(1) = 3$. In each problem, find $h'(1)$ for the indicated function h.

17. $h(x) = f(x)g(x)$ **18.** $h(x) = \dfrac{f(x)}{g(x)}$

19. $h(x) = \dfrac{g(x)}{f(x)}$ **20.** $h(x) = f(x)f(x)$

21. $h(x) = \dfrac{1}{f(x)}$ **22.** $h(x) = \dfrac{1}{g(x)}$

B *In Problems 23–30, find the indicated derivatives and simplify.*

23. $f'(x)$ for $f(x) = (2x + 1)(x^2 - 3x)$
24. y' for $y = (x^3 + 2x^2)(3x - 1)$
25. $\dfrac{dy}{dx}$ for $y = (2x - x^2)(5x + 2)$
26. $\dfrac{d}{dx}[(3 - x^3)(x^2 - x)]$
27. y' for $y = \dfrac{5x - 3}{x^2 + 2x}$
28. $f'(x)$ for $f(x) = \dfrac{3x^2}{2x - 1}$
29. $\dfrac{d}{dx} \dfrac{x^2 - 3x + 1}{x^2 - 1}$
30. $\dfrac{dy}{dx}$ for $y = \dfrac{x^4 - x^3}{3x - 1}$

In Problems 31–34, find $f'(x)$ and find the equation of the line tangent to the graph of f at $x = 2$.

31. $f(x) = (1 + 3x)(5 - 2x)$
32. $f(x) = (7 - 3x)(1 + 2x)$

33. $f(x) = \dfrac{x - 8}{3x - 4}$

34. $f(x) = \dfrac{2x - 5}{2x - 3}$

In Problems 35–42, find the value(s) of x (to two decimal places) where $f'(x) = 0$.

35. $f(x) = (2x - 15)(x^2 + 18)$
36. $f(x) = (2x - 3)(x^2 - 6)$
37. $f(x) = (x^2 - 5)(x^2 - x)$
38. $f(x) = (x^2 - 3)(x^2 - 2x)$
39. $f(x) = \dfrac{x}{x^2 + 1}$
40. $f(x) = \dfrac{x}{x^2 + 9}$
41. $f(x) = \dfrac{x^3 + 12x + 1}{x^2 + 1}$
42. $f(x) = \dfrac{x^3 + 14x - 1}{x^2 + 1}$

In Problems 43–46, find $f'(x)$ two ways; by using the product or quotient rule and by simplifying first.

43. $f(x) = x^3(x^4 - 1)$ **44.** $f(x) = x^4(x^3 - 1)$

45. $f(x) = \dfrac{x^3 + 9}{x^3}$ **46.** $f(x) = \dfrac{x^4 + 4}{x^4}$

C *In Problems 47–58, find each derivative and simplify.*

47. $f'(x)$ for $f(x) = (2x^4 - 3x^3 + x)(x^2 - x + 5)$
48. $\dfrac{dy}{dx}$ for $y = (x^2 - 3x + 1)(x^3 + 2x^2 - x)$
49. $\dfrac{d}{dx} \dfrac{3x^2 - 2x + 3}{4x^2 + 5x - 1}$
50. y' for $y = \dfrac{x^3 - 3x + 4}{2x^2 + 3x - 2}$
51. $\dfrac{dy}{dx}$ for $y = 9x^{1/3}(x^3 + 5)$
52. $\dfrac{d}{dx}[(4x^{1/2} - 1)(3x^{1/3} + 2)]$
53. $f'(x)$ for $f(x) = \dfrac{6\sqrt[3]{x}}{x^2 - 3}$

54. y' for $y = \dfrac{2\sqrt{x}}{x^2 - 3x + 1}$

55. $\dfrac{d}{dx} \dfrac{x^3 - 2x^2}{\sqrt[3]{x^2}}$

56. $\dfrac{dy}{dx}$ for $y = \dfrac{x^2 - 3x + 1}{\sqrt[4]{x}}$

57. $f'(x)$ for $f(x) = \dfrac{(2x^2 - 1)(x^2 + 3)}{x^2 + 1}$

58. y' for $y = \dfrac{2x - 1}{(x^3 + 2)(x^2 - 3)}$

Problems 59–62 refer to a function of the form $f(x) = [u(x)]^n$, where $u(x)$ is a differentiable function.

59. Use the product rule to show that if $n = 2$, then $f'(x) = 2u(x)u'(x)$.

60. Use the product rule and the result from Problem 59 to show that if $n = 3$, then $f'(x) = 3[u(x)]^2u'(x)$.

61. Based on the results in Problems 59 and 60, write a formula for $f'(x)$ for any n.

62. Use the quotient rule to find $f'(x)$ if $n = -1$. Does this agree with your formula in Problem 61?

▶ APPLICATIONS

Business & Economics

63. *Sales analysis.* The total sales S (in thousands of CD's) for a compact disk are given by

$$S(t) = \frac{90t^2}{t^2 + 50}$$

where t is the number of months since the release of the CD.
(A) Find $S'(t)$.

(B) Find $S(10)$ and $S'(10)$. Write a brief verbal interpretation of these results and illustrate graphically.
(C) Use the results from part (B) to estimate the total sales after 11 months.

64. *Sales analysis.* A communications company has installed a cable television system in a city. The total number N (in thousands) of subscribers t months after the installation of the system is given by

$$N(t) = \frac{200t}{t + 5}$$

(A) Find $N'(t)$.

(B) Find $N(15)$ and $N'(15)$. Write a brief verbal interpretation of these results and illustrate graphically.
(C) Use the results from part (B) to estimate the total number of subscribers after 16 months.

65. *Price–demand equation.* According to classical economic theory, the demand x for a quantity in a free market decreases as the price p increases (see the figure). Suppose

that the number x of CD players people are willing to buy per week from a retail chain at a price of $\$p$ is given by

$$x = \frac{4{,}000}{0.1p + 1} \qquad 10 \leq p \leq 70$$

Figure for 65 and 66

(A) Find the demand and the instantaneous rate of change of demand with respect to price when the price is $\$40$. Write a brief verbal interpretation of these results.
(B) Use the results from part (A) to estimate the demand if the price is increased to $\$41$.

66. *Price–supply equation.* Also according to classical economic theory, the supply x for a quantity in a free market increases as the price p increases (see the figure). Suppose that the number x of CD players a retail chain is willing to sell per week at a price of $\$p$ is given by

$$x = \frac{100p}{0.1p + 1} \qquad 10 \leqslant p \leqslant 70$$

(A) Find the supply and the instantaneous rate of change of supply with respect to price when the price is $40. Write a brief verbal interpretation of these results.

(B) Use the results from part (A) to estimate the supply if the price is increased to $41.

Life Sciences

67. *Medicine.* A drug is injected into the bloodstream of a patient through her right arm. The concentration of the drug (in milligrams per cubic centimeter) in the bloodstream of the left arm t hours after the injection is given by

$$C(t) = \frac{0.14t}{t^2 + 1}$$

(A) Find $C'(t)$.

(B) Find $C'(0.5)$ and $C'(3)$, and interpret the results verbally and graphically.

68. *Drug sensitivity.* One hour after x milligrams of a particular drug are given to a person, the change in body temperature $T(x)$, in degrees Fahrenheit, is given approximately by

$$T(x) = x^2 \left(1 - \frac{x}{9} \right) \qquad 0 \leqslant x \leqslant 7$$

The rate at which T changes with respect to the size of the dosage x, $T'(x)$, is called the *sensitivity* of the body to the dosage.

(A) Find $T'(x)$, using the product rule.

(B) Find $T'(3)$ and $T'(6)$, and illustrate graphically.

Social Sciences

69. *Learning.* In the early days of quantitative learning theory (around 1917), L. L. Thurstone found that a given person successfully accomplished $N(x)$ acts after x practice acts, as given by

$$N(x) = \frac{100x + 200}{x + 32}$$

(A) Find the instantaneous rate of change of learning, $N'(x)$, with respect to the number of practice acts x.

(B) Find $N'(4)$ and $N'(68)$, and illustrate graphically.

SECTION 8-6 ■ # Chain Rule: Power Form

■ CHAIN RULE: POWER RULE
■ COMBINING RULES OF DIFFERENTIATION

In this section we develop a rule for differentiating powers of functions—a special case of the very important *chain rule,* which we will return to in Chapter 10. Also, for the first time, we will encounter some product forms that cannot be simplified by multiplication and must be differentiated by the product rule.

■ CHAIN RULE: POWER RULE

We have already made extensive use of the power rule,

$$\frac{d}{dx} x^n = nx^{n-1} \tag{1}$$

Now we want to generalize this rule so that we can differentiate functions of the form $[u(x)]^n$, where $u(x)$ is a differentiable function. Is rule (1) still valid if we replace x with a function $u(x)$?

Let $u(x) = 2x^2$ and $f(x) = [u(x)]^3 = 8x^6$. Which of the following is $f'(x)$?

(A) $3[u(x)]^2$ (B) $3[u'(x)]^2$ (C) $3[u(x)]^2u'(x)$

The calculations in Explore–Discuss 1 show that we cannot generalize the power rule simply by replacing x with $u(x)$ in (1).

How can we find a formula for the derivative of $[u(x)]^n$, where $u(x)$ is an arbitrary differentiable function? Let's begin by considering the derivatives of $[u(x)]^2$ and $[u(x)]^3$ to see if a general pattern emerges. Since $[u(x)]^2 = u(x)u(x)$, we use the product rule to write

$$\frac{d}{dx}[u(x)]^2 = \frac{d}{dx}[u(x)u(x)]$$
$$= u(x)u'(x) + u(x)u'(x)$$
$$= 2u(x)u'(x) \tag{2}$$

Since $[u(x)]^3 = [u(x)]^2u(x)$, we now use the product rule and the result in (2) to write

$$\frac{d}{dx}[u(x)]^3 = \frac{d}{dx}\{[u(x)]^2u(x)\}$$
$$= [u(x)]^2\frac{d}{dx}u(x) + u(x)\frac{d}{dx}[u(x)]^2$$
$$= [u(x)]^2u'(x) + u(x)[2u(x)u'(x)]$$
$$= 3[u(x)]^2u'(x)$$

Use (2) to substitute for $\frac{d}{dx}[u(x)]^2$.

Continuing in this fashion, it can be shown that

$$\frac{d}{dx}[u(x)]^n = n[u(x)]^{n-1}u'(x) \qquad n \text{ a positive integer} \tag{3}$$

Using more advanced techniques, formula (3) can be established for all real numbers n. Thus, we have the **general power rule:**

General Power Rule

If $u(x)$ is a differentiable function, n is any real number, and
$$y = f(x) = [u(x)]^n$$
then
$$f'(x) = n[u(x)]^{n-1}u'(x)$$
This rule is often written more compactly as
$$y' = nu^{n-1}u' \qquad \text{or} \qquad \frac{d}{dx}u^n = nu^{n-1}\frac{du}{dx} \qquad \text{where } u = u(x)$$

The general power rule is a special case of a very important and useful differentiation rule called the **chain rule.** In essence, the chain rule will enable us to differentiate a composition form $f[g(x)]$ if we know how to differentiate $f(x)$ and $g(x)$. We defer a complete discussion of the chain rule until Chapter 10.

EXAMPLE 1 ➤ **Differentiating Power Forms** Find $f'(x)$:

(A) $f(x) = (3x + 1)^4$ (B) $f(x) = (x^3 + 4)^7$

(C) $f(x) = \dfrac{1}{(x^2 + x + 4)^3}$ (D) $f(x) = \sqrt{3 - x}$

Solution (A) $f(x) = (3x + 1)^4$ Let $u = 3x + 1$, $n = 4$.

$$f'(x) = 4(3x + 1)^3(3x + 1)'$$ $nu^{n-1}\dfrac{du}{dx}$

$$= 4(3x + 1)^3 3$$ $\dfrac{du}{dx} = 3$
$$= 12(3x + 1)^3$$

(B) $f(x) = (x^3 + 4)^7$ Let $u = (x^3 + 4)$, $n = 7$.

$$f'(x) = 7(x^3 + 4)^6(x^3 + 4)'$$ $nu^{n-1}\dfrac{du}{dx}$

$$= 7(x^3 + 4)^6 3x^2$$ $\dfrac{du}{dx} = 3x^2$
$$= 21x^2(x^3 + 4)^6$$

(C) $f(x) = \dfrac{1}{(x^2 + x + 4)^3} = (x^2 + x + 4)^{-3}$ Let $u = x^2 + x + 4$, $n = -3$.

$$f'(x) = -3(x^2 + x + 4)^{-4}(x^2 + x + 4)'$$ $nu^{n-1}\dfrac{du}{dx}$

$$= -3(x^2 + x + 4)^{-4}(2x + 1)$$ $\dfrac{du}{dx} = 2x + 1$

$$= \dfrac{-3(2x + 1)}{(x^2 + x + 4)^4}$$

(D) $f(x) = \sqrt{3 - x} = (3 - x)^{1/2}$ Let $u = 3 - x$, $n = \frac{1}{2}$.

$$f'(x) = \dfrac{1}{2}(3 - x)^{-1/2}(3 - x)'$$ $nu^{n-1}\dfrac{du}{dx}$

$$= \dfrac{1}{2}(3 - x)^{-1/2}(-1)$$ $\dfrac{du}{dx} = -1$

$$= -\dfrac{1}{2(3 - x)^{1/2}} \quad \text{or} \quad -\dfrac{1}{2\sqrt{3 - x}}$$ ◀

MATCHED PROBLEM 1 ➤ Find $f'(x)$: (A) $f(x) = (5x + 2)^3$ (B) $f(x) = (x^4 - 5)^5$

(C) $f(x) = \dfrac{1}{(x^2 + 4)^2}$ (D) $f(x) = \sqrt{4 - x}$ ◀

Notice that we used two steps to differentiate each function in Example 1. First, we applied the general power rule; then we found du/dx. As you gain experience with the general power rule, you may want to combine these two steps. If you do this, be certain to multiply by du/dx. For example,

$$\frac{d}{dx}(x^5 + 1)^4 = 4(x^5 + 1)^3 5x^4 \quad \textit{Correct}$$

$$\frac{d}{dx}(x^5 + 1)^4 \neq 4(x^5 + 1)^3 \quad \textit{du/dx = 5x^{4} is missing}$$

If we let $u(x) = x$, then $du/dx = 1$, and the general power rule reduces to the (ordinary) power rule discussed in Section 8-4. Compare the following:

$$\frac{d}{dx}x^n = nx^{n-1} \quad \textit{Yes—power rule}$$

$$\frac{d}{dx}u^n = nu^{n-1}\frac{du}{dx} \quad \textit{Yes—general power rule}$$

$$\frac{d}{dx}u^n \neq nu^{n-1} \quad \textit{Unless u(x) = x + k so that du/dx = 1}$$

■ COMBINING RULES OF DIFFERENTIATION

The following examples illustrate the use of the general power rule in combination with other rules of differentiation.

EXAMPLE 2 ➤ **Tangent Lines** Find $f'(x)$ and the equation of the line tangent to the graph of f at $x = 2$ for $f(x) = x^2\sqrt{2x + 12}$. Check with a graphing utility.

Solution
$$f(x) = x^2\sqrt{2x + 12}$$
$$= x^2(2x + 12)^{1/2} \quad \textit{Apply the product rule.}$$

$$f'(x) = x^2\frac{d}{dx}(2x + 12)^{1/2} + (2x + 12)^{1/2}\frac{d}{dx}x^2 \quad \textit{Use the general power rule to differentiate (2x + 12)^{1/2} and the ordinary power rule to differentiate x^{2}.}$$

$$= x^2\left[\frac{1}{2}(2x + 12)^{-1/2}\right](2) + (2x + 12)^{1/2}(2x)$$

$$= \frac{x^2}{\sqrt{2x + 12}} + 2x\sqrt{2x + 12}$$

$$f'(2) = \frac{4}{\sqrt{16}} + 4\sqrt{16} = 1 + 16 = 17$$

$$f(2) = 4\sqrt{16} = 16$$

$$(x_1, y_1) = (2, f(2)) = (2, 16) \quad \textit{Point}$$
$$m = f'(2) = 17 \quad \textit{Slope}$$
$$y - 16 = 17(x - 2) \quad \textit{y - y\textsubscript{1} = m(x - x\textsubscript{1})}$$
$$y = 17x - 18 \quad \textit{Tangent line}$$

Figure 1A checks the values of $f(2)$ and $f'(2)$, and Figure 1B shows the graphs of $y = f(x)$ and the tangent line at $x = 2$. Note that we graphed the equation $y = 17x - 18$ in order to check our work. We did not use the built-in routine for graphing tangent lines. This routine would produce the same graph, but it would not tell us if we have the correct equation for the tangent line.

(A)

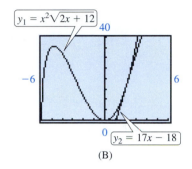

(B)

FIGURE 1

MATCHED PROBLEM 2 ▶ Find $f'(x)$ and the equation of the line tangent to the graph of f at $x = 3$ for $f(x) = x\sqrt{15 - 2x}$. Check with a graphing utility. ◀

EXAMPLE 3 ▶ **Alternate Methods of Differentiation** Let $f(x) = \dfrac{2x^2 + x}{(x^2 + 1)^2}$.

(A) Use the quotient rule to find $f'(x)$.
(B) Write f as a product, and use the product rule to find $f'(x)$.
(C) Find the value(s) of x where the tangent line to the graph of f is horizontal. Round answers to two decimal places.

Solution (A) Use the quotient rule:

$$f'(x) = \frac{(x^2 + 1)^2 \dfrac{d}{dx}(2x^2 + x) - (2x^2 + x)\dfrac{d}{dx}(x^2 + 1)^2}{[(x^2 + 1)^2]^2}$$

$$= \frac{(x^2 + 1)^2(4x + 1) - (2x^2 + x)2(x^2 + 1)2x}{(x^2 + 1)^4} \qquad (4)$$

$$= \frac{(x^2 + 1)[(x^2 + 1)(4x + 1) - 4x(2x^2 + x)]}{(x^2 + 1)^4}$$

$$= \frac{4x^3 + x^2 + 4x + 1 - 8x^3 - 4x^2}{(x^2 + 1)^3}$$

$$= \frac{-4x^3 - 3x^2 + 4x + 1}{(x^2 + 1)^3} \qquad (5)$$

(B) Rewrite f and use the product rule:

$$f(x) = \frac{2x^2 + x}{(x^2 + 1)^2} = (2x^2 + x)(x^2 + 1)^{-2}$$

$$f'(x) = (2x^2 + x)\frac{d}{dx}(x^2 + 1)^{-2} + (x^2 + 1)^{-2}\frac{d}{dx}(2x^2 + x)$$

$$= (2x^2 + x)(-2)(x^2 + 1)^{-3}2x + (x^2 + 1)^{-2}(4x + 1) \qquad (6)$$

$$= \frac{-4x(2x^2 + x)}{(x^2 + 1)^3} + \frac{4x + 1}{(x^2 + 1)^2}$$

$$= \frac{-4x(2x^2 + x)}{(x^2 + 1)^3} + \frac{(4x + 1)(x^2 + 1)}{(x^2 + 1)^3}$$

$$= \frac{-4x(2x^2 + x) + (4x + 1)(x^2 + 1)}{(x^2 + 1)^3}$$

$$= \frac{-8x^3 - 4x^2 + 4x^3 + 4x + x^2 + 1}{(x^2 + 1)^3}$$

$$= \frac{-4x^3 - 3x^2 + 4x + 1}{(x^2 + 1)^3}$$

Note that this final simplified form is the same as (5) in part (A), as we would expect.

(C) We must solve the equation $f'(x) = 0$ to locate the horizontal tangent lines for the graph of f. We could use either intermediate form (4) or intermediate form (6) to solve this equation, but using the final simplified form (5) gives some additional insight into the solution. Since a fraction is 0 when the numerator is 0 and the denominator is not, we see that the values of x where the tangent line is horizontal are the solutions of the cubic equation

$$-4x^3 - 3x^2 + 4x + 1 = 0 \qquad (7)$$

Using a root approximation routine on a graphing utility (Fig. 2), we find that equation (7) has three solutions (rounded to two decimal places): -1.35, -0.22, and 0.83. (Recall from Section 2-1 that a cubic polynomial can have at most three roots.) Thus, the graph of f has horizontal tangent lines at $x \approx -1.35$, $x \approx -0.22$, and $x \approx 0.83$.

FIGURE 2
$y_1 = -4x^3 - 3x^2 + 4x + 1$

MATCHED PROBLEM 3 ➤ Repeat Example 3 for:

$$f(x) = \frac{x^2 - x}{(x^2 + 2)^2}$$ ◀

explore–discuss 2

> Refer to Example 3, where we used simplified form (5) for f' to locate the horizontal tangent lines for $f(x) = (2x^2 + x)/(x^2 + 1)^2$.
>
> (A) Use either unsimplified form (4) or unsimplified form (6) to solve the equation $f'(x) = 0$. Without referring to the solution in Example 3, can you be certain that you have found all the solutions?
> (B) Find a viewing window that clearly shows the graph of f and the three horizontal tangent lines. In the next chapter we will see that locating horizontal tangent lines is an important and useful step in the analysis of the graph of a function.

As Example 3C and Explore–Discuss 2 illustrate, using the simplified form of a derivative often gives additional insight into the solution of a problem. In Example 3C, using form (5) to solve $f'(x) = 0$ assures us that we have found all the solutions. If a simplified form for the derivative of a quotient is desired, it is usually easier to use the quotient rule than to convert to a product. (Compare the algebraic simplifications in Example 3A with those in Example 3B.) There is one special case where converting to a product is the preferred method—when working with a fraction whose numerator is a constant.

EXAMPLE 4 ➤ **Alternate Methods of Differentiation** Find $f'(x)$ two ways for:

$$f(x) = \frac{4}{(x^2 + 9)^3}$$

Solution **Method 1.** Use the quotient rule:

$$f'(x) = \frac{(x^2 + 9)^3 \dfrac{d}{dx} 4 - 4 \dfrac{d}{dx}(x^2 + 9)^3}{[(x^2 + 9)^3]^2}$$

$$= \frac{(x^2 + 9)^3(0) - 4[3(x^2 + 9)^2(2x)]}{(x^2 + 9)^6}$$

$$= \frac{-24x(x^2 + 9)^2}{(x^2 + 9)^6} = \frac{-24x}{(x^2 + 9)^4}$$

Method 2.   Rewrite as a product, and use the general power rule:

$$f(x) = \frac{4}{(x^2 + 9)^3} = 4(x^2 + 9)^{-3}$$

$$f'(x) = 4(-3)(x^2 + 9)^{-4}(2x)$$

$$= \frac{-24x}{(x^2 + 9)^4}$$

Which method do you prefer?

◀

MATCHED PROBLEM 4 ▶ Find $f'(x)$ two ways for: $f(x) = \dfrac{5}{(x^3 + 1)^2}$

◀

Answers to Matched Problems   1. (A) $15(5x + 2)^2$ (B) $20x^3(x^4 - 5)^4$ (C) $-4x/(x^2 + 4)^3$
(D) $-1/(2\sqrt{4 - x})$

2. $f'(x) = \sqrt{15 - 2x} - \dfrac{x}{\sqrt{15 - 2x}}; y = 2x + 3$

3. (A) $f'(x) = \dfrac{(x^2 + 2)^2(2x - 1) - (x^2 - x)(4x)(x^2 + 2)}{(x^2 + 2)^4}$

$$= \frac{-2x^3 + 3x^2 + 4x - 2}{(x^2 + 2)^3}$$

(B) $f'(x) = (x^2 - x)(-4x)(x^2 + 2)^{-3} + (x^2 + 2)^{-2}(2x - 1)$

$$= \frac{-2x^3 + 3x^2 + 4x - 2}{(x^2 + 2)^3}$$

(C) $x \approx -1.11, x \approx 0.41, x \approx 2.20$

4. $\dfrac{-30x^2}{(x^3 + 1)^3}$

EXERCISE   8-6

*The answers to many of the problems in this exercise set
contain both an unsimplified form and a simplified form of a
derivative. When checking your work, first check that you
applied the rules correctly, and then check that you performed
the algebraic simplification correctly.*

A   *In Problems 1–6, replace the* **?** *with an expression that
will make the indicated equation valid.*

1. $\dfrac{d}{dx}(3x + 4)^4 = 4(3x + 4)^3$**?**

2. $\dfrac{d}{dx}(5 - 2x)^6 = 6(5 - 2x)^5$**?**

3. $\dfrac{d}{dx}(4 - 2x^2)^3 = 3(4 - 2x^2)^2$**?**

4. $\dfrac{d}{dx}(3x^2 + 7)^5 = 5(3x^2 + 7)^4$**?**

5. $\dfrac{d}{dx}(1 + 2x + 3x^2)^7 = 7(1 + 2x + 3x^2)^6$**?**

6. $\dfrac{d}{dx}(4 - 3x - 2x^2)^8 = 8(4 - 3x - 2x^2)^7$**?**

In Problems 7–18, find f'(x) using the general power rule and simplify.

7. $f(x) = (2x + 5)^3$

8. $f(x) = (3x - 7)^5$

9. $f(x) = (5 - 2x)^4$

10. $f(x) = (9 - 5x)^2$

11. $f(x) = (3x^2 + 5)^5$

12. $f(x) = (5x^2 - 3)^6$

13. $f(x) = (x^3 - 2x^2 + 2)^8$

14. $f(x) = (2x^2 + x + 1)^7$

15. $f(x) = (2x - 5)^{1/2}$

16. $f(x) = (4x + 3)^{1/2}$

17. $f(x) = (x^4 + 1)^{-2}$

18. $f(x) = (x^5 + 2)^{-3}$

In Problems 19–22, find f'(x) and the equation of the line tangent to the graph of f at the indicated value of x. Find the value(s) of x where the tangent line is horizontal.

19. $f(x) = (2x - 1)^3$; $x = 1$

20. $f(x) = (3x - 1)^4$; $x = 1$

21. $f(x) = (4x - 3)^{1/2}$; $x = 3$

22. $f(x) = (2x + 8)^{1/2}$; $x = 4$

B *In Problems 23–40, find dy/dx using the general power rule and simplify.*

23. $y = 3(x^2 - 2)^4$

24. $y = 2(x^3 + 6)^5$

25. $y = 2(x^2 + 3x)^{-3}$

26. $y = 3(x^3 + x^2)^{-2}$

27. $y = \sqrt{x^2 + 8}$

28. $y = \sqrt[3]{3x - 7}$

29. $y = \sqrt[3]{3x + 4}$

30. $y = \sqrt{2x - 5}$

31. $y = (x^2 - 4x + 2)^{1/2}$

32. $y = (2x^2 + 2x - 3)^{1/2}$

33. $y = \dfrac{1}{2x + 4}$

34. $y = \dfrac{1}{3x - 7}$

35. $y = \dfrac{1}{(x^3 + 4)^5}$

36. $y = \dfrac{1}{(x^2 - 3)^6}$

37. $y = \dfrac{1}{4x^2 - 4x + 1}$

38. $y = \dfrac{1}{2x^2 - 3x + 1}$

39. $y = \dfrac{4}{\sqrt{x^2 - 3x}}$

40. $y = \dfrac{3}{\sqrt[3]{x - x^2}}$

In Problems 41–46, find f'(x), and find the equation of the line tangent to the graph of f at the indicated value of x.

41. $f(x) = x(4 - x)^3$; $x = 2$

42. $f(x) = x^2(1 - x)^4$; $x = 2$

43. $f(x) = \dfrac{x}{(2x - 5)^3}$; $x = 3$

44. $f(x) = \dfrac{x^4}{(3x - 8)^2}$; $x = 4$

45. $f(x) = x\sqrt{2x + 2}$; $x = 1$

46. $f(x) = x\sqrt{x - 6}$; $x = 7$

In Problems 47–58, find the value(s) of x (to two decimal places) where the tangent line is horizontal.

47. $f(x) = x^2(x - 5)^3$

48. $f(x) = x^3(x - 7)^4$

49. $f(x) = \sqrt{x^2 - 8x + 20}$

50. $f(x) = \sqrt{x^2 + 4x + 5}$

51. $f(x) = \sqrt{x^4 - 4x^3 + 6x + 10}$

52. $f(x) = \sqrt{x^4 - 5x^2 + x + 9}$

53. $f(x) = (x^2 + 2x - 15)\sqrt{x^2 + 1}$

54. $f(x) = (x^2 - 3x - 20)\sqrt{x^2 + 1}$

55. $f(x) = \dfrac{x}{(2x + 5)^2}$

56. $f(x) = \dfrac{x - 1}{(x - 3)^2}$

57. $f(x) = \dfrac{x^3 - x + 1}{(x^2 + 1)^2}$

58. $f(x) = \dfrac{x^3 - 2x + 2}{(x^2 + 1)^2}$

C *In Problems 59–70, find each derivative and simplify.*

59. $\dfrac{d}{dx}[3x(x^2 + 1)^3]$

60. $\dfrac{d}{dx}[2x^2(x^3 - 3)^4]$

61. $\dfrac{d}{dx}\dfrac{(x^3 - 7)^4}{2x^3}$

62. $\dfrac{d}{dx}\dfrac{3x^2}{(x^2 + 5)^3}$

63. $\dfrac{d}{dx}[(2x - 3)^2(2x^2 + 1)^3]$

64. $\dfrac{d}{dx}[(x^2 - 1)^3(x^2 - 2)^2]$

65. $\dfrac{d}{dx}(4x^2\sqrt{x^2 - 1})$

66. $\dfrac{d}{dx}(3x\sqrt{2x^2 + 3})$

67. $\dfrac{d}{dx}\dfrac{2x}{\sqrt{x - 3}}$

68. $\dfrac{d}{dx}\dfrac{x^2}{\sqrt{x^2 + 1}}$

69. $\dfrac{d}{dx}\sqrt{(2x - 1)^3(x^2 + 3)^4}$

70. $\dfrac{d}{dx}\sqrt{\dfrac{4x + 1}{2x^2 + 1}}$

▶ APPLICATIONS

Business & Economics

71. *Marginal cost.* The total cost (in hundreds of dollars) of producing *x* calculators per day is

$$C(x) = 10 + \sqrt{2x + 16} \qquad 0 \le x \le 50$$

(see the figure).

Figure for 71

(A) Find the marginal cost at a production level of *x* calculators.

(B) Find $C'(24)$ and $C'(42)$, and interpret the results verbally and graphically.

72. *Marginal cost.* The total cost (in hundreds of dollars) of producing *x* cameras per week is

$$C(x) = 6 + \sqrt{4x + 4} \qquad 0 \le x \le 30$$

(A) Find the marginal cost at a production level of *x* cameras.

(B) Find $C'(15)$ and $C'(24)$, and interpret the results verbally and graphically.

73. *Price–supply equation.* The number *x* of stereo speakers a retail chain is willing to sell per week at a price of $p is given by

$$x = 80\sqrt{p + 25} - 400 \qquad 20 \le p \le 100$$

Find the supply and the instantaneous rate of change of supply with respect to price when the price is $75. Write a brief verbal interpretation of these results. (See the figure at the top of the next column.)

74. *Price–demand equation.* The number *x* of stereo speakers people are willing to buy per week from a retail chain

Figure for 73 and 74

at a price of $p is given by

$$x = 1,000 - 60\sqrt{p + 25} \qquad 20 \le p \le 100$$

(see the figure).
Find the demand and the instantaneous rate of change of demand with respect to price when the price is $75. Write a brief verbal interpretation of these results.

75. *Compound interest.* If $1,000 is invested at an annual interest rate *r* compounded monthly, the amount in the account at the end of 4 years is given by

$$A = 1,000(1 + \tfrac{1}{12}r)^{48}$$

Find the rate of change of the amount *A* with respect to the interest rate *r*.

76. *Compound interest.* If $100 is invested at an annual interest rate *r* compounded semiannually, the amount in the account at the end of 5 years is given by

$$A = 100(1 + \tfrac{1}{2}r)^{10}$$

Find the rate of change of the amount *A* with respect to the interest rate *r*.

Life Sciences

77. *Bacteria growth.* The number *y* of bacteria in a certain colony after *x* days is given approximately by

$$y = (3 \times 10^6)\left[1 - \frac{1}{\sqrt[3]{(x^2 - 1)^2}}\right]$$

Find dy/dx.

78. *Pollution.* A small lake in a resort area became contaminated with harmful bacteria because of excessive septic tank seepage. After treating the lake with a bactericide, the Department of Public Health estimated the bacteria concentration (number per cubic centimeter) after t days to be given by

$$C(t) = 500(8 - t)^2 \qquad 0 \le t \le 7$$

(A) Find $C'(t)$ using the general power rule.

(B) Find $C'(1)$ and $C'(6)$, and interpret the results verbally and graphically.

Social Sciences

79. *Learning.* In 1930, L. L. Thurstone developed the following formula to indicate how learning time T depends on the length of a list n:

$$T = f(n) = \frac{c}{k}\, n\sqrt{n - a}$$

where a, c, and k are empirical constants. Suppose that for a particular person, time T (in minutes) for learning a list of length n is

$$T = f(n) = 2n\sqrt{n - 2}$$

(A) Find dT/dn.

(B) Find $f'(11)$ and $f'(27)$, and interpret the results verbally and graphically.

SECTION 8-7 ■ # Marginal Analysis in Business and Economics

■ MARGINAL COST, REVENUE, AND PROFIT
■ APPLICATION
■ MARGINAL AVERAGE COST, REVENUE, AND PROFIT

■ ## MARGINAL COST, REVENUE, AND PROFIT

One important use of calculus in business and economics is in *marginal analysis*. We introduced the concept of *marginal cost* earlier. There is no reason to stop there. Economists also talk about *marginal revenue* and *marginal profit*. Recall that the word "marginal" refers to an instantaneous rate of change—that is, a derivative. Thus, we define the following:

Marginal Cost, Revenue, and Profit

If x is the number of units of a product produced in some time interval, then

$$\text{Total cost} = C(x) \qquad\qquad \text{Total revenue} = R(x)$$

$$\textbf{Marginal cost} = C'(x) \qquad\qquad \textbf{Marginal revenue} = R'(x)$$

$$\text{Total profit} = P(x) = R(x) - C(x)$$

$$\textbf{Marginal profit} = P'(x) = R'(x) - C'(x)$$
$$= (\text{Marginal revenue}) - (\text{Marginal cost})$$

Marginal cost (or revenue or profit) is the instantaneous rate of change of cost (or revenue or profit) relative to production at a given production level.

It is important to remember that whenever we refer to a cost function $C(x)$ it is understood that $C(x)$ represents the *total cost* of producing x items. To find the exact cost of producing a particular item, we use the difference of two successive values of $C(x)$:

$$\text{Total cost of producing } x + 1 \text{ items} = C(x + 1)$$
$$\text{Total cost of producing } x \text{ items} = C(x)$$
$$\text{Exact cost of producing the } (x + 1)\text{st item} = C(x + 1) - C(x)$$

As we noted in Section 8-4, the marginal cost function can be used to approximate this exact cost. To see why, we return to the definition of a derivative:

$$C'(x) = \lim_{h \to 0} \frac{C(x + h) - C(x)}{h} \qquad \textcolor{blue}{\textit{Marginal cost}}$$

$$C'(x) \approx \frac{C(x + h) - C(x)}{h} \qquad h \neq 0$$

$$C'(x) \approx C(x + 1) - C(x) \qquad h = 1$$

Thus, we see that the marginal cost $C'(x)$ approximates $C(x + 1) - C(x)$, the exact cost of producing the $(x + 1)$st item. These observations are summarized below and illustrated in Figure 1.

FIGURE 1
$C'(x) \approx C(x + 1) - C(x)$

Marginal Cost and Exact Cost

If $C(x)$ is the total cost of producing x items, then the marginal cost function approximates the exact cost of producing the $(x + 1)$st item:

Marginal Cost   **Exact Cost**

$$C'(x) \approx C(x + 1) - C(x)$$

Similar interpretations can be made for total revenue functions and total profit functions.

EXAMPLE 1 ➤ **Cost Analysis** A company manufactures fuel tanks for automobiles. The total weekly cost (in dollars) of producing x tanks is given by

$$C(x) = 10,000 + 90x - 0.05x^2$$

(A) Find the marginal cost function.
(B) Find $C'(500)$, and discuss the various interpretations of this result.
(C) Find the exact cost of producing the 501st tank, and discuss the relationship between this result and the marginal cost found in part (B).

Solution (A) $C'(x) = 90 - 0.1x$
(B) $C'(500) = 90 - 0.1(500) = \$40$

The standard interpretation of a derivative certainly applies to marginal cost functions. Thus, at a production level of 500 tanks per week, total production costs are increasing at the rate of $40 per tank. The special interpretation of a marginal cost function also applies. That is, $40 is the approximate cost of producing the 501st tank.

(C) $C(501) = 10,000 + 90(501) - 0.05(501)^2$

$\qquad = \$42,539.95$ *Total cost of producing 501 tanks per week*

$\quad\ C(500) = 10,000 + 90(500) - 0.05(500)^2$

$\qquad = \$42,500.00$ *Total cost of producing 500 tanks per week*

$\quad\ C(501) - C(500) = 42,539.95 - 42,500.00$

$\qquad\qquad\qquad\quad = \$39.95$ *Exact cost of producing the 501st tank*

Comparing this result with the marginal cost found in part (B), we see that the marginal cost does provide a good approximation to the exact cost of producing the 501st tank.

See Figure 2 for a check of these calculations.   ◀

FIGURE 2
$y_1 = 10,000 + 90x - 0.05x^2$

MATCHED PROBLEM 1 ▶ A company manufactures automatic transmissions for automobiles. The total weekly cost (in dollars) of producing x transmissions is given by

$$C(x) = 50,000 + 600x - 0.75x^2$$

(A) Find the marginal cost function.

(B) Find $C'(200)$, and discuss the various interpretations of this result.

(C) Find the exact cost of producing the 201st transmission, and discuss the relationship between this result and the marginal cost found in part (B).   ◀

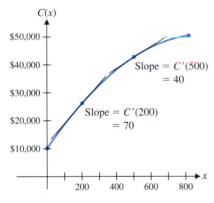

FIGURE 3
$C(x) = 10,000 + 90x - 0.05x^2$

It is instructive to compute both the approximate cost $C'(x)$ and the exact cost $C(x + 1) - C(x)$ for comparison purposes. However, in actual practice the marginal cost is used much more frequently than the exact cost. One reason for this is that the marginal cost is easily visualized when examining the graph of the total cost function. Figure 3 shows the graph of the cost function discussed in Example 1 with tangent lines added at $x = 200$ and $x = 500$. The graph clearly shows that as production increases, the slope of the tangent line decreases. Thus, the cost of producing the next tank also decreases, a desirable characteristic of a total cost function. We will have much more to say about graphical analysis in the next chapter.

■ APPLICATION

We now want to discuss how price, demand, revenue, cost, and profit are tied together in typical applications. Although either price or demand can be used as the independent variable in a price–demand equation, it is common practice to use demand as the independent variable when marginal revenue, cost, and profit are also involved.

explore – discuss 1

TABLE 1

DEMAND	PRICE
x	p
3,000	$7
6,000	$4

The market research department of a company used test marketing to determine the demand for a new radio (Table 1).

(A) Assuming that the relationship between price p and demand x is linear, find the price–demand equation and write the result in the form $x = f(p)$. Graph the equation and find the domain of f. Discuss the effect of price increases and decreases on demand.

(B) Solve the equation found in part (A) for p, obtaining an equation of the form $p = g(x)$. Graph this equation and find the domain of g. Discuss the effect of price increases and decreases on demand.

EXAMPLE 2 ▶

Production Strategy The market research department of a company recommends that the company manufacture and market a new transistor radio. After suitable test marketing, the research department presents the following **price–demand equation:**

$$x = 10,000 - 1,000p \quad \text{x is demand at price p} \tag{1}$$

Or, solving (1) for p,

$$p = 10 - 0.001x \tag{2}$$

where x is the number of radios retailers are likely to buy at $\$p$ per radio.

The financial department provides the following **cost function.**

$$C(x) = 7,000 + 2x \tag{3}$$

where $\$7,000$ is the estimate of fixed costs (tooling and overhead) and $\$2$ is the estimate of variable costs per radio (materials, labor, marketing, transportation, storage, etc.).

(A) Find the domain of the function defined by the price–demand equation in (2).
(B) Find the marginal cost function $C'(x)$ and interpret.
(C) Find the revenue function as a function of x, and find its domain.
(D) Find the marginal revenue at $x = 2,000$, $5,000$, and $7,000$. Interpret these results.
(E) Graph the cost function and the revenue function in the same coordinate system, find the intersection points of these two graphs, and interpret the results.
(F) Find the profit function and its domain, and sketch its graph.
(G) Find the marginal profit at $x = 1,000$, $4,000$, and $6,000$. Interpret these results.

Solution (A) Since price p and demand x must be nonnegative, we have $x \geq 0$ and

$$p = 10 - 0.001x \geq 0$$
$$10 \geq 0.001x$$
$$10,000 \geq x$$

Thus, the permissible values of x are $0 \leq x \leq 10,000$.

(B) The marginal cost is $C'(x) = 2$. Since this is a constant, it costs an additional \$2 to produce one more radio at any production level.

(C) The **revenue** is the amount of money R received by the company for manufacturing and selling x radios at \$$p$ per radio and is given by

$$R = (\text{Number of radios sold})(\text{Price per radio}) = xp$$

In general, the revenue R can be expressed as a function of p by using equation (1) or as a function of x by using equation (2). As we mentioned earlier, when using marginal functions, we will always use the number of items x as the independent variable. Thus, the **revenue function** is

$$R(x) = xp = x(10 - 0.001x) \quad \text{Using equation (2)} \tag{4}$$
$$= 10x - 0.001x^2$$

Since equation (2) is defined only for $0 \leqslant x \leqslant 10{,}000$, it follows that the domain of the revenue function is $0 \leqslant x \leqslant 10{,}000$.

(D) The **marginal revenue** is

$$R'(x) = 10 - 0.002x$$

For production levels of $x = 2{,}000$, $5{,}000$, and $7{,}000$, we have

$$R'(2{,}000) = 6 \qquad R'(5{,}000) = 0 \qquad R'(7{,}000) = -4$$

This means that at production levels of 2,000, 5,000, and 7,000, the respective approximate changes in revenue per unit change in production are \$6, \$0, and $-\$4$. That is, at the 2,000 output level, revenue increases as production increases; at the 5,000 output level, revenue does not change with a ''small'' change in production; and at the 7,000 output level, revenue decreases with an increase in production.

(E) The intersection points of the graphs of $R(x)$ and $C(x)$ may be located using a built-in intersection routine on a graphing utility (Fig. 4). These intersection points are called the **break-even points,** because revenue equals cost at these production levels—the company neither makes nor loses money, but just breaks even. Referring to Figure 4, we see that the break-even points are $(1{,}000, 9{,}000)$ and $(7{,}000, 21{,}000)$. Furthermore, we see that cost is greater than revenue for production levels between 0 and 1,000 and also between 7,000 and 10,000.

FIGURE 4
$y_1 = 7{,}000 + 2x$
$y_2 = 10x - 0.001x^2$

Consequently, the company incurs a loss at these levels. On the other hand, for production levels between 1,000 and 7,000, revenue is greater than cost, so the company makes a profit. These results are illustrated graphically in Figure 5.

FIGURE 5

(F) The **profit function** is

$$P(x) = R(x) - C(x)$$
$$= (10x - 0.001x^2) - (7,000 + 2x)$$
$$= -0.001x^2 + 8x - 7,000$$

The domain of the cost function is $x \geq 0$ and the domain of the revenue function is $0 \leq x \leq 10,000$. Thus, the domain of the profit function is the set of x values for which both functions are defined; that is, $0 \leq x \leq 10,000$. The graph of the profit function is shown in Figure 6. Notice that the x coordinates of the break-even points in Figure 5 are the x intercepts of the profit function. Furthermore, the intervals where cost is greater than revenue and where revenue is greater than cost correspond, respectively, to the intervals where profit is negative and the intervals where profit is positive.

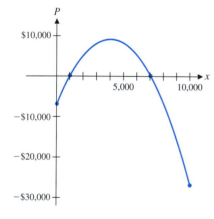

FIGURE 6

(G) The **marginal profit** is

$$P'(x) = -0.002x + 8$$

For production levels of 1,000, 4,000, and 6,000, we have

$$P'(1,000) = 6 \qquad P'(4,000) = 0 \qquad P'(6,000) = -4$$

This means that at production levels of 1,000, 4,000, and 6,000, the respective approximate changes in profit per unit change in production are $6, $0, and −$4. That is, at the 1,000 output level, profit will be increased if production is increased; at the 4,000 output level, profit does not change for "small" changes in production; and at the 6,000 output level, profits will decrease if production is increased. It seems the best production level to produce a maximum profit is 4,000. ◄

Example 2 warrants careful study, since a number of important ideas in economics and calculus are involved. In the next chapter, we will develop a systematic procedure for finding the production level (and, using the demand equation, the selling price) that will maximize profit.

MATCHED PROBLEM 2 ➤ Refer to the revenue and profit equations in Example 2.

(A) Find $R'(3,000)$ and $R'(6,000)$, and interpret the results.
(B) Find $P'(2,000)$ and $P'(7,000)$, and interpret the results. ◄

explore – discuss 2

Let

$$C(x) = 12,000 + 5x \quad \text{and} \quad R(x) = 9x - 0.002x^2$$

Explain why $\neq$ is used below. Then find the correct expression for the profit function.

$$P(x) = R(x) - C(x) \neq 9x - 0.002x^2 - 12,000 + 5x$$

■ MARGINAL AVERAGE COST, REVENUE, AND PROFIT

Sometimes, it is desirable to carry out marginal analysis relative to **average cost (cost per unit), average revenue (revenue per unit), and average profit (profit per unit).** The relevant definitions are summarized in the following box:

Marginal Average Cost, Revenue, and Profit

If x is the number of units of a product produced in some time interval, then

Cost per unit: Average cost $= \overline{C}(x) = \dfrac{C(x)}{x}$

Marginal average cost $= \overline{C}'(x) = \dfrac{d}{dx}\overline{C}(x)$

Revenue per unit: Average revenue $= \overline{R}(x) = \dfrac{R(x)}{x}$

Marginal average revenue $= \overline{R}'(x) = \dfrac{d}{dx}\overline{R}(x)$

Profit per unit: Average profit $= \overline{P}(x) = \dfrac{P(x)}{x}$

Marginal average profit $= \overline{P}'(x) = \dfrac{d}{dx}\overline{P}(x)$

EXAMPLE 3 ➤ **Cost Analysis**   A small machine shop manufactures drill bits used in the petroleum industry. The shop manager estimates that the total daily cost (in dollars) of producing x bits is

$$C(x) = 1{,}000 + 25x - 0.1x^2$$

(A) Find $\overline{C}(x)$ and $\overline{C}'(x)$.
(B) Find $\overline{C}(10)$ and $\overline{C}'(10)$, and interpret.
(C) Use the results in part (B) to estimate the average cost per bit at a production level of 11 bits per day.

Solution   (A) $\displaystyle \overline{C}(x) = \frac{C(x)}{x} = \frac{1{,}000 + 25x - 0.1x^2}{x}$

$$= \frac{1{,}000}{x} + 25 - 0.1x \qquad \textcolor{blue}{\text{Average cost function}}$$

$$\overline{C}'(x) = \frac{d}{dx}\,\overline{C}(x) = -\frac{1{,}000}{x^2} - 0.1 \qquad \textcolor{blue}{\text{Marginal average cost function}}$$

(B) $\displaystyle \overline{C}(10) = \frac{1{,}000}{10} + 25 - 0.1(10) = \$124$

$$\overline{C}'(10) = -\frac{1{,}000}{10^2} - 0.1 = -\$10.10$$

At a production level of 10 bits per day, the average cost of producing a bit is $124, and this cost is decreasing at the rate of $10.10 per bit.

(C)   If production is increased by 1 bit, then the average cost per bit will decrease by approximately $10.10. Thus, the average cost per bit at a production level of 11 bits per day is approximately $124 − $10.10 = $113.90.
See Figure 7 for a check of these calculations.

FIGURE 7
$y_1 = 1{,}000 + 25x - 0.1x^2$
$y_2 = y_1/x$

MATCHED PROBLEM 3 ➤ Consider the cost function for the production of radios from Example 2:

$$C(x) = 7{,}000 + 2x$$

(A) Find $\overline{C}(x)$ and $\overline{C}'(x)$.

(B) Find $\overline{C}(100)$ and $\overline{C}'(100)$, and interpret.

(C) Use the results in part (B) to estimate the average cost per radio at a production level of 101 radios. ◄

---

explore – discuss 3

A student produced the following solution to Matched Problem 3:

$$C(x) = 7,000 + 2x \quad \text{Cost}$$

$$C'(x) = 2 \qquad\qquad \text{Marginal cost}$$

$$\frac{C'(x)}{x} = \frac{2}{x} \qquad\qquad \text{"Average" of the marginal cost}$$

Explain why the last function is not the same as the marginal average cost function.

Caution

1. The marginal average cost function must be computed by first finding the average cost function and then finding its derivative. As Explore–Discuss 3 illustrates, reversing the order of these two steps produces a different function that does not have any useful economic interpretations.

2. Recall that the marginal cost function has two interpretations: the usual interpretation of any derivative as an instantaneous rate of change, and the special interpretation as an approximation to the exact cost of the $(x + 1)$st item. This special interpretation does not apply to the marginal average cost function. Referring to Example 3, it would be incorrect to interpret $\overline{C}'(10) = -\$10.10$ to mean that the average cost of the next bit is approximately $-\$10.10$. In fact, the phrase "average cost of the next bit" does not even make sense. Averaging is a concept applied to a collection of items, not to a single item.

These remarks also apply to revenue and profit functions.

Answers to Matched Problems

1. (A) $C'(x) = 600 - 1.5x$

   (B) $C'(200) = 300$. At a production level of 200 transmissions, total costs are increasing at the rate of $300 per transmission. Also, the approximate cost of producing the 201st transmission is $300.

   (C) $C(201) - C(200) = \$299.25$. The marginal cost from part (B) provides a good approximation to this exact cost.

2. (A) $R'(3,000) = 4$. At a production level of 3,000, a unit increase in production will increase revenue by approx. $4.

   $R'(6,000) = -2$. At a production level of 6,000, a unit increase in production will decrease revenue by approx. $2.

(B) $P'(2,000) = 4$. At a production level of 2,000, a unit increase in production will increase profit by approx. $4.
  $P'(7,000) = -6$. At a production level of 7,000, a unit increase in production will decrease profit by approx. $6.

3. (A) $\overline{C}(x) = \dfrac{7,000}{x} + 2; \ \overline{C}'(x) = -\dfrac{7,000}{x^2}$

  (B) $\overline{C}(100) = \$72; \ \overline{C}'(100) = -\$0.70$. At a production level of 100 radios, the average cost per radio is $72, and this average cost is decreasing at a rate of $0.70 per radio.
  (C) Approx. $71.30.

EXERCISE 8-7

 APPLICATIONS

Business & Economics

1. *Cost analysis.* The total cost (in dollars) of producing x food processors is

   $$C(x) = 2,000 + 50x - 0.5x^2$$

   (A) Find the exact cost of producing the 21st food processor.
   (B) Use the marginal cost to approximate the cost of producing the 21st food processor.

2. *Cost analysis.* The total cost (in dollars) of producing x electric guitars is

   $$C(x) = 1,000 + 100x - 0.25x^2$$

   (A) Find the exact cost of producing the 51st guitar.
   (B) Use the marginal cost to approximate the cost of producing the 51st guitar.

3. *Cost analysis.* The total cost (in dollars) of manufacturing x auto body frames is

   $$C(x) = 60,000 + 300x$$

(A) Find the average cost per unit if 500 frames are produced.

(B) Find the marginal average cost at a production level of 500 units, and interpret the results.

(C) Use the results from parts (A) and (B) to estimate the average cost per frame if 501 frames are produced.

4. *Cost analysis.* The total cost (in dollars) of printing x dictionaries is

   $$C(x) = 20,000 + 10x$$

(A) Find the average cost per unit if 1,000 dictionaries are produced.

(B) Find the marginal average cost at a production level of 1,000 units, and interpret the results.

(C) Use the results from parts (A) and (B) to estimate the average cost per dictionary if 1,001 dictionaries are produced.

5. *Revenue analysis.* The total revenue (in dollars) from the sale of x clock radios is

   $$R(x) = 100x - 0.025x^2$$

   Evaluate the marginal revenue at the given values of x, and interpret the results verbally and graphically.
   (A) $x = 1,600$ (B) $x = 2,500$

6. *Revenue analysis.* The total revenue (in dollars) from the sale of x steam irons is

   $$R(x) = 50x - 0.05x^2$$

Evaluate the marginal revenue at the given values of x, and interpret the results verbally and graphically.

(A) $x = 400$ (B) $x = 650$

7. *Profit analysis.* The total profit (in dollars) from the sale of x skateboards is

$$P(x) = 30x - 0.5x^2 - 250$$

(A) Find the exact profit from the sale of the 26th skateboard.

(B) Use the marginal profit to approximate the profit from the sale of the 26th skateboard.

8. *Profit analysis.* The total profit (in dollars) from the sale of x portable stereos is

$$P(x) = 22x - 0.1x^2 - 400$$

(A) Find the exact profit from the sale of the 41st stereo.

(B) Use the marginal profit to approximate the profit from the sale of the 41st stereo.

9. *Profit analysis.* The total profit (in dollars) from the sale of x video cassettes is

$$P(x) = 5x - 0.005x^2 - 450$$

Evaluate the marginal profit at the given values of x, and interpret the results verbally and graphically.

(A) $x = 450$ (B) $x = 750$

10. *Profit analysis.* The total profit (in dollars) from the sale of x cameras is

$$P(x) = 12x - 0.02x^2 - 1,000$$

Evaluate the marginal profit at the given values of x, and interpret the results verbally and graphically.

(A) $x = 200$ (B) $x = 350$

11. *Profit analysis.* The total profit (in dollars) from the sale of x lawn mowers is

$$P(x) = 30x - 0.03x^2 - 750$$

(A) Find the average profit per mower if 50 mowers are produced.

(B) Find the marginal average profit at a production level of 50 mowers and interpret.

(C) Use the results from parts (A) and (B) to estimate the average profit per mower if 51 mowers are produced.

12. *Profit analysis.* The total profit (in dollars) from the sale of x charcoal grills is

$$P(x) = 20x - 0.02x^2 - 320$$

(A) Find the average profit per grill if 40 grills are produced.

(B) Find the marginal average profit at a production level of 40 grills and interpret.

(C) Use the results from parts (A) and (B) to estimate the average profit per grill if 41 grills are produced.

13. *Revenue, cost, and profit.* The price–demand equation and the cost function for the production of table saws are given, respectively, by

$$p = 200 - \frac{x}{30} \quad \text{and} \quad C(x) = 72{,}000 + 60x$$

where x is the number of saws that can be sold at a price of \$$p$ per saw and $C(x)$ is the total cost (in dollars) of producing x saws.

(A) Find the marginal cost.

(B) Find the revenue function in terms of x.

(C) Find the marginal revenue.

(D) Find $R'(1{,}500)$ and $R'(4{,}500)$, and interpret the results.

(E) Graph the cost function and the revenue function on the same coordinate system for $0 \leqslant x \leqslant 6{,}000$. Find the break-even points, and indicate regions of loss and profit.

(F) Find the profit function in terms of x.

(G) Find the marginal profit.

(H) Find $P'(1{,}500)$ and $P'(3{,}000)$, and interpret the results.

14. *Revenue, cost, and profit.* The price–demand equation and the cost function for the production of television sets are given, respectively, by

$$p = 300 - \frac{x}{30} \quad \text{and} \quad C(x) = 150{,}000 + 30x$$

where x is the number of sets that can be sold at a price of \$$p$ per set and $C(x)$ is the total cost (in dollars) of producing x sets.

(A) Find the marginal cost.

(B) Find the revenue function in terms of x.

(C) Find the marginal revenue.

(D) Find $R'(3,000)$ and $R'(6,000)$, and interpret the results.

(E) Graph the cost function and the revenue function on the same coordinate system for $0 \leq x \leq 9,000$. Find the break-even points, and indicate regions of loss and profit.

(F) Find the profit function in terms of x.

(G) Find the marginal profit.

(H) Find $P'(1,500)$ and $P'(4,500)$, and interpret the results.

15. *Revenue, cost, and profit.* A company is planning to manufacture and market a new two-slice electric toaster. After conducting extensive market surveys, the research department provides the following estimates: a weekly demand of 200 toasters at a price of $16 per toaster and a weekly demand of 300 toasters at a price of $14 per toaster. The financial department estimates that weekly fixed costs will be $1,400 and variable costs (cost per unit) will be $4.

(A) Assume that the price–demand equation is linear. Use the research department's estimates to find the price–demand equation.

(B) Find the revenue function in terms of x.

(C) Assume that the cost function is linear. Use the financial department's estimates to find the cost function.

(D) Graph the cost function and the revenue function on the same coordinate system for $0 \leq x \leq 1,000$. Find the break-even points, and indicate regions of loss and profit.

(E) Find the profit function in terms of x.

(F) Evaluate the marginal profit at $x = 250$ and $x = 475$, and interpret the results.

16. *Revenue, cost, and profit.* The company in Problem 15 is also planning to manufacture and market a four-slice toaster. For this toaster, the research department's estimates are a weekly demand of 300 toasters at a price of $25 per toaster and a weekly demand of 400 toasters at a price of

$20. The financial department's estimates are fixed weekly costs of $5,000 and variable costs of $5 per toaster. Assume the price–demand equation and cost function are linear (see Problem 15, A and C).

(A) Use the research department's estimates to find the price–demand equation.

(B) Find the revenue function in terms of x.

(C) Use the financial department's estimates to find the cost function in terms of x.

(D) Graph the cost function and the revenue function on the same coordinate system for $0 \leq x \leq 800$. Find the break-even points, and indicate regions of loss and profit.

(E) Find the profit function in terms of x.

(F) Evaluate the marginal profit at $x = 325$ and $x = 425$, and interpret the results.

17. *Revenue, cost, and profit.* The total cost and the total revenue (in dollars) for the production and sale of x ski jackets are given, respectively, by

$$C(x) = 24x + 21,900 \quad \text{and} \quad R(x) = 200x - 0.2x^2$$

$$0 \leq x \leq 1,000$$

(A) Find the value of x where the graph of $R(x)$ has a horizontal tangent line.

(B) Find the profit function $P(x)$.

(C) Find the value of x where the graph of $P(x)$ has a horizontal tangent line.

(D) Graph $C(x)$, $R(x)$, and $P(x)$ on the same coordinate system for $0 \leq x \leq 1,000$. Find the break-even points. Find the x intercepts for the graph of $P(x)$.

18. *Revenue, cost, and profit.* The total cost and the total revenue (in dollars) for the production and sale of x hair dryers are given, respectively, by

$$C(x) = 5x + 2,340 \quad \text{and} \quad R(x) = 40x - 0.1x^2$$

$$0 \leq x \leq 400$$

(A) Find the value of x where the graph of $R(x)$ has a horizontal tangent line.

(B) Find the profit function $P(x)$.

(C) Find the value of x where the graph of $P(x)$ has a horizontal tangent line.

(D) Graph $C(x)$, $R(x)$, and $P(x)$ on the same coordinate system for $0 \leq x \leq 400$. Find the break-even points. Find the x intercepts for the graph of $P(x)$.

19. *Revenue, cost, and profit.* The price–demand equation and the cost function for the production of garden hoses by a certain company are given by

$$p = 20 - \sqrt{x} \quad \text{and} \quad C(x) = 500 + 2x$$

where x is the number of garden hoses that can be sold at a price of $\$p$ per hose and $C(x)$ is the total cost (in dollars) of producing x garden hoses.

(A) Find the profit function $P(x)$.

(B) What restrictions must be placed on the domain of $P(x)$? Why?

(C) Determine the production levels that will result in a profit and those that will result in a loss. Illustrate graphically.

20. *Revenue, cost, and profit.* The price–demand equation and the cost function for the production of hand-woven silk scarves by a certain mill are given by

$$p = 60 - 2\sqrt{x} \qquad \text{and} \qquad C(x) = 3,000 + 5x$$

where x is the number of scarves that can be sold at a price of $\$p$ per scarf and $C(x)$ is the total cost (in dollars) of producing x scarves.

(A) Find the profit function $P(x)$.

(B) What restrictions must be placed on the domain of $P(x)$? Why?

(C) Determine the production levels that will result in a profit and those that will result in a loss. Illustrate graphically.

21. *Break-even analysis.* Table 2 contains price–demand and total cost data for the production of overhead projectors, where p is the wholesale price (in dollars) of a projector for an annual demand of x projectors and C is the total cost (in dollars) of producing x projectors.

TABLE 2

x	p ($)	C ($)
3,190	581	1,130,000
4,570	405	1,241,000
5,740	181	1,410,000
5,330	85	1,620,000

(A) Find a quadratic regression equation for the price–demand data using x as the independent variable.

(B) Find a linear regression equation for the cost data using x as the independent variable. Use this equation to estimate the fixed costs and the variable costs per projector. Round answers to the nearest dollar.

(C) Find the break-even points. Round answers to the nearest integer.

(D) Find the price range for which the company will make a profit. Round answers to the nearest dollar.

22. *Break-even analysis.* Table 3 contains price–demand and total cost data for the production of treadmills, where p is the wholesale price (in dollars) of a treadmill for an annual demand of x treadmills and C is the total cost (in dollars) of producing x treadmills.

TABLE 3

x	p ($)	C ($)
2,910	1,435	3,650,000
3,415	1,280	3,870,000
4,645	1,125	4,190,000
7,330	910	4,380,000

(A) Find a linear regression equation for the price–demand data using x as the independent variable.

(B) Find a linear regression equation for the cost data using x as the independent variable. Use this equation to estimate the fixed costs and the variable costs per treadmill. Round answers to the nearest dollar.

(C) Find the break-even points. Round answers to the nearest integer.

(D) Find the price range for which the company will make a profit. Round answers to the nearest dollar.

CHAPTER 8
GROUP ACTIVITY

Minimal Average Cost

If $C(x)$ is the total cost of producing x items, the marginal cost function $C'(x)$ gives the approximate cost of the next item produced, while the average cost function $\overline{C}(x)$ gives the average cost per item for the items already produced. Thus, $C'(x)$ looks forward to the next item, while $\overline{C}(x)$ looks backward at all the items produced thus far. Given this difference in viewpoint, it is somewhat surprising that there is an important relationship between these two functions. As we will see, information gained from comparing the values of $C'(x)$ and $\overline{C}(x)$ can help determine the production level that minimizes average cost.

(A) The total cost (in dollars) of producing x items is given by

$$C(x) = 0.01x^2 + 40x + 3{,}600$$

Find $C'(x)$ and $\overline{C}(x)$, and complete Table 1.

(B) Repeat part (A) for

$$C(x) = 0.00016x^3 - 0.12x^2 + 30x + 10{,}000$$

(C) Examine the values in the tables from parts (A) and (B), and write a brief verbal description of the behavior of each function. Does each average cost function appear to have a minimum value? What is the minimal value, and where does it occur? What relationship do you observe between the minimum average cost and the marginal cost at the production level that minimizes average cost?

(D) Confirm your observations in part (C) by examining the graphs of $C'(x)$ and $\overline{C}(x)$.

(E) The following statements can help justify the relationship you observed in part (C). In each case, fill in the blank with "increase" or "decrease" and justify your choice.
1. If $C'(x) < \overline{C}(x)$ (that is, the cost of the next item is less than the average cost of the items already produced), then increasing production by 1 item will _____ the average cost.
2. If $C'(x) > \overline{C}(x)$ (that is, the cost of the next item is more than the average cost of the items already produced), then increasing production by 1 item will _____ the average cost.

(F) Discuss the validity of the following statement for an arbitrary cost function $C(x)$:

If the minimum value of $\overline{C}(x)$ occurs at a production level x, then $C'(x) = \overline{C}(x)$ at that production level.

(G) We used a quadratic function and a cubic function in parts (A) and (B) to illustrate the relationship between $C'(x)$ and $\overline{C}(x)$. But linear cost functions are one of the most important types. To see why we did not choose a linear cost function, try to parallel the above development for the cost function

$$C(x) = 30x + 12{,}000$$

Do any of your findings contradict the statements in parts (E) and (F)?

TABLE 1

x	$\overline{C}(x)$	$C'(x)$
100		
200		
300		
400		
500		
600		
700		
800		
900		
1,000		

C H A P T E R 8 R E V I E W ■ Important Terms and Symbols

8-1 *Rate of Change and Slope.* Average rate of change; difference quotient; average velocity; instantaneous velocity; instantaneous rate of change; secant line; slope of a secant line; slope of a graph; slope of a tangent line

$$\frac{f(a + h) - f(a)}{h}; \quad \lim_{h \to 0} \frac{f(a + h) - f(a)}{h}$$

8-2 *Limits.* Limit; absolute value function; one-sided limits; limit from the left; left-hand limit; limit from the right; right-hand limit; properties of limits; limit of a polynomial function; 0/0 indeterminate form

$$\lim_{x \to c} f(x); \quad \lim_{x \to c^-} f(x); \quad \lim_{x \to c^+} f(x)$$

8-3 *The Derivative.* Derivative; differentiation; differentiable; interpretations of the derivative; numerical differentiation; symbolic differentiation; nondifferentiable; continuous

$$f'(x) = \lim_{h \to 0} \frac{f(x + h) - f(x)}{h}$$

8-4 *Derivatives of Constants, Power Forms, and Sums.* Derivative notation; derivative of a constant function rule; power rule; constant times a function rule; sum and difference rule; marginal cost function

$$f'(x); \quad y'; \quad \frac{dy}{dx}$$

8-5 *Derivatives of Products and Quotients.* Product rule; quotient rule

8-6 *Chain Rule: Power Form.* General power rule; chain rule; combining rules of differentiation

8-7 *Marginal Analysis in Business and Economics.* Marginal cost; marginal revenue; marginal profit; exact cost; price–demand equation; cost function; revenue function; marginal revenue; break-even points; profit function; marginal profit; average cost; marginal average cost; average revenue; marginal average revenue; average profit; marginal average profit

$$C'(x); \quad \overline{C}(x); \quad \overline{C}'(x); \quad R'(x); \quad \overline{R}(x); \quad \overline{R}'(x);$$
$$P'(x); \quad \overline{P}(x); \quad \overline{P}'(x)$$

C H A P T E R 8 ■ Summary of Rules of Differentiation

$$\frac{d}{dx} k = 0$$

$$\frac{d}{dx} x^n = n x^{n-1}$$

$$\frac{d}{dx} k f(x) = k f'(x)$$

$$\frac{d}{dx} [u(x) \pm v(x)] = u'(x) \pm v'(x)$$

$$\frac{d}{dx} [F(x)S(x)] = F(x)S'(x) + S(x)F'(x)$$

$$\frac{d}{dx} \frac{T(x)}{B(x)} = \frac{B(x)T'(x) - T(x)B'(x)}{[B(x)]^2}$$

$$\frac{d}{dx} [u(x)]^n = n[u(x)]^{n-1} u'(x)$$

CHAPTER   8   ■   Review Exercise

Work through all the problems in this chapter review and check your answers in the back of the book. Answers to all review problems are there along with section numbers in italics to indicate where each type of problem is discussed. Where weaknesses show up, review appropriate sections in the text.

Many of the problems in this exercise set ask you to find a derivative. Most of the answers to these problems contain both an unsimplified form and a simplified form of the derivative. When checking your work, first check that you applied the rules correctly, and then check that you performed the algebraic simplification correctly.

A

1. Find the indicated quantities for $y = f(x) = 2x^2 + 5$:
    (A) The change in y if x changes from 1 to 3
    (B) The average rate of change of y with respect to x if x changes from 1 to 3
    (C) The slope of the secant line through the points $(1, f(1))$ and $(3, f(3))$ on the graph of $y = f(x)$
    (D) The instantaneous rate of change of y with respect to x at $x = 1$
    (E) The slope of the line tangent to the graph of $y = f(x)$ at $x = 1$
    (F) $f'(1)$

2. Use the graph of f shown below to approximate $f'(-1)$ and $f'(1)$ to the nearest integer.

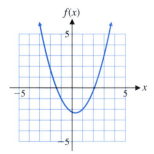

Figure for 2

3. If $\lim_{x \to 1} f(x) = 2$ and $\lim_{x \to 1} g(x) = 4$, find:

    (A) $\lim_{x \to 1} (5f(x) + 3g(x))$ (B) $\lim_{x \to 1} [f(x)g(x)]$

    (C) $\lim_{x \to 1} \dfrac{g(x)}{f(x)}$

In Problems 4–6, using the given graph of f to estimate the indicated limits and function values.

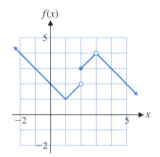

Figure for 4–6

4. (A) $\lim_{x \to 1^-} f(x)$ (B) $\lim_{x \to 1^+} f(x)$

    (C) $\lim_{x \to 1} f(x)$ (D) $f(1)$

5. (A) $\lim_{x \to 2^-} f(x)$ (B) $\lim_{x \to 2^+} f(x)$

    (C) $\lim_{x \to 2} f(x)$ (D) $f(2)$

6. (A) $\lim_{x \to 3^-} f(x)$ (B) $\lim_{x \to 3^+} f(x)$

    (C) $\lim_{x \to 3} f(x)$ (D) $f(3)$

7. Use the following expression and the definition of derivative to find $f'(x)$:

    $$f(x + h) - f(x) = 3x^2h + 3xh^2 + h^3 + 2xh$$

8. Given $f(5) = 4$, $f'(5) = -1$, $g(5) = 2$, and $g'(5) = -3$, find $h'(5)$ for each of the following functions:
    (A) $h(x) = 2f(x) + 3g(x)$ (B) $h(x) = f(x)g(x)$

    (C) $h(x) = \dfrac{f(x)}{g(x)}$ (D) $h(x) = [f(x)]^2$

9. Replace the **?** in the equation below with an expression that makes the equation valid.

    $$\frac{d}{dx} (3x^2 + 4x + 1)^5 = 5(3x^2 + 4x + 1)^4 \underline{\ \ ?\ \ }$$

In Problems 10–19, find $f'(x)$ and simplify.

10. $f(x) = 3x^4 - 2x^2 + 1$ 11. $f(x) = 2x^{1/2} - 3x$

12. $f(x) = 5$ 13. $f(x) = \dfrac{1}{2x^2} + \dfrac{x^2}{2}$

14. $f(x) = (2x - 1)(3x + 2)$

15. $f(x) = (x^2 - 1)(x^3 - 3)$

16. $f(x) = \dfrac{2x}{x^2 + 2}$ **17.** $f(x) = \dfrac{1}{3x + 2}$

18. $f(x) = (2x - 3)^3$ **19.** $f(x) = (x^2 + 2)^{-2}$

B

20. Let $f(x) = 0.5x^2 - 5$.
(A) Find the slope of the secant line through $(2, f(2))$ and $(4, f(4))$.
(B) Find the slope of the secant line through $(2, f(2))$ and $(2 + h, f(2 + h))$, $h \neq 0$.
(C) Find the slope of the tangent line at $x = 2$.

21. Sketch a possible graph of a function $y = f(x)$ passing through the indicated points and having the indicated slope:

POINT	SLOPE
$(-3, -3)$	-2
$(-2, -4)$	0
$(0, 0)$	4
$(2, 4)$	0
$(3, 3)$	-2

In Problems 22–29, find the indicated derivative and simplify.

22. $\dfrac{dy}{dx}$ for $y = 3x^4 - 2x^{-3} + 5$

23. y' for $y = (2x^2 - 3x + 2)(x^2 + 2x - 1)$

24. $f'(x)$ for $f(x) = \dfrac{2x - 3}{(x - 1)^2}$

25. y' for $y = 2\sqrt{x} + \dfrac{4}{\sqrt{x}}$

26. $\dfrac{d}{dx}[(x^2 - 1)(2x + 1)^2]$

27. $\dfrac{d}{dx}\sqrt[3]{x^3 - 5}$

28. $\dfrac{dy}{dx}$ for $y = \dfrac{3x^2 + 4}{x^2}$

29. $\dfrac{d}{dx}\dfrac{(x^2 + 2)^4}{2x - 3}$

30. For $y = f(x) = x^2 + 4$, find:
(A) The slope of the graph at $x = 1$

(B) The equation of the tangent line at $x = 1$ in the form $y = mx + b$

31. Repeat Problem 30 for $f(x) = x^3(x + 1)^2$.

In Problems 32–38, find the value(s) of x (to two decimal places) where the tangent line is horizontal.

32. $f(x) = 10x - x^2$ **33.** $f(x) = (x + 3)(x^2 - 45)$

34. $f(x) = x^4 - x^3 - 4x^2 + 5x$

35. $f(x) = x^2(2x - 15)^3$

36. $f(x) = \dfrac{x}{x^2 + 4}$

37. $f(x) = \dfrac{x^3 + 20x + 1}{x^2 + 2}$

38. $f(x) = \dfrac{x^2 + x}{(x^2 + 1)^2}$

39. If an object moves along the y axis (scale in feet) so that it is at $y = f(x) = 16x^2 - 4x$ at time x (in seconds), find:
(A) The instantaneous velocity function
(B) The velocity at time $x = 3$ seconds

40. An object moves along the y axis (scale in feet) so that at time x (in seconds) it is at $y = f(x) = 96x - 16x^2$. Find:
(A) The instantaneous velocity function
(B) The time(s) when the velocity is 0

41. Complete the following table for $f(x) = 5^x$, and estimate $f'(0)$ to two decimal places.

h	-0.1	-0.01	-0.001	$\rightarrow 0 \leftarrow$	0.001	0.01	0.1
$\dfrac{f(h) - f(0)}{h}$	?	?	?	$\rightarrow ? \leftarrow$	?	?	?

42. Graph $f(x) = 4^{-x}$, zoom in on the graph near $x = 0$, and use numerical differentiation to approximate $f'(0)$ to two decimal places.

43. Let $f(x) = x^3$, $g(x) = (x - 4)^3$, and $h(x) = (x + 3)^3$.
(A) How are the graphs of f, g, and h related? Illustrate your conclusion by graphing f, g, and h on the same coordinate axes.
(B) How would you expect the graphs of the derivatives of these functions to be related? Illustrate your conclusion by graphing f', g', and h' on the same coordinate axes.

44. Let $f(x)$ be a differentiable function and let k be a nonzero constant. For each function g, write a brief verbal description of the relationship between the graphs of f and g. Do the same for the graphs of f' and g'.

(A) $g(x) = f(x + k)$ (B) $g(x) = f(x) + k$

In Problems 45–54, find each limit, if it exists.

45. $\lim\limits_{x \to 3} \dfrac{2x - 3}{x + 5}$

46. $\lim\limits_{x \to 3} (2x^2 - x + 1)$

47. $\lim\limits_{x \to 0} \dfrac{2x}{3x^2 - 2x}$

48. $\lim\limits_{x \to 3} \dfrac{x - 3}{x^2 - 9}$

49. $\lim\limits_{x \to 4^-} \dfrac{|x - 4|}{x - 4}$

50. $\lim\limits_{x \to 4^+} \dfrac{|x - 4|}{x - 4}$

51. $\lim\limits_{x \to 4} \dfrac{|x - 4|}{x - 4}$

52. $\lim\limits_{h \to 0} \dfrac{[(2 + h)^2 - 1] - [2^2 - 1]}{h}$

53. $\lim\limits_{h \to 0} \dfrac{f(2 + h) - f(2)}{h}$ for $f(x) = x^2 + 4$

54. $\lim\limits_{h \to 0} \dfrac{f(x + h) - f(x)}{h}$ for $f(x) = \dfrac{1}{x + 2}$

55. Let

$$f(x) = \frac{x^3 - 4x^2 - 4x + 16}{|x^2 - 4|}$$

Graph f and use zoom and trace to investigate the left- and right-hand limits at the indicated values of c.

(A) $c = -2$ (B) $c = 0$ (C) $c = 2$

In Problems 56 and 57, use the definition of the derivative and the two step-process to find $f'(x)$.

56. $f(x) = x^2 - x$ **57.** $f(x) = \sqrt{x} - 3$

C *Problems 58–61 refer to the function f in the figure. Determine whether f is differentiable at the indicated value of x.*

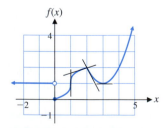

Figure for 58–61

58. $x = 0$ **59.** $x = 1$ **60.** $x = 2$ **61.** $x = 3$

In Problems 62–65, find $f'(x)$ and simplify.

62. $f(x) = (x - 4)^4(x + 3)^3$

63. $f(x) = \dfrac{x^5}{(2x + 1)^4}$

64. $f(x) = \dfrac{\sqrt{x^2 - 1}}{x}$

65. $f(x) = \dfrac{x}{\sqrt{x^2 + 4}}$

66. The domain of the power function $f(x) = x^{1/5}$ is the set of all real numbers. Find the domain of the derivative $f'(x)$. Discuss the nature of the graph of $y = f(x)$ for any x values excluded from the domain of $f'(x)$.

67. Let

$$f(x) = \frac{x^3 + m|x|}{x}$$ where m is a constant

Put your graphing utility in the dot mode, and use it to investigate the indicated limits.

(A) Find: $\lim\limits_{x \to 0^-} f(x)$ and $\lim\limits_{x \to 0^+} f(x)$ for $m = -1$

(B) Find: $\lim\limits_{x \to 0^-} f(x)$ and $\lim\limits_{x \to 0^+} f(x)$ for $m = 2$

(C) Find m so that: $\lim\limits_{x \to 0^-} f(x) = \lim\limits_{x \to 0^+} f(x)$

(D) Write a brief verbal description of each graph. How does the graph in (C) differ from the graphs in (A) and (B)?

68. Let $f(x) = 1 - |x - 1|,\ 0 \leqslant x \leqslant 2$ (see the figure).

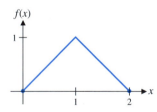

Figure for 68

(A) $\lim\limits_{h \to 0^-} \dfrac{f(1 + h) - f(1)}{h} = ?$

(B) $\lim\limits_{h \to 0^+} \dfrac{f(1 + h) - f(1)}{h} = ?$

(C) $\lim\limits_{h \to 0} \dfrac{f(1 + h) - f(1)}{h} = ?$

(D) Does $f'(1)$ exist?

▶ APPLICATIONS

Business & Economics

69. *Cost analysis.* The total cost (in dollars) of producing x television sets is

$$C(x) = 10,000 + 200x - 0.1x^2$$

(A) Find the exact cost of producing the 101st television set.

(B) Use the marginal cost to approximate the cost of producing the 101st television set.

70. *Cost analysis.* The total cost (in dollars) of producing x bicycles is

$$C(x) = 5,000 + 40x + 0.05x^2$$

(A) Find the total cost and the marginal cost at a production level of 100 bicycles, and interpret the results verbally and graphically.

(B) Find the average cost and the marginal average cost at a production level of 100 bicycles, and interpret the results verbally and graphically.

71. *Cost analysis.* The total cost (in dollars) of producing x laser printers per week is shown in the figure. Which is greater, the approximate cost of producing the 201st printer or the approximate cost of producing the 601st printer? Does this graph represent a manufacturing process that is becoming more efficient or less efficient as production levels increase? Explain.

Figure for 71

72. *Cost analysis.* Let

$$p = 25 - 0.01x \quad \text{and} \quad C(x) = 2x + 9,000$$
$$0 \leqslant x \leqslant 2,500$$

be the price–demand equation and the cost function, respectively, for the manufacture of umbrellas.

(A) Find the marginal cost, average cost, and marginal average cost functions.

(B) Express the revenue in terms of x, and find the marginal revenue, average revenue, and marginal average revenue functions.

(C) Find the profit, marginal profit, average profit, and marginal average profit functions.

(D) Find the break-even point(s).

(E) Evaluate the marginal profit at $x = 1,000$, $1,150$, and $1,400$, and interpret the results.

(F) Graph $R = R(x)$ and $C = C(x)$ on the same coordinate system, and locate regions of profit and loss.

73. *Employee training.* A company producing computer components has established that on the average, a new employee can assemble $N(t)$ components per day after t days of on-the-job training, as given by

$$N(t) = \frac{40t}{t + 2}$$

(A) Find the average rate of change of $N(t)$ from 3 days to 6 days.

(B) Find the instantaneous rate of change of $N(t)$ at 3 days.

74. *Sales analysis.* Past sales records for a swimming pool manufacturer indicate that the total number of swimming pools, N (in thousands), sold during a year are given by

$$N(t) = t\sqrt{4 + t}$$

where t is the number of months since the beginning of the year. Find $N(5)$ and $N'(5)$, and interpret verbally and graphically.

75. *Compound interest.* If $5,000 is invested in an account that earns 7% compounded annually, then the amount in the account after t years is given by

$$A(t) = 5,000(1.07)^t$$

(A) Find $A(10)$ and use numerical differentiation to approximate $A'(10)$ to the nearest dollar.

(B) Write a brief verbal interpretation of the results in part (A) and illustrate graphically.

76. *Natural gas consumption.* The data in Table 1 give the United States consumption of natural gas in trillions of cubic feet.

TABLE 1

YEAR	NATURAL GAS CONSUMPTION
1960	12.7
1965	16.0
1970	21.9
1975	20.1
1980	19.7
1985	17.2
1990	18.6

(A) Let x represent time (in years), with $x = 0$ corresponding to 1960, and let y represent the corresponding U.S. consumption of natural gas. Enter the data set in a graphing utility and find a cubic regression equation for the data.

(B) If $y = N(x)$ denotes the regression equation found in part (A), find $N(35)$ and $N'(35)$, and write a brief verbal interpretation of these results.

77. *Break-even analysis.* Table 2 contains price–demand and total cost data from a bakery for the production of kringles (a Danish pastry), where p is the price (in dollars) of a kringle for a daily demand of x kringles and C is the total cost (in dollars) of producing x kringles.

TABLE 2

x	p ($)	C ($)
125	9	740
140	8	785
170	7	850
200	6	900

(A) Find a linear regression equation for the price–demand data using x as the independent variable.

(B) Find a linear regression equation for the cost data using x as the independent variable. Use this equation to estimate the fixed costs and the variable costs per kringle.

(C) Find the break-even points.

(D) Find the price range for which the bakery will make a profit.

In all answers, round dollar amounts to the nearest cent.

Life Sciences

78. *Pollution.* A sewage treatment plant disposes of its effluent through a pipeline that extends 1 mile toward the center of a large lake. The concentration of effluent $C(x)$, in parts per million, x meters from the end of the pipe is given approximately by

$$C(x) = 500(x + 1)^{-2}$$

What is the instantaneous rate of change of concentration at 9 meters? At 99 meters?

79. *Medicine.* The body temperature (in degrees Fahrenheit) of a patient t hours after being given a fever-reducing drug is given by

$$F(t) = 98 + \frac{4}{\sqrt{t + 1}}$$

Find $F(3)$ and $F'(3)$. Write a brief verbal interpretation of these results and illustrate graphically.

Social Sciences

80. *Learning.* If a person learns N items in t hours, as given by

$$N(t) = 20\sqrt{t}$$

find the rate of learning after:
(A) 1 hour   (B) 4 hours

81. *Population.* The number of married couples in the United States can be described by the function

$$M(t) = 40.7(1.01)^t$$

where M is number of married couples (in millions) and t is time in years since 1960.

(A) Find $M(50)$, and use numerical differentiation to approximate $M'(50)$. Round both quantities to one decimal place.

(B) Write a brief verbal interpretation of the results in part (A) and illustrate graphically.

(C) Use the results in part (B) to estimate the number of married couples in 2011 and in 2012.

CHAPTER 9

Graphing and Optimization

Historically, one of the primary reasons for studying calculus was the development of powerful tools for sketching graphs of functions. In fact, prior to the very recent advent of graphing utilities, hand-sketching was the only way most students could produce graphs. Now things have changed. An accurate graph of a function specified by an equation can be produced quickly and easily on a graphing utility. Does this mean that the old tools are no longer relevant? On the contrary, efficient use of graphing technology still requires a person at the helm, directing the electronic device, and this person still needs a sound foundation in graphing tools and analysis.

In the first four sections of this chapter, we will develop the graphing concepts and tools necessary to satisfy the following objectives:

1. Analyze and discuss the important features of graphs of functions specified by equations. This will involve algebraic, numeric, and graphical techniques, and a graphing utility where appropriate.
2. Analyze, discuss, and sketch by hand the graphs of functions that are not specified by equations. This will involve utilizing tables, charts, verbal descriptions, or other graphs.

In the last section, we will use these graphing tools to solve optimization problems, one of the most important applications of calculus to real-world problems.

A brief review of Section 2-1 would be helpful before beginning this chapter. In particular, we will make frequent use of the following important property of polynomials:

The graph of a polynomial function of positive degree n can have at most $(n - 1)$ turning points and at most n zeros or x intercepts.

SECTION   9-1   ■   # Continuity and Graphs

■ CONTINUITY
■ CONTINUITY PROPERTIES
■ INFINITE LIMITS
■ SIGN CHARTS

In this section we return to the limit concept and use it to study an important property of functions called *continuity*. An understanding of continuity is essential for sketching and analyzing graphs. We will also see that continuity provides a simple and efficient method for solving inequalities, a tool that we will use extensively in later sections.

■ CONTINUITY

Compare the graphs shown in Figure 1, which were discussed in Examples 2–4 in Section 8-2. Notice that two of the graphs are broken; that is, they cannot be drawn without lifting a pen off the paper. Informally, a function is *continuous over an interval* if its graph over the interval can be drawn without removing a pen from the paper. A function whose graph is broken (disconnected) at $x = c$ is said to be *discontinuous* at $x = c$. Function f (Fig. 1A) is continuous for all x. Function g (Fig. 1B) is discontinuous at $x = 2$, but is continuous over any interval that does not include 2. Function h (Fig. 1C) is discontinuous at $x = 0$, but is continuous over any interval that does not include 0.

(A) $\lim_{x \to 2} f(x) = 4$
$f(2) = 4$

(B) $\lim_{x \to 2} g(x) = 4$
$g(2)$ is not defined

(C) $\lim_{x \to 0} h(x)$ does not exist
$h(0)$ is not defined

F I G U R E   1

Most graphs of natural phenomena are continuous, whereas many graphs in business and economics applications have discontinuities. Figure 2A illustrates temperature variation over a 24 hour period—a continuous phenomenon. Figure 2B illustrates warehouse inventory over a 1 week period—a discontinuous phenomenon.

(A) Temperature for a 24 hour period

(B) Inventory in a warehouse during 1 week

F I G U R E 2

explore – discuss 1

(A) Write a brief verbal description of the temperature variation illustrated in Figure 2A, including estimates of the high and low temperatures during this period and the times at which they occurred.

(B) Write a brief verbal description of the changes in inventory illustrated in Figure 2B, including estimates of the changes in inventory and the times at which these changes occurred.

The preceding discussion leads to the following formal definition of continuity:

Continuity

A function f is **continuous at the point $x = c$** if

1. $\lim\limits_{x \to c} f(x)$ exists   2. $f(c)$ exists   3. $\lim\limits_{x \to c} f(x) = f(c)$

A function is **continuous on the open interval* (a, b)** if it is continuous at each point on the interval.

_____

* When used in this context, (a, b) represents the set of real numbers satisfying $a < x < b$.

If one or more of the three conditions in the definition fails, then the function is **discontinuous** at $x = c$.

e x p l o r e – d i s c u s s 2

Sketch a graph of a function that is discontinuous at a point because it fails to satisfy condition 1 in the definition of continuity. Repeat for conditions 2 and 3.

EXAMPLE 1 ➤ **Continuity of a Function Defined by a Graph**   Use the definition of continuity to discuss the continuity of the function whose graph is shown in Figure 3.

Solution   We begin by identifying the points of discontinuity. Examining the graph, we see breaks and/or holes in the graph at $x = -4, -2, 1$, and 3. Now we must determine which conditions in the definition of continuity are not satisfied at each of these points. In each case, we find the value of the function and the limit of the function at the point in question.

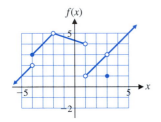

FIGURE 3

Discontinuity at $x = -4$:

$$\lim_{x \to -4^-} f(x) = 2$$ Since the one-sided limits are different, the limit does not exist (Section 8-2).

$$\lim_{x \to -4^+} f(x) = 3$$

$$\lim_{x \to -4} f(x) \text{ does not exist}$$

$$f(-4) = 3$$

Thus, f is not continuous at $x = -4$ because condition 1 is not satisfied.

Discontinuity at $x = -2$:

$$\lim_{x \to -2^-} f(x) = 5$$ The hole at $(-2, 5)$ indicates that 5 is not the value of f at -2. Since there is no solid dot elsewhere on the vertical line

$$\lim_{x \to -2^+} f(x) = 5$$ $x = -2$, $f(-2)$ is not defined.

$$\lim_{x \to -2} f(x) = 5$$

$$f(-2) \text{ does not exist}$$

Thus, f is not continuous at $x = -2$ because condition 2 is not satisfied.

Discontinuity at $x = 1$:

$$\lim_{x \to 1^-} f(x) = 4$$

$$\lim_{x \to 1^+} f(x) = 1$$

$$\lim_{x \to 1} f(x) \text{ does not exist}$$

$$f(1) \text{ does not exist}$$

This time, f is not continuous at $x = 1$ because both conditions 1 and 2 are not satisfied.

Discontinuity at $x = 3$:

$$\lim_{x \to 3^-} f(x) = 3 \quad \text{\color{blue}The solid dot at (3, 1) indicates that } f(3) = 1.$$

$$\lim_{x \to 3^+} f(x) = 3$$

$$\lim_{x \to 3} f(x) = 3$$

$$f(3) = 1$$

Conditions 1 and 2 are satisfied, but f is not continuous at $x = 3$ because condition 3 is not satisfied.

Having identified and discussed all points of discontinuity, we can now conclude that f is continuous except at $x = -4, -2, 1,$ and 3. ◄

MATCHED PROBLEM 1 ➤ Use the definition of continuity to discuss the continuity of the function whose graph is shown in Figure 4.

$f(x)$

FIGURE 4 ◄

For functions defined by equations, it is also important to be able to locate points of discontinuity by examining the equation.

EXAMPLE 2 ➤ **Continuity of Functions Defined by Equations**   Using the definition of continuity, discuss the continuity of each function at the indicated point(s).

(A) $f(x) = x + 2$ at $x = 2$

(B) $g(x) = \dfrac{x^2 - 4}{x - 2}$ at $x = 2$

(C) $h(x) = \dfrac{|x|}{x}$ at $x = 0$ and at $x = 1$

Solution   (A) f is continuous at $x = 2$, since

$$\lim_{x \to 2} f(x) = 4 = f(2)$$

(See Fig. 1A.)

(B) g is not continuous at $x = 2$, since $g(2) = 0/0$ is not defined. (See Fig. 1B.)

(C) h is not continuous at $x = 0$, since $h(0) = |0|/0$ is not defined; also, $\lim_{x \to 0} h(x)$ does not exist.

h is continuous at $x = 1$, since

$$\lim_{x \to 1} \frac{|x|}{x} = 1 = h(1)$$

(See Fig. 1C.)   ◄

MATCHED PROBLEM 2 ► Using the definition of continuity, discuss the continuity of each function at the indicated point(s).

(A) $f(x) = x + 1$ at $x = 1$ (B) $g(x) = \dfrac{x^2 - 1}{x - 1}$ at $x = 1$

(C) $h(x) = \dfrac{x - 1}{|x - 1|}$ at $x = 1$ and at $x = 0$ ◄

explore – discuss 3

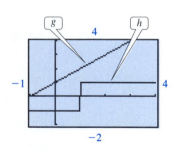

FIGURE 5

The graphing utility graphs of

$$g(x) = \frac{x^2 - 1}{x - 1} \qquad \text{and} \qquad h(x) = \frac{x - 1}{|x - 1|}$$

shown in Figure 5 seem to indicate that these functions are continuous at $x = 1$. But the results in Matched Problem 2 show that both functions are discontinuous at $x = 1$. Resolve this apparent contradiction by changing the window variables Xmin and Xmax so that the graphing utility shows the correct behavior of each function at $x = 1$.

We can also talk about one-sided continuity, just as we talked about one-sided limits. For example, a function is said to be **continuous on the right** at $x = c$ if $\lim_{x \to c^+} f(x) = f(c)$ and **continuous on the left** at $x = c$ if $\lim_{x \to c^-} f(x) = f(c)$. A function is **continuous on the closed interval [a, b]** if it is continuous on the open interval (a, b) and is continuous on the right at a and continuous on the left at b.

Figure 6A illustrates a function that is continuous on the closed interval $[-1, 1]$. Figure 6B illustrates a function that is continuous on a half-closed interval $[0, \infty)$.

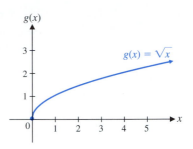

(A) f is continuous on the
closed interval $[-1, 1]$

(B) g is continuous on the
half-closed interval $[0, \infty)$

FIGURE 6
Continuity on closed and half-closed intervals

■ CONTINUITY PROPERTIES

Functions have some useful **general continuity properties:**

> **If two functions are continuous on the same interval, then their sum, difference, product, and quotient are continuous on the same interval, except for values of x that make a denominator 0.**

These properties, along with Theorem 1 below, enable us to determine intervals of continuity for some important classes of functions without having to look at their graphs or use the three conditions in the definition.

theorem 1

Continuity Properties of Some Specific Functions

(A) A constant function $f(x) = k$, where k is a constant, is continuous for all x.
  $f(x) = 7$ is continuous for all x.

(B) For n a positive integer, $f(x) = x^n$ is continuous for all x.
  $f(x) = x^5$ is continuous for all x.

(C) A polynomial function is continuous for all x.
  $2x^3 - 3x^2 + x - 5$ is continuous for all x.

(D) A rational function is continuous for all x except those values that make a denominator 0.
  $\dfrac{x^2 + 1}{x - 1}$ is continuous for all x except x = 1, a value that makes the denominator 0.

(E) For n an odd positive integer greater than 1, $\sqrt[n]{f(x)}$ is continuous wherever $f(x)$ is continuous.
  $\sqrt[3]{x^2}$ is continuous for all x.

(F) For n an even positive integer, $\sqrt[n]{f(x)}$ is continuous wherever $f(x)$ is continuous and nonnegative.
  $\sqrt[4]{x}$ is continuous on the interval $[0, \infty)$.

Notice that Theorem 1C follows from 1A, 1B, and the general continuity properties stated above. Also, note that Theorem 1D follows from 1C and the general continuity properties, since a rational function is a function that can be expressed as the quotient of two polynomials.

EXAMPLE 3 ▶ **Using Continuity Properties**   Using Theorem 1 and the general properties of continuity, determine where each function is continuous.

(A) $f(x) = x^2 - 2x + 1$ (B) $f(x) = \dfrac{x}{(x + 2)(x - 3)}$

(C) $f(x) = \sqrt[3]{x^2 - 4}$ (D) $f(x) = \sqrt{x - 2}$

Solution   (A) Since f is a polynomial function, f is continuous for all x.
(B) Since f is a rational function, f is continuous for all x except -2 and 3 (values that make the denominator 0).
(C) The polynomial function $x^2 - 4$ is continuous for all x. Since $n = 3$ is odd, f is continuous for all x.
(D) The polynomial function $x - 2$ is continuous for all x and nonnegative for $x \geqslant 2$. Since $n = 2$ is even, f is continuous for $x \geqslant 2$, or on the interval $[2, \infty)$.
◀

MATCHED PROBLEM 3 ▶ Using Theorem 1 and the general properties of continuity, determine where each function is continuous.

(A) $f(x) = x^4 + 2x^2 + 1$ (B) $f(x) = \dfrac{x^2}{(x + 1)(x - 4)}$

(C) $f(x) = \sqrt{x - 4}$ (D) $f(x) = \sqrt[3]{x^3 + 1}$ ◀

■ **INFINITE LIMITS**

A function is discontinuous at any point c where $\lim_{x \to c} f(x)$ does not exist. For example, if the one-sided limits are different at $x = c$, then the limit does not exist and the function is discontinuous at $x = c$ (see Fig. 3). Another situation where a limit may fail to exist involves functions whose values become very large as x approaches c. The special symbol ∞ is often used to describe this type of behavior. To illustrate this case, consider the functions

$$f(x) = \frac{1}{x} \quad \text{and} \quad g(x) = \frac{1}{x^2}$$

The graph of f in Figure 7A indicates that the values of $f(x)$ are very large negative numbers if x is near 0 on the left and very large positive numbers if x is near 0 on the right. In neither case does the limit exist. However, it is convenient to use the special symbol ∞ to describe the nature of the graph at $x = 0$. Thus, we write

$$\lim_{x \to 0^-} f(x) = -\infty \quad \text{and} \quad \lim_{x \to 0^+} f(x) = \infty$$

Since these two statements represent different types of behavior, we cannot write a single limit statement to describe the nature of the graph at $x = 0$.

(A) $y_1 = f(x) = \dfrac{1}{x}$ (B) $\displaystyle\lim_{x\to 0^-} \dfrac{1}{x} = -\infty$ (C) $\displaystyle\lim_{x\to 0^+} \dfrac{1}{x} = \infty$

FIGURE 7

The graph of g in Figure 8A shows that the values of $g(x)$ are very large positive numbers when x is near 0 on either side of 0. Once again, the limit of $g(x)$ as x approaches 0 does not exist, but we can describe the behavior of the graph of g near 0 by writing

$$\lim_{x\to 0} g(x) = \infty$$

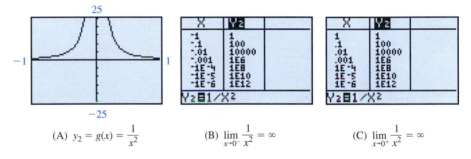

(A) $y_2 = g(x) = \dfrac{1}{x^2}$ (B) $\displaystyle\lim_{x\to 0^-} \dfrac{1}{x^2} = \infty$ (C) $\displaystyle\lim_{x\to 0^+} \dfrac{1}{x^2} = \infty$

FIGURE 8

For both functions f and g, the line $x = 0$ (the vertical axis) is a *vertical asymptote*. These ideas are summarized in the following box.

Vertical Asymptotes

If the limit of a function f fails to exist as x approaches c from the left because the values of $f(x)$ are becoming very large positive numbers (or very large negative numbers), we say that*

$$\lim_{x\to c^-} f(x) = \infty \quad (\text{or} -\infty)$$

---

* The precise definition of this limit statement is as follows: $\lim_{x\to c^-} f(x) = \infty$ if for each $N > 0$, there exists a $d > 0$ such that $f(x) > N$ whenever $c - d < x < c$. Similar statements can be made for limits from the right, unrestricted limits, and limits involving very large negative numbers.

(Continued)

656 CHAPTER 9 GRAPHING AND OPTIMIZATION

Vertical Asymptotes *(Continued)*

If this happens as x approaches c from the right, we say that

$$\lim_{x \to c^+} f(x) = \infty \quad (\text{or} - \infty)$$

If both one-sided limits exhibit the same behavior, we say that

$$\lim_{x \to c} f(x) = \infty \quad (\text{or} - \infty)$$

If any of the above hold, the line $x = c$ is a **vertical asymptote** for the graph of $y = f(x)$.

EXAMPLE 4 ➤ **Limits at Points of Discontinuity** For the function

$$f(x) = \frac{1 - x}{x^4 - x^2}$$

use ∞ and $-\infty$, as appropriate, to describe the behavior at each point of discontinuity, and identify all vertical asymptotes.

Solution First, we factor the denominator and identify the points of discontinuity:

$$x^4 - x^2 = x^2(x^2 - 1) = x^2(x - 1)(x + 1)$$

Thus, f is discontinuous at $x = -1$, 0, and 1. We use a numerical approach to investigate the behavior of $f(x)$ near each of these discontinuities.

Behavior at $x = -1$ [values of $f(x)$ rounded to the nearest integer]:

TABLE 1

x	-1.01	-1.001	$-1.0001 \to$	-1	$\leftarrow -0.9999$	-0.999	-0.99
$f(x)$	98	998	$9{,}998 \to \infty$	$-\infty \leftarrow -10{,}002$		$-1{,}002$	-102

Examining the values of $f(x)$ near $x = -1$ (Table 1), we see that the values of $f(x)$ are large positive numbers for x near -1 on the left and large negative numbers for x near -1 on the right. Thus, we must use one-sided limits to describe the behavior at $x = -1$:

$$\lim_{x \to -1^-} \frac{1 - x}{x^4 - x^2} = \infty \quad \text{and} \quad \lim_{x \to -1^+} \frac{1 - x}{x^4 - x^2} = -\infty$$

This shows that the line $x = -1$ is a vertical asymptote for the graph of $y = f(x)$.

Behavior at $x = 0$ [values of $f(x)$ rounded to the nearest integer]:

TABLE 2

x	-0.01	-0.001	$\to 0 \leftarrow$	0.001	0.01
$f(x)$	$-10{,}101$	$-1{,}001{,}001 \to -\infty \leftarrow -999{,}001$			$-9{,}901$

Since the values of $f(x)$ for x near 0, on either side of 0, are very large negative numbers (Table 2), we can use a single limit statement to describe the behavior at $x = 0$:

$$\lim_{x \to 0} \frac{1 - x}{x^4 - x^2} = -\infty$$

Thus, the line $x = 0$ (the y axis) is a vertical asymptote for the graph of $y = f(x)$.

Behavior at $x = 1$ [values of $f(x)$ rounded to three decimal places]:

TABLE 3

x	0.9	0.99	$0.999 \to$	1	$\leftarrow 1.001$	1.01	1.1
$f(x)$	-0.650	-0.513	$-0.501 \to -0.5 \leftarrow -0.499$			-0.488	-0.394

The values in Table 3 suggest that $\lim_{x \to 1} f(x)$ exists. We confirm this by using algebraic simplification (notice that this limit is an indeterminate form):

$$\lim_{x \to 1} \frac{1 - x}{x^4 - x^2} = \lim_{x \to 1} \frac{1 - x}{x^2(x + 1)(x - 1)} \qquad \frac{0}{0} \text{ indeterminate form}$$

$$= \lim_{x \to 1} \frac{-1}{x^2(x + 1)} \qquad \frac{1 - x}{x - 1} = -1, x \neq 1$$

$$= -0.5$$

Since this limit exists, there is no vertical asymptote at $x = 1$.

The graph of $y = f(x)$ (Fig. 9) illustrates the behavior indicated by all these limit statements.

$$\lim_{x \to -1^-} f(x) = \infty \qquad \lim_{x \to 0} f(x) = -\infty$$

$$\lim_{x \to -1^+} f(x) = -\infty \qquad \lim_{x \to 1} f(x) = -0.5$$

We will have much more to say about sketching and analyzing graphs involving asymptotes in Section 9-4. ◄

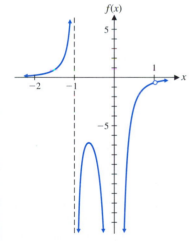

FIGURE 9

$f(x) = \dfrac{1 - x}{x^4 - x^2}$

MATCHED PROBLEM 4 ► For the function

$$f(x) = \frac{x - 3}{x^2 - 4x + 3}$$

use ∞ and $-\infty$, as appropriate, to describe the behavior at each point of discontinuity, and identify all vertical asymptotes. ◄

Caution

Figure 10A shows the graph of the function f from Example 4 on a graphing utility. It appears that the graphing utility has also drawn the vertical asymptote at $x = -1$, but this is not the case. As we saw in Example 4, points close to -1 on the left have large positive y coordinates, while points close to -1 on the right have large negative y coordinates. For the given x range on this particular graphing utility, there are no points on the screen with x coordinate -1. The graphing utility simply connected the last point to the left of -1 with the first point to the right of -1. Since these points are not visible on the screen, this gives the appearance of a vertical asymptote.

Figure 10B shows the graph of the same function with a much larger y range. Now these two points are visible and, in fact, the graph appears to be continuous at $x = -1$, which we know is not true. When you graph functions with vertical asymptotes on a graphing utility, you should proceed as we did in Example 4 to identify the asymptotes first. You cannot depend on the graphing utility to identify asymptotes.

(A)

(B)

FIGURE 10

■ SIGN CHARTS

FIGURE 11
$y_1 = 2x - 3$

Knowing where a function is positive or negative can be very useful in the analysis of functions, as we will see in the following sections. This information is frequently displayed in the form of a **sign chart**, a specially labeled line graph that shows at a glance where the function is positive and where it is negative. For example, from the graph of the linear function $f(x) = 2x - 3$ shown in Figure 11, we see that $f(x)$ is negative to the left of the x intercept ($x = 1.5$) and positive to the right of the x intercept. This information is summarized in the following sign chart for $f(x) = 2x - 3$:

The sign chart for the linear function $f(x) = 2x - 3$ was easy to construct because we are very familiar with the properties of a line. For more complicated functions, we can use continuity properties as an aid in constructing sign charts. The properties we will need are discussed below.

Suppose a function f is continuous over the interval $(1, 8)$ and $f(x) \neq 0$ for any x in $(1, 8)$. Also suppose $f(2) = 5$, a positive number. Is it possible for $f(x)$ to be negative for any x in $(1, 8)$? The answer is "no." If $f(7)$ were -3, for example, as shown in Figure 12, how would it be possible to join the points $(2, 5)$ and $(7, -3)$ with the graph of a continuous function without crossing the x axis between 1 and 8 at least once? [Crossing the x axis would violate our assumption that $f(x) \neq 0$ for any x in $(1, 8)$.] Thus, we conclude that $f(x)$ must be positive for all x in $(1, 8)$. If $f(2)$ were negative, then, using the same type of reasoning, $f(x)$ would have to be negative over the whole interval $(1, 8)$.

In general, **if f is continuous and $f(x) \neq 0$ on the interval (a, b), then $f(x)$ cannot change sign on (a, b).** This is the essence of Theorem 2.

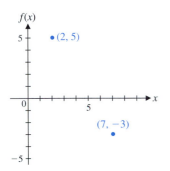

f(x)
• (2, 5)
(7, −3)

FIGURE 12

theorem 2

Sign Properties on an Interval (a, b)

If f is continuous on (a, b) and $f(x) \neq 0$ for all x in (a, b), then either $f(x) > 0$ for all x in (a, b) or $f(x) < 0$ for all x in (a, b).

Theorem 2 provides the basis for an effective method of constructing sign charts. Example 5 illustrates the process.

EXAMPLE 5 ➤ Constructing Sign Charts   Construct a sign chart for

$$f(x) = \frac{x + 1}{x - 2}$$

and use it to solve the following inequalities:

(A) $f(x) > 0$ (B) $f(x) \leq 0$

Solution   The given rational function is discontinuous at $x = 2$, and $f(x) = 0$ at $x = -1$. Figure 13 shows a graph of f.

We begin the construction of the sign chart by plotting $x = 2$ and $x = -1$, which we call *partition numbers,* on a real number line:

8
−5   9
−6

FIGURE 13
Graph of f

We use an open dot at 2 and the abbreviation ND to note the fact that $f(x)$ is *not defined* at $x = 2$. The partition numbers 2 and -1 determine three open intervals: $(-\infty, -1)$, $(-1, 2)$, and $(2, \infty)$. Since we formed these intervals by excluding the points of discontinuity and the zeros for f, we can conclude that the function f is continuous and nonzero on each of these intervals. From Theorem 2, we know that $f(x)$ does not change sign on any of these intervals, including those portions of the intervals that are not visible in Figure 13. Thus, we can find the sign of $f(x)$ on each of these intervals by examining the graph of f over any interval that contains all the partition numbers. Using the graph in Figure 13, we complete the sign chart for $f(x)$.

Sign chart for $f(x) = \dfrac{x + 1}{x - 2}$:

(A) From the sign chart, the solution of the inequality

$$f(x) > 0 \qquad \text{is} \qquad \begin{array}{ll} x < -1 \text{ or } x > 2 & \text{\color{blue}{Inequality notation}} \\ (-\infty, -1) \cup (2, \infty) & \text{\color{blue}{Interval notation}} \end{array}$$

(B) To solve $f(x) \leq 0$, we proceed as in part (A) and then check each partition number where f is defined to see if f satisfies the given inequality. Since f is defined at $x = -1$ and $f(-1) = 0$ satisfies the inequality $f(x) \leq 0$, we have

$$f(x) \leq 0 \qquad \text{for} \qquad \begin{array}{ll} -1 \leq x < 2 & \text{\color{blue}{Inequality notation}} \\ [-1, 2) & \text{\color{blue}{Interval notation}} \end{array} \qquad \blacktriangleleft$$

MATCHED PROBLEM 5 ➤ Construct a sign chart for

$$f(x) = \frac{x^2 - 1}{x - 3}$$

and use it to solve:

(A) $f(x) < 0$ (B) $f(x) \geq 0$ $\qquad\qquad\qquad\blacktriangleleft$

In general, given a function f, we will call all values x such that f is discontinuous at x or $f(x) = 0$ **partition numbers. These numbers determine open intervals where $f(x)$ does not change sign.** Once the partition numbers have been determined, it is an easy matter to use the graph of the function to construct a sign chart for the function.

EXAMPLE 6 ➤ **Constructing Sign Charts with a Graphing Utility**   Construct a sign chart for $f(x) = x^3 - 4x^2 + x + 4$. Round partition numbers to two decimal places.

Solution

The graph of f in Figure 14 shows three points where $f(x) = 0$. Since $f(x)$ is a cubic polynomial, there are no other solutions to the equation $f(x) = 0$. We use a root approximation routine to find these solutions correct to two decimal places (details omitted):

$$x = -0.81, \quad x = 1.47, \quad \text{and} \quad x = 3.34$$

These are the only partition numbers for f. Plotting these partition numbers on a number line and using Figure 14 to determine the sign of f over each interval formed by the partition numbers produces the sign chart shown below.

Sign chart for $f(x) = x^3 - 4x^2 + x + 4$:

FIGURE 14
$y_1 = x^3 - 4x^2 + x + 4$

Another approach to forming a sign chart is to graph $f(x)/|f(x)|$ as shown in Figure 15. This figure conveys the same information as the sign chart. The portions of this graph that lie below the x axis correspond to the minus signs in the sign chart, and the portions that lie above the x axis correspond to the plus signs in the sign chart. ◀

In some applications of sign charts, we will want to form the complete chart showing the partition numbers and the intervals they determine. However, in many cases, it will suffice to work with either the graph of the function (Fig. 14) or a graphing utility representation of a sign chart (Fig. 15).

FIGURE 15
$y_2 = y_1/|y_1|$

MATCHED PROBLEM 6 ▶

Let $f(x) = -x^3 + 4x^2 - 2x - 2$. Construct a completely labeled sign chart for $f(x)$, and then find a graphing utility representation of this chart. Round partition numbers to two decimal places. ◀

Answers to Matched Problems

1. f is not continuous at $x = -3, -1, 2,$ and 4.

   $x = -3$: $\lim_{x \to -3} f(x) = 3$, but $f(-3)$ does not exist

   $x = -1$: $f(-1) = 1$, but $\lim_{x \to -1} f(x)$ does not exist

   $x = 2$: $\lim_{x \to 2} f(x) = 5$, but $f(2) = 3$

   $x = 4$: $\lim_{x \to 4} f(x)$ does not exist, and $f(4)$ does not exist

2. (A) f is continuous at $x = 1$, since $\lim_{x \to 1} f(x) = 2 = f(1)$.

   (B) g is not continuous at $x = 1$, since $g(1)$ is not defined.

   (C) h is not continuous at $x = 1$ for two reasons: $h(1)$ does not exist and $\lim_{x \to 1} h(x)$ does not exist.

   h is continuous at $x = 0$, since $\lim_{x \to 0} h(x) = -1 = h(0)$.

3. (A) Since f is a polynomial function, f is continuous for all x.

(B) Since f is a rational function, f is continuous for all x except -1 and 4 (values that make the denominator 0).

(C) The polynomial function $x - 4$ is continuous for all x and nonnegative for $x \geq 4$. Since $n = 2$ is even, f is continuous for $x \geq 4$, or on the interval $[4, \infty)$.

(D) The polynomial function $x^3 + 1$ is continuous for all x. Since $n = 3$ is odd, f is continuous for all x.

4. f has a vertical asymptote at $x = 1$, since $\lim_{x \to 1^-} f(x) = -\infty$ and $\lim_{x \to 1^+} f(x) = \infty$; f is discontinuous at $x = 3$, since $f(3)$ does not exist, but there is no vertical asymptote at $x = 3$, since $\lim_{x \to 3} f(x) = 0.5$.

5.

(A) $-\infty < x < -1$ or $1 < x < 3$; $(-\infty, -1) \cup (1, 3)$

(B) $-1 \leq x \leq 1$ or $3 < x < \infty$; $[-1, 1] \cup (3, \infty)$

6.

EXERCISE 9-1

A *In Problems 1–8, sketch a possible graph of a function that satisfies the given conditions at $x = 1$, and discuss the continuity of f at $x = 1$.*

1. $f(1) = 2$ and $\lim_{x \to 1} f(x) = 2$

2. $f(1) = -2$ and $\lim_{x \to 1} f(x) = 2$

3. $f(1) = 2$ and $\lim_{x \to 1} f(x) = -2$

4. $f(1) = -2$ and $\lim_{x \to 1} f(x) = -2$

5. $f(1) = -2$, $\lim_{x \to 1^-} f(x) = -2$, and $\lim_{x \to 1^+} f(x) = -2$

6. $f(1) = 2$, $\lim_{x \to 1^-} f(x) = 2$, and $\lim_{x \to 1^+} f(x) = 2$

7. $f(1) = -2$, $\lim_{x \to 1^-} f(x) = 2$, and $\lim_{x \to 1^+} f(x) = -2$

8. $f(1) = 2$, $\lim_{x \to 1^-} f(x) = 2$, and $\lim_{x \to 1^+} f(x) = -2$

In Problems 9–12, sketch a possible graph of a function that is continuous for all x except x = 1 and satisfies the given conditions at x = 1.

9. $\lim\limits_{x \to 1} f(x) = -\infty$ **10.** $\lim\limits_{x \to 1} f(x) = \infty$

11. $\lim\limits_{x \to 1^-} f(x) = -\infty$ and $\lim\limits_{x \to 1^+} f(x) = \infty$

12. $\lim\limits_{x \to 1^-} f(x) = \infty$ and $\lim\limits_{x \to 1^+} f(x) = -\infty$

Use Theorem 1 to determine where each function in Problems 13–18 is continuous.

13. $f(x) = 2x - 3$ **14.** $g(x) = 3 - 5x$

15. $h(x) = \dfrac{2}{x - 5}$ **16.** $k(x) = \dfrac{x}{x + 3}$

17. $g(x) = \dfrac{x - 5}{(x - 3)(x + 2)}$

18. $F(x) = \dfrac{1}{x(x + 7)}$

B   Problems 19–24 refer to the function f shown in the graph. Use the graph to estimate limits as outlined below.

Figure for 19–24

For each value of c:

(A)   Find $\lim\limits_{x \to c^-} f(x)$, $\lim\limits_{x \to c^+} f(x)$, $\lim\limits_{x \to c} f(x)$, and $f(c)$.

(B)   Is f continuous at x = c? Explain.

19. $c = 0$ **20.** $c = -1$ **21.** $c = 1$

22. $c = 2$ **23.** $c = -2$ **24.** $c = 0.5$

25. Given the following function f:

$$f(x) = \begin{cases} 2 & \text{if } x \text{ is an integer} \\ 1 & \text{if } x \text{ is not an integer} \end{cases}$$

(A) Graph f.   (B) $\lim\limits_{x \to 2} f(x) = ?$ (C) $f(2) = ?$

(D) Is f continuous at x = 2?
(E) Where is f discontinuous?

26. Given the following function g:

$$g(x) = \begin{cases} -1 & \text{if } x \text{ is an even integer} \\ 1 & \text{if } x \text{ is not an even integer} \end{cases}$$

(A) Graph g.   (B) $\lim\limits_{x \to 1} g(x) = ?$ (C) $g(1) = ?$

(D) Is g continuous at x = 1?
(E) Where is g discontinuous?

In Problems 27–36, use $-\infty$ or ∞ where appropriate to describe the behavior at each point of discontinuity, and identify all vertical asymptotes.

27. $f(x) = \dfrac{1}{x + 3}$ **28.** $g(x) = \dfrac{x}{4 - x}$

29. $h(x) = \dfrac{x^2 + 4}{x^2 - 4}$ **30.** $k(x) = \dfrac{x^2 - 9}{x^2 + 9}$

31. $F(x) = \dfrac{x^2 - 4}{x^2 + 4}$ **32.** $G(x) = \dfrac{x^2 + 9}{9 - x^2}$

33. $H(x) = \dfrac{x^2 - 2x - 3}{x^2 - 4x + 3}$ **34.** $K(x) = \dfrac{x^2 + 2x - 3}{x^2 - 4x + 3}$

35. $T(x) = \dfrac{8x - 16}{x^4 - 8x^3 + 16x^2}$

36. $S(x) = \dfrac{6x + 9}{x^4 + 6x^3 + 9x^2}$

In Problems 37–48, construct a completely labeled sign chart for f(x). Round partition numbers to two decimal places. Then solve the indicated inequality. Express answers in both inequality and interval notation.

37. $f(x) = x^2 - x - 12; f(x) < 0$

38. $f(x) = x^2 - 2x - 8; f(x) < 0$

39. $f(x) = x^2 - 10x + 21; f(x) > 0$

40. $f(x) = x^2 + 7x + 10; f(x) > 0$

41. $f(x) = x^3 - 3x^2 - 2x + 5; f(x) < 0$

42. $f(x) = x^3 + 3x^2 - 4x - 8; f(x) < 0$

43. $f(x) = x^4 - 6x^2 + 3x + 5; f(x) > 0$

44. $f(x) = x^4 - 4x^2 - 2x + 2; f(x) > 0$

45. $f(x) = \dfrac{x^2 + 5x}{x - 3}; f(x) > 0$

46. $f(x) = \dfrac{x - 4}{x^2 + 2x}; f(x) < 0$

47. $f(x) = \dfrac{3 + 6x - x^3}{x - 1}; f(x) \le 0$

48. $f(x) = \dfrac{x^3 - 5x + 1}{x + 1}; f(x) \ge 0$

Use Theorem 1 to determine where each function in Problems 49–56 is continuous. Express the answer in interval notation.

49. $F(x) = 2x^8 - 3x^4 + 5$

50. $h(x) = \dfrac{x^4 - 3x + 5}{x^2 + 2x}$

51. $g(x) = \sqrt{x - 5}$

52. $f(x) = \sqrt{3 - x}$

53. $K(x) = \sqrt[3]{x - 5}$

54. $H(x) = \sqrt[3]{3 - x}$

55. $f(x) = \dfrac{x^2 - 1}{x^2 - 3x + 2}$

56. $k(x) = \dfrac{x^2 - 4}{x^2 + x - 2}$

In Problems 57–62, graph f and locate all points of discontinuity.

57. $f(x) = \begin{cases} 1 + x & \text{if } x < 1 \\ 5 - x & \text{if } x \geq 1 \end{cases}$

58. $f(x) = \begin{cases} x^2 & \text{if } x \leq 1 \\ 2x & \text{if } x > 1 \end{cases}$

59. $f(x) = \begin{cases} 1 + x & \text{if } x \leq 2 \\ 5 - x & \text{if } x > 2 \end{cases}$

60. $f(x) = \begin{cases} x^2 & \text{if } x \leq 2 \\ 2x & \text{if } x > 2 \end{cases}$

61. $f(x) = \begin{cases} -x & \text{if } x < 0 \\ 1 & \text{if } x = 0 \\ x & \text{if } x > 0 \end{cases}$

62. $f(x) = \begin{cases} 1 & \text{if } x < 0 \\ 0 & \text{if } x = 0 \\ 1 + x & \text{if } x > 0 \end{cases}$

C

63. Use the graph of the function g to answer the following questions:

(A) Is g continuous on the open interval $(-1, 2)$?

(B) Is g continuous from the right at $x = -1$? That is, does $\lim_{x \to -1^+} g(x) = g(-1)$?

(C) Is g continuous from the left at $x = 2$? That is, does $\lim_{x \to 2^-} g(x) = g(2)$?

(D) Is g continuous on the closed interval $[-1, 2]$?

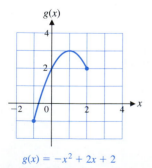

$g(x) = -x^2 + 2x + 2$

Figure for 63

64. Use the graph of the function f to answer the following questions:

(A) Is f continuous on the open interval $(0, 3)$?

(B) Is f continuous from the right at $x = 0$? That is, does $\lim_{x \to 0^+} f(x) = f(0)$?

(C) Is f continuous from the left at $x = 3$? That is, does $\lim_{x \to 3^-} f(x) = f(3)$?

(D) Is f continuous on the closed interval $[0, 3]$?

$f(x) = x^2 - 2x + 2$

Figure for 64

*Problems 65 and 66 refer to the **greatest integer function,** which is denoted by $[\![x]\!]$ and is defined as follows:*

$$[\![x]\!] = \text{Greatest integer} \leq x$$

For example,

$$[\![-3.6]\!] = \text{Greatest integer} \leq -3.6 = -4$$
$$[\![2]\!] = \text{Greatest integer} \leq 2 = 2$$
$$[\![2.5]\!] = \text{Greatest integer} \leq 2.5 = 2$$

The graph of $f(x) = [\![x]\!]$ is shown. There, we can see that

$$\begin{aligned} [\![x]\!] = -2 & \quad \text{for} & -2 \leq x < -1 \\ [\![x]\!] = -1 & \quad \text{for} & -1 \leq x < 0 \\ [\![x]\!] = 0 & \quad \text{for} & 0 \leq x < 1 \\ [\![x]\!] = 1 & \quad \text{for} & 1 \leq x < 2 \\ [\![x]\!] = 2 & \quad \text{for} & 2 \leq x < 3 \end{aligned}$$

and so on.

$f(x) = [\![x]\!]$

Figure for 65 and 66

65. (A) Is f continuous from the right at $x = 0$?
(B) Is f continuous from the left at $x = 0$?
(C) Is f continuous on the open interval $(0, 1)$?
(D) Is f continuous on the closed interval $[0, 1]$?
(E) Is f continuous on the half-closed interval $[0, 1)$?

66. (A) Is f continuous from the right at $x = 2$?
(B) Is f continuous from the left at $x = 2$?
(C) Is f continuous on the open interval $(1, 2)$?
(D) Is f continuous on the closed interval $[1, 2]$?
(E) Is f continuous on the half-closed interval $[1, 2)$?

In Problems 67–70, sketch a possible graph for a function f that is continuous for all real numbers and satisfies the given conditions. Find the x intercepts for f.

67. $f(x) < 0$ on $(-\infty, -5)$ and $(2, \infty)$; $f(x) > 0$ on $(-5, 2)$.

68. $f(x) > 0$ on $(-\infty, -4)$ and $(3, \infty)$; $f(x) < 0$ on $(-4, 3)$.

69. $f(x) < 0$ on $(-\infty, -6)$ and $(-1, 4)$; $f(x) > 0$ on $(-6, -1)$ and $(4, \infty)$.

70. $f(x) > 0$ on $(-\infty, -3)$ and $(2, 7)$; $f(x) < 0$ on $(-3, 2)$ and $(7, \infty)$.

71. The function $f(x) = 2/(1 - x)$ satisfies $f(0) = 2$ and $f(2) = -2$. Is f equal to 0 anywhere on the interval $(-1, 3)$? Does this contradict Theorem 2? Explain.

72. The function $f(x) = 6/(x - 4)$ satisfies $f(2) = -3$ and $f(7) = 2$. Is f equal to 0 anywhere on the interval $(0, 9)$? Does this contradict Theorem 2? Explain.

73. The function f is continuous and never 0 on the interval $(0, 4)$, and continuous and never 0 on the interval $(4, 8)$. Also, $f(2) = 3$ and $f(6) = -3$. Discuss the validity of the following statement and illustrate your conclusions with graphs: Either $f(4) = 0$ or f is discontinuous at $x = 4$.

74. The function f is continuous and never 0 on the interval $(-3, 1)$, and continuous and never 0 on the interval $(1, 4)$. Also, $f(-2) = -3$ and $f(3) = 4$. Discuss the validity of the following statement and illustrate your conclusions with graphs: Either $f(1) = 0$ or f is discontinuous at $x = 1$.

▶ **APPLICATIONS**

Business & Economics

75. *Postal rates.* First-class postage in 1995 was \$0.32 for the first ounce (or any fraction thereof) and \$0.23 for each additional ounce (or fraction thereof) up to 11 ounces. If $P(x)$ is the amount of postage for a letter weighing x ounces, then we can write

$$P(x) = \begin{cases} \$0.32 & \text{if } 0 < x \leq 1 \\ \$0.55 & \text{if } 1 < x \leq 2 \\ \$0.78 & \text{if } 2 < x \leq 3 \\ \text{and so on} \end{cases}$$

(A) Graph P for $0 < x \leq 5$.
(B) Find $\lim_{x \to 4.5} P(x)$ and $P(4.5)$.

(C) Find $\lim_{x \to 4} P(x)$ and $P(4)$.
(D) Is P continuous at $x = 4.5$? At $x = 4$?

76. *Telephone rates.* A person placing a station-to-station call on Saturday from San Francisco to New York is charged \$0.30 for the first minute (or any fraction thereof) and \$0.20 for each additional minute (or fraction thereof). If the length of a call is x minutes, then the long-distance charge $R(x)$ is

$$R(x) = \begin{cases} \$0.30 & \text{if } 0 < x \leq 1 \\ \$0.50 & \text{if } 1 < x \leq 2 \\ \$0.70 & \text{if } 2 < x \leq 3 \\ \text{and so on} \end{cases}$$

(A) Graph R for $0 < x \leq 6$.
(B) Find $\lim_{x \to 2.5} R(x)$ and $R(2.5)$.
(C) Find $\lim_{x \to 2} R(x)$ and $R(2)$.
(D) Is R continuous at $x = 2.5$? At $x = 2$?

77. *Pricing.* An office products firm sells custom-printed pencils for companies to use for promotional purposes. The minimum order is 150 pencils, and discounts are given for volume purchases, as shown in the table. If x is the number of pencils ordered, then the price per pencil is \$0.49 for $150 \leq x < 250$, \$0.39 for $250 \leq x < 500$, and so on.

(A) Let $y = p(x)$ represent the price per pencil. Graph $y = p(x)$ for $150 \leq x \leq 1{,}500$.

(B) Identify the discontinuities of p and discuss the behavior at each discontinuity.

(C) Let $y = C(x)$ be the total cost for an order of x pencils. Graph $y = C(x)$ for $150 \leq x \leq 1{,}500$.

(D) Identify the discontinuities of C and discuss the behavior at each discontinuity.

Table for 77

QUANTITY ORDERED	150	250	500	1,000 or more
PRICE PER PENCIL	$0.49	$0.39	$0.29	$0.24

78. *Pricing.* The office products firm in Problem 77 also sells custom-printed pens. The minimum order is 200 pens, and discounts are given for volume purchases, as shown in the table. Let x represent the number of pens ordered.

(A) Let $y = p(x)$ represent the price per pen. Graph $y = p(x)$ for $200 \leq x \leq 900$.

(B) Identify the discontinuities of p and discuss the behavior at each discontinuity.

(C) Let $y = C(x)$ be the total cost for an order of x pens. Graph $y = C(x)$ for $200 \leq x \leq 900$.

(D) Identify the discontinuities of C and discuss the behavior at each discontinuity.

Table for 78

QUANTITY ORDERED	200	300	500	700 or more
PRICE PER PEN	$1.19	$1.14	$1.04	$0.93

79. *Income.* A personal computer salesperson receives a base salary of $1,000 per month and a commission of 5% of all sales over $10,000 during the month. If the monthly sales are $20,000 or more, the salesperson is given an additional $500 bonus. Let $E(s)$ represent the person's earnings during the month as a function of the monthly sales s.

(A) Graph $E(s)$ for $0 \leq s \leq 30{,}000$.

(B) Find $\lim_{s \to 10{,}000} E(s)$ and $E(10{,}000)$.

(C) Find $\lim_{s \to 20{,}000} E(s)$ and $E(20{,}000)$.

(D) Is E continuous at $s = 10{,}000$? At $s = 20{,}000$?

80. *Equipment rental.* An office equipment rental and leasing company rents typewriters for $10 per day (and any fraction

thereof) or for $50 per 7 day week. Let $C(x)$ be the cost of renting a typewriter for x days.

(A) Graph $C(x)$ for $0 \leq x \leq 10$.

(B) Find $\lim_{x \to 4.5} C(x)$ and $C(4.5)$.

(C) Find $\lim_{x \to 8} C(x)$ and $C(8)$.

(D) Is C continuous at $x = 4.5$? At $x = 8$?

Life Sciences

81. *Animal supply.* A medical laboratory raises its own rabbits. The number of rabbits $N(t)$ available at any time t depends on the number of births and deaths. When a birth or death occurs, the function N generally has a discontinuity, as shown in the figure.

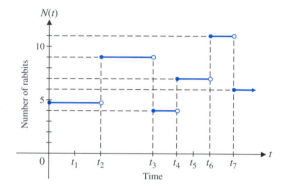

Figure for 81

(A) Where is the function N discontinuous?

(B) $\lim_{t \to t_5} N(t) = ?$; $N(t_5) = ?$

(C) $\lim_{t \to t_3} N(t) = ?$; $N(t_3) = ?$

Social Sciences

82. *Learning.* The graph might represent the history of a particular person learning the material on limits and continuity in

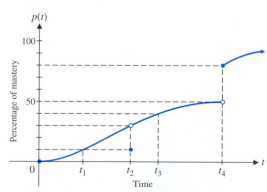

Figure for 82

this book. At time t_2, the student's mind goes blank during a quiz. At time t_4, the instructor explains a concept particularly well, and suddenly, a big jump in understanding takes place.

(A) Where is the function p discontinuous?
(B) $\lim_{t \to t_1} p(t) = \mathbf{?}$; $p(t_1) = \mathbf{?}$
(C) $\lim_{t \to t_2} p(t) = \mathbf{?}$; $p(t_2) = \mathbf{?}$
(D) $\lim_{t \to t_4} p(t) = \mathbf{?}$; $p(t_4) = \mathbf{?}$

S E C T I O N   9-2   ■   First Derivative and Graphs

- INCREASING AND DECREASING FUNCTIONS
- LOCAL EXTREMA
- FIRST-DERIVATIVE TEST
- ANALYZING GRAPHS

Since the derivative is associated with the slope of the graph of a function at a point, we might expect that it is also associated with other properties of a graph. As we will see in this and the next section, the derivative can tell us a great deal about the shape of the graph of a function. In addition, this investigation will lead to methods for finding absolute maximum and minimum values of functions. Manufacturing companies can use these methods to find production levels that will minimize cost or maximize profit. Pharmacologists can use them to find levels of drug dosages that will produce maximum sensitivity to a drug. And so on.

■ Increasing and Decreasing Functions

explore – discuss 1

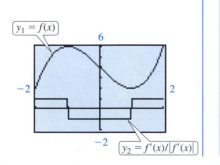

FIGURE 1

Figure 1 shows the graphs of $y_1 = f(x)$ and $y_2 = f'(x)/|f'(x)|$, where

$$f(x) = x^3 - 3x + 4$$

and

$$f'(x) = 3x^2 - 3 = 3(x + 1)(x - 1)$$

Thus, the graph of y_2 represents a sign chart for $f'(x)$. Discuss the relationship between the graph of f and the sign of $f'(x)$ over each interval where $f'(x)$ has constant sign. Also, describe the behavior of the graph of f at each partition number for $f'(x)$.

Graphs of functions generally have *rising* and *falling* sections (as we scan the graphs from left to right). Referring to the graph of $f(x) = x^3 - 3x + 4$ in Figure 1, we see that on the interval $(-2, -1)$, the graph of f is rising; that is, $f(x)$ is *increasing*,* and the slope of the graph is positive [$f'(x) > 0$]. On the other hand, on the

---

* Formally, we say that $f(x)$ is **increasing** on an interval (a, b) if $f(x_2) > f(x_1)$ whenever $a < x_1 < x_2 < b$; and f is **decreasing** on (a, b) if $f(x_2) < f(x_1)$ whenever $a < x_1 < x_2 < b$.

interval $(-1, 1)$, the graph of f is falling; that is, $f(x)$ is *decreasing,* and the slope of the graph is negative $[f'(x) < 0]$. Finally, on the interval $(1, 2)$, once again the graph of f is rising, $f(x)$ is increasing, and $f'(x) > 0$. At $x = -1$ and $x = 1$, the slope of the graph is 0 $[f'(x) = 0]$ and the tangent lines are horizontal.

In general, if $f'(x) > 0$ (is positive) on the interval (a, b), then $f(x)$ increases ($\nearrow$) and the graph of f rises as we move from left to right over the interval; if $f'(x) < 0$ (is negative) on an interval (a, b), then $f(x)$ decreases ($\searrow$) and the graph of f falls as we move from left to right over the interval. We summarize these important results in the box.

Increasing and Decreasing Functions

For the interval (a, b):

$f'(x)$	$f(x)$	GRAPH OF f	EXAMPLES
$+$	Increases $\nearrow$	Rises $\nearrow$	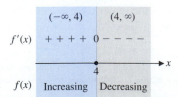
$-$	Decreases $\searrow$	Falls $\searrow$	

EXAMPLE 1 ➤ **Finding Intervals Where a Function Is Increasing or Decreasing** Given the function $f(x) = 8x - x^2$:

(A) Which values of x correspond to horizontal tangent lines?
(B) For which values of x is $f(x)$ increasing? Decreasing?
(C) Illustrate the results in parts (A) and (B) with a graph.

Solution (A) $f'(x) = 8 - 2x = 0$

$$x = 4$$

FIGURE 2
$f'(x) = 8 - 2x$

Thus, a horizontal tangent line exists at $x = 4$ only (see Fig. 2).

(B) We will construct a sign chart for $f'(x)$ to determine which values of x make $f'(x) > 0$ and which values make $f'(x) < 0$. Recall from Section 9-1 that the partition numbers for a function are the points where the function is 0 or discontinuous. Thus, when constructing a sign chart for $f'(x)$, we must locate all points where $f'(x) = 0$ or $f'(x)$ is discontinuous. From part (A) we know that $f'(x) = 8 - 2x = 0$ at $x = 4$. Since $f'(x) = 8 - 2x$ is a polynomial, it is continuous for all x. Thus, 4 is the only partition number. We construct a sign chart for the intervals $(-\infty, 4)$ and $(4, \infty)$, using the graph of f' in Figure 2 to determine the sign over each interval:

	$(-\infty, 4)$	$(4, \infty)$
$f'(x)$	$+ + + +$ 0	$- - - -$
		4 $\rightarrow x$
$f(x)$	Increasing	Decreasing

Thus, $f(x)$ is increasing on $(-\infty, 4)$ and decreasing on $(4, \infty)$.

(C)

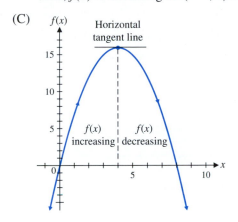

MATCHED PROBLEM 1 ▶ Repeat Example 1 for $f(x) = x^2 - 6x + 10$. ◀

As Example 1 illustrates, sign charts will play an important role in using the derivative to analyze the graph of a function f. The partition numbers for f' are central to the construction of these sign charts and also to the analysis of the graph of $y = f(x)$. We already know that if $f'(c) = 0$, then the graph of $y = f(x)$ will have a horizontal tangent line at $x = c$. But the partition numbers for f' also include the numbers c where $f'(c)$ does not exist. There are two possibilities at this type of number: $f(c)$ does not exist, or $f(c)$ exists, but the slope of the tangent line at $x = c$ is undefined.

Critical Values of f

The values of x in the domain of f where $f'(x) = 0$ or $f'(x)$ does not exist are called the **critical values** of f. (The critical values of f are always partition numbers for f', but f' may have partition numbers that are not critical values.)

It is important to understand that although f' may not be defined at a critical value c, f must be defined at c.

Critical values of a function f are always in the domain of f.

Example 2 will illustrate the relationship between critical values and partition numbers.

EXAMPLE 2 ▶ **Partition Numbers and Critical Values** For each function, find the partition numbers for f', the critical values for f, and determine the intervals where f is increasing and those where f is decreasing.

(A) $f(x) = 1 + x^3$ (B) $f(x) = (1 - x)^{1/3}$ (C) $f(x) = \dfrac{1}{x - 2}$

Solution   (A) $f(x) = 1 + x^3$ $f'(x) = 3x^2 = 0$

$$x = 0$$

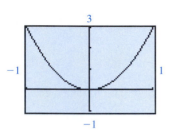

FIGURE 3
$f'(x) = 3x^2$

The only partition number for f' is $x = 0$ (see Fig. 3). Since 0 is in the domain of f, $x = 0$ is also the only critical value for f.

Sign chart for $f'(x) = 3x^2$ (partition number is 0):

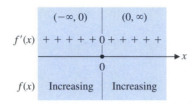

The sign chart indicates that $f(x)$ is increasing on $(-\infty, 0)$ and $(0, \infty)$. Since f is continuous at $x = 0$, it follows that $f(x)$ is increasing for all x. The graph of f is shown in Figure 4.

(B) $f(x) = (1 - x)^{1/3}$ $f'(x) = -\dfrac{1}{3}(1 - x)^{-2/3} = \dfrac{-1}{3(1 - x)^{2/3}}$

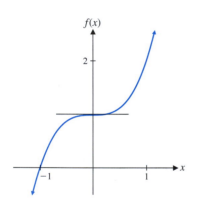

FIGURE 4

To find partition numbers for f', we note that f' is continuous for all x except for values of x for which the denominator is 0; that is, $f'(1)$ does not exist and f' is discontinuous at $x = 1$ (see Fig. 5). Since the numerator is the constant -1, $f'(x) \neq 0$ for any value of x. Thus, $x = 1$ is the only partition number for f'. Since 1 is in the domain of f, $x = 1$ is also the only critical value of f. As before, we use the abbreviation ND to note the fact that $f'(x)$ is not defined at $x = 1$.

Sign chart for $f'(x) = -1/[3(1 - x)^{2/3}]$ (partition number is 1):

FIGURE 5
$f'(x) = \dfrac{-1}{3[(1 - x)^2]^{1/3}}$*

The sign chart indicates that f is decreasing on $(-\infty, 1)$ and $(1, \infty)$. Since f is continuous at $x = 1$, it follows that $f(x)$ is decreasing for all x. Thus, **a continuous function can be decreasing (or increasing) on an interval containing values of x where $f'(x)$ does not exist.** The graph of f is shown in Figure 6. Notice that the undefined derivative at $x = 1$ results in a vertical tangent line at $x = 1$. In general, **a vertical tangent will occur at $x = c$ if f is continuous at $x = c$ and $|f'(x)|$ becomes larger and larger as x approaches c.**

---

* Note the special way we wrote $(1 - x)^{2/3}$. Many graphing utilities will not produce correct graphs for functions involving fractional exponents. In most cases, writing $x^{p/q}$ as $(x^p)^{1/q}$ should rectify the problem.

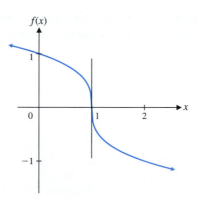

FIGURE 6

(C) $f(x) = \dfrac{1}{x - 2}$ $f'(x) = \dfrac{-1}{(x - 2)^2}$

To find the partition numbers for f', note that $f'(x) \neq 0$ for any x and f' is not defined at $x = 2$ (see Fig. 7). Thus, $x = 2$ is the only partition number for f'. However, $x = 2$ is *not* in the domain of f. Consequently, $x = 2$ is not a critical value of f. This function has no critical values.

Sign chart for $f'(x) = -1/(x - 2)^2$ (partition number is 2):

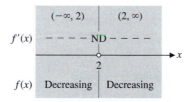

Thus, f is decreasing on $(-\infty, 2)$ and $(2, \infty)$. See the graph of f in Figure 8.

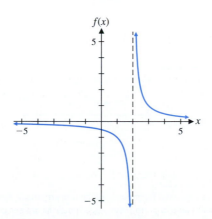

FIGURE 8

FIGURE 7

$f'(x) = \dfrac{-1}{(x - 2)^2}$

MATCHED PROBLEM 2 ➤   For each function, find the partition numbers for f', the critical values for f, and determine the intervals where f is increasing and those where f is decreasing.

(A) $f(x) = 1 - x^3$ (B) $f(x) = (1 + x)^{1/3}$ (C) $f(x) = \dfrac{1}{x}$

e x p l o r e – d i s c u s s   2

A student examined the sign chart in Example 2C and concluded that $f(x) = 1/(x - 2)$ is decreasing for all x except $x = 2$. But $f(1) = -1 < f(3) = 1$, which seems to indicate that f is increasing. Discuss the difference between the correct answer in Example 2C and the student's answer. Explain why the student's description of where f is decreasing is unacceptable.

Caution

Example 2C illustrates two important ideas.

1. Do not assume all partition numbers for the derivative f' are critical values of the function f. A partition number must also be in the domain of f in order to be a critical value.*

2. The values where a function is increasing or decreasing must always be expressed in terms of open intervals that are subsets of the domain of the function.

■ LOCAL EXTREMA

When the graph of a continuous function changes from rising to falling, a high point, or *local maximum*, occurs; and when the graph changes from falling to rising, a low point, or *local minimum*, occurs. In Figure 9, high points occur at c_3 and c_6, and low

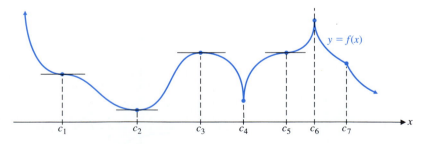

FIGURE 9

_____
* There do exist functions with the property that a partition number for f' is in the domain of f but is not a critical value of f. However, such functions are not considered in this text.

points occur at c_2 and c_4. In general, we call $f(c)$ a **local maximum** if there exists an interval (m, n) containing c such that

$$f(x) \leq f(c) \qquad \text{for all } x \text{ in } (m, n)$$

Note that this inequality need only hold for values of x near c; hence, the use of the term *local*.

The quantity $f(c)$ is called a **local minimum** if there exists an interval (m, n) containing x such that

$$f(x) \geq f(c) \qquad \text{for all } x \text{ in } (m, n)$$

The quantity $f(c)$ is called a **local extremum** if it is either a local maximum or a local minimum. A point on a graph where a local extremum occurs is also called a **turning point.** Thus, in Figure 9, we see that local maxima occur at c_3 and c_6, local minima occur at c_2 and c_4, and all four values produce local extrema. Also note that the local maximum $f(c_3)$ is not the highest point on the graph in Figure 9. Later in this chapter we will consider the problem of finding the highest and lowest points on a graph. For now, we are concerned only with locating local extrema.

How can we locate local maxima and minima if we are given the equation for a function and not its graph? Since a function must change from increasing to decreasing at a local maximum and from decreasing to increasing at a local minimum, we would expect local extrema to occur at points where the derivative changes sign. But this can occur only at critical values. Thus, we have the following theorem:

theorem 1

Existence of Local Extrema

If f is continuous on the interval (a, b) and $f(c)$ is a local extremum, then either $f'(c) = 0$ or $f'(c)$ does not exist (is not defined).

Theorem 1 states that a local extremum can occur only at a critical value, but it does not imply that every critical value produces a local extremum. In Figure 9, c_1 and c_5 are critical values (the slope is 0), but the function does not have a local maximum or local minimum at either of these values.

Our strategy for finding local extrema is now clear. We find all critical values for f and test each one to see if it produces a local maximum, a local minimum, or neither.

■ FIRST-DERIVATIVE TEST

If $f'(x)$ exists on both sides of a critical value c, then the sign of $f'(x)$ can be used to determine whether the point $(c, f(c))$ is a local maximum, a local minimum, or neither. The various possibilities are summarized in the box and illustrated in Figure 10.

First-Derivative Test for Local Extrema

Let c be a critical value of f [$f(c)$ defined and either $f'(c) = 0$ or $f'(c)$ not defined]. Construct a sign chart for $f'(x)$ close to and on either side of c.

SIGN CHART	$f(c)$
$f'(x)$ $\quad$ $-\,-\,-$ $\quad$ $+\,+\,+$ $\xleftarrow{\quad\quad m \quad\quad c \quad\quad n \quad}\to x$ $f(x)$ $\quad$ Decreasing $\quad$ Increasing	$f(c)$ is a local minimum. If $f'(x)$ changes from negative to positive at c, then $f(c)$ is a local minimum.
$f'(x)$ $\quad$ $+\,+\,+$ $\quad$ $-\,-\,-$ $\xleftarrow{\quad\quad m \quad\quad c \quad\quad n \quad}\to x$ $f(x)$ $\quad$ Increasing $\quad$ Decreasing	$f(c)$ is a local maximum. If $f'(x)$ changes from positive to negative at c, then $f(c)$ is a local maximum.
$f'(x)$ $\quad$ $-\,-\,-$ $\quad$ $-\,-\,-$ $\xleftarrow{\quad\quad m \quad\quad c \quad\quad n \quad}\to x$ $f(x)$ $\quad$ Decreasing $\quad$ Decreasing	$f(c)$ is not a local extremum. If $f'(x)$ does not change sign at c, then $f(c)$ is neither a local maximum nor a local minimum.
$f'(x)$ $\quad$ $+\,+\,+$ $\quad$ $+\,+\,+$ $\xleftarrow{\quad\quad m \quad\quad c \quad\quad n \quad}\to x$ $f(x)$ $\quad$ Increasing $\quad$ Increasing	$f(c)$ is not a local extremum. If $f'(x)$ does not change sign at c, then $f(c)$ is neither a local maximum nor a local minimum.

EXAMPLE 3 ➤ **Locating Local Extrema**   Given $f(x) = x^3 + 3x^2 - 20x - 20$:

(A) Find the critical values of f, correct to two decimal places.

(B) Construct a sign chart for $f'(x)$ and use it to locate local extrema. Describe the graph of f verbally.

(C) Illustrate the results from parts (A) and (B) graphically.

Solution   (A) Find all numbers x in the domain of f where $f'(x) = 0$ or $f'(x)$ does not exist. First, find $f'(x)$:

$$f'(x) = 3x^2 + 6x - 20 = 0 \qquad\qquad\qquad (1)$$

FIGURE 10
Local extrema

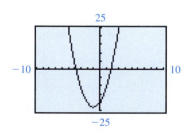

FIGURE 11
$f'(x) = 3x^2 + 6x - 20$

The quadratic formula can be used to find the exact solutions to this equation. However, decimal approximations are generally more useful than exact values when analyzing graphs. Thus, we use root approximation techniques on a graphing utility to solve (1). The graph in Figure 11 shows that f' has one zero near $x = -4$ and another near $x = 2$. Since a quadratic polynomial can have at most two zeros, this graph shows all the solutions to equation (1). Using a root approximation routine (details omitted), we find that the solutions to (1) (to two decimal places) are

$$x = -3.77 \quad \text{and} \quad x = 1.77$$

Since $f'(x)$ exists for all x, the critical values are $x = -3.77$ and $x = 1.77$.
(B) Partition numbers for $f'(x)$ are $x = -3.77$ and $x = 1.77$ (which also happen to

be critical values for f). We construct a sign chart for $f'(x)$ using the graph of f' in Figure 11 to determine the sign of $f'(x)$ for each interval in the chart:

The sign chart indicates that $f(x)$ increases on $(-\infty, -3.77)$, has a local maximum at $x = -3.77$, decreases on $(-3.77, 1.77)$, has a local minimum at $x = 1.77$, and increases on $(1.77, \infty)$. These facts are summarized in the following table:

x	$f'(x)$	$f(x)$	GRAPH OF f
$(-\infty, -3.77)$	+	Increasing	Rising
$x = -3.77$	0	Local maximum	Horizontal tangent
$(-3.77, 1.77)$	−	Decreasing	Falling
$x = 1.77$	0	Local minimum	Horizontal tangent
$(1.77, \infty)$	+	Increasing	Rising

(C) Figure 12A shows the graph of $f(x)$ on a graphing utility, and Figure 12B shows a sketch of this graph that illustrates the information from parts (A) and (B). Note that we used trace to determine a y range for Figure 12A that will display the local extrema. We also added a background grid to aid in sketching the graph.

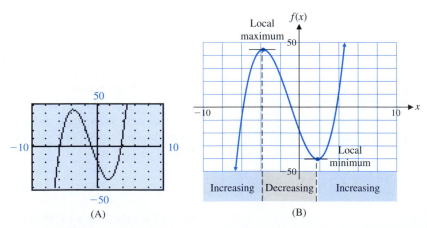

FIGURE 12
$f(x) = x^3 + 3x^2 - 20x - 20$

MATCHED PROBLEM 3 ▶ Repeat Example 3 for $f(x) = -x^3 + 3x^2 + 5x - 7$. ◀

Caution

Refer to the function f in Example 3. If you graph f and f' in the same viewing window (Fig. 13A), be certain not to confuse the zeros of $f'(x)$ with the zeros of $f(x)$. To avoid this possible error, graph f' alone and find its zeros, as we did in Figure 11. If you want to display information about the sign of $f'(x)$ in the same window as the graph of f, use the expression $f'(x)/|f'(x)|$, or in this case, we multiplied by 10 and used $10f'(x)/|f'(x)|$ to separate the horizontal line segments from the x axis. See Figure 13B.

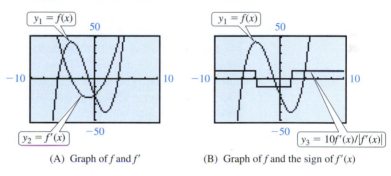

(A) Graph of f and f' (B) Graph of f and the sign of $f'(x)$

FIGURE 13
$f(x) = x^3 + 3x^2 - 20x - 20$

Most graphing utilities have built-in routines labeled *maximum* and *minimum*. These routines are designed to find *absolute* maxima and minima, not *local* maxima and minima. (We will discuss absolute extrema in Section 9-5.) Although it is possible to use these built-in routines to find local extrema, they may produce misleading results. Furthermore, they will not locate a horizontal tangent where the graph does not change direction, as in Example 2A. Thus, the preferred method for locating local extrema is to find the critical values and apply the first-derivative test.

How can you tell if you have found all the local extrema for a function? In general, this can be a difficult question to answer. However, in the case of a polynomial function, there is an easily determined upper limit on the number of local extrema. Since the local extrema are the x intercepts of the derivative, this limit is a consequence of the number of x intercepts of a polynomial. The relevant information is summarized in the following theorem, which is stated without proof:

theorem 2

Local Extrema for Polynomial Functions

If $f(x) = a_n x^n + a_{n-1} x^{n-1} + \cdots + a_1 x + a_0$, $a_n \neq 0$, is an nth-degree polynomial, then f has at most $n - 1$ local extrema.

Theorem 2 does not guarantee that every nth-degree polynomial has exactly $n - 1$ local extrema; it says only that there can never be more than $n - 1$ local extrema. For example, the third-degree polynomial in Example 3 has two local extrema, while the third-degree polynomial in Example 2A does not have any.

■ ANALYZING GRAPHS

In addition to providing information for hand sketching graphs, the derivative is also an important tool for analyzing graphs and discussing the interplay between a function and its rate of change. The next two examples illustrate this process in the context of some applications to economics.

EXAMPLE 4 ➤ **Agricultural Exports and Imports**   Over the past several decades, the United States has exported more agricultural products than it has imported, maintaining a positive balance of trade in this area. However, the trade balance fluctuated considerably during this period. The graph in Figure 14 approximates the rate of change of the trade balance over a 15 year period, where $B(t)$ is the trade balance (in billions of dollars) and t is time (in years).

(A) Write a brief verbal description of the graph of $y = B(t)$, including a discussion of any local extrema.
(B) Sketch a possible graph of $y = B(t)$.

Solution   (A) The graph of the derivative $y = B'(t)$ contains the same essential information as a sign chart. That is, we see that $B'(t)$ is positive on $(0, 4)$, 0 at $t = 4$, negative on $(4, 12)$, 0 at $t = 12$, and positive on $(12, 15)$. Hence, the trade balance increases for the first 4 years to a local maximum, decreases for the next 8 years to a local minimum, and then increases for the final 3 years.

(B) Without additional information concerning the actual values of $y = B(t)$, we cannot produce an accurate graph. However, we can sketch a possible graph that illustrates the important features, as shown in Figure 15. The absence of a scale on the vertical axis is a consequence of the lack of information about the values of $B(t)$.

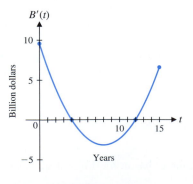

FIGURE 14
Rate of change of the balance of trade

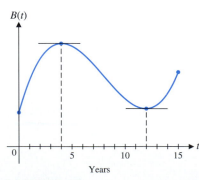

FIGURE 15
Balance of trade

MATCHED PROBLEM 4 ➤ The graph in Figure 16 approximates the rate of change of the U.S. share of the total world production of motor vehicles over a 20 year period, where $S(t)$ is the U.S. share (as a percentage) and t is time (in years).

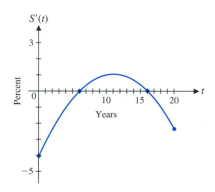

FIGURE 16

(A) Write a brief verbal description of the graph of $y = S(t)$, including a discussion of any local extrema.

(B) Sketch a possible graph of $y = S(t)$.

EXAMPLE 5 ➤ **Revenue Analysis** The graph of the total revenue $R(x)$ (in dollars) from the sale of x bookcases is shown in Figure 17.

(A) Write a brief verbal description of the graph of the marginal revenue function $y = R'(x)$, including a discussion of any x intercepts.

(B) Sketch a possible graph of $y = R'(x)$.

Solution (A) The graph of $y = R(x)$ indicates that $R(x)$ increases on $(0, 550)$, has a local maximum at $x = 550$, and decreases on $(550, 1,000)$. Consequently, the marginal revenue function $R'(x)$ must be positive on $(0, 550)$, 0 at $x = 550$, and negative on $(550, 1,000)$.

(B) A possible graph of $y = R'(x)$ illustrating the information summarized in part (A) is shown in Figure 18.

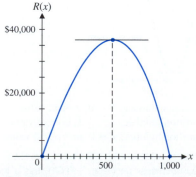

FIGURE 17
Revenue

FIGURE 18
Marginal revenue

MATCHED PROBLEM 5 ➤ The graph of the total revenue $R(x)$ (in dollars) from the sale of x desks is shown in Figure 19.

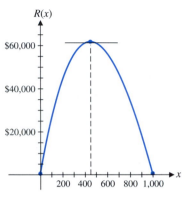

FIGURE 19

(A) Write a brief verbal description of the graph of the marginal revenue function $y = R'(x)$, including a discussion of any x intercepts.
(B) Sketch a possible graph of $y = R'(x)$.   ◀

Comparing Examples 4 and 5, we see that we were able to obtain more information about the function from the graph of its derivative (Example 4), than we were when the process was reversed (Example 5). In the next section we will introduce some ideas that will enable us to extract additional information about the derivative from the graph of the function.

Answers to Matched Problems

1. (A) Horizontal tangent line at $x = 3$.
   (B) Decreasing on $(-\infty, 3)$; increasing on $(3, \infty)$
   (C) $f(x)$

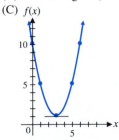

2. (A) Partition number: $x = 0$, critical value: $x = 0$, decreasing for all x
   (B) Partition number: $x = -1$, critical value: $x = -1$, increasing for all x
   (C) Partition number: $x = 0$, no critical values, decreasing on $(-\infty, 0)$ and $(0, \infty)$

3. (A) Critical values: $x = -0.63$ and $x = 2.63$
   (B) $f(x)$ decreases on $(-\infty, -0.63)$, has a local minimum at $x = -0.63$, increases on $(-0.63, 2.63)$, has a local maximum at $x = 2.63$, and decreases on $(2.63, \infty)$

(C)

4. (A) The U.S. share of the world market decreases for 6 years to a local minimum, increases for the next 10 years to a local maximum, and then decreases for the final 4 years.

(B) $S(t)$

5. (A) The marginal revenue is positive on $(0, 450)$, 0 at $x = 450$, and negative on $(450, 1,000)$.

(B) $R'(x)$

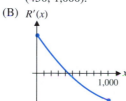

EXERCISE 9-2

A *Problems 1–8 refer to the graph of $y = f(x)$.*

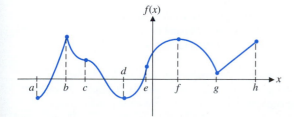

Figure for 1–8

1. Identify the intervals over which $f(x)$ is increasing.

2. Identify the intervals over which $f(x)$ is decreasing.

3. Identify the intervals over which $f'(x) < 0$.

4. Identify the intervals over which $f'(x) > 0$.

5. Identify the x coordinates of the points where $f'(x) = 0$.

6. Identify the x coordinates of the points where $f'(x)$ does not exist.

7. Identify the x coordinates of the points where $f(x)$ has a local maximum.

8. Identify the x coordinates of the points where $f(x)$ has a local minimum.

In Problems 9 and 10, $f(x)$ is continuous on $(-\infty, \infty)$ and has critical values at $x = a, b, c,$ and d. Use the sign chart for $f'(x)$ to determine whether f has a local maximum, a local minimum, or neither at each critical value.

9.

10.

In Problems 11–18, $f(x)$ is continuous on $(-\infty, \infty)$. Use the given information to sketch the graph of f.

11.

x	-2	-1	0	1	2
$f(x)$	-1	1	2	3	1

12.

x	-2	-1	0	1	2
$f(x)$	1	3	2	1	-1

13.

x	-2	-1	0	2	4
$f(x)$	2	1	2	1	0

14.

x	-2	-1	0	2	3
$f(x)$	-3	0	2	-1	0

15. $f(-2) = 4, f(0) = 0, f(2) = -4;$
$f'(-2) = 0, f'(0) = 0, f'(2) = 0;$
$f'(x) > 0$ on $(-\infty, -2)$ and $(2, \infty)$;
$f'(x) < 0$ on $(-2, 0)$ and $(0, 2)$

16. $f(-2) = -1, f(0) = 0, f(2) = 1;$
$f'(-2) = 0, f'(2) = 0;$
$f'(x) > 0$ on $(-\infty, -2), (-2, 2),$ and $(2, \infty)$

17. $f(-1) = 2, f(0) = 0, f(1) = -2;$
$f'(-1) = 0, f'(1) = 0, f'(0)$ is not defined;
$f'(x) > 0$ on $(-\infty, -1)$ and $(1, \infty)$;
$f'(x) < 0$ on $(-1, 0)$ and $(0, 1)$

18. $f(-1) = 2, f(0) = 0, f(1) = 2;$
$f'(-1) = 0, f'(1) = 0, f'(0)$ is not defined;
$f'(x) > 0$ on $(-\infty, -1)$ and $(0, 1)$;
$f'(x) < 0$ on $(-1, 0)$ and $(1, \infty)$

B Problems 19–24 involve functions f_1–f_6 and their derivatives g_1–g_6. Use the graphs shown in figures (A) and (B) to match each function f_i with its derivative g_j.

19. f_1 **20.** f_2 **21.** f_3
22. f_4 **23.** f_5 **24.** f_6

Figure (A) for 19-24

Figure (B) for 19-24

In Problems 25–30, find the intervals where $f(x)$ is increasing, the intervals where $f(x)$ is decreasing, and sketch the graph. Add horizontal tangent lines.

25. $f(x) = 4 + 8x - x^2$

26. $f(x) = 2x^2 - 8x + 9$

27. $f(x) = x^3 - 3x + 1$

28. $f(x) = x^3 - 12x + 2$

29. $f(x) = 10 - 12x + 6x^2 - x^3$

30. $f(x) = x^3 + 3x^2 + 3x$

In Problems 31–48, find the intervals where $f(x)$ is increasing, the intervals where $f(x)$ is decreasing, and the local extrema. Round critical values to two decimal places.

31. $f(x) = 4 + 10x - x^2$ **32.** $f(x) = 5 + 8x - 2x^2$

33. $f(x) = 2x^3 + 4$ **34.** $f(x) = 2 - 3x^3$

35. $f(x) = 2 - 6x - 2x^3$ **36.** $f(x) = x^3 + 9x + 7$

37. $f(x) = x^3 - 12x + 8$ **38.** $f(x) = 3x - x^3$

39. $f(x) = x^3 - 3x^2 - 24x + 7$

40. $f(x) = x^3 + 3x^2 - 9x + 5$

41. $f(x) = 2x^2 - x^4$ **42.** $f(x) = x^4 - 8x^2 + 3$

43. $f(x) = x^4 - 2x^2 + 3x$

44. $f(x) = x^4 - x^2 - 4x$

45. $f(x) = x^4 - 3x^3 + 2x$

46. $f(x) = x^4 + 3x^3 - 3x$

47. $f(x) = 0.25x^4 + 2x^3 + 2.5x^2 - 11x$

48. $f(x) = 0.25x^4 - x^3 - 2x^2 + 9x + 9$

In Problems 49–52, use the given graph of $y = f'(x)$ to find the intervals where f is increasing, the intervals where f is decreasing, and the local extrema. Sketch a possible graph for $y = f(x)$.

49.

50.

51.

52.

In Problems 53–56, use the given graph of $y = f(x)$ to find the intervals where $f'(x) > 0$, the intervals where $f'(x) < 0$, and the values of x for which $f'(x) = 0$. Sketch a possible graph for $y = f'(x)$.

53.

54.

55.

56.

61. $f(x) = 1 + \dfrac{1}{x} + \dfrac{1}{x^2}$

62. $f(x) = 3 - \dfrac{4}{x} - \dfrac{2}{x^2}$

63. $f(x) = \dfrac{x^2}{x - 2}$

64. $f(x) = \dfrac{x^2}{x + 1}$

65. $f(x) = x^4(x - 6)^2$

66. $f(x) = x^3(x - 5)^2$

67. $f(x) = 3(x - 2)^{2/3} + 4$

68. $f(x) = 6(4 - x)^{2/3} + 4$

69. $f(x) = 2\sqrt{x} - x,\ x > 0$

70. $f(x) = x - 4\sqrt{x},\ x > 0$

C *In Problems 57–70, find the critical values, the intervals where $f(x)$ is increasing, the intervals where $f(x)$ is decreasing, and the local extrema. Do not graph.*

57. $f(x) = \dfrac{x - 1}{x + 2}$

58. $f(x) = \dfrac{x + 2}{x - 3}$

59. $f(x) = x + \dfrac{4}{x}$

60. $f(x) = \dfrac{9}{x} + x$

71. Let $f(x) = x^3 + kx$, where k is a constant. Discuss the number of local extrema and the shape of the graph of f if:
(A) $k > 0$ (B) $k < 0$ (C) $k = 0$

72. Let $f(x) = x^4 + kx^2$, where k is a constant. Discuss the number of local extrema and the shape of the graph of f if:
(A) $k > 0$ (B) $k < 0$ (C) $k = 0$

▶ **APPLICATIONS**

Business & Economics

73. *Profit analysis.* The graph of the total profit $P(x)$ (in dollars) from the sale of x cordless electric screwdrivers is shown in the figure.

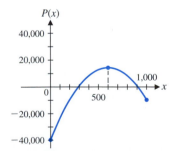

Figure for 73

74. *Revenue analysis.* The graph of the total revenue $R(x)$ (in dollars) from the sale of x cordless electric screwdrivers is shown in the figure.

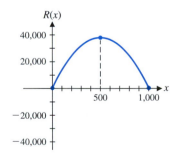

Figure for 74

(A) Write a brief verbal description of the graph of the marginal profit function $y = P'(x)$, including a discussion of any x intercepts.

(B) Sketch a possible graph of $y = P'(x)$.

(A) Write a brief verbal description of the graph of the marginal revenue function $y = R'(x)$, including a discussion of any x intercepts.

(B) Sketch a possible graph of $y = R'(x)$.

75. *Price analysis.* The graph in the figure approximates the rate of change of the price of bacon over a 70 month period, where $B(t)$ is the price of a pound of sliced bacon (in dollars) and t is time (in months).

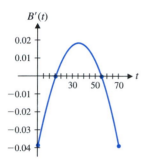

Figure for 75

(A) Write a brief verbal description of the graph of $y = B(t)$, including a discussion of any local extrema.

(B) Sketch a possible graph of $y = B(t)$.

76. *Price analysis.* The graph in the figure approximates the rate of change of the price of eggs over a 70 month period, where $E(t)$ is the price of a dozen eggs (in dollars) and t is time (in months).

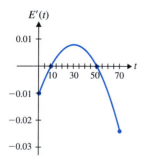

Figure for 76

(A) Write a brief verbal description of the graph of $y = E(t)$, including a discussion of any local extrema.

(B) Sketch a possible graph of $y = E(t)$.

77. *Average cost.* A manufacturer incurs the following costs in producing x toasters in one day for $0 < x < 150$: fixed costs, \$320; unit production cost, \$20 per toaster; equip-ment maintenance and repairs, $x^2/20$ dollars. Thus, the cost of manufacturing x toasters in one day is given by

$$C(x) = \frac{x^2}{20} + 20x + 320 \qquad 0 < x < 150$$

(A) What is the average cost, $\overline{C}(x)$, per toaster if x toasters are produced in one day?

(B) Find the critical values for $\overline{C}(x)$, the intervals where the average cost per toaster is decreasing, the intervals where the average cost per toaster is increasing, and the local extrema.

78. *Average cost.* A manufacturer incurs the following costs in producing x blenders in one day for $0 < x < 200$: fixed costs, \$450; unit production cost, \$60 per blender; equip-ment maintenance and repairs, $x^2/18$ dollars.

(A) What is the average cost, $\overline{C}(x)$, per blender if x blenders are produced in one day?

(B) Find the critical values for $\overline{C}(x)$, the intervals where the average cost per blender is decreasing, the intervals where the average cost per blender is increasing, and the local extrema.

79. *Marginal analysis.* Show that profit will be increasing over production intervals (a, b) for which marginal revenue is greater than marginal cost. [*Hint:* $P(x) = R(x) - C(x)$]

80. *Marginal analysis.* Show that profit will be decreasing over production intervals (a, b) for which marginal revenue is less than marginal cost.

Life Sciences

81. *Medicine.* A drug is injected into the bloodstream of a pa-tient through the right arm. The concentration of the drug in the bloodstream of the left arm t hours after the injection is approximated by

$$C(t) = \frac{0.14t}{t^2 + 1} \qquad 0 < t < 24$$

Find the critical values for $C(t)$, the intervals where the concentration of the drug is increasing, the intervals where the concentration of the drug is decreasing, and the local extrema.

82. *Medicine.* The concentration $C(t)$, in milligrams per cubic centimeter, of a particular drug in a patient's bloodstream is given by

$$C(t) = \frac{0.16t}{t^2 + 4t + 4} \qquad 0 < t < 12$$

where t is the number of hours after the drug is taken orally. Find the critical values for $C(t)$, the intervals where the concentration of the drug is increasing, the intervals where the concentration of the drug is decreasing, and the local extrema.

Social Sciences

83. *Politics.* Public awareness of a Congressional candidate before and after a successful campaign was approximated by

$$P(t) = \frac{8.4t}{t^2 + 49} + 0.1 \qquad 0 < t < 24$$

where t is time (in months) after the campaign started and $P(t)$ is the fraction of people in the Congressional district who could recall the candidate's (and later, Congressman's) name. Find the critical values for $P(t)$, the time intervals where the fraction is increasing, the time intervals where the fraction is decreasing, and the local extrema.

SECTION 9-3 ■ Second Derivative and Graphs

- CONCAVITY
- INFLECTION POINTS
- SECOND-DERIVATIVE TEST
- ANALYZING GRAPHS

In the preceding section, we saw that the derivative can be used to determine when a graph is rising and falling. Now we want to see what the *second derivative* (the derivative of the derivative) can tell us about the shape of a graph.

■ CONCAVITY

Consider the functions

$$f(x) = x^2 \qquad \text{and} \qquad g(x) = \sqrt{x}$$

for x in the interval $(0, \infty)$. Since

$$f'(x) = 2x > 0 \qquad \text{for } 0 < x < \infty$$

and

$$g'(x) = \frac{1}{2\sqrt{x}} > 0 \qquad \text{for } 0 < x < \infty$$

both functions are increasing on $(0, \infty)$.

(A) Discuss the difference in the shapes of the graphs of f and g shown in Figure 1.

(B) Complete the following table and discuss the relationship between the values of the derivatives of f and g and the shapes of their graphs.

x	0.25	0.5	0.75	1
$f'(x)$				
$g'(x)$				

(A) $f(x) = x^2$

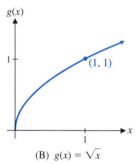

(B) $g(x) = \sqrt{x}$

FIGURE 1

We use the term *concave upward* to describe a graph that opens upward and *concave downward* to describe a graph that opens downward. Thus, the graph of f in Figure 1A is concave upward, and the graph of g in Figure 1B is concave downward. Finding a mathematical formulation of concavity will help us sketch and analyze graphs.

It will be instructive to examine the slopes of f and g at various points on their graphs (see Fig. 2). We can make two observations about each graph. Looking at the graph of f in Figure 2A, we see that $f'(x)$ (the slope of the tangent line) is *increasing* and that the graph lies *above* each tangent line. Looking at Figure 2B, we see that $g'(x)$ is *decreasing* and that the graph lies *below* each tangent line.

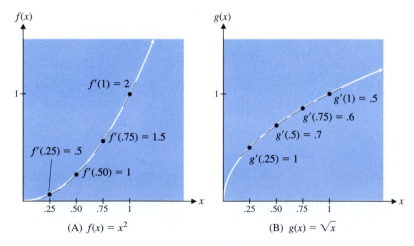

(A) $f(x) = x^2$ (B) $g(x) = \sqrt{x}$

FIGURE 2

With these ideas in mind, we state the general definition of concavity:

The graph of a function f is concave upward on the interval (a, b) if $f'(x)$ is *increasing* on (a, b) and is concave downward on the interval (a, b) if $f'(x)$ is *decreasing* on (a, b).

Geometrically, the graph is concave upward on (a, b) if it lies above its tangent lines in (a, b) and is concave downward on (a, b) if it lies below its tangent lines in (a, b).

How can we determine when $f'(x)$ is increasing or decreasing? In the preceding section, we used the derivative of a function to determine when that function is increasing or decreasing. Thus, to determine when the function $f'(x)$ is increasing or decreasing, we use the derivative of $f'(x)$. The derivative of the derivative of a function is called the *second derivative* of the function. Various notations for the second derivative are given in the following box:

Second Derivative

For $y = f(x)$, the **second derivative** of f, provided it exists, is

$$f''(x) = \frac{d}{dx} f'(x)$$

Other notations for $f''(x)$ are

$$\frac{d^2y}{dx^2} \qquad y''$$

Returning to the functions f and g discussed at the beginning of this section, we have

$$f(x) = x^2 \qquad\qquad g(x) = \sqrt{x} = x^{1/2}$$

$$f'(x) = 2x \qquad\qquad g'(x) = \frac{1}{2} x^{-1/2} = \frac{1}{2\sqrt{x}}$$

$$f''(x) = \frac{d}{dx} 2x = 2 \qquad g''(x) = \frac{d}{dx} \frac{1}{2} x^{-1/2} = -\frac{1}{4} x^{-3/2} = -\frac{1}{4\sqrt{x^3}}$$

For $x > 0$, we see that $f''(x) > 0$; thus, $f'(x)$ is increasing and the graph of f is concave upward (see Fig. 2A). For $x > 0$, we also see that $g''(x) < 0$; thus, $g'(x)$ is decreasing and the graph of g is concave downward (see Fig. 2B). These ideas are summarized in the following box:

Concavity

For the interval (a, b):

$f''(x)$	$f'(x)$	GRAPH OF $y = f(x)$	EXAMPLES
$+$	Increasing	Concave upward	
$-$	Decreasing	Concave downward	

Be careful not to confuse concavity with falling and rising. As Figure 3 illustrates, a graph that is concave upward on an interval may be falling, rising, or both falling and rising on that interval. A similar statement holds for a graph that is concave downward.

$$f''(x) > 0 \text{ over } (a, b)$$
Concave upward

(A) $f'(x)$ is negative and increasing. Graph of f is falling.

(B) $f'(x)$ increases from negative to positive. Graph of f falls, then rises.

(C) $f'(x)$ is positive and increasing. Graph of f is rising.

$$f''(x) < 0 \text{ over } (a, b)$$
Concave downward

(D) $f'(x)$ is positive and decreasing. Graph of f is rising.

(E) $f'(x)$ decreases from positive to negative. Graph of f rises, then falls.

(F) $f'(x)$ is negative and decreasing. Graph of f is falling.

FIGURE 3
Concavity

EXAMPLE 1 ➤ **Determining Concavity of a Graph** Let $f(x) = x^3$. Describe the relationships among the sign of $f''(x)$, the increasing/decreasing properties of $f'(x)$, and the concavity of the graph of f.

Solution   We begin by graphing $f(x) = x^3, f'(x) = 3x^2$, and $f''(x) = 6x$ (Fig. 4).

(A) $f''(x) = 6x$

(B) $f'(x) = 3x^2$

(C) $f(x) = x^3$

FIGURE 4

A table provides a convenient way to describe the relationships among these three graphs:

x	$f''(x)$	$f'(x)$	$f(x)$
$(-\infty, 0)$	$-$	Decreasing	Concave downward
$(0, \infty)$	$+$	Increasing	Concave upward

◄

◄

MATCHED PROBLEM 1 ► Repeat Example 1 for $f(x) = 1 - x^3$.   ◄

■ INFLECTION POINTS

The graph of $f(x) = x^3$ in Example 1 changes from concave downward to concave upward at the point $(0, 0)$ (Fig. 4C). This point is called an *inflection point.*

explore – discuss 2

Discuss the relationship between the change in concavity of each of the following functions at $x = 0$ and the second derivative at and near 0.

(A) $f(x) = x^5$ (B) $g(x) = x^{4/3}$ (C) $h(x) = x^4$

In general, an **inflection point** is a point on the graph of the function where the concavity changes (from upward to downward or from downward to upward). In order for the concavity to change at a point, $f''(x)$ must change sign at that point. But in Section 9-1, we saw that the partition numbers identify the points where a function can change sign. Thus, we have the following theorem:

theorem 1

Inflection Points

If $y = f(x)$ is continuous on (a, b) and has an inflection point at $x = c$, then either $f''(c) = 0$ or $f''(c)$ does not exist.

Note that inflection points can occur only at partition numbers of f'', but not every partition number of f'' produces an inflection point. Two additional requirements must be satisfied for an inflection point:

A partition number c for f'' produces an inflection point for the graph of f only if:
1. $f''(x)$ changes sign at c
2. c is in the domain of f

Figure 5 illustrates several typical cases.

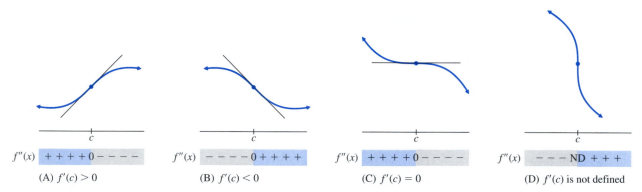

$f''(x)$ $+ + + + 0 - - - -$ $f''(x)$ $- - - - 0 + + + +$ $f''(x)$ $+ + + + 0 - - - -$ $f''(x)$ $- - - \text{ND} + + +$

(A) $f'(c) > 0$ (B) $f'(c) < 0$ (C) $f'(c) = 0$ (D) $f'(c)$ is not defined

FIGURE 5
Inflection points

If $f'(c)$ exists and $f''(x)$ changes sign at $x = c$, then the tangent line at an inflection point $(c, f(c))$ will always lie below the graph on the side that is concave upward and above the graph on the side that is concave downward (see Figs. 5A, B, and C).

EXAMPLE 2 ➤ **Locating Inflection Points**   Find the x coordinates (rounded to two decimal places) of the inflection points, the intervals where the graph is concave upward, and the intervals where the graph is concave downward for

$$f(x) = 0.25x^4 - 3x^3 + 9x^2 + 12x - 5$$

Solution   We begin by finding the second derivative:

$$f(x) = 0.25x^4 - 3x^3 + 9x^2 + 12x - 5$$
$$f'(x) = x^3 - 9x^2 + 18x + 12$$
$$f''(x) = 3x^2 - 18x + 18$$

Next, we investigate the graphs of f and f''. The portion of the graph of f visible in a standard viewing window (Fig. 6A) appears to be concave upward. However, the graph of f'' (Fig. 6B) shows that f'' changes sign twice, once near $x = 1$ and again

(A) $y = f(x)$ (B) $y = f''(x)$ (C) $y = f(x)$

FIGURE 6

near $x = 5$. Since $f''(x)$ is a second-degree polynomial, these are the only points where $f''(x)$ changes sign, and there must be two inflection points that are not visible in the standard viewing window. We use trace to determine the new viewing window for the graph of f shown in Figure 6C.

Now we can see that the graph of f does indeed have two inflection points. To locate these inflection points, we must solve the equation $f''(x) = 0$. Applying a root approximation routine (details omitted), we find that

$$f''(x) = 3x^2 - 18x + 18 = 0 \qquad \text{at} \quad x = 1.27 \quad \text{and} \quad x = 4.73$$

A sign chart for $f''(x)$ provides a convenient way to summarize the information about the graph of f obtained from f'':

Thus, we see that the graph of f is concave upward on $(-\infty, 1.27)$ and $(4.73, \infty)$, concave downward on $(1.27, 4.73)$, and has inflection points at $x = 1.27$ and $x = 4.73$. These results are illustrated on the graph of f in Figure 7.

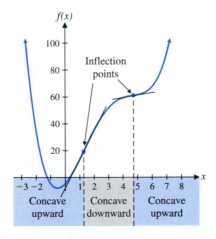

FIGURE 7

MATCHED PROBLEM 2 ➤ Find the x coordinates (rounded to two decimal places) of the inflection points, the intervals where the graph is concave upward, and the intervals where the graph is concave downward for

$$f(x) = 0.25x^4 + 3.5x^3 + 14x^2$$

It is important to remember that the partition numbers of f'' are only candidates for inflection points. The function f must be defined at $x = c$, and the second derivative must change sign at $x = c$ in order for the graph to have an inflection point at $x = c$. For example, consider

$$f(x) = x^4 \qquad\qquad g(x) = \frac{1}{x}$$

$$f'(x) = 4x^3 \qquad\qquad g'(x) = -\frac{1}{x^2}$$

$$f''(x) = 12x^2 \qquad\qquad g''(x) = \frac{2}{x^3}$$

In each case, $x = 0$ is a partition number for the second derivative, but neither graph has an inflection point at $x = 0$. Function f does not have an inflection point at $x = 0$ because $f''(x)$ does not change sign at $x = 0$ (see Fig. 8A). Function g does not have an inflection point at $x = 0$ because $g(0)$ is not defined (see Fig. 8B).

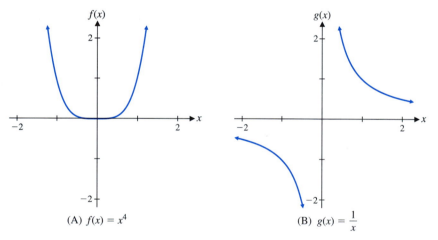

(A) $f(x) = x^4$　　　　　　(B) $g(x) = \frac{1}{x}$

FIGURE 8

■ SECOND-DERIVATIVE TEST

Now we want to see how the second derivative can be used to find local extrema. Suppose f is a function satisfying $f'(c) = 0$ and $f''(c) > 0$. First, note that if $f''(c) > 0$, then it follows from the properties of limits* that $f''(x) > 0$ in some interval (m, n) containing c. Thus, the graph of f must be concave upward in this interval. But this implies that $f'(x)$ is increasing in this interval. Since $f'(c) = 0, f'(x)$ must change from negative to positive at $x = c$, and $f(c)$ is a local minimum (see Fig. 9). Reasoning in the same fashion, we conclude that if $f'(c) = 0$ and $f''(c) < 0$, then

---

* Actually, we are assuming that $f''(x)$ is continuous in an interval containing c. It is very unlikely that we will encounter a function for which $f''(c)$ exists but $f''(x)$ is not continuous in an interval containing c.

$f(c)$ is a local maximum. Of course, it is possible that both $f'(c) = 0$ and $f''(c) = 0$. In this case, the second derivative cannot be used to determine the shape of the graph around $x = c$; $f(c)$ may be a local minimum, a local maximum, or neither.

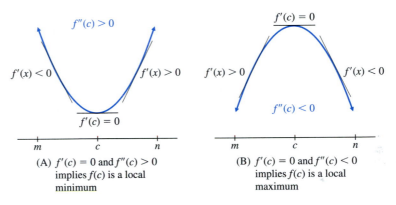

(A) $f'(c) = 0$ and $f''(c) > 0$ implies $f(c)$ is a local minimum

(B) $f'(c) = 0$ and $f''(c) < 0$ implies $f(c)$ is a local maximum

FIGURE 9
The second derivative and local extrema

Second-Derivative Test for Local Maxima and Minima

Let c be a critical value for $f(x)$.

$f'(c)$	$f''(c)$	GRAPH OF f IS	$f(c)$	EXAMPLE
0	+	Concave upward	Local minimum	$\smile$
0	−	Concave downward	Local maximum	$\frown$
0	0	?	Test fails	

The first-derivative test must be used whenever $f''(c) = 0$ or $f''(c)$ does not exist.

EXAMPLE 3 ➤ **Testing for Local Extrema** Use the second-derivative test to find the local maxima and minima for

$$f(x) = x^3 - 6x^2 + 9x + 1$$

Solution Take first and second derivatives and find critical values:

$$f(x) = x^3 - 6x^2 + 9x + 1$$
$$f'(x) = 3x^2 - 12x + 9 = 3(x - 1)(x - 3)$$
$$f''(x) = 6x - 12$$

Critical values are $x = 1$ and $x = 3$.

$$f''(1) = -6 < 0 \quad \text{f has a local maximum at x = 1.}$$
$$f''(3) = 6 > 0 \quad \text{f has a local minimum at x = 3.}$$

MATCHED PROBLEM 3 ➤ Use the second-derivative test to find the local maxima and minima for

$$f(x) = x^3 - 9x^2 + 24x - 10$$

_____

Caution

Do not assume that $f''(c) = 0$ implies $f(c)$ is not a local extremum. For example, if $f(x) = x^4$, then $f''(0) = 0$, but $f(0)$ is a local minimum (see Fig. 8). Therefore:

The first-derivative test *must* be used whenever $f''(c) = 0$ or $f''(c)$ does not exist.

■ ANALYZING GRAPHS

In the next two examples, we will combine increasing/decreasing properties with concavity properties to analyze the graph of a function.

EXAMPLE 4 ➤ **Analyzing a Graph** Figure 10 shows the graph of the derivative of a function f. Use this graph to discuss the graph of f. Include a sketch of a possible graph of f.

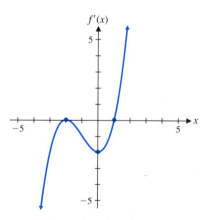

FIGURE 10

Solution The sign of the derivative determines where the original function is increasing and decreasing, and the increasing/decreasing properties of the derivative determine the concavity of the original function. The relevant information obtained from the graph of f' is summarized in Table 1, and a possible graph of f is shown in Figure 11.

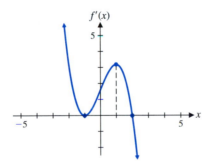

$f(x)$

FIGURE 11

TABLE 1

x	$f'(x)$ (Fig. 10)	$f(x)$ (Fig. 11)
$-\infty < x < -2$	Negative and increasing	Decreasing and concave upward
$x = -2$	Local maximum	Inflection point
$-2 < x < 0$	Negative and decreasing	Decreasing and concave downward
$x = 0$	Local minimum	Inflection point
$0 < x < 1$	Negative and increasing	Decreasing and concave upward
$x = 1$	x intercept	Local minimum
$1 < x < \infty$	Positive and increasing	Increasing and concave upward

◀

MATCHED PROBLEM 4 ▶ Figure 12 shows the graph of the derivative of a function f. Use this graph to discuss the graph of f. Include a sketch of a possible graph of f.

$f'(x)$

FIGURE 12

◀

Refer to Table 1 in the solution for Example 4. Note that the inflection points of f correspond to the local extrema of f'. Theorem 2 states that this is always the case, as long as f' is continuous at $x = c$.

theorem 2

> Let f be a function with a continuous first derivative at $x = c$. Then f has an inflection point at $x = c$ if and only if f' has a local extremum at $x = c$.

▶ **EXAMPLE 5** ▶ **Maximum Rate of Change** Using past records, a company estimates that it will sell $N(x)$ items after spending $\$x$ thousand on advertising, as given by

$$N(x) = 2,000 - 2x^3 + 60x^2 - 450x \qquad 5 \leqslant x \leqslant 15$$

When is the rate of change of sales with respect to advertising expenditures increasing? Decreasing? What is the maximum rate of change? Graph N and N' on the same coordinate system and interpret.

Solution The rate of change of sales with respect to advertising expenditures is

$$N'(x) = -6x^2 + 120x - 450 = -6(x - 5)(x - 15)$$

To determine when this rate is increasing and decreasing, we find $N''(x)$, the derivative of $N'(x)$:

$$N''(x) = -12x + 120 = 12(10 - x)$$

The graphs of N'', N', and N are shown in Figure 13, and the information obtained by analyzing the graphs of $N'(x)$ and $N''(x)$ is summarized in Table 2.

(A) $N''(x)$ (B) $N'(x)$ (C) $N(x)$

TABLE 2

x	$N''(x)$	$N'(x)$	$N(x)$
$5 < x < 10$	$+$	Positive and increasing	Increasing and concave upward
$x = 10$	0	Local maximum	Inflection point
$10 < x < 15$	$-$	Positive and decreasing	Increasing and concave downward

Thus, we see that $N'(x)$, the rate of change of sales, is increasing on $(5, 10)$ and decreasing on $(10, 15)$. Furthermore, the maximum rate of change, $N'(10) = 150$, corresponds to the inflection point of N at $x = 10$ (see Fig. 14). This value of x where

FIGURE 14

the rate of change of sales reaches its maximum value is referred to as the **point of diminishing returns.** Additional amounts spent on advertising beyond this point will still increase sales, but by smaller and smaller amounts. ◀

MATCHED PROBLEM 5 ▶ Repeat Example 5 for $N(x) = 5,000 - x^3 + 60x^2 - 900x$, $10 \leq x \leq 30$. ◀

Answers to Matched Problems

1.

x	$f''(x)$	$f'(x)$	$f(x)$
$(-\infty, 0)$	$+$	Increasing	Concave upward
$(0, \infty)$	$-$	Decreasing	Concave downward

2. The graph of f is concave upward on $(-\infty, -5.21)$ and $(-1.79, \infty)$, concave downward on $(-5.21, -1.79)$, and has inflection points at $x = -5.21$ and $x = -1.79$.

3. $f(2)$ is a local maximum; $f(4)$ is a local minimum

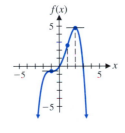

4.

x	$f'(x)$	$f(x)$
$-\infty < x < -1$	Positive and decreasing	Increasing and concave downward
$x = -1$	Local minimum	Inflection point
$-1 < x < 1$	Positive and increasing	Increasing and concave upward
$x = 1$	Local maximum	Inflection point
$1 < x < 2$	Positive and decreasing	Increasing and concave downward
$x = 2$	x intercept	Local maximum
$2 < x < \infty$	Negative and decreasing	Decreasing and concave downward

5. $N'(x)$ is increasing on $(10, 20)$, decreasing on $(20, 30)$; maximum rate of change is $N'(20) = 300$; $x = 20$ is point of diminishing returns

EXERCISE 9-3

A   *Problems 1–8 refer to the graph of $y = f(x)$.*

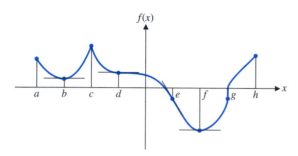

$f(x)$

Figure for 1–8

1. Identify intervals over which the graph of f is concave upward.

2. Identify intervals over which the graph of f is concave downward.

3. Identify intervals over which $f''(x) < 0$.

4. Identify intervals over which $f''(x) > 0$.

5. Identify intervals over which $f'(x)$ is increasing.

6. Identify intervals over which $f'(x)$ is decreasing.

7. Identify x coordinates of inflection points.

8. Identify x coordinates of local extrema for $f'(x)$.

In Problems 9–12, match the indicated conditions with one of the graphs shown in the figure.

$f(x)$ $f(x)$ $f(x)$ $f(x)$

(A)   (B)   (C)   (D)

Figure for 9–12

9. $f'(x) > 0$ and $f''(x) > 0$ on (a, b)
10. $f'(x) > 0$ and $f''(x) < 0$ on (a, b)
11. $f'(x) < 0$ and $f''(x) > 0$ on (a, b)
12. $f'(x) < 0$ and $f''(x) < 0$ on (a, b)

In Problems 13–18, describe the graph of f at the given point relative to the existence of a local maximum or minimum with

one of the following phrases: "Local maximum," "Local minimum," "Neither," or "Unable to determine from the given information." Assume that $f(x)$ is continuous on $(-\infty, \infty)$.

13. $(2, f(2))$ if $f'(2) = 0$ and $f''(2) > 0$
14. $(4, f(4))$ if $f'(4) = 1$ and $f''(4) < 0$
15. $(-3, f(-3))$ if $f'(-3) = 0$ and $f''(-3) = 0$
16. $(-1, f(-1))$ if $f'(-1) = 0$ and $f''(-1) < 0$
17. $(6, f(6))$ if $f'(6) = 1$ and $f''(6)$ does not exist
18. $(5, f(5))$ if $f'(5) = 0$ and $f''(5)$ does not exist

In Problems 19–26, $f(x)$ is continuous on $(-\infty, \infty)$. Use the given information to sketch the graph of f.

19.

x	-4	-2	-1	0	2	4
$f(x)$	0	3	1.5	0	-1	-3

$f'(x)$

$f''(x)$

20.

x	-4	-2	-1	0	2	4
$f(x)$	0	-2	-1	0	1	3

$f'(x)$

$f''(x)$

21.

x	-3	0	1	2	4	5
$f(x)$	-4	0	2	1	-1	0

22.

x	-4	-2	0	2	4	6
$f(x)$	0	3	0	-2	0	3

23. $f(0) = 2,\ f(1) = 0,\ f(2) = -2;\ f'(0) = 0,\ f'(2) = 0;$ $f'(x) > 0$ on $(-\infty, 0)$ and $(2, \infty);\ f'(x) < 0$ on $(0, 2);$ $f''(1) = 0; f''(x) > 0$ on $(1, \infty); f'(x) < 0$ on $(-\infty, 1)$

24. $f(-2) = -2, f(0) = 1, f(2) = 4; f'(-2) = 0, f'(2) = 0;$ $f'(x) > 0$ on $(-2, 2); f'(x) < 0$ on $(-\infty, -2)$ and $(2, \infty);$ $f''(0) = 0; f''(x) > 0$ on $(-\infty, 0); f'(x) < 0$ on $(0, \infty)$

25. $f(-1) = 0,\ f(0) = -2,\ f(1) = 0;\ f'(0) = 0,\ f'(-1)$ and $f'(1)$ are not defined; $f'(x) > 0$ on $(0, 1)$ and $(1, \infty);$ $f'(x) < 0$ on $(-\infty, -1)$ and $(-1, 0); f''(-1)$ and $f''(1)$ are not defined; $f''(x) > 0$ on $(-1, 1); f''(x) < 0$ on $(-\infty, -1)$ and $(1, \infty)$

26. $f(0) = -2, f(1) = 0, f(2) = 4; f'(0) = 0, f'(2) = 0, f'(1)$ is not defined; $f'(x) > 0$ on $(0, 1)$ and $(1, 2); f'(x) < 0$ on $(-\infty, 0)$ and $(2, \infty); f''(1)$ is not defined; $f''(x) > 0$ on $(-\infty, 1); f''(x) < 0$ on $(1, \infty)$

In Problems 27–36, find the indicated derivative for each function.

27. $f''(x)$ for $f(x) = x^3 - 2x^2 - 1$
28. $g''(x)$ for $g(x) = x^4 - 3x^2 + 5$
29. d^2y/dx^2 for $y = 2x^5 - 3$
30. d^2y/dx^2 for $y = 3x^4 - 7x$
31. y'' for $y = 1 - 2x + x^3$
32. y'' for $y = 3x^2 - x^3$
33. y'' for $y = (x^2 - 1)^3$
34. y'' for $y = (x^2 + 4)^4$
35. $f''(x)$ for $f(x) = 3x^{-1} + 2x^{-2} + 5$
36. $f''(x)$ for $f(x) = x^2 - x^{1/3}$

B *In Problems 37–48, find all local maxima and minima using the second-derivative test whenever it applies. If the second-derivative test fails, use the first-derivative test.*

37. $f(x) = 2x^2 - 8x + 6$ **38.** $f(x) = 6x - x^2 + 4$
39. $f(x) = 2x^3 - 3x^2 - 12x - 5$
40. $f(x) = 2x^3 + 3x^2 - 12x - 1$
41. $f(x) = 3 - x^3 + 3x^2 - 3x$
42. $f(x) = x^3 + 6x^2 + 12x + 2$
43. $f(x) = x^4 - 8x^2 + 10$
44. $f(x) = x^4 - 18x^2 + 50$
45. $f(x) = x^6 + 3x^4 + 2$ **46.** $f(x) = 4 - x^6 - 6x^4$
47. $f(x) = x + \dfrac{16}{x}$ **48.** $f(x) = x + \dfrac{25}{x}$

In Problems 49–60, find the intervals where the graph of f is concave upward, the intervals where the graph is concave downward, and the inflection points. Round all x coordinates to two decimal places.

49. $f(x) = x^3 - 18x^2 + 10x - 11$
50. $f(x) = x^3 + 24x^2 + 15x - 12$
51. $f(x) = x^4 - 24x^2 + 10x - 5$
52. $f(x) = x^4 + 6x^2 + 9x + 11$
53. $f(x) = -x^4 + 4x^3 + 2x^2 + 4x - 9$
54. $f(x) = -x^4 + 2x^3 + 6x^2 + 5x + 10$
55. $f(x) = 0.25x^4 - 2x^3 + 1.5x^2 + 21x + 15$
56. $f(x) = 0.25x^4 - 3.5x^3 + 12x^2 + 21x$
57. $f(x) = x^5 + 2x^4 + 4x^2 - 5$
58. $f(x) = x^5 - 3x^4 + x^3 - x^2 + 10$
59. $f(x) = x^5 - 3x^4 - x^3 + 7x^2 - 2$
60. $f(x) = x^5 - 2x^4 - 3x^3 + 4x^2 + 4x + 5$

In Problems 61–68, find local maxima, local minima, and inflection points. Sketch the graph of each function. Include tangent lines at each local extreme point and inflection point.

61. $f(x) = x^3 - 6x^2 + 16$
62. $f(x) = x^3 - 9x^2 + 15x + 10$
63. $f(x) = x^3 + x + 2$
64. $f(x) = 1 - 3x - x^3$
65. $f(x) = (2 - x)^3 + 1$
66. $f(x) = (1 + x)^3 - 1$

67. $f(x) = x^3 - 12x$
68. $f(x) = 27x - x^3$

In Problems 69–72, use the graph of $y = f'(x)$ to discuss the graph of $y = f(x)$. Organize your conclusions in a table (see Example 4, page 696), and sketch a possible graph of $y = f(x)$.

69.

70.

71.

72.

In Problems 73–76, assume that f, f', and f'' are continuous for all real numbers.

73. Explain how you can locate inflection points for the graph of $y = f(x)$ by examining the graph of $y = f'(x)$.

74. Explain how you can determine where $f'(x)$ is increasing or decreasing by examining the graph of $y = f(x)$.

75. Explain how you can locate local maxima and minima for the graph of $y = f'(x)$ by examining the graph of $y = f(x)$.

76. Explain how you can locate local maxima and minima for the graph of $y = f(x)$ by examining the graph of $y = f'(x)$.

C   *Find the inflection points in Problems 77–80.*

77. $f(x) = \dfrac{1}{x^2 + 12}$

78. $f(x) = \dfrac{x^2}{x^2 + 12}$

79. $f(x) = \dfrac{x}{x^2 + 12}$

80. $f(x) = \dfrac{x^3}{x^2 + 12}$

 APPLICATIONS

Business & Economics

81. *Inflation.* One commonly used measure of inflation is the annual rate of change of the Consumer Price Index (CPI). A newspaper headline proclaims that the rate of change of inflation for consumer prices is increasing. What does this say about the shape of the graph of the CPI?

82. *Inflation.* Another commonly used measure of inflation is the annual rate of change of the Producers Price Index (PPI). A government report states that the rate of change of inflation for producer prices is decreasing. What does this say about the shape of the graph of the PPI?

83. *Cost analysis.* A company manufactures a variety of lighting fixtures at different locations. The total cost $C(x)$ (in dollars) of producing x desk lamps per week at plant A is shown in the figure. Discuss the shape of the graph of the marginal cost function $C'(x)$ and interpret in terms of the efficiency of the production process at this plant.

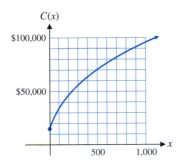

Figure for 83
Production costs at plant A

84. *Cost analysis.* The company in Problem 83 produces the same lamp at another plant. The total cost $C(x)$ (in dollars) of producing x desk lamps per week at plant B is shown in the figure. Discuss the shape of the graph of the marginal cost function $C'(x)$ and interpret in terms of the efficiency of the production process at plant B. Compare the production processes at these two plants.

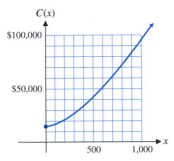

Figure for 84
Production costs at plant B

85. *Revenue.* The marketing research department for a computer company used a large city to test market their new product. They found that the relationship between price p (dollars per unit) and the demand x (units per week) was given approximately by

$$p = 1,296 - 0.12x^2 \qquad 0 < x < 80$$

Thus, the weekly revenue can be approximated by

$$R(x) = xp = 1,296x - 0.12x^3 \qquad 0 < x < 80$$

(A) Find the local extrema for the revenue function.
(B) Over which intervals is the graph of the revenue function concave upward? Concave downward?

86. *Profit.* Suppose the cost equation for the company in Problem 85 is

$$C(x) = 830 + 396x$$

(A) Find the local extrema for the profit function.
(B) Over which intervals is the graph of the profit function concave upward? Concave downward?

87. *Advertising.* A company estimates that it will sell $N(x)$ units of a product after spending $\$x$ thousand on advertising, as given by

$$N(x) = -3x^3 + 225x^2 - 3,600x + 17,000$$

$$10 \leqslant x \leqslant 40$$

(A) When is the rate of change of sales $N'(x)$ increasing? Decreasing?
(B) Find the inflection points for the graph of N.
(C) Graph N and N' on the same coordinate system.
(D) What is the maximum rate of change of sales?

88. *Advertising.* A company estimates that it will sell $N(x)$ units of a product after spending $\$x$ thousand on advertising, as given by

$$N(x) = -2x^3 + 90x^2 - 750x + 2,000 \qquad 5 \leqslant x \leqslant 25$$

(A) When is the rate of change of sales $N'(x)$ increasing? Decreasing?
(B) Find the inflection points for the graph of N.
(C) Graph N and N' on the same coordinate system.
(D) What is the maximum rate of change of sales?

89. *Advertising.* An automobile dealer uses television advertising to promote car sales. Using past records, the dealer arrived at the data in the table, where x is the number of ads placed monthly and y is the number of cars sold that month.

NUMBER OF ADS	NUMBER OF CARS
x	y
10	325
12	339
20	417
30	546
35	615
40	682
50	795

(A) Enter the data in a graphing utility and find a cubic regression equation for the number of cars sold monthly as a function of the number of ads.

(B) How many ads should the dealer place each month to maximize the rate of change of sales with respect to the number of ads, and how many cars can the dealer expect to sell with this number of ads? Round answers to the nearest integer.

90. *Advertising.* A music store uses radio advertising to promote sales of CD's. The store manager used past records to determine the data in the table, where x is the number of ads placed monthly and y is the number of CD's sold that month.

NUMBER OF ADS	NUMBER OF CD'S
x	y
10	345
14	488
20	746
30	1,228
40	1,671
50	1,955

(A) Enter the data in a graphing utility and find a cubic regression equation for the number of CD's sold monthly as a function of the number of ads.

(B) How many ads should the store manager place each month to maximize the rate of change of sales with respect to the number of ads, and how many CD's can the manager expect to sell with this number of ads? Round answers to the nearest integer.

Life Sciences

91. *Population growth—bacteria.* A drug that stimulates reproduction is introduced into a colony of bacteria. After t minutes, the number of bacteria is given approximately by

$$N(t) = 1,000 + 30t^2 - t^3 \qquad 0 \leqslant t \leqslant 20$$

(A) When is the rate of growth $N'(t)$ increasing? Decreasing?

(B) Find the inflection points for the graph of N.

(C) Sketch the graphs of N and N' on the same coordinate system.

(D) What is the maximum rate of growth?

92. *Drug sensitivity.* One hour after x milligrams of a particular drug are given to a person, the change in body temperature $T(x)$, in degrees Fahrenheit, is given by

$$T(x) = x^2 \left(1 - \frac{x}{9} \right) \qquad 0 \leqslant x \leqslant 6$$

The rate at which $T(x)$ changes with respect to the size of the dosage x, $T'(x)$, is called the *sensitivity* of the body to the dosage.

(A) When is $T'(x)$ increasing? Decreasing?

(B) Where does the graph of T have inflection points?

(C) Sketch the graphs of T and T' on the same coordinate system.

(D) What is the maximum value of $T'(x)$?

Social Sciences

93. *Learning.* The time T (in minutes) it takes a person to learn a list of length n is

$$T(n) = 0.08n^3 - 1.2n^2 + 6n \qquad n \geqslant 0$$

(A) When is the rate of change of T with respect to the length of the list increasing? Decreasing?

(B) Where does the graph of T have inflection points? Graph T and T' on the same coordinate system.

(C) What is the minimum value of $T'(n)$?

SECTION 9-4 ■ **Graphical Analysis: Unified and Extended**

- LIMITS AT INFINITY
- VERTICAL ASYMPTOTES
- SUMMARY OF GRAPH PROPERTIES
- ANALYZING GRAPHS
- APPLICATION

In this section we will apply, in a systematic way, all the graphing concepts discussed in the preceding three sections. Before we do this, we need to discuss the behavior of a graph as x increases or decreases without bound and also at certain points of discontinuity.

■ LIMITS AT INFINITY

An important element in the analysis of the graph of a function is the behavior of the function as x increases or decreases without bound. Recall from Section 9-1 that we used the special symbol ∞ to describe limits that increase or decrease without bound. We will write $x \to \infty$ to indicate that x is increasing without bound and $x \to -\infty$ to indicate that x is decreasing without bound.

We begin by considering power functions of the form x^p and $1/x^p$, where p is a positive real number.

explore–discuss 1

(A) Complete the following table:

x	100	1,000	10,000	100,000	1,000,000
x^2					
$1/x^2$					

(B) Describe verbally the behavior of x^2 as x increases without bound. Then use limit notation to describe this behavior.

(C) Repeat part (B) for $1/x^2$.

If p is a positive real number, then x^p increases as x increases, and it can be shown that there is no upper bound on the values of x^p. We indicate this by writing

$$x^p \to \infty \quad \text{as} \quad x \to \infty \qquad \text{or} \qquad \lim_{x \to \infty} x^p = \infty$$

Since the reciprocals of very large numbers are very small numbers, it follows that $1/x^p$ approaches 0 as x increases without bound. We indicate this behavior by writing

$$\frac{1}{x^p} \to 0 \quad \text{as} \quad x \to \infty \qquad \text{or} \qquad \lim_{x \to \infty} \frac{1}{x^p} = 0$$

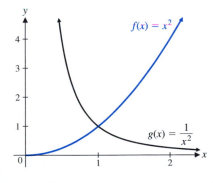

FIGURE 1

$\lim_{x \to \infty} f(x) = \infty$

$\lim_{x \to \infty} g(x) = 0$

Figure 1 illustrates this behavior for $f(x) = x^2$ and $g(x) = 1/x^2$.

Limits of power forms as x decreases without bound behave in a similar manner, with two important differences. First, if x is negative, then x^p is not a real number for all values of p. For example, $x^{1/2} = \sqrt{x}$ is not a real number for negative values of x. Second, if x^p represents a real number for all real x, then it may approach ∞ or $-\infty$, depending on the value of p. For example,

$$\lim_{x \to -\infty} x^2 = \infty \qquad \text{but} \qquad \lim_{x \to -\infty} x^3 = -\infty$$

For the function g in Figure 1, the line $y = 0$ (the x axis) is called a *horizontal asymptote*. In general, a line $y = b$ is a **horizontal asymptote** for the graph of

$y = f(x)$ if $f(x)$ approaches b as x either decreases without bound or increases without bound. Symbolically, $y = b$ is a horizontal asymptote if either

$$\lim_{x \to -\infty} f(x) = b \qquad \text{or} \qquad \lim_{x \to \infty} f(x) = b$$

In the first case, the graph of f will be close to the horizontal line $y = b$ for large (in absolute value) negative x (see Fig. 2). In the second case, the graph will be close to the horizontal line $y = b$ for large positive x.

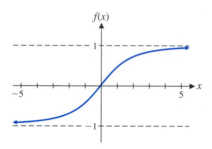

FIGURE 2

$$\lim_{x \to -\infty} f(x) = -1$$

$$\lim_{x \to \infty} f(x) = 1$$

Theorem 1 summarizes the various possibilities for limits of power functions as x increases or decreases without bound.

theorem 1

Limits at Infinity for Power Functions

If p is a positive real number and k is any real constant, then

1. $\displaystyle\lim_{x \to -\infty} \frac{k}{x^p} = 0$ 2. $\displaystyle\lim_{x \to \infty} \frac{k}{x^p} = 0$

3. $\displaystyle\lim_{x \to -\infty} kx^p = \pm\infty$ 4. $\displaystyle\lim_{x \to \infty} kx^p = \pm\infty$

provided that x^p names a real number for negative values of x. The limits in 3 and 4 will be either $-\infty$ or ∞, depending on k and p.

It is important to understand that the symbol ∞ does not represent an actual number that x is approaching, but is used only to indicate that x is increasing with no upper limit on its size. And, as we discussed in Section 9-1, when we say that a limit is equal to ∞, we are using ∞ to describe the behavior of a limit that does not exist. Finally, we note that **limit properties 1–6, listed in Theorem 1 in Section 8-2, are valid if we replace the statement $x \to c$ with $x \to \infty$ or $x \to -\infty$.**

In Section 2-1, we noted that the highest-degree term in a polynomial dominates all other terms combined. Now we can use the ideas discussed above to gain additional insight into this type of behavior.

EXAMPLE 1 ➤ **Limit of a Polynomial Function at Infinity** Let

$$p(x) = 2x^3 - x^2 - 7x + 3$$

Discuss the limit of $p(x)$ as x approaches ∞ or $-\infty$.

Solution We begin with a graphical approach. Figure 3 clearly shows that for points distant from the origin, the graph of $y = p(x)$ is very similar to the graph of $y = 2x^3$, the

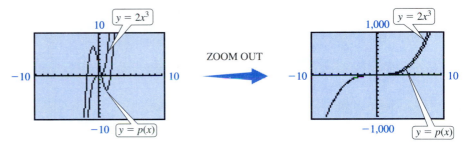

FIGURE 3
$p(x) = 2x^3 - x^2 - 7x + 3$

term of highest degree in $p(x)$. We confirm this observation algebraically by factoring out $2x^3$ and writing $p(x)$ as

$$p(x) = 2x^3\left(1 - \frac{1}{2x} - \frac{7}{2x^2} + \frac{3}{2x^3}\right)$$

Using Theorem 1 and limit properties from Section 8-2, we have

$$\lim_{x\to\infty}\left(1 - \frac{1}{2x} - \frac{7}{2x^2} + \frac{3}{2x^3}\right) = 1 - 0 - 0 + 0 = 1$$

Thus, for large values of x,

$$\left(1 - \frac{1}{2x} - \frac{7}{2x^2} + \frac{3}{2x^3}\right) \approx 1$$

and

$$p(x) = 2x^3\left(1 - \frac{1}{2x} - \frac{7}{2x^2} + \frac{3}{2x^3}\right) \approx 2x^3$$

Since $2x^3 \to \infty$ as $x \to \infty$, we can conclude that

$$\lim_{x\to\infty}(2x^3 - x^2 - 7x + 3) = \lim_{x\to\infty}2x^3\left(1 - \frac{1}{2x} - \frac{7}{2x^2} + \frac{3}{2x^3}\right) = \infty$$

Similarly, $2x^3 \to -\infty$ as $x \to -\infty$ implies that

$$\lim_{x \to -\infty} (2x^3 - x^2 - 7x + 3) = \lim_{x \to -\infty} 2x^3 \left(1 - \frac{1}{2x} - \frac{7}{2x^2} + \frac{3}{2x^3} \right) = -\infty$$

This confirms our earlier observation that the behavior of $p(x)$ for large values is the same as the behavior of the highest-degree term, $2x^3$. ◀

MATCHED PROBLEM 1 ➤ Let $p(x) = -4x^4 + 2x^3 + 3x$. Discuss the limit of $p(x)$ as x approaches ∞ and as x approaches $-\infty$. ◀

The ideas introduced in Example 1 are generalized to any polynomial in Theorem 2.

theorem 2

> **Limits at Infinity for Polynomial Functions**
>
> If
>
> $$p(x) = a_n x^n + a_{n-1} x^{n-1} + \cdots + a_1 x + a_0 \qquad a_n \neq 0, \quad n \geq 1$$
>
> then
>
> $$\lim_{x \to \infty} p(x) = \lim_{x \to \infty} a_n x^n = \pm\infty$$
>
> and
>
> $$\lim_{x \to -\infty} p(x) = \lim_{x \to -\infty} a_n x^n = \pm\infty$$
>
> Each limit will be either $-\infty$ or ∞, depending on a_n and n.

A polynomial of degree 0 is a constant function, $p(x) = a_0$, and its limit as x approaches ∞ or $-\infty$ is the number a_0. For any polynomial of degree 1 or greater, Theorem 2 states that the limit as x approaches ∞ or $-\infty$ cannot be equal to a number. Thus, it follows that **polynomials of degree 1 or greater never have horizontal asymptotes.**

explore – discuss 2

Write a verbal description of the relationship between a polynomial and its highest-degree term as stated in Theorem 2.

Since a rational function is the ratio of two polynomials, it is not surprising that reciprocals of powers of x can be used to analyze limits at infinity for rational functions. For example, consider the rational function

$$f(x) = \frac{2x^2 - 5x + 7}{3x^2 - 4x + 8}$$

FIGURE 4
Graph of $y = f(x)$

From Figure 4, it appears that the graph of f is approaching a limit as $x \to \infty$ and as $x \to -\infty$. We could obtain good estimates of these limits by graphing f over a longer interval on the x axis. However, using algebraic techniques leads to some general results that are very easy to apply. We begin by factoring the highest-degree term out of the numerator and the denominator:

$$f(x) = \frac{2x^2\left(1 - \dfrac{5}{2x} + \dfrac{7}{2x^2}\right)}{3x^2\left(1 - \dfrac{4}{3x} + \dfrac{8}{3x^2}\right)} = \frac{2}{3} \cdot \frac{1 - \dfrac{5}{2x} + \dfrac{7}{2x^2}}{1 - \dfrac{4}{3x} + \dfrac{8}{3x^2}}$$

Using Theorem 1, we have

$$\lim_{x \to \infty} f(x) = \lim_{x \to \infty} \left(\frac{2}{3} \cdot \frac{1 - \dfrac{5}{2x} + \dfrac{7}{2x^2}}{1 - \dfrac{4}{3x} + \dfrac{8}{3x^2}}\right) = \frac{2}{3} \cdot \frac{1 - 0 + 0}{1 - 0 + 0} = \frac{2}{3}$$

and

$$\lim_{x \to -\infty} f(x) = \lim_{x \to -\infty} \left(\frac{2}{3} \cdot \frac{1 - \dfrac{5}{2x} + \dfrac{7}{2x^2}}{1 - \dfrac{4}{3x} + \dfrac{8}{3x^2}}\right) = \frac{2}{3} \cdot \frac{1 - 0 + 0}{1 - 0 + 0} = \frac{2}{3}$$

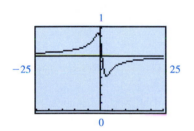

FIGURE 5
Graph of $y = f(x)$ and $y = \frac{2}{3}$

Thus, for this rational function, the behavior as x approaches infinity is determined by the ratio of the highest-degree term in the numerator to the highest-degree term in the denominator (see Fig. 5). Theorem 3 generalizes this to any rational function and lists the three possible outcomes. This provides us with a useful tool for analyzing the behavior of rational functions as x approaches infinity.

theorem 3

Limits at Infinity and Horizontal Asymptotes for Rational Functions

If

$$f(x) = \frac{a_m x^m + a_{m-1}x^{m-1} + \cdots + a_1 x + a_0}{b_n x^n + b_{n-1}x^{n-1} + \cdots + b_1 x + b_0} \qquad a_m \neq 0, \quad b_n \neq 0$$

then

$$\lim_{x \to \infty} f(x) = \lim_{x \to \infty} \frac{a_m x^m}{b_n x^n} \qquad \text{and} \qquad \lim_{x \to -\infty} f(x) = \lim_{x \to -\infty} \frac{a_m x^m}{b_n x^n}$$

There are three possible cases for these limits:

1. If $m < n$, then $\lim\limits_{x \to \infty} f(x) = \lim\limits_{x \to -\infty} f(x) = 0$, and the line $y = 0$ (the x axis) is a horizontal asymptote for $f(x)$.

(Continued)

> **Limits at Infinity and Horizontal Asymptotes for Rational Functions**   *(Continued)*
>
> 2. If $m = n$, then $\lim\limits_{x \to \infty} f(x) = \lim\limits_{x \to -\infty} f(x) = a_m/b_n$, and the line $y = a_m/b_n$ is a horizontal asymptote for $f(x)$.
> 3. If $m > n$, then each limit will be ∞ or $-\infty$, depending on m, n, a_m, and b_n, and $f(x)$ does not have a horizontal asymptote.

Notice in cases 1 and 2 of Theorem 3 that the limit is the same if x approaches ∞ or $-\infty$. Thus, **a rational function can have at most one horizontal asymptote.**

e x p l o r e – d i s c u s s 3

> Case 1 in Theorem 3 can be stated verbally as follows:
>
> If the degree of the numerator of a rational function is less than the degree of the denominator, then the x axis is a horizontal asymptote.
>
> Write similar verbal statements for cases 2 and 3.

E X A M P L E 2 ➤ **Horizontal Asymptotes for Rational Functions**   Find all horizontal asymptotes for each function:

(A) $f(x) = \dfrac{5x^3 - 2x^2 + 1}{4x^3 + 2x - 7}$ (B) $f(x) = \dfrac{3x^4 - x^2 + 1}{8x^6 - 10}$

(C) $f(x) = \dfrac{2x^5 - x^3 - 1}{6x^3 + 2x^2 - 7}$

Solution   (A) $\dfrac{a_m x^m}{b_n x^n} = \dfrac{5x^3}{4x^3} = \dfrac{5}{4}$

The line $y = \frac{5}{4}$ is a horizontal asymptote for $f(x)$ (Theorem 3, case 2).

(B) $\dfrac{a_m x^m}{b_n x^n} = \dfrac{3x^4}{8x^6} = \dfrac{3}{8x^2}$

The line $y = 0$ (the x axis) is a horizontal asymptote for $f(x)$ (Theorem 3, case 1).

(C) $\dfrac{a_m x^m}{b_n x^n} = \dfrac{2x^5}{6x^3} = \dfrac{x^2}{3}$

Thus, $f(x)$ does not have a horizontal asymptote (Theorem 3, case 3). ◀

MATCHED PROBLEM 2 ➤ Find all horizontal asymptotes for each function:

(A) $f(x) = \dfrac{4x^3 - 5x + 8}{2x^4 - 7}$ (B) $f(x) = \dfrac{5x^6 + 3x}{2x^5 - x - 5}$

(C) $f(x) = \dfrac{2x^3 - x + 7}{4x^3 + 3x^2 - 100}$ ◀

■ VERTICAL ASYMPTOTES

Theorem 3 adds an important tool for locating horizontal asymptotes to our mathematical toolbox. In Section 9-1 we discussed another type of asymptote—a vertical asymptote. Now we look further into this subject.

explore–discuss 4

(A) Figure 6 shows the graphs of two functions f and g on a graphing utility. Use the graphs to discuss the limit of each function as x approaches 1. Does either function appear to have a vertical asymptote at $x = 1$?

(A) $y = f(x)$ (B) $y = g(x)$

FIGURE 6

(B) The functions we used to produce the graphs in Figure 6 are

$$f(x) = \frac{1}{x - 1} \qquad \text{and} \qquad g(x) = \frac{2x - 2}{x^2 - 2x + 1.01}$$

Use the equations to discuss the limit of each function as x approaches 1. Does either function have a vertical asymptote at $x = 1$?

Graphing utilities are useful tools for analyzing graphs of functions. However, when vertical asymptotes are involved, graphs can be misleading, as illustrated in Figure 6. Theorem 4 provides a simple and effective algebraic tool for locating vertical asymptotes.

theorem 4

Locating Vertical Asymptotes

Let $f(x) = n(x)/d(x)$, where both n and d are continuous at $x = c$. If, at $x = c$, the denominator $d(x)$ is 0 and the numerator $n(x)$ is not 0, then the line $x = c$ is a vertical asymptote for the graph of f.

[*Note:* Since a rational function is a ratio of two polynomial functions and polynomial functions are continuous for all real numbers, this theorem includes rational functions as a special case.]

If $f(x) = n(x)/d(x)$ and both $n(c) = 0$ and $d(c) = 0$, then the limit of $f(x)$ as x approaches c involves an indeterminate form and Theorem 4 does not apply:

$$\lim_{x \to c} f(x) = \lim_{x \to c} \frac{n(x)}{d(x)} \quad \frac{0}{0} \text{ indeterminate form}$$

Algebraic simplification is often useful in this situation.

EXAMPLE 3 ➤ **Locating Vertical Asymptotes**   Find all vertical asymptotes for

$$f(x) = \frac{x^2 + x - 2}{x^2 - 1}$$

Solution   Let $n(x) = x^2 + x - 2$ and $d(x) = x^2 - 1$. Factoring the denominator, we see that

$$d(x) = x^2 - 1 = (x - 1)(x + 1)$$

Since $d(-1) = 0$ and $n(-1) = -2 \neq 0$, Theorem 4 tells us that the line $x = -1$ is a vertical asymptote. On the other hand, $d(1) = 0$ but $n(1) = 0$ also, so Theorem 4 does not apply at $x = 1$. We use algebraic simplification to investigate the behavior of the function at $x = 1$:

$$\lim_{x \to 1} f(x) = \lim_{x \to 1} \frac{x^2 + x - 2}{x^2 - 1}$$

$$= \lim_{x \to 1} \frac{(x - 1)(x + 2)}{(x - 1)(x + 1)}$$

$$= \lim_{x \to 1} \frac{x + 2}{x + 1}$$

$$= \frac{3}{2}$$

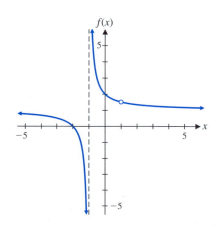

FIGURE 7
$f(x) = \dfrac{x^2 + x - 2}{x^2 - 1}$

Since the limit exists as x approaches 1, f does not have a vertical asymptote at $x = 1$ (see Fig. 7).   ◄

MATCHED PROBLEM 3 ➤   Find all vertical asymptotes for

$$f(x) = \frac{x - 3}{x^2 - 4x + 3}$$

◀

■ **SUMMARY OF GRAPH PROPERTIES**

We now have powerful tools that we can use to investigate the graph of a function. Given the equation for the function, we can use these tools in conjunction with a graphing utility to perform a detailed analysis of the graph of the function. Given information about the function in other forms, such as a graph of the derivative or a sign chart, we can use these tools to verbally describe the graph and to produce a hand sketch of the graph. We organize these tools in the next box for convenient reference.

Properties of the Graph of $y = f(x)$

Part 1.   *Analyze f(x):*

(A) Find the domain of f. [The domain of f is the set of all real numbers x that produce real values for $f(x)$.]
(B) Find intercepts. [The y intercept is $f(0)$, if it exists; the x intercepts are the solutions to $f(x) = 0$, if they exist.]
(C) Find asymptotes and behavior at infinity. [Use Theorems 1–4, if they apply; otherwise, calculate limits at points of discontinuity and as x increases and decreases with bound.]

Part 2.   *Analyze f'(x):*   Find any critical values for $f(x)$ and any partition numbers for $f'(x)$. [Remember, every critical value for $f(x)$ is also a partition number for $f'(x)$, but some partition numbers for $f'(x)$ may not be critical values for $f(x)$.] Use the sign of $f'(x)$ to determine the intervals where $f(x)$ is increasing and decreasing, and to find the local maxima and minima.

Part 3.   *Analyze f''(x):*   Use the sign of $f''(x)$ to determine where the graph of f is concave upward and concave downward, and to find any inflection points.

Parts 2 and 3 in the box refer to analyzing the sign of a derivative. We have discussed a number of different ways you can perform this analysis. To review these methods, consider the function $f(x) = x^2$ with derivative $f'(x) = 2x$. Figure 8A (at the top of the next page) shows the graph of f', Figure 8B shows a graphing utility representation of the sign chart for f', and Figure 8C shows a conventional sign chart. All three representations convey the same information about the graph of f. We will continue to use the conventional sign chart in the text, because it provides a conve-

(A) Graph of $f'(x)$

(B) Graph of $\dfrac{f'(x)}{|f'(x)|}$

(C) Sign chart

FIGURE 8

nient interpretation and summary of the relevant graph properties. When solving problems, you should use the method that you feel works best for you.

■ ANALYZING GRAPHS

We now consider examples where we will find all the graph information for the given functions by following the outline given in the preceding box.

EXAMPLE 4 ➤ **Analyzing a Graph** Find all the pertinent graph information for the function

$$f(x) = 0.1x^4 - x^3 - 5x^2 + 25x + 35$$

Round any approximate values to two decimal places.

Solution   Part 1. *Analyze $f(x)$:* $f(x) = 0.1x^4 - x^3 - 5x^2 + 25x + 35$

(A) Domain: All real x

(B) y intercept: $f(0) = 35$
x intercept: Solve $f(x) = 0$. Since the y intercept is 35, we choose a scale for the y axis that will show this intercept when we graph $y = f(x)$ (Fig. 9). The graph shows three x intercepts, which are easily located using a root approximation routine (details omitted):

$$x = -5.45, -1.19, 4.41 \quad \text{x intercepts}$$

(C) Asymptotes and behavior at infinity: Since $f(x)$ is a polynomial function, there are no horizontal or vertical asymptotes. The graph in Figure 9 seems to indicate that the values of $f(x)$ are decreasing without bound as x approaches ∞. But since f is a polynomial with highest-degree term $0.1x^4$, the values of $f(x)$ must be increasing without bound both as x approaches $-\infty$ and as x approaches ∞. Thus, there must be a point to the right of those visible in Figure 9 at which $f(x)$ begins to increase and the graph must cross the x axis a fourth time. To investigate this further, we trace to the right in Figure 9 looking for a point where the values of $f(x)$ begin to increase (Fig. 10A). Then we continue to the right, looking for a fourth x intercept (Fig. 10B). Having discovered this fourth intercept, we must amend our earlier results in part (B):

$$x = -5.45, -1.19, 4.41, 12.23 \quad \text{x intercepts}$$

FIGURE 9
$f(x) = 0.1x^4 - x^3 - 5x^2 + 25x + 35$

Adjusting the y range to show the local minimum discovered in Figure 10A produces the graph shown in Figure 10C.

(A) $f(x)$ begins to increase (B) $x = 12.23$ is the fourth intercept (C) Final graph

FIGURE 10

Since $f(x)$ is a fourth-degree polynomial, it can have at most four x intercepts and at most three local extrema. Thus, we are assured that the graph in Figure 10C shows all the important graph properties of $f(x)$.

FIGURE 11
Graph of $f'(x)$

Part 2. *Analyze* $f'(x)$*:* $f'(x) = 0.4x^3 - 3x^2 - 10x + 25$
To find the critical values for $f(x)$, we graph $f'(x)$ (Fig. 11) and use root approximation techniques to solve $f'(x) = 0$ (details omitted):

Critical values for $f(x)$: $-3.72, 1.78, 9.45$

Partition numbers for $f'(x)$: $-3.72, 1.78, 9.45$

Sign chart for $f'(x)$:

We see that $f(x)$ is decreasing on $(-\infty, -3.72)$ and $(1.78, 9.45)$, increasing on $(-3.72, 1.78)$ and $(9.45, \infty)$, has local minima at $x = -3.72$ and 9.45, and a local maximum at $x = 1.78$.

FIGURE 12
Graph of $f''(x)$

Part 3. *Analyze* $f''(x)$*:* $f''(x) = 1.2x^2 - 6x - 10$
To find the partition numbers for $f''(x)$, we graph $f''(x)$ (Fig. 12) and use root approximation techniques to solve $f''(x) = 0$ (details omitted):

Partition numbers for $f''(x)$: $-1.32, 6.32$

Sign chart for $f''(x)$:

Thus, $f(x)$ is concave upward on $(-\infty, -1.32)$ and $(6.32, \infty)$, concave down-ward on $(-1.32, 6.32)$, and has inflection points at $x = -1.32$ and $x = 6.32$. ◄

MATCHED PROBLEM 4 ➤ Find all the pertinent graph information for the function

$$f(x) = 0.1x^4 + x^3 - 4x^2 - 25x + 25$$

Round any approximate values to two decimal places. ◄

EXAMPLE 5 ➤ **Analyzing a Graph**   Find all the pertinent graph information for the function

$$f(x) = \frac{x - 1}{x - 2}$$

and illustrate with a sketch.

Solution   **Part 1.** *Analyze $f(x)$:* $f(x) = \dfrac{x - 1}{x - 2}$

(A) Domain: All real x, except $x = 2$

(B) y intercept: $f(0) = \dfrac{0 - 1}{0 - 2} = \dfrac{1}{2}$

x intercepts: Since a fraction is 0 when its numerator is 0 and the denominator is not 0, the x intercept is $x = 1$.

(C) Horizontal asymptote: $\dfrac{a_m x^m}{b_n x^n} = \dfrac{x}{x} = 1$

Thus, the line $y = 1$ is a horizontal asymptote.

Vertical asymptote: The denominator is 0 for $x = 2$, and the numerator is not 0 for this value. Therefore, the line $x = 2$ is a vertical asymptote.

Part 2. *Analyze $f'(x)$:* $f'(x) = \dfrac{(x - 2)(1) - (x - 1)(1)}{(x - 2)^2} = \dfrac{-1}{(x - 2)^2}$

Critical values for $f(x)$:   None

Partition number for $f'(x)$: $x = 2$

FIGURE 13
Graph of f'

Sign chart for $f'(x)$ (see Fig. 13):

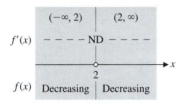

Thus, $f(x)$ is decreasing on $(-\infty, 2)$ and $(2, \infty)$. There are no local extrema.

Part 3. *Analyze $f''(x)$:* $f''(x) = \dfrac{2}{(x-2)^3}$

Partition number for $f''(x)$: 2

Sign chart for $f''(x)$ (see Fig. 14):

FIGURE 14
Graph of f''

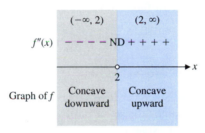

Thus, the graph of f is concave downward on $(-\infty, 2)$ and concave upward on $(2, \infty)$. Since $f(2)$ is not defined, there is no inflection point at $x = 2$, even though $f''(x)$ changes sign at $x = 2$.

Figure 15A shows a graphing utility graph of $y = f(x)$. To illustrate the graph properties, we insert intercepts and asymptotes, and then sketch the graph in Figure 15B, using Figure 15A as an aid.

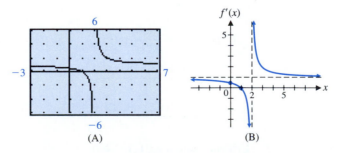

FIGURE 15
Graph of f

MATCHED PROBLEM 5 ▶ Find all the pertinent graph information for the function

$$f(x) = \frac{2x}{1 - x}$$

and illustrate with a sketch. ◀

■ APPLICATION

▶ EXAMPLE 6 ▶ **Average Cost** The total weekly cost (in dollars) of producing x CD players is given by

$$C(x) = 16{,}000 + 25x + 0.1x^2$$

Find the pertinent graph information for the average cost function. Graph the average cost function and the marginal cost function in the same viewing window, find the point of intersection, and interpret these results.

Solution The average cost function is

$$\overline{C}(x) = \frac{16{,}000 + 25x + 0.1x^2}{x} = \frac{16{,}000}{x} + 25 + 0.1x$$

Part 1. *Analyze* $\overline{C}(x)$:

(A) Domain: Since negative values of x do not make sense and $\overline{C}(0)$ is not defined, the domain is the set of positive real numbers.
(B) Intercepts: None
(C) Behavior at infinity and asymptotes:

$$\frac{a_m x^m}{b_n x^n} = \frac{0.1x^2}{x} = 0.1x \to \infty \qquad \text{as} \qquad x \to \infty$$

Thus, there is no horizontal asymptote.

Vertical asymptote: The line $x = 0$ is a vertical asymptote, since the denominator is 0 and the numerator is not 0 for $x = 0$.

Part 2. *Analyze* $\overline{C}'(x)$:

$$\overline{C}'(x) = -\frac{16{,}000}{x^2} + 0.1$$

$$= \frac{0.1x^2 - 16{,}000}{x^2}$$

$$= \frac{0.1(x - 400)(x + 400)}{x^2}$$

Critical value for $\overline{C}(x)$:   400

Partition numbers for $\overline{C}'(x)$:   0 and 400

Sign chart for $\overline{C}'(x)$ (details omitted):

Thus, $\overline{C}(x)$ is decreasing on $(0, 400)$, increasing on $(400, \infty)$, and has a local minimum at $x = 400$.

Part 3. *Analyze* $\overline{C}''(x)$: $\overline{C}''(x) = \dfrac{32{,}000}{x^3}$

$\overline{C}''(x)$ is positive for all positive x. Therefore, the graph of $y = \overline{C}(x)$ is concave upward on $(0, \infty)$.

FIGURE 16

The marginal cost function is $C'(x) = 25 + 0.2x$. To select a viewing window for the graphs of $y = \overline{C}(x)$ and $y = C'(x)$, we note that the value of $\overline{C}(x)$ at its local minimum is $\overline{C}(400) = 105$. Choosing the window parameters so that this point is near the center of the screen produces the graph in Figure 16. The graphs intersect at $(400, 105)$. Thus, we have the following interpretation:

The minimum average cost occurs when the average cost is equal to the marginal cost.

In this case, the minimum average cost is $105 when 400 CD players are produced.
◀

MATCHED PROBLEM 6 ➤ The total weekly cost (in dollars) of producing x food processors is $C(x) = 1{,}600 + 0.25x^2$. Find the pertinent graph information for the average cost function. Graph the average cost function and the marginal cost function in the same viewing window, find the point of intersection, and find the minimum average cost.   ◀

The relationship between average cost and marginal cost illustrated in Example 6 is true for most, but not all, cost functions. See the Chapter 8 Group Activity for a more detailed treatment of this important principle of economics.

Answers to Matched Problems

1. $p(x) = -4x^4 \left(1 - \dfrac{1}{2x} - \dfrac{3}{4x^3} \right) \approx -4x^4$ for x large (in absolute value);

   $\lim\limits_{x \to -\infty} p(x) = -\infty$, $\lim\limits_{x \to \infty} p(x) = -\infty$

2. (A) $y = 0$ (B) No horizontal asymptote   (C) $y = \frac{1}{2}$

3. $x = 1$

4. Domain: $(-\infty, \infty)$; y intercept: $f(0) = 25$; x intercepts: $-11.42, -4.68, 0.90, 5.19$

   Asymptotes and behavior at infinity: There are no horizontal or vertical asymptotes; $\lim\limits_{x \to \pm\infty} f(x) = \infty$

Increasing on $(-8.95, -2.01)$ and $(3.47, \infty)$; decreasing on $(-\infty, -8.95)$ and $(-2.01, 3.47)$

Local maximum at $x = -2.01$; local minima at $x = -8.95$ and $x = 3.47$

Concave upward on $(-\infty, -6.09)$ and $(1.09, \infty)$; concave downward on $(-6.09, 1.09)$

Inflection points at $x = -6.09$ and $x = 1.09$

5. Domain: All real x, except $x = 1$; y intercept: $f(0) = 0$; x intercept: 0

Horizontal asymptote: $y = -2$; vertical asymptote: $x = 1$

Increasing on $(-\infty, 1)$ and $(1, \infty)$

Concave upward on $(-\infty, 1)$; concave downward on $(1, \infty)$

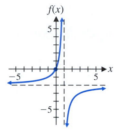

6. Domain: $(0, \infty)$; intercepts: None

Behavior at infinity: $\lim\limits_{x \to \infty} \overline{C}(x) = \infty$; vertical asymptote: $x = 0$

Decreasing on $(0, 80)$; increasing on $(80, \infty)$

Local minimum at $x = 80$

Concave upward on $(0, \infty)$

Intersection point: $(80, 40)$; minimum average cost is $40 at $x = 80$

EXERCISE 9-4

A *Problems 1–14 refer to the graph of $y = f(x)$ shown in the figure.*

Figure for 1–14

1. Identify the intervals over which $f'(x) < 0$.

2. Identify the intervals over which $f'(x) > 0$.

3. Identify the intervals over which $f(x)$ is increasing.

4. Identify the intervals over which $f(x)$ is decreasing.

5. Identify the x coordinate(s) of the point(s) where $f(x)$ has a local maximum.

6. Identify the x coordinate(s) of the point(s) where $f(x)$ has a local minimum.

7. Identify the intervals over which $f''(x) < 0$.

8. Identify the intervals over which $f''(x) > 0$.

9. Identify the intervals over which the graph of f is concave upward.

10. Identify the intervals over which the graph of *f* is concave downward.

11. Identify the *x* coordinate(s) of the inflection point(s).

12. Identify the horizontal asymptote(s).

13. Identify the vertical asymptote(s).

14. Identify the *x* and *y* intercepts.

In Problems 15–24, use the given information to sketch the graph of f. Assume that f is continuous on its domain and that all intercepts are included in the table of values.

15. Domain: All real *x*

x	-4	-2	0	2	4
$f(x)$	0	1	0	-1	0

16. Domain: All real *x*

x	-4	-2	-1	0	1	2	4
$f(x)$	0	3	2	1	2	3	0

17. Domain: All real x; $\lim\limits_{x \to \pm\infty} f(x) = 2$

x	-4	-2	0	2	4
$f(x)$	0	-2	0	-2	0

18. Domain: All real x; $\lim\limits_{x \to -\infty} f(x) = -3$; $\lim\limits_{x \to \infty} f(x) = 3$

x	-2	-1	0	1	2
$f(x)$	0	2	0	-2	0

19. Domain: All real x, except $x = -2$; $\lim\limits_{x \to -2^-} f(x) = \infty$; $\lim\limits_{x \to -2^+} f(x) = -\infty$; $\lim\limits_{x \to \infty} f(x) = 1$

x	-4	0	4	6
$f(x)$	0	0	3	2

20. Domain: All real x, except $x = 1$; $\lim\limits_{x \to 1^-} f(x) = \infty$;

$\lim\limits_{x \to 1^+} f(x) = \infty$; $\lim\limits_{x \to \infty} f(x) = -2$

x	-4	-2	0	2
$f(x)$	0	-2	0	0

21. Domain: All real x, except $x = -1$; $f(-3) = 2$, $f(-2) = 3$, $f(0) = -1$, $f(1) = 0$; $f'(x) > 0$ on $(-\infty, -1)$ and $(-1, \infty)$; $f''(x) > 0$ on $(-\infty, -1)$; $f''(x) < 0$ on $(-1, \infty)$; vertical asymptote: $x = -1$; horizontal asymptote: $y = 1$

22. Domain: All real x, except $x = 1$; $f(0) = -2$, $f(2) = 0$; $f'(x) < 0$ on $(-\infty, 1)$ and $(1, \infty)$; $f''(x) < 0$ on $(-\infty, 1)$; $f''(x) > 0$ on $(1, \infty)$; vertical asymptote: $x = 1$; horizontal asymptote: $y = -1$

23. Domain: All real x, except $x = -2$ and $x = 2$; $f(-3) = -1$, $f(0) = 0$, $f(3) = 1$; $f'(x) < 0$ on $(-\infty, -2)$ and $(2, \infty)$; $f'(x) > 0$ on $(-2, 2)$; $f''(x) < 0$ on $(-\infty, -2)$ and $(-2, 0)$; $f''(x) > 0$ on $(0, 2)$ and $(2, \infty)$; vertical asymptotes: $x = -2$ and $x = 2$; horizontal asymptote: $y = 0$

24. Domain: All real x, except $x = -1$ and $x = 1$; $f(-2) = 1$, $f(0) = 0$, $f(2) = 1$; $f'(x) > 0$ on $(-\infty, -1)$ and $(0, 1)$; $f'(x) < 0$ on $(-1, 0)$ and $(1, \infty)$; $f''(x) > 0$ on $(-\infty, -1)$, $(-1, 1)$, and $(1, \infty)$; vertical asymptotes: $x = -1$ and $x = 1$; horizontal asymptote: $y = 0$

B *In Problems 25–34, evaluate the indicated limit. Use $-\infty$ or ∞ where appropriate.*

25. $\lim\limits_{x \to \infty} (4x^3 + 2x - 9)$

26. $\lim\limits_{x \to \infty} (-2x^5 + 5x^3 + 3)$

27. $\lim\limits_{x \to -\infty} (-3x^6 + 9x^5 + 4)$

28. $\lim\limits_{x \to -\infty} (7x^4 - 4x^3 + 7x)$

29. $\lim\limits_{x \to \infty} \dfrac{4x + 7}{5x - 9}$

30. $\lim\limits_{x \to \infty} \dfrac{2 - 3x^3}{7 + 4x^3}$

31. $\lim\limits_{x \to \infty} \dfrac{5x^2 + 11}{7x - 2}$

32. $\lim\limits_{x \to \infty} \dfrac{5x + 11}{7x^3 - 2}$

33. $\lim\limits_{x \to -\infty} \dfrac{7x^4 - 14x^2}{6x^5 + 3}$

34. $\lim\limits_{x \to -\infty} \dfrac{4x^7 - 8x}{6x^4 + 9x^2}$

In Problems 35–48, find all horizontal and vertical asymptotes.

35. $f(x) = \dfrac{2x}{x + 2}$

36. $f(x) = \dfrac{3x + 2}{x - 4}$

37. $f(x) = \dfrac{x^2 + 1}{x^2 - 1}$

38. $f(x) = \dfrac{x^2 - 1}{x^2 + 2}$

39. $f(x) = \dfrac{x^3}{x^2 + 6}$

40. $f(x) = \dfrac{x}{x^2 - 4}$

41. $f(x) = \dfrac{x}{x^2 + 4}$

42. $f(x) = \dfrac{x^2 + 9}{x}$

43. $f(x) = \dfrac{x^2}{x - 3}$

44. $f(x) = \dfrac{x + 5}{x^2}$

45. $f(x) = \dfrac{2x^2 + 3x - 2}{x^2 - x - 2}$

46. $f(x) = \dfrac{x^2 + 7x + 12}{2x^2 + 5x - 12}$

47. $f(x) = \dfrac{2x^2 - 5x + 2}{x^2 - x - 2}$

48. $f(x) = \dfrac{x^2 - x - 12}{2x^2 + 5x - 12}$

In Problems 49–60, find all the pertinent graph information and illustrate graphically.

49. $f(x) = x^2 - 6x + 5$

50. $f(x) = 3 + 2x - x^2$

51. $f(x) = x^3 - 6x^2$

52. $f(x) = 3x^2 - x^3$

53. $f(x) = (x + 4)(x - 2)^2$

54. $f(x) = (2 - x)(x + 1)^2$

55. $f(x) = 8x^3 - 2x^4$

56. $f(x) = x^4 - 4x^3$

57. $f(x) = \dfrac{x + 3}{x - 3}$

58. $f(x) = \dfrac{2x - 4}{x + 2}$

59. $f(x) = \dfrac{x}{x - 2}$

60. $f(x) = \dfrac{2 + x}{3 - x}$

61. Theorem 2 states that

$$\lim\limits_{x \to \infty} (a_n x^n + a_{n-1} x^{n-1} + \cdots + a_0) = \pm\infty$$

What conditions must n and a_n satisfy for the limit to be ∞? For the limit to be $-\infty$?

62. Theorem 2 also states that

$$\lim\limits_{x \to -\infty} (a_n x^n + a_{n-1} x^{n-1} + \cdots + a_0) = \pm\infty$$

What conditions must n and a_n satisfy for the limit to be ∞? For the limit to be $-\infty$?

63. Let $p(x) = x^3 - 2x^2$.

(A) Find $\lim\limits_{x \to \infty} p'(x)$ and $\lim\limits_{x \to \infty} p''(x)$. Describe the shape of the graph of $y = p(x)$ for large positive values of x.

(B) Find $\lim\limits_{x \to -\infty} p'(x)$ and $\lim\limits_{x \to -\infty} p''(x)$. Describe the shape of the graph of $y = p(x)$ for large (in absolute value) negative values of x.

64. Repeat Problem 63 for $p(x) = x^4 - 2x^3$.

73. $f(x) = \dfrac{1}{1 + x^2}$ **74.** $f(x) = \dfrac{x^2}{1 + x^2}$

75. $f(x) = (x^2 + 3)(9 - x^2)$

76. $f(x) = (x^2 - 1)(x^2 - 5)$

C *In Problems 65–84, find all the pertinent graph information. Round any approximate values to two decimal places*

77. $f(x) = x^4 - 5x^3 + 3x^2 + 8x - 5$

78. $f(x) = x^4 + 2x^3 - 5x^2 - 4x + 4$

65. $f(x) = x^3 - x$ **66.** $f(x) = x^3 + x$

79. $f(x) = x^4 - 21x^3 + 100x^2 + 20x + 100$

67. $f(x) = -x^3 + 10x^2 + 30x + 50$

80. $f(x) = x^4 - 12x^3 + 28x^2 + 76x - 50$

68. $f(x) = -x^3 - 10x^2 + 40x + 50$

81. $f(x) = -x^4 - x^3 + 2x^2 - 2x + 3$

69. $f(x) = x + \dfrac{1}{x}$ **70.** $f(x) = x - \dfrac{1}{x}$

82. $f(x) = -x^4 + x^3 + x^2 + 6$

83. $f(x) = 0.1x^5 + 0.3x^4 - 4x^3 - 5x^2 + 40x + 30$

71. $f(x) = \dfrac{x}{x^2 - 4}$ **72.** $f(x) = \dfrac{1}{x^2 - 4}$

84. $f(x) = x^5 + 4x^4 - 7x^3 - 20x^2 + 20x - 20$

▶ **A P P L I C A T I O N S**

Business & Economics

85. *Pollution.* In Silicon Valley (California), a number of computer-related manufacturing firms were found to be contaminating underground water supplies with toxic chemicals stored in leaking underground containers. A water quality control agency ordered the companies to take immediate corrective action and to contribute to a monetary pool for testing and cleanup of the underground contamination. Suppose the required monetary pool (in millions of dollars) for the testing and cleanup is estimated to be given by

$$P(x) = \frac{2x}{1 - x} \qquad 0 \le x < 1$$

where x is the percentage (expressed as a decimal fraction) of the total contaminant removed.
(A) Where is $P(x)$ increasing? Decreasing?
(B) Where is the graph of P concave upward? Downward?
(C) Find any horizontal and vertical asymptotes.
(D) Find the x and y intercepts.
(E) Sketch a graph of P.

86. *Employee training.* A company producing computer components has established that on the average a new employee can assemble $N(t)$ components per day after t days of on-the-job training, as given by

$$N(t) = \frac{100t}{t + 9} \qquad t \ge 0$$

(A) Where is $N(t)$ increasing? Decreasing?
(B) Where is the graph of N concave upward? Downward?

(C) Find any horizontal and vertical asymptotes.
(D) Find the intercepts.
(E) Sketch a graph of N.

87. *Replacement time.* An office copier has an initial price of $3,200. A maintenance/service contract costs $300 for the first year and increases $100 per year thereafter. It can be shown that the total cost of the copier (in dollars) after n years is given by

$$C(n) = 3,200 + 250n + 50n^2$$

(A) Write an expression for the average cost per year, $\overline{C}(n)$, for n years.
(B) When is the average cost per year minimum? (This is frequently referred to as the **replacement time** for this piece of equipment.)

88. *Construction costs.* The management of a manufacturing plant wishes to add a fenced-in rectangular storage yard of 20,000 square feet, using the plant building as one side of the yard (see the figure). If x is the distance (in feet) from the building to the fence parallel to the building, then show that the length of the fence required for the yard is given by

$$L(x) = 2x + \frac{20,000}{x} \qquad x > 0$$

Figure for 88

What are the dimensions of the rectangle requiring the least amount of fencing?

89. *Average and marginal costs.* The total daily cost (in dollars) of producing x park benches is given by

$$C(x) = 1,000 + 5x + 0.1x^2$$

(A)   Summarize the pertinent graph information for the average cost function.

(B)   Find the point of intersection of the graph of the average cost function and the marginal cost function, and interpret the results.

90. *Average and marginal costs.* The total daily cost (in dollars) of producing x picnic tables is given by

$$C(x) = 500 + 2x + 0.2x^2$$

(A)   Summarize the pertinent graph information for the average cost function.

(B)   Find the point of intersection of the graph of the average cost function and the marginal cost function, and interpret the results.

91. *Minimizing average costs.* The data in the table give the total daily costs y (in dollars) of producing x pepperoni pizzas at various production levels.

NUMBER OF PIZZAS x	TOTAL COSTS ($) y
50	395
100	475
150	640
200	910
250	1,140
300	1,450

(A)   Enter the data in a graphing utility and find a quadratic regression equation for the total costs.

(B)   Use the regression equation from part (A) to find the minimum average cost (to the nearest cent) and the corresponding production level (to the nearest integer).

92. *Minimizing average costs.* The data in the table give the total daily costs y (in dollars) of producing x deluxe pizzas at various production levels.

NUMBER OF PIZZAS x	TOTAL COSTS ($) y
50	595
100	755
150	1,110
200	1,380
250	1,875
300	2,410

(A)   Enter the data in a graphing utility and find a quadratic regression equation for the total costs.

(B)   Use the regression equation from part (A) to find the minimum average cost (to the nearest cent) and the corresponding production level (to the nearest integer).

Life Sciences

93. *Medicine.* A drug is injected into the bloodstream of a patient through her right arm. The concentration of the drug in the bloodstream of the left arm t hours after the injection is given by

$$C(t) = \frac{0.14t}{t^2 + 1}$$

Summarize the pertinent graph information and illustrate with a graph.

94. *Physiology.* In a study on the speed of muscle contraction in frogs under various loads, researchers W. O. Fems and J. Marsh found that the speed of contraction decreases with increasing loads. More precisely, they found that the relationship between speed of contraction S (in centimeters per second) and load w (in grams) is given approximately by

$$S(w) = \frac{26 + 0.06w}{w} \qquad w \geq 5$$

Summarize the pertinent graph information and illustrate with a graph.

Social Sciences

95. *Psychology—retention.* An experiment on retention is conducted in a psychology class. Each student in the class is given 1 day to memorize the same list of 30 special characters. The lists are turned in at the end of the day, and for each succeeding day for 30 days each student is asked to turn in a list of as many of the symbols as can be recalled. Averages are taken, and it is found that

$$N(t) = \frac{5t + 20}{t} \qquad t \geq 1$$

provides a good approximation of the average number of symbols, $N(t)$, retained after t days. Summarize the pertinent graph information and illustrate with a graph.

SECTION   9-5   ■   Optimization; Absolute Maxima and Minima

- ■ ABSOLUTE MAXIMA AND MINIMA
- ■ APPLICATIONS

We are now ready to consider one of the most important applications of the derivative, namely, the use of derivatives to find the *absolute maximum* or *minimum* value of a function. As we mentioned earlier, an economist may be interested in the price or production level of a commodity that will bring a maximum profit; a doctor may be interested in the time it takes for a drug to reach its maximum concentration in the bloodstream after an injection; and a city planner might be interested in the location of heavy industry in a city to produce minimum pollution in residential and business areas. Before we launch an attack on problems of this type, which are called *optimization problems,* we have to say a few words about the procedures needed to find absolute maximum and absolute minimum values of functions. We have most of the tools we need from the previous sections.

■ ABSOLUTE MAXIMA AND MINIMA

First, what do we mean by *absolute maximum* and *absolute minimum?* We say that $f(c)$ is an **absolute maximum** of f if

$$f(c) \geq f(x)$$

for all x in the domain of f. Similarly, $f(c)$ is called an **absolute minimum** of f if

$$f(c) \leq f(x)$$

for all x in the domain of f. Figure 1 illustrates some typical examples.

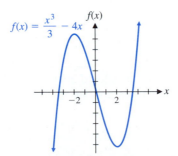

(A) No absolute maximum or minimum
   One local maximum at $x = -2$
   One local minimum at $x = 2$

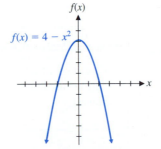

(B) Absolute maximum at $x = 0$
   No absolute minimum

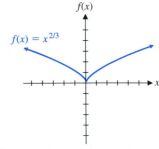

(C) Absolute minimum at $x = 0$
   No absolute maximum

FIGURE 1

Functions f, g, and h, along with their graphs, are shown in Figure 2.

(A) $f(x) = 1 - |x|$ (B) $g(x) = x - \dfrac{1}{x}$ (C) $h(x) = x - \dfrac{|x|}{x}$

FIGURE 2

(A) Which of these functions are continuous on $[-1, 1]$?

(B) Find the absolute maximum and the absolute minimum of each function on $[-1, 1]$, if they exist, and the corresponding values of x that produce these absolute extrema.

(C) Suppose that p is continuous on $[-1, 1]$ and satisfies $p(-1) = 0$ and $p(1) = 0$. Sketch a possible graph for p. Does the function you graphed have an absolute maximum? An absolute minimum? Can you modify your sketch so that p does not have an absolute maximum or an absolute minimum on $[-1, 1]$?

In many practical problems, the domain of a function is restricted because of practical or physical considerations. If the domain is restricted to some closed interval, as is often the case, then Theorem 1 can be proved.

Extreme Value Theorem

A function f that is continuous on a closed interval $[a, b]$ (see Section 9-1) has both an absolute maximum value and an absolute minimum value on that interval.

It is important to understand that the absolute maximum and minimum values depend on both the function f and the interval $[a, b]$. Figure 3 illustrates four cases.

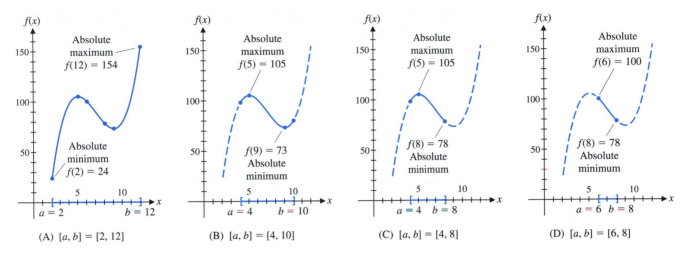

(A) $[a, b] = [2, 12]$ (B) $[a, b] = [4, 10]$ (C) $[a, b] = [4, 8]$ (D) $[a, b] = [6, 8]$

FIGURE 3
Absolute extrema for $f(x) = x^3 - 21x^2 + 135x - 170$ for various closed intervals

In all four cases illustrated in Figure 3, the absolute maximum value and absolute minimum value both occur at a critical value or an end point. In general:

Absolute extrema (if they exist) must always occur at critical values or at end points.

Thus, to find the absolute maximum or minimum value of a continuous function on an interval, we simply identify the critical values and any included end points, and then examine the graph to locate the highest or lowest point on the graph over the given interval.

EXAMPLE 1 ➤ Finding Absolute Extrema   Find the absolute maximum and minimum values of

$$f(x) = x^3 - 9x^2 + 15x$$

on each of the following intervals:

(A) $[0, 4]$ (B) $[0, 8]$ (C) $[0, \infty)$

Solution   The function is continuous for all values of x. The derivative is given by

$$f'(x) = 3x^2 - 18x + 15 = 3(x - 1)(x - 5)$$

Thus, $x = 1$ and $x = 5$ are the critical values. Examining the graph of f in Figure 4, we see that f has a local maximum at $x = 1$ and a local minimum at $x = 5$.

(A) The highest point on the interval $[0, 4]$ occurs at $x = 1$ and the lowest point at $x = 4$ (see Fig. 4). Thus:

Absolute maximum: $f(1) = 7$ *x = 1 is a critical value.*
Absolute minimum: $f(4) = -20$ *x = 4 is an end point.*

FIGURE 4
Graph of f

(B) On the interval [0, 8], the highest point occurs at $x = 8$ and the lowest point at $x = 5$ (see Fig. 4). Thus:

Absolute maximum: $f(8) = 56$ $x = 8$ is an end point.
Absolute minimum: $f(5) = -25$ $x = 5$ is a critical value.

(C) On the interval [0, ∞), the lowest point occurs at $x = 5$, but there is no highest point since $f(x) \rightarrow \infty$ as $x \rightarrow \infty$ (see Fig. 4). Thus:

Absolute maximum: Does not exist
Absolute minimum: $f(5) = -25$ $x = 5$ is a critical value.   ◀

MATCHED PROBLEM 1 ➤   Find the absolute maximum and minimum values of

$$f(x) = 2x^3 - 15x^2 + 24x$$

on each of the following intervals:

(A) [0, 2]   (B) [2, 5]   (C) [5, ∞)   ◀

■ APPLICATIONS

Now we want to solve some applied problems that involve absolute extrema. Before beginning, we outline in the next box the steps to follow in solving this type of problem. The first step is the most difficult one. The techniques used to solve optimization problems are best illustrated through a series of examples.

A Strategy for Solving Applied Optimization Problems

Step 1.   Introduce variables and a function f, including the domain I of f, and then construct a mathematical model of the form

Maximize (or minimize) $f(x)$ on the interval I

Step 2.   Find the absolute maximum (or minimum) value of $f(x)$ on the interval I and the value(s) of x where this occurs.

Step 3.   Use the solution to the mathematical model to answer the questions asked in the problem.

▶ EXAMPLE 2 ➤   **Maximize Revenue and Profit**   A company manufactures and sells x transistor radios per week. If the weekly cost and price–demand equations are

$$C(x) = 5{,}000 + 2x$$
$$p = 10 - 0.001x \qquad 0 \leqslant x \leqslant 10{,}000$$

find the following for each week:

(A) The maximum revenue
(B) The maximum profit, the production level that will realize the maximum profit,

and the price that the company should charge for each radio to realize the maximum profit

Solution (A) The revenue received for selling x radios at $\$p$ per radio is

$$R(x) = xp$$
$$= x(10 - 0.001x)$$
$$= 10x - 0.001x^2$$

Thus, the mathematical model is

$$\text{Maximize} \quad R(x) = 10x - 0.001x^2 \qquad 0 \leq x \leq 10,000$$
$$R'(x) = 10 - 0.002x$$
$$10 - 0.002x = 0$$
$$x = 5,000 \qquad \text{Only critical value}$$

Examining the graph of R shown in Figure 5 over the interval [0, 10,000], we see that the maximum value occurs at the critical value $x = 5,000$. Thus, the maximum revenue is

$$\text{Max } R(x) = R(5,000) = \$25,000$$

FIGURE 5
Graph of R

(B) Profit = Revenue − Cost

$$P(x) = R(x) - C(x)$$
$$= 10x - 0.001x^2 - 5,000 - 2x$$
$$= 8x - 0.001x^2 - 5,000$$

The mathematical model is

$$\text{Maximize} \quad P(x) = 8x - 0.001x^2 - 5,000 \qquad 0 \leq x \leq 10,000$$
$$P'(x) = 8 - 0.002x$$
$$8 - 0.002x = 0$$
$$x = 4,000 \qquad \text{Only critical value}$$

From the graph of P (Fig. 6), we conclude that

$$\text{Max } P(x) = P(4,000) = \$11,000$$

Using the price–demand equation with $x = 4,000$, we find

$$p = 10 - 0.001(4,000) = \$6$$

Thus, a maximum profit of $\$11,000$ per week is realized when 4,000 radios are produced weekly and sold for $\$6$ each. Notice that this is not the same level of production that produces the maximum revenue. ◀

FIGURE 6
Graph of P

All the results in Example 2 are illustrated in Figure 7 (at the top of the next page). We also note that profit is maximum when

$$P'(x) = R'(x) - C'(x) = 0$$

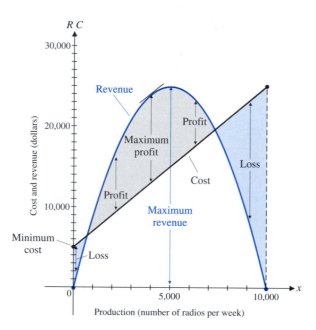

FIGURE 7

that is, when the marginal revenue is equal to the marginal cost (the rate of increase in revenue is the same as the rate of increase in cost at the 4,000 output level—notice that the slopes of the two curves are the same at this point).

MATCHED PROBLEM 2 ➤ Repeat Example 2 for

$$C(x) = 90,000 + 30x$$

$$p = 300 - \frac{x}{30} \qquad 0 \le x \le 9,000$$

◀

EXAMPLE 3 ➤ **Maximizing Profit** The government has decided to tax the company in Example 2 $2 for each radio produced. Taking into account this additional cost, how many radios should the company manufacture each week in order to maximize its weekly profit? What is the maximum weekly profit? How much should the company charge for the radios to realize the maximum weekly profit?

Solution   The tax of $2 per unit changes the company's cost equation:

$$C(x) = \text{Original cost} + \text{Tax}$$
$$= 5,000 + 2x + 2x$$
$$= 5,000 + 4x$$

The new profit function is

$$P(x) = R(x) - C(x)$$
$$= 10x - 0.001x^2 - 5,000 - 4x$$
$$= 6x - 0.001x^2 - 5,000$$

Thus, we must solve the following:

$$\text{Maximize} \quad P(x) = 6x - 0.001x^2 - 5{,}000 \qquad 0 \leqslant x \leqslant 10{,}000$$
$$P'(x) = 6 - 0.002x$$
$$6 - 0.002x = 0$$
$$x = 3{,}000$$

Examining the graph of P (Fig. 8), we find that

$$\text{Max } P(x) = P(3{,}000) = \$4{,}000$$

FIGURE 8
Graph of P

Using the price–demand equation with $x = 3{,}000$, we find

$$p = 10 - 0.001(3{,}000) = \$7$$

Thus, the company's maximum profit is $4,000 when 3,000 radios are produced and sold weekly at a price of $7.

Even though the tax caused the company's cost to increase by $2 per radio, the price that the company should charge to maximize its profit increases by only $1. The company must absorb the other $1 with a resulting decrease of $7,000 in maximum profit. ◀

MATCHED PROBLEM 3 ➤ Repeat Example 3 if

$$C(x) = 90{,}000 + 30x$$
$$p = 300 - \frac{x}{30} \qquad 0 \leqslant x \leqslant 9{,}000$$

and the government decides to tax the company $20 for each unit produced. Compare the results with the results in Matched Problem 2B. ◀

▶ EXAMPLE 4 ➤ **Maximizing Yield** A walnut grower estimates from past records that if 20 trees are planted per acre, each tree will average 60 pounds of nuts per year. If for each additional tree planted per acre (up to 15) the average yield per tree drops 2 pounds, how many trees should be planted to maximize the yield per acre? What is the maximum yield?

Solution Let x be the number of additional trees planted per acre. Then

$$20 + x = \text{Total number of trees per acre}$$
$$60 - 2x = \text{Yield per tree}$$
$$\text{Yield per acre} = (\text{Total number of trees per acre})(\text{Yield per tree})$$
$$Y(x) = (20 + x)(60 - 2x)$$
$$= 1{,}200 + 20x - 2x^2 \qquad 0 \leqslant x \leqslant 15$$

Thus, we must solve the following:

$$\text{Maximize} \quad Y(x) = 1{,}200 + 20x - 2x^2 \qquad 0 \leqslant x \leqslant 15$$

$$Y'(x) = 20 - 4x$$

$$20 - 4x = 0$$

$$x = 5$$

The graph of Y in Figure 9 shows that

$$\text{Max } Y(x) = Y(5) = 1{,}250 \text{ pounds per acre}$$

Thus, a maximum yield of 1,250 pounds of nuts per acre is realized if 25 trees are planted per acre.   ◀

FIGURE 9
Graph of Y

MATCHED PROBLEM 4 ▶ Repeat Example 4 starting with 30 trees per acre and a reduction of 1 pound per tree for each additional tree planted.   ◀

explore – discuss 2

In Example 4, letting x be the number of *additional* trees planted per acre produced a simple and direct solution to the problem. However, this is not the most obvious choice for a variable. Suppose we proceed as follows:

Let x be the total number of trees planted per acre and let y be the yield per tree. Then the yield per acre is given by xy.

(A) Find y when $x = 20$ and when $x = 21$. Find the equation of the line through these two points.
(B) Use the equation from part (A) to express the yield per acre in terms of either x or y, and use this expression to solve Example 4.
(C) Compare this method of solution to the one used in Example 4 with respect to ease of comprehension and ease of computation.

EXAMPLE 5 ▶ Maximizing Area   A farmer wants to construct a rectangular pen next to a barn 60 feet long, using all of the barn as part of one side of the pen. Find the dimensions of the pen with the largest area that the farmer can build if:

(A) 160 feet of fencing material is available
(B) 250 feet of fencing material is available

Solution   (A) We begin by constructing and labeling the figure shown in the margin on the facing page. The area of the pen is

$$A = (x + 60)y$$

Before we can maximize the area, we must determine a relationship between x and y in order to express A as a function of one variable. In this case, x and y are related to the total amount of available fencing material:

$$x + y + 60 + x + y = 160$$
$$2x + 2y = 100$$
$$y = 50 - x$$

Thus,

$$A(x) = (x + 60)(50 - x)$$

Now we need to determine the permissible values of x; that is, the domain of the function A. Since the farmer wants to use all of the barn as part of one side of the pen, x cannot be negative. Since y is the other dimension of the pen, y cannot be negative. Thus,

$$y = 50 - x \geqslant 0$$
$$50 \geqslant x$$

F I G U R E   10

Graph of A

The domain of A is [0, 50]. Thus, we must solve the following:

$$\text{Maximize} \quad A(x) = (x + 60)(50 - x) \qquad 0 \leqslant x \leqslant 50$$
$$A(x) = 3,000 - 10x - x^2$$
$$A'(x) = -10 - 2x$$
$$-10 - 2x = 0$$
$$x = -5$$

Since $x = -5$ is not in the interval [0, 50], there are no critical values in the interval. $A(x)$ is continuous on [0, 50], so the absolute maximum must occur at one of the end points. Examining the graph of A (Fig. 10), we see that the maximum occurs at $x = 0$. Thus,

$$\text{Max } A(x) = A(0) = 3,000$$

If $x = 0$, then $y = 50$. Thus, the dimensions of the pen with largest area are 60 feet by 50 feet.

(B) If 250 feet of fencing material is available, then

$$x + y + 60 + x + y = 250$$
$$2x + 2y = 190$$
$$y = 95 - x$$

The model becomes

$$\text{Maximize} \quad A(x) = (x + 60)(95 - x) \qquad 0 \leqslant x \leqslant 95$$
$$A(x) = 5,700 + 35x - x^2$$
$$A'(x) = 35 - 2x$$
$$35 - 2x = 0$$
$$x = \tfrac{35}{2} = 17.5 \quad \textit{The only critical value}$$

From Figure 11, we see that

$$\text{Max } A(x) = A(17.5) = 6,006.25$$
$$y = 95 - 17.5 = 77.5$$

F I G U R E   11

Graph of A

This time, the dimensions of the pen with the largest area are 77.5 feet by 77.5 feet.

60 ft

77.5 ft   77.5 ft

MATCHED PROBLEM 5 ➤ Repeat Example 5 if the barn is 80 feet long.

➤ **EXAMPLE 6** ➤ **Inventory Control**   A recording company anticipates that there will be a demand for 20,000 copies of a certain CD during the following year. It costs the company $0.50 to store a CD for 1 year. Each time it must make additional CD's, it costs $200 to set up the equipment. How many CD's should the company make during each production run in order to minimize its total storage and set-up costs?

Solution   This type of problem is called an **inventory control problem.** One of the basic assumptions made in such problems is that the demand is uniform. For example, if there are 250 working days in a year, then the daily demand would be 20,000 ÷ 250 = 80 CD's. The company could decide to produce all 20,000 CD's at the beginning of the year. This would certainly minimize the set-up costs, but would result in very large storage costs. At the other extreme, it could produce 80 CD's each day. This would minimize the storage costs, but would result in very large set-up costs. Somewhere between these two extremes is the optimal solution that will minimize the total storage and set-up costs. Let

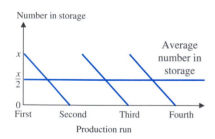

Number in storage

Average number in storage

First   Second   Third   Fourth

Production run

FIGURE 12

x = Number of CD's manufactured during each production run
y = Number of production runs

It is easy to see that the total set-up cost for the year is $200y$, but what is the total storage cost? If the demand is uniform, then the number of CD's in storage between production runs will decrease from x to 0, and the average number in storage each day is $x/2$. This result is illustrated in Figure 12.

Since it costs $0.50 to store a CD for 1 year, the total storage cost is $0.5(x/2) = 0.25x$ and the total cost is

$$\text{Total cost} = \text{Set-up cost} + \text{Storage cost}$$
$$C = 200y + 0.25x$$

In order to write the total cost C as a function of one variable, we must find a relationship between x and y. If the company produces x CD's in each of y production runs, then the total number of CD's produced is xy. Thus,

$$xy = 20,000$$
$$y = \frac{20,000}{x}$$

Certainly, x must be at least 1 and cannot exceed 20,000. Thus, we must solve the following:

$$\text{Minimize} \quad C(x) = 200 \left(\frac{20,000}{x} \right) + 0.25x \qquad 1 \leqslant x \leqslant 20,000$$

$$C(x) = \frac{4,000,000}{x} + 0.25x$$

$$C'(x) = -\frac{4,000,000}{x^2} + 0.25$$

In the preceding examples, we found critical values algebraically. This time, we will use a graphing utility to solve the equation $C'(x) = 0$. Using trace to determine Ymin and Ymax for the graph of C', and then using a root approximation routine (Fig. 13), we see that the only critical value is $x = 4,000$. Using trace again to adjust Ymin and Ymax for the graph of C produces Figure 14.

FIGURE 13
Graph of C'

FIGURE 14
Graph of C

Thus,

$$\text{Min } C(x) = C(4,000) = 2,000$$

$$y = \frac{20,000}{4,000} = 5$$

The company will minimize its total cost by making 4,000 CD's five times during the year. ◀

MATCHED PROBLEM 6 ➤ Repeat Example 6 if it costs $250 to set up a production run and $0.40 to store a CD for 1 year. ◀

Answers to Matched Problems

1. (A) Absolute maximum: $f(1) = 11$; absolute minimum: $f(0) = 0$
   (B) Absolute maximum: $f(2) = 4$; absolute minimum: $f(4) = -16$
   (C) Absolute maximum: Does not exist; absolute minimum: $f(5) = -5$
2. (A) Max $R(x) = R(4,500) = \$675,000$
   (B) Max $P(x) = P(4,050) = \$456,750$; $p = \$165$
3. Max $P(x) = P(3,750) = \$378,750$; $p = \$175$; price increases $10, profit decreases $78,000
4. Max $Y(x) = Y(15) = 2,025$ lb/acre
5. (A) 80 ft by 40 ft (B) 82.5 ft by 82.5 ft
6. Make 5,000 CD's four times during the year

EXERCISE   9-5

A   *Problems 1–8 refer to the graph of y = f(x). Find the absolute maximum and minimum over the indicated interval.*

Figure for 1–8

1. [0, 10]   **2.** [1, 9]   **3.** [0, 8]
4. [2, 10]   **5.** [4, 6]   **6.** [8, 10]
7. [1, 5]   **8.** [6, 8]

In Problems 9–14, find the absolute maximum and minimum, if either exists, for each function.

9. $f(x) = x^2 - 4x + 5$ **10.** $f(x) = x^2 + 6x + 7$
11. $f(x) = 10 + 8x - x^2$ **12.** $f(x) = 6 - 8x - x^2$
13. $f(x) = 1 - x^3$ **14.** $f(x) = 1 - x^4$

B   *In Problems 15–18, find the indicated extremum of each function, if it exists.*

15.   Absolute maximum value of

$$f(x) = 24 - 2x - \frac{8}{x} \qquad x > 0$$

16.   Absolute minimum value of

$$f(x) = 3x + \frac{27}{x} \qquad x > 0$$

17.   Absolute minimum value of

$$f(x) = 5 + 3x + \frac{12}{x^2} \qquad x > 0$$

18.   Absolute maximum value of

$$f(x) = 10 - 2x - \frac{27}{x^2} \qquad x > 0$$

In Problems 19–22, find the absolute maximum and minimum, if either exists, for each function on the indicated intervals.

19. $f(x) = x^3 - 6x^2 + 9x - 6$
    (A) [-1, 5]   (B) [-1, 3]   (C) [2, 5]

20. $f(x) = 2x^3 - 3x^2 - 12x + 24$
    (A) [-3, 4]   (B) [-2, 3]   (C) [-2, 1]

21. $f(x) = (x - 1)(x - 5)^3 + 1$
    (A) [0, 3]   (B) [1, 7]   (C) [3, 6]

22. $f(x) = x^4 - 8x^2 + 16$
    (A) [-1, 3]   (B) [0, 2]   (C) [-3, 4]

C   *Preliminary word problems:*

23.   How would you divide a 10 inch line so that the product of the two lengths is a maximum?

24.   What quantity should be added to 5 and subtracted from 5 in order to produce the maximum product of the results?

25.   Find two numbers whose difference is 30 and whose product is a minimum.

26.   Find two positive numbers whose sum is 60 and whose product is a maximum.

27.   Find the dimensions of a rectangle with perimeter 100 centimeters that has maximum area. Find the maximum area.

28.   Find the dimensions of a rectangle of area 225 square centimeters that has the least perimeter. What is the perimeter?

Problems 29–32 refer to a rectangular area enclosed by a fence that costs $B per foot. Discuss the existence of a solution and the economical implications of each optimization problem.

29.   Given a fixed area, minimize the cost of the fencing.

30.   Given a fixed area, maximize the cost of the fencing.

31.   Given a fixed amount to spend on fencing, maximize the enclosed area.

32.   Given a fixed amount to spend on fencing, minimize the enclosed area.

▶ APPLICATIONS

Business & Economics

33. *Maximum revenue and profit.* A company manufactures and sells x television sets per month. The monthly cost and price–demand equations are

$$C(x) = 72,000 + 60x$$

$$p = 200 - \frac{x}{30} \qquad 0 \leq x \leq 6,000$$

(A) Find the maximum revenue.

(B) Find the maximum profit, the production level that will realize the maximum profit, and the price the company should charge for each television set.

(C) If the government decides to tax the company $5 for each set it produces, how many sets should the company manufacture each month in order to maximize its profit? What is the maximum profit? What should the company charge for each set?

34. *Maximum revenue and profit.* Repeat Problem 33 for

$$C(x) = 60,000 + 60x$$

$$p = 200 - \frac{x}{50} \qquad 0 \leq x \leq 10,000$$

35. *Car rental.* A car rental agency rents 200 cars per day at a rate of $30 per day. For each $1 increase in rate, 5 fewer cars are rented. At what rate should the cars be rented to produce the maximum income? What is the maximum income?

36. *Rental income.* A 300 room hotel in Las Vegas is filled to capacity every night at $80 a room. For each $1 increase in rent, 3 fewer rooms are rented. If each rented room costs $10 to service per day, how much should the management charge for each room to maximize gross profit? What is the maximum gross profit?

37. *Agriculture.* A commercial cherry grower estimates from past records that if 30 trees are planted per acre, each tree will yield an average of 50 pounds of cherries per season. If for each additional tree planted per acre (up to 20), the average yield per tree is reduced by 1 pound, how many trees should be planted per acre to obtain the maximum yield per acre? What is the maximum yield?

38. *Agriculture.* A commercial pear grower must decide on the optimum time to have fruit picked and sold. If the pears are picked now, they will bring 30¢ per pound, with each tree

yielding an average of 60 pounds of salable pears. If the average yield per tree increases 6 pounds per tree per week for the next 4 weeks, but the price drops 2¢ per pound per week, when should the pears be picked to realize the maximum return per tree? What is the maximum return?

39. *Manufacturing.* A candy box is to be made out of a piece of cardboard that measures 8 by 12 inches. Squares of equal size will be cut out of each corner, and then the ends and sides will be folded up to form a rectangular box. What size square should be cut from each corner to obtain a maximum volume?

40. *Packaging.* A parcel delivery service will deliver a package only if the length plus girth (distance around) does not exceed 108 inches.

Figure for 40

(A) Find the dimensions of a rectangular box with square ends that satisfies the delivery service's restriction and has maximum volume. What is the maximum volume?

(B) Find the dimensions (radius and height) of a cylindrical container that meets the delivery service's requirement and has maximum volume. What is the maximum volume?

41. *Construction costs.* A fence is to be built to enclose a rectangular area of 800 square feet. The fence along three sides is to be made of material that costs $6 per foot. The material for the fourth side costs $18 per foot. Find the dimensions of the rectangle that will allow the most economical fence to be built.

42. *Construction costs.* The owner of a retail lumber store wants to construct a fence to enclose an outdoor storage area adjacent to the store, using all of the store as part of one side of the area (see the figure). Find the dimensions that will enclose the largest area if:

Figure for 42

(A) 240 feet of fencing material is used.
(B) 400 feet of fencing material is used.

43. *Inventory control.* A publishing company sells 50,000 copies of a certain book each year. It costs the company $1 to store a book for 1 year. Each time it must print additional copies, it costs the company $1,000 to set up the presses. How many books should the company produce during each printing in order to minimize its total storage and set-up costs?

44. *Operational costs.* The cost per hour for fuel to run a train is $v^2/4$ dollars, where v is the speed of the train in miles per hour. (Note that the cost goes up as the square of the speed.) Other costs, including labor, are $300 per hour. How fast should the train travel on a 360 mile trip to minimize the total cost for the trip?

45. *Construction costs.* A freshwater pipeline is to be run from a source on the edge of a lake to a small resort community on an island 5 miles off-shore, as indicated in the figure.

Figure for 45

(A) If it costs 1.4 times as much to lay the pipe in the lake as it does on land, what should x be (in miles) to minimize the total cost of the project?
(B) If it costs only 1.1 times as much to lay the pipe in the lake as it does on land, what should x be to minimize the total cost of the project? [*Note:* Compare with Problem 50.]

46. *Manufacturing costs.* A manufacturer wants to produce cans that will hold 12 ounces (approximately 22 cubic inches) in the form of a right circular cylinder. Find the dimensions (radius of an end and height) of the can that will use the smallest amount of material. Assume the circular ends are cut out of squares, with the corner portions wasted, and the sides are made from rectangles, with no waste.

Life Sciences

47. *Bacteria control.* A recreational swimming lake is treated periodically to control harmful bacteria growth. Suppose t days after a treatment, the concentration of bacteria per cubic centimeter is given by

$$C(t) = 30t^2 - 240t + 500 \qquad 0 \leq t \leq 8$$

How many days after a treatment will the concentration be minimal? What is the minimum concentration?

48. *Drug concentration.* The concentration $C(t)$, in milligrams per cubic centimeter, of a particular drug in a patient's bloodstream is given by

$$C(t) = \frac{0.16t}{t^2 + 4t + 4}$$

where t is the number of hours after the drug is taken. How many hours after the drug is given will the concentration be maximum? What is the maximum concentration?

49. *Laboratory management.* A laboratory uses 500 white mice each year for experimental purposes. It costs $4 to feed a mouse for 1 year. Each time mice are ordered from a supplier, there is a service charge of $10 for processing the order. How many mice should be ordered each time in order to minimize the total cost of feeding the mice and of placing the orders for the mice?

50. *Bird flights.* Some birds tend to avoid flights over large bodies of water during daylight hours. (It is speculated that more energy is required to fly over water than land, because air generally rises over land and falls over water during the day.) Suppose an adult bird with this tendency is taken from its nesting area on the edge of a large lake to an island 5 miles off-shore and is then released (see the figure).
(A) If it takes 1.4 times as much energy to fly over water as land, how far up-shore (x, in miles) should the bird head in order to minimize the total energy expended in returning to the nesting area?
(B) If it takes only 1.1 times as much energy to fly over water as land, how far up-shore should the bird head in order to minimize the total energy expended in returning to the nesting area? [*Note:* Compare with Problem 45.]

Figure for 50

Figure for 52

51. *Botany.* If it is known from past experiments that the height (in feet) of a given plant after t months is given approximately by

$$H(t) = 4t^{1/2} - 2t \qquad 0 \leqslant t \leqslant 2$$

how long, on the average, will it take a plant to reach its maximum height? What is the maximum height?

52. *Pollution.* Two heavy industrial areas are located 10 miles apart, as indicated in the figure. If the concentration of particulate matter (in parts per million) decreases as the reciprocal of the square of the distance from the source, and area A_1 emits eight times the particulate matter as A_2, then the concentration of particulate matter at any point between the two areas is given by

$$C(x) = \frac{8k}{x^2} + \frac{k}{(10-x)^2} \qquad 0.5 \leqslant x \leqslant 9.5, \quad k > 0$$

How far from A_1 will the concentration of particulate matter between the two areas be at a minimum?

Social Sciences

53. *Politics.* In a newly incorporated city, it is estimated that the voting population (in thousands) will increase according to

$$N(t) = 30 + 12t^2 - t^3 \qquad 0 \leqslant t \leqslant 8$$

where t is time in years. When will the rate of increase be most rapid?

54. *Learning.* A large grocery chain found that, on the average, a checker can memorize $P\%$ of a given price list in x continuous hours, as given approximately by

$$P(x) = 96x - 24x^2 \qquad 0 \leqslant x \leqslant 3$$

How long should a checker plan to take to memorize the maximum percentage? What is the maximum?

CHAPTER 9
GROUP ACTIVITY

Maximizing Profit

A company manufactures and sells x air conditioners per month. The monthly cost and price–demand equations are

$$C(x) = 180x + 20,000$$
$$p = 220 - 0.001x \qquad 0 \leqslant x \leqslant 100,000$$

(A) How many air conditioners should the company manufacture each month in order to maximize its monthly profit? What is the maximum monthly profit, and what should the company charge for each air conditioner to realize the maximum monthly profit?

(B) Repeat part (A) if the government decides to tax the company at the rate of $18 per air conditioner produced. How much revenue will the government receive from the tax on these air conditioners?

(C) Repeat part (A) if the government raises the tax to $23 per air conditioner. Discuss the effect of this tax increase on the government's tax revenue.

(D) Repeat part (A) if the government sets the tax rate at $$t$ per air conditioner. What value of t will maximize the government's tax revenue? What is the government's maximum tax revenue?

CHAPTER 9 REVIEW ■ **Important Terms and Symbols**

9-1 *Continuity and Graphs.* Continuous at a point; continuous on an open interval; discontinuous at a point; continuous on the right; continuous on the left; continuous on a closed interval; general continuity properties; continuity properties of specific functions; vertical asymptote; sign chart; partition number

$$\lim_{x \to c} f(x) = f(c) \text{ if } f \text{ is continuous at } x = c;$$

$$\lim_{x \to c} f(x) = \infty; \quad \lim_{x \to c} f(x) = -\infty$$

9-2 *First Derivative and Graphs.* Increasing and decreasing functions; rising and falling graphs; critical values; local extremum; local maximum; local minimum; turning point; first-derivative test for local extrema

9-3 *Second Derivative and Graphs.* Concave upward; concave downward; second derivative; concavity and the second derivative; inflection point; second-derivative test for local maxima and minima; point of diminishing returns

$$f''(x); \quad \frac{d^2y}{dx^2}; \quad y''$$

9-4 *Graphical Analysis: Unified and Extended.* Horizontal asymptote; limits at infinity for power functions, polynomials, and rational functions; locating vertical asymptotes; graph properties [analyze $f(x)$ to find domain, intercepts, behavior at infinity and asymptotes; analyze $f'(x)$ to find increasing and decreasing regions, local extrema; analyze $f''(x)$ to find concave upward and concave downward regions, inflection points]

9-5 *Optimization; Absolute Maxima and Minima.* Absolute maximum; absolute minimum; absolute extrema of a function on a closed interval; optimization problems; inventory control

---

CHAPTER 9 ■ **Review Exercise**

Work through all the problems in this chapter review and check your answers in the back of the book. Answers to all review problems are there along with section numbers in italics to indicate where each type of problem is discussed. Where weaknesses show up, review appropriate sections in the text.

A *Problems 1–8 refer to the graph of $y = f(x)$. Identify the points or intervals on the x axis that produce the indicated behavior.*

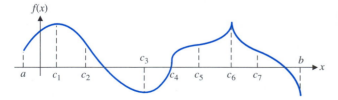

Figure for 1–8

1. $f(x)$ is increasing
2. $f'(x) < 0$
3. Graph of f is concave downward
4. Local minima
5. Absolute maxima
6. $f'(x)$ appears to be 0

7. $f'(x)$ does not exist
8. Inflection points

In Problems 9–11, use the graph of the function f shown in the figure to answer each question.

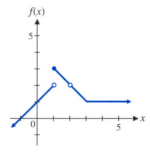

Figure for 9–11

9. (A) $\lim_{x \to 1} f(x) = $ **?** (B) $f(1) = $ **?**
   (C) Is f continuous at $x = 1$?
10. (A) $\lim_{x \to 2} f(x) = $ **?** (B) $f(2) = $ **?**
    (C) Is f continuous at $x = 2$?
11. (A) $\lim_{x \to 3} f(x) = $ **?** (B) $f(3) = $ **?**
    (C) Is f continuous at $x = 3$?

In Problems 12 and 13, use the given information to sketch the graph of f. Assume that f is continuous on its domain and that all intercepts are included in the given information.

12. Domain: All real x

x	-3	-2	-1	0	2	3
$f(x)$	0	3	2	0	-3	0

13. Domain: All real x; $f(-2) = 1$, $f(0) = 0$, $f(2) = 1$; $f'(0) = 0$; $f'(x) < 0$ on $(-\infty, 0)$; $f'(x) > 0$ on $(0, \infty)$; $f''(-2) = 0, f''(2) = 0; f''(x) < 0$ on $(-\infty, -2)$ and $(2, \infty)$; $f''(x) > 0$ on $(-2, 2)$; $\lim_{x \to -\infty} f(x) = 2$; $\lim_{x \to \infty} f(x) = 2$

14. Find $f''(x)$ for $f(x) = x^4 + 5x^3$.

15. Find y'' for $y = 3x + \dfrac{4}{x}$.

B *Problems 16 and 17 refer to the function f described in the figure.*

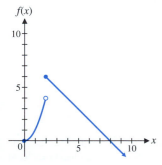

Figure for 16 and 17

$$f(x) = \begin{cases} x^2 & \text{if } 0 \le x < 2 \\ 8 - x & \text{if } x \ge 2 \end{cases}$$

16. (A) $\lim_{x \to 2^-} f(x) = $ **?**   (B) $\lim_{x \to 2^+} f(x) = $ **?**   (C) $\lim_{x \to 2} f(x) = $ **?**
(D) $f(2) = $ **?**   (E) Is f continuous at $x = 2$?

17. (A) $\lim_{x \to 5^-} f(x) = $ **?**   (B) $\lim_{x \to 5^+} f(x) = $ **?**   (C) $\lim_{x \to 5} f(x) = $ **?**
(D) $f(5) = $ **?**   (E) Is f continuous at $x = 5$?

In Problems 18–20, solve each inequality. Express the answer in interval notation. Round any approximate values to two decimal places.

18. $x^2 - x < 12$

19. $\dfrac{x - 5}{x^2 + 3x} > 0$

20. $x^3 + x^2 - 4x - 2 > 0$

Problems 21–24 refer to the function

$$f(x) = x^3 - 18x^2 + 81x$$

21. Using $f(x)$:
(A) Determine the domain of f.
(B) Find any intercepts for the graph of f.
(C) Discuss the behavior at infinity and find any asymptotes.

22. Using $f'(x)$:
(A) Find critical values for $f(x)$.
(B) Find partition numbers for $f'(x)$.
(C) Find intervals over which $f(x)$ is increasing; decreasing.
(D) Find any local maxima and minima.

23. Using $f''(x)$:
(A) Find intervals over which the graph of f is concave upward; concave downward.
(B) Find any inflection points.

24. Illustrate the results of Problems 21–23 with a graph.

Problems 25–28 refer to the function

$$y = f(x) = \frac{3x}{x + 2}$$

25. Using $f(x)$:
(A) Determine the domain of f.
(B) Find any intercepts for the graph of f.
(C) Discuss the behavior at infinity and find any asymptotes.

26. Using $f'(x)$:
(A) Find critical values for $f(x)$.
(B) Find partition numbers for $f'(x)$.
(C) Find intervals over which $f(x)$ is increasing; decreasing.
(D) Find any local maxima and minima.

27. Using $f''(x)$:
(A) Find intervals over which the graph of f is concave upward; concave downward.
(B) Find any inflection points.

28. Illustrate the results of Problems 25–27 with a graph.

29. Use the graph of $y = f'(x)$ to discuss the graph of $y = f(x)$. Organize your conclusions in a table (see Example 4, page 697). Sketch a possible graph for $y = f(x)$.

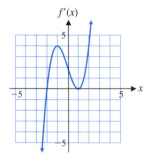

$f'(x)$

Figure for 29 and 30

30. Refer to the graph of $y = f'(x)$ in Problem 29. Which of the following could be the graph of $y = f''(x)$?

(A)

$f''(x)$

(B)

$f''(x)$

(C)

$f''(x)$

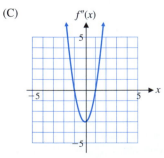

31. Find the absolute maximum and absolute minimum, if either exists, for

$$f(x) = x^3 - 6x^2 - 15x + 12 \qquad x \geq 0$$

32. Find the absolute maximum and absolute minimum, if either exists, for

$$y = f(x) = x^3 - 12x + 12 \qquad -3 \leq x \leq 5$$

33. Find the absolute minimum, if it exists, for

$$y = f(x) = x^2 + \frac{16}{x^2} \qquad x > 0$$

In Problems 34–38, determine where f is continuous. Express the answer in interval notation.

34. $f(x) = 2x^2 - 3x + 1$ **35.** $f(x) = \dfrac{1}{x + 5}$

36. $f(x) = \dfrac{x - 3}{x^2 - x - 6}$ **37.** $f(x) = \sqrt{x - 3}$

38. $f(x) = \sqrt[3]{1 - x^2}$

In Problems 39–48, evaluate the indicated limit. Use $-\infty$ or ∞ where appropriate.

39. $\displaystyle\lim_{x \to 2^-} \frac{x}{2 - x}$ **40.** $\displaystyle\lim_{x \to 2^+} \frac{x}{2 - x}$

41. $\displaystyle\lim_{x \to 2} \frac{x}{2 - x}$ **42.** $\displaystyle\lim_{x \to 2} \frac{x}{(2 - x)^2}$

43. $\displaystyle\lim_{x \to \infty} (-3x^5)$ **44.** $\displaystyle\lim_{x \to -\infty} (3x^2 - 4x^3)$

45. $\displaystyle\lim_{x \to \infty} (4x^6 - 2x^5)$ **46.** $\displaystyle\lim_{x \to \infty} \frac{6x^3 + 4x^2 + 5}{3x^3 + 2x + 7}$

47. $\displaystyle\lim_{x \to \infty} \frac{6x^4 + 4x^2 + 5}{3x^3 + 2x + 7}$ **48.** $\displaystyle\lim_{x \to \infty} \frac{6x^3 + 4x^2 + 5}{3x^4 + 2x + 7}$

Find horizontal and vertical asymptotes, if they exist, in Problems 49 and 50.

49. $f(x) = \dfrac{x}{x^2 + 9}$ **50.** $f(x) = \dfrac{x^3}{x^2 - 9}$

51. Let $y = f(x)$ be a polynomial function with local minima at $x = a$ and $x = b$, $a < b$. Must f have at least one local maximum between a and b? Justify your answer.

52. The derivative of $f(x) = x^{-1}$ is $f'(x) = -x^{-2}$. Since $f'(x) < 0$ for $x \neq 0$, is it correct to say that $f(x)$ is decreasing for all x except $x = 0$? Explain.

53. Discuss the difference between a partition number for $f'(x)$ and a critical value for $f(x)$, and illustrate with examples.

C

54. Find the absolute maximum for $f'(x)$ if

$$f(x) = 6x^2 - x^3 + 8$$

Graph f and f' on the same coordinate system for $0 \leqslant x \leqslant 4$.

55. Find two positive numbers whose product is 400 and whose sum is a minimum. What is the minimum sum?

In Problems 56–58, summarize the pertinent graph information. Round any approximate values to two decimal places.

56. $f(x) = (x - 1)^3(x + 3)$

57. $f(x) = x^4 + x^3 - 4x^2 - 3x + 4$

58. $f(x) = 0.25x^4 - 5x^3 + 31x^2 - 70x$

▶ A P P L I C A T I O N S

Business & Economics

59. *Pricing.* An office products firm gives volume discounts on purchases of three-ring binders, as shown in the table.

QUANTITY ORDERED	1	12	24 or more
PRICE PER BINDER	$1.99	$1.76	$1.49

(A) Let $y = p(x)$ represent the price per binder if x binders are ordered. Graph $y = p(x)$ for $1 \leqslant x \leqslant 30$.
(B) Identify the discontinuities of p and discuss the behavior at each discontinuity.
(C) Let $y = C(x)$ be the total cost for an order of x binders. Graph $y = C(x)$ for $1 \leqslant x \leqslant 30$.
(D) Identify the discontinuities of C and discuss the behavior at each discontinuity.

60. *Price analysis.* The graph in the figure approximates the rate of change of the price of tomatoes over a 60 month period, where $p(t)$ is the price of a pound of tomatoes and t is time (in months).

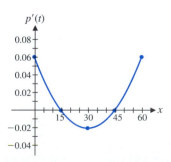

Figure for 60

(A) Write a brief verbal description of the graph of $y = p(t)$, including a discussion of local extrema and inflection points.
(B) Sketch a possible graph of $y = p(t)$.

61. *Maximum revenue and profit.* A company manufactures and sells x electric stoves per month. The monthly cost and price–demand equations are

$$C(x) = 350x + 50,000$$
$$p = 500 - 0.025x \qquad 0 \leqslant x \leqslant 20,000$$

(A) Find the maximum revenue.
(B) How many stoves should the company manufacture each month to maximize its profit? What is the maximum monthly profit? How much should the company charge for each stove?
(C) If the government decides to tax the company $20 for each stove it produces, how many stoves should the company manufacture each month to maximize its profit? What is the maximum monthly profit? How much should the company charge for each stove?

62. *Construction.* A fence is to be built to enclose a rectangular area. The fence along three sides is to be made of material that costs $5 per foot. The material for the fourth side costs $15 per foot.

(A) If the area is 5,000 square feet, find the dimensions of the rectangle that will allow the most economical fence to be built.
(B) If $3,000 is available for the fencing, find the dimensions of the rectangle that will enclose the most area.

63. *Rental income.* A 200 room hotel in Fresno is filled to capacity every night at a rate of $40 per room. For each $1 increase in the nightly rate, 4 fewer rooms are rented. If each rented room costs $8 a day to service, how much should the management charge per room in order to maximize gross profit? What is the maximum gross profit?

64. *Inventory control.* A computer store sells 7,200 boxes of floppy disks annually. It costs the store $0.20 to store a box of disks for 1 year. Each time it reorders disks, the store must pay a $5.00 service charge for processing the order. How many times during the year should the store order disks in order to minimize the total storage and reorder costs?

65. *Average cost.* The total cost (in dollars) of producing x garbage disposals per day is given by

$$C(x) = 4,000 + 10x + 0.1x^2$$

Find the point of intersection of the graph of the average cost function and the marginal cost function, and interpret the results.

66. *Advertising.* A chain of appliance stores uses television ads to promote the sales of refrigerators. Analyzing past records produced the data in the table, where x is the number of ads placed monthly and y is the number of refrigerators sold that month.

NUMBER OF ADS	NUMBER OF REFRIGERATORS
x	y
10	271
20	427
25	526
30	629
45	887
48	917

(A) Enter the data in a graphing utility, set the utility to display two decimal places, and find a cubic regression equation for the number of refrigerators sold monthly as a function of the number of ads.

(B) How many ads should be placed each month to maximize the rate of change of sales with respect to the number of ads, and how many refrigerators can be expected to be sold with this number of ads? Round answers to the nearest integer.

Life Sciences

67. *Bacteria control.* If t days after a treatment, the bacteria count per cubic centimeter in a body of water is given by

$$C(t) = 20t^2 - 120t + 800 \qquad 0 \leq t \leq 9$$

in how many days will the count be a minimum?

Social Sciences

68. *Politics.* In a new suburb, it is estimated that the number of registered voters will grow according to

$$N = 10 + 6t^2 - t^3 \qquad 0 \leq t \leq 5$$

where t is time in years and N is in thousands. When will the rate of increase be maximum?

CHAPTER 10

Additional Derivative Topics

In this chapter we complete our discussion of derivatives by first looking at the differentiation of forms that involve the exponential and logarithmic functions and then considering some additional topics and applications involving all the different types of functions we have encountered thus far. You will probably find it helpful to review some of the important properties of the exponential and logarithmic functions given in Chapter 2 before proceeding further.

S E C T I O N   10-1   ■ **The Constant *e* and Continuous Compound Interest**

■ THE CONSTANT *e*
■ CONTINUOUS COMPOUND INTEREST

In Chapter 2, both the exponential function with base *e* and continuous compound interest were introduced informally. Now, with limit concepts at our disposal, we can give precise definitions of *e* and continuous compound interest.

■ THE CONSTANT *e*

The special irrational number *e* is a particularly suitable base for both exponential and logarithmic functions. The reasons for choosing this number as a base will become clear as we develop differentiation formulas for the exponential function e^x and the natural logarithmic function ln *x*.

In precalculus treatments (Chapter 2), the number *e* is informally defined as an irrational number that can be approximated by the expression $[1 + (1/n)]^n$ by taking *n* sufficiently large. Now we will use the limit concept to formally define *e* as either of the following two limits:

The Number e

$$e = \lim_{n \to \infty} \left(1 + \frac{1}{n} \right)^n \qquad \text{or, alternately,} \qquad e = \lim_{s \to 0} (1 + s)^{1/s}$$

$$e = 2.718\ 281\ 828\ 459...$$

FIGURE 1
$y_1 = (1 + x)^{1/x}$

We will use both these limit forms. [*Note:* If $s = 1/n$, then as $n \to \infty$, $s \to 0$.]

The proof that the indicated limits exist and represent an irrational number between 2 and 3 is not easy and is omitted here. Many people reason (incorrectly) that the limits are 1, since "$(1 + s)$ approaches 1 as $s \to 0$, and 1 to any power is 1." A little experimentation with a graphing utility will convince you otherwise. To obtain evidence that $\lim_{s \to 0}(1 + s)^{1/s}$ exists and equals 2.718..., we graph $f(x) = (1 + x)^{1/x}$ and examine the y coordinates as we trace along the curve near $x = 0$ (see Fig. 1). Note that the y coordinates are greater than 2.718... to the left of 0 and less than 2.718... to the right of 0. Try zooming in several times about the point $(0, e)$. Note that the function is discontinuous at 0.

Exactly who discovered e is still being debated. It is named after the great mathematician Leonard Euler (1707–1783), who computed e to twenty-three decimal places using $[1 + (1/n)]^n$.

■ CONTINUOUS COMPOUND INTEREST

Now we will see how e appears quite naturally in the important application of compound interest. Let us start with simple interest, move on to compound interest, and then on to continuous compound interest.

If a principal P is borrowed at an annual rate of r,* then after t years at simple interest the borrower will owe the lender an amount A given by

$$A = P + Prt = P(1 + rt) \quad \text{Simple interest} \tag{1}$$

On the other hand, if interest is compounded n times a year, then the borrower will owe the lender an amount A given by

$$A = P \left(1 + \frac{r}{n} \right)^{nt} \quad \text{Compound interest} \tag{2}$$

where r/n is the interest rate per compounding period and nt is the number of compounding periods. Suppose P, r, and t in (2) are held fixed and n is increased. Will the amount A increase without bound, or will it tend to approach some limiting value?

Let us perform a calculator experiment before we attack the general limit problem. If $P = \$100$, $r = 0.06$, and $t = 2$ years, then

---

* If r is the interest rate written as a decimal, then $100r\%$ is the rate using %. For example, if $r = 0.12$, then we have $100r\% = 100(0.12)\% = 12\%$. The expressions 0.12 and 12% are therefore equivalent. Unless stated otherwise, all formulas in this book use r in decimal form.

$$A = 100 \left(1 + \frac{0.06}{n}\right)^{2n}$$

We compute A for several values of n in Table 1. The biggest gain appears in the first step; then the gains slow down as n increases. In fact, it appears that A might be tending to approach \$112.75 as n gets larger and larger.

TABLE 1

COMPOUNDING FREQUENCY	n	$A = 100 \left(1 + \frac{0.06}{n}\right)^{2n}$
Annually	1	\$112.3600
Semiannually	2	112.5509
Quarterly	4	112.6493
Weekly	52	112.7419
Daily	365	112.7486
Hourly	8,760	112.7496

explore – discuss 1

(A) Suppose \$1,000 is deposited in a savings account that earns 6% simple interest. How much will be in the account after 2 years?

(B) Suppose \$1,000 is deposited in a savings account that earns compound interest at a rate of 6% per year. How much will be in the account after 2 years if interest is compounded annually? Semiannually? Quarterly? Weekly?

(C) How frequently must interest be compounded at the 6% rate in order to have \$1,150 in the account after 2 years?

Now we turn back to the general problem for a moment. Keeping P, r, and t fixed in equation (2), we compute the following limit and observe an interesting and useful result:

$$\lim_{n \to \infty} P \left(1 + \frac{r}{n}\right)^{nt} = P \lim_{n \to \infty} \left(1 + \frac{r}{n}\right)^{(n/r)rt}$$

Insert r/r in the exponent and let $s = r/n$. Note that $n \to \infty$ implies $s \to 0$.

$$= P \lim_{s \to 0}[(1 + s)^{1/s}]^{rt}$$

Use the limit property given in the footnote below.*

$$= P[\lim_{s \to 0}(1 + s)^{1/s}]^{rt}$$

$$\lim_{s \to 0}(1 + s)^{1/s} = e$$

$$= Pe^{rt}$$

The resulting formula is called the **continuous compound interest formula,** a very important and widely used formula in business and economics.

---

* The following new limit property is used: If $\lim_{x \to c} f(x)$ exists, then $\lim_{x \to c}[f(x)]^p = [\lim_{x \to c} f(x)]^p$, provided the last expression names a real number.

Continuous Compound Interest

$$A = Pe^{rt}$$

where

P = Principal
r = Annual nominal interest rate compounded continuously
t = Time in years
A = Amount at time t

▷ **EXAMPLE 1** ➤ **Computing Continuously Compounded Interest** If $100 is invested at an annual nominal rate of 6% compounded continuously, what amount will be in the account after 2 years? How much interest will be earned?

Solution
$$A = Pe^{rt}$$
$$= 100e^{(0.06)(2)} \quad \text{6\% is equivalent to } r = 0.06.$$
$$\approx \$112.7497$$

(Compare this result with the values calculated in Table 1.) The interest earned is $112.7497 − $100 = $12.7497. ◀

MATCHED PROBLEM 1 ➤ What amount (to the nearest cent) will an account have after 5 years if $100 is invested at an annual nominal rate of 8% compounded annually? Semiannually? Continuously? ◀

▷ **EXAMPLE 2** ➤ **Graphing the Growth of an Investment** If $100 is invested at 12% compounded continuously,* graph the amount in the account relative to time for a period of 10 years. What amount will be in the account after 10 years?

Solution Figure 2 shows the graph of $A = 100e^{0.12t}$, $0 \le t \le 10$, on a graphing utility. We evaluate the function at $t = 10$ to obtain $332.01 for the amount in the account after 10 years.

FIGURE 2
$y_1 = 100e^{0.12x}$ ◀

---

* Following common usage, we will often write ''at 12% compounded continuously,'' understanding that this means ''at an annual nominal rate of 12% compounded continuously.''

MATCHED PROBLEM 2 ▶ If $5,000 is invested at 20% compounded continuously, graph the amount in the account relative to time for a period of 10 years. What amount will be in the account after 10 years? ◀

▶ EXAMPLE 3 ▶ Computing Growth Time   How long will it take an investment of $5,000 to grow to $8,000 if it is invested at 12% compounded continuously?

Solution   We locate the intersection point of the graphs of $y_1 = 5{,}000e^{0.12x}$ and $y_2 = 8{,}000$, as shown in Figure 3, and find that $x = t \approx 3.92$ years.

FIGURE 3 ◀

MATCHED PROBLEM 3 ▶ How long will it take an investment of $10,000 to grow to $15,000 if it is invested at 9% compounded continuously? ◀

▶ EXAMPLE 4 ▶ Computing Doubling Time   How long will it take money to double if it is invested at 18% compounded continuously?

Solution   Since doubling time is independent of the amount invested, we may assume that $P = 1$ and $A = 2$. Figure 4 shows that $y_1 = e^{0.18x}$ and $y_2 = 2$ intersect at $x \approx 3.85$. Therefore, the doubling time is approximately 3.85 years.

FIGURE 4 ◀

MATCHED PROBLEM 4 ▶ How long will it take money to triple if it is invested at 12% compounded continuously? ◀

explore–discuss 2

You are considering three options for investing $10,000: at 7% compounded annually, at 6% compounded monthly, and at 5% compounded continuously.

(A) Which option would be the best for investing $10,000 for 8 years?
(B) How long would you need to invest your money for the third option to be the best?

Answers to Matched Problems   1. $146.93; $148.02; $149.18
2. $y_1 = 5,000e^{0.2x}$ 3. 4.51 yr   4. 9.16 yr

EXERCISE   10-1

A   *Use a calculator to evaluate A to the nearest cent in Problems 1 and 2.*

1. $A = \$1,000e^{0.1t}$ for $t = 2, 5,$ and 8
2. $A = \$5,000e^{0.08t}$ for $t = 1, 4,$ and 10

B   *In Problems 3–8, solve for t or r to two decimal places.*

3. $2 = e^{0.06t}$ **4.** $2 = e^{0.03t}$ **5.** $3 = e^{0.1t}$
6. $3 = e^{0.25t}$ **7.** $2 = e^{5r}$ **8.** $3 = e^{10r}$

C   *In Problems 9 and 10, complete each table to five decimal places using a calculator.*

9.

n	$[1 + (1/n)]^n$
10	2.593 74
100	
1,000	
10,000	
100,000	
1,000,000	
10,000,000	
↓	↓
∞	$e = 2.718\ 281\ 828\ 459...$

10.

s	$(1 + s)^{1/s}$
0.01	2.704 81
−0.01	
0.001	
−0.001	
0.000 1	
−0.000 1	
0.000 01	
−0.000 01	
↓	↓
0	$e = 2.718\ 281\ 828\ 459...$

11. It can be shown that the number e satisfies the inequality

$$\left(1 + \frac{1}{n}\right)^n < e < \left(1 + \frac{1}{n}\right)^{n+1}   \qquad n \geq 1$$

Illustrate this graphically by graphing $y_1 = (1 + 1/n)^n$, $y_2 = 2.718\ 281\ 828 \approx e$, and $y_3 = (1 + 1/n)^{n+1}$ in the same viewing window for $1 \leq n \leq 20$.

12. It can be shown that

$$e^s = \lim_{n\to\infty} \left(1 + \frac{s}{n}\right)^n$$

for any real number s. Illustrate this graphically for $s = 2$ by graphing $y_1 = (1 + 2/n)^n$ and $y_2 = 7.389\ 056\ 099 \approx e^2$ in the same viewing window for $1 \leq n \leq 50$.

13. Use trace and zoom on a graphing utility to investigate

$$\lim_{n \to \infty} (1 + n)^{1/n}$$

Do you think the limit exists? If so, what do you think it is?

14. Use trace and zoom on a graphing utility to investigate

$$\lim_{n \to 0} \left(1 + \frac{1}{n}\right)^{n}$$

Do you think the limit, or a one-sided limit, exists? If so, what do you think the limit, or one-sided limit, is?

▶ APPLICATIONS

Business & Economics

15. *Continuous compound interest.* If $20,000 is invested at an annual nominal rate of 12% compounded continuously, how much will it be worth in 8.5 years?

16. *Continuous compound interest.* Assume $1 had been invested at an annual nominal rate of 4% compounded continuously at the time of the birth of Christ. What would be the value of the account in solid gold Earths in the year 2000? Assume that the Earth weighs approximately 2.11×10^{26} ounces and that gold will be worth $1,000 an ounce in the year 2000. What would be the value of the account in dollars at simple interest?

17. *Present value.* A note will pay $20,000 at maturity 10 years from now. How much should you be willing to pay for the note now if money is worth 7% compounded continuously?

18. *Present value.* A note will pay $50,000 at maturity 5 years from now. How much should you be willing to pay for the note now if money is worth 8% compounded continuously?

19. *Continuous compound interest.* An investor bought stock for $20,000. Four years later, the stock was sold for $30,000. If interest is compounded continuously, what annual nominal rate of interest did the original $20,000 investment earn?

20. *Continuous compound interest.* A family paid $40,000 cash for a house. Ten years later, they sold the house for $100,000. If interest is compounded continuously, what annual nominal rate of interest did the original $40,000 investment earn?

21. *Present value.* Solving $A = Pe^{rt}$ for P, we obtain

$$P = Ae^{-rt}$$

which is the present value of the amount A due in t years if money earns interest at an annual nominal rate r compounded continuously.

(A) Graph $P = 10,000e^{-0.08t}$, $0 \le t \le 50$.

(B) $\lim_{t \to \infty} 10,000e^{-0.08t} = ?$ [Guess, using part (A).]

[*Conclusion:* The longer the duration of time until the amount A is due, the smaller its present value, as we would expect.]

22. *Present value.* Referring to Problem 21, in how many years will the $10,000 have to be due in order for its present value to be $5,000?

23. *Doubling time.* How long will it take money to double if it is invested at 25% compounded continuously?

24. *Doubling time.* How long will it take money to double if it is invested at 5% compounded continuously?

25. *Doubling rate.* At what nominal rate compounded continuously must money be invested to double in 5 years?

26. *Doubling rate.* At what nominal rate compounded continuously must money be invested to double in 3 years?

27. *Growth time.* A man with $20,000 to invest decides to diversify his investments by placing $10,000 in an account that earns 7.2% compounded continuously and $10,000 in an account that earns 8.4% compounded annually. Determine how long it will take for his total investment in the two accounts to grow to $35,000.

28. *Growth time.* A woman invests $5,000 in an account that earns 8.8% compounded continuously and $7,000 in an account that earns 9.6% compounded annually. Determine how long it will take for her total investment in the two accounts to grow to $20,000.

29. *Doubling times.*
(A) Show that the doubling time t (in years) at an annual rate r compounded continuously is given by

$$t = \frac{\ln 2}{r}$$

(B) Graph the doubling-time equation from part (A) for $0.02 \leqslant r \leqslant 0.30$. Are these restrictions on r reasonable? Explain.
(C) Determine the doubling times (in years, to two decimal places) for $r = 5\%, 10\%, 15\%, 20\%, 25\%$, and 30%.

30. *Doubling rates.*
(A) Show that the rate r that doubles an investment at continuously compounded interest in t years is given by

$$r = \frac{\ln 2}{t}$$

(B) Graph the doubling-rate equation from part (A) for $1 \leqslant t \leqslant 20$. Are these restrictions on t reasonable? Explain.
(C) Determine the doubling rates for $t = 2, 4, 6, 8, 10$, and 12 years.

Life Sciences

31. *Radioactive decay.* A mathematical model for the decay of radioactive substances is given by

$$Q = Q_0 e^{rt}$$

where

Q_0 = Amount of the substance at time $t = 0$
r = Continuous compound rate of decay
t = Time in years
Q = Amount of the substance at time t

If the continuous compound rate of decay of radium per year is $r = -0.000\ 433\ 2$, how long will it take a certain amount of radium to decay to half the original amount? (This period of time is the *half-life* of the substance.)

32. *Radioactive decay.* The continuous compound rate of decay of carbon-14 per year is $r = -0.000\ 123\ 8$. How long will it take a certain amount of carbon-14 to decay to half the original amount? (Use the radioactive decay model in Problem 31.)

33. *Radioactive decay.* A cesium isotope has a half-life of 30 years. What is the continuous compound rate of decay? (Use the radioactive decay model in Problem 31.)

34. *Radioactive decay.* A strontium isotope has a half-life of 90 years. What is the continuous compound rate of decay? (Use the radioactive decay model in Problem 31.)

Social Sciences

35. *World population.* A mathematical model for world population growth over short periods of time is given by

$$P = P_0 e^{rt}$$

where

P_0 = Population at time $t = 0$
r = Continuous compound rate of growth
t = Time in years
P = Population at time t

How long will it take the world population to double if it continues to grow at its current continuous compound rate of 2% per year?

36. *World population.* Repeat Problem 35 under the assumption that the world population is growing at a continuous compound rate of 1% per year.

37. *Population growth.* Some underdeveloped nations have population doubling times of 20 years. At what continuous compound rate is the population growing? (Use the population growth model in Problem 35.)

38. *Population growth.* Some developed nations have population doubling times of 120 years. At what continuous compound rate is the population growing? (Use the population growth model in Problem 35.)

39. *World population.* If the world population is now 5 billion (5×10^9) people and if it continues to grow at a continuous compound rate of 2% per year, how long will it be before there is only 1 square yard of land per person? (The Earth has approximately 1.68×10^{14} square yards of land.)

SECTION   10-2   ■   ## Derivatives of Logarithmic and Exponential Functions

■ DERIVATIVE FORMULAS FOR ln x AND e^x
■ GRAPHING TECHNIQUES
■ APPLICATION

In this section, we discuss derivative formulas for ln x and e^x. Out of all the possible choices for bases for the logarithmic and exponential functions, $\log_b x$ and b^x, it turns out (as we will see in this and the next section) that the simplest derivative formulas occur when the base b is chosen to be e.

explore – discuss 1

(A) Draw tangent lines to the graph of $f(x) = 2^x$ for several values of x, and estimate the slopes of those tangent lines (see Fig. 1). Is the graph of the derivative above or below the graph of $f(x) = 2^x$? Confirm your answer by graphing both $f(x)$ and its numerical derivative (see Fig. 2).

FIGURE 1
$y_1 = 2^x$

FIGURE 2
$y_1 = 2^x$
$y_2 = n\text{Deriv}(y_1, x, x)$

(B) Draw tangent lines to the graph of $g(x) = 3^x$ for several values of x, and estimate the slopes of those tangent lines. Is the graph of the derivative above or below the graph of $g(x) = 3^x$? Confirm your answer by graphing both $g(x)$ and its numerical derivative.

Explore–Discuss 1 led to the discovery that for $f(x) = 2^x$, the graphs of $f(x)$ and $f'(x)$ look much the same, with $f'(x)$ always below $f(x)$. Similarly, for $g(x) = 3^x$, the graphs of $g(x)$ and $g'(x)$ look much the same, with $g'(x)$ always above $g(x)$. These observations suggest that there may be an exponential function with base between 2 and 3 that is its own derivative. In this section we will show that the exponential function with base $e = 2.718. . .$ has exactly this property—that is,

$$\frac{d}{dx} e^x = e^x$$

We will also obtain the derivative formula

$$\frac{d}{dx} \ln x = \frac{1}{x}$$

for the natural logarithm function.

■ DERIVATIVE FORMULAS FOR ln x AND e^x

We are now ready to derive a formula for the derivative of

$$f(x) = \ln x = \log_e x \qquad x > 0$$

using the definition of the derivative

$$f'(x) = \lim_{h \to 0} \frac{f(x + h) - f(x)}{h}$$

and the two-step process discussed in Section 8-3.

Step 1. Simplify the difference quotient first:

$$\frac{f(x + h) - f(x)}{h} = \frac{\ln(x + h) - \ln x}{h}$$

$$= \frac{1}{h} [\ln(x + h) - \ln x] \qquad \text{Use ln } A - \text{ln } B = \text{ln } \frac{A}{B}.$$

$$= \frac{1}{h} \ln \frac{x + h}{x} \qquad\qquad \text{Multiply by 1} = x/x \text{ to change form.}$$

$$= \frac{x}{x} \cdot \frac{1}{h} \ln \frac{x + h}{x}$$

$$= \frac{1}{x} \left[\frac{x}{h} \ln\left(1 + \frac{h}{x}\right) \right] \qquad \text{Use } p \text{ ln } A = \text{ln } A^p.$$

$$= \frac{1}{x} \ln\left(1 + \frac{h}{x}\right)^{x/h}$$

Step 2. Find the limit: Let $s = h/x$. For x fixed, if $h \to 0$, then $s \to 0$. Thus,

$$\frac{d}{dx} \ln x = \lim_{h \to 0} \frac{f(x + h) - f(x)}{h}$$

$$= \lim_{h \to 0} \left[\frac{1}{x} \ln\left(1 + \frac{h}{x}\right)^{x/h} \right] \qquad \text{Let } s = h/x. \text{ Note that } h \to 0 \text{ implies } s \to 0.$$

$$= \frac{1}{x} \lim_{s \to 0} [\ln(1 + s)^{1/s}]$$ *Use the new limit property given in the footnote below.**

$$= \frac{1}{x} \ln [\lim_{s \to 0} (1 + s)^{1/s}]$$ *Use the definition of e.*

$$= \frac{1}{x} \ln e$$ *$\ln e = \log_e e = 1$*

$$= \frac{1}{x}$$

Thus,

$$\frac{d}{dx} \ln x = \frac{1}{x}$$

In the next section, we will show that, in general,

$$\frac{d}{dx} \log_b x = \frac{1}{\ln b} \left(\frac{1}{x} \right)$$

which is a somewhat more complicated result than the above—unless $b = e$.

In the process of finding the derivative of e^x, we will use (without proof) the fact that

$$\lim_{h \to 0} \left(\frac{e^h - 1}{h} \right) = 1$$

---

explore – discuss 2

Graph $(e^h - 1)/h$ for $-1 \leqslant h \leqslant 1$. Zoom in on the graph several times about the point $(0, 1)$. Do the results make it seem reasonable to conclude that $\lim_{h \to 0} (e^h - 1)/h = 1$? Is $(e^h - 1)/h$ discontinuous at $h = 0$? Discuss.

We now apply the two-step process to the exponential function $f(x) = e^x$.

Step 1.   Simplify the difference quotient first:

$$\frac{f(x + h) - f(x)}{h} = \frac{e^{x+h} - e^x}{h}$$ *Use $e^{a+b} = e^a e^b$.*

$$= \frac{e^x e^h - e^x}{h}$$ *Factor out e^x.*

$$= e^x \left(\frac{e^h - 1}{h} \right)$$

---

* The following new limit property is used: If $\lim_{x \to c} f(x)$ exists and is positive, then $\lim_{x \to c} [\ln f(x)] = \ln [\lim_{x \to c} f(x)]$.

Step 2.   Compute the limit of the result in step 1:

$$\frac{d}{dx} e^x = \lim_{h \to 0} \frac{f(x + h) - f(x)}{h}$$

$$= \lim_{h \to 0} e^x \left(\frac{e^h - 1}{h} \right)$$

$$= e^x \lim_{h \to 0} \left(\frac{e^h - 1}{h} \right) \qquad \text{Use the assumed limit given above.}$$

$$= e^x \cdot 1 = e^x$$

Thus,

$$\frac{d}{dx} e^x = e^x$$

In the next section, we will show that

$$\frac{d}{dx} b^x = b^x \ln b$$

which is, again, a somewhat more complicated result than the above—unless $b = e$.

The two results just obtained explain why e^x is so widely used that it is some-times referred to as *the* exponential function. These two new and important deriva-tive formulas are restated in the box for reference.

Derivatives of the Natural Logarithmic and Exponential Functions

$$\frac{d}{dx} \ln x = \frac{1}{x} \qquad \frac{d}{dx} e^x = e^x$$

These new derivative formulas can be combined with the rules of differentiation discussed in Chapter 8 to differentiate a wide variety of functions.

EXAMPLE 1   ➤   Finding Derivatives   Find $f'(x)$ for:

(A) $f(x) = 2e^x + 3 \ln x$ 　　　　　　　　(B) $f(x) = \dfrac{e^x}{x^3}$

(C) $f(x) = (\ln x)^4$ 　　　　　　　　　　　(D) $f(x) = \ln x^4$

Solution   (A) $f'(x) = \boxed{2 \dfrac{d}{dx} e^x + 3 \dfrac{d}{dx} \ln x}$

$$= 2e^x + 3 \left(\frac{1}{x} \right) = 2e^x + \frac{3}{x}$$

(B) $f'(x) = \dfrac{x^3 \dfrac{d}{dx} e^x - e^x \dfrac{d}{dx} x^3}{(x^3)^2}$ Quotient rule

$$= \frac{x^3 e^x - e^x 3x^2}{x^6} = \frac{x^2 e^x (x-3)}{x^6} = \frac{e^x(x-3)}{x^4}$$

(C) $\dfrac{d}{dx} (\ln x)^4 = 4(\ln x)^3 \dfrac{d}{dx} \ln x$ Power rule for functions

$$= 4(\ln x)^3 \left(\frac{1}{x}\right) = \frac{4(\ln x)^3}{x}$$

(D) $\dfrac{d}{dx} \ln x^4 = \dfrac{d}{dx} (4 \ln x)$ Property of logarithms

$$= 4\left(\frac{1}{x}\right) = \frac{4}{x}$$

◄

MATCHED PROBLEM 1 ➤ Find $f'(x)$ for:

(A) $f(x) = 4 \ln x - 5e^x$ (B) $f(x) = x^2 e^x$
(C) $f(x) = \ln x^3$ (D) $f(x) = (\ln x)^3$

◄

Caution

$$\frac{d}{dx} e^x \neq xe^{x-1} \qquad \frac{d}{dx} e^x = e^x$$

The power rule cannot be used to differentiate the exponential function. The power rule applies to exponential forms x^n where the exponent is a constant and the base is a variable. In the exponential form e^x, the base is a constant and the exponent is a variable.

■ **GRAPHING TECHNIQUES**

Using the techniques discussed in Chapter 9, we can use first and second derivatives to gain useful information about the graphs of $y = \ln x$ and $y = e^x$. With the derivative formulas given above, we can construct Table 1.

TABLE 1

$\ln x$		e^x	
$y = \ln x$	$x > 0$	$y = e^x$	$-\infty < x < \infty$
$y' = 1/x > 0$	$x > 0$	$y' = e^x > 0$	$-\infty < x < \infty$
$y'' = -1/x^2 < 0$	$x > 0$	$y'' = e^x > 0$	$-\infty < x < \infty$

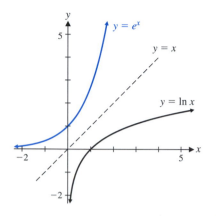

FIGURE 3
e^x is continuous on $(-\infty, \infty)$
$\ln x$ is continuous on $(0, \infty)$

From the table, we can see that both functions are increasing throughout their respective domains, the graph of $y = \ln x$ is always concave downward, and the graph of $y = e^x$ is always concave upward. It can be shown that the y axis is a vertical asymptote for the graph of $y = \ln x$ ($\lim_{x \to 0^+} \ln x = -\infty$), and the x axis is a horizontal asymptote for the graph of $y = e^x$ ($\lim_{x \to -\infty} e^x = 0$). Both equations are graphed in Figure 3.

Notice that if we fold the page along the dashed line $y = x$, the two graphs match exactly (see Section 2-3). Also notice that both graphs are unbounded as $x \to \infty$. Comparing each graph with the graph of $y = x$ (the dashed line), we conclude that $\ln x$ grows more slowly than x and e^x grows more rapidly than x. In fact, the limits in the following box can be established:

Limit Properties of Logarithmic and Exponential Functions

$$\lim_{x \to \infty} \frac{\ln x}{x^p} = 0, \quad p > 0 \qquad \text{and} \qquad \lim_{x \to \infty} \frac{x^p}{e^x} = 0, \quad p > 0 \qquad (1)$$

These limits indicate that $\ln x$ grows more slowly than any positive power of x, and e^x grows more rapidly than any positive power of x.

Now we will apply graphing techniques to a more complicated function.

EXAMPLE 2   ➤   **Analyzing a Graph**   Find all the pertinent graph information for the function

$$f(x) = 2 + \frac{50 \ln x}{e^x}$$

following the strategy discussed in Section 9-4, and illustrate with a sketch. Round any approximate values to two decimal places.

Solution   **Part 1.**   *Analyze f(x):* $f(x) = 2 + \dfrac{50 \ln x}{e^x}$

(A) Domain: All $x > 0$
(B) y intercepts: None, since 0 is not in the domain of f.
    x intercept: The graph of f shown in Figure 4 indicates a single x intercept, which we locate using a root approximation routine: $x = 0.91$
(C) Asymptotes and behavior at infinity: We have not developed general limit techniques (as we did for rational functions in Theorem 3 of Section 9-4) for determining the behavior at infinity for functions involving exponential and logarithmic forms. Instead, we use the power of the graphing utility to draw graphs of f, compute several values of $f(x)$, locate asymptotes, and determine the behavior at infinity. The graph of f shown in Figure 4 suggests that the line $y = 2$ is a horizontal asymptote. This is confirmed by the computation of several values of $f(x)$ as $x \to \infty$ (see Fig. 5). [It is unnecessary to compute values of $f(x)$ as $x \to -\infty$, because f is undefined for $x < 0$.] For this particular function f, we

FIGURE 4
$y_1 = 2 + \dfrac{50 \ln x}{e^x}$

```
10→X:Y1
     2.005226860
20→X:Y1
     2.000000309
100→X:Y1
     2.000000000
```

```
.1→X:Y1
     -102.1732575
.01→X:Y1
     -225.9673989
.001→X:Y1
     -343.0425488
```

can also conclude that the line $y = 2$ is a horizontal asymptote by using the limit properties of the logarithmic and exponential functions:

$$\lim_{x \to \infty} f(x) = \lim_{x \to \infty} \left(2 + \frac{50 \ln x}{e^x} \right) \qquad \text{Multiply the fraction by } x/x = 1.$$

$$= \lim_{x \to \infty} \left(2 + \frac{50 \ln x}{x} \cdot \frac{x}{e^x} \right)$$

$$= 2 + 50 \left(\lim_{x \to \infty} \frac{\ln x}{x} \right) \left(\lim_{x \to \infty} \frac{x}{e^x} \right) \qquad \text{Use limit properties (1).}$$

$$= 2 + 50(0)(0) = 2$$

The graph of f shown in Figure 4 also suggests that the y axis is a vertical asymptote. This is confirmed by the computation of several values of $f(x)$ as $x \to 0^+$ (see Fig. 6). [We do not need to consider the limit of $f(x)$ as $x \to 0^-$, because f is undefined for $x < 0$.] Once again, we can use the limit properties to confirm our graphical observations:

$$\lim_{x \to 0^+} f(x) = \lim_{x \to 0^+} \left(2 + \frac{50 \ln x}{e^x} \right)$$

$$= 2 + 50 \frac{\displaystyle\lim_{x \to 0^+} \ln x}{\displaystyle\lim_{x \to 0^+} e^x} \qquad \begin{array}{l} \lim\limits_{x \to 0^+} e^x = e^0 = 1, \text{ since } e^x \text{ is} \\[4pt] \text{continuous at } O. \end{array}$$

$$= 2 + 50 \frac{\displaystyle\lim_{x \to 0^+} \ln x}{1} = -\infty \qquad \lim\limits_{x \to 0^+} \ln x = -\infty$$

Part 2.   *Analyze $f'(x)$:*

$$f'(x) = \frac{e^x \left(\dfrac{50}{x} \right) - (50 \ln x)e^x}{e^{2x}}$$

$$= \frac{\dfrac{50}{x} - 50 \ln x}{e^x}$$

To find the critical values for $f(x)$, we graph $f'(x)$ (Fig. 7) and use root approximation techniques to solve $f'(x) = 0$.

Graph of $f'(x)$

Critical value for $f(x)$:   1.76

Partition numbers for $f'(x)$:   0, 1.76

Sign chart for $f'(x)$:

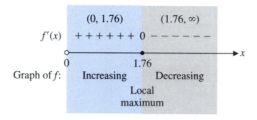

We see that $f(x)$ is increasing on $(0, 1.76)$, decreasing on $(1.76, \infty)$, and has a local maximum at $x = 1.76$.

Part 3.   *Analyze $f''(x)$:*

$$f''(x) = \frac{e^x\left(\dfrac{-50}{x^2} - \dfrac{50}{x}\right) - \left(\dfrac{50}{x} - 50 \ln x\right)e^x}{e^{2x}}$$

$$= \frac{\dfrac{-50}{x^2} - \dfrac{100}{x} + 50 \ln x}{e^x}$$

FIGURE 8
Graph of $f''(x)$

To find the partition numbers for $f''(x)$, we graph $f''(x)$, as shown in Figure 8, and use root approximation techniques to solve $f''(x) = 0$:

Partition numbers for $f''(x)$:   0, 2.55

Sign chart for $f''(x)$:

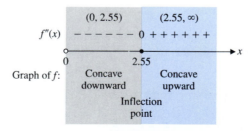

Thus, $f(x)$ is concave downward on $(0, 2.55)$, concave upward on $(2.55, \infty)$, and has an inflection point at $x = 2.55$.

To illustrate the pertinent graph information, we indicate asymptotes, intercepts, local extrema, and inflection points on the graph of f shown in Figure 9.

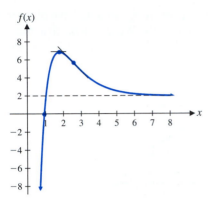

FIGURE 9
Graph of f

◀

MATCHED PROBLEM 2 ➤ Find all the pertinent graph information for the function $f(x) = x \ln x$ and illustrate with a sketch.

◀

■ **APPLICATION**

▶

EXAMPLE 3 ➤ **Maximizing Profit** The market research department of a chain of pet stores test marketed their aquarium pumps (as well as other items) in several of their stores in a test city. They found that the weekly demand for aquarium pumps is given approximately by

$$p = 12 - 2 \ln x \qquad 0 < x < 90$$

where x is the number of pumps sold each week and $\$p$ is the price of one pump. If each pump costs the chain \$3, how should it be priced in order to maximize the weekly profit?

Solution Although we want to find the price that maximizes the weekly profit, it will be easier to first find the number of pumps that will maximize the weekly profit. The revenue equation is

$$R(x) = xp = 12x - 2x \ln x$$

The cost equation is

$$C(x) = 3x$$

and the profit equation is

$$
\begin{aligned}
P(x) &= R(x) - C(x) \\
&= 12x - 2x \ln x - 3x \\
&= 9x - 2x \ln x
\end{aligned}
$$

Thus, we must solve the following:

$$\text{Maximize} \quad P(x) = 9x - 2x \ln x \qquad 0 < x < 90$$

We approximate the maximum value of $y_1 = 9x - 2x \ln x$ for $0 < x < 90$ as indicated in Figure 10. We conclude that the maximum weekly profit occurs when $x \approx 33.115$ and $p \approx 12 - 2 \ln 33.115 \approx \$5$.

FIGURE 10 ◀

MATCHED PROBLEM 3 ➤ Repeat Example 3 if each pump costs the chain \$3.50. ◀

Answers to Matched Problems

1. (A) $(4/x) - 5e^x$
   (B) $xe^x(x + 2)$
   (C) $3/x$
   (D) $3(\ln x)^2/x$

2. Domain: $(0, \infty)$
   y intercept: None $[f(0)$ is not defined$]$
   x intercept: $x = 1$
   Increasing on (e^{-1}, ∞)
   Decreasing on $(0, e^{-1})$
   Local minimum at $x = e^{-1} \approx 0.368$
   Concave upward on $(0, \infty)$

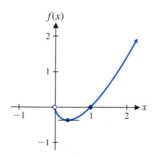

3. Maximum profit occurs for $x = e^{3.25} \approx 26$ and $p = \$5.50$.

EXERCISE 10-2

For many of the problems in this exercise set, the answers in the back of the book include both an unsimplified form and a simplified form. When checking your work, first check that you applied the rules correctly, and then check that you performed the algebraic simplification correctly.

A   *In Problems 1–32, find $f'(x)$ and simplify.*

1. $f(x) = 6e^x - 7 \ln x$
2. $f(x) = 4e^x + 5 \ln x$
3. $f(x) = 2x^e + 3e^x$
4. $f(x) = 4e^x - ex^e$
5. $f(x) = \ln x^5$
6. $f(x) = (\ln x)^5$
7. $f(x) = (\ln x)^2$
8. $f(x) = \ln x^2$

B

9. $f(x) = x^4 \ln x$ 10. $f(x) = x^3 \ln x$
11. $f(x) = x^3 e^x$ 12. $f(x) = x^4 e^x$
13. $f(x) = \dfrac{e^x}{x^2 + 9}$ 14. $f(x) = \dfrac{e^x}{x^2 + 4}$
15. $f(x) = \dfrac{\ln x}{x^4}$ 16. $f(x) = \dfrac{\ln x}{x^3}$
17. $f(x) = (x + 2)^3 \ln x$ 18. $f(x) = (x - 1)^2 \ln x$
19. $f(x) = (x + 1)^3 e^x$ 20. $f(x) = (x - 2)^3 e^x$
21. $f(x) = \dfrac{x^2 + 1}{e^x}$ 22. $f(x) = \dfrac{x + 1}{e^x}$
23. $f(x) = x(\ln x)^3$ 24. $f(x) = x(\ln x)^2$
25. $f(x) = (4 - 5e^x)^3$ 26. $f(x) = (5 - \ln x)^4$
27. $f(x) = \sqrt{1 + \ln x}$ 28. $f(x) = \sqrt{1 + e^x}$
29. $f(x) = xe^x - e^x$ 30. $f(x) = x \ln x - x$
31. $f(x) = 2x^2 \ln x - x^2$ 32. $f(x) = x^2 e^x - 2xe^x + 2e^x$

In Problems 33–36, find the equation of the line tangent to the graph of $y = f(x)$ at the indicated value of x.

33. $f(x) = e^x$; $x = 1$
34. $f(x) = e^x$; $x = 2$
35. $f(x) = \ln x$; $x = e$
36. $f(x) = \ln x$; $x = 1$

37.   A student claims that the tangent line to the graph of $f(x) = e^x$ at $x = 3$ passes through the point $(2, 0)$ (see the figure). Is she correct? Will the tangent line at $x = 4$ pass through $(3, 0)$? Explain.

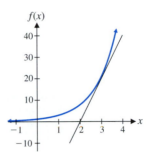

Figure for 37

38.   A student claims that the tangent line to the graph of $g(x) = \ln x$ at $x = 3$ passes through the origin (see the figure). Is he correct? Will the tangent line at $x = 4$ pass through the origin? Explain.

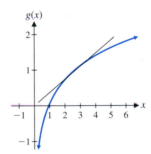

Figure for 38

C   *In Problems 39–44, find the indicated extremum of each function for x > 0.*

39.   Absolute maximum value of

$$f(x) = 4x - x \ln x$$

40.   Absolute minimum value of

$$f(x) = x \ln x - 3x$$

41.   Absolute minimum value of

$$f(x) = \frac{e^x}{x}$$

42.   Absolute maximum value of

$$f(x) = \frac{x^2}{e^x}$$

43. Absolute maximum value of

$$f(x) = \frac{1 + 2 \ln x}{x}$$

44. Absolute minimum value of

$$f(x) = \frac{1 - 5 \ln x}{x}$$

In Problems 45–48, locate horizontal and vertical asymptotes by graphing f and computing several appropriate values of f(x). Discuss the behavior of f at infinity and at each vertical asymptote.

45. $f(x) = 2 + 5e^x - e^{2x}$

46. $f(x) = 3 - 10e^{-x} + e^{-3x}$

47. $f(x) = \dfrac{x^6}{e^x \ln x} - 5$

48. $f(x) = 7 + \dfrac{6 \ln x}{\sqrt{x - 2}}$

In Problems 49–52, locate horizontal and vertical asymptotes using limit properties of the logarithmic and exponential functions. Discuss the behavior of f at infinity and at each vertical asymptote.

49. $f(x) = 3 + 5x^2e^{-x} + 2x^4e^{-x}$

50. $f(x) = 8 - 3xe^x + x^2e^x$

51. $f(x) = 6 + 2 \ln x - 4x^3e^{-x}$

52. $f(x) = 2 - 3 \ln x + x^2e^{-x}$

In Problems 53–60, find all the pertinent graph information and illustrate with a sketch. Round any approximate values to two decimal places.

53. $f(x) = 1 - e^x$
54. $f(x) = 1 - \ln x$
55. $f(x) = x - \ln x$
56. $f(x) = e^x - x$
57. $f(x) = (3 - x)e^x$
58. $f(x) = (x - 2)e^x$

59. $f(x) = x^2 \ln x$
60. $f(x) = \dfrac{\ln x}{x}$

In Problems 61–64, approximate the critical values of f(x) to two decimal places and find the intervals where f(x) is increasing, the intervals where f(x) is decreasing, and the local extrema.

61. $f(x) = e^x - 2x^2$
62. $f(x) = e^x + x^2$
63. $f(x) = 20 \ln x - e^x$
64. $f(x) = x^2 - 3x \ln x$

65. Find the points of intersection of $f(x) = e^x$ and $g(x) = x^4$. Note that there are three points of intersection and that e^x is greater than x^4 for large values of x.

66. Find the points of intersection of $f(x) = e^x$ and $h(x) = x^5$. Note that there are two points of intersection and that e^x is greater than x^5 for large values of x.

67. Find the points of intersection of $f(x) = \ln x$ and $g(x) = x^{1/4}$. Note that $\ln x$ is less than $x^{1/4}$ for large values of x.

68. Find the points of intersection of $f(x) = \ln x$ and $h(x) = x^{1/5}$. Note that $\ln x$ is less than $x^{1/5}$ for large values of x.

69. A student graphs the function $f(x) = 3e^{-x} + \ln x$ on a graphing utility (see the figure) and conjectures that f is increasing for $x > 0$. Explain why the conjecture is false, and find the intervals on which f is increasing and decreasing.

Figure for 69: $y_1 = 3e^{-x} + \ln x$

70. A student graphs the function $f(x) = e^x - x^2 + 10x^4$ on a graphing utility (see the figure) and conjectures that f is always concave upward. Explain why the conjecture is false, and find the intervals on which f is concave upward and concave downward.

Figure for 70: $y_1 = e^x - x^2 + 10x^4$

▶ APPLICATIONS

Business & Economics

71. *Maximum profit.* A national food service runs food concessions for sporting events throughout the country. Their marketing research department chose a particular football stadium to test market a new jumbo hot dog. It was found that the demand for the new hot dog is given approximately by

$$p = 5 - \ln x \qquad 5 \leqslant x \leqslant 50$$

where x is the number of hot dogs (in thousands) that can be sold during one game at a price of $\$p$. If the concessionaire pays $\$1$ for each hot dog, how should the hot dogs be priced to maximize the profit per game?

72. *Maximum profit.* On a national tour of a rock band, the demand for T-shirts is given by

$$p = 15 - 4 \ln x \qquad 1 \leqslant x \leqslant 40$$

where x is the number of T-shirts (in thousands) that can be sold during a single concert at a price of $\$p$. If the shirts cost the band $\$5$ each, how should they be priced in order to maximize the profit per concert?

73. *Maximum profit.* A regional chain of department stores has collected the data in the table at the top of the next column, showing weekly sales of a certain brand of jeans. The same jeans have been offered at various prices ranging from the regular price of $\$35.99$ to the lowest sale price of $\$23.99$. Each pair of this brand costs the chain $\$20.00$. Use logarithmic regression ($p = a + b \ln x$) to find the price (to the nearest cent) that will maximize profit.

PAIRS OF JEANS	PRICE PER PAIR ($)
x	p
21,543	23.99
14,029	25.99
12,130	27.99
9,169	29.99
6,964	31.99
5,506	33.99
4,187	35.99

74. *Maximum profit.* A mail order company specializes in the sale of athletic shoes. Its market research team has determined the weekly demand shown in the table for a particular model of basketball shoe when priced at various levels. The company is able to purchase this model from the manufacturer for $\$62$ per pair. Use logarithmic regression ($p = a + b \ln x$) to find the price (to the nearest cent) that will maximize profit.

DEMAND	PRICE PER PAIR ($)
x	p
4,321	74
3,064	79
2,499	84
2,047	89
1,823	94
1,781	99

75. *Minimum average cost.* The cost of producing x units of a product is given by

$$C(x) = 600 + 100x - 100 \ln x \qquad x \geqslant 1$$

Find the minimum average cost.

76. *Minimum average cost.* The cost of producing x units of a product is given by

$$C(x) = 1,000 + 200x - 200 \ln x \qquad x \geqslant 1$$

Find the minimum average cost.

77. *Maximizing revenue.* A cosmetic company is planning the introduction and promotion of a new lipstick line. The marketing research department, after test marketing the new line in a carefully selected large city, found that the demand in that city is given approximately by

$$p = 10e^{-x} \qquad 0 \leq x \leq 2$$

where x thousand lipsticks were sold per week at a price of p dollars each.

(A) At what price will the weekly revenue $R(x) = xp$ be maximum? What is the maximum weekly revenue in the test city?

(B) Graph R for $0 \leq x \leq 2$.

78. *Maximizing revenue.* Repeat Problem 77 using the demand equation $p = 12e^{-x}, 0 \leq x \leq 2$.

Life Sciences

79. *Blood pressure.* An experiment was set up to find a relationship between weight and systolic blood pressure in normal children. Using hospital records for 5,000 normal children, it was found that the systolic blood pressure was given approximately by

$$P(x) = 17.5(1 + \ln x) \qquad 10 \leq x \leq 100$$

where $P(x)$ is measured in millimeters of mercury and x is measured in pounds. What is the rate of change of blood pressure with respect to weight at the 40 pound weight level? At the 90 pound weight level?

80. *Blood pressure.* Graph the systolic blood pressure equation in Problem 79.

81. *Drug concentration.* The concentration of a drug in the bloodstream t hours after injection is given approximately by

$$C(t) = 4.35e^{-t} \qquad 0 \leq t \leq 5$$

where $C(t)$ is concentration in milligrams per milliliter.

(A) What is the rate of change of concentration after 1 hour? After 4 hours?

(B) Graph C.

82. *Water pollution.* The use of iodine crystals is a popular way of making small quantities of nonpotable water safe to drink. Crystals placed in a 1 ounce bottle of water will dissolve until the solution is saturated. After saturation, half of this solution is poured into a quart container of nonpotable water, and after about an hour, the water is usually safe to drink. The half empty 1 ounce bottle is then refilled to be used again in the same way. Suppose the concentration of iodine in the 1 ounce bottle t minutes after the crystals are introduced can be approximated by

$$C(t) = 250(1 - e^{-t}) \qquad t \geq 0$$

where $C(t)$ is the concentration of iodine in micrograms per milliliter.

(A) What is the rate of change of the concentration after 1 minute? After 4 minutes?

(B) Graph C for $0 \leq t \leq 5$.

Social Sciences

83. *Psychology—stimulus/response.* In psychology, the Weber–Fechner law for stimulus response is

$$R = k \ln\left(\frac{S}{S_0}\right)$$

where R is the response, S is the stimulus, and S_0 is the lowest level of stimulus that can be detected. Find dR/dS.

84. *Psychology—learning.* A mathematical model for the average of a group of people learning to type is given by

$$N(t) = 10 + 6 \ln t \qquad t \geq 1$$

where $N(t)$ is the number of words per minute typed after t hours of instruction and practice (2 hours per day, 5 days per week). What is the rate of learning after 10 hours of instruction and practice? After 100 hours?

SECTION   10-3 ■ ## Chain Rule: General Form

■ COMPOSITE FUNCTIONS
■ CHAIN RULE
■ GENERALIZED DERIVATIVE RULES
■ OTHER LOGARITHMIC AND EXPONENTIAL FUNCTIONS

explore – discuss 1

(A) We have shown that the function $f(x) = e^x$ is its own derivative. Use the graph of $g(x) = e^{x^2}$ in Figure 1 to explain why $g(x)$ is not its own derivative. Then confirm your explanation by graphing the numerical derivative of $g(x)$.

(B) Use the graph of $h(x) = 2xe^{x^2}$ in Figure 2 to explain why $h(x)$ could be the derivative of $g(x)$. [We will show in this section that $h(x)$ is indeed the derivative of $g(x)$.]

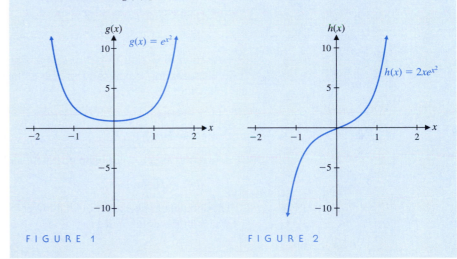

FIGURE 1   FIGURE 2

In Section 8-6, we introduced the power form of the chain rule:

$$\frac{d}{dx}[u(x)]^n = n[u(x)]^{n-1}u'(x) \quad \text{For example,}$$

$$\frac{d}{dx}(x^2 - 3)^5 = 5(x^2 - 3)^4 \frac{d}{dx}(x^2 - 3)$$

$$= 10x(x^2 - 3)^4$$

This general power rule is a special case of one of the most important derivative rules of all—the *chain rule*—which will enable us to determine the derivatives of some fairly complicated functions in terms of derivatives of more elementary functions.

Suppose you were asked to find the derivative of

$$m(x) = \ln(2x + 1) \quad \text{or} \quad n(x) = e^{3x^2 - 1}$$

We have formulas for computing derivatives of $\ln x$ and e^x, and polynomial functions in general, but not in the indicated combinations. The chain rule is used to compute derivatives of functions that are *compositions* of more elementary functions whose derivatives are known. We therefore start this section with a brief review of *composite functions*.

■ COMPOSITE FUNCTIONS

Let us look at function m more closely:

$$m(x) = \ln(2x + 1)$$

The function m is a combination of the natural logarithm function and a linear function. To see this more clearly, let

$$y = f(u) = \ln u \qquad \text{and} \qquad u = g(x) = 2x + 1$$

Then we can express y as a function of x as follows:

$$y = f(u) = f[g(x)] = \ln(2x + 1) = m(x)$$

The function m is said to be the *composite* of the two simpler functions f and g. (Loosely speaking, we can think of m as a function of a function.) In general, we have the following:

Composite Functions

A function m is a **composite** of functions f and g if

$$m(x) = f[g(x)]$$

The domain of m is the set of all numbers x such that x is in the domain of g and $g(x)$ is in the domain of f.

EXAMPLE 1 ➤ Composite Functions   Let $f(u) = e^u$, $g(x) = 3x^2 + 1$, and $m(v) = v^{3/2}$. Find:

(A) $f[g(x)]$ (B) $g[f(u)]$ (C) $m[g(x)]$

Solution   (A) $f[g(x)] = e^{g(x)} = e^{3x^2 + 1}$
(B) $g[f(u)] = 3[f(u)]^2 + 1 = 3(e^u)^2 + 1 = 3e^{2u} + 1$
(C) $m[g(x)] = [g(x)]^{3/2} = (3x^2 + 1)^{3/2}$ ◀

MATCHED PROBLEM 1 ➤   Let $f(u) = \ln u$, $g(x) = 2x^3 + 4$, and $m(v) = v^{-5}$. Find:

(A) $f[g(x)]$ (B) $g[f(u)]$ (C) $m[g(x)]$ ◀

EXAMPLE 2 ➤   Composite Functions   Write each function as a composition of the natural logarithm or exponential function and a polynomial.

(A) $y = \ln(x^3 - 2x^2 + 1)$ (B) $y = e^{x^2 + 4}$

Solution (A) Let

$$y = f(u) = \ln u$$
$$u = g(x) = x^3 - 2x^2 + 1$$

Check: $y = f[g(x)] = \ln[g(x)] = \ln(x^3 - 2x^2 + 1)$

(B) Let

$$y = f(u) = e^u$$
$$u = g(x) = x^2 + 4$$

Check: $y = f[g(x)] = e^{g(x)} = e^{x^2 + 4}$ ◀

MATCHED PROBLEM 2 ▶ Repeat Example 2 for:

(A) $y = e^{2x^3 + 7}$ (B) $y = \ln(x^4 + 10)$ ◀

■ CHAIN RULE

The word *chain* in the name *chain rule* comes from the fact that a function formed by composition (such as those in Example 1) involves a chain of functions — that is, a function of a function. The *chain rule* will enable us to compute the derivative of a composite function in terms of the derivatives of the functions making up the composition.

Suppose

$$y = m(x) = f[g(x)]$$

is a composite of f and g, where

$$y = f(u) \qquad \text{and} \qquad u = g(x)$$

We would like to express the derivative dy/dx in terms of the derivatives of f and g. From the definition of a derivative (see Section 8-3), we have

$$\frac{dy}{dx} = \lim_{h \to 0} \frac{m(x + h) - m(x)}{h} \qquad \text{Substitute}$$

$$m(x + h) = f[g(x + h)]$$
$$\text{and } m(x) = f[g(x)]$$

$$= \lim_{h \to 0} \frac{f[g(x + h)] - f[g(x)]}{h} \qquad \text{Multiply by}$$

$$1 = \frac{g(x + h) - g(x)}{g(x + h) - g(x)}$$

$$= \lim_{h \to 0} \left[\frac{f[g(x + h)] - f[g(x)]}{h} \cdot \frac{g(x + h) - g(x)}{g(x + h) - g(x)} \right]$$

$$= \lim_{h \to 0} \left[\frac{f[g(x + h)] - f[g(x)]}{g(x + h) - g(x)} \cdot \frac{g(x + h) - g(x)}{h} \right] \qquad (1)$$

We recognize the second factor in (1) as the difference quotient for $g(x)$. To interpret the first factor as the difference quotient for $f(u)$, we let $k = g(x + h) - g(x)$. Since $u = g(x)$, we can write

$$u + k\;\boxed{= g(x) + g(x + h) - g(x)} = g(x + h)$$

Substituting in (1), we now have

$$\frac{dy}{dx} = \lim_{h \to 0} \left[\frac{f(u + k) - f(u)}{k} \cdot \frac{g(x + h) - g(x)}{h} \right] \tag{2}$$

If we assume that $k = [g(x + h) - g(x)] \to 0$ as $h \to 0$, then we can find the limit of each difference quotient in (2):

$$\frac{dy}{dx} = \left[\lim_{k \to 0} \frac{f(u + k) - f(u)}{k} \right]\left[\lim_{h \to 0} \frac{g(x + h) - g(x)}{h} \right]$$

$$= f'(u)g'(x)$$

$$= \frac{dy}{du}\frac{du}{dx}$$

This result is correct under rather general conditions, and is called the *chain rule*, but our "derivation" is superficial, because it ignores a number of hidden problems. Since a formal proof of the chain rule is beyond the scope of this book, we simply state it as follows:

Chain Rule

If $y = f(u)$ and $u = g(x)$, define the composite function

$$y = m(x) = f[g(x)]$$

Then

$$\frac{dy}{dx} = \frac{dy}{du}\frac{du}{dx} \qquad \text{provided } \frac{dy}{du} \text{ and } \frac{du}{dx} \text{ exist}$$

Or, equivalently,

$$m'(x) = f'[g(x)]g'(x) \qquad \text{provided } f'[g(x)] \text{ and } g'(x) \text{ exist}$$

EXAMPLE 3   ▶   **Using the Chain Rule**   Find dy/dx, given:

(A) $y = \ln(x^2 - 4x + 2)$ (B) $y = e^{2x^3 + 5}$ (C) $y = (3x^2 + 1)^{3/2}$

Solution   (A) Let $y = \ln u$ and $u = x^2 - 4x + 2$. Then

$$\boxed{\begin{aligned} \frac{dy}{dx} &= \frac{dy}{du}\frac{du}{dx} \\ &= \frac{1}{u}(2x - 4) \end{aligned}}^{*}$$

$$= \frac{1}{x^2 - 4x + 2}(2x - 4) \quad \text{Since } u = x^2 - 4x + 2$$

$$= \frac{2x - 4}{x^2 - 4x + 2}$$

(B) Let $y = e^u$ and $u = 2x^3 + 5$. Then

$$\boxed{\begin{aligned} \frac{dy}{dx} &= \frac{dy}{du}\frac{du}{dx} \\ &= e^u(6x^2) \end{aligned}}$$

$$= 6x^2 e^{2x^3 + 5} \quad \text{Since } u = 2x^3 + 5$$

(C) We have two methods:

Method 1. *Chain rule — general form:* Let $y = u^{3/2}$ and $u = 3x^2 + 1$. Then

$$\boxed{\begin{aligned} \frac{dy}{dx} &= \frac{dy}{du}\frac{du}{dx} \\ &= \tfrac{3}{2}u^{1/2}(6x) \end{aligned}}$$

$$= \tfrac{3}{2}(3x^2 + 1)^{1/2}(6x) \quad \text{Since } u = 3x^2 + 1$$

$$= 9x(3x^2 + 1)^{1/2} \quad \text{or} \quad 9x\sqrt{3x^2 + 1}$$

Method 2. *Chain rule — power form (general power rule):*

$$\frac{d}{dx}(3x^2 + 1)^{3/2} = \tfrac{3}{2}(3x^2 + 1)^{1/2}\frac{d}{dx}(3x^2 + 1) \qquad \frac{d}{dx}[u(x)]^n = n[u(x)]^{n-1}\frac{d}{dx}u(x)$$

$$= \tfrac{3}{2}(3x^2 + 1)^{1/2}(6x)$$

$$= 9x(3x^2 + 1)^{1/2} \quad \text{or} \quad 9x\sqrt{3x^2 + 1} \qquad \blacktriangleleft$$

---

* After some experience with the chain rule, the steps in the dashed boxes are usually done mentally.

The general power rule stated in Section 8-6 can be derived using the chain rule as follows: Given $y = [u(x)]^n$, let $y = v^n$ and $v = u(x)$. Then

$$\frac{dy}{dx} = \frac{dy}{dv}\frac{dv}{dx}$$

$$= nv^{n-1}\frac{d}{dx}u(x)$$

$$= n[u(x)]^{n-1}\frac{d}{dx}u(x) \quad \text{Since } v = u(x)$$

MATCHED PROBLEM 3 ➤   Find dy/dx, given:

(A) $y = e^{3x^4 + 6}$ (B) $y = \ln(x^2 + 9x + 4)$ (C) $y = (2x^3 + 4)^{-5}$
                                                          (Use two methods.)   ◀

The chain rule can be extended to compositions of three or more functions. For example, if $y = f(w)$, $w = g(u)$, and $u = h(x)$, then

$$\frac{dy}{dx} = \frac{dy}{dw}\frac{dw}{du}\frac{du}{dx}$$

EXAMPLE 4 ➤   **Using the Chain Rule**   For $y = h(x) = e^{1 + (\ln x)^2}$, find dy/dx.

Solution   Note that h is of the form $y = e^w$, where $w = 1 + u^2$ and $u = \ln x$. Thus,

$$\frac{dy}{dx} = \frac{dy}{dw}\frac{dw}{du}\frac{du}{dx}$$

$$= e^w(2u)\left(\frac{1}{x}\right)$$

$$= e^{1 + u^2}(2u)\left(\frac{1}{x}\right) \quad \text{Since } w = 1 + u^2$$

$$= e^{1 + (\ln x)^2}(2 \ln x)\left(\frac{1}{x}\right) \quad \text{Since } u = \ln x$$

$$= \frac{2}{x}(\ln x)e^{1 + (\ln x)^2}$$ ◀

MATCHED PROBLEM 4 ➤   For $y = h(x) = [\ln(1 + e^x)]^3$, find dy/dx.   ◀

■ GENERALIZED DERIVATIVE RULES

In practice, it is not necessary to introduce additional variables when using the chain rule, as we did in Examples 3 and 4. Instead, the chain rule can be used to extend the derivative rules for specific functions to general derivative rules for compositions. This is what we did above when we showed that the general power rule is a consequence of the chain rule. The same technique can be applied to functions of the form

$y = e^{f(x)}$ and $y = \ln[f(x)]$ (see Problems 67 and 68 in Exercise 10-3). The results are summarized in the following box:

General Derivative Rules

$$\frac{d}{dx}[f(x)]^n = n[f(x)]^{n-1}f'(x) \tag{3}$$

$$\frac{d}{dx}\ln[f(x)] = \frac{1}{f(x)}f'(x) \tag{4}$$

$$\frac{d}{dx}e^{f(x)} = e^{f(x)}f'(x) \tag{5}$$

For power, natural logarithm, or exponential forms, we can either use the chain rule discussed earlier or these special differentiation formulas, which are based on the chain rule. Use whichever is easier for you. In Example 5, we will use the general derivative rules.

EXAMPLE 5 ➤ **Using General Derivative Rules**

(A) $\dfrac{d}{dx} e^{2x} = e^{2x} \dfrac{d}{dx} 2x$ Using (5)

$\qquad\quad = e^{2x}(2) = 2e^{2x}$

(B) $\dfrac{d}{dx} \ln(x^2 + 9) = \dfrac{1}{x^2 + 9} \dfrac{d}{dx}(x^2 + 9)$ Using (4)

$\qquad\qquad\qquad = \dfrac{1}{x^2 + 9} 2x = \dfrac{2x}{x^2 + 9}$

(C) $\dfrac{d}{dx}(1 + e^{x^2})^3 = 3(1 + e^{x^2})^2 \dfrac{d}{dx}(1 + e^{x^2})$ Using (3)

$\qquad\qquad\qquad = 3(1 + e^{x^2})^2 e^{x^2} \dfrac{d}{dx} x^2$ Using (5)

$\qquad\qquad\qquad = 3(1 + e^{x^2})^2 e^{x^2}(2x)$

$\qquad\qquad\qquad = 6xe^{x^2}(1 + e^{x^2})^2$ ◀

MATCHED PROBLEM 5 ➤ Find:

(A) $\dfrac{d}{dx} \ln(x^3 + 2x)$ (B) $\dfrac{d}{dx} e^{3x^2 + 2}$ (C) $\dfrac{d}{dx}(2 + e^{-x^2})^4$ ◀

■ **OTHER LOGARITHMIC AND EXPONENTIAL FUNCTIONS**

In most applications involving logarithmic or exponential functions, the number e is the preferred base. However, there are situations where it is convenient to use a base other than e. Derivatives of $y = \log_b x$ and $y = b^x$ can be obtained by expressing these functions in terms of the natural logarithmic and exponential functions.

We begin by finding a relationship between $\log_b x$ and $\ln x$ for any base $b, b > 0$ and $b \neq 1$. Some of you may prefer to remember the process, and others the formula.

$$y = \log_b x \quad \text{\textit{Change to exponential form.}}$$
$$b^y = x \quad \text{\textit{Take the natural logarithm of both sides.}}$$
$$\ln b^y = \ln x \quad \text{\textit{Recall that } } \ln b^y = y \ln b.$$
$$y \ln b = \ln x \quad \text{\textit{Solve for y.}}$$
$$y = \frac{1}{\ln b} \ln x$$

Thus,

$$\log_b x = \frac{1}{\ln b} \ln x \quad \text{\textit{Change-of-base formula*}} \tag{6}$$

Differentiating both sides of (6), we have

$$\frac{d}{dx} \log_b x = \frac{1}{\ln b} \frac{d}{dx} \ln x = \frac{1}{\ln b} \left(\frac{1}{x} \right)$$

---

e x p l o r e – d i s c u s s 2

(A) The graphs of $f(x) = \log_2 x$ and $g(x) = \log_4 x$ are shown in Figure 3. Which graph belongs to which function?

FIGURE 3

(B) Graph $f'(x)$ and $g'(x)$.
(C) The function $f(x)$ is related to $g(x)$ in the same way that $f'(x)$ is related to $g'(x)$. What is that relationship?

---

* Equation (6) is a special case of the **general change-of-base formula** for logarithms (which can be derived in the same way): $\log_b x = (\log_a x)/(\log_a b)$.

EXAMPLE 6 ➤ Differentiating Logarithmic Functions   Find $f'(x)$ for:

(A) $f(x) = \log_2 x$ (B) $f(x) = \log(1 + x^3)$

Solution   (A) $f(x) = \log_2 x = \dfrac{1}{\ln 2} \ln x$ Using (6)

$$f'(x) = \frac{1}{\ln 2}\left(\frac{1}{x}\right)$$

(B) $f(x) = \log(1 + x^3)$ Recall that $\log r = \log_{10} r$.

$$= \frac{1}{\ln 10}\ln(1 + x^3)$$ Using (6)

$$f'(x) = \frac{1}{\ln 10}\left(\frac{1}{1+x^3}\,3x^2\right) = \frac{1}{\ln 10}\left(\frac{3x^2}{1+x^3}\right)$$ ◄

MATCHED PROBLEM 6 ➤ Find $f'(x)$ for:

(A) $f(x) = \log x$ (B) $f(x) = \log_3(x + x^2)$ ◄

Now we want to find a relationship between b^x and e^x for any base b, $b > 0$ and $b \neq 1$.

$$y = b^x$$ Take the natural logarithm of both sides.

$$\ln y = \ln b^x$$

$$= x \ln b$$ If $\ln A = B$, then $A = e^B$.

$$y = e^{x \ln b}$$

Thus,

$$b^x = e^{x \ln b} \tag{7}$$

Differentiating both sides of (7), we have

$$\frac{d}{dx}b^x = e^{x \ln b}\ln b = b^x \ln b$$

EXAMPLE 7 ➤ Differentiating Exponential Functions   Find $f'(x)$ for:

(A) $f(x) = 2^x$ (B) $f(x) = 10^{x^5 + x}$

Solution   (A) $f(x) = 2^x = e^{x \ln 2}$ Using (7)

$$f'(x) = e^{x \ln 2}\ln 2 = 2^x \ln 2$$

(B) $f(x) = 10^{x^5 + x} = e^{(x^5 + x)\ln 10}$ Using (7)

$$f'(x) = e^{(x^5 + x)\ln 10}(5x^4 + 1)\ln 10$$
$$= 10^{x^5 + x}(5x^4 + 1)\ln 10$$ ◄

MATCHED PROBLEM 7 ➤ Find $f'(x)$ for:

(A) $f(x) = 5^x$ (B) $f(x) = 4^{x^2 + 3x}$ ◄

Answers to Matched Problems

1. (A) $\ln(2x^3 + 4)$ (B) $2(\ln u)^3 + 4$ (C) $(2x^3 + 4)^{-5}$
2. (A) $y = f(u) = e^u; u = g(x) = 2x^3 + 7$
   (B) $y = f(u) = \ln u; u = g(x) = x^4 + 10$
3. (A) $12x^3e^{3x^4+6}$ (B) $\dfrac{2x + 9}{x^2 + 9x + 4}$ (C) $-30x^2(2x^3 + 4)^{-6}$
4. $\dfrac{3e^x[\ln(1 + e^x)]^2}{1 + e^x}$
5. (A) $\dfrac{3x^2 + 2}{x^3 + 2x}$ (B) $6xe^{3x^2+2}$ (C) $-8xe^{-x^2}(2 + e^{-x^2})^3$
6. (A) $\dfrac{1}{\ln 10}\left(\dfrac{1}{x}\right)$ (B) $\dfrac{1}{\ln 3}\left(\dfrac{1 + 2x}{x + x^2}\right)$
7. (A) $5^x \ln 5$ (B) $4^{x^2+3x}(2x + 3)\ln 4$

EXERCISE 10-3

A   *Write each composite function in Problems 1–6 in the form $y = f(u)$ and $u = g(x)$.*

1. $y = (2x + 5)^3$
2. $y = (3x - 7)^5$
3. $y = \ln(2x^2 + 7)$
4. $y = \ln(x^2 - 2x + 5)$
5. $y = e^{x^2-2}$
6. $y = e^{3x^3+5x}$

In Problems 7–12, express y in terms of x. Use the chain rule to find dy/dx, and then express dy/dx in terms of x.

7. $y = u^2; u = 2 + e^x$
8. $y = u^3; u = 3 - \ln x$
9. $y = e^u; u = 2 - x^4$
10. $y = e^u; u = x^6 + 5x^2$
11. $y = \ln u; u = 4x^5 - 7$
12. $y = \ln u; u = 2 + 3x^4$

In Problems 13–46, find the derivative and simplify. (For some of these problems, the answers in the back of the book include both an unsimplified form and a simplified form. When checking your work, first check that you applied the rules correctly, and then check that you performed the algebraic simplification correctly.)

13. $\dfrac{d}{dx}\ln(x - 3)$
14. $\dfrac{d}{dw}\ln(w + 100)$
15. $\dfrac{d}{dt}\ln(3 - 2t)$
16. $\dfrac{d}{dy}\ln(4 - 5y)$
17. $\dfrac{d}{dx}3e^{2x}$
18. $\dfrac{d}{dy}2e^{3y}$
19. $\dfrac{d}{dt}2e^{-4t}$
20. $\dfrac{d}{dr}6e^{-3r}$

B

21. $\dfrac{d}{dx}100e^{-0.03x}$
22. $\dfrac{d}{dt}1,000e^{0.06t}$
23. $\dfrac{d}{dx}\ln(x + 1)^4$
24. $\dfrac{d}{dx}\ln(x + 1)^{-3}$
25. $\dfrac{d}{dx}(2e^{2x} - 3e^x + 5)$
26. $\dfrac{d}{dt}(1 + e^{-t} - e^{-2t})$
27. $\dfrac{d}{dx}e^{3x^2-2x}$
28. $\dfrac{d}{dx}e^{x^3-3x^2+1}$
29. $\dfrac{d}{dt}\ln(t^2 + 3t)$
30. $\dfrac{d}{dx}\ln(x^3 - 3x^2)$
31. $\dfrac{d}{dx}\ln(x^2 + 1)^{1/2}$
32. $\dfrac{d}{dx}\ln(x^4 + 5)^{3/2}$
33. $\dfrac{d}{dt}[\ln(t^2 + 1)]^4$
34. $\dfrac{d}{dw}[\ln(w^3 - 1)]^2$
35. $\dfrac{d}{dx}(e^{2x} - 1)^4$
36. $\dfrac{d}{dx}(e^{x^2} + 3)^5$

37. $\dfrac{d}{dx}\dfrac{e^{2x}}{x^2+1}$

38. $\dfrac{d}{dx}\dfrac{e^{x+1}}{x+1}$

39. $\dfrac{d}{dx}(x^2+1)e^{-x}$

40. $\dfrac{d}{dx}(1-x)e^{2x}$

41. $\dfrac{d}{dx}(e^{-x}\ln x)$

42. $\dfrac{d}{dx}\dfrac{\ln x}{e^x+1}$

43. $\dfrac{d}{dx}\dfrac{1}{\ln(1+x^2)}$

44. $\dfrac{d}{dx}\dfrac{1}{\ln(1-x^3)}$

45. $\dfrac{d}{dx}\sqrt[3]{\ln(1-x^2)}$

46. $\dfrac{d}{dt}\sqrt[5]{\ln(1-t^5)}$

C

47. Let $f(x)=\ln x$, $g(x)=e^{-x}$, and $h(x)=f(g(x))$. Which of f, g, and h are increasing? Which are decreasing? Explain your reasoning, and then confirm your answers by graphing each function.

48. Let $f(x)=e^{-x}$, $g(x)=\ln(1/x)$, and $h(x)=f(g(x))$. Which of f, g, and h are increasing? Which are decreasing? Explain your reasoning, and then confirm your answers by graphing each function.

49. Let $f(x)=e^x$, $g(x)=x^2$, and $h(x)=f(g(x))$. Which of f, g, and h increases the most rapidly for $x>0$? The least rapidly? Explain your reasoning, and then confirm your answers by graphing each function.

50. Let $f(x)=-\ln x$, $g(x)=e^{-x}$, and $h(x)=f(g(x))$. Which of f, g, and h decreases the most rapidly for $x>0$? The least rapidly? Explain your reasoning, and then confirm your answers by graphing each function.

In Problems 51–56, apply the graphing strategy discussed in Section 9-4 to f, state the pertinent information, and sketch the graph.

51. $f(x)=1-e^{-x}$

52. $f(x)=2-3e^{-2x}$

53. $f(x)=\ln(1-x)$

54. $f(x)=\ln(2x+4)$

55. $f(x)=e^{-0.5x^2}$

56. $f(x)=\ln(x^2+4)$

In Problems 57 and 58, express y in terms of x. Use the chain rule to find dy/dx, and express dy/dx in terms of x.

57. $y=1+w^2$; $w=\ln u$; $u=2+e^x$

58. $y=\ln w$; $w=1+e^u$; $u=x^2$

In Problems 59–66, find each derivative.

59. $\dfrac{d}{dx}\log_2(3x^2-1)$

60. $\dfrac{d}{dx}\log(x^3-1)$

61. $\dfrac{d}{dx}10^{x^2+x}$

62. $\dfrac{d}{dx}8^{1-2x^2}$

63. $\dfrac{d}{dx}\log_3(4x^3+5x+7)$

64. $\dfrac{d}{dx}\log_5(5^{x^2-1})$

65. $\dfrac{d}{dx}2^{x^3-x^2+4x+1}$

66. $\dfrac{d}{dx}10^{\ln x}$

67. Use the chain rule to derive the formula:

$$\dfrac{d}{dx}\ln[f(x)]=\dfrac{1}{f(x)}f'(x)$$

68. Use the chain rule to derive the formula:

$$\dfrac{d}{dx}e^{f(x)}=e^{f(x)}f'(x)$$

69. (A) Use the graphs of $f(x)=\ln(x^2+1)$ and $g(x)=1/(x^2+1)$ (see the figure) to explain why $g(x)$ is not the derivative of $f(x)$.

(B) Compute $f'(x)$ and sketch its graph.

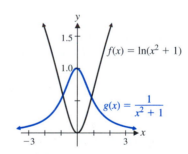

Figure for 69

70. (A) How many times is the chain rule used in computing the derivative of $f(x)=e^{\sqrt{3x}}$?

(B) Construct a function $g(x)$ such that the chain rule is used three times in computing its derivative.

71. Suppose a student reasons that the functions $f(x)=\ln[5(x^2+3)^4]$ and $g(x)=4\ln(x^2+3)$ must have the same derivative, since he has entered $f(x)$, $g(x)$, $f'(x)$, and $g'(x)$ into a graphing utility, but only three graphs appear (see the figure). Is his reasoning correct? Are $f'(x)$ and $g'(x)$ the same function? Explain.

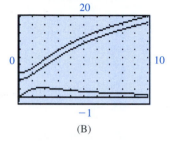

(A)   (B)

Figure for 71

72. Suppose a student reasons that the functions $f(x) = (x + 1) \ln(x + 1) - x$ and $g(x) = (x + 1)^{1/3}$ must have

the same derivative, since she has entered $f(x)$, $g(x)$, $f'(x)$, and $g'(x)$ into a graphing utility, but only three graphs appear (see the figure). Is her reasoning correct? Are $f'(x)$ and $g'(x)$ the same function? Explain.

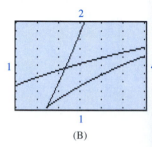

(A)   (B)

Figure for 72

▶ APPLICATIONS

Business & Economics

73. *Maximum revenue.* Suppose the price–demand equation for x units of a commodity is determined from empirical data to be

$$p = 100e^{-0.05x}$$

where x units are sold per day at a price of $\$p$ each. Find the production level and price that maximize revenue. What is the maximum revenue?

74. *Maximum revenue.* Repeat Problem 73 using the price–demand equation

$$p = 10e^{-0.04x}$$

75. *Maximum profit.* Refer to Problem 73. If the daily fixed cost is $400 and the cost per unit is $6, use approximation techniques to find the production level and the price that maximize profit. What is the maximum profit? [*Hint:* Graph $y = P(x)$ and $y = P'(x)$ in the same viewing window.]

76. *Maximum profit.* Refer to Problem 74. If the daily fixed cost is $30 and the cost per unit is $0.70, use approximation techniques to find the production level and the price that maximize profit. What is the maximum profit? [*Hint:* Graph $y = P(x)$ and $y = P'(x)$ in the same viewing window.]

77. *Salvage value.* The salvage value S (in dollars) of a company airplane after t years is estimated to be given by

$$S(t) = 300,000e^{-0.1t}$$

What is the rate of depreciation (in dollars per year) after 1 year? 5 years? 10 years?

78. *Resale value.* The resale value R (in dollars) of a company car after t years is estimated to be given by

$$R(t) = 20,000e^{-0.15t}$$

What is the rate of depreciation (in dollars per year) after 1 year? 2 years? 3 years?

79. *Promotion and maximum profit.* A recording company has produced a new compact disk featuring a very popular recording group. Before launching a national sales campaign, the marketing research department chose to test market the CD in a bellwether city. Their interest is in determining the length of a sales campaign that will maximize total profits. From empirical data, the research department estimates that the proportion of a target group of 50,000 persons buying the CD after t days of television promotion is given by $1 - e^{-0.03t}$. If $4 is received for each CD sold, then the total revenue after t days of promotion will be approximated by

$$R(t) = (4)(50,000)(1 - e^{-0.03t}) \qquad t \geq 0$$

Television promotion costs are

$$C(t) = 4{,}000 + 3{,}000t \qquad t \geq 0$$

(A) How many days of television promotion should be used to maximize total profit? What is the maximum total profit? What percentage of the target market will have purchased the CD when the maximum profit is reached?

(B) Graph the profit function.

80. *Promotion and maximum profit.* Repeat Problem 79 using the revenue equation

$$R(t) = (3)(60{,}000)(1 - e^{-0.04t})$$

Life Sciences

81. *Blood pressure and age.* A research group using hospital records developed the following approximate mathematical model relating systolic blood pressure and age:

$$P(x) = 40 + 25 \ln(x + 1) \qquad 0 \leq x \leq 65$$

where $P(x)$ is pressure measured in millimeters of mercury and x is age in years. What is the rate of change of pressure at the end of 10 years? At the end of 30 years? At the end of 60 years?

82. *Biology.* A yeast culture at room temperature (68 °F) is placed in a refrigerator maintaining a constant temperature of 38 °F. After t hours, the temperature T of the culture is given approximately by

$$T = 30e^{-0.58t} + 38 \qquad t \geq 0$$

What is the rate of change of temperature of the culture at the end of 1 hour? At the end of 4 hours?

83. *Bacterial growth.* A single cholera bacterium divides every 0.5 hour to produce two complete cholera bacteria. If we

start with a colony of 5,000 bacteria, then after t hours there will be

$$A(t) = 5{,}000 \cdot 2^{2t}$$

bacteria. Find $A'(t)$, $A'(1)$, and $A'(5)$, and interpret the results.

84. *Bacterial growth.* Repeat Problem 83 for a starting colony of 1,000 bacteria where a single bacterium divides every 0.25 hour.

Social Sciences

85. *Sociology.* Daniel Lowenthal, a sociologist at Columbia University, made a 5 year study on the sale of popular records relative to their position in the top 20. He found that the average number of sales $N(n)$ of the nth ranking record was given approximately by

$$N(n) = N_1 e^{-0.09(n-1)} \qquad 1 \leq n \leq 20$$

where N_1 was the number of sales of the top record on the list at a given time. Graph N for $N_1 = 1{,}000{,}000$ records.

86. *Political science.* Thomas W. Casstevens, a political scientist at Oakland University, has studied legislative turnover. He (with others) found that the number $N(t)$ of continuously serving members of an elected legislative body remaining t years after an election is given approximately by a function of the form

$$N(t) = N_0 e^{-ct}$$

In particular, for the 1965 election for the U.S. House of Representatives, it was found that

$$N(t) = 434e^{-0.0866t}$$

What is the rate of change after 2 years? After 10 years?

S E C T I O N   10-4   ■   Implicit Differentiation

- ■ SPECIAL FUNCTION NOTATION
- ■ IMPLICIT DIFFERENTIATION

▨ SPECIAL FUNCTION NOTATION

The equation

$$y = 2 - 3x^2 \tag{1}$$

defines a function f with y as a dependent variable and x as an independent variable. Using function notation, we would write

$$y = f(x) \qquad \text{or} \qquad f(x) = 2 - 3x^2$$

In order to reduce to a minimum the number of symbols involved in a discussion, we will often write equation (1) in the form

$$y = 2 - 3x^2 = y(x)$$

where y is *both* a dependent variable and a function symbol. This is a convenient notation and no harm is done as long as one is aware of the double role of y. Other examples are

$$x = 2t^2 - 3t + 1 = x(t)$$
$$z = \sqrt{u^2 - 3u} = z(u)$$
$$r = \frac{1}{(s^2 - 3s)^{2/3}} = r(s)$$

This type of notation will simplify much of the discussion and work that follows.

Until now we have considered functions involving only one independent variable. There is no reason to stop there. The concept can be generalized to functions involving two or more independent variables, and this will be done in detail in Chapter 13. For now, we will "borrow" the notation for a function involving two independent variables. For example,

$$F(x, y) = x^2 - 2xy + 3y^2 - 5$$

specifies a function F involving two independent variables.

■ IMPLICIT DIFFERENTIATION

Consider the equation

$$3x^2 + y - 2 = 0 \tag{2}$$

and the equation obtained by solving (2) for y in terms of x,

$$y = 2 - 3x^2 \tag{3}$$

Both equations define the same function using x as the independent variable and y as the dependent variable. For (3), we can write

$$y = f(x)$$

where

$$f(x) = 2 - 3x^2 \tag{4}$$

and we have an **explicit** (clearly stated) rule that enables us to determine y for each value of x. On the other hand, the y in equation (2) is the same y as in equation (3), and equation (2) **implicitly** gives (implies though does not plainly express) y as a function of x. Thus, we say that equations (3) and (4) define the function f explicitly, and equation (2) defines f implicitly.

The direct use of an equation that defines a function implicitly to find the derivative of the dependent variable with respect to the independent variable is called **implicit differentiation.** Let us differentiate (2) implicitly and (3) directly, and compare results.

Starting with

$$3x^2 + y - 2 = 0$$

we think of y as a function of x—that is, $y = y(x)$—and write

$$3x^2 + y(x) - 2 = 0$$

Then we differentiate both sides with respect to x:

$$\frac{d}{dx}[3x^2 + y(x) - 2] = \frac{d}{dx}0 \quad \text{Since } y \text{ is a function of } x, \text{ but is not}$$
$$\frac{d}{dx}3x^2 + \frac{d}{dx}y(x) - \frac{d}{dx}2 = 0 \quad \begin{array}{l}\text{explicitly given, we simply write}\\ \frac{d}{dx}y(x) = y' \text{ to indicate its derivative.}\end{array}$$
$$6x + y' - 0 = 0$$

Now, we solve for y':

$$y' = -6x$$

Note that we get the same result if we start with equation (3) and differentiate directly:

$$y = 2 - 3x^2$$
$$y' = -6x$$

Why are we interested in implicit differentiation? In general, why do we not solve for y in terms of x and differentiate directly? The answer is that there are many equations of the form

$$F(x, y) = 0 \tag{5}$$

that are either difficult or impossible to solve for y explicitly in terms of x (try it for $x^2y^5 - 3xy + 5 = 0$ or for $e^y - y = 3x$, for example). But it can be shown that, under fairly general conditions on F, equation (5) will define one or more functions where y is a dependent variable and x is an independent variable. To find y' under these conditions, we differentiate (5) implicitly.

explore – discuss 1

(A) How many tangent lines are there to the graph in Figure 1 when $x = 0$? When $x = 1$? When $x = 2$? When $x = 4$? When $x = 6$?

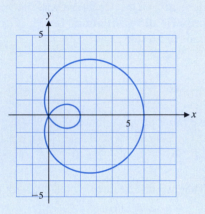

FIGURE 1

(B) Sketch the tangent lines referred to in part (A) and estimate each of their slopes.
(C) Explain why the graph in Figure 1 is not the graph of a function.

EXAMPLE 1 ➤ Differentiating Implicitly   Given

$$F(x, y) = x^2 + y^2 - 25 = 0 \tag{6}$$

find y' and the slope of the graph at $x = 3$.

Solution   We start with the graph of $x^2 + y^2 - 25 = 0$ (a circle, as shown in Figure 2) so that we can interpret our results geometrically. From the graph, it is clear that equation (6) does not define a function. But with a suitable restriction on the variables, equation (6) can define two or more functions. For example, the upper half and the lower half of the circle each define a function. A point on each half-circle that corresponds to $x = 3$ is found by substituting $x = 3$ into (6) and solving for y:

$$x^2 + y^2 - 25 = 0$$
$$(3)^2 + y^2 = 25$$
$$y^2 = 16$$
$$y = \pm 4$$

Thus, the point $(3, 4)$ is on the upper half-circle and the point $(3, -4)$ is on the lower half-circle. We will use these results in a moment. We now differentiate (6) implicitly, treating y as a function of x; that is, $y = y(x)$:

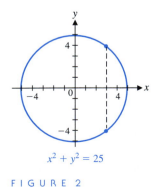

$x^2 + y^2 = 25$

FIGURE 2

$$x^2 + y^2 - 25 = 0$$

$$x^2 + [y(x)]^2 - 25 = 0$$

$$\frac{d}{dx}\{x^2 + [y(x)]^2 - 25\} = \frac{d}{dx}0$$

$$\frac{d}{dx}x^2 + \frac{d}{dx}[y(x)]^2 - \frac{d}{dx}25 = 0 \qquad \text{Use the chain rule.}$$

$$2x + 2[y(x)]^{2-1}y'(x) - 0 = 0$$

$$2x + 2yy' = 0 \qquad \text{Solve for } y' \text{ in terms of } x \text{ and } y.$$

$$y' = -\frac{2x}{2y}$$

$$y' = -\frac{x}{y} \qquad \text{Leave the answer in terms of } x \text{ and } y.$$

We have found y' without first solving $x^2 + y^2 - 25 = 0$ for y in terms of x. And by leaving y' in terms of x and y, we can use $y' = -x/y$ to find y' for *any* point on the graph of $x^2 + y^2 - 25 = 0$ (except where $y = 0$). In particular, for $x = 3$, we found that $(3, 4)$ and $(3, -4)$ are on the graph; thus, the slope of the graph at $(3, 4)$ is

$$y'|_{(3,4)} = -\tfrac{3}{4} \qquad \text{The slope of the graph at } (3, 4)$$

and the slope at $(3, -4)$ is

$$y'|_{(3,-4)} = -\tfrac{3}{-4} = \tfrac{3}{4} \qquad \text{The slope of the graph at } (3, -4)$$

The symbol

$$\boldsymbol{y'|_{(a,b)}}$$

is used to indicate that we are evaluating y' at $x = a$ and $y = b$.

The results are interpreted geometrically on the original graph as shown in Figure 3. ◀

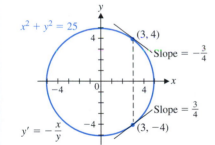

FIGURE 3

In Example 1, the fact that y' is given in terms of both x and y is not a great disadvantage. We have only to make certain that when we want **to evaluate y' for a particular value of x and y, say, (x_0, y_0), the ordered pair must satisfy the original equation.**

MATCHED PROBLEM 1 ▶ Graph $x^2 + y^2 - 169 = 0$, find y' by implicit differentiation, and find the slope of the graph when $x = 5$. ◀

EXAMPLE 2 ▶ Differentiating Implicitly   Find the equation(s) of the tangent line(s) to the graph of

$$y - xy^2 + x^2 + 1 = 0 \tag{7}$$

at the point(s) where $x = 1$.

Solution   We first find y when $x = 1$:

$$y - xy^2 + x^2 + 1 = 0$$
$$y - (1)y^2 + (1)^2 + 1 = 0$$
$$y - y^2 + 2 = 0$$
$$y^2 - y - 2 = 0$$
$$(y - 2)(y + 1) = 0$$
$$y = -1, 2$$

Thus, there are two points on the graph of (7) where $x = 1$; namely, $(1, -1)$ and $(1, 2)$. We next find the slope of the graph at these two points by differentiating (7) implicitly:

$$y - xy^2 + x^2 + 1 = 0$$

$$\frac{d}{dx} y - \frac{d}{dx} xy^2 + \frac{d}{dx} x^2 + \frac{d}{dx} 1 = \frac{d}{dx} 0 \qquad \text{Use the product rule and the}$$

$$\text{chain rule for } \frac{d}{dx} xy^2.$$

$$y' - (x \cdot 2yy' + y^2) + 2x = 0$$

$$y' - 2xyy' - y^2 + 2x = 0 \qquad \text{Solve for } y' \text{ by getting all terms}$$

$$y' - 2xyy' = y^2 - 2x \qquad \text{involving } y' \text{ on one side.}$$

$$(1 - 2xy)y' = y^2 - 2x$$

$$y' = \frac{y^2 - 2x}{1 - 2xy}$$

Now, find the slope at each point:

$$y'|_{(1,-1)} = \frac{(-1)^2 - 2(1)}{1 - 2(1)(-1)} = \frac{1 - 2}{1 + 2} = \frac{-1}{3} = -\frac{1}{3}$$

$$y'|_{(1,2)} = \frac{(2)^2 - 2(1)}{1 - 2(1)(2)} = \frac{4 - 2}{1 - 4} = \frac{2}{-3} = -\frac{2}{3}$$

Equation of the tangent line at $(1, -1)$:   Equation of the tangent line at $(1, 2)$:

$$y - y_1 = m(x - x_1) \qquad\qquad\qquad\qquad y - y_1 = m(x - x_1)$$
$$y + 1 = -\tfrac{1}{3}(x - 1) \qquad\qquad\qquad\quad y - 2 = -\tfrac{2}{3}(x - 1)$$
$$y + 1 = -\tfrac{1}{3}x + \tfrac{1}{3} \qquad\qquad\qquad\quad y - 2 = -\tfrac{2}{3}x + \tfrac{2}{3}$$
$$y = -\tfrac{1}{3}x - \tfrac{2}{3} \qquad\qquad\qquad\qquad\quad y = -\tfrac{2}{3}x + \tfrac{8}{3}$$

In general, it is difficult to graph an equation of the form $F(x, y) = 0$. We can, however, graph the equation $y - xy^2 + x^2 + 1 = 0$ by first using the quadratic formula to solve for y in terms of x, and then graphing the two resulting functions (see Fig. 4). Note, as indicated by Figure 4, that we can easily confirm our computations of the slopes of the tangent lines when $x = 1$.

(A) Slope at $(1, 2)$ (B) Slope at $(1, -1)$

FIGURE 4

$$y_1 = \frac{1 + \sqrt{1 + 4x(x^2 + 1)}}{2x}; \; y_2 = \frac{1 - \sqrt{1 + 4x(x^2 + 1)}}{2x}$$

MATCHED PROBLEM 2 ➤   Repeat Example 2 for $x^2 + y^2 - xy - 7 = 0$ at $x = 1$.

explore – discuss 2

> The slopes of the tangent lines to $y^2 + 3xy + 4x = 9$ when $x = 0$ can be found in either of the following ways: (1) by differentiating the equation implicitly; or (2) by solving for y explicitly in terms of x (using the quadratic formula), and then computing the derivative. Which of the two methods is more efficient? Explain.

EXAMPLE 3 ➤   **Differentiating Implicitly**   Find x' for $x = x(t)$ defined implicitly by

$$t \ln x = xe^t - 1$$

and evaluate x' at $(t, x) = (0, 1)$.

Solution   It is important to remember that x is the dependent variable and t is the independent variable. Therefore, we differentiate both sides of the equation with respect to t (using product and chain rules where appropriate), and then solve for x':

$$t \ln x = xe^t - 1 \qquad \text{Differentiate implicitly with respect to } t.$$

$$\frac{d}{dt}(t \ln x) = \frac{d}{dt}(xe^t) - \frac{d}{dt} 1$$

$$t\,\frac{x'}{x} + \ln x = xe^t + x'e^t \qquad \text{Clear fractions.}$$

$$x\left(t\,\frac{x'}{x}\right) + x \ln x = x \cdot xe^t + xe^t x' \qquad x \neq 0$$

$$tx' + x \ln x = x^2 e^t + xe^t x' \qquad \text{Solve for } x'.$$

$$tx' - xe^t x' = x^2 e^t - x \ln x \qquad \text{Factor out } x'.$$

$$(t - xe^t)x' = x^2 e^t - x \ln x$$

$$x' = \frac{x^2 e^t - x \ln x}{t - xe^t}$$

Now, we evaluate x' at $(t, x) = (0, 1)$, as requested:

$$x'|_{(0,1)} = \frac{(1)^2 e^0 - 1 \ln 1}{0 - 1 e^0}$$

$$= \frac{1}{-1} = -1$$

◀

MATCHED PROBLEM 3 ▶ Find x' for $x = x(t)$ defined implicitly by

$$1 + x \ln t = t e^x$$

and evaluate x' at $(t, x) = (1, 0)$.

◀

Answers to Matched Problems

1. $y' = -x/y$; when $x = 5$, $y = \pm 12$, thus, $y'|_{(5,12)} = -\frac{5}{12}$ and $y'|_{(5,-12)} = \frac{5}{12}$

2. $y' = \dfrac{y - 2x}{2y - x}$; $y = \frac{4}{5}x - \frac{14}{5}$, $y = \frac{1}{5}x + \frac{14}{5}$

3. $x' = \dfrac{te^x - x}{t \ln t - t^2 e^x}$; $x'|_{(1,0)} = -1$

EXERCISE 10-4

In Problems 1–18, find y' without solving for y in terms of x (use implicit differentiation). Evaluate y' at the indicated point.

A

1. $y - 3x^2 + 5 = 0$; $(1, -2)$
2. $3x^4 + y - 2 = 0$; $(1, -1)$
3. $y^2 - 3x^2 + 8 = 0$; $(2, 2)$
4. $3y^2 + 2x^3 - 14 = 0$; $(1, 2)$
5. $y^2 + y - x = 0$; $(2, 1)$
6. $2y^3 + y^2 - x = 0$; $(3, 1)$

B

7. $xy - 6 = 0$; $(2, 3)$
8. $3xy - 2x - 2 = 0$; $(2, 1)$
9. $2xy + y + 2 = 0$; $(-1, 2)$
10. $2y + xy - 1 = 0$; $(-1, 1)$
11. $x^2 y - 3x^2 - 4 = 0$; $(2, 4)$
12. $2x^3 y - x^3 + 5 = 0$; $(-1, 3)$
13. $e^y = x^2 + y^2$; $(1, 0)$ 14. $x^2 - y = 4e^y$; $(2, 0)$
15. $x^3 - y = \ln y$; $(1, 1)$ 16. $\ln y = 2y^2 - x$; $(2, 1)$
17. $x \ln y + 2y = 2x^3$; $(1, 1)$
18. $xe^y - y = x^2 - 2$; $(2, 0)$

In Problems 19 and 20, find x' for $x = x(t)$ defined implicitly by the given equation. Evaluate x' at the indicated point.

19. $x^2 - t^2 x + t^3 + 11 = 0$; $(-2, 1)$
20. $x^3 - tx^2 - 4 = 0$; $(-3, -2)$

In Problems 21–24, find the equation(s) of the tangent line(s) to the graphs of the indicated equations at the point(s) with the given value of x.

21. $xy - x - 4 = 0$; $x = 2$
22. $3x + xy + 1 = 0$; $x = -1$
23. $y^2 - xy - 6 = 0$; $x = 1$
24. $xy^2 - y - 2 = 0$; $x = 1$

25. If $xe^y = 1$, find y' in two ways: first by implicit differentiation, then by solving for y explicitly in terms of x. Which method do you prefer? Explain.

26. Explain the difficulty in solving $x^3 + y + xe^y = 1$ for y as an explicit function of x. Find the slope of the tangent line to the graph of the equation at the point $(0, 1)$.

C *In Problems 27–34, find y' and the slope of the tangent line to the graph of each equation at the indicated point.*

27. $(1 + y)^3 + y = x + 7$; $(2, 1)$
28. $(y - 3)^4 - x = y$; $(-3, 4)$
29. $(x - 2y)^3 = 2y^2 - 3$; $(1, 1)$
30. $(2x - y)^4 - y^3 = 8$; $(-1, -2)$
31. $\sqrt{7 + y^2} - x^3 + 4 = 0$; $(2, 3)$
32. $6\sqrt{y^3 + 1} - 2x^{3/2} - 2 = 0$; $(4, 2)$
33. $\ln(xy) = y^2 - 1$; $(1, 1)$
34. $e^{xy} - 2x = y + 1$; $(0, 0)$

In Problems 35–38, round all approximate values to two decimal places.

35. Find the slopes of the tangent lines at the points on the graph of the equation $y^2 - 9x^2 + 8x - 5 = 0$ where $x = 2$. Confirm your answers by graphing the equation.

36. Find the slopes of the tangent lines at the points on the graph of the equation $3y^2 - x^2 + xy + 5 = 0$ where $x = -3$. Confirm your answers by graphing the equation.

37. Find the equation(s) of the tangent line(s) at the point(s) on the graph of the equation

$$y^3 - xy - x^3 = 2$$

where $x = 1$.

38. Refer to the equation in Problem 37. Find the equation(s) of the tangent line(s) at the point(s) on the graph where $y = -1$.

▶ **APPLICATIONS**

Business & Economics

For the demand equations in Problems 39–42, find the rate of change of p with respect to x by differentiating implicitly (x is the number of items that can be sold at a price of $p.)

39. $x = p^2 - 2p + 1,000$

40. $x = p^3 - 3p^2 + 200$

41. $x = \sqrt{10,000 - p^2}$

42. $x = \sqrt[3]{1,500 - p^3}$

Life Sciences

43. *Biophysics.* In biophysics, the equation

$$(L + m)(V + n) = k$$

is called the *fundamental equation of muscle contraction*, where m, n, and k are constants, and V is the velocity of the shortening of muscle fibers for a muscle subjected to a load of L. Find dL/dV using implicit differentiation.

44. *Biophysics.* In Problem 43, find dV/dL using implicit differentiation.

SECTION 10-5 ■ **Related Rates**

The workers in a union are concerned that the rate at which wages are increasing is lagging behind the rate of increase in the company's profits. An automobile dealer wants to predict how badly an anticipated increase in interest rates will decrease his rate of sales. An investor is studying the connection between the rate of increase in the Dow Jones Average and the rate of increase in the Gross National Product over the past 50 years.

In each of these situations there are two quantities—wages and profits in the first instance, for example—that are changing with respect to time. We would like to discover the precise relationship between the rates of increase (or decrease) of the two quantities. We will begin our discussion of such *related rates* by considering some familiar situations in which the two quantities are distances, and the two rates are velocities.

EXAMPLE 1 ➤ **Related Rates and Motion** A 26 foot ladder is placed against a wall (Fig. 1). If the top of the ladder is sliding down the wall at 2 feet per second, at what rate is the bottom of the ladder moving away from the wall when the bottom of the ladder is 10 feet away from the wall?

FIGURE 1

Solution Many people reason that since the ladder is of constant length, the bottom of the ladder will move away from the wall at the same rate that the top of the ladder is moving down the wall. This is not the case, as we will see.

At any moment in time, let x be the distance of the bottom of the ladder from the wall, and let y be the distance of the top of the ladder on the wall (see Fig. 1). Both x and y are changing with respect to time and can be thought of as functions of time; that is, $x = x(t)$ and $y = y(t)$. Furthermore, x and y are related by the Pythagorean relationship:

$$x^2 + y^2 = 26^2 \tag{1}$$

Differentiating (1) implicitly with respect to time t, and using the chain rule where appropriate, we obtain

$$2x\frac{dx}{dt} + 2y\frac{dy}{dt} = 0 \tag{2}$$

The rates dx/dt and dy/dt are related by equation (2); hence, this type of problem is referred to as a **related rates problem.**

Now our problem is to find dx/dt when $x = 10$ feet, given that $dy/dt = -2$ (y is decreasing at a constant rate of 2 feet per second). We have all the quantities we need in equation (2) to solve for dx/dt, except y. When $x = 10$, y can be found using (1):

$$10^2 + y^2 = 26^2$$
$$y = \sqrt{26^2 - 10^2} = 24 \text{ feet}$$

Substitute $dy/dt = -2$, $x = 10$, and $y = 24$ into (2); then solve for dx/dt:

$$2(10)\frac{dx}{dt} + 2(24)(-2) = 0$$

$$\frac{dx}{dt} = \frac{-2(24)(-2)}{2(10)} = 4.8 \text{ feet per second}$$

Thus, the bottom of the ladder is moving away from the wall at a rate of 4.8 feet per second.

An alternative method is to solve (2) for dx/dt, and then substitute $y = \sqrt{26^2 - x^2}$ and $dy/dt = -2$:

$$2x\frac{dx}{dt} + 2y\frac{dy}{dt} = 0$$

$$\frac{dx}{dt} = \frac{-y}{x}\frac{dy}{dt}$$

$$= \frac{2\sqrt{26^2 - x^2}}{x}$$

Now we graph dx/dt and evaluate at $x = 10$, again concluding that $dx/dt = 4.8$ feet per second when the bottom of the ladder is 10 feet away from the wall (see Fig. 2). Note that the rate at which the bottom of the ladder is moving away from the wall decreases as the ladder falls. ◄

FIGURE 2
$$y_1 = \frac{2\sqrt{26^2 - x^2}}{x}$$

MATCHED PROBLEM 1 ➤ A 26 foot ladder is placed against a wall (Fig. 1). If the bottom of the ladder is moving away from the wall at 3 feet per second, at what rate is the top moving down when the top of the ladder is 24 feet up the wall? ◀

e x p l o r e – d i s c u s s 1

(A) For which values of x and y in Example 1 is dx/dt equal to 2 (that is, the same rate at which the ladder is sliding down the wall)?
(B) When is dx/dt greater than 2? Less than 2?

Suggestions for Solving Related Rates Problems

Step 1. Sketch a figure if helpful.

Step 2. Identify all relevant variables, including those whose rates are given and those whose rates are to be found.

Step 3. Express all given rates and rates to be found as derivatives.

Step 4. Find an equation connecting the variables in step 2.

Step 5. Implicitly differentiate the equation found in step 4, using the chain rule where appropriate, and substitute in all given values.

Step 6. Solve for the derivative that will give the unknown rate.

EXAMPLE 2 ➤ **Related Rates and Motion** Suppose two motor boats leave from the same point at the same time. If one travels north at 15 miles per hour and the other travels east at 20 miles per hour, how fast will the distance between them be changing after 2 hours?

Solution First, draw a picture, as shown in Figure 3.

All variables, x, y, and z, are changing with time. Hence, they can be thought of as functions of time; $x = x(t)$, $y = y(t)$, and $z = z(t)$, given implicitly. It now makes sense to take derivatives of each variable with respect to time. From the Pythagorean theorem,

$$z^2 = x^2 + y^2 \tag{3}$$

We also know that

$$\frac{dx}{dt} = 20 \text{ miles per hour} \quad \text{and} \quad \frac{dy}{dt} = 15 \text{ miles per hour}$$

FIGURE 3

We would like to find dz/dt at the end of 2 hours; that is, when $x = 40$ miles and $y = 30$ miles. To do this, we differentiate both sides of (3) with respect to t and solve for dz/dt:

$$2z \frac{dz}{dt} = 2x \frac{dx}{dt} + 2y \frac{dy}{dt} \qquad (4)$$

We have everything we need except z. When $x = 40$ and $y = 30$, we find z from (3) to be 50. Substituting the known quantities into (4), we obtain

$$2(50) \frac{dz}{dt} = 2(40)(20) + 2(30)(15)$$

$$\frac{dz}{dt} = 25 \text{ miles per hour}$$

Thus, the boats will be separating at a rate of 25 miles per hour. ◀

MATCHED PROBLEM 2 ▶ Repeat Example 2 for the situation at the end of 3 hours. ◀

EXAMPLE 3 ▶ **Related Rates and Motion** Suppose a point is moving along the graph of $x^2 + y^2 = 25$ (Fig. 4). When the point is at $(-3, 4)$, its x coordinate is increasing at the rate of 0.4 unit per second. How fast is the y coordinate changing at that moment?

Solution Since both x and y are changing with respect to time, we can think of each as a function of time:

$$x = x(t) \qquad \text{and} \qquad y = y(t)$$

but restricted so that

$$x^2 + y^2 = 25 \qquad (5)$$

Our problem is now to find dy/dt, given $x = -3$, $y = 4$, and $dx/dt = 0.4$. Implicitly differentiating both sides of (5) with respect to t, we have

$$x^2 + y^2 = 25$$

$$2x \frac{dx}{dt} + 2y \frac{dy}{dt} = 0 \qquad \text{Divide both sides by 2.}$$

$$x \frac{dx}{dt} + y \frac{dy}{dt} = 0 \qquad \text{Substitute } x = -3, y = 4, \text{ and}$$
$$\qquad\qquad\qquad\qquad dx/dt = 0.4, \text{ and solve for } dy/dt.$$

$$(-3)(0.4) + 4 \frac{dy}{dt} = 0$$

$$\frac{dy}{dt} = 0.3 \text{ unit per second} \qquad ◀$$

FIGURE 4

$x^2 + y^2 = 25$

$(-3, 4)$

MATCHED PROBLEM 3 ▶ A point is moving on the graph of $y^3 = x^2$. When the point is at $(-8, 4)$, its y coordinate is decreasing at 2 units per second. How fast is the x coordinate changing at that moment? ◀

▶

EXAMPLE 4 ➤ Related Rates and Business   Suppose that for a company manufacturing transistor radios, the cost, revenue, and profit equations are given by

$$C = 5,000 + 2x \quad \text{Cost equation}$$
$$R = 10x - 0.001x^2 \quad \text{Revenue equation}$$
$$P = R - C \quad \text{Profit equation}$$

where the production output in 1 week is x radios. If production is increasing at the rate of 500 radios per week when production is 2,000 radios, find the rate of increase in:

(A) Cost   (B) Revenue   (C) Profit

Solution   If production x is a function of time (it must be, since it is changing with respect to time), then C, R, and P must also be functions of time. These functions are implicitly (rather than explicitly) given. Letting t represent time in weeks, we differentiate both sides of each of the three equations above with respect to t, and then substitute $x = 2,000$ and $dx/dt = 500$ to find the desired rates.

(A) $C = 5,000 + 2x$ Think: $C = C(t)$ and $x = x(t)$.

$$\frac{dC}{dt} = \frac{d}{dt}(5,000) + \frac{d}{dt}(2x) \quad \text{Differentiate both sides with respect to } t.$$

$$\frac{dC}{dt} = 0 + 2\frac{dx}{dt} = 2\frac{dx}{dt}$$

Since $dx/dt = 500$ when $x = 2,000$,

$$\frac{dC}{dt} = 2(500) = \$1,000 \text{ per week}$$

Cost is increasing at a rate of $1,000 per week.

(B) $R = 10x - 0.001x^2$

$$\frac{dR}{dt} = \frac{d}{dt}(10x) - \frac{d}{dt}0.001x^2$$

$$\frac{dR}{dt} = 10\frac{dx}{dt} - 0.002x\frac{dx}{dt}$$

$$\frac{dR}{dt} = (10 - 0.002x)\frac{dx}{dt}$$

Since $dx/dt = 500$ when $x = 2,000$,

$$\frac{dR}{dt} = [10 - 0.002(2,000)](500) = \$3,000 \text{ per week}$$

Revenue is increasing at a rate of $3,000 per week.

(C) $P = R - C$

$$\frac{dP}{dt} = \frac{dR}{dt} - \frac{dC}{dt}$$

$= \$3,000 - \$1,000$ Results from parts (A) and (B)

$= \$2,000$ per week

Profit is increasing at a rate of $2,000 per week.   ◄

MATCHED PROBLEM 4 ➤   Repeat Example 4 for a production level of 6,000 radios per week.   ◄

explore–discuss 2

(A) In Example 4 suppose that $x(t) = 500t + 500$. Find the time and production level at which the profit is maximized.

(B) Suppose that $x(t) = t^2 + 492t + 16$. Find the time and production level at which the profit is maximized.

(C) Explain why it is unnecessary to know a formula for $x(t)$ in order to determine the production level at which the profit is maximized.

Answers to Matched Problems   1. $dy/dt = -1.25$ ft/sec   2. $dz/dt = 25$ mi/hr   3. $dx/dt = 6$ units/sec
4. (A) $dC/dt = \$1,000$/wk   (B) $dR/dt = -\$1,000$/wk   (C) $dP/dt = -\$2,000$/wk

EXERCISE 10-5

A   *In Problems 1–6, assume $x = x(t)$ and $y = y(t)$. Find the indicated rate, given the other information.*

1. $y = 2x^2 - 1$; $dy/dt = ?$, $dx/dt = 2$ when $x = 30$
2. $y = 2x^{1/2} + 3$; $dy/dt = ?$, $dx/dt = 8$ when $x = 4$
3. $x^2 + y^2 = 25$; $dy/dt = ?$, $dx/dt = -3$ when $x = 3$ and $y = 4$
4. $y^2 + x = 11$; $dx/dt = ?$, $dy/dt = -2$ when $x = 2$ and $y = 3$
5. $x^2 + xy + 2 = 0$; $dy/dt = ?$, $dx/dt = -1$ when $x = 2$ and $y = -3$
6. $y^2 + xy - 3x = -3$; $dx/dt = ?$, $dy/dt = -2$ when $x = 1$ and $y = 0$

B

7. A point is moving on the graph of $xy = 36$. When the point is at $(4, 9)$, its x coordinate is increasing at 4 units per second. How fast is the y coordinate changing at that moment?

8. A point is moving on the graph of $4x^2 + 9y^2 = 36$. When the point is at $(3, 0)$, its y coordinate is decreasing at 2 units per second. How fast is its x coordinate changing at that moment?

9. A boat is being pulled toward a dock as indicated in the figure. If the rope is being pulled in at 3 feet per second, how fast is the distance between the dock and the boat decreasing when it is 30 feet from the dock?

Figure for 9 and 10

10. Refer to Problem 9. Suppose the distance between the boat and the dock is decreasing at 3.05 feet per second. How fast is the rope being pulled in when the boat is 10 feet from the dock?

11. A rock is thrown into a still pond and causes a circular ripple. If the radius of the ripple is increasing at 2 feet per second, how fast is the area changing when the radius is 10 feet? [Use $A = \pi R^2$, $\pi \approx 3.14$.]

12. Refer to Problem 11. How fast is the circumference of a circular ripple changing when the radius is 10 feet? [Use $C = 2\pi R$, $\pi \approx 3.14$.]

13. The radius of a spherical balloon is increasing at the rate of 3 centimeters per minute. How fast is the volume changing when the radius is 10 centimeters? [Use $V = \frac{4}{3}\pi R^3$, $\pi \approx 3.14$.]

14. Refer to Problem 13. How fast is the surface area of the sphere increasing? [Use $S = 4\pi R^2$, $\pi \approx 3.14$.]

15. Boyle's law for enclosed gases states that if the volume is kept constant, then the pressure P and temperature T are related by the equation

$$\frac{P}{T} = k$$

where k is a constant. If the temperature is increasing at 3 Kelvins per hour, what is the rate of change of pressure when the temperature is 250 K and the pressure is 500 pounds per square inch?

16. Boyle's law for enclosed gases states that if the temperature is kept constant, then the pressure P and volume V of the gas are related by the equation

$$VP = k$$

where K is a constant. If the volume is decreasing by 5 cubic inches per second, what is the rate of change of the pressure when the volume is 1,000 cubic inches and the pressure is 40 pounds per square inch?

17. A 10 foot ladder is placed against a vertical wall. Suppose the bottom slides away from the wall at a constant rate of 3 feet per second. How fast is the top sliding down the wall (negative rate) when the bottom is 6 feet from the wall? [*Hint:* Use the Pythagorean theorem: $a^2 + b^2 = c^2$, where c is the length of the hypotenuse of a right triangle and a and b are the lengths of the two shorter sides.]

18. A weather balloon is rising vertically at the rate of 5 meters per second. An observer is standing on the ground 300

meters from the point where the balloon was released. At what rate is the distance between the observer and the balloon changing when the balloon is 400 meters high?

19. Refer to Problem 17. Let x denote the distance from the bottom of the ladder to the wall, and let y denote the distance from the top of the ladder to the ground. Graph dy/dt as a function of x, and evaluate it at $x = 4, 6,$ and 8. Does dy/dt have a maximum or minimum value? Discuss.

20. Refer to Problem 18. Let y denote the distance from the balloon to the ground, and let z denote the distance from the balloon to the observer. Graph dz/dt as a function of y, and evaluate it at $y = 300, 600,$ and 900. Does dz/dt ever exceed 5 meters per second? Explain.

C

21. A street light is on top of a 20 foot pole. A person who is 5 feet tall walks away from the pole at the rate of 5 feet per second. At what rate is the tip of the person's shadow moving away from the pole when he is 20 feet from the pole?

22. Refer to Problem 21. At what rate is the person's shadow growing when he is 20 feet from the pole?

23. Helium is pumped into a spherical balloon at a constant rate of 4 cubic feet per second. How fast is the radius increasing after 1 minute? After 2 minutes? Is there any time at which the radius is increasing at a rate of 100 feet per second? Explain.

24. A point is moving along the x axis at a constant rate of 5 units per second. At which point is its distance from $(0, 1)$ increasing at a rate of 2 units per second? At 4 units per second? At 5 units per second? At 10 units per second? Explain.

25. A point is moving on the graph of $y = e^x + x + 1$ in such a way that its x coordinate is always increasing at a rate of 3 units per second. How fast is the y coordinate changing when the point crosses the x axis?

26. A point is moving on the graph of $x^3 + y^2 = 1$ in such a way that its y coordinate is always increasing at a rate of 2 units per second. At which point(s) is the x coordinate increasing at a rate of 1 unit per second?

▶ **APPLICATIONS**

Business & Economics

27. *Cost, revenue, and profit rates.* Suppose that for a company manufacturing calculators, the cost, revenue, and profit equations are given by

$$C = 90,000 + 30x$$

$$R = 300x - \frac{x^2}{30}$$

$$P = R - C$$

where the production output in 1 week is x calculators. If production is increasing at a rate of 500 calculators per week when production output is 6,000 calculators, find the rate of increase (decrease) in:
(A) Cost　　(B) Revenue　　(C) Profit

28. *Cost, revenue, and profit rates.* Repeat Problem 27 for

$$C = 72,000 + 60x$$

$$R = 200x - \frac{x^2}{30}$$

$$P = R - C$$

where production is increasing at a rate of 500 calculators per week at a production level of 1,500 calculators.

29. *Advertising.* A retail store estimates that weekly sales s and weekly advertising costs x (both in dollars) are related by

$$s = 60,000 - 40,000e^{-0.0005x}$$

The current weekly advertising costs are $2,000, and these costs are increasing at the rate of $300 per week. Find the current rate of change of sales.

30. *Advertising.* Repeat Problem 29 for

$$s = 50,000 - 20,000e^{-0.0004x}$$

31. *Price–demand.* The price p (in dollars) and demand x for a product are related by

$$2x^2 + 5xp + 50p^2 = 80,000$$

(A) If the price is increasing at a rate of $2 per month when the price is $30, find the rate of change of the demand.

(B) If the demand is decreasing at a rate of 6 units per month when the demand is 150 units, find the rate of change of the price.

32. *Price–demand.* Repeat Problem 31 for

$$x^2 + 2xp + 25p^2 = 74,500$$

Life Sciences

33. *Pollution.* An oil tanker aground on a reef is leaking oil that forms a circular oil slick about 0.1 foot thick (see the figure). To estimate the rate dV/dt (in cubic feet per minute) at which the oil is leaking from the tanker, it was found that the radius of the slick was increasing at 0.32 foot per minute ($dR/dt = 0.32$) when the radius R was 500 feet. Find dV/dt, using $\pi \approx 3.14$.

$$A = \pi R^2$$
$$V = 0.1\,A$$

Figure for 33

Social Sciences

34. *Learning.* A person who is new on an assembly line performs an operation in T minutes after x performances of the operation, as given by

$$T = 6\left(1 + \frac{1}{\sqrt{x}}\right)$$

If $dx/dt = 6$ operations per hour, where t is time in hours, find dT/dt after 36 performances of the operation.

Elasticity of Demand

We normally expect that by lowering the price of a commodity we will increase demand, and that by raising the price we will decrease demand. To measure the sensitivity of demand to changes in price, we define the **elasticity of demand,** denoted by $E(p)$, to be the limit of the relative change in demand x divided by the relative change in price p. Therefore, if $x = f(p)$ is the price–demand equation, then

$$E(p) = \lim_{h \to 0} \frac{\dfrac{f(p+h) - f(p)}{f(p)}}{\dfrac{h}{p}}$$

$$= \lim_{h \to 0} \left[\frac{p}{f(p)} \cdot \frac{f(p+h) - f(p)}{h} \right]$$

$$= \frac{p}{f(p)} f'(p)$$

Since an increase in price normally produces a decrease in demand, we expect $E(p)$ to be negative. If $E(p) = -1$, a 10% increase in price produces a 10% decrease in demand, and we say that demand has **unit elasticity.** If $E(p) = -4$, a 10% increase in price produces a 40% decrease in demand; we express such sensitivity of demand to price by saying that demand is **elastic** whenever $E(p) < -1$. If $E(p) = -0.25$, a 10% increase in price produces only a 2.5% decrease in demand; we express such insensitivity to price by saying that demand is **inelastic** whenever $E(p) > -1$.

(A) Consider the price–demand equation

$$x = f(p) = 10{,}000 - 500p \qquad 0 \leqslant p \leqslant 20$$

Calculate the elasticity of demand at $p = 4, 7, 10, 13,$ and 16, and describe each case as elastic, inelastic, or of unit elasticity.

(B) Graph the elasticity of demand $E(p)$ and the revenue $R(p)$ as functions of price, and graph the line $y = -1$. Use the graphs to explain the connection between elasticity and the increasing/decreasing behavior of the revenue function.

(C) Repeat part (B) for the price–demand equation

$$x = f(p) = \sqrt{144 - 2p} \qquad 0 \leqslant p \leqslant 72$$

(D) Suppose that $x = g(p)$ is any price–demand equation for which $g(p)$ is differentiable. Let the associated revenue function be $R(p) = xp = g(p)p$. Show that $R'(p) = g(p)[E(p) + 1]$. Explain why revenue increases with price if and only if demand is inelastic.

C H A P T E R   10   R E V I E W   ■   Important Terms and Symbols

10-1 *The Constant e and Continuous Compound Interest.* Definition of e; continuous compound interest

$$A = Pe^{rt}$$

10-2 *Derivatives of Logarithmic and Exponential Functions.* Derivative formulas for the natural logarithmic and exponential functions; graph properties of $y = \ln x$ and $y = e^x$

10-3 *Chain Rule: General Form.* Composite functions; chain rule; general derivative formulas; derivative formulas for $y = \log_b x$ and $y = b^x$

10-4 *Implicit Differentiation.* Special function notation; function explicitly defined; function implicitly defined; implicit differentiation

$$y = f(x); \quad y = y(x); \quad F(x, y) = 0; \quad y'|_{(a, b)}$$

10-5 *Related Rates.* Related rates

$$x = x(t); \quad y = y(t)$$

C H A P T E R   10   ■   Additional Rules of Differentiation

$$\frac{d}{dx} \ln x = \frac{1}{x}$$

$$\frac{d}{dx} e^x = e^x$$

$$\frac{d}{dx} b^x = \frac{d}{dx} e^{x \ln b} = e^{x \ln b} \ln b = b^x \ln b$$

$$\frac{d}{dx} \ln[f(x)] = \frac{1}{f(x)} f'(x)$$

$$\frac{d}{dx} e^{f(x)} = e^{f(x)} f'(x)$$

$$\frac{d}{dx} [f(x)]^n = n[f(x)]^{n-1} f'(x)$$

$$\frac{d}{dx} \log_b x = \frac{1}{\ln b} \frac{d}{dx} \ln x = \frac{1}{\ln b} \left(\frac{1}{x} \right)$$

$$\frac{dy}{dx} = \frac{dy}{du} \frac{du}{dx}, \quad \frac{dy}{dx} = \frac{dy}{dw} \frac{dw}{du} \frac{du}{dx}, \quad \text{and so on}$$

C H A P T E R   10   ■   Review Exercise

Work through all the problems in this chapter review and check your answers in the back of the book. Answers to all review problems are there along with section numbers in italics to indicate where each type of problem is discussed. Where weaknesses show up, review appropriate sections in the text.

A

1. Use a calculator to evaluate $A = 2,000e^{0.09t}$ to the nearest cent for $t = 5$, 10, and 20.

Find the indicated derivatives in Problems 2–4.

2. $\dfrac{d}{dx} (2 \ln x + 3e^x)$

3. $\dfrac{d}{dx} e^{2x-3}$

4. y' for $y = \ln(2x + 7)$

5. Let $y = \ln u$ and $u = 3 + e^x$.
(A) Express y in terms of x.
(B) Use the chain rule to find dy/dx, and then express dy/dx in terms of x.

6. Find y' for $y = y(x)$ defined implicitly by the equation $2y^2 - 3x^3 - 5 = 0$, and evaluate at $(x, y) = (1, 2)$.

7. For $y = 3x^2 - 5$, where $x = x(t)$ and $y = y(t)$, find dy/dt if $dx/dt = 3$ when $x = 12$.

B

8. Graph $y = 100e^{-0.1x}$.

9. Use trace and zoom on a graphing utility to investigate

$$\lim_{n \to \infty} \left(1 + \frac{2}{n} \right)^n$$

Do you think the limit exists? If so, what would you guess it is?

10. Let $f(x) = \ln(x^2 + 1)$, $g(x) = 1/(x + 1)$, and $h(x) = f(g(x))$. Which of f, g, and h are increasing for $x > 0$? Which are decreasing? Explain your reasoning, and then confirm your answers by graphing each function.

Find the indicated derivatives in Problems 11–16.

11. $\dfrac{d}{dz}[(\ln z)^7 + \ln z^7]$

12. $\dfrac{d}{dx}(x^6 \ln x)$

13. $\dfrac{d}{dx}\dfrac{e^x}{x^6}$

14. y' for $y = \ln(2x^3 - 3x)$

15. $f'(x)$ for $f(x) = e^{x^3 - x^2}$

16. dy/dx for $y = e^{-2x} \ln 5x$

17. Find the equation of the line tangent to the graph of $y = f(x) = 1 + e^{-x}$ at $x = 0$. At $x = -1$.

18. Find y' for $y = y(x)$ defined implicitly by the equation $x^2 - 3xy + 4y^2 = 23$, and find the slope of the graph at $(-1, 2)$.

19. Find x' for $x = x(t)$ defined implicitly by $x^3 - 2t^2x + 8 = 0$, and evaluate at $(t, x) = (-2, 2)$.

20. Find y' for $y = y(x)$ defined implicitly by $x - y^2 = e^y$, and evaluate at $(1, 0)$.

21. Find y' for $y = y(x)$ defined implicitly by $\ln y = x^2 - y^2$, and evaluate at $(1, 1)$.

22. Find the slopes of the tangent lines at the points on the graph of the equation $x^3 + y^2 - 1 = 0$ where $x = -2$. Confirm your answers by graphing the equation.

23. A point is moving on the graph of $y^2 - 4x^2 = 12$ so that its x coordinate is decreasing at 2 units per second when $(x, y) = (1, 4)$. Find the rate of change of the y coordinate.

24. A 17 foot ladder is placed against a wall. If the foot of the ladder is pushed toward the wall at 0.5 foot per second, how fast is the top rising when the foot of the ladder is 8 feet from the wall?

25. Water from a water heater is leaking onto a floor. A circular pool is created whose area is increasing at the rate of 24 square inches per minute. How fast is the radius R of the pool increasing when the radius is 12 inches? $[A = \pi R^2]$

C *In Problems 26–29, find the absolute maximum value of $f(x)$ for $x > 0$.*

26. $f(x) = 11x - 2x \ln x$

27. $f(x) = 10xe^{-2x}$

28. $f(x) = 3x - x^2 + e^{-x}$

29. $f(x) = \dfrac{\ln x}{e^x}$

In Problems 30 and 31, apply the graphing strategy discussed in Section 9-4 to f, summarize the pertinent information, and sketch the graph.

30. $f(x) = 5 - 5e^{-x}$

31. $f(x) = x^3 \ln x$

32. Let $y = w^3$, $w = \ln u$, and $u = 4 - e^x$.
(A) Express y in terms of x.
(B) Use the chain rule to find dy/dx, and then express dy/dx in terms of x.

Find the indicated derivatives in Problems 33–35.

33. y' for $y = 5^{x^2 - 1}$

34. $\dfrac{d}{dx} \log_5(x^2 - x)$

35. $\dfrac{d}{dx} \sqrt{\ln(x^2 + x)}$

36. Find y' for $y = y(x)$ defined implicitly by the equation $e^{xy} = x^2 + y + 1$, and evaluate at $(0, 0)$.

37. A rock is thrown into a still pond and causes a circular ripple. Suppose the radius is increasing at a constant rate of 3 feet per second. Show that the area does not increase at a constant rate. When is the rate of increase of the area the smallest? The largest? Explain.

38. A point moves along the graph of $y = x^3$ in such a way that its y coordinate is increasing at a constant rate of 5 units per second. Does the x coordinate ever increase at a faster rate than the y coordinate? Explain.

▶ APPLICATIONS

Business & Economics

39. *Doubling time.* How long will it take money to double if it is invested at 5% interest compounded:
(A) Annually?
(B) Continuously?

40. *Continuous compound interest.* If $100 is invested at 10% interest compounded continuously, the amount (in dollars) at the end of t years is given by

$$A = 100e^{0.1t}$$

Find $A'(t)$, $A'(1)$, and $A'(10)$.

41. *Marginal analysis.* If the price–demand equation for x units of a commodity is

$$p(x) = 1,000e^{-0.02x}$$

find the marginal revenue equation.

42. *Maximum revenue.* For the price–demand equation in Problem 41, find the production level and price per unit that produces the maximum revenue. What is the maximum revenue?

43. *Maximum revenue.* Graph the revenue function from Problems 41 and 42 for $0 \le x \le 100$.

44. *Minimum average cost.* The cost of producing x units of a product is given by

$$C(x) = 200 + 50x - 50 \ln x \qquad x \ge 1$$

Find the minimum average cost.

45. *Demand equation.* Given the demand equation

$$x = \sqrt{5,000 - 2p^3}$$

find the rate of change of p with respect to x by implicit differentiation (x is the number of items that can be sold at a price of $\$p$ per item).

46. *Rate of change of revenue.* A company is manufacturing a new video game and can sell all it manufactures. The revenue (in dollars) is given by

$$R = 36x - \frac{x^2}{20}$$

where the production output in 1 day is x games. If production is increasing at 10 games per day when production is 250 games per day, find the rate of increase in revenue.

Life Sciences

47. *Drug concentration.* The concentration of a drug in the bloodstream t hours after injection is given approximately by

$$C(t) = 5e^{-0.3t}$$

where $C(t)$ is concentration in milligrams per milliliter. What is the rate of change of concentration after 1 hour? After 5 hours?

48. *Wound healing.* A circular wound on an arm is healing at the rate of 45 square millimeters per day (the area of the wound is decreasing at this rate). How fast is the radius R of the wound decreasing when $R = 15$ millimeters? $[A = \pi R^2]$

Social Sciences

49. *Psychology—learning.* In a computer assembly plant, a new employee, on the average, is able to assemble

$$N(t) = 10(1 - e^{-0.4t})$$

units after t days of on-the-job training.
(A) What is the rate of learning after 1 day? After 5 days?
(B) Graph N for $0 \le t \le 10$.

50. *Learning.* A new worker on the production line performs an operation in T minutes after x performances of the operation, as given by

$$T = 2\left(1 + \frac{1}{x^{3/2}}\right)$$

If, after performing the operation 9 times, the rate of improvement is $dx/dt = 3$ operations per hour, find the rate of improvement in time dT/dt in performing each operation.

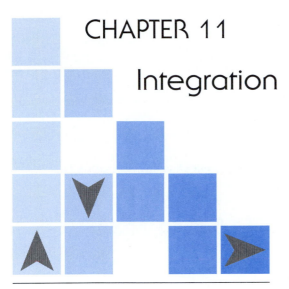

CHAPTER 11

Integration

The last three chapters dealt with differential calculus. We now begin the development of the second main part of calculus, called *integral calculus*. Two types of integrals will be introduced, the *indefinite integral* and the *definite integral,* each quite different from the other. But through the remarkable *fundamental theorem of calculus,* we will show that not only are the two integral forms intimately related, but both are intimately related to differentiation.

S E C T I O N   11-1   ■   ## Antiderivatives and Indefinite Integrals

- ANTIDERIVATIVES
- A GEOMETRIC–NUMERIC LOOK AT ANTIDERIVATIVES
- ANTIDERIVATIVES AND INDEFINITE INTEGRALS: ALGEBRAIC FORMS
- ANTIDERIVATIVES AND INDEFINITE INTEGRALS: EXPONENTIAL AND LOGARITHMIC FORMS
- APPLICATIONS

Many operations in mathematics have reverses—compare addition and subtraction, multiplication and division, and powers and roots. We now know how to find the derivatives of many functions. The reverse operation, *antidifferentiation* (the reconstruction of a function from its derivative) will receive our attention in this and the next two sections. A function F is an **antiderivative** of a function f if $F'(x) = f(x)$. We will look at the problem geometrically, numerically, and algebraically, and we will use an algebraic approach to develop special antiderivative formulas in much the same way as we developed derivative formulas.

■ ANTIDERIVATIVES

(A) Find three antiderivatives of $2x$.

(B) How many antiderivatives of $2x$ exist, and how are they related to each other?

(C) What notation would you use to represent all antiderivatives of $2x$?

What is an antiderivative of x^2?

$\dfrac{x^3}{3}$ is an antiderivative of x^2

since

$$\frac{d}{dx}\left(\frac{x^3}{3}\right) = x^2$$

Note also that

$$\frac{d}{dx}\left(\frac{x^3}{3} + 2\right) = x^2 \qquad \frac{d}{dx}\left(\frac{x^3}{3} - \pi\right) = x^2 \qquad \frac{d}{dx}\left(\frac{x^3}{3} + \sqrt{5}\right) = x^2$$

Hence,

$$\frac{x^3}{3} + 2 \qquad \frac{x^3}{3} - \pi \qquad \frac{x^3}{3} + \sqrt{5}$$

are also antiderivatives of x^2, since each has x^2 as a derivative. In fact, it appears that

$$\frac{x^3}{3} + C \qquad \text{for any real number } C$$

is an antiderivative of x^2, since

$$\frac{d}{dx}\left(\frac{x^3}{3} + C\right) = x^2$$

Thus, antidifferentiation of a given function does not lead to a unique function, but to a whole family of functions.

Does the expression

$$\frac{x^3}{3} + C \qquad \text{with } C \text{ any real number}$$

include all antiderivatives of x^2? Theorem 1 (which we state without proof) indicates that the answer is yes.

theorem 1

On Antiderivatives

If the derivatives of two functions are equal on an open interval (a, b), then the functions can differ by at most a constant. Symbolically: If F and G are differentiable functions on the interval (a, b) and $F'(x) = G'(x)$ for all x in (a, b), then $F(x) = G(x) + k$ for some constant k.

■ A GEOMETRIC–NUMERIC LOOK AT ANTIDERIVATIVES

explore–discuss 2

Given $f'(x) = 1$:

(A) What does $f'(x) = 1$ tell us about graphs of any antiderivative function f?
(B) Sketch, in the same coordinate system, graphs of three different antiderivatives $y = f(x)$ satisfying $f(0) = -2$, $f(0) = 0$, and $f(0) = 2$. What are the geometric relationships among the graphs?

TABLE 1
Properties of the Graph of $y = f(x)$ Determined from the Graph of $y = f'(x)$

$f'(x)$	GRAPH OF $y = f(x)$
Positive	Increasing
Negative	Decreasing
Increasing	Concave upward
Decreasing	Concave downward

We can visualize certain characteristics of an antiderivative by using slopes. If we are given the graph of a derivative function $y = f'(x)$, we can sketch a possible graph of an antiderivative function f showing its characteristic shape. Recall the properties listed in Table 1.

EXAMPLE 1 ➤ **Analyzing the Graph of $y = f(x)$ Given the Graph of $y = f'(x)$**   Given the graph of $y = f'(x)$ in Figure 1, sketch possible graphs of three antiderivative functions f such that $f(0) = -1$, $f(0) = 0$, and $f(0) = 2$, all using the same set of coordinate axes.

Solution   Make up a table summarizing the information obtained from the graph of $y = f'(x)$ (Table 2).

FIGURE 1

TABLE 2

x	$f'(x)$	GRAPH OF $y = f(x)$
$0 < x < 1$	Positive and decreasing	Increasing and concave downward
$x = 1$	x intercept	Local maximum
$1 < x < 3$	Negative and decreasing	Decreasing and concave downward
$x = 3$	Local minimum	Inflection point
$3 < x < 5$	Negative and increasing	Decreasing and concave upward
$x = 5$	x intercept	Local minimum
$5 < x < 6$	Positive and increasing	Increasing and concave upward

If $f(x)$ is one antiderivative of $f'(x)$, since all antiderivatives are given by $y = f(x) + C$, **the graph of any antiderivative function can be obtained from another by a vertical translation.** We know where each of the three graphs of $y = f(x)$ start and the general shape over the interval $(0, 6)$. Figure 2 shows possible graphs of three antiderivative functions having the required characteristics.

FIGURE 2 ◀

MATCHED PROBLEM 1 ▶ Given the graph of $y = f'(x)$ in Figure 3, sketch possible graphs of three antiderivative functions f such that $f(0) = -1, f(0) = 0$, and $f(0) = 1$, all using the same set of coordinate axes. ◀

FIGURE 3

■ ANTIDERIVATIVES AND INDEFINITE INTEGRALS: ALGEBRAIC FORMS

Theorem 1 states that if the derivatives of two functions are equal, then the functions differ by at most a constant. We use the symbol

$$\int f(x)\,dx$$

called the **indefinite integral,** to represent the family of all antiderivatives of $f(x)$, and write

$$\int f(x)\,dx = F(x) + C \qquad \text{if} \qquad F'(x) = f(x)$$

The symbol $\int$ is called an **integral sign,** and the function $f(x)$ is called the **integrand.** The symbol dx indicates that the antidifferentiation is performed with respect to the variable x. (We will have more to say about the symbols $\int$ and dx later in this chapter.) The arbitrary constant C is called the **constant of integration.** Referring to the preceding discussion, we can write

$$\int x^2\,dx = \frac{x^3}{3} + C \qquad \text{since} \qquad \frac{d}{dx}\left(\frac{x^3}{3} + C\right) = x^2$$

Of course, variables other than x can be used in indefinite integrals. For example,

$$\int t^2 \, dt = \frac{t^3}{3} + C \qquad \text{since} \qquad \frac{d}{dt}\left(\frac{t^3}{3} + C\right) = t^2$$

or

$$\int u^2 \, du = \frac{u^3}{3} + C \qquad \text{since} \qquad \frac{d}{du}\left(\frac{u^3}{3} + C\right) = u^2$$

The fact that indefinite integration and differentiation are reverse operations, except for the addition of the constant of integration, can be expressed symbolically as

$$\frac{d}{dx}\left[\int f(x) \, dx\right] = f(x) \qquad \text{The derivative of the indefinite integral of f(x) is f(x).}$$

and

$$\int F'(x) \, dx = F(x) + C \qquad \text{The indefinite integral of the derivative of F(x) is F(x) + C.}$$

Just as with differentiation, we can develop formulas and special properties that will enable us to find indefinite integrals of many frequently encountered functions. To start, we list some formulas that can be established using the definitions of antiderivative and indefinite integral, and the properties of derivatives considered in Chapter 8.

Indefinite Integral Formulas and Properties

For k and C constants:

1. $\int k \, dx = kx + C$

2. $\displaystyle\int x^n \, dx = \frac{x^{n+1}}{n+1} + C \qquad n \neq -1$

3. $\int kf(x) \, dx = k \int f(x) \, dx$

4. $\int [f(x) \pm g(x)] \, dx = \int f(x) \, dx \pm \int g(x) \, dx$

We will establish formula 2 and property 3 here (the others may be shown to be true in a similar manner). To establish formula 2, we simply differentiate the right side to obtain the integrand on the left side. Thus,

$$\frac{d}{dx}\left(\frac{x^{n+1}}{n+1} + C\right) = \frac{(n+1)x^n}{n+1} + 0 = x^n \qquad n \neq -1$$

(Notice that formula 2 cannot be used when $n = -1$; that is, when the integrand is x^{-1} or $1/x$. The indefinite integral of $x^{-1} = 1/x$ will be considered later in this section.)

To establish property 3, let F be a function such that $F'(x) = f(x)$. Then

$$k \int f(x) \, dx = k \int F'(x) \, dx = k[F(x) + C_1] = kF(x) + kC_1$$

and since $[kF(x)]' = kF'(x) = kf(x)$, we have

$$\int kf(x)\, dx = \int kF'(x)\, dx = kF(x) + C_2$$

But $kF(x) + kC_1$ and $kF(x) + C_2$ describe the same set of functions, since C_1 and C_2 are arbitrary real numbers. Thus, property 3 is established.

Caution

It is important to remember that property 3 states that **a constant factor can be moved across an integral sign. A variable factor cannot be moved across an integral sign:**

Constant Factor	Variable Factor
$\int 5x^{1/2}\, dx = 5 \int x^{1/2}\, dx$	$\int x x^{1/2}\, dx \neq x \int x^{1/2}\, dx$

Now let us put the formulas and properties to use.

EXAMPLE 2 ➤ Using Indefinite Integral Properties and Formulas

(A) $\int 5\, dx = 5x + C$

(B) $\displaystyle\int x^4\, dx = \frac{x^{4+1}}{4+1} + C = \frac{x^5}{5} + C$

(C) $\displaystyle\int 5t^7\, dt = 5 \int t^7\, dt = 5\frac{t^8}{8} + C = \frac{5}{8}t^8 + C$

(D) $\int (4x^3 + 2x - 1)\, dx$ $= \int 4x^3\, dx + \int 2x\, dx - \int dx$ *Property 4 can be extended to the sum and difference of an arbitrary number of functions.*

$= 4 \int x^3\, dx + 2 \int x\, dx - \int dx$

$= \dfrac{4x^4}{4} + \dfrac{2x^2}{2} - x + C$

$= x^4 + x^2 - x + C$

(E) $\displaystyle\int \frac{3\, dx}{x^2} = \int 3x^{-2}\, dx = \frac{3x^{-2+1}}{-2+1} + C = -3x^{-1} + C$

(F) $\displaystyle\int 5\sqrt[3]{u^2}\, du = 5 \int u^{2/3}\, du = 5\frac{u^{(2/3)+1}}{\frac{2}{3}+1} + C$

$= 5\dfrac{u^{5/3}}{\frac{5}{3}} + C = 3u^{5/3} + C$ ◄

To check any of the results in Example 2, we differentiate the final result to obtain the integrand in the original indefinite integral. When you evaluate an indefinite integral, do not forget to include the arbitrary constant C.

MATCHED PROBLEM 2 ➤ Find each of the following:

(A) $\int dx$ (B) $\int 3t^4 \, dt$ (C) $\int (2x^5 - 3x^2 + 1) \, dx$

(D) $\int 4\sqrt[5]{w^3} \, dw$ (E) $\int \left(2x^{2/3} - \dfrac{3}{x^4} \right) dx$ ◀

EXAMPLE 3 ➤ Using Indefinite Integral Properties and Formulas

(A) $\displaystyle\int \frac{x^3 - 3}{x^2} \, dx = \int \left(\frac{x^3}{x^2} - \frac{3}{x^2} \right) dx$

$$= \int (x - 3x^{-2}) \, dx$$

$$= \int x \, dx - 3 \int x^{-2} \, dx$$

$$= \frac{x^{1+1}}{1+1} - 3 \frac{x^{-2+1}}{-2+1} + C$$

$$= \tfrac{1}{2}x^2 + 3x^{-1} + C$$

(B) $\displaystyle\int \left(\frac{2}{\sqrt[3]{x}} - 6\sqrt{x} \right) dx = \int (2x^{-1/3} - 6x^{1/2}) \, dx$

$$= 2 \int x^{-1/3} \, dx - 6 \int x^{1/2} \, dx$$

$$= 2\frac{x^{(-1/3)+1}}{-\frac{1}{3}+1} - 6\frac{x^{(1/2)+1}}{\frac{1}{2}+1} + C$$

$$= 2\frac{x^{2/3}}{\frac{2}{3}} - 6\frac{x^{3/2}}{\frac{3}{2}} + C$$

$$= 3x^{2/3} - 4x^{3/2} + C$$ ◀

MATCHED PROBLEM 3 ➤ Find each indefinite integral.

(A) $\displaystyle\int \frac{x^4 - 8x^3}{x^2} \, dx$ (B) $\displaystyle\int \left(8\sqrt[3]{x} - \frac{6}{\sqrt{x}} \right) dx$ ◀

■ ANTIDERIVATIVES AND INDEFINITE INTEGRALS: EXPONENTIAL AND LOGARITHMIC FORMS

We now give indefinite integral formulas for e^x and $1/x$. (Recall that the form $x^{-1} = 1/x$ is not covered by formula 2, given earlier.)

Indefinite Integral Formulas

5. $\int e^x \, dx = e^x + C$ 6. $\displaystyle\int \frac{1}{x} \, dx = \ln |x| + C$ $x \ne 0$

Formula 5 follows immediately from the derivative formula for the exponential function discussed in the last chapter. Because of the absolute value, formula 6 does not follow directly from the derivative formula for the natural logarithm function. Let us show that

$$\frac{d}{dx} \ln |x| = \frac{1}{x} \qquad x \neq 0$$

We consider two cases, $x > 0$ and $x < 0$:

Case 1. $x > 0$:

$$\frac{d}{dx} \ln |x| = \frac{d}{dx} \ln x \quad \text{Since } |x| = x \text{ for } x > 0$$

$$= \frac{1}{x}$$

Case 2. $x < 0$:

$$\frac{d}{dx} \ln |x| = \frac{d}{dx} \ln(-x) \qquad \text{Since } |x| = -x \text{ for } x < 0$$

$$= \frac{1}{-x} \frac{d}{dx}(-x)$$

$$= \frac{-1}{-x} = \frac{1}{x}$$

Thus,

$$\frac{d}{dx} \ln |x| = \frac{1}{x} \qquad x \neq 0$$

and hence,

$$\int \frac{1}{x} \, dx = \ln |x| + C \qquad x \neq 0$$

What about the indefinite integral of $\ln x$? We postpone a discussion of $\int \ln x \, dx$ until Section 12-3, where we will be able to find it using a technique called *integration by parts*.

EXAMPLE 4 ▶ Exponential and Logarithmic Forms

$$\int \left(2e^x + \frac{3}{x} \right) dx = 2 \int e^x \, dx + 3 \int \frac{1}{x} \, dx$$

$$= 2e^x + 3 \ln |x| + C \qquad \blacktriangleleft$$

MATCHED PROBLEM 4 ▶ Find: $\displaystyle \int \left(\frac{5}{x} - 4e^x \right) dx$ ◀

[*Note:* More general forms of exponential and logarithmic functions are considered in Section 11-2.]

■ APPLICATIONS

Let us now consider some applications of the indefinite integral to see why we are interested in finding antiderivatives of functions.

EXAMPLE 5 ➤ Curves   Find the equation of the curve that passes through (2, 5) if its slope is given by $dy/dx = 2x$ at any point x.

Solution   We are interested in finding a function $y = f(x)$ such that

$$\frac{dy}{dx} = 2x \tag{1}$$

and

$$y = 5 \qquad \text{when} \qquad x = 2 \tag{2}$$

If $dy/dx = 2x$, then

$$\begin{aligned} y &= \int 2x \, dx \\ &= x^2 + C \end{aligned} \tag{3}$$

Since $y = 5$ when $x = 2$, we determine the *particular value of C* so that

$$5 = 2^2 + C$$

Thus, $C = 1$, and

$$y = x^2 + 1$$

is the *particular antiderivative* out of all those possible from (3) that satisfies both (1) and (2). See Figure 4. ◀

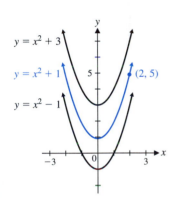

$y = x^2 + 3$

$y = x^2 + 1$

$y = x^2 - 1$

(2, 5)

FIGURE 4
$y = x^2 + C$

Note how Example 5 differs from Example 1, where we reconstructed a possible graph of a function using the graph of its derivative. In Example 5 we actually found an equation for the antiderivative whose graph passes through (2, 5). Consequently, a precise graph of the antiderivative could be produced.

explore – discuss 3

Graph the derivative function $y = f'(x) = 2x$ in Example 5 and use the information from the graph to confirm the shape of the graphs of the antiderivative functions shown in Figure 4.

MATCHED PROBLEM 5 ➤ Find the equation of the curve that passes through (2, 6) if its slope is given by $dy/dx = 3x^2$ at any point x. ◀

In certain situations, it is easier to determine the rate at which something happens than how much of it has happened in a given length of time (for example, population growth rates, business growth rates, rate of healing of a wound, rates of learning or forgetting). If a rate function (derivative) is given and we know the value of the dependent variable for a given value of the independent variable, then—if the rate function is not too complicated—we can often find the original function by integration.

EXAMPLE 6 ➤ **Cost Function**   If the marginal cost of producing x units is given by

$$C'(x) = 0.3x^2 + 2x$$

and the fixed cost is \$2,000, find the cost function $C(x)$ and the cost of producing 20 units.

Solution   Recall that marginal cost is the derivative of the cost function and that fixed cost is cost at a 0 production level. Thus, the mathematical problem is to find $C(x)$ given

$$C'(x) = 0.3x^2 + 2x \qquad C(0) = 2,000$$

We now find the indefinite integral of $0.3x^2 + 2x$ and determine the arbitrary integration constant using $C(0) = 2,000$:

$$C'(x) = 0.3x^2 + 2x$$
$$C(x) = \int (0.3x^2 + 2x)\, dx$$
$$= 0.1x^3 + x^2 + K \qquad \text{Since } C \text{ represents the cost, we use } K \text{ for the}$$
$$\text{constant of integration.}$$

But

$$C(0) = (0.1)0^3 + 0^2 + K = 2,000$$

Thus, $K = 2,000$, and the particular cost function is

$$C(x) = 0.1x^3 + x^2 + 2,000$$

We now find $C(20)$, the cost of producing 20 units:

$$C(20) = (0.1)20^3 + 20^2 + 2,000$$
$$= \$3,200$$

See Figure 5 for a geometric representation.   ◄

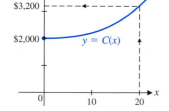

FIGURE 5

MATCHED PROBLEM 6 ➤ Find the revenue function $R(x)$ when the marginal revenue is

$$R'(x) = 400 - 0.4x$$

and no revenue results at a 0 production level. What is the revenue at a production level of 1,000 units?   ◄

EXAMPLE 7 ➤ **Advertising**   An FM radio station is launching an aggressive advertising campaign in order to increase the number of daily listeners. The station currently has 27,000 daily listeners, and management expects the number of daily listeners, $S(t)$, to grow at the rate of

$$S'(t) = 60t^{1/2}$$

listeners per day, where t is the number of days since the campaign began. How long should the campaign last if the station wants the number of daily listeners to grow to 41,000?

Solution   We must solve the equation $S(t) = 41,000$ for t, given that

$$S'(t) = 60t^{1/2} \quad \text{and} \quad S(0) = 27,000$$

First, we use integration to find $S(t)$:

$$S(t) = \int 60t^{1/2} \, dt$$

$$= 60 \frac{t^{3/2}}{\frac{3}{2}} + C$$

$$= 40t^{3/2} + C$$

Since

$$S(0) = 40(0)^{3/2} + C = 27,000$$

we have $C = 27,000$, and

$$S(t) = 40t^{3/2} + 27,000$$

Now we use a graphing utility to solve the equation $S(t) = 41,000$. From Figure 6, we see that the advertising campaign should last approximately 50 days. ◄

50,000

0   60

Intersection
X=49.664419 Y=41000

0

FIGURE 6
$y_1 = 40x^{1.5} + 27,000$
$y_2 = 41,000$

MATCHED PROBLEM 7 ➤   The current monthly circulation of the magazine *Computing News* is 640,000 copies. Due to competition from a new magazine in the same field, the monthly circulation of *Computing News*, $C(t)$, is expected to decrease at the rate of

$$C'(t) = -6,000t^{1/3}$$

copies per month, where t is the time in months since the new magazine began publication. How long will it take for the circulation of *Computing News* to decrease to 460,000 copies per month? ◄

_____

Caution

1. $\displaystyle \int e^x \, dx \neq \frac{e^{x+1}}{x+1} + C$

The power rule only applies to power functions of the form x^n where the exponent n is a real constant not equal to -1 and the base x is the variable. The function e^x is an exponential function with variable exponent x and constant base e. The correct form for this problem is

$$\int e^x\, dx = e^x + C$$

2. $\displaystyle \int x(x^2 + 2)\, dx \neq \frac{x^2}{2}\left(\frac{x^3}{3} + 2x\right) + C$

The integral of a product is not equal to the product of the integrals. The correct form for this problem is

$$\int x(x^2 + 2)\, dx = \int (x^3 + 2x)\, dx = \frac{x^4}{4} + x^2 + C$$

3. Not all elementary functions have elementary antiderivatives in closed form. But finding one when it exists can markedly simplify the solution of some problems.

Answers to Matched Problems   1.

2. (A) $x + C$ (B) $\frac{3}{5}t^5 + C$ (C) $(x^6/3) - x^3 + x + C$ (D) $\frac{5}{2}w^{8/5} + C$
(E) $\frac{6}{5}x^{5/3} + x^{-3} + C$
3. (A) $\frac{1}{3}x^3 - 4x^2 + C$ (B) $6x^{4/3} - 12x^{1/2} + C$ 4. $5 \ln |x| - 4e^x + C$
5. $y = x^3 - 2$ 6. $R(x) = 400x - 0.2x^2; R(1,000) = \$200,000$
7. $t = (40)^{3/4} \approx 16$ mo

EXERCISE   11-1

A   *In Problems 1–16, find each indefinite integral. (Check by differentiating.)*

1. $\int 7\, dx$

2. $\int \pi\, dx$

3. $\int x^6\, dx$

4. $\int x^3\, dx$

5. $\int 8t^3\, dt$

6. $\int 10t^4\, dt$

7. $\int (2u + 1)\, du$

8. $\int (1 - 2u)\, du$

9. $\int (3x^2 + 2x - 5)\, dx$

10. $\int (2 + 4x - 6x^2)\, dx$

11. $\int (s^4 - 8s^5)\, ds$

12. $\int (t^5 + 6t^3)\, dt$

13. $\int 3e^t\, dt$

14. $\int 2e^t\, dt$

15. $\int 2z^{-1}\, dz$

16. $\int \dfrac{3}{s}\, ds$

Each graph in Problems 17–20 is the graph of a derivative function $y = f'(x)$. For each, sketch possible graphs of three antiderivative functions f satisfying $f(0) = -2$, $f(0) = 0$, and $f(0) = 2$, respectively, all on the same coordinate system.

17.

18.

19. $f'(x)$

20. $f'(x)$

33.

34.

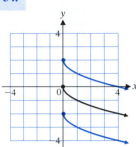

In Problems 21–30, find all the antiderivatives for each derivative.

21. $\dfrac{dy}{dx} = 200x^4$

22. $\dfrac{dx}{dt} = 42t^5$

23. $\dfrac{dP}{dx} = 24 - 6x$

24. $\dfrac{dy}{dx} = 3x^2 - 4x^3$

25. $\dfrac{dy}{du} = 2u^5 - 3u^2 - 1$

26. $\dfrac{dA}{dt} = 3 - 12t^3 - 9t^5$

27. $\dfrac{dy}{dx} = e^x + 3$

28. $\dfrac{dy}{dx} = x - e^x$

29. $\dfrac{dx}{dt} = 5t^{-1} + 1$

30. $\dfrac{du}{dv} = \dfrac{4}{v} + \dfrac{v}{4}$

In Problems 35–56, find each indefinite integral. (Check by differentiation.)

35. $\int 6x^{1/2}\, dx$

36. $\int 8t^{1/3}\, dt$

37. $\int 8x^{-3}\, dx$

38. $\int 12u^{-4}\, du$

39. $\int \dfrac{du}{\sqrt{u}}$

40. $\int \dfrac{dt}{\sqrt[3]{t}}$

41. $\int \dfrac{dx}{4x^3}$

42. $\int \dfrac{6\, dm}{m^2}$

43. $\int \dfrac{du}{2u^5}$

44. $\int \dfrac{dy}{3y^4}$

45. $\int \left(3x^2 - \dfrac{2}{x^2} \right) dx$

46. $\int \left(4x^3 + \dfrac{2}{x^3} \right) dx$

47. $\int \left(10x^4 - \dfrac{8}{x^5} - 2 \right) dx$

48. $\int \left(\dfrac{6}{x^4} - \dfrac{2}{x^3} + 1 \right) dx$

49. $\int \left(3\sqrt{x} + \dfrac{2}{\sqrt{x}} \right) dx$

50. $\int \left(\dfrac{2}{\sqrt[3]{x}} - \sqrt[3]{x^2} \right) dx$

51. $\int \left(\sqrt[3]{x^2} - \dfrac{4}{x^3} \right) dx$

52. $\int \left(\dfrac{12}{x^5} - \dfrac{1}{\sqrt[3]{x^2}} \right) dx$

53. $\int \dfrac{e^x - 3x}{4}\, dx$

54. $\int \dfrac{e^x - 3x^2}{2}\, dx$

55. $\int (2z^{-3} + z^{-2} + z^{-1})\, dz$

56. $\int (3x^{-2} - x^{-1})\, dx$

 B Which sets of graphs in Problems 31–34 could be the graphs of three antiderivative functions from a family of antiderivative functions? Explain.

31.

32.

Each graph in Problems 57–60 is the graph of a derivative function y = f'(x). For each, sketch possible graphs of three antiderivative functions f satisfying f(0) = −2, f(0) = 0, and f(0) = 2, respectively, all on the same coordinate system.

57.

58.

59.

60.

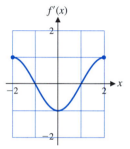

In Problems 61–70, find the particular antiderivative of each derivative that satisfies the given condition.

61. $\dfrac{dy}{dx} = 2x - 3$; $y(0) = 5$

62. $\dfrac{dy}{dx} = 5 - 4x$; $y(0) = 20$

63. $C'(x) = 6x^2 - 4x$; $C(0) = 3{,}000$

64. $R'(x) = 600 - 0.6x$; $R(0) = 0$

65. $\dfrac{dx}{dt} = \dfrac{20}{\sqrt{t}}$; $x(1) = 40$

66. $\dfrac{dR}{dt} = \dfrac{100}{t^2}$; $R(1) = 400$

67. $\dfrac{dy}{dx} = 2x^{-2} + 3x^{-1} - 1$; $y(1) = 0$

68. $\dfrac{dy}{dx} = 3x^{-1} + x^{-2}$; $y(1) = 1$

69. $\dfrac{dx}{dt} = 4e^t - 2$; $x(0) = 1$

70. $\dfrac{dy}{dt} = 5e^t - 4$; $y(0) = -1$

71. Find the equation of the curve that passes through (2, 3) if its slope is given by

$$\frac{dy}{dx} = 4x - 3$$

for each x.

72. Find the equation of the curve that passes through (1, 3) if its slope is given by

$$\frac{dy}{dx} = 12x^2 - 12x$$

for each x.

C *In Problems 73–78, find each indefinite integral.*

73. $\displaystyle\int \frac{2x^4 - x}{x^3}\, dx$

74. $\displaystyle\int \frac{x^{-1} - x^4}{x^2}\, dx$

75. $\displaystyle\int \frac{x^5 - 2x}{x^4}\, dx$

76. $\displaystyle\int \frac{1 - 3x^4}{x^2}\, dx$

77. $\displaystyle\int \frac{x^2 e^x - 2x}{x^2}\, dx$

78. $\displaystyle\int \frac{1 - xe^x}{x}\, dx$

For each derivative in Problems 79–84, find an antiderivative that satisfies the given condition.

79. $\dfrac{dM}{dt} = \dfrac{t^2 - 1}{t^2}$; $M(4) = 5$

80. $\dfrac{dR}{dx} = \dfrac{1 - x^4}{x^3}$; $R(1) = 4$

81. $\dfrac{dy}{dx} = \dfrac{5x + 2}{\sqrt[3]{x}}$; $y(1) = 0$

82. $\dfrac{dx}{dt} = \dfrac{\sqrt{t^3} - t}{\sqrt{t^3}}$; $x(9) = 4$

83. $p'(x) = -\dfrac{10}{x^2}$; $p(1) = 20$

84. $p'(x) = \dfrac{10}{x^3}$; $p(1) = 15$

▶ APPLICATIONS

Business & Economics

85. *Cost function.* The marginal average cost for producing x digital sports watches is given by

$$\overline{C}'(x) = -\frac{1,000}{x^2} \qquad \overline{C}(100) = 25$$

where $\overline{C}(x)$ is the average cost in dollars. Find the average cost function and the cost function. What are the fixed costs?

86. *Paper and paperboard production.* In spite of the prediction of a paperless computerized office, paper and paperboard production in the United States has steadily increased. In 1990 the production was 80.3 million short tons, and since 1970 production has been growing at a rate given by

$$f'(t) = 0.048t + 0.95$$

where t is years after 1970 (data from American Paper Institute, Inc.). Noting that $f(20) = 80.3$, find $f(t)$. Also find $f(0)$ and $f(30)$, production levels for the years 1970 and 2000.

87. *Manufacturing costs.* The graph of the marginal cost function from the manufacturing of x thousand watches per month [where cost $C(x)$ is in thousands of dollars per month] is given in the figure.

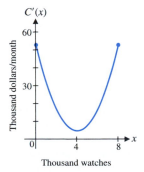

Figure for 87

(A) Using the graph shown, verbally describe the shape of the graph of the cost function $C(x)$ as x increases from 0 to 8,000 watches per month.

(B) Given the equation of the marginal cost function

$$C'(x) = 3x^2 - 24x + 53$$

find the cost function if monthly fixed costs at 0 output are \$30,000. What is the cost for manufacturing 4,000 watches per month? 8,000 watches per month?

(C) Graph the cost function for $0 \leqslant x \leqslant 8$. [Check the shape of the graph relative to the analysis in part (A).]

(D) Why do you think that the graph of the cost function is steeper at both ends than in the middle?

88. *Revenue.* The graph of the marginal revenue function from the sale of x digital sports watches is given in the figure.

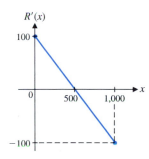

Figure for 88

(A) Using the graph shown, verbally describe the shape of the graph of the revenue function $R(x)$ as x increases from 0 to 1,000.

(B) Find the equation of the marginal revenue function (the linear function shown in the figure).

(C) Find the equation of the revenue function that satisfies $R(0) = 0$. Graph the revenue function over the interval $[0, 1,000]$. [Check the shape of the graph relative to the analysis in part (A).]

(D) Find the price–demand equation and determine the price when the demand is 700 units.

89. *Sales analysis.* The monthly sales of a particular personal computer are expected to decline at the rate of

$$S'(t) = -25t^{2/3}$$

computers per month, where t is time in months and $S(t)$ is the number of computers sold each month. The company plans to stop manufacturing this computer when the monthly sales reach 800 computers. If the monthly sales now ($t = 0$) are 2,000 computers, find $S(t)$. How long (to the nearest month) will the company continue to manufacture this computer?

90. *Sales analysis.* The rate of change of the monthly sales of a new home video game cartridge is given by

$$S'(t) = 500t^{1/4} \qquad S(0) = 0$$

where t is the number of months since the game was released and $S(t)$ is the number of cartridges sold each month. Find $S(t)$. When (to the nearest month) will the monthly sales reach 20,000 cartridges?

91. *Sales analysis.* Repeat Problem 89 if $S'(t) = -25t^{2/3} - 70$ and all other information remains the same.

92. *Sales analysis.* Repeat Problem 90 if $S'(t) = 500t^{1/4} + 300$ and all other information remains the same.

93. *Labor costs and learning.* A defense contractor is starting production on a new missile control system. On the basis of data collected while assembling the first 16 control systems, the production manager obtained the following function describing the rate of labor use:

$$g(x) = 2,400x^{-1/2}$$

where $g(x)$ is the number of labor-hours required to assemble the xth unit of the control system. For example, after assembling 16 units, the rate of assembly is 600 labor-hours per unit, and after assembling 25 units, the rate of assembly is 480 labor-hours per unit. The more units assembled, the more efficient the process because of learning. If 19,200 labor-hours are required to assemble the first 16 units, how many labor-hours, $L(x)$, will be required to assemble the first x units? The first 25 units?

94. *Labor costs and learning.* If the rate of labor use in Problem 93 is

$$g(x) = 2,000x^{-1/3}$$

and if the first 8 control units require 12,000 labor-hours, how many labor-hours, $L(x)$, will be required for the first x control units? The first 27 control units?

Life Sciences

95. *Weight–height.* For an average person, the rate of change of weight W (in pounds) with respect to height h (in inches) is given approximately by

$$\frac{dW}{dh} = 0.0015h^2$$

Find $W(h)$ if $W(60) = 108$ pounds. Also, find the weight for a person who is 5 feet 10 inches tall.

96. *Wound healing.* If the area A of a healing wound changes at a rate given approximately by

$$\frac{dA}{dt} = -4t^{-3} \qquad 1 \le t \le 10$$

where t is time in days and $A(1) = 2$ square centimeters, what will the area of the wound be in 10 days?

Social Sciences

97. *Urban growth.* The rate of growth of the population, $N(t)$, of a newly incorporated city t years after incorporation is estimated to be

$$\frac{dN}{dt} = 400 + 600\sqrt{t} \qquad 0 \le t \le 9$$

If the population was 5,000 at the time of incorporation, find the population 9 years later.

98. *Learning.* A beginning high school language class was chosen for an experiment in learning. Using a list of 50 words, the experiment involved measuring the rate of vocabulary memorization at different times during a continuous 5 hour study session. It was found that the average rate of learning for the whole class was inversely proportional to the time spent studying and was given approximately by

$$V'(t) = \frac{15}{t} \qquad 1 \le t \le 5$$

If the average number of words memorized after 1 hour of study was 15 words, what was the average number of words learned after t hours of study for $1 \le t \le 5$? After 4 hours of study? (Round answer to the nearest whole number.)

SECTION 11-2 ■ **Integration by Substitution**

- REVERSING THE CHAIN RULE
- INTEGRATION BY SUBSTITUTION
- ADDITIONAL SUBSTITUTION TECHNIQUES
- APPLICATION

Many of the indefinite integral formulas introduced in the preceding section are based on corresponding derivative formulas studied earlier. We now consider indefinite integral formulas and procedures based on the chain rule for differentiation.

■ REVERSING THE CHAIN RULE

Recall the chain rule:

$$\frac{d}{dx} f[g(x)] = f'[g(x)]g'(x)$$

The expression on the right is formed by taking the derivative of the outside function f and multiplying it by the derivative of the inside function g. If we recognize an integrand as a chain-rule form $f'[g(x)]g'(x)$, then we can easily find an antiderivative and its indefinite integral:

Reversing the Chain Rule

$$\int f'[g(x)]g'(x)\, dx = f[g(x)] + C \qquad (1)$$

---

explore – discuss 1

(A) Which of the following has e^{x^3-1} as an antiderivative?

$$x^2 e^{x^3-1} \qquad 3x^2 e^{x^3-1} \qquad 3x e^{x^3-1}$$

(B) Which of the following would have e^{x^3-1} as an antiderivative if it were multiplied by a constant factor? A variable factor?

$$3x e^{x^3-1} \qquad x^2 e^{x^3-1}$$

We are interested in finding the indefinite integral:

$$\int 3x^2 e^{x^3-1}\, dx \qquad (2)$$

The integrand appears to be the chain-rule form $e^{g(x)}g'(x)$, which is the derivative of $e^{g(x)}$. Since

$$\frac{d}{dx} e^{x^3-1} = 3x^2 e^{x^3-1}$$

it follows that

$$\int 3x^2 e^{x^3-1}\, dx = e^{x^3-1} + C \tag{3}$$

How does the following indefinite integral differ from (2)?

$$\int x^2 e^{x^3-1}\, dx \tag{4}$$

It is missing the constant factor 3. That is, $x^2 e^{x^3-1}$ is within a constant factor of being the derivative of e^{x^3-1}. But because a constant factor can be moved across the integral sign, this causes us little trouble in finding the indefinite integral of $x^2 e^{x^3-1}$. We introduce the constant factor 3, and at the same time multiply by $\frac{1}{3}$ and move the $\frac{1}{3}$ factor outside the integral sign. This is equivalent to multiplying the integrand in (4) by 1.

$$\int x^2 e^{x^3-1}\, dx = \int \frac{3}{3} x^2\, e^{x^3-1}\, dx$$

$$= \frac{1}{3}\int 3x^2 e^{x^3-1}\, dx = \frac{1}{3} e^{x^3-1} + C \tag{5}$$

The derivative of the right side of (5) is the integrand of the indefinite integral (4). You should check this.

How does the following indefinite integral differ from (2)?

$$\int 3x e^{x^3-1}\, dx \tag{6}$$

It is missing a variable factor x. This is more serious. As tempting as it might be, we *cannot* adjust (6) by introducing the variable factor x and moving $1/x$ outside the integral sign, as we did with the constant 3 in (5). If we could move $1/x$ across the integral sign, what would stop us from moving the whole integrand across the integral sign? Then, indefinite integration would become a trivial exercise and would not give us the results we want—antiderivatives of the integrand.

Summary of the Above Integral Forms

$\int 3x^2 e^{x^3-1} dx = e^{x^3-1} + C$ Integrand is a chain-rule form.

$\int x^2 e^{x^3-1}\, dx = \dfrac{1}{3} e^{x^3-1} + C$ Integrand can be adjusted to a chain-rule form.

$\int 3x e^{x^3-1}\, dx = ?$ Integrand cannot be adjusted to a chain-rule form.

Caution

A constant factor can be moved across an integral sign, but a variable factor cannot.

There is nothing wrong with educated guessing when looking for an antiderivative of a given function, and you are encouraged to do so. You have only to check the result by differentiation. And if you are right, you go on your way; if you are wrong, you simply try another approach.

In Section 10-3, we saw that the chain rule extends the derivative formulas for x^n, e^x, and $\ln x$ to derivative formulas for $[f(x)]^n$, $e^{f(x)}$, and $\ln[f(x)]$. The chain rule can also be used to extend the indefinite integral formulas discussed in Section 11-1. Some general formulas are summarized in the following box:

General Indefinite Integral Formulas

1. $\displaystyle \int [f(x)]^n f'(x)\, dx = \frac{[f(x)]^{n+1}}{n+1} + C \qquad n \neq -1$

2. $\int e^{f(x)} f'(x)\, dx = e^{f(x)} + C$

3. $\displaystyle \int \frac{1}{f(x)} f'(x)\, dx = \ln|f(x)| + C$

Each formula can be verified by using the chain rule to show that the derivative of the function on the right is the integrand on the left. For example,

$$\frac{d}{dx}[e^{f(x)} + C] = e^{f(x)} f'(x)$$

verifies formula 2.

EXAMPLE 1 ➤ Reversing the Chain Rule

(A) $\displaystyle \int (3x+4)^{10} 3\, dx = \frac{(3x+4)^{11}}{11} + C$ Formula 1 with $f(x) = 3x + 4$ and $f'(x) = 3$

Check: $\displaystyle \frac{d}{dx} \frac{(3x+4)^{11}}{11} = 11 \frac{(3x+4)^{10}}{11} \frac{d}{dx}(3x+4) = (3x+4)^{10} 3$

(B) $\int e^{x^2} 2x\, dx = e^{x^2} + C$ Formula 2 with $f(x) = x^2$ and $f'(x) = 2x$

Check: $\displaystyle \frac{d}{dx} e^{x^2} = e^{x^2} \frac{d}{dx} x^2 = e^{x^2} 2x$

(C) $\displaystyle \int \frac{1}{1+x^3} 3x^2\, dx = \ln|1 + x^3| + C$ Formula 3 with $f(x) = 1 + x^3$ and $f'(x) = 3x^2$

Check: $\displaystyle \frac{d}{dx} \ln|1 + x^3| = \frac{1}{1+x^3} \frac{d}{dx}(1 + x^3) = \frac{1}{1+x^3} 3x^2$ ◀

MATCHED PROBLEM 1 ➤   Find each indefinite integral.

(A) $\int (2x^3 - 3)^{20} 6x^2 \, dx$ (B) $\int e^{5x} 5 \, dx$ (C) $\int \frac{1}{4 + x^2} 2x \, dx$ ◄

■ INTEGRATION BY SUBSTITUTION

The key step in using formulas 1, 2, and 3 is recognizing the form of the integrand. Some people find it difficult to identify $f(x)$ and $f'(x)$ in these formulas and prefer to use a *substitution* to simplify the integrand. The *method of substitution,* which we now discuss, becomes increasingly useful as one progresses in studies of integration.

We start by introducing the idea of the *differential*. We represented the derivative by the symbol dy/dx taken as a whole. We now define dy and dx as two separate quantities with the property that their ratio is still equal to $f'(x)$:

Differentials

If $y = f(x)$ defines a differentiable function, then:

1. The **differential dx** of the independent variable x is an arbitrary real number.
2. The **differential dy** of the dependent variable y is defined as the product of $f'(x)$ and dx—that is, as

$$dy = f'(x) \, dx$$

Differentials involve mathematical subtleties that are treated carefully in advanced mathematics courses. Here, we are interested in them mainly as a bookkeeping device to aid in the process of finding indefinite integrals. We can always check the results by differentiating.

EXAMPLE 2 ➤   Differentials

(A) If $y = f(x) = x^2$, then   (B) If $u = g(x) = e^{3x}$, then

$$dy = f'(x) \, dx = 2x \, dx \qquad\qquad du = g'(x) \, dx = 3e^{3x} \, dx$$

(C) If $w = h(t) = \ln(4 + 5t)$, then

$$dw = h'(t) \, dt = \frac{5}{4 + 5t} \, dt$$ ◄

MATCHED PROBLEM 2 ➤   (A) Find dy for $y = f(x) = x^3$.   (B) Find du for $u = h(x) = \ln(2 + x^2)$.
(C) Find dv for $v = g(t) = e^{-5t}$.

The **method of substitution** is developed through the following examples.

EXAMPLE 3 ➤ Using Substitution   Find $\int (x^2 + 2x + 5)^5(2x + 2)\ dx$.

Solution   If

$$u = x^2 + 2x + 5$$

then the differential of u is

$$du = (2x + 2)\ dx$$

Notice that du is one of the factors in the integrand. Substitute u for $x^2 + 2x + 5$ and du for $(2x + 2)\ dx$ to obtain

$$\int (x^2 + 2x + 5)^5(2x + 2)\ dx = \int u^5\ du$$

$$= \frac{u^6}{6} + C$$

$$= \frac{1}{6}(x^2 + 2x + 5)^6 + C \quad \text{Since } u = x^2 + 2x + 5$$

Check

$$\frac{d}{dx}\frac{1}{6}(x^2 + 2x + 5)^6 = \frac{1}{6}(6)(x^2 + 2x + 5)^5\frac{d}{dx}(x^2 + 2x + 5)$$

$$= (x^2 + 2x + 5)^5(2x + 2) \qquad ◄$$

MATCHED PROBLEM 3 ➤ Find $\int (x^2 - 3x + 7)^4(2x - 3)\ dx$ by substitution.   ◄

The substitution method is also called the **change-of-variable method,** since u replaces the variable x in the process. Substituting $u = f(x)$ and $du = f'(x)\ dx$ in formulas 1, 2, and 3 produces the general indefinite integral formulas in the following box:

General Indefinite Integral Formulas

4. $\displaystyle\int u^n\ du = \frac{u^{n+1}}{n+1} + C \qquad n \neq -1$

5. $\int e^u\ du = e^u + C$

6. $\displaystyle\int \frac{1}{u}\ du = \ln |u| + C$

These formulas are valid if u is an independent variable or if u is a function of another variable and du is its differential with respect to that variable.

The substitution method for evaluating certain indefinite integrals is outlined in the following box:

> **Integration by Substitution**
> _____
>
> **Step 1.** Select a substitution that appears to simplify the integrand. In partic-
> ular, try to select u so that du is a factor in the integrand.
>
> **Step 2.** Express the integrand entirely in terms of u and du, completely elimi-
> nating the original variable and its differential.
>
> **Step 3.** Evaluate the new integral, if possible.
>
> **Step 4.** Express the antiderivative found in step 3 in terms of the original
> variable.

EXAMPLE 4 ➤ **Using Substitution**   Use a substitution to find the following:

$$\text{(A) } \int (3x + 4)^6 3 \, dx \qquad \text{(B) } \int e^{t^2} 2t \, dt$$

Solution   (A) If we let $u = 3x + 4$, then $du = 3 \, dx$, and

$$\int (3x + 4)^6 3 \, dx = \int u^6 \, du \qquad \text{Use formula 4.}$$

$$= \frac{u^7}{7} + C$$

$$= \frac{(3x + 4)^7}{7} + C \quad \text{Since } u = 3x + 4$$

$$\text{Check: } \quad \frac{d}{dx} \frac{(3x + 4)^7}{7} = \frac{7(3x + 4)^6}{7} \frac{d}{dx}(3x + 4) = (3x + 4)^6 3$$

(B) If we let $u = t^2$, then $du = 2t \, dt$, and

$$\int e^{t^2} 2t \, dt = \int e^u \, du \quad \text{Use formula 5.}$$
$$= e^u + C$$
$$= e^{t^2} + C \quad \text{Since } u = t^2$$

$$\text{Check: } \quad \frac{d}{dt} e^{t^2} = e^{t^2} \frac{d}{dt} t^2 = e^{t^2} 2t$$

◀

MATCHED PROBLEM 4 ➤ Use a substitution to find each indefinite integral.

$$\text{(A) } \int (2x^3 - 3)^4 6x^2 \, dx \qquad \text{(B) } \int e^{5w} 5 \, dw$$

◀

___
Caution

Integration by substitution is an effective procedure for some indefinite integrals, but not all. Substitution is not helpful for $\int e^{x^2}\, dx$ or $\int (\ln x)\, dx$, for example.

■ ADDITIONAL SUBSTITUTION TECHNIQUES

In order to use the substitution method, **the integrand must be expressed entirely in terms of u and du.** In some cases, the integrand will have to be modified before making a substitution and using one of the integration formulas. Example 5 illustrates this process.

EXAMPLE 5 ➤ Substitution Techniques Integrate:

$$(A)\ \int \frac{1}{4x + 7}\, dx \qquad (B)\ \int te^{-t^2}\, dt \qquad (C)\ \int 4x^2 \sqrt{x^3 + 5}\, dx$$

Solution (A) If $u = 4x + 7$, then $du = 4\, dx$. We are missing a factor of 4 in the integrand to match formula 6 exactly. Recalling that a constant factor can be moved across an integral sign, we proceed as follows:

$$\int \frac{1}{4x + 7}\, dx = \int \frac{1}{4x + 7}\frac{4}{4}\, dx$$

$$= \frac{1}{4} \int \frac{1}{4x + 7}\, 4\, dx \qquad \text{Substitute } u = 4x + 7 \text{ and } du = 4\, dx.$$

$$= \frac{1}{4} \int \frac{1}{u}\, du \qquad \text{Use formula 6.}$$

$$= \tfrac{1}{4} \ln |u| + C$$
$$= \tfrac{1}{4} \ln |4x + 7| + C \qquad \text{Since } u = 4x + 7$$

Check: $\dfrac{d}{dx} \dfrac{1}{4} \ln |4x + 7| = \dfrac{1}{4} \dfrac{1}{4x + 7} \dfrac{d}{dx}(4x + 7) = \dfrac{1}{4}\dfrac{1}{4x + 7} 4 = \dfrac{1}{4x + 7}$

(B) If $u = -t^2$, then $du = -2t\, dt$. Proceed as in part (A):

$$\int te^{-t^2}\, dt = \int e^{-t^2}\frac{-2}{-2} t\, dt$$

$$= -\frac{1}{2} \int e^{-t^2}(-2t)\, dt \quad \text{Substitute } u = -t^2 \text{ and } du = -2t\, dt.$$

$$= -\frac{1}{2} \int e^u\, du \qquad \text{Use formula 5.}$$

$$= -\tfrac{1}{2}e^u + C$$
$$= -\tfrac{1}{2}e^{-t^2} + C \qquad \text{Since } u = -t^2$$

Check: $\dfrac{d}{dt}(-\tfrac{1}{2}e^{-t^2}) = -\tfrac{1}{2}e^{-t^2}\dfrac{d}{dt}(-t^2) = -\tfrac{1}{2}e^{-t^2}(-2t) = te^{-t^2}$

(C) $\int 4x^2 \sqrt{x^3 + 5}\ dx = 4 \int \sqrt{x^3 + 5}\ (x^2)\ dx$ *Move the 4 across the integral*
sign and proceed as before.

$$= 4 \int \sqrt{x^3 + 5}\ \frac{3}{3}\ (x^2)\ dx$$

$$= \frac{4}{3} \int \sqrt{x^3 + 5}(3x^2)\ dx \qquad \textit{Substitute } u = x^3 + 5 \textit{ and}$$
du = 3x² dx.

$$= \frac{4}{3} \int \sqrt{u}\ du$$

$$= \frac{4}{3} \int u^{1/2}\ du \qquad\qquad \textit{Use formula 4.}$$

$$= \frac{4}{3}\frac{u^{3/2}}{\frac{3}{2}} + C$$

$$= \tfrac{8}{9}u^{3/2} + C$$
$$= \tfrac{8}{9}(x^3 + 5)^{3/2} + C \qquad \textit{Since } u = x^3 + 5$$

Check: $\dfrac{d}{dx}[\tfrac{8}{9}(x^3 + 5)^{3/2}] = \tfrac{4}{3}(x^3 + 5)^{1/2}\dfrac{d}{dx}(x^3 + 5)$

$$= \tfrac{4}{3}(x^3 + 5)^{1/2}3x^2 = 4x^2\sqrt{x^3 + 5} \qquad \blacktriangleleft$$

MATCHED PROBLEM 5 ➤   Integrate:

(A) $\int e^{-3x}\ dx$ (B) $\displaystyle\int \frac{x}{x^2 - 9}\ dx$ (C) $\int 5t^2(t^3 + 4)^{-2}\ dt$ $\blacktriangleleft$

Even if it is not possible to find a substitution that makes an integrand match one of the integration formulas exactly, a substitution may sufficiently simplify the integrand so that other techniques can be used.

EXAMPLE 6 ➤   **Substitution Techniques**   Find: $\displaystyle\int \frac{x}{\sqrt{x + 2}}\ dx$

Solution   Proceeding as before, if we let $u = x + 2$, then $du = dx$ and

$$\int \frac{x}{\sqrt{x + 2}}\ dx = \int \frac{x}{\sqrt{u}}\ du$$

Notice that this substitution is not yet complete, because we have not expressed the integrand entirely in terms of u and du. As we noted earlier, only a constant factor can be moved across an integral sign, so we cannot move x outside the integral sign (as much as we would like to). Instead, we must return to the original substitution, solve for x in terms of u, and use the resulting equation to complete the substitution:

$$u = x + 2 \qquad \textit{Solve for x in terms of u.}$$
$$u - 2 = x \qquad \textit{Substitute this expression for x.}$$

Thus,

$$\int \frac{x}{\sqrt{x+2}}\, dx = \int \frac{u-2}{\sqrt{u}}\, du \qquad \text{Simplify the integrand.}$$

$$= \int \frac{u-2}{u^{1/2}}\, du$$

$$= \int (u^{1/2} - 2u^{-1/2})\, du$$

$$\boxed{= \int u^{1/2}\, du - 2 \int u^{-1/2}\, du}$$

$$= \frac{u^{3/2}}{\frac{3}{2}} - 2\frac{u^{1/2}}{\frac{1}{2}} + C$$

$$= \tfrac{2}{3}(x+2)^{3/2} - 4(x+2)^{1/2} + C \quad \text{Since } u = x + 2$$

Check: $\dfrac{d}{dx}[\tfrac{2}{3}(x+2)^{3/2} - 4(x+2)^{1/2}] = (x+2)^{1/2} - 2(x+2)^{-1/2}$

$$= \frac{x+2}{(x+2)^{1/2}} - \frac{2}{(x+2)^{1/2}}$$

$$= \frac{x}{(x+2)^{1/2}}$$

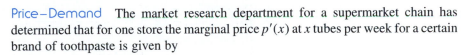

MATCHED PROBLEM 6 ➤ Find: $\int x\sqrt{x+1}\, dx$

■ APPLICATION

EXAMPLE 7 ➤ **Price–Demand** The market research department for a supermarket chain has determined that for one store the marginal price $p'(x)$ at x tubes per week for a certain brand of toothpaste is given by

$$p'(x) = -0.015e^{-0.01x}$$

(A) Find the price–demand equation if the weekly demand is 50 tubes when the price of a tube is \$2.35.

(B) Find the weekly demand when the price of a tube is \$1.89.

Solution (A) $p(x) = \int -0.015e^{-0.01x}\, dx$

$$= -0.015 \int e^{-0.01x}\, dx$$

$$= -0.015 \int e^{-0.01x} \frac{-0.01}{-0.01}\, dx$$

$$= \frac{-0.015}{-0.01} \int e^{-0.01x}(-0.01)\, dx \quad \text{Substitute } u = -0.01x \text{ and } du = -0.01\, dx.$$

$$= 1.5 \int e^u\, du$$

$$= 1.5e^u + C$$

$$= 1.5e^{-0.01x} + C \qquad \text{Since } u = -0.01x$$

We find C by noting that

$$p(50) = 1.5e^{-0.01(50)} + C = \$2.35$$
$$C = \$2.35 - 1.5e^{-0.5} \quad \text{Use a calculator.}$$
$$C = \$2.35 - 0.91$$
$$C = \$1.44$$

Thus,

$$p(x) = 1.5e^{-0.01x} + 1.44$$

(B) To find the demand when the price is $1.89, we use a graphing utility to solve $p(x) = \$1.89$ for x. From Figure 1, we see that the weekly demand is 120 tubes when the price is $1.89 per tube.   ◀

FIGURE 1
$y_1 = 1.5e^{-0.01x} + 1.44$
$y_2 = 1.89$

MATCHED PROBLEM 7 ▶ The marginal price $p'(x)$ at a supply level of x tubes per week for a certain brand of toothpaste is given by

$$p'(x) = 0.001e^{0.01x}$$

(A) Find the price–supply equation if the supplier is willing to supply 100 tubes per week at a price of $1.65 each.
(B) How many tubes would the supplier be willing to supply at a price of $1.98 each?   ◀

explore – discuss 2

In each of the following examples explain why $\neq$ is used; then work the problem correctly using either an appropriate substitution or another method.

1. $\displaystyle \int (x^2 + 3)^2 \, dx = \int (x^2 + 3)^2 \, \frac{2x}{2x} \, dx$

$$\neq \frac{1}{2x} \int (x^2 + 3)^2 2x \, dx$$

2. $\displaystyle \int \frac{1}{10x + 3} \, dx = \int \frac{1}{u} \, dx \qquad u = 10x + 3$

$$\neq \ln|u| + C$$

We conclude with two final cautions (the first was stated earlier, but is worth repeating):

Caution

1. A variable cannot be moved across an integral sign!
2. An integral must be expressed entirely in terms of u and du before applying integration formulas 4, 5, and 6.

Answers to Matched Problems

1. (A) $\frac{1}{21}(2x^3 - 3)^{21} + C$ (B) $e^{5x} + C$
   (C) $\ln|4 + x^2| + C$ or $\ln(4 + x^2) + C$, since $4 + x^2 > 0$

2. (A) $dy = 3x^2\,dx$ (B) $du = \dfrac{2x}{2 + x^2}\,dx$ (C) $dv = -5e^{-5t}\,dt$

3. $\frac{1}{5}(x^2 - 3x + 7)^5 + C$
4. (A) $\frac{1}{5}(2x^3 - 3)^5 + C$ (B) $e^{5w} + C$
5. (A) $-\frac{1}{3}e^{-3x} + C$ (B) $\frac{1}{2}\ln|x^2 - 9| + C$ (C) $-\frac{5}{3}(t^3 + 4)^{-1} + C$
6. $\frac{2}{5}(x + 1)^{5/2} - \frac{2}{3}(x + 1)^{3/2} + C$ 7. (A) $p(x) = 0.1e^{0.01x} + 1.38$ (B) 179 tubes

---

EXERCISE 11-2

In Problems 1–38, find each indefinite integral, and check the result by differentiating.

A

1. $\int (x^2 - 4)^5 2x\,dx$
2. $\int (x^3 + 1)^4 3x^2\,dx$
3. $\int e^{4x}4\,dx$
4. $\int e^{-3x}(-3)\,dx$
5. $\int \dfrac{1}{2t + 3}2\,dt$
6. $\int \dfrac{1}{5t - 7}5\,dt$

B

7. $\int (3x - 2)^7\,dx$
8. $\int (5x + 3)^9\,dx$
9. $\int (x^2 + 3)^7 x\,dx$
10. $\int (x^3 - 5)^4 x^2\,dx$
11. $\int 10e^{-0.5t}\,dt$
12. $\int 4e^{0.01t}\,dt$
13. $\int \dfrac{1}{10x + 7}\,dx$
14. $\int \dfrac{1}{100 - 3x}\,dx$
15. $\int xe^{2x^2}\,dx$
16. $\int x^2 e^{4x^3}\,dx$
17. $\int \dfrac{x^2}{x^3 + 4}\,dx$
18. $\int \dfrac{x}{x^2 - 2}\,dx$
19. $\int \dfrac{t}{(3t^2 + 1)^4}\,dt$
20. $\int \dfrac{t^2}{(t^3 - 2)^5}\,dt$
21. $\int \dfrac{x^2}{(4 - x^3)^2}\,dx$
22. $\int \dfrac{x}{(5 - 2x^2)^5}\,dx$
23. $\int x\sqrt{x + 4}\,dx$
24. $\int x\sqrt{x - 9}\,dx$
25. $\int \dfrac{x}{\sqrt{x - 3}}\,dx$
26. $\int \dfrac{x}{\sqrt{x + 5}}\,dx$
27. $\int x(x - 4)^9\,dx$
28. $\int x(x + 6)^8\,dx$
29. $\int e^{2x}(1 + e^{2x})^3\,dx$
30. $\int e^{-x}(1 - e^{-x})^4\,dx$
31. $\int \dfrac{1 + x}{4 + 2x + x^2}\,dx$
32. $\int \dfrac{x^2 - 1}{x^3 - 3x + 7}\,dx$

33. $\int (2x + 1)e^{x^2 + x + 1}\,dx$
34. $\int (x^2 + 2x)e^{x^3 + 3x^2}\,dx$
35. $\int (e^x - 2x)^3(e^x - 2)\,dx$
36. $\int (x^2 - e^x)^4(2x - e^x)\,dx$
37. $\int \dfrac{x^3 + x}{(x^4 + 2x^2 + 1)^4}\,dx$
38. $\int \dfrac{x^2 - 1}{(x^3 - 3x + 7)^2}\,dx$

In Problems 39–44, imagine that the indicated "solutions" were given to you by a student whom you are tutoring in this class.

(A) How would you have the student check each solution?
(B) Is the solution right or wrong? If the solution is wrong, explain what is wrong and how it can be corrected.
(C) Show a correct solution for each incorrect solution, and check the result by differentiation.

39. $\int \dfrac{1}{2x - 3}\,dx = \ln|2x - 3| + C$

40. $\int \dfrac{x}{x^2 + 5}\,dx = \ln|x^2 + 5| + C$

41. $\int x^3 e^{x^4}\,dx = e^{x^4} + C$

42. $\int e^{4x - 5}\,dx = e^{4x - 5} + C$

43. $\int 2(x^2 - 2)^2\,dx = \dfrac{(x^2 - 2)^2}{3x} + C$

44. $\int (-10x)(x^2 - 3)^{-4}\,dx = (x^2 - 3)^{-5} + C$

C *In Problems 45–56, find each indefinite integral, and check the result by differentiating.*

45. $\int x\sqrt{3x^2 + 7}\, dx$

46. $\int x^2\sqrt{2x^3 + 1}\, dx$

47. $\int x(x^3 + 2)^2\, dx$

48. $\int x(x^2 + 2)^2\, dx$

49. $\int x^2(x^3 + 2)^2\, dx$

50. $\int (x^2 + 2)^2\, dx$

51. $\int \dfrac{x^3}{\sqrt{2x^4 + 3}}\, dx$

52. $\int \dfrac{x^2}{\sqrt{4x^3 - 1}}\, dx$

53. $\int \dfrac{(\ln x)^3}{x}\, dx$

54. $\int \dfrac{e^x}{1 + e^x}\, dx$

55. $\int \dfrac{1}{x^2}e^{-1/x}\, dx$

56. $\int \dfrac{1}{x\ln x}\, dx$

In Problems 57–62, find the antiderivative of each derivative.

57. $\dfrac{dx}{dt} = 7t^2(t^3 + 5)^6$

58. $\dfrac{dm}{dn} = 10n(n^2 - 8)^7$

59. $\dfrac{dy}{dt} = \dfrac{3t}{\sqrt{t^2 - 4}}$

60. $\dfrac{dy}{dx} = \dfrac{5x^2}{(x^3 - 7)^4}$

61. $\dfrac{dp}{dx} = \dfrac{e^x + e^{-x}}{(e^x - e^{-x})^2}$

62. $\dfrac{dm}{dt} = \dfrac{\ln(t - 5)}{t - 5}$

Use substitution techniques to derive the integration formulas in Problems 63 and 64. Then check your work by differentiation.

63. $\int e^{au}\, du = \dfrac{1}{a}e^{au} + C, \quad a \neq 0$

64. $\int \dfrac{1}{au + b}\, du = \dfrac{1}{a}\ln|au + b| + C, \quad a \neq 0$

▶ **APPLICATIONS**

Business & Economics

65. *Price–demand equation.* The marginal price for a weekly demand of x bottles of baby shampoo in a drug store is given by

$$p'(x) = \dfrac{-6{,}000}{(3x + 50)^2}$$

(A) Find the price–demand equation if the weekly demand is 150 when the price of a bottle of shampoo is $4.

(B) What is the weekly demand when the price is $2.50?

66. *Price–supply equation.* The marginal price at a supply level of x bottles of baby shampoo per week is given by

$$p'(x) = \dfrac{300}{(3x + 25)^2}$$

(A) Find the price–supply equation if the distributor of the shampoo is willing to supply 75 bottles a week at a price of $1.60 per bottle.

(B) How many bottles would the supplier be willing to supply at a price of $1.75 per bottle?

67. *Cost function.* The weekly marginal cost of producing x pairs of tennis shoes is given by

$$C'(x) = 12 + \dfrac{500}{x + 1}$$

where $C(x)$ is cost in dollars. If the fixed costs are $2,000 per week, find the cost function. What is the average cost per pair of shoes if 1,000 pairs of shoes are produced each week?

68. *Revenue function.* The weekly marginal revenue from the sale of x pairs of tennis shoes is given by

$$R'(x) = 40 - 0.02x + \dfrac{200}{x + 1} \qquad R(0) = 0$$

where $R(x)$ is revenue in dollars. Find the revenue function. Find the revenue from the sale of 1,000 pairs of shoes.

69. *Marketing.* An automobile company is ready to introduce a new line of cars with a national sales campaign. After test marketing the line in a carefully selected city, the marketing research department estimates that sales (in millions of dollars) will increase at the monthly rate of

$$S'(t) = 10 - 10e^{-0.1t} \qquad 0 \leqslant t \leqslant 24$$

t months after the national campaign has started.

(A) What will be the total sales, $S(t)$, t months after the beginning of the national campaign if we assume 0 sales at the beginning of the campaign?

(B) What are the estimated total sales for the first 12 months of the campaign?

(C) When (to the nearest month) will the estimated total sales reach $100 million?

70. *Marketing.* Repeat Problem 69 if the monthly rate of increase in sales is found to be approximated by

$$S'(t) = 20 - 20e^{-0.05t} \qquad 0 \le t \le 24$$

71. *Oil production.* Using data from the first 3 years of production as well as geological studies, the management of an oil company estimates that oil will be pumped from a producing field at a rate given by

$$R(t) = \frac{100}{t + 1} + 5 \qquad 0 \le t \le 20$$

where $R(t)$ is the rate of production (in thousands of barrels per year) t years after pumping begins. How many barrels of oil, $Q(t)$, will the field produce the first t years if $Q(0) = 0$? How many barrels will be produced the first 9 years?

72. *Oil production.* In Problem 71, if the rate is found to be

$$R(t) = \frac{120t}{t^2 + 1} + 3 \qquad 0 \le t \le 20$$

how many barrels of oil, $Q(t)$, will the field produce the first t years if $Q(0) = 0$? How many barrels will be produced the first 5 years?

Life Sciences

73. *Biology.* A yeast culture is growing at the rate of $W'(t) = 0.2e^{0.1t}$ grams per hour. If the starting culture weighs 2 grams, what will be the weight of the culture, $W(t)$, after t hours? After 8 hours?

74. *Medicine.* The rate of healing for a skin wound (in square centimeters per day) is approximated by $A'(t) = -0.9e^{-0.1t}$. If the initial wound has an area of 9 square centimeters, what will its area, $A(t)$, be after t days? After 5 days?

75. *Pollution.* A contaminated lake is treated with a bactericide. The rate of decrease in harmful bacteria t days after the treatment is given by

$$\frac{dN}{dt} = -\frac{2{,}000t}{1 + t^2} \qquad 0 \le t \le 10$$

where $N(t)$ is the number of bacteria per milliliter of water. If the initial count was 5,000 bacteria per milliliter, find $N(t)$ and then find the bacteria count after 10 days.

76. *Pollution.* An oil tanker aground on a reef is losing oil and producing an oil slick that is radiating outward at a rate given approximately by

$$\frac{dR}{dt} = \frac{60}{\sqrt{t + 9}} \qquad t \ge 0$$

where R is the radius (in feet) of the circular slick after t minutes. Find the radius of the slick after 16 minutes if the radius is 0 when $t = 0$.

Social Sciences

77. *Learning.* In a particular business college, it was found that an average student enrolled in an advanced typing class progressed at a rate of $N'(t) = 6e^{-0.1t}$ words per minute per week, t weeks after enrolling in a 15 week course. If at the beginning of the course a student could type 40 words per minute, how many words per minute, $N(t)$, would the student be expected to type t weeks into the course? After completing the course?

78. *Learning.* In the same business college, it was also found that an average student enrolled in a beginning shorthand class progressed at a rate of $N'(t) = 12e^{-0.06t}$ words per minute per week, t weeks after enrolling in a 15 week course. If at the beginning of the course a student could take dictation in shorthand at 0 words per minute, how many words per minute, $N(t)$, would the student be expected to handle t weeks into the course? After completing the course?

79. *College enrollment.* The projected rate of increase in enrollment in a new college is estimated by

$$\frac{dE}{dt} = 5{,}000(t + 1)^{-3/2} \qquad t \ge 0$$

where $E(t)$ is the projected enrollment in t years. If enrollment is 2,000 now ($t = 0$), find the projected enrollment 15 years from now.

S E C T I O N   11-3   ■   # Differential Equations—Growth and Decay

- DIFFERENTIAL EQUATIONS AND SLOPE FIELDS
- CONTINUOUS COMPOUND INTEREST REVISITED
- EXPONENTIAL GROWTH LAW
- POPULATION GROWTH; RADIOACTIVE DECAY; LEARNING
- A COMPARISON OF EXPONENTIAL GROWTH PHENOMENA

In the previous section, we considered equations of the form

$$\frac{dy}{dx} = 6x^2 - 4x \qquad p'(x) = -400e^{-0.04x}$$

These are examples of *differential equations*. In general, an equation is a **differential equation** if it involves an unknown function and one or more of its derivatives. Other examples of differential equations are

$$\frac{dy}{dx} = ky \qquad y'' - xy' + x^2 = 5 \qquad \frac{dy}{dx} = 2xy$$

The first and third equations are called **first-order** equations because each involves a first derivative, but no higher derivative. The second equation is called a **second-order** equation because it involves a second derivative, and no higher derivatives. Finding solutions to different types of differential equations (functions that satisfy the equation) is the subject matter for whole books and courses on this topic. Here, we will consider only a few very special but very important first-order equations that have immediate and significant applications.

We start by looking at some first-order equations geometrically—in terms of *slope fields*. We then reconsider continuous compound interest as modeled by a first-order differential equation, and from this we will be able to generalize the approach to a wide variety of other types of growth phenomena.

■ DIFFERENTIAL EQUATIONS AND SLOPE FIELDS

We introduce the concept of *slope field* through an example. Suppose we are given the first-order differential equation

$$\frac{dy}{dx} = 0.2y \tag{1}$$

A function f is a solution of (1) if $y = f(x)$ satisfies equation (1) for all values of x in the domain of f. Geometrically interpreted, equation (1) gives us the slope of a solution curve that passes through the point (x, y). For example, if $y = f(x)$ is a solution of (1) that passes through the point $(0, 2)$, then the slope of f at $(0, 2)$ is given by

$$\frac{dy}{dx} = 0.2(2) = 0.4$$

We indicate this by drawing a short segment of the tangent line at the point (0, 2), as shown in Figure 1A. This procedure is repeated for points (− 3, 1) and (2, 3), as also shown in Figure 1A. Assuming that the graph of *f* passes through all three points, we sketch an approximate graph of *f* in Figure 1B.

(A)

(B)

FIGURE 1

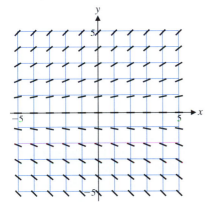

FIGURE 2

If we continue the process of drawing tangent line segments at each point in the grid in Figure 1—a task easily handled by computers but not by hand—we obtain a *slope field*. A slope field for differential equation (1), drawn by a computer, is shown in Figure 2. In general, a **slope field** for a first-order differential equation is obtained by drawing tangent line segments determined by the equation at each point in a grid. In a more advanced treatment of the subject, one can find out a lot about the shape and behavior of solution curves of first-order differential equations by looking at slope fields. Our interests are more modest here.

explore–discuss 1

(A) In Figure 1A (or a copy), draw tangent line segments for a solution curve of differential equation (1) that passes through (− 3, − 1), (0, − 2), and (2, − 3).

(B) In Figure 1B (or a copy), sketch an approximate graph of the solution curve that passes through these three points. (Repeat the tangent line segments first.)

(C) Of all the elementary functions discussed in the first two chapters, make a conjecture as to what type of function appears to be a solution to differential equation (1).

In Explore–Discuss 1, if you guessed that solutions to (1) are exponential functions, you are to be congratulated. We now show that

$$y = Ce^{0.2x} \qquad\qquad (2)$$

is a solution to (1) for any real number C. To do this, we substitute $y = Ce^{0.2x}$ into (1) to see if the left side is equal to the right side for all real x:

$$\frac{dy}{dx} = 0.2y$$

Left side: $\quad \dfrac{dy}{dx} = \dfrac{d}{dx}(Ce^{0.2x}) = 0.2Ce^{0.2x}$

Right side: $\quad 0.2y = 0.2Ce^{0.2x}$

Thus, (2) is a solution of (1) for C any real number. [Later in this section we will show how to find (2) directly from (1).]

Which values of C will produce solution curves that pass through $(0, 2)$ and $(0, -2)$, respectively? Substituting the coordinates of each point into (2) and solving for C (a task left the reader), we obtain

$$y = 2e^{0.2x} \qquad \text{and} \qquad y = -2e^{0.2x} \qquad\qquad (3)$$

as can be easily checked. The graphs of equations (3) are shown in Figure 3 and confirm the results in Figure 1B.

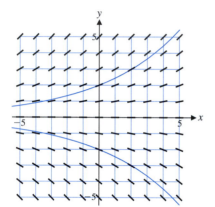

FIGURE 3

e x p l o r e – d i s c u s s   2

(A) In Figure 3 (or a copy), sketch in an approximate solution curve that passes through $(0, 3)$ and one that passes through $(0, -3)$.

(B) Use a graphing utility to graph $y = Ce^{0.2x}$ for $C = -4, -3, -2, 2, 3, 4$, all in the same viewing window. Notice how these solution curves follow the flow of the tangent line segments in the slope field in Figure 2.

As indicated above, drawing slope fields by hand is not a task for human beings —a 20 by 20 grid would require drawing 400 tangent line segments! Repetitive tasks of this type are what computers are for. A few problems in Exercise 11-3 involve interpreting slope fields, not drawing them.

■ CONTINUOUS COMPOUND INTEREST REVISITED

Let P be the initial amount of money deposited in an account, and let A be the amount in the account at any time t. Instead of assuming that the money in the account earns a particular rate of interest, suppose we say that the rate of growth of the amount of money in the account at any time t is proportional to the amount present at that time. Since dA/dt is the rate of growth of A with respect to t, we have

$$\frac{dA}{dt} = rA \qquad A(0) = P \qquad A, P > 0 \tag{4}$$

where r is an appropriate constant. We would like to find a function $A = A(t)$ that satisfies these conditions. Multiplying both sides of equation (4) by $1/A$, we obtain

$$\frac{1}{A}\frac{dA}{dt} = r$$

Now we integrate each side with respect to t:

$$\int \frac{1}{A}\frac{dA}{dt}\,dt = \int r\,dt \qquad \frac{dA}{dt}\,dt = A'(t)\,dt = dA$$

$$\int \frac{1}{A}\,dA = \int r\,dt$$

$$\ln|A| = rt + C \qquad |A| = A, \text{ since } A > 0$$

$$\ln A = rt + C$$

We convert this last equation into the equivalent exponential form

$$A = e^{rt + C} \qquad \textit{Definition of logarithmic function:}$$
$$\qquad\qquad y = \ln x \text{ if and only if } x = e^y$$
$$= e^C e^{rt} \qquad \textit{Property of exponents:} \quad b^m b^n = b^{m+n}$$

Since $A(0) = P$, we evaluate $A(t) = e^C e^{rt}$ at $t = 0$ and set it equal to P:

$$A(0) = e^C e^0 = e^C = P$$

Hence, $e^C = P$, and we can rewrite $A = e^C e^{rt}$ in the form

$$A = Pe^{rt}$$

This is the same continuous compound interest formula obtained in Section 10-1, where the principal P is invested at an annual nominal rate r compounded continuously for t years.

■ EXPONENTIAL GROWTH LAW

In general, if the rate of change with respect to time of a quantity Q is proportional to the amount present and $Q(0) = Q_0$, then proceeding in exactly the same way as above, we obtain the following:

Exponential Growth Law

If $\dfrac{dQ}{dt} = rQ$ and $Q(0) = Q_0$,   then $Q = Q_0 e^{rt}$,

where

  $Q_0 =$ Amount at $t = 0$
  $r =$ Continuous compound growth rate (expressed as a decimal)
  $t =$ Time
  $Q =$ Quantity at time t

The constant r in the exponential growth law is sometimes called the **growth constant,** or the **growth rate.** The last term can be misleading, since the rate of growth of Q with respect to time is dQ/dt, not r. Notice that if $r < 0$, then $dQ/dt < 0$ and Q is decreasing. This type of growth is called **exponential decay.**

Once we know that the rate of growth of something is proportional to the amount present, then we know it has exponential growth and we can use the results summarized in the box without having to solve the differential equation each time. The exponential growth law applies not only to money invested at interest compounded continuously, but also to many other types of problems — population growth, radioactive decay, natural resource depletion, and so on.

■ POPULATION GROWTH; RADIOACTIVE DECAY; LEARNING

The world population is growing at an ever-increasing rate, as illustrated in Figure 4. **Population growth** over certain periods of time often can be approximated by the exponential growth law described above.

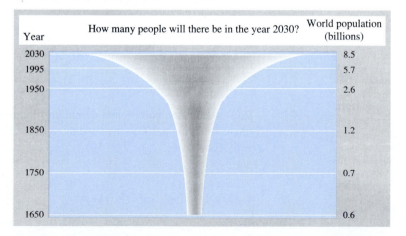

Year	How many people will there be in the year 2030?	World population (billions)
2030		8.5
1995		5.7
1950		2.6
1850		1.2
1750		0.7
1650		0.6

FIGURE 4
The population explosion
Source: World Bank, 1994; *World Almanac,* 1994

E X A M P L E 1 ➤   **Population Growth**   India had a population of about 0.9 billion people in 1995 ($t = 0$) and a growth rate of 1.3% per year, which we will assume is compounded continuously. Let P represent the population (in billions) t years after 1995.

(A) Find an equation that represents India's population growth after 1995, assuming the 1.3% growth rate continues.

(B) What is the estimated population (to the nearest tenth of a billion) for India in the year 2030?

Solution   (A) The exponential growth law applies, and we have

$$\frac{dP}{dt} = 0.013P \qquad P(0) = 0.9$$

Thus,

$$P = 0.9e^{0.013t} \tag{5}$$

(B) See the graph of equation (5) in Figure 5. Using equation (5), we can estimate the population in India in 2030 ($t = 35$):

$$P = 0.9e^{0.013(35)} = 1.4 \text{ billion people}$$

FIGURE 5

Years after 1995

M A T C H E D P R O B L E M 1 ➤   Assuming the same continuous compound growth rate as in Example 1, what will India's population be (to the nearest tenth of a billion) in the year 2012?

E X A M P L E 2 ➤   **Population Growth**   If the exponential growth law applies to Canada's population growth, at what continuous compound growth rate will the population double over the next 100 years?

Solution   The problem is to find r, given $P = 2P_0$ and $t = 100$:

$$P = P_0 e^{rt}$$

$$2P_0 = P_0 e^{100r}$$

$$2 = e^{100r} \qquad \text{Take the natural logarithm of both sides and}$$

$$100r = \ln 2 \qquad \text{reverse the equation.}$$

$$r = \frac{\ln 2}{100}$$

$$\approx 0.0069 \quad \text{or} \quad 0.69\%$$

FIGURE 6
$y_1 = e^{100x}$; $y_2 = 2$

Figure 6 shows a graphical solution of the equation $2 = e^{100r}$.   ◄

MATCHED PROBLEM 2 ▶   If the exponential growth law applies to population growth in Nigeria, find the doubling time (to the nearest year) of the population if it continues to grow at 2.1% per year compounded continuously.   ◄

We now turn to another type of exponential growth—**radioactive decay.** In 1946, Willard Libby (who later received a Nobel prize in chemistry) found that as long as a plant or animal is alive, radioactive carbon-14 is maintained at a constant level in its tissues. Once the plant or animal is dead, however, the radioactive carbon-14 diminishes by radioactive decay at a rate proportional to the amount present. Thus,

$$\frac{dQ}{dt} = rQ \qquad Q(0) = Q_0$$

and we have another example of the exponential growth law. The continuous compound rate of decay for radioactive carbon-14 has been found to be 0.000 123 8; thus, $r = -0.000\ 123\ 8$, since decay implies a negative continuous compound growth rate.

▶   EXAMPLE 3 ▶   Archaeology   A piece of human bone was found at an archaeological site in Africa. If 10% of the original amount of radioactive carbon-14 was present, estimate the age of the bone (to the nearest 100 years).

Solution   Using the exponential growth law for

$$\frac{dQ}{dt} = -0.000\ 123\ 8Q \qquad Q(0) = Q_0$$

we find that

$$Q = Q_0 e^{-0.0001238t}$$

and our problem is to find t so that $Q = 0.1Q_0$ (since the amount of carbon-14 present now is 10% of the amount present, Q_0, at the death of the person).

Thus,

$$0.1Q_0 = Q_0 e^{-0.0001238t}$$

$$0.1 = e^{-0.0001238t}$$

$$\ln 0.1 = \ln e^{-0.0001238t}$$

$$t = \frac{\ln 0.1}{-0.000\ 123\ 8} \approx 18{,}600 \text{ years}$$

See Figure 7 for a graphical solution.

FIGURE 7
$y_1 = e^{-0.0001238x}; y_2 = 0.1$

MATCHED PROBLEM 3 ▶ Estimate the age of the bone in Example 3 (to the nearest 100 years) if 50% of the original amount of carbon-14 is present. ◀

In learning certain skills such as typing and swimming, a mathematical model often used is one that assumes there is a maximum skill attainable, say, M, and the rate of improving is proportional to the difference between that achieved, y, and that attainable, M. Mathematically,

$$\frac{dy}{dt} = k(M - y) \qquad y(0) = 0$$

We solve this type of problem using the same technique that was used to obtain the exponential growth law. First, multiply both sides of the first equation by $1/(M - y)$ to obtain

$$\frac{1}{M - y} \frac{dy}{dt} = k$$

and then integrate each side with respect to t:

$$\int \frac{1}{M - y} \frac{dy}{dt}\, dt = \int k\, dt$$

$$-\int \frac{1}{M - y} \left(-\frac{dy}{dt} \right) dt = \int k\, dt \qquad \text{Substitute } u = M - y \text{ and}$$

$$-\int \frac{1}{u}\, du = \int k\, dt \qquad \qquad du = -dy = -\frac{dy}{dt}\, dt.$$

$$-\ln|u| = kt + C \qquad \text{Substitute } M - y = u.$$

$$-\ln(M - y) = kt + C \qquad \text{Absolute value signs are not required. (Why?)}$$

$$\ln(M - y) = -kt - C$$

Change this last equation to equivalent exponential form:

$$M - y = e^{-kt-C}$$

$$M - y = e^{C}e^{-kt}$$

$$y = M - e^{-C}e^{-kt}$$

Now, $y(0) = 0$; hence,

$$y(0) = M - e^{-C}e^{0} = 0$$

Solving for e^{-C}, we obtain

$$e^{-C} = M$$

and our final solution is

$$y = M - Me^{-kt} = M(1 - e^{-kt})$$

EXAMPLE 4 ▶   **Learning**   For a particular person who is learning to swim, it is found that the distance y (in feet) the person is able to swim in 1 minute after t hours of practice is given approximately by

$$y = 50(1 - e^{-0.04t})$$

(A) Graph the equation and $y = 50$ in the same viewing window, using $0 \leqslant t \leqslant 75$ and $0 \leqslant y \leqslant 60$.

(B) Find the rate of improvement (to two decimal places) after 10 hours of practice.

Solution (A)

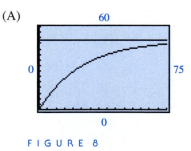

FIGURE 8

(B) Rate of improvement $= y'(10) = 1.34$ feet per hour of practice (see Fig. 9).

FIGURE 9

MATCHED PROBLEM 4 ▶   In Example 4, what is the rate of improvement (to two decimal places) after 50 hours of practice?

■ A COMPARISON OF EXPONENTIAL GROWTH PHENOMENA

The graphs and equations given in Table 1 compare several widely used growth models. These are divided basically into two groups: unlimited growth and limited growth. Following each equation and graph is a short (and necessarily incomplete) list of areas in which the models are used. This only touches on a subject that has been extensively developed and which you are likely to encounter in greater depth in the future.

TABLE 1
Exponential Growth

DESCRIPTION	MODEL	SOLUTION	GRAPH	USES
Unlimited growth: Rate of growth is proportional to the amount present	$\dfrac{dy}{dt} = ky$ $k, t > 0$ $y(0) = c$	$y = ce^{kt}$		• Short-term population growth (people, bacteria, etc.) • Growth of money at continuous compound interest • Price–supply curves
Exponential decay: Rate of growth is proportional to the amount present	$\dfrac{dy}{dt} = -ky$ $k, t > 0$ $y(0) = c$	$y = ce^{-kt}$		• Depletion of natural resources • Radioactive decay • Light absorption in water • Price–demand curves • Atmospheric pressure (t is altitude)
Limited growth: Rate of growth is proportional to the difference between the amount present and a fixed limit	$\dfrac{dy}{dt} = k(M - y)$ $k, t > 0$ $y(0) = 0$	$y = M(1 - e^{-kt})$		• Sales fads (for example, skateboards) • Depreciation of equipment • Company growth • Learning
Logistic growth: Rate of growth is proportional to the amount present and to the difference between the amount present and a fixed limit	$\dfrac{dy}{dt} = ky(M - y)$ $k, t > 0$ $y(0) = \dfrac{M}{1 + c}$	$y = \dfrac{M}{1 + ce^{-kMt}}$		• Long-term population growth • Epidemics • Sales of new products • Rumor spread • Company growth

Answers to Matched Problems   1. 1.1 billion people   2. 33 yr   3. 5,600 yr   4. 0.27 ft/hr

EXERCISE   11-3

A   *In Problems 1–6, find the general or particular solution, as indicated, for each differential equation.*

1. $\dfrac{dy}{dx} = e^{0.5x}$

2. $\dfrac{dy}{dx} = \dfrac{2}{x}$

3. $\dfrac{dy}{dx} = x^2 - x;\ y(0) = 0$

4. $\dfrac{dy}{dx} = \sqrt{x};\ y(0) = 0$

5. $\dfrac{dy}{dx} = -2xe^{-x^2};\ y(0) = 3$

6. $\dfrac{dy}{dx} = e^{x-3};\ y(3) = -5$

B   *Problems 7–12 refer to the following slope fields:*

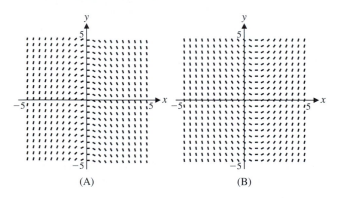

(A)   (B)

7.   Which slope field is associated with the differential equation $dy/dx = x - 1$? Briefly justify your answer.

8.   Which slope field is associated with the differential equation $dy/dx = -x$? Briefly justify your answer.

9.   Solve the differential equation $dy/dx = x - 1$, and find the particular solution that passes through $(0, -2)$.

10.   Solve the differential equation $dy/dx = -x$, and find the particular solution that passes through $(0, 3)$.

11.   Graph the particular solution found in Problem 9 in the appropriate figure above (or a copy).

12.   Graph the particular solution found in Problem 10 in the appropriate figure above (or a copy).

In Problems 13–18, find the general or particular solution, as indicated, for each differential equation.

13. $\dfrac{dy}{dx} = -0.8y$

14. $\dfrac{dy}{dx} = 0.6y$

15. $\dfrac{dy}{dx} = 0.07y;\ y(0) = 1,000$

16. $\dfrac{dy}{dx} = -0.004y;\ y(0) = 5$

17. $\dfrac{dx}{dt} = -x$

18. $\dfrac{dy}{dt} = y$

C   *Problems 19–24 refer to the following slope fields:*

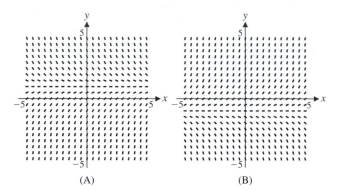

(A)   (B)

19.   Which slope field is associated with the differential equation $dy/dx = 1 - y$? Briefly justify your answer.

20.   Which slope field is associated with the differential equation $dy/dx = y + 1$? Briefly justify your answer.

21.   Show that $y = 1 - Ce^{-x}$ is a solution of the differential equation $dy/dx = 1 - y$ for any real number C. Find the particular solution that passes through $(0, 0)$.

22.   Show that $y = Ce^x - 1$ is a solution of the differential equation $dy/dx = y + 1$ for any real number C. Find the particular solution that passes through $(0, 0)$.

23.   Graph the particular solution found in Problem 21 in the appropriate figure above (or a copy).

24.   Graph the particular solution found in Problem 22 in the appropriate figure above (or a copy).

25.   Graph $y = 1 - Ce^{-x}$ for $C = -2, -1, 1$, and 2 for $-5 \leqslant x \leqslant 5$, $-5 \leqslant y \leqslant 5$, all in the same viewing window. Observe how the solution curves go with the flow of the tangent line segments in the corresponding slope field shown above.

26.   Graph $y = Ce^x - 1$ for $C = -2, -1, 1$, and 2 for $-5 \leqslant x \leqslant 5$, $-5 \leqslant y \leqslant 5$, all in the same viewing window. Observe how the solution curves go with the flow of the tangent line segments in the corresponding slope field shown above.

In Problems 27–34, graph the given examples of the various cases in Table 1.

27. Unlimited growth:
$y = 1,000e^{0.08t}$
$0 \leqslant t \leqslant 15$
$0 \leqslant y \leqslant 3,500$

28. Unlimited growth:
$y = 5,250e^{0.12t}$
$0 \leqslant t \leqslant 10$
$0 \leqslant y \leqslant 20,000$

29. Exponential decay:
$p = 100e^{-0.05x}$
$0 \leqslant x \leqslant 30$
$0 \leqslant p \leqslant 100$

30. Exponential decay:
$p = 1,000e^{-0.08x}$
$0 \leqslant x \leqslant 40$
$0 \leqslant p \leqslant 1,000$

31. Limited growth:
$N = 100(1 - e^{-0.05t})$
$0 \leqslant t \leqslant 100$
$0 \leqslant N \leqslant 100$

32. Limited growth:
$N = 1,000(1 - e^{-0.07t})$
$0 \leqslant t \leqslant 70$
$0 \leqslant N \leqslant 1,000$

33. Logistic growth:
$$N = \frac{1,000}{1 + 999e^{-0.4t}}$$
$0 \leqslant t \leqslant 40$
$0 \leqslant N \leqslant 1,000$

34. Logistic growth:
$$N = \frac{400}{1 + 99e^{-0.4t}}$$
$0 \leqslant t \leqslant 30$
$0 \leqslant N \leqslant 400$

▶ **APPLICATIONS**

Business & Economics

35. *Continuous compound interest.* Find the amount A in an account after t years if

$$\frac{dA}{dt} = 0.08A \quad \text{and} \quad A(0) = 1,000$$

36. *Continuous compound interest.* Find the amount A in an account after t years if

$$\frac{dA}{dt} = 0.12A \quad \text{and} \quad A(0) = 5,250$$

37. *Continuous compound interest.* Find the amount A in an account after t years if

$$\frac{dA}{dt} = rA \quad A(0) = 8,000 \quad A(2) = 9,020$$

38. *Continuous compound interest.* Find the amount A in an account after t years if

$$\frac{dA}{dt} = rA \quad A(0) = 5,000 \quad A(5) = 7,460$$

39. *Price–demand.* The marginal price dp/dx at x units of demand per week is proportional to the price p. There is no weekly demand at a price of \$100 per unit [$p(0) = 100$], and there is a weekly demand of 5 units at a price of \$77.88 per unit [$p(5) = 77.88$].
(A) Find the price–demand equation.
(B) At a demand of 10 units per week, what is the price (to the nearest cent)?
(C) At a price of \$45, what is the demand (to the nearest unit)?

40. *Price–supply.* The marginal price dp/dx at x units of supply per day is proportional to the price p. There is no supply at a price of \$10 per unit [$p(0) = 10$], and there is a daily supply of 50 units at a price of \$12.84 per unit [$p(50) = 12.84$].
(A) Find the price–supply equation.
(B) At a supply of 100 units per day, what is the price (to the nearest cent)?
(C) At a price of \$38, what is the supply (to the nearest unit)?

41. *Advertising.* A company is trying to expose a new product to as many people as possible through television advertising. Suppose the rate of exposure to new people is proportional to the number of those who have not seen the product out of L possible viewers. No one is aware of the product at the start of the campaign, and after 10 days 40% of L are aware of the product. Mathematically,

$$\frac{dN}{dt} = k(L - N) \quad N(0) = 0 \quad N(10) = 0.4L$$

(A) Solve the differential equation.
(B) What percent of L will have been exposed after 5 days of the campaign?
(C) How many days will it take to expose 80% of L?

42. *Advertising.* Suppose the differential equation for Problem 41 is

$$\frac{dN}{dt} = k(L - N) \quad N(0) = 0 \quad N(10) = 0.1L$$

(A) Interpret $N(10) = 0.1L$ verbally.
(B) Solve the differential equation.
(C) How many days will it take to expose 50% of L?

Life Sciences

43. *Biology.* For relatively clear bodies of water, light intensity is reduced according to

$$\frac{dI}{dx} = -kI \qquad I(0) = I_0$$

where I is the intensity of light at x feet below the surface. For the Sargasso Sea off the West Indies, $k = 0.009\ 42$. Find I in terms of x, and find the depth at which the light is reduced to half of that at the surface.

44. *Blood pressure.* It can be shown under certain assumptions that blood pressure P in the largest artery in the human body (the aorta) changes between beats with respect to time t according to

$$\frac{dP}{dt} = -aP \qquad P(0) = P_0$$

where a is a constant. Find $P = P(t)$ that satisfies both conditions.

45. *Drug concentration.* A single injection of a drug is administered to a patient. The amount Q in the body then decreases at a rate proportional to the amount present, and for a particular drug the rate is 4% per hour. Thus,

$$\frac{dQ}{dt} = -0.04Q \qquad Q(0) = Q_0$$

where t is time in hours.
(A) If the initial injection is 3 milliliters [$Q(0) = 3$], find $Q = Q(t)$ satisfying both conditions.
(B) How many milliliters (to two decimal places) are in the body after 10 hours?
(C) How many hours (to two decimal places) will it take for only 1 milliliter of the drug to be left in the body?

46. *Simple epidemic.* A community of 1,000 individuals is assumed to be homogeneously mixed. One individual who has just returned from another community has influenza. Assume the home community has not had influenza shots and all are susceptible. One mathematical model for an influenza epidemic assumes that influenza tends to spread at a rate in direct proportion to the number who have it, N, and to the number who have not yet contracted it—in this case, $1,000 - N$. Mathematically,

$$\frac{dN}{dt} = kN(1,000 - N) \qquad N(0) = 1$$

where N is the number of people who have contracted influenza after t days. For $k = 0.0004$, it can be shown that $N(t)$

is given by

$$N(t) = \frac{1,000}{1 + 999e^{-0.4t}}$$

(A) How many people have contracted influenza after 10 days? After 20 days?
(B) How many days will it take until half the community has contracted influenza?
(C) Find $\lim_{t\to\infty} N(t)$.

47. *Nuclear accident.* One of the dangerous radioactive isotopes detected after the Chernobyl nuclear accident in 1986 was cesium-137. If 93.3% of the cesium-137 emitted during the accident is still present 3 years later, find the continuous compound rate of decay of this isotope.

48. *Insecticides.* Many countries have banned the use of the insecticide DDT because of its long-term adverse effects. Five years after a particular country stopped using DDT, the amount of DDT in the ecosystem had declined to 75% of the amount present at the time of the ban. Find the continuous compound rate of decay of DDT.

Social Sciences

49. *Archaeology.* A skull from an ancient tomb was discovered and was found to have 5% of the original amount of radioactive carbon-14 present. Estimate the age of the skull. (See Example 3.)

50. *Learning.* For a particular person learning to type, it was found that the number of words per minute, N, the person was able to type after t hours of practice was given approximately by

$$N = 100(1 - e^{-0.02t})$$

See Table 1 (limited growth) for a characteristic graph. What is the rate of improvement after 10 hours of practice? After 40 hours of practice?

51. *Small group analysis.* In a study on small group dynamics, sociologists Stephan and Mischler found that, when the members of a discussion group of 10 were ranked according to the number of times each participated, the number of times $N(k)$ the kth-ranked person participated was given approximately by

$$N(k) = N_1 e^{-0.11(k-1)} \qquad 1 \le k \le 10$$

where N_1 is the number of times the 1st-ranked person participated in the discussion. If, in a particular discussion group of 10 people, $N_1 = 180$, estimate how many times the 6th-ranked person participated. The 10th-ranked person.

52. *Perception.* One of the oldest laws in mathematical psychology is the Weber–Fechner law (discovered in the middle of the nineteenth century). It concerns a person's sensed perception of various strengths of stimulation involving weights, sound, light, shock, taste, and so on. One form of the law states that the rate of change of sensed sensation S with respect to stimulus R is inversely proportional to the strength of the stimulus R. Thus,

$$\frac{dS}{dR} = \frac{k}{R}$$

where k is a constant. If we let R_0 be the threshold level at which the stimulus R can be detected (the least amount of sound, light, weight, and so on, that can be detected), then it is appropriate to write

$$S(R_0) = 0$$

Find a function S in terms of R that satisfies the above conditions.

53. *Rumor spread.* A group of 400 parents, relatives, and friends are waiting anxiously at Kennedy Airport for a student charter flight to return after a year in Europe. It is stormy and the plane is late. A particular parent thought he had heard that the plane's radio had gone out and related this news to some friends, who in turn passed it on to others, and so on. Sociologists have studied rumor propagation and have found that a rumor tends to spread at a rate in direct proportion to the number who have heard it, x, and to the number who have not, $P - x$, where P is the total population. Mathematically, for our case, $P = 400$ and

$$\frac{dx}{dt} = 0.001x(400 - x) \qquad x(0) = 1$$

where t is time (in minutes). From this, it can be shown that

$$x(t) = \frac{400}{1 + 399e^{-0.4t}}$$

See Table 1 (logistic growth) for a characteristic graph.

(A) How many people have heard the rumor after 5 minutes? 20 minutes?

(B) Find $\lim_{t \to \infty} x(t)$.

54. *Rumor spread.* In Problem 53, how long (to the nearest minute) will it take for half of the group of 400 to have heard the rumor?

SECTION 11-4 ■ A Geometric–Numeric Introduction to the Definite Integral

- ■ AREA
- ■ RATE, AREA, AND DISTANCE
- ■ RATE, AREA, AND TOTAL CHANGE

The first three sections of this chapter were focused on the *indefinite integral*—the set of all antiderivatives of a given function. In this and the next section we introduce a related form called the *definite integral*. Our approach in this section will be intuitive and informal. In the next section these concepts will be made more precise.

We introduce the *definite integral* by considering the area bounded by the graph of a function f, the x axis, and the vertical lines $x = a$ and $x = b$, as shown in Figure 1. If $f(x)$ is positive, then we refer to this area as the area under the graph of $y = f(x)$ from $x = a$ to $x = b$.

We will denote the shaded area in Figure 1 by the **definite integral symbol:**

$$\int_a^b f(x)\, dx = \left(\begin{array}{c}\text{Area under the graph}\\ \text{from } x = a \text{ to } x = b\end{array}\right)$$

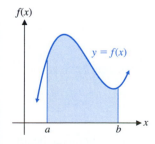

$f(x)$

$y = f(x)$

a b x

FIGURE 1
Area under the graph of $y = f(x)$

This symbol is precisely defined in Section 11-5, where we will also see how the definite integral of a function over an interval $[a, b]$ is related to the indefinite integral (antiderivative) of the function through the fundamental theorem of calculus. To keep things simple at the start, we restrict $f(x)$ to positive values, but this restriction will be relaxed shortly. We also assume all functions referred to are continuous over any interval under consideration.

Surprisingly, out of this analysis of area we will be able to find the total change in a function over an interval from its derivative. For example, given a rate function (derivative) or a table of rate values, we will be able to determine the distance that a steel ball dropped from a bridge travels in a given time interval, or the additional costs resulting from an increase in production.

■ AREA

How do we find the shaded area in Figure 2? That is, how do we find the area bounded by the graph of $f(x) = 0.25x^2 + 1$, the x axis, and the vertical lines $x = 1$ and $x = 5$? [This cumbersome description is usually shortened to "the area under the graph of $f(x) = 0.25x^2 + 1$ from $x = 1$ to $x = 5$."] Our standard geometric area formulas do not apply directly, but the formula for the area of a rectangle can be used indirectly. To see how, we look at a method of approximating the area under the graph by using rectangles that will give us any accuracy desired, which is quite different from finding the area exactly. Our first area approximation is made by dividing the interval [1, 5] on the x axis into four equal parts of length

$$\Delta x = \frac{5 - 1}{4} = 1$$

(It is customary to denote the length of the subintervals by Δx, which is read "delta x," since Δ is the Greek letter delta.) We then place a rectangle on each subinterval with a height determined by the function evaluated at the left end point of the subinterval (see Fig. 3).

Summing the areas of the rectangles in Figure 3, we obtain a **left sum** of four rectangles, denoted by L_4, as follows:

$$L_4 = f(1) \cdot 1 + f(2) \cdot 1 + f(3) \cdot 1 + f(4) \cdot 1$$
$$= 1.25 + 2.00 + 3.25 + 5 = 11.5$$

From Figure 3, since $f(x)$ is increasing, it is clear that the left sum L_4 underestimates the area, and we can write

$$11.5 = L_4 < \int_1^5 (0.25x^2 + 1)\, dx = \text{Area}$$

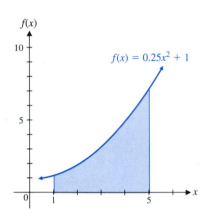

FIGURE 2
What is the shaded area?

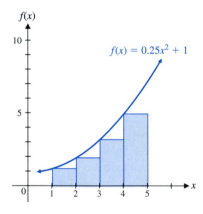

FIGURE 3
Left rectangles

explore – discuss 1

If $f(x)$ were decreasing over the interval [1, 5], would the left sum L_4 over- or underestimate the actual area under the curve? Explain.

Now suppose we use the right end point of each subinterval to obtain the height of the rectangle placed on top of it. Superimposing this result on top of Figure 3, we obtain Figure 4.

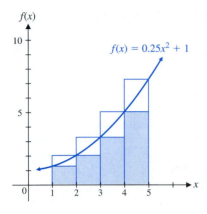

FIGURE 4
Left and right rectangles

Summing the areas of the higher rectangles in Figure 4, we obtain the **right sum** of the four rectangles, denoted by R_4, as follows (compare R_4 below with L_4 above and note that R_4 can be obtained from L_4 by deleting one rectangle area and adding one more):

$$R_4 = f(2) \cdot 1 + f(3) \cdot 1 + f(4) \cdot 1 + f(5) \cdot 1$$
$$= 2.00 + 3.25 + 5.00 + 7.25 = 17.5$$

From Figure 4, since $f(x)$ is increasing, it is clear that the right sum R_4 overestimates the area, and we conclude that the actual area is between 11.5 and 17.5. That is,

$$11.5 = L_4 < \int_1^5 (0.25x^2 + 1)\, dx < R_4 = 17.5 \tag{1}$$

Since the actual area lies between the left estimate and the right estimate, the average should be even closer:

$$\text{Average} = \frac{L_4 + R_4}{2} = \frac{11.5 + 17.5}{2} = 14.5$$

explore – discuss 2

If $f(x)$ in Figure 4 were decreasing over the interval $[1, 5]$, would the right sum R_4 over- or underestimate the actual area under the curve? Explain.

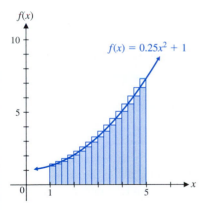

FIGURE 5

The first approximation of the area under the curve in (1) is fairly coarse, but the method outlined can be continued with increasingly accurate results by dividing the interval $[1, 5]$ into more and more equal subintervals. Of course, this is not a job for hand calculation, but a job computers are designed to do.* Figure 5 shows left and right rectangle approximations for 16 equal subdivisions.
For this case,

$$\Delta x = \frac{5 - 1}{16} = 0.25$$

$$L_{16} = f(1) \cdot \Delta x + f(1.25) \cdot \Delta x + \cdots + f(4.75) \cdot \Delta x$$
$$= 13.59$$

$$R_{16} = f(1.25) \cdot \Delta x + f(1.50) \cdot \Delta x + \cdots + f(5) \cdot \Delta x$$
$$= 15.09$$

Thus, we now know that the area under the curve is between 13.59 and 15.09. That is,

$$13.59 = L_{16} < \int_1^5 (0.25x^2 + 1)\, dx < R_{16} = 15.09 \tag{2}$$

$$\text{Average} = \frac{L_{16} + R_{16}}{2} = \frac{13.59 + 15.09}{2} = 14.34$$

* The computer software that accompanies this text will perform these calculations (see the Preface).

For 100 equal subdivisions, computer calculations give us

$$14.214 = L_{100} < \int_1^5 (0.25x^2 + 1)\, dx < R_{100} = 14.454 \tag{3}$$

$$\text{Average} = \frac{L_{100} + R_{100}}{2} = \frac{14.214 + 14.454}{2} = 14.334$$

For an approximation process to be useful, we need an estimate of the maximum possible error produced by the process. We define the **error in an approximation** to be the absolute value of the difference between the approximated value and the actual value.

e x p l o r e – d i s c u s s   3

If a and b are the coordinates of two points on a real number line, then the distance between these two points is given by $|b - a|$. The following inequalities involving distance on a real number line are very useful for estimating errors in approximations.

1. If x is between a and b on a number line, then

$$|x - a| \leq |b - a| \qquad \text{and} \qquad |x - b| \leq |b - a|$$

2. If $A = \dfrac{a + b}{2}$, then

$$|x - A| \leq \frac{1}{2}|b - a|$$

Interpret each of these inequalities both verbally and geometrically on a real number line.

A function is **monotone** over an interval $[a, b]$ if it is either increasing over $[a, b]$ or decreasing over $[a, b]$. The following important observation about monotone functions forms the basis for estimating the error when a left sum or a right sum is used to approximate an area.

If f is a monotone function, then the area under the graph of $f(x)$ always lies between the left sum L_n and the right sum R_n for any integer n.

Consider a positive monotone function $y = f(x)$, as shown in Figure 6. If $\int_a^b f(x)\, dx$ is the actual area under the graph of $y = f(x)$ from $x = a$ to $x = b$, then this area lies between L_n and R_n. Thus, approximating this area with either L_n or R_n produces an error that is less than the absolute value of the difference between the left sum and the right sum, $|R_n - L_n|$. This difference is just the sum of the areas of the white rectangles shown in Figure 6. (Think about this.)

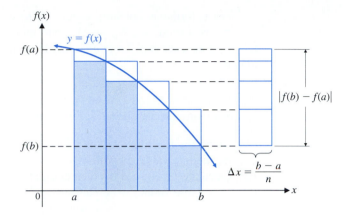

F I G U R E 6
$|R_n - L_n| = |f(b) - f(a)|\Delta x$

Thus,

$$\text{Error} \leq |R_n - L_n| = |f(b) - f(a)|\Delta x = |f(b) - f(a)|\frac{b - a}{n}$$

If the approximation is the average of left and right sums, then the bound for the error is cut in half. These results are summarized in the following box for convenient reference:

Error Bounds for Left and Right Sums and Their Average (Monotone Functions)

If $f(x)$ is monotonic on the interval $[a, b]$ and

$$I = \int_a^b f(x)\, dx, \quad L_n = \text{Left sum},$$

$$R_n = \text{Right sum}, \quad \text{and} \quad A_n = \frac{L_n + R_n}{2}$$

then the following error bounds hold:

$$|I - L_n| \leq |f(b) - f(a)|\frac{b - a}{n}$$

$$|I - R_n| \leq |f(b) - f(a)|\frac{b - a}{n}$$

$$|I - A_n| \leq |f(b) - f(a)|\frac{b - a}{2n}$$

These formulas not only enable us to compute error bounds for a particular approximation, L_n, R_n, or A_n, but more importantly, they can be used to tell us in advance how large n should be to produce an error less than a specified amount.

EXAMPLE 1 ➤ Approximating Areas   Given the function $f(x) = 9 - 0.25x^2$, we are interested in approximating the area under $y = f(x)$ from $x = 2$ to $x = 5$.

(A) Graph the function over the interval $[0, 6]$; then draw left and right rectangles for the interval $[2, 5]$ with $n = 6$.
(B) Calculate L_6, R_6, and A_6; then calculate error bounds for each.
(C) How large should n be chosen for L_n, R_n, and A_n for the approximation of $\int_2^5 (9 - 0.25x^2)\, dx$ to be within 0.05 of the true value?

Solution   (A) $\Delta x = 0.5$:

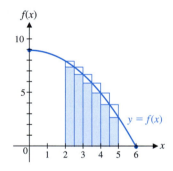

(B) $L_6 = f(2) \cdot \Delta x + f(2.5) \cdot \Delta x + f(3) \cdot \Delta x + f(3.5) \cdot \Delta x + f(4) \cdot \Delta x$
$\qquad\quad + f(4.5) \cdot \Delta x$
$\qquad = 18.53$

$R_6 = f(2.5) \cdot \Delta x + f(3) \cdot \Delta x + f(3.5) \cdot \Delta x + f(4) \cdot \Delta x + f(4.5) \cdot \Delta x$
$\qquad\quad + f(5) \cdot \Delta x$
$\qquad = 15.91$

$A_6 = \dfrac{L_6 + R_6}{2} = 17.22$

Error bound for L_6 and R_6:

$$\text{Error} \leq |f(5) - f(2)| \frac{5 - 2}{6} = |2.75 - 8|(0.5) = 2.625$$

Error bound for A_6:

$$\text{Error} \leq \frac{2.625}{2} = 1.3125$$

(C) For L_n and R_n, find n such that Error $\leqslant 0.05$:

$$|f(b) - f(a)| \frac{b-a}{n} \leqslant 0.05$$

$$|2.75 - 8| \frac{3}{n} \leqslant 0.05$$

$$15.75 \leqslant 0.05n$$

$$n \geqslant \frac{15.75}{0.05} = 315$$

For A_n, find n such that Error $\leqslant 0.05$:

$$|f(b) - f(a)| \frac{b-a}{2n} \leqslant 0.05$$

$$|2.75 - 8| \frac{3}{2n} \leqslant 0.05$$

$$7.875 \leqslant 0.05n$$

$$n \geqslant \frac{7.875}{0.05} = 157.5$$

$$n \geqslant 158 \qquad \text{Round up to the next integer.} \quad \blacktriangleleft$$

MATCHED PROBLEM 1 ➤ Given the function $f(x) = 8 - 0.5x^2$, we are interested in approximating the area under $y = f(x)$ from $x = 1$ to $x = 3$.

(A) Graph the function over the interval $[0, 4]$; then draw left and right rectangles for the interval $[1, 3]$ with $n = 4$.

(B) Calculate L_4, R_4, and A_4; then calculate error bounds for each.

(C) How large should n be chosen for L_n, R_n, and A_n for the approximation of $\int_2^5 (8 - 0.5x^2)\, dx$ to be within 0.5 of the true value? $\quad \blacktriangleleft$

If f is not monotonic, then the area may not lie between the left sum and the right sum, and our error estimates cannot be applied to the entire interval. However, we can usually divide the interval $[a, b]$ into subintervals such that f is monotonic on each, and then proceed as above (see Fig. 7).

The discussion until now assumed that the graph of $y = f(x)$ is above the x axis; that is, $f(x)$ is positive on the interval $[a, b]$. If part (or all) of the graph lies below the x axis, then for some (or all) values of x, $f(x)$ will be negative and $f(x)\,\Delta x$ will be the negative of the area of a rectangle lying below the x axis (see Fig. 8, next page). In this case, the left sum and the right sum, L_n and R_n, are the sums of the areas of the rectangles above the x axis and the negatives of the areas of the rectangles below the x axis. It is customary to refer to these as **signed areas**, since the positive quantities represent areas and the negative quantities represent negatives of areas.

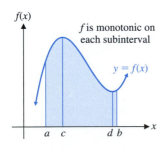

$f(x)$

f is monotonic on each subinterval

$y = f(x)$

$a \quad c \qquad d\ b$

x

FIGURE 7
A nonmonotonic function

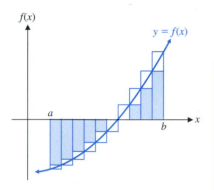

$f(x)$

$y = f(x)$

a

b

x

FIGURE 8
Signed areas

This observation leads to the following expanded interpretation of the *definite integral symbol:*

Definite Integral Symbol for Functions with Negative Values

If $f(x)$ is positive for some values of x on $[a, b]$ and negative for others, then the **definite integral symbol**

$$\int_a^b f(x)\, dx$$

represents the cumulative sum of the signed areas between the graph of $y = f(x)$ and the x axis where the areas above the x axis are counted positively and the areas below the x axis are counted negatively (see Fig. 9, where A and B are actual areas of the indicated regions).

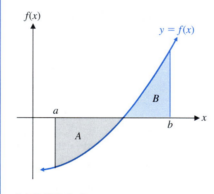

$f(x)$

$y = f(x)$

B

a

b

x

A

FIGURE 9
$$\int_a^b f(x)\, dx = -A + B$$

If a function $f(x)$ is monotonic and has both negative and positive values on the interval $[a, b]$, as in Figures 8 and 9, then the process for approximating $\int_a^b f(x)\, dx$ by L_n, R_n, or A_n, including the use of the associated error bound formulas, proceeds unchanged from that used above when $f(x)$ was restricted to positive quantities.

■ RATE, AREA, AND DISTANCE

We now turn our attention to a seemingly unrelated problem involving distance, rate, and time. In our introduction to the derivative in Chapter 8, we started with the concept

$$\text{Average rate} = \frac{\text{Total distance}}{\text{Elapsed time}} \tag{4}$$

and then developed the concept of instantaneous rate as the limit of a difference quotient:

$$\frac{ds}{dt} = \lim_{h \to 0} \frac{s(t + h) - s(t)}{h}$$

$$= \lim_{\Delta t \to 0} \frac{s(t + \Delta t) - s(t)}{\Delta t}$$

where s is the position of a moving object at time t. (The *increment notation* Δt is sometimes used in place of h in the difference quotient, and is convenient for our purpose here.) Now we are going to reverse the process. That is, we will start with a rate function, either in the form of an equation or a table, and work back to distance. Central to the process is equation (4) in the form

$$\text{Distance} = (\text{Rate})(\text{Time}) \tag{5}$$

A concrete example should make the process clear.

From physics it can be shown (neglecting air resistance) that a small steel ball dropped from a bridge (Fig. 10) will have an instantaneous rate of descent given approximately by

$$r(t) = 32t$$

where $r(t)$ is the rate in feet per second at the end of t seconds. (Note that the rate at which the ball falls increases with time due to gravity, as you would expect.) How far will the ball fall between the first and fourth second? To answer this question, we look at the area under the curve $r(t) = 32t$ from $t = 1$ to $t = 4$. But how do we get distance from area? To make the connection, we divide the interval [1, 4] into three equal parts and estimate the area under the curve using L_3 and R_3, as shown in Figure 11.

$$\text{Area of first left rectangle} = r(1)\,\Delta t$$

But

$$r(1)\,\Delta t = (\text{Rate at end of 1st second}) \times (1 \text{ second})$$
$$= \text{Distance traveled from } t = 1 \text{ to } t = 2 \text{ at a constant rate of } r(1)$$

Since $r(t)$ is increasing over the interval [1, 2], $r(1)$ is the minimum rate on this interval. Thus, $r(1)\,\Delta t$ is an underestimate of the actual distance the steel ball travels from $t = 1$ to $t = 2$.

$$\text{Area of first right rectangle} = r(2)\,\Delta t$$

But

$$r(2)\,\Delta t = (\text{Rate at end of 2nd second}) \times (1 \text{ second})$$
$$= \text{Distance traveled from } t = 1 \text{ to } t = 2 \text{ at a constant rate of } r(2)$$

Since $r(t)$ is increasing over the interval [1, 2], $r(2)$ is the maximum rate on this interval. Thus, $r(2)\,\Delta t$ is an overestimate of the actual distance the steel ball travels from $t = 1$ to $t = 2$.

FIGURE 10
Steel ball dropped from a bridge

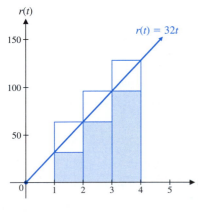

FIGURE 11
$\Delta x = 1$

The same analysis applies to the second and third rectangles, and we conclude:

$$L_3 \leq \left(\begin{array}{c} \text{Distance traveled} \\ \text{from } t = 1 \text{ to } t = 4 \end{array} \right) \leq R_3$$

Proceeding in the same way for arbitrary n, we can write

$$L_n \leq \left(\begin{array}{c} \text{Distance traveled} \\ \text{from } t = 1 \text{ to } t = 4 \end{array} \right) \leq R_n$$

This suggests that the actual area under the rate function curve $r(t) = 32t$ from $t = 1$ to $t = 4$ represents the exact distance the steel ball travels from $t = 1$ to $t = 4$; that is,

$$\int_1^4 r(t) \, dt = \left(\begin{array}{c} \text{Distance traveled} \\ \text{from } t = 1 \text{ to } t = 4 \end{array} \right)$$

In general, it can be shown that:

Rate, Area, and Distance

If $r = r(t)$ is a positive rate function for an object moving on a line, then

$$\int_a^b r(t) \, dt = \left(\begin{array}{c} \text{Net distance traveled} \\ \text{from } t = a \text{ to } t = b \end{array} \right)$$
$$= \left(\begin{array}{c} \text{Total change in position} \\ \text{from } t = a \text{ to } t = b \end{array} \right)$$

EXAMPLE 2 ➤ Distance and Area

(A) Using the rate function $r(t) = 32t$ from the above discussion, estimate the distance the ball falls from $t = 1$ to $t = 4$ using A_6. Calculate an error bound for this estimate.

(B) How large should n be so that the error in using A_n is not greater than 0.5 foot?

Solution (A) $\Delta t = \dfrac{b - a}{n} = \dfrac{4 - 1}{6} = 0.5$

$L_6 = r(1) \Delta t + r(1.5) \Delta t + r(2) \Delta t + r(2.5) \Delta t + r(3) \Delta t + r(3.5) \Delta t$
$= 216$

$R_6 = r(1.5) \Delta t + r(2) \Delta t + r(2.5) \Delta t + r(3) \Delta t + r(3.5) \Delta t + r(4) \Delta t$
$= 264$

$A_6 = \dfrac{216 + 264}{2} = 240$ feet   Approximate distance traveled from
$t = 1$ to $t = 4$

Error $\leq |r(4) - r(1)| \dfrac{4 - 1}{2 \cdot 6} = 24$ feet

Thus,

$$\begin{pmatrix} \text{Distance traveled} \\ \text{from } t = 1 \text{ to } t = 4 \end{pmatrix} = \int_1^4 r(t)\, dt = 240 \pm 24 \text{ feet}$$

[*Note:* $\int_1^4 r(t)\, dt = 240 \pm 24$ feet is a common and convenient way of representing $240 - 24 \leqslant \int_1^4 r(t)\, dt \leqslant 240 + 24$, and we will use this form where appropriate.]

(B) Solve the error bound inequality:

$$|r(4) - r(1)| \frac{4 - 1}{2n} \leqslant 0.5$$

$$\frac{144}{n} \leqslant 0.5$$

$$n \geqslant 288 \qquad \blacktriangleleft$$

MATCHED PROBLEM 2 ▶ (A) Using the rate function $r(t) = 32t$, approximate the distance traveled from $t = 2$ to $t = 5$ using A_6. Calculate an error bound for this estimate.

(B) How large should n be so that the error in using A_n is not greater than 1 foot? ◀

■ RATE, AREA, AND TOTAL CHANGE

Recall from Chapter 8 that marginal cost (or revenue or profit) is the instantaneous rate of change of cost (or revenue or profit) relative to production at a given production level. We now investigate how total change in cost is related to marginal cost and area through a concrete example.

A company in Florida manufactures cruising power boats. Marginal costs at various production levels are given in Table 1, where x is the number of boats produced per month and the marginal cost $C'(x)$ is in thousands of dollars.

TABLE 1
Marginal Costs

x	5	10	15	20	25
$C'(x)$	24	20	17	15	14

We are interested in estimating the additional cost (total change in cost) in going from a production level of 5 boats per month to 25 boats per month. We make this estimate by graphing the values in Table 1 along with left and right rectangles, as shown in Figure 12.

We assume that a continuous decreasing curve representing the marginal cost function passes through each of the Table 1 points as indicated, but we do not know the equation of this curve. (This situation occurs very frequently in real-world applications.) Nevertheless, we can still estimate the total change in cost by proceeding as we did above in finding distance, given a rate.

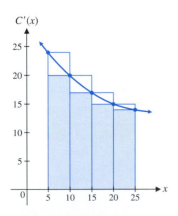

FIGURE 12
$\Delta x = 5$

Area of first left rectangle $= C'(5)\, \Delta x$

But

$$C'(5)\, \Delta x = \left(\begin{array}{c} \text{Approximate cost of producing} \\ \text{the next boat at an output} \\ \text{level of 5 boats} \end{array} \right) \times 5$$

$$= \text{Approximate additional cost of increasing}$$
$$\text{production from 5 to 10 boats per month}$$

Since $C'(x)$ is decreasing over the interval $[5, 25]$, $C'(5)$ is the maximum rate on this interval. Thus, $C'(5)\, \Delta x$ is an overestimate of the additional cost of increasing production from 5 to 10 boats per month.

Area of first right rectangle $= C'(10)\, \Delta x$

But

$$C'(10)\, \Delta x = \left(\begin{array}{c} \text{Approximate cost of producing} \\ \text{the next boat at an output} \\ \text{level of 10 boats} \end{array} \right) \times 5$$

$$= \text{Approximate additional cost of increasing}$$
$$\text{production from 5 to 10 boats per month}$$

Since $C'(x)$ is decreasing over the interval $[5, 25]$, $C'(10)$ is the minimum rate on this interval. Thus, $C'(10)\, \Delta x$ is an underestimate of the additional cost of increasing production from 5 to 10 boats per month.

The same analysis applies to the second, third, and fourth rectangles, and we conclude

$$R_4 \leq \left(\begin{array}{c} \text{Additional cost of increasing} \\ \text{production from 5 to 25 boats} \\ \text{per month} \end{array} \right) \leq L_4$$

This suggests that the actual area under the graph of the marginal cost function $y = C'(x)$—if we had the whole graph—would represent the additional cost of increasing production from 5 to 25 boats per month, and we would be able to write

$$\int_5^{25} C'(x)\, dx = \left(\begin{array}{c} \text{Additional cost of increasing} \\ \text{production from 5 to 25 boats} \\ \text{per month} \end{array} \right)$$

In general, it can be shown that:

Rate, Area, and Total Change

If $y = F'(x)$ is a rate function (derivative), then the cumulated sum of the signed areas between the graph of $y = F'(x)$ and the x axis from $x = a$ to $x = b$ represents the total net change in $F(x)$ from $x = a$ to $x = b$. Symbolically,

$$\int_a^b F'(x)\, dx = \left(\begin{array}{c} \text{Total net change in } F(x) \\ \text{from } x = a \text{ to } x = b \end{array} \right)$$

Informally, a **definite integral** produces the total change in a function from its rate of change. We are very close to a landmark result in mathematics, the *fundamental theorem of calculus*, which will be discussed in Section 11-5.

EXAMPLE 3 ➤ **Change in Cost and Area** Refer to Table 1 and Figure 12 above, and use A_4 to estimate the additional cost (total change in cost) in going from a production level of 5 boats per month to 25 boats per month. Calculate an error bound for this estimate.

Solution $\Delta x = 5$

$$L_4 = C'(5) \, \Delta x + C'(10) \, \Delta x + C'(15) \, \Delta x + C'(20) \, \Delta x$$
$$= \$380 \text{ thousand} \quad \textit{Overestimate}$$

$$R_4 = C'(10) \, \Delta x + C'(15) \, \Delta x + C'(20) \, \Delta x + C'(25) \, \Delta x$$
$$= \$330 \text{ thousand} \quad \textit{Underestimate}$$

$$A_4 = \frac{380 + 330}{2} = \$355 \text{ thousand}$$

Error bound estimate for A_4:

$$\text{Error} \leq |C'(5) - C'(25)| \frac{25 - 5}{2 \cdot 4} = \$25 \text{ thousand}$$

Thus, the additional cost (total change in cost) in going from a production level of 5 boats per month to 25 boats per month is

$$\int_5^{25} C'(x) \, dx = \$355{,}000 \pm \$25{,}000$$

◀

MATCHED PROBLEM 3 ➤ Refer to Table 1 and Figure 12 above, and use A_2 to estimate the additional cost (total change in cost) in going from a production level of 10 boats per month to 20 boats per month. Calculate an error bound for this estimate. ◀

Answers to Matched Problems   1. (A) $\Delta x = 0.5$:

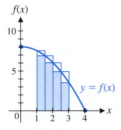

(B) $L_4 = 12.625, R_4 = 10.625, A_4 = 11.625$; Error for L_4 and $R_4 = 2$; A_4 error bound $= 1$
(C) $n > 16$ for L_n and R_n; $n > 8$ for A_n

2. (A) $\left(\begin{array}{c} \text{Distance traveled} \\ \text{from } t = 2 \text{ to } t = 5 \end{array} \right) = 336 \pm 24 \text{ ft}$ (B) $n \geq 144$

3. $\left(\begin{array}{c} \text{Additional cost of increasing} \\ \text{production from 10 to 20 boats} \\ \text{per month} \end{array} \right) = \$172{,}500 \pm \$12{,}500$

EXERCISE 11-4

A *Problems 1–10 involve estimating the area under the curves in figures (A)–(D) from x = 1 to x = 4. For each figure, divide the interval [1, 4] into three equal subintervals.*

(A)

(B)

(C)

(D)

Figure for 1–10

1. Draw in left and right rectangles for figures (A) and (B).

2. Draw in left and right rectangles for figures (C) and (D).

3. Using the results of Problem 1, compute L_3, R_3, and A_3 for figure (A) and for figure (B).

4. Using the results of Problem 2, compute L_3, R_3, and A_3 for figure (C) and for figure (D).

5. Replace the question marks with L_3 and R_3 as appropriate. Explain your choice.

$$? \leq \int_1^4 f(x)\, dx \leq ? \qquad ? \leq \int_1^4 g(x)\, dx \leq ?$$

6. Replace the question marks with L_3 and R_3 as appropriate. Explain your choice.

$$? \leq \int_1^4 u(x)\, dx \leq ? \qquad ? \leq \int_1^4 v(x)\, dx \leq ?$$

7. Compute error bounds for L_3, R_3, and A_3 found in Problem 3 for both figures.

8. Compute error bounds for L_3, R_3, and A_3 found in Problem 4 for both figures.

9. Using results from the preceding problems, interpret the following in geometric language:

$$\int_1^4 f(x)\, dx = 16.5 \pm 3.5$$

10. Using results from the preceding problems, interpret the following in geometric language:

$$\int_1^4 u(x)\, dx = 10.5 \pm 3.5$$

11. An arrow is shot vertically into the air with an initial rate of 128 feet per second. Neglecting air resistance, its rate of ascent $r(t)$ (in feet per second) at the end of t seconds is shown in the figure.

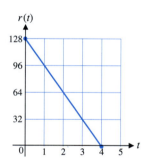

Figure for 11

(A) Estimate the height the arrow reaches using left and right sums and their average over four equal subdivisions. (Read values directly from the graph.) Calculate error bounds for these estimates.

(B) Explain how the areas of the rectangles used in the left or right sum approximations in part (A) are related to distance.

(C) How many equal subdivisions on the interval [0, 4] should be used to have A_n approximate the height the arrow reaches within 1 foot?

12. Referring to Problem 11, use left and right sums and their average using two equal subdivisions to estimate the distance the arrow travels during the first 2 seconds of flight. Estimate error bounds for each estimate.

B   *Problems 13 and 14 refer to the figure showing two parcels of land along a river.*

Figure for 13 and 14

13. You are interested in purchasing both parcels of land shown in the figure and wish to make a quick check on their combined area. There is no equation for the river frontage, so you use the average of the left and right sums of rectangles covering the area. The 1,000 foot baseline is divided into ten equal parts. At the end of each subinterval, a measurement is made from the baseline to the river, and the results are tabulated (Table 2). Let x be the distance from the left end of the baseline and let $h(x)$ be the distance from the baseline to the river at x. Estimate the combined area of both parcels using A_{10}, and calculate an error bound for this estimate. How many subdivisions of the baseline would be required so that the error in using A_n would not exceed 2,500 square feet?

TABLE 2

x	0	100	200	300	400	500
$h(x)$	0	183	235	245	260	286

x	600	700	800	900	1,000
$h(x)$	322	388	453	489	500

14. Estimate the separate area of each parcel in Problem 13 using A_5 for each, and calculate an error bound for each estimate. The baseline length of each parcel is 500 feet.

15. A car is traveling at 75 miles per hour (110 feet per second), and the driver slams on the breaks to stop. During the 7 seconds it takes to stop, the speed of the car at the end of each second is recorded by a special instrument as follows:

t	0	1	2	3	4	5	6	7
$r(t)$	110	85	63	45	29	16	5	0

where $r(t)$ is the speed (in feet per second) at the end of t seconds, $0 \leq t \leq 7$.
   (A) Estimate the distance required to stop using the average of left and right sums over seven equal subdivisions. Calculate an error bound for this estimate.
   (B) How many equal subdivisions should be used on the interval [0, 7] to have A_n approximate the stopping distance within 5 feet?

16. In Problem 15, estimate the distance traveled during the first 3 seconds after applying the brakes using the average of left and right sums over three equal subdivisions. Calculate an error bound for this estimate. How many equal subdivisions should be used on the interval [0, 3] to have A_n approximate the distance within 10 feet?

17. The figure shows the graph of a function that is not monotonic over the interval [0, 5]. We wish to estimate the area under the curve over this interval.

Figure for 17

   (A) Find a point P on the x axis that divides [0, 5] into two parts such that $f(x)$ is monotonic over each.
   (B) Divide the interval [0, 5] into five equal subintervals, and sketch left and right rectangles over the interval.
   (C) Describe the behavior of left and right rectangles to the left and to the right of P relative to under- and overestimating the true area.
   (D) Find numbers N_1 and N_2 by combining appropriate left and right rectangle sums so that

$$N_1 \leq \int_0^5 f(x)\, dx \leq N_2$$

18. Find numbers N_1 and N_2 by combining appropriate left and right rectangle sums from Problem 17 so that

$$N_1 \leq \int_1^4 f(x)\, dx \leq N_2$$

C   *Problems 19 and 20 refer to the figure.*

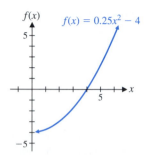

$f(x) = 0.25x^2 - 4$

Figure for 19 and 20

19. Approximate $\int_2^5 (0.25x^2 - 4)\, dx$ using L_6, R_6, and A_6. Compute error bounds for each. (Round all values to two decimal places.) Describe in geometric terms what the definite integral over the interval $[2, 5]$ represents.

20. Approximate $\int_1^6 (0.25x^2 - 4)\, dx$ using L_5, R_5, and A_5. Compute error bounds for each. (Round all values to two decimal places.) Describe in geometric terms what the definite integral over the interval $[1, 6]$ represents.

The left sum L_n or the right sum R_n is used to approximate each definite integral in Problems 21–24 to the indicated accuracy. How large must n be chosen in each case? (Each function is monotonic over the indicated interval.)

21. $\int_1^2 x^x\, dx = R_n \pm 0.05$ **22.** $\int_0^3 2^x\, dx = L_n \pm 0.5$

23. $\int_0^2 e^{-x^2}\, dx = L_n \pm 0.005$ **24.** $\int_0^1 e^{t^2}\, dt = R_n \pm 0.05$

25. If f is monotonic on the interval $[a, b]$, give an argument why we are justified in writing

$$\int_a^b f(x)\, dx = \lim_{n \to \infty} L_n$$

26. If f is monotonic on the interval $[a, b]$, give an argument why we are justified in writing

$$\int_a^b f(x)\, dx = \lim_{n \to \infty} R_n$$

▶ APPLICATIONS

Business & Economics

27. *Cost.* A company manufactures mountain bikes. The research department produced the marginal cost graph shown, where $C'(x)$ is in dollars and x is the number of bikes produced per month. Estimate the increase in cost going from a production level of 300 bikes per month to 900 bikes per month. Use the average of the left and right sums over two equal subintervals, and compute an error bound for this estimate.

Figure for 27 and 28

28. *Cost.* Referring to Problem 27, estimate the increase in cost going from a production level of 0 bikes per month to 600 bikes per month using left and right sums over two equal subintervals. Replace the question marks with the values of L_2 or R_2 as appropriate:

$$? \leq \int_0^{600} C'(x)\, dx \leq ?$$

29. *Continuous compound interest.* Ten thousand dollars is deposited in an account where the rate of change of the amount in the account is given by

$$A'(t) = 800e^{0.08t}$$

t years after the initial deposit. By how much will the account change from the end of the second year to the end of the sixth year? Estimate the amount by using left and right sums over four equal subintervals. (Calculate all quantities to the nearest dollar.) Replace the question marks with the values of L_4 or R_4 as appropriate:

$$? \leq \int_2^6 800e^{0.08t}\, dt \leq ?$$

30. *Continuous compound interest.* One thousand dollars is deposited in an account where the rate of change of the amount in the account is given by

$$A'(t) = 50e^{0.05t}$$

t years after the initial deposit. By how much will the account change from the end of the first year to the end of the fifth year? Estimate the amount by using the average of the left sum and the right sum over four equal subintervals. Calculate an error bound for this estimate. (Calculate all quantities to the nearest dollar.)

31. *Employee training.* A company producing computer components has established that, on the average, a new employee can assemble $N(t)$ components per day after t days of on-the-job training, as indicated in the following table (a new employee's productivity continuously increases with time on the job):

t	0	20	40	60	80	100	120
$N(t)$	10	51	68	76	81	84	86

Use the average of left and right sums to estimate the total number of units produced by a new employee over the first 60 days. Over the second 60 days. Use three equal subintervals for each. Calculate an error bound for each estimate.

32. *Employee training.* For a new employee in Problem 31, use left and right sums to estimate the number of units produced over the time interval [20, 100]. Replace the question marks with the values of L_4 or R_4 as appropriate:

$$? \leqslant \int_{20}^{100} N(t)\,dt \leqslant ?$$

33. *Revenue.* The research department for a market chain in a city established the following price–demand equation for a premium beer sold by the six-pack:

$$p = 8 - \frac{x}{50} \qquad 0 \leqslant x \leqslant 300 \qquad \textit{Price–demand}$$

where x is the number of six-packs sold per day at a price of p dollars each. From the price–demand equation we obtain the revenue and marginal revenue equations and their graphs shown in the figure.

(A) Interpret $\int_{100}^{200} R'(x)\,dx$ geometrically and relative to change in revenue.

(B) Approximate $\int_{100}^{200} R'(x)\,dx$ using the average of the left and right sums over four equal subintervals. Calculate an error bound for this estimate.

(C) Evaluate $R(200) - R(100)$. How does this quantity relate to part (A)?

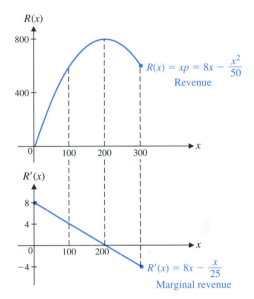

Figure for 33 and 34

34. *Revenue.* Refer to Problem 33.

(A) Interpret $\int_{100}^{300} R'(x)\,dx$ geometrically and relative to change in revenue. Guess what the value should be from the graph.

(B) Approximate $\int_{100}^{300} R'(x)\,dx$ using the average of left and right sums over four equal subintervals. Calculate an error bound for this estimate.

(C) Evaluate $R(300) - R(100)$. How does this quantity relate to part (A)?

Life Sciences

35. *Medicine.* The rate of healing $A'(t)$ (in square centimeters per day) for a certain type of abrasive skin wound is given approximately by Table 3, where $A(t)$ is the area of the wound in square centimeters after t days of healing.

TABLE 3

t	0	1	2	3	4	5	6
$A'(t)$	.90	.81	.74	.67	.60	.55	.49

t	7	8	9	10
$A'(t)$	.45	.40	.36	.33

(A) Approximate the area of the wound that will be healed in the first 5 days of healing using the left and right sums over five equal subintervals.

(B) Replace the question marks with values of L_5 and R_5 as appropriate:

$$? \leqslant \int_0^5 A'(t)\, dt \leqslant ?$$

36. *Medicine.* Refer to Problem 35. Approximate the area of the wound that will be healed in the second 5 days of healing. Use the average of the left and right sums over five equal subintervals. Calculate an error bound for this estimate.

Social Sciences

37. *Learning.* During a special study on learning, a psychologist found that, on the average, the rate of learning a list of special symbols in a code, $N'(x)$, after x days of practice was given approximately by the following table values:

x	0	2	4	6	8	10	12
$N'(x)$	29	26	23	21	19	17	15

$N(x)$ is the number of special symbols learned after x days of practice. Approximate the number of code symbols learned from the sixth to the twelfth day using the average of the left and right sums over three equal subintervals. Calculate an error bound for this estimate.

38. *Learning.* For the data in Problem 37, approximate the number of code symbols learned during the first 6 days of practice using left and right sums over three equal subintervals. Replace the question marks with values of L_3 and R_3 as appropriate:

$$? \leqslant \int_0^6 N'(x)\, dx \leqslant ?$$

SECTION   11-5   ■ **Definite Integral as a Limit of a Sum; Fundamental Theorem of Calculus**

- ■ DEFINITE INTEGRAL AS A LIMIT OF A SUM
- ■ NUMERICAL INTEGRATION
- ■ RECOGNIZING A DEFINITE INTEGRAL—AVERAGE VALUE
- ■ FUNDAMENTAL THEOREM OF CALCULUS (FTC)

The previous section presented an intuitive and informal introduction to the definite integral and the fundamental theorem of calculus. There, we used the definite integral symbol $\int_a^b f(x)\, dx$ to represent the cumulative sums of signed areas between the graph of $y = f(x)$ and the x axis from $x = a$ to $x = b$. See Figure 1, where A, B, and C represent actual areas of the respective regions.

Signed area is not the only interpretation of a definite integral. In the previous section, we also discovered that if $r(t)$ represents the velocity of an object moving on a line at time t, then $\int_a^b r(t)\, dt$ represents the net change in position from $t = a$ to $t = b$. And if $f(x) = C'(x)$, the marginal cost for producing x items, then $\int_a^b C'(x)\, dx$ represents the total change in cost going from a production level of a units to b units. Other applications of the definite integral include finding volumes, arc lengths, useful lives of vending machines, income distributions, resource depletion, income from a continuous income stream, and much more.

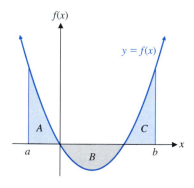

FIGURE 1
$$\int_a^b f(x)\, dx = A - B + C$$

■ **DEFINITE INTEGRAL AS A LIMIT OF A SUM**

We now make some of the concepts in the preceding section more precise. We start by giving a precise description of left and right sums: If we partition the interval $[a, b]$ into n equal subintervals of length $\Delta x = (b - a)/n$ with end points $a =$

$x_0, x_1, \ldots, x_n = b$, then, using **summation notation:**

$$L_n = \textbf{Left sum} \quad = \sum_{k=1}^{n} f(x_{k-1})\, \Delta x$$

$$= f(x_0)\, \Delta x + f(x_1)\, \Delta x + \cdots + f(x_{n-1})\, \Delta x$$

$$R_n = \textbf{Right sum} = \sum_{k=1}^{n} f(x_k)\, \Delta x$$

$$= f(x_1)\, \Delta x + f(x_2)\, \Delta x + \cdots + f(x_n)\, \Delta x$$

We are now ready to give a general definition of the *definite integral* in a form that not only covers the forms above and in the preceding section as special cases, but also opens the concept up to many other interpretations and applications.

Definition of a Definite Integral

Let f be a continuous function defined on the closed interval $[a, b]$, and let

1. $a = x_0 < x_1 < \cdots < x_{n-1} < x_n = b$
2. $\Delta x_k = x_k - x_{k-1} \quad$ for $k = 1, 2, \ldots, n$
3. $\Delta x_k \to 0 \quad$ as $n \to \infty$
4. $x_{k-1} \le c_k \le x_k \quad$ for $k = 1, 2, \ldots, n$

Then

$$\int_a^b f(x)\, dx = \lim_{n \to \infty} \sum_{k=1}^{n} f(c_k)\, \Delta x_k$$

$$= \lim_{n \to \infty} [f(c_1)\, \Delta x_1 + f(c_2)\, \Delta x_2 + \cdots + f(c_n)\, \Delta x_n]$$

is called a **definite integral** of f from a to b. The **integrand** is $f(x)$, the **lower limit** is a, and the **upper limit** is b.

In the definition of a definite integral, we divide the closed interval $[a, b]$ into n subintervals of arbitrary length in such a way that the length of each subinterval $\Delta x_k = x_k - x_{k-1}$ tends to 0 as n increases without bound. From each of the n subintervals we then select a point c_k and form the sum

$$\sum_{k=1}^{n} f(c_k)\, \Delta x_k = f(c_1)\, \Delta x_1 + f(c_2)\, \Delta x_2 + \cdots + f(c_n)\, \Delta x_n$$

which is called a **Riemann sum** [named after the celebrated German mathematician Georg Riemann (1826–1866)].

Under the conditions stated in the definition, it can be shown that the limit of the Riemann sum always exists, and it is a real number. The limit is independent of the nature of the subdivisions of $[a, b]$ as long as condition 3 holds, and it is independent of the choice of c_k as long as condition 4 holds.

The left and right sums discussed above are particularly simple Riemann sums with subintervals all the same length and c_k chosen in a regular manner as an end point of a subinterval. Either of these two sums, as well as others we will discuss below, can be used to approximate a definite integral. Generally, the larger n is, the better the approximation.

The next box lists several properties of the definite integral, which follow directly from the definition. Note that properties 3 and 4 parallel properties given earlier in Section 11-1 for the indefinite integral.

Definite Integral Properties

1. $\displaystyle\int_a^a f(x)\, dx = 0$ $\qquad\qquad\qquad\qquad\qquad\qquad \displaystyle\int_5^5 x^2\, dx = 0$

2. $\displaystyle\int_a^b f(x)\, dx = -\int_b^a f(x)\, dx$ $\qquad\qquad\quad \displaystyle\int_1^5 x^2\, dx = -\int_5^1 x^2\, dx$

3. $\displaystyle\int_a^b kf(x)\, dx = k\int_a^b f(x)\, dx,\quad k \text{ a constant}$ $\quad \displaystyle\int_2^7 3x^2\, dx = 3\int_2^7 x^2\, dx$

4. $\displaystyle\int_a^b [f(x) \pm g(x)]\, dx = \int_a^b f(x)\, dx \pm \int_a^b g(x)\, dx$

$\qquad\qquad\qquad\qquad\qquad\qquad\qquad \displaystyle\int_2^6 [x^{-1} - x^2]\, dx = \int_2^6 x^{-1}\, dx - \int_2^6 x^2\, dx$

5. $\displaystyle\int_a^b f(x)\, dx = \int_a^c f(x)\, dx + \int_c^b f(x)\, dx$ $\quad \displaystyle\int_{-1}^4 \sqrt[3]{x}\, dx = \int_{-1}^2 \sqrt[3]{x}\, dx + \int_2^4 \sqrt[3]{x}\, dx$

■ NUMERICAL INTEGRATION

It is important to remember that irrespective of its interpretation in any applied problem:

> **A definite integral can always be interpreted geometrically in terms of accumulated signed areas, just as a derivative can always be interpreted geometrically in terms of slope.**

Numerical integration is any numerical process that is used to approximate a definite integral (which is different from finding it exactly). A numerical method may be any method that approximates the accumulated signed areas bounded by the graph of the integrand and the x axis from $x = a$ to $x = b$, and may or may not be a Riemann sum relative to the integrand $f(x)$. The left and right sums (discussed above) and their average are examples of numerical integration. The left and right sums are Riemann sums relative to $f(x)$, but their average is not.

Many numerical integration techniques exist. We will add the *trapezoidal sum* and the *midpoint sum* to the three already discussed. The Group Activity at the end of this chapter presents another technique, called *Simpson's rule,* and compares its effectiveness relative to the methods considered here. We will end this section by discussing the use and limitations of the built-in numerical integration routine that is now available in most graphing calculators and many computer programs.

For a numerical integration technique to be useful, we need to have an idea of how closely it approximates the definite integral. For the left and right sums we summarize the following from the last section:

Left and Right Sums

$$I = \int_a^b f(x)\,dx \qquad L_n = \textbf{Left sum} \qquad R_n = \textbf{Right sum}$$

(A) $f(x)$ *is increasing:*

$$L_n \leq \int_a^b f(x)\,dx \leq R_n$$

R_n overestimates ⟶

L_n underestimates

(B) $f(x)$ *is decreasing:*

$$R_n \leq \int_a^b f(x)\,dx \leq L_n$$

L_n overestimates

R_n underestimates ⟶

(C) $f(x)$ *is a constant function:*

$$L_n = R_n = \int_a^b f(x)\,dx$$

(D) $f(x)$ *has both increasing and decreasing portions on [a, b]:* Divide interval $[a, b]$ into decreasing and increasing portions. Apply (A) or (B) to each subinterval as appropriate, and then combine the results.

explore – discuss 1

Give a geometric and verbal argument that when $f(x)$ is a constant function, L_n and R_n give exact results (zero error).

The average of L_n and R_n should be closer to the exact value than either L_n or R_n. Referring to Figure 2, we can see that the average A_n is generally a better estimate:

$$A_k = \frac{L_k + R_k}{2} \qquad \text{Average of areas of kth left and right rectangles}$$

$$= \frac{f(x_{k-1})\,\Delta x + f(x_k)\,\Delta x}{2}$$

$$= \frac{f(x_{k-1}) + f(x_k)}{2}\,\Delta x \qquad \text{Area of kth trapezoid (see Appendix A, Table I)}$$

Thus, the average sum A_n is the same as the sum of trapezoids fitted to the curve as shown in Figure 2B. From Figure 2 it is clear that the area of the trapezoid is closer to

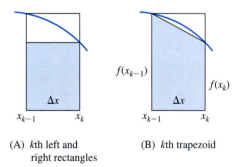

(A) kth left and right rectangles

(B) kth trapezoid

FIGURE 2
kth left and right rectangles and trapezoid

the area under the curve than the area of either the left or right rectangle. We also note that since the curve is concave down, the trapezoid underestimates the actual area. If the curve were concave up, then the trapezoid would overestimate the actual area (draw a picture). This discussion leads to the *trapezoidal sum,* or *rule,* for approximating definite integrals. If we partition the interval $[a, b]$ into n equal subintervals of length $\Delta x = (b - a)/n$ with end points $a = x_0, x_1, \ldots, x_n = b$, then we have the following:

Trapezoidal Sum (Rule):

$$T_n = A_n = \sum_{k=1}^{n} \frac{f(x_{k-1}) + f(x_k)}{2} \, \Delta x$$

$$= \frac{f(x_0) + f(x_1)}{2} \, \Delta x + \frac{f(x_1) + f(x_2)}{2} \, \Delta x + \cdots + \frac{f(x_{n-1}) + f(x_n)}{2} \, \Delta x$$

$$= [f(x_0) + 2f(x_1) + 2f(x_2) + \cdots + 2f(x_{n-1}) + f(x_n)] \frac{\Delta x}{2}$$

Figure 3 compares the use of the trapezoidal sum T_4 and the left and right sums L_4 and R_4 in approximating the definite integral $\int_0^4 [-0.25(x-4)^2 + 4] \, dx$.

 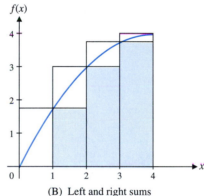

(A) Trapezoidal sum   (B) Left and right sums

FIGURE 3
Trapezoidal sum versus left and right sums

Another way of balancing the error resulting from the use of a left or right sum is to use the height of a rectangle at the midpoint of each subinterval. This leads to the *midpoint sum, M_n*. If we partition the interval $[a, b]$ into n equal subintervals of length $\Delta x = (b - a)/n$ with end points $a = x_0, x_1, \ldots, x_n = b$, then we have the following:

Midpoint Sum:

$$M_n = \sum_{k=1}^{n} f\left(\frac{x_{k-1} + x_k}{2}\right) \Delta x \quad \text{A Riemann sum}$$

$$= f\left(\frac{x_0 + x_1}{2}\right) \Delta x + f\left(\frac{x_1 + x_2}{2}\right) \Delta x + \cdots + f\left(\frac{x_{n-1} + x_n}{2}\right) \Delta x$$

Figure 4 illustrates a kth midpoint rectangle and compares it geometrically with a kth trapezoid. At a glance, it is not clear which is the better approximation. However, using simple geometry, we can show that the kth midpoint rectangle has the

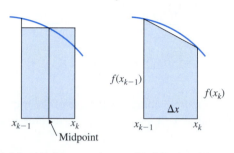

(A) kth midpoint rectangle (B) kth trapezoid

FIGURE 4

same area as another trapezoid whose top is tangent to the curve at the point of intersection of the curve and the vertical line that passes through the midpoint of the base. Figure 5A shows the tangent trapezoid (which is not the same trapezoid shown in Fig. 4B). Figure 5B shows the tangent trapezoid and the midpoint rectangle. Since the two shaded triangles at the top of the figure are congruent, they have the same area. So, in forming the tangent trapezoid, we are just taking the shaded triangle from the upper right corner of the midpoint rectangle, rotating it, and placing it on top of the upper left part of the rectangle.

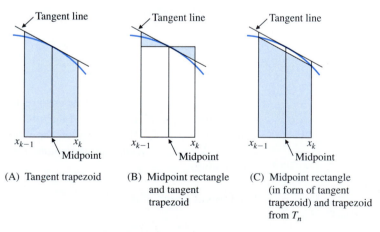

(A) Tangent trapezoid (B) Midpoint rectangle and tangent trapezoid (C) Midpoint rectangle (in form of tangent trapezoid) and trapezoid from T_n

FIGURE 5

Why did we go to all this trouble? Figure 5C shows the reason. The figure shows the relationship between the midpoint rectangle (using the equivalent tangent trapezoid) and the trapezoid related to T_n. The following important results are apparent: For a given value of n, M_n generally produces a smaller error than T_n. If the curve is concave downward, M_n overestimates and T_n underestimates; if the curve is concave upward, M_n underestimates and T_n overestimates (draw a picture). A summary of these results is shown in the following box.

Trapezoidal Sums and Midpoint Sums

$$I = \int_a^b f(x)\, dx \qquad T_n = \textbf{Trapezoidal sum} \qquad M_n = \textbf{Midpoint sum}$$

(A) *Graph of $y = f(x)$ is concave down:*

$$T_n \leq \int_a^b f(x)\, dx \leq M_n$$

Tangent line

M_n overestimates

T_n underestimates →

Midpoint

(B) *Graph of $y = f(x)$ is concave up:*

$$M_n \leq \int_a^b f(x)\, dx \leq T_n$$

T_n overestimates

Tangent line

M_n underestimates →

Midpoint

(C) *$f(x)$ is either a constant or linear function:*

$$M_n = T_n = \int_a^b f(x)\, dx$$

(D) *Graph of $y = f(x)$ has both concave up and concave down parts:* Find inflection points and divide the interval $[a, b]$ accordingly. Use (A) or (B) on each subinterval as appropriate, and then combine results.

explore – discuss 2

Give a geometric and verbal argument that when $f(x)$ is a constant or linear function, T_n and M_n give exact results (zero error).

We return to an example considered in the last section and compare an approximation using L_n and R_n with an approximation using T_n and M_n. First, we state some important assumptions:

Important Assumptions

In all the applications in this book dealing with data sets such as that in Table 1, Example 1 (below), unless stated to the contrary, we assume that a unique continuous function f exists whose graph passes through all the data points and has the increasing/decreasing and concavity properties of the graph of the data set. [We say that the graph of a data set with more than two points is concave up (down) if there exists a smooth curve through the data points that is concave up (down).] These assumptions are reasonable for all the applications we will consider.

EXAMPLE 1

Left and Right Sums versus Trapezoidal and Midpoint Sums As discussed in Section 11-4, a company in Florida manufactures cruising power boats. Marginal costs at various production levels are given in Table 1, where x is the number of boats produced per month and the marginal cost $C'(x)$ is in thousands of dollars.

We are interested in estimating the additional cost (total change in cost) in going from a production level of 5 boats per month to 25 boats per month. The total change in cost is given by

$$I = \int_5^{25} C'(x) \, dx$$

where $y = C'(x)$ is a unique continuous function that includes the data points in Table 1 on its graph.

(A) Plot the data points in a graphing utility and determine whether the graph of $y = C'(x)$ is decreasing or increasing and whether it is concave up or down. Indicate which of L_n, R_n, T_n and M_n underestimate I and which overestimate I.

(B) Find L_4 and R_4, and replace the question marks with either as appropriate:

$$? \leq \int_5^{25} C'(x) \, dx \leq ?$$

(C) Find T_2 and M_2, and replace the question marks with either as appropriate:

$$? \leq \int_5^{25} C'(x) \, dx \leq ?$$

TABLE 1
Marginal Costs

x	5	10	15	20	25
$C'(x)$	24	20	17	15	14

Solution

(A) Referring to Figure 6 and our assumptions above, we conclude that the graph of $y = C'(x)$ is decreasing and is concave up on the interval $[5, 25]$. Thus, L_n and T_n overestimate I, and R_n and M_n underestimate I.

(B) From Table 1,

$$L_4 = (24 + 20 + 17 + 15)5 = \$380 \text{ thousand} \quad \textit{Overestimate}$$

$$R_4 = (20 + 17 + 15 + 14)5 = \$330 \text{ thousand} \quad \textit{Underestimate}$$

Thus,

$$R_4 = \$330 \text{ thousand} \leq \int_5^{25} C'(x) \, dx \leq \$380 \text{ thousand} = L_4$$

(C) From Table 1,

$$T_2 = (24 + 2 \cdot 17 + 14)\tfrac{10}{2} = \$360 \text{ thousand} \quad \textit{Overestimate}$$

$$M_2 = (20 + 15)10 = \$350 \text{ thousand} \quad \textit{Underestimate}$$

Thus,

$$M_2 = \$350 \text{ thousand} \leq \int_5^{25} C'(x) \, dx \leq \$360 \text{ thousand} = T_2 \qquad \blacktriangleleft$$

Note that even with just two subdivisions, the trapezoidal and midpoint estimates are much better than the left and right estimates with four subdivisions. And with no more computation!

MATCHED PROBLEM 1 ➤ Repeat Example 1 with Table 1 replaced by Table 2:

TABLE 2
Marginal Costs

x	5	10	15	20	25
$C'(x)$	27	26	24	21	17

$\blacktriangleleft$

Any of the approximation sums described above can be made more accurate by taking n larger or by increasing the number of data points. Over regions where a function is monotonic or does not change in concavity, we can tell how close our approximation is to the true value. If an approximation is not close enough, we increase n until the desired accuracy is achieved. (There are general formulas for error bounds for each of the approximation sums discussed above, but they involve the first or second derivative of the integrand, which is not very helpful if we do not have an equation for the integrand.)

Unless stated to the contrary, we will use the numerical integration methods described above when we have a table of data points representing specific values of a continuous integrand or if we have a continuous graph of data without an explicit equation representing an integrand. Both situations are frequently encountered in real-world applications.

30

0   30

0

FIGURE 6

If the integrand in a definite integral is represented by a continuous function given explicitly by an equation, then a **numerical integration routine,** preprogrammed in many graphing calculators and certain mathematical software packages, yields accurate results very quickly. (Consult the user's manual for your own graphing calculator or computer for specific details.) The following format is used in several popular graphing utilities, and this is the format that will be illustrated and used in this book:

Format for Numerical Integration Routine

---

fnInt(expression,variable,lower,upper)*

---

* Default error setting is $\pm (1 \times 10^{-5})$

From now on, unless stated to the contrary, *numerical integration routine* will refer to a built-in routine in a graphing calculator or computer software. The other numerical integration processes discussed above will be referred to by name or symbol.

EXAMPLE 2 ➤ Using a Numerical Integration Routine   Approximate each of the following definite integrals (to three decimal places) using a numerical integration routine:

(A) $\int_{0}^{8} [-0.25(x-4)^2 + 4]\, dx$ (B) $\int_{-1.5}^{2} x^3\, dx$ (C) $\int_{-1}^{2} e^{-x^2}\, dx$

Solution   (A) $\int_{0}^{8} [-0.25(x-4)^2 + 4]\, dx = 21.333$

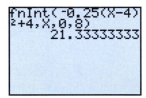

FIGURE 7

(B) $\int_{-1.5}^{2} x^3\, dx = 2.734$

FIGURE 8

(C) $\int_{-1}^{2} e^{-x^2} dx = 1.629$

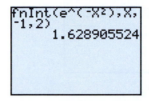

```
fnInt(e^(-X²),X,
-1,2)
          1.628905524
```

FIGURE 9 ◀

MATCHED PROBLEM 2 ➤ Approximate each of the following definite integrals (to three decimal places) using a numerical integration routine:

(A) $\int_{-1}^{4} (x^2 - 2x) \, dx$ ◀ (B) $\int_{-3}^{8} \sqrt[3]{x} \, dx$ (C) $\int_{1.2}^{10} \frac{x}{\ln x} \, dx$ ◀

Another approach to problems involving integrands that are not given in equation form (such as in Example 1) is to fit a regression curve to the data set, and then use a numerical integration routine. The process is often quick and easy, but suffers from not having a convenient error estimate for the result. This approach is illustrated in Example 3.

▶ EXAMPLE 3 ➤ Using a Regression Equation for a Data Set

(A) Fit a quadratic regression equation (see Section 2-4) to the data set in Example 1, and graph it and the data set in the same viewing window.

(B) Use the quadratic regression equation from part (A) and fnInt to approximate $\int_{5}^{25} C'(x) \, dx$ to the nearest thousand dollars. (Compare the results with those in Example 1.)

Solution (A) See Figure 10.

(A)

(B)

(C)

FIGURE 10

(B) Total change in cost $= \int_5^{25} C'(x)\, dx =$ \$353 thousand (Fig. 11).

MATCHED PROBLEM 3 ▶   (A) Fit a quadratic regression equation (see Section 2-4) to the data set in Matched Problem 1, and graph it and the data set in the same viewing window.

(B) Use the quadratic regression equation from part (A) and fnInt to approximate $\int_5^{25} C'(x)\, dx$ to the nearest thousand dollars. (Compare the results with those in Matched Problem 1.)   ◀

■ RECOGNIZING A DEFINITE INTEGRAL—AVERAGE VALUE

Suppose the temperature F (in degrees Fahrenheit) in the middle of a small shallow lake from 8 AM ($t = 0$) to 6 PM ($t = 10$) during a day in June is given approximately as shown in Figure 12.

How can we compute the average temperature from 8 AM to 6 PM, the average of a continuous function over the interval [0, 10]? We know that the average of a finite number of values $a_1, a_2, \ldots, a_n$ is given by

$$\text{Average} = \frac{a_1 + a_2 + \cdots + a_n}{n}$$

Thus, it would seem reasonable to divide the time interval [0, 10] into n equal subintervals, compute the temperature at a point in each subinterval, and then use the average of these values as an approximation of the average value of the continuous function $y = F(t)$ over [0, 10]. We would expect the approximations to improve as n increases, and we are led to define the limit of the average of n values as $n \to \infty$ to be average temperature over the time interval [0, 10]. To compute this limit we proceed as follows:

$$\begin{pmatrix} \text{Average} \\ \text{temperature} \\ \text{for } n \text{ values} \end{pmatrix} = \frac{1}{n}\,[F(t_1) + F(t_2) + \cdots + F(t_n)] \qquad (1)$$

Form (1) looks something like a Riemann sum, but we are missing Δt. If we let $\Delta t = (b - a)/n$, then (1) can be written in the form (which you should confirm):

$$\begin{pmatrix} \text{Average} \\ \text{temperature} \\ \text{for } n \text{ values} \end{pmatrix} = \frac{1}{b - a}\,[F(t_1)\,\Delta t + F(t_2)\,\Delta t + \cdots + F(t_n)\,\Delta t]$$

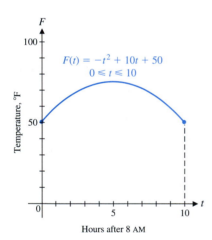

$F(t) = -t^2 + 10t + 50$
$0 \leqslant t \leqslant 10$

Hours after 8 AM

FIGURE 12

The expression within the square brackets is a Riemann sum, and its limit as $n \to \infty$ is the definite integral of $F(t)$ over the interval $[0, 10]$. Thus, using a numerical integration routine, the average temperature (to the nearest degree) is given by

$$\begin{pmatrix} \text{Average} \\ \text{temperature over} \\ [a, b] = [0, 10] \end{pmatrix} = \frac{1}{b - a} \int_a^b F(t)\, dt$$

$$= \frac{1}{10} \int_0^{10} (-t^2 + 10t + 50)\, dt$$

$$= 67°\text{F}$$

Proceeding as above for an arbitrary continuous function f over an interval $[a, b]$, we obtain the following general formula:

Average Value of a Continuous Function f Over $[a, b]$

$$\frac{1}{b - a} \int_a^b f(x)\, dx$$

■ FUNDAMENTAL THEOREM OF CALCULUS (FTC)

We are now ready to present a very important result in mathematics, **the fundamental theorem of calculus,** which we abbreviate by **FTC.** We start by returning to the marginal cost discussed above.

If $C'(x)$ represents marginal cost at an output of x units, then $\int_a^b C'(x)\, dx$ represents the total change in cost as production changes from $x = a$ units to $x = b$ units (see Fig. 13). We can also get the total change in cost by using $C(b) - C(a)$, where $C(x)$ is the cost function, an antiderivative of the marginal cost function $C'(x)$. Thus, it appears that we can write

$$\int_a^b C'(x)\, dx = C(b) - C(a)$$

It turns out that this remarkable result is true under rather general conditions, which we now state:

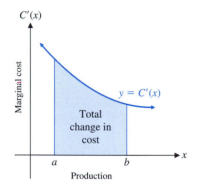

FIGURE 13
$\int_a^b C'(x)\, dx = $ Total change in cost

Fundamental Theorem of Calculus (FTC)

If f is a continuous function on the closed interval $[a, b]$ and F is *any* antiderivative of f, then

$$\int_a^b f(x)\, dx = F(x) \Big|_a^b = F(b) - F(a) \qquad F'(x) = f(x)$$

(The definite integral of a rate gives net change.)

Why is the fundamental theorem *fundamental?* Because it makes a connection between the derivative and the integral, the two basic concepts in calculus. If we can find an antiderivative of $f(x)$, then we can evaluate $\int_a^b f(x)\,dx$, the limit of a Riemann sum over the interval $[a, b]$, exactly by evaluating the antiderivative of $f(x)$ at the end points a and b and taking the difference. The symbol $F(x)|_a^b$ is used to represent the **net change** in $F(x)$ from $x = a$ to $x = b$, and is included as a convenient intermediate step in the evaluation of a definite integral by the fundamental theorem. It is important to remember that:

Any antiderivative of $f(x)$ can be used in the fundamental theorem. One generally chooses the simplest antiderivative by letting $C = 0$, since any value of C will drop out when computing the difference $F(b) - F(a)$.

e x p l o r e – d i s c u s s 3

It can be shown that $F'(x) = f(x)$, where

$$F(x) = \int_a^x f(t)\,dt \qquad a \leqslant x \leqslant b$$

Assuming this result, use properties of a definite integral to show that

$$\int_a^b f(t)\,dt = F(b) - F(a)$$

We now use the FTC to evaluate

$$\int_0^5 \frac{x}{x^2 + 10}\,dx$$

to three decimal places. The simplest antiderivative of the integrand is $\frac{1}{2}\ln|x^2 + 10|$, which can be replaced by $\frac{1}{2}\ln(x^2 + 10)$, since $x^2 + 10 > 0$ for $0 \leqslant x \leqslant 5$. Thus,

$$\int_0^5 \frac{x}{x^2 + 10}\,dx = \frac{1}{2}\ln(x^2 + 10)\,\bigg|_0^5 \qquad \text{By the FTC}$$

$$= \frac{1}{2}\ln 35 - \frac{1}{2}\ln 10 = 0.626$$

Historically, the FTC was the main tool used in evaluating definite integrals. However, now we have a new tool—applying a numerical integration routine (fnInt), we get:

$$\int_0^5 \frac{x}{x^2 + 10}\,dx = 0.626$$

A numerical integration routine gives results even when the FTC cannot be used—for example, when the integrand does not have an elementary antiderivative or it is too difficult to find. The integrands in the following two definite integrals (from Example 2 and Matched Problem 2) do not have elementary antiderivatives, but are easy to approximate using a numerical integration routine:

$$\int_{-1}^{2} e^{-x^2}\, dx \qquad \int_{1.2}^{10} \frac{x}{\ln x}\, dx$$

Thus:

> **Unless directed to the contrary, if the integrand in a definite integral is given by an equation representing a continuous function over constant limits of integration, use a numerical integration routine (fnInt) to approximate the integral.**

But what if one of the limits of integration is not a constant? For example, consider the following function defined by a definite integral:

$$A(x) = \int_{1}^{x}(t^2 - 4t + 5)\, dt \qquad 1 \leq x \leq 5$$

Geometrically, A is an area function that gives the area under the graph of $y = t^2 - 4t + 5$ from $t = 1$ to $t = x$ (see Fig. 14).

To write an explicit equation for the area function $A(x)$ not involving the definite integral, we have to use the FTC as follows (a numerical integration routine is not helpful here):

$$A(x) = \int_{1}^{x}(t^2 - 4t + 5)\, dt = \left(\frac{t^3}{3} - 2t^2 + 5t\right)\Big|_{1}^{x}$$

$$= \left(\frac{x^3}{3} - 2x^2 + 5x\right) - \left(\frac{1^3}{3} - 2\cdot 1^2 + 5\cdot 1\right)$$

$$= \frac{x^3}{3} - 2x^2 + 5x - \frac{10}{3}$$

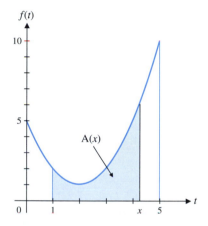

f(*t*)

A(x)

FIGURE 14

The FTC is used in this way in many practical problems. The next example is a case in point.

EXAMPLE 4 ▶ **Change in Profit** A company manufactures and wholesales x hand calculators per day. The daily marginal profit (in dollars) is given by

$$P'(x) = -0.002x + 8$$

The company is currently manufacturing 2,000 calculators per day, but is planning to increase production.

(A) Find a function $F(x)$ that represents the change in profit from a production level of 2,000 calculators a day to x thousand calculators per day—that is, $2{,}000 \leq x \leq 6{,}000$.

(B) Graph $y = F(x)$, and find the production level that produces the maximum increase in profit. Verbally describe the results.

Solution   (A) For $2{,}000 \le x \le 6{,}000$, the change in profit function is given by

$$F(x) = \int_{2,000}^{x} P'(t)\, dt \quad \text{Use the FTC.}$$

$$= \int_{2,000}^{x} (-0.002t + 8)\, dt = (-0.001t^2 + 8t)\Big|_{2,000}^{x}$$

$$= (-0.001x^2 + 8x) - [-0.001(2{,}000)^2 + 8(2{,}000)]$$

$$= -0.001x^2 + 8x - 12{,}000 \qquad 2{,}000 \le x \le 6{,}000$$

(B) We use the maximum routine on a graphing utiity, as shown in the margin. Thus, by increasing production from 2,000 to 4,000 calculators per day, the company will realize the greatest gain in daily profit—a gain of $4,000 per day. (The gain in profit increases and then decreases as production goes up, because the price of the calculator must be reduced in order to sell the increased production. This fact is hidden in the marginal profit equation.) ◀

MATCHED PROBLEM 4 ▶   For the company in Example 4, the daily marginal revenue (in dollars) is given by

$$R'(x) = -0.002x + 10$$

The company is currently manufacturing 2,000 calculators per day, but is planning to increase production.

(A) Find a function $G(x)$ that represents the total change in revenue for $2{,}000 \le x \le 6{,}000$.

(B) Graph $y = G(x)$, and find the production level that produces the maximum increase in revenue. Verbally describe the results. ◀

Answers to Matched Problems   1. (A) The graph of $y = C'(x)$ is decreasing and concave down; L_n and M_n overestimate I; R_n and T_n underestimate I

(B) $R_4 = \$440$ thousand $\le \int_5^{25} C'(x)\, dx \le \$490$ thousand $= L_4$

(C) $T_2 = \$460$ thousand $\le \int_5^{25} C'(x)\, dx \le \$470$ thousand $= M_2$

2. (A) 6.667   (B) 8.755   (C) 30.157

3. (A)

 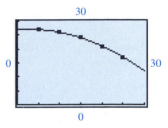

(B) Total change in cost $= \int_5^{25} C'(x)\, dx =$ \$467 thousand

```
fnInt(Y₁,X,5,25)
          466.6666667
```

4. (A) $G(x) = -0.001x^2 + 10x - 16{,}000, \quad 2{,}000 \le x \le 6{,}000$

(B) By increasing production from 2,000 to 5,000 calculators per day, the company will realize the greatest gain in daily revenue, a gain of \$9,000 per day. (Note the difference between this result and the answer to Example 4B.)

EXERCISE 11-5

A *Problems 1–6 refer to the following figure with the indicated areas:*

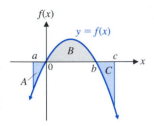

$f(x)$

$y = f(x)$

Area $A = 2.33$
Area $B = 10.67$
Area $C = 5.63$

Figure for 1–6

1. $\int_a^0 f(x)\, dx = \,?$

2. $\int_b^c f(x)\, dx = \,?$

3. $\int_a^b f(x)\, dx = \,?$

4. $\int_0^c f(x)\, dx = \,?$

5. $\int_0^b \dfrac{f(x)}{10}\, dx = \,?$

6. $\int_0^b 10 f(x)\, dx = \,?$

Evaluate each definite integral in Problems 7–14 to three decimal places using the FTC, and confirm each answer with a numerical integration routine.

7. $\int_2^3 2x\, dx$

8. $\int_1^2 3x^2\, dx$

9. $\int_0^4 (3x^2 - 4)\, dx$

10. $\int_0^2 (6x^2 - 2x)\, dx$

11. $\displaystyle\int_{0}^{1} e^{2x}\, dx$

12. $\displaystyle\int_{-1}^{1} e^{5x}\, dx$

13. $\displaystyle\int_{1}^{3.5} 2x^{-1}\, dx$

14. $\displaystyle\int_{1}^{2} \frac{dx}{x}$

In Problems 15–20, write each function in a form that is free of any definite integrals.

15. $F(x) = \displaystyle\int_{2}^{x} 2t\, dt$

16. $G(x) = \displaystyle\int_{1}^{x} 3t^2\, dt$

17. $g(x) = \displaystyle\int_{0}^{x} (3u^2 - 4)\, du$

18. $f(x) = \displaystyle\int_{0}^{x} (6u^2 - 2u)\, du$

19. $r(t) = \displaystyle\int_{0}^{t} e^{2x}\, dx$

20. $g(t) = \displaystyle\int_{-1}^{t} e^{5x}\, dx$

21. A plot of land along a river is shown in the figure. To estimate its area A, measurements (in meters) were made as indicated. Estimate the area of the plot using a trapezoidal sum and a midpoint sum in the form $T_2 \leq A \leq M_2$ or $M_2 \leq A \leq T_2$, as appropriate. Calculate T_2 and M_2 to the nearest square meter.

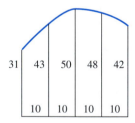

Figure for 21

22. Another plot of land along the same river as in Problem 21 is shown in the figure. Estimate the area A of the plot using a trapezoidal sum and a midpoint sum in the form $T_2 \leq A \leq M_2$ or $M_2 \leq A \leq T_2$, as appropriate. Calculate T_2 and M_2 to the nearest square meter.

Figure for 22

B *Problems 23–26 refer to the figure for Problems 1–6 and the indicated areas.*

23. $\displaystyle\int_{b}^{0} f(x)\, dx = ?$

24. $\displaystyle\int_{0}^{a} f(x)\, dx = ?$

25. $\displaystyle\int_{c}^{0} f(x)\, dx = ?$

26. $\displaystyle\int_{b}^{a} f(x)\, dx = ?$

Evaluate each definite integral in Problems 27–34 (to three decimal places) using a numerical integration routine.

27. $\displaystyle\int_{2}^{3} 12(x^2 - 4)^5 x\, dx$

28. $\displaystyle\int_{0}^{1} 32(x^2 + 1)^7 x\, dx$

29. $\displaystyle\int_{-5}^{10} e^{-0.05t}\, dt$

30. $\displaystyle\int_{-10}^{25} e^{-0.01t}\, dt$

31. $\displaystyle\int_{-1}^{7} \frac{u}{\sqrt{u+2}}\, du$

32. $\displaystyle\int_{0}^{3} s\sqrt{s+1}\, ds$

33. $\displaystyle\int_{-1}^{2} 0.2e^{-0.3x^2}\, dx$

34. $\displaystyle\int_{2}^{3} \frac{x+1}{\ln(x+2)}\, dx$

In Problems 35–40, write each function in a form that is free of any definite integrals.

35. $f(x) = \displaystyle\int_{2}^{x} 12(t^2 - 4)^5 t\, dt$

36. $g(x) = \displaystyle\int_{0}^{x} 32(t^2 + 1)^7 t\, dt$

37. $G(x) = \displaystyle\int_{5}^{x} e^{0.05t}\, dt$

38. $F(x) \displaystyle\int_{10}^{x} e^{0.01t}\, dt$

39. $m(x) = \displaystyle\int_{2}^{x} \frac{t}{t^2 - 9}\, dt$

40. $n(x) = \displaystyle\int_{0}^{x} \frac{s}{\sqrt{s^2 + 1}}\, ds$

41. Each of the following integrals represents a single function, a set of functions, or a real number. Explain which one each integral represents.

(A) $\displaystyle\int_{a}^{b} f(x)\, dx$

(B) $\displaystyle\int_{a}^{x} f(t)\, dt, \; a \leq x \leq b$

(C) $\displaystyle\int f(x)\, dx$

42. Explain the difference between a definite integral and an indefinite integral in terms of real numbers and functions.

In Problems 43–50:

(A) *Find the average value of each function over the indicated interval.*

(B) *Graph the function and its average value over the indicated interval in the same viewing window.*

43. $f(x) = 500 - 50x$; $[0, 10]$

44. $g(x) = 2x + 7$; $[0, 5]$

45. $f(t) = 3t^2 - 2t$; $[-1, 2]$

46. $g(t) = 4t - 3t^2$; $[-2, 2]$

47. $f(x) = \sqrt[3]{x}$; $[1, 8]$

48. $g(x) = \sqrt{x + 1}$; $[3, 8]$

49. $f(x) = 4e^{-0.2x}$; $[0, 10]$

50. $f(x) = 64e^{0.08x}$; $[0, 10]$

51. In Example 2 in Section 11-4 we used the rate function $r(t) = 32t$ and left and right sums to estimate the distance that a steel ball dropped from a bridge falls over the time interval $[1, 4]$. Find this distance by evaluating a definite integral.

52. Repeat Problem 51 for the time interval $[2, 5]$.

53. The sums L_n, R_n, T_n, and M_n are used to approximate the definite integral $\int_a^b f(x)\,dx$, where the graph of f is shown in the figure. The values of the approximations, 6.63, 5.89, 5.78, 6.01, got mixed up. Associate an appropriate value with each approximation.

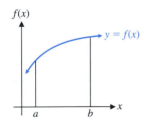

$f(x)$

$y = f(x)$

a b x

Figure for 53

54. The sums L_n, R_n, T_n, and M_n are used to approximate the definite integral $\int_a^b g(s)\,ds$, where the graph of g is shown in the figure. The values of the approximations, 2.74, 3.02, 2.83, 2.66, got mixed up. Associate an appropriate value with each approximation.

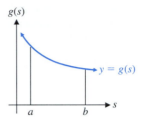

$g(s)$

$y = g(s)$

a b s

Figure for 54

55. A rocket blasts off vertically from a launching pad and explodes 8 seconds after lift-off. A black box is recovered that has recorded the velocity, $v(t)$, at the end of each second of flight, as shown in Table 3.

TABLE 3

t (sec)	0	1	2	3	4	5	6	7	8
$v(t)$ (ft/sec)	0	21	53	89	132	194	264	340	428

(A) Graph the data set from Table 3 in a graphing utility. (Save the data set in the graphing utility for Problem 57.)

(B) Represent H, the height of the rocket at the time of the explosion, symbolically by a definite integral. Replace the question marks in

$$? \leq H \leq ?$$

with T_4 and M_4, as appropriate, and explain why you made the choice. [*Hint:* Use part (A).]

(C) Estimate the height H at the time of the explosion using the form $T_4 \leq H \leq M_4$ or $M_4 \leq H \leq T_4$, as appropriate. Calculate T_4 and M_4 to the nearest foot.

56. The manufacturer of a powerful new sports car tested its acceleration from a standing start. Speed, $r(t)$, was recorded on a special instrument at the end of each of the first 6 seconds, as shown in Table 4.

TABLE 4

t (sec)	0	1	2	3	4	5	6
$r(t)$ (mi/hr)	0	5	12	20	30	44	60
$r(t)$ (ft/sec)	0	7.335	17.604	29.34	44.01	64.548	88.02

(A) Graph the data set represented by the first and third rows of Table 4 in a graphing utility. (Save the data set in the graphing utility for Problem 58.)

(B) Represent D, the distance (in feet) that the car travels during the first 6 seconds, symbolically by a definite integral. Replace the question marks in

$$? \leq D \leq ?$$

with T_3 and M_3, as appropriate, and explain why you made the choice. [*Hint:* Use part (A).]

(C) Estimate the distance D that the car travels in the first 6 seconds using the form $T_3 \leq D \leq M_3$ or $M_3 \leq D \leq T_3$, as appropriate. Calculate T_3 and M_3 to the nearest foot.

57. Refer to Problem 55.

(A) Find a quadratic regression equation for the data set in Table 3, and plot it and the data set in the same viewing window.

(B) Use the regression equation and fnInt to estimate (to the nearest foot) the height of the rocket at the time of the explosion.

58. Refer to Problem 56.

(A) Find a quadratic regression equation for the data set represented by rows 1 and 3 in Table 4, and plot it and the data set in the same viewing window.

(B) Use the regression equation and fnInt to estimate (to the nearest foot) the distance that the car traveled by the end of the sixth second.

C *Write Problems 59–62 in the form $\int_a^b f(x)\,dx$ and evaluate using the fundamental theorem of calculus.*

59. $\lim\limits_{n\to\infty} [(1 - c_1^2)\,\Delta x + (1 - c_2^2)\,\Delta x + \cdots + (1 + c_n^2)\,\Delta x]$,

where $\Delta x = \dfrac{5 - 2}{n}$ and $c_k = 2 + k\dfrac{3}{n}$, $k = 1, 2, \ldots, n$.

60. $\lim\limits_{n\to\infty} [(c_1^2 - 3)\,\Delta x + (c_2^2 - 3)\,\Delta x + \cdots + (c_n^2 - 3)\,\Delta x]$,

where $\Delta x = \dfrac{10 - 0}{n}$ and $c_k = 0 + k\dfrac{10}{n}$,

$k = 1, 2, \ldots, n$.

61. $\lim\limits_{n\to\infty} [(3c_1^2 - 2c_1 + 3)\,\Delta x + (3c_2^2 - 2c_2 + 3)\,\Delta x +$

$\cdots + (3c_n^2 - 2c_n + 3)\,\Delta x]$, where $\Delta x = \dfrac{12 - 2}{n}$ and

$c_k = 2 + k\dfrac{10}{n}$, $k = 1, 2, \ldots, n$.

62. $\lim\limits_{n\to\infty} [(4c_1^3 + 3c_1^2 - 5)\,\Delta x + (4c_2^3 + 3c_2^2 - 5)\,\Delta x +$

$\cdots + (4c_n^3 + 3c_n^2 - 5)\,\Delta x]$, where $\Delta x = \dfrac{2 - 1}{n}$ and

$c_k = 1 + k\dfrac{1}{n}$, $k = 1, 2, \ldots, n$.

63. If $\int_0^x f(t)\,dt = e^{2x} - 1$, find $f(0)$.

64. If $\int_0^x f(t)\,dt = x^3 - 2x^2$, find $f(3)$.

Problems 65 and 66 refer to the following discussion: In a more advanced treatment of exponential and logarithmic functions, the natural logarithmic function is defined in terms of a definite integral as follows:

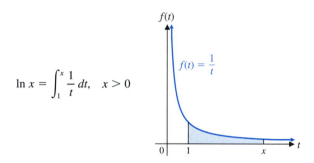

$$\ln x = \int_1^x \frac{1}{t}\,dt, \quad x > 0$$

Round all calculations to four decimal places.

65. (A) Approximate

$$\ln 2 = \int_1^2 \frac{1}{t}\,dt$$

using the midpoint sum M_5 and the trapezoidal sum T_5, and write the answer in the form

$$M_5 \leq \ln 2 = \int_1^2 \frac{1}{t}\,dt \leq T_5 \quad \text{or}$$

$$T_5 \leq \ln 2 = \int_1^2 \frac{1}{t}\,dt \leq M_5$$

as appropriate.

(B) Approximate

$$\ln 2 = \int_1^2 \frac{1}{t}\,dt$$

using a numerical integration routine.

(C) Evaluate ln 2 directly on a calculator.

66. (A) Approximate

$$\ln 1.6 = \int_1^{1.6} \frac{1}{t}\,dt$$

using the midpoint sum M_3 and the trapezoidal sum T_3, and write the answer in the form

$$M_3 \leqslant \ln 1.6 = \int_1^{1.6} \frac{1}{t}\, dt \leqslant T_3 \quad \text{or}$$

$$T_3 \leqslant \ln 1.6 = \int_1^{1.6} \frac{1}{t}\, dt \leqslant M_3$$

as appropriate.

(B) Approximate

$$\ln 1.6 = \int_1^{1.6} \frac{1}{t}\, dt$$

using a numerical integration routine.

(C) Evaluate ln 1.6 directly on a calculator.

 APPLICATIONS

Business & Economics

67. *Employee training.* A company producing computer components has established that, on the average, a new employee can assemble $N(t)$ components per day after t days of on-the-job training, as indicated in Table 5. (A new employee's productivity usually increases with time on the job up to a leveling-off point.) We are interested in estimating the number of units produced during the first 120 days by a new employee.

TABLE 5

t	0	20	40	60	80	100	120
$N(t)$	10	51	68	76	81	84	85

(A) Graph the data set in a graphing utility.

(B) If T_3 and M_3 are used for the estimate, indicate which underestimates and which overestimates the new employee's production, and how you know.

(C) Estimate the new employee's production in the first 120 days on the job, and write the answer in the form

$$T_3 \leqslant \int_0^{120} N(t)\ dt \leqslant M_3 \quad \text{or}$$

$$M_3 \leqslant \int_0^{120} N(t)\ dt \leqslant T_3$$

as appropriate.

68. *Employee training.* Repeat Problem 67 for the first 80 days using T_2 and M_2.

69. *Cost.* A company manufactures mountain bikes. The research department produced the following marginal cost function:

$$C'(x) = 500 - \frac{x}{3} \qquad 0 \leqslant x \leqslant 900$$

where $C'(x)$ is in dollars and x is the number of bikes produced per month. Compute the increase in cost going from a production level of 300 bikes per month to 900 bikes per month. Set up a definite integral and evaluate.

70. *Cost.* Referring to Problem 69, compute the increase in cost going from a production level of 0 bikes per month to 600 bikes per month. Set up a definite integral and evaluate.

71. *Average cash reserves.* Suppose cash reserves (in thousands of dollars) are approximated by

$$C(x) = 1 + 12x - x^2 \qquad 0 \leqslant x \leqslant 12$$

where x is the number of months after the first of the year.
(A) What is the average cash reserve for the first quarter?
(B) Graph the cash reserve function and its average for the first quarter in the same viewing window.

72. *Average cash reserves.* Repeat Problem 71 for the second quarter.

73. *Change in revenue.* A company manufactures and wholesales garden hoses. The company has a current production level of 400 hoses per day and wishes to increase production. The marginal revenue (in dollars per hose) at a production level of x hoses per day is $R'(x) = 30 - 0.9\sqrt{x}$.
(A) Find a function $F(x)$ that represents the change in revenue from a production level of 400 hoses per day to x hoses per day for $400 \leqslant x \leqslant 2{,}000$.

(B) Graph $y = F(x)$ and find the production level (to the nearest unit) that produces the maximum increase in revenue (to the nearest dollar). Verbally describe the results.

74. *Change in revenue.* A company manufactures and wholesales clock radios. The company has a current production level of 1,000 clock radios per week and wishes to increase production. The marginal revenue (in dollars per clock radio) at a production level of x clock radios per week is $R'(x) = 100 - 0.05x$.

(A) Find a function $G(x)$ that represents the change in revenue from a production level of 1,000 clock radios per week to x clock radios per week for $1,000 \leqslant x \leqslant 3,000$.

(B) Graph $y = G(x)$ and find the production level (to the nearest unit) that produces the maximum increase in revenue (to the nearest dollar). Verbally describe the results.

75. *Salvage value.* A new piece of industrial equipment will depreciate in value rapidly at first, then less rapidly as time goes on. Suppose the rate (in dollars per year) at which the book value of a new milling machine changes is given approximately by

$$V'(t) = 500(t - 12) \qquad 0 \leqslant t \leqslant 10$$

where $V(t)$ is the value of the machine after t years. What is the total loss in value of the machine in the first 5 years? In the second 5 years? Set up appropriate integrals and solve.

76. *Maintenance costs.* Maintenance costs for an apartment house generally increase as the building gets older. From past records, a managerial service determines that the rate of increase in maintenance costs (in dollars per year) for a particular apartment complex is given approximately by

$$M'(x) = 90x^2 + 5,000$$

where x is the age of the apartment complex in years, and $M(x)$ is the total (accumulated) cost of maintenance for x years. Write a definite integral that will give the total costs from the end of the second year to the end of the seventh year after the apartment complex was built, and evaluate it.

77. *Average cost.* The total cost (in dollars) of manufacturing x auto body frames is

$$C(x) = 60,000 + 300x$$

(A) Find the average cost per frame if 500 frames are produced. [*Hint:* Recall that $\overline{C}(x)$ is the average cost per unit.]

(B) Find the average value of the cost function over the interval [0, 500].

(C) Discuss the difference between the answers in parts (A) and (B).

78. *Average cost.* The total cost (in dollars) of printing x dictionaries is $C(x) = 20,000 + 10x$.

(A) Find the average cost per dictionary if 1,000 dictionaries are produced.

(B) Find the average value of the cost function over the interval [0, 1,000].

(C) Discuss the difference between the answers in parts (A) and (B).

79. *Useful life.* An amusement company maintains records for each video game it installs in an arcade. Suppose that $C(t)$ and $R(t)$ represent the total accumulated costs and revenues (in thousands of dollars), respectively, t years after a particular game has been installed. Also,

$$C'(t) = 2 \qquad R'(t) = 9e^{-0.5t}$$

The value of t for which $C'(t) = R'(t)$ is called the **useful life** of the game.

(A) Find the useful life of the game (to the nearest year).

(B) Express the total profit accumulated during the useful life of the game as a definite integral, and evaluate it.

80. *Useful life.* Repeat Problem 79 if $C'(t) = 1$ and $R'(t) = 7.5e^{-0.5t}$.

81. *Useful life.* Refer to Problem 79. The total accumulated costs $C(t)$ and revenues $R(t)$ (in thousands of dollars) for a coin-operated photocopying machine satisfy

$$C'(t) = 0.091t \quad \text{and} \quad R'(t) = 5te^{-t^2}$$

where t is the time in years. Find the useful life of the machine (to the nearest year). What is the total profit (to the nearest dollar) accumulated during the useful life of the machine?

82. *Useful life.* Refer to Problem 79. The total accumulated costs $C(t)$ and revenues $R(t)$ (in thousands of dollars) for a coal mine satisfy

$$C'(t) = 3 \quad \text{and} \quad R'(t) = 15e^{-0.1t}$$

where t is the number of years the mine has been in operation. Find the useful life of the mine (to the nearest year). What is the total profit (to the nearest dollar) accumulated during the useful life of the mine?

83. *Cost.* The marginal cost at various levels of output per month for a company manufacturing watches is shown in Table 6. The output x is given in thousands of units per month, and $C(x)$ is in thousands of dollars per month.

TABLE 6

x	0	1	2	3	4	5	6	7	8
$C'(x)$	58	30	18	9	5	7	17	33	51

(A) Graph the data set in a graphing utility. (Save the data set in the utility for Problem 85.)

(B) Which of the two sums, trapezoidal and midpoint, overestimates and which underestimates the marginal cost?

(C) The company now has a production level of 2 thousand watches per month and wishes to increase it to 8 thousand. Estimate the increased cost using T_3 and M_3, and write the answer in the form

$$T_3 \leqslant \int_2^8 C'(x) \, dx \leqslant M_3 \quad \text{or}$$

$$M_3 \leqslant \int_2^8 C'(x) \, dx \leqslant T_3$$

as appropriate. Calculate T_3 and M_3 to the nearest thousand dollars.

84. *Cost.* Refer to Problem 83. Suppose the company now has a production level of 1 thousand watches per month and wishes to increase it to 7 thousand. Estimate the increased cost using T_3 and M_3, and write the answer in the form

$$T_3 \leqslant \int_1^7 C'(x) \, dx \leqslant M_3 \quad \text{or}$$

$$M_3 \leqslant \int_1^7 C'(x) \, dx \leqslant T_3$$

as appropriate. Calculate T_3 and M_3 to the nearest thousand dollars.

85. *Cost.*
(A) Find a quadratic regression equation for the data in Table 6 (Problem 83), and graph it and the data set in the same viewing window.
(B) Use the regression equation from part (A) and an appropriate definite integral to approximate the increased cost (to the nearest thousand dollars) in going from a production level of 2 thousand watches per month to 8 thousand watches per month.

86. *Cost.*
(A) Find a cubic regression equation for the data in Table 6 (Problem 83), and graph it and the data set in the same viewing window.
(B) Use the regression equation from part (A) and an appropriate definite integral to approximate the increased cost (to the nearest thousand dollars) in going from a production level of 1 thousand watches per month to 7 thousand watches per month.

87. *Oil production.* Using data from the first 3 years of production as well as geological studies, the management of an oil company estimates that oil will be pumped from a producing field at a rate given by

$$R(t) = \frac{100}{t+1} + 5 \qquad 0 \leqslant t \leqslant 20$$

where $R(t)$ is the rate of production (in thousands of barrels per year) t years after pumping begins. Approximately how many barrels of oil (to the nearest thousand barrels) will the field produce during the first 10 years of production? From the end of the 10th year to the end of the 20th year?

88. *Oil production.* In Problem 87, if the rate is found to be

$$R(t) = \frac{120t}{t^2+1} + 3 \qquad 0 \leqslant t \leqslant 20$$

approximately how many barrels of oil will the field produce (to the nearest thousand barrels) during the first 5 years of production? From the end of the 5th year to the end of the 10th year?

89. *Oil production.* Using the rate function in Problem 87, find a function $Q(x)$ that represents the total amount of oil produced by the field from year $t = 0$ to year $t = x$ for $0 \leqslant x \leqslant 20$.

90. *Oil production.* Using the rate function in Problem 88, find a function $S(x)$ that represents the total amount of oil produced by the field from year $t = 0$ to year $t = x$ for $0 \leqslant x \leqslant 20$.

Life Sciences

91. *Medicine.* A drug is injected into the bloodstream of a patient through her right arm. The concentration of the drug in the bloodstream of the left arm t hours after the injection is given by

$$C(t) = \frac{0.14t}{t^2+1}$$

What is the average concentration of the drug in the bloodstream of the left arm during the first hour after the injection? During the first 2 hours after the injection? Round answers to four decimal places.

92. *Temperature.* If the temperature $C(t)$ in an aquarium is made to change according to

$$C(t) = t^3 - 2t + 10 \qquad 0 \leqslant t \leqslant 2$$

(in degrees Celsius) over a 2 hour period, what is the average temperature over this period?

93. *Medicine.* The rate of healing $A'(t)$ (in square centimeters per day) for a certain type of abrasive skin wound is given approximately by Table 7 (next page), where $A(t)$ is the area of the wound in square centimeters after t days of healing.

TABLE 7

t	0	1	2	3	4	5
$A'(t)$	1.08	0.96	0.85	0.75	0.66	0.58

t	6	7	8	9	10
$A'(t)$	0.51	0.45	0.40	0.36	0.33

(A) Plot the data set in a graphing utility.

(B) Approximate the area of the wound (in square centimeters) that will be healed in the first 6 days of healing. Use M_3 and T_3, and express the answer in the form

$$? \leq \int_0^6 A'(t)\, dt \leq ? $$

where the question marks are replaced by M_3 and T_3, as appropriate. Compute M_3 and T_3 to two decimal places.

94. *Medicine.* Refer to Problem 93. Approximate the area of the wound (in square centimeters) that will be healed from the end of day 6 to the end of day 10. Use M_2 and T_2, and express the answer in the form

$$? \leq \int_6^{10} A'(t)\, dt \leq ? $$

where the question marks are replaced by M_2 and T_2, as appropriate. Compute M_2 and T_2 to two decimal places.

95. *Medicine.* Find an exponential regression equation ($y = ab^x$) for the data in Table 7 (Problem 93), and use the equa-

tion and an appropriate definite integral to estimate the area (in square centimeters) of healing in the first 6 days.

96. *Medicine.* Find a quadratic regression equation for the data in Table 7 (Problem 93), and use the equation and an appropriate definite integral to estimate the area (in square centimeters) of healing from the end of the 6th day to the end of the 10th day.

Social Sciences

97. *Urban growth.* The rate of growth of the population, $N(t)$, of a newly incorporated city t years after incorporation is estimated to be

$$ N'(t) = 400 + 600\sqrt{t} \qquad 0 \leq t \leq 9 $$

How much will the population increase by the end of 7 years after incorporation?

98. *Learning.* A beginning high school language class was chosen for an experiment in learning. Using a list of 100 words, the experiment involved measuring the rate of vocabulary memorization at different times during a continuous 5 hour study session. It was found that the average rate of learning for the whole class was inversely proportional to the time spent studying and was given approximately by

$$ V'(t) = \frac{30}{t} \qquad 1 \leq t \leq 5 $$

How many words will be memorized, on the average, by the class from the end of the first hour to the end of the fifth hour?

CHAPTER 11
GROUP ACTIVITY

Simpson's Rule

Introduction to Simpson's Rule

In Sections 11-4 and 11-5 we investigated the use of left, right, trapezoidal, and midpoint sums as numerical integration devices for approximating definite integrals. For increasing or decreasing functions, we found that the errors from the use of left and right sums were about the same in absolute value and of opposite sign (Fig. 1).

$$ L_n \leq \int_a^b f(x)\, dx \leq R_n $$

R_n overestimates

L_n underestimates

FIGURE 1

It seemed reasonable to take their average, which would balance out the error, and we found that their average produced the trapezoidal sum:

$$T_n = \frac{L_n + R_n}{2} \quad \text{Trapezoidal sum}$$

The midpoint sum also balanced out the error. We found that if the graph of the integrand was concave up or concave down, then the errors from the use of the midpoint sum and the trapezoidal sum were of opposite sign (Fig. 2). (Recall that the area of a midpoint sum rectangle is the same as the area of a corresponding tangent line trapezoid.) It also appears from Figure 2, and it can be proved in general, that the trapezoidal sum error is about double the midpoint sum error.

$$T_n \leq \int_a^b f(x)\, dx \leq M_n$$

Tangent line

M_n overestimates

T_n underestimates

Midpoint

FIGURE 2

This suggests that a weighted average of the two estimates, with the midpoint sum being counted double the trapezoidal sum, might be an even better estimate than either separately. This weighted average is called **Simpson's rule,** and is given symbolically as follows:

$$S_{2n} = \frac{2M_n + T_n}{3} \quad \text{Simpson's rule} \tag{1}$$

You may wonder why we use a subscript of $2n$ for the Simpson sum and n for the midpoint and trapezoidal sums. Looking back in Section 11-5, you will find that when we used both the midpoint sum and the trapezoidal sum on the same interval $[a, b]$, we divided the interval into $2n$ equal subdivisions, evaluated the integrand at each of these points, and then used n equal subdivisions for each of the sums. When we combine the two sums in Simpson's rule, equation (1), we are using integrand values at each of the $2n$ points on $[a, b]$. Thus, **Simpson's rule always requires an even number of equal subdivisions of $[a, b]$.**

Problem 1. For $\Delta x = (b - a)/4$, show, starting with equation (1) in the form

$$S_4 = \frac{2M_2 + T_2}{3}$$

that

$$S_4 = [f(x_0) + 4f(x_1) + 2f(x_2) + 4f(x_3) + f(x_4)]\frac{\Delta x}{3}$$

Error Comparisons

One way to compare the effectiveness of a numerical integration technique is to find the error produced for a given value of n and to see what happens to the error as n is increased. We do this for the five sums discussed above, L_n, R_n, M_n, T_n, and S_n, by using these sums to approximate a definite integral of known value and looking at the corresponding errors.

Problem 2. Given the definite integral

$$I = \int_4^{12} \frac{dx}{x} = 1.099$$

use the values of the integrand in Table 1 to complete Table 2.

TABLE 1
Integrand Values

x	4	5	6	7	8	9	10	11	12
$1/x$	0.250	0.200	0.167	0.143	0.125	0.111	0.100	0.091	0.083

TABLE 2
Error

	$L_n - I$	$R_n - I$	$M_n - I$	$T_n - I$	$S_{2n} - I$
$n = 2$	0.401				0.002
$n = 4$			-0.009		

Table 2 suggests that the relationships between error and the change in n shown in Table 3 exist. This can be confirmed in general.

TABLE 3
Error and Change in n

CHANGE IN n	FACTOR BY WHICH ERROR IS CHANGED		
	L_n, R_n	M_n, T_n	S_n
$2n$	$\frac{1}{2}$	$\frac{1}{4}$	$\frac{1}{16}$
$10n$	$\frac{1}{10}$ Adds one more digit of accuracy	$\frac{1}{10^2}$ Adds two more digits of accuracy	$\frac{1}{10^4}$ Adds four more digits of accuracy

We have already noted that L_n and R_n give exact results if the integrand is a constant function. Also, M_n and T_n give exact results for constant and linear functions. What about Simpson's rule? Simpson's rule gives exact results if the integrand is any polynomial of degree 3 or less. It should now be clear why Simpson's rule is so popular. However, Simpson's rule is generally not used in numerical integration routines. More powerful algorithms exist. For example, the Gauss–Kronrod algorithm is used in a number of popular graphing utilities and gives exact results for polynomials of degree 5 or less.

Problem 3.   Use Simpson's rule to approximate the definite integral

$$I = \int_2^{10} \frac{x}{\ln x}\, dx$$

for $2n = 4$ and for $2n = 8$. Compute S_{2n} to three decimal places. Find I (to three decimal places) using a numerical integration routine, and compute the error, $|I - S_{2n}|$, for the two estimates above.

Application

We now turn to an example where a numerical integration routine cannot be used.

Problem 4.   A company manufactures and wholesales a popular brand of sunglasses. Their financial research department collected the data in Table 4, showing marginal cost $C'(x)$ (in dollars) at selected production levels of x pairs of sunglasses per hour.

TABLE 4
Marginal Costs

x	50	100	150	200	250	300	350	400	450
$C'(x)$	21.80	16.95	15.31	14.50	14.00	13.66	13.42	13.25	13.11

(A) If the company is now producing 50 pairs of sunglasses per hour and increases production to 450 pairs per hour, estimate the total increase in costs per hour using L_n and R_n with an appropriate value of n. How are L_n and R_n related to the actual increase in cost?

(B) Use Simpson's rule to estimate the increase in cost in part (A).

CHAPTER   11   REVIEW   ■   Important Terms and Symbols

11-1 *Antiderivatives and Indefinite Integrals.* Antiderivative; indefinite integral; integral sign; integrand; constant of integration

11-2 *Integration by Substitution.* Reversing the chain rule; general integral formulas; differential; integration by substitution

11-3 *Differential Equations—Growth and Decay.* Differential equation; first-order; second-order; slope field; continuous compound interest; exponential growth law; population growth; radioactive decay; unlimited growth; limited growth; logistic growth

$$\frac{dQ}{dt} = rQ; \quad Q = Q_0 e^{rt}$$

11-4 *A Geometric–Numeric Introduction to the Definite Integral.* Area under a curve; definite integral symbol; left rectangles; right rectangles; left sum; right sum; monotone functions; error bound for left and right sum; signed areas; rate, area, and distance; rate, area, and total change

$$\int_a^b f(x)\, dx; \quad L_n; \quad R_n; \quad A_n$$

11-5 *Definite Integral as a Limit of a Sum; Fundamental Theorem of Calculus.* Left sum; right sum; Riemann sum; definite integral as a limit of a Riemann sum; integrand; lower limit; upper limit; properties of definite integrals; numerical integration; left, right, midpoint, and trapezoid sums; numerical integration routine; average value of a continuous function; fundamental theorem of calculus

$$L_n; \quad R_n; \quad M_n; \quad T_n;$$

$$\int_a^b f(x)\, dx; \quad F(x) \Big|_a^b = F(b) - F(a);$$

$$\sum_{k=1}^n f(c_k)\, \Delta x_k; \quad \frac{1}{b-a} \int_a^b f(x)\, dx$$

CHAPTER   11   ■   Integration Formulas and Properties

$\int k\, dx = kx + C$

$\int kf(x)\, dx = k \int f(x)\, dx$

$\int [f(x) \pm g(x)]\, dx = \int f(x)\, dx \pm \int g(x)\, dx$

$\displaystyle \int u^n\, du = \frac{u^{n+1}}{n+1} + C \qquad n \neq -1$

$\int e^u\, du = e^u + C$

$\displaystyle \int e^{au}\, du = \frac{1}{a} e^{au} + C \qquad a \neq 0$

$\displaystyle \int \frac{1}{u}\, du = \ln |u| + C \qquad u \neq 0$

$\displaystyle \int_a^a f(x)\, dx = 0$

$\displaystyle \int_a^b f(x)\, dx = - \int_b^a f(x)\, dx$

$\displaystyle \int_a^b kf(x)\, dx = k \int_a^b f(x)\, dx \qquad k \text{ a constant}$

$\displaystyle \int_a^b [f(x) \pm g(x)]\, dx = \int_a^b f(x)\, dx \pm \int_a^b g(x)\, dx$

$\displaystyle \int_a^b f(x)\, dx = \int_a^c f(x)\, dx + \int_c^b f(x)\, dx$

CHAPTER   11   ■   Review Exercise

Work through all the problems in this chapter review and check your answers in the back of the book. Answers to all review problems are there along with section numbers in italics to indicate where each type of problem is discussed. Where weaknesses show up, review appropriate sections in the text.

A   *Find each integral in Problems 1–6.*

1. $\int (3t^2 - 2t)\, dt$

2. $\int_2^5 (2x - 3)\, dx$

3. $\int (3t^{-2} - 3)\, dt$

4. $\int_1^x t\, dt$

5. $\int e^{-0.5x} \, dx$

6. $\int_1^5 \frac{2}{u} \, du$

7. Find a function $y = f(x)$ that satisfies both conditions:

$$\frac{dy}{dx} = 3x^2 - 2 \qquad f(0) = 4$$

8. From the graph of $y = f'(x)$ shown in the figure, verbally describe the shape of the graph of an antiderivative function f. How would the graph of one antiderivative of $f'(x)$ differ from another?

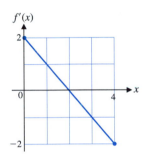

$f'(x)$

Figure for 8 and 9

9. From the graph of $y = f'(x)$ shown in the figure, sketch possible graphs of three antiderivative functions f such that $f(0) = -2$, $f(0) = 0$, and $f(0) = 2$, all on the same set of coordinate axes.

10. Find all antiderivatives of:

(A) $\dfrac{dy}{dx} = 8x^3 - 4x - 1$

(B) $\dfrac{dx}{dt} = e^t - 4t^{-1}$

11. Approximate $I = \int_1^5 (x^2 + 1) \, dx$ using T_2 and M_2, and write the answer in the form

$$? \leqslant \int_1^5 (x^2 + 1) \, dx \leqslant ?$$

with the question marks replaced by T_2 and M_2, as appropriate. Explain how you know which to use.

12. Evaluate the integral I in Problem 11, and find the actual errors $(M_2 - I)$ and $(T_2 - I)$. Perform all calculations to three decimal places.

13. Use the values in Table 1 and left and right sums, L_5 and R_5, to approximate $\int_7^{17} f(x) \, dx$. Write the answer in the form

$$? \leqslant \int_7^{17} f(x) \, dx \leqslant ?$$

with the question marks replaced by L_5 and R_5, as appropriate. Perform all calculations to one decimal place. Explain how you know which of L_5 and R_5 overestimates and which underestimates the integral.

TABLE 1

x	7	9	11	13	15	17
$f(x)$	3.4	3.3	2.8	2.3	1.5	0.5

14. Find the average value of $f(x) = 6x^2 + 2x$ over the interval $[-1, 2]$.

15. Describe a rectangle that would have the same area as the area under the graph of $f(x) = 6x^2 + 2x$ from $x = -1$ to $x = 2$ (see Problem 14).

B *Use the graph and actual areas of the indicated regions in the figure to evaluate the integrals in Problems 16–23:*

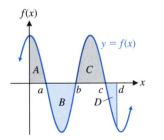

$f(x)$

Area $A = 1$
Area $B = 2$
Area $C = 2$
Area $D = 0.6$

Figure for 16–23

16. $\displaystyle\int_a^b 5f(x) \, dx$

17. $\displaystyle\int_b^c \frac{f(x)}{5} \, dx$

18. $\displaystyle\int_b^d f(x) \, dx$

19. $\displaystyle\int_a^c f(x) \, dx$

20. $\displaystyle\int_0^d f(x) \, dx$

21. $\displaystyle\int_b^a f(x) \, dx$

22. $\displaystyle\int_c^b f(x) \, dx$

23. $\displaystyle\int_d^0 f(x) \, dx$

24. For the graph of $y = f'(x)$ shown, verbally describe the shape of the graph of an antiderivative function f. How would the graph of one antiderivative of $f'(x)$ differ from another?

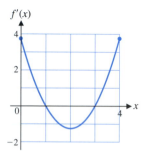

Figure for 24 and 25

25. For the graph of $y = f'(x)$ shown, sketch possible graphs of three antiderivative functions f such that $f(0) = -1$, $f(0) = 0$, and $f(0) = 1$, all on the same set of coordinate axes.

Problems 26–31 refer to the slope field shown in the figure.

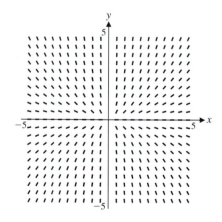

Figure for 26–31

26. (A) For $dy/dx = (2y)/x$, what is the slope of a solution curve at $(2, 1)$? At $(-2, -1)$?
(B) For $dy/dx = (2x)/y$, what is the slope of a solution curve at $(2, 1)$? At $(-2, -1)$?

27. Is the slope field shown in the figure for $dy/dx = (2x)/y$ or for $dy/dx = (2y)/x$? Explain.

28. Show that $y = Cx^2$ is a solution of $dy/dx = (2y)/x$ for any real number C.

29. Referring to Problem 28, find the particular solution of $dy/dx = (2y)/x$ that passes through $(2, 1)$. Through $(-2, -1)$.

30. Graph the two particular solutions found in Problem 29 in the slope field shown (or a copy).

31. Graph in the same viewing window $y = Cx^2$ for $C = -2$, -1, 1, and 2 using $-5 \leqslant x \leqslant 5$ and $-5 \leqslant y \leqslant 5$.

Find each integral in Problems 32–42.

32. $\int \sqrt[3]{6x - 5}\ dx$

33. $\int_0^x 10(2t - 1)^4\ dt$

34. $\int \left(\dfrac{2}{x^2} - 2xe^{x^2} \right) dx$

35. $\int_0^4 x\sqrt{x^2 + 4}\ dx$

36. $\int_1^x (e^{-2t} + t^{-1})\ dt$

37. $\int_0^{10} 10e^{-0.02x}\ dx$

38. $\int_0^3 \dfrac{x}{1 + x^2}\ dx$

39. $\int_0^3 \dfrac{x}{(1 + x^2)^2}\ dx$

40. $\int x^3(2x^4 + 5)^5\ dx$

41. $\int \dfrac{e^{-x}}{e^{-x} + 3}\ dx$

42. $\int \dfrac{e^x}{(e^x + 2)^2}\ dx$

43. Find a function $y = f(x)$ that satisfies both conditions:
$$\dfrac{dy}{dx} = 3x^{-1} - x^{-2} \qquad f(1) = 5$$

44. Find the equation of the curve that passes through $(2, 10)$ if its slope is given by
$$\dfrac{dy}{dx} = 6x + 1$$
for each x.

45. The sums L_n, R_n, T_n, and M_n are used to approximate the definite integral $\int_a^b f(x)\ dx$, where the graph of f is shown in the figure. Suppose the values of the approximations, 5.11, 4.83, 5.01, 4.79, got mixed up. Associate an appropriate value with each approximation.

Figure for 45

46. The velocity $v(t)$ (in meters per second) of a woman running a 100 meter race was recorded by special radar instruments at the end of each of the first 8 seconds, as shown in Table 2. We are interested in estimating the distance she ran by the end of the eighth second.

TABLE 2

t	0	1	2	3	4	5	6	7	8
$v(t)$	0	3.51	6.29	8.79	10.15	11.14	11.57	11.91	11.95

(A) Symbolically represent the distance by an appropriate definite integral. Explain how you can tell which of T_4 and M_4 over- or underestimate the integral.

(B) Estimate the total distance the woman ran in the first 8 seconds. Write the answer in the form

$$? \leq \int_0^8 v(t) \, dt \leq ?$$

using T_4 and M_4 in place of the question marks, as appropriate. Calculate T_4 and M_4 to two decimal places.

47. Find a cubic regression equation for the data set in Table 2 (Problem 46), and use this equation to estimate the distance the woman ran in the first 8 seconds (to two decimal places).

48. If $F(x) = \int_0^x e^{-0.25t} \, dt$, find $F'(x)$.

49. (A) Find the average value of $f(x) = 3\sqrt{x}$ over the interval $[1, 9]$.
(B) Graph $f(x) = 3\sqrt{x}$ and its average over the interval $[1, 9]$ in the same coordinate system.

C Find each integral in Problems 50–54.

50. $\displaystyle \int \frac{(\ln x)^2}{x} \, dx$

51. $\displaystyle \int x(x^3 - 1)^2 \, dx$

52. $\displaystyle \int \frac{x}{\sqrt{6 - x}} \, dx$

53. $\displaystyle \int_0^7 x\sqrt{16 - x} \, dx$

54. $\displaystyle \int_{-1}^x t(t + 1)^4 \, dt$

55. Find a function $y = f(x)$ that satisfies both conditions:

$$\frac{dy}{dx} = 9x^2 e^{x^3} \qquad f(0) = 2$$

56. Solve the differential equation:

$$\frac{dN}{dt} = 0.06N \qquad N(0) = 800 \qquad N > 0$$

Problems 57–59 involve estimating the value of the definite integral

$$I = \int_1^2 e^{-x^2} \, dx$$

(Recall, we mentioned that this integrand does not have an elementary antiderivative.)

57. Graph $f(x) = e^{-x^2}$ over the interval $[1, 2]$. If M_n and T_n are used to estimate I, which overestimates and which underestimates I, and why?

58. Use M_5 and T_5 to estimate I, and write the answer in the form

$$? \leq \int_1^2 e^{-x^2} \, dx \leq ?$$

with the question marks replaced by M_5 and T_5, as appropriate. Compute M_5 and T_5 to three decimal places.

59. Evaluate I (to three decimal places) using a numerical integration routine. Evaluate the errors $(M_5 - I)$ and $(T_5 - I)$.

60. Write the following in the form $\int_a^b f(x) \, dx$, and evaluate to three decimal places:

$$\lim_{n \to \infty} \left[\left(\frac{1}{c_1 - 2} \right) \Delta x + \left(\frac{1}{c_2 - 2} \right) \Delta x + \cdots + \left(\frac{1}{c_n - 2} \right) \Delta x \right]$$

where

$$\Delta x = \frac{9 - 4}{n} \qquad \text{and} \qquad c_k = 4 + k\frac{5}{n},$$

$$k = 1, 2, \ldots, n$$

61. If $\int_1^x f(t) \, dt = \ln x + x - 1$, find $f(1)$.

Graph Problems 62–65 and identify each as unlimited growth, exponential decay, limited growth, or logistic growth:

62. $N = 50(1 - e^{-0.07t}); \ 0 \leq t \leq 80, 0 \leq N \leq 60$

63. $p = 500e^{-0.03x}; \ 0 \leq x \leq 100, 0 \leq p \leq 500$

64. $A = 200e^{0.08t}; \ 0 \leq t \leq 20, 0 \leq A \leq 1,000$

65. $N = \dfrac{100}{1 + 9e^{-0.3t}}; \ 0 \leq t \leq 25, 0 \leq N \leq 100$

▶ **A P P L I C A T I O N S**

Business & Economics

66. *Cost.* A company manufactures downhill skis. The research department produced the marginal cost graph shown in the figure, where $C'(x)$ is in dollars and x is the number of pairs of skis produced per week. Estimate the increase in cost going from a production level of 200 to 600 pairs of skis per week. Use left and right sums over two equal subintervals. Replace the question marks with the values of L_2 and R_2 as appropriate:

$$? \leqslant \int_{200}^{600} C'(x)\, dx \leqslant ?$$

Figure for 66

67. *Cost.* Explain how the increase in production cost is related to the left or right rectangles used in Problem 66.

68. *Cost.* Assuming that the marginal cost function in Problem 66 is linear, find its equation, and write a definite integral that represents the increase in costs going from a production level of 200 to 600 pairs of skis per week. Evaluate the definite integral.

69. *Profit and production.* The weekly marginal profit for an output of x units is given approximately by

$$P'(x) = 150 - \frac{x}{10} \qquad 0 \leqslant x \leqslant 40$$

What is the total change in profit for a production change from 10 units per week to 40 units? Set up a definite integral and evaluate it.

70. *Profit function.* If the marginal profit for producing x units per day is given by

$$P'(x) = 100 - 0.02x \qquad P(0) = 0$$

where $P(x)$ is the profit in dollars, find the profit function P and the profit on 10 units of production per day.

71. *Resource depletion.* An oil well starts out producing oil at the rate of 60,000 barrels of oil per year, but the production

rate is expected to decrease by 4,000 barrels per year. Thus, if $P(t)$ is the total production (in thousands of barrels) in t years, then

$$P'(t) = f(t) = 60 - 4t \qquad 0 \leqslant t \leqslant 15$$

Write a definite integral that will give the total production after 15 years of operation and evaluate it.

72. *Inventory.* Suppose the inventory of a certain item t months after the first of the year is given approximately by

$$I(t) = 10 + 36t - 3t^2 \qquad 0 \leqslant t \leqslant 12$$

What is the average inventory for the second quarter of the year?

73. *Price–supply.* Given the price–supply function

$$p = S(x) = 8(e^{0.05x} - 1)$$

find the average price (in dollars) over the supply interval [40, 50].

74. *Employee training.* A company producing high-quality home sound systems found that, on the average, a new employee can assemble $N(t)$ components per day after t days of on-the-job training, as indicated in Table 3.

TABLE 3

t	0	5	10	15	20	25	30
$N(t)$	3	9	14	18	21	23	24

(A) Symbolically represent, by means of an appropriate definite integral I, the number of units an average new employee can be expected to assemble during the first 30 days of employment.

(B) If the midpoint sum M_n and the trapezoidal sum T_n are used to estimate I, which will overestimate and which will underestimate I, and how can you tell?

(C) Estimate I using T_3 and M_3, and write the answer in the form

$$? \leqslant I \leqslant ?$$

with the question marks replaced by T_3 and M_3, as appropriate.

(D) Find a quadratic regression curve for the data set, and estimate I by integrating this function over the interval [0, 30] using a numerical integration routine. Round answer to the nearest unit.

75. *Useful life.* The total accumulated costs $C(t)$ and revenues $R(t)$ (in thousands of dollars), respectively, for a coal mine satisfy

$$C'(t) = 3 \qquad \text{and} \qquad R'(t) = 20e^{-0.1t}$$

where t is the number of years the mine has been in operation. Find the useful life of the mine to the nearest year. What is the total profit accumulated during the useful life of the mine?

76. *Marketing.* The market research department for an automobile company estimates that the sales (in millions of dollars) of a new automobile will increase at the monthly rate of

$$S'(t) = 4e^{-0.08t} \qquad 0 \leq t \leq 24$$

t months after the introduction of the automobile. What will be the total sales $S(t)$, t months after the automobile is introduced if we assume that there were 0 sales at the time the automobile entered the marketplace? What are the estimated total sales during the first 12 months after the introduction of the automobile? How long will it take for the total sales to reach $40 million?

77. *Change in revenue.* A company manufactures and wholesales a sports watch at a current production level of 1,000 watches per week and wishes to increase production. The marginal revenue is $R'(x) = 120 - 0.06x$, where x is the number of watches produced per week and revenues are in dollars.

(A) Find a function $G(x)$ that represents the change in revenue from a production level of 1,000 watches per week to x watches per week for $1,000 \leq x \leq 3,000$.

(B) Find the production level (to the nearest unit) that produces the maximum increase in revenue (to the nearest dollar). Verbally describe the results.

Life Sciences

78. *Pollution.* In an industrial area, the concentration $C(t)$ of a particulate matter in the atmosphere (in parts per million) during a 12 hour period is given in the figure.

(A) Represent the average concentration over the 12 hour period symbolically in terms of an appropriate definite integral I.

(B) If the midpoint sum M_n and the trapezoidal sum T_n are used to estimate I, which overestimates and which underestimates I, and why?

(C) Approximate I using T_6 and M_6 (take function values from the graph) and write the answer in the form

$$? \leq I \leq ?$$

with the question marks replaced by appropriate quantities (to one decimal place) involving T_6 and M_6. Interpret verbally.

Figure for 78

79. *Pollution.* An environmental protection agency estimates that the rate of seepage of toxic chemicals from a waste dump (in gallons per year) is given by

$$R(t) = \frac{1{,}000}{(1 + t)^2}$$

where t is time in years since the discovery of the seepage. Find the total amount of toxic chemicals that seep from the dump during the first 4 years after the seepage is discovered.

80. *Population.* According to the World Bank, the population in the Americas (north, central, and south) in 1995 was about 770 million, and was growing at the rate of about 1% compounded continuously.

(A) If the population continues to grow at this rate, what is the estimated population in the Americas for the year 2030?

(B) At the indicated growth rate, how long will it take the population in the Americas to double?

Social Sciences

81. *Archaeology.* The continuous compound rate of decay for carbon-14 is $r = -0.000\,123\,8$. A piece of animal bone found at an archaeological site contains 4% of the original amount of carbon-14. Estimate the age of the bone.

82. *Learning.* In a particular business college, it was found that an average student enrolled in a typing class progressed at a rate of $N'(t) = 7e^{-0.1t}$ words per minute t weeks after enrolling in a 15 week course. If at the beginning of the course a student could type 25 words per minute, how many words per minute, $N(t)$, would the student be expected to type t weeks into the course? After completing the course?

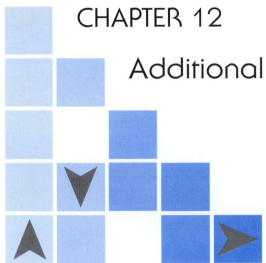

CHAPTER 12

Additional Integration Topics

This chapter contains additional topics on integration. Since they are essentially independent of one another, they may be taken up in any order, and certain sections may be omitted if desired.

SECTION 12-1 ■ # Area between Curves

■ AREA BETWEEN A CURVE AND THE x AXIS
■ AREA BETWEEN TWO CURVES
■ APPLICATION: INCOME DISTRIBUTION

In the last chapter we found that the definite integral $\int_a^b f(x)\,dx$ represents the sum of the signed areas between the graph of $y = f(x)$ and the x axis from $x = a$ to $x = b$, where the areas above the x axis are counted positively and the areas below the x axis are counted negatively (see Fig. 1). In this section we are interested in using the definite integral to find the actual area between a curve and the x axis or the actual area between two curves. These areas are always nonnegative quantities—**area measure is never negative.**

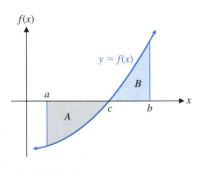

FIGURE 1

$$\int_a^b f(x)\,dx = -A + B$$

■ AREA BETWEEN A CURVE AND THE x AXIS

In Figure 1, A represents the area between $y = f(x)$ and the x axis from $x = a$ to $x = c$, and B represents the area between $y = f(x)$ and the x axis from $x = c$ to $x = b$. Both A and B are positive quantities. Since $f(x) \geqslant 0$ on the interval $[c, b]$,

$$\int_c^b f(x)\,dx = B$$

893

And since $f(x) \leq 0$ on the interval $[a, c]$,

$$\int_a^c f(x)\, dx = -A$$

or

$$A = -\int_a^c f(x)\, dx = \int_a^c [-f(x)]\, dx$$

Thus:

> **The area between the graph of a negative function and the x axis is equal to the definite integral of the negative of the function.**

explore – discuss 1

Sketch a graph of a function f such that $f(x) \leq 0$ over the interval $[1, 5]$. (No equation is necessary.) Sketch a graph of $y = -f(x)$ over the same interval in the same coordinate system. Explain how these figures relate to the above discussion.

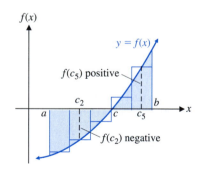

FIGURE 2
Area and Riemann sums

All the above interpretation of the definite integral relative to area, as we saw in the last chapter, is based on the definition of the definite integral as the limit of a Riemann sum:

$$\int_a^b f(x)\, dx = \lim_{n \to \infty} \sum_{k=1}^n f(c_k)\, \Delta x_k$$

Figure 2 shows a particular Riemann sum, the midpoint sum M_5, over the interval $[a, b]$. The product $f(c_k)\, \Delta x_k$ is negative for any rectangle on $[a, c]$ and positive for any rectangle on $[c, b]$. Thus, $f(c_k)\, \Delta x_k$ represents the negative of the area of a rectangle on $[a, c]$, and $-f(c_k)\, \Delta x_k$ represents the actual area of the rectangle. Consequently, for the interval $[a, c]$, where $f(x) \leq 0$,

$$\text{Area} = \lim_{n \to \infty} \sum_{k=1}^n [-f(c_k)]\, \Delta x_k = \int_a^c [-f(x)]\, dx$$

But for the interval $[c, b]$, where $f(x) \geq 0$,

$$\text{Area} = \lim_{n \to \infty} \sum_{k=1}^n f(c_k)\, \Delta x_k = \int_c^b f(x)\, dx$$

In summary:

Area between a Curve and the x Axis

For a function f continuous over $[a, b]$, the area between $y = f(x)$ and the x axis from $x = a$ to $x = b$ can be found using definite integrals as follows:

For $f(x) \geqslant 0$ over $[a, b]$:   Area $= \displaystyle\int_a^b f(x)\, dx$

For $f(x) \leqslant 0$ over $[a, b]$:   Area $= \displaystyle\int_a^b [-f(x)]\, dx$

If $f(x)$ is positive for some values of x and negative for others on an interval (as in Fig. 1), the area between the graph of f and the x axis can be obtained by dividing the interval into subintervals over which f is always positive or always negative, finding the area over each subinterval, and then summing these areas.

EXAMPLE 1 ➤ Area between a Curve and the x Axis   Find the area bounded by $f(x) = 6x - x^2$ and $y = 0$ for $1 \leqslant x \leqslant 4$.

Solution   First, sketch a graph of the region with the aid of a graphing utility (Fig. 3). (The solution of every area problem should begin with a graph of the region in question. After a little experience, a glance at a graphing utility graph should be sufficient to establish relationships.)

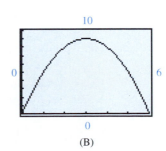

(A)   (B)

FIGURE 3

Since $f(x) \geqslant 0$ on $[1, 4]$,

$$A = \int_1^4 (6x - x^2)\, dx = 24$$

◄

[*Note:* Unless stated to the contrary, from now on we will use the numerical integration routine (fnInt) to evaluate definite integrals.]

MATCHED PROBLEM 1 ➤ Find the area (to three decimal places) bounded by $f(x) = x^2 + 1$ and $y = 0$ for $-1 \leqslant x \leqslant 3$. ◄

EXAMPLE 2 ➤ **Area between a Curve and the x Axis**　Find the area (to three decimal places) between the graph of $f(x) = x^2 - 2x$ and the x axis over the indicated intervals:

(A) $[1, 2]$　(B) $[-1, 1]$

Solution　We begin by graphing the area regions in Figure 4.

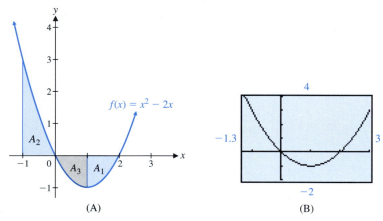

(A)　　　　(B)

FIGURE 4

(A) From the graph, we see that $f(x) \leqslant 0$ for $1 \leqslant x \leqslant 2$, so we integrate $-f(x)$:

$$A_1 = \int_1^2 [-f(x)]\, dx = \int_1^2 (2x - x^2)\, dx = 0.667$$

(B) Since the graph shows that $f(x) \geqslant 0$ on $[-1, 0]$ and $f(x) \leqslant 0$ on $[0, 1]$, the computation of this area requires two integrals:

$$A = A_2 + A_3 = \int_{-1}^0 f(x)\, dx + \int_0^1 [-f(x)]\, dx$$
$$= \int_{-1}^0 (x^2 - 2x)\, dx + \int_0^1 (2x - x^2)\, dx$$
$$= 1.333 + 0.667 = 2$$
◄

MATCHED PROBLEM 2 ➤ Find the area (to three decimal places) between the graph of $f(x) = x^2 - 9$ and the x axis over the indicated intervals:

(A) $[0, 2]$　(B) $[2, 4]$ ◄

■ AREA BETWEEN TWO CURVES

Consider the area bounded by $y = f(x)$ and $y = g(x)$, where $f(x) \geqslant g(x) \geqslant 0$, for $a \leqslant x \leqslant b$, as indicated in Figure 5.

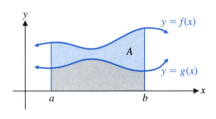

$$\left(\begin{matrix} \text{Area } A \text{ between} \\ f(x) \text{ and } g(x) \end{matrix} \right) = \left(\begin{matrix} \text{Area} \\ \text{under } f(x) \end{matrix} \right) - \left(\begin{matrix} \text{Area} \\ \text{under } g(x) \end{matrix} \right)$$

Areas are from x = a to x = b above the x axis.

$$= \int_a^b f(x)\, dx - \int_a^b g(x)\, dx$$

Use definite integral property 4 (Section 11-5).

$$= \int_a^b [f(x) - g(x)]\, dx$$

FIGURE 5

It can be shown that the above result does not require $f(x)$ or $g(x)$ to remain positive over the interval $[a, b]$. A more general result is stated in the box:

Area between Two Curves

If f and g are continuous and $f(x) \geqslant g(x)$ over the interval $[a, b]$, then the area bounded by $y = f(x)$ and $y = g(x)$ for $a \leqslant x \leqslant b$ is given exactly by

$$A = \int_a^b [f(x) - g(x)]\, dx$$

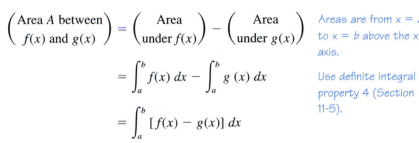

---

e x p l o r e – d i s c u s s   2

A Riemann sum for the integral representing the area between the graphs of $y = f(x)$ and $y = g(x)$ has the form

$$\sum_{k=1}^{n} [f(c_k) - g(c_k)]\, \Delta x_k$$

If $f(x) \geqslant g(x)$, then each term in this sum represents the area of a rectangle with height $f(c_k) - g(c_k)$ and width Δx. Discuss the relationship between these rectangles and the area between the graphs of $y = f(x)$ and $y = g(x)$.

EXAMPLE 3 ➤   **Area between Two Curves**   Find the area (to three decimal places) bounded by $f(x) = \frac{1}{2}x + 3$, $g(x) = -x^2 + 1$, $x = -2$, and $x = 1$.

Solution   First, graph the area region, as shown in Figure 6.

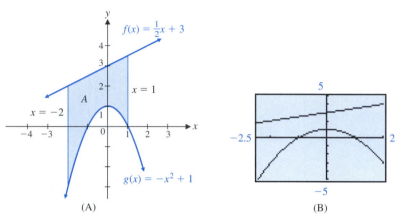

(A)   (B)

FIGURE 6

Observe from the graph that $f(x) \geqslant g(x)$ for $-2 \leqslant x \leqslant 1$, so

$$A = \int_{-2}^{1} [f(x) - g(x)] \, dx = \int_{-2}^{1} [(\tfrac{1}{2}x + 3) - (-x^2 + 1)] \, dx = 8.250 \quad \blacktriangleleft$$

MATCHED PROBLEM 3 ➤   Find the area (to three decimal places) bounded by $f(x) = x^2 - 1$, $g(x) = -0.5x - 3$, $x = -1$, and $x = 2$.   ◀

EXAMPLE 4 ➤   **Area between Two Curves**   Find the area (to three decimal places) bounded by $f(x) = 5 - x^2$ and $g(x) = 2 - 2x$.

Solution   First, graph the area region, as shown in Figure 7.

(A)   (B)   (C)

FIGURE 7

Since the problem does not include any limits on the values of x, we must determine the appropriate values from the graph. These are found by locating the intersection points (which we do graphically using a built-in routine in a graphing utility as shown in Figures 7B and 7C).

We will perform the integration over the interval $[-1, 3]$. Note that the area between the graphs for $x < -1$ is unbounded on the left, and the area between the graphs for $x > 3$ is unbounded on the right. Figure 7 shows that $f(x) \geqslant g(x)$ over the interval $[-1, 3]$. Thus,

$$A = \int_{-1}^{3} [f(x) - g(x)] \, dx = \int_{-1}^{3} [(5 - x^2) - (2 - 2x)] \, dx = 10.667 \qquad \blacktriangleleft$$

MATCHED PROBLEM 4 ➤ Find the area (to three decimal places) bounded by $f(x) = 6 - x^2$ and $g(x) = x$.

$\blacktriangleleft$

EXAMPLE 5 ➤ Area between Two Curves Find the area (to three decimal places) bounded by $f(x) = x^2 - x$ and $g(x) = 2x$ for $-2 \leqslant x \leqslant 3$.

Solution The area regions are shown in Figure 8.

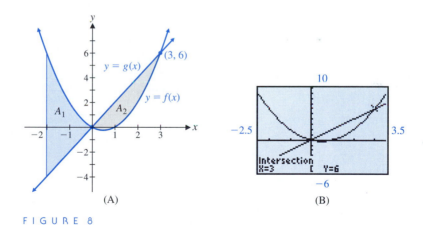

FIGURE 8

Note that $f(x) \geqslant g(x)$ on the interval $[-2, 0]$, but $g(x) \geqslant f(x)$ on the interval $[0, 3]$. Thus, two integrals are required to compute this area:

$$A_1 = \int_{-2}^{0} [f(x) - g(x)] \, dx = \int_{-2}^{0} [(x^2 - x) - (2x)] \, dx = 8.667$$

$$A_2 = \int_{0}^{3} [g(x) - f(x)] \, dx = \int_{0}^{3} [2x - (x^2 - x)] \, dx = 4.500$$

The total area between the graphs is

$$A = A_1 + A_2 = 8.667 + 4.500 = 13.167 \qquad \blacktriangleleft$$

MATCHED PROBLEM 5 ➤   Find the area (to three decimal places) bounded by $f(x) = 2x^2$ and $g(x) = 4 - 2x$ for $-2 \leqslant x \leqslant 2$.   ◄

▷

■ APPLICATION: INCOME DISTRIBUTION

The U.S. Bureau of the Census compiles and analyzes a great deal of data having to do with the distribution of income among families in the United States. For 1992 the Bureau reported that the lowest 20% of families received 4% of all family income, and the top 20% received 45%. Table 1 and Figure 9 give a detailed picture of the distribution of family income in 1992.

T A B L E 1 **Family Income Distribution in the United States in 1992**

INCOME LEVEL	x	y
Under $17,000	0.20	0.04
Under $30,000	0.40	0.15
Under $44,000	0.60	0.31
Under $64,000	0.80	0.55

Source: U.S. Bureau of the Census

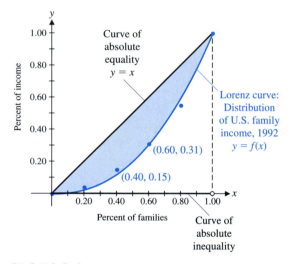

FIGURE 9
Lorenz chart

The graph of $y = f(x)$ through the data points in Figure 9 is called a **Lorenz curve** and is generally found using *regression analysis,* a technique of fitting a particular elementary function to a data set over a given interval. The variable x **represents the cumulative percentage of families at or below a given income level** and y **represents the cumulative percentage of total family income received.** For example, data point (0.40, 0.15) in Table 1 or on the Lorenz curve in Figure 9 indicates that the bottom 40% of families (those with incomes under $30,000) receive 15% of the total income for all families; data point (0.60, 0.31) indicates that the bottom 60% of families receive 31% of the total income for all families; and so on.

Absolute equality of income would occur if the area between the Lorenz curve and $y = x$ were 0. In this case, the Lorenz curve would be $y = x$ and all families would receive equal shares of the total income. That is, 5% of the families would receive 5% of the income, 20% of the families would receive 20% of the income, 65% of the families would receive 65% of the income, and so on. The maximum possible area between a Lorenz curve and $y = x$ is $\frac{1}{2}$, the area of the triangle below $y = x$. In this case, we would have **absolute inequality**—all the income would be in the hands of one family and the rest would have none. In actuality, Lorenz curves lie between these two extremes. But as the shaded area increases, the greater the inequality of income distribution.

The ratio of the area bounded by $y = x$ and the Lorenz curve $y = f(x)$ to the area of the triangle under the line $y = x$ from $x = 0$ to $x = 1$, is called the **index of income concentration.** The area bounded by $y = x$ and $y = f(x)$ is given by $\int_0^1 [x - f(x)]\, dx$, and the area of the triangle below $y = x$ is $\frac{1}{2}$. Thus, we have the following:

Index of Income Concentration

If $y = f(x)$ is the equation of a Lorenz curve, then

$$\text{Index of income concentration} = 2 \int_0^1 [x - f(x)]\, dx$$

The index of income concentration is always a number between 0 and 1:

A measure of 0 indicates absolute equality—all individuals share equally in the income. A measure of 1 indicates absolute inequality—one individual has all the income and the rest have none.

The closer the index is to 0, the closer the income is to being equally distributed. The closer the index is to 1, the closer the income is to being concentrated in a few hands. The index of income concentration is used to compare income distributions at various points in time, between different groups of people, before and after taxes are paid, between different countries, and so on.

EXAMPLE 6 ➤ Distribution of Income The Lorenz curve for the distribution of income in a certain country in 1990 is given by $f(x) = x^{2.6}$. Economists predict that the Lorenz curve for the country in the year 2010 will be given by $g(x) = x^{1.8}$. Find the index of income concentration (to three decimal places) for each curve, and interpret the results.

Solution   The Lorenz curves are shown in Figure 10.

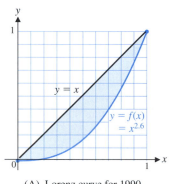
(A) Lorenz curve for 1990

(B) Projected Lorenz curve
for 2010

FIGURE 10

The index of income concentration in 1990 is (see Figure 10A)

$$2 \int_0^1 [x - f(x)]\, dx = 2 \int_0^1 [x - x^{2.6}]\, dx = 0.444$$

The projected index of income concentration in 2010 is (see Figure 10B)

$$2 \int_0^1 [x - g(x)]\, dx = 2 \int_0^1 [x - x^{1.8}]\, dx = 0.286$$

If this projection is correct, the index of income concentration will decrease, and income will be more equally distributed in the year 2010 than in 1990. ◄

MATCHED PROBLEM 6 ►   Repeat Example 6 if the projected Lorenz curve in the year 2010 is given by $g(x) = x^{3.8}$. ◄

Answers to Matched Problems
1. $A = \int_{-1}^3 (x^2 + 1)\, dx = 13.333$
2. (A) $A = \int_0^2 (9 - x^2)\, dx = 15.333$
   (B) $A = \int_2^3 (9 - x^2)\, dx + \int_3^4 (x^2 - 9)\, dx = 6$
3. $A = \int_{-1}^2 [(x^2 - 1) - (-0.5x - 3)]\, dx = 9.750$
4. $A = \int_{-3}^2 [(6 - x^2) - x]\, dx = 20.833$
5. $A = \int_{-2}^1 [(4 - 2x) - 2x^2]\, dx + \int_1^2 [2x^2 - (4 - 2x)]\, dx = 12.667$
6. Index of income concentration = 0.583; income will be less equally distributed in 2010.

EXERCISE 12-1

A *Set up definite integrals in Problems 1–4 that represent the indicated shaded area in figures (A)–(D).*

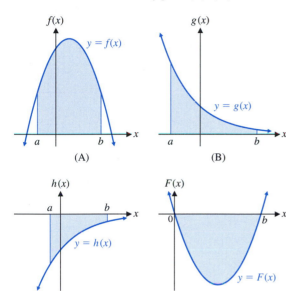

(A)

(B)

(C)

(D)

1. Shaded area in figure (B)
2. Shaded area in figure (A)
3. Shaded area in figure (C)
4. Shaded area in figure (D)

Problems 5 and 6 refer to figure (C) above.

5. Explain why $\int_a^b h(x)\, dx$ does not represent the area between the graph of $y = h(x)$ and the x axis from $x = a$ to $x = b$ in figure (C).

6. Explain why $\int_a^b [-h(x)]\, dx$ represents the area between the graph of $y = h(x)$ and the x axis from $x = a$ to $x = b$ in figure (C).

In Problems 7–18, find the area bounded by the graphs of the indicated equations over the given intervals. Compute answers to three decimal places.

7. $y = 3x^2;\ y = 0,\ 1 \leqslant x \leqslant 2$
8. $y = 4x^3;\ y = 0,\ 1 \leqslant x \leqslant 2$
9. $y = -2x - 1;\ y = 0,\ 0 \leqslant x \leqslant 4$
10. $y = 2x - 4;\ y = 0,\ -2 \leqslant x \leqslant 1$
11. $y = x^2 + 2;\ y = 0,\ -1 \leqslant x \leqslant 0$

12. $y = 3x^2 + 1;\ y = 0,\ -2 \leqslant x \leqslant 0$
13. $y = x^2 - 4;\ y = 0,\ -1 \leqslant x \leqslant 2$
14. $y = 3x^2 - 12;\ y = 0,\ -2 \leqslant x \leqslant 1$
15. $y = e^x;\ y = 0,\ -1 \leqslant x \leqslant 2$
16. $y = e^{-x};\ y = 0,\ -2 \leqslant x \leqslant 1$
17. $y = -1/t;\ y = 0,\ 0.5 \leqslant t \leqslant 1$
18. $y = -1/t;\ y = 0,\ 0.1 \leqslant t \leqslant 1$

B *Set up definite integrals in Problems 19–26 that represent the indicated shaded areas in figures (A) and (B) over the given intervals.*

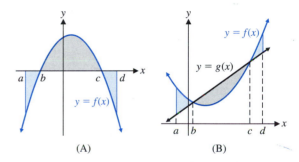

(A)

(B)

19. Over interval $[a, b]$ in figure (A)
20. Over interval $[c, d]$ in figure (A)
21. Over interval $[b, d]$ in figure (A)
22. Over interval $[a, c]$ in figure (A)
23. Over interval $[c, d]$ in figure (B)
24. Over interval $[a, b]$ in figure (B)
25. Over interval $[a, c]$ in figure (B)
26. Over interval $[b, d]$ in figure (B)

Problems 27 and 28 refer to figures (A) and (B) above.

27. Referring to figure (B), explain how you would use definite integrals and the functions f and g to find the area bounded by the two functions from $x = a$ to $x = d$.

28. Referring to figure (A), explain how you would use definite integrals to find the area between the graph of $y = f(x)$ and the x axis from $x = a$ to $x = d$.

In Problems 29–48, find the area bounded by the graphs of the indicated equations over the given intervals (when stated). Compute answers to three decimal places.

29. $y = -x;\ y = 0,\ -2 \leqslant x \leqslant 1$
30. $y = -x + 1;\ y = 0,\ -1 \leqslant x \leqslant 2$

31. $y = x^2 - 4; y = 0, 0 \leq x \leq 3$
32. $y = 4 - x^2; y = 0, 0 \leq x \leq 4$
33. $y = 4 - x^2; y = 0, -3 \leq x \leq 4$
34. $y = x^2 - 4; y = 0, -4 \leq x \leq 3$
35. $y = -2x + 8; y = 12; -1 \leq x \leq 2$
36. $y = 2x + 6; y = 3; -1 \leq x \leq 2$
37. $y = 3x^2; y = 12$
38. $y = x^2; y = 9$
39. $y = 4 - x^2; y = -5$
40. $y = x^2 - 1; y = 3$
41. $y = x^2 + 1; y = 2x - 2; -1 \leq x \leq 2$
42. $y = x^2 - 1; y = x - 2; -2 \leq x \leq 1$

43. $y = e^{0.5x}; y = -\dfrac{1}{x}; 1 \leq x \leq 2$

44. $y = \dfrac{1}{x}; y = -e^x; 0.5 \leq x \leq 1$

45. $y = 3 - 5x - 2x^2; y = 2x^2 + 3x - 2$
46. $y = 3 - 2x^2; y = 2x^2 - 4x$

47. $y = -0.5x + 2.25; y = \dfrac{1}{x}$

48. $y = x - 4.25; y = -\dfrac{1}{x}$

C *In Problems 49–62, find the area bounded by the graphs of the indicated equations over the given intervals (when stated). Compute intersection points and answers to three decimal places.*

49. $y = 10 - 2x; y = 4 + 2x; 0 \leq x \leq 4$
50. $y = 3x; y = x + 5; 0 \leq x \leq 5$
51. $y = x^3; y = 4x$
52. $y = x^3 + 1; y = x + 1$
53. $y = x^3 - 3x^2 - 9x + 12; y = x + 12$
54. $y = x^3 - 6x^2 + 9x; y = x$
55. $y = x^4 - 4x^2 + 1; y = x^2 - 3$
56. $y = x^4 - 6x^2; y = 4x^2 - 9$
57. $y = x^3 - x^2 + 2; y = -x^3 + 8x - 2$
58. $y = 2x^3 + 2x^2 - x; y = -2x^3 - 2x^2 + 2x$
59. $y = e^{-x}; y = 3 - 2x$
60. $y = 2 - (x + 1)^2; y = e^{x+1}$
61. $y = e^x; y = 5x - x^3$
62. $y = 2 - e^x; y = x^3 + 3x^2$

▶ APPLICATIONS

In the following applications it is helpful to look at graphs to get a clearer understanding of a problem and to interpret results.

Business & Economics

63. *Oil production.* Using data from the first 3 years of production as well as geological studies, the management of an oil company estimates that oil will be pumped from a producing field at a rate given by

$$R(t) = \frac{100}{t + 10} + 10 \qquad 0 \leq t \leq 15$$

where $R(t)$ is the rate of production (in thousands of barrels per year) t years after pumping begins. Find the area (to the nearest integer) between the graph of R and the t axis over the interval $[5, 10]$ and interpret the results.

64. *Oil production.* In Problem 63, if the rate is found to be

$$R(t) = \frac{100t}{t^2 + 25} + 4 \qquad 0 \leq t \leq 25$$

find the area (to the nearest integer) between the graph of R and the t axis over the interval $[5, 15]$ and interpret the results.

65. *Useful life.* An amusement company maintains records for each video game it installs in an arcade. Suppose that $C(t)$ and $R(t)$ represent the total accumulated costs and revenues (in thousands of dollars), respectively, t years after a particular game has been installed. If

$$C'(t) = 2 \qquad \text{and} \qquad R'(t) = 9e^{-0.3t}$$

find the area (to three decimal places) between the graphs of C' and R' over the interval on the t axis from 0 to the useful life of the game and interpret the results.

66. *Useful life.* Repeat Problem 65 if

$$C'(t) = 2t \qquad \text{and} \qquad R'(t) = 5te^{-0.1t^2}$$

67. *Income distribution.* As part of a study of the effects of World War II on the economy of the United States, an economist used data from the U.S. Bureau of the Census to produce the following Lorenz curves for distribution of income in the United States in 1935 and in 1947:

$$f(x) = x^{2.4} \quad \text{Lorenz curve for 1935}$$

$$g(x) = x^{1.6} \quad \text{Lorenz curve for 1947}$$

Find the index of income concentration (to three decimal places) for each Lorenz curve and interpret the results.

68. *Income distribution.* Using data from the U.S. Bureau of the Census, an economist produced the following Lorenz curves for distribution of income in the United States in 1962 and in 1972:

$$f(x) = \tfrac{3}{10}x + \tfrac{7}{10}x^2 \quad \text{Lorenz curve for 1962}$$

$$g(x) = \tfrac{1}{2}x + \tfrac{1}{2}x^2 \quad \text{Lorenz curve for 1972}$$

Find the index of income concentration (to three decimal places) for each Lorenz curve and interpret the results.

69. *Distribution of wealth.* Lorenz curves also can be used to provide a relative measure of the distribution of the total assets of a country. Using data in a report by the U.S. Congressional Joint Economic Committee, an economist produced the following Lorenz curves for the distribution of total assets in the United States in 1963 and in 1983:

$$f(x) = x^{10} \quad \text{Lorenz curve for 1963}$$

$$g(x) = x^{12} \quad \text{Lorenz curve for 1983}$$

Find the index of income concentration (to three decimal places) for each Lorenz curve and interpret the results.

70. *Income distribution.* The government of a small country is planning sweeping changes in the tax structure in order to provide a more equitable distribution of income. The Lorenz curves for the current income distribution and for the projected income distribution after enactment of the tax changes are given below. Find the index of income concentration (to three decimal places) for each Lorenz curve. Will the proposed changes provide a more equitable income distribution? Explain.

$$f(x) = x^{2.3} \quad \text{Current Lorenz curve}$$

$$g(x) = 0.4x + 0.6x^2 \quad \text{Projected Lorenz curve after changes in tax laws}$$

Life Sciences

71. *Biology.* A yeast culture is growing at a rate of $W'(t) = 0.3e^{0.1t}$ grams per hour. Find the area (to two decimal places) between the graph of W' and the t axis over the interval $[0, 10]$ and interpret the results.

72. *Natural resource depletion.* The instantaneous rate of change of the demand for lumber in the United States since 1970 ($t = 0$) in billions of cubic feet per year is estimated to be given by

$$Q'(t) = 12 + 0.006t^2 \quad 0 \leq t \leq 50$$

Find the area (to two decimal places) between the graph of Q' and the t axis over the interval $[15, 20]$ and interpret the results.

Social Sciences

73. *Learning.* A beginning high school language class was chosen for an experiment on learning. Using a list of 50 words, the experiment involved measuring the rate of vocabulary memorization at different times during a continuous 5 hour study session. It was found that the average rate of learning for the whole class was inversely proportional to the time spent studying and was given approximately by

$$V'(t) = \frac{15}{t} \quad 1 \leq t \leq 5$$

Find the area (to the nearest integer) between the graph of V' and the t axis over the interval $[2, 4]$ and interpret the results.

74. *Learning.* Repeat Problem 73 if $V'(t) = 13/t^{1/2}$ and the interval is changed to $[1, 4]$.

SECTION 12-2 ■ Applications in Business and Economics

- PROBABILITY DENSITY FUNCTIONS
- CONTINUOUS INCOME STREAM
- FUTURE VALUE OF A CONTINUOUS INCOME STREAM
- CONSUMERS' AND PRODUCERS' SURPLUS

This section contains a number of important applications of the definite integral from business and economics. Included are three independent topics: probability density functions, continuous income streams, and consumers' and producers' surplus. Any of the three may be covered as time and interests dictate, and in any order.

■ PROBABILITY DENSITY FUNCTIONS

We will now take a brief look at the use of the definite integral to determine probabilities. Our approach will be intuitive and informal. A more formal treatment of the subject requires the use of the special "improper" integral form $\int_{-\infty}^{\infty} f(x)\,dx$, which we have not defined or discussed at this point.

Suppose an experiment is designed in such a way that any real number x on the interval $[c, d]$ is a possible outcome. For example, x may represent an IQ score, the height of a person in inches, or the life of a light bulb in hours. Technically, we refer to x as a *continuous random variable*.

In certain situations it is possible to find a function f with x as an independent variable such that the function f can be used to determine the probability that the outcome x of an experiment will be in the interval $[c, d]$. Such a function, called a **probability density function,** must satisfy the following three conditions (see Fig. 1):

1. $f(x) \geqslant 0$ for all real x
2. The area under the graph of $f(x)$ over the interval $(-\infty, \infty)$ is exactly 1.
3. If $[c, d]$ is a subinterval of $(-\infty, \infty)$, then

$$\text{Probability}(c \leqslant x \leqslant d) = \int_{c}^{d} f(x)\,dx$$

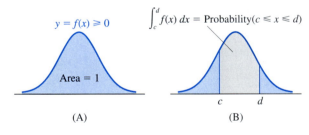

$y = f(x) \geqslant 0$

Area $= 1$

$\int_{c}^{d} f(x)\,dx = \text{Probability}(c \leqslant x \leqslant d)$

(A)

(B)

FIGURE 1
Probability density function

EXAMPLE 1 ➤   **Duration of Telephone Calls**   Suppose the length of telephone calls (in minutes) in a public telephone booth is a continuous random variable with probability density function shown in Figure 2.

$$f(t) = \begin{cases} \frac{1}{4}e^{-t/4} & \text{if } t \geqslant 0 \\ 0 & \text{otherwise} \end{cases}$$

(A) Determine the probability (to two decimal places) that a call selected at random will last between 2 and 3 minutes.

(B) Find b (to two decimal places) so that the probability of a call selected at random lasting between 2 and b minutes is .5.

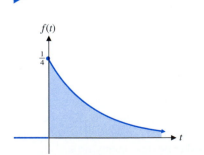

$f(t)$

$\frac{1}{4}$

t

FIGURE 2

Solution   (A) Probability$(2 \leqslant t \leqslant 3) = \int_2^3 \tfrac{1}{4} e^{-t/4} \, dt$

$$= (-e^{-t/4})\big|_2^3$$
$$= -e^{-3/4} + e^{-1/2} \approx .13$$

(B) We want to find b such that Probability$(2 \leqslant t \leqslant b) = .5$; that is,

$$\int_2^b \tfrac{1}{4} e^{-t/4} \, dt = .5 \tag{1}$$

Using the FTC on the left side of (1) produces

$$\int_2^b \tfrac{1}{4} e^{-t/4} \, dt = -e^{-t/4}\big|_2^b = -e^{-b/4} + e^{-1/2}$$

which leads to the exponential equation

$$-e^{-b/4} + e^{-1/2} = .5$$

From Figure 3 we see that $b = 8.96$ minutes. ◄

F I G U R E   3
$y_1 = -e^{-x/4} + e^{-1/2}$
$y_2 = 0.5$

MATCHED PROBLEM 1 ➤   (A) In Example 1, find the probability (to two decimal places) that a call selected at random will last 4 minutes or less.
(B) Find b (to two decimal places) so that the probability of a call selected at random lasting b minutes or less is .9. ◄

One of the most important probability density functions is the **normal probability density function** defined below and graphed in Figure 4.

$$f(x) = \frac{1}{\sigma\sqrt{2\pi}} e^{-(x-\mu)^2/2\sigma^2} \qquad \begin{array}{l} \mu \text{ is the mean.} \\ \sigma \text{ is the standard deviation.} \end{array}$$

F I G U R E   4
Normal curve

It can be shown (but not easily) that the area under the normal curve in Figure 4 over the interval $(-\infty, \infty)$ is exactly 1. Since $\int e^{-x^2} \, dx$ is nonintegrable in terms of elementary functions (that is, the antiderivative cannot be expressed as a finite combination of simple functions), probabilities such as

$$\text{Probability}(c \leqslant x \leqslant d) = \frac{1}{\sigma\sqrt{2\pi}} \int_c^d e^{-(x-\mu)^2/2\sigma^2} \, dx$$

have generally been determined by making an appropriate substitution in the integrand and then using a table of areas under the standard normal curve (that is, the normal curve with $\mu = 0$ and $\sigma = 1$). But now with the readily available numerical integration routines in many graphing calculators, the integral can be approximated very quickly and accurately. For example, if $\mu = 0$ and $\sigma = 1$, then (to three decimal places)

$$\text{Probability}(0 \leqslant x \leqslant 2) = \frac{1}{\sqrt{2\pi}} \int_0^2 e^{-x^2/2} \, dx = .477$$

■ CONTINUOUS INCOME STREAM

We start with a simple example having an obvious solution and generalize the concept to examples having less obvious solutions.

Suppose an aunt has established a trust that pays you $2,000 a year for 10 years. What is the total amount you will receive from the trust by the end of the tenth year? Since there are 10 payments of $2,000 each, you will receive

$$10 \times \$2,000 = \$20,000$$

We now look at the same problem from a different point of view, a point of view that will be useful in more complex problems. Let us assume that the income stream is continuous at a rate of $2,000 per year. In Figure 5 the area under the graph of $f(t) = 2,000$ from 0 to t represents the income accumulated t years after the start. For example, for $t = \frac{1}{4}$ year, the income would be $\frac{1}{4}(2,000) = \$500$; for $t = \frac{1}{2}$ year, the income would be $\frac{1}{2}(2,000) = \$1,000$; for $t = 1$ year, the income would be $1(2,000) = \$2,000$; for $t = 5.3$ years, the income would be $5.3(2,000) = \$10,600$; and for $t = 10$ years, the income would be $10(2,000) = \$20,000$. The total income over a 10 year period—that is, the area under the graph of $f(t) = 2,000$ from 0 to 10—is also given by the definite integral

$$\int_{0}^{10} 2,000 \, dt = 2,000t \Big|_{0}^{10} = 2,000(10) - 2,000(0) = \$20,000$$

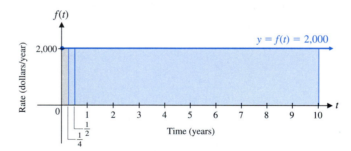

FIGURE 5
Continuous income stream

We now apply the idea of a continuous income stream to a less obvious problem.

EXAMPLE 2 ▶   **Continuous Income Stream**   The rate of change of the income produced by a vending machine located at an airport is given by

$$f(t) = 5,000e^{0.04t}$$

where t is time in years since the installation of the machine. Find the total income (to the nearest dollar) produced by the machine during the first 5 years of operation.

Solution   The area under the graph of the rate of change function from 0 to 5 represents the total change in income over the first 5 years (Fig. 6), and hence is given by a definite integral:

$$\text{Total income} = \int_0^5 5{,}000 e^{0.04t}\, dt = \$27{,}675$$

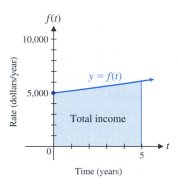

FIGURE 6
Continuous income stream

Thus, the vending machine produces a total income of $27,675 during the first 5 years of operation.   ◄

MATCHED PROBLEM 2 ➤   Referring to Example 2, find the total income (to the nearest dollar) produced during the second 5 years of operation.   ◄

     In reality, income from a vending machine is not usually received as a single payment at the end of each year, even though the rate is given as a yearly rate. Income is usually collected on a daily or weekly basis. In problems of this type it is convenient to assume that income is actually received in a **continuous stream;** that is, we assume that income is a continuous function of time and the rate of change is an instantaneous rate. The rate of change is called the **rate of flow** of the continuous income stream. In general, we have the following:

Total Income for a Continuous Income Stream

If $f(t)$ is the rate of flow of a continuous income stream, then the **total income** produced during the time period from $t = a$ to $t = b$ is

$$\text{Total income} = \int_a^b f(t)\, dt$$

■ FUTURE VALUE OF A CONTINUOUS INCOME STREAM

In Section 10-1, we discussed the continuous compound interest formula

$$A = Pe^{rt}$$

where P is the principal (or present value), A is the amount (or future value), r is the annual rate of continuous compounding (expressed as a decimal), and t is time in years. For example, if money is worth 12% compounded continuously, then the future value of a $10,000 investment in 5 years is (to the nearest dollar)

$$A = 10,000e^{0.12(5)} = \$18,221$$

Now we want to apply the future value concept to the income produced by a continuous income stream. Suppose $f(t)$ is the rate of flow of a continuous income stream, and the income produced by this continuous income stream is invested as soon as it is received at a rate r, compounded continuously. We already know how to find the total income produced after T years, but how can we find the total of the income produced and the interest earned by this income? Since the income is received in a continuous flow, we cannot just use the formula $A = Pe^{rt}$. This formula is valid only for a single deposit P, not for a continuous flow of income. Instead, we use a Riemann sum approach that will allow us to apply the formula $A = Pe^{rt}$ repeatedly. To begin, we divide the time interval $[0, T]$ into n equal subintervals of length Δt and choose an arbitrary point c_k in each subinterval, as illustrated in Figure 7.

The total income produced during the time period from $t = t_{k-1}$ to $t = t_k$ is equal to the area under the graph of $f(t)$ over this subinterval and is approximately equal to $f(c_k)\Delta t$, the area of the shaded rectangle in Figure 7. The income received during this time period will earn interest for approximately $T - c_k$ years. Thus, using the future value formula $A = Pe^{rt}$ with $P = f(c_k)\Delta t$ and $t = T - c_k$, the future value of the income produced during the time period from $t = t_{k-1}$ to $t = t_k$ is approximately equal to

$$f(c_k)\Delta te^{(T-c_k)r}$$

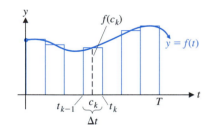

FIGURE 7

The total of these approximate future values over n subintervals is then

$$f(c_1)\Delta te^{(T-c_1)r} + f(c_2)\Delta te^{(T-c_2)r} + \cdots + f(c_n)\Delta te^{(T-c_n)r} = \sum_{k=1}^{n} f(c_k)e^{r(T-c_k)}\Delta t$$

This has the form of a Riemann sum, and the limit of this sum is a definite integral. (See the definition of definite integral in Section 11-5.) Thus, the *future value, FV,* of the income produced by the continuous income stream is given by

$$FV = \int_0^T f(t)e^{r(T-t)}\, dt$$

Since r and T are constants, we also can write

$$FV = \int_0^T f(t)e^{rT}e^{-rt}\, dt = e^{rT}\int_0^T f(t)e^{-rt}\, dt \tag{2}$$

This last form is preferable, since the integral is usually easier to evaluate than the first form.

Future Value of a Continuous Income Stream

If $f(t)$ is the rate of flow of a continuous income stream, $0 \leqslant t \leqslant T$, and if the income is continuously invested at a rate r compounded continuously, then the **future value, FV,** at the end of T years is given by

$$FV = \int_0^T f(t)e^{r(T-t)}\, dt = e^{rT}\int_0^T f(t)e^{-rt}\, dt$$

The future value of a continuous income stream is the total value of all money produced by the continuous income stream (income and interest) at the end of T years.

We return to the trust set up for you by your aunt. Suppose the $2,000 per year you receive from the trust is invested as soon as it is received at 8% compounded continuously. We consider the trust income a continuous income stream with a flow rate of $2,000 per year. What is its future value (to the nearest dollar) by the end of the 10th year? Using the definite integral for future value from the box, we have

$$FV = e^{rT}\int_0^T f(t)e^{-rt}\, dt$$

$$FV = e^{0.08(10)}\int_0^{10} 2,000e^{-0.08t}\, dt \quad r = 0.08,\ T = 10,\ f(t) = 2,000$$

$$= \$30,639$$

Thus, at the end of 10 years you will have received $30,639, including interest. How much is interest? Since you received $20,000 in income from the trust, the interest is the difference between the future value and income. Thus,

$$\$30,639 - \$20,000 = \$10,639$$

is the interest earned by the income received from the trust over the 10 year period. We now apply the same analysis to Example 2, the slightly more involved vending machine problem.

EXAMPLE 3 ➤ **Future Value of a Continuous Income Stream** Using the continuous income rate of flow for the vending machine in Example 2,

$$f(t) = 5,000e^{0.04t}$$

find the future value of this income stream at 12% compounded continuously for 5 years, and find the total interest earned. Compute answers to the nearest dollar.

Solution   Using the formula

$$FV = e^{rT} \int_0^T f(t)e^{-rt} \, dt$$

with $r = 0.12$, $T = 5$, and $f(t) = 5{,}000e^{0.04t}$, we have

$$FV = e^{0.12(5)} \int_0^5 5{,}000e^{0.04t}e^{-0.12t} \, dt$$

$$= 5{,}000e^{0.6} \int_0^5 e^{-0.08t} \, dt = \$37{,}545$$

Thus, the future value of the income stream at 12% compounded continuously at the end of 5 years is $37,545.

In Example 2, we saw that the total income produced by this vending machine over a 5 year period was $27,675. The difference between the future value and income is interest. Thus,

$$\$37{,}545 - \$27{,}675 = \$9{,}870$$

is the interest earned by the income produced by the vending machine during the 5 year period.   ◀

MATCHED PROBLEM 3 ➤   Repeat Example 3 if the interest rate is 9% compounded continuously.   ◀

■ CONSUMERS' AND PRODUCERS' SURPLUS

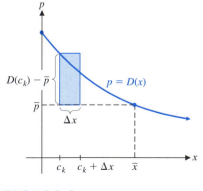

FIGURE 8

Let $p = D(x)$ be the price–demand equation for a product, where x is the number of units of the product that consumers will purchase at a price of $\$p$ per unit. Suppose $\bar{p}$ is the current price and $\bar{x}$ is the number of units that can be sold at that price. The price–demand curve in Figure 8 shows that if the price is higher than $\bar{p}$, then the demand x is less than $\bar{x}$, but some consumers are still willing to pay the higher price. Consumers who are willing to pay more than $\bar{p}$ but are still able to buy the product at $\bar{p}$ have saved money. We want to determine the total amount saved by all the consumers who are willing to pay a price higher than $\bar{p}$ for this product.

To do this, consider the interval $(c_k, \ c_k + \Delta x]$, where $c_k + \Delta x < \bar{x}$. If the price remained constant over this interval, then the savings on each unit would be the difference between $D(c_k)$, the price consumers are willing to pay, and $\bar{p}$, the price they actually pay. Since Δx represents the number of units purchased by consumers over the interval, the total savings to consumers over this interval is approximately equal to

$$[D(c_k) - \bar{p}]\Delta x \quad \text{(Savings per unit)} \times \text{(Number of units)}$$

which is the area of the shaded rectangle shown in Figure 8. If we divide the interval $[0, \bar{x}]$ into n equal subintervals, then the total savings to consumers is approximately equal to

$$[D(c_1) - \bar{p}]\Delta x + [D(c_2) - \bar{p}]\Delta x + \cdots + [D(c_n) - \bar{p}]\Delta x = \sum_{k=1}^{n} [D(c_k) - \bar{p}]\Delta x$$

which we recognize as a Riemann sum for the following integral:

$$\int_{0}^{\bar{x}} [D(x) - \bar{p}]\, dx$$

Thus, we define the *consumers' surplus* to be this integral.

Consumers' Surplus

If $(\bar{x}, \bar{p})$ is a point on the graph of the price–demand equation $p = D(x)$ for a particular product, then the **consumers' surplus, CS,** at a price level of $\bar{p}$ is

$$CS = \int_{0}^{\bar{x}} [D(x) - \bar{p}]\, dx$$

which is the area between $p = \bar{p}$ and $p = D(x)$ from $x = 0$ to $x = \bar{x}$.

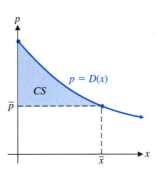

The consumers' surplus represents the total savings to consumers who are willing to pay more than $\bar{p}$ for the product but are still able to buy the product for $\bar{p}$.

EXAMPLE 4 ➤ Consumers' Surplus Find the consumers' surplus at a price level of $8 for the price–demand equation

$$p = D(x) = 20 - 0.05x$$

Solution   **Step 1.** Find $\bar{x}$, the demand when the price is $\bar{p} = 8$. From Figure 9, we see that $\bar{x} = 240$ when $\bar{p} = 8$.

Step 2. Sketch a graph showing the consumers' surplus region (Fig. 10):

FIGURE 9
$y_1 = 20 - 0.05x$
$y_2 = 8$

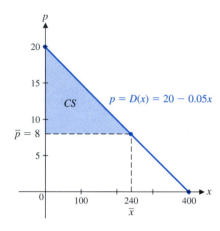

FIGURE 10

Step 3. Find the consumers' surplus (the shaded area in the graph):

$$CS = \int_0^{\bar{x}} [D(x) - \bar{p}]\, dx$$

$$= \int_0^{240} (20 - 0.05x - 8)\, dx$$

$$= \int_0^{240} (12 - 0.05x)\, dx$$

$$= \$1{,}440$$

Thus, the total savings to consumers who are willing to pay a higher price for the product is $1,440.   ◄

MATCHED PROBLEM 4 ➤   Repeat Example 4 for a price level of $4.   ◄

If $p = S(x)$ is the price–supply equation for a product, $\bar{p}$ is the current price, and $\bar{x}$ is the current supply, then some suppliers are still willing to supply some units at a lower price than $\bar{p}$. The additional money that these suppliers gain from the higher price is called the *producers' surplus* and can be expressed in terms of a definite integral (proceeding as we did for the consumers' surplus).

Producers' Surplus
___

If $(\bar{x}, \bar{p})$ is a point on the graph of the price–supply equation $p = S(x)$, then the **producers' surplus, PS,** at a price level of $\bar{p}$ is

$$PS = \int_0^{\bar{x}} [\bar{p} - S(x)]\, dx$$

which is the area between $p = \bar{p}$ and $p = S(x)$ from $x = 0$ to $x = \bar{x}$.

The producers' surplus represents the total gain to producers who are willing to supply units at a lower price than $\bar{p}$ but are still able to supply units at $\bar{p}$.

EXAMPLE 5 ➤ **Producers' Surplus** Find the producers' surplus (to the nearest dollar) at a price level of $20 for the price–supply equation

$$p = S(x) = 2 + 0.0002x^2$$

Solution **Step 1.** Find $\bar{x}$, the supply when the price is $\bar{p} = 20$. From Figure 11, we see that $\bar{x} = 300$ when $\bar{p} = 20$.

FIGURE 11
$y_1 = 2 + 0.0002x^2$
$y_2 = 20$

Step 2. Sketch a graph (Fig. 12).

Step 3. Find the producers' surplus (the shaded area in the graph):

$$PS = \int_0^{\bar{x}} [\bar{p} - S(x)]\, dx = \int_0^{300} [20 - (2 + 0.0002x^2)]\, dx$$

$$= \int_0^{300} (18 - 0.0002x^2)\, dx$$

$$= \$3,600$$

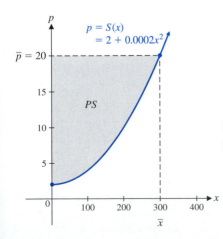

FIGURE 12

Thus, the total gain to producers who are willing to supply units at a lower price is $3,600.   ◀

MATCHED PROBLEM 5 ▶ Repeat Example 5 for a price level of $4.   ◀

In a free competitive market, the price of a product is determined by the relationship between supply and demand. If $p = D(x)$ and $p = S(x)$ are the price–demand and price–supply equations, respectively, for a product and if $(\bar{x}, \bar{p})$ is the point of intersection of these equations, then $\bar{p}$ is called the **equilibrium price** and $\bar{x}$ is called the **equilibrium quantity.** If the price stabilizes at the equilibrium price $\bar{p}$, then this is the price level that will determine both the consumers' surplus and the producers' surplus.

▶ EXAMPLE 6 ▶ **Equilibrium Price and Consumers' and Producers' Surplus**   Find the equilibrium price and then find the consumers' surplus and producers' surplus at the equilibrium price level (round answers to the nearest dollar) if

$$p = D(x) = 20 - 0.05x \qquad \text{and} \qquad p = S(x) = 2 + 0.0002x^2$$

Solution   Step 1.   Plot the price–demand and supply–demand equations in the same viewing window and find the point of intersection, which is the equilibrium point. From Figure 13 we see that the equilibrium price is $\bar{p} = \$10$ and the equilibrium quantity is $\bar{x} = 200$.

Step 2.   Sketch a graph showing the consumers' and producers' surplus (Fig. 14) using Figure 13 as an aid.

FIGURE 13
$y_1 = 20 - 0.05x$
$y_2 = 2 + 0.0002x^2$

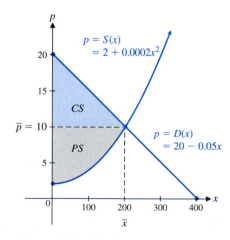

FIGURE 14

Step 3. Find the consumers' surplus:

$$CS = \int_0^{\bar{x}} [D(x) - \bar{p}] \, dx$$

$$= \int_0^{200} (20 - 0.05x - 10) \, dx$$

$$= \int_0^{200} (10 - 0.05x) \, dx$$

$$= \$1,000$$

Step 4. Find the producers' surplus:

$$PS = \int_0^{\bar{x}} [\bar{p} - S(x)] \, dx$$

$$= \int_0^{200} [10 - (2 + 0.0002x^2)] \, dx$$

$$= \int_0^{200} (8 - 0.0002x^2) \, dx$$

$$= \$1,067 \qquad \blacktriangleleft$$

MATCHED PROBLEM 6 ➤ Repeat Example 6 for

$$p = D(x) = 25 - 0.001x^2 \qquad \text{and} \qquad p = S(x) = 5 + 0.1x \qquad \blacktriangleleft$$

Answers to Matched Problems   1. (A) .63   (B) 9.21 min   2. \$33,803   3. $FV = \$34,691$; interest = \$7,016
4. \$2,560   5. \$133   6. $\bar{p} = 15$; $CS = \$667$; $PS = \$500$

EXERCISE 12-2

▶ **APPLICATIONS**

Business & Economics

Unless stated to the contrary, compute all monetary answers to the nearest dollar and all probabilities to two decimal places.

1. The life expectancy (in years) of a certain brand of clock radio is a continuous random variable with probability density function

$$f(x) = \begin{cases} 2/(x + 2)^2 & \text{if } x \geq 0 \\ 0 & \text{otherwise} \end{cases}$$

(A) Find the probability that a randomly selected clock radio lasts at most 6 years.

(B) Find the probability that a randomly selected clock radio lasts from 6 to 12 years.

(C) Graph $y = f(x)$ for [0, 12] and show the shaded region for part (A).

2. The shelf-life (in years) of a certain brand of flashlight batteries is a continuous random variable with probability density function

$$f(x) = \begin{cases} 1/(x + 1)^2 & \text{if } x \geq 0 \\ 0 & \text{otherwise} \end{cases}$$

(A) Find the probability that a randomly selected battery has a shelf-life of 3 years or less.

(B) Find the probability that a randomly selected battery has a shelf-life of from 3 to 9 years.

(C) Graph $y = f(x)$ for [0, 10] and show the shaded region for part (A).

3. In Problem 1, find d so that the probability of a randomly selected clock radio lasting d years or less is .8.

4. In Problem 2, find d so that the probability of a randomly selected battery lasting d years or less is .5.

5. A manufacturer guarantees a product for 1 year. The time to failure of the product after it is sold is given by the probability density function

$$f(t) = \begin{cases} .01e^{-.01t} & \text{if } t \geq 0 \\ 0 & \text{otherwise} \end{cases}$$

where t is time in months. What is the probability that a buyer chosen at random will have a product failure:

(A) During the warranty period?

(B) During the second year after purchase?

6. In a certain city, the daily use of water (in hundreds of gallons) per household is a continuous random variable with probability density function

$$f(x) = \begin{cases} .15e^{-.15x} & \text{if } x \geq 0 \\ 0 & \text{otherwise} \end{cases}$$

Find the probability that a household chosen at random will use:

(A) At most 400 gallons of water per day

(B) Between 300 and 600 gallons of water per day

7. In Problem 5, what is the probability that the product will last at least 1 year? [*Hint:* Recall that the total area under the probability density function curve is 1.]

8. In Problem 6, what is the probability that a household will use more than 400 gallons of water per day? [See the hint in Problem 7.]

9. Find the total income produced by a continuous income stream in the first 5 years if the rate of flow is $f(t) = 2,500$.

10. Find the total income produced by a continuous income stream in the first 10 years if the rate of flow is $f(t) = 3,000$.

11. Interpret the results in Problem 9 with both a graph and a verbal description of the graph.

12. Interpret the results in Problem 10 with both a graph and a verbal description of the graph.

13. Find the total income produced by a continuous income stream in the first 3 years if the rate of flow is $f(t) = 400e^{0.05t}$.

14. Find the total income produced by a continuous income stream in the first 2 years if the rate of flow is $f(t) = 600e^{0.06t}$.

15. Interpret the results in Problem 13 with both a graph and a verbal description of the graph.

16. Interpret the results in Problem 14 with both a graph and a verbal description of the graph.

17. Starting at age 25, you deposit $2,000 a year into an IRA account for retirement. Treat the yearly deposits into the account as a continuous income stream. If money in the account earns 5% compounded continuously, how much will be in the account 40 years later when you retire at age 65? How much of the final amount is interest?

18. Suppose in Problem 17 that you start the IRA deposits at age 30, but the account earns 6% compounded continuously. Treat the yearly deposits into the account as a continuous income stream. How much will be in the account 35 years later when you retire at age 65? How much of the final amount is interest?

19. Find the future value at 10% interest compounded continuously for 4 years for the continuous income stream with rate of flow $f(t) = 1,500e^{-0.02t}$.

20. Find the future value at 7% interest compounded continuously for 6 years for the continuous income stream with rate of flow $f(t) = 2,000e^{0.06t}$.

21. Compute the interest earned in Problem 19.

22. Compute the interest earned in Problem 20.

23. An investor is presented with a choice of two investments, an established clothing store and a new computer store. Each choice requires the same initial investment and each produces a continuous income stream of 10% compounded continuously. The rate of flow of income from the clothing store is $f(t) = 12,000$, and the rate of flow of income from the computer store is expected to be $g(t) = 10,000e^{0.05t}$. Compare the future values of these investments to determine which is the better choice over the next 5 years.

24. Refer to Problem 23. Which investment is the better choice over the next 10 years?

25. An investor has $10,000 to invest in either a bond that matures in 5 years or a business that will produce a continuous stream of income over the next 5 years with rate of flow $f(t) = 2,000$. If both the bond and the continuous income stream earn 8% compounded continuously, which is the better investment?

26. Refer to Problem 25. Which is the better investment if the rate of flow of the income from the business is $f(t) = 3,000$?

27. A business is planning to purchase a piece of equipment that will produce a continuous stream of income for 8 years with rate of flow $f(t) = 9,000$. If the continuous income stream earns 12% compounded continuously, what single deposit into an account earning the same interest rate will produce the same future value as the continuous income stream? (This deposit is called the **present value** of the continuous income stream.)

28. Refer to Problem 27. Find the present value of a continuous income stream at 8% compounded continuously for 12 years if the rate of flow is $f(t) = 1,000e^{0.03t}$.

29. Find the future value at a rate r compounded continuously for T years for a continuous income stream with rate of flow $f(t) = k$, where k is a constant.

30. Find the future value at a rate r compounded continuously for T years for a continuous income stream with rate of flow $f(t) = ke^{ct}$, where c and k are constants, $c \neq r$.

31. Find the consumers' surplus at a price level of $\bar{p} = \$150$ for the price–demand equation

$$p = D(x) = 400 - 0.05x$$

32. Find the consumers' surplus at a price level of $\bar{p} = \$120$ for the price–demand equation

$$p = D(x) = 200 - 0.02x$$

33. Interpret the results in Problem 31 with both a graph and a verbal description of the graph.

34. Interpret the results in Problem 32 with both a graph and a verbal description of the graph.

35. Find the producers' surplus at a price level of $\bar{p} = \$67$ for the price–supply equation

$$p = S(x) = 10 + 0.1x + 0.0003x^2$$

36. Find the producers' surplus at a price level of $\bar{p} = \$55$ for the price–supply equation

$$p = S(x) = 15 + 0.1x + 0.003x^2$$

37. Interpret the results in Problem 35 with both a graph and a verbal description of the graph.

38. Interpret the results in Problem 36 with both a graph and a verbal description of the graph.

In Problems 39–46, find the consumers' surplus and the producers' surplus at the equilibrium price level for the given price–demand and price–supply equations. Include a graph that identifies the consumers' surplus and the producers' surplus. Round all values to the nearest integer.

39. $p = D(x) = 50 - 0.1x; p = S(x) = 11 + 0.05x$
40. $p = D(x) = 25 - 0.004x^2; p = S(x) = 5 + 0.004x^2$
41. $p = D(x) = 80e^{-0.001x}; p = S(x) = 30e^{0.001x}$
42. $p = D(x) = 185e^{-0.005x}; p = S(x) = 25e^{0.005x}$
43. $p = D(x) = 80 - 0.04x; p = S(x) = 30e^{0.001x}$
44. $p = D(x) = 190 - 0.2x; p = S(x) = 25e^{0.005x}$
45. $p = D(x) = 80e^{-0.001x}; p = S(x) = 15 + 0.0001x^2$
46. $p = D(x) = 185e^{-0.005x}; p = S(x) = 20 + 0.002x^2$

S E C T I O N   12-3   ■   **Integration by Parts**

In Section 11-1, we said we would return later to the indefinite integral

$$\int \ln x \, dx$$

since none of the integration techniques considered up to that time could be used to find an antiderivative for ln x. We will now develop a very useful technique, called *integration by parts,* that will enable us to find not only the above integral, but also many others, including integrals such as

$$\int x \ln x \, dx \qquad \text{and} \qquad \int xe^x \, dx$$

The technique of integration by parts is also used to derive many integration formulas that are tabulated in mathematical handbooks. Some of these handbook formulas are discussed in the next section.

The method of integration by parts is based on the product formula for derivatives. If f and g are differentiable functions, then

$$\frac{d}{dx}\,[f(x)g(x)] = f(x)g'(x) + g(x)f'(x)$$

which can be written in the equivalent form

$$f(x)g'(x) = \frac{d}{dx}\,[f(x)g(x)] - g(x)f'(x)$$

Integrating both sides, we obtain

$$\int f(x)g'(x)\,dx = \int \frac{d}{dx}\,[f(x)g(x)]\,dx - \int g(x)f'(x)\,dx$$

The first integral to the right of the equal sign is $f(x)g(x) + C$. (Why?) We will leave out the constant of integration for now, since we can add it after integrating the second integral to the right of the equal sign. So we have

$$\int f(x)g'(x)\,dx = f(x)g(x) - \int g(x)f'(x)\,dx$$

This equation can be transformed into a more convenient form by letting $u = f(x)$ and $v = g(x)$; then $du = f'(x)\,dx$ and $dv = g'(x)\,dx$. Making these substitutions, we obtain the **integration by parts formula:**

Integration by Parts Formula

$$\int u\,dv = uv - \int v\,du$$

This formula can be very useful when the integral on the left is difficult or impossible to integrate using standard formulas. If u and dv are chosen with care—this is the crucial part of the process—then the integral on the right side may be easier to integrate than the one on the left. The formula provides us with another tool that is helpful in many, but not all cases. We are able to easily check the results by differentiating to get the original integrand, a good habit to get into. Several examples will demonstrate the use of the formula.

EXAMPLE 1 ➤ **Integration by Parts**   Find $\int xe^x\,dx$ using integration by parts and check the result.

Solution   First, write the integration by parts formula:

$$\int u\,dv = uv - \int v\,du \tag{1}$$

Now try to identify u and dv in $\int xe^x \, dx$ so that the integral $\int v \, du$ on the right side of (1) is easier to integrate than $\int u \, dv = \int xe^x \, dx$ on the left side. There are essentially two reasonable choices in selecting u and dv in $\int xe^x \, dx$:

Choice 1   Choice 2

$u \quad dv \qquad\qquad u \quad dv$

$\int x \ e^x \, dx \qquad \int e^x \ x \, dx$

We will pursue choice 1 and leave choice 2 for you to explore (see Explore–Discuss 1 following this example).

From choice 1, $u = x$ and $dv = e^x \, dx$. Looking at formula (1), we need du and v to complete the right side. It is convenient to proceed with the following arrangement: Let

$$u = x \qquad\qquad dv = e^x \, dx$$

then

$$du = dx \qquad \int dv = \int e^x \, dx$$
$$v = e^x$$

Any constant may be added to v, but we will always choose 0 for simplicity. The general arbitrary constant of integration will be added at the end of the process.

Substituting these results in formula (1), we obtain

$$\int u \, dv = uv - \int v \, du$$
$$\int xe^x \, dx = xe^x - \int e^x \, dx \qquad \text{The right integral is easy to integrate.}$$
$$= xe^x - e^x + C \qquad \text{Now add the arbitrary constant } C.$$

Check $\qquad \dfrac{d}{dx}(xe^x - e^x + C) = xe^x + e^x - e^x = xe^x$ ◀

---

explore – discuss 1

Pursue choice 2 in Example 1 using the integration by parts formula, and explain why this choice does not work out.

MATCHED PROBLEM 1 ▶   Find: $\int xe^{2x} \, dx$ ◀

EXAMPLE 2 ▶   Integration by Parts   Find: $\int x \ln x \, dx$

Solution   As before, we have essentially two choices in choosing u and dv:

Choice 1   Choice 2

$u \qquad dv \qquad\qquad u \qquad dv$

$\int x \ \ln x \, dx \qquad \int \ln x \ x \, dx$

Choice 1 is rejected, since we do not as yet know how to find an antiderivative of $\ln x$. So we move to choice 2 and choose $u = \ln x$ and $dv = x\, dx$; then proceed as in Example 1. Let

$$u = \ln x \qquad\qquad dv = x\, dx$$

then

$$du = \frac{1}{x}\, dx \qquad\qquad \int dv = \int x\, dx$$

$$v = \frac{x^2}{2}$$

Substitute these results into the integration by parts formula:

$$\int u\, dv = uv - \int v\, du$$

$$\int x \ln x\, dx = (\ln x)\left(\frac{x^2}{2}\right) - \int \left(\frac{x^2}{2}\right)\left(\frac{1}{x}\right) dx$$

$$= \frac{x^2}{2} \ln x - \int \frac{x}{2}\, dx \qquad\qquad \text{An easy integral to evaluate}$$

$$= \frac{x^2}{2} \ln x - \frac{x^2}{4} + C$$

Check $\quad \dfrac{d}{dx}\left(\dfrac{x^2}{2} \ln x - \dfrac{x^2}{4} + C\right) = x \ln x + \left(\dfrac{x^2}{2} \cdot \dfrac{1}{x}\right) - \dfrac{x}{2} = x \ln x$ ◄

MATCHED PROBLEM 2 ➤ Find: $\int x \ln 2x\, dx$ ◄

As you should have discovered in Explore–Discuss 1, some choices for u and dv will lead to integrals that are more complicated than the original integral. This does not mean there is an error in the calculations or the integration by parts formula. It simply means that the particular choice of u and dv does not change the problem into one we can solve. When this happens, we must look for a different choice of u and dv. In some problems, it is possible that no choice will work. These observations and some guidelines for selecting u and dv are summarized in the box.

Integration by Parts: Selection of u and dv

For $\int u\, dv = uv - \int v\, du$:

1. The product $u\, dv$ must equal the original integrand.
2. It must be possible to integrate dv (preferably by using standard formulas or simple substitutions).

Integration by Parts: Selection of u and dv (Continued)

---

3. The new integral $\int v\, du$ should not be any more involved than the original integral $\int u\, dv$.
4. For integrals involving $x^p e^{ax}$, try

$$u = x^p \qquad \text{and} \qquad dv = e^{ax}\, dx$$

5. For integrals involving $x^p(\ln x)^q$, try

$$u = (\ln x)^q \qquad \text{and} \qquad dv = x^p\, dx$$

In some cases, repeated use of the integration by parts formula will lead to the evaluation of the original integral. The next example provides an illustration of such a case.

EXAMPLE 3 ➤ **Repeated Use of Integration by Parts** Find: $\int x^2 e^{-x}\, dx$

Solution   Following suggestion 4 in the box, we choose

$$u = x^2 \qquad\qquad dv = e^{-x}\, dx$$

Then,

$$du = 2x\, dx \qquad\qquad v = -e^{-x}$$

and

$$
\begin{aligned}
\int x^2 e^{-x}\, dx &= x^2(-e^{-x}) - \int (-e^{-x})2x\, dx \\
&= -x^2 e^{-x} + 2 \int x e^{-x}\, dx
\end{aligned}
\tag{2}
$$

The new integral is not one we can evaluate by standard formulas, but it is simpler than the original integral. Applying the integration by parts formula to it will produce an even simpler integral. For the integral $\int x e^{-x}\, dx$, we choose

$$u = x \qquad\qquad dv = e^{-x}\, dx$$

Then,

$$du = dx \qquad\qquad v = -e^{-x}$$

and

$$
\begin{aligned}
\int x e^{-x}\, dx &= x(-e^{-x}) - \int (-e^{-x})\, dx \\
&= -x e^{-x} + \int e^{-x}\, dx \\
&= -x e^{-x} - e^{-x} \qquad\qquad \textit{Choose 0 for the constant.}
\end{aligned}
\tag{3}
$$

Substituting (3) into (2), we have

$$
\begin{aligned}
\int x^2 e^{-x}\, dx &= -x^2 e^{-x} + 2(-x e^{-x} - e^{-x}) + C \qquad \textit{Add an arbitrary constant here.} \\
&= -x^2 e^{-x} - 2x e^{-x} - 2e^{-x} + C
\end{aligned}
$$

Check $\quad \dfrac{d}{dx}(-x^2 e^{-x} - 2xe^{-x} - 2e^{-x} + C) = x^2 e^{-x} - 2xe^{-x} + 2xe^{-x} - 2e^{-x} + 2e^{-x}$

$$= x^2 e^{-x} \qquad \blacktriangleleft$$

MATCHED PROBLEM 3 ➤ Find: $\int x^2 e^{2x}\, dx$ $\qquad \blacktriangleleft$

EXAMPLE 4 ➤ **Using Integration by Parts** Find $F(x) = \int_1^x \ln t\, dt$, $x \geq 1$, and interpret the result geometrically.

Solution First, we will find $\int \ln t\, dt$, and then return to the definite integral. Following suggestion 5 in the box (with $p = 0$), we choose

$$u = \ln t \qquad dv = dt$$

Then,

$$du = \dfrac{1}{t}\, dt \qquad v = t$$

Hence,

$$\int \ln t\, dt = (\ln t)(t) - \int (t)\dfrac{1}{t}\, dt$$

$$= t \ln t - t + C$$

Note that this is the important result we mentioned at the beginning of this section. Now, we have

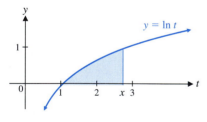

$$F(x) = \int_1^x \ln t\, dt = (t \ln t - t)\,\Big|_1^x$$

$$= (x \ln x - x) - (1 \ln 1 - 1)$$

$$= x \ln x - x + 1 \qquad x \geq 1$$

The integral represents the area under the curve $y = \ln t$ from $t = 1$ to $t = x$, as shown in Figure 1. $\qquad \blacktriangleleft$

FIGURE 1

MATCHED PROBLEM 4 ➤ Find: $G(x) = \int_1^x \ln 3t\, dt$, $x \geq 1$ $\qquad \blacktriangleleft$

explore – discuss 2

Try using the integration by parts formula for $\int e^{x^2}\, dx$ and explain why it doesn't solve the problem.

Answers to Matched Problems

1. $\dfrac{x}{2} e^{2x} - \dfrac{1}{4} e^{2x} + C$ $\quad$ 2. $\dfrac{x^2}{2} \ln 2x - \dfrac{x^2}{4} + C$

3. $\dfrac{x^2}{2} e^{2x} - \dfrac{x}{2} e^{2x} + \dfrac{1}{4} e^{2x} + C$ $\quad$ 4. $G(x) = x \ln 3x - x - \ln 3 + 1$, $x \geq 1$

EXERCISE 12-3

A *In Problems 1–4, integrate using the method of integration by parts. Assume x > 0 whenever the natural logarithm function is involved.*

1. $\int xe^{3x}\, dx$

2. $\int xe^{4x}\, dx$

3. $\int x^2 \ln x\, dx$

4. $\int x^3 \ln x\, dx$

B *Problems 5–18 are mixed—some require integration by parts and others can be solved using techniques we have considered earlier. Integrate as indicated, assuming x > 0 whenever the natural logarithm function is involved.*

5. $\int xe^{-x}\, dx$

6. $\int (x - 1)e^{-x}\, dx$

7. $\int xe^{x^2}\, dx$

8. $\int xe^{-x^2}\, dx$

9. $\int_0^x (t - 3)e^t\, dt,\ x \geq 0$

10. $\int_0^x (t + 1)e^t\, dt,\ x \geq 0$

11. $\int_1^x \ln 2t\, dt,\ x \geq 1$

12. $\int_1^x \ln\left(\dfrac{t}{2}\right) dt,\ x \geq 1$

13. $\int \dfrac{2x}{x^2 + 1}\, dx$

14. $\int \dfrac{x^2}{x^3 + 5}\, dx$

15. $\int \dfrac{\ln x}{x}\, dx$

16. $\int \dfrac{e^x}{e^x + 1}\, dx$

17. $\int \sqrt{x} \ln x\, dx$

18. $\int \dfrac{\ln x}{\sqrt{x}}\, dx$

In Problems 19–22, illustrate each integral graphically and describe what each integral represents in terms of areas.

19. Problem 9

20. Problem 10

21. Problem 11

22. Problem 12

C *Problems 23–40 are mixed—some may require use of the integration by parts formula along with techniques we have considered earlier; others may require repeated use of the integration by parts formula. Assume g(x) > 0 whenever ln g(x) is involved.*

23. $\int x^2 e^x\, dx$

24. $\int x^3 e^x\, dx$

25. $\int xe^{ax}\, dx,\ a \neq 0$

26. $\int \ln ax\, dx,\ a > 0$

27. $\int_1^x \dfrac{\ln t}{t^2}\, dt,\ x \geq 1$

28. $\int_1^x t^3 e^{t^2}\, dt,\ x \geq 1$

29. $\int_0^x \ln(t + 4)\, dt,\ x \geq 0$

30. $\int_0^x \ln(4 - t)\, dt,\ x \geq 0$

31. $\int xe^{x-2}\, dx$

32. $\int xe^{x+1}\, dx$

33. $\int x \ln(1 + x^2)\, dx$

34. $\int x \ln(1 + x)\, dx$

35. $\int e^x \ln(1 + e^x)\, dx$

36. $\int \dfrac{\ln(1 + \sqrt{x})}{\sqrt{x}}\, dx$

37. $\int (\ln x)^2\, dx$

38. $\int x(\ln x)^2\, dx$

39. $\int (\ln x)^3\, dx$

40. $\int x(\ln x)^3\, dx$

▶ **APPLICATIONS**

Business & Economics

41. *Sales analysis.* The monthly sales of a particular personal computer are expected to decline at the rate of

$$S'(t) = -4te^{0.1t}$$

computers per month, where t is time in months and $S(t)$ is the number of computers sold each month. The company plans to stop manufacturing this computer when the monthly sales fall to 800 computers. If the monthly sales now ($t = 0$) are 2,000 computers, find $S(t)$. How long (to the nearest month) will the company continue to manufacture this computer?

42. *Sales analysis.* The rate of change of the monthly sales of a new home video game cartridge is given by

$$S'(t) = 350 \ln(x + 1) \qquad S(0) = 0$$

where t is the number of months since the game was released and $S(t)$ is the number of cartridges sold each month. Find $S(t)$. When (to the nearest month) will the monthly sales reach 15,000 cartridges?

43. *Revenue function.* The marginal revenue from the sales of x hair driers per week is given by

$$R'(x) = 65 - 6 \ln(x + 1) \qquad R(0) = 0$$

where $R(x)$ is the revenue in dollars. Find the revenue function and the production level (to the nearest unit) for a revenue of $20,000 per week. What is the weekly revenue (to the nearest dollar) at a production level of 1,000 hair driers per week?

44. *Cost function.* The marginal cost of producing x pairs of tennis shoes per hour is given by

$$C'(x) = 12 + xe^{-0.05x} \qquad C(10) = \$500$$

where $C(x)$ is cost in dollars. Find the cost function and the production level (to the nearest unit) for a cost of $1,500 per hour. What is the cost per hour (to the nearest dollar) at a production level of 100 pairs of shoes per hour?

45. *Profit.* If the marginal profit (in millions of dollars per year) is given by

$$P'(t) = 2t - te^{-t}$$

find a function that gives the total profit earned over the first x years of operation by use of an appropriate definite integral.

46. *Production.* An oil field is estimated to produce oil at a rate of $R(t)$ thousand barrels per month t months from now, as given by

$$R(t) = 10e^{-0.1t}$$

Find a function that gives the total production in the first x years of operation by use of an appropriate definite integral.

47. *Profit.* Interpret Problem 45 with both a graph and a verbal description of the graph.

48. *Production.* Interpret Problem 46 with both a graph and a verbal description of the graph.

Life Sciences

49. *Medicine.* After a person takes a pill, the drug contained in the pill is assimilated into the bloodstream. The rate of assimilation (in milligrams per minute) t minutes after taking the pill is

$$C'(t) = te^{-0.2t} \qquad C(0) = 0$$

where $C(t)$ is the concentration in milligrams. Find $C(t)$. How long (to the nearest minute) will it take to have 15 milligrams of the drug in the bloodstream? How many milligrams (to one decimal place) will be in the bloodstream after 20 minutes?

50. *Biology.* A yeast culture is growing at the rate of

$$W'(t) = \sqrt{t + 1}\, \ln(t + 1)$$

grams per hour. If the starting culture weighs 2 grams, what will be the weight of the culture, $W(t)$, after t hours? How long (to the nearest hour) will it take the culture to weigh 60 grams? How much (to the nearest gram) will the culture weigh after 10 hours?

Social Sciences

51. *Learning.* In a particular business college, it was found that an average student enrolled in an advanced typing class progressed at a rate of

$$N'(t) = (t + 6)e^{-0.25t}$$

words per minute per week, t weeks after enrolling in a 15 week course. If, at the beginning of the course, a student could type 40 words per minute, how many words per minute, $N(t)$, would the student be expected to type t weeks into the course? How long (to the nearest week) should it take the student to achieve the 70 word per minute level? How many words per minute should the student be able to type by the end of the course?

52. *Learning.* In the same business college, it was also found that an average student enrolled in a beginning shorthand class progressed at a rate of

$$N'(t) = (t + 10)e^{-0.1t}$$

words per minute per week, t weeks after enrolling in a 15 week course. If, at the beginning of the course, a student could take dictation in shorthand at 0 words per minute, how many words per minute, $N(t)$, would the student be expected to handle t weeks into the course? How long (to the nearest week) should it take the student to achieve 90 words per minute? How many words per minute should the student be able to handle by the end of the course?

SECTION   12-4   ■   ## Integration Using Tables

- USING A TABLE OF INTEGRALS
- SUBSTITUTION AND INTEGRAL TABLES
- REDUCTION FORMULAS
- APPLICATION

A **table of integrals** is a list of integration formulas used to evaluate integrals. People who frequently evaluate complex integrals may refer to tables that contain hundreds of formulas. Tables of this type are included in mathematical handbooks available in most college book stores. Table II of Appendix A contains a short list of integral formulas illustrating the types found in more extensive tables. Some of these formulas can be derived using the integration techniques discussed earlier, while others require techniques we have not considered. However, it is possible to verify each formula by differentiating the right side.

■ USING A TABLE OF INTEGRALS

The formulas in Table II (and in larger integral tables) are organized by categories, such as "Integrals Involving $a + bu$," "Integrals Involving $\sqrt{u^2 - a^2}$," and so on. The variable u is the variable of integration. All other symbols represent constants. To use a table to evaluate an integral, you must first find the category that most closely agrees with the form of the integrand and then find a formula in that category that can be made to match the integrand exactly by assigning values to the constants in the formula. The following examples illustrate this process.

EXAMPLE 1 ➤ **Integration Using Tables** Use Table II to find: $\displaystyle\int \frac{x}{(5 + 2x)(4 - 3x)}\, dx$

Solution Since the integrand

$$f(x) = \frac{x}{(5 + 2x)(4 - 3x)}$$

is a rational function involving terms of the form $a + bu$ and $c + du$, we examine formulas 15–20 in Table II to see if any of the integrands in these formulas can be made to match $f(x)$ exactly. Comparing the integrand in formula 16 with $f(x)$, we see that this integrand will match $f(x)$ if we let $a = 5$, $b = 2$, $c = 4$, and $d = -3$. Letting $u = x$ and substituting for a, b, c, and d in formula 16, we have

$$\int \frac{u}{(a + bu)(c + du)}\, du = \frac{1}{ad - bc}\left(\frac{a}{b}\ln|a + bu| - \frac{c}{d}\ln|c + du|\right) \qquad \text{Formula 16}$$

$$\int \frac{x}{(5 + 2x)(4 - 3x)}\, dx = \frac{1}{5\cdot(-3) - 2\cdot 4}\left(\frac{5}{2}\ln|5 + 2x| - \frac{4}{-3}\ln|4 - 3x|\right) + C$$

$$a \quad b \quad c \quad d \qquad\qquad a\cdot d - b\cdot c = 5\cdot(-3) - 2\cdot 4 = -23$$
$$= -\tfrac{5}{46}\ln|5 + 2x| - \tfrac{4}{69}\ln|4 - 3x| + C$$

Notice that the constant of integration C is not included in any of the formulas in Table II. However, you must still include C in all antiderivatives. ◄

MATCHED PROBLEM 1 ➤ Use Table II to find: $\displaystyle\int \frac{1}{(5 + 3x)^2(1 + x)}\, dx$ ◄

EXAMPLE 2 ➤ **Integration Using Tables** Find

$$\int_3^x \frac{1}{t\sqrt{25 - t^2}}\, dt$$

using tables and the FTC.

Solution   First, we use Table II to find

$$\int \frac{1}{t\sqrt{25 - t^2}} \, dt$$

Since the integrand involves the expression $\sqrt{25 - t^2}$, we examine formulas 29–31 and select formula 29 with $a^2 = 25$ and $a = 5$:

$$\int \frac{1}{u\sqrt{a^2 - u^2}} \, du = -\frac{1}{a} \ln \left| \frac{a + \sqrt{a^2 - u^2}}{u} \right| \qquad \text{Formula 29}$$

$$\int \frac{1}{t\sqrt{25 - t^2}} \, dt = -\frac{1}{5} \ln \left| \frac{5 + \sqrt{25 - t^2}}{t} \right| + C$$

Thus,

$$\int_3^x \frac{1}{t\sqrt{25 - t^2}} \, dt = -\frac{1}{5} \ln \left| \frac{5 + \sqrt{25 - t^2}}{t} \right| \Bigg|_3^x$$

$$= -\frac{1}{5} \ln \left| \frac{5 + \sqrt{25 - x^2}}{x} \right| + \frac{1}{5} \ln 3$$

◀

MATCHED PROBLEM 2 ▶   Find $\displaystyle\int_6^x \frac{1}{t^2\sqrt{100 - t^2}} \, dt$ using tables and the FTC.   ◀

■ SUBSTITUTION AND INTEGRAL TABLES

As Examples 1 and 2 illustrate, if the integral we want to evaluate can be made to match one in the table exactly, then evaluating the indefinite integral consists of simply substituting the correct values of the constants into the formula. What happens if we cannot match an integral with one of the formulas in the table? In many cases, a substitution will change the given integral into one that corresponds to a table entry. The following examples illustrate several frequently used substitutions.

EXAMPLE 3 ▶   Integration Using Substitution and Tables   Find: $\displaystyle\int \frac{x^2}{\sqrt{16x^2 - 25}} \, dx$

Solution   In order to relate this integral to one of the formulas involving $\sqrt{u^2 - a^2}$ (formulas 40–45), we observe that if $u = 4x$, then

$$u^2 = 16x^2 \qquad \text{and} \qquad \sqrt{16x^2 - 25} = \sqrt{u^2 - 25}$$

Thus, we will use the substitution $u = 4x$ to change this integral into one that appears in the table:

$$\int \frac{x^2}{\sqrt{16x^2 - 25}} \, dx = \frac{1}{4} \int \frac{\frac{1}{16}u^2}{\sqrt{u^2 - 25}} \, du \qquad \begin{array}{l} \text{Substitution:} \\ u = 4x, \ du = 4 \ dx \\ x = \frac{1}{4}u \end{array}$$

$$= \frac{1}{64} \int \frac{u^2}{\sqrt{u^2 - 25}} \, du$$

This last integral can be evaluated by using formula 44 with $a = 5$:

$$\int \frac{u^2}{\sqrt{u^2 - a^2}} \, du = \frac{1}{2}(u\sqrt{u^2 - a^2} + a^2 \ln|u + \sqrt{u^2 - a^2}|) \qquad \text{Formula 44}$$

$$\int \frac{x^2}{\sqrt{16x^2 - 25}} \, dx = \frac{1}{64} \int \frac{u^2}{\sqrt{u^2 - 25}} \, du \qquad \text{Use formula 44 with } a = 5.$$

$$= \frac{1}{128}(u\sqrt{u^2 - 25} + 25 \ln|u + \sqrt{u^2 - 25}|) + C \qquad \text{Substitute } u = 4x.$$

$$= \frac{1}{128}(4x\sqrt{16x^2 - 25} + 25 \ln|4x + \sqrt{16x^2 - 25}|) + C \qquad \blacktriangleleft$$

MATCHED PROBLEM 3 ➤ Find: $\int \sqrt{9x^2 - 16} \, dx$ ◀

EXAMPLE 4 ➤ Integration Using Substitution and Tables Find: $\displaystyle\int \frac{x}{\sqrt{x^4 + 1}} \, dx$

Solution None of the formulas in the table involve fourth powers; however, if we let $u = x^2$, then

$$\sqrt{x^4 + 1} = \sqrt{u^2 + 1}$$

and this form does appear in formulas 32–39. Thus, we substitute $u = x^2$:

$$\int \frac{1}{\sqrt{x^4 + 1}} \, x \, dx = \frac{1}{2} \int \frac{1}{\sqrt{u^2 + 1}} \, du \qquad \begin{array}{l} \text{Substitution:} \\ u = x^2, \, du = 2x \, dx \end{array}$$

We recognize the last integral as formula 36 with $a = 1$:

$$\int \frac{1}{\sqrt{u^2 + a^2}} \, du = \ln|u + \sqrt{u^2 + a^2}| \qquad \text{Formula 36}$$

$$\int \frac{x}{\sqrt{x^4 + 1}} \, dx = \frac{1}{2} \int \frac{1}{\sqrt{u^2 + 1}} \, du \qquad \text{Use formula 36 with } a = 1.$$

$$= \frac{1}{2} \ln|u + \sqrt{u^2 + 1}| + C \qquad \text{Substitute } u = x^2.$$

$$= \frac{1}{2} \ln|x^2 + \sqrt{x^4 + 1}| + C \qquad \blacktriangleleft$$

MATCHED PROBLEM 4 ➤ Find: $\int x\sqrt{x^4 + 1} \, dx$ ◀

■ REDUCTION FORMULAS

EXAMPLE 5 ➤ Using Reduction Formulas Use Table II to find: $\int x^2 e^{3x} \, dx$

Solution Since the integrand involves the function e^{3x}, we examine formulas 46–48 and conclude that formula 47 can be used for this problem. Letting $u = x$, $n = 2$, and $a = 3$ in formula 47, we have

$$\int u^n e^{au} \, du = \frac{u^n e^{au}}{a} - \frac{n}{a} \int u^{n-1} e^{au} \, du \qquad \text{Formula 47}$$

$$\int x^2 e^{3x} \, dx = \frac{x^2 e^{3x}}{3} - \frac{2}{3} \int x e^{3x} \, dx$$

Notice that the expression on the right still contains an integral, but the exponent of x has been reduced by 1. Formulas of this type are called **reduction formulas** and are designed to be applied repeatedly until an integral that can be evaluated is obtained. Applying formula 47 to $\int xe^{3x}\, dx$ with $n = 1$, we have

$$\int x^2 e^{3x}\, dx = \frac{x^2 e^{3x}}{3} - \frac{2}{3}\left(\frac{xe^{3x}}{3} - \frac{1}{3}\int e^{3x}\, dx\right)$$

$$= \frac{x^2 e^{3x}}{3} - \frac{2xe^{3x}}{9} + \frac{2}{9}\int e^{3x}\, dx$$

This last expression contains an integral that is easy to evaluate:

$$\int e^{3x}\, dx = \tfrac{1}{3}e^{3x}$$

After making a final substitution and adding a constant of integration, we have

$$\int x^2 e^{3x}\, dx = \frac{x^2 e^{3x}}{3} - \frac{2xe^{3x}}{9} + \frac{2}{27}e^{3x} + C \qquad \blacktriangleleft$$

MATCHED PROBLEM 5 ➤ Use Table II to find: $\int (\ln x)^2\, dx$ ◀

■ APPLICATION

▶ EXAMPLE 6 ➤ Marginal Cost A company manufactures cycling gloves. It has fixed costs of $50 per hour, and the marginal cost is given by

$$C'(x) = \frac{40 + 18x}{1 + 2x}$$

where $C(x)$ is the total hourly cost at an output of x pairs of gloves per hour. Find $C(x)$. How many pairs of gloves are produced per hour at a total cost of $155 per hour? What is the hourly cost (to the nearest cent) when 10 pairs of gloves are produced per hour?

Solution
$$C(x) = \int \frac{40 + 18x}{1 + 2x}\, dx$$
$$= 9x + 15.5 \ln(1 + 2x) + K \quad \text{Use Formula 20}$$

Since $C(0) = 50$, $K = 50$. Thus,

$$C(x) = 9x + 15.5 \ln(1 + 2x) + 50$$

From Figure 1 we see that $x = 7$ when $C(x) = 155$, so 7 pairs of gloves are produced per hour at a total cost of $155 per hour. From Figure 2, we see that $C(10) = \$187.19$; this is the hourly cost when 10 pairs of gloves are produced.

FIGURE 1
$y_1 = 9x + 15.5 \ln(1 + 2x) + 50$
$y_2 = 155$

FIGURE 2
$y_1 = 9x + 15.5 \ln(1 + 2x) + 50$

MATCHED PROBLEM 6 ➤ Repeat Example 6 if fixed costs are $45 per hour and the marginal cost is given by

$$C'(x) = \frac{50 + 24x}{1 + 3x}$$

explore – discuss 1

Use algebraic manipulation, including algebraic long division, on the integrand in Example 6 to show that

$$\frac{40 + 18x}{1 + 2x} = 9 + \frac{31}{1 + 2x}$$

Use this result to find the indefinite integral in Example 6 without resorting to table formulas.

Answers to Matched Problems

1. $\frac{1}{2}\left(\frac{1}{5 + 3x}\right) + \frac{1}{4}\ln\left|\frac{1 + x}{5 + 3x}\right| + C$ 2. $-\frac{\sqrt{100 - x^2}}{100x} + \frac{1}{75}$

3. $\frac{1}{6}(3x\sqrt{9x^2 - 16} - 16\ln|3x + \sqrt{9x^2 - 16}|) + C$

4. $\frac{1}{4}(x^2\sqrt{x^4 + 1} + \ln|x^2 + \sqrt{x^4 + 1}|) + C$ 5. $x(\ln x)^2 - 2x \ln x + 2x + C$

6. $C(x) = 8x + 14\ln(1 + 3x) + 45$; 8 pairs; \$173.08/hr

EXERCISE 12-4

A *Use Table II to find each indefinite integral in Problems 1–10.*

1. $\displaystyle\int \frac{1}{x(1 + x)}\,dx$

2. $\displaystyle\int \frac{1}{x^2(1 + x)}\,dx$

3. $\displaystyle\int \frac{1}{(3 + x)^2(5 + 2x)}\,dx$

4. $\displaystyle\int \frac{x}{(5 + 2x)^2(2 + x)}\,dx$

5. $\displaystyle\int \frac{x}{\sqrt{16 + x}}\,dx$

6. $\displaystyle\int \frac{1}{x\sqrt{16 + x}}\,dx$

7. $\displaystyle\int \frac{1}{x\sqrt{x^2 + 4}}\,dx$

8. $\displaystyle\int \frac{1}{x^2\sqrt{x^2 - 16}}\,dx$

9. $\int x^2 \ln x\,dx$

10. $\int x^3 \ln x\,dx$

Evaluate each definite integral in Problems 11–16 (to four decimal places) using tables and the FTC. Check with a numerical integration routine.

11. $\displaystyle\int_1^3 \frac{x^2}{3+x}\,dx$

12. $\displaystyle\int_2^6 \frac{x}{(6+x)^2}\,dx$

13. $\displaystyle\int_0^7 \frac{1}{(3+x)(1+x)}\,dx$

14. $\displaystyle\int_0^7 \frac{x}{(3+x)(1+x)}\,dx$

15. $\displaystyle\int_0^4 \frac{1}{\sqrt{x^2+9}}\,dx$

16. $\displaystyle\int_4^5 \sqrt{x^2-16}\,dx$

B *In Problems 17–28, use substitution techniques and Table II to find each indefinite integral.*

17. $\displaystyle\int \frac{\sqrt{4x^2+1}}{x^2}\,dx$

18. $\displaystyle\int x^2\sqrt{9x^2-1}\,dx$

19. $\displaystyle\int \frac{x}{\sqrt{x^4-16}}\,dx$

20. $\displaystyle\int x\sqrt{x^4-16}\,dx$

21. $\displaystyle\int x^2\sqrt{x^6+4}\,dx$

22. $\displaystyle\int \frac{x^2}{\sqrt{x^6+4}}\,dx$

23. $\displaystyle\int \frac{1}{x^3\sqrt{4-x^4}}\,dx$

24. $\displaystyle\int \frac{\sqrt{x^4+4}}{x}\,dx$

25. $\displaystyle\int \frac{e^x}{(2+e^x)(3+4e^x)}\,dx$

26. $\displaystyle\int \frac{e^x}{(4+e^x)^2(2+e^x)}\,dx$

27. $\displaystyle\int \frac{\ln x}{x\sqrt{4+\ln x}}\,dx$

28. $\displaystyle\int \frac{1}{x\ln x\sqrt{4+\ln x}}\,dx$

C *In Problems 29–34, use Table II to find each indefinite integral.*

29. $\displaystyle\int x^2 e^{5x}\,dx$

30. $\displaystyle\int x^2 e^{-4x}\,dx$

31. $\displaystyle\int x^3 e^{-x}\,dx$

32. $\displaystyle\int x^3 e^{2x}\,dx$

33. $\displaystyle\int (\ln x)^3\,dx$

34. $\displaystyle\int (\ln x)^4\,dx$

Problems 35–42 are mixed—some require the use of Table II and others can be solved using techniques we considered earlier.

35. $\displaystyle\int_3^x t\sqrt{t^2-9}\,dt,\ x \geqslant 3$

36. $\displaystyle\int_3^x t^2\sqrt{t^2-9}\,dt,\ x \geqslant 3$

37. $\displaystyle\int_2^x \frac{1}{t^2-1}\,dt,\ x \geqslant 2$

38. $\displaystyle\int_2^x \frac{2t}{(t^2-1)^2}\,dt,\ x \geqslant 2$

39. $\displaystyle\int \frac{x+1}{x^2+2x}\,dx$

40. $\displaystyle\int \frac{x+1}{x^2+x}\,dx$

41. $\displaystyle\int \frac{x+1}{x^2+3x}\,dx$

42. $\displaystyle\int \frac{x^2+1}{x^2+3x}\,dx$

 APPLICATIONS

Use Table II to evaluate indefinite integrals wherever appropriate.

Business & Economics

43. *Cost.* A company manufactures downhill skis. It has fixed costs of $25,000, and the marginal cost is given by

$$C'(x) = \frac{250+10x}{1+0.05x}$$

where $C(x)$ is the total cost at an output of x pairs of skis. Find the cost function $C(x)$ and determine the production level (to the nearest unit) that produces a cost of $150,000. What is the cost (to the nearest dollar) for a production level of 850 skis?

44. *Cost.* A company manufactures a portable CD player. It has fixed costs of $11,000 per week, and the marginal cost given by

$$C'(x) = \frac{65+20x}{1+0.4x}$$

where $C(x)$ is the total cost per week at an output of x players per week. Find the cost function $C(x)$ and determine the production level (to the nearest unit) that produces a cost of $52,000 per week. What is the cost (to the nearest dollar) for a production level of 700 players per week?

45. *Profit.* The marginal profit for a small car agency in selling x cars per week is given by

$$P'(x) = x\sqrt{2+3x}$$

where $P(x)$ is the profit in dollars. The agency's profit on the sale of only 1 car per week is $-\$2,000$ (that is, a loss of $2,000). Find the profit function and the number of cars that must be sold (to the nearest unit) to produce a profit of $13,000 per week. How much will the weekly profit be (to the nearest dollar) if 10 cars are sold per week?

46. *Revenue.* The marginal revenue for a company manufacturing and selling x solar powered calculators per week is given by

$$R'(x) = \frac{x}{\sqrt{1 + 2x}} \qquad R(0) = 0$$

where $R(x)$ is the revenue in dollars. Find the revenue function and the number of calculators that must be sold (to the nearest unit) to produce $10,000 in revenue per week. How much will the weekly revenue be (to the nearest dollar) if 1,000 calculators are sold per week?

47. *Marketing.* After test marketing a new high-fiber cereal, the market research department of a major cereal producer estimates that monthly sales (in millions of dollars) will grow at the monthly rate of

$$S'(t) = \frac{4t^2}{(1 + 2t)^2}$$

t months after the cereal is introduced. If we assume 0 sales at the time the cereal is first introduced, find the total sales, $S(t)$, t months after the cereal is introduced. Find the total sales (to the nearest tenth of a million dollars) during the first 2 years this cereal is on the market. How long (to the nearest month) will it take sales to reach $12 million?

48. *Marketing.* After test marketing a new shampoo, the market research department estimates that monthly sales (in millions of dollars) will grow at the monthly rate of

$$S'(t) = \frac{t^2}{(1 + t)^2}$$

t months after the shampoo is introduced. If we assume 0 sales at the time the shampoo is first introduced, find the total sales, $S(t)$, t months after the shampoo is introduced. Find the total sales (to the nearest tenth of a million dollars) during the first 18 months this shampoo is on the market. How long (to the nearest month) will it take sales to reach $7 million?

Life Sciences

49. *Pollution.* An oil tanker aground on a reef is losing oil and producing an oil slick that is radiating outward at a rate given approximately by

$$\frac{dR}{dt} = \frac{100}{\sqrt{t^2 + 9}} \qquad t \geq 0$$

where R is the radius (in feet) of the circular slick after t minutes. Find the radius, $R(t)$, t minutes after the slick starts to form if the radius is 0 when $t = 0$. Find the radius (to the nearest foot) of the slick after 4 minutes. How long (to the nearest minute) will it take the slick to have a radius of 300 feet?

50. *Pollution.* An oil tanker aground on a reef is losing oil and producing an oil slick that is radiating outward at a rate given approximately by

$$\frac{dR}{dt} = \frac{80}{\sqrt{t^2 + 4}} \qquad t \geq 0$$

where R is the radius (in feet) of the circular slick after t minutes. Find the radius, $R(t)$, t minutes after the slick starts to form if the radius is 0 when $t = 0$. Find the radius (to the nearest foot) of the slick after 5 minutes. How long (to the nearest minute) will it take the slick to have a radius of 250 feet?

Social Sciences

51. *Learning.* A person learns N items at a rate given approximately by

$$N'(t) = \frac{60}{\sqrt{t^2 + 25}} \qquad t \geq 0$$

where t is the number of hours of continuous study. Determine the total number of items learned in the first t hours of continuous study, assuming $N(0) = 0$. How many items are learned in the first 5 hours of continuous study? How long (to the nearest tenth of an hour) will it take to learn 65 items?

CHAPTER 12
GROUP ACTIVITY

Analysis of Income Concentration from Raw Data

We start with raw data on income distribution supplied in table form by the U.S. Bureau of the Census, as given in Table 1. From the raw data in the table, we wish to compare income distribution among Whites and income distribution among Blacks in the United States in 1992. The approach will be numeric, geometric, and symbolic. We will first organize in table form the data that correspond to data points for a Lorenz curve. We will then find Lorenz curves of the form $f(x) = x^p$ for each set of data points. We will interpret the income distribution geometrically by graphing the Lorenz curves and $y = x$ for both sets of data. Finally, we compute the index of income concentration for Blacks and for Whites, and interpret the results. (See the discussion of Lorenz curves in Section 12-1 for relevant background material.)

TABLE 1
Income Distribution by Population Fifths

FAMILIES, 1992 RACE	UPPER LIMIT OF EACH FIFTH*				
	Lowest	*Second*	*Third*	*Fourth*	*Top 5%†*
TOTAL	$16,960	$30,000	$44,200	$64,300	$106,509
WHITE	19,000	32,000	46,250	66,252	109,900
BLACK	7,531	15,609	26,800	44,200	75,619

FAMILIES, 1992 RACE	PERCENTAGE DISTRIBUTION OF TOTAL INCOME					
	Lowest fifth	*Second fifth*	*Third fifth*	*Fourth fifth*	*Highest fifth*	*Top 5%*
TOTAL	4.4	10.5	16.5	24.0	44.6	17.6
WHITE	4.9	10.9	16.7	23.7	43.8	17.3
BLACK	3.0	8.2	15.0	25.0	48.8	18.5

* The highest fifth does not have an upper limit.
† Lower limit for top 5%.
Source: U.S. Bureau of the Census, U.S. Department of Commerce

Part 1. *Numeric analysis:* Complete Tables 2 and 3 using the data in Table 1. Round income levels to the nearest thousand dollars and represent percents in decimal form to two decimal places. Remember that x represents the cumulative percentage of families in a given category and y represents the corresponding cumulative percentage of income received by these families. Verbally describe the meaning of the last two lines in Tables 2 and 3 after they are completed.

Part 2. *Geometric analysis:*

(A) Sketch separate graphs on suitable graph paper for each table by plotting the data points from the x and y columns in Tables 2 and 3. Also, graph the line $y = x$ over the interval [0, 1] in each graph.

TABLE 2
Black Family Income Distribution in 1992

INCOME LEVEL	x	y
Under $8,000	0.20	0.03
Under		0.11
Under		
Under $44,000	0.80	

TABLE 3
White Family Income Distribution in 1992

INCOME LEVEL	x	y
Under $19,000	0.20	0.05
Under		0.16
Under		
Under $66,000	0.80	

(B) Find p (to one decimal place) so that the graph of the function $f(x) = x^p$ goes through the point $(0.20, 0.03)$ in Table 2. Plot this curve on the corresponding graph, $0 \le x \le 1$. Also, find q (to one decimal place) so that the function $g(x) = x^q$ goes through the point $(0.20, 0.05)$ in Table 3. Plot this curve on the corresponding graph, $0 \le x \le 1$. Repeat this process for each of the remaining points in each table. Compare all the graphs based on Table 2 and select the value of p that best fits the data (by eye). Do the same for all the graphs based on Table 3.

(C) From the final graphs chosen in part (B) can you draw any conclusions about whether income is distributed more equitably among Blacks or Whites? Verbally support your conclusions.

Part 3. *Symbolic analysis:* Use the values of p and q you determined in part 2B to compute the index of income concentration for Blacks and for Whites, and interpret.

C H A P T E R 12 R E V I E W ■ **Important Terms and Symbols**

12-1 *Area between Curves.* Area between a curve and the x axis; area between two curves; distribution of income; Lorenz curve; index of income concentration; perfect or absolute equality; perfect or absolute inequality

12-2 *Applications in Business and Economics.* Probability density function; normal probability density function; continuous income stream; rate of flow; total income; future value of a continuous income stream; consumers' surplus; producers' surplus; equilibrium price; equilibrium quantity

12-3 *Integration by Parts.* Integration by parts; selection of u and dv

$$\int u\, dv = uv - \int v\, du$$

12-4 *Integration Using Tables.* Table of integrals; substitution and integral tables; reduction formulas

---

C H A P T E R 12 ■ **Review Exercise**

Work through all the problems in this chapter review and check your answers in the back of the book. Answers to all review problems are there along with section numbers in italics to indicate where each type of problem is discussed. Where weaknesses show up, review appropriate sections in the text.

Compute all numerical answers to three decimal places unless directed otherwise.

A *In Problems 1–3, set up definite integrals that represent the shaded areas in the figure over the indicated intervals.*

1. Interval $[a, b]$
2. Interval $[b, c]$
3. Interval $[a, c]$

4. Sketch a graph of the area between the graphs of $y = \ln x$ and $y = 0$ over the interval $[0.5, e]$, and find the area.

In Problems 5–8, evaluate each integral.

5. $\int xe^{4x}\, dx$

6. $\int x \ln x\, dx$

7. $\int \dfrac{1}{x(1 + x)^2}\, dx$

8. $\int \dfrac{1}{x^2\sqrt{1 + x}}\, dx$

B *In Problems 9–12, set up definite integrals that represent the shaded areas in the figure over the indicated intervals.*

9. Interval $[a, b]$
10. Interval $[b, c]$
11. Interval $[b, d]$
12. Interval $[a, d]$

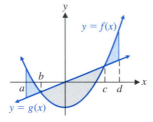

13. Sketch a graph of the area bounded by the graphs of $y = x^2 - 6x + 9$ and $y = 9 - x$, and find the area.

In Problems 14–19, evaluate each integral.

14. $\displaystyle\int_0^x te^t\, dt,\ x \geqslant 0$

15. $\displaystyle\int_0^x \dfrac{t^2}{\sqrt{t^2 + 16}}\, dt,\ x \geqslant 0$

16. $\displaystyle\int \sqrt{9x^2 - 49}\, dx$

17. $\displaystyle\int te^{-0.5t}\, dt$

18. $\displaystyle\int x^2 \ln x\, dx$

19. $\displaystyle\int \dfrac{1}{1 + 2e^x}\, dx$

20. Sketch a graph of the area bounded by the indicated graphs and find the area. Approximate intersection points and area to two decimal places.
(A) $y = x^3 - 6x^2 + 9x;\ y = x$
(B) $y = x^3 - 6x^2 + 9x;\ y = x + 1$

C *In Problems 21–28, evaluate each integral.*

21. $\displaystyle\int \dfrac{(\ln x)^2}{x}\, dx$

22. $\int x(\ln x)^2\, dx$

23. $\displaystyle\int \dfrac{x}{\sqrt{x^2 - 36}}\, dx$

24. $\displaystyle\int \dfrac{x}{\sqrt{x^4 - 36}}\, dx$

25. $\displaystyle\int_1^x t \ln 2t\, dt,\ x \geqslant 1$

26. $\int (\ln x)^2\, dx$

27. $\int xe^{-2x^2}\, dx$

28. $\int x^2 e^{-2x}\, dx$

▶ **APPLICATIONS**

Business & Economics

29. *Product warranty.* A manufacturer warrants a product for parts and labor for 1 year, and for parts only for a second year. The time to a failure of the product after it is sold is given by the probability density function

$$f(t) = \begin{cases} 0.21e^{-0.21t} & \text{if } t \geqslant 0 \\ 0 & \text{otherwise} \end{cases}$$

What is the probability that a buyer chosen at random will have a product failure:
(A) During the first year of warranty?
(B) During the second year of warranty?

30. *Product warranty.* Graph the probability density function for Problem 29 over the interval $[0, 3]$, interpret part (B) of Problem 29 geometrically, and verbally describe the geometric representation.

31. *Continuous income stream.* The rate of flow (in dollars per year) of a continuous income stream for a 5 year period is given by

$$f(t) = 2{,}500e^{0.05t} \qquad 0 \leqslant t \leqslant 5$$

(A) Graph $y = f(t)$ over $[0, 5]$ and shade in the area that represents the total income received from the end of the first year to the end of the fourth year.
(B) Find the total income received, to the nearest dollar, from the end of the first year to the end of the fourth year.

32. *Future value of a continuous income stream.* The continuous income stream in Problem 31 is invested as it is received at 15% compounded continuously.
(A) Find the future value (to the nearest dollar) at the end of the 5 year period.
(B) Find the interest earned (to the nearest dollar) during this 5 year period.

33. *Income distribution.* An economist produced the following Lorenz curves for the current income distribution and the projected income distribution 10 years from now in a certain country:

$$f(x) = 0.1x + 0.9x^2 \quad \text{Current Lorenz curve}$$

$$g(x) = x^{1.5} \quad \text{Projected Lorenz curve}$$

(A) Graph $y = x$ and the current Lorenz curve on one set of coordinate axes for [0, 1], and graph $y = x$ and the projected Lorenz curve on another set of coordinate axes over the same interval.

(B) Looking at the areas bounded by the Lorenz curves and $y = x$, can you say that the income will be more or less equitably distributed 10 years from now?

(C) Compute the index of income concentration (to one decimal place) for the current and projected curves. Now what can you say about the distribution of income 10 years from now? More equitable or less?

34. *Consumers' and producers' surplus.* Find the consumers' surplus and the producers' surplus at the equilibrium price level for each pair of price–demand and price–supply equations. Include a graph that identifies the consumers' surplus and the producers' surplus. Round all values to the nearest integer.
(A) $p = D(x) = 70 - 0.2x; p = S(x) = 13 + 0.0012x^2$
(B) $p = D(x) = 70 - 0.2x; p = S(x) = 13e^{0.006x}$

Life Sciences

35. *Medicine.* A hospital patient receives a drug intravenously. Starting with 100 milliliters of the drug in the system, the amount is gradually increased at the rate

$$Q'(t) = \frac{60t}{(t+1)^2(t+2)}$$

where t is time in hours. Find $Q(t)$, the amount of the drug in the system t hours later. How many milliliters of the drug (to two decimal places) will be in the system 6 hours later? How long (to the nearest one-hundredth of an hour) will it take to have 110 milliliters of the drug in the system?

36. *Medicine.* In Problem 35, the quantity of the drug in the system approaches a fixed quantity as time increases. What (to two decimal places) is this fixed quantity?

37. *Medicine.* For a particular doctor, the length of time (in hours) spent with a patient per office visit has the probability density function

$$f(t) = \begin{cases} \dfrac{\frac{4}{3}}{(t+1)^2} & \text{if } 0 \le t \le 3 \\ 0 & \text{otherwise} \end{cases}$$

(A) What is the probability that this doctor will spend less than 1 hour with a randomly selected patient?
(B) What is the probability that this doctor will spend more than 1 hour with a randomly selected patient?

38. *Medicine.* Illustrate part (B) in Problem 37 geometrically. Describe the geometric interpretation verbally.

Social Sciences

39. *Politics.* The rate of change of the voting population of a city with respect to time t (in years) is estimated to be

$$N'(t) = \frac{3{,}000t}{\sqrt{1+2t}}$$

If the voting population is now 350,000, what will be the voting population, $N(t)$, t years from now? What will the voting population be (to the nearest thousand) 5 years from now? How long (to the nearest year) will it take the voting population to reach 380,000?

40. *Psychology.* Rats were trained to go through a maze by rewarding them with a food pellet upon successful completion. After the seventh successful run, it was found that the probability density function for length of time (in minutes) until success on the eighth trial is given by

$$f(t) = \begin{cases} .5e^{-.5t} & \text{if } t \ge 0 \\ 0 & \text{otherwise} \end{cases}$$

What is the probability that a rat selected at random after seven successful runs will take 2 or more minutes to complete the eighth run successfully? [Recall that the area under a probability density function curve from $-\infty$ to ∞ is 1.]

CHAPTER 13

Multivariable Calculus

SECTION 13-1 ■ **Functions of Several Variables**

- FUNCTIONS OF TWO OR MORE INDEPENDENT VARIABLES
- EXAMPLES OF FUNCTIONS OF SEVERAL VARIABLES
- THREE-DIMENSIONAL COORDINATE SYSTEMS

■ FUNCTIONS OF TWO OR MORE INDEPENDENT VARIABLES

In Section 1-1, we introduced the concept of a function with one independent variable. Now we will broaden the concept to include functions with more than one independent variable. We start with an example.

A small manufacturing company produces a standard type of surfboard and no other products. If fixed costs are \$500 per week and variable costs are \$70 per board produced, then the weekly cost function is given by

$$C(x) = 500 + 70x \tag{1}$$

where x is the number of boards produced per week. The cost function is a function of a single independent variable x. For each value of x from the domain of C there exists exactly one value of $C(x)$ in the range of C.

Now, suppose the company decides to add a high-performance competition board to its line. If the fixed costs for the competition board are \$200 per week and the variable costs are \$100 per board, then the cost function (1) must be modified to

$$C(x, y) = 700 + 70x + 100y \tag{2}$$

where $C(x, y)$ is the cost for weekly output of x standard boards and y competition boards. Equation (2) is an example of a function with two independent variables, x and y. Of course, as the company expands its product line even further, its weekly

939

cost function must be modified to include more and more independent variables, one for each new product produced.

In general, an equation of the form

$$z = f(x, y)$$

describes a **function of two independent variables** if for each permissible ordered pair (x, y), there is one and only one value of z determined by $f(x, y)$. The variables x and y are **independent variables,** and the variable z is a **dependent variable.** The set of all ordered pairs of permissible values of x and y is the **domain** of the function, and the set of all corresponding values $f(x, y)$ is the **range** of the function. Unless otherwise stated, we will assume that the domain of a function specified by an equation of the form $z = f(x, y)$ is the set of all ordered pairs of real numbers (x, y) such that $f(x, y)$ is also a real number. It should be noted, however, that certain conditions in practical problems often lead to further restrictions of the domain of a function.

We can similarly define functions of three independent variables, $w = f(x, y, z)$; of four independent variables, $u = f(w, x, y, z)$; and so on. In this chapter, we will primarily concern ourselves with functions of two independent variables.

EXAMPLE 1 ➤ **Evaluating a Function of Two Independent Variables**   For the cost function $C(x, y) = 700 + 70x + 100y$ described earlier, find $C(10, 5)$.

Solution
$$C(10, 5) = 700 + 70(10) + 100(5)$$
$$= \$1,900$$ ◄

MATCHED PROBLEM 1 ➤ Find $C(20, 10)$ for the cost function in Example 1.   ◄

EXAMPLE 2 ➤ **Evaluating a Function of Three Independent Variables**   For $f(x, y, z) = 2x^2 - 3xy + 3z + 1$, find $f(3, 0, -1)$.

Solution
$$f(3, 0, -1) = 2(3)^2 - 3(3)(0) + 3(-1) + 1$$
$$= 18 - 0 - 3 + 1 = 16$$ ◄

MATCHED PROBLEM 2 ➤ Find $f(-2, 2, 3)$ for f in Example 2.   ◄

EXAMPLE 3 ➤ **Revenue, Cost, and Profit Functions**   The surfboard company discussed at the beginning of this section has determined that the demand equations for the two types of boards they produce are given by

$$p = 210 - 4x + y$$
$$q = 300 + x - 12y$$

where p is the price of the standard board, q is the price of the competition board, x is the weekly demand for standard boards, and y is the weekly demand for competition boards.

(A) Find the weekly revenue function $R(x, y)$, and evaluate $R(20, 10)$.

(B) If the weekly cost function is

$$C(x, y) = 700 + 70x + 100y$$

find the weekly profit function $P(x, y)$, and evaluate $P(20, 10)$.

Solution   (A)

$$\text{Revenue} = \begin{pmatrix} \text{Demand for} \\ \text{standard} \\ \text{boards} \end{pmatrix} \times \begin{pmatrix} \text{Price of a} \\ \text{standard} \\ \text{board} \end{pmatrix} + \begin{pmatrix} \text{Demand for} \\ \text{competition} \\ \text{boards} \end{pmatrix} \times \begin{pmatrix} \text{Price of a} \\ \text{competition} \\ \text{board} \end{pmatrix}$$

$$\begin{aligned}
R(x, y) &= xp + yq \\
&= x(210 - 4x + y) + y(300 + x - 12y) \\
&= 210x + 300y - 4x^2 + 2xy - 12y^2
\end{aligned}$$

$$\begin{aligned}
R(20, 10) &= 210(20) + 300(10) - 4(20)^2 + 2(20)(10) - 12(10)^2 \\
&= \$4,800
\end{aligned}$$

(B)   Profit = Revenue − Cost

$$\begin{aligned}
P(x, y) &= R(x, y) - C(x, y) \\
&= 210x + 300y - 4x^2 + 2xy - 12y^2 - 700 - 70x - 100y \\
&= 140x + 200y - 4x^2 + 2xy - 12y^2 - 700
\end{aligned}$$

$$\begin{aligned}
P(20, 10) &= 140(20) + 200(10) - 4(20)^2 + 2(20)(10) - 12(10)^2 - 700 \\
&= \$1,700 \quad \blacktriangleleft
\end{aligned}$$

MATCHED PROBLEM 3 ➤   Repeat Example 3 if the demand and cost equations are given by

$$\begin{aligned}
p &= 220 - 6x + y \\
q &= 300 + 3x - 10y \\
C(x, y) &= 40x + 80y + 1,000
\end{aligned} \quad \blacktriangleleft$$

■ EXAMPLES OF FUNCTIONS OF SEVERAL VARIABLES

A number of concepts we have already considered can be thought of in terms of functions of two or more variables. We list a few of these below.

Area of a rectangle $A(x, y) = xy$

Volume of a box $V(x, y, z) = xyz$

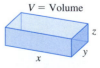

Volume of a right $V(r, h) = \pi r^2 h$
circular cylinder

Simple interest	$A(P, r, t) = P(1 + rt)$	A = Amount P = Principal r = Annual rate t = Time in years
Compound interest	$A(P, r, t, n) = P\left(1 + \dfrac{r}{n}\right)^{nt}$	A = Amount P = Principal r = Annual rate t = Time in years n = Compound periods per year
IQ	$Q(M, C) = \dfrac{M}{C}(100)$	Q = IQ = Intelligence quotient M = MA = Mental age C = CA = Chronological age
Resistance for blood flow in a vessel (Poiseuille's law)	$R(L, r) = k\dfrac{L}{r^4}$	R = Resistance L = Length of vessel r = Radius of vessel k = Constant

EXAMPLE 4 ▶ **Package Design**   A company uses a box with a square base and an open top for one of its products (see the figure). If x is the length (in inches) of each side of the base and y is the height (in inches), find the total amount of material $M(x, y)$ required to construct one of these boxes, and evaluate $M(5, 10)$.

Solution

$$\text{Area of base} = x^2$$
$$\text{Area of one side} = xy$$
$$\text{Total material} = (\text{Area of base}) + 4(\text{Area of one side})$$
$$M(x, y) = x^2 + 4xy$$
$$M(5, 10) = (5)^2 + 4(5)(10)$$
$$= 225 \text{ square inches} \quad ◀$$

MATCHED PROBLEM 4 ▶   For the box in Example 4, find the volume $V(x, y)$, and evaluate $V(5, 10)$.   ◀

The next example concerns the **Cobb–Douglas production function,**

$$f(x, y) = kx^m y^n$$

where k, m, and n are positive constants with $m + n = 1$. Economists use this function to describe the number of units $f(x, y)$ produced from the utilization of x units of labor and y units of capital (for equipment such as tools, machinery, buildings, and so on). Cobb–Douglas production functions are also used to describe the productivity of a single industry, of a group of industries producing the same product, or even of an entire country.

EXAMPLE 5 ➤ Productivity   The productivity of a steel manufacturing company is given approximately by the function

$$f(x, y) = 10x^{0.2}y^{0.8}$$

with the utilization of x units of labor and y units of capital. If the company uses 3,000 units of labor and 1,000 units of capital, how many units of steel will be produced?

Solution   The number of units of steel produced is given by

$$f(3,000, 1,000) = 10(3,000)^{0.2}(1,000)^{0.8} \quad \textit{Use a calculator.}$$
$$\approx 12,457 \text{ units}$$

MATCHED PROBLEM 5 ➤ Refer to Example 5. Find the steel production if the company uses 1,000 units of labor and 2,000 units of capital.

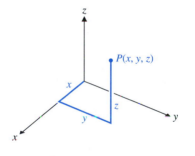

FIGURE 1
Rectangular coordinate system

■ THREE-DIMENSIONAL COORDINATE SYSTEMS

We now take a brief look at some graphs of functions of two independent variables. Since functions of the form $z = f(x, y)$ involve two independent variables, x and y, and one dependent variable, z, we need a *three-dimensional coordinate system* for their graphs. A **three-dimensional coordinate system** is formed by three mutually perpendicular number lines intersecting at their origins (see Fig. 1). In such a system, every ordered **triplet of numbers (x, y, z)** can be associated with a unique point, and conversely.

EXAMPLE 6 ➤ Three-Dimensional Coordinates   Locate $(-3, 5, 2)$ in a rectangular coordinate system.

Solution

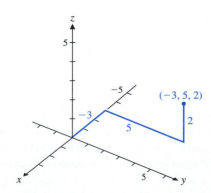

MATCHED PROBLEM 6 ➤ Find the coordinates of the corners A, C, G, and D of the rectangular box shown in the figure.

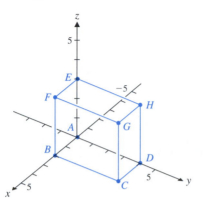

explore – discuss 1

Imagine that you are facing the front of a classroom whose rectangular walls meet at right angles. Suppose that the point of intersection of the floor, front wall, and left side wall is the origin of a three-dimensional coordinate system in which every point in the room has nonnegative coordinates. Then the plane $z = 0$ (or, equivalently, the xy plane) can be described as "the floor," and the plane $z = 2$ can be described as "the plane parallel to, but 2 units above, the floor." Give similar descriptions of the following planes:

(A) $x = 0$ (B) $x = 3$ (C) $y = 0$
(D) $y = 4$ (E) $x = -1$

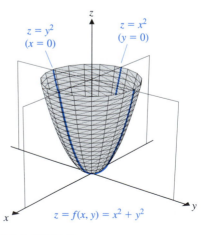

$z = y^2$
$(x = 0)$

$z = x^2$
$(y = 0)$

$z = f(x, y) = x^2 + y^2$

FIGURE 2
Paraboloid

What does the graph of $z = x^2 + y^2$ look like? If we let $x = 0$ and graph $z = 0^2 + y^2 = y^2$ in the yz plane, we obtain a parabola; if we let $y = 0$ and graph $z = x^2 + 0^2 = x^2$ in the xz plane, we obtain another parabola. It can be shown that the graph of $z = x^2 + y^2$ is either one of these parabolas rotated around the z axis (see Fig. 2). This cup-shaped figure is a *surface* and is called a **paraboloid.**

In general, the graph of any function of the form $z = f(x, y)$ is called a **surface.** The graph of such a function is the graph of all ordered triplets of numbers (x, y, z) that satisfy the equation. Graphing functions of two independent variables is often a very difficult task, and the general process will not be dealt with in this book. We present only a few simple graphs to suggest extensions of earlier geometric interpretations of the derivative and local maxima and minima to functions of two variables. Note that $z = f(x, y) = x^2 + y^2$ appears (see Fig. 2) to have a local minimum at $(x, y) = (0, 0)$. Figure 3 shows a local maximum at $(x, y) = (0, 0)$.

Figure 4 shows a point at $(x, y) = (0, 0)$, called a **saddle point,** which is neither a local minimum nor a local maximum. Note that if $x = 0$, then the saddle point is a local minimum, and if $y = 0$, then the saddle point is a local maximum. More will be said about local maxima and minima in Section 13-3.

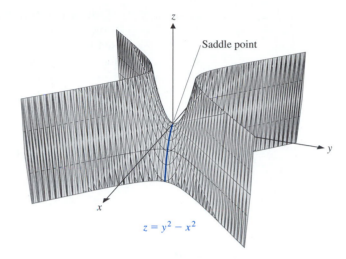

FIGURE 3
Local maximum: $f(0, 0) = 4$

FIGURE 4
Saddle point at $(0, 0, 0)$

---

explore – discuss 2

(A) Let $f(x, y) = x^2 + y^2$. The cross-section of the surface $z = f(x, y)$ by the plane $x = 2$ is the graph of $z = f(2, y) = 4 + y^2$, which is a parabola (see Fig. 5). Explain why each cross-section of $z = f(x, y)$ by a plane parallel to the yz plane is a parabola that opens upward. Explain why each cross-section of $z = f(x, y)$ by a plane parallel to the xz plane is a parabola that opens upward.

FIGURE 5

(B) Let $g(x, y) = y^2 - x^2$. Explain why each cross-section of $z = g(x, y)$ by a plane parallel to the yz plane is a parabola that opens upward (see Fig. 4). Explain why each cross-section of $z = f(x, y)$ by a plane parallel to the xz plane is a parabola that opens downward.

Some graphing utilities are designed to draw graphs (like those of Figs. 2, 3, and 4) of functions of two independent variables. Others, like the graphing calculator used for the displays in this text, are designed to draw graphs of functions of just a single independent variable. When using the latter type of calculator, we can graph cross-sections by planes parallel to the xz plane or yz plane to gain insight into the graph of a function of two independent variables.

EXAMPLE 7 ➤ **Graphing Cross-Sections**

(A) Describe the cross-sections of $f(x, y) = 2x^2 + y^2$ in the planes $y = 0$, $y = 1$, $y = 2$, $y = 3$, and $y = 4$.

(B) Describe the cross-sections of $f(x, y) = 2x^2 + y^2$ in the planes $x = 0$, $x = 1$, $x = 2$, $x = 3$, and $x = 4$.

Solution (A) The cross-section of $f(x, y) = 2x^2 + y^2$ by the plane $y = 0$ is the graph of the function $f(x, 0) = 2x^2$ in this plane. We can examine the shape of this cross-section by graphing $y_1 = 2x^2$ on a graphing utility (Fig. 6). Similarly, the graphs of $y_2 = f(x, 1) = 2x^2 + 1$, $y_3 = f(x, 2) = 2x^2 + 4$, $y_4 = f(x, 3) = 2x^2 + 9$, and $y_5 = f(x, 4) = 2x^2 + 16$ show the shapes of the other four cross-sections (see Fig. 6). Each of these is a parabola that opens upward. Note the correspondence between the graphs in Figure 6 and the actual cross-sections of $f(x, y) = 2x^2 + y^2$ shown in Figure 7.

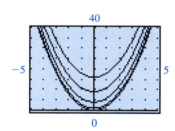

FIGURE 6
$y_1 = 2x^2$
$y_2 = 2x^2 + 1$
$y_3 = 2x^2 + 4$
$y_4 = 2x^2 + 9$
$y_5 = 2x^2 + 16$

FIGURE 7

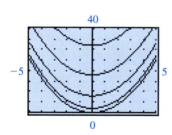

FIGURE 8
$y_1 = x^2$
$y_2 = 2 + x^2$
$y_3 = 8 + x^2$
$y_4 = 18 + x^2$
$y_5 = 32 + x^2$

(B) The five cross-sections are represented by the graphs of the functions $f(0, y) = y^2$, $f(1, y) = 2 + y^2$, $f(2, y) = 8 + y^2$, $f(3, y) = 18 + y^2$, and $f(4, y) = 32 + y^2$. These five functions are graphed in Figure 8. (Note that changing the name of the independent variable from y to x for graphing purposes does not affect the graph displayed.) Each of the five cross-sections is a parabola that opens upward. ◄

MATCHED PROBLEM 7 ➤ (A) Describe the cross-sections of $g(x, y) = y^2 - x^2$ in the planes $y = 0$, $y = 1$, $y = 2$, $y = 3$, and $y = 4$.

(B) Describe the cross-sections of $g(x, y) = y^2 - x^2$ in the planes $x = 0$, $x = 1$, $x = 2$, $x = 3$, and $x = 4$. ◄

Answers to Matched Problems
1. $3,100   2. 30
3. (A) $R(x, y) = 220x + 300y - 6x^2 + 4xy - 10y^2$; $R(20, 10) = \$4,800$
   (B) $P(x, y) = 180x + 220y - 6x^2 + 4xy - 10y^2 - 1,000$; $P(20, 10) = \$2,200$
4. $V(x, y) = x^2 y$; $V(5, 10) = 250$ in.^{3} 5. 17,411 units
6. $A(0, 0, 0)$; $C(2, 4, 0)$; $G(2, 4, 3)$; $D(0, 4, 0)$
7. (A) Each cross-section is a parabola that opens downward.
   (B) Each cross-section is a parabola that opens upward.

EXERCISE 13-1

A *For the functions*

$f(x, y) = 10 + 2x - 3y$ *and* $g(x, y) = x^2 - 3y^2$

find the indicated values in Problems 1–8.

1. $f(0, 0)$ 2. $f(2, 1)$ 3. $f(-3, 1)$
4. $f(2, -7)$ 5. $g(0, 0)$ 6. $g(0, -1)$
7. $g(2, -1)$ 8. $g(-1, 2)$

B *Find the indicated value of the given function in Problems 9–22.*

9. $A(2, 3)$ for $A(x, y) = xy$
10. $V(2, 4, 3)$ for $V(x, y, z) = xyz$
11. $Q(12, 8)$ for $Q(M, C) = \dfrac{M}{C}(100)$
12. $T(50, 17)$ for $T(V, x) = \dfrac{33V}{x + 33}$
13. $V(2, 4)$ for $V(r, h) = \pi r^2 h$
14. $S(4, 2)$ for $S(x, y) = 5x^2 y^3$
15. $R(1, 2)$ for
    $R(x, y) = -5x^2 + 6xy - 4y^2 + 200x + 300y$
16. $P(2, 2)$ for
    $P(x, y) = -x^2 + 2xy - 2y^2 - 4x + 12y + 5$
17. $R(6, 0.5)$ for $R(L, r) = 0.002\dfrac{L}{r^4}$
18. $L(2,000, 50)$ for $L(w, v) = (1.25 \times 10^{-5})wv^2$
19. $A(100, 0.06, 3)$ for $A(P, r, t) = P + Prt$
20. $A(10, 0.04, 3, 2)$ for $A(P, r, t, n) = P\left(1 + \dfrac{r}{n}\right)^{tn}$
21. $A(100, 0.08, 10)$ for $A(P, r, t) = Pe^{rt}$
22. $A(1,000, 0.06, 8)$ for $A(P, r, t) = Pe^{rt}$
23. Let $F(x, y) = x^2 + e^x y - y^2$. Find all values of x such that $F(x, 2) = 0$.
24. Let $G(a, b, c) = a^3 + b^3 + c^3 - (ab + ac + bc) - 6$. Find all values of b such that $G(2, b, 1) = 0$.

C

25. For the function $f(x, y) = x^2 + 2y^2$, find:
    $$\frac{f(x + h, y) - f(x, y)}{h}$$
26. For the function $f(x, y) = x^2 + 2y^2$, find:
    $$\frac{f(x, y + k) - f(x, y)}{k}$$
27. For the function $f(x, y) = 2xy^2$, find:
    $$\frac{f(x + h, y) - f(x, y)}{h}$$
28. For the function $f(x, y) = 2xy^2$, find:
    $$\frac{f(x, y + k) - f(x, y)}{k}$$
29. Find the coordinates of E and F in the figure for Matched Problem 6 (in the text).
30. Find the coordinates of B and H in the figure for Matched Problem 6 (in the text).
31. Let $f(x, y) = x^2$.
    (A) Explain why the cross-sections of the surface $z = f(x, y)$ by planes parallel to $y = 0$ are parabolas.
    (B) Describe the cross-sections of the surface by the planes $x = 0$, $x = 1$, and $x = 2$.
    (C) Describe the surface $z = f(x, y)$.
32. Let $f(x, y) = \sqrt{4 - y^2}$.
    (A) Explain why the cross-sections of the surface $z = f(x, y)$ by planes parallel to $x = 0$ are semicircles of radius 2.
    (B) Describe the cross-sections of the surface by the planes $y = 0$, $y = 2$, and $y = 3$.
    (C) Describe the surface $z = f(x, y)$.

33. Let $f(x, y) = \sqrt{36 - x^2 - y^2}$.
  (A) Describe the cross-sections of the surface $z = f(x, y)$ by the planes $y = 1$, $y = 2$, $y = 3$, $y = 4$, and $y = 5$.
  (B) Describe the cross-sections of the surface $z = f(x, y)$ by the planes $x = 0$, $x = 1$, $x = 2$, $x = 3$, $x = 4$, and $x = 5$.
  (C) Describe the surface $z = f(x, y)$.

34. Let $f(x, y) = 100 + 10x + 25y - x^2 - 5y^2$.
  (A) Describe the cross-sections of the surface $z = f(x, y)$ by the planes $y = 0$, $y = 1$, $y = 2$, and $y = 3$.
  (B) Describe the cross-sections of the surface $z = f(x, y)$ by the planes $x = 0$, $x = 1$, $x = 2$, and $x = 3$.
  (C) Describe the surface $z = f(x, y)$.

▶ APPLICATIONS

Business & Economics

35. *Cost function.* A small manufacturing company produces two models of a surfboard: a standard model and a competition model. If the standard model is produced at a variable cost of $70 each, the competition model at a variable cost of $100 each, and the total fixed costs per month are $2,000, then the monthly cost function is given by

$$C(x, y) = 2,000 + 70x + 100y$$

where x and y are the numbers of standard and competition models produced per month, respectively. Find $C(20, 10)$, $C(50, 5)$, and $C(30, 30)$.

36. *Advertising and sales.* A company spends $x thousand per week on newspaper advertising and $y thousand per week on television advertising. Its weekly sales are found to be given by

$$S(x, y) = 5x^2y^3$$

Find $S(3, 2)$ and $S(2, 3)$.

37. *Revenue function.* A supermarket sells two brands of coffee: brand A at $p per pound and brand B at $q per pound. The daily demand equations for brands A and B are, respectively,

$$x = 200 - 5p + 4q$$
$$y = 300 + 2p - 4q$$

(both in pounds). Find the daily revenue function $R(p, q)$. Evaluate $R(2, 3)$ and $R(3, 2)$.

38. *Revenue, cost, and profit functions.* A company manufactures ten-speed and three-speed bicycles. The weekly demand and cost equations are

$$p = 230 - 9x + y$$
$$q = 130 + x - 4y$$
$$C(x, y) = 200 + 80x + 30y$$

where $p is the price of a ten-speed bicycle, $q is the price of a three-speed bicycle, x is the weekly demand for ten-speed bicycles, y is the weekly demand for three-speed bicycles, and $C(x, y)$ is the cost function. Find the weekly revenue function $R(x, y)$ and the weekly profit function $P(x, y)$. Evaluate $R(10, 15)$ and $P(10, 15)$.

39. *Productivity.* The Cobb–Douglas production function for a petroleum company is given by

$$f(x, y) = 20x^{0.4}y^{0.6}$$

where x is the utilization of labor and y is the utilization of capital. If the company uses 1,250 units of labor and 1,700 units of capital, how many units of petroleum will be produced?

40. *Productivity.* The petroleum company in Problem 39 is taken over by another company that decides to double both the units of labor and the units of capital utilized in the production of petroleum. Use the Cobb–Douglas production function given in Problem 39 to find the amount of petroleum that will be produced by this increased utilization of labor and capital. What is the effect on productivity of doubling both the units of labor and the units of capital?

41. *Future value.* At the end of each year, $2,000 is invested into an IRA earning 9% compounded annually. How much will be in the account at the end of 30 years? Use the annuity formula

$$F(P, i, n) = P\frac{(1 + i)^n - 1}{i}$$

where

P = Periodic payment

i = Rate per period

n = Number of payments (periods)

$F = FV$ = Future value

42. *Package design.* The packaging department in a company has been asked to design a rectangular box with no top and a partition down the middle (see the figure). If x, y, and z are the dimensions (in inches), find the total amount of material $M(x, y, z)$ used in constructing one of these boxes, and evaluate $M(10, 12, 6)$.

Figure for 42

Life Sciences

43. *Marine biology.* In using scuba diving gear, a marine biologist estimates the time of a dive according to the equation

$$T(V, x) = \frac{33V}{x + 33}$$

where

T = Time of dive in minutes

V = Volume of air, at sea level pressure, compressed into tanks

x = Depth of dive in feet

Find $T(70, 47)$ and $T(60, 27)$.

44. *Blood flow.* Poiseuille's law states that the resistance, R, for blood flowing in a blood vessel varies directly as the length of the vessel, L, and inversely as the fourth power of its radius, r. Stated as an equation,

$$R(L, r) = k\frac{L}{r^4} \qquad k \text{ a constant}$$

Find $R(8, 1)$ and $R(4, 0.2)$.

45. *Physical anthropology.* Anthropologists, in their study of race and human genetic groupings, often use an index called the *cephalic index.* The cephalic index, C, varies directly as the width, W, of the head, and inversely as the length, L, of the head (both viewed from the top). In terms of an equation,

$$C(W, L) = 100\,\frac{W}{L}$$

where

W = Width in inches

L = Length in inches

Find $C(6, 8)$ and $C(8.1, 9)$.

Social Sciences

46. *Safety research.* Under ideal conditions, if a person driving a car slams on the brakes and skids to a stop, the length of the skid marks (in feet) is given by the formula

$$L(w, v) = kwv^2$$

where

k = Constant

w = Weight of car in pounds

v = Speed of car in miles per hour

For $k = 0.000\ 013\ 3$, find $L(2,000, 40)$ and $L(3,000, 60)$.

47. *Psychology.* The intelligence quotient (IQ) is defined to be the ratio of mental age (MA), as determined by certain tests, and chronological age (CA), multiplied by 100. Stated as an equation,

$$Q(M, C) = \frac{M}{C} \cdot 100$$

where

Q = IQ M = MA C = CA

Find $Q(12, 10)$ and $Q(10, 12)$.

Partial Derivatives

■ PARTIAL DERIVATIVES
■ SECOND-ORDER PARTIAL DERIVATIVES

■ PARTIAL DERIVATIVES

We know how to differentiate many kinds of functions of one independent variable and how to interpret the results. What about functions with two or more independent variables? Let us return to the surfboard example considered at the beginning of the chapter.

For the company producing only the standard board, the cost function was

$$C(x) = 500 + 70x$$

Differentiating with respect to x, we obtain the marginal cost function

$$C'(x) = 70$$

Since the marginal cost is constant, \$70 is the change in cost for a 1 unit increase in production at any output level.

For the company producing two types of boards, a standard model and a competition model, the cost function was

$$C(x, y) = 700 + 70x + 100y$$

Now suppose we differentiate with respect to x, holding y fixed, and denote this by $C_x(x, y)$; or suppose we differentiate with respect to y, holding x fixed, and denote this by $C_y(x, y)$. Differentiating in this way, we obtain

$$C_x(x, y) = 70 \qquad C_y(x, y) = 100$$

Each of these is called a **partial derivative,** and, in this example, each represents marginal cost. The first is the change in cost due to a 1 unit increase in production of the standard board with the production of the competition model held fixed. The second is the change in cost due to a 1 unit increase in production of the competition board with the production of the standard board held fixed.

In general, if $z = f(x, y)$, then the **partial derivative of f with respect to x,** denoted by $\partial z/\partial x, f_x$, or $f_x(x, y)$, is defined by

$$\frac{\partial z}{\partial x} = \lim_{h \to 0} \frac{f(x + h, y) - f(x, y)}{h}$$

provided the limit exists. We recognize this as the ordinary derivative of f with respect to x, holding y constant. Thus, we are able to continue to use all the derivative rules and properties discussed in Chapters 8–10 for partial derivatives.

Similarly, the **partial derivative of f with respect to y,** denoted by $\partial z/\partial y, f_y$, or $f_y(x, y)$, is defined by

$$\frac{\partial z}{\partial y} = \lim_{k \to 0} \frac{f(x, y + k) - f(x, y)}{k}$$

which is the ordinary derivative with respect to y, holding x constant.

Parallel definitions and interpretations hold for functions with three or more independent variables.

EXAMPLE 1 ➤ **Partial Derivatives** For $z = f(x, y) = 2x^2 - 3x^2y + 5y + 1$, find:

(A) $\partial z/\partial x$ (B) $f_x(2, 3)$

Solution   (A) $z = 2x^2 - 3x^2y + 5y + 1$

Differentiating with respect to x, holding y constant (that is, treating y as a constant), we obtain

$$\frac{\partial z}{\partial x} = 4x - 6xy$$

(B) $f(x, y) = 2x^2 - 3x^2y + 5y + 1$

First, differentiate with respect to x. From part (A) we have

$$f_x(x, y) = 4x - 6xy$$

Then evaluate at $(2, 3)$:

$$f_x(2, 3) = 4(2) - 6(2)(3) = -28$$

Alternatively, substitute $y = 3$ into $f(x, y)$ and graph the function $f(x, 3) = -7x^2 + 16$, which represents the cross-section of the surface $z = f(x, y)$ by the plane $y = 3$. Then determine the slope of the tangent line when $x = 2$. Again, we conclude that $f_x(2, 3) = -28$ (see Fig. 1).   ◀

FIGURE 1
$y_1 = -7x^2 + 16$

MATCHED PROBLEM 1 ➤ For f in Example 1, find:

(A) $\partial z/\partial y$ (B) $f_y(2, 3)$ ◀

EXAMPLE 2 ➤ **Partial Derivatives Using the Chain Rule** For $z = f(x, y) = e^{x^2 + y^2}$, find:

(A) $\partial z/\partial x$ (B) $f_y(2, 1)$

Solution   (A) Using the chain rule [thinking of $z = e^u$, $u = u(x)$; y is held constant], we obtain

$$\frac{\partial z}{\partial x} = e^{x^2 + y^2} \frac{\partial(x^2 + y^2)}{\partial x}$$

$$= 2xe^{x^2 + y^2}$$

(B) $f_y(x, y) = e^{x^2 + y^2} \frac{\partial(x^2 + y^2)}{\partial y} = 2ye^{x^2 + y^2}$

$$f_y(2, 1) = 2(1)e^{(2)^2 + (1)^2}$$
$$= 2e^5$$ ◀

MATCHED PROBLEM 2 ➤ For $z = f(x, y) = (x^2 + 2xy)^5$, find:

(A) $\partial z/\partial y$ (B) $f_x(1, 0)$ ◀

▶   EXAMPLE 3 ▶   Profit   The profit function for the surfboard company in Example 3 in Section 13-1 was

$$P(x, y) = 140x + 200y - 4x^2 + 2xy - 12y^2 - 700$$

Find $P_x(15, 10)$ and $P_x(30, 10)$, and interpret the results.

Solution

$$P_x(x, y) = 140 - 8x + 2y$$
$$P_x(15, 10) = 140 - 8(15) + 2(10) = 40$$
$$P_x(30, 10) = 140 - 8(30) + 2(10) = -80$$

At a production level of 15 standard and 10 competition boards per week, increasing the production of standard boards by 1 unit and holding the production of competition boards fixed at 10 will increase profit by approximately $40. At a production level of 30 standard and 10 competition boards per week, increasing the production of standard boards by 1 unit and holding the production of competition boards fixed at 10 will decrease profit by approximately $80.   ◀

MATCHED PROBLEM 3 ▶   For the profit function in Example 3, find $P_y(25, 10)$ and $P_y(25, 15)$, and interpret the results.   ◀

explore – discuss 1

Let $P(x, y)$ be the profit function of Example 3 and Matched Problem 3.

(A) Assume the production of competition boards remains fixed at 10. Which production level of standard boards will yield a maximum profit? Calculate that maximum profit.

(B) Assume the production of standard boards remains fixed at 25. Which production level of competition boards will yield a maximum profit? Calculate that maximum profit.

▶   EXAMPLE 4 ▶   Productivity   The productivity of a major computer manufacturer is given approximately by the Cobb–Douglas production function

$$f(x, y) = 15x^{0.4}y^{0.6}$$

with the utilization of x units of labor and y units of capital. The partial derivative $f_x(x, y)$ represents the rate of change of productivity with respect to labor and is called the **marginal productivity of labor.** The partial derivative $f_y(x, y)$ represents the rate of change of productivity with respect to capital and is called the **marginal productivity of capital.** If the company is currently utilizing 4,000 units of labor and 2,500 units of capital, find the marginal productivity of labor and the marginal productivity of capital. For the greatest increase in productivity, should the management of the company encourage increased use of labor or increased use of capital?

Solution

$$f_x(x, y) = 6x^{-0.6}y^{0.6}$$
$$f_x(4,000, 2,500) = 6(4,000)^{-0.6}(2,500)^{0.6}$$
$$\approx 4.53 \qquad \text{Marginal productivity of labor}$$
$$f_y(x, y) = 9x^{0.4}y^{-0.4}$$
$$f_y(4,000, 2,500) = 9(4,000)^{0.4}(2,500)^{-0.4}$$
$$\approx 10.86 \qquad \text{Marginal productivity of capital}$$

At the current level of utilization of 4,000 units of labor and 2,500 units of capital, each 1 unit increase in labor utilization (keeping capital utilization fixed at 2,500 units) will increase production by approximately 4.53 units, and each 1 unit increase in capital utilization (keeping labor utilization fixed at 4,000 units) will increase production by approximately 10.86 units. Thus, the management of the company should encourage increased use of capital. ◀

MATCHED PROBLEM 4 ➤ The productivity of an airplane manufacturing company is given approximately by the Cobb–Douglas production function

$$f(x, y) = 40x^{0.3}y^{0.7}$$

(A) Find $f_x(x, y)$ and $f_y(x, y)$.
(B) If the company is currently using 1,500 units of labor and 4,500 units of capital, find the marginal productivity of labor and the marginal productivity of capital.
(C) For the greatest increase in productivity, should the management of the company encourage increased use of labor or increased use of capital? ◀

Partial derivatives have simple geometric interpretations, as indicated in Figure 2. If we hold x fixed, say, $x = a$, then $f_y(a, y)$ is the slope of the curve obtained by intersecting the plane $x = a$ with the surface $z = f(x, y)$. A similar interpretation is given to $f_x(x, b)$.

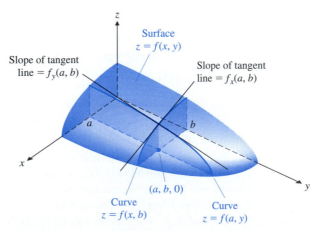

FIGURE 2

Let $f(x, y) = x^2y - 2xy^2 + 3$ and consider the surface $z = f(x, y)$.

(A) The cross-section of the surface by the plane $x = 2$ is the graph of $f(2, y) = 4y - 4y^2 + 3$, which is a parabola. Use an ordinary derivative to find the slope of the tangent line to this parabola when $y = \frac{1}{2}$.

(B) Use partial derivatives to confirm your answer to part (A).

(C) Explain why the cross-section of the surface by the plane $y = 1$ is also a parabola. Use an ordinary derivative to find the slope of the tangent line to this parabola when $x = 3$.

(D) Use partial derivatives to confirm your answer to part (C).

■ SECOND-ORDER PARTIAL DERIVATIVES

The function

$$z = f(x, y) = x^4y^7$$

has two **first-order partial derivatives,**

$$\frac{\partial z}{\partial x} = f_x = f_x(x, y) = 4x^3y^7 \qquad \text{and} \qquad \frac{\partial z}{\partial y} = f_y = f_y(x, y) = 7x^4y^6$$

Each of these partial derivatives, in turn, has two partial derivatives called **second-order partial derivatives** of $z = f(x, y)$. Generalizing the various notations we have for first-order partial derivatives, the four second-order partial derivatives of $z = f(x, y) = x^4y^7$ are written as

Equivalent notations

$$f_{xx} = f_{xx}(x, y) = \frac{\partial^2 z}{\partial x^2} = \frac{\partial}{\partial x}\left(\frac{\partial z}{\partial x}\right) = \frac{\partial}{\partial x}(4x^3y^7) = 12x^2y^7$$

$$f_{xy} = f_{xy}(x, y) = \frac{\partial^2 z}{\partial y\,\partial x} = \frac{\partial}{\partial y}\left(\frac{\partial z}{\partial x}\right) = \frac{\partial}{\partial y}(4x^3y^7) = 28x^3y^6$$

$$f_{yx} = f_{yx}(x, y) = \frac{\partial^2 z}{\partial x\,\partial y} = \frac{\partial}{\partial x}\left(\frac{\partial z}{\partial y}\right) = \frac{\partial}{\partial x}(7x^4y^6) = 28x^3y^6$$

$$f_{yy} = f_{yy}(x, y) = \frac{\partial^2 z}{\partial y^2} = \frac{\partial}{\partial y}\left(\frac{\partial z}{\partial y}\right) = \frac{\partial}{\partial y}(7x^4y^6) = 42x^4y^5$$

In the mixed partial derivative $\partial^2 z/\partial y\,\partial x = f_{xy}$, we started with $z = f(x, y)$ and first differentiated with respect to x (holding y constant). Then we differentiated with respect to y (holding x constant). In the other mixed partial derivative, $\partial^2 z/\partial x\,\partial y = f_{yx}$, the order of differentiation was reversed; however, the final result was the same—that is, $f_{xy} = f_{yx}$. Although it is possible to find functions for which $f_{xy} \neq f_{yx}$, such functions rarely occur in applications involving partial derivatives. Thus, for all the functions in this text, we will assume that $f_{xy} = f_{yx}$.

In general, we have the following definitions:

Second-Order Partial Derivatives

If $z = f(x, y)$, then

$$f_{xx} = f_{xx}(x, y) = \frac{\partial^2 z}{\partial x^2} = \frac{\partial}{\partial x}\left(\frac{\partial z}{\partial x}\right)$$

$$f_{xy} = f_{xy}(x, y) = \frac{\partial^2 z}{\partial y\, \partial x} = \frac{\partial}{\partial y}\left(\frac{\partial z}{\partial x}\right)$$

$$f_{yx} = f_{yx}(x, y) = \frac{\partial^2 z}{\partial x\, \partial y} = \frac{\partial}{\partial x}\left(\frac{\partial z}{\partial y}\right)$$

$$f_{yy} = f_{yy}(x, y) = \frac{\partial^2 z}{\partial y^2} = \frac{\partial}{\partial y}\left(\frac{\partial z}{\partial y}\right)$$

EXAMPLE 5 ➤ **Second-Order Partial Derivatives** For $z = f(x, y) = 3x^2 - 2xy^3 + 1$, find:

(A) $\dfrac{\partial^2 z}{\partial x\, \partial y}$, $\dfrac{\partial^2 z}{\partial y\, \partial x}$ (B) $\dfrac{\partial^2 z}{\partial x^2}$ (C) $f_{yx}(2, 1)$

Solution (A) First differentiate with respect to y and then with respect to x:

$$\frac{\partial z}{\partial y} = -6xy^2 \qquad \frac{\partial^2 z}{\partial x\, \partial y} = \frac{\partial}{\partial x}\left(\frac{\partial z}{\partial y}\right) = \frac{\partial}{\partial x}(-6xy^2) = -6y^2$$

First differentiate with respect to x and then with respect to y:

$$\frac{\partial z}{\partial x} = 6x - 2y^3 \qquad \frac{\partial^2 z}{\partial y\, \partial x} = \frac{\partial}{\partial y}\left(\frac{\partial z}{\partial x}\right) = \frac{\partial}{\partial y}(6x - 2y^3) = -6y^2$$

(B) Differentiate with respect to x twice:

$$\frac{\partial z}{\partial x} = 6x - 2y^3 \qquad \frac{\partial^2 z}{\partial x^2} = \frac{\partial}{\partial x}\left(\frac{\partial z}{\partial x}\right) = 6$$

(C) First find $f_{yx}(x, y)$; then evaluate at $(2, 1)$. Again, remember that f_{yx} means to differentiate with respect to y first and then with respect to x. Thus,

$$f_y(x, y) = -6xy^2 \qquad f_{yx}(x, y) = -6y^2$$

and

$$f_{yx}(2, 1) = -6(1)^2 = -6$$

◀

MATCHED PROBLEM 5 ➤ For $z = f(x, y) = x^3 y - 2y^4 + 3$, find:

(A) $\dfrac{\partial^2 z}{\partial y\, \partial x}$ (B) $\dfrac{\partial^2 z}{\partial y^2}$ (C) $f_{xy}(2, 3)$ (D) $f_{yx}(2, 3)$

◀

Answers to Matched Problems

1. (A) $\partial z/\partial y = -3x^2 + 5$ (B) $f_y(2, 3) = -7$
2. (A) $10x(x^2 + 2xy)^4$ (B) 10
3. $P_y(25, 10) = 10$: at a production level of $x = 25$ and $y = 10$, increasing y by 1 unit and holding x fixed at 25 will increase profit by approx. \$10; $P_y(25, 15) = -110$: at a production level of $x = 25$ and $y = 15$, increasing y by 1 unit and holding x fixed at 25 will decrease profit by approx. \$110
4. (A) $f_x(x, y) = 12x^{-0.7}y^{0.7}$; $f_y(x, y) = 28x^{0.3}y^{-0.3}$
   (B) Marginal productivity of labor ≈ 25.89; marginal productivity of capital ≈ 20.14
   (C) Labor
5. (A) $3x^2$ (B) $-24y^2$ (C) 12 (D) 12

EXERCISE   13-2

A   *For $z = f(x, y) = 10 + 3x + 2y$, find each of the following:*

1. $\partial z/\partial x$
2. $\partial z/\partial y$
3. $f_y(1, 2)$
4. $f_x(1, 2)$

For $z = f(x, y) = 3x^2 - 2xy^2 + 1$, find each of the following:

5. $\partial z/\partial y$
6. $\partial z/\partial x$
7. $f_x(2, 3)$
8. $f_y(2, 3)$

For $S(x, y) = 5x^2y^3$, find each of the following:

9. $S_x(x, y)$
10. $S_y(x, y)$
11. $S_y(2, 1)$
12. $S_x(2, 1)$

B   *For $C(x, y) = x^2 - 2xy + 2y^2 + 6x - 9y + 5$, find each of the following:*

13. $C_x(x, y)$
14. $C_y(x, y)$
15. $C_x(2, 2)$
16. $C_y(2, 2)$
17. $C_{xy}(x, y)$
18. $C_{yx}(x, y)$
19. $C_{xx}(x, y)$
20. $C_{yy}(x, y)$

For $z = f(x, y) = e^{2x + 3y}$, find each of the following:

21. $\dfrac{\partial z}{\partial x}$
22. $\dfrac{\partial z}{\partial y}$
23. $\dfrac{\partial^2 z}{\partial x\, \partial y}$
24. $\dfrac{\partial^2 z}{\partial y\, \partial x}$
25. $f_{xy}(1, 0)$
26. $f_{yx}(0, 1)$
27. $f_{xx}(0, 1)$
28. $f_{yy}(1, 0)$

In Problems 29–38, find $f_x(x, y)$ and $f_y(x, y)$ for each function f.

29. $f(x, y) = (x^2 - y^3)^3$
30. $f(x, y) = \sqrt{2x - y^2}$
31. $f(x, y) = (3x^2y - 1)^4$
32. $f(x, y) = (3 + 2xy^2)^3$
33. $f(x, y) = \ln(x^2 + y^2)$
34. $f(x, y) = \ln(2x - 3y)$

35. $f(x, y) = y^2 e^{xy^2}$
36. $f(x, y) = x^3 e^{x^2y}$
37. $f(x, y) = \dfrac{x^2 - y^2}{x^2 + y^2}$
38. $f(x, y) = \dfrac{2x^2y}{x^2 + y^2}$

39. (A) Let $f(x, y) = y^3 + 4y^2 - 5y + 3$. Show that $\partial f/\partial x = 0$.
    (B) Explain why there are an infinite number of functions $g(x, y)$ such that $\partial g/\partial x = 0$.

40. (A) Find an example of a function $f(x, y)$ such that $\partial f/\partial x = 3$ and $\partial f/\partial y = 2$.
    (B) How many such functions are there? Explain.

In Problems 41–46, find $f_{xx}(x, y)$, $f_{xy}(x, y)$, $f_{yx}(x, y)$, and $f_{yy}(x, y)$ for each function f.

41. $f(x, y) = x^2y^2 + x^3 + y$
42. $f(x, y) = x^3y^3 + x + y^2$
43. $f(x, y) = \dfrac{x}{y} - \dfrac{y}{x}$
44. $f(x, y) = \dfrac{x^2}{y} - \dfrac{y^2}{x}$
45. $f(x, y) = xe^{xy}$
46. $f(x, y) = x \ln(xy)$

C

47. For
$$P(x, y) = -x^2 + 2xy - 2y^2 - 4x + 12y - 5$$
find values of x and y such that
$$P_x(x, y) = 0 \quad \text{and} \quad P_y(x, y) = 0$$
simultaneously.

48. For
$$C(x, y) = 2x^2 + 2xy + 3y^2 - 16x - 18y + 54$$
find values of x and y such that
$$C_x(x, y) = 0 \quad \text{and} \quad C_y(x, y) = 0$$
simultaneously.

49. For

$$F(x, y) = x^3 - 2x^2y^2 - 2x - 4y + 10$$

find all values of x and y such that

$$F_x(x, y) = 0 \quad \text{and} \quad F_y(x, y) = 0$$

simultaneously.

50. For

$$G(x, y) = x^2 \ln y - 3x - 2y + 1$$

find all values of x and y such that

$$G_x(x, y) = 0 \quad \text{and} \quad G_y(x, y) = 0$$

simultaneously.

51. Let $f(x, y) = 4 - 3x + y - x^2 - 2y^2$.
  (A) Find the maximum value of $f(x, y)$ when $y = 1$.
  (B) Explain why the answer to part (A) is not the maximum value of the function $f(x, y)$.

52. Let $g(x, y) = ye^x - xy^2 + 1$.
  (A) Approximate the minimum value of $g(x, y)$ when $y = 2$.
  (B) Explain why the answer to part (A) is not the minimum value of the function $g(x, y)$.

In Problems 53 and 54, show that the function f satisfies
$f_{xx}(x, y) + f_{yy}(x, y) = 0.$

53. $f(x, y) = \ln(x^2 + y^2)$

54. $f(x, y) = x^3 - 3xy^2$

55. For $f(x, y) = x^2 + 2y^2$, find:

  (A) $\displaystyle \lim_{h \to 0} \frac{f(x + h, y) - f(x, y)}{h}$

  (B) $\displaystyle \lim_{k \to 0} \frac{f(x, y + k) - f(x, y)}{k}$

56. For $f(x, y) = 2xy^2$, find:

  (A) $\displaystyle \lim_{h \to 0} \frac{f(x + h, y) - f(x, y)}{h}$

  (B) $\displaystyle \lim_{k \to 0} \frac{f(x, y + k) - f(x, y)}{k}$

▶ APPLICATIONS

Business & Economics

57. *Profit function.* A firm produces two types of calculators, x of type A and y of type B each week. The weekly revenue and cost functions (in dollars) are

$$R(x, y) = 80x + 90y + 0.04xy - 0.05x^2 - 0.05y^2$$

$$C(x, y) = 8x + 6y + 20,000$$

Find $P_x(1,200, 1,800)$ and $P_y(1,200, 1,800)$, and interpret the results.

58. *Advertising and sales.* A company spends $\$x$ per week on newspaper advertising and $\$y$ per week on television advertising. Its weekly sales were found to be given by

$$S(x, y) = 10x^{0.4}y^{0.8}$$

Find $S_x(3,000, 2,000)$ and $S_y(3,000, 2,000)$, and interpret the results.

59. *Demand equations.* A supermarket sells two brands of coffee, brand A at $\$p$ per pound and brand B at $\$q$ per pound. The daily demand equations for brands A and B are, respectively,

$$x = 200 - 5p + 4q$$

$$y = 300 + 2p - 4q$$

Find $\partial x/\partial p$ and $\partial y/\partial p$, and interpret the results.

60. *Revenue and profit functions.* A company manufactures ten-speed and three-speed bicycles. The weekly demand and cost functions are

$$p = 230 - 9x + y$$

$$q = 130 + x - 4y$$

$$C(x, y) = 200 + 80x + 30y$$

where $\$p$ is the price of a ten-speed bicycle, $\$q$ is the price of a three-speed bicycle, x is the weekly demand for ten-speed bicycles, y is the weekly demand for three-speed bicycles, and $C(x, y)$ is the cost function. Find $R_x(10, 5)$ and $P_x(10, 5)$, and interpret the results.

61. *Productivity.* The productivity of a certain third-world country is given approximately by the function

$$f(x, y) = 10x^{0.75}y^{0.25}$$

with the utilization of x units of labor and y units of capital.
(A) Find $f_x(x, y)$ and $f_y(x, y)$.
(B) If the country is now using 600 units of labor and 100 units of capital, find the marginal productivity of labor and the marginal productivity of capital.
(C) For the greatest increase in the country's productivity, should the government encourage increased use of labor or increased use of capital?

62. *Productivity.* The productivity of an automobile manufacturing company is given approximately by the function

$$f(x, y) = 50\sqrt{xy} = 50x^{0.5}y^{0.5}$$

with the utilization of x units of labor and y units of capital.
(A) Find $f_x(x, y)$ and $f_y(x, y)$.
(B) If the company is now using 250 units of labor and 125 units of capital, find the marginal productivity of labor and the marginal productivity of capital.
(C) For the greatest increase in the company's productivity, should the management encourage increased use of labor or increased use of capital?

*Problems 63–66 refer to the following: If a decrease in demand for one product results in an increase in demand for another product, then the two products are said to be **competitive, or substitute, products**. (Real whipping cream and imitation whipping cream are examples of competitive, or substitute, products.) If a decrease in demand for one product results in a decrease in demand for another product, then the two products are said to be **complementary products**. (Fishing boats and outboard motors are examples of complementary products.) Partial derivatives can be used to test whether two products are competitive, complementary, or neither. We start with demand functions for two products where the demand for either depends on the prices for both:*

$x = f(p, q)$ Demand function for product A

$y = g(p, q)$ Demand function for product B

The variables x and y represent the number of units demanded of products A and B, respectively, at a price p for 1 unit of product A and a price q for 1 unit of product B. Normally, if the price of A increases while the price of B is held constant, then the demand for A will decrease; that is, $f_p(p, q) < 0$. Then, if A and B are competitive products, the demand for B will increase; that is, $g_p(p, q) > 0$. Similarly, if the price of B increases while the price of A is held constant, then the demand for B will decrease; that is, $g_q(p, q) < 0$. And if A and B are

competitive products, then the demand for A will increase; that is, $f_q(p, q) > 0$. Reasoning similarly for complementary products, we arrive at the following test:

Test for Competitive and Complementary Products

PARTIAL DERIVATIVES				PRODUCTS A AND B
$f_q(p, q) > 0$	and	$g_p(p, q) > 0$		Competitive (Substitute)
$f_q(p, q) < 0$	and	$g_p(p, q) < 0$		Complementary
$f_q(p, q) \geq 0$	and	$g_p(p, q) \leq 0$		Neither
$f_q(p, q) \leq 0$	and	$g_p(p, q) \geq 0$		Neither

Use this test in Problems 63–66 to determine whether the indicated products are competitive, complementary, or neither.

63. *Product demand.* The weekly demand equations for the sale of butter and margarine in a supermarket are

$$x = f(p, q) = 8,000 - 0.09p^2 + 0.08q^2 \quad \text{Butter}$$

$$y = g(p, q) = 15,000 + 0.04p^2 - 0.3q^2 \quad \text{Margarine}$$

64. *Product demand.* The daily demand equations for the sale of brand A coffee and brand B coffee in a supermarket are

$$x = f(p, q) = 200 - 5p + 4q \quad \text{Brand A coffee}$$

$$y = g(p, q) = 300 + 2p - 4q \quad \text{Brand B coffee}$$

65. *Product demand.* The monthly demand equations for the sale of skis and ski boots in a sporting goods store are

$$x = f(p, q) = 800 - 0.004p^2 - 0.003q^2 \quad \text{Skis}$$

$$y = g(p, q) = 600 - 0.003p^2 - 0.002q^2 \quad \text{Ski boots}$$

66. *Product demand.* The monthly demand equations for the sale of tennis rackets and tennis balls in a sporting goods store are

$$x = f(p, q) = 500 - 0.5p - q^2 \quad \text{Tennis rackets}$$

$$y = g(p, q) = 10,000 - 8p - 100q^2 \quad \text{Tennis balls (cans)}$$

Life Sciences

67. *Medicine.* The following empirical formula relates the surface area A (in square inches) of an average human body to its weight w (in pounds) and its height h (in inches):

$$A = f(w, h) = 15.64w^{0.425}h^{0.725}$$

Knowing the surface area of a human body is useful, for example, in studies pertaining to hypothermia (heat loss due to exposure).

(A) Find $f_w(w, h)$ and $f_h(w, h)$.

(B) For a 65 pound child who is 57 inches tall, find $f_w(65, 57)$ and $f_h(65, 57)$, and interpret the results.

68. *Blood flow.* Poiseuille's law states that the resistance, R, for blood flowing in a blood vessel varies directly as the length of the vessel, L, and inversely as the fourth power of its radius, r. Stated as an equation,

$$R(L, r) = k\frac{L}{r^4} \qquad k \text{ a constant}$$

Find $R_L(4, 0.2)$ and $R_r(4, 0.2)$, and interpret the results.

Social Sciences

69. *Physical anthropology.* Anthropologists, in their study of race and human genetic groupings, often use the cephalic index, C, which varies directly as the width, W, of the head, and inversely as the length, L, of the head (both viewed from the top). In terms of an equation,

$$C(W, L) = 100\frac{W}{L}$$

where

$$W = \text{Width in inches} \qquad L = \text{Length in inches}$$

Find $C_W(6, 8)$ and $C_L(6, 8)$, and interpret the results.

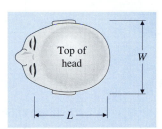

70. *Safety research.* Under ideal conditions, if a person driving a car slams on the brakes and skids to a stop, the length of the skid marks (in feet) is given by the formula

$$L(w, v) = kwv^2$$

where

$$k = \text{Constant}$$
$$w = \text{Weight of car in pounds}$$
$$v = \text{Speed of car in miles per hour}$$

For $k = 0.000\ 013\ 3$, find $L_w(2,500, 60)$ and $L_v(2,500, 60)$, and interpret the results.

SECTION 13-3 ■ ## Maxima and Minima

We are now ready to undertake a brief but useful analysis of local maxima and minima for functions of the type $z = f(x, y)$. Basically, we are going to extend the second-derivative test developed for functions of a single independent variable. To start, we assume that all second-order partial derivatives exist for the function f in some circular region in the xy plane. This guarantees that the surface $z = f(x, y)$ has no sharp points, breaks, or ruptures. In other words, we are dealing only with smooth surfaces with no edges (like the edge of a box); or breaks (like an earthquake fault); or sharp points (like the bottom point of a golf tee). See Figure 1 (page 960).

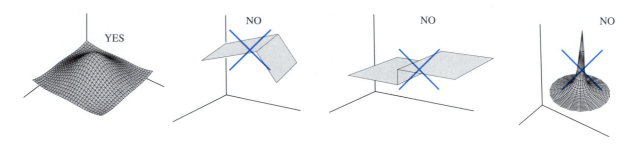

FIGURE 1

In addition, we will not concern ourselves with boundary points or absolute maxima–minima theory. In spite of these restrictions, the procedure we are now going to describe will help us solve a large number of useful problems.

What does it mean for $f(a, b)$ to be a local maximum or a local minimum? We say that $f(a, b)$ **is a local maximum** if there exists a circular region in the domain of f with (a, b) as the center, such that

$$f(a, b) \geq f(x, y)$$

for all (x, y) in the region. Similarly, we say that $f(a, b)$ **is a local minimum** if there exists a circular region in the domain of f with (a, b) as the center, such that

$$f(a, b) \leq f(x, y)$$

for all (x, y) in the region. Figure 2A illustrates a local maximum, Figure 2B a local minimum, and Figure 2C a **saddle point,** which is neither a local maximum nor a local minimum.

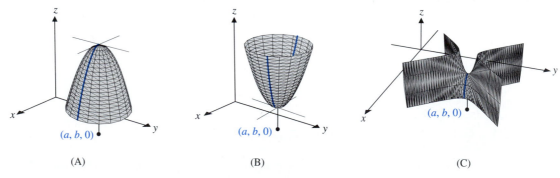

(A)   (B)   (C)

FIGURE 2

What happens to $f_x(a, b)$ and $f_y(a, b)$ if $f(a, b)$ is a local minimum or a local maximum and the partial derivatives of f exist in a circular region containing (a, b)? Figure 2 suggests that $f_x(a, b) = 0$ and $f_y(a, b) = 0$, since the tangent lines to the given curves are horizontal. Theorem 1 indicates that our intuitive reasoning is correct.

theorem 1

Local Extrema and Partial Derivatives

Let $f(a, b)$ be a local extremum (a local maximum or a local minimum) for the function f. If both f_x and f_y exist at (a, b), then

$$f_x(a, b) = 0 \qquad \text{and} \qquad f_y(a, b) = 0 \tag{1}$$

The converse of this theorem is false. That is, if $f_x(a, b) = 0$ and $f_y(a, b) = 0$, then $f(a, b)$ may or may not be a local extremum; for example, the point $(a, b, f(a, b))$ may be a saddle point (see Fig. 2C).

Theorem 1 gives us *necessary* (but not *sufficient*) conditions for $f(a, b)$ to be a local extremum. We thus find all points (a, b) such that $f_x(a, b) = 0$ and $f_y(a, b) = 0$ and test these further to determine whether $f(a, b)$ is a local extremum or a saddle point. Points (a, b) such that conditions (1) hold are called **critical points.**

explore–discuss 1

(A) Let $f(x, y) = y^2 + 1$. Explain why $f(x, y)$ has a local minimum at every point on the x axis. Verify that every point on the x axis is a critical point. Explain why the graph of $z = f(x, y)$ could be described as a "trough."

(B) Let $g(x, y) = x^3$. Show that every point on the y axis is a critical point. Explain why no point on the y axis is a local extremum. Explain why the graph of $z = g(x, y)$ could be described as a "slide."

The next theorem, using second-derivative tests, gives us *sufficient* conditions for a critical point to produce a local extremum or a saddle point. (As was the case with Theorem 1, we state this theorem without proof.)

theorem 2

Second-Derivative Test for Local Extrema

Given:

1. $z = f(x, y)$
2. $f_x(a, b) = 0$ and $f_y(a, b) = 0$ [(a, b) is a critical point]
3. All second-order partial derivatives of f exist in some circular region containing (a, b) as a center
4. $A = f_{xx}(a, b)$, $B = f_{xy}(a, b)$, $C = f_{yy}(a, b)$

Then:

Case 1.   If $AC - B^2 > 0$ and $A < 0$, then $f(a, b)$ is a local maximum.

Case 2.   If $AC - B^2 > 0$ and $A > 0$, then $f(a, b)$ is a local minimum.

Case 3.   If $AC - B^2 < 0$, then f has a saddle point at (a, b).

Case 4.   If $AC - B^2 = 0$, the test fails.

FIGURE 3

To illustrate the use of Theorem 2, we will first find the local extremum for a very simple function whose solution is almost obvious: $z = f(x, y) = x^2 + y^2 + 2$. From the function f itself and its graph (Fig. 3), it is clear that a local minimum is found at $(0, 0)$. Let us see how Theorem 2 confirms this observation.

Step 1. Find critical points: Find (x, y) such that $f_x(x, y) = 0$ and $f_y(x, y) = 0$ simultaneously:

$$f_x(x, y) = 2x = 0 \qquad\qquad f_y(x, y) = 2y = 0$$
$$x = 0 \qquad\qquad\qquad\qquad y = 0$$

The only critical point is $(a, b) = (0, 0)$.

Step 2. Compute $A = f_{xx}(0, 0)$, $B = f_{xy}(0, 0)$, and $C = f_{yy}(0, 0)$:

$$f_{xx}(x, y) = 2 \qquad \text{thus} \qquad A = f_{xx}(0, 0) = 2$$
$$f_{xy}(x, y) = 0 \qquad \text{thus} \qquad B = f_{xy}(0, 0) = 0$$
$$f_{yy}(x, y) = 2 \qquad \text{thus} \qquad C = f_{yy}(0, 0) = 2$$

Step 3. Evaluate $AC - B^2$ and try to classify the critical point $(0, 0)$ using Theorem 2:

$$AC - B^2 = (2)(2) - (0)^2 = 4 > 0 \qquad \text{and} \qquad A = 2 > 0$$

Therefore, case 2 in Theorem 2 holds. That is, $f(0, 0) = 2$ is a local minimum.

We will now use Theorem 2 in the following examples to analyze extrema without the aid of graphs.

EXAMPLE 1 ➤ **Finding Local Extrema** Use Theorem 2 to find local extrema for:

$$f(x, y) = -x^2 - y^2 + 6x + 8y - 21$$

Solution **Step 1.** Find critical points: Find (x, y) such that $f_x(x, y) = 0$ and $f_y(x, y) = 0$ simultaneously:

$$f_x(x, y) = -2x + 6 = 0 \qquad\qquad f_y(x, y) = -2y + 8 = 0$$
$$x = 3 \qquad\qquad\qquad\qquad y = 4$$

The only critical point is $(a, b) = (3, 4)$.

Step 2. Compute $A = f_{xx}(3, 4)$, $B = f_{xy}(3, 4)$, and $C = f_{yy}(3, 4)$:

$$f_{xx}(x, y) = -2 \qquad \text{thus} \qquad A = f_{xx}(3, 4) = -2$$
$$f_{xy}(x, y) = 0 \qquad \text{thus} \qquad B = f_{xy}(3, 4) = 0$$
$$f_{yy}(x, y) = -2 \qquad \text{thus} \qquad C = f_{yy}(3, 4) = -2$$

Step 3. Evaluate $AC - B^2$ and try to classify the critical point $(3, 4)$ using Theorem 2:

$$AC - B^2 = (-2)(-2) - (0)^2 = 4 > 0 \qquad \text{and} \qquad A = -2 < 0$$

Therefore, case 1 in Theorem 2 holds. That is, $f(3, 4) = 4$ is a local maximum. ◀

MATCHED PROBLEM 1 ➤ Use Theorem 2 to find local extrema for: $f(x, y) = x^2 + y^2 - 10x - 2y + 36$

EXAMPLE 2 ➤ **Finding Local Extrema—Multiple Critical Points** Use Theorem 2 to find local extrema for:

$$f(x, y) = x^3 + y^3 - 6xy$$

Solution **Step 1.** Find critical points for $f(x, y) = x^3 + y^3 - 6xy$:

$$f_x(x, y) = 3x^2 - 6y = 0 \qquad \textit{Solve for y.}$$
$$6y = 3x^2$$
$$y = \tfrac{1}{2}x^2 \qquad\qquad\qquad (2)$$
$$f_y(x, y) = 3y^2 - 6x = 0$$
$$3y^2 = 6x \qquad \textit{Use (2) to eliminate y.}$$
$$3(\tfrac{1}{2}x^2)^2 = 6x$$
$$\tfrac{3}{4}x^4 = 6x \qquad \textit{Solve for x.}$$
$$3x^4 - 24x = 0$$
$$3x(x^3 - 8) = 0$$
$$x = 0 \quad \text{or} \quad x = 2$$
$$y = 0 \qquad\qquad y = \tfrac{1}{2}(2)^2 = 2$$

The critical points are $(0, 0)$ and $(2, 2)$.

Since there are two critical points, steps 2 and 3 must be performed twice.

Test (0, 0) **Step 2.** Compute $A = f_{xx}(0, 0)$, $B = f_{xy}(0, 0)$, and $C = f_{yy}(0, 0)$:

$$f_{xx}(x, y) = 6x \qquad \text{thus} \qquad A = f_{xx}(0, 0) = 0$$
$$f_{xy}(x, y) = -6 \qquad \text{thus} \qquad B = f_{xy}(0, 0) = -6$$
$$f_{yy}(x, y) = 6y \qquad \text{thus} \qquad C = f_{yy}(0, 0) = 0$$

Step 3. Evaluate $AC - B^2$ and try to classify the critical point $(0, 0)$ using Theorem 2:

$$AC - B^2 = (0)(0) - (-6)^2 = -36 < 0$$

Therefore, case 3 in Theorem 2 applies. That is, f has a saddle point at $(0, 0)$.

Now we will consider the second critical point, $(2, 2)$.

Test (2, 2) **Step 2.** Compute $A = f_{xx}(2, 2)$, $B = f_{xy}(2, 2)$, and $C = f_{yy}(2, 2)$:

$$f_{xx}(x, y) = 6x \qquad \text{thus} \qquad A = f_{xx}(2, 2) = 12$$
$$f_{xy}(x, y) = -6 \qquad \text{thus} \qquad B = f_{xy}(2, 2) = -6$$
$$f_{yy}(x, y) = 6y \qquad \text{thus} \qquad C = f_{yy}(2, 2) = 12$$

Step 3. Evaluate $AC - B^2$ and try to classify the critical point $(2, 2)$ using Theorem 2:

$$AC - B^2 = (12)(12) - (-6)^2 = 108 > 0 \qquad \text{and} \qquad A = 12 > 0$$

Thus, case 2 in Theorem 2 applies, and $f(2, 2) = -8$ is a local minimum.

Our conclusions may be confirmed geometrically by graphing cross-sections of the function f. The cross-sections of f by the planes $y = 0$, $x = 0$, $y = x$, and $y = -x$ [each of these planes contains $(0, 0)$] are represented by the graphs of the functions $f(x, 0) = x^3$, $f(0, y) = y^3$, $f(x, x) = 2x^3 - 6x^2$, and $f(x, -x) = 6x^2$, respectively, as shown in Figure 4A (note that the first two functions have the same graph). The cross-sections of f by the planes $y = 2$, $x = 2$, $y = x$, and $y = 4 - x$ [each of these planes contains $(2, 2)$] are represented by the graphs of the functions $f(x, 2) = x^3 - 12x + 8$, $f(2, y) = y^3 - 12y + 8$, $f(x, x) = 2x^3 - 6x^2$, and $f(x, 4 - x) = x^3 + (4 - x)^3 + 6x^2 - 24x$, respectively, as shown in Figure 4B (the first two functions have the same graph). Figure 4B illustrates the fact that since f has a local minimum at $(2, 2)$, each of the cross-sections of f through $(2, 2)$ has a local minimum of -8 at $(2, 2)$. Figure 4A, on the other hand, indicates that some cross-sections of f through $(0, 0)$ have a local minimum, some a local maximum, and some neither one, at $(0, 0)$.

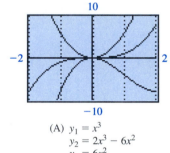

(A) $y_1 = x^3$
$y_2 = 2x^3 - 6x^2$
$y_3 = 6x^2$

(B) $y_1 = x^3 - 12x + 8$
$y_2 = 2x^3 - 6x^2$
$y_3 = x^3 + (4 - x)^3 + 6x^2 - 24x$

FIGURE 4

MATCHED PROBLEM 2 ➤   Use Theorem 2 to find local extrema for: $f(x, y) = x^3 + y^2 - 6xy$

explore – discuss 2

Let $f(x, y) = x^4 + y^2 + 3$, $g(x, y) = 10 - x^2 - y^4$, and $h(x, y) = x^3 + y^2$.

(A) Show that each function has only $(0, 0)$ as a critical point.
(B) Explain why $f(x, y)$ has a minimum at $(0, 0)$, $g(x, y)$ has a maximum at $(0, 0)$, and $h(x, y)$ has neither a minimum nor a maximum at $(0, 0)$.
(C) Are the results of part (B) consequences of Theorem 2? Explain.

EXAMPLE 3 ➤   **Profit**   Suppose the surfboard company discussed earlier has developed the yearly profit equation

$$P(x, y) = -2x^2 + 2xy - y^2 + 10x - 4y + 107$$

where x is the number (in thousands) of standard surfboards produced per year, y is the number (in thousands) of competition surfboards produced per year, and P is profit (in thousands of dollars). How many of each type of board should be produced per year to realize a maximum profit? What is the maximum profit?

Solution   Step 1.   Find critical points:

$$P_x(x, y) = -4x + 2y + 10 = 0$$
$$P_y(x, y) = 2x - 2y - 4 = 0$$

Solving this system, we obtain $(3, 1)$ as the only critical point.

Step 2.   Compute $A = P_{xx}(3, 1)$, $B = P_{xy}(3, 1)$, and $C = P_{yy}(3, 1)$:

$P_{xx}(x, y) = -4$	thus	$A = P_{xx}(3, 1) = -4$
$P_{xy}(x, y) = 2$	thus	$B = P_{xy}(3, 1) = 2$
$P_{yy}(x, y) = -2$	thus	$C = P_{yy}(3, 1) = -2$

Step 3.   Evaluate $AC - B^2$ and try to classify the critical point $(3, 1)$ using Theorem 2:

$$AC - B^2 = (-4)(-2) - (2)^2 = 8 - 4 = 4 > 0 \qquad \text{and} \qquad A = -4 < 0$$

Therefore, case 1 in Theorem 2 applies. That is, $P(3, 1) = \$120{,}000$ is a local maximum. This is obtained by producing 3,000 standard boards and 1,000 competition boards per year. ◀

MATCHED PROBLEM 3 ▶   Repeat Example 3 with: $P(x, y) = -2x^2 + 4xy - 3y^2 + 4x - 2y + 77$ ◀

▶   EXAMPLE 4 ▶   **Package Design**   The packaging department in a company has been asked to design a rectangular box with no top and a partition down the middle. The box must have a volume of 48 cubic inches. Find the dimensions that will minimize the amount of material used to construct the box.

FIGURE 5

Solution   Refer to Figure 5. The amount of material used in constructing this box is

	Base	Front, back	Sides, partition	
$M =$	xy	$+ \; 2xz$	$+ \quad 3yz$	(3)

The volume of the box is

$$V = xyz = 48 \tag{4}$$

Since Theorem 2 applies only to functions with two independent variables, we must use (4) to eliminate one of the variables in (3):

$$M = xy + 2xz + 3yz \qquad \text{Substitute } z = 48/xy.$$

$$= xy + 2x\left(\frac{48}{xy}\right) + 3y\left(\frac{48}{xy}\right)$$

$$= xy + \frac{96}{y} + \frac{144}{x}$$

Thus, we must find the minimum value of

$$M(x, y) = xy + \frac{96}{y} + \frac{144}{x} \qquad x > 0 \quad \text{and} \quad y > 0$$

Step 1.   Find critical points:

$$M_x(x, y) = y - \frac{144}{x^2} = 0$$

$$y = \frac{144}{x^2} \tag{5}$$

$$M_y(x, y) = x - \frac{96}{y^2} = 0$$

$$x = \frac{96}{y^2} \qquad \text{Solve for } y^2.$$

$$y^2 = \frac{96}{x} \qquad \text{Use (5) to eliminate } y \text{ and solve for } x.$$

$$\left(\frac{144}{x^2}\right)^2 = \frac{96}{x}$$

$$\frac{20{,}736}{x^4} = \frac{96}{x} \qquad \text{Multiply both sides by } x^4/96 \\ \qquad\qquad (\text{recall, } x > 0).$$

$$x^3 = \frac{20{,}736}{96} = 216$$

$$x = 6 \qquad \text{Use (5) to find } y.$$

$$y = \frac{144}{36} = 4$$

Thus, $(6, 4)$ is the only critical point.

Step 2.   Compute $A = M_{xx}(6, 4)$, $B = M_{xy}(6, 4)$, and $C = M_{yy}(6, 4)$:

$$M_{xx}(x, y) = \frac{288}{x^3} \qquad \text{thus} \qquad A = M_{xx}(6, 4) = \tfrac{288}{216} = \tfrac{4}{3}$$

$$M_{xy}(x, y) = 1 \qquad \text{thus} \qquad B = M_{xy}(6, 4) = 1$$

$$M_{yy}(x, y) = \frac{192}{y^3} \qquad \text{thus} \qquad C = M_{yy}(6, 4) = \tfrac{192}{64} = 3$$

Step 3.   Evaluate $AC - B^2$ and try to classify the critical point $(6, 4)$ using Theorem 2:

$$AC - B^2 = (\tfrac{4}{3})(3) - (1)^2 = 3 > 0 \qquad \text{and} \qquad A = \tfrac{4}{3} > 0$$

Therefore, case 2 in Theorem 2 applies; $M(x, y)$ has a local minimum at $(6, 4)$. If $x = 6$ and $y = 4$, then

$$z = \frac{48}{xy} = \frac{48}{(6)(4)} = 2$$

Thus, the dimensions that will require the minimum amount of material are 6 inches by 4 inches by 2 inches.   ◀

MATCHED PROBLEM 4 ➤ If the box in Example 4 must have a volume of 384 cubic inches, find the dimensions that will require the least amount of material. ◄

Answers to Matched Problems
1. $f(5, 1) = 10$ is a local minimum
2. f has a saddle point at $(0, 0)$; $f(6, 18) = -108$ is a local minimum
3. Local maximum for $x = 2$ and $y = 1$; $P(2, 1) = \$80,000$
4. 12 in. by 8 in. by 4 in.

EXERCISE 13-3

Find local extrema in Problems 1–20 using Theorem 2.

A

1. $f(x, y) = 6 - x^2 - 4x - y^2$
2. $f(x, y) = 3 - x^2 - y^2 + 6y$
3. $f(x, y) = x^2 + y^2 + 2x - 6y + 14$
4. $f(x, y) = x^2 + y^2 - 4x + 6y + 23$

B

5. $f(x, y) = xy + 2x - 3y - 2$
6. $f(x, y) = x^2 - y^2 + 2x + 6y - 4$
7. $f(x, y) = -3x^2 + 2xy - 2y^2 + 14x + 2y + 10$
8. $f(x, y) = -x^2 + xy - 2y^2 + x + 10y - 5$
9. $f(x, y) = 2x^2 - 2xy + 3y^2 - 4x - 8y + 20$
10. $f(x, y) = 2x^2 - xy + y^2 - x - 5y + 8$

C

11. $f(x, y) = e^{xy}$
12. $f(x, y) = x^2y - xy^2$
13. $f(x, y) = x^3 + y^3 - 3xy$
14. $f(x, y) = 2y^3 - 6xy - x^2$
15. $f(x, y) = 2x^4 + y^2 - 12xy$
16. $f(x, y) = 16xy - x^4 - 2y^2$
17. $f(x, y) = x^3 - 3xy^2 + 6y^2$

18. $f(x, y) = 2x^2 - 2x^2y + 6y^3$
19. $f(x, y) = y^3 + 2x^2y^2 - 3x - 2y + 8$
20. $f(x, y) = x \ln y + x^2 - 4x - 5y + 3$

21. Explain why $f(x, y) = x^2$ has an infinite number of local extrema.

22. (A) Find the local extrema of the functions $f(x, y) = x + y$, $g(x, y) = x^2 + y^2$, and $h(x, y) = x^3 + y^3$.
   (B) Discuss the local extrema of the function $k(x) = x^n + y^n$, where n is a positive integer.

23. (A) Show that $(0, 0)$ is a critical point for the function $f(x, y) = x^4e^y + x^2y^4 + 1$, but that the second-derivative test for local extrema fails.
   (B) Use cross-sections, as in Example 2, to decide whether f has a local maximum, a local minimum, or a saddle point at $(0, 0)$.

24. (A) Show that $(0, 0)$ is a critical point for the function $g(x, y) = e^{xy^2} + x^2y^3 + 2$, but that the second-derivative test for local extrema fails.
   (B) Use cross-sections, as in Example 2, to decide whether g has a local maximum, a local minimum, or a saddle point at $(0, 0)$.

▶ APPLICATIONS

Business & Economics

25. *Product mix for maximum profit.* A firm produces two types of calculators, x thousand of type A and y thousand of type B per year. If the revenue and cost equations for the year are (in millions of dollars)

$$R(x, y) = 2x + 3y$$

$$C(x, y) = x^2 - 2xy + 2y^2 + 6x - 9y + 5$$

determine how many of each type of calculator should be produced per year to maximize profit. What is the maximum profit?

26. *Automation–labor mix for minimum cost.* The annual labor and automated equipment cost (in millions of dollars) for a company's production of television sets is given by

$$C(x, y) = 2x^2 + 2xy + 3y^2 - 16x - 18y + 54$$

where x is the amount spent per year on labor and y is the amount spent per year on automated equipment (both in millions of dollars). Determine how much should be spent on each per year to minimize this cost. What is the minimum cost?

27. *Maximizing profit.* A department store sells two brands of inexpensive calculators. The store pays $6 for each brand A calculator and $8 for each brand B calculator. The research department has estimated the following weekly demand equations for these two competitive products:

$$x = 116 - 30p + 20q \quad \text{Demand equation for brand A}$$
$$y = 144 + 16p - 24q \quad \text{Demand equation for brand B}$$

where p is the selling price for brand A and q is the selling price for brand B.

(A) Determine the demands x and y when $p = \$10$ and $q = \$12$; when $p = \$11$ and $q = \$11$.

(B) How should the store price each calculator to maximize weekly profits? What is the maximum weekly profit? [*Hint:* $C = 6x + 8y$, $R = px + qy$, and $P = R - C$.]

28. *Maximizing profit.* A store sells two brands of color print film. The store pays $2 for each roll of brand A film and $3 for each roll of brand B film. A consulting firm has estimated the following daily demand equations for these two competitive products:

$$x = 75 - 40p + 25q \quad \text{Demand equation for brand A}$$
$$y = 80 + 20p - 30q \quad \text{Demand equation for brand B}$$

where p is the selling price for brand A and q is the selling price for brand B.

(A) Determine the demands x and y when $p = \$4$ and $q = \$5$; when $p = \$4$ and $q = \$4$.

(B) How should the store price each brand of film to maximize daily profits? What is the maximum daily profit? [*Hint:* $C = 2x + 3y$, $R = px + qy$, and $P = R - C$.]

29. *Minimizing cost.* A satellite television reception station is to be located at $P(x, y)$ so that the sum of the squares of the distances from P to the three towns A, B, and C is minimum (see the figure). Find the coordinates of P. This location will minimize the cost of providing satellite cable television for all three towns.

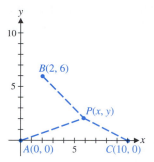

Figure for 29

30. *Minimizing cost.* Repeat Problem 29, replacing the coordinates of B with $B(6, 9)$ and the coordinates of C with $C(9, 0)$.

31. *Minimum material.* A rectangular box with no top and two parallel partitions (see the figure) is to be made to hold a volume of 64 cubic inches. Find the dimensions that will require the least amount of material.

Figure for 31

32. *Minimum material.* A rectangular box with no top and two intersecting partitions (see the figure) is to be made to hold a volume of 72 cubic inches. What should its dimensions be in order to use the least amount of material in its construction?

Figure for 32

33. *Maximum volume.* A mailing service states that a rectangular package shall have the sum of the length and girth not to exceed 120 inches (see the figure). What are the dimensions

of the largest (in volume) mailing carton that can be constructed meeting these restrictions?

Length

Girth

Figure for 33

34. *Maximum shipping volume.* A shipping box is reinforced with steel bands in all three directions, as indicated in the figure. A total of 150 inches of steel tape are to be used, with 6 inches of waste because of a 2 inch overlap in each direction. Find the dimensions of the box with maximum volume that can be taped as indicated.

Figure for 34

S E C T I O N   13-4   ■   Maxima and Minima Using Lagrange Multipliers

- FUNCTIONS OF TWO INDEPENDENT VARIABLES
- FUNCTIONS OF THREE INDEPENDENT VARIABLES

■ FUNCTIONS OF TWO INDEPENDENT VARIABLES

We will now consider a particularly powerful method of solving a certain class of maxima–minima problems. The method is due to Joseph Louis Lagrange (1736–1813), an eminent eighteenth century French mathematician, and it is called the **method of Lagrange multipliers.** We introduce the method through an example; then we will formalize the discussion in the form of a theorem.

A rancher wants to construct two feeding pens of the same size along an existing fence (see Fig. 1). If the rancher has 720 feet of fencing materials available, how long should x and y be in order to obtain the maximum total area? What is the maximum area?

The total area is given by

$$f(x, y) = xy$$

which can be made as large as we like, providing there are no restrictions on x and y. But there are restrictions on x and y, since we have only 720 feet of fencing. That is, x and y must be chosen so that

$$3x + y = 720$$

This restriction on x and y, called a **constraint,** leads to the following maxima–minima problem:

Existing fence

FIGURE 1

$$\text{Maximize} \quad f(x, y) = xy \tag{1}$$

$$\text{Subject to} \quad 3x + y = 720 \quad \text{or} \quad 3x + y - 720 = 0 \tag{2}$$

This problem is a special case of a general class of problems of the form

$$\text{Maximize (or Minimize)} \quad z = f(x, y) \tag{3}$$

$$\text{Subject to} \quad\quad\quad g(x, y) = 0 \tag{4}$$

Of course, we could try to solve (4) for y in terms of x, or for x in terms of y, then substitute the result into (3), and use methods developed in Section 9-5 for functions of a single variable. But what if (4) were more complicated than (2), and solving for one variable in terms of the other was either very difficult or impossible? In the method of Lagrange multipliers, we will work with $g(x, y)$ directly and avoid having to solve (4) for one variable in terms of the other. In addition, the method generalizes to functions of arbitrarily many variables subject to one or more constraints.

Now, to the method. We form a new function F, using functions f and g in (3) and (4), as follows:

$$F(x, y, \lambda) = f(x, y) + \lambda g(x, y) \tag{5}$$

where λ (the Greek letter lambda) is called a **Lagrange multiplier.** Theorem 1 gives the basis for the method.

theorem 1

The Method of Lagrange Multipliers for Functions of Two Variables

---

Any local maxima or minima of the function $z = f(x, y)$ subject to the constraint $g(x, y) = 0$ will be among those points (x_0, y_0) for which (x_0, y_0, λ_0) is a solution to the system

$$F_x(x, y, \lambda) = 0$$
$$F_y(x, y, \lambda) = 0$$
$$F_\lambda(x, y, \lambda) = 0$$

where $F(x, y, \lambda) = f(x, y) + \lambda g(x, y)$, provided all the partial derivatives exist.

We now solve the fence problem using the method of Lagrange multipliers.

Step 1. Formulate the problem in the form of equations (3) and (4):

Maximize $f(x, y) = xy$

Subject to $g(x, y) = 3x + y - 720 = 0$

Step 2. Form the function F, introducing the Lagrange multiplier λ:

$$F(x, y, \lambda) = f(x, y) + \lambda g(x, y)$$
$$= xy + \lambda(3x + y - 720)$$

Step 3. Solve the system $F_x = 0$, $F_y = 0$, $F_\lambda = 0$. (The solutions are called **critical points** for F.)

$$F_x = y + 3\lambda = 0$$
$$F_y = x + \lambda = 0$$
$$F_\lambda = 3x + y - 720 = 0$$

From the first two equations, we see that

$$y = -3\lambda$$
$$x = -\lambda$$

Substitute these values for x and y into the third equation and solve for λ:

$$-3\lambda - 3\lambda = 720$$
$$-6\lambda = 720$$
$$\lambda = -120$$

Thus,

$$y = -3(-120) = 360 \text{ feet}$$
$$x = -(-120) = 120 \text{ feet}$$

and $(x_0, y_0, \lambda_0) = (120, 360, -120)$ is the only critical point for F.

Step 4. According to Theorem 1, if the function $f(x, y)$, subject to the constraint $g(x, y) = 0$, has a local maximum or minimum, it must occur at $x = 120$, $y = 360$. Although it is possible to develop a test similar to Theorem 2 in Section 13-3 to determine the nature of this local extremum, we will not do so. [Note that Theorem 2 cannot be applied to $f(x, y)$ at $(120, 360)$, since this point is not a critical point of the unconstrained function $f(x, y)$.] We will simply assume that the maximum value of $f(x, y)$ must occur for $x = 120$, $y = 360$. Thus,

$$\text{Max } f(x, y) = f(120, 360)$$
$$= (120)(360) = 43{,}200 \text{ square feet}$$

The key steps in applying the method of Lagrange multipliers are listed in the following box:

Method of Lagrange Multipliers—Key Steps

Step 1. Formulate the problem in the form

Maximize (or Minimize) $z = f(x, y)$

Subject to $g(x, y) = 0$

Step 2. Form the function F:

$F(x, y, \lambda) = f(x, y) + \lambda g(x, y)$

Step 3. Find the critical points for F; that is, solve the system

$F_x(x, y, \lambda) = 0$
$F_y(x, y, \lambda) = 0$
$F_\lambda(x, y, \lambda) = 0$

Step 4. If (x_0, y_0, λ_0) is the only critical point of F, then we assume that (x_0, y_0) will always produce the solution to the problems we consider. If F has more than one critical point, then we evaluate $z = f(x, y)$ at (x_0, y_0) for each critical point (x_0, y_0, λ_0) of F. For the problems we consider, we assume that the largest of these values is the maximum value of $f(x, y)$, subject to the constraint $g(x, y) = 0$, and the smallest is the minimum value of $f(x, y)$, subject to the constraint $g(x, y) = 0$.

EXAMPLE 1 ➤ Minimization Subject to a Constraint   Minimize $f(x, y) = x^2 + y^2$ subject to $x + y = 10$.

Solution   Step 1.   Minimize $f(x, y) = x^2 + y^2$
Subject to $g(x, y) = x + y - 10 = 0$

Step 2. $F(x, y, \lambda) = x^2 + y^2 + \lambda(x + y - 10)$

Step 3. $F_x = 2x + \lambda = 0$
$F_y = 2y + \lambda = 0$
$F_\lambda = x + y - 10 = 0$

From the first two equations, $x = -\lambda/2$ and $y = -\lambda/2$. Substituting these into the third equation, we obtain

$$-\frac{\lambda}{2} - \frac{\lambda}{2} = 10$$

$$-\lambda = 10$$

$$\lambda = -10$$

The only critical point is $(x_0, y_0, \lambda_0) = (5, 5, -10)$.

FIGURE 2
$h(x) = x^2 + (10 - x)^2$

MATCHED PROBLEM 1 ➤

Step 4. Since $(5, 5, -10)$ is the only critical point for F, we conclude that (see step 4 in the box)

$$\text{Min } f(x, y) = f(5, 5) = (5)^2 + (5)^2 = 50$$

Since $g(x, y)$ has a relatively simple form, an alternative to the method of Lagrange multipliers is to solve $g(x, y) = 0$ for y, and then substitute into $f(x, y)$ to obtain the function $h(x) = f(x, 10 - x) = x^2 + (10 - x)^2$ in the single variable x. Then minimize h (see Fig. 2). Again, we conclude that $\text{Min } f(x, y) = f(5, 5) = 50$. This technique depends on being able to solve the constraint for one of the two variables, and thus is not always available as an alternative to the method of Lagrange multipliers. ◀

Maximize $f(x, y) = 25 - x^2 - y^2$ subject to $x + y = 4$. ◀

Figures 3 and 4 illustrate the results obtained in Example 1 and Matched Problem 1, respectively.

FIGURE 3

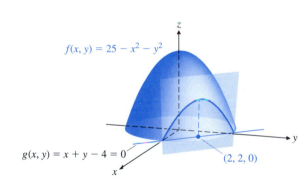

FIGURE 4

explore – discuss 1

FIGURE 5

Consider the problem of minimizing $f(x, y) = 3x^2 + 5y^2$ subject to the constraint $g(x, y) = 2x + 3y - 6 = 0$.

(A) Compute the value of $f(x, y)$ when x and y are integers, $0 \leq x \leq 3$, $0 \leq y \leq 2$. Record your answers next to the points (x, y) (see Fig. 5).
(B) Graph the constraint $g(x, y) = 0$.
(C) Estimate the minimum value of f on the basis of your graph and the computations from part (A).
(D) Use the method of Lagrange multipliers to solve the minimization problem.

▶

E X A M P L E   2   ➤   **Productivity**   The Cobb–Douglas production function for a new product is given by

$$N(x, y) = 16x^{0.25}y^{0.75}$$

where x is the number of units of labor and y is the number of units of capital required to produce $N(x, y)$ units of the product. Each unit of labor costs \$50 and each unit of capital costs \$100. If \$500,000 has been budgeted for the production of this product, how should this amount be allocated between labor and capital in order to maximize production? What is the maximum number of units that can be produced?

Solution   The total cost of using x units of labor and y units of capital is $50x + 100y$. Thus, the constraint imposed by the \$500,000 budget is

$$50x + 100y = 500,000$$

Step 1.   Maximize $N(x, y) = 16x^{0.25}y^{0.75}$
Subject to $g(x, y) = 50x + 100y - 500,000 = 0$

Step 2. $F(x, y, \lambda) = 16x^{0.25}y^{0.75} + \lambda(50x + 100y - 500,000)$

Step 3. $F_x = 4x^{-0.75}y^{0.75} + 50\lambda = 0$
$F_y = 12x^{0.25}y^{-0.25} + 100\lambda = 0$
$F_\lambda = 50x + 100y - 500,000 = 0$

From the first two equations,

$$\lambda = -\tfrac{2}{25}x^{-0.75}y^{0.75} \qquad \text{and} \qquad \lambda = -\tfrac{3}{25}x^{0.25}y^{-0.25}$$

Thus,

$$-\tfrac{2}{25}x^{-0.75}y^{0.75} = -\tfrac{3}{25}x^{0.25}y^{-0.25} \qquad \text{Multiply both sides by } x^{0.75}y^{0.25}.$$
$$-\tfrac{2}{25}y = -\tfrac{3}{25}x \qquad\qquad \text{(We can assume } x \neq 0 \text{ and } y \neq 0.)$$
$$y = \tfrac{3}{2}x$$

Now, substitute for y in the third equation and solve for x:

$$50x + 100(\tfrac{3}{2}x) - 500,000 = 0$$
$$200x = 500,000$$
$$x = 2,500$$

Thus,

$$y = \tfrac{3}{2}(2,500) = 3,750$$

and

$$\lambda = -\tfrac{2}{25}(2,500)^{-0.75}(3,750)^{0.75} \approx -0.1084$$

The only critical point of F is $(2,500, 3,750, -0.1084)$.

Step 4.   Since F has only one critical point, we conclude that maximum productivity occurs when 2,500 units of labor and 3,750 units of capital are used (see step 4 in the box). Thus,

$$\text{Max } N(x, y) = N(2{,}500, 3{,}750)$$
$$= 16(2{,}500)^{0.25}(3{,}750)^{0.75}$$
$$\approx 54{,}216 \text{ units}$$

The negative of the value of the Lagrange multiplier found in step 3 is called the **marginal productivity of money** and gives the approximate increase in production for each additional dollar spent on production. In Example 2, increasing the production budget from $500,000 to $600,000 would result in an approximate increase in production of

$$0.1084(100{,}000) = 10{,}840 \text{ units}$$

Note that simplifying the constraint equation

$$50x + 100y - 500{,}000 = 0$$

to

$$x + 2y - 10{,}000 = 0$$

before forming the function $F(x, y, \lambda)$ would make it difficult to interpret $-\lambda$ correctly. Thus, **in marginal productivity problems, the constraint equation should not be simplified.**

MATCHED PROBLEM 2 ► The Cobb–Douglas production function for a new product is given by

$$N(x, y) = 20x^{0.5}y^{0.5}$$

where x is the number of units of labor and y is the number of units of capital required to produce $N(x, y)$ units of the product. Each unit of labor costs $40 and each unit of capital costs $120.

(A) If $300,000 has been budgeted for the production of this product, how should this amount be allocated in order to maximize production? What is the maximum production?

(B) Find the marginal productivity of money in this case, and estimate the increase in production if an additional $40,000 is budgeted for production. ◄

explore – discuss 2

Consider the problem of maximizing $f(x, y) = 4 - x^2 - y^2$ subject to the constraint $g(x, y) = y - x^2 + 1 = 0$.

(A) Explain why $f(x, y) = 3$ whenever (x, y) is a point on the circle of radius 1 centered at the origin. What is the value of $f(x, y)$ when (x, y) is a point on the circle of radius 2 centered at the origin? On the circle of radius 3 centered at the origin? (See Fig. 6.)

(B) Explain why some points on the parabola $y - x^2 + 1 = 0$ lie inside the circle $x^2 + y^2 = 1$.

(C) In light of part (B), would you guess that the maximum value of $f(x, y)$ subject to the constraint is greater than 3? Explain.

(D) Use Lagrange multipliers to solve the maximization problem.

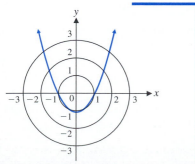

FIGURE 6

■ FUNCTIONS OF THREE INDEPENDENT VARIABLES

We have indicated that the method of Lagrange multipliers can be extended to functions with arbitrarily many independent variables with one or more constraints. We now state a theorem for functions with three independent variables and one constraint, and consider an example that will demonstrate the advantage of the method of Lagrange multipliers over the method used in Section 13-3.

theorem 2

The Method of Lagrange Multipliers for Functions of Three Variables

Any local maxima or minima of the function $w = f(x, y, z)$ subject to the constraint $g(x, y, z) = 0$ will be among the set of points (x_0, y_0, z_0) for which $(x_0, y_0, z_0, \lambda_0)$ is a solution to the system

$$F_x(x, y, z, \lambda) = 0$$
$$F_y(x, y, z, \lambda) = 0$$
$$F_z(x, y, z, \lambda) = 0$$
$$F_\lambda(x, y, z, \lambda) = 0$$

where $F(x, y, z, \lambda) = f(x, y, z) + \lambda g(x, y, z)$, provided all the partial derivatives exist.

▶ EXAMPLE 3 ➤ **Package Design** A rectangular box with an open top and one partition is to be constructed from 162 square inches of cardboard (Fig. 7). Find the dimensions that will result in a box with the largest possible volume.

Solution We must maximize

$$V(x, y, z) = xyz$$

subject to the constraint that the amount of material used is 162 square inches. Thus, x, y, and z must satisfy

$$xy + 2xz + 3yz = 162$$

Step 1. Maximize $V(x, y, z) = xyz$
          Subject to $g(x, y, z) = xy + 2xz + 3yz - 162 = 0$

Step 2. $F(x, y, z, \lambda) = xyz + \lambda(xy + 2xz + 3yz - 162)$

FIGURE 7

Step 3. $F_x = yz + \lambda(y + 2z) = 0$
         $F_y = xz + \lambda(x + 3z) = 0$
         $F_z = xy + \lambda(2x + 3y) = 0$
         $F_\lambda = xy + 2xz + 3yz - 162 = 0$

From the first two equations, we can write

$$\lambda = \frac{-yz}{y + 2z} \qquad \lambda = \frac{-xz}{x + 3z}$$

Eliminating λ, we have

$$\frac{-yz}{y + 2z} = \frac{-xz}{x + 3z}$$
$$-xyz - 3yz^2 = -xyz - 2xz^2$$
$$3yz^2 = 2xz^2 \qquad \textit{We can assume } z \neq 0.$$
$$3y = 2x$$
$$x = \tfrac{3}{2}y$$

From the second and third equations,

$$\lambda = \frac{-xz}{x + 3z} \qquad\qquad \lambda = \frac{-xy}{2x + 3y}$$

Eliminating λ, we have

$$\frac{-xz}{x + 3z} = \frac{-xy}{2x + 3y}$$
$$-2x^2z - 3xyz = -x^2y - 3xyz$$
$$2x^2z = x^2y \qquad \textit{We can assume } x \neq 0.$$
$$2z = y$$
$$z = \tfrac{1}{2}y$$

Substituting $x = \tfrac{3}{2}y$ and $z = \tfrac{1}{2}y$ in the fourth equation, we have

$$(\tfrac{3}{2}y)y + 2(\tfrac{3}{2}y)(\tfrac{1}{2}y) + 3y(\tfrac{1}{2}y) - 162 = 0$$
$$\tfrac{3}{2}y^2 + \tfrac{3}{2}y^2 + \tfrac{3}{2}y^2 = 162$$
$$y^2 = 36 \qquad \textit{We can assume } y > 0.$$
$$y = 6$$
$$x = \tfrac{3}{2}(6) = 9 \qquad \textit{Using } x = \tfrac{3}{2}y$$
$$z = \tfrac{1}{2}(6) = 3 \qquad \textit{Using } z = \tfrac{1}{2}y$$

and, finally,

$$\lambda = \frac{-(6)(3)}{6 + 2(3)} = -\frac{3}{2} \qquad \textit{Using } \lambda = \frac{-yz}{y + 2z}$$

Thus, the only critical point of F with x, y, and z all positive is $(9, 6, 3, -\tfrac{3}{2})$.

Step 4. The box with maximum volume has dimensions 9 inches by 6 inches by 3 inches. ◄

MATCHED PROBLEM 3 ➤ A box of the same type as described in Example 3 is to be constructed from 288 square inches of cardboard. Find the dimensions that will result in a box with the largest possible volume. ◄

Suppose we had decided to solve Example 3 by the method used in Section 13-3. First, we would have to solve the material constraint for one of the variables, say, z:

$$z = \frac{162 - xy}{2x + 3y}$$

Then we would eliminate z in the volume function and maximize

$$V(x, y) = xy\,\frac{162 - xy}{2x + 3y}$$

Using the method of Lagrange multipliers allows us to avoid the formidable task of finding the partial derivatives of V.

Answers to Matched Problems

1. Max $f(x, y) = f(2, 2) = 17$ (see Fig. 4)
2. (A) 3,750 units of labor and 1,250 units of capital;
   Max $N(x, y) = N(3,750, 1,250) \approx 43,301$ units
   (B) Marginal productivity of money ≈ 0.1443; increase in production $\approx 5,774$ units
3. 12 in. by 8 in. by 4 in.

EXERCISE   13-4

Use the method of Lagrange multipliers in Problems 1–14.

A

1. Maximize $f(x, y) = 2xy$
   Subject to $x + y = 6$

2. Minimize $f(x, y) = 6xy$
   Subject to $y - x = 6$

3. Minimize $f(x, y) = x^2 + y^2$
   Subject to $3x + 4y = 25$

4. Maximize $f(x, y) = 25 - x^2 - y^2$
   Subject to $2x + y = 10$

B

5. Find the maximum and minimum of $f(x, y) = 2xy$ subject to $x^2 + y^2 = 18$.

6. Find the maximum and minimum of $f(x, y) = x^2 - y^2$ subject to $x^2 + y^2 = 25$.

7. Maximize the product of two numbers if their sum must be 10.

8. Minimize the product of two numbers if their difference must be 10.

C

9. Minimize $f(x, y, z) = x^2 + y^2 + z^2$
   Subject to $2x - y + 3z = -28$

10. Maximize $f(x, y, z) = xyz$
    Subject to $2x + y + 2z = 120$

11. Maximize and minimize $f(x, y, z) = x + y + z$
    Subject to $x^2 + y^2 + z^2 = 12$

12. Maximize and minimize $f(x, y, z) = 2x + 4y + 4z$
    Subject to $x^2 + y^2 + z^2 = 9$

13. Maximize $f(x, y) = y + xy^2$
    Subject to $x + y^2 = 1$

14. Maximize and minimize $f(x, y) = x + e^y$
    Subject to $x^2 + y^2 = 1$

15. Consider the problem of maximizing $f(x, y)$ subject to $g(x, y) = 0$, where $g(x, y) = y - 5$. Explain how the maximization problem can be solved without using the method of Lagrange multipliers.

16. Consider the problem of minimizing $f(x, y)$ subject to $g(x, y) = 0$, where $g(x, y) = 4x - y + 3$. Explain how the minimization problem can be solved without using the method of Lagrange multipliers.

17. Consider the problem of maximizing $f(x, y) = e^{-(x^2 + y^2)}$ subject to the constraint $g(x, y) = x^2 + y - 1 = 0$.
    (A) Solve the constraint equation for y, and then substitute into $f(x, y)$ to obtain a function $h(x)$ in the single variable x. Graph h and solve the original maximization problem by maximizing h.
    (B) Confirm your answer using the method of Lagrange multipliers.

18. Consider the problem of minimizing $f(x, y) = x^2 + 2y^2$ subject to the constraint $g(x, y) = ye^x - 1 = 0$.
    (A) Solve the constraint equation for y, and then substitute into $f(x, y)$ to obtain a function $h(x)$ in the single variable x. Graph h and solve the original minimization problem by minimizing h.
    (B) Confirm your answer using the method of Lagrange multipliers.

▶ APPLICATIONS

Business & Economics

19. *Budgeting for least cost.* A manufacturing company produces two models of a television set, x units of model A and y units of model B per week, at a cost (in dollars) of

$$C(x, y) = 6x^2 + 12y^2$$

If it is necessary (because of shipping considerations) that

$$x + y = 90$$

how many of each type of set should be manufactured per week to minimize cost? What is the minimum cost?

20. *Budgeting for maximum production.* A manufacturing firm has budgeted $60,000 per month for labor and materials. If $x thousand is spent on labor and $y thousand is spent on materials, and if the monthly output (in units) is given by

$$N(x, y) = 4xy - 8x$$

how should the $60,000 be allocated to labor and materials in order to maximize N? What is the maximum N?

21. *Productivity.* A consulting firm for a manufacturing company arrived at the following Cobb–Douglas production function for a particular product:

$$N(x, y) = 50x^{0.8}y^{0.2}$$

where x is the number of units of labor and y is the number of units of capital required to produce $N(x, y)$ units of the product. Each unit of labor costs $40 and each unit of capital costs $80.

(A) If $400,000 is budgeted for production of the product, determine how this amount should be allocated to maximize production, and find the maximum production.

(B) Find the marginal productivity of money in this case, and estimate the increase in production if an additional $50,000 is budgeted for the production of this product.

22. *Productivity.* The research department for a manufacturing company arrived at the following Cobb–Douglas production function for a particular product:

$$N(x, y) = 10x^{0.6}y^{0.4}$$

where x is the number of units of labor and y is the number of units of capital required to produce $N(x, y)$ units of the product. Each unit of labor costs $30 and each unit of capital costs $60.

(A) If $300,000 is budgeted for production of the product, determine how this amount should be allocated to maximize production, and find the maximum production.

(B) Find the marginal productivity of money in this case, and estimate the increase in production if an additional $80,000 is budgeted for the production of this product.

23. *Maximum volume.* A rectangular box with no top and two intersecting partitions is to be constructed from 192 square inches of cardboard (see the figure). Find the dimensions that will maximize the volume.

Figure for 23

24. *Maximum volume.* A mailing service states that a rectangular package shall have the sum of the length and girth not to exceed 120 inches (see the figure). What are the dimensions of the largest (in volume) mailing carton that can be constructed meeting these restrictions?

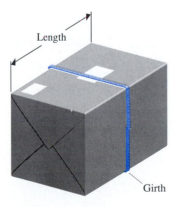

Length

Girth

Figure for 24

Life Sciences

25. *Agriculture.* Three pens of the same size are to be built along an existing fence (see the figure at the top of the next page). If 400 feet of fencing are available, what length should x and y be to produce the maximum total area? What is the maximum area?

Figure for 25

26. *Diet and minimum cost.* A group of guinea pigs is to receive 25,600 calories per week. Two available foods produce $200xy$ calories for a mixture of x kilograms of type M food and y kilograms of type N food. If type M costs \$1 per kilogram and type N costs \$2 per kilogram, how much of each type of food should be used to minimize weekly food costs? What is the minimum cost? [*Note:* $x \geqslant 0, y \geqslant 0$]

S E C T I O N   13-5   ■ Method of Least Squares

■ LEAST SQUARES APPROXIMATION
■ APPLICATIONS

■ LEAST SQUARES APPROXIMATION

Regression analysis is the process of fitting an elementary function to a set of data points using the **method of least squares.** The mechanics of using regression techniques were introduced in Section 2-4. Now, using the optimization techniques of Section 13-3, we are able to develop and explain the mathematical foundation of the method of least squares. We begin with **linear regression,** the process of finding the equation of the line that is the "best" approximation to a set of data points.

A manufacturer wants to approximate the cost function for a product. The value of the cost function has been determined for certain levels of production, as listed in Table 1. Although these points do not all lie on a line (see Fig. 1), they are very close to being linear. The manufacturer would like to approximate the cost function by a linear function; that is, determine values a and b so that the line

$$y = ax + b$$

is, in some sense, the "best" approximation to the cost function.

T A B L E 1

| NUMBER OF UNITS | COST |
x, in hundreds	y, in thousands of dollars
2	4
5	6
6	7
9	8

F I G U R E 1

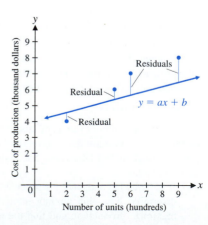

F I G U R E 2

What do we mean by "best"? Since the line $y = ax + b$ will not go through all four points, it is reasonable to examine the differences between the y coordinates of the points listed in the table and the y coordinates of the corresponding points on the line. Each of these differences is called the **residual** at that point (see Fig. 2). For example, at $x = 2$, the point from Table 1 is $(2, 4)$ and the point on the line is $(2, 2a + b)$, so the residual is

$$4 - (2a + b) = 4 - 2a - b$$

All the residuals are listed in Table 2.

Our criterion for the "best" approximation is the following: Determine the values of a and b that *minimize the sum of the squares* of the residuals. The resulting line is called the **least squares line,** or the **regression line.** To this end, we minimize

$$F(a, b) = (4 - 2a - b)^2 + (6 - 5a - b)^2 + (7 - 6a - b)^2 + (8 - 9a - b)^2$$

TABLE 2

x	y	$ax + b$	RESIDUAL
2	4	$2a + b$	$4 - 2a - b$
5	6	$5a + b$	$6 - 5a - b$
6	7	$6a + b$	$7 - 6a - b$
9	8	$9a + b$	$8 - 9a - b$

Step 1. Find critical points:

$$\begin{aligned} F_a(a, b) &= 2(4 - 2a - b)(-2) + 2(6 - 5a - b)(-5) \\ &\quad + 2(7 - 6a - b)(-6) + 2(8 - 9a - b)(-9) \\ &= -304 + 292a + 44b = 0 \end{aligned}$$

$$\begin{aligned} F_b(a, b) &= 2(4 - 2a - b)(-1) + 2(6 - 5a - b)(-1) \\ &\quad + 2(7 - 6a - b)(-1) + 2(8 - 9a - b)(-1) \\ &= -50 + 44a + 8b = 0 \end{aligned}$$

After dividing each equation by 2, we solve the system

$$146a + 22b = 152$$
$$22a + 4b = 25$$

obtaining $(a, b) = (0.58, 3.06)$ as the only critical point.

Step 2. Compute $A = F_{aa}(a, b)$, $B = F_{ab}(a, b)$, and $C = F_{bb}(a, b)$:

$$\begin{array}{lll} F_{aa}(a, b) = 292 & \text{thus} & A = F_{aa}(0.58, 3.06) = 292 \\ F_{ab}(a, b) = 44 & \text{thus} & B = F_{ab}(0.58, 3.06) = 44 \\ F_{bb}(a, b) = 8 & \text{thus} & C = F_{bb}(0.58, 3.06) = 8 \end{array}$$

Step 3. Evaluate $AC - B^2$ and try to classify the critical point (a, b) using Theorem 2 in Section 13-3:

$$AC - B^2 = (292)(8) - (44)^2 = 400 > 0 \quad \text{and} \quad A = 292 > 0$$

Therefore, case 2 in Theorem 2 applies, and $F(a, b)$ has a local minimum at the critical point $(0.58, 3.06)$.

Thus, the least squares line for the given data is

$$y = 0.58x + 3.06 \quad \text{\textit{Least squares line}}$$

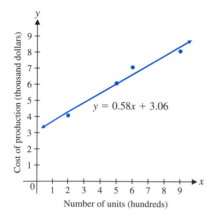

y = 0.58x + 3.06

FIGURE 3

The sum of the squares of the residuals is minimized for this choice of a and b (see Fig. 3).

This linear function can now be used by the manufacturer to estimate any of the quantities normally associated with the cost function—such as costs, marginal costs, average costs, and so on. For example, the cost of producing 2,000 units is approximately

$$y = (0.58)(20) + 3.06 = 14.66 \quad \text{or} \quad \$14,660$$

The marginal cost function is

$$\frac{dy}{dx} = 0.58$$

The average cost function is

$$\bar{y} = \frac{0.58x + 3.06}{x}$$

In general, if we are given a set of n points $(x_1, y_1), (x_2, y_2), \ldots, (x_n, y_n)$, we want to determine the line $y = ax + b$ for which the sum of the squares of the residuals is minimized. Using summation notation, the sum of the squares of the residuals is given by

$$F(a, b) = \sum_{k=1}^{n} (y_k - ax_k - b)^2$$

Note that in this expression the variables are a and b, and the x_k and y_k are all known values. To minimize $F(a, b)$, we thus compute the partial derivatives with respect to a and b and set them equal to 0:

$$F_a(a, b) = \sum_{k=1}^{n} 2(y_k - ax_k - b)(-x_k) = 0$$

$$F_b(a, b) = \sum_{k=1}^{n} 2(y_k - ax_k - b)(-1) = 0$$

Dividing each equation by 2 and simplifying, we see that the coefficients a and b of the least squares line $y = ax + b$ must satisfy the following system of *normal equations*:

$$\left(\sum_{k=1}^{n} x_k^2 \right) a + \left(\sum_{k=1}^{n} x_k \right) b = \sum_{k=1}^{n} x_k y_k$$

$$\left(\sum_{k=1}^{n} x_k \right) a + nb = \sum_{k=1}^{n} y_k$$

Solving this system for a and b produces the formulas given in the box.

Least Squares Approximation

For a set of n points $(x_1, y_1), (x_2, y_2), \ldots, (x_n, y_n)$, the coefficients of the least squares line $y = ax + b$ are the solutions of the system of **normal equations**

$$\left(\sum_{k=1}^{n} x_k^2\right)a + \left(\sum_{k=1}^{n} x_k\right)b = \sum_{k=1}^{n} x_k y_k$$

$$\left(\sum_{k=1}^{n} x_k\right)a + nb = \sum_{k=1}^{n} y_k \tag{1}$$

and are given by the formulas

$$a = \frac{n\left(\sum\limits_{k=1}^{n} x_k y_k\right) - \left(\sum\limits_{k=1}^{n} x_k\right)\left(\sum\limits_{k=1}^{n} y_k\right)}{n\left(\sum\limits_{k=1}^{n} x_k^2\right) - \left(\sum\limits_{k=1}^{n} x_k\right)^2} \tag{2}$$

$$b = \frac{\sum\limits_{k=1}^{n} y_k - a\left(\sum\limits_{k=1}^{n} x_k\right)}{n} \tag{3}$$

TABLE 3

x_k	y_k	$x_k y_k$	x_k^2
2	4	8	4
5	6	30	25
6	7	42	36
9	8	72	81
Totals 22	25	152	146

For the data of Table 1 we tabulate the sums required for the normal equations and their solution in Table 3.

The normal equations (1) are then

$$146a + 22b = 152$$
$$22a + 4b = 25$$

The solution to the normal equations given by equations (2) and (3) is

$$a = \frac{4(152) - (22)(25)}{4(146) - (22)^2} = 0.58$$

$$b = \frac{25 - 0.58(22)}{4} = 3.06$$

Note that Table 3 thus provides a convenient format for the computation of step 1 above.

Many graphing utilities have a linear regression feature that solves the system of normal equations obtained by setting the partial derivatives of the sum of squares of the residuals equal to 0. Therefore, in practice, we simply enter the given data points and use the linear regression feature (as in Section 2-4) to determine the line $y = ax + b$ that best fits the data (see Fig. 4, at the top of the next page). There is no need to compute partial derivatives, or even to tabulate sums (as in Table 3).

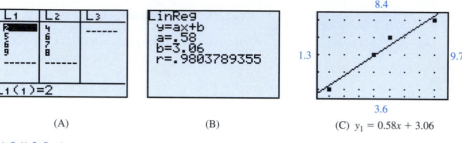

(A)   (B)   (C) $y_1 = 0.58x + 3.06$

FIGURE 4

explore – discuss 1

(A) Plot the four points $(0, 0)$, $(0, 1)$, $(10, 0)$, and $(10, 1)$. Which line would you guess "best" fits these four points? Use formulas (2) and (3) to test your conjecture.

(B) Plot the four points $(0, 0)$, $(0, 10)$, $(1, 0)$, and $(1, 10)$. Which line would you guess "best" fits these four points? Use formulas (2) and (3) to test your conjecture.

(C) If either of your conjectures was wrong, explain how your reasoning was mistaken.

The method of least squares can also be applied to find the quadratic equation $y = ax^2 + bx + c$ that best fits a set of data points. In this case, the sum of the squares of the residuals is a function of three variables:

$$F(a, b, c) = \sum_{k=1}^{n} (y_k - ax_k^2 - bx_k - c)^2$$

There are now three partial derivatives to compute and set equal to 0:

$$F_a(a, b, c) = \sum_{k=1}^{n} 2(y_k - ax_k^2 - bx_k - c)(-x_k^2) = 0$$

$$F_b(a, b, c) = \sum_{k=1}^{n} 2(y_k - ax_k^2 - bx_k - c)(-x_k) = 0$$

$$F_c(a, b, c) = \sum_{k=1}^{n} 2(y_k - ax_k^2 - bx_k - c)(-1) = 0$$

The resulting set of three linear equations in the three variables a, b, and c is called the *set of normal equations for quadratic regression*. A quadratic regression feature on a calculator is designed to solve such normal equations after the given set of points has been entered. Figure 5 illustrates the computation for the data of Table 1.

(A) (B) (C) $y_1 = -0.0417x^2 + 1.0383x + 2.06$

FIGURE 5

explore – discuss 2

(A) Use the graphs in Figure 4 and 5 to predict which technique, linear regression or quadratic regression, yields the smaller sum of squares of the residuals for the data of Table 1. Explain.

(B) Confirm your prediction by computing the sum of squares of the residuals in each case.

The method of least squares can similarly be applied to the other regression equations considered in Section 2-4 (cubic, quartic, logarithmic, exponential, and power regression equations). Details are explored in some of the exercises at the end of this section.

■ APPLICATIONS

▶ EXAMPLE 1 ➤ Educational Testing Table 4 lists the midterm and final examination scores for 10 students in a calculus course.

TABLE 4

MIDTERM	FINAL	MIDTERM	FINAL
49	61	78	77
53	47	83	81
67	72	85	79
71	76	91	93
74	68	99	99

(A) Use formulas (1), (2), and (3) to find the normal equations and the least squares line for the data given in Table 4.

(B) Use the linear regression feature on a graphing utility to find and graph the least squares line.

(C) Use the least squares line to predict the final examination score for a student who scored 95 on the midterm examination.

Solution   (A) Table 5 shows a convenient way to compute all the sums in the formulas for a and b. Thus,

TABLE 5

x_k	y_k	$x_k y_k$	x_k^2
49	61	2,989	2,401
53	47	2,491	2,809
67	72	4,824	4,489
71	76	5,396	5,041
74	68	5,032	5,476
78	77	6,006	6,084
83	81	6,723	6,889
85	79	6,715	7,225
91	93	8,463	8,281
99	99	9,801	9,801
Totals 750	753	58,440	58,496

$$\sum_{k=1}^{10} x_k = 750 \qquad \sum_{k=1}^{10} y_k = 753 \qquad \sum_{k=1}^{10} x_k y_k = 58{,}440 \qquad \sum_{k=1}^{10} x_k^2 = 58{,}496$$

and the normal equations are

$$58{,}496a + 750b = 58{,}440$$
$$750a + 10b = 753$$

Using formulas (2) and (3),

$$a = \frac{10(58{,}440) - (750)(753)}{10(58{,}496) - (750)^2} = \frac{19{,}650}{22{,}460} \approx 0.875$$

$$b = \frac{753 - 0.875(750)}{10} = 9.675$$

The least squares line is given (approximately) by

$$y = 0.875x + 9.675$$

(B) We enter the data and use the linear regression feature, as shown in Figure 6.

(A)

(B)

(C)

FIGURE 6

(C) If $x = 95$, then $y = 0.875(95) + 9.683 \approx 92.8$ is the predicted score on the final exam. This is also indicated above in Figure 6C. If we assume the exam score must be an integer, we would predict a score of 93. ◄

MATCHED PROBLEM 1 ➤ Repeat Example 1 for the scores listed in Table 6.

TABLE 6

MIDTERM	FINAL	MIDTERM	FINAL
54	50	84	80
60	66	88	95
75	80	89	85
76	68	97	94
78	71	99	86

◄

EXAMPLE 2 ➤ **Energy Consumption**   The use of fuel oil for home heating in the United States has declined steadily for several decades. Table 7 lists the percentage of occupied housing units in the United States that were heated by fuel oil for various years between 1960 and 1991. Use the data in the table and linear regression to predict the percentage of occupied housing units in the United States that will be heated by fuel oil in the year 2000.

TABLE 7
Occupied Housing Units Heated by Fuel Oil

YEAR	PERCENT	YEAR	PERCENT
1960	32.4	1985	14.1
1970	26.0	1987	14.0
1975	22.5	1989	13.3
1980	18.1	1991	12.3
1983	14.9		

Solution   We enter the data, with $x = 0$ representing 1960, $x = 10$ representing 1970, etc., and use the linear regression feature, as shown in Figure 7.

(A)

(B)

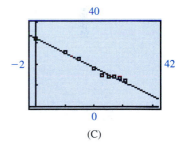

(C)

FIGURE 7

Figure 7 indicates that the least squares line is $y = -0.68x + 32.26$. The estimated percentage of occupied housing units heated by fuel oil in the year 2000 (corresponding to $x = 40$) is thus $-0.68(40) + 32.26 = 5.06\%$.   ◀

MATCHED PROBLEM 2 ➤   In 1950, coal was still a major source of fuel for home energy consumption, and the percentage of occupied housing units heated by fuel oil was only 22.1. Add the data for 1950 to the data for Example 2, and compute the new least squares line and the new estimate for the percentage of occupied housing units heated by fuel oil in the year 2000. Discuss the discrepancy between the two estimates.   ◀

explore–discuss 3

T A B L E 8

t (hr)	A (mg)
0	50.0
5	46.4
10	43.1
15	39.9
20	37.1

The data in Table 8 give the amount A (in milligrams) of the radioactive isotope gallium-67, used in the diagnosis of malignant tumors, at time t (in hours).

(A) Since the data describe radioactive decay, we would expect the relationship between A and t to be exponential—that is, $A = ae^{bt}$ for some constants a and b. Show that if $A = ae^{bt}$, then the relationship between $\ln A$ and t is linear.

(B) Compute $\ln A$ for each data point and find the line that "best" fits the data $(t, \ln A)$.

(C) Determine the "best" values of a and b.

(D) Confirm your results by using the exponential regression feature on a graphing utility.

Answers to Matched Problems

1. (A) $y = 0.85x + 9.47$
   (B)

   (C) 90.3

2. $y = -0.40x + 25.75$; 9.75%

EXERCISE   13-5

A *In Problems 1–6, find the least squares line. Graph the data and the least squares line.*

1.
x	y
1	1
2	3
3	4
4	3

2.
x	y
1	-2
2	-1
3	3
4	5

3.
x	y
1	8
2	5
3	4
4	0

4.
x	y
1	20
2	14
3	11
4	3

5.
x	y
1	3
2	4
3	5
4	6

6.
x	y
1	2
2	3
3	3
4	2

B *In Problems 7–12, find the least squares line and use it to estimate y for the indicated value of x.*

7.

x	y
0	10
5	22
10	31
15	46
20	51

Estimate y when x = 25.

8.

x	y
−5	60
0	50
5	30
10	20
15	15

Estimate y when x = 20.

9.

x	y
−1	14
1	12
3	8
5	6
7	5

Estimate y when x = 2.

10.

x	y
2	−4
6	0
10	8
14	12
18	14

Estimate y when x = 15.

11.

x	y	x	y
0.5	25	9.5	12
2	22	11	11
3.5	21	12.5	8
5	21	14	5
6.5	18	15.5	1

Estimate y when x = 8.

12.

x	y	x	y
0	−15	12	11
2	−9	14	13
4	−7	16	19
6	−7	18	25
8	−1	20	33

Estimate y when x = 10.

C

13. To find the coefficients of the parabola

$$y = ax^2 + bx + c$$

that is the "best" fit for the points (1, 2), (2, 1), (3, 1), and (4, 3), minimize the sum of the squares of the residuals

$$F(a, b, c) = (a + b + c - 2)^2$$
$$+ (4a + 2b + c - 1)^2$$
$$+ (9a + 3b + c - 1)^2$$
$$+ (16a + 4b + c - 3)^2$$

by solving the system of normal equations

$$F_a(a, b, c) = 0 \quad F_b(a, b, c) = 0 \quad F_c(a, b, c) = 0$$

for a, b, and c. Graph the points and the parabola.

14. Repeat Problem 13 for the points (−1, −2), (0, 1), (1, 2), and (2, 0).

Problems 15 and 16 refer to the system of normal equations and the formulas for a and b.

15. Verify formulas (2) and (3) by solving the system of normal equations (1) for a and b.

16. If

$$\bar{x} = \frac{1}{n} \sum_{k=1}^{n} x_k \quad \text{and} \quad \bar{y} = \frac{1}{n} \sum_{k=1}^{n} y_k$$

are the averages of the x and y coordinates, respectively, show that the point $(\bar{x}, \bar{y})$ satisfies the equation of the least squares line $y = ax + b$.

17. (A) Suppose that n = 5 and that the x coordinates of the data points (x_1, y_1), (x_2, y_2), . . . , (x_n, y_n) are −2, −1, 0, 1, 2. Show that system (1) implies that

$$a = \frac{\sum x_k y_k}{\sum x_k^2}$$

and that b is equal to the average of the values of y_k.

(B) Show that the conclusion of part (A) holds whenever the average of the x coordinates of the data points is 0.

18. (A) Give an example of a set of six data points such that half of the points lie above the least squares line and half lie below.

(B) Give an example of a set of six data points such that just one of the points lies above the least squares line and five lie below.

19. (A) Find and graph the linear and quadratic functions that best fit the data points (0, 1.3), (1, 0.6), (2, 1.5), (3, 3.6), and (4, 7.4).

(B) Which of the two functions best fits the data? Explain.

20. (A) Find and graph the linear, quadratic, and logarithmic functions that best fit the data points (1, 3.2), (2, 4.2), (3, 4.7), (4, 5.0), and (5, 5.3).

(B) Which of the three functions best fits the data? Explain.

21. Describe the normal equations for cubic regression. How many equations are there? What are the variables? What techniques could be used to solve them?

22. Describe the normal equations for quartic regression. How many equations are there? What are the variables? What techniques could be used to solve them?

APPLICATIONS

Business & Economics

23. *Production.* Data for passenger car production in Mexico for the years 1985–1991 are given in the table.

Passenger Car Production in Mexico

YEAR	THOUSANDS PER MONTH	YEAR	THOUSANDS PER MONTH
1985	23.8	1989	37.9
1986	16.5	1990	51.2
1987	19.0	1991	61.1
1988	29.0		

(A) Find the least squares line for the data using $x = 0$ for 1985.

(B) Use the least squares line to estimate the monthly production (in thousands) in the year 1998.

24. *Purchasing.* Data for the purchase of U.S. auto parts by Japanese car makers (for their cars made in United States) are given in the table.

Purchase of U.S. Auto Parts by Japanese Car Makers

YEAR	BILLION DOLLARS	YEAR	BILLION DOLLARS
1986	2.1	1991	8.5
1987	2.5	1992	11.2
1988	3.9	1993	12.9
1989	5.6	1994	17.0
1990	7.1		

(A) Find the least squares line for the data using $x = 0$ for 1986.

(B) Use the least squares line to estimate the purchase of auto parts by Japanese car makers (for their cars made in the United States) in the year 2000.

25. *Maximizing profit.* The market research department for a drug store chain chose two summer resort areas to test market a new sun screen lotion packaged in 4 ounce plastic bottles. After a summer of varying the selling price and recording the monthly demand, the research department arrived at the demand table given at the top of the next column, where y is the number of bottles purchased per month (in thousands) at x dollars per bottle.

x	y
5.0	2.0
5.5	1.8
6.0	1.4
6.5	1.2
7.0	1.1

(A) Find a demand equation using the method of least squares.

(B) If each bottle of sun screen costs the drug store chain $4, how should it be priced to achieve a maximum monthly profit? [*Hint:* Use the result of part (A), with $C = 4y$, $R = xy$, and $P = R - C$.]

26. *Maximizing profit.* A market research consultant for a supermarket chain chose a large city to test market a new brand of mixed nuts packaged in 8 ounce cans. After a year of varying the selling price and recording the monthly demand, the consultant arrived at the demand table given below, where y is the number of cans purchased per month (in thousands) at x dollars per can.

x	y
4.0	4.2
4.5	3.5
5.0	2.7
5.5	1.5
6.0	0.7

(A) Find a demand equation using the method of least squares.

(B) If each can of nuts costs the supermarket chain $3, how should it be priced to achieve a maximum monthly profit?

Life Sciences

27. *Medicine.* If a person dives into cold water, a neural reflex response automatically shuts off blood circulation to the skin and muscles and reduces the pulse rate. A medical research team conducted an experiment using a group of ten 2-year-olds. A child's face was placed momentarily in cold water, and the corresponding reduction in pulse rate was recorded. The data for the average reduction in heart rate for each temperature are summarized in the table.

WATER TEMPERATURE (°F)	PULSE RATE REDUCTION
50	15
55	13
60	10
65	6
70	2

(A) If T is water temperature (in degrees Fahrenheit) and P is pulse rate reduction (in beats per minute), use the method of least squares to find a linear equation relating T and P.

(B) Use the equation found in part (A) to find P when $T = 57$.

28. *Biology.* In biology there is an approximate rule, called the *bioclimatic rule for temperate climates*, that has been known for a couple of hundred years. This rule states that in spring and early summer, periodic phenomena such as blossoming of flowers, appearance of insects, and ripening of fruit usually come about 4 days later for each 500 feet of altitude. Stated as a formula,

$$d = 8h \qquad 0 \leqslant h \leqslant 4$$

where d is the change in days and h is the altitude (in thousands of feet). To test this rule, an experiment was set up to record the difference in blossoming time of the same type of apple tree at different altitudes. A summary of the results is given in the table.

h	d
0	0
1	7
2	18
3	28
4	33

(A) Use the method of least squares to find a linear equation relating h and d. Does the bioclimatic rule, $d = 8h$, appear to be approximately correct?

(B) How much longer will it take this type of apple tree to blossom at 3.5 thousand feet than at sea level? [Use the linear equation found in part (A).]

Social Sciences

29. *Political science.* Association of economic class and party affiliation did not start with Roosevelt's New Deal; it goes back to the time of Andrew Jackson (1767–1845). Paul Lazarsfeld of Columbia University published an article in the November 1950 issue of *Scientific American* in which he discusses statistical investigations of the relationships between economic class and party affiliation. The data in the table are taken from this article.

(A) If A represents the average assessed value per person in a given ward in 1836 and D represents the percentage of people in that ward voting Democratic in 1836, use the method of least squares to find a linear equation relating A and D.

(B) If the average assessed value per person in a ward had been $300, what is the predicted percentage of people in that ward that would have voted Democratic?

Political Affiliations, 1836

WARD	AVERAGE ASSESSED VALUE PER PERSON (IN $100)	DEMOCRATIC VOTES (%)
12	1.7	51
3	2.1	49
1	2.3	53
5	2.4	36
2	3.6	65
11	3.7	35
10	4.7	29
4	6.2	40
6	7.1	34
9	7.4	29
8	8.7	20
7	11.9	23

30. *Education.* The table lists the high school grade-point averages (GPA's) of 10 students, along with their grade-point averages after one semester of college.

HIGH SCHOOL GPA	COLLEGE GPA	HIGH SCHOOL GPA	COLLEGE GPA
2.0	1.5	3.0	2.3
2.2	1.5	3.1	2.5
2.4	1.6	3.3	2.9
2.7	1.8	3.4	3.2
2.9	2.1	3.7	3.5

(A) Find the least squares line for the data.

(B) Estimate the college GPA for a student with a high school GPA of 3.5.

(C) Estimate the high school GPA necessary for a college GPA of 2.7.

31. *Olympic Games.* The table gives the winning heights in the pole vault in the Olympic Games from 1896 to 1992.
(A) Find the least squares line for the data using $x = 0$ for 1896.
(B) Estimate the winning height in the pole vault in the Olympic Games of 2008.

Olympic Pole Vault Winning Height

YEAR	HEIGHT (FT)	YEAR	HEIGHT (FT)
1896	10.81	1952	14.93
1900	10.82	1956	14.96
1904	11.50	1960	15.43
1906	11.60	1964	16.73
1908	12.17	1968	17.71
1912	12.96	1972	18.04
1920	13.46	1976	18.04
1924	12.96	1980	18.96
1928	13.78	1984	18.85
1932	14.16	1988	18.35
1936	14.27	1992	19.02
1948	14.10		

32. *Tuition.* The table gives resident and nonresident tuition for a random sample of state universities.

Tuition at State Universities (1994)

RESIDENT	NONRESIDENT	RESIDENT	NONRESIDENT
$2,290	$ 7,480	$2,540	$11,332
4,365	14,069	2,904	10,603
3,876	9,648	4,618	9,664
1,900	5,600	2,835	6,735
2,376	5,810	3,328	7,580
2,187	5,367	2,088	7,052
2,799	8,292	1,798	5,970
1,750	4,941	1,679	6,438

(A) Find the least squares line for the data.
(B) Estimate the nonresident tuition at a state university where the resident tuition is $4,000.

SECTION 13-6 ■ **Double Integrals Over Rectangular Regions**

■ INTRODUCTION
■ DEFINITION OF THE DOUBLE INTEGRAL
■ AVERAGE VALUE OVER RECTANGULAR REGIONS
■ VOLUME AND DOUBLE INTEGRALS

■ **INTRODUCTION**

We have generalized the concept of differentiation to functions with two or more independent variables. How can we do the same with integration, and how can we interpret the results? Let us first look at the operation of antidifferentiation. We can antidifferentiate a function of two or more variables with respect to one of the variables by treating all the other variables as though they were constants. Thus, this operation is the reverse operation of partial differentiation, just as ordinary antidifferentiation is the reverse operation of ordinary differentiation. We write $\int f(x, y)\, dx$ to indicate that we are to antidifferentiate $f(x, y)$ with respect to x, holding y fixed; we write $\int f(x, y)\, dy$ to indicate that we are to antidifferentiate $f(x, y)$ with respect to y, holding x fixed.

EXAMPLE 1 ➤ Partial Antidifferentiation Evaluate:

(A) $\int (6xy^2 + 3x^2)\, dy$ (B) $\int (6xy^2 + 3x^2)\, dx$

Solution   (A) Treating x as a constant and using the properties of antidifferentiation from Section 11-1, we have

$$\int (6xy^2 + 3x^2)\, dy = \int 6xy^2\, dy + \int 3x^2\, dy$$

$$= 6x \int y^2\, dy + 3x^2 \int dy$$

$$= 6x \left(\frac{y^3}{3} \right) + 3x^2(y) + C(x)$$

$$= 2xy^3 + 3x^2y + C(x)$$

The dy tells us we are looking for the antiderivative of $6xy^2 + 3x^2$ with respect to y only, holding x constant.

Notice that the constant of integration actually can be *any function of x alone,* since, for any such function

$$\frac{\partial}{\partial y} C(x) = 0$$

Check: We can verify that our answer is correct by using partial differentiation:

$$\frac{\partial}{\partial y}[2xy^3 + 3x^2y + C(x)] = 6xy^2 + 3x^2 + 0$$

$$= 6xy^2 + 3x^2$$

(B) Now we treat y as a constant:

$$\int (6xy^2 + 3x^2)\, dx = \int 6xy^2\, dx + \int 3x^2\, dx$$

$$= 6y^2 \int x\, dx + 3 \int x^2\, dx$$

$$= 6y^2 \left(\frac{x^2}{2} \right) + 3 \left(\frac{x^3}{3} \right) + E(y)$$

$$= 3x^2y^2 + x^3 + E(y)$$

This time, the antiderivative contains an arbitrary function $E(y)$ of y alone.

Check: $\dfrac{\partial}{\partial x}[3x^2y^2 + x^3 + E(y)] = 6xy^2 + 3x^2 + 0$

$$= 6xy^2 + 3x^2$$

MATCHED PROBLEM 1 ➤   Evaluate:

(A) $\int (4xy + 12x^2y^3)\, dy$ (B) $\int (4xy + 12x^2y^3)\, dx$

Now that we have extended the concept of antidifferentiation to functions with two variables, we also can evaluate definite integrals of the form

$$\int_a^b f(x, y)\, dx \qquad \text{or} \qquad \int_c^d f(x, y)\, dy$$

EXAMPLE 2 ➤   **Evaluating a Partial Antiderivative**   Evaluate, substituting the limits of integration in y if dy is used and in x if dx is used:

(A) $\displaystyle\int_0^2 (6xy^2 + 3x^2)\,dy$

(B) $\displaystyle\int_0^1 (6xy^2 + 3x^2)\,dx$

Solution   (A) From Example 1A, we know that $\int (6xy^2 + 3x^2)\,dy = 2xy^3 + 3x^2y + C(x)$. According to properties of the definite integral for a function of one variable, we can use any antiderivative to evaluate the definite integral. Thus, choosing $C(x) = 0$, we have

$$\int_0^2 (6xy^2 + 3x^2)\,dy = (2xy^3 + 3x^2y)\,\Big|_{y=0}^{y=2}$$
$$= [2x(2)^3 + 3x^2(2)] - [2x(0)^3 + 3x^2(0)]$$
$$= 16x + 6x^2$$

(B) From Example 1B, we know that $\int (6xy^2 + 3x^2)\,dx = 3x^2y^2 + x^3 + E(y)$. Thus, choosing $E(y) = 0$, we have

$$\int_0^1 (6xy^2 + 3x^2)\,dx = (3x^2y^2 + x^3)\,\Big|_{x=0}^{x=1}$$
$$= [3y^2(1)^2 + (1)^3] - [3y^2(0)^2 + (0)^3]$$
$$= 3y^2 + 1 \qquad ◄$$

MATCHED PROBLEM 2 ➤   Evaluate:

(A) $\displaystyle\int_0^1 (4xy + 12x^2y^3)\,dy$ (B) $\displaystyle\int_0^3 (4xy + 12x^2y^3)\,dx$ ◄

Notice that integrating and evaluating a definite integral with integrand $f(x, y)$ with respect to y produces a function of x alone (or a constant). Likewise, integrating and evaluating a definite integral with integrand $f(x, y)$ with respect to x produces a function of y alone (or a constant). Each of these results, involving at most one variable, can now be used as an integrand in a second definite integral.

EXAMPLE 3 ➤   **Evaluating Iterated Integrals**   Evaluate:

(A) $\displaystyle\int_0^1 \left[\int_0^2 (6xy^2 + 3x^2)\,dy \right] dx$

(B) $\displaystyle\int_0^2 \left[\int_0^1 (6xy^2 + 3x^2)\,dx \right] dy$

Solution   (A) Example 2A showed that

$$\int_0^2 (6xy^2 + 3x^2)\, dy = 16x + 6x^2$$

Thus,

$$\int_0^1 \left[\int_0^2 (6xy^2 + 3x^2)\, dy \right] dx = \int_0^1 (16x + 6x^2)\, dx$$

$$= (8x^2 + 2x^3)\ \Big|_{x=0}^{x=1}$$

$$= [8(1)^2 + 2(1)^3] - [8(0)^2 + 2(0)^3] = 10$$

(B) Example 2B showed that

$$\int_0^1 (6xy^2 + 3x^2)\, dx = 3y^2 + 1$$

Thus,

$$\int_0^2 \left[\int_0^1 (6xy^2 + 3x^2)\, dx \right] dy = \int_0^2 (3y^2 + 1)\, dy$$

$$= (y^3 + y)\ \Big|_{y=0}^{y=2}$$

$$= [(2)^3 + 2] - [(0)^3 + 0] = 10$$

A numerical integration routine can be used as an alternative to the fundamental theorem of calculus (FTC) to evaluate the last integrals in parts (A) and (B), $\int_0^1 (16x + 6x^2)\, dx$ and $\int_0^2 (3y^2 + 1)\, dy$, since the integrand in each case is a function of a single variable (see Fig. 1).   ◄

FIGURE 1

MATCHED PROBLEM 3 ➤   Evaluate:

(A) $\int_0^3 \left[\int_0^1 (4xy + 12x^2y^3)\, dy \right] dx$

(B) $\int_0^1 \left[\int_0^3 (4xy + 12x^2y^3)\, dx \right] dy$ ◄

■ DEFINITION OF THE DOUBLE INTEGRAL

Notice that the answers in Examples 3A and 3B are identical. This is not an accident. In fact, it is this property that enables us to define the *double integral,* as follows:

Double Integral

The **double integral** of a function $f(x, y)$ over a rectangle

$$R = \{(x, y) | a \leqslant x \leqslant b, \quad c \leqslant y \leqslant d\}$$

is

$$\iint\limits_{R} f(x, y)\, dA$$

$$= \int_{a}^{b} \left[\int_{c}^{d} f(x, y)\, dy \right] dx$$

$$= \int_{c}^{d} \left[\int_{a}^{b} f(x, y)\, dx \right] dy$$

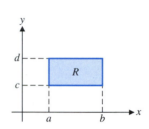

In the double integral $\iint_{R} f(x, y)\, dA$, $f(x, y)$ is called the **integrand** and R is called the **region of integration.** The expression dA indicates that this is an integral over a two-dimensional region. The integrals

$$\int_{a}^{b} \left[\int_{c}^{d} f(x, y)\, dy \right] dx \qquad \text{and} \qquad \int_{c}^{d} \left[\int_{a}^{b} f(x, y)\, dx \right] dy$$

are referred to as **iterated integrals** (the brackets are often omitted), and the order in which dx and dy are written indicates the order of integration. This is not the most general definition of the double integral over a rectangular region; however, it is equivalent to the general definition for all the functions we will consider.

EXAMPLE 4 ➤ Evaluating a Double Integral   Evaluate:

$$\iint\limits_{R} (x + y)\, dA \qquad \text{over} \qquad R = \{(x, y) | 1 \leqslant x \leqslant 3, \quad -1 \leqslant y \leqslant 2\}$$

Solution   Region R is illustrated in Figure 2. We can choose either order of iteration. As a check, we will evaluate the integral both ways:

$$\iint\limits_{R} (x + y)\, dA = \int_{1}^{3} \int_{-1}^{2} (x + y)\, dy\, dx$$

$$= \int_{1}^{3} \left[\left(xy + \frac{y^2}{2} \right) \Big|_{y=-1}^{y=2} \right] dx$$

$$= \int_{1}^{3} [(2x + 2) - (-x + \tfrac{1}{2})]\, dx$$

$$= \int_{1}^{3} (3x + \tfrac{3}{2})\, dx = 15$$

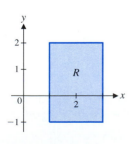

FIGURE 2

$$\iint\limits_R (x + y)\, dA = \int_{-1}^{2} \int_{1}^{3} (x + y)\, dx\, dy$$

$$= \int_{-1}^{2} \left[\left(\frac{x^2}{2} + xy \right) \Big|_{x=1}^{x=3} \right] dy$$

$$= \int_{-1}^{2} [(\tfrac{9}{2} + 3y) - (\tfrac{1}{2} + y)]\, dy$$

$$= \int_{-1}^{2} (4 + 2y)\, dy = 15 \quad \blacktriangleleft$$

MATCHED PROBLEM 4 ➤ Evaluate both ways:

$$\iint\limits_R (2x - y)\, dA \quad \text{over} \quad R = \{(x, y)| -1 \leqslant x \leqslant 5, \ 2 \leqslant y \leqslant 4\} \quad \blacktriangleleft$$

EXAMPLE 5 ➤ **The Double Integral of an Exponential Function** Evaluate to two decimal places:

$$\iint\limits_R 2xe^{x^2+y}\, dA \quad \text{over} \quad R = \{(x, y)|0 \leqslant x \leqslant 1, \ -1 \leqslant y \leqslant 1\}$$

Solution Region R is illustrated in Figure 3.

$$\iint\limits_R 2xe^{x^2+y}\, dA = \int_{-1}^{1} \int_{0}^{1} 2xe^{x^2+y}\, dx\, dy$$

$$= \int_{-1}^{1} \left[(e^{x^2+y}) \Big|_{x=0}^{x=1} \right] dy$$

$$= \int_{-1}^{1} (e^{1+y} - e^y)\, dy \approx 4.04 \quad \blacktriangleleft$$

FIGURE 3

MATCHED PROBLEM 5 ➤ Evaluate to two decimal places:

$$\iint\limits_R \frac{x}{y^2} e^{x/y}\, dA \quad \text{over} \quad R = \{(x, y)|0 \leqslant x \leqslant 1, \ 1 \leqslant y \leqslant 2\} \quad \blacktriangleleft$$

■ **AVERAGE VALUE OVER RECTANGULAR REGIONS**

In Section 11-5, the average value of a function $f(x)$ over an interval $[a, b]$ was defined as

$$\frac{1}{b - a} \int_{a}^{b} f(x)\, dx$$

This definition is easily extended to functions of two variables over rectangular regions, as shown in the next box. Notice that the denominator in the expression given in the box, $(b - a)(d - c)$, is simply the area of the rectangle R.

Average Value Over Rectangular Regions

The **average value** of the function $f(x, y)$ over the rectangle

$$R = \{(x, y) | a \leqslant x \leqslant b, \quad c \leqslant y \leqslant d\}$$

is

$$\frac{1}{(b - a)(d - c)} \iint_R f(x, y) \, dA$$

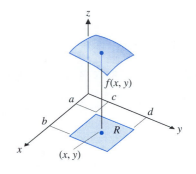

EXAMPLE 6 ➤ **Average Value**   Find the average value of $f(x, y) = 4 - \frac{1}{2}x - \frac{1}{2}y$ over the rectangle $R = \{(x, y) | 0 \leqslant x \leqslant 2, \quad 0 \leqslant y \leqslant 2\}$.

Solution   Region R is illustrated in Figure 4.

$$\frac{1}{(b - a)(d - c)} \iint_R f(x, y) \, dA = \frac{1}{(2 - 0)(2 - 0)} \iint_R \left(4 - \frac{1}{2}x - \frac{1}{2}y \right) dA$$

$$= \frac{1}{4} \int_0^2 \int_0^2 \left(4 - \frac{1}{2}x - \frac{1}{2}y \right) dy \, dx$$

$$= \frac{1}{4} \int_0^2 \left[\left(4y - \frac{1}{2}xy - \frac{1}{4}y^2 \right) \Big|_{y=0}^{y=2} \right] dx$$

$$= \frac{1}{4} \int_0^2 (7 - x) \, dx = 3$$

FIGURE 4

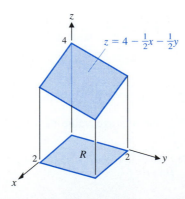

FIGURE 5

Figure 5 illustrates the surface $z = f(x, y)$, and the result obtained (3) is the average of the z values over the region R. ◀

MATCHED PROBLEM 6 ▶ Find the average value of $f(x, y) = x + 2y$ over the rectangle $R = \{(x, y)|0 \leqslant x \leqslant 2, \quad 0 \leqslant y \leqslant 1\}$. ◀

explore – discuss 1

(A) Which of the two functions, $f(x, y) = 4 - x^2 - y^2$ or $g(x, y) = 4 - x - y$, would you guess has the greater average value over the rectangle $R = \{(x, y)|0 \leqslant x \leqslant 1, \quad 0 \leqslant y \leqslant 1\}$? Explain.

(B) Use double integrals to check the correctness of your guess in part (A).

■ VOLUME AND DOUBLE INTEGRALS

One application of the definite integral of a function with one variable is the calculation of areas, so it is not surprising that the definite integral of a function of two variables can be used to calculate volumes of solids.

Volume Under a Surface

If $f(x, y) \geqslant 0$ over a rectangle $R = \{(x, y)|a \leqslant x \leqslant b, \quad c \leqslant y \leqslant d\}$, then the volume of the solid formed by graphing f over the rectangle R is given by

$$V = \iint\limits_{R} f(x, y) \, dA$$

A proof of the statement in the box is left to a more advanced text.

EXAMPLE 7 ▶ **Volume** Find the volume of the solid under the graph of $f(x, y) = 1 + x^2 + y^2$ over the rectangle $R = \{(x, y)|0 \leqslant x \leqslant 1, \quad 0 \leqslant y \leqslant 1\}$.

Solution   Figure 6 shows the region R and Figure 7 illustrates the volume under consideration.

$$V = \iint\limits_R (1 + x^2 + y^2) \, dA$$

$$= \int_0^1 \int_0^1 (1 + x^2 + y^2) \, dx \, dy$$

$$= \int_0^1 \left[(x + \tfrac{1}{3}x^3 + xy^2) \Big|_{x=0}^{x=1} \right] dy$$

$$= \int_0^1 (\tfrac{4}{3} + y^2) \, dy$$

$$= \tfrac{5}{3} \text{ cubic units}$$

FIGURE 6

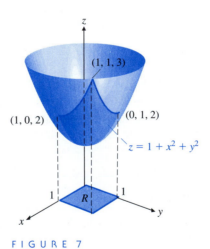

FIGURE 7

MATCHED PROBLEM 7 ▶ Find the volume of the solid under the graph of $f(x, y) = 1 + x + y$ over the rectangle $R = \{(x, y) | 0 \leq x \leq 1, \quad 0 \leq y \leq 2\}$.

explore – discuss 2

Consider the solid under the graph of $f(x, y) = 4 - y^2$ and above the rectangle $R = \{(x, y) | 0 \leq x \leq 3, \quad 0 \leq y \leq 2\}$.

(A) Explain why each cross-section of the solid by a plane parallel to the yz plane has the same area, and compute that area.

(B) Compute the areas of the cross-sections of the solid by the planes $y = 0$, $y = \tfrac{1}{2}$, and $y = 1$.

(C) Compute the volume of the solid in two different ways.

Answers to Matched Problems
1. (A) $2xy^2 + 3x^2y^4 + C(x)$ (B) $2x^2y + 4x^3y^3 + E(y)$
2. (A) $2x + 3x^2$ (B) $18y + 108y^3$ 3. (A) 36   (B) 36
4. 12   5. 0.42   6. 2   7. 5 cubic units

EXERCISE   13-6

A   *In Problems 1–6, find each antiderivative. Then use the antiderivative to evaluate the definite integral.*

1. (A) $\displaystyle\int 12x^2y^3\,dy$ (B) $\displaystyle\int_0^1 12x^2y^3\,dy$

2. (A) $\displaystyle\int 12x^2y^3\,dx$ (B) $\displaystyle\int_{-1}^2 12x^2y^3\,dx$

3. (A) $\displaystyle\int (4x + 6y + 5)\,dx$ (B) $\displaystyle\int_{-2}^3 (4x + 6y + 5)\,dx$

4. (A) $\displaystyle\int (4x + 6y + 5)\,dy$ (B) $\displaystyle\int_1^4 (4x + 6y + 5)\,dy$

5. (A) $\displaystyle\int \frac{x}{\sqrt{y + x^2}}\,dx$ (B) $\displaystyle\int_0^2 \frac{x}{\sqrt{y + x^2}}\,dx$

6. (A) $\displaystyle\int \frac{x}{\sqrt{y + x^2}}\,dy$ (B) $\displaystyle\int_1^5 \frac{x}{\sqrt{y + x^2}}\,dy$

B   *In Problems 7–12, evaluate each iterated integral. (See the indicated problem for the evaluation of the inner integral.)*

7. $\displaystyle\int_{-1}^2 \int_0^1 12x^2y^3\,dy\,dx$ 8. $\displaystyle\int_0^1 \int_{-1}^2 12x^2y^3\,dx\,dy$

   (see Problem 1)   (see Problem 2)

9. $\displaystyle\int_1^4 \int_{-2}^3 (4x + 6y + 5)\,dx\,dy$

   (see Problem 3)

10. $\displaystyle\int_{-2}^3 \int_1^4 (4x + 6y + 5)\,dy\,dx$

   (see Problem 4)

11. $\displaystyle\int_1^5 \int_0^2 \frac{x}{\sqrt{y + x^2}}\,dx\,dy$ 12. $\displaystyle\int_0^2 \int_1^5 \frac{x}{\sqrt{y + x^2}}\,dy\,dx$

   (see Problem 5)   (see Problem 6)

Use both orders of iteration to evaluate each double integral in Problems 13–16.

13. $\displaystyle\iint_R xy\,dA; R = \{(x, y)|0 \leqslant x \leqslant 2,\ \ 0 \leqslant y \leqslant 4\}$

14. $\displaystyle\iint_R \sqrt{xy}\,dA; R = \{(x, y)|1 \leqslant x \leqslant 4,\ \ 1 \leqslant y \leqslant 9\}$

15. $\displaystyle\iint_R (x + y)^5\,dA; R = \{(x, y)|-1 \leqslant x \leqslant 1,\ \ 1 \leqslant y \leqslant 2\}$

16. $\displaystyle\iint_R xe^y\,dA; R = \{(x, y)|-2 \leqslant x \leqslant 3,\ \ 0 \leqslant y \leqslant 2\}$

In Problems 17–20, find the average value of each function over the given rectangle.

17. $f(x, y) = (x + y)^2$;
    $R = \{(x, y)|1 \leqslant x \leqslant 5,\ \ -1 \leqslant y \leqslant 1\}$
18. $f(x, y) = x^2 + y^2; R = \{(x, y)|-1 \leqslant x \leqslant 2,\ \ 1 \leqslant y \leqslant 4\}$
19. $f(x, y) = x/y; R = \{(x, y)|1 \leqslant x \leqslant 4,\ \ 2 \leqslant y \leqslant 7\}$
20. $f(x, y) = x^2y^3; R = \{(x, y)|-1 \leqslant x \leqslant 1,\ \ 0 \leqslant y \leqslant 2\}$

In Problems 21–24, find the volume of the solid under the graph of each function over the given rectangle.

21. $f(x, y) = 2 - x^2 - y^2$;
    $R = \{(x, y)|0 \leqslant x \leqslant 1,\ \ 0 \leqslant y \leqslant 1\}$
22. $f(x, y) = 5 - x; R = \{(x, y)|0 \leqslant x \leqslant 5,\ \ 0 \leqslant y \leqslant 5\}$
23. $f(x, y) = 4 - y^2; R = \{(x, y)|0 \leqslant x \leqslant 2,\ \ 0 \leqslant y \leqslant 2\}$
24. $f(x, y) = e^{-x-y}; R = \{(x, y)|0 \leqslant x \leqslant 1,\ \ 0 \leqslant y \leqslant 1\}$

C   *Evaluate each double integral in Problems 25–28. Select the order of integration carefully—each problem is easy to do one way and difficult the other.*

25. $\displaystyle\iint_R xe^{xy}\,dA; R = \{(x, y)|0 \leqslant x \leqslant 1,\ \ 1 \leqslant y \leqslant 2\}$

26. $\displaystyle\iint_R xye^{x^2y}\,dA; R = \{(x, y)|0 \leqslant x \leqslant 1,\ \ 1 \leqslant y \leqslant 2\}$

27. $\displaystyle\iint_R \frac{2y + 3xy^2}{1 + x^2}\,dA;$

   $R = \{(x, y)|0 \leqslant x \leqslant 1,\ \ -1 \leqslant y \leqslant 1\}$

28. $\displaystyle\iint\limits_{R} \frac{2x + 2y}{1 + 4y + y^2}\, dA;$

$R = \{(x, y)|1 \leqslant x \leqslant 3, \quad 0 \leqslant y \leqslant 1\}$

29. Show that $\int_0^2 \int_0^2 (1 - y)\, dx\, dy = 0$. Does the double integral represent the volume of a solid? Explain.

30. (A) Find the average values of the functions $f(x, y) = x + y$, $g(x, y) = x^2 + y^2$, and $h(x, y) = x^3 + y^3$ over the rectangle

$$R = \{(x, y)|0 \leqslant x \leqslant 1, \quad 0 \leqslant y \leqslant 1\}$$

(B) Does the average value of $k(x, y) = x^n + y^n$ over the rectangle $R_1 = \{(x, y)|0 \leqslant x \leqslant 1, \quad 0 \leqslant y \leqslant 1\}$ increase or decrease as n increases? Explain.

(C) Does the average value of $k(x, y) = x^n + y^n$ over the rectangle $R_2 = \{(x, y)|0 \leqslant x \leqslant 2, \quad 0 \leqslant y \leqslant 2\}$ increase or decrease as n increases? Explain.

31. Let $f(x, y) = x^3 + y^2 - e^{-x} - 1$.
(A) Find the average value of $f(x, y)$ over the rectangle $R = \{(x, y)|-2 \leqslant x \leqslant 2, \quad -2 \leqslant y \leqslant 2\}$.
(B) Graph the set of all points (x, y) in R for which $f(x, y) = 0$.
(C) For which points (x, y) in R is $f(x, y)$ greater than 0? Less than 0? Explain.

32. Find the dimensions of the square S centered at the origin for which the average value of $f(x, y) = x^2 e^y$ over S is equal to 100.

▶ APPLICATIONS

Business & Economics

33. *Multiplier principle.* Suppose Congress enacts a one-time-only 10% tax rebate that is expected to infuse $\$y$ billion, $5 \leqslant y \leqslant 7$, into the economy. If every individual and corporation is expected to spend a proportion x, $0.6 \leqslant x \leqslant 0.8$, of each dollar received, then by the **multiplier principle** in economics, the total amount of spending S (in billions of dollars) generated by this tax rebate is given by

$$S(x, y) = \frac{y}{1 - x}$$

What is the average total amount of spending for the indicated ranges of the values of x and y? Set up a double integral and evaluate.

34. *Multiplier principle.* Repeat Problem 33 if $6 \leqslant y \leqslant 10$ and $0.7 \leqslant x \leqslant 0.9$.

35. *Cobb–Douglas production function.* If an industry invests x thousand labor-hours, $10 \leqslant x \leqslant 20$, and $\$y$ million, $1 \leqslant y \leqslant 2$, in the production of N thousand units of a certain item, then N is given by

$$N(x, y) = x^{0.75}y^{0.25}$$

What is the average number of units produced for the indicated ranges of x and y? Set up a double integral and evaluate.

36. *Cobb–Douglas production function.* Repeat Problem 35 for

$$N(x, y) = x^{0.5}y^{0.5}$$

where $10 \leqslant x \leqslant 30$ and $1 \leqslant y \leqslant 3$.

Life Sciences

37. *Population distribution.* In order to study the population distribution of a certain species of insects, a biologist has constructed an artificial habitat in the shape of a rectangle 16 feet long and 12 feet wide. The only food available to the insects in this habitat is located at its center. The biologist has determined that the concentration C of insects per square foot at a point d units from the food supply (see the figure) is given approximately by

$$C = 10 - \tfrac{1}{10}d^2$$

What is the average concentration of insects throughout the habitat? Express C as a function of x and y, set up a double integral, and evaluate.

Figure for 37

38. *Population distribution.* Repeat Problem 37 for a square habitat that measures 12 feet on each side, where the insect concentration is given by

$$C = 8 - \tfrac{1}{10}d^2$$

39. *Pollution.* A heavy industrial plant located in the center of a small town emits particulate matter into the atmosphere.

Suppose the concentration of particulate matter (in parts per million) at a point d miles from the plant (see the figure) is given by

$$C = 100 - 15d^2$$

If the boundaries of the town form a rectangle 4 miles long and 2 miles wide, what is the average concentration of particulate matter throughout the city? Express C as a function of x and y, set up a double integral, and evaluate.

Figure for 39

40. *Pollution.* Repeat Problem 39 if the boundaries of the town form a rectangle 8 miles long and 4 miles wide and the concentration of particulate matter is given by

$$C = 100 - 3d^2$$

Social Sciences

41. *Safety research.* Under ideal conditions, if a person driving a car slams on the brakes and skids to a stop, the length of the skid marks (in feet) is given by the formula

$$L = 0.000\ 013\ 3xy^2$$

where x is the weight of the car (in pounds) and y is the speed of the car (in miles per hour). What is the average length of the skid marks for cars weighing between 2,000 and 3,000 pounds and traveling at speeds between 50 and 60 miles per hour? Set up a double integral and evaluate.

42. *Safety research.* Repeat Problem 41 for cars weighing between 2,000 and 2,500 pounds and traveling at speeds between 40 and 50 miles per hour.

43. *Psychology.* The intelligence quotient Q for an individual with mental age x and chronological age y is given by

$$Q(x, y) = 100\,\frac{x}{y}$$

In a group of sixth graders, the mental age varies between 8 and 16 years and the chronological age varies between 10 and 12 years. What is the average intelligence quotient for this group? Set up a double integral and evaluate.

44. *Psychology.* Repeat Problem 43 for a group with mental ages between 6 and 14 years and chronological ages between 8 and 10 years.

CHAPTER 13
GROUP ACTIVITY

z = f(x, y) = 4 − x² − y²

FIGURE 1
Paraboloid

Double Integrals Over More General Regions

The definition of the double integral can be extended to nonrectangular regions. Consider, for example, the solid under the graph of the paraboloid $f(x, y) = 4 - x^2 - y^2$ and above the xy plane (see Fig. 1).

Since the intersection of $f(x, y)$ with the xy plane is the circle $x^2 + y^2 = 4$, to compute the volume of the solid we evaluate $\iint_C f(x, y)\, dA$ over the circular region C bounded by $x^2 + y^2 = 4$. We write

$$C = \{(x, y)\mid -\sqrt{4 - x^2} \leqslant y \leqslant \sqrt{4 - x^2},\quad -2 \leqslant x \leqslant 2\}$$

Then

$$\iint_C f(x, y)\, dA = \int_{-2}^{2} \int_{-\sqrt{4-x^2}}^{\sqrt{4-x^2}} (4 - x^2 - y^2)\, dy\, dx$$

$$= \int_{-2}^{2} \left[\left(4y - x^2 y - \frac{y^3}{3} \right) \Big|_{-\sqrt{4-x^2}}^{\sqrt{4-x^2}} \right] dx$$

$$= \int_{-2}^{2} \frac{4}{3} (4 - x^2)^{3/2}\, dx \approx 25.13$$

More generally, a region R in the xy plane is a **regular x region** if there exist functions $f(x)$ and $g(x)$ and numbers a and b such that

$$R = \{(x, y) \,|\, g(x) \leqslant y \leqslant f(x), \quad a \leqslant x \leqslant b\}$$

The double integral of a function $F(x, y)$ over a regular x region R is defined by

$$\iint\limits_{R} F(x, y)\, dA = \int_a^b \left[\int_{g(x)}^{f(x)} F(x, y)\, dy \right] dx$$

Similarly, interchanging the roles of x and y, a region S in the xy plane is a **regular y region** if there exist functions $h(y)$ and $k(y)$ and numbers c and d such that

$$S = \{(x, y) \,|\, h(y) \leqslant x \leqslant k(y), \quad c \leqslant y \leqslant d\}$$

The double integral of a function $F(x, y)$ over a regular y region S is defined by

$$\iint\limits_{S} F(x, y)\, dA = \int_c^d \left[\int_{h(y)}^{k(y)} F(x, y)\, dx \right] dy$$

(A) Graph the region R bounded by the graphs of $y = -x$, $y = x^2$, and $x = 1$ for $x \geqslant 0$, and explain why R is a regular x region.

(B) Evaluate $\iint_R 3xy^2\, dA$ over the region R of part (A).

(C) Graph the region S bounded by the graphs of $y = \sqrt{x}$, $x + y = 2$, and $y = 0$, and explain why S is a regular y region.

(D) Evaluate $\iint_S (y - 4x)\, dA$ over the region S of part (C).

(E) Explain why the region T bounded by the graphs of $y = x$ and $y = \frac{1}{2}x^2$ is both a regular x region and a regular y region.

(F) Evaluate $\iint_T 4x^3y\, dA$ in two different ways.

(G) What proportion of the volume of a right circular cylinder of height 4, with base the circular region C inside $x^2 + y^2 = 4$, lies inside the paraboloid $f(x, y) = 4 - x^2 - y^2$?

C H A P T E R   1 3   R E V I E W   ■   Important Terms and Symbols

13-1 *Functions of Several Variables.* Functions of two independent variables; functions of several independent variables; Cobb–Douglas production function; three-dimensional coordinate system; triplet of numbers (x, y, z); surface; paraboloid; saddle point

$$z = f(x, y); \quad w = f(x, y, z)$$

13-2 *Partial Derivatives.* Partial derivative of f with respect to x; partial derivative of f with respect to y; marginal pro-

ductivity of labor; marginal productivity of capital; second-order partial derivatives

$$\frac{\partial z}{\partial x}; \quad \frac{\partial z}{\partial y}; \quad f_x(x, y); \quad f_y(x, y);$$

$$\frac{\partial^2 z}{\partial x^2} = f_{xx}(x, y); \quad \frac{\partial^2 z}{\partial x\, \partial y} = f_{yx}(x, y); \quad \frac{\partial^2 z}{\partial y\, \partial x} = f_{xy}(x, y);$$

$$\frac{\partial^2 z}{\partial y^2} = f_{yy}(x, y)$$

13-3 *Maxima and Minima.* Local maximum; local minimum; saddle point; critical point; second-derivative test

13-4 *Maxima and Minima Using Lagrange Multipliers.* Constraint; Lagrange multiplier; critical point; method of Lagrange multipliers for functions of two variables; marginal productivity of money; method of Lagrange multipliers for functions of three variables

13-5 *Method of Least Squares.* Least squares approximation; regression analysis; method of least squares; linear regression; residual; least squares line; regression line; normal equations

13-6 *Double Integrals Over Rectangular Regions.* Double integral; integrand; region of integration; iterated integral;

average value over rectangular region; volume under a surface

$$\iint\limits_R f(x, y)\, dA = \int_a^b \left[\int_c^d f(x, y)\, dy \right] dx$$

$$= \int_c^d \left[\int_a^b f(x, y)\, dx \right] dy;$$

$$\frac{1}{(b - a)(d - c)} \iint\limits_R f(x, y)\, dA;$$

$$V = \iint\limits_R f(x, y)\, dA$$

C H A P T E R   13   ■   **Review Exercise**

Work through all the problems in this chapter review and check your answers in the back of the book. Answers to all review problems are there along with section numbers in italics to indicate where each type of problem is discussed. Where weaknesses show up, review appropriate sections in the text.

A

1. For $f(x, y) = 2{,}000 + 40x + 70y$, find $f(5, 10)$, $f_x(x, y)$, and $f_y(x, y)$.

2. For $z = x^3 y^2$, find $\partial^2 z / \partial x^2$ and $\partial^2 z / \partial x\, \partial y$.

3. Evaluate: $\int (6xy^2 + 4y)\, dy$

4. Evaluate: $\int (6xy^2 + 4y)\, dx$

5. Evaluate: $\int_0^1 \int_0^1 4xy\, dy\, dx$

B

6. For $f(x, y) = 3x^2 - 2xy + y^2 - 2x + 3y - 7$, find $f(2, 3)$, $f_x(x, y)$, and $f_y(2, 3)$.

7. For $f(x, y) = -4x^2 + 4xy - 3y^2 + 4x + 10y + 81$, find

$$[f_{xx}(2, 3)][f_{yy}(2, 3)] - [f_{xy}(2, 3)]^2$$

8. If $f(x, y) = x + 3y$ and $g(x, y) = x^2 + y^2 - 10$, find the critical points of $F(x, y, \lambda) = f(x, y) + \lambda g(x, y)$.

9. Use the least squares line for the data in the table to estimate y when $x = 10$.

x	y
2	12
4	10
6	7
8	3

10. For $R = \{(x, y)| -1 \leqslant x \leqslant 1, \quad 1 \leqslant y \leqslant 2\}$, evaluate the following in two ways:

$$\iint\limits_R (4x + 6y)\, dA$$

C

11. For $f(x, y) = e^{x^2 + 2y}$, find f_x, f_y, and f_{xy}.

12. For $f(x, y) = (x^2 + y^2)^5$, find f_x and f_{xy}.

13. Find all critical points and test for extrema for

$$f(x, y) = x^3 - 12x + y^2 - 6y$$

14. Use Lagrange multipliers to maximize $f(x, y) = xy$ subject to $2x + 3y = 24$.

15. Use Lagrange multipliers to minimize $f(x, y, z) = x^2 + y^2 + z^2$ subject to $2x + y + 2z = 9$.

16. Find the least squares line for the data in the table.

x	y	x	y
10	50	60	80
20	45	70	85
30	50	80	90
40	55	90	90
50	65	100	110

17. Find the average value of $f(x, y) = x^{2/3}y^{1/3}$ over the rectangle

$$R = \{(x, y)| -8 \leqslant x \leqslant 8, \quad 0 \leqslant y \leqslant 27\}$$

18. Find the volume of the solid under the graph of $z = 3x^2 + 3y^2$ over the rectangle

$$R = \{(x, y)|0 \leqslant x \leqslant 1, \quad -1 \leqslant y \leqslant 1\}$$

19. Without doing any computation, predict the average value of $f(x, y) = x + y$ over the rectangle $R = \{(x, y)| -10 \leqslant x \leqslant 10, \quad -10 \leqslant y \leqslant 10\}$. Then check the correctness of your prediction by evaluating a double integral.

20. (A) Find the dimensions of the square S centered at the origin such that the average value of

$$f(x, y) = \frac{e^x}{y + 10}$$

over S is equal to 5.

(B) Is there a square centered at the origin over which

$$f(x, y) = \frac{e^x}{y + 10}$$

has average value 0.05? Explain.

▶ APPLICATIONS

Business & Economics

21. *Maximizing profit.* A company produces x units of product A and y units of product B (both in hundreds per month). The monthly profit equation (in thousands of dollars) is found to be

$$P(x, y) = -4x^2 + 4xy - 3y^2 + 4x + 10y + 81$$

(A) Find $P_x(1, 3)$ and interpret the results.

(B) How many of each product should be produced each month to maximize profit? What is the maximum profit?

22. *Minimizing material.* A rectangular box with no top and six compartments (see the figure) is to have a volume of 96 cubic inches. Find the dimensions that will require the least amount of material.

Figure for 22

23. *Profit.* A company's annual profits (in millions of dollars) over a 5 year period are given in the table. Use the least squares line to estimate the profit for the sixth year.

YEAR	PROFIT
1	2
2	2.5
3	3.1
4	4.2
5	4.3

24. *Productivity.* The Cobb–Douglas production function for a product is

$$N(x, y) = 10x^{0.8}y^{0.2}$$

where x is the number of units of labor and y is the number of units of capital required to produce N units of the product.

(A) Find the marginal productivity of labor and the marginal productivity of capital at $x = 40$ and $y = 50$. For the greatest increase in productivity, should management encourage increased use of labor or increased use of capital?

(B) If each unit of labor costs $100, each unit of capital costs $50, and $10,000 is budgeted for production of this product, use the method of Lagrange multipliers to determine the allocations of labor and capital that will maximize the number of units produced and find the maximum production. Find the marginal productivity of money and approximate the increase in production that would result from an increase of $2,000 in the amount budgeted for production.

(C) If $50 \leq x \leq 100$ and $20 \leq y \leq 40$, find the average number of units produced. Set up a definite integral and evaluate.

Life Sciences

25. *Marine biology.* The function used for timing dives with scuba gear is

$$T(V, x) = \frac{33V}{x + 33}$$

where T is the time of the dive in minutes, V is the volume of air (in cubic feet, at sea level pressure) compressed into tanks, and x is the depth of the dive in feet. Find $T_x(70, 17)$ and interpret the results.

26. *Pollution.* A heavy industrial plant located in the center of a small town emits particulate matter into the atmosphere. Suppose the concentration of particulate matter (in parts per million) at a point d miles from the plant is given by

$$C = 100 - 24d^2$$

If the boundaries of the town form a square 4 miles long and 4 miles wide, what is the average concentration of particulate matter throughout the town? Express C as a function of x and y, set up a double integral, and evaluate.

Social Sciences

27. *Sociology.* Joseph Cavanaugh, a sociologist, found that the number of long-distance telephone calls, n, between two cities in a given period of time varied (approximately) jointly as the populations P_1 and P_2 of the two cities, and varied inversely as the distance, d, between the two cities. In terms of an equation for a time period of 1 week,

$$n(P_1, P_2, d) = 0.001 \frac{P_1 P_2}{d}$$

Find $n(100{,}000, 50{,}000, 100)$.

28. *Education.* At the beginning of the semester, students in a foreign language course are given a proficiency exam. The same exam is given at the end of the semester. The results for 5 students are given in the table. Use the least squares line to estimate the score on the second exam for a student who scored 40 on the first exam.

FIRST EXAM	SECOND EXAM
30	60
50	75
60	80
70	85
90	90

29. *Population density.* The table gives the population per square mile in the United States for the years 1900–1990.

U.S. Population Density

YEAR	POPULATION (PER SQUARE MILE)	YEAR	POPULATION (PER SQUARE MILE)
1900	25.6	1950	50.7
1910	31.0	1960	50.6
1920	35.6	1970	57.4
1930	41.2	1980	64.0
1940	44.2	1990	70.3

(A) Find the least squares line for the data.

(B) Use linear regression to estimate the population density in the United States in the year 2000.

(C) Use quadratic regression and exponential regression to make the estimate of part (B).

30. *Life expectancy.* The table gives life expectancies for males and females in a sample of Central and South American countries.

Life Expectancies for Central and South American Countries

MALES	FEMALES	MALES	FEMALES
62.30	67.50	70.15	74.10
68.05	75.05	62.93	66.58
72.40	77.04	68.43	74.88
63.39	67.59	66.68	72.80
55.11	59.43		

(A) Find the least squares line for the data.

(B) Use linear regression to estimate the life expectancy of a female in a Central or South American country in which the life expectancy for males is 60 years.

(C) Use quadratic and logarithmic regression to make the estimate of part (B).

31. It is reasonable to conjecture from the data given in the table that many Japanese women tend to leave the work force to marry and have children, but then reenter the work force when the children are grown.

(A) Explain why you might expect cubic regression to provide a better fit to the data than linear or quadratic regression.

(B) Investigate your expectation by plotting the data points and graphing the curve of best fit using linear, quadratic, and cubic regression.

Women in the Work Force in Japan (1993)

AGE	PERCENTAGE OF WOMEN EMPLOYED	AGE	PERCENTAGE OF WOMEN EMPLOYED
22	75	47	72
27	64	52	66
32	52	57	57
37	61	62	40
42	70		

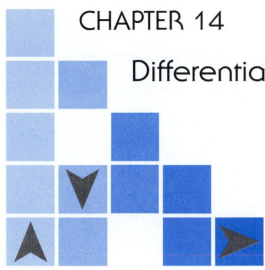

CHAPTER 14

Differential Equations

In Section 11-3 we considered *differential equations* of the form

$$\frac{dy}{dt} = ky, \qquad \frac{dy}{dt} = k(M - y), \qquad \text{and} \qquad \frac{dy}{dt} = ky(M - y)$$

Each of these equations establishes a relationship between a quantity y and its rate of growth dy/dt. Their solutions provide models for several types of exponential growth laws (see Table 1, page 837). Many different relationships in business, economics, and the sciences can be stated in terms of differential equations. Unfortunately, there is no single method that will solve all the differential equations that may be encountered—even in very simple applications. In this chapter, after discussing some basic concepts in the first section, we will consider methods for solving several types of differential equations that have significant applications.

SECTION 14-1 ■ **Basic Concepts**

- SOLUTIONS OF DIFFERENTIAL EQUATIONS
- IMPLICIT SOLUTIONS
- APPLICATION

In this section we review the basic concepts introduced in Section 11-3 and discuss in more detail what is meant by a solution to a differential equation, including explicit and implicit representations. Specific techniques for solving differential equations will be discussed in the next two sections.

■ SOLUTIONS OF DIFFERENTIAL EQUATIONS

A **differential equation** is an equation involving an unknown function, usually denoted by y, and one or more of its derivatives. For example,

$$y' = 0.2xy \tag{1}$$

is a differential equation. Since only the first derivative of the unknown function y appears in this equation, it is called a **first-order** differential equation. In general, the **order** of a differential equation is the highest derivative of the unknown function present in the function.

Notice that we used y' rather than dy/dx to represent the derivative of y with respect to x in equation (1). This is customary practice in the study of differential equations. Unless indicated otherwise, we will assume that y is a function of the independent variable x and that y' refers to the derivative of y with respect to x.

explore – discuss 1

Consider the differential equation

$$\frac{dy}{dx} = 2x \tag{2}$$

(A) Which of the following functions satisfy this equation?

$$y = x^2 \qquad y = 2x^2 \qquad y = x^2 + 2$$

(B) Can you find other functions that satisfy equation (2)? What form do all these functions have?

(C) Discuss possible solution methods for any differential equation of the form

$$\frac{dy}{dx} = f(x)$$

Figure 1 shows a *slope field* for equation (1). Recall from Section 11-3 that the **slope field** for a differential equation is obtained by drawing tangent line segments determined by the equation at each point in a grid. This provides a geometric interpretation of the equation that indicates the general shape of solutions to the equation. Due to the large number of line segments, slope fields are usually generated by computer.

Now consider the function

$$y = 2e^{0.1x^2}$$

whose derivative is

$$y' = 2e^{0.1x^2}(0.2x) = 0.4xe^{0.1x^2}$$

Substituting for y and y' in equation (1) gives

$$y' = 0.2xy \tag{1}$$

$$0.4xe^{0.1x^2} = 0.2x(2e^{0.1x^2})$$

$$0.4xe^{0.1x^2} = 0.4xe^{0.1x^2}$$

FIGURE 1
Slope field for $y' = 0.2xy$

which is certainly true for all values of x. This shows that the function $y = 2e^{0.1x^2}$ is a **solution** of equation (1). But this function is not the only solution. In fact, if C is any constant, substituting $y = Ce^{0.1x^2}$ and $y' = 0.2xCe^{0.1x^2}$ in equation (1) yields the identity

$$y' = 0.2xy$$
$$0.2xCe^{0.1x^2} = 0.2xCe^{0.1x^2}$$

It turns out that all solutions of $y' = 0.2xy$ can be obtained from $y = Ce^{0.1x^2}$ by assigning C appropriate values; hence, $y = Ce^{0.1x^2}$ is called the **general solution** of equation (1). The collection of all functions of the form $y = Ce^{0.1x^2}$ is called the **family of solutions** of equation (1). The function $y = 2e^{0.1x^2}$, obtained by letting $C = 2$ in the general solution, is called a **particular solution** of equation (1). Other particular solutions are

$$
\begin{array}{ll}
y = -2e^{0.1x^2} & C = -2 \\
y = -e^{0.1x^2} & C = -1 \\
y = 0 & C = 0 \\
y = e^{0.1x^2} & C = 1
\end{array}
$$

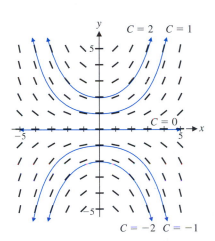

FIGURE 2
Particular solutions of $y' = 0.2xy$

These particular solutions are graphed in Figure 2, along with the slope field for equation (1). Notice how the particular solutions conform to the shapes indicated by the slope field.

Remark: Recognizing General Solutions
Most, but not all, first-order differential equations have general solutions that involve one arbitrary constant. Whenever we find a function, such as $y = Ce^{0.1x^2}$, that satisfies a given differential equation for any value of the constant C, we will assume this function is the general solution.

EXAMPLE 1 ➤ Verifying the Solution of a Differential Equation   Show that

$$y = Cx^2 + 1$$

is the general solution of the differential equation

$$xy' = 2y - 2$$

On the same set of axes, graph the particular solutions obtained by letting $C = -2$, -1, 0, 1, and 2.

Solution   Substituting $y = Cx^2 + 1$ and $y' = 2Cx$ in the differential equation, we have

$$
\begin{aligned}
xy' &= 2y - 2 \\
x(2Cx) &= 2(Cx^2 + 1) - 2 \\
2Cx^2 &= 2Cx^2 + 2 - 2 \\
2Cx^2 &= 2Cx^2
\end{aligned}
$$

which shows that $y = Cx^2 + 1$ is the general solution. The particular solutions corresponding to $C = -2, -1, 0, 1,$ and 2 are graphed in Figure 3.

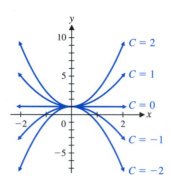

FIGURE 3
Particular solutions of $xy' = 2y - 2$

MATCHED PROBLEM 1 ➤   Show that

$$y = Cx + 1$$

is the general solution of the differential equation

$$xy' = y - 1$$

On the same set of axes, graph the particular solutions obtained by letting $C = -2, -1, 0, 1,$ and 2. ◀

In many applications, we will be interested in finding a particular solution $y(x)$ that satisfies an **initial condition** of the form $y(x_0) = y_0$. The value of the constant C in the general solution must then be selected so that this initial condition is satisfied; that is, so that the solution curve (from the family of solution curves) passes through (x_0, y_0).

EXAMPLE 2 ➤   **Finding a Particular Solution**   Using the general solution in Example 1, find a particular solution of the differential equation $xy' = 2y - 2$ that satisfies the indicated initial condition, if such a solution exists.

(A) $y(1) = 3$ (B) $y(0) = 3$ (C) $y(0) = 1$

Solution   (A) *Initial condition $y(1) = 3$.* From Example 1, the general solution of the differential equation is

$$y = Cx^2 + 1$$

Substituting $x = 1$ and $y = 3$ in this general solution yields

$$3 = C(1)^2 + 1$$
$$= C + 1$$
$$C = 2$$

Thus, the particular solution satisfying the initial condition $y(1) = 3$ is

$$y = 2x^2 + 1$$

See the graph labeled $C = 2$ in Figure 3.

(B) *Initial condition $y(0) = 3$.* Substituting $x = 0$ and $y = 3$ in the general solution yields

$$3 = C(0)^2 + 1$$

No matter what value of C we select, $C(0)^2 = 0$ and this equation reduces to

$$3 = 1$$

which is *never* valid. Thus, we must conclude that there is no particular solution of this differential equation that will satisfy the initial condition $y(0) = 3$.

(C) *Initial condition $y(0) = 1$.* Substituting $x = 0$ and $y = 1$ in the general solution, we have

$$1 = C(0)^2 + 1$$
$$1 = 1$$

This equation is valid for *all* values of C. Thus, all solutions of this differential equation satisfy the initial condition $y(0) = 1$ (see Fig. 3). ◀

Remark: Nature of Particular Solutions

The situation illustrated in Example 2A is the most common. Most first-order differential equations you will encounter will have exactly one particular solution that satisfies a given initial condition. However, we must always be aware of the possibility that a differential equation may not have any particular solutions that satisfy a given initial condition, or the possibility that it may have many particular solutions satisfying a given initial condition.

MATCHED PROBLEM 2 ➤ Using the general solution in Matched Problem 1, find a particular solution of the differential equation $xy' = y - 1$ that satisfies the indicated initial condition, if such a solution exists.

(A) $y(1) = 2$ (B) $y(0) = 2$ (C) $y(0) = 1$ ◀

■ IMPLICIT SOLUTIONS

The solution method we will discuss in the next section always produces an implicitly defined solution of a differential equation. In most of the differential equations we will consider, an explicit form of the solution can then be found by solving this implicit equation for the dependent variable in terms of the independent variable.

Let y be a function defined implicitly by the equation

$$xy = C \qquad \text{where } C \text{ is a constant} \tag{3}$$

(A) Differentiate equation (3) implicitly to obtain a differential equation involving y and y'.

(B) Solve equation (3) for y, and verify that the resulting function satisfies the differential equation you found in part (B).

EXAMPLE 3 ➤ **Verifying an Implicit Solution**   If y is defined implicitly by the equation

$$y^3 + e^y - x^4 = C \tag{4}$$

show that y satisfies the differential equation

$$(3y^2 + e^y)y' = 4x^3$$

Solution   We use implicit differentiation to show that y satisfies the given differential equation:

$$y^3 + e^y - x^4 = C$$

$$\frac{d}{dx}(y^3 + e^y - x^4) = \frac{d}{dx}C \qquad \text{Remember, } \frac{d}{dx}x^n = nx^{n-1} \text{ and } \frac{d}{dx}e^x = e^x,$$

$$\boxed{\frac{d}{dx}y^3 + \frac{d}{dx}e^y - \frac{d}{dx}x^4 = 0} \qquad \begin{array}{l} \text{but } \frac{d}{dx}y^n = ny^{n-1}y' \text{ and } \frac{d}{dx}e^y = e^yy' \\ \text{since } y \text{ is a function of } x. \end{array}$$

$$3y^2y' + e^yy' - 4x^3 = 0$$

$$(3y^2 + e^y)y' = 4x^3$$

Since the last equation is the given differential equation, our calculations show that any function y defined implicitly by equation (4) is a solution of this differential equation. We cannot find an explicit formula for y, since none exists in terms of finite combinations of elementary functions.   ◄

MATCHED PROBLEM 3 ➤ If y is defined implicitly by the equation

$$y + e^{y^2} - x^2 = C$$

show that y satisfies the differential equation

$$(1 + 2ye^{y^2})y' = 2x$$ ◄

EXAMPLE 4 ➤ **Finding an Explicit Solution**   If y is defined implicitly by the equation

$$y^2 - x^2 = C$$

show that y satisfies the differential equation

$$yy' = x$$

Find an explicit expression for the particular solution that satisfies the initial condition $y(0) = 2$.

Solution   Using implicit differentiation, we have

$$y^2 - x^2 = C$$

$$\frac{d}{dx}(y^2 - x^2) = \frac{d}{dx}C$$

$$2yy' - 2x = 0$$

$$2yy' = 2x$$

$$yy' = x$$

which shows that y satisfies the given differential equation. Substituting $x = 0$ and $y = 2$ in $y^2 - x^2 = C$, we have

$$y^2 - x^2 = C$$

$$(2)^2 - (0)^2 = C$$

$$4 = C$$

Thus, the particular solution satisfying $y(0) = 2$ is a solution of the equation

$$y^2 - x^2 = 4$$

or

$$y^2 = 4 + x^2$$

Solving the last equation for y yields two explicit solutions,

$$y_1(x) = \sqrt{4 + x^2} \quad \text{and} \quad y_2(x) = -\sqrt{4 + x^2}$$

The first of these two solutions satisfies

$$y_1(0) = \sqrt{4 + (0)^2} = \sqrt{4} = 2$$

while the second satisfies

$$y_2(0) = -\sqrt{4 + (0)^2} = -\sqrt{4} = -2$$

Thus, the particular solution of the differential equation $yy' = x$ that satisfies the initial condition $y(0) = 2$ is

$$y(x) = \sqrt{4 + x^2}$$

Check $y'(x) = \dfrac{1}{2}(4 + x^2)^{-1/2}2x$ $\qquad\qquad yy' = x$

$$= \frac{x}{\sqrt{4 + x^2}} \qquad\qquad \sqrt{4 + x^2}\left(\frac{x}{\sqrt{4 + x^2}}\right) \overset{?}{=} x$$

$$x \overset{\checkmark}{=} x$$

MATCHED PROBLEM 4 ▶ If y is defined implicitly by the equation

$$y^2 - x = C$$

show that y satisfies the differential equation

$$2yy' = 1$$

Find an explicit expression for the particular solution that satisfies the initial condition $y(0) = 3$. ◀

■ APPLICATION

In economics, the price of a product is often studied over a period of time, and so it is natural to view price as a function of time. Let $p(t)$ be the price of a particular product at time t. If $p(t)$ approaches a limiting value $\bar{p}$ as t approaches infinity, then the price for this product is said to be **dynamically stable** and $\bar{p}$ is referred to as the **equilibrium price.** (Later in this chapter, this definition of equilibrium price will be related to the one given in Section 12-2.) In order to study the behavior of price as a function of time, economists often assume that the price satisfies a differential equation. This approach is illustrated in the next example.

▶ EXAMPLE 5 ▶ **Dynamic Price Stability**   The price $p(t)$ of a product is assumed to satisfy the differential equation

$$\frac{dp}{dt} = 10 - 0.5p \qquad t \geq 0$$

(A)   Show that

$$p(t) = 20 - Ce^{-0.5t}$$

is the general solution of this differential equation, and evaluate

$$\bar{p} = \lim_{t \to \infty} p(t)$$

(B)   Find the three particular solutions that satisfy the initial conditions $p(0) = 40$, $p(0) = 10$, and $p(0) = 20$.

(C)   Discuss the long-term behavior of the price of this product.

Solution   (A) $p(t) = 20 - Ce^{-0.5t}$
$$p'(t) = 0.5Ce^{-0.5t}$$

Substituting in the given differential equation, we have

$$\frac{dp}{dt} = 10 - 0.5p$$
$$0.5Ce^{-0.5t} = 10 - 0.5(20 - Ce^{-0.5t})$$
$$= 10 - 10 + 0.5Ce^{-0.5t}$$
$$= 0.5Ce^{-0.5t}$$

which shows that $p(t) = 20 - Ce^{-0.5t}$ is the general solution of this differential equation. To find the equilibrium price, we must evaluate $\lim_{t \to \infty} p(t)$.

$$
\begin{aligned}
\bar{p} &= \lim_{t \to \infty} p(t) \\
&= \lim_{t \to \infty} (20 - Ce^{-0.5t}) \\
&= 20 - C \lim_{t \to \infty} e^{-0.5t} \qquad \lim_{t \to \infty} e^{-0.5t} = 0 \\
&= 20 - C \cdot 0 \\
&= 20
\end{aligned}
$$

(B) We must evaluate the constant C for each of the indicated initial conditions. In each case, we will make use of the equation

$$p(0) = 20 - Ce^0 = 20 - C$$

$p(0) = 40$	$p(0) = 10$	$p(0) = 20$
$20 - C = 40$	$20 - C = 10$	$20 - C = 20$
$C = -20$	$C = 10$	$C = 0$
$p(t) = 20 + 20e^{-0.5t}$	$p(t) = 20 - 10e^{-0.5t}$	$p(t) = 20$

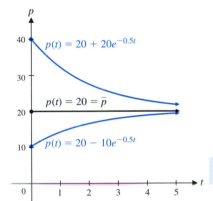

$p(t) = 20 + 20e^{-0.5t}$

$p(t) = 20 = \bar{p}$

$p(t) = 20 - 10e^{-0.5t}$

FIGURE 4

Particular solutions of $\dfrac{dp}{dt} = 10 - 0.5p$

(C) Graphing the three particular solutions gives us some insight into their long-term behavior (Fig. 4). The equilibrium price $\bar{p}$ does not depend on the constant C and consequently does not depend on the initial value of the price function $p(0)$ (see Fig. 4). If $p(0) > \bar{p}$, then the price decreases and approaches $\bar{p}$ as a limit. If $p(0) < \bar{p}$, then the price increases and approaches $\bar{p}$ as a limit. If $p(0) = \bar{p}$, then the price remains constant for all t. ◀

MATCHED PROBLEM 5 ▶ (A) Show that the price function

$$p(t) = 25 - Ce^{-0.2t}$$

is the general solution of the differential equation

$$\frac{dp}{dt} = 5 - 0.2p \qquad t \geq 0$$

Find the equilibrium price $\bar{p}$.

(B) Find the three particular solutions that satisfy the initial conditions $p(0) = 40$, $p(0) = 5$, and $p(0) = 25$.

(C) Discuss the long-term behavior of the price of this product. ◀

Answers to Matched Problems

1.

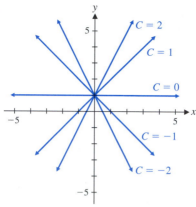

2. (A) $y = x + 1$
   (B) No particular solution exists.
   (C) $y = Cx + 1$ for any C

3. $\dfrac{d}{dx}y + \dfrac{d}{dx}e^{y^2} - \dfrac{d}{dx}x^2 = \dfrac{d}{dx}C$

   $y' + e^{y^2}2yy' - 2x = 0$

   $(1 + 2ye^{y^2})y' = 2x$

4. $y(x) = \sqrt{9 + x}$

5. (A) $\bar{p} = 25$
   (B) $p(t) = 25 + 15e^{-0.2t}$; $p(t) = 25 - 20e^{-0.2t}$; $p(t) = 25$
   (C) If $p(0) > \bar{p}$, then the price decreases and approaches $\bar{p}$ as a limit. If $p(0) < \bar{p}$, then the price increases and approaches $\bar{p}$ as a limit. If $p(0) = \bar{p}$, then the price remains constant for all t.

EXERCISE 14-1

A *In Problems 1–4, show that the given function is the general solution of the indicated differential equation.*

1. $y = Cx^2$; $xy' = 2y$

2. $y = C(x - 1)^2$; $(x - 1)y' = 2y$

3. $y = \dfrac{C}{x}$; $xy' = -y$

4. $y = \dfrac{C}{x + 1}$; $(x + 1)y' = -y$

In Problems 5–8, determine which of the slope fields (A)–(D) shown at the right is associated with the indicated differential equation. Briefly justify your answer.

5. $xy' = 2y$ (From Problem 1)
6. $(x - 1)y' = 2y$ (From Problem 2)
7. $xy' = -y$ (From Problem 3)
8. $(x + 1)y' = -y$ (From Problem 4)

In Problems 9–12, use the appropriate slope field—(A), (B), (C), or (D), given for Problems 5–8 (or a copy)—to graph the particular solutions obtained by letting $C = -2, -1, 0, 1$, and 2.

9. $y = Cx^2$ (From Problem 1)
10. $y = C(x - 1)^2$ (From Problem 2)

(A)

(B)

(C)

(D)

11. $y = \dfrac{C}{x}$ (From Problem 3)

12. $y = \dfrac{C}{x + 1}$ (From Problem 4)

B *In Problems 13–18, show that the given function y is the general solution of the indicated differential equation. Find a particular solution satisfying the given initial condition.*

13. $y = Ce^x - 5x - 5;\ y' = y + 5x;\ y(0) = 2$
14. $y = Ce^{-x} + 2x - 2;\ y' = 2x - y;\ y(0) = 3$
15. $y = e^x + Ce^{2x};\ y' = 2y - e^x;\ y(0) = -1$
16. $y = e^x + Ce^{-x};\ y' = 2e^x - y;\ y(0) = 1$
17. $y = x + \dfrac{C}{x};\ xy' = 2x - y;\ y(2) = 3$
18. $y = Cx + \dfrac{2}{x};\ x^2 y' = xy - 4;\ y(2) = -3$

If y is defined implicitly by the given equation, use implicit differentiation in Problems 19–22 to show that y satisfies the indicated differential equation.

19. $y^3 + xy - x^3 = C;\ (3y^2 + x)y' = 3x^2 - y$
20. $y + x^3 y^3 - 2x = C;\ (1 + 3x^3 y^2)y' = 2 - 3x^2 y^3$
21. $xy + e^{y^2} - x^2 = C;\ (x + 2ye^{y^2})y' = 2x - y$
22. $y + e^{xy} - x = C;\ (1 + xe^{xy})y' = 1 - ye^{xy}$

If y is defined implicitly by the given equation, use implicit differentiation in Problems 23–26 to show that y satisfies the indicated differential equation. Find an explicit expression for the particular solution that satisfies the given initial condition.

23. $y^2 + x^2 = C;\ yy' = -x;\ y(0) = 3$
24. $y^2 + 2x^2 = C;\ yy' = -2x;\ y(0) = -4$
25. $\ln(2 - y) = x + C;\ y' = y - 2;\ y(0) = 1$
26. $\ln(5 - y) = 2x + C;\ y' = 2(y - 5);\ y(0) = 2$

In Problems 27–30, use the general solution y of the differential equation to find a particular solution that satisfies the indicated initial condition. Discuss the behavior of these solutions as $x \to \infty$.

27. $y = 2 + Ce^{-x};\ y' = 2 - y$
   (A) $y(0) = 1$ (B) $y(0) = 2$ (C) $y(0) = 3$
28. $y = 4 + Ce^{-x};\ y' = 4 - y$
   (A) $y(0) = 2$ (B) $y(0) = 4$ (C) $y(0) = 5$

29. $y = 2 + Ce^x;\ y' = y - 2$
   (A) $y(0) = 1$ (B) $y(0) = 2$ (C) $y(0) = 3$
30. $y = 4 + Ce^x;\ y' = y - 4$
   (A) $y(0) = 2$ (B) $y(0) = 4$ (C) $y(0) = 5$

C *In Problems 31 and 32, use the general solution y of the differential equation to find a particular solution that satisfies the indicated initial condition. Describe the behavior of these solutions as $x \to \infty$.*

31. $y = \dfrac{10}{1 + Ce^{-x}};\ y' = 0.1y(10 - y)$
   (A) $y(0) = 1$ (B) $y(0) = 10$ (C) $y(0) = 20$
   [*Hint:* The particular solution in part (A) has an inflection point at $x = \ln 9$.]

32. $y = \dfrac{5}{1 + Ce^{-x}};\ y' = 0.2y(5 - y)$
   (A) $y(0) = 1$ (B) $y(0) = 5$ (C) $y(0) = 10$
   [*Hint:* The particular solution in part (A) has an inflection point at $x = \ln 4$.]

In Problems 33 and 34, use the general solution y of the differential equation to find a particular solution that satisfies the indicated initial condition.

33. $y = Cx^3 + 2;\ xy' = 3y - 6$
   (A) $y(0) = 2$ (B) $y(0) = 0$ (C) $y(1) = 1$
34. $y = Cx^4 + 1;\ xy' = 4y - 4$
   (A) $y(0) = 1$ (B) $y(0) = 0$ (C) $y(1) = 2$

In Problems 35 and 36, use window dimensions Xmin = −5, Xmax = 5, Ymin = −5, and Ymax = 5.

35. Given that $y = x + Ce^{-x}$ is the general solution of the differential equation $y' + y = 1 + x$:
   (A) In the same viewing window, graph the particular solutions obtained by letting $C = 0, 1, 2$, and 3.

   (B) What do the graphs of the solutions for $C > 0$ have in common?
   (C) In the same viewing window, graph the particular solutions obtained by letting $C = 0, -1, -2$, and -3.

   (D) What do the graphs of the solutions for $C < 0$ have in common?

36. Repeat Problem 35, given that $y = x + (C/x)$ is the general solution of the differential equation $xy' = 2x - y$.

▶ APPLICATIONS

Business & Economics

37. *Price stability.* The price $p(t)$ of a product is assumed to satisfy the differential equation

$$\frac{dp}{dt} = 0.5 - 0.1p$$

(A)   Show that

$$p(t) = 5 - Ce^{-0.1t}$$

is the general solution of this differential equation, and evaluate

$$\bar{p} = \lim_{t \to \infty} p(t)$$

(B)   Find the particular solutions that satisfy the initial conditions $p(0) = 1$ and $p(0) = 10$.

(C)   Discuss the long-term behavior of the price of this product.

38. *Price stability.* The price $p(t)$ of a product is assumed to satisfy the differential equation

$$\frac{dp}{dt} = 0.8 - 0.2p$$

(A)   Show that

$$p(t) = 4 - Ce^{-0.2t}$$

is the general solution of this differential equation, and evaluate

$$\bar{p} = \lim_{t \to \infty} p(t)$$

(B)   Find the particular solutions that satisfy the initial conditions $p(0) = 2$ and $p(0) = 8$.

(C)   Discuss the long-term behavior of the price of this product.

39. *Continuous compound interest.* If money is deposited at the continuous rate of $200 per year into an account earning 8% compounded continuously, then the amount A in the account after t years satisfies the differential equation

$$\frac{dA}{dt} = 0.08A + 200$$

(A)   Show that

$$A = Ce^{0.08t} - 2,500$$

is the general solution of this differential equation.

(B)   Find the particular solutions satisfying $A(0) = 0$ and $A(0) = 1,000$.

(C)   Compare the long-term behavior of the two particular solutions in part (B). Discuss the effect of the value of $A(0)$ on the amount in the account.

40. *Continuous compound interest.* If money is withdrawn at the continuous rate of $500 per year from an account earning 10% compounded continuously, then the amount A in the account after t years satisfies the differential equation

$$\frac{dA}{dt} = 0.1A - 500$$

(A)   Show that

$$A = 5,000 + Ce^{0.1t}$$

is the general solution of this differential equation.

(B)   Find the particular solutions satisfying $A(0) = 4,000$ and $A(0) = 6,000$.

(C)   Compare the long-term behavior of the two particular solutions in part (B). Discuss the effect of the value of $A(0)$ on the amount in the account.

Life Sciences

41. *Population growth—Verhulst growth law.* The number $N(t)$ of bacteria in a culture at time t is assumed to satisfy the Verhulst growth law

$$\frac{dN}{dt} = 100 - 0.5N$$

(A)   Show that

$$N(t) = 200 - Ce^{-0.5t}$$

is the general solution of this differential equation, and evaluate

$$\bar{N} = \lim_{t \to \infty} N(t)$$

where $\bar{N}$ is the equilibrium size of the population.

(B)   Find the particular solutions that satisfy $N(0) = 50$ and $N(0) = 300$.

(C)   Discuss the long-term behavior of this population.

42. *Population growth—logistic growth.* The population $N(t)$ of a certain species of animal in a controlled habitat at time t is assumed to satisfy the logistic growth law

$$\frac{dN}{dt} = \frac{1}{500} N(1,000 - N)$$

(A) Show that

$$N(t) = \frac{1,000}{1 + Ce^{-2t}}$$

is the general solution of this differential equation, and evaluate

$$\overline{N} = \lim_{t \to \infty} N(t)$$

(B) Find the particular solutions that satisfy $N(0) = 200$ and $N(0) = 2,000$.

(C) Discuss the long-term behavior of this population.

Social Sciences

43. *Rumor spread—Gompertz growth law.* The rate of propagation of a rumor is assumed to satisfy the Gompertz growth law

$$\frac{dN}{dt} = Ne^{-0.5t}$$

where $N(t)$ is the number of individuals who have heard the rumor at time t.

(A) Show that $N(t) = Ce^{-2e^{-0.5t}}$ is the general solution of this differential equation, and evaluate

$$\overline{N} = \lim_{t \to \infty} N(t)$$

(B) Find the particular solutions that satisfy $N(0) = 100$ and $N(0) = 200$.

(C) Discuss the effect of the value of $N(0)$ on the long-term propagation of this rumor.

SECTION 14-2 ■ **Separation of Variables**

- SEPARATION OF VARIABLES
- EXPONENTIAL GROWTH
- LIMITED GROWTH
- LOGISTIC GROWTH

■ **SEPARATION OF VARIABLES**

In this section we will develop a technique called *separation of variables*, which can be used to solve differential equations that can be expressed in the form

$$f(y)y' = g(x) \tag{1}$$

We have already used this technique informally in Section 11-3 to derive some important exponential growth laws. We will now present a more formal development of this method.

e x p l o r e – d i s c u s s 1

Use algebraic manipulation to express each of the following equations in the form $f(y)y' = g(x)$, or explain why it is not possible to do so.

(A) $y' = x^2y$ (B) $y' = x^2 + xy$ (C) $y' = x^2y + xy$

The solution of differential equations by the method of separating the variables is based on the substitution formula for indefinite integrals. If $y = y(x)$ is a differentiable function of x, then

$$\int f(y)\, dy = \int f[y(x)]y'(x)\, dx \qquad \text{Substitution formula for indefinite integrals} \qquad (2)$$

If $y(x)$ is also the solution of (1), then $y(x)$ and $y'(x)$ must satisfy (1). That is,

$$f[y(x)]y'(x) = g(x)$$

Substituting $g(x)$ for $f[y(x)]y'(x)$ in (2), we have

$$\int f(y)\, dy = \int g(x)\, dx$$

Thus, the solution $y(x)$ of (1) is given implicitly by this equation. Of course, for this method to be useful, we must be able to evaluate both of these indefinite integrals. This discussion is summarized in Theorem 1:

t h e o r e m   1

> **Separation of Variables**
>
> The solution of the differential equation
>
> $$f(y)y' = g(x)$$
>
> is given implicitly by the equation
>
> $$\int f(y)\, dy = \int g(x)\, dx$$

Several examples will demonstrate the use of this method.

EXAMPLE 1 ➤ **Using Separation of Variables**   Solve: $y' = 2xy^2$

Solution
$$y' = 2xy^2 \qquad \text{Multiply by } 1/y^2 \text{ to separate the variables.}$$

$$\frac{1}{y^2}y' = 2x \qquad \text{Convert to an equation involving indefinite integrals.}$$

$$\int \frac{1}{y^2}\, dy = \int 2x\, dx \qquad \text{Evaluate each indefinite integral.}$$

$$-\frac{1}{y} + C_1 = x^2 + C_2 \qquad \text{Combine the two constants of integration into a single arbitrary constant.}$$

$$-\frac{1}{y} = x^2 + C \qquad \text{Solve for } y.$$

$$y = -\frac{1}{x^2 + C} \qquad \text{General solution of } y' = 2xy^2$$

Check $y' = \dfrac{d}{dx}\left(-\dfrac{1}{x^2 + C}\right) = \dfrac{2x}{(x^2 + C)^2}$

$$y' = 2xy^2$$ Substitute $y' = 2x/(x^2 + C)^2$ and
$y = -1/(x^2 + C)$.

$$\dfrac{2x}{(x^2 + C)^2} \overset{?}{=} 2x\left(-\dfrac{1}{x^2 + C}\right)^2$$

$$\dfrac{2x}{(x^2 + C)^2} \overset{\checkmark}{=} \dfrac{2x}{(x^2 + C)^2}$$

This verifies that our general solution is correct. (You should develop the habit of checking the solution of each differential equation you solve, as we have done here. From now on, we leave it to you to check most of the examples worked in the text.) ◀

MATCHED PROBLEM 1 ▶ Solve: $y' = 4x^3y^2$ ◀

In some cases the general solution obtained by the technique of separation of variables does not include all the solutions to a differential equation. For example, the constant function $y = 0$ also satisfies the differential equation in Example 1. (Verify this.) Yet the solution $y = 0$ cannot be obtained from the expression

$$y = -\dfrac{1}{x^2 + C}$$

for any choice of the constant C. Solutions of this type are referred to as **singular solutions** and are usually discussed in more advanced courses. We will not attempt to find the singular solutions of any of the differential equations we consider.

EXAMPLE 2 ▶ Using Separation of Variables Find the general solution of

$$(1 + x^2)y' = 2x(y - 1)$$

Then find the particular solution that satisfies the initial condition $y(0) = 3$.

Solution First, we find the general solution:

$$(1 + x^2)y' = 2x(y - 1)$$ Multiply by $1/(1 + x^2)$ and $1/(y - 1)$ to separate the variables.

$$\dfrac{y'}{y - 1} = \dfrac{2x}{1 + x^2}$$ Convert to an equation involving indefinite integrals.

$$\int \dfrac{dy}{y - 1} = \int \dfrac{2x\,dx}{1 + x^2}$$ Evaluate each indefinite integral.

$$\ln|y - 1| = \ln(1 + x^2) + C$$

where C is an arbitrary constant. Notice that the two constants of integration always can be combined to form a single arbitrary constant. Also note that we can use $1 + x^2$ instead of $|1 + x^2|$, since $1 + x^2$ is always positive. In order to solve for y, we convert this last equation to exponential form:

$$|y - 1| = e^{\ln(1 + x^2) + C}$$
$$= e^C e^{\ln(1 + x^2)} \qquad \text{Use the property } e^{\ln r} = r.$$
$$= e^c(1 + x^2)$$

It can be shown that if we replace e^c with an arbitrary constant K, then we can omit the absolute value signs on the left side of the last equation.* The resulting equation can then be used to find the general solution to the original differential equation. Thus,

$$y - 1 = K(1 + x^2)$$
$$y = 1 + K(1 + x^2) \qquad \text{General solution}$$

To find the particular solution that satisfies $y(0) = 3$, we substitute $x = 0$ and $y = 3$ in the general solution and solve for K:

$$3 = 1 + K(1 + 0)$$
$$K = 2$$
$$y = 1 + 2(1 + x^2) = 3 + 2x^2 \qquad \text{Particular solution} \qquad \blacktriangleleft$$

MATCHED PROBLEM 2 ➤ Find the general solution of

$$(2 + x^4)y' = 4x^3(y - 3)$$

Then find the particular solution that satisfies the initial condition $y(0) = 5$. ◄

---

explore – discuss 2

Let y be a positive quantity that is changing with respect to time t, such as the price of a product, the amount of a drug in the bloodstream, or the population of a country. If y satisfies the differential equation

$$\frac{dy}{dt} = ky \qquad k \text{ a constant}$$

then we say that the rate of change of y with respect to time t is proportional to y. If $k > 0$, y is increasing (growing); if $k = 0$, y is constant; and if $k < 0$, y is decreasing (declining or decaying). For each of the following differential equations, write a brief verbal description of the rate of change of y and discuss the possible increasing/decreasing properties of y.

$$\text{(A)} \quad \frac{dy}{dt} = k \qquad \text{(B)} \quad \frac{dy}{dt} = k(100 - y) \qquad \text{(C)} \quad \frac{dy}{dt} = ky(100 - y)$$

---

* In any problem involving separation of variables, you may assume that the equations

$$e^{\ln|f(y)|} = e^C g(x) \qquad \text{and} \qquad f(y) = Kg(x)$$

are equivalent (both C and K are arbitrary constants). Justification of this assumption involves properties of the absolute value and exponential functions.

■ EXPONENTIAL GROWTH

We now return to the study of exponential growth laws first begun in Section 11-3. This time we will place more emphasis on determining the relevant growth law for a particular application and on using separation of variables to solve the corresponding differential equation in each problem. We begin with the familiar exponential growth law.

Exponential Growth Law

If the rate of change with respect to time t of a quantity y is proportional to the amount present, then y satisfies the differential equation

$$\frac{dy}{dt} = ky$$

Exponential growth includes both the case where y is increasing and the case where y is decreasing, or decaying.

EXAMPLE 3 ➤ **Product Analysis**   A certain brand of mothballs evaporate at a rate proportional to their volume, losing half their volume every 4 weeks. If the volume of each mothball is initially 15 cubic centimeters and a mothball becomes ineffective when its volume reaches 1 cubic centimeter, how long will these mothballs be effective? Round answer to one decimal place.

Solution   The volume of each mothball is decaying at a rate proportional to its volume. If V is the volume of a mothball after t weeks, then

$$\frac{dV}{dt} = kV$$

Since the initial volume is 15 cubic centimeters, we know that $V(0) = 15$. After 4 weeks, the volume will be half the original volume, so $V(4) = 7.5$. Summarizing these requirements, we have the following exponential decay model:

$$\frac{dV}{dt} = kV$$

$$V(0) = 15 \qquad V(4) = 7.5$$

We want to determine the value of t that satisfies the equation $V(t) = 1$. First, we use separation of variables to find the general solution of the differential equation:

$$\frac{dV}{dt} = kV$$

$$\frac{1}{V}\frac{dV}{dt} = k$$

$$\int \frac{dV}{V} = \int k \, dt$$

$$\ln V = kt + C \qquad\qquad \textit{We can write } \ln V \textit{ in place of } \ln|V|, \textit{ since } V > 0.$$

$$V = e^{kt+C} = e^{C}e^{kt} = Ae^{kt} \qquad \textit{General solution}$$

where $A = e^{c}$ is a positive constant. Now we use the initial condition to determine the value of the constant A:

$$V(0) = Ae^{0} = A = 15$$

$$V(t) = 15e^{kt}$$

Substituting $t = 4$ in this last form for $V(t)$, we see that k must satisfy

$$15e^{4k} = 7.5$$

Graphing both sides of this equation and finding the intersection point (Fig. 1), we see that $k \approx -0.1733$. Thus, the particular solution is

$$V(t) = 15e^{-0.1733t}$$

To determine how long the mothballs will be effective, we solve the equation $V(t) = 1$ for t by graphing both sides and finding the intersection point (Fig. 2). Thus, the mothballs will be effective for approximately 15.6 weeks.

FIGURE 1
$y_1 = 15e^{4x}; y_2 = 7.5$
$k \approx 0.1733$

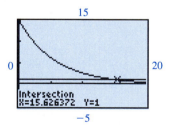

FIGURE 2
$y_1 = 15e^{-0.1733x}; y_2 = 1$
$t \approx 15.6$ weeks

◀

MATCHED PROBLEM 3 ▶ Repeat Example 3 if the mothballs lose half their volume every 5 weeks.   ◀

■ LIMITED GROWTH

In certain situations there is an upper limit (or a lower limit), say M, on the values a variable can assume. This limiting value leads to the *limited growth law*.

> ### Limited Growth Law
>
> If the rate of change with respect to time t of a quantity y is proportional to the difference between y and a limiting value M, then y satisfies the differential equation
>
> $$\frac{dy}{dt} = k(M - y)$$

When we speak of limited growth, we will include both the case where y increases and approaches M from below and the case where y decreases and approaches M from above.

EXAMPLE 4 ▶ **Sales Growth** The annual sales of a new company are expected to grow at a rate proportional to the difference between the sales and an upper limit of $20 million. The sales are 0 initially and $4 million for the second year of operation.

(A) What should the company expect the sales to be (to the nearest tenth of a million dollars) during the 10th year?

(B) In what year should the sales be expected to reach $15 million?

Solution If S is the annual sales (in millions of dollars) during year t, then the model for this problem is

$$\frac{dS}{dt} = k(20 - S) \tag{3}$$

$$S(0) = 0 \qquad S(2) = 4$$

This is a limited growth model. For part (A) we want to find $S(10)$, and for part (B) we want to solve $S(t) = 15$ for t. First, separating the variables in (3) and integrating both sides, we obtain

$$\int \frac{dS}{20 - S} = \int k \, dt$$

$$-\ln(20 - S) = kt + C \qquad \text{We can write } -\ln(20 - S) \text{ in place}$$
$$\ln(20 - S) = -kt - C \qquad \text{of } -\ln|20 - S|, \text{ since } 0 < S < 20.$$
$$20 - S = e^{-kt - C} = e^{-C}e^{-kt} = Ae^{-kt} \quad A = e^{-C}$$
$$S = 20 - Ae^{-kt} \qquad \text{General solution}$$

Applying the condition $S(0) = 0$, we have

$$S(0) = 20 - A = 0$$
$$A = 20$$

Thus,

$$S(t) = 20 - 20e^{-kt}$$

Applying the condition $S(2) = 4$ produces the following equation:

$$20 - 20e^{-2k} = 4$$

Solving this graphically (Fig. 3), we find that $k \approx 0.1116$, and the particular solution is (see Fig. 4)

$$S(t) = 20 - 20e^{-0.1116t}$$

(A) $S(10) = 20 - 20e^{-1.116} \approx \$13.4$ million
(B) Solving the equation $S(t) = 15$ graphically (Fig. 4), we find that the sales will exceed \$15 million during the 13th year.

FIGURE 3
$y_1 = 20 - 20e^{-2x}$; $y_2 = 4$
$k \approx 0.1116$

FIGURE 4
$y_1 = 20 - 20e^{-0.1116x}$; $y_2 = 15$
$t \approx 12.4$

◄

MATCHED PROBLEM 4 ➤ Repeat Example 4 if the sales during the 2nd year are \$3 million. ◄

■ LOGISTIC GROWTH

If a quantity first begins to grow exponentially but then starts to approach a limiting value, it is said to exhibit *logistic growth*. More formally, we have:

Logistic Growth Law

If the rate of change with respect to time t of a quantity y is proportional to both the amount present and the difference between y and a limiting value M, then y satisfies the differential equation

$$\frac{dy}{dt} = ky(M - y)$$

Theoretically, functions satisfying logistic growth laws can be increasing or decreasing, just as was the case for the exponential and limited growth laws. However, decreasing functions are seldom encountered in actual practice.

EXAMPLE 5 ➤ **Population Growth** In a study of ciliate protozoans, it has been shown that the rate of growth of the number of *Paramecium caudatum* in a medium with fixed volume is proportional to the product of the number present and the difference between an upper limit of 375 and the number present. Suppose the medium initially contains 25 paramecia. After 1 hour there are 125 paramecia. How many paramecia are present after 2 hours?

Solution If P is the number of paramecia in the medium at time t, then the model for this problem is the following logistic growth model:

$$\frac{dP}{dt} = kP(375 - P)$$

$$P(0) = 25 \qquad P(1) = 125$$

We want to find $P(2)$. First, we separate the variables and convert to an equation involving indefinite integrals:

$$\frac{1}{P(375 - P)}\frac{dP}{dt} = k$$

$$\int \frac{1}{P(375 - P)}\,dP = \int k\,dt \tag{4}$$

The integral on the left side of (4) can be evaluated either by using an algebraic identity* or by using formula 9 in Table II. We will use formula 9 with $u = P$, $a = 375$, and $b = -1$:

$$\int \frac{1}{u(a + bu)}\,du = \frac{1}{a}\ln\left|\frac{u}{a + bu}\right| \qquad \text{\color{blue}Formula 9}$$

$$\int \frac{1}{P(375 - P)}\,dP = \frac{1}{375}\ln\left|\frac{P}{375 - P}\right| \qquad \text{\color{blue}Since $0 < P < 375$, the absolute value signs can be omitted.}$$

$$= \frac{1}{375}\ln\left(\frac{P}{375 - P}\right)$$

Returning to equation (4), we have

$$\frac{1}{375}\ln\left(\frac{P}{375 - P}\right) = \int k\,dt = kt + D \qquad \text{\color{blue}D is a constant.}$$

$$\ln\left(\frac{P}{375 - P}\right) = 375kt + 375D$$

$$\frac{P}{375 - P} = e^{375kt + 375D} = e^{Bt}e^{C} \qquad \text{\color{blue}$B = 375k$, $C = 375D$}$$

---

* If you do not wish to use Table II to evaluate integrals, you can use the algebraic identity

$$\frac{1}{x(a - x)} = \frac{1}{a}\left[\frac{(a - x) + x}{x(a - x)}\right] = \frac{1}{a}\left(\frac{1}{x} + \frac{1}{a - x}\right)$$

to evaluate this integral and in all logistic growth problems in Exercise 14-2.

$$P = 375e^{Bt}e^C - Pe^{Bt}e^C$$

$$P(e^{Bt}e^C + 1) = 375e^{Bt}e^C$$

$$P = \frac{375e^{Bt}e^C}{e^{Bt}e^C + 1}$$ Multiply numerator and denominator by $e^{-Bt}e^{-C}$ and let $A = e^{-C}$.

$$= \frac{375}{1 + Ae^{-Bt}}$$ General solution

From the condition $P(0) = 25$, we get the equation

$$\frac{375}{1 + A} = 25$$

Figure 5 shows that $A = 14$. Substituting $A = 14$ in the general solution and applying the condition $P(1) = 125$ produces the following equation:

$$\frac{375}{1 + 14e^{-B}} = 125$$

From Figure 6, we see that $B \approx 1.9459$, and we have

$$P(t) = \frac{375}{1 + 14e^{-1.9459t}}$$ Particular solution

To determine the population after 2 hours, we evaluate $P(2)$ (Fig. 7) and find that there will be approximately 292 paramecia after 2 hours. The S-shaped curve in Figure 7 is typical of logistic growth functions. Notice that the upper limit of 375 is a horizontal asymptote.

FIGURE 5
$y_1 = \dfrac{375}{1 + x}$; $y_2 = 25$
$A = 14$

FIGURE 6
$y_1 = \dfrac{375}{1 + 14e^{-x}}$; $y_2 = 125$
$B \approx 1.9459$

FIGURE 7
$y_1 = \dfrac{375}{1 + 14e^{-1.9459x}}$; $y_2 = 375$
Logistic growth   ◀

MATCHED PROBLEM 5 ▶ Repeat Example 5 for 15 paramecia initially and 150 paramecia after 1 hour. ◀

Answers to Matched Problems

1. $y = -1/(x^4 + C)$
2. General solution: $y = 3 + K(2 + x^4)$; particular solution: $y = 5 + x^4$
3. 19.5 weeks
4. (A) \$11.13 million   (B) 17th year ($t \approx 17.06$ yr)
5. 343 paramecia

EXERCISE 14-2

A *In Problems 1–4, write a differential equation that describes the rate of change of the indicated quantity.*

1. The annual sales y of a company are increasing at the rate of $100,000 per year.

2. The annual sales y of a company are increasing at a rate proportional to the annual sales.

3. The fish population y in a lake is growing at a rate proportional to the difference between the population and an upper limit of 10,000 fish.

4. In a community of 100,000, the number of people y who have contracted an infectious disease is growing at a rate proportional to the product of the number of people who have contracted the disease and the number who have not.

In Problems 5–8, write a verbal description of the rate of change of the given quantity satisfying the indicated differential equation and discuss the increasing/decreasing properties of the quantity.

5. The annual sales y (in millions of dollars) of a company satisfy $dy/dt = 0.1y$ and $y(0) = 2$.

6. The annual sales y (in millions of dollars) of a company satisfy $dy/dt = 0.1(8 - y)$ and $y(0) = 2$.

7. The number of people y who have heard a rumor satisfies $dy/dt = 0.2y(5,000 - y)$ and $y(0) = 1$.

8. The amount of a drug y (in milliliters) in a patient's bloodstream satisfies $dy/dt = -0.2y$ and $y(0) = 5$.

In Problems 9–18, find the general solution for each differential equation. Then find the particular solution satisfying the initial condition.

9. $y' = 1$; $y(0) = 2$

10. $y' = 2x$; $y(0) = 4$

11. $y' = \dfrac{1}{x^{1/2}}$; $y(1) = -2$

12. $y' = 3x^{1/2}$; $y(0) = 7$

13. $y' = y$; $y(0) = 10$

14. $y' = y - 10$; $y(0) = 15$

15. $y' = 25 - y$; $y(0) = 5$

16. $y' = 3x^2 y$; $y(0) = \frac{1}{2}$

17. $y' = \dfrac{y}{x}$; $y(1) = 5$; $x > 0$

18. $y' = \dfrac{y}{x^2}$; $y(-1) = 2e$

B *In Problems 19–28, find the general solution for each differential equation. Then find the particular solution satisfying the initial condition.*

19. $y' = \dfrac{1}{y^2}$; $y(1) = 3$

20. $y' = \dfrac{x^2}{y^2}$; $y(0) = 2$

21. $y' = ye^x$; $y(0) = 3e$

22. $y' = -y^2 e^x$; $y(0) = \frac{1}{2}$

23. $y' = \dfrac{e^x}{e^y}$; $y(0) = \ln 2$

24. $y' = y^2(2x + 1)$; $y(0) = -\frac{1}{5}$

25. $y' = xy + x$; $y(0) = 2$

26. $y' = (2x + 4)(y - 3)$; $y(0) = 2$

27. $y' = (2 - y)^2 e^x$; $y(0) = 1$

28. $y' = \dfrac{x^{1/2}}{(y - 5)^2}$; $y(1) = 7$

In Problems 29–34, find the general solution for each differential equation. Do not attempt to find an explicit expression for the solution.

29. $y' = \dfrac{1 + x^2}{1 + y^2}$

30. $y' = \dfrac{6x - 9x^2}{2y + 4}$

31. $xyy' = (1 + x^2)(1 + y^2)$

32. $(xy^2 - x)y' = x^2 y - x^2$

33. $x^2 e^y y' = x^3 + x^3 e^y$

34. $y' = \dfrac{xe^x}{\ln y}$

C *In Problems 35–40, find an explicit expression for the particular solution for each differential equation.*

35. $xyy' = \ln x$; $y(1) = 1$

36. $y' = \dfrac{xe^{x^2}}{y}$; $y(0) = 2$

37. $xy' = x\sqrt{y} + 2\sqrt{y}$; $y(1) = 4$

38. $y' = x(x - 1)^{1/2}(y - 1)^{1/2}$; $y(1) = 1$

39. $yy' = xe^{-y^2}$; $y(0) = 1$

40. $yy' = x(1 + y^2)$; $y(0) = 1$

41. In many applications involving limited growth, the limiting value M is not known in advance and must be determined from data. Suppose that the rate of change with respect to time t of a quantity y is proportional to the difference between y and an unknown upper limit M, and that $y = 0$ when $t = 0$. Using the limited growth law, it follows that

$$y = M(1 - e^{-kt}) \qquad M > 0, k > 0$$

(A) Using the data given in the table, find two equations that M and k must satisfy, and solve each equation for M.

t	1	1.5
y	3	4

(B) Use a graphing utility and the equations from part (A) to approximate M to one decimal place.

42. Refer to Problem 41. Suppose that the rate of change with respect to time of a quantity y is proportional to the difference between y and an unknown lower limit M, and that

$y = 10$ when $t = 0$. Using the limited growth law, it follows that

$$y = M + (10 - M)e^{-kt} \qquad M > 0, k > 0$$

(A) Using the data given in the table, find two equations that M and k must satisfy, and solve each equation for M.

t	1	5
y	9	7

(B) Use a graphing utility and the equations from part (A) to approximate M to one decimal place.

 APPLICATIONS

Business & Economics

43. *Continuous compound interest.* If $5,000 is invested at 12% compounded continuously, how much will be in the account at the end of 10 years?

44. *Continuous compound interest.* If a sum of money is invested at 9% compounded continuously, how long will it take for the sum to double?

45. *Advertising.* A company is using radio advertising to introduce a new product to a community of 100,000 people. Suppose the rate at which people learn about the new product is proportional to the number who have not yet heard of it.
    (A) If no one is aware of the product at the start of the advertising campaign and after 7 days 20,000 people are aware of the product, how long will it take for 50,000 people to become aware of the product?

    (B) Suppose the company is dissatisfied with the result in part (A) and wants to decrease to 14 days the amount of time it takes for 50,000 people to become aware of their product. In order for this to happen, how many people must become aware of the product during the first 7 days?

46. *Advertising.* Prior to the beginning of an advertising campaign, 10% of the potential users of a certain brand are aware of the brand name. After the first week of the cam-

paign, 20% of the consumers are aware of the brand name. Assume that the percentage of informed consumers is growing at a rate proportional to the product of the percentage of informed consumers and the percentage of uninformed consumers.
    (A) What percentage of consumers will be aware of the brand name after 5 weeks of advertising?

    (B) Suppose the company is dissatisfied with the result in part (A) and increases the intensity of the advertising campaign so that 25% of the consumers are aware of the brand name after the first week. What effect does this have on the percentage of consumers that will be aware of the brand name after 5 weeks of advertising?

47. *Product analysis.* A company wishes to analyze a new room deodorizer. The active ingredient evaporates at a rate proportional to the amount present. Half of the ingredient evaporates in the first 30 days after the deodorizer is installed. If the deodorizer becomes ineffective after 90% of the active ingredient has evaporated, how long will one of these deodorizers remain effective?

48. *Natural resources.* According to the U.S. Department of Agriculture Forest Service, the total consumption of wood in the United States was 11.6 billion cubic feet in 1960 and 13.4 billion cubic feet in 1970. If the consumption is increasing at a rate proportional to the total consumption, how much wood will be used in the year 2000?

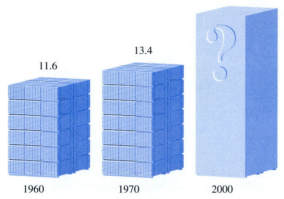

1960   1970   2000

49. *Sales growth.* The annual sales of a new company are expected to grow at a rate proportional to the difference between the sales and an upper limit of $5 million. If the sales are 0 initially and $1 million during the fourth year of operation, when will the sales reach $4 million?

50. *Sales analysis.* The annual sales of a company have declined from $8 million 2 years ago to $6 million today. If the annual sales continue to decline at a rate proportional to the difference between the annual sales and a lower limit of $3 million, find the annual sales 3 years from now.

51. *Sales analysis.* A new company has 0 sales initially, sales of $2 million during the first year, and sales of $5 million during the third year. If the annual sales S are assumed to be growing at a rate proportional to the difference between the sales and an unknown upper limit M, then by the limited growth law,

$$S(t) = M(1 - e^{-kt}) \qquad M > 0, k > 0$$

where t is time (in years) and $S(t)$ represents sales (in millions of dollars). Use approximation techniques to find k to one decimal place and M to the nearest million. (See Problem 41.)

52. *Sales analysis.* Refer to Problem 51. Approximate k to one decimal place and M to the nearest million if the sales during the third year are $4 million and all other information is unchanged.

Newton's law of cooling states that the rate of change of the temperature of an object is proportional to the difference between the temperature of the object and the temperature of the surrounding medium. Use this law to formulate models for Problems 53–56, and then solve using the techniques discussed in this section.

53. *Manufacturing.* As part of a manufacturing process, a metal bar is to be heated in an oven until its temperature reaches 500°F. The oven is maintained at a constant temperature of 800°F. Before it is placed in the oven, the temperature of the bar is 80°F. After 2 minutes in the oven, the temperature of the bar is 200°F. How long should the bar be left in the oven?

54. *Manufacturing.* The next step in the manufacturing process described in Problem 53 calls for the heated bar to be cooled in a vat of water until its temperature reaches 100°F. The water in the vat is maintained at a constant temperature of 50°F. If the temperature of the bar is 500°F when it is first placed in the water and the bar has cooled to 400°F after 5 minutes in the water, how long should the bar be left in the water?

Life Sciences

55. *Food preparation.* A pie is removed from an oven where the temperature is 325°F and placed in a freezer with a constant temperature of 25°F. After 1 hour in the freezer, the temperature of the pie is 225°F. What is the temperature of the pie after 4 hours in the freezer?

56. *Food preparation.* A roast is taken from a freezer where the temperature is 20°F and placed in an oven with a constant temperature of 350°F. After 1 hour in the oven, the temperature of the roast is 185°F. What is the temperature of the roast after 3 hours in the oven?

57. *Population growth.* A culture of bacteria is growing at a rate proportional to the number present. The culture initially contains 100 bacteria. After 1 hour there are 140 bacteria in the culture.
(A) How many bacteria will be present after 5 hours?
(B) When will the culture contain 1,000 bacteria?

58. *Population growth.* A culture of bacteria is growing in a medium that can support a maximum of 1,100 bacteria. The rate of change of the number of bacteria is proportional to the product of the number present and the difference between 1,100 and the number present. The culture initially contains 100 bacteria. After 1 hour there are 140 bacteria.
(A) How many bacteria are present after 5 hours?
(B) When will the culture contain 1,000 bacteria?

59. *Simple epidemic.* An influenza epidemic has spread throughout a community of 50,000 people at a rate proportional to the product of the number of people who have been infected and the number who have not been infected. If 100 individuals were infected initially and 500 were infected 10 days later:
(A) How many people will be infected after 20 days?
(B) When will half the community be infected?

60. *Ecology.* A fish population in a large lake is declining at a rate proportional to the difference between the population and a lower limit of 5,000 fish.

(A) If the population has declined from 15,000 fish 3 years ago to 10,000 today, find the population 6 years from now.

(B) Suppose the fish population 6 years from now turns out to be 8,000, indicating that the lake might be able to support more than the original lower limit of 5,000 fish. Assuming that the population is still proportional to the difference between the population and an unknown lower limit *M*, approximate *M* to the nearest hundred.

Social Sciences

61. *Sensory perception.* A person is subjected to a physical stimulus that has a measurable magnitude, but the intensity of the resulting sensation is difficult to measure. If *s* is the magnitude of the stimulus and *I(s)* is the intensity of sensation, experimental evidence suggests that

$$\frac{dI}{ds} = k\,\frac{I}{s}$$

for some constant *k*. Express *I* as a function of *s*.

62. *Learning.* The number of words per minute, *N*, a person can type increases with practice. Suppose the rate of change of *N* is proportional to the difference between *N* and an upper limit of 140. It is reasonable to assume that a beginner cannot type at all. Thus, $N = 0$ when $t = 0$. If a person can type 35 words per minute after 10 hours of practice:

(A) How many words per minute can that individual type after 20 hours of practice?

(B) How many hours must that individual practice to be able to type 105 words per minute?

63. *Rumor spread.* A rumor spreads through a population of 1,000 people at a rate proportional to the product of the number who have heard it and the number who have not heard it. If 5 people initiated a rumor and 10 people had heard it after 1 day:

(A) How many people will have heard the rumor after 7 days?

(B) How long will it take for 850 people to hear the rumor?

S E C T I O N   14-3   ■   # First-Order Linear Differential Equations

■ THE PRODUCT RULE REVISITED
■ SOLUTION OF FIRST-ORDER LINEAR DIFFERENTIAL EQUATIONS
■ APPLICATIONS

■ THE PRODUCT RULE REVISITED

The solution method we will discuss in this section is based in part on the product rule. Recall that if *F* and *S* are two differentiable functions, then

$$(FS)' = FS' + F'S$$

In Chapter 8, we started with *F* and *S* and used the product rule to find $(FS)'$. Now we want to use the rule in the reverse direction. That is, we want to be able to recognize when a sum of two terms is the derivative of a product.

explore – discuss 1

Match each sum on the left with the derivative of a product on the right, where y is an unknown differentiable function of x.

(A) $x^2y' + 2xy$ (1) $(e^{x^2}y)'$

(B) $e^{-2x}y' - 2e^{-2x}y$ (2) $\left(\dfrac{1}{x^2}\,y\right)'$

(C) $\dfrac{1}{x^2}\,y' - \dfrac{2}{x^3}\,y$ (3) $(e^{-2x}y)'$

(D) $e^{x^2}y' + 2xe^{x^2}y$ (4) $(x^2y)'$

In general, if $m(x)y' + n(x)y$ is the derivative of a product, how must $m(x)$ and $n(x)$ be related?

■ SOLUTION OF FIRST-ORDER LINEAR DIFFERENTIAL EQUATIONS

A differential equation that can be expressed in the form

$$y' + f(x)y = g(x)$$

is called a **first-order linear differential equation.**

For example,

$$y' + \frac{2}{x}\,y = x \tag{1}$$

is a first-order linear differential equation with $f(x) = 2/x$ and $g(x) = x$. This equation cannot be solved by the method of separation of variables. (Try to separate the variables to convince yourself that this is true.) Instead, we will change the form of the equation by multiplying both sides by x^2:

$$x^2y' + 2xy = x^3 \tag{2}$$

How was x^2 chosen? We will discuss that below. Let us first see how this choice leads to a solution of the problem.

Since $(x^2)' = 2x$, the left side of (2) is the derivative of a product:

$$x^2y' + 2xy = (x^2y)'$$

Thus, we can write equation (2) as

$$(x^2y)' = x^3$$

Now we can integrate both sides:

$$\int (x^2y)'\,dx = \int x^3\,dx$$

$$x^2y = \frac{x^4}{4} + C$$

Solving for y, we obtain the general solution

$$y = \frac{x^2}{4} + \frac{C}{x^2}$$

The function x^2, which we used to transform the original equation into one we could solve as illustrated, is called an *integrating factor*. It turns out that there is a specific formula for determining the integrating factor for any first-order linear differential equation. Furthermore, this integrating factor can then be used to find the solution of the differential equation, just as we used x^2 to find the solution to (1). The formula for the integrating factor and a step-by-step summary of the solution process are given in the box.

Solving First-Order Linear Differential Equations

Step 1. Write the equation in the **standard form:**

$$y' + f(x)y = g(x)$$

Step 2. Compute the **integrating factor:**

$$I(x) = e^{\int f(x)\,dx}$$

[When evaluating $\int f(x)\,dx$, choose 0 for the constant of integration.]

Step 3. Multiply both sides of the standard form by the integrating factor $I(x)$. The left side should now be in the form $[I(x)y]'$:

$$[I(x)y]' = I(x)g(x)$$

Step 4. Integrate both sides:

$$I(x)y = \int I(x)g(x)\,dx$$

[When evaluating $\int I(x)g(x)\,dx$, include an arbitrary constant of integration]:

Step 5. Solve for y to obtain the **general solution:**

$$y = \frac{1}{I(x)} \int I(x)g(x)\,dx$$

EXAMPLE 1 ➤ Using an Integrating Factor   Solve: $2xy' + y = 10x^2$

Solution   Step 1. Multiply both sides by $1/(2x)$ to obtain the standard form:

$$y' + \frac{1}{2x}y = 5x \quad f(x) = \frac{1}{2x} \text{ and } g(x) = 5x$$

Step 2. Find the integrating factor:

$$I(x) = e^{\int f(x)\, dx}$$
$$= e^{\int [1/(2x)]\, dx} \qquad \textit{Assume } x > 0 \textit{ and choose 0 for the constant of integration.}$$
$$= e^{(1/2)\ln x} \qquad \textit{r ln t = ln } t^r$$
$$= e^{\ln x^{1/2}} \qquad e^{\ln r} = r, r > 0$$
$$= x^{1/2} \qquad \textit{Integrating factor}$$

Step 3. Multiply both sides of the standard form by the integrating factor:

$$x^{1/2}\left(y' + \frac{1}{2x}\,y\right) = x^{1/2}(5x)$$
$$x^{1/2}y' + \tfrac{1}{2}x^{-1/2}y = 5x^{3/2} \qquad \textit{The left side should have the form } [I(x)y]'.$$
$$(x^{1/2}y)' = 5x^{3/2}$$

Step 4. Integrate both sides:

$$\int (x^{1/2}y)'\, dx = \int 5x^{3/2}\, dx \qquad \textit{Include an arbitrary constant of integration on the right side.}$$

$$x^{1/2}y = 2x^{5/2} + C$$

Step 5. Solve for y:

$$y = \frac{1}{x^{1/2}}(2x^{5/2} + C)$$

$$= 2x^2 + \frac{C}{x^{1/2}} \qquad \textit{General solution}$$

◀

MATCHED PROBLEM 1 ▶ Solve: $xy' + 3y = 4x$ ◀

In Example 1, notice that we assumed $x > 0$ to avoid introducing absolute value signs in the integrating factor. Many of the problems in this section will require the evaluation of expressions of the form $e^{\int [h'(x)/h(x)]\, dx}$. In order to avoid the complications caused by the introduction of absolute value signs, we will assume that the domain of $h(x)$ has been restricted so that $h(x) > 0$. This will simplify the solution process. We state the following familiar formulas for convenient reference:

Basic Formulas Involving the Natural Logarithm Function

If the domain of $h(x)$ is restricted so that $h(x) > 0$, then

$$\int \frac{h'(x)}{h(x)}\, dx = \ln h(x) \qquad \text{and} \qquad e^{\ln h(x)} = h(x)$$

If a first-order linear differential equation is written in standard form, then multiplying both sides of the equation by its integrating factor will always convert the left side of the equation into the derivative of $I(x)y$. Thus, it is possible to omit steps 3 and 4, and proceed directly to step 5. This approach is illustrated in the next example. You decide which is easier to use—the step-by-step procedure or the formula in step 5.

EXAMPLE 2 ➤   Using an Integrating Factor   Find the particular solution of the equation

$$y' + 2xy = 4x$$

satisfying the initial condition $y(0) = 5$.

Solution   Since the equation is already in standard form, we begin by finding the integrating factor:

$$I(x) = e^{\int 2x\, dx} = e^{x^2} \quad f(x) = 2x$$

Proceeding directly to step 5, we have

$$y = \frac{1}{I(x)} \int I(x)g(x)\, dx \quad g(x) = 4x$$

$$= \frac{1}{e^{x^2}} \int e^{x^2}(4x)\, dx$$

$$= e^{-x^2} \int 4xe^{x^2}\, dx \qquad \text{Let } u = x^2,\, du = 2x\, dx.$$
$$= e^{-x^2}(2e^{x^2} + C)$$
$$= 2 + Ce^{-x^2} \qquad \text{General solution}$$

Substituting $x = 0$ and $y = 5$ in the general solution, we have

$$5 = 2 + C$$
$$C = 3$$
$$y = 2 + 3e^{-x^2} \quad \text{Particular solution}$$ ◀

MATCHED PROBLEM 2 ➤   Find the particular solution of $y' + 3x^2y = 9x^2$ satisfying the initial condition $y(0) = 7$.   ◀

Caution

1. When integrating both sides of an equation such as

$$(x^{1/2}y)' = 5x^{3/2}$$

remember that

$$x^{1/2}y = \int 5x^{3/2}\, dx \neq 2x^{5/2} \quad \text{``}+ C\text{'' is missing in the antiderivative.}$$

Remember: A constant of integration must be included when evaluating $\int I(x)g(x)\, dx$**.** If you omit this constant, you will not be able to find the general solution of the differential equation. See step 4 of Example 1 for the correct procedure.

2. $\dfrac{1}{e^{x^2}} \displaystyle\int 4xe^{x^2}\, dx \neq \dfrac{1}{e^{x^3}} \displaystyle\int 4xe^{x^3}\, dx = \displaystyle\int 4x\, dx$

Just as a variable factor cannot be moved across the integral sign, **a variable factor outside the integral sign cannot be used to cancel a factor inside the integral sign.** See Example 2 for the correct procedure.

■ APPLICATIONS

If P is the initial amount deposited into an account earning $100r\%$ compounded continuously, and A is the amount in the account after t years, then we saw in Section 11-3 that A satisfies the exponential growth equation

$$\dfrac{dA}{dt} = rA \qquad A(0) = P$$

Now suppose that money is continuously withdrawn from this account at a rate of $\$m$ per year. Then the amount A in the account at time t must satisfy

$$\left(\begin{array}{c}\text{Rate of change} \\ \text{of amount } A\end{array}\right) = \left(\begin{array}{c}\text{Rate of growth} \\ \text{from continuous} \\ \text{compounding}\end{array}\right) - \left(\begin{array}{c}\text{Rate of} \\ \text{withdrawal}\end{array}\right)$$

$$\dfrac{dA}{dt} \qquad = \qquad rA \qquad - \qquad m$$

or

$$\dfrac{dA}{dt} - rA = -m \tag{3}$$

explore–discuss 2

How does equation (3) change if instead of making continuous withdrawals, money is continuously deposited into the account at the rate of $\$m$ per year?

▶ EXAMPLE 3 ▶ **Continuous Compound Interest** An initial deposit of $10,000 is made into an account earning 8% compounded continuously. Money is then continuously withdrawn at a constant rate of $1,000 a year until the account is depleted. Find the amount in the account at any time t. When will the amount be 0? What is the total amount withdrawn from this account?

Solution The amount A in the account at any time t must satisfy [see equation (3)]

$$\dfrac{dA}{dt} - 0.08A = -1,000 \qquad A(0) = 10,000$$

The integrating factor for this equation is

$$I(t) = e^{\int -0.08\, dt} = e^{-0.08t} \quad f(t) = -0.08$$

Multiplying both sides of the differential equation by $I(t)$ and following the step-by-step procedure, we have

$$e^{-0.08t}\frac{dA}{dt} - 0.08e^{-0.08t}A = -1{,}000e^{-0.08t}$$

$$(e^{-0.08t}A)' = -1{,}000e^{-0.08t}$$

$$e^{-0.08t}A = \int -1{,}000e^{-0.08t}\, dt$$

$$= 12{,}500e^{-0.08t} + C$$

$$A = 12{,}500 + Ce^{0.08t} \qquad \textit{General solution}$$

Applying the initial condition $A(0) = 10{,}000$ yields

$$A(0) = 12{,}500 + C = 10{,}000$$

$$C = -2{,}500$$

$$A(t) = 12{,}500 - 2{,}500e^{0.08t} \qquad \textit{Amount in the account at any time t}$$

To determine when the amount in the account is 0, we graph A and use a built-in root approximation routine (Fig. 1). Thus, the account is depleted after 20.118 years. Since money is being withdrawn at the rate of \$1,000 per year, the total amount withdrawn is

$$1{,}000(20.118) = \$20{,}118$$

FIGURE 1
$y_1 = 12{,}500 - 2{,}500e^{0.08x}$

MATCHED PROBLEM 3 ▶ Repeat Example 3 if the account earns 5% compounded continuously. ◀

▶ EXAMPLE 4 ▶ **Equilibrium Price** In economics, the supply S and the demand D for a commodity often can be considered as functions of both the price, $p(t)$, and the rate of change of the price, $p'(t)$. (Thus, S and D are ultimately functions of time t.) The **equilibrium price at time t** is the solution of the equation $S = D$. If $p(t)$ is the solution of this equation, then the **long-range equilibrium price** is

$$\bar{p} = \lim_{t \to \infty} p(t)$$

For example, if

$$D = 50 - 2p(t) + 2p'(t) \qquad S = 20 + 4p(t) + 5p'(t)$$

and $p(0) = 15$, then the equilibrium price at time t is the solution of the equation

$$50 - 2p(t) + 2p'(t) = 20 + 4p(t) + 5p'(t)$$

This simplifies to

$$p'(t) + 2p(t) = 10 \quad f(t) = 2 \text{ and } g(t) = 10$$

which is a first-order linear equation with integrating factor

$$I(t) = e^{\int 2\,dt} = e^{2t}$$

Proceeding directly to step 5, we have

$$p(t) = \frac{1}{e^{2t}} \int 10e^{2t}\,dt$$

$$= e^{-2t}(5e^{2t} + C)$$

$$p(t) = 5 + Ce^{-2t} \qquad \text{\color{blue}General solution}$$

$$p(0) = 5 + C = 15$$

$$C = 10$$

$$p(t) = 5 + 10e^{-2t} \qquad \text{\color{blue}Equilibrium price at time } t \text{ (see Fig. 2)}$$

$$\bar{p} = \lim_{t \to \infty}(5 + 10e^{-2t}) = 5 \quad \text{\color{blue}Long-range equilibrium price}$$

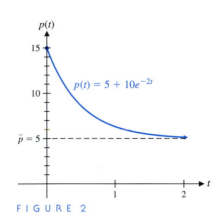

FIGURE 2

MATCHED PROBLEM 4 ▶ If $D = 70 + 2p(t) + 2p'(t)$, $S = 30 + 6p(t) + 3p'(t)$, and $p(0) = 25$, find the equilibrium price at time t, and find the long-range equilibrium price.

◀ EXAMPLE 5 ▶ **Pollution Control** A company has a 1,000 gallon holding tank which is used to control the release of pollutants into a sewage system. Initially, the tank contains 500 gallons of water. Each gallon of water contains 2 pounds of pollutants. Additional polluted water containing 5 pounds of pollutants per gallon is pumped into the tank at the rate of 100 gallons per hour and is thoroughly mixed with the water already present in the tank. At the same time, the uniformly mixed water in the tank is released into the sewage system at a rate of 50 gallons per hour (see Fig. 3). This process continues for 5 hours. At the end of this 5 hour period, determine:

(A) The total amount of pollutants in the tank.
(B) The rate (in pounds per gallon) at which pollutants are being released into the sewage system.

Water enters tank
at 100 gallons per hour

Water level
is rising

Water leaves tank
at 50 gallons per hour

FIGURE 3

Solution Let $p(t)$ be the total amount (in pounds) of pollutants in the tank t hours after this process begins. Since the tank initially contains 500 gallons of water and each gallon of water contains 2 pounds of pollutants,

$$p(0) = 2 \cdot 500 = 1{,}000 \quad \text{Initial amount of pollutants in the tank}$$

Since polluted water is entering and leaving the tank at different rates and with different concentrations of pollutants, the rate of change of the amount of pollutants in the tank will depend on the rate at which pollutants enter the tank and the rate at which they leave the tank:

$$\begin{pmatrix} \text{Rate of change} \\ \text{of pollutants} \end{pmatrix} = \begin{pmatrix} \text{Rate pollutants} \\ \text{enter the tank} \end{pmatrix} - \begin{pmatrix} \text{Rate pollutants} \\ \text{leave the tank} \end{pmatrix}$$

Finding expressions for the two rates on the right side of this equation will produce a differential equation involving $p'(t)$.

Since water containing 5 pounds of pollutants per gallon is entering the tank at a constant rate of 100 gallons per hour, pollutants are entering the tank at a constant rate of

$$\begin{pmatrix} \text{5 pounds} \\ \text{per gallon} \end{pmatrix} \times \begin{pmatrix} \text{100 gallons} \\ \text{per hour} \end{pmatrix} = 500 \text{ pounds per hour}$$

How fast are the pollutants leaving the tank? Since the total amount of pollutants in the tank is increasing, the rate at which the pollutants leave the tank will depend on the amount of pollutants in the tank at time t and the amount of water in the tank at time t. Since 100 gallons of water enter the tank each hour and only 50 gallons leave each hour, the amount of water in the tank increases at the rate of 50 gallons per hour. Thus, the amount of water in the tank at time t is

$$\begin{pmatrix} \text{Initial amount} \\ \text{of water} \end{pmatrix} + \begin{pmatrix} \text{Gallons} \\ \text{per hour} \end{pmatrix} \times \begin{pmatrix} \text{Number} \\ \text{of hours} \end{pmatrix}$$
$$500 \qquad + \qquad 50t$$

The amount of pollutants in each gallon of water at any time t is

$$\frac{p(t)}{500 + 50t} \quad \frac{\begin{pmatrix} \text{Total amount} \\ \text{of pollutants} \end{pmatrix}}{\begin{pmatrix} \text{Total amount} \\ \text{of water} \end{pmatrix}} = \text{Pollutants per gallon}$$

Since water is leaving the tank at the rate of 50 gallons per hour, the rate at which the pollutants are leaving the tank is

$$\frac{50p(t)}{500 + 50t} = \frac{p(t)}{10 + t}$$

Thus, the rate of change of $p(t)$ must satisfy

$$\begin{pmatrix} \text{Rate of change} \\ \text{of pollutants} \end{pmatrix} = \begin{pmatrix} \text{Rate pollutants} \\ \text{enter the tank} \end{pmatrix} - \begin{pmatrix} \text{Rate pollutants} \\ \text{leave the tank} \end{pmatrix}$$

$$p'(t) \qquad\qquad = \qquad\quad 500 \qquad\quad - \qquad\quad \frac{p(t)}{10 + t}$$

This gives the following model for this problem:

$$p'(t) = 500 - \frac{p(t)}{10 + t} \qquad p(0) = 1{,}000$$

or

$$p'(t) + \frac{1}{10 + t} p(t) = 500 \qquad\qquad \text{First-order linear equation with } f(t) = 1/(10 + t) \text{ and } g(t) = 500$$

$$I(t) = e^{\int dt/(10 + t)} = e^{\ln(10 + t)} = 10 + t \qquad \text{Integrating factor}$$

$$p(t) = \frac{1}{10 + t} \int 500(10 + t)\, dt \qquad\qquad \text{Proceeding directly to step 5}$$

$$= \frac{1}{10 + t} [250(10 + t)^2 + C]$$

$$p(t) = 250(10 + t) + \frac{C}{10 + t} \qquad\qquad \text{General solution}$$

$$1{,}000 = 250(10) + \frac{C}{10} \qquad\qquad \text{Initial condition: } p(0) = 1{,}000$$

$$\frac{C}{10} = -1{,}500$$

$$C = -15{,}000$$

$$p(t) = 250(10 + t) - \frac{15{,}000}{10 + t} \qquad\qquad \text{Particular solution (see Fig. 4)}$$

$p(t)$

$p(t) = 250(10 + t) - \dfrac{15{,}000}{10 + t}$

5,000

2,750

1,000

0   5   10 t

FIGURE 4

(A) To find the total amount of pollutants after 5 hours, we evaluate $p(5)$:

$$p(5) = 250(15) - \frac{15{,}000}{15} = 2{,}750 \text{ pounds}$$

(B) After 5 hours, the tank contains 750 gallons of water. The rate at which pollutants are being released into the sewage system is

$$\frac{2{,}750}{750} \approx 3.67 \text{ pounds per gallon}$$

MATCHED PROBLEM 5 ▶ Repeat Example 5 if water is released from the tank at the rate of 75 gallons per hour.

◀

Answers to Matched Problems

1. $I(x) = x^3$; $y = x + (C/x^3)$ 2. $I(x) = e^{x^3}$; $y = 3 + 4e^{-x^3}$
3. $A(t) = 20{,}000 - 10{,}000e^{0.05t}$; $A = 0$ when $t = (\ln 2)/0.05 \approx 13.863$ yr; total withdrawals = \$13,863
4. $p(t) = 10 + 15e^{-4t}$; $\bar{p} = 10$
5. $p(t) = 125(20 + t) - \dfrac{12{,}000{,}000}{(20 + t)^3}$ (A) 2,357 lb   (B) Approx. 3.77 lb/gal

---

EXERCISE   14-3

In all problems, assume $h(x) > 0$ whenever $\ln h(x)$ is involved.

A   *In Problems 1–8, replace each **?** with a function of x that will make the integrand equal to the derivative of a product, and then find the antiderivative. Choose 0 for the constant of integration.*

1. $\int (x^3 y' + \textbf{?}\, y)\, dx$
2. $\int (e^{2x} y' + \textbf{?}\, y)\, dx$
3. $\int (e^{-3x} y' + \textbf{?}\, y)\, dx$
4. $\int (x^{1/2} y' + \textbf{?}\, y)\, dx$
5. $\int (\textbf{?}\, y' + 4x^3 y)\, dx$
6. $\int (\textbf{?}\, y' + 0.2e^{0.2x} y)\, dx$
7. $\int (\textbf{?}\, y' - 0.5e^{-0.5x} y)\, dx$
8. $\int \left(\textbf{?}\, y' - \dfrac{1}{x^2}\, y \right) dx$

In Problems 9–20, find the integrating factor, the general solution, and the particular solution satisfying the given initial condition.

9. $y' + 2y = 4$; $y(0) = 1$
10. $y' - 3y = 3$; $y(0) = -1$
11. $y' + y = e^{-2x}$; $y(0) = 3$
12. $y' - 2y = e^{3x}$; $y(0) = 2$
13. $y' - y = 2e^x$; $y(0) = -4$
14. $y' + 4y = 3e^{-4x}$; $y(0) = 5$
15. $y' + y = 9x^2 e^{-x}$; $y(0) = 2$
16. $y' - 3y = 6\sqrt{x}\, e^{3x}$; $y(0) = -2$
17. $xy' + y = 2x$; $y(1) = 1$
18. $xy' - 3y = 4x$; $y(1) = 1$
19. $xy' + 2y = 10x^3$; $y(2) = 8$
20. $xy' - y = \dfrac{9}{x^2}$; $y(2) = 3$

B   *In Problems 21–30, find the integrating factor $I(x)$ for each equation, and then find the general solution.*

21. $y' + xy = 5x$
22. $y' - 2xy = 6x$
23. $y' - 2y = 4x$
24. $y' + y = x^2$
25. $xy' + y = xe^x$
26. $xy' + 2y = xe^{3x}$
27. $xy' + y = x \ln x$
28. $xy' - y = x \ln x$
29. $2xy' + 3y = 20x$
30. $2xy' - y = 6x^2$

In Problems 31–34, suppose that the indicated "solutions" were given to you by a student whom you are tutoring in this class.

(A) *How would you have the student check each solution?*

(B) *Is the solution right or wrong? If the solution is wrong, explain what is wrong and how it can be corrected.*

(C) *Show a correct solution for each incorrect solution, and check the result.*

31.   Equation: $y' + \dfrac{1}{x} y = (x + 1)^2$

Solution: $y = \dfrac{1}{x} \displaystyle\int x(x + 1)^2\, dx$

$= \dfrac{1}{3}(x + 1)^3 + C$

32.   Equation: $y' + \dfrac{2}{x} y = x - 1$

Solution: $y = \dfrac{1}{x^2} \displaystyle\int x^2(x - 1)\, dx$

$= \dfrac{1}{2}(x - 1)^2 + C$

33.   Equation: $y' + 3y = e^{-x}$

Solution: $y = \dfrac{1}{e^{3x}} \displaystyle\int e^{2x}\, dx$

$= \dfrac{1}{2} e^{-x}$

34.   Equation: $y' - 2y = e^x$

Solution: $y = e^{2x} \displaystyle\int e^{-x}\, dx$

$= -e^x$

C   *In Problems 35–40, find the general solution two ways. First, use an integrating factor and then use separation of variables.*

35. $y' = \dfrac{1 - y}{x}$
36. $y' = \dfrac{y + 2}{x + 1}$
37. $y' = \dfrac{2x + 2xy}{1 + x^2}$
38. $y' = \dfrac{4x + 2xy}{4 + x^2}$
39. $y' = 2x(y + 1)$
40. $y' = 3x^2(y + 2)$

41. Use an integrating factor to find the general solution of the unlimited growth model,

$$\frac{dy}{dt} = ky$$

[*Hint:* Remember, the antiderivative of the constant function 0 is an arbitrary constant C.]

42. Use an integrating factor to find the general solution of the limited growth model,

$$\frac{dy}{dt} = k(L - y)$$

 ▶ A P P L I C A T I O N S

Business & Economics

43. *Continuous compound interest.* An initial deposit of $20,000 is made into an account that earns 4% compounded continuously. Money is then withdrawn at a constant rate of $4,000 a year until the amount in the account is 0. Find the amount in the account at any time t. When is the amount 0? What is the total amount withdrawn from the account?

44. *Continuous compound interest.* An initial deposit of $50,000 is made into an account that earns 10% compounded continuously. Money is then withdrawn at a constant rate of $6,000 a year until the amount in the account is 0. Find the amount in the account at any time t. When is the amount 0? What is the total amount withdrawn from the account?

45. *Continuous compound interest.* An initial deposit of $P is made into an account that earns 5% compounded continuously. Money is then withdrawn at a constant rate of $1,500 a year. After 10 years of continuous withdrawals, the amount in the account is 0. Find the initial deposit P.

46. *Continuous compound interest.* An initial deposit of $P is made into an account that earns 8% compounded continuously. Money is then withdrawn at a constant rate of $3,000 a year. After 5 years of continuous withdrawals, the amount in the account is 0. Find the initial deposit P.

47. *Continuous compound interest.* An initial deposit of $7,000 is made into an account earning 8% compounded continuously. Thereafter, money is deposited into the account at a constant rate of $2,000 per year. Find the amount in this account at any time t. How much is in this account after 5 years?

48. *Continuous compound interest.* An initial deposit of $10,000 is made into an account earning 6% compounded continuously. Thereafter, money is deposited into the account at a constant rate of $6,000 per year. Find the amount in this account at any time t. How much is in this account after 10 years?

49. *Continuous compound interest.* An initial investment of $10,000 is made in a mutual fund. Additional investments are made at the rate of $1,000 per year. After 10 years the fund has a value of $35,000. What continuous compound rate of interest has this fund earned over this 10 year period? Express the answer as a percentage, correct to two decimal places.

50. *Continuous compound interest.* A trust fund is established by depositing $100,000 into an account. Withdrawals are made at the rate of $20,000 per year. After 5 years, the trust fund contains $20,000. What continuous compound rate of interest has this fund earned over this 5 year period? Express the answer as a percentage, correct to two decimal places.

51. *Supply–demand.* The supply S and demand D for a certain commodity satisfy the equations

$$S = 35 - 2p(t) + 3p'(t) \text{ and } D = 95 - 5p(t) + 2p'(t)$$

If $p(0) = 30$, find the equilibrium price at time t and the long-range equilibrium price.

52. *Supply–demand.* The supply S and demand D for a certain commodity satisfy the equations

$$S = 70 - 3p(t) + 2p'(t) \text{ and } D = 100 - 5p(t) + p'(t)$$

If $p(0) = 5$, find the equilibrium price at time t and the long-range equilibrium price.

Life Sciences

53. *Pollution.* A 1,000 gallon holding tank contains 200 gallons of water. Initially, each gallon of water in the tank contains 2 pounds of pollutants. Water containing 3 pounds of pollutants per gallon enters the tank at a rate of 75 gallons per hour, and the uniformly mixed water is released from the tank at a rate of 50 gallons per hour. How many pounds of pollutants are in the tank after 2 hours? At what rate (in pounds per gallon) are the pollutants being released after 2 hours? (See the figure at the top of the next page.)

Water enters tank at 75 gallons per hour

Water leaves tank at 50 gallons per hour

54. *Pollution.* Rework Problem 53 if water is entering the tank at the rate of 100 gallons per hour.

55. *Pollution.* Rework Problem 53 if water is entering the tank at the rate of 50 gallons per hour.

56. *Pollution.* Rework Problem 53 if water is entering the tank at the rate of 150 gallons per hour.

57. *Pollution.* Refer to Problem 53. When will the tank contain 1,000 pounds of pollutants? Round answer to one decimal place.

58. *Pollution.* Refer to Problem 56. When will the tank contain 1,500 pounds of pollutants? Round answer to one decimal place.

In a recent article in the College Mathematics Journal *(Jan. 1987, 18:1), Arthur Segal proposes the following model for weight loss or gain:*

$$\frac{dw}{dt} + 0.005w = \frac{1}{3,500} C$$

where $w(t)$ is a person's weight (in pounds) after t days of consuming exactly C calories per day. Use this model to solve Problems 59–62.

59. *Weight loss.* A person weighing 160 pounds goes on a diet of 2,100 calories per day. How much will this person weigh after 30 days on this diet? How long will it take this person to lose 10 pounds? Find $\lim_{t \to \infty} w(t)$ and interpret the results.

60. *Weight loss.* A person weighing 200 pounds goes on a diet of 2,800 calories per day. How much will this person weigh after 90 days on this diet? How long will it take this person to lose 25 pounds? Find $\lim_{t \to \infty} w(t)$ and interpret the results.

61. *Weight loss.* A person weighing 130 pounds would like to lose 5 pounds during a 30 day period. How many calories per day should this person consume to reach this goal?

62. *Weight loss.* A person weighing 175 pounds would like to lose 10 pounds during a 45 day period. How many calories per day should this person consume to reach this goal?

Social Sciences

In 1960, William K. Estes proposed the following model for measuring a student's performance in the classroom:

$$\frac{dk}{dt} + \ell k = \lambda \ell$$

where $k(t)$ is the student's knowledge after t weeks (expressed as a percentage and measured by performance on examinations), ℓ is a constant called the coefficient of learning and representing the student's ability to learn (expressed as a percentage and determined by IQ or some similar general intelligence predictor), and λ is a constant representing the fraction of available time the student spends performing helpful acts that should increase knowledge of the subject (studying, going to class, and so on). Use this model to solve Problems 63 and 64.

63. *Learning theory.* Students enrolled in a beginning Spanish class are given a pretest the first day of class in order to determine their initial knowledge of the subject. The results of the pretest, the coefficient of learning, and the fraction of time spent performing helpful acts for two students in the class are given in the table. Use the Estes model to predict the knowledge of each student after 6 weeks in the class.

	SCORE ON PRETEST	COEFFICIENT OF LEARNING	FRACTION OF HELPFUL ACTS
STUDENT A	0.1 (10%)	0.8	0.9
STUDENT B	0.4 (40%)	0.8	0.7

64. *Learning theory.* Refer to Problem 63. When will both students have the same level of knowledge? Round answer to one decimal place.

Euler's Method

In this chapter we developed techniques for finding the solutions to certain types of differential equations. But many differential equations do not have solutions that can be expressed in terms of a finite combination of familiar functions. In these cases it is necessary to approximate the solutions. Here, we will consider a method of approximating the solution to any **initial-value problem** of the form

$$y' = f(x, y) \qquad y(a) = C \qquad a \leqslant x \leqslant b \qquad (1)$$

To avoid confusion between the actual solution of an initial-value problem and an approximation to that solution, we will call $y = y(x)$ the **exact solution** of (1) if

$$y'(x) = f(x, y(x)) \quad \text{for } a \leqslant x \leqslant b \qquad \text{and} \qquad y(a) = C$$

Note that an exact solution is always a particular solution. General solutions involving arbitrary constants cannot be approximated numerically.

The technique for approximating solutions to initial-value problems that we will discuss is called **Euler's method.** This method is based on the following observation about the slope of a line: If m is the slope of the line segment through the points (x_0, y_0) and (x_1, y_1), and $\Delta x = x_1 - x_0$ is the change in x, then $y_1 = y_0 + m\Delta x$ (see Fig. 1).

FIGURE 1

To introduce Euler's method, we consider the following initial-value problem:

$$y' = y \qquad y(0) = 1 \qquad (2)$$

Of course, this is a very familiar equation whose exact solution is $y = e^x$. Studying this simple equation will enable us to compare the approximate solution with the exact solution. We want to approximate the solution to (2) for $0 \leqslant x \leqslant 1$. To do this we divide the interval $[0, 1]$ into five equal subintervals, each of length $\Delta x = 0.2$, and find approximate values of the solution at the end points of each of these intervals. Recall from our earlier discussion of slope fields that at a given point in the plane, the equation $y' = y$ gives the slope of the tangent line to a solution through that point (see Fig. 2).

FIGURE 2
Slope field for $y' = y$

FIGURE 3
Approximating $y(0.2)$

The slope at the given initial point $(0, y(0)) = (0, 1)$ is $y' = 1$. We draw a line with slope 1 starting at $(0, 1)$ and ending at $(0.2, y_1)$ (see Fig. 3). From our earlier observation about slope, we conclude that

$$y_1 = 1 + 0.2(1) = 1.2 \approx y(0.2)$$

(A) Find the slope at $(0.2, 1.2)$, draw a line segment with this slope starting at $(0.2, 1.2)$ and ending at $(0.4, y_2)$, and use y_2 to approximate $y(0.4)$.

(B) Repeat part (A) to find approximate values for $y(0.6)$, $y(0.8)$, and $y(1)$.

Figure 4 shows the line segments connecting the approximate values of $y(x)$, and Figure 5 compares this graph with the graph of the exact solution.

Exact solution: $y = e^x$

Approximate solution

FIGURE 4

FIGURE 5

The ideas introduced above are summarized in the following box.

Euler's Method

If $y(x)$ is the (exact) solution to the initial-value problem

$$y' = f(x, y) \qquad y(x_0) = y_0$$

$x_0, x_1, \ldots, x_n$ is the sequence of values defined by

$$x_k = x_{k-1} + \Delta x \qquad k = 1, 2, \ldots, n$$

and $y_1, y_2, \ldots, y_n$ is the sequence of values defined by

$$y_k = y_{k-1} + f(x_{k-1}, y_{k-1})\Delta x \qquad k = 1, 2, \ldots, n \qquad (3)$$

then

$$y(x_k) \approx y_k \qquad k = 1, 2, \ldots, n$$

(C) Formula (3) in Euler's method is a **recursion formula;** that is, each approximate value is calculated in terms of the preceding one. Verify that this formula gives the same approximate values you found in parts (A) and (B).

(D) Discuss the relationship between formula (3) and the slope of a line segment as illustrated in Figure 1.

(E) The distance between successive x values, Δx, is called the **step size.** We have used equal step sizes. Will this procedure work if the step sizes are not equal?

Now let's consider the following initial value problem:

$$y' = y + 7x - 3 \qquad y(0) = 1 \qquad (4)$$

(F) Find the exact solution of (4).

(G) Approximate the solution for $0 \leqslant x \leqslant 1$ using Euler's method with a step size of $\Delta x = 0.2$. Construct a table comparing the approximate values with the exact values.

(H) Repeat part (G) with a step size of $\Delta x = 0.1$.

(I) To compare the exact solution from part (F) and the approximate solution from part (G) on a graphing utility, enter the values of x_k and the approximate values of y_k as a data set in the statistics editor. Then enter the exact solution in the function editor, and graph the data set and the function in the same viewing window. Repeat for the approximate values from part (H). Computing and graphing approximate values is easily automated by writing a short program. Table 1 lists programs for several graphing calculators, and Figure 6 illustrates the use of one of these programs for the data set from part (G). In Figure 6A, note that the function $f(x, y)$ was entered in capital letters preceded by quotation marks. The TI-85 program also requires the quotation marks, but the function must be entered in lowercase letters.

TABLE 1

Graphing Calculator Programs for Euler's Method

TI-82/TI-83	TI-85
$\{0\} \to L_1$	$\{0\} \to xStat$
$\{0\} \to L_2$	$\{0\} \to yStat$
Input "X0? ",X	Input "X0: ",x
Input "Y0? ",Y	Input "Y0: ",y
Input "DELTA X? ",D	Input "DELTA x: ",D
Input "NBR OF STEPS? ",N	Input "NUMBER OF STEPS: ",N
Input "F(X,Y)? ",Y_0	Input "f(x,y): ",Y
FnOff 0	ST ▶ Eq(Y,y9)
$0 \to I$	FnOff 9
Lb1 A	$0 \to I$
$X \to L_1(I+1)$	Lb1 A
$Y \to L_2(I+1)$	$x \to x$ Stat(I+1)
$X+Y_0*D \to Y$	$y \to yStat(I+1)$
$X+D \to X$	$y+y9*D \to y$
$I+1 \to I$	$X+D \to x$
If I>N	$I+1 \to I$
Stop	If I>N
Goto A	Stop
	Goto A

(A) Input to program

(B) Output from program

(C) Graph of exact and
approximate solutions

FIGURE 6

Using the TI-82/TI-83 program for Euler's method

(J) The distance between the exact and approximate solutions represents the error in the approximation. Based on the tables and graphs you have constructed, what happens to this error each time you compute the next approximate value? What happens to this error if you decrease the step size?

So far we have considered only equations whose exact solution is known. Now let's consider an equation whose exact solution cannot be found by any of the methods we have discussed.

$$y' = \frac{1}{x} + \frac{1}{y} \qquad y(1) = 1$$

(K) Use Euler's method with a step size of 0.2 to approximate $y(2)$.
(L) Repeat part (K) with smaller step sizes until you believe you have approximated $y(2)$ correctly to one decimal place.

C H A P T E R   14   R E V I E W   ■   **Important Terms and Symbols**

14-1   *Basic Concepts.* Differential equation; first-order differential equation; order; slope field; solution; general solution; family of solutions; particular solution; initial condition; implicit solution; dynamically stable price; equilibrium price

14-2   *Separation of Variables.* Separation of variables; singular solution; exponential growth law; limited growth law; logistic growth law

$$\int f(y)\,dy = \int g(x)\,dx; \quad \frac{dy}{dt} = ky; \quad \frac{dy}{dt} = k(M - y);$$

$$\frac{dy}{dt} = ky(M - y)$$

14-3   *First-Order Linear Differential Equations.* First-order linear differential equation; standard form; integrating factor; general solution; equilibrium price at time t; long-range equilibrium price

$$y' + f(x)y = g(x); \quad I(x) = e^{\int f(x)\,dx};$$

$$y = \frac{1}{I(x)} \int I(x)g(x)\,dx$$

C H A P T E R   14   ■   **Review Exercise**

Work through all the problems in this chapter review and check your answers in the back of the book. Answers to all review problems are there along with section numbers in italics to indicate where each type of problem is discussed. Where weaknesses show up, review appropriate sections in the text.

A   *In Problems 1 and 2, show that the given function is the general solution of the indicated differential equation.*

1. $y = C\sqrt{x};\ 2xy' = y$ **2.** $y = 1 + Ce^{-x/3};\ 3y' + y = 1$

In Problems 3 and 4, determine which of the following slope fields is associated with the indicated differential equation. Briefly justify your answer.

(A)   (B)

3. $2xy' = y$ (From Problem 1)

4. $3y' + y = 1$ (From Problem 2)

In Problems 5 and 6, use the appropriate slope field—(A) or (B), given for Problems 3 and 4 (or a copy)—to graph the particular solutions obtained by letting C = −2, −1, 0, 1, and 2.

5. $y = C\sqrt{x}$ (From Problem 1)

6. $y = 1 + Ce^{-x/3}$ (From Problem 2)

In Problems 7 and 8, write a differential equation that describes the rate of change of the indicated quantity.

7. The price y of a product is decreasing at a rate proportional to the difference between the price and a lower limit of $5.

8. The price y of a product is increasing at a rate proportional to the price.

In Problems 9 and 10, write a verbal description of the rate of change of the given quantity satisfying the indicated differential equation and discuss the increasing/decreasing properties of the quantity.

9. The number of people y who have contracted an infectious disease satisfies $dy/dt = 0.2y(50,000 − y)$ and $y(0) = 1$.

10. The amount of radioactive material y (in grams) at the site of a nuclear accident satisfies $dy/dt = −0.001y$ and $y(0) = 100$.

In Problems 11–16, find the general solution.

11. $y' = -\dfrac{4y}{x}$

12. $y' = -\dfrac{4y}{x} + x$

13. $y' = 3x^2y^2$

14. $y' = 2y - e^x$

15. $y' = \dfrac{5}{x}y + x^6$

16. $y' = \dfrac{3 + y}{2 + x}, x > -2$

B *In Problems 17–24, find the particular solution that satisfies the given condition.*

17. $y' = 10 - y; y(0) = 0$

18. $y' + y = x; y(0) = 0$

19. $y' = 2ye^{-x}; y(0) = 1$

20. $y' = \dfrac{2x - y}{x + 4}; y(0) = 1$

21. $y' = \dfrac{x}{y + 4}; y(0) = 0$

22. $y' + \dfrac{2}{x}y = \ln x; y(1) = 2$

23. $yy' = \dfrac{x(1 + y^2)}{1 + x^2}; y(0) = 1$

24. $y' + 2xy = 2e^{-x^2}; y(0) = 1$

C

25. Solve the following differential equation two ways. First, use an integrating factor and then use separation of variables.

$$xy' - 4y = 8$$

26. Give an example of an equation that can be solved using an integrating factor but cannot be solved using separation of variables. Solve the equation.

27. Give an example of an equation that can be solved using separation of variables but cannot be solved using an integrating factor. Solve the equation.

28. Find the general solution of the following differential equation and then find the particular solutions satisfying the indicated initial conditions.

$$xy' - 5y = -10$$

(A) $y(0) = 2$ (B) $y(0) = 0$ (C) $y(1) = 1$

29. Find an explicit expression for the particular solution of the differential equation $yy' = x$ satisfying the initial condition $y(0) = 4$.

30. Given that $y = x^3 + Cx$ is the general solution of the differential equation $xy' - y = 2x^3$:

(A) In the same viewing window, graph the particular solutions obtained by letting $C = 0, 1, 2$, and 3.

(B) What do the graphs of the solutions for $C > 0$ have in common?

(C) In the same viewing window, graph the particular solutions obtained by letting $C = 0, -1, -2$, and −3.

(D) What do the graphs of the solutions for $C < 0$ have in common?

31. The rate of change with respect to time t of a quantity y is proportional to the difference between y and an unknown upper limit M, and $y = 0$ when $t = 0$. Using the limited growth law, it follows that

$$y = M(1 - e^{-kt}) \qquad M > 0, k > 0$$

(A) Using the data given in the table, find two equations that M and k must satisfy, and solve each equation for M.

t	2	5
y	4	7

(B) Use a graphing utility and the equations from part (A) to approximate M to one decimal place.

▶ APPLICATIONS

Business & Economics

32. *Depreciation.* A refrigerator costs $500 when it is new. The value of the refrigerator depreciates to $25 over a 20 year period. If the rate of change of the value is proportional to the value, find the value of the refrigerator 5 years after it was purchased.

33. *Sales growth.* The rate of growth of the annual sales of a new company is proportional to the difference between the annual sales S and an upper limit of $200,000. Assume that $S(0) = 0$.

(A) If the sales are $50,000 for the first year, how long will it take for the annual sales to reach $150,000? Round answer to the nearest year.

(B) Suppose the company is dissatisfied with the results in part (A) and wants the annual sales to reach $150,000 after just 3 years. In order for this to happen, what must the sales be for the first year? Round answer to the nearest thousand dollars.

34. *Price stability.* The supply S and the demand D for a certain commodity satisfy the equations

$$S = 100 + p + p' \quad \text{and} \quad D = 200 - p - p'$$

(A) Find the equilibrium price p at any time t.

(B) Find the long-range equilibrium price $\bar{p}$.

(C) Graph the particular solutions that satisfy the initial conditions $p(0) = 75$ and $p(0) = 25$.

(D) Discuss the long-term behavior of the price of this commodity.

35. *Continuous compound interest.* An initial deposit of $60,000 is made into an account that earns 5% compounded continuously. Money is then withdrawn at a constant rate of $5,000 per year until the amount in the account is 0. Find the amount in the account at any time t. When is the amount 0? What is the total amount withdrawn from the account?

36. *Continuous compound interest.* An initial investment of $15,000 is made in a mutual fund. Additional investments are made at the rate of $2,000 per year. After 10 years the fund has a value of $70,000. What continuous compound rate of interest has this fund earned over this 10 year period? Express the answer as a percentage, correct to two decimal places.

Life Sciences

37. *Crop yield.* The yield per acre, $y(t)$ (in bushels per acre), of a corn crop satisfies the equation

$$\frac{dy}{dt} = 100 + e^{-t} - y$$

If $y(0) = 0$, find y at any time t.

38. *Pollution.* A 1,000 gallon holding tank contains 100 gallons of unpolluted water. Water containing 2 pounds of pollutant per gallon is pumped into the tank at the rate of 75 gallons per hour. The uniformly mixed water is released from the tank at 50 gallons per hour.

(A) Find the total amount of pollutants in the tank after 2 hours. Round answer to one decimal place.

(B) When will the tank contain 700 pounds of pollutants? Round answer to one decimal place.

39. *Ecology.* A bird population on an island is declining at a rate proportional to the difference between the population and a lower limit of 200 birds.

(A) If the population has declined from 1,000 birds 5 years ago to 500 today, find the population 4 years from now.

(B) Suppose the bird population 4 years from now turns out to be 400, indicating that the island might be able to support more than the original lower limit of 200 birds. Assuming that the population is still proportional to the difference between the population and an unknown lower limit M, approximate M to the nearest integer.

Social Sciences

40. *Rumor spread.* A single individual starts a rumor in a community of 200 people. The rumor spreads at a rate proportional to the number of people who have not yet heard the rumor. After 2 days, 10 people have heard the rumor.

(A) How many people will have heard the rumor after 5 days?

(B) How long will it take for the rumor to spread to 100 people?

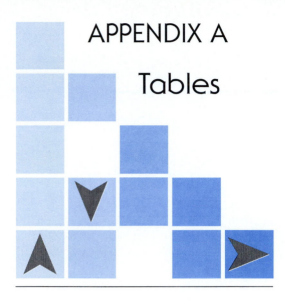

APPENDIX A

Tables

TABLE I
Basic Geometric Formulas

- ### 1. SIMILAR TRIANGLES

(A) Two triangles are similar if two angles of one triangle have the same measure as two angles of the other.

(B) If two triangles are similar, their corresponding sides are proportional:

$$\frac{a}{a'} = \frac{b}{b'} = \frac{c}{c'}$$

- ### 2. PYTHAGOREAN THEOREM

$$c^2 = a^2 + b^2$$

- ### 3. RECTANGLE

$A = ab$ Area

$P = 2a + 2b$ Perimeter

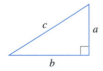

■ ## 4. PARALLELOGRAM

h = Height

$A = ah = ab \sin \theta$ Area

$P = 2a + 2b$ Perimeter

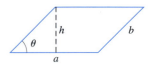

■ ## 5. TRIANGLE

h = Height

$A = \frac{1}{2}hc$ Area

$P = a + b + c$ Perimeter

$s = \frac{1}{2}(a + b + c)$ Semiperimeter

$A = \sqrt{s(s - a)(s - b)(s - c)}$ Area—Heron's formula

■ ## 6. TRAPEZOID

Base a is parallel to base b.

h = Height

$A = \frac{1}{2}(a + b)h$ Area

■ ## 7. CIRCLE

R = Radius

D = Diameter

$D = 2R$

$A = \pi R^2 = \frac{1}{4}\pi D^2$ Area

$C = 2\pi R = \pi D$ Circumference

$\dfrac{C}{D} = \pi$ For all circles

$\pi \approx 3.141\ 59$

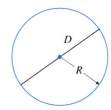

■ ## 8. RECTANGULAR SOLID

$V = abc$ Volume

$T = 2ab + 2ac + 2bc$ Total surface area

TABLE I   BASIC GEOMETRIC FORMULAS   **1057**

T A B L E I *(Continued)*

9. RIGHT CIRCULAR CYLINDER

R = Radius of base

h = Height

$V = \pi R^2 h$ Volume

$S = 2\pi R h$ Lateral surface area

$T = 2\pi R(R + h)$ Total surface area

10. RIGHT CIRCULAR CONE

R = Radius of base

h = Height

s = Slant height

$V = \frac{1}{3}\pi R^2 h$ Volume

$S = \pi R s = \pi R\sqrt{R^2 + h^2}$ Lateral surface area

$T = \pi R(R + s) = \pi R(R + \sqrt{R^2 + h^2})$ Total surface area

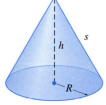

11. SPHERE

R = Radius

D = Diameter

$D = 2R$

$V = \frac{4}{3}\pi R^3 = \frac{1}{6}\pi D^3$ Volume

$S = 4\pi R^2 = \pi D^2$ Surface area

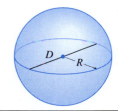

TABLE II
Integration Formulas

[*Note:* The constant of integration is omitted for each integral, but must be included in any particular application of a formula. The variable u is the variable of integration; all other symbols represent constants.]

■ INTEGRALS INVOLVING u^n

1. $\displaystyle\int u^n \, du = \frac{u^{n+1}}{n+1}, \quad n \neq -1$

2. $\displaystyle\int u^{-1} \, du = \int \frac{1}{u} \, du = \ln|u|$

■ INTEGRALS INVOLVING $a + bu, \quad a \neq 0$ AND $b \neq 0$

3. $\displaystyle\int \frac{1}{a+bu} \, du = \frac{1}{b} \ln|a+bu|$

4. $\displaystyle\int \frac{u}{a+bu} \, du = \frac{u}{b} - \frac{a}{b^2} \ln|a+bu|$

5. $\displaystyle\int \frac{u^2}{a+bu} \, du = \frac{(a+bu)^2}{2b^3} - \frac{2a(a+bu)}{b^3} + \frac{a^2}{b^3} \ln|a+bu|$

6. $\displaystyle\int \frac{u}{(a+bu)^2} \, du = \frac{1}{b^2} \left(\ln|a+bu| + \frac{a}{a+bu} \right)$

7. $\displaystyle\int \frac{u^2}{(a+bu)^2} \, du = \frac{(a+bu)}{b^3} - \frac{a^2}{b^3(a+bu)} - \frac{2a}{b^3} \ln|a+bu|$

8. $\displaystyle\int u(a+bu)^n \, du = \frac{(a+bu)^{n+2}}{(n+2)b^2} - \frac{a(a+bu)^{n+1}}{(n+1)b^2}, \quad n \neq -1, -2$

9. $\displaystyle\int \frac{1}{u(a+bu)} \, du = \frac{1}{a} \ln \left| \frac{u}{a+bu} \right|$

10. $\displaystyle\int \frac{1}{u^2(a+bu)} \, du = -\frac{1}{au} + \frac{b}{a^2} \ln \left| \frac{a+bu}{u} \right|$

11. $\displaystyle\int \frac{1}{u(a+bu)^2} \, du = \frac{1}{a(a+bu)} + \frac{1}{a^2} \ln \left| \frac{u}{a+bu} \right|$

12. $\displaystyle\int \frac{1}{u^2(a+bu)^2} \, du = -\frac{a+2bu}{a^2u(a+bu)} + \frac{2b}{a^3} \ln \left| \frac{a+bu}{u} \right|$

■ INTEGRALS INVOLVING $a^2 - u^2, \quad a > 0$

13. $\displaystyle\int \frac{1}{u^2-a^2} \, du = \frac{1}{2a} \ln \left| \frac{u-a}{u+a} \right|$

14. $\displaystyle\int \frac{1}{a^2-u^2} \, du = \frac{1}{2a} \ln \left| \frac{u+a}{u-a} \right|$

TABLE II INTEGRATION FORMULAS **1059**

TABLE II *(Continued)*

- **INTEGRALS INVOLVING** $(a + bu)$ **AND** $(c + du)$, $b \neq 0$, $d \neq 0$, **AND** $ad - bc \neq 0$

15. $\displaystyle\int \frac{1}{(a + bu)(c + du)}\, du = \frac{1}{ad - bc} \ln\left|\frac{c + du}{a + bu}\right|$

16. $\displaystyle\int \frac{u}{(a + bu)(c + du)}\, du = \frac{1}{ad - bc}\left(\frac{a}{b}\ln|a + bu| - \frac{c}{d}\ln|c + du|\right)$

17. $\displaystyle\int \frac{u^2}{(a + bu)(c + du)}\, du = \frac{1}{bd}u - \frac{1}{ad - bc}\left(\frac{a^2}{b^2}\ln|a + bu| - \frac{c^2}{d^2}\ln|c + du|\right)$

18. $\displaystyle\int \frac{1}{(a + bu)^2(c + du)}\, du = \frac{1}{ad - bc}\frac{1}{a + bu} + \frac{d}{(ad - bc)^2}\ln\left|\frac{c + du}{a + bu}\right|$

19. $\displaystyle\int \frac{u}{(a + bu)^2(c + du)}\, du = -\frac{a}{b(ad - bc)}\frac{1}{a + bu} - \frac{c}{(ad - bc)^2}\ln\left|\frac{c + du}{a + bu}\right|$

20. $\displaystyle\int \frac{a + bu}{c + du}\, du = \frac{bu}{d} + \frac{ad - bc}{d^2}\ln|c + du|$

- **INTEGRALS INVOLVING** $\sqrt{a + bu}$, $a \neq 0$ **AND** $b \neq 0$

21. $\displaystyle\int \sqrt{a + bu}\, du = \frac{2\sqrt{(a + bu)^3}}{3b}$

22. $\displaystyle\int u\sqrt{a + bu}\, du = \frac{2(3bu - 2a)}{15b^2}\sqrt{(a + bu)^3}$

23. $\displaystyle\int u^2\sqrt{a + bu}\, du = \frac{2(15b^2u^2 - 12abu + 8a^2)}{105b^3}\sqrt{(a + bu)^3}$

24. $\displaystyle\int \frac{1}{\sqrt{a + bu}}\, du = \frac{2\sqrt{a + bu}}{b}$

25. $\displaystyle\int \frac{u}{\sqrt{a + bu}}\, du = \frac{2(bu - 2a)}{3b^2}\sqrt{a + bu}$

26. $\displaystyle\int \frac{u^2}{\sqrt{a + bu}}\, du = \frac{2(3b^2u^2 - 4abu + 8a^2)}{15b^3}\sqrt{a + bu}$

27. $\displaystyle\int \frac{1}{u\sqrt{a + bu}}\, du = \frac{1}{\sqrt{a}}\ln\left|\frac{\sqrt{a + bu} - \sqrt{a}}{\sqrt{a + bu} + \sqrt{a}}\right|, \quad a > 0$

28. $\displaystyle\int \frac{1}{u^2\sqrt{a + bu}}\, du = -\frac{\sqrt{a + bu}}{au} - \frac{b}{2a\sqrt{a}}\ln\left|\frac{\sqrt{a + bu} - \sqrt{a}}{\sqrt{a + bu} + \sqrt{a}}\right|, \quad a > 0$

- **INTEGRALS INVOLVING** $\sqrt{a^2 - u^2}$, $a > 0$

29. $\displaystyle\int \frac{1}{u\sqrt{a^2 - u^2}}\, du = -\frac{1}{a}\ln\left|\frac{a + \sqrt{a^2 - u^2}}{u}\right|$

TABLE II *(Continued)*

30. $\displaystyle\int \frac{1}{u^2\sqrt{a^2 - u^2}}\,du = -\frac{\sqrt{a^2 - u^2}}{a^2 u}$

31. $\displaystyle\int \frac{\sqrt{a^2 - u^2}}{u}\,du = \sqrt{a^2 - u^2} - a\ln\left|\frac{a + \sqrt{a^2 - u^2}}{u}\right|$

■ INTEGRALS INVOLVING $\sqrt{u^2 + a^2}$, $a > 0$

32. $\displaystyle\int \sqrt{u^2 + a^2}\,du = \frac{1}{2}\left(u\sqrt{u^2 + a^2} + a^2\ln|u + \sqrt{u^2 + a^2}|\right)$

33. $\displaystyle\int u^2\sqrt{u^2 + a^2}\,du = \frac{1}{8}\left[u(2u^2 + a^2)\sqrt{u^2 + a^2} - a^4\ln|u + \sqrt{u^2 + a^2}|\right]$

34. $\displaystyle\int \frac{\sqrt{u^2 + a^2}}{u}\,du = \sqrt{u^2 + a^2} - a\ln\left|\frac{a + \sqrt{u^2 + a^2}}{u}\right|$

35. $\displaystyle\int \frac{\sqrt{u^2 + a^2}}{u^2}\,du = -\frac{\sqrt{u^2 + a^2}}{u} + \ln|u + \sqrt{u^2 + a^2}|$

36. $\displaystyle\int \frac{1}{\sqrt{u^2 + a^2}}\,du = \ln|u + \sqrt{u^2 + a^2}|$

37. $\displaystyle\int \frac{1}{u\sqrt{u^2 + a^2}}\,du = \frac{1}{a}\ln\left|\frac{u}{a + \sqrt{u^2 + a^2}}\right|$

38. $\displaystyle\int \frac{u^2}{\sqrt{u^2 + a^2}}\,du = \frac{1}{2}\left(u\sqrt{u^2 + a^2} - a^2\ln|u + \sqrt{u^2 + a^2}|\right)$

39. $\displaystyle\int \frac{1}{u^2\sqrt{u^2 + a^2}}\,du = -\frac{\sqrt{u^2 + a^2}}{a^2 u}$

■ INTEGRALS INVOLVING $\sqrt{u^2 - a^2}$, $a > 0$

40. $\displaystyle\int \sqrt{u^2 - a^2}\,du = \frac{1}{2}\left(u\sqrt{u^2 - a^2} - a^2\ln|u + \sqrt{u^2 - a^2}|\right)$

41. $\displaystyle\int u^2\sqrt{u^2 - a^2}\,du = \frac{1}{8}\left[u(2u^2 - a^2)\sqrt{u^2 - a^2} - a^4\ln|u + \sqrt{u^2 - a^2}|\right]$

42. $\displaystyle\int \frac{\sqrt{u^2 - a^2}}{u^2}\,du = -\frac{\sqrt{u^2 - a^2}}{u} + \ln|u + \sqrt{u^2 - a^2}|$

43. $\displaystyle\int \frac{1}{\sqrt{u^2 - a^2}}\,du = \ln|u + \sqrt{u^2 - a^2}|$

44. $\displaystyle\int \frac{u^2}{\sqrt{u^2 - a^2}}\,du = \frac{1}{2}\left(u\sqrt{u^2 - a^2} + a^2\ln|u + \sqrt{u^2 - a^2}|\right)$

TABLE II   INTEGRATION FORMULAS   **1061**

T A B L E II   *(Continued)*

45. $\displaystyle\int \frac{1}{u^2\sqrt{u^2-a^2}}\,du = \frac{\sqrt{u^2-a^2}}{a^2 u}$

■ INTEGRALS INVOLVING e^{au}, $a \neq 0$

46. $\displaystyle\int e^{au}\,du = \frac{e^{au}}{a}$

47. $\displaystyle\int u^n e^{au}\,du = \frac{u^n e^{au}}{a} - \frac{n}{a}\int u^{n-1} e^{au}\,du$

48. $\displaystyle\int \frac{1}{c + de^{au}}\,du = \frac{u}{c} - \frac{1}{ac}\ln|c + de^{au}|,\quad c \neq 0$

■ INTEGRALS INVOLVING $\ln u$

49. $\displaystyle\int \ln u\,du = u\ln u - u$

50. $\displaystyle\int \frac{\ln u}{u}\,du = \frac{1}{2}(\ln u)^2$

51. $\displaystyle\int u^n \ln u\,du = \frac{u^{n+1}}{n+1}\ln u - \frac{u^{n+1}}{(n+1)^2},\quad n \neq -1$

52. $\displaystyle\int (\ln u)^n\,du = u(\ln u)^n - n\int (\ln u)^{n-1}\,du$

■ INTEGRALS INVOLVING TRIGONOMETRIC FUNCTIONS OF au, $a \neq 0$

53. $\displaystyle\int \sin au\,du = -\frac{1}{a}\cos au$

54. $\displaystyle\int \cos au\,du = \frac{1}{a}\sin au$

55. $\displaystyle\int \tan au\,du = -\frac{1}{a}\ln|\cos au|$

56. $\displaystyle\int \cot au\,du = \frac{1}{a}\ln|\sin au|$

57. $\displaystyle\int \sec au\,du = \frac{1}{a}\ln|\sec au + \tan au|$

58. $\displaystyle\int \csc au\,du = \frac{1}{a}\ln|\csc au - \cot au|$

59. $\displaystyle\int (\sin au)^2\, du = \frac{u}{2} - \frac{1}{4a} \sin 2au$

60. $\displaystyle\int (\cos au)^2\, du = \frac{u}{2} + \frac{1}{4a} \sin 2au$

61. $\displaystyle\int (\sin au)^n\, du = -\frac{1}{an} (\sin au)^{n-1} \cos au + \frac{n-1}{n} \int (\sin au)^{n-2}\, du, \quad n \neq 0$

62. $\displaystyle\int (\cos au)^n\, du = \frac{1}{an} \sin au(\cos au)^{n-1} + \frac{n-1}{n} \int (\cos au)^{n-2}\, du, \quad n \neq 0$

ANSWERS

■ CHAPTER 1

Exercise 1-1

1.

x	y
-3	-5
-2	-4
-1	-3
0	-2
1	-1
2	0
3	1

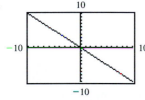

3.

x	y
-3	-5
-2	0
-1	3
0	4
1	3
2	0
3	-5

5.

x	y
-3	8
-2	3
-1	0
0	-1
1	0
2	3
3	8

7. (A) -7 (B) -8 (C) 0 (D) -4 (E) $-7, -2, 7$ (F) $-8, 0, 6$ **9.** Inside   **11.** Outside   **13.** Inside
15. Xmin $= -5$, Xmax $= 4$, Ymin $= -2$, Ymax $= 7$ **17.** Xmin $= -8$, Xmax $= 12$, Ymin $= -5$, Ymax $= 7$
19.

21.

23.

25.

x	0	2	4	6
y	9	1	1	9

27.

x	-5	-3	-1	1	3	5
y	4.5	-1.3	-3.9	-3.3	0.5	7.5

29.

x	− 3	− 1	1	3	5
y	10.5	− 2.5	4.5	7.5	− 17.5

31. (A) 2.25   (B) − 1.75
33. (A) 2.10   (B) 0.35 and 1.53
35. − 0.57, 0.76, 2.31
37. ± 6.32, ± 2.24
39. 2.23607

41. Mathematical coordinates of a geometric point in a rectangular coordinate system represent a unique point that has no dimensions. A pixel, on the other hand, is a rectangle with dimensions, and its screen coordinates, usually taken at the center of the rectangle, represent the infinitely many points within the rectangle.

43. A 0.89 in. square or a 2.40 in. square can be cut out. Dimensions for smaller square: $0.89 \times 9.22 \times 6.72$ in. Dimensions for larger square: $2.40 \times 6.20 \times 3.70$ in.

45. A 0.93 in. square or a 3.92 in. square can be cut out. Dimensions for smaller square: $0.93 \times 10.14 \times 10.61$ in. Dimensions for larger square: $3.92 \times 4.16 \times 6.12$ in.

47. 8,815 cases

Exercise 1-2

1. Function   **3.** Not a function   **5.** Function   **7.** Function   **9.** Not a function   **11.** Function   **13.** 4   **15.** − 5   **17.** − 6
19. − 2   **21.** − 12   **23.** − 1   **25.** − 6   **27.** 12   **29.** $\frac{3}{4}$ **31.** $y = 0$ **33.** $y = 4$ **35.** $x = -5, 0, 4$ **37.** $x = -6$
39. All real numbers   **41.** All real numbers except − 4   **43.** All real numbers except − 4 and 1   **45.** All real numbers, except − 3
47. $x \leqslant 7$ **49.** $x < 7$
51. $f(2) = 0$, and 0 is a number; therefore, $f(2)$ exists. On the other hand, $f(3)$ is not defined, since the denominator assumes a value of 0; therefore, we say that $f(3)$ does not exist.
53. $g(x) = 2x^3 - 5$ **55.** $G(x) = 2\sqrt{x} - x^2$ **57.** Function f multiplies the domain element by 2 and subtracts 3 from the result.
59. Function F multiplies the cube of the domain element by 3 and subtracts twice the square root of the domain element from the result.
61. A function with domain R **63.** A function with domain R **65.** Not a function; for example, when $x = 1$, $y = \pm 3$
67. A function with domain all real numbers except $x = 4$ **69.** Not a function; for example, when $x = 4$, $y = \pm 3$ **71.** 4   **73.** $h + 2$

75. $h - 1$ **77.** 4   **79.** $8a + 4h - 7$ **81.** $3a^2 + 3ah + h^2$ **83.** $\dfrac{1}{\sqrt{a + h} + \sqrt{a}}$ **85.** $P(w) = 2w + \dfrac{50}{w}, w > 0$

87. $A(l) = l(50 - l), 0 \leqslant l \leqslant 50$ **89.**

$p(8) = 71$ dollars per chip; $p(11) = 53$ dollars per chip

91. (A) $R(x) = xp(x) = x(119 - 6x)$ million dollars; Domain: $1 \leqslant x \leqslant 15$ (C)

(B)

x (million)	R(x) (million $)
1	113
3	303
6	498
9	585
12	564
15	435

93. (A) $P(x) = R(x) - C(x) = x(119 - 6x) - (234 + 23x)$ million dollars; Domain: $1 \leqslant x \leqslant 15$ (C)

(B)

x (million)	P(x) (million $)
1	−144
3	0
6	126
9	144
12	54
15	−144

95. (A) $V(x) = x(8 - 2x)(12 - 2x)$ (B) $0 \leqslant x \leqslant 4$ (C)

x	V(x)
1	60
2	64
3	36

(D)

97. $v = \dfrac{75 - w}{15 + w}$; 1.9032 cm/sec

Exercise 1-3

1. Domain: All real numbers; Range: All real numbers   **3.** Domain: All real numbers; Range: $(-\infty, 0]$ **5.** Domain: $[0, \infty)$; Range: $(-\infty, 0]$

7. Domain: All real numbers; Range: All real numbers   **9.**   **11.**

13.   **15.**   **17.**   **19.**

21. The graph of $g(x) = -|x + 3|$ is the graph of $y = |x|$ reflected in the x axis and shifted 3 units to the left.
23. The graph of $f(x) = (x - 4)^2 - 3$ is the graph of $y = x^2$ shifted 4 units to the right and 3 units down.
25. The graph of $f(x) = 7 - \sqrt{x}$ is the graph of $y = \sqrt{x}$ reflected in the x axis and shifted 7 units up.
27. The graph of $h(x) = -3|x|$ is the graph of $y = |x|$ reflected in the x axis and vertically expanded by a factor of 3.
29. The graph of the basic function $y = x^2$ is shifted 2 units to the left and 3 units down. Equation: $y = (x + 2)^2 - 3$.
31. The graph of the basic function $y = x^2$ is reflected in the x axis and shifted 3 units to the right and 2 units up.
Equation: $y = 2 - (x - 3)^2$.
33. The graph of the basic function $y = \sqrt{x}$ is reflected in the x axis and shifted 4 units up. Equation: $y = 4 - \sqrt{x}$.
35. The graph of the basic function $y = x^3$ is shifted 2 units to the left and 1 unit down. Equation: $y = (x + 2)^3 - 1$.
37. $g(x) = \sqrt{x - 2} - 3$ **39.** $g(x) = -|x + 3|$ **41.** $g(x) = -(x - 2)^3 - 1$

43. The basic graph of the function $y = |x|$ is reflected in the x axis and has a vertical contraction by a factor of 0.5. Equation: $y = -0.5|x|$.
45. The graph of the basic function $y = x^2$ is reflected in the x axis and vertically expanded by a factor of 2. Equation: $y = -2x^2$.
47. The graph of the basic function $y = \sqrt[3]{x}$ is reflected in the x axis and vertically expanded by a factor of 3. Equation: $y = -3\sqrt[3]{x}$.
49. (A) The graph of the basic function $y = \sqrt{x}$ is reflected in the x axis, vertically expanded by a factor of 4, and shifted up 115 units.
(B)

51. (A) The graph of the basic function $y = x^3$ is vertically compressed by a factor of 0.00048 and shifted right 500 units and up 60,000 units.
(B)

53. (A) The graph of the basic function $y = x$ is vertically expanded by a factor of 5.5 and shifted down 220 units.
(B)

55. (A) The graph of the basic function $y = \sqrt{x}$ is vertically expanded by a factor of 7.08.
(B)

Exercise 1-4

1. (D)   **3.** (C); slope is 0   **5.**

7.

9. Slope $= 2$; y intercept $= -3$
11. Slope $= -\frac{2}{3}$; y intercept $= 2$
13. $y = -2x + 4$
15. $y = -\frac{3}{5}x + 3$

17.

19.

21.

23. $y = -3x + 5$; $m = -3$
25. $y = -\frac{2}{3}x + 4$; $m = -\frac{2}{3}$

27. (A)

(B) x intercept: 3.5; y intercept: -4.2
(C) x intercept: 3.5; y intercept: -4.2
(D) $x > 3.5$ or $(3.5, \infty)$

29. $x = 3$; $y = -5$ **31.** $x = -1$; $y = -3$ **33.** $y + 1 = -3(x - 4)$; $y = -3x + 11$
35. $y + 5 = \frac{2}{3}(x + 6)$; $y = \frac{2}{3}x - 1$ **37.** $y = 0x - 5$, or $y = -5$ **39.** $\frac{1}{3}$ **41.** $-\frac{1}{5}$
43. Not defined   **45.** 0   **47.** $(y - 3) = \frac{1}{3}(x - 1)$; $x - 3y = -8$
49. $(y + 2) = -\frac{1}{5}(x + 5)$; $x + 5y = -15$ **51.** $x = 2$ **53.** $y = 3$
55. A linear function   **57.** Not a function   **59.** A constant function

61. (A)

(B) Varying C produces a family of parallel lines. This is verified by observing that varying C does not change the slope of the lines but changes the intercepts.

63. (A) $130; $220

(B)

(C) 6

65. (A) $C(x) = 180x + 200$

(B) $2,360

(C)

67. (A)

(B) $p(x) = -\frac{1}{60}x + 200$

(C) $p(3,000) = 150

69. $0.2x + 0.1y = 20$

71. (A) 64 g; 35 g

(B)

(C) $-\frac{1}{5}$

Exercise 1-5

1. (A), (C), (E), (F)   **3.** (A) m (B) g (C) f (D) n

5. (A) x intercepts: 1, 3; y intercept: -3
(B) Vertex: (2, 1)
(C) Maximum: 1
(D) Range: $y \le 1$ or $(-\infty, 1]$
(E) Increasing interval: $x \le 2$ or $(-\infty, 2]$
(F) Decreasing interval: $x \ge 2$ or $[2, \infty)$

7. (A) x intercepts: $-3, -1$; y intercept: 3
(B) Vertex: $(-2, -1)$
(C) Minimum: -1
(D) Range: $y \ge -1$ or $[-1, \infty)$
(E) Increasing interval: $x \ge -2$ or $[-2, \infty)$
(F) Decreasing interval: $x \le -2$ or $(-\infty, -2]$

9. (A) x intercepts: 1, 3; y intercept: -3
(B) Vertex: (2, 1)
(C) Maximum: 1
(D) Range: $y \le 1$ or $(-\infty, 1]$

11. (A) x intercepts: $-3, -1$; y intercept: 3   (B) Vertex: $(-2, -1)$ (C) Minimum: -1 (D) Range: $y \ge -1$ or $[-1, \infty)$

13. $y = -[x - (-2)]^2 + 5$ or $y = -(x + 2)^2 + 5$ **15.** $y = (x - 1)^2 - 3$

17.

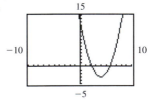

(A) x intercepts: 2.3, 5.7; y intercept: 13
(B) Vertex: $(4, -3)$
(C) Minimum: -3
(D) Range: $y \ge -3$ or $[-3, \infty)$

19.

(A) x intercepts: $-6.2, 0.2$; y intercept: 1
(B) Vertex: $(-3, 10)$
(C) Maximum: 10
(D) Range: $y \le 10$ or $(-\infty, 10]$

21.

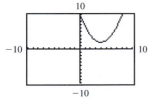

(A) x intercepts: none; y intercept: 10
(B) Vertex: (4, 2)
(C) Minimum: 2
(D) Range: $y \ge 2$ or $[2, \infty)$

23. $f(x) = (x - 4)^2 - 3$
(A) x intercepts: 2.3, 5.7; y intercept: 13
(B) Vertex: $(4, -3)$
(C) Minimum: -3
(D) Range: $y \ge -3$ or $[-3, \infty)$

25. $M(x) = -(x + 3)^2 + 10$
(A) x intercepts: $-6.2, 0.2$; y intercept: 1
(B) Vertex: $(-3, 10)$
(C) Maximum: 10
(D) Range: $y \le 10$ or $(-\infty, 10]$

27. $G(x) = 0.5(x - 4)^2 + 2$
(A) x intercepts: none; y intercept: 10
(B) Vertex: (4, 2)
(C) Minimum: 2
(D) Range: $y \ge 2$ or $[2, \infty)$

29. $x = -5.37, 0.37$ **31.** $-1.37 < x < 2.16$ **33.** $x \le -0.74$ or $x \ge 4.19$ **35.** The vertex of the parabola is on the x axis.

37.

39.

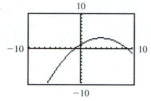

(A) *x* intercepts: − 3.08, 9.08; *y* intercept: − 7
(B) Vertex: (3, − 9.25)
(C) Minimum: − 9.25
(D) Range: $y \geq -9.25$ or $[-9.25, \infty)$

41. $g(x) = 0.25(x - 3)^2 - 9.25$
(A) *x* intercepts: − 3.08, 9.08; *y* intercept: − 7
(B) Vertex: (3, − 9.25)
(C) Minimum: − 9.25
(D) Range: $y \geq -9.25$ or $[-9.25, \infty)$

(A) *x* intercepts: − 1.10, 9.10; *y* intercept: 1.20
(B) Vertex: (4, 3.12)
(C) Maximum: 3.12
(D) Range: $y \leq 3.12$ or $(-\infty, 3.12]$

43. $f(x) = -0.12(x - 4)^2 + 3.12$
(A) *x* intercepts: − 1.10, 9.10; *y* intercept: 1.20
(B) Vertex: (4, 3.12)
(C) Maximum: 3.12
(D) Range: $y \leq 3.12$ or $(-\infty, 3.12]$

45.

47.

(A) 1.64, 7.61
(B) $1.64 < x < 7.61$
(C) $0 \leq x < 1.64$ or $7.61 < x \leq 10$

(A) 1.10, 5.57
(B) $1.10 < x < 5.57$
(C) $0 \leq x < 1.10$ or $5.57 < x \leq 8$

49. $f(x) = x^2 + 1$ and $g(x) = -(x - 4)^2 - 1$ are two examples. The graphs do not cross the *x* axis.

51. (A)

(B) 9.917 million chips (9,917,000 chips); 590.042 million dollars ($590,042,000)
(C) Same as B
(D) $59

53. (A)

(B) Break even at 3.000 million (3,000,000) and 13.000 million chip (13,000,000) production
(C) Loss: $1 \leq x < 3$ or $13 < x \leq 15$; Profit: $3 < x < 13$

55. (A) $x = 0.14$ cm
(B) $x = 0.14$ cm

Chapter 1 Review Exercise

1. *(1-1)*

2. *(1-1)*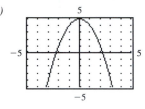

3. Xmin $= -9$; Xmax $= 8$; Ymin $= -10$; Ymax $= 5$ *(1-1)*

4. (A) Not a function (B) A function (C) A function
  (D) Not a function *(1-2)*

5. (A) -2 (B) -8 (C) 0 (D) Not defined *(1-2)*

6. (A) $y = 4$ (B) $x = 0$ (C) $y = 1$ (D) $x = -1$ or 1
  (E) $y = -2$ (F) $y = -5$ or 5 *(1-2)*

7. (A)   (B)   (C)   (D)

(1-3)

8. (A) n (B) g (C) m; slope is zero (D) f; slope is not defined *(1-4)*   **9.** $y = -\frac{2}{3}x + 6$ *(1-4)*

10. Vertical line: $x = -6$; Horizontal line: $y = 5$ *(1-4)*   **12.** (B), (C), (D), (F) *(1-5)*

11. x intercept $= 9$; y intercept $= -6$; Slope $= \frac{2}{3}$ *(1-4)*   **13.** (A) g (B) m (C) n (D) f *(1-3, 1-5)*

14. (A) x intercepts: -4, 0; y intercept: 0
  (B) Vertex: $(-2, -4)$
  (C) Minimum: -4
  (D) Range: $y \geq -4$ or $[-4, \infty)$
  (E) Increasing on $[-2, \infty)$
  (F) Decreasing on $(-\infty, -2]$ *(1-5)*

15. Linear functions: (A), (C), (E), (F); Constant function: (D) *(1-4)*

16. (A) All real numbers except $x = -2$ and 3
  (B) $x < 5$ *(1-2)*

17. Function g multiplies a domain element by 2 and then subtracts 3 times the square root of the domain element from the result. *(1-2)*

18. The graph of $x = -3$ is a vertical line with x intercept -3, and the graph of $y = 2$ is a horizontal line with y intercept 2. *(1-4)*

19. *(1-1)*

20. $x = -4.52, 1.43, 3.08$ *(1-1)*

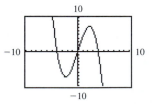

21. -2 *(1-2)*   **22.** $2a + h - 3$ *(1-2)*

23. The graph of function m is the graph of $y = |x|$ reflected in the x axis and shifted to the right 4 units. *(1-3)*

24. The graph of function g is the graph of $y = x^3$ vertically contracted by a factor of 0.3 and shifted up 3 units. *(1-3)*

25. The graph of $y = x^2$ is vertically expanded by a factor of 2, reflected in the x axis, and shifted to the left 3 units. Equation: $y = -2(x + 3)^2$. *(1-3)*

26. $f(x) = 2\sqrt{x+3} - 1$ *(1-3)*   **27.** (A) $y = -\frac{2}{3}x$ (B) $y = 3$ *(1-4)*   **28.** (A) $3x + 2y = 1$ (B) $y = 5$ (C) $x = -2$ *(1-4)*
29. $y = -(x-4)^2 + 3$ *(1-3, 1-5)*

30.

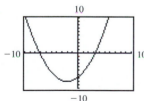

(A) x intercepts: -0.4, 8.4; y intercept: 1.2
(B) Vertex: $(4.0, 7.6)$
(C) Maximum: 7.6
(D) Range: $x \leqslant 7.6$ or $(-\infty, 7.6]$ *(1-5)*

31. $f(x) = -0.4(x-4)^2 + 7.6$ (A) x intercepts: -0.4, 8.4; y intercept: 1.2 (B) Vertex: $(4.0, 7.6)$
(C) Maximum: 7.6 (D) Range: $x \leqslant 7.6$ or $(-\infty, 7.6]$ *(1-5)*
32. $-0.74 < x < 4.19$ *(1-5)*
33. The graph of $y = \sqrt[3]{x}$ is vertically expanded by a factor of 2, reflected in the x axis, and shifted 1 unit left and 1 unit down.
Equation: $y = -2\sqrt[3]{x+1} - 1$. *(1-3)*
34. The graphs appear to be perpendicular to each other. (It can be shown that if the slopes of two slant lines are the negative reciprocals of each other, then the two lines are perpendicular.) *(1-4)*
35. (A) $\dfrac{1}{\sqrt{x+h}+\sqrt{x}}$ (B) $\dfrac{-1}{x(x+h)}$ *(1-3)*
36.

(A) x intercepts: -7.2, 3.2; y intercept: -6.9
(B) Vertex: $(-2, -8.1)$ (C) Minimum: -8.1
(D) Range: $x \geqslant -8.1$ or $[-8.1, \infty)$
(E) Decreasing: $(-\infty, -2]$; Increasing: $[-2, \infty)$ *(1-5)*

37. $G(x) = 0.3(x+2)^2 - 8.1$ (A) x intercepts: -7.2, 3.2; y intercept: -6.9 (B) Vertex: $(-2, -8.1)$ (C) Minimum: -8.1
(D) Range: $x \geqslant -8.1$ or $[-8.1, \infty)$ (E) Decreasing: $(-\infty, -2]$; Increasing: $[-2, \infty)$ *(1-5)*
38. 1.4422 *(1-1)*   **39.** (A) $x = 2.39$ (B) $x < 2.39$ *(1-1)*
40. (A) $V(t) = -1,250t + 12,000, \ 0 \leqslant t \leqslant 8$

(B) $V(5) = \$5,750$ *(1-4)*

41. (A)

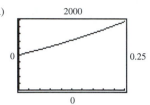

(B) $r = 0.1447$ or 14.47% compounded annually *(1-1, 1-3)*

42. (A) $R = 1.6C$
(B) $192
(C) $110
(D) 1.6; the slope indicates the change in retail price per unit change in cost. *(1-4)*

43. (A) $C = 84,000 + 15x$; $R = 50x$

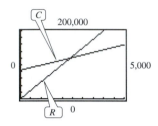

(B) $R = C$ at $x = 2,400$ units; $R < C$ for $0 \le x < 2,400$;
$R > C$ for $x > 2,400$
(C) $R = C$ at $x = 2,400$ units; $R < C$ for $0 \le x < 2,400$;
$R > C$ for $x > 2,400$ *(1-4)*

44. (A)

(B) $R = C$ for $x = 4.686$ thousand units (4,686 units) and for $x = 27.314$ thousand units (27,314 units); $R < C$ for $1 \le x < 4.686$ or $27.314 < x \le 40$; $R > C$ for $4.686 < x < 27.314$.
(C) Maximum revenue is 500 thousand dollars ($500,000). This occurs at an output of 20 thousand units (20,000 units). At this output, the wholesale price is $p(20) = \$25$. *(1-4, 1-5)*

45. (A) $P(x) = R(x) - C(x) = x(50 - 1.25x) - (160 + 10x)$

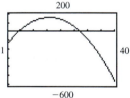

(B) $P = 0$ for $x = 4.686$ thousand units (4,686 units) and for $x = 27.314$ thousand units (27,314 units); $P < 0$ for $1 \le x < 4.686$ or $27.314 < x \le 40$; $P > 0$ for $4.686 < x < 27.314$.
(C) Maximum profit is 160 thousand dollars ($160,000). This occurs at an output of 16 thousand units (16,000 units). At this output, the wholesale price is $p(16) = \$30$. *(1-5)*

46. (A) $A(x) = -\frac{3}{2}x^2 + 420x$
(B) Domain: $0 \le x \le 280$
(C) Maximum combined area is 29,400 ft. This occurs for $x = 140$ ft and $y = 105$ ft. *(1-5)*

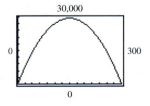

47. (A) $P(x) = 15x + 20$
(B) 95
(C)

(D) 15 *(1-5)*

48. (A) 1 lb; 3 lb
(B)

(C) $\frac{1}{30}$ *(1-5)*

■ CHAPTER 2

Exercise 2-1

1. (A) 2 (B) 1 (C) 2 (D) 0 (E) 1 (F) 1   **3.** (A) 5 (B) 4 (C) 5 (D) 1 (E) 1 (F) 1
5. (A) 6 (B) 5 (C) 6 (D) 0 (E) 1 (F) 1   **7.** (A) 3 (B) 4 (C) Negative   **9.** (A) 4 (B) 5 (C) Negative
11. (A) 0 (B) 1 (C) Negative   **13.** (A) 5 (B) 6 (C) Positive
15. (A) x intercept: -2; y intercept: -1 (B) Domain: all real numbers except 2 (C) Vertical asymptote: $x = 2$; horizontal asymptote: $y = 1$
(D)   (E)

17. (A) x intercept: 0; y intercept: $\tilde{0}$ (B) Domain: all real numbers except -2 (C) Vertical asymptote: $x = -2$; horizontal asymptote: $y = 3$
(D)   (E)

19. (A) x intercept: 2; y intercept: -1 (B) Domain: all real numbers except 4 (C) Vertical asymptote: $x = 4$; horizontal asymptote: $y = -2$
(D)   (E)

21. The graph will look more and more like the graph of $y = 2x^4$.   **23.** The graph will look more and more like the graph of $y = -x^5$.
25. (A)

$y = 2x^4$

$y = 2x^4 - 5x^2 + x + 2$

(B)

$y = 2x^4$

$y = 2x^4 - 5x^2 + x + 2$

27. (A)

$$y = -x^5 \qquad\qquad y = -x^5 + 4x^3 - 4x + 1$$

(B)

$$y = -x^5 \qquad\qquad y = -x^5 + 4x^3 - 4x + 1$$

29. (A) Maximum number of x intercepts is 3; maximum number of turning points is 2.

(B) x intercept: $x = -1.77$; turning points at $x = -0.82, 0.82$

31. (A) Maximum number of x intercepts is 4; maximum number of turning points is 3.

(B) x intercepts: $x = -0.33, 1.54$; turning point at $x = 0.91$

33. (A) x intercept: 0; y intercept: 0

(B) Vertical asymptotes: $x = -2, x = 3$; Horizontal asymptote: $y = 2$

(C)

(D)

35. (A) x intercept: $\pm\sqrt{3}$; y intercept: $-\frac{2}{3}$.

(B) Vertical asymptotes: $x = -3, x = 3$; Horizontal asymptote: $y = -2$

(C)

(D)

37. (A) x intercept: 0; y intercept: 0   (B) Vertical asymptotes: $x = -3, x = 2$; Horizontal asymptote: $y = 0$

(C)

(D)

39. $f(x) = x^2 - x - 2$ **41.** $f(x) = 4x - x^3$

43. (A) $C(x) = 180x + 200$

(B) $\overline{C}(x) = \dfrac{180x + 200}{x}$

(C) $\overline{C}(x)$

(D) \$180 per board

45. (A) $\overline{C}(n) = \dfrac{2{,}500 + 175n + 25n^2}{n}$

(B) 10 yr; \$675.00 per yr

47. (A) $\overline{C}(x) = \dfrac{0.00048(x - 500)^3 + 60{,}000}{x}$

(B)

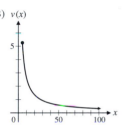

(C) 750 cases per month; \$90 per case

49. (A) 0.06 cm/sec   (B) $v(x)$

Exercise 2-2

1.

3.

5.

7.

9.

11.

13. 4^{6xy} **15.** e **17.** $8e^{3.6t}$

19. The graph of g is a reflection of the graph f in the x axis.

21. The graph of g is the graph of f shifted 1 unit to the left.

23. The graph of g is the graph of f shifted 1 unit up.

25. The graph of g is the graph of f vertically expanded by a factor of 2 and shifted to the left 2 units.

27.

29.

31.

33.

35.

37. The top curve belongs to f; the bottom curve belongs to g.
39. The top curve belongs to g; the bottom curve belongs to f.
41. $x = 1$ **43.** $x = -1, 6$ **45.** $x = 3$ **47.** $x = 3$ **49.** $x = -3, 0$

51.

53.

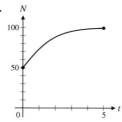

55. (A) \$2,633.56   **57.** (A) \$11,871.65   **59.** \$9,217
     (B) \$7,079.54   (B) \$20,427.93

61. (A) \$10,850.88   **63.** \$28,847.49   **65.** N approaches 2 as t increases without bound.   **67.** (A) 10%
     (B) \$10,838.29   (B) 1%

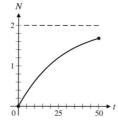

69. (A) $N = 40,000e^{0.21t}$ **71.** (A) $P = 5.7e^{0.0114t}$
     (B) 215,000; 613,000   (B) 6.8 billion; 8.5 billion
     (C)   (C)

Exercise 2-3

1. $27 = 3^3$ **3.** $10^0 = 1$ **5.** $8 = 4^{3/2}$ **7.** $\log_7 49 = 2$ **9.** $\log_4 8 = \frac{3}{2}$ **11.** $\log_b A = u$ **13.** 0 **15.** 1 **17.** 1 **19.** 3
21. -3 **23.** 3 **25.** $\log_b P - \log_b Q$ **27.** $5 \log_b L$ **29.** $\log_b p - \log_b q - \log_b r - \log_b s$ **31.** $x = 9$ **33.** $y = 2$ **35.** $b = 10$
37. $x = 2$ **39.** $y = -2$ **41.** $b = 100$ **43.** $5 \log_b x - 3 \log_b y$ **45.** $\frac{1}{3} \log_b N$ **47.** $2 \log_b x + \frac{1}{3} \log_b y$ **49.** $\log_b 50 - 0.2t \log_b 2$
51. $\log_b P + t \log_b(1 + r)$ **53.** $\log_e 100 - 0.01t$ **55.** $x = 2$ **57.** $x = 8$ **59.** $x = 7$ **61.** No solution
63.

65. The graph of $y = \log_2(x - 2)$ is the graph of $y = \log_2 x$ shifted to the right 2 units.
67. Domain: $(-1, \infty)$; Range: All real numbers
69. (A) 3.547 43 (B) -2.160 32 (C) 5.626 29 (D) -3.197 04
71. (A) 13.4431 (B) 0.0089 (C) 16.0595 (D) 0.1514
73. 1.0792 **75.** 1.4595 **77.** 30.6589 **79.** 18.3559

81. Increasing: $(0, \infty)$

83. Decreasing: $(0, 1]$
Increasing: $[1, \infty)$

85. Increasing: $(-2, \infty)$

87. Increasing: $(0, \infty)$

89. The calculator interprets log 13/7 as $\dfrac{\log 13}{7}$, not as $\log \frac{13}{7}$. **91.** Because $b^0 = 1$ for any permissible base b ($b > 0$, $b \neq 1$).

93. $y = c \cdot 10^{0.8x}$ **95.** $x > \sqrt{x} > \ln x$ for $1 < x \leq 16$ **97.** 12 yr **99.** 9.87 yr; 9.80 yr **101.** 8.664% compounded continuously
105. 901 yr

Exercise 2-4

1. (A)

(B) 9 million
3. (A)

(B) $5.4 million

5. (A)   (B) $\bar{x} = 195; \bar{p} = \$10.09$

7. (A)

(B) 0.233

9. (A)

(B) 91 mph

Chapter 2   Review Exercise

1. $v = \ln u$ *(2-3)*   **2.** $y = \log x$ *(2-3)*   **3.** $M = e^N$ *(2-3)*   **4.** $u = 10^v$ *(2-3)*   **5.** 5^{2x} *(2-2)*   **6.** e^{2u^2} *(2-2)*   **7.** $x = 9$ *(2-3)*   **8.** $x = 6$ *(2-3)*
9. $x = 4$ *(2-3)*   **10.** $x = 2.157$ *(2-3)*   **11.** $x = 13.128$ *(2-3)*   **12.** $x = 1,273.503$ *(2-3)*   **13.** $x = 0.318$ *(2-3)*
14. (A) 3   (B) 2   (C) 3   (D) 1   (E) 1   (F) 1 *(2-1)*   **15.** (A) 4   (B) 3   (C) 4   (D) 0   (E) 1   (F) 1 *(2-1)*
16. (A) 2   (B) 3   (C) Positive *(2-1)*   **17.** (A) 3   (B) 4   (C) Negative *(2-1)*
18. (A) x intercept: -4; y intercept: -2 (B) All real numbers, except $x = 2$ (C) Vertical asymptote: $x = 2$; Horizontal asymptote: $y = 1$
(D)   (E) 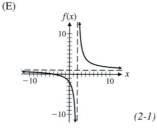 *(2-1)*

19. (A) x intercept: $\frac{4}{3}$; y intercept: -2 (B) All real numbers, except $x = -2$ (C) Vertical asymptote: $x = -2$; Horizontal asymptote: $y = 3$
(D)   (E) *(2-1)*

20. $x = 8$ *(2-3)*   **21.** $x = 3$ *(2-3)*   **22.** $x = 3$ *(2-2)*   **23.** $x = -1, 3$ *(2-2)*   **24.** $x = 0, \frac{3}{2}$ *(2-2)*   **25.** $x = -2$ *(2-3)*   **26.** $x = \frac{1}{2}$ *(2-3)*

27. $x = 27$ *(2-3)*   **28.** $x = 13.3113$ *(2-3)*   **29.** $x = 158.7552$ *(2-3)*   **30.** $x = 0.0097$ *(2-3)*   **31.** $x = 1.4359$ *(2-3)*   **32.** $x = 1.4650$ *(2-3)*

33. $x = 92.1034$ *(2-3)*   **34.** $x = 9.0065$ *(2-3)*   **35.** $x = 2.1081$ *(2-3)*   **36.** $x = 2.8074$ *(2-3)*   **37.** $x = -1.0387$ *(2-3)*

38. They look very much alike *(2-1)*   **39.** *(2-1)* (A)   (B)

40. $1 + 2e^x - e^{-x}$ *(2-2)*   **41.** $2e^{-2x} - 2$ *(2-2)*

42. Increasing: $[-2, 4]$ *(2-2)*   **43.** Decreasing: $[0, \infty)$ *(2-2)*   **44.** Increasing: $(-1, 10]$ *(2-3)*

45. $\log 10^\pi = \pi$ and $10^{\log\sqrt{2}} = \sqrt{2}$; $\ln e^\pi = \pi$ and $e^{\ln\sqrt{2}} = \sqrt{2}$ *(2-3)*   **46.** $x = 2$ *(2-3)*   **47.** $x = 2$ *(2-3)*   **48.** $x = 1$ *(2-3)*   **49.** $x = 300$ *(2-3)*

50. $y = ce^{-5t}$ *(2-3)*   **51.** If $\log_1 x = y$, then $1^y = x$; that is, $1 = x$ for all positive real numbers x, which is not possible. *(2-3)*

52. \$10,263.65 *(2-2)*   **53.** \$10,272.17 *(2-2)*   **54.** 8 yr *(2-2, 2-3)*   **55.** 6.93 yr *(2-2, 2-3)*

56. (A) $C(x) = 40x + 300$; $\overline{C}(x) = \dfrac{40x + 300}{x}$ **57.** (A) $\overline{C}(x) = \dfrac{20x^3 - 360x^2 + 2,300x - 1,000}{x}$

(B) $\overline{C}(x)$

(C) $y = 40$ (D) \$40 per pair *(2-1)*

(B)

(C) 8.667 thousand cases (8,667)   (D) \$567 per case *(2-1)*

58. (A)

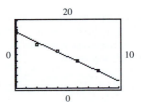

(B) 5 million *(2-4)*

59. (A)

(B) \$10 *(2-4)*

60. $\bar{p} = \$8.38$; $\bar{x} = 5.7$ million   **61.** (A) $N = 2^{2t}$ or $N = 4^t$ (B) 15 days *(2-2, 2-3)*   **62.** $k = 0.009\ 42$; 489 ft *(2-2, 2-3)*

(2-4)

63. (A)

(B) 2,521 bacteria *(2-4)*

64. 23.4 yr *(2-2, 2-3)*   **65.** 23.1 yr *(2-2, 2-3)*

■ CHAPTER 3

Exercise 3-1

1. 0.095; $\frac{1}{6}$ yr   **3.** 18%; $\frac{5}{12}$ yr   **5.** $I = \$20$ **7.** $r = 0.08$ or 8%   **9.** $A = \$112$ **11.** $P = \$888.89$ **13.** $r = I/Pt$ **15.** $P = A/(1 + rt)$
17. The graphs are linear, all with y intercept \$1,000; their slopes are 40, 80, and 120, respectively.   **19.** \$140   **21.** \$9.23   **23.** \$7,685.00
25. 10.125%   **27.** 18%   **29.** \$1,680; 36%   **31.** 9.126%   **33.** \$9,693.91   **35.** 13.538%   **37.** 21.335%   **39.** 11.161%

Exercise 3-2

1. $A = \$112.68$ **3.** $A = \$3,433.50$ **5.** $P = \$2,419.99$ **7.** $P = \$7,351.04$ **9.** 0.75% per month   **11.** 1.75% per quarter
13. 9.6% compounded monthly   **15.** 9% compounded semiannually   **17.** (A) \$126.25; \$26.25 (B) \$126.90; \$26.90 (C) \$127.05; \$27.05
19. (A) \$7,147.51 (B) \$10,217.39
21. All three graphs are increasing, curve upward, and have the same y intercept; the greater the interest rate, the greater the increase. The amounts at the end of 8 years are \$1,376.40, \$1,892.46, and \$2,599.27, respectively.
23. (A) \$6,755.64 (B) \$4,563.87   **25.** (A) 10.38% (B) 12.68%   **27.** $5\frac{1}{2}$ yr   **29.** $n \approx 12$ **31.** (A) $7\frac{1}{4}$ yr (B) 6 yr   **33.** \$22,702.60
35. \$196,993.25   **37.** \$14.26/ft²/mo   **39.** 18 yr
41. 9% compounded monthly, since its effective rate is 9.38%; the effective rate of 9.3% compounded annually is 9.3%
43. (A) In 1998, 222 years after the signing, it would be worth \$76,139.26.   (C)
     (B) If interest were compounded monthly, daily, or continuously, it would
          be worth \$77,409.05, \$78,033.73, or \$78,055.09, respectively.

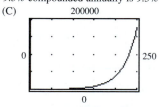

45. 2 yr, 10 mo   **47.** \$328,791.70   **49.** 4.952 yr; 4.959 yr   **51.** 18 quarters   **53.** 7.14%
55. \$5,935.34   **57.** 9.08%   **59.** (A) 8.60% (B) 8.60% (C) 8.57%   **61.** 14.48%   **63.** 7.95%

Exercise 3-3

1. $FV = \$13,435.19$ **3.** $FV = \$60,401.98$ **5.** $PMT = \$123.47$ **7.** $PMT = \$310.62$ **9.** $n = 17$
11. Value: \$30,200.99; interest: \$10,200.99   **13.** \$20,931.01   **15.** \$331.46   **17.** \$625.28

19. First year: $50.76; second year: $168.09; third year: $296.42   **21.** (A) $413,092   (B) $393,965   **23.** $177.46; $1,481.92
25. 3 yr, 5 mo   **27.** 11.58%   **29.** 7.13%   **31.** After 11 quarterly payments   **33.** (A) $290.66   (B) 337 mo   (C) 10.4%

Exercise 3-4

1. $PV = \$3,458.41$ **3.** $PV = \$4,606.09$ **5.** $PMT = \$199.29$ **7.** $PMT = \$586.01$ **9.** $n = 29$ **11.** $109,421.92
13. $14,064.67; $16,800.00   **15.** (A) $36.59/mo; $58.62 interest   (B) $38.28/mo; $89.04 interest   **17.** $273.69/mo; $7,705.68 interest
19. Amortization schedule:

PAYMENT NUMBER	PAYMENT	INTEREST	UNPAID BALANCE REDUCTION	UNPAID BALANCE
0				$5,000.00
1	$ 758.05	$ 225.00	$ 533.05	4,466.95
2	758.05	201.01	557.04	3,909.91
3	758.05	175.95	582.10	3,327.81
4	758.05	149.75	608.30	2,719.51
5	758.05	122.38	635.67	2,083.84
6	758.05	93.77	664.28	1,419.56
7	758.05	63.88	694.17	725.39
8	758.03	32.64	725.39	0.00
Totals	$6,064.38	$1,064.38	$5,000.00	

21. First year interest = $625.07; Second year interest = $400.91; Third year interest = $148.46
23. $85,846.38; $128,153.62   **25.** $143.85/mo; $904.80
27. Monthly payment = $841.39
(A) $70,952.33   (B) $55,909.02   (C) $36,813.32
29. (A) Monthly payment = $395.04; Total interest = $64,809.60
(B) 114 mo, or 9.5 yr; Interest saved = $38,375.04
31. $29,799

33. All three graphs are decreasing, curve downward, and have the same x intercept; the unpaid balances are always in the ratio $2:3:4$. The monthly payments are $402.31, $603.47, and $804.62, with total interest amounting to $94,831.60, $142,249.20, and $189,663.20, respectively.
35. 14.45%   **37.** 10.21%   **39.** (A) $446,774.43   (B) $1,241.04   (C) 14.7%

Chapter 3   Review Exercise

1. $A = \$104.50$ *(3-1)*   **2.** $P = \$800$ *(3-1)*   **3.** $t = 0.75$ yr, or 9 mo *(3-1)*   **4.** $r = 6\%$ *(3-1)*   **5.** $A = \$1,393.68$ *(3-2)*
6. $P = \$3,193.50$ *(3-2)*   **7.** $FV = \$69,770.03$ *(3-3)*   **8.** $PMT = \$115.00$ *(3-3)*   **9.** $PV = \$33,944.27$ *(3-4)*   **10.** $PMT = \$166.07$ *(3-4)*
11. $n \approx 16$ *(3-2)*   **12.** $n \approx 41$ *(3-3)*   **13.** $3,350.00; $350.00 *(3-1)*   **14.** $27,551.32 *(3-2)*   **15.** $9,422.24 *(3-2)*
16. (A) $115,573.86   (B) $359.64   (C) $171,228.80 *(3-4)*   **17.** $164,402 *(3-2)*
18. 9% compounded quarterly, since its effective rate is 9.31%, while the effective rate of 9.25% compounded annually is 9.25% *(3-2)*
19. $27,971.23; $8,771.23 *(3-3)*   **20.** $11.64 *(3-1)*   **21.** $10,210.25 *(3-2)*   **22.** $6,268.21 *(3-2)*   **23.** 15% *(3-1)*
24. $10,318.91; $2,281.09 *(3-4)*   **25.** 2 yr, 3 mo *(3-2)*   **26.** 5 yr, 10 mo; 3 yr, 11 mo *(3-2)*   **27.** (A) $571,499   (B) $1,973,277 *(3-3)*
28. 10.45% *(3-2)*   **29.** 20% *(3-1)*   **30.** $10,988.22; $12,000 *(3-4)*   **31.** $99.85/mo; $396.36 *(3-4)*   **32.** $526.28/mo *(3-3)*
33. 6.93 yr; 7.27 yr *(3-2)*   **34.** $175.28; $2,516.80 *(3-4)*   **35.** 5 yr, 10 mo *(3-3)*   **36.** 18 yr *(3-4)*   **37.** 28.8% *(3-1)*
38. Amortization schedule:

PAYMENT NUMBER	PAYMENT	INTEREST	UNPAID BALANCE REDUCTION	UNPAID BALANCE
0				$1,000.00
1	$ 265.82	$25.00	$ 240.82	759.18
2	265.82	18.98	246.84	512.34
3	265.82	12.81	253.01	259.33
4	265.81	6.48	259.33	0.00
Totals	$1,063.27	$63.27	$1,000.00	

(3-4)

39. 1 yr, 1 mo *(3-2)*   **40.** $55,347.48; $185,830.24 *(3-3)*
41. West Lake S & L: 9.800%; Security S & L: 9.794% *(3-2)*
42. 8.24% *(3-1)*   **43.** 43 *(3-3)*
44. (A) $1,435.63   (B) $74,397.48   (C) $11,625.04 *(3-4)*
45. The certificate would be worth $81,041.86 when the 360th payment is made. By reducing the principal the loan would be paid off in 238 months. If the monthly payment were then invested at 8.75% compounded monthly, it would be worth $105,167.89 at the time of the 360th payment. *(3-2, 3-3, 3-4)*
46. The lower rate would save $15,485.03 in interest payments. *(3-4)*
47. $3,176.14 *(3-2)*   **48.** 8.37% *(3-2)*   **49.** $4,744.73 *(3-1)*

50. $6,697.11 *(3-4)*   **51.** 9.38% *(3-2)*   **52.** (A) $398,807   (B) $374,204 *(3-3)*
53. (A) $746.79/mo; $896.76/mo   (B) $73,558.78; $41,482.19 *(3-4)*   **54.** $19,239 *(3-4)*   **55.** 33.517% *(3-4)*
56. (A) 10.741%   (B) 15 yr; 41 yr *(3-3)*

■ CHAPTER 4

Exercise 4-1

1. Graph (B); no solution   **3.** Graph (A); $x = -3$, $y = 1$ **5.** $x = 3$, $y = 2$ **7.** $x = 2$, $y = 4$ **9.** No solution (parallel lines)
11. $x = 4$, $y = 5$ **13.** $x = 1$, $y = 4$ **15.** $u = 2$, $v = -3$ **17.** $m = 8$, $n = 6$ **19.** $x = 1$, $y = -5$ **21.** No solution (inconsistent)
23. $x = -\frac{4}{3}$, $y = 1$ **25.** Infinitely many solutions (dependent)   **27.** $x = 4{,}000$, $y = 280$ **29.** $x = \frac{4}{7}$, $y = \frac{3}{7}$ **31.** $x = 1.1$, $y = 0.3$
33. $x = -\frac{5}{4}$, $y = \frac{5}{3}$ **35.** $(2.41, 1.12)$ **37.** $(-3.31, -2.24)$ **39.**

41.

43.

45. (A) $(20, -24)$
(B) $(-4, 6)$
(C) No solution

47. (A) Equilibrium price = \$6.50; Equilibrium quantity = 500
(B)

49. (A) $p = 0.001q + 0.15$
(B) $p = -0.002q + 1.74$
(C) Equilibrium price = \$0.68;
Equilibrium quantity = 530
(D)

51. (A) For $x = 120$ units,
Cost = \$216,000 = Revenue
(B)

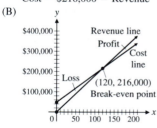

53. (A) For $x = 1{,}920$ tapes,
Cost = \$38,304 = Revenue
(B)

55. Mix A: 80 g; mix B: 60 g   **57.** (A)

(B) $d = 141$ cm (approx.)   (C) Vacillate

Exercise 4-2

1. A is 2×3; C is 1×3 **3.** C **5.** B **7.** $a_{12} = -4$; $a_{23} = -5$ **9.** $-1, 8, 0$ **11.** $\begin{bmatrix} 4 & -6 & | & -8 \\ 1 & -3 & | & 2 \end{bmatrix}$

13. $\begin{bmatrix} -4 & 12 & | & -8 \\ 4 & -6 & | & -8 \end{bmatrix}$ **15.** $\begin{bmatrix} 1 & -3 & | & 2 \\ 8 & -12 & | & -16 \end{bmatrix}$ **17.** $\begin{bmatrix} 1 & -3 & | & 2 \\ 0 & 6 & | & -16 \end{bmatrix}$ **19.** $\begin{bmatrix} 1 & -3 & | & 2 \\ 2 & 0 & | & -12 \end{bmatrix}$ **21.** $\begin{bmatrix} 1 & -3 & | & 2 \\ 3 & -3 & | & -10 \end{bmatrix}$

23. $x_1 = 3, x_2 = 2$; each pair of lines has the same intersection point.

25. $x_1 = 3, x_2 = 1$
27. $x_1 = 2, x_2 = 1$
29. $x_1 = 2, x_2 = 4$
31. No solution
33. $x_1 = 1, x_2 = 4$

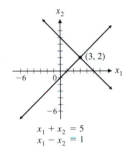

$x_1 + x_2 = 5$
$x_1 - x_2 = 1$

$x_1 + x_2 = 5$
$-2x_2 = -4$

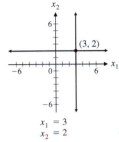

$x_1 + x_2 = 5$
$x_2 = 2$

$x_1 = 3$
$x_2 = 2$

35. Infinitely many solutions: $x_2 = s$, $x_1 = 2s - 3$ for any real number s **37.** Infinitely many solutions: $x_2 = s$, $x_1 = \frac{1}{2}s + \frac{1}{2}$ for any real number s
39. $x_1 = -1, x_2 = 3$ **41.** No solution   **43.** Infinitely many solutions: $x_2 = t$, $x_1 = \frac{3}{2}t + 2$ for any real number t **45.** $x_1 = 2, x_2 = -1$
47. $x_1 = 2, x_2 = -1$ **49.** $x_1 = 1.1, x_2 = 0.3$ **51.** $x_1 = -23.125, x_2 = 7.8125$ **53.** $x_1 = 3.225, x_2 = -6.9375$

Exercise 4-3

1. Yes   **3.** No   **5.** No   **7.** Yes   **9.** $x_1 = -2, x_2 = 3, x_3 = 0$ **11.** $x_1 = 2t + 3, x_2 = -t - 5, x_3 = t$ for t any real number.

13. No solution   **15.** $x_1 = 2s + 3t - 5, x_2 = s, x_3 = -3t + 2, x_4 = t$ for s and t any real numbers   **17.** $\begin{bmatrix} 1 & 0 & | & -7 \\ 0 & 1 & | & 3 \end{bmatrix}$

19. $\begin{bmatrix} 1 & 0 & 0 & | & -5 \\ 0 & 1 & 0 & | & 4 \\ 0 & 0 & 1 & | & -2 \end{bmatrix}$ **21.** $\begin{bmatrix} 1 & 0 & 2 & | & -\frac{5}{3} \\ 0 & 1 & -2 & | & \frac{1}{3} \\ 0 & 0 & 0 & | & 0 \end{bmatrix}$ **23.** $x_1 = -2, x_2 = 3, x_3 = 1$ **25.** $x_1 = 0, x_2 = -2, x_3 = 2$

27. $x_1 = 2t + 3, x_2 = t - 2, x_3 = t$ for t any real number   **29.** $x_1 = 1, x_2 = 2$ **31.** No solution
33. $x_1 = t - 1, x_2 = 2t + 2, x_3 = t$ for t any real number   **35.** $x_1 = -2s + t + 1, x_2 = s, x_3 = t$ for s and t any real numbers
37. No solution   **39.** $x_1 = 2.5t - 4, x_2 = t, x_3 = -5$ for t any real number   **41.** $x_1 = 1, x_2 = -2, x_3 = 1$
43. (A) Dependent system with two parameters and an infinite number of solutions   (B) Dependent system with one parameter and an infinite number of solutions   (C) Independent system with a unique solution   (D) Impossible
45. $x_1 = 2s - 3t + 3, x_2 = s + 2t + 2, x_3 = s, x_4 = t$ for s and t any real numbers   **47.** $x_1 = -0.5, x_2 = 0.2, x_3 = 0.3, x_4 = -0.4$
49. $x_1 = 2s - 1.5t + 1, x_2 = s, x_3 = -t + 1.5, x_4 = 0.5t - 0.5, x_5 = t$ for s and t any real numbers
51. 20 one-person boats, 220 two-person boats, 100 four-person boats
53. $(t - 80)$ one-person boats, $(-2t + 420)$ two-person boats, t four-person boats, $80 \leq t \leq 210$, t an integer
55. No solution; no production schedule will use all the labor-hours in all departments
57. $(5t - 29)$ 6,000 gal cars, $(-6t + 53)$ 8,000 gal cars, and t 18,000 gal cars, where $t = 6, 7,$ or 8 **59.** 57.65%
61. 8 oz food A, 2 oz food B, 4 oz food C   **63.** No solution   **65.** 8 oz food A, $(-2t + 10)$ oz food B, t oz food C, $0 \leq t \leq 5$
67. $(10 - t)$ barrels of mix A, $(t - 5)$ barrels of mix B, $(25 - 2t)$ barrels of mix C, and t barrels of mix D, where t is an integer, $5 \leq t \leq 10$
69. Company A: 10 hr; company B: 15 hr
71. (A) 6th St. and Washington Ave.: $x_1 + x_2 = 1,200$; 6th St. and Lincoln Ave.: $x_2 + x_3 = 1,000$; 5th St. and Lincoln Ave.: $x_3 + x_4 = 1,300$
    (B) $x_1 = 1,500 - t, x_2 = t - 300, x_3 = 1,300 - t,$ and $x_4 = t$, where $300 \leq t \leq 1,300$ (C) 1,300; 300
    (D) Washington Ave.: 500; 6th St.: 700; Lincoln Ave.: 300

Exercise 4-4

1. $\begin{bmatrix} -1 & 0 \\ 5 & -3 \end{bmatrix}$ **3.** $\begin{bmatrix} 2 & 0 & 3 \\ 7 & 7 & -5 \end{bmatrix}$ **5.** Not defined   **7.** $\begin{bmatrix} 5 & -7 \\ -5 & 2 \\ 0 & 4 \end{bmatrix}$ **9.** $\begin{bmatrix} 5 & -10 & 0 & 20 \\ -15 & 10 & -5 & 30 \end{bmatrix}$ **11.** $[10]$

13. $\begin{bmatrix} 5 \\ -3 \end{bmatrix}$ **15.** $\begin{bmatrix} 2 & 4 \\ 1 & -5 \end{bmatrix}$ **17.** $\begin{bmatrix} 1 & -5 \\ -2 & -4 \end{bmatrix}$ **19.** $[-7]$ **21.** $\begin{bmatrix} -15 & 6 \\ -20 & 8 \end{bmatrix}$ **23.** $[11]$ **25.** $\begin{bmatrix} 3 & -2 & -4 \\ 6 & -4 & -8 \\ -9 & 6 & 12 \end{bmatrix}$

27. $\begin{bmatrix} -12 & 12 & 18 \\ 20 & -18 & -6 \end{bmatrix}$ **29.** Not defined **31.** $\begin{bmatrix} 11 & 2 \\ 4 & 27 \end{bmatrix}$ **33.** $\begin{bmatrix} 6 & 4 \\ 0 & -3 \end{bmatrix}$ **35.** $\begin{bmatrix} -1.3 & -0.7 \\ -0.2 & -0.5 \\ 0.1 & 1.1 \end{bmatrix}$

37. $\begin{bmatrix} -66 & 69 & 39 \\ 92 & -18 & -36 \end{bmatrix}$ **39.** Not defined **41.** $\begin{bmatrix} -18 & 48 \\ 54 & -34 \end{bmatrix}$ **43.** $\begin{bmatrix} -26 & -15 & -25 \\ -4 & -18 & 4 \\ 2 & 43 & -19 \end{bmatrix}$ **45.** $a = -1, b = 1, c = 3, d = -5$

47. $x = 2, y = -3$ **49.** $x = 3, y = 2$ **51.** $a = 3, b = 4, c = 1, d = 2$ **53.** All are true. **55.**

	Guitar	Banjo	
	$33	$26	Materials
	$57	$77	Labor

57.

	Basic car	Air	AM/FM radio	Cruise control
Model A	$1,180	$82	$54	$30
Model B	$1,075	$98	$81	$30
Model C	$1,715	$106	$108	$36

(Markup)

59. (A) $11.80 (B) $30.30 (C) MN gives the labor costs per boat at each plant.

(D) $MN = $

	Plant I	Plant II	
	$11.80	$13.80	One-person boat
	$18.50	$21.60	Two-person boat
	$26.00	$30.30	Four-person boat

61. (A) $A^2 = \begin{bmatrix} 0 & 0 & 2 & 0 & 0 \\ 1 & 0 & 0 & 0 & 1 \\ 0 & 1 & 0 & 2 & 0 \\ 1 & 0 & 0 & 0 & 1 \\ 0 & 0 & 1 & 0 & 0 \end{bmatrix}$ There is one way to travel from Baltimore to Atlanta with one intermediate connection; there are two ways to travel from Atlanta to Chicago with one intermediate connection. In general, the elements in A^2 indicate the number of different ways to travel from the ith city to the jth city with one intermediate connection.

(B) $A^3 = \begin{bmatrix} 2 & 0 & 0 & 0 & 2 \\ 0 & 1 & 0 & 2 & 0 \\ 0 & 0 & 3 & 0 & 0 \\ 0 & 1 & 0 & 2 & 0 \\ 1 & 0 & 0 & 0 & 1 \end{bmatrix}$ There is one way to travel from Denver to Baltimore with two intermediate connections; there are two ways to travel from Atlanta to El Paso with two intermediate connections. In general, the elements in A^3 indicate the number of different ways to travel from the ith city to the jth city with two intermediate connections.

(C) $A + A^2 + A^3 + A^4 = \begin{bmatrix} 2 & 3 & 2 & 5 & 2 \\ 1 & 1 & 4 & 2 & 1 \\ 4 & 1 & 3 & 2 & 4 \\ 1 & 1 & 4 & 2 & 1 \\ 1 & 1 & 1 & 3 & 1 \end{bmatrix}$ It is possible to travel from any origin to any destination with at most three intermediate connections.

63. (A) 70 g (B) 30 g
(C) MN gives the amount (in grams) of protein, carbohydrate, and fat in 20 oz of each mix.

(D) $MN = $

	Mix X	Mix Y	Mix Z	
	70	60	50	Protein
	380	360	340	Carbohydrate
	50	40	30	Fat

65. (A) $2,650 (B) $4,400
(C) NM gives the total cost per town.

(D) $NM = \begin{bmatrix} \$2,650 \\ \$4,400 \end{bmatrix}$ Berkeley Oakland (Cost per town)

(E) $[1 \quad 1]N = [3,000 \quad 1,300 \quad 13,000]$

	Telephone call	House call	Letter

(F) $N\begin{bmatrix} 1 \\ 1 \\ 1 \end{bmatrix} = \begin{bmatrix} 6,500 \\ 10,800 \end{bmatrix}$ Berkeley Oakland (Total contacts)

Exercise 4-5

1. $\begin{bmatrix} 2 & -3 \\ 4 & 5 \end{bmatrix}$ **3.** $\begin{bmatrix} 2 & -3 \\ 4 & 5 \end{bmatrix}$ **5.** $\begin{bmatrix} -2 & 1 & 3 \\ 2 & 4 & -2 \\ 5 & 1 & 0 \end{bmatrix}$ **7.** $\begin{bmatrix} -2 & 1 & 3 \\ 2 & 4 & -2 \\ 5 & 1 & 0 \end{bmatrix}$ **15.** $\begin{bmatrix} -1 & 0 \\ -3 & 1 \end{bmatrix}$ **17.** $\begin{bmatrix} 3 & -2 \\ -1 & 1 \end{bmatrix}$

19. $\begin{bmatrix} 7 & -3 \\ -2 & 1 \end{bmatrix}$ **21.** $\begin{bmatrix} -5 & -12 & 3 \\ -2 & -4 & 1 \\ 2 & 5 & -1 \end{bmatrix}$ **23.** $\begin{bmatrix} 6 & -2 & -1 \\ -5 & 2 & 1 \\ -3 & 1 & 1 \end{bmatrix}$ **25.** $\begin{bmatrix} -2 & -3 \\ 3 & 4 \end{bmatrix}$ **27.** Does not exist

29. $\begin{bmatrix} 1.5 & -0.5 \\ -2 & 1 \end{bmatrix}$ **31.** $\begin{bmatrix} 1 & 2 & 2 \\ -2 & -3 & -4 \\ -1 & -2 & -1 \end{bmatrix}$ **33.** Does not exist **35.** $\begin{bmatrix} -3 & -2 & 1.5 \\ 4 & 3 & -2 \\ 3 & 2 & -1.25 \end{bmatrix}$ **37.** $\begin{bmatrix} -1.75 & -0.375 & 0.5 \\ -5.5 & -1.25 & 1 \\ 0.5 & 0.25 & 0 \end{bmatrix}$

41. M^{-1} exists if and only if all the elements on the main diagonal are nonzero. **43.** $\begin{bmatrix} 0.5 & -0.3 & 0.85 & -0.25 \\ 0 & 0.1 & 0.05 & -0.25 \\ -1 & 0.9 & -1.55 & 0.75 \\ -0.5 & 0.4 & -1.3 & 0.5 \end{bmatrix}$

45. $\begin{bmatrix} 1.75 & 5.25 & 8.75 & -1 & -18.75 \\ 1.25 & 3.75 & 6.25 & -1 & -13.25 \\ -4.75 & -13.25 & -22.75 & 3 & 48.75 \\ -1.375 & -4.625 & -7.875 & 1 & 16.375 \\ 3.25 & 8.75 & 15.25 & -2 & -32.25 \end{bmatrix}$ **47.** 36 44 59 86 61 82 68 95 25 37 49 64 63 81 47 66 43 62

49. GONE WITH THE WIND
51. 27 19 82 9 89 50 114 101 70 123 50 100 75 72 89 63 96 99 69 99 52 105 106 56 97 59 127 110 81 132
53. THE GREATEST SHOW ON EARTH

Exercise 4-6

1. $3x_1 + x_2 = 5$ **3.** $-3x_1 + x_2 = 3$ **5.** $\begin{bmatrix} 3 & -4 \\ 2 & 1 \end{bmatrix}\begin{bmatrix} x_1 \\ x_2 \end{bmatrix} = \begin{bmatrix} 1 \\ 5 \end{bmatrix}$ **7.** $\begin{bmatrix} 1 & -3 & 2 \\ -2 & 3 & 0 \\ 1 & 1 & 4 \end{bmatrix}\begin{bmatrix} x_1 \\ x_2 \\ x_3 \end{bmatrix} = \begin{bmatrix} -3 \\ 1 \\ -2 \end{bmatrix}$
$\quad\ 2x_1 - x_2 = -4$ $\quad\ 2x_1 + x_3 = -4$
$\quad\quad\quad\quad -x_1 + 3x_2 - 2x_3 = 2$

9. $x_1 = -8, x_2 = 2$ **11.** $x_1 = 0, x_2 = 4$ **13.** (A) $x_1 = -3, x_2 = 2$ **15.** (A) $x_1 = 17, x_2 = -5$ **17.** (A) $x_1 = 1, x_2 = 0, x_3 = 0$
$\quad\quad\quad\quad\quad\quad\quad\quad\quad\quad\quad\quad\quad\quad\quad\quad$ (B) $x_1 = -1, x_2 = 2$ (B) $x_1 = 7, x_2 = -2$ (B) $x_1 = -7, x_2 = -2, x_3 = 3$
$\quad\quad\quad\quad\quad\quad\quad\quad\quad\quad\quad\quad\quad\quad\quad\quad$ (C) $x_1 = -8, x_2 = 3$ (C) $x_1 = 24, x_2 = -7$ (C) $x_1 = 17, x_2 = 5, x_3 = -7$

19. (A) $x_1 = 8, x_2 = -6, x_3 = -2$ **21.** $X = (A - B)^{-1}C$ **27.** (A) $x_1 = 1, x_2 = 0$
$\quad$ (B) $x_1 = -6, x_2 = 6, x_3 = 2$ **23.** $X = (A + I)^{-1}C$ $\quad$ (B) $x_1 = -2,000, x_2 = 1,000$
$\quad$ (C) $x_1 = 20, x_2 = -16, x_3 = -10$ **25.** $X = (A + B)^{-1}(C + D)$ $\quad$ (C) $x_1 = 2,001, x_2 = -1,000$
29. $x_1 = 18.2, x_2 = 27.9\ x_3 = -15.2$ **31.** $x_1 = 24, x_2 = 5, x_3 = -2, x_4 = 15$
33. Concert 1: 6,000 \$4 tickets and 4,000 \$8 tickets; Concert 2: 5,000 \$4 tickets and 5,000 \$8 tickets; Concert 3: 3,000 \$4 tickets and 7,000 \$8 tickets
35. Order 1: 280 hr at plant A and 25 hr at plant B; Order 2: 160 hr at plant A and 150 hr at plant B; Order 3: 80 hr at plant A and 225 hr at plant B
37. President: \$56,600, executive vice president: \$47,000; associative vice president: \$37,400; assistant vice president: \$27,900
39. A: \$2,927,000; B: \$3,372,000; C: \$3,675,000; D: \$3,926,000
41. (A) Freezers: \$346,000; Refrigerators: \$254,000; Accounting: \$85,000; Maintenance: \$127,000
$\quad$ (B) Sum of direct costs = \$600,000; Sum of total costs = \$346,000 + \$254,000 = \$600,000; the direct costs of the service departments are
$\quad\quad$ distributed among the production departments.
43. Diet 1: 60 oz mix A and 80 oz mix B; Diet 2: 20 oz mix A and 60 oz mix B; Diet 3: 0 oz mix A and 100 oz mix B

Exercise 4-7

1. 40¢ from A; 20¢ from E **3.** $X = \begin{bmatrix} x_1 \\ x_2 \end{bmatrix} = \begin{bmatrix} 12.4 \\ 7.2 \end{bmatrix}$ **5.** 20¢ from A; 10¢ from B; 10¢ from E

7. Agriculture: \$18 billion; Building: \$15.6 billion; Energy: \$22.4 billion **9.** $\begin{bmatrix} 24 \\ 46 \end{bmatrix}$ **11.** $\begin{bmatrix} 38.6 \\ 18 \\ 17.4 \end{bmatrix}$

13. (A) Agriculture: \$80 million; Manufacturing: \$64 million
$\quad$ (B) The final demand for agriculture increases to \$54 million and the final demand for manufacturing decreases to \$38 million.

15. The total output of the energy sector should be 75% of the total output of the mining sector.   **17.** Each element should be between 0 and 1, inclusive.
19. Coal: $28 billion; Steel: $26 billion   **21.** Agriculture: $148 million; Tourism: $146 million
23. Agriculture: $40.1 billion; Manufacturing: $29.4 billion; Energy: $34.4 billion
25. Year 1: Agriculture: $65 billion; Energy: $83 billion; Labor: $71 billion; Manufacturing: $88 billion.
    Year 2: Agriculture: $81 billion; Energy: $97 billion; Labor: $83 billion; Manufacturing: $99 billion.
    Year 3: Agriculture: $117 billion; Energy: $124 billion; Labor: $106 billion; Manufacturing: $120 billion.

Chapter 4   Review Exercise

1. $x = 4, y = 4$ *(4-1)*   **2.** $x = 4$ $y = 4$ *(4-1)*   **3.** $\begin{bmatrix} 3 & 3 \\ 4 & 2 \end{bmatrix}$ *(4-4)*   **4.** Not defined *(4-4)*   **5.** $\begin{bmatrix} -3 & 0 \\ 1 & -1 \end{bmatrix}$ *(4-4)*   **6.** $\begin{bmatrix} 4 & 3 \\ 7 & 4 \end{bmatrix}$ *(4-4)*

7. Not defined *(4-4)*   **8.** $\begin{bmatrix} 5 \\ 5 \end{bmatrix}$ *(4-4)*   **9.** $\begin{bmatrix} 2 & 3 \\ 4 & 6 \end{bmatrix}$ *(4-4)*   **10.** [8] *(4-4)*   **11.** Not defined *(4-4)*   **12.** $\begin{bmatrix} -2 & 3 \\ 3 & -4 \end{bmatrix}$ *(4-5)*

13. $x_1 = 9, x_2 = -11$ *(4-1)*   **14.** $x_1 = 9, x_2 = -11$ *(4-2)*   **15.** $x_1 = 9, x_2 = -11; x_1 = 16, x_2 = -19; x_1 = -2, x_2 = 4$ *(4-6)*

16. Not defined *(4-4)*   **17.** $\begin{bmatrix} 10 & -8 \\ 4 & 6 \end{bmatrix}$ *(4-4)*   **18.** $\begin{bmatrix} -2 & 8 \\ 8 & 6 \end{bmatrix}$ *(4-4)*   **19.** $\begin{bmatrix} -2 & -1 & -3 \\ 4 & 2 & 6 \\ 6 & 3 & 9 \end{bmatrix}$ *(4-4)*   **20.** [9] *(4-4)*

21. $\begin{bmatrix} 10 & -5 & 1 \\ -1 & -4 & -5 \\ 1 & -7 & -2 \end{bmatrix}$ *(4-4)*   **22.** $\begin{bmatrix} -\frac{5}{2} & 2 & -\frac{1}{2} \\ 1 & -1 & 1 \\ \frac{1}{2} & 0 & -\frac{1}{2} \end{bmatrix}$ *(4-5)*   **23.** (A) $x_1 = 2, x_2 = 1, x_3 = -1$
(B) $x_1 = -5t - 12, x_2 = 3t + 7, x_3 = t$ for t any real number *(4-3)*

24. $x_1 = 2, x_2 = 1, x_3 = -1; x_1 = 1, x_2 = -2, x_3 = 1; x_1 = -1, x_2 = 2, x_3 = -2$ *(4-6)*   **25.** $X = \begin{bmatrix} 48 \\ 56 \end{bmatrix}$ *(4-7)*

26. $x = 3.46, y = 1.69$ *(4-1)*   **27.** $\begin{bmatrix} -0.9 & -0.1 & 5 \\ 0.8 & 0.2 & -4 \\ 0.1 & -0.1 & 0 \end{bmatrix}$ *(4-5)*   **28.** $x_1 = 1,400, x_2 = 3,200, x_3 = 2,400$ *(4-6)*

29. $x_1 = 1,400, x_2 = 3,200, x_3 = 2,400$ *(4-3)*   **30.** $X = \begin{bmatrix} 81 \\ 49 \\ 62 \end{bmatrix}$ *(4-7)*   **31.** (A) Unique solution
(B) Either no solution or an
infinite number of solutions *(4-6)*

32. (A) Unique solution
(B) No solution
(C) Infinite number of solutions *(4-3)*
33. (B) is the only correct solution. *(4-6)*

34. (A) $C = 243,000 + 22.45x; R = 59.95x$
(B) $x = 6,480$ machines; $R = C = \$388,476$
(C) Profit occurs if $x > 6,480$;
loss occurs if $x < 6,480$.

35. $0.01x_1 + 0.02x_2 = 4.5$
$0.02x_1 + 0.05x_2 = 10$
$x_1 = 250$ tons of ore A,
$x_2 = 100$ tons of ore B *(4-3)*

36. (A) $\overset{X}{\begin{bmatrix} x_1 \\ x_2 \end{bmatrix}} = \overset{A^{-1}}{\begin{bmatrix} 500 & -200 \\ -200 & 100 \end{bmatrix}} \begin{bmatrix} 4.5 \\ 10 \end{bmatrix} = \begin{bmatrix} 250 \\ 100 \end{bmatrix}$ (B) $\overset{X}{\begin{bmatrix} x_1 \\ x_2 \end{bmatrix}} = \overset{A^{-1}}{\begin{bmatrix} 500 & -200 \\ -200 & 100 \end{bmatrix}} \begin{bmatrix} 2.3 \\ 5 \end{bmatrix} = \begin{bmatrix} 150 \\ 40 \end{bmatrix}$ *(4-6)*
    $x_1 = 250$ tons of ore A, $x_2 = 100$ tons of ore B $x_1 = 150$ tons of ore A, $x_2 = 40$ tons of ore B

37. 1 model A truck, 6 model B trucks, and 5 model C trucks; or 3 model A trucks, 3 model B trucks, and 6 model C trucks; or 5 model A trucks and 7 model C trucks. *(4-3)*

38. (A) Elements in *MN* give the cost of materials for each alloy from each supplier.

Supplier A   Supplier B

(B) $MN = \begin{bmatrix} \$7,620 & \$7,530 \\ \$13,880 & \$13,930 \end{bmatrix} \begin{matrix} \text{Alloy 1} \\ \text{Alloy 2} \end{matrix}$

Supplier A   Supplier B

(C) $[1 \quad 1]MN = [\$21,500 \quad \$21,460]$ Total material costs *(4-4)*

39. (A) \$6.35

(B) Elements in *MN* give the total labor costs for each calculator at each plant.

CA   TX

(C) $MN = \begin{bmatrix} \$3.65 & \$3.00 \\ \$6.35 & \$5.20 \end{bmatrix} \begin{matrix} \text{Model A} \\ \text{Model B} \end{matrix}$ *(4-4)*

40. \$2,000 at 5%; \$3,000 at 10% *(4-3)*   **41.** \$2,000 at 5%; \$3,000 at 10% *(4-6)*

42. Concert 1: 5,000 \$8 tickets, 15,000 \$12 tickets, and 5,000 \$20 tickets;
Concert 2: 7,500 \$8 tickets, 10,000 \$12 tickets, and 7,500 \$20 tickets;
Concert 3: 10,000 \$8 tickets, 5,000 \$12 tickets, and 10,000 \$20 tickets *(4-6)*

43. (A) Agriculture: \$80 billion; Fabrication: \$60 billion (B) Agriculture: \$135 billion; Fabrication: \$145 billion *(4-7)*

44. GRAPHING UTILITY *(4-5)*   **45.** (A) 1st & Elm: $x_1 + x_4 = 1,300$ (B) $x_1 = 1,300 - t, x_2 = 900 - t, x_3 = t - 200,$
2nd & Elm: $x_1 - x_2 = 400$ $x_4 = t$, where $200 \leq t \leq 900$
2nd & Oak: $x_2 + x_3 = 700$ (C) 900; 200
1st & Oak: $x_3 - x_4 = -200$ (D) Elm St.: 800; 2nd St.: 400; Oak St.: 300 *(4-3)*

■ CHAPTER 5

Exercise 5-1

1.

3.

5.

7.

9.

11.

13. IV   **15.** I   **17.**

19.

21. IV; (8, 0), (18, 0), (6, 4)   **23.** I; (0, 16), (6, 4), (18, 0)   **25.** Bounded

27. Bounded

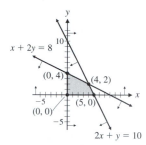

$x + 2y = 8$
$(0, 4)$
$(4, 2)$
$(5, 0)$
$(0, 0)$
$2x + y = 10$

29. Unbounded

$x + 2y = 8$
$(0, 10)$
$(8, 0)$
$(4, 2)$
$2x + y = 10$

31. Bounded

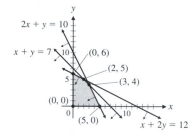

$2x + y = 10$
$x + y = 7$
$(0, 6)$
$(2, 5)$
$(3, 4)$
$(0, 0)$
$(5, 0)$
$x + 2y = 12$

33. Unbounded

$(0, 16)$
$(4, 8)$
$(10, 2)$
$(14, 0)$
$2x + y = 16$
$x + y = 12$
$x + 2y = 14$

35. Bounded

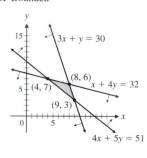

$3x + y = 30$
$(8, 6)$
$x + 4y = 32$
$(4, 7)$
$(9, 3)$
$4x + 5y = 51$

37. Empty

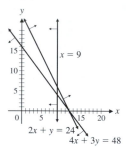

$x = 9$
$2x + y = 24$
$4x + 3y = 48$

39. Unbounded

$2x - y = 4$
$(8, 12)$
$(4, 4)$
$x - y = 0$
$(0, 0)$
$x = 8$

41. Bounded

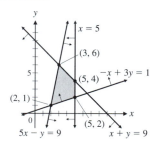

$x = 5$
$(3, 6)$
$(5, 4)$
$-x + 3y = 1$
$(2, 1)$
$5x - y = 9$
$(5, 2)$
$x + y = 9$

43. Bounded

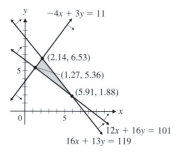

$-4x + 3y = 11$
$(2.14, 6.53)$
$(1.27, 5.36)$
$(5.91, 1.88)$
$12x + 16y = 101$
$16x + 13y = 119$

45. (A) $3x + 4y = 36$ and $3x + 2y = 30$ intersect at $(8, 3)$; $3x + 4y = 36$ and $x = 0$ intersect at $(0, 9)$; $3x + 4y = 36$ and $y = 0$ intersect at $(12, 0)$; $3x + 2y = 30$ and $x = 0$ intersect at $(0, 15)$; $3x + 2y = 30$ and $y = 0$ intersect at $(10, 0)$; $x = 0$ and $y = 0$ intersect at $(0, 0)$
(B) $(8, 3)$, $(0, 9)$, $(10, 0)$, $(0, 0)$

47. $6x + 4y \leqslant 108$
$x + y \leqslant 24$
$x \geqslant 0$
$y \geqslant 0$

$(6, 18)$

49. (A) All production schedules in the feasible region that are on the graph of $50x + 60y = 1,100$ will result in a profit of $1,100.
(B) There are many possible choices. For example, producing 5 trick skis and 15 slalom skis will produce a profit of $1,150. All the production schedules in the feasible region that are on the graph of $50x + 60y = 1,150$ will result in a profit of $1,150.

51. $20x + 10y \geqslant 460$
$30x + 30y \geqslant 960$
$5x + 10y \geqslant 220$
$x \geqslant 0$
$y \geqslant 0$

$(14, 18)$
$(20, 12)$

53. $10x + 20y \leqslant 800$
$20x + 10y \leqslant 640$
$x \geqslant 0$
$y \geqslant 0$

$(16, 32)$

Exercise 5-2

1. 16   **3.** 84   **5.** 32   **7.** 36   **9.** Max $P = 30$ at $x_1 = 4$ and $x_2 = 2$
11. Min $z = 14$ at $x_1 = 4$ and $x_2 = 2$; no max
13. Max $P = 260$ at $x_1 = 2$ and $x_2 = 5$ **15.** Min $z = 140$ at $x_1 = 14$ and $x_2 = 0$; no max
17. Min $P = 20$ at $x_1 = 0$ and $x_2 = 2$; Max $P = 150$ at $x_1 = 5$ and $x_2 = 0$
19. Feasible region empty; no optimal solutions
21. Min $P = 140$ at $x_1 = 3$ and $x_2 = 8$; Max $P = 260$ at $x_1 = 8$ and $x_2 = 10$, at $x_1 = 12$ and $x_2 = 2$, or at any point on the line segment from (8, 10) to (12, 2)
23. Max $P = 26,000$ at $x_1 = 400$ and $x_2 = 600$ **25.** Max $P = 5,507$ at $x_1 = 6.62$ and $x_2 = 4.25$
27. (A) $2a < b$ (B) $\frac{1}{4}a < b < 2a$ (C) $b < \frac{1}{4}a$ (D) $b = 2a$ (E) $b = \frac{1}{4}a$
29. (A) Max profit = \$780 when 6 trick skis and 18 slalom skis are produced.
  (B) Max profit decreases to \$720 when 18 trick skis and no slalom skis are produced.
  (C) Max profit increases to \$1,080 when no trick skis and 24 slalom skis are produced.
31. (A) Plant A: 5 days; Plant B: 4 days; min cost \$8,600
  (B) Plant A: 10 days; Plant B: 0 days; min cost \$6,000
  (C) Plant A: 0 days; Plant B: 10 days; min cost \$8,000
33. 7 buses, 15 vans; min cost \$9,900
35. (A) Max $P =$ \$450 when 750 gallons are produced using the old process exclusively.
  (B) Max $P =$ \$380 when 400 gallons are produced using the old process and 700 gallons are produced using the new process.
  (C) Max $P =$ \$288 when 1,440 gallons are produced using the new process exclusively.
37. (A) 150 bags brand A, 100 bags brand B; max nitrogen 1,500 lb
  (B) 0 bags brand A, 250 bags brand B; min nitrogen 750 lb
39. 20 yd³ A, 12 yd³ B; \$1,020   **41.** 48; 16 mice, 32 rats

Exercise 5-3

1. (A) 2 (B) 2 basic and 3 nonbasic variables
  (C) 2 linear equations with 2 variables
3. (A) 5 (B) 4 (C) 5 basic and 4 nonbasic variables
  (D) 5 linear equations with 5 variables

5.

	NONBASIC	BASIC	FEASIBLE?
(A)	x_1, x_2	s_1, s_2	Yes
(B)	x_1, s_1	x_2, s_2	Yes
(C)	x_1, s_2	x_2, s_1	No
(D)	x_2, s_1	x_1, s_2	No
(E)	x_2, s_2	x_1, s_1	Yes
(F)	s_1, s_2	x_1, x_2	Yes

7.

	x_1	x_2	s_1	s_2	FEASIBLE?
(A)	0	0	50	40	Yes
(B)	0	50	0	-60	No
(C)	0	20	30	0	Yes
(D)	25	0	0	15	Yes
(E)	40	0	-30	0	No
(F)	20	10	0	0	Yes

9.

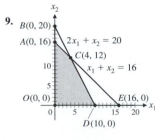

$$x_1 + x_2 + s_1 \quad\quad\quad = 16$$
$$2x_1 + x_2 \quad\quad + s_2 = 20$$

x_1	x_2	s_1	s_2	INTERSECTION POINT	FEASIBLE?
0	0	16	20	O	Yes
0	16	0	4	A	Yes
0	20	-4	0	B	No
16	0	0	-12	E	No
10	0	6	0	D	Yes
4	12	0	0	C	Yes

11.

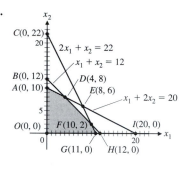

$$2x_1 + x_2 + s_1 \qquad\qquad = 22$$
$$x_1 + x_2 \qquad + s_2 \qquad = 12$$
$$x_1 + 2x_2 \qquad\qquad + s_3 = 20$$

x_1	x_2	s_1	s_2	s_3	INTERSECTION POINT	FEASIBLE?
0	0	22	12	20	O	Yes
0	22	0	-10	-24	C	No
0	12	10	0	-4	B	No
0	10	12	2	0	A	Yes
11	0	0	1	9	G	Yes
12	0	-2	0	8	H	No
20	0	-18	-8	0	I	No
10	2	0	0	6	F	Yes
8	6	0	-2	0	E	No
4	8	6	0	0	D	Yes

Exercise 5-4

1. (A) Basic: x_2, s_1, P; nonbasic: x_1, s_2 (B) $x_1 = 0, x_2 = 12, s_1 = 15, s_2 = 0, P = 20$ (C) Additional pivot required

3. (A) Basic: x_2, x_3, s_3, P; nonbasic: x_1, s_1, s_2 (B) $x_1 = 0, x_2 = 15, x_3 = 5, s_1 = 0, s_2 = 0, s_3 = 12, P = 45$ (C) No solution

5.

Enter ↓ (x_1)

$$\begin{array}{c} \text{Exit} \rightarrow s_1 \\ s_2 \\ P \end{array}
\begin{bmatrix} x_1 & x_2 & s_1 & s_2 & P & \\ 1 & 4 & 1 & 0 & 0 & 4 \\ 3 & 5 & 0 & 1 & 0 & 24 \\ \hline -8 & -5 & 0 & 0 & 1 & 0 \end{bmatrix}$$

$$\begin{array}{c} x_1 \\ \sim\ s_2 \\ P \end{array}
\begin{bmatrix} 1 & 4 & 1 & 0 & 0 & 4 \\ 0 & -7 & -3 & 1 & 0 & 12 \\ \hline 0 & 27 & 8 & 0 & 1 & 32 \end{bmatrix}$$

7.

Enter ↓ (x_1)

$$\begin{array}{c} \text{Exit} \rightarrow x_2 \\ s_2 \\ s_3 \\ P \end{array}
\begin{bmatrix} x_1 & x_2 & s_1 & s_2 & s_3 & P & \\ 2 & 1 & 1 & 0 & 0 & 0 & 4 \\ 3 & 0 & 1 & 1 & 0 & 0 & 8 \\ 0 & 0 & 2 & 0 & 1 & 0 & 2 \\ \hline -4 & 0 & -3 & 0 & 0 & 1 & 5 \end{bmatrix}$$

$$\begin{array}{c} x_1 \\ \sim\ s_2 \\ s_3 \\ P \end{array}
\begin{bmatrix} 1 & \frac{1}{2} & \frac{1}{2} & 0 & 0 & 0 & 2 \\ 0 & -\frac{3}{2} & -\frac{1}{2} & 1 & 0 & 0 & 2 \\ 0 & 0 & 2 & 0 & 1 & 0 & 2 \\ \hline 0 & 2 & -1 & 0 & 0 & 1 & 13 \end{bmatrix}$$

9. (A)
$$\begin{aligned} 2x_1 + x_2 + s_1 \qquad\qquad &= 10 \\ x_1 + 3x_2 \qquad + s_2 \qquad &= 10 \\ -15x_1 - 10x_2 \qquad\qquad + P &= 0 \end{aligned}$$

(B)

Enter ↓ (x_1)

$$\begin{array}{c} \text{Exit} \rightarrow s_1 \\ s_2 \\ P \end{array}
\begin{bmatrix} x_1 & x_2 & s_1 & s_2 & P & \\ 2 & 1 & 1 & 0 & 0 & 10 \\ 1 & 3 & 0 & 1 & 0 & 10 \\ \hline -15 & -10 & 0 & 0 & 1 & 0 \end{bmatrix}$$

(C) Max $P = 80$ at $x_1 = 4$ and $x_2 = 2$

11. (A)
$$\begin{aligned} 2x_1 + x_2 + s_1 \qquad\qquad &= 10 \\ x_1 + 3x_2 \qquad + s_2 \qquad &= 10 \\ -30x_1 - x_2 \qquad\qquad + P &= 0 \end{aligned}$$

(B)

Enter ↓ (x_1)

$$\begin{array}{c} \text{Exit} \rightarrow s_1 \\ s_2 \\ P \end{array}
\begin{bmatrix} x_1 & x_2 & s_1 & s_2 & P & \\ 2 & 1 & 1 & 0 & 0 & 10 \\ 1 & 3 & 0 & 1 & 0 & 10 \\ \hline -30 & -1 & 0 & 0 & 1 & 0 \end{bmatrix}$$

(C) Max $P = 150$ at $x_1 = 5$ and $x_2 = 0$

13. Max $P = 260$ at $x_1 = 2$ and $x_2 = 5$

15. No optimal solution exists.

17. Max $P = 7$ at $x_1 = 3$ and $x_2 = 5$

19. Max $P = 58$ at $x_1 = 12, x_2 = 0$, and $x_3 = 2$

21. Max $P = 17$ at $x_1 = 4, x_2 = 3$, and $x_3 = 0$

23. Max $P = 22$ at $x_1 = 1, x_2 = 6$, and $x_3 = 0$

25. Max $P = 26,000$ at $x_1 = 400$ and $x_2 = 600$

27. Max $P = 450$ at $x_1 = 0, x_2 = 180$, and $x_3 = 30$

29. Max $P = 88$ at $x_1 = 24$ and $x_2 = 8$

31. Choosing either column produces the same optimal solution: Max $P = 13$ at $x_1 = 3$ and $x_2 = 10$.

33. Choosing Column 1: Max $P = 60$ at $x_1 = 12$, $x_2 = 8$, and $x_3 = 0$. Choosing Column 2: Max $P = 60$ at $x_1 = 0$, $x_2 = 20$, and $x_3 = 0$.

35. Let: $x_1 = $ Number of A components $\qquad$ Maximize $\quad P = 7x_1 + 8x_2 + 10x_3$
$\quad x_2 = $ Number of B components $\qquad$ Subject to $\quad 2x_1 + 3x_2 + 2x_3 \leq 1{,}000$
$\quad x_3 = $ Number of C components $\qquad\qquad\qquad x_1 + \ x_2 + 2x_3 \leq \ 800$
$\qquad\qquad\qquad\qquad\qquad\qquad\qquad\qquad\qquad\qquad x_1, x_2, x_3 \geq 0$

$\quad$ 200 A components, 0 B components, and 300 C components; max profit is \$4,400

37. Let: $x_1 = $ Amount invested in government bonds $\qquad$ Maximize $\quad P = 0.08x_1 + 0.13x_2 + 0.15x_3$
$\quad x_2 = $ Amount invested in mutual funds $\qquad$ Subject to $\qquad x_1 + x_2 + x_3 \leq 100{,}000$
$\quad x_3 = $ Amount invested in money market funds $\qquad\qquad\qquad -x_1 + x_2 + x_3 \leq 0$
$\qquad\qquad\qquad\qquad\qquad\qquad\qquad\qquad\qquad\qquad x_1, x_2, x_3 \geq 0$

$\quad$ \$50,000 in government bonds, \$0 in mutual funds, and \$50,000 in money market funds; max return is \$11,500

39. Let: $x_1 = $ Number of ads placed in daytime shows $\qquad$ Maximize $\quad P = 14{,}000x_1 + 24{,}000x_2 + 18{,}000x_3$
$\quad x_2 = $ Number of ads placed in prime-time shows $\qquad$ Subject to $\qquad\quad x_1 + \qquad x_2 + \qquad x_3 \leq 15$
$\quad x_3 = $ Number of ads placed in late-night shows $\qquad\qquad\qquad 1{,}000x_1 + 2{,}000x_2 + 1{,}500x_3 \leq 20{,}000$
$\qquad\qquad\qquad\qquad\qquad\qquad\qquad\qquad\qquad\qquad x_1, x_2, x_3 \geq 0$

$\quad$ 10 daytime ads, 5 prime-time ads, and 0 late-night ads; max number of potential customers is 260,000

41. (A) Let: $x_1 = $ Number of colonial houses $\qquad$ Maximize $\quad P = 20{,}000x_1 + 18{,}000x_2 + 24{,}000x_3$
$\quad x_2 = $ Number of split-level houses $\qquad$ Subject to $\qquad \frac{1}{2}x_1 + \qquad \frac{1}{2}x_2 + \qquad x_3 \leq 30$
$\quad x_3 = $ Number of ranch houses $\qquad\qquad\qquad 60{,}000x_1 + 60{,}000x_2 + 80{,}000x_3 \leq 3{,}200{,}000$
$\qquad\qquad\qquad\qquad\qquad\qquad\qquad\qquad 4{,}000x_1 + \ 3{,}000x_2 + \ 4{,}000x_3 \leq 180{,}000$
$\qquad\qquad\qquad\qquad\qquad\qquad\qquad\qquad\qquad\qquad x_1, x_2, x_3 \geq 0$

$\quad$ 20 colonial, 20 split-level, and 10 ranch houses; max profit is \$1,000,000

(B) 0 colonial, 40 split-level, and 10 ranch houses; max profit is \$960,000; 20,000 labor-hours are not used

(C) 45 colonial, 0 split-level, and 0 ranch houses; max profit is \$1,125,000; 7.5 acres of land and \$500,000 of capital are not used

43. (A) Let: $x_1 = $ Number of boxes of assortment I $\qquad$ Maximize $\quad P = 4x_1 + 3x_2 + 5x_3$
$\quad x_2 = $ Number of boxes of assortment II $\qquad$ Subject to $\qquad 4x_1 + 12x_2 + 8x_3 \leq 4{,}800$
$\quad x_3 = $ Number of boxes of assortment III $\qquad\qquad\qquad 4x_1 + \ 4x_2 + 8x_3 \leq 4{,}000$
$\qquad\qquad\qquad\qquad\qquad\qquad\qquad\qquad 12x_1 + \ 4x_2 + 8x_3 \leq 5{,}600$
$\qquad\qquad\qquad\qquad\qquad\qquad\qquad\qquad\qquad\qquad x_1, x_2, x_3 \geq 0$

$\quad$ 200 boxes of assortment I, 100 boxes of assortment II, and 350 boxes of assortment III; max profit is \$2,850

(B) 100 boxes of assortment I, 0 boxes of assortment II, and 550 boxes of assortment III; max profit is \$3,150; 200 fruit-filled candies are not used

(C) 0 boxes of assortment I, 50 boxes of assortment II, and 675 boxes of assortment III; max profit is \$3,525; 400 fruit-filled candies are not used

45. Let: $x_1 = $ Number of grams of food A $\qquad$ Maximize $\quad P = 3x_1 + 4x_2 + 5x_3$
$\quad x_2 = $ Number of grams of food B $\qquad$ Subject to $\qquad x_1 + 3x_2 + 2x_3 \leq 30$
$\quad x_3 = $ Number of grams of food C $\qquad\qquad\qquad 2x_1 + \ x_2 + 2x_3 \leq 24$
$\qquad\qquad\qquad\qquad\qquad\qquad\qquad\qquad\qquad\qquad x_1, x_2, x_3 \geq 0$

$\quad$ 0 g food A, 3 g food B, and 10.5 g food C; max protein is 64.5 units

47. Let: $x_1 = $ Number of undergraduate students $\qquad$ Maximize $\quad P = 18x_1 + 25x_2 + 30x_3$
$\quad x_2 = $ Number of graduate students $\qquad$ Subject to $\qquad x_1 + \qquad x_2 + \qquad x_3 \leq 20$
$\quad x_3 = $ Number of faculty members $\qquad\qquad\qquad 100x_1 + 150x_2 + 200x_3 \leq 3{,}200$
$\qquad\qquad\qquad\qquad\qquad\qquad\qquad\qquad\qquad\qquad x_1, x_2, x_3 \geq 0$

$\quad$ 0 undergraduate students, 16 graduate students, and 4 faculty members; max number of interviews is 520

Exercise 5-5

1. (A) Maximize $\quad P = 4y_1 + 5y_2$ $\qquad$ **(B)** $\quad y_1 + 2y_2 + x_1 \qquad\qquad = 8$ $\qquad$ **(C)**
$\qquad$ Subject to $\quad y_1 + 2y_2 \leq 8$ $\qquad\qquad\qquad\quad 3y_1 + \ y_2 \qquad + x_2 \qquad = 9$
$\qquad\qquad\qquad\qquad 3y_1 + \ y_2 \leq 9$ $\qquad\qquad\qquad -4y_1 - 5y_2 \qquad\qquad\quad + P = 0$
$\qquad\qquad\qquad\qquad\qquad y_1, y_2 \geq 0$

$$\begin{array}{cccccc|c} y_1 & y_2 & x_1 & x_2 & P & \\ 1 & 2 & 1 & 0 & 0 & 8 \\ 3 & 1 & 0 & 1 & 0 & 9 \\ -4 & -5 & 0 & 0 & 1 & 0 \end{array}$$

3. (A) Max $P = 121$ at $y_1 = 3$ and $y_2 = 5$ $\quad$ **(B)** Min $C = 121$ at $x_1 = 1$ and $x_2 = 2$

5. (A) Maximize $\quad P = 13y_1 + 12y_2$ $\quad$ **(B)** Min $C = 26$ at $x_1 = 0$ $\qquad$ **7. (A)** Maximize $\quad P = 15y_1 + 8y_2$ $\quad$ **(B)** Min $C = 54$ at $x_1 = 6$
$\qquad$ Subject to $\quad 4y_1 + 3y_2 \leq 9$ $\qquad$ and $x_2 = 13$ $\qquad\qquad\qquad$ Subject to $\quad 2y_1 + \ y_2 \leq 7$ $\qquad\qquad$ and $x_2 = 1$
$\qquad\qquad\qquad\quad y_1 + \ y_2 \leq 2$ $\qquad\qquad\qquad\qquad\qquad\qquad\qquad\qquad\qquad 3y_1 + 2y_2 \leq 12$
$\qquad\qquad\qquad\qquad\quad y_1, y_2 \geq 0$ $\qquad\qquad\qquad\qquad\qquad\qquad\qquad\qquad\qquad\qquad y_1, y_2 \geq 0$

9. (A) Maximize $P = 8y_1 + 4y_2$ **(B)** Min $C = 32$ at $x_1 = 0$ **11. (A)** Maximize $P = 6y_1 + 4y_2$ **(B)** No optimal solution exists.
  Subject to $2y_1 - 2y_2 \leq 11$ and $x_2 = 8$ Subject to $-3y_1 + y_2 \leq 7$
  $\quad\quad\quad y_1 + 3y_2 \leq 4$ $\quad\quad\quad\quad y_1 - 2y_2 \leq 9$
  $\quad\quad\quad\quad y_1, y_2 \geq 0$ $\quad\quad\quad\quad\quad y_1, y_2 \geq 0$

13. Min $C = 24$ at $x_1 = 8$ and $x_2 = 0$ **15.** Min $C = 20$ at $x_1 = 0$ and $x_2 = 4$ **17.** Min $C = 140$ at $x_1 = 14$ and $x_2 = 0$
19. Min $C = 44$ at $x_1 = 6$ and $x_2 = 2$ **21.** Min $C = 43$ at $x_1 = 0$, $x_2 = 1$, and $x_3 = 3$ **23.** No optimal solution exists.
25. 2 variables and 4 problem constraints   **27.** 2 constraints and any number of variables
29. No; the dual problem is not a standard maximization problem.   **31.** Yes; multiply both sides of the inequality by -1.
33. Min $C = 44$ at $x_1 = 0$, $x_2 = 3$, and $x_3 = 5$ **35.** Min $C = 166$ at $x_1 = 0$, $x_2 = 12$, $x_3 = 20$, and $x_4 = 3$
37. (A) Let: $x_1 = $ Numbers of hours the Cedarburg plant is operated   Minimize $C = 70x_1 + 75x_2 + 90x_3$
  $\quad\quad\quad x_2 = $ Number of hours the Grafton plant is operated   Subject to $20x_1 + 10x_2 + 20x_3 \geq 300$
  $\quad\quad\quad x_3 = $ Number of hours the West Bend plant is operated $\quad\quad\quad\quad 10x_1 + 20x_2 + 20x_3 \geq 200$
  $\quad\quad\quad\quad\quad\quad\quad\quad\quad\quad\quad\quad\quad\quad\quad\quad\quad\quad\quad\quad\quad\quad\quad x_1, x_2, x_3 \geq 0$
  Cedarburg plant 10 hr per day, West Bend plant 5 hr per day, Grafton plant not used; min cost is $1,150
  (B) West Bend plant 15 hr per day, Cedarburg and Grafton plants not used; min cost is $1,350
  (C) Grafton plant 10 hr per day, West Bend plant 10 hr per day, Cedarburg plant not used; min cost is $1,650

39. Let: $x_1 = $ Number of single-sided drives ordered from Associated Electronics   Minimize $C = 250x_1 + 350x_2 + 290x_3 + 320x_4$
  $\quad\quad\quad x_2 = $ Number of double-sided drives ordered from Associated Electronics   Subject to $x_1 + x_2 \quad\quad\quad\quad\quad \leq 1,000$
  $\quad\quad\quad x_3 = $ Number of single-sided drives ordered from Digital Drives $\quad\quad\quad\quad\quad\quad\quad\quad x_3 + x_4 \leq 2,000$
  $\quad\quad\quad x_4 = $ Number of double-sided drives ordered from Digital Drives $\quad\quad\quad\quad x_1 \quad\quad + x_3 \quad\quad \geq 1,200$
  $\quad\quad\quad\quad\quad\quad\quad\quad\quad\quad\quad\quad\quad\quad\quad\quad\quad\quad\quad\quad\quad\quad\quad\quad x_2 + \quad\quad x_4 \geq 1,600$
  $\quad\quad\quad\quad\quad\quad\quad\quad\quad\quad\quad\quad\quad\quad\quad\quad\quad\quad\quad\quad\quad\quad\quad\quad\quad x_1, x_2, x_3, x_4 \geq 0$

  1,000 single-sided drives from Associated Electronics, 200 single-sided and 1,600 double-sided drives from Digital Drives; min cost is $820,000

41. Let: $x_1 = $ Number of ounces of food L Minimize $C = 20x_1 + 24x_2 + 18x_3$
  $\quad\quad\quad x_2 = $ Number of ounces of food M Subject to $20x_1 + 10x_2 + 10x_3 \geq 300$
  $\quad\quad\quad x_3 = $ Number of ounces of food N $\quad\quad\quad\quad 10x_1 + 10x_2 + 10x_3 \geq 200$
  $\quad\quad\quad\quad\quad\quad\quad\quad\quad\quad\quad\quad\quad\quad\quad\quad\quad\quad 10x_1 + 15x_2 + 10x_3 \geq 240$
  $\quad\quad\quad\quad\quad\quad\quad\quad\quad\quad\quad\quad\quad\quad\quad\quad\quad\quad\quad\quad\quad\quad x_1, x_2, x_3 \geq 0$

  10 oz L, 8 oz M, 2 oz N; min cholesterol intake is 428 units

43. Let: $x_1 = $ Number of students bused from North Division to Central   Minimize $C = 5x_1 + 2x_2 + 3x_3 + 4x_4$
  $\quad\quad\quad x_2 = $ Number of students bused from North Division to Washington   Subject to $x_1 + x_2 \quad\quad\quad\quad\quad \geq 300$
  $\quad\quad\quad x_3 = $ Number of students bused from South Division to Central $\quad\quad\quad\quad\quad\quad\quad\quad x_3 + x_4 \geq 500$
  $\quad\quad\quad x_4 = $ Number of students bused from South Division to Washington $\quad\quad\quad\quad x_1 \quad\quad + x_3 \quad\quad \leq 400$
  $\quad\quad\quad\quad\quad\quad\quad\quad\quad\quad\quad\quad\quad\quad\quad\quad\quad\quad\quad\quad\quad\quad\quad\quad x_2 \quad\quad + x_4 \leq 500$
  $\quad\quad\quad\quad\quad\quad\quad\quad\quad\quad\quad\quad\quad\quad\quad\quad\quad\quad\quad\quad\quad\quad\quad\quad\quad x_1, x_2, x_3, x_4 \geq 0$

  300 students bused from North Division to Washington, 400 from South Division to Central, and 100 from South Division to Washington; min cost is $2,200

Exercise 5-6

1. (A) Maximize $P = 5x_1 + 2x_2 - Ma_1$ **(B)**
  Subject to $x_1 + 2x_2 + s_1 \quad\quad\quad\quad = 12$
  $\quad\quad\quad\quad x_1 + x_2 \quad\quad - s_2 + a_1 = 4$
  $\quad\quad\quad\quad\quad x_1, x_2, s_1, s_2, a_1 \geq 0$

	x_1	x_2	s_1	s_2	a_1	P	
	1	2	1	0	0	0	12
	1	1	0	-1	1	0	4
	$-M-5$	$-M-2$	0	M	0	1	$-4M$

  (C) $x_1 = 12$, $x_2 = 0$, $s_1 = 0$, $s_2 = 8$, $a_1 = 0$, $P = 60$ **(D)** Max $P = 60$ at $x_1 = 12$ and $x_2 = 0$

3. (A) Maximize $P = 3x_1 + 5x_2 - Ma_1$ **(B)**
  Subject to $2x_1 + x_2 + s_1 \quad\quad = 8$
  $\quad\quad\quad\quad x_1 + x_2 \quad\quad + a_1 = 6$
  $\quad\quad\quad\quad\quad x_1, x_2, s_1, a_1 \geq 0$

	x_1	x_2	s_1	a_1	P	
	2	1	1	0	0	8
	1	1	0	1	0	6
	$-M-3$	$-M-5$	0	0	1	$-6M$

  (C) $x_1 = 0$, $x_2 = 6$, $s_1 = 2$, $a_1 = 0$, $P = 30$ **(D)** Max $P = 30$ at $x_1 = 0$ and $x_2 = 6$

5. (A) Maximize $P = 4x_1 + 3x_2 - Ma_1$ **(B)**
  Subject to $-x_1 + 2x_2 + s_1 \quad\quad\quad\quad = 2$
  $\quad\quad\quad\quad x_1 + x_2 \quad\quad - s_2 + a_1 = 4$
  $\quad\quad\quad\quad\quad x_1, x_2, s_1, s_2, a_1 \geq 0$

	x_1	x_2	s_1	s_2	a_1	P	
	-1	2	1	0	0	0	2
	1	1	0	-1	1	0	4
	$-M-4$	$-M-3$	0	M	0	1	$-4M$

  (C) No optimal solution exists.   **(D)** No optimal solution exists.

7. (A) Maximize $P = 5x_1 + 10x_2 - Ma_1$
Subject to
$$x_1 + x_2 + s_1 \qquad = 3$$
$$2x_1 + 3x_2 \qquad - s_2 + a_1 = 12$$
$$x_1, x_2, s_1, s_2, a_1 \geqslant 0$$

(B)

x_1	x_2	s_1	s_2	a_1	P	
1	1	1	0	0	0	3
2	3	0	−1	1	0	12
$-2M - 5$	$-3M - 10$	0	M	0	1	$-12M$

(C) $x_1 = 0, x_2 = 3, s_1 = 0, s_2 = 0, a_1 = 3, P = -3M + 30$
(D) No optimal solution exists.

9. Min $P = 1$ at $x_1 = 3$ and $x_2 = 5$; Max $P = 16$ at $x_1 = 8$ and $x_2 = 0$
11. Max $P = 44$ at $x_1 = 2$ and $x_2 = 8$
13. No optimal solution exists. **15.** Min $C = -9$ at $x_1 = 0$, $x_2 = \frac{7}{4}$, and $x_3 = \frac{3}{4}$
17. Max $P = 32$ at $x_1 = 0$, $x_2 = 4$, and $x_3 = 2$
19. Max $P = 65$ at $x_1 = \frac{35}{2}$, $x_2 = 0$, and $x_3 = \frac{15}{2}$ **21.** Max $P = 120$ at $x_1 = 20$, $x_2 = 0$, and $x_3 = 20$
23. Problem 5—Unbounded feasible region: Problem 7—Empty feasible region:

25. Min $C = -30$ at $x_1 = 0$, $x_2 = \frac{3}{4}$, and $x_3 = 0$ **27.** Max $P = 17$ at $x_1 = \frac{49}{5}$, $x_2 = 0$, and $x_3 = \frac{22}{5}$
29. Min $C = \frac{135}{2}$ at $x_1 = \frac{15}{4}$, $x_2 = \frac{3}{4}$, and $x_3 = 0$ **31.** Max $P = 372$ at $x_1 = 28$, $x_2 = 4$, and $x_3 = 0$
33. Let: x_1 = Number of 16k modules manufactured daily   Maximize $P = 18x_1 + 30x_2$
x_2 = Number of 64k modules manufactured daily   Subject to $10x_1 + 15x_2 \leqslant 2{,}200$
$$2x_1 + 4x_2 \leqslant 500$$
$$x_1 \qquad \geqslant 50$$
$$x_1, x_2 \geqslant 0$$

130 of the 16k modules and 60 of the 64k modules; max profit is \$4,140
35. Let: x_1 = Number of ads placed in the *Sentinel*   Minimize $C = 200x_1 + 200x_2 + 100x_3$
x_2 = Number of ads placed in the *Journal*   Subject to $x_1 + x_2 + x_3 \leqslant 10$
x_3 = Number of ads placed in the *Tribune* $2{,}000x_1 + 500x_2 + 1{,}500x_3 \geqslant 16{,}000$
$$x_1, x_2, x_3 \geqslant 0$$

2 ads in the *Sentinel*, 0 ads in the *Journal*, 8 ads in the *Tribune*; min cost is \$1,200
37. Let: x_1 = Number of bottles of brand A Minimize $C = 0.6x_1 + 0.4x_2 + 0.9x_3$
x_2 = Number of bottles of brand B Subject to $10x_1 + 10x_2 + 20x_3 \geqslant 100$
x_3 = Number of bottles of brand C $2x_1 + 3x_2 + 4x_3 \leqslant 24$
$$x_1, x_2, x_3 \geqslant 0$$

0 bottles of A, 4 bottles of B, 3 bottles of C; min cost is \$4.30
39. Let: x_1 = Number of cubic yards of mix A Maximize $P = 12x_1 + 16x_2 + 8x_3$
x_2 = Number of cubic yards of mix B Subject to $12x_1 + 8x_2 + 16x_3 \leqslant 700$
x_3 = Number of cubic yards of mix C $16x_1 + 8x_2 + 16x_3 \geqslant 800$
$$x_1, x_2, x_3 \geqslant 0$$

25 yd³ A, 50 yd³ B, 0 yd³ C; max is 1,100 lb
41. Let: x_1 = Number of car frames produced at the Milwaukee plant   Maximize $P = 50x_1 + 70x_2 + 50x_3 + 70x_4$
x_2 = Number of truck frames produced at the Milwaukee plant   Subject to $x_1 \qquad + x_3 \qquad \leqslant 250$
x_3 = Number of car frames produced at the Racine plant $x_2 \qquad + x_4 \leqslant 350$
x_4 = Number of truck frames produced at the Racine plant $x_1 + x_2 \qquad\qquad \leqslant 300$
$$x_3 + x_4 \leqslant 200$$
$$150x_1 + 200x_2 \qquad\qquad\qquad \leqslant 50{,}000$$
$$135x_3 + 180x_4 \leqslant 35{,}000$$
$$x_1, x_2, x_3, x_4 \geqslant 0$$

43. Let: x_1 = Number of barrels of A used in regular gasoline
x_2 = Number of barrels of A used in premium gasoline
x_3 = Number of barrels of B used in regular gasoline
x_4 = Number of barrels of B used in premium gasoline
x_5 = Number of barrels of C used in regular gasoline
x_6 = Number of barrels of C used in premium gasoline

Maximize $P = 10x_1 + 18x_2 + 8x_3 + 16x_4 + 4x_5 + 12x_6$
Subject to
$$x_1 + x_2 \leq 40{,}000$$
$$x_3 + x_4 \leq 25{,}000$$
$$x_5 + x_6 \leq 15{,}000$$
$$x_1 + x_3 + x_5 \geq 30{,}000$$
$$x_2 + x_4 + x_6 \geq 25{,}000$$
$$-5x_1 + 5x_3 + 15x_5 \geq 0$$
$$-15x_2 - 5x_4 + 5x_6 \geq 0$$
$$x_1, x_2, x_3, x_4, x_5, x_6 \geq 0$$

45. Let: x_1 = Number of ounces of food L
x_2 = Number of ounces of food M
x_3 = Number of ounces of food N

Minimize $C = 0.4x_1 + 0.6x_2 + 0.8x_3$
Subject to
$$30x_1 + 10x_2 + 30x_3 \geq 400$$
$$10x_1 + 10x_2 + 10x_3 \geq 200$$
$$10x_1 + 30x_2 + 20x_3 \geq 300$$
$$8x_1 + 4x_2 + 6x_3 \leq 150$$
$$60x_1 + 40x_2 + 50x_3 \leq 900$$
$$x_1, x_2, x_3 \geq 0$$

47. Let: x_1 = Number of students from town A enrolled in school I
x_2 = Number of students from town A enrolled in school II
x_3 = Number of students from town B enrolled in school I
x_4 = Number of students from town B enrolled in school II
x_5 = Number of students from town C enrolled in school I
x_6 = Number of students from town C enrolled in school II

Minimize $C = 4x_1 + 8x_2 + 6x_3 + 4x_4 + 3x_5 + 9x_6$
Subject to
$$x_1 + x_2 = 500$$
$$x_3 + x_4 = 1{,}200$$
$$x_5 + x_6 = 1{,}800$$
$$x_1 + x_3 + x_5 \leq 2{,}000$$
$$x_2 + x_4 + x_6 \leq 2{,}000$$
$$x_1 + x_3 + x_5 \geq 1{,}400$$
$$x_2 + x_4 + x_6 \geq 1{,}400$$
$$x_1 \leq 300$$
$$x_2 \leq 300$$
$$x_3 \leq 720$$
$$x_4 \leq 720$$
$$x_5 \leq 1{,}080$$
$$x_6 \leq 1{,}080$$
$$x_1, x_2, x_3, x_4, x_5, x_6 \geq 0$$

Chapter 5   Review Exercise

1. Bounded *(5-1)*

2. Unbounded *(5-1)*

3. Max $P = 24$ at $x_1 = 4$ and $x_2 = 0$ *(5-2)*

4. $2x_1 + x_2 + s_1 = 8$
$x_1 + 2x_2 + s_2 = 10$ *(5-3)*

5. 2 basic and 2 nonbasic variables *(5-3)*

6.

x_1	x_2	s_1	s_2	FEASIBLE?
0	0	8	10	Yes
0	8	0	−6	No
0	5	3	0	Yes
4	0	0	6	Yes
10	0	−12	0	No
2	4	0	0	Yes

(5-3)

7. $\text{Exit} \rightarrow$

Enter ↓

$$\begin{array}{c} \\ s_1 \\ s_2 \\ P \end{array} \begin{bmatrix} x_1 & x_2 & s_1 & s_2 & P & \\ 2 & 1 & 1 & 0 & 0 & 8 \\ 1 & 2 & 0 & 1 & 0 & 10 \\ \hdashline -6 & -2 & 0 & 0 & 1 & 0 \end{bmatrix}$$ *(5-4)*

8. Max $P = 24$ at $x_1 = 4$ and $x_2 = 0$ *(5-4)*

9. Basic variables: x_2, s_2, s_3, P; nonbasic variables: x_1, x_3, s_1

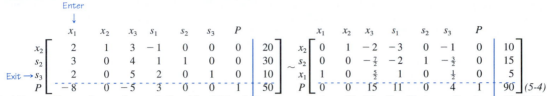

$$
\begin{array}{c}
\text{Enter} \\
\downarrow
\end{array}
$$

	x_1	x_2	x_3	s_1	s_2	s_3	P	
x_2	2	1	3	-1	0	0	0	20
s_2	3	0	4	1	1	0	0	30
Exit → s_3	2	0	5	2	0	1	0	10
P	-8	0	-5	3	0	0	1	50

$\sim$

	x_1	x_2	x_3	s_1	s_2	s_3	P	
x_2	0	1	-2	-3	0	-1	0	10
s_2	0	0	$-\tfrac{7}{2}$	-2	1	$-\tfrac{3}{2}$	0	15
x_1	1	0	$\tfrac{5}{2}$	1	0	$\tfrac{1}{2}$	0	5
P	0	0	15	11	0	4	1	90	*(5-4)*

10. (A) $x_1 = 0, x_2 = 2, s_1 = 0, s_2 = 5, P = 12$; additional pivoting required   **11.** Min $C = 40$ at $x_1 = 0$ and $x_2 = 20$ *(5-2)*
(B) $x_1 = 0, x_2 = 0, s_1 = 0, s_2 = 7, P = 22$; no optimal solution exists
(C) $x_1 = 6, x_2 = 0, s_1 = 15, s_2 = 0, P = 10$; optimal solution *(5-4)*

12. Maximize $P = 15y_1 + 20y_2$
Subject to $y_1 + 2y_2 \le 5$
$\qquad\quad 3y_1 + \;\; y_2 \le 2$
$\qquad\qquad\; y_1, y_2 \ge 0$ *(5-5)*

13. $\begin{aligned}
y_1 + \;\; 2y_2 + x_1 \qquad\qquad &= 5 \\
3y_1 + \;\;\; y_2 \qquad\; + x_2 \quad\;\; &= 2 \\
-15y_1 - 20y_2 \qquad\qquad\; + P &= 0
\end{aligned}$ *(5-5)*

14.
y_1	y_2	x_1	x_2	P	
1	2	1	0	0	5
3	1	0	1	0	2
-15	-20	0	0	1	0	*(5-5)*

15. Max $P = 40$ at $y_1 = 0$ and $y_2 = 2$ *(5-4)*   **16.** Min $C = 40$ at $x_1 = 0$ and $x_2 = 20$ *(5-5)*   **17.** Max $P = 26$ at $x_1 = 2$ and $x_2 = 5$ *(5-2)*
18. Max $P = 26$ at $x_1 = 2$ and $x_2 = 5$ *(5-4)*   **19.** Min $C = 51$ at $x_1 = 9$ and $x_2 = 3$ *(5-2)*   **20.** Maximize $P = 10y_1 + 15y_2 + 3y_3$
21. Min $C = 51$ at $x_1 = 9$ and $x_2 = 3$ *(5-5)* $\qquad\qquad\qquad\qquad\qquad\qquad\qquad\qquad$ Subject to $y_1 + \;\; y_2 \qquad\quad \le 3$
22. No optimal solution exists. *(5-4)* $\qquad\qquad\qquad\qquad\qquad\qquad\qquad\qquad\qquad\qquad\quad$ $y_1 + 2y_2 + y_3 \le 8$
23. Max $P = 23$ at $x_1 = 4, x_2 = 1$, and $x_3 = 0$ *(5-4)* $\qquad\qquad\qquad\qquad\qquad\qquad\qquad\qquad\qquad$ $y_1, y_2, y_3 \ge 0$ *(5-5)*
24. (A) Modified problem:
$\quad$ Maximize $P = x_1 + 3x_2 - Ma_1$
$\quad$ Subject to $x_1 + \;\; x_2 - s_1 + a_1 \qquad\;\; = 6$
$\qquad\qquad\;\;\; x_1 + 2x_2 \qquad\qquad + s_2 = 8$
$\qquad\qquad\;\;\; x_1, x_2, s_1, s_2, a_1 \ge 0$
(B) Preliminary simplex tableau: $\qquad\qquad$ Initial simplex tableau:

x_1	x_2	s_1	a_1	s_2	P	
1	1	-1	1	0	0	6
1	2	0	0	1	0	8
-1	-3	0	M	0	1	0

x_1	x_2	s_1	a_1	s_2	P	
1	1	-1	1	0	0	6
1	2	0	0	1	0	8
$-M-1$	$-M-3$	M	0	0	1	$-6M$

(C) $x_1 = 4, x_2 = 2, s_1 = 0, a_1 = 0, s_2 = 0, P = 10$
(D) Since $a_1 = 0$, the solution to the original problem is Max $P = 10$ at $x_1 = 4$ and $x_2 = 2$. *(5-6)*

25. (A) Modified problem:
$\quad$ Maximize $P + x_1 + x_2 - Ma_1$
$\quad$ Subject to $x_1 + \;\; x_2 - s_1 + a_1 \qquad\;\; = 5$
$\qquad\qquad\;\;\; x_1 + 2x_2 \qquad\qquad + s_2 = 4$
$\qquad\qquad\;\;\; x_1, x_2, s_1, s_2, a_1 \ge 0$
(B) Preliminary simplex tableau: $\qquad\qquad$ Initial simplex tableau:

x_1	x_2	s_1	a_1	s_2	P	
1	1	-1	1	0	0	5
1	2	0	0	1	0	4
-1	-1	0	M	0	1	0

x_1	x_2	s_1	a_1	s_2	P	
1	1	-1	1	0	0	5
1	2	0	0	1	0	4
$-M-1$	$-M-1$	M	0	0	1	$-5M$

(C) $x_1 = 4, x_2 = 0, s_1 = 0, s_2 = 0, a_1 = 1, P = -M + 4$ (D) Since $a_1 \ne 0$, the original problem has no solution. *(5-6)*

26. Maximize $P = 2x_1 + 3x_2 + x_3 - Ma_1 - Ma_2$
$\quad$ Subject to $\quad x_1 - 3x_2 + \;\; x_3 + s_1 \qquad\qquad\qquad = 7$
$\qquad\qquad\quad\;\; x_1 + \;\; x_2 - 2x_3 \qquad - s_2 + a_1 \qquad\; = 2$
$\qquad\qquad\;\; 3x_1 + 2x_2 - \;\; x_3 \qquad\qquad\qquad + a_2 = 4$
$\qquad\qquad\quad\;\; x_1, x_2, x_3, s_1, s_2, a_1, a_2 \ge 0$ *(5-6)*

27. The geometric method solves maximization and minimization problems involving two decision variables. If the feasible region is bounded, there are no restrictions on the coefficients or constants. If the feasible region is unbounded and the coefficients of the objective function are positive, then a minimization problem has a solution, but a maximization does not. *(5-2)*

28. The basic simplex method with slack variables solves standard maximization problems involving $\le$ constraints with nonnegative constants on the right side. *(5-4)*

29. The dual method solves minimization problems with positive coefficients in the objective function. *(5-5)*

30. The big M method solves any linear programming problem. *(5-6)*

31. Max $P = 36$ at $x_1 = 6$, $x_2 = 8$ *(5-2)*

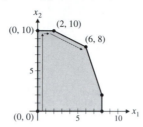

32. Min $C = 15$ at $x_1 = 3$ and $x_2 = 3$ *(5-5)*

33. Min $C = 15$ at $x_1 = 3$ and $x_2 = 3$ *(5-6)*

34. Min $C = 9,960$ at $x_1 = 0$, $x_2 = 240$, $x_3 = 400$, and $x_4 = 60$ *(5-5)*

35. (A) Let: x_1 = Number of regular sails Maximize $P = 100x_1 + 200x_2$
   x_2 = Number of competition sails Subject to $2x_1 + 3x_2 \leq 150$
   $4x_1 + 10x_2 \leq 380$
   $x_1, x_2 \geq 0$
   Max $P = \$8,500$ when 45 regular and 20 competition sails are produced.
   (B) Max profit increases to $\$9,880$ when 38 competition and no regular sails are produced.
   (C) Max profit decreases to $\$7,500$ when no competition and 75 regular sails are produced. *(5-2)*

36. (A) Let: x_1 = Amount invested in oil stock Maximize $P = 0.12x_1 + 0.09x_2 + 0.05x_3$
   x_2 = Amount invested in steel stock Subject to $x_1 + x_2 + x_3 \leq 150,000$
   x_3 = Amount invested in government bonds $x_1 \leq 50,000$
   $x_1 + x_2 - x_3 \leq 25,000$
   $x_1, x_2, x_3 \geq 0$
   Max return is $\$12,500$ when $\$50,000$ is invested in oil stock, $\$37,500$ is invested in steel stock, and $\$62,500$ in government bonds.
   (B) Max return is $\$13,625$ when $\$87,500$ is invested in steel stock and $\$62,500$ in government bonds. *(5-4)*

37. Let: x_1 = Number of motors shipped from factory A to plant X Minimize $C = 5x_1 + 8x_2 + 9x_3 + 7x_4$
   x_2 = Number of motors shipped from factory A to plant Y Subject to $x_1 + x_2 \leq 1,500$
   x_3 = Number of motors shipped from factory B to plant X $x_3 + x_4 \leq 1,000$
   x_4 = Number of motors shipped from factory B to plant Y $x_1 + x_3 \geq 900$
   $x_2 + x_4 \geq 1,200$
   $x_1, x_2, x_3, x_4 \geq 0$
   Min $C = \$13,100$ when 900 motors are shipped from factory A to plant X, 200 motors are shipped from factory A to plant Y, and 1,000 motors are shipped from factory B to plant Y *(5-5)*

38. Let: x_1 = Number of pounds of long-grain rice used in brand A Maximize $P = 0.8x_1 + 0.5x_2 - 1.9x_3 - 2.2x_4$
   x_2 = Number of pounds of long-grain rice used in brand B Subject to $0.1x_1 - 0.9x_3 \leq 0$
   x_3 = Number of pounds of wild rice used in brand A $0.05x_2 - 0.95x_4 \leq 0$
   x_4 = Number of pounds of wild rice used in brand B $x_1 + x_2 \leq 8,000$
   $x_3 + x_4 \leq 500$
   $x_1, x_2, x_3, x_4 \geq 0$
   Max profit is $\$3,350$ when 1,350 lb long-grain rice and 150 lb wild rice are used to produce 1,500 lb brand A, and 6,650 lb long-grain rice and 350 lb wild rice are used to produce 7,000 lb brand B. *(5-4)*

39. (A) Let: x_1 = Number of grams of mix A Minimize $C = 0.04x_1 + 0.09x_2$
   x_2 = Number of grams of mix B Subject to $2x_1 + 5x_2 \geq 850$
   $2x_1 + 4x_2 \geq 800$
   $4x_1 + 5x_2 \geq 1,150$
   $x_1, x_2 \geq 0$
   Min $C = \$16.50$ when 300 g mix A and 50 g mix B are used.
   (B) The minimum cost decreases to $\$13.00$ when 100 g mix A and 150 g mix B are used.
   (C) The minimum cost increases to $\$17.00$ when 425 g mix A and no mix B are used. *(5-2)*

■ CHAPTER 6

Exercise 6-1

1. 115   **3.** 300   **5.** 165   **7.** 210   **9.** 95   **11.** 260

13. (A) 4 ways:

Coin 1 Outcome	Coin 2 Outcome	Combined Outcomes
H	H	(H, H)
	T	(H, T)
T	H	(T, H)
	T	(T, T)

(B) $2 \cdot 2 = 4$ ways

15. (A) 12 combined outcomes:

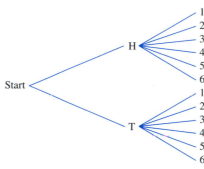

(B) $2 \cdot 6 = 12$ combined outcomes

17. $n(A \cap B') = 60$, $n(A \cap B) = 20$, $n(A' \cap B) = 30$, $n(A' \cap B') = 90$ **19.** $n(A \cap B') = 5$, $n(A \cap B) = 20$, $n(A' \cap B) = 35$, $n(A' \cap B') = 40$

21.

	A	A'	Totals
B	30	60	90
B'	40	70	110
Totals	70	130	200

23.

	A	A'	Totals
B	20	35	55
B'	25	20	45
Totals	45	55	100

25. $5 \cdot 3 \cdot 4 \cdot 2 = 120$
27. $6 \cdot 5 \cdot 4 \cdot 3 = 360$; $6 \cdot 6 \cdot 6 \cdot 6 = 1,296$; $6 \cdot 5 \cdot 5 \cdot 5 = 750$
29. $10 \cdot 9 \cdot 8 \cdot 7 \cdot 6 = 30,240$; $10 \cdot 10 \cdot 10 \cdot 10 \cdot 10 = 100,000$;
  $10 \cdot 9 \cdot 9 \cdot 9 \cdot 9 = 65,610$
31. $26 \cdot 26 \cdot 26 \cdot 10 \cdot 10 \cdot 10 = 17,576,000$;
  $26 \cdot 25 \cdot 24 \cdot 10 \cdot 9 \cdot 8 = 11,232,000$

33. No, the same 8 combined choices are available either way.   **35.** 14   **37.** 14

39. (A) 6 combined outcomes:

41. 12   **43.** (A) 1,010   (B) 190   (C) 270   **45.** 1,570
47. (A) 102   (B) 689   (C) 1,470   (D) 1,372

(B) $3 \cdot 2 = 6$

49. (A) 12 classifications:

51. 2,905

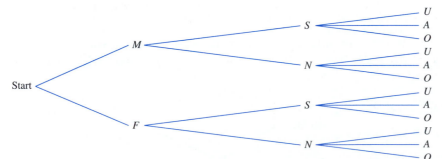

(B) $2 \cdot 2 \cdot 3 = 12$

Exercise 6-2

1. 24   **3.** 9   **5.** 990   **7.** 10   **9.** 35   **11.** 1   **13.** 60   **15.** 6,497,400   **17.** 10   **19.** 270,725
21. $P_{10,3} = 10 \cdot 9 \cdot 8 = 720$ **23.** $C_{7,3} = 35$; $P_{7,3} = 210$
25. The factorial function $x!$ grows much faster than the exponential function 3^x, which in turn grows much faster than the cubic function x^3.
27. $C_{13,5} = 1,287$ **29.** $C_{13,5}C_{13,2} = 100,386$ **31.** $C_{8,3}C_{10,4}C_{7,2} = 246,960$ **33.** (A) $C_{8,2} = 28$ (B) $C_{8,3} = 56$ (C) $C_{8,4} = 70$
35. $P_{5,2} = 20$; $P_{5,3} = 60$; $P_{5,4} = 120$; $P_{5,5} = 120$ **37.** (A) $P_{8,5} = 6,720$ (B) $C_{8,5} = 56$ (C) $2 \cdot C_{6,4} = 30$
39. For many calculators $k = 69$, but your calculator may be different.   **41.** (A) $C_{24,3} = 2,024$ (B) $C_{19,3} = 969$
43. (A) $C_{30,10} = 30,045,015$ (B) $C_{8,2}C_{12,5}C_{10,3} = 2,661,120$
45. (A) $C_{6,3}C_{5,2} = 200$ (B) $C_{6,4}C_{5,1} = 75$ (C) $C_{6,5} = 6$ (D) $C_{11,5} = 462$ (E) $C_{6,4}C_{5,1} + C_{6,5} = 81$
47. 336; 512   **49.** $P_{4,2} = 12$

Exercise 6-3

1. Occurrence of E is certain.   **3.** $\frac{1}{2}$
5. (A) Reject; no probability can be negative (B) Reject; $P(J) + P(G) + P(P) + P(S) \neq 1$ (C) Acceptable
7. $P(J) + P(P) = .56$ **9.** $\frac{1}{8}$ **11.** $1/P_{10,3} \approx .0014$ **13.** $C_{26,5}/C_{52,5} \approx .025$ **15.** $C_{12,5}/C_{52,5} \approx .000\ 305$
17. $S = \{$All days in a year, 365, excluding leap year$\}$; $\frac{1}{365}$, assuming each day is as likely as any other day for a person to be born
19. $1/P_{5,5} = 1/5! = .008\ 33$ **21.** $\frac{1}{36}$ **23.** $\frac{5}{36}$ **25.** $\frac{1}{6}$ **27.** $\frac{7}{9}$ **29.** 0   **31.** $\frac{1}{3}$ **33.** $\frac{2}{9}$ **35.** $\frac{2}{3}$ **37.** $\frac{1}{4}$ **39.** $\frac{1}{4}$ **41.** $\frac{3}{4}$
43. $\frac{1}{9}$ **45.** $\frac{1}{3}$ **47.** $\frac{1}{9}$ **49.** $\frac{4}{9}$ **51.** $C_{16,5}/C_{52,5} \approx .001\ 68$ **53.** $48/C_{52,5} \approx .000\ 018\ 5$ **55.** $4/C_{52,5} \approx .000\ 001\ 5$
57. $C_{4,2}C_{4,3}/C_{52,5} \approx .000\ 009$ **59.** (A) $1/P_{12,4} \approx .000\ 084$ (B) $1/12^4 \approx .000\ 048$
61. (A) $C_{6,3}C_{5,2}/C_{11,5} \approx .433$ (B) $C_{6,4}C_{5,1}/C_{11,5} \approx .162$ (C) $C_{6,5}C_{11,5} \approx .013$ (D) $(C_{6,4}C_{5,1} + C_{6,5})/C_{11,5} \approx .175$
63. (A) $1/P_{8,3} \approx .0030$ (B) $1/8^3 \approx .0020$ **65.** (A) $P_{6,2}/P_{11,2} \approx .273$ (B) $(C_{5,3} + C_{6,1}C_{5,2})/C_{11,3} \approx .424$

Exercise 6-4

1. .997   **3.** (1), $\frac{1}{2}$ **5.** (2), $\frac{7}{10}$ **7.** .4   **9.** .25   **11.** .05   **13.** .2   **15.** .6   **17.** .65   **19.** $\frac{1}{4}$ **21.** $\frac{11}{36}$
23. (A) $\frac{3}{5}$; $\frac{3}{5}$ (B) $\frac{1}{3}$; $\frac{3}{7}$ (C) $\frac{2}{3}$; $\frac{3}{2}$ (D) $\frac{11}{4}$; $\frac{9}{11}$ **25.** (A) $\frac{3}{11}$ (B) $\frac{13}{18}$ (C) $\frac{4}{5}$ or .8 (D) .49   **27.** 1 : 1   **29.** 7 : 1   **31.** 2 : 1   **33.** 1 : 2
35. (A) $\frac{1}{8}$ (B) \$8   **37.** (A) .31; $\frac{31}{69}$ (B) .6; $\frac{3}{2}$ **39.** $\frac{11}{26}$; $\frac{11}{15}$ **41.** $\frac{7}{13}$; $\frac{7}{6}$ **43.** .78   **45.** $\frac{250}{1,000} = .25$
47. Either events A, B, and C are mutually exclusive, or events A and B are not mutually exclusive and the other pairs of events are mutually exclusive.
49. There are fewer calculator steps, and, in addition, 365! produces an overflow error on many calculators, while $P_{365,n}$ does not produce an overflow error for many values of n.
51. $P(E) = 1 - \dfrac{12!}{(12-n)!\,12^n}$ **55.** (A) $P(C \cup S) = P(C) + P(S) - P(C \cap S) = .45 + .75 - .35 = .85$ (B) $P(C' \cap S') = .15$
57. (A) $P(M_1 \cup A) = P(M_1) + P(A) - P(M_1 \cap A) = .2 + .3 - .05 = .45$ **59.** $P(K' \cap D') = .9$ **61.** .83
    (B) $P[(M_2 \cap A') \cup (M_3 \cap A')] = P(M_2 \cap A') + P(M_3 \cap A') = .2 + .35 = .55$
63. $P(A \cap S) = \frac{50}{1,000} = .05$ **65.** (A) $P(U \cup N) = .22$; $\frac{11}{39}$ (B) $P[(D \cap A) \cup (R \cap A)] = .3$; $\frac{7}{3}$ **67.** $1 - C_{15,3}/C_{20,3} \approx .6$

Exercise 6-5

1. .50   **3.** .20   **5.** .10   **7.** .06   **9.** .50   **11.** .30   **13.** Independent   **15.** Dependent   **17.** (A) $\frac{1}{2}$ (B) $2(\frac{1}{2})^8 \approx .007\ 81$
19. (A) $\frac{1}{4}$ **21.** (A) .18   **23.** (A) Independent and not mutually exclusive   **25.** $(\frac{1}{2})(\frac{1}{2}) = \frac{1}{4}$; $\frac{1}{2} + \frac{1}{2} - \frac{1}{4} = \frac{3}{4}$
    (B) Dependent   (B) .26   (B) Dependent and mutually exclusive.
27. (A) $(\frac{1}{4})(\frac{13}{51}) \approx .0637$ (B) $(\frac{1}{4})(\frac{1}{4}) = .0625$ **29.** (A) $\frac{3}{13}$ (B) Independent   **31.** (A) Dependent (B) Independent

33. (A) Start, (B) Start

35. (A) $\frac{24}{49}$ (B) $\frac{11}{21}$ **37.** $\frac{5}{18}$ **39.** (A) .167 (B) .25 (C) .25   **43.** $P(A|A) = P(A \cap A)/P(A) = P(A)/P(A) = 1$

45. $P(A)P(B) \neq 0 = P(A \cap B)$

47. (A)

	H	S	B	Totals
Y	.400	.180	.020	.600
N	.150	.120	.130	.400
Totals	.550	.300	.150	1.000

(B) $P(Y|H) = \dfrac{.400}{.550} \approx .727$ (C) $P(Y|B) = \dfrac{.020}{.150} \approx .133$

(D) $P(S) = .300; P(S|Y) = .300$ (E) $P(H) = .550; P(H|Y) \approx .667$

(F) $P(B \cap N) = .130$ (G) Yes   (H) No   (I) No

49. (A) .167   (B) .25   (C) .25

51. (A)

	C	C'	Totals
R	.06	.44	.50
R'	.02	.48	.50
Totals	.08	.92	1.00

(B) Dependent

(C) $P(C|R) = .12$ and $P(C) = .08$; since $P(C|R) > P(C)$, the red dye should be banned.

(D) $P(C|R) = .04$ and $P(C) = .08$; since $P(C|R) < P(C)$, the red dye should not be banned, since it appears to prevent cancer.

53. (A)

	A	B	C	Totals
F	.130	.286	.104	.520
F'	.120	.264	.096	.480
Totals	.250	.550	.200	1.000

(B) $P(A|F) = \dfrac{.130}{.520} = .250; P(A|F') = \dfrac{.120}{.480} = .250$

(C) $P(C|F) = \dfrac{.104}{.520} = .200; P(C|F') = \dfrac{.096}{.480} = .200$

(D) $P(A) = .250$ (E) $P(B) = .550; P(B|F') = .550$ (F) $P(F \cap C) = .104$

(G) No; A, B and C are independent of F and F'.

Exercise 6-6

1. $(.6)(.8) = .48$ **3.** $(.6)(.8) + (.4)(.3) = .60$ **5.** .80   **7.** .417   **9.** .375   **11.** .222   **13.** .50   **15.** .278   **17.** .125

19. .50   **21.** .375   **23.** Start   **25.** .25   **27.** .333   **29.** .50   **31.** .745

33. M and U are independent, which follows from any one of the following:
$P(M|U) = c = P(M); P(U|M) = a = P(U); P(M \cap U) = ca = P(M)P(U).$

35. .235   **37.** $\dfrac{P(U_1 \cap R)}{P(R)} + \dfrac{P(U_1' \cap R)}{P(R)} = \dfrac{P(U_1 \cap R) + P(U_1' \cap R)}{P(R)} = \dfrac{P(R)}{P(R)} = 1$

39. .913; .226   **41.** .091; .545; .364   **43.** .667; .000 412

45. .231; .036   **47.** .941; .0588

Exercise 6-7

1. $E(X) = -.1$ **3.** Probability distribution: **5.** Payoff table: **7.** Payoff table: **9.** $-\$0.50$

11. $-\$0.036; \$0.036$

x_i	0	1	2
p_i	$\frac{1}{4}$	$\frac{1}{2}$	$\frac{1}{4}$

$E(X) = 1$

x_i	\$1	$-\$1$
p_i	$\frac{1}{2}$	$\frac{1}{2}$

$E(X) = 0$; game is fair

x_i	$-\$3$	$-\$2$	$-\$1$	\$0	\$1	\$2
p_i	$\frac{1}{6}$	$\frac{1}{6}$	$\frac{1}{6}$	$\frac{1}{6}$	$\frac{1}{6}$	$\frac{1}{6}$

$E(X) = -50¢$; game is not fair

13. $40. Let $x =$ amount you should lose if a 6 turns up. Set up a payoff table; then set the expected value of the game equal to zero and solve for x.

15. Win $1   **17.** A_2; $210   **19.** Payoff table:

x_i	$35	$-$1
p_i	$\frac{1}{38}$	$\frac{37}{38}$

$$E(X) = -5.26¢$$

21. .002   **23.** Payoff table:

x_i	$499	$99	$19	$4	$-$1
p_i	.0002	.0006	.001	.004	.9942

$$E(X) = -80¢$$

25. (A)

x_i	0	1	2
p_i	$\frac{7}{15}$	$\frac{7}{15}$	$\frac{1}{15}$

(B) .60

27. (A)

x_i	$-$5	$195	$395	$595
p_i	.985	.0149	.000 059 9	.000 000 06

(B) $E(X) \approx -$2$

29. Payoff table:

x_i	$4,850	$-$150
p_i	.01	.99

$$E(X) = -$100$$

31. Site A, with $E(X) = 3.6 million
33. 1.54
35. For A_1, $E(X) = 4, and for A_2, $E(X) = 4.80; A_2 is better

Chapter 6   Review Exercise

1. (A) 12 combined outcomes:

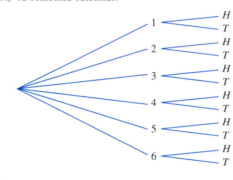

(B) $6 \cdot 2 = 12$ *(6-1)*

2. (A) 65 (B) 75 (C) 35 (D) 105 (E) 150
(F) 85 (G) 115 (H) 45 *(6-1)*
3. 15; 30 *(6-2)*   **4.** $6 \cdot 5 \cdot 4 \cdot 3 \cdot 2 \cdot 1 = 720$ *(6-1)*
5. $P_{6,6} = 6! = 720$ *(6-2)*   **6.** $C_{13,5}/C_{52,5} \approx .0005$ *(6-3)*
7. $1/P_{15,2} \approx .0048$ *(6-3)*
8. $1/P_{10,3} \approx .0014$; $1/C_{10,3} \approx .0083$ *(6-3)*
9. .05 *(6-3)*   **10.** Payoff table:

x_i	$-$2	$-$1	$0	$1	$2
p_i	$\frac{1}{5}$	$\frac{1}{5}$	$\frac{1}{5}$	$\frac{1}{5}$	$\frac{1}{5}$

$E(X) = 0$; game is fair *(6-7)*

11. (A) .7 (B) .6 *(6-4)*   **12.** $P(R \cup G) = .8$; odds for $R \cup G$ are 8 to 2 *(6-4)*   **13.** $\frac{5}{11} \approx .455$ *(6-4)*   **14.** .27 *(6-5)*   **15.** .20 *(6-5)*
16. .02 *(6-5)*   **17.** .03 *(6-5)*   **18.** .15 *(6-5)*   **19.** .1304 *(6-5)*   **20.** .1 *(6-5)*   **21.** No, since $P(T|Z) \neq P(T)$ *(6-5)*
22. Yes, since $P(S \cap X) = P(S)P(X)$ *(6-5)*   **23.** .4 *(6-5)*   **24.** .2 *(6-5)*   **25.** .3 *(6-5)*   **26.** .08 *(6-5)*   **27.** .18 *(6-5)*
28. .26 *(6-5)*   **29.** .31 *(6-6)*   **30.** .43 *(6-6)*
31. (A) $\frac{10}{32}$ (B) $\frac{1}{4}$ (C) As the sample in part (A) is increased in size, approximate empirical probabilities should approach the theoretical probabilities. *(6-3)*
32. Payoff table:

x_i	$5	$-$4	$2
p_i	.25	.5	.25

$E(X) = -25¢$; game is not fair *(6-7)*

33. (A) $\frac{1}{3}$ (B) $\frac{2}{9}$ *(6-5)*   **34.** (A) $\frac{2}{13}$; 2 to 11 (B) $\frac{4}{13}$; 4 to 9 (C) $\frac{12}{13}$; 12 to 1 *(6-4)*
35. (A) 1 to 8 (B) $8 *(6-4)*
36. (A) P(2 heads) $= .21$, P(1 head) $= .48$, P(0 heads) $= .31$
(B) P(2 heads) $= .25$, P(1 head) $= .50$, P(0 heads) $= .25$
(C) 2 heads, 250; 1 head, 500; 0 heads, 250 *(6-3, 6-7)*
37. 5 children, 15 grandchildren, and 30 great grandchildren, for a total of 50 descendants *(6-1, 6-2)*
38. $\frac{1}{2}$; since the coin has no memory, the 10th toss is independent of the preceding 9 tosses. *(6-5)*

39. (A)

x_i	2	3	4	5	6	7	8	9	10	11	12
p_i	$\frac{1}{36}$	$\frac{2}{36}$	$\frac{3}{36}$	$\frac{4}{36}$	$\frac{5}{36}$	$\frac{6}{36}$	$\frac{5}{36}$	$\frac{4}{36}$	$\frac{3}{36}$	$\frac{2}{36}$	$\frac{1}{36}$

(B) $E(X) = 7$ *(6-7)*

40. $A = \{(1, 3), (2, 2), (3, 1), (2, 6), (3, 5), (4, 4), (5, 3), (6, 2), (6, 6)\}$;
$B = \{(1, 5), (2, 4), (3, 3), (4, 2), (5, 1), (6, 6)\}$; $P(A) = \frac{1}{4}$; $P(B) = \frac{1}{6}$;
$P(A \cap B) = \frac{1}{36}$; $P(A \cup B) = \frac{7}{18}$ *(6-4)*
41. (1) Probability of an event cannot be negative; (2) sum of probabilities of simple events must be 1; (3) probability of an event cannot be greater than 1. *(6-3)*

42.

	A	A'	Totals
B	15	30	45
B'	35	20	55
Totals	50	50	100

(6-4)

43. (A) .6 (B) $\frac{5}{6}$ *(6-5)*   **44.** (A) $\frac{1}{13}$ (B) Independent *(6-5)*   **45.** 336; 512; 392 *(6-1)*
46. (A) $P_{6,3} = 120$ (B) $C_{5,2} = 10$ *(6-2)*   **47.** (A) $\frac{6}{25}$ (B) $\frac{3}{10}$ *(6-5)*   **48.** Part (B) *(6-5)*
49. (A) 1.2 (B) 1.2 *(6-7)*   **50.** (A) $\frac{3}{5}$ (B) $\frac{1}{3}$ (C) $\frac{7}{15}$ (D) $\frac{9}{14}$ (E) $\frac{5}{8}$ (F) $\frac{3}{10}$ *(6-5, 6-6)*
51. No *(6-5)*   **52.** (A) $C_{13,5}/C_{52,5}$ (B) $C_{13,3} \cdot C_{13,2}/C_{52,5}$ *(6-3)*   **53.** $C_{8,2}/C_{10,4} = \frac{2}{15}$ *(6-3)*
54. $N_1 \cdot N_2 \cdot N_3$ *(6-1)*

55. Events S and H are mutually exclusive. Hence, $P(S \cap H) = 0$, while $P(S) \neq 0$ and $P(H) \neq 0$. Therefore, $P(S \cap H) \neq P(S)P(H)$, which implies S and H are dependent. *(6-5)*

56. (A) .350 (B) $\frac{3}{8} = .375$ (C) 375 *(6-3)*   **57.** $-.0172$; .0172; no *(6-7)*

58. (A) $P_{10,3} = 720$ (B) $P_{6,3}/P_{10,3} = \frac{1}{6}$ (C) $C_{10,3} = 120$ (D) $(C_{6,3} + C_{6,2} \cdot C_{4,1})/C_{10,3} = \frac{2}{3}$ *(6-1, 6-2, 6-3)*

59. 33 *(6-1)*   **60.** $1 - C_{7,3}/C_{10,3} = \frac{17}{24}$ *(6-3)*   **61.** $\frac{12}{51} \approx .235$ *(6-5)*   **62.** $\frac{12}{51} \approx .235$ *(6-5)*

63. (A)

x_i	2	3	4	5	6
p_i	$\frac{9}{36}$	$\frac{12}{36}$	$\frac{10}{36}$	$\frac{4}{36}$	$\frac{1}{36}$

(B) $E(X) = \frac{10}{3}$ *(6-7)*

64. $E(X) \approx -\$0.167$; no; $\$(10/3) \approx \$3.33$ *(6-7)*

65. $2^5 = 32$; 6 *(6-1)*

66. (A) $\frac{1}{4}$; 1 to 3 (B) $\$3$ *(6-4, 6-6)*

67. $1 - 10!/(5!10^5) \approx .70$ *(6-4)*

68. $P_{5,5} = 120$ *(6-1)*

69. (A) 610 (B) 390 (C) 270 *(6-1)*

70. (A) .8 (B) .2 (C) .5 *(6-1, 6-4)*   **71.** $P(A \cap P) = P(A)P(P|A) = .34$ *(6-5)*

72. (A) $P(A) = .290$; $P(B) = .290$; $P(A \cap B) = .100$; $P(A|B) = .345$; $P(B|A) = .345$

(B) No, since $P(A \cap B) \neq P(A)P(B)$.

(C) $P(C) = .880$; $P(D) = .120$; $P(C \cap D) = 0$; $P(C|D) = 0$; $P(D|C) = 0$

(D) Yes, since $C \cap D = \emptyset$; dependent, since $P(C \cap D) = 0$ and $P(C)P(D) \neq 0$ *(6-5)*

73. A: $E(X) = \$7.6$ million; B: $E(X) = \$7.8$ million; plan B *(6-7)*

74. Payoff table:

x_i	$\$270$	$-\$30$
p_i	.08	.92

$E(X) = -\$6$ *(6-7)*

75. $1 - (C_{10,4}/C_{12,4}) \approx .576$ *(6-2, 6-4)*

76. (A)

x_i	0	1	2
p_i	$\frac{12}{22}$	$\frac{9}{22}$	$\frac{1}{22}$

(B) $E(X) = \frac{1}{2}$ *(6-7)*

77. .955 *(6-6)*   **78.** $\frac{6}{7} \approx .857$ *(6-6)*

■ CHAPTER 7

Exercise 7-1

     A   *B*

1. $S_1 = [.8 \quad .2]$; the probability of being in state A after one trial is .8, and the probability of being in state B after one trial is .2.

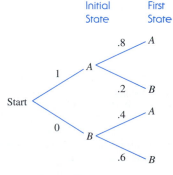

A: $(1)(.8) + (0)(.4) = .8$
B: $(1)(.2) + (0)(.6) = .2$

     A   *B*

3. $S_1 = [.6 \quad .4]$; the probability of being in state A after one trial is .6, and the probability of being in state B after one trial is .4.

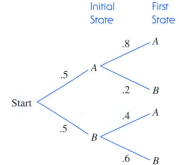

A: $(.5)(.8) + (.5)(.4) = .6$
B: $(.5)(.2) + (.5)(.6) = .4$

     A   *B*

5. $S_2 = [.72 \quad .28]$; the probability of being in state A after two trials is .72, and the probability of being in state B after two trials is .28.

     A   *B*

7. $S_2 = [.64 \quad .36]$; the probability of being in state A after two trials is .64, and the probability of being in state B after two trials is .36.

9.

$A\ [.4\ .6]$
$B\ [.7\ .3]$

11.

$A\ [.1\ .4\ .5]$
$B\ [.5\ .2\ .3]$
$C\ [.7\ .2\ .1]$

13. $a = .5, b = .6, c = .7$ **15.** $a = .7, b = 1, c = .2$ **17.**

$A\ [.3\ .7]$
$B\ [.9\ .1]$

19.

$A\ [.6\ .1\ .3]$
$B\ [.2\ .3\ .5]$
$C\ [0\ 0\ 1]$

21. .35   **23.** .212   **25.** $S_2 = [.43\ \ .35\ \ .22]$; the probabilities of going from state A to states A, B, and C in two trials

27. $S_3 = [.212\ \ .298\ \ .49]$; the probabilities of going from state C to states A, B, and C in three trials

29. $S_4 = [.425\ \ .575]$ **31.** $S_4 = [.452\ \ .152\ \ .396]$

35. (A) $\begin{bmatrix} .0154 & .3534 & .0153 & .6159 \\ 0 & 1 & 0 & 0 \\ .0102 & .2962 & .0103 & .6833 \\ 0 & 0 & 0 & 1 \end{bmatrix}$ (B) .6159 (C) .2962 (D) 0

39. (A) $[.25\ \ .75]$ (B) $[.25\ \ .75]$ (C) $[.25\ \ .75]$ (D) $[.25\ \ .75]$
(E) All the state matrices appear to approach the same matrix, $S = [.25\ \ .75]$, regardless of the values in the initial-state matrix.

41. $Q = \begin{bmatrix} .25 & .75 \\ .25 & .75 \end{bmatrix}$; the rows of Q are the same as the matrix S from Problem 39.

43. (A) R = Rain R' = No rain (B) $\begin{bmatrix} .4 & .6 \\ .06 & .94 \end{bmatrix}$ (C) Saturday: .196; Sunday: .12664

45. (A) (B) $\begin{bmatrix} .8 & .2 \\ .2 & .8 \end{bmatrix}$ (C) 32%; 39.2%

47. (A) N = National Property
U = United Family
O = Other companies

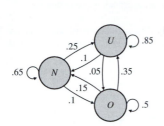

$$\begin{array}{c} \\ N \\ (B) \;\; U \\ O \end{array} \begin{array}{ccc} N & U & O \\ \begin{bmatrix} .65 & .25 & .1 \\ .1 & .85 & .05 \\ .15 & .35 & .5 \end{bmatrix} \end{array}$$

(C) 38.5%; 32%
(D) 45%; 53.65%

49. (A) B = Beginning agent
I = Intermediate agent
T = Terminated agent
Q = Qualified agent

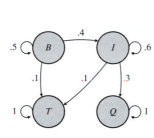

$$\begin{array}{c} \\ B \\ (B) \;\; I \\ T \\ Q \end{array} \begin{array}{cccc} B & I & T & Q \\ \begin{bmatrix} .5 & .4 & .1 & 0 \\ 0 & .6 & .1 & .3 \\ 0 & 0 & 1 & 0 \\ 0 & 0 & 0 & 1 \end{bmatrix} \end{array}$$

(C) .12; .3612

51. (A)
$$\begin{array}{c} \\ HMO \\ PPO \\ FFS \end{array} \begin{array}{ccc} HMO & PPO & FFS \\ \begin{bmatrix} .8 & .15 & .05 \\ .2 & .7 & .1 \\ .25 & .3 & .45 \end{bmatrix} \end{array}$$
(B) HMO: 34.75%; PPO: 37%; FFS: 28.25%   (C) HMO: 42.2625%; PPO: 39.5875%; FFS: 18.15%

53. (A)
$$\begin{array}{c} \\ H \\ R \end{array} \begin{array}{cc} H & R \\ \begin{bmatrix} .88 & .12 \\ .05 & .95 \end{bmatrix} \end{array}$$
H = Homeowner
R = Renter
(B) H: 42.433%   (C) H: 40.219%

Exercise 7-2

1. Regular   **3.** Regular   **5.** Not regular   **7.** Regular   **9.** Not regular   **11.** Regular

13. $S = [.4 \quad .6], \bar{P} = \begin{bmatrix} .4 & .6 \\ .4 & .6 \end{bmatrix}$ **15.** $S = [.375 \quad .625], \bar{P} = \begin{bmatrix} .375 & .625 \\ .375 & .625 \end{bmatrix}$ **17.** $S = [.3 \quad .5 \quad .2], \bar{P} = \begin{bmatrix} .3 & .5 & .2 \\ .3 & .5 & .2 \\ .3 & .5 & .2 \end{bmatrix}$

19. $S = [.6 \quad .24 \quad .16], \bar{P} = \begin{bmatrix} .6 & .24 & .16 \\ .6 & .24 & .16 \\ .6 & .24 & .16 \end{bmatrix}$ **21.** $S = [.3553 \quad .6447]$ **23.** $S = [.3636 \quad .4091 \quad .2273]$

25. (A)

(B) $\begin{array}{c} \\ Red \\ Blue \end{array} \begin{array}{cc} Red & Blue \\ \begin{bmatrix} .4 & .6 \\ .2 & .8 \end{bmatrix} \end{array}$

(C) $[.25 \quad .75]$; in the long run, the red urn will be selected 25% of the time and the blue urn 75% of the time.

27. (A) The state matrices alternate between [.2 .8] and [.8 .2]; hence, they do not approach any one matrix.
(B) The state matrices are all equal to S_0; hence, S_0 is a stationary matrix.
(C) The powers of P alternate between P and I, the 2×2 identity; hence, they do not approach a limiting matrix.
(D) Parts (B) and (C) of Theorem 1 are not valid for this matrix. Since P is not regular, this is not a contradiction.
29. (A) Since P is not regular, it may have more than one stationary matrix.
(B) [.5 0 .5] is another stationary matrix.
(C) P has an infinite number of stationary matrices.

31. $\bar{P} = \begin{bmatrix} 1 & 0 & 0 \\ .25 & 0 & .75 \\ 0 & 0 & 1 \end{bmatrix}$; each row of $\bar{P}$ is a stationary matrix for P.   **33.** 72.5%

35. (A) $S_1 = [.516\ .484]$; $S_2 = [.577\ .423]$
(B)

YEAR	DATA (%)	MODEL (%)
1970	43.3	43.3
1980	51.5	51.6
1990	57.5	57.7

(C) 74.1%

37. GTT: 25%; NCJ: 25%; Dash: 50%
39. Poor: 20%; satisfactory: 40%; preferred: 40%
41. 51%
43. Stationary matrix = [.25 .50 .25]
45. (A) [.25 .75] (B) 42.5%; 51.25%
(C) 60% rapid transit; 40% automobile

47. (A) $S_1 = [.334\ .666]$; $S_2 = [.343\ .657]$
(B)

YEAR	DATA (%)	MODEL (%)
1970	30.9	30.9
1980	33.3	33.4
1990	34.4	34.3

(C) 35%

Exercise 7-3

1. B, C **3.** A, D **5.** B is an absorbing state; absorbing chain   **7.** C is an absorbing state; not an absorbing chain

9. $\begin{array}{c} \\ B \\ A \\ C \end{array}\begin{array}{c} B\ \ \ A\ \ \ C \\ \begin{bmatrix} 1 & 0 & 0 \\ .5 & .2 & .3 \\ .1 & .5 & .4 \end{bmatrix}\end{array}$

11. $\begin{array}{c} \\ B \\ D \\ A \\ C \end{array}\begin{array}{c} B\ \ \ D\ \ \ A\ \ \ C \\ \begin{bmatrix} 1 & 0 & 0 & 0 \\ 0 & 1 & 0 & 0 \\ .4 & .1 & .3 & .2 \\ .4 & .3 & 0 & .3 \end{bmatrix}\end{array}$

13. $\begin{array}{c} \\ B \\ C \\ A \end{array}\begin{array}{c} B\ \ \ C\ \ \ A \\ \begin{bmatrix} 1 & 0 & 0 \\ 0 & 1 & 0 \\ .3 & .5 & .2 \end{bmatrix}\end{array}$

15. $\begin{array}{c} \\ B \\ D \\ A \\ C \end{array}\begin{array}{c} B\ \ \ D\ \ \ A\ \ \ C \\ \begin{bmatrix} 1 & 0 & 0 & 0 \\ 0 & 1 & 0 & 0 \\ .2 & .4 & .1 & .3 \\ .2 & .1 & .5 & .2 \end{bmatrix}\end{array}$

17. $\bar{P} = \begin{array}{c} \\ A \\ B \\ C \end{array}\begin{array}{c} A\ \ \ B\ \ \ C \\ \begin{bmatrix} 1 & 0 & 0 \\ 0 & 1 & 0 \\ .2 & .8 & 0 \end{bmatrix}\end{array}$; $P(C \text{ to } A) = .2$; $P(C \text{ to } B) = .8$. It will take an average of 2 trials to go from C to either A or B.

19. $\bar{P} = \begin{array}{c} \\ A \\ B \\ C \end{array}\begin{array}{c} A\ \ \ B\ \ \ C \\ \begin{bmatrix} 1 & 0 & 0 \\ 1 & 0 & 0 \\ 1 & 0 & 0 \end{bmatrix}\end{array}$; $P(B \text{ to } A) = 1$; $P(C \text{ to } A) = 1$. It will take an average of 4 trials to go from B to A and an average of 3 trials to go from C to A.

21. $\bar{P} = \begin{array}{c} \\ A \\ B \\ C \\ D \end{array}\begin{array}{c} A\ \ \ \ B\ \ \ \ C\ \ \ D \\ \begin{bmatrix} 1 & 0 & 0 & 0 \\ 0 & 1 & 0 & 0 \\ .36 & .64 & 0 & 0 \\ .44 & .56 & 0 & 0 \end{bmatrix}\end{array}$; $P(C \text{ to } A) = .36$; $P(C \text{ to } B) = .64$; $P(D \text{ to } A) = .44$; $P(D \text{ to } B) = .56$. It will take an average of 3.2 trials to go from C to either A or B and an average of 2.8 trials to go from D to either A or B.

23. (A) [.2 .8 0] (B) [.26 .74 0]   **25.** (A) [1 0 0] (B) [1 0 0]
27. (A) [.44 .56 0 0] (B) [.36 .64 0 0] (C) [.408 .592 0 0] (D) [.384 .616 0 0]

29. $\begin{array}{c} \\ A \\ B \\ C \\ D \end{array}\begin{array}{c} A\ \ \ \ \ B\ \ \ \ C\ \ \ D \\ \begin{bmatrix} 1 & 0 & 0 & 0 \\ 0 & 1 & 0 & 0 \\ .6375 & .3625 & 0 & 0 \\ .7375 & .2625 & 0 & 0 \end{bmatrix}\end{array}$

31. $\begin{array}{c} \\ A \\ B \\ C \\ D \\ E \end{array}\begin{array}{c} A\ \ \ \ \ \ B\ \ \ \ \ C\ \ \ D\ \ E \\ \begin{bmatrix} 1 & 0 & 0 & 0 & 0 \\ 0 & 1 & 0 & 0 & 0 \\ .0875 & .9125 & 0 & 0 & 0 \\ .1875 & .8125 & 0 & 0 & 0 \\ .4375 & .5625 & 0 & 0 & 0 \end{bmatrix}\end{array}$

33. $\begin{array}{c} \\ A \\ B \\ C \\ D \end{array}\begin{array}{c} A\ \ \ \ \ B\ \ \ \ C\ \ \ D \\ \begin{bmatrix} 0 & .52 & 0 & .48 \\ 0 & 1 & 0 & 0 \\ 0 & .36 & 0 & .64 \\ 0 & 0 & 0 & 1 \end{bmatrix}\end{array}$

39. (A) 75% (B) 12.5% (C) 7.5 months
43. (A) 91.52% (B) 4.96% (C) 6.32 days

41. (A) Company A: 30%, company B: 15%, company C: 55% (B) 5 yr
45. (A) .375 (B) 1.75 exits

Chapter 7 Review Exercise

1. $S_1 = [.32 \quad .68]$, $S_2 = [.328 \quad .672]$. The probability of being in state A after one trial is .32 and after two trials is .328; the probability of being in state B after one trial is .68 and after two trials is .672 *(7-1)*

2. State A is absorbing; chain is absorbing. *(7-2, 7-3)*   **3.** No absorbing states; chain is regular. *(7-2, 7-3)*

4. No absorbing states; chain is neither. *(7-2, 7-3)*

5.
$$\begin{array}{c} \\ A \\ B \\ C \end{array} \begin{array}{ccc} A & B & C \\ \left[\begin{array}{ccc} 0 & 1 & 0 \\ .1 & 0 & .9 \\ 0 & 1 & 0 \end{array}\right] \end{array}$$

No absorbing states, chain is neither. *(7-1, 7-2, 7-3)*

6.
$$\begin{array}{c} \\ A \\ B \\ C \end{array} \begin{array}{ccc} A & B & C \\ \left[\begin{array}{ccc} 0 & 1 & 0 \\ .1 & .2 & .7 \\ 0 & 0 & 1 \end{array}\right] \end{array}$$

C is absorbing, chain is absorbing. *(7-1, 7-2, 7-3)*

7.
$$\begin{array}{c} \\ A \\ B \\ C \end{array} \begin{array}{ccc} A & B & C \\ \left[\begin{array}{ccc} 0 & 0 & 1 \\ .1 & .2 & .7 \\ 0 & 1 & 0 \end{array}\right] \end{array}$$

No absorbing states, chain is regular. *(7-1, 7-2, 7-3)*

8.
$$\begin{array}{c} \\ A \\ B \\ C \\ D \end{array} \begin{array}{cccc} A & B & C & D \\ \left[\begin{array}{cccc} .3 & .2 & 0 & .5 \\ 0 & 1 & 0 & 0 \\ 0 & 0 & .2 & .8 \\ 0 & 0 & .3 & .7 \end{array}\right] \end{array}$$

B is absorbing, chain is neither. *(7-1, 7-2, 7-3)*

9.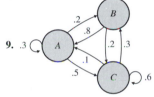
$$\begin{array}{c} \\ A \\ B \\ C \end{array} \begin{array}{ccc} A & B & C \\ \left[\begin{array}{ccc} .3 & .2 & .5 \\ .8 & 0 & .2 \\ .1 & .3 & .6 \end{array}\right] \end{array} \quad (7-1)$$

10. (A) .3 (B) .675 *(7-1)*

11. $S = [.25 \quad .75]$; $\bar{P} = \begin{array}{c} \\ A \\ B \end{array} \begin{array}{cc} A & B \\ \left[\begin{array}{cc} .25 & .75 \\ .25 & .75 \end{array}\right] \end{array} \quad (7-2)$

12. $S = [.4 \quad .48 \quad .12]$; $\bar{P} = \begin{array}{c} \\ A \\ B \\ C \end{array} \begin{array}{ccc} A & B & C \\ \left[\begin{array}{ccc} .4 & .48 & .12 \\ .4 & .48 & .12 \\ .4 & .48 & .12 \end{array}\right] \end{array} \quad (7-2)$

13. $\begin{array}{c} \\ A \\ B \\ C \end{array} \begin{array}{ccc} A & B & C \\ \left[\begin{array}{ccc} 1 & 0 & 0 \\ 0 & 1 & 0 \\ .75 & .25 & 0 \end{array}\right] \end{array}$; $P(C \text{ to } A) = .75$; $P(C \text{ to } B) = .25$. It takes an average of 2.5 trials to go from C to an absorbing state. *(7-3)*

14. $\begin{array}{c} \\ A \\ B \\ C \\ D \end{array} \begin{array}{cccc} A & B & C & D \\ \left[\begin{array}{cccc} 1 & 0 & 0 & 0 \\ 0 & 1 & 0 & 0 \\ .2 & .8 & 0 & 0 \\ .3 & .7 & 0 & 0 \end{array}\right] \end{array}$; $P(C \text{ to } A) = .2$; $P(C \text{ to } B) = .8$; $P(D \text{ to } A) = .3$; $P(D \text{ to } B) = .7$. It takes an average of 2 trials to go from C to an absorbing state and an average of 3 trials to go from D to an absorbing state. *(7-3)*

19. $\begin{array}{c} \\ B \\ D \\ A \\ C \end{array} \begin{array}{cccc} B & D & A & C \\ \left[\begin{array}{cccc} 1 & 0 & 0 & 0 \\ 0 & 1 & 0 & 0 \\ .1 & .1 & .6 & .2 \\ .2 & .2 & .3 & .3 \end{array}\right] \end{array} \quad (7-3)$

20. (A) $[.1 \quad .4 \quad .5]$ (B) $[.1 \quad .4 \quad .5]$ *(7-3)*   **21.** (A) $[.25 \quad .75 \quad 0]$ (B) $[.55 \quad .45 \quad 0]$ *(7-3)*

22. (A)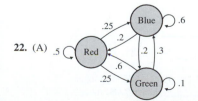

(B) $\begin{array}{c} \\ R \\ B \\ G \end{array} \begin{array}{ccc} R & B & G \\ \left[\begin{array}{ccc} .5 & .25 & .25 \\ .2 & .6 & .2 \\ .6 & .3 & .1 \end{array}\right] \end{array}$ (C) Regular (D) $\begin{array}{c} \\ R \\ B \\ G \end{array} \begin{array}{ccc} R & B & G \\ \left[\begin{array}{ccc} .4 & .4 & .2 \\ .4 & .4 & .2 \\ .4 & .4 & .2 \end{array}\right] \end{array}$ In the long run, the red urn will be selected 40% of the time, the blue urn 40% of the time, and the green urn 20% of the time. *(7-2)*

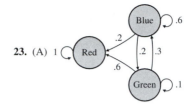

23. (A) 1

(B)
$$\begin{array}{cc} & \begin{array}{ccc} R & B & G \end{array} \\ \begin{array}{c} R \\ B \\ G \end{array} & \begin{bmatrix} 1 & 0 & 0 \\ .2 & .6 & .2 \\ .6 & .3 & .1 \end{bmatrix} \end{array}$$
(C) Absorbing

(D)
$$\begin{array}{cc} & \begin{array}{ccc} R & B & G \end{array} \\ \begin{array}{c} R \\ B \\ G \end{array} & \begin{bmatrix} 1 & 0 & 0 \\ 1 & 0 & 0 \\ 1 & 0 & 0 \end{bmatrix} \end{array}$$

Once the red urn is selected, the blue and green urns will never be selected again. It will take an average of 3.67 trials to reach the red urn from the blue urn and an average of 2.33 trials to reach the red urn from the green urn. *(7-3)*

25. No such chain exists. *(7-2, 7-3)*   **26.** No such chain exists. *(7-2, 7-3)*   **27.** No such chain exists. *(7-2)*

28. $S = [1 \quad 0 \quad 0]$ and $S' = [0 \quad 1 \quad 0]$ are both stationary matrices for $P = \begin{array}{cc} & \begin{array}{ccc} A & B & C \end{array} \\ \begin{array}{c} A \\ B \\ C \end{array} & \begin{bmatrix} 1 & 0 & 0 \\ 0 & 1 & 0 \\ .6 & .3 & .1 \end{bmatrix} \end{array}$ *(7-3)*

29. $P = \begin{array}{cc} & \begin{array}{cc} A & B \end{array} \\ \begin{array}{c} A \\ B \end{array} & \begin{bmatrix} 0 & 1 \\ 1 & 0 \end{bmatrix} \end{array}$ *(7-2, 7-3)*   **30.** No such chain exists. *(7-2)*   **31.** No such chain exists. *(7-3)*

32. (A) .7   **(B)** $\begin{array}{cc} & \begin{array}{cc} X & X' \end{array} \\ \begin{array}{c} X \\ X' \end{array} & \begin{bmatrix} .7 & .3 \\ .5 & .5 \end{bmatrix} \end{array}$ **(C)** $\begin{array}{c} \begin{array}{cc} X & X' \end{array} \\ [.2 \quad .8] \end{array}$

(D) $\begin{array}{c} \begin{array}{cc} X & X' \end{array} \\ [.54 \quad .46] \end{array}$; 54% of the consumers will purchase brand X on the next purchase.   **(E)** $\begin{array}{c} \begin{array}{cc} X & X' \end{array} \\ [.625 \quad .375] \end{array}$ **(F)** 62.5% *(7-2)*

33. (A) Brand *A*: 24%, brand *B*: 32%, brand *C*: 44%   **(B)** 4 yr *(7-3)*

34. (A) $S_1 = \begin{array}{c} \begin{array}{cc} V & V' \end{array} \\ [.585 \quad .415] \end{array}$; $S_2 = \begin{array}{c} \begin{array}{cc} V & V' \end{array} \\ [.728 \quad .272] \end{array}$ **(B)**

YEAR	DATA (%)	MODEL (%)
1984	10.6	10.6
1988	58.0	58.5
1992	72.5	72.8

(C) 79% *(7-2)*

35. (A) 63.75%   **(B)** 15%   **(C)** 8.75 yr *(7-3)*   **36.** $\bar{P} = \begin{array}{cc} & \begin{array}{ccc} \text{Red} & \text{Pink} & \text{White} \end{array} \\ \begin{array}{c} \text{Red} \\ \text{Pink} \\ \text{White} \end{array} & \begin{bmatrix} 1 & 0 & 0 \\ 1 & 0 & 0 \\ 1 & 0 & 0 \end{bmatrix} \end{array}$ *(7-3)*

37. (A) $S_1 = \begin{array}{c} \begin{array}{cc} F & F' \end{array} \\ [.022 \quad .978] \end{array}$; $S_2 = \begin{array}{c} \begin{array}{cc} F & F' \end{array} \\ [.0175 \quad .9825] \end{array}$ **(B)**

YEAR	DATA (%)	MODEL (%)
1980	2.8	2.8
1985	2.3	2.2
1990	1.9	1.75

(C) 0.4% *(7-2)*

■ CHAPTER 8

Exercise 8-1

1. 45   **3.** 15   **5.** 15   **7.**

h	-0.1	-0.01	$-0.001 \rightarrow 0 \leftarrow 0.001$	0.01	0.1
$\dfrac{f(1 + h) - f(1)}{h}$	5.7	5.97	$5.997 \rightarrow 6 \leftarrow 6.003$	6.03	6.3

9. 6
11. 6 ft/sec
13. 6
15. 100 ft/sec
17. 80 ft/sec

19. (A) 4   (D)
(B) 3
(C) 2

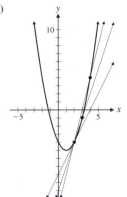

21. Slope at $x = -1$ is 1; slope at $x = 3$ is -2.
23. Slope at $x = -3$ is -5; slope at $x = -1$ is 0; slope at $x = 1$ is -1; slope at $x = 3$ is 4.

25.

27.

29. 8   **31.** $\frac{1}{4}$ **33.**

x	-3	-2	-1	0	1	2	3
Slope	-3	-2	-1	0	1	2	3

Slope function: $y = x$

35. The slope of the line is m. The slope of the graph at any point on the graph is also m.

37. $\dfrac{f(1 + h) - f(1)}{h} = 6 + 3h \to 6$ as $h \to 0$

39.

h	-0.1	-0.01	$-0.001 \to$	0	$\leftarrow 0.001$	0.01	0.1
$\dfrac{f(0 + h) - f(0)}{h}$	-1	-1	-1	$\to -1 \ne 1 \leftarrow$ 1		1	1

The slope of the graph is not defined at $(0, 0)$.

41. (A) -0.12 hr/yr   (B) $0.37/yr

43. (A)

L1	L2	L3
0	37.1	------
5	36.1	
10	35.3	
15	34.9	
20	34.5	
------	------	

L₁={0,5,10,15,2...

```
QuadReg
y=ax²+bx+c
a=.0045714286
b=-.2194285714
c=37.08857143
```

(B) -0.13 weekly hr/yr
45. (A) At a production level of 1,000 car seats, the revenue is $35,000 and is increasing at the rate of $10 per seat.
(B) At a production level of 1,300 car seats, the revenue is $35,750 and is decreasing at the rate of $5 per seat.
47. In 1990, the annual production was 11,360,000 metric tons and was decreasing at the rate of 130,000 metric tons per year.
49. (A) $67.5 billion/yr   (B) $1.7 billion/yr
51. (A)

L1	L2	L3
0	476.9	------
1	526.2	
2	583.6	
3	652.4	
4	728.6	
------	------	

L₁={0,1,2,3,4}

```
QuadReg
y=ax²+bx+c
a=4.657142857
b=44.33142857
c=476.9342857
```

(B) $62.96 billion/yr
53. In 1990, the number of male infant deaths per 100,000 births was 9.8 and was decreasing at the rate of 0.42 death per 100,000 births per year.

Exercise 8-2

1.

3.

5.

7. (A) 2 (B) 2 (C) 2 (D) 2

9. (A) 1 (B) 2 (C) Does not exist (D) 2   **11.** (A) 1 (B) 2 (C) Does not exist (D) Does not exist
13. (A) 1 (B) 1 (C) 1 (D) 3   **15.** -4 **17.** 36   **19.** $\frac{5}{9}$ **21.** $\sqrt{5}$ **23.** 1.4   **25.** 47   **27.** $\frac{5}{3}$ **29.** 243   **31.** 3
33. -6 **35.** 1   **37.** -1 **39.** Does not exist   **41.** 1   **43.** 0.5   **45.** -1 **47.** -5 **49.** $\frac{2}{3}$ **51.** 0   **53.** 3   **55.** 4
57. 0   **59.** $1/(2\sqrt{2})$ **61.** Does not exist   **63.** Slope $=4$

65. Slope $=\frac{1}{4}$

67. 80 ft/sec   **69.** $-1, 0, 2$ **71.** (A) The limit does not exist. The values of $1/x$ are large negative numbers when x is close to 0 on the left.
  (B) The limit does not exist. The values of $1/x$ are large positive numbers when x is close to 0 on the right.
73. (A) $\lim_{x\to 1^-} f(x) = 1$; $\lim_{x\to 1^+} f(x) = -1$ (B) $\lim_{x\to 1^-} f(x) = -3$; $\lim_{x\to 1^+} f(x) = 3$ (C) $m = -1$ **75.** $2a$ **77.** $1/(2\sqrt{a})$
79. (A) $-1, 1, 3$ (B) $2a - 3$
  (C) The slopes are the same as in part (A). In part (A), each value of a required a new limit operation. In part (B), a single limit operation
  produces the slope for all values of a.
81. 10   **83.** 0.693   **85.** 2.718   **87.** Typical values of n are 95 on a TI-81, 94 on a TI-82, and 126 on a TI-85.
89. $\lim_{x\to -2^-} f(x) = \lim_{x\to -2^+} f(x) = 2$ **91.** $\lim_{x\to 2^-} f(x) = -4$, $\lim_{x\to 2^+} f(x) = 4$ **93.** $\lim_{x\to -3^-} f(x) = -3$, $\lim_{x\to -3^+} f(x) = 3$
  $\lim_{x\to 2^-} f(x) = \lim_{x\to 2^+} f(x) = 2$ $\lim_{x\to 3^-} f(x) = -3$, $\lim_{x\to 3^+} f(x) = 3$

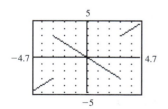

95. (A) At a production level of 900 units, the revenue is $99,000 and is
    increasing at the rate of $20 per jigsaw.

(B) At a production level of 1,200 units, the revenue is $96,000 and is
    decreasing at the rate of $40 per jigsaw.

97. In 1990, the revolving-credit debt was $233.1 billion and was increasing at the rate of $23.8 billion per year.

99. The alveolar pressure at 9,000 ft is 70.3% of that at sea level and is decreasing at the rate of 2.94% per 1000 ft..

101. The secondary school enrollment in 1995 was 17.2 million and was increasing at the rate of 0.56 million students per year.

Exercise 8-3

1. (A) 1; slope of the secant line through $(0, f(0))$ and $(1, f(1))$
  (B) $2 + h$; slope of the secant line through $(1 + h, f(1 + h))$ and $(1, f(1))$
  (C) 2; slope of the tangent line at $(1, f(1))$
3. $4x - 3$ **5.** $3x^2 - 2x$ **7.** $f'(x) = 2; f'(1) = 2, f'(2) = 2, f'(3) = 2$ **9.** $f'(x) = -2x; f'(1) = -2, f'(2) = -4, f'(3) = -6$
11. $f'(-3) = -1, f'(3) = 2; f'(-1) = 0$ **13.** $f'(-5) = -3, f'(1) = 1, f'(7) = -3; f'(-2) = 0, f'(4) = 0$
15. (A) 5 (B) $3 + h$ (C) 3 (D) $y = 3x - 1$ **17.** (A) 5 m/sec (B) $3 + h$ m/sec (C) 3 m/sec
19. $f'(x) = 6 - 2x; f'(1) = 4, f'(2) = 2, f'(3) = 0$ **21.** $f'(x) = 1/(2\sqrt{x}); f'(1) = \frac{1}{2}, f'(2) = 1/(2\sqrt{2}), f'(3) = 1/(2\sqrt{3})$
23. $f'(x) = 1/x^2; f'(1) = 1, f'(2) = \frac{1}{4}, f'(3) = \frac{1}{9}$ **25.** Yes   **27.** No   **29.** No   **31.** Yes
33. (A) $f'(x) = 2x - 4$ **35.** $v = f'(x) = 8x - 2$; 6 ft/sec, 22 ft/sec, 38 ft/sec
  (B) $-4, 0, 4$ **37.**   **39.**

41. 0.69   **43.** 0.71
45. (A) The graphs of g and h are vertical translations of the graph of f. All three functions should have the same derivative. (B) $2x$
47. (A) The slope of the graph of f is 0 at any point on the graph.   **49.** f is nondifferentiable at $x = 1$

51. No   **53.** No   **57.** (A) $2 - 2x$ (B) 2 (C) -2

59. (A) $S'(t) = 1/\sqrt{t + 10}$
(B) $S(15) = 10$; $S'(15) = 0.2$. After 15 months, the total sales are $10 million and are increasing at the rate of $0.2 million, or $200,000, per month.
(C) The estimated total sales are $10.2 million after 16 months and $10.4 million after 17 months.

61. (A) $A(5) = 133.82$; $A'(5) = 7.80$
(B) After 5 years, the original $100 investment has grown to $133.82 and is continuing to grow at the rate of $7.80 per year.

63. (A) In March, the price of the stock was $80 and was increasing at the rate of $5 per month.
(B) The stock reached its highest price in April when the rate of change was 0.

65. (A)

(B) In 1995, 14.5% of the households had black-and-white television sets, and this number was decreasing at the rate of 3.3% per year.

67. (A) $P'(t) = 12 - 2t$
(B) $P(3) = 107$; $P'(3) = 6$. After 3 hr, the ozone level is 107 ppb and is increasing at the rate of 6 ppb/hr.

69. (A) $P(30) = 47.1$; $P'(30) = 0.9$
(B) In 2010, there will be 47.1 million people aged 65 or older, and the number in this group will be growing at the rate of 900,000/yr.
(C) The estimated population is 48 million in 2011 and 48.9 million in 2012.

Exercise 8-4

1. 0   **3.** 0   **5.** $12x^{11}$ **7.** 1   **9.** $-7x^{-8} = -7/x^8$ **11.** $\frac{5}{2}x^{3/2}$ **13.** $-5x^{-6} = -5/x^6$ **15.** $8x^3$ **17.** $2x^5$ **19.** $x^4/3$

21. 12   **23.** 2   **25.** 9   **27.** $-10x^{-6} = -10/x^6$ **29.** $-16x^{-5} = -16/x^5$ **31.** $x^{-3} = 1/x^3$ **33.** $-x^{-2/3} = -1/x^{2/3}$ **35.** $4x - 3$

37. $15x^4 - 6x^2$ **39.** $-12x^{-5} - 4x^{-3} = \dfrac{-12}{x^5} - \dfrac{4}{x^3}$ **41.** $-\dfrac{1}{2}x^{-2} + 2x^{-4} = \dfrac{-1}{2x^2} + \dfrac{2}{x^4}$ **43.** $2x^{-1/3} - \dfrac{5}{3}x^{-2/3} = \dfrac{2}{x^{1/3}} - \dfrac{5}{3x^{2/3}}$

45. $-\dfrac{9}{5}x^{-8/5} + 3x^{-3/2} = \dfrac{-9}{5x^{8/5}} + \dfrac{3}{x^{3/2}}$ **47.** $-\dfrac{1}{3}x^{-4/3} = \dfrac{-1}{3x^{4/3}}$ **49.** $-6x^{-3/2} + 6x^{-3} + 1 = \dfrac{-6}{x^{3/2}} + \dfrac{6}{x^3} + 1$

51. (A) $f'(x) = 6 - 2x$ (B) $f'(2) = 2$; $f'(4) = -2$ (C) $y = 2x + 4$; $y = -2x + 16$ (D) $x = 3$

53. (A) $f'(x) = 12x^3 - 12x$ (B) $f'(2) = 72$; $f'(4) = 720$ (C) $y = 72x - 127$; $y = 720x - 2,215$ (D) $x = -1, 0, 1$

55. (A) $v = f'(x) = 176 - 32x$ (B) $f'(0) = 176$ ft/sec; $f'(3) = 80$ ft/sec   (C) 5.5 sec

57. (A) $v = f'(x) = 3x^2 - 18x + 15$ (B) $f'(0) = 15$ ft/sec; $f'(3) = -12$ ft/sec   (C) $x = 1$ sec, 5 sec

59. $x = 2.78$ **61.** $x = -1.12, x = 1.79$ **63.** $x = -1.49, x = 0.48, x = 1.77$ **65.** $x = 0, x = 3$

67. $x = -2.07$, $x = -0.80$, $x = 0.92$, $x = 1.95$ **69.** $x = -b/(2a)$
71. (A) $x^3 + x$ (B) x^3 (C) $x^3 - x$ **73.** $-20x^{-2} = -20/x^2$ **75.** $2x - 3 - 10x^{-3} = 2x - 3 - (10/x^3)$
79. The domain of $f'(x)$ is all real numbers except $x = 0$. At $x = 0$, the graph of $y = f(x)$ is smooth, but it has a vertical tangent.
81. (A) $C'(x) = 60 - (x/2)$
 (B) $C'(60) = \$30/\text{racket}$. At a production level of 60 rackets, the rate of change of total cost relative to production is \$30 per racket; thus, the cost of producing 1 more racket at this level of production is approx. \$30.
 (C) \$29.75; the marginal cost of \$30 per racket found in part (B) is a close approximation to this value.
 (D) $C'(80) = \$20$ per racket. At a production level of 80 rackets, the rate of change of total cost relative to production is \$20 per racket; thus, the cost of producing 1 more racket at this level of production is approx. \$20.

83. The approximate cost of producing the 101st oven is greater than that of the 401st oven. Since these marginal costs are decreasing, the manufacturing process is becoming more efficient.
85. (A) $N'(x) = 3{,}780/x^2$
 (B) $N'(10) = 37.8$. At the \$10,000 level of advertising, sales are increasing at the rate of 37.8 boats per \$1,000 spent on advertising.
   $N'(20) = 9.45$. At the \$20,000 level of advertising, sales are increasing at the rate of 9.45 boats per \$1,000 spent on advertising.
87. (A) -1.37 beats/min (B) -0.58 beat/min   **89.** (A) 25 items/hr (B) 8.33 items/hr

Exercise 8-5

1. $2x^3(2x) + (x^2 - 2)(6x^2) = 10x^4 - 12x^2$ **3.** $(x - 3)(2) + (2x - 1)(1) = 4x - 7$ **5.** $\dfrac{(x - 3)(1) - x(1)}{(x - 3)^2} = \dfrac{-3}{(x - 3)^2}$

7. $\dfrac{(x - 2)(2) - (2x + 3)(1)}{(x - 2)^2} = \dfrac{-7}{(x - 2)^2}$ **9.** $(x^2 + 1)(2) + (2x - 3)(2x) = 6x^2 - 6x + 2$ **11.** $\dfrac{(2x - 3)(2x) - (x^2 + 1)(2)}{(2x - 3)^2} = \dfrac{2x^2 - 6x - 2}{(2x - 3)^2}$

13. $(x^2 + 2)2x + (x^2 - 3)2x = 4x^3 - 2x$ **15.** $\dfrac{(x^2 - 3)2x - (x^2 + 2)2x}{(x^2 - 3)^2} = \dfrac{-10x}{(x^2 - 3)^2}$ **17.** 8   **19.** 1   **21.** $\frac{1}{8}$

23. $(2x + 1)(2x - 3) + (x^2 - 3x)(2) = 6x^2 - 10x - 3$ **25.** $(2x - x^2)(5) + (5x + 2)(2 - 2x) = -15x^2 + 16x + 4$

27. $\dfrac{(x^2 + 2x)(5) - (5x - 3)(2x + 2)}{(x^2 + 2x)^2} = \dfrac{-5x^2 + 6x + 6}{(x^2 + 2x)^2}$ **29.** $\dfrac{(x^2 - 1)(2x - 3) - (x^2 - 3x + 1)(2x)}{(x^2 - 1)^2} = \dfrac{3x^2 - 4x + 3}{(x^2 - 1)^2}$

31. $f'(x) = (1 + 3x)(-2) + (5 - 2x)(3)$; $y = -11x + 29$ **33.** $f'(x) = \dfrac{(3x - 4)(1) - (x - 8)(3)}{(3x - 4)^2}$; $y = 5x - 13$

35. $x = 2$, $x = 3$ **37.** $x = -1.49$, $x = 0.48$, $x = 1.77$ **39.** $x = -1$, $x = 1$ **41.** $x = -2.51$, $x = -1.48$, $x = 1.12$, $x = 2.87$

43. $7x^6 - 3x^2$ **45.** $-27x^{-4} = -\dfrac{27}{x^4}$ **47.** $(2x^4 - 3x^3 + x)(2x - 1) + (x^2 - x + 5)(8x^3 - 9x^2 + 1) = 12x^5 - 25x^4 + 52x^3 - 42x^2 - 2x + 5$

49. $\dfrac{(4x^2 + 5x - 1)(6x - 2) - (3x^2 - 2x + 3)(8x + 5)}{(4x^2 + 5x - 1)^2} = \dfrac{23x^2 - 30x - 13}{(4x^2 + 5x - 1)^2}$ **51.** $9x^{1/3}(3x^2) + (x^3 + 5)(3x^{-2/3}) = \dfrac{30x^3 + 15}{x^{2/3}}$

53. $\dfrac{(x^2 - 3)(2x^{-2/3}) - 6x^{1/3}(2x)}{(x^2 - 3)^2} = \dfrac{-10x^2 - 6}{(x^2 - 3)^2 x^{2/3}}$ **55.** $x^{-2/3}(3x^2 - 4x) + (x^3 - 2x^2)(-\frac{2}{3}x^{-5/3}) = -\frac{8}{3}x^{1/3} + \frac{7}{3}x^{4/3}$

57. $\dfrac{(x^2 + 1)[(2x^2 - 1)(2x) + (x^2 + 3)(4x)] - (2x^2 - 1)(x^2 + 3)(2x)}{(x^2 + 1)^2} = \dfrac{4x^5 + 8x^3 + 16x}{(x^2 + 1)^2}$ **61.** $f'(x) = n[u(x)]^{n-1}u'(x)$

63. (A) $S'(t) = \dfrac{(t^2 + 50)(180t) - 90t^2(2t)}{(t^2 + 50)^2} = \dfrac{9{,}000t}{(t^2 + 50)^2}$

(B) $S(10) = 60$; $S'(10) = 4$. After 10 months, the total sales are 60,000 CD's, and sales are increasing at the rate of 4,000 CD's per month.

(C) Approx. 64,000 CD's

65. (A) $x = 800$; $dx/dp = -16$. At a price level of $40, the demand is 800 CD players per week, and demand is decreasing at the rate of 16 players per dollar.

(B) Approx. 784 CD players

67. (A) $C'(t) = \dfrac{(t^2 + 1)(0.14) - 0.14t(2t)}{(t^2 + 1)^2} = \dfrac{0.14 - 0.14t^2}{(t^2 + 1)^2}$

(B) $C'(0.5) = 0.0672$. After 0.5 hr, concentration is increasing at the rate of 0.0672 mg/cc/hr. $C'(3) = -0.0112$. After 3 hr, concentration is decreasing at the rate of 0.0112 mg/cc/hr.

69. (A) $N'(x) = \dfrac{(x + 32)(100) - (100x + 200)}{(x + 32)^2} = \dfrac{3{,}000}{(x + 32)^2}$

(B) $N'(4) = 2.31$; $N'(68) = 0.30$

Exercise 8-6

1. 3   **3.** $(-4x)$ **5.** $(2 + 6x)$ **7.** $6(2x + 5)^2$ **9.** $-8(5 - 2x)^3$ **11.** $30x(3x^2 + 5)^4$ **13.** $8(x^3 - 2x^2 + 2)^7(3x^2 - 4x)$

15. $(2x - 5)^{-1/2} = \dfrac{1}{(2x - 5)^{1/2}}$ **17.** $-8x^3(x^4 + 1)^{-3} = \dfrac{-8x^3}{(x^4 + 1)^3}$ **19.** $f'(x) = 6(2x - 1)^2$; $y = 6x - 5$; $x = \frac{1}{2}$

21. $f'(x) = 2(4x - 3)^{-1/2} = \dfrac{2}{(4x - 3)^{1/2}}$; $y = \dfrac{2}{3}x + 1$; none   **23.** $24x(x^2 - 2)^3$ **25.** $-6(x^2 + 3x)^{-4}(2x + 3) = \dfrac{-6(2x + 3)}{(x^2 + 3x)^4}$

27. $x(x^2 + 8)^{-1/2} = \dfrac{x}{(x^2 + 8)^{1/2}}$ **29.** $(3x + 4)^{-2/3} = \dfrac{1}{(3x + 4)^{2/3}}$ **31.** $(x^2 - 4x + 2)^{-1/2}(x - 2) = \dfrac{x - 2}{(x^2 - 4x + 2)^{1/2}}$

33. $-2(2x + 4)^{-2} = \dfrac{-2}{(2x + 4)^2}$ **35.** $-15x^2(x^3 + 4)^{-6} = \dfrac{-15x^2}{(x^3 + 4)^6}$ **37.** $(-8x + 4)(4x^2 - 4x + 1)^{-2} = \dfrac{-4}{(2x - 1)^3}$

39. $-2(2x - 3)(x^2 - 3x)^{-3/2} = \dfrac{-2(2x - 3)}{(x^2 - 3x)^{3/2}}$ **41.** $f'(x) = (4 - x)^3 - 3x(4 - x)^2 = 4(4 - x)^2(x - 1);\ y = -16x + 48$

43. $f'(x) = \dfrac{(2x - 5)^3 - 6x(2x - 5)^2}{(2x - 5)^6} = \dfrac{-4x - 5}{(2x - 4)^4};\ y = -17x + 54$ **45.** $f'(x) = (2x + 2)^{1/2} + x(2x + 2)^{-1/2} = \dfrac{3x + 2}{(2x + 2)^{1/2}};\ y = \tfrac{5}{2}x - \tfrac{1}{2}$

47. $x = 0,\ x = 2,\ x = 5$ **49.** $x = 4$ **51.** $x = -0.64,\ x = 0.83,\ x = 2.81$ **53.** $x = -2.90,\ x = 0.16,\ x = 1.41$ **55.** $x = 2.5$

57. $x = -2.71,\ x = -0.19,\ x = 1.00,\ x = 1.90$

59. $18x^2(x^2 + 1)^2 + 3(x^2 + 1)^3 = 3(x^2 + 1)^2(7x^2 + 1)$ **61.** $\dfrac{24x^5(x^3 - 7)^3 - (x^3 - 7)^4 6x^2}{4x^6} = \dfrac{3(x^3 - 7)^3(3x^3 + 7)}{2x^4}$

63. $(2x - 3)^2[12x(2x^2 + 1)^2] + (2x^2 + 1)^3[4(2x - 3)] = 4(2x^2 + 1)^2(2x - 3)(8x^2 - 9x + 1)$

65. $4x^3(x^2 - 1)^{-1/2} + 8x(x^2 - 1)^{1/2} = \dfrac{12x^3 - 8x}{(x^2 - 1)^{1/2}}$ **67.** $\dfrac{(x - 3)^{1/2}(2) - x(x - 3)^{-1/2}}{x - 3} = \dfrac{x - 6}{(x - 3)^{3/2}}$

69. $\tfrac{1}{2}[(2x - 1)^3(x^2 + 3)^4]^{-1/2}[8x(2x - 1)^3(x^2 + 3)^3 + 6(x^2 + 3)^4(2x - 1)^2] = (2x - 1)^{1/2}(x^2 + 3)(11x^2 - 4x + 9)$

71. (A) $C'(x) = (2x + 16)^{-1/2} = \dfrac{1}{(2x + 16)^{1/2}}$

(B) $C'(24) = \tfrac{1}{8}$, or \$12.50. At a production level of 24 calculators, total cost is increasing at the rate of \$12.50 per calculator; also, the cost of producing the 25th calculator is approx. \$12.50.
$C'(42) = \tfrac{1}{10}$, or \$10.00. At a production level of 42 calculators, total cost is increasing at the rate of \$10.00 per calculator; also, the cost of producing the 43rd calculator is approx. \$10.00.

73. $x = 400$ and $dx/dp = 4$. At a price of \$75, the supply is 400 speakers per week, and supply is increasing at the rate of 4 speakers per dollar.

75. $4{,}000(1 + \tfrac{1}{12}r)^{47}$ **77.** $\dfrac{(4 \times 10^6)x}{(x^2 - 1)^{5/3}}$

79. (A) $f'(n) = n(n - 2)^{-1/2} + 2(n - 2)^{1/2}$

(B) $f'(11) = \tfrac{29}{3} = 9.67$. When the list contains 11 items, the learning time is increasing at the rate of 9.67 min per item.
$f'(27) = \tfrac{77}{5} = 15.4$. When the list contains 27 items, the learning time is increasing at the rate of 15.4 min per item.

Exercise 8-7

1. (A) \$29.50   (B) \$30

3. (A) \$420   (B) $\overline{C}'(500) = -0.24$. At a production level of 500 frames, average cost is decreasing at the rate of 24¢ per frame.
(C) Approx. \$419.76

5. (A) $R'(1,600) = 20$. At a production level of 1,600 radios, revenue is increasing at the rate of $20 per radio.

(B) $R'(2,500) = -25$. At a production level of 2,500 radios, revenue is decreasing at the rate of $25 per radio.

7. (A) $4.50 (B) $5

9. (A) $P'(450) = 0.5$. At a production level of 450 cassettes, profit is increasing at the rate of 50¢ per cassette.

(B) $P'(750) = -2.5$. At a production level of 750 cassettes, profit is decreasing at the rate of $2.50 per cassette.

11. (A) $13.50

(B) $\overline{P}'(50) = \$0.27$. At a production level of 50 mowers, the average profit per mower is increasing at the rate of $0.27 per mower.

(C) Approx. $13.77

13. (A) $C'(x) = 60$ (B) $R(x) = 200x - (x^2/30)$ (C) $R'(x) = 200 - (x/15)$

(D) $R'(1,500) = 100$. At a production level of 1,500 saws, revenue is increasing at the rate of $100 per saw.
$R'(4,500) = -100$. At a production level of 4,500 saws, revenue is decreasing at the rate of $100 per saw.

(E) Break-even points: (600, 108,000) and (3,600, 288,000) (F) $P(x) = -(x^2/30) + 140x - 72,000$

(G) $P'(x) = -(x/15) + 140$

(H) $P'(1,500) = 40$. At a production level of 1,500 saws, profit is increasing at the rate of $40 per saw.
$P'(3,000) = -60$. At a production level of 3,000 saws, profit is decreasing at the rate of $60 per saw.

15. (A) $p = 20 - 0.02x$ (B) $R(x) = 20x - 0.02x^2$ (C) $C(x) = 4x + 1,400$

(D) Break-even points: (100, 1,800) and (700, 4,200) (E) $P(x) = 16x - 0.02x^2 - 1,400$

(F) $P'(250) = 6$. At a production level of 250 toasters, profit is increasing at the rate of $6 per toaster.
$P'(475) = -3$. At a production level of 475 toasters, profit is decreasing at the rate of $3 per toaster.

17. (A) $x = 500$ (B) $P(x) = 176x - 0.2x^2 - 21,900$
(C) $x = 440$
(D) Break-even points: (150, 25,500) and (730, 39,420); x intercepts
for $P(x)$: $x = 150$ and $x = 730$

19. (A) $P(x) = 18x - x^{3/2} - 500$
(B) $0 \leqslant x \leqslant 400$
(C) A profit is realized for $44 < x < 258$, and a loss is incurred for
$0 \leqslant x < 44$ and $258 < x \leqslant 400$.

21. (A)   (B) Fixed costs ≈ \$721,680;
variable costs ≈ \$121

(C) (713, 807,703), (5,423, 1,376,227)
(D) $\$254 \leqslant p \leqslant \$1,133$

```
QuadReg
y=ax²+bx+c
a=1.4101002ᴇ-5
b=-.2732556676
c=1320.924694
```

```
LinReg
y=ax+b
a=120.7047281
b=721680.1282
r=.9934384133
```

Chapter 8 Review Exercise

1. (A) 16 (B) 8 (C) 8 (D) 4 (E) 4 (F) 4 *(8-1, 8-3, 8-4)*   **2.** $f'(-1) \approx -2; f'(1) \approx 1$ *(8-1, 8-3)*   **3.** (A) 22 (B) 8 (C) 2 *(8-2)*
4. (A) 1 (B) 1 (C) 1 (D) 1 *(8-2)*   **5.** (A) 2 (B) 3 (C) Does not exist (D) 3 *(8-2)*
6. (A) 4 (B) 4 (C) 4 (D) Does not exist *(8-2)*   **7.** $3x^2 + 2x$ *(8-3)*   **8.** (A) -11 (B) -14 (C) $\frac{5}{2}$ (D) -8 *(8-4, 8-5, 8-6)*

9. $(6x + 4)$ *(8-6)*   **10.** $12x^3 - 4x$ *(8-4)*   **11.** $x^{-1/2} - 3 = \dfrac{1}{x^{1/2}} - 3$ *(8-4)*   **12.** 0 *(8-4)*   **13.** $-x^{-3} + x = \dfrac{-1}{x^3} + x$ *(8-4)*

14. $(2x - 1)(3) + (3x + 2)(2) = 12x + 1$ *(8-5)*   **15.** $(x^2 - 1)(3x^2) + (x^3 - 3)(2x) = 5x^4 - 3x^2 - 6x$ *(8-5)*

16. $\dfrac{(x^2 + 2)2 - 2x(2x)}{(x^2 + 2)^2} = \dfrac{4 - 2x^2}{(x^2 + 2)^2}$ *(8-5)*   **17.** $(-1)(3x + 2)^{-2}(3) = \dfrac{-3}{(3x + 2)^2}$ *(8-6)*   **18.** $3(2x - 3)^2(2) = 6(2x - 3)^2$ *(8-6)*

19. $-2(x^2 + 2)^{-3}(2x) = \dfrac{-4x}{(x^2 + 2)^3}$ *(8-6)*   **20.** (A) 3 (B) $2 + 0.5h$ (C) 2 *(8-1, 8-3)*

21. *(8-1)*

22. $12x^3 + 6x^{-4} = 12x^3 + \dfrac{6}{x^4}$ *(8-4)*

23. $(2x^2 - 3x + 2)(2x + 2) + (x^2 + 2x - 1)(4x - 3) = 8x^3 + 3x^2 - 12x + 7$ *(8-5)*

24. $\dfrac{(x - 1)^2(2) - (2x - 3)(2)(x - 1)}{(x - 1)^4} = \dfrac{4 - 2x}{(x - 1)^3}$ *(8-5)*

25. $x^{-1/2} - 2x^{-3/2} = \dfrac{1}{x^{1/2}} - \dfrac{2}{x^{3/2}}$ *(8-4)*

26. $(x^2 - 1)[2(2x + 1)(2)] + (2x + 1)^2(2x) = 2(2x + 1)(4x^2 + x - 2)$ *(8-5, 8-6)*

27. $\dfrac{1}{3}(x^3 - 5)^{-2/3}(3x^2) = \dfrac{x^2}{(x^3 - 5)^{2/3}}$ *(8-6)*

28. $-8x^{-3} = \dfrac{-8}{x^3}$ *(8-4)*   **29.** $\dfrac{(2x - 3)(4)(x^2 + 2)^3(2x) - (x^2 + 2)^4(2)}{(2x - 3)^2} = \dfrac{2(x^2 + 2)^3(7x^2 - 12x - 2)}{(2x - 3)^2}$ *(8-5, 8-6)*

30. (A) $m = f'(1) = 2$ (B) $y = 2x + 3$ *(8-3, 8-4)*   **31.** (A) $m = f'(1) = 16$ (B) $y = 16x - 12$ *(8-3, 8-5)*   **32.** $x = 5$ *(8-4)*
33. $x = -5, x = 3$ *(8-5)*   **34.** $x = -1.37, x = 0.60, x = 1.52$ *(8-4)*   **35.** $x = 0, x = 3, x = \frac{15}{2}$ *(8-5)*   **36.** $x = -2, x = 2$ *(8-5)*

37. $x = -2.96$, $x = -2.20$, $x = 1.85$, $x = 3.31$ *(8-5)*   **38.** $x = -1.89$, $x = -0.36$, $x = 0.74$ *(8-5, 8-6)*
39. (A) $v = f'(x) = 32x - 4$ (B) $f'(3) = 92$ ft/sec *(8-4)*   **40.** (A) $v = f'(x) = 96 - 32x$ (B) $x = 3$ sec *(8-4)*
41.   **42.** -1.39 *(8-1, 8-3)*

h	-0.1	-0.01	$-0.001 \to 0$	$\leftarrow 0.001$	0.01	0.1
$\dfrac{f(h) - f(0)}{h}$	1.49	1.60	$1.61 \to 1.61 \leftarrow 1.61$		1.62	1.75

1.61 *(8-1, 8-3)*

43. (A) The graph of g is the graph of f shifted 4 units to the right, and the graph of h is the graph of f shifted 3 units to the left:

(B) The graph of g' is the graph of f' shifted 4 units to the right, and the graph of h' is the graph of f' shifted 3 units to the left:

(8-4)

44. (A) The graph of g is a horizontal translation of the graph of f, and the graph of g' is a horizontal translation of the graph of f'.
 (B) The graph of g is a vertical translation of the graph of f, and the graph of g' is the same as the graph of f'. *(8-3)*
45. $\frac{3}{8}$ *(8-2)*   **46.** 16 *(8-2)*   **47.** -1 *(8-2)*   **48.** $\frac{1}{6}$ *(8-2)*   **49.** -1 *(8-2)*   **50.** 1 *(8-2)*   **51.** Does not exist *(8-2)*   **52.** 4 *(8-2)*

53. 4 *(8-2)*   **54.** $\dfrac{-1}{(x + 2)^2}$ *(8-2)*

55. (A) $\lim_{x \to -2^-} f(x) = -6$; $\lim_{x \to -2^+} f(x) = 6$; $\lim_{x \to -2} f(x)$ does not exist   (B) $\lim_{x \to 0} f(x) = 4$
 (C) $\lim_{x \to 2^-} f(x) = 2$; $\lim_{x \to 2^+} f(x) = -2$; $\lim_{x \to 2} f(x)$ does not exist *(8-2)*
56. $2x - 1$ *(8-3)*   **57.** $1/(2\sqrt{x})$ *(8-3)*   **58.** No *(8-3)*   **59.** No *(8-3)*   **60.** No *(8-3)*   **61.** Yes *(8-3)*

62. $(x - 4)^4(3)(x + 3)^2 + (x + 3)^3(4)(x - 4)^3 = 7x(x - 4)^3(x + 3)^2$ *(8-5, 8-6)*   **63.** $\dfrac{(2x + 1)^4(5x^4) - x^5(4)(2x + 1)^3(2)}{(2x + 1)^8} = \dfrac{x^4(2x + 5)}{(2x + 1)^5}$ *(8-5, 8-6)*

64. $\dfrac{x(\frac{1}{2})(x^2 - 1)^{-1/2}(2x) - (x^2 - 1)^{1/2}}{x^2} = \dfrac{1}{x^2(x^2 - 1)^{1/2}}$ *(8-5, 8-6)*   **65.** $\dfrac{(x^2 + 4)^{1/2} - x(\frac{1}{2})(x^2 + 4)^{-1/2}(2x)}{x^2 + 4} = \dfrac{4}{(x^2 + 4)^{3/2}}$ *(8-5, 8-6)*

66. The domain of $f'(x)$ is all real numbers except $x = 0$. At $x = 0$, the graph of $y = f(x)$ is smooth, but it has a vertical tangent. *(8-3)*
67. (A) $\lim_{x \to 0^-} f(x) = 1$; $\lim_{x \to 0^+} f(x) = -1$ (B) $\lim_{x \to 0^-} f(x) = -2$; $\lim_{x \to 0^+} f(x) = 2$ (C) $m = 0$ *(8-2)*
68. (A) 1 (B) -1 (C) Does not exist   (D) No *(8-3)*   **69.** (A) \$179.90   (B) \$180 *(8-7)*
70. (A) $C(100) = 9,500$; $C'(100) = 50$. At a production level of 100 bicycles, the total cost is \$9,500, and cost is increasing at the rate of \$50 per bicycle.

(B) $\overline{C}(100) = 95$; $\overline{C}'(100) = -0.45$. At a production level of 100 bicycles, the average cost is \$95, and average cost is decreasing at a rate of \$0.45 per bicycle. *(8-7)*

71. The approximate cost of producing the 201st printer is greater than that of the 601st printer. Since these marginal costs are decreasing, the manufacturing process is becoming more efficient. *(8-7)*

72. (A) $C'(x) = 2; \overline{C}(x) = 2 + \dfrac{9,000}{x}; \overline{C}'(x) = \dfrac{-9,000}{x^2}$

(B) $R(x) = xp = 25x - 0.01x^2; R'(x) = 25 - 0.02x; \overline{R}(x) = 25 - 0.01x; \overline{R}'(x) = -0.01$

(C) $P(x) = R(x) - C(x) = 23x - 0.01x^2 - 9,000; P'(x) = 23 - 0.02x; \overline{P}(x) = 23 - 0.01x - \dfrac{9,000}{x}; \overline{P}'(x) = -0.01 + \dfrac{9,000}{x^2}$

(D) (500, 10,000) and (1,800, 12,600)

(E) $P'(1,000) = 3$. Profit is increasing at the rate of \$3 per umbrella.
$P'(1,150) = 0$. Profit is flat.
$P'(1,400) = -5$. Profit is decreasing at the rate of \$5 per umbrella.

(F)

73. (A) 2 components/day (B) 3.2 components/day *(8-5)*

74. $N(5) = 15, N'(5) = 3.8333$, after 5 months, the total sales are 15,000 pools, and sales are increasing at the rate of 3,833 pools per month. *(8-6)*

75. (A) $A(10) = \$9,836; A'(10) \approx \$665$

(B) After 10 yr, the amount in the account is \$9,836, and it is growing at the rate of \$665/yr. *(8-3)*

76. (A)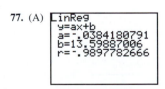

(B) In 1995 consumption was 20.5 trillion ft³ of natural gas and this number was increasing at the rate of 0.8 trillion ft³/yr. *(8-3)*

77. (A)
```
LinReg
y=ax+b
a=-.0384180791
b=13.59887006
r=-.9897782666
```

(B) Fixed costs: \$484.21; variable costs per kringle: \$2.11

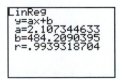

(C) (51, 591.15), (248, 1,007.62)

(D) \$4.07 < p < \$11.64 *(8-3)*

78. $C'(9) = -1$ ppm/m; $C'(99) = -0.001$ ppm/m *(8-6)*

79. $F(3) = 100; F'(3) = -0.25.$ After 3 hr, the body temperature is 100°F, and the temperature is decreasing at the rate of 0.25°F/hr. *(8-6)*

80. (A) 10 items/hr   (B) 5 items/hr *(8-4)*   **81.** (A) $M(50) \approx 66.9; M'(50) \approx 0.7$

(B) In 2010, there will be 66.9 million married couples, and this number will be growing at the rate of 0.7 million = 700,000 couples per year.

(C) 2011: 67.6 million couples; 2012: 68.3 million couples *(8-3)*

■ CHAPTER 9

Exercise 9-1

1. f is continuous at $x = 1$, since $\lim_{x \to 1} f(x) = f(1)$

3. f is discontinuous at $x = 1$, since $\lim_{x \to 1} f(x) \neq f(1)$

5. f is continuous at $x = 1$, since $\lim_{x \to 1} f(x) = f(1)$

7. f is discontinuous at $x = 1$, since $\lim_{x \to 1} f(x)$ does not exist

9.

11.

13. All x **15.** All x, except $x = 5$ **17.** All x, except $x = -2$ and $x = 3$

19. (A) $\lim_{x\to 0^-} f(x) = 1$, $\lim_{x\to 0^+} f(x) = 1$, $\lim_{x\to 0} f(x) = 1$, $f(0) = 1$ (B) Yes, all three conditions are satisfied.

21. (A) $\lim_{x\to 1^-} f(x) = 2$, $\lim_{x\to 1^+} f(x) = 1$, $\lim_{x\to 1} f(x)$ does not exist, $f(1) = 1$ (B) No, because $\lim_{x\to 1} f(x)$ does not exist.

23. (A) $\lim_{x\to -2^-} f(x) = 1$, $\lim_{x\to -2^+} f(x) = 1$, $\lim_{x\to -2} f(x) = 1$, $f(-2) = 3$ (B) No, because $\lim_{x\to -2} f(x) \neq f(-2)$.

25. (A)

(B) 1
(C) 2
(D) No
(E) All integers

27. f is discontinuous at $x = -3$; $\lim_{x\to -3^-} f(x) = -\infty$ and $\lim_{x\to -3^+} f(x) = \infty$; the line $x = -3$ is a vertical asymptote.

29. h is discontinuous at $x = -2$; $\lim_{x\to -2^-} h(x) = \infty$ and $\lim_{x\to -2^+} h(x) = -\infty$; the line $x = -2$ is a vertical asymptote; h is discontinuous at $x = 2$; $\lim_{x\to 2^-} h(x) = -\infty$ and $\lim_{x\to 2^+} h(x) = \infty$; the line $x = 2$ is a vertical asymptote.

31. F is continuous for all real numbers x and has no vertical asymptotes.

33. H is discontinuous at $x = 1$; $\lim_{x\to 1^-} H(x) = -\infty$ and $\lim_{x\to 1^+} H(x) = \infty$; the line $x = 1$ is a vertical asymptote; H is discontinuous at $x = 3$; $\lim_{x\to 3} H(x) = 2$, but $H(3)$ does not exist; there is no vertical asymptote at $x = 3$.

35. T is discontinuous at $x = 0$; $\lim_{x\to 0^-} T(x) = -\infty$ and $\lim_{x\to 0^+} T(x) = -\infty$; the line $x = 0$ (the y axis) is a vertical asymptote; T is discontinuous at $x = 4$; $\lim_{x\to 4^-} T(x) = \infty$ and $\lim_{x\to 4^+} T(x) = \infty$; the line $x = 4$ is a vertical asymptote.

37. $-3 < x < 4$; $(-3, 4)$ **39.** $x < 3$ or $x > 7$; $(-\infty, 3) \cup (7, \infty)$

41. $x < -1.33$ or $1.20 < x < 3.13$; $(-\infty, -1.33) \cup (1.20, 3.13)$

43. $x < -2.53$ or $x > -0.72$; $(-\infty, -2.53) \cup (-0.72, \infty)$

45. $-5 < x < 0$ or $x > 3$; $(-5, 0) \cup (3, \infty)$

47. $x \leqslant -2.15$ or $-0.52 \leqslant x < 1$ or $x \geqslant 2.67$; $(-\infty, -2.15] \cup [-0.52, 1) \cup [2.67, \infty)$

49. $(-\infty, \infty)$ **51.** $[5, \infty)$ **53.** $(-\infty, \infty)$ **55.** $(-\infty, 1), (1, 2), (2, \infty)$

57. Discontinuous at $x = 1$

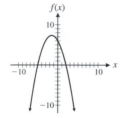

59. Continuous for all x

61. Discontinuous at $x = 0$

63. (A) Yes
(B) Yes
(C) Yes
(D) Yes

65. (A) Yes
(B) No
(C) Yes
(D) No
(E) Yes

67. x intercepts: $x = -5, 2$

69. x intercepts: $x = -6, -1, 4$

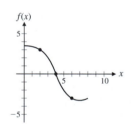

71. No, but this does not contradict Theorem 2, since f is discontinuous at $x = 1$.

73. The following sketches illustrate that either condition is possible. Theorem 2 implies that one of these two conditions must occur.

75. (A)

(B) $\lim_{x \to 4.5} P(x) = \$1.24$; $P(4.5) = \$1.24$
(C) $\lim_{x \to 4} P(x)$ does not exist; $P(4) = \$1.01$
(D) Continuous at $x = 4.5$; not continuous at $x = 4$

77. (A)

(B) p is discontinuous at $x = 250$, 500, and 1,000. In each case, the limit from the left is greater than the limit from the right, reflecting the corresponding drop in price at these order quantities.

(C)

(D) C is discontinuous at $x = 150$, 250, 500, and 1,000. In each case, the limit from the left is greater than the limit from the right, reflecting savings to the customer due to the corresponding drop in price at these order quantities.

79. (A)

(B) $\lim_{s \to 10,000} E(s) = \$1,000$; $E(10,000) = \$1,000$
(C) $\lim_{s \to 20,000} E(s)$ does not exist; $E(20,000) = \$2,000$
(D) Yes; no

81. (A) t_2, t_3, t_4, t_6, t_7 (B) $\lim_{t \to t_5} N(t) = 7$; $N(t_5) = 7$ (C) $\lim_{t \to t_3} N(t)$ does not exist; $N(t_3) = 4$

Exercise 9-2

1. (a, b); (d, f); (g, h) **3.** (b, c); (c, d); (f, g) **5.** c, d, f **7.** b, f
9. Local maximum at $x = a$; local minimum at $x = c$; no local extrema at $x = b$ and $x = d$

11.   **13.**   **15.**   **17.**

19. g_4 **21.** g_6 **23.** g_2
25. Increasing on $(-\infty, 4)$ **27.** Increasing on $(-\infty, -1)$, $(1, \infty)$ **29.** Decreasing for all x
  Decreasing on $(4, \infty)$ Decreasing on $(-1, 1)$ Horizontal tangent at $x = 2$
  Horizontal tangent at $x = 4$ Horizontal tangents at $x = -1, 1$

31. Increasing on $(-\infty, 5)$; decreasing on $(5, \infty)$; local maximum at $x = 5$ **33.** Increasing for all x; no local extrema
35. Decreasing for all x; no local extrema
37. Increasing on $(-\infty, -2)$ and $(2, \infty)$; decreasing on $(-2, 2)$; local maximum at $x = -2$; local minimum at $x = 2$
39. Increasing on $(-\infty, -2)$ and $(4, \infty)$; decreasing on $(-2, 4)$; local maximum at $x = -2$; local minimum at $x = 4$
41. Increasing on $(-\infty, -1)$ and $(0, 1)$; decreasing on $(-1, 0)$ and $(1, \infty)$; local maxima at $x = -1$ and $x = 1$; local minimum at $x = 0$
43. Increasing on $(-1.26, \infty)$; decreasing on $(-\infty, -1.26)$; local minimum at $x = -1.26$
45. Increasing on $(-0.43, 0.54)$ and $(2.14, \infty)$; decreasing on $(-\infty, -0.43)$ and $(0.54, 2.14)$; local maximum at $x = 0.54$; local minima at $x = -0.43$
and $x = 2.14$
47. Increasing on $(-4.17, -2.78)$ and $(0.95, \infty)$; decreasing on $(-\infty, -4.17)$ and $(-2.78, 0.95)$; local maximum at $x = -2.78$; local minima at
$x = -4.17$ and $x = 0.95$
49. Increasing on $(-1, 2)$; decreasing on $(-\infty, -1)$ and $(2, \infty)$;
  local minimum at $x = -1$; local maximum at $x = 2$

51. Increasing on $(-1, 2)$ and $(2, \infty)$; decreasing on $(-\infty, -1)$;
  local minimum at $x = -1$

53. $f'(x) > 0$ on $(-\infty, -1)$ and $(3, \infty)$; $f'(x) < 0$ on $(-1, 3)$;
   $f'(x) = 0$ at $x = -1$ and $x = 3$

55. $f'(x) > 0$ on $(-2, 1)$ and $(3, \infty)$; $f'(x) < 0$ on $(-\infty, -2)$ and $(1, 3)$;
   $f'(x) = 0$ at $x = -2$, $x = 1$, and $x = 3$

57. No critical values; increasing on $(-\infty, -2)$ and $(-2, \infty)$; no local extrema
59. Critical values: $x = -2$, $x = 2$; increasing on $(-\infty, -2)$ and $(2, \infty)$; decreasing on $(-2, 0)$ and $(0, 2)$; local maximum at $x = -2$; local minimum at $x = 2$
61. Critical value: $x = -2$; increasing on $(-2, 0)$; decreasing on $(-\infty, -2)$ and $(0, \infty)$; local minimum at $x = -2$
63. Critical values: $x = 0$, $x = 4$; increasing on $(-\infty, 0)$ and $(4, \infty)$; decreasing on $(0, 2)$ and $(2, 4)$; local maximum at $x = 0$; local minimum at $x = 4$
65. Critical values: $x = 0$, $x = 4$, $x = 6$; increasing on $(0, 4)$ and $(6, \infty)$; decreasing on $(-\infty, 0)$ and $(4, 6)$; local maximum at $x = 4$; local minima at $x = 0$ and $x = 6$
67. Critical value: $x = 2$; increasing on $(2, \infty)$; decreasing on $(-\infty, 2)$; local minimum at $x = 2$
69. Critical value: $x = 1$; increasing on $(0, 1)$; decreasing on $(1, \infty)$; local maximum at $x = 1$
71. (A) There are no critical values and no local extrema. The function is increasing for all x.
   (B) There are two critical values, $x = \pm\sqrt{-k/3}$. The function increases on $(-\infty, -\sqrt{-k/3})$ to a local maximum at $x = -\sqrt{-k/3}$, decreases on $(-\sqrt{-k/3}, \sqrt{-k/3})$ to a local minimum at $x = \sqrt{-k/3}$, and increases on $(\sqrt{-k/3}, \infty)$.
   (C) The only critical value is $x = 0$. There are no local extrema. The function is increasing for all x.
73. (A) The marginal profit is positive on $(0, 600)$, 0 at $x = 600$, and   (B) $P'(x)$
   negative on $(600, 1{,}000)$.

75. (A) The price decreases for the first 15 months to a local minimum,   (B) $B(t)$
   increases for the next 40 months to a local maximum, and then
   decreases for the remaining 15 months.

77. (A) $\overline{C}(x) = \dfrac{x}{20} + 20 + \dfrac{320}{x}$ (B) Critical value: $x = 80$; decreasing for $0 < x < 80$; increasing for $80 < x < 150$; local minimum at $x = 80$

79. $P(x)$ is increasing over (a, b) if $P'(x) = R'(x) - C'(x) > 0$ over (a, b); that is, if $R'(x) > C'(x)$ over (a, b).
81. Critical value: $t = 1$; increasing for $0 < t < 1$; decreasing for $1 < t < 24$; local maximum at $t = 1$
83. Critical value: $t = 7$; increasing for $0 < t < 7$; decreasing for $7 < t < 24$; local maximum at $t = 7$

Exercise 9-3

1. $(a, c), (c, d), (e, g)$ **3.** $(d, e), (g, h)$ **5.** $(a, c), (c, d), (e, g)$ **7.** d, e, g **9.** (C)   **11.** (D)

13. Local minimum   **15.** Unable to determine   **17.** Neither

19.   **21.**   **23.**   **25.**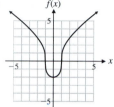

27. $6x - 4$ **29.** $40x^3$ **31.** $6x$ **33.** $24x^2(x^2 - 1) + 6(x^2 - 1)^2 = 6(x^2 - 1)(5x^2 - 1)$ **35.** $6x^{-3} + 12x^{-4}$

37. $f(2) = -2$ is a local minimum   **39.** $f(-1) = 2$ is a local maximum; $f(2) = -25$ is a local minimum   **41.** No local extrema

43. $f(-2) = -6$ is a local minimum; $f(0) = 10$ is a local maximum; $f(2) = -6$ is a local minimum   **45.** $f(0) = 2$ is a local minimum

47. $f(-4) = -8$ is a local maximum; $f(4) = 8$ is a local minimum

49. Concave upward on $(6, \infty)$; concave downward on $(-\infty, 6)$; inflection point at $x = 6$

51. Concave upward on $(-\infty, -2)$ and $(2, \infty)$; concave downward on $(-2, 2)$; inflection points at $x = -2$ and $x = 2$

53. Concave upward on $(-0.15, 2.15)$; concave downward on $(-\infty, -0.15)$ and $(2.15, \infty)$; inflection points at $x = -0.15$ and $x = 2.15$

55. Concave upward on $(-\infty, 0.27)$ and $(3.73, \infty)$; concave downward on $(0.27, 3.73)$; inflection points at $x = 0.27$ and $x = 3.73$

57. Concave upward on $(-1.40, \infty)$; concave downward on $(-\infty, -1.40)$; inflection point at $x = -1.40$

59. Concave upward on $(-0.61, 0.66)$ and $(1.74, \infty)$; concave downward on $(-\infty, -0.61)$ and $(0.66, 1.74)$; inflection points at $x = -0.61$, $x = 0.66$, and $x = 1.74$

61. Local maximum at $x = 0$ **63.** Inflection point at $x = 0$ **65.** Inflection point at $x = 2$ **67.** Local maximum at $x = -2$
Local minimum at $x = 4$ Local minimum at $x = 2$
Inflection point at $x = 2$ Inflection point at $x = 0$

69.

x	$f'(x)$	$f(x)$
$-\infty < x < -1$	Positive and decreasing	Increasing and concave downward
$x = -1$	x intercept	Local maximum
$-1 < x < 0$	Negative and decreasing	Decreasing and concave downward
$x = 0$	Local minimum	Inflection point
$0 < x < 2$	Negative and increasing	Decreasing and concave upward
$x = 2$	Local maximum	Inflection point
$2 < x < \infty$	Negative and decreasing	Decreasing and concave downward

71.

x	$f'(x)$	$f(x)$
$-\infty < x < -2$	Negative and increasing	Decreasing and concave upward
$x = -2$	Local maximum	Inflection point
$-2 < x < 0$	Negative and decreasing	Decreasing and concave downward
$x = 0$	Local minimum	Inflection point
$0 < x < 2$	Negative and increasing	Decreasing and concave upward
$x = 2$	Local maximum	Inflection point
$2 < x < \infty$	Negative and decreasing	Decreasing and concave downward

73. If $f'(x)$ has a local extremum at $x = c$, then $f''(x)$ must change from increasing to decreasing or from decreasing to increasing at $x = c$. Thus, the graph of $y = f(x)$ must change concavity at $x = c$, and there must be an inflection point at $x = c$.
75. If there is an inflection point on the graph of $y = f(x)$ at $x = c$, then $f(x)$ must change concavity at $x = c$. Consequently, $f'(x)$ must change from increasing to decreasing or from decreasing to increasing at $x = c$, and $x = c$ is a local extremum for $f'(x)$.
77. Inflection points at $x = -2$ and $x = 2$ **79.** Inflection points at $x = -6$, $x = 0$, and $x = 6$ **81.** The graph of the CPI is concave upward.
83. The graph of $y = C'(x)$ is positive and decreasing. Since marginal costs are decreasing, the production process is becoming more efficient as production increases.
85. (A) Local maximum at $x = 60$ (B) Concave downward on the whole interval $(0, 80)$
87. (A) Increasing on $(10, 25)$;   **89.** (A)   **91.** (A) Increasing on $(0, 10)$;   **93.** (A) Increasing on $(5, \infty)$;
decreasing on $(25, 40)$ decreasing on $(10, 20)$ decreasing on $(0, 5)$
(B) Inflection point at $x = 25$ (B) Inflection point at $t = 10$ (B) Inflection point at $n = 5$
(C)   (C)

(B) 32 ads to sell 574 cars per month

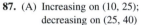

(D) Max $N'(x) = N'(25) = 2{,}025$ (D) $N'(10) = 300$

(C) $T'(5) = 0$

Exercise 9-4

1. $(-\infty, b), (0, e), (e, g)$ **3.** $(b, d), (d, 0), (g, \infty)$ **5.** $x = 0$ **7.** $(-\infty, a), (d, e), (h, \infty)$ **9.** $(a, d), (e, h)$
11. $x = a, x = h$ **13.** $x = d, x = e$
15.   **17.**   **19.**

21.   **23.**

25. ∞ **27.** $-\infty$ **29.** $\frac{4}{5}$ **31.** ∞ **33.** 0
35. Horizontal asymptote: $y = 2$; vertical asymptote: $x = -2$
37. Horizontal asymptote: $y = 1$; vertical asymptotes: $x = -1$ and $x = 1$
39. No horizontal or vertical asymptotes
41. Horizontal asymptote: $y = 0$; no vertical asymptotes
43. No horizontal asymptote; vertical asymptote: $x = 3$
45. Horizontal asymptote: $y = 2$; vertical asymptotes: $x = -1$ and $x = 2$
47. Horizontal asymptote: $y = 2$; vertical asymptote: $x = -1$

49. Domain: All real numbers
y intercept: 5; x intercepts: 1,5
Decreasing on $(-\infty, 3)$
Increasing on $(3, \infty)$
Local minimum at $x = 3$
Concave upward on $(-\infty, \infty)$

51. Domain: All real numbers
y intercept: 0; x intercepts: 0, 6
Increasing on $(-\infty, 0)$ and $(4, \infty)$
Decreasing on $(0, 4)$
Local maximum at $x = 0$
Local minimum at $x = 4$
Concave upward on $(2, \infty)$
Concave downward on $(-\infty, 2)$
Inflection point at $x = 2$

53. Domain: All real numbers
y intercept: 16; x intercepts: $-4, 2$
Increasing on $(-\infty, -2)$ and $(2, \infty)$
Decreasing on $(-2, 2)$
Local maximum at $x = -2$
Local minimum at $x = 2$
Concave upward on $(0, \infty)$
Concave downward on $(-\infty, 0)$
Inflection point at $x = 0$

55. Domain: All real numbers
y intercept: 0; x intercepts: 0, 4
Increasing on $(-\infty, 3)$
Decreasing on $(3, \infty)$
Local maximum at $x = 3$
Concave upward on $(0, 2)$
Concave downward on $(-\infty, 0)$ and $(2, \infty)$
Inflection points at $x = 0$ and $x = 2$

57. Domain: All real numbers, except 3
y intercept: -1; x intercept: -3
Horizontal asymptote: $y = 1$
Vertical asymptote: $x = 3$
Decreasing on $(-\infty, 3)$ and $(3, \infty)$
Concave upward on $(3, \infty)$
Concave downward on $(-\infty, 3)$

59. Domain: All real numbers, except 2
y intercept: 0; x intercept: 0
Horizontal asymptote: $y = 1$
Vertical asymptote: $x = 2$
Decreasing on $(-\infty, 2)$ and $(2, \infty)$
Concave downward on $(-\infty, 2)$
Concave upward on $(2, \infty)$

61. For any $n \geqslant 1$, the limit will be ∞ if $a > 0$ and $-\infty$ if $a < 0$.

63. (A) $\lim_{x \to \infty} p'(x) = \infty$; $\lim_{x \to \infty} p''(x) = \infty$. The graph of $y = p(x)$ is
increasing and concave upward for large positive values of x.
(B) $\lim_{x \to -\infty} p'(x) = \infty$; $\lim_{x \to -\infty} p''(x) = -\infty$. The graph of $y = p(x)$ is increasing and concave downward for large negative values of x.

65. Domain: All real numbers
y intercept: 0; x intercepts: $-1, 0, 1$
$\lim_{x \to -\infty} f(x) = -\infty$; $\lim_{x \to \infty} f(x) = \infty$
Increasing on $(-\infty, -\sqrt{3}/3)$ and $(\sqrt{3}/3, \infty)$
Decreasing on $(-\sqrt{3}/3, \sqrt{3}/3)$
Local maximum at $x = -\sqrt{3}/3$
Local minimum at $x = \sqrt{3}/3$
Concave downward on $(-\infty, 0)$
Concave upward on $(0, \infty)$
Inflection point at $x = 0$

67. Domain: All real numbers
y intercept: 50; x intercept: 12.68
$\lim_{x \to -\infty} f(x) = \infty$; $\lim_{x \to \infty} f(x) = -\infty$
Increasing on $(-1.26, 7.93)$
Decreasing on $(-\infty, -1.26)$ and $(7.93, \infty)$
Local maximum at $x = 7.93$
Local minimum at $x = -1.26$
Concave upward on $(-\infty, \frac{10}{3})$
Concave downward on $(\frac{10}{3}, \infty)$
Inflection point at $x = \frac{10}{3}$

69. Domain: All real numbers, except 0
Vertical asymptote: $x = 0$
$\lim_{x \to -\infty} f(x) = -\infty$; $\lim_{x \to \infty} f(x) = \infty$
Increasing on $(-\infty, -1)$ and $(1, \infty)$
Decreasing on $(-1, 0)$ and $(0, 1)$
Local maximum at $x = -1$
Local minimum at $x = 1$
Concave upward on $(0, \infty)$
Concave downward on $(-\infty, 0)$

71. Domain: All real numbers, except ± 2
y intercept: 0; x intercept: 0
Horizontal asymptote: $y = 0$
Vertical asymptotes: $x = -2, x = 2$
Decreasing on $(-\infty, -2), (-2, 2)$, and $(2, \infty)$
Concave upward on $(-2, 0)$ and $(2, \infty)$
Concave downward on $(-\infty, -2)$ and $(0, 2)$
Inflection point at $x = 0$

73. Domain: All real numbers
y intercept: 1
Horizontal asymptote: $y = 0$
Increasing on $(-\infty, 0)$
Decreasing on $(0, \infty)$
Local maximum at $x = 0$
Concave upward on $(-\infty, -\sqrt{3}/3)$ and $(\sqrt{3}/3, \infty)$
Concave downward on $(-\sqrt{3}/3, \sqrt{3}/3)$
Inflection points at $x = -\sqrt{3}/3$ and $x = \sqrt{3}/3$

75. Domain: All real numbers
y intercept: 27; x intercepts: − 3, 3
$\lim_{x \to \pm\infty} f(x) = -\infty$
Increasing on $(-\infty, -\sqrt{3})$ and $(0, \sqrt{3})$
Decreasing on $(-\sqrt{3}, 0)$ and $(\sqrt{3}, \infty)$
Local maxima at $x = -\sqrt{3}$ and $x = \sqrt{3}$
Local minimum at $x = 0$
Concave upward on $(-1, 1)$
Concave downward on $(-\infty, -1)$ and $(1, \infty)$
Inflection points at $x = -1$ and $x = 1$

77. Domain: All real numbers
x intercepts: − 1.18, 0.61, 1.87, 3.71; y intercept: − 5
$\lim_{x \to \pm\infty} f(x) = \infty$
Decreasing on $(-\infty, -0.53)$ and $(1.24, 3.04)$
Increasing on $(-0.53, 1.24)$ and $(3.04, \infty)$
Local minima at $x = -0.53$ and $x = 3.04$
Local maximum at $x = 1.24$
Concave upward on $(-\infty, 0.22)$ and $(2.28, \infty)$
Concave downward on $(0.22, 2.28)$
Inflection points at $x = 0.22$ and $x = 2.28$

79. Domain: All real numbers
y intercept: 100; x intercepts: 8.01, 13.36
$\lim_{x \to \pm\infty} f(x) = \infty$
Increasing on $(-0.10, 4.57)$ and $(11.28, \infty)$
Decreasing on $(-\infty, -0.10)$ and $(4.57, 11.28)$
Local maximum at $x = 4.57$
Local minima at $x = -0.10$ and $x = 11.28$
Concave upward on $(-\infty, 1.95)$ and $(8.55, \infty)$
Concave downward on $(1.95, 8.55)$
Inflection points at $x = 1.95$ and $x = 8.55$

81. Domain: All real numbers
x intercepts: − 2.40, 1.16; y intercept: 30
$\lim_{x \to \pm\infty} f(x) = -\infty$
Increasing on $(-\infty, -1.58)$
Decreasing on $(-1.58, \infty)$
Local maximum at $x = -1.58$
Concave downward on $(-\infty, -0.88)$ and $(0.38, \infty)$
Concave upward on $(-0.88, 0.38)$
Inflection points at $x = -0.88$ and $x = 0.38$

83. Domain: All real numbers
x intercepts: − 6.68, − 3.64, − 0.72; y intercept: 30
$\lim_{x \to -\infty} f(x) = -\infty$; $\lim_{x \to \infty} f(x) = \infty$
Decreasing on $(-5.59, -2.27)$ and $(1.65, 3.82)$
Increasing on $(-\infty, -5.59)$, $(-2.27, 1.65)$, and $(3.82, \infty)$
Local minima at $x = -2.27$ and $x = 3.82$
Local maxima at $x = -5.59$ and $x = 1.65$
Concave upward on $(-4.31, -0.40)$ and $(2.91, \infty)$
Concave downward on $(-\infty, -4.31)$ and $(-0.40, 2.91)$
Inflection points at $x = -4.31$, $x = -0.40$, and $x = 2.91$

85. (A) Increasing on $(0, 1)$
(B) Concave upward on $(0, 1)$
(C) $x = 1$ is a vertical asymptote
(D) The origin is both an x and a y intercept
(E)

87. (A) $\overline{C}(n) = \dfrac{3{,}200}{n} + 250 + 50n$

(B) 8 yr

89. (A) Domain: $x > 0$
Vertical asymptote: $x = 0$
$\lim_{x \to \infty} f(x) = \infty$
Increasing on $(100, 0)$
Decreasing on $(0, 100)$
Local minimum at $x = 100$
Concave upward on $(0, \infty)$
(B) $(100, 25)$; the minimum average cost of $25 occurs at $x = 100$, the point where marginal cost equals average cost

91. (A)

```
QuadReg
y=ax²+bx+c
a=.0100714286
b=.7835714286
c=316
```

(B) Minimum average cost is $4.35 when 177 pizzas are produced daily.

93. Domain: $t \geq 0$
t intercept: $t = 0$
y intercept: $y = 0$
Horizontal asymptote: $y = 0$
Increasing on $(0, 1)$
Decreasing on $(1, \infty)$
Local maximum at $t = 1$
Concave upward on $(\sqrt{3}, \infty)$
Concave downward on $(0, \sqrt{3})$
Inflection point at $x = \sqrt{3}$

95. Horizontal asymptote: $y = 5$
Decreasing on $(1, \infty)$
Concave upward on $(1, \infty)$

Exercise 9-5

1. Min $f(x) = f(0) = 1$; Max $f(x) = f(10) = 9$ **3.** Min $f(x) = f(0) = 1$; Max $f(x) = f(3) = 8$ **5.** Min $f(x) = f(6) = 3$; Max $f(x) = f(4) = 7$
7. Min $f(x) = f(1) = f(5) = 5$; Max $f(x) = f(3) = 8$ **9.** Min $f(x) = f(2) = 1$; no maximum   **11.** Max $f(x) = f(4) = 26$; no minimum
13. No absolute extrema exist.   **15.** Max $f(x) = f(2) = 16$ **17.** Min $f(x) = f(2) = 14$
19. (A) Max $f(x) = f(5) = 14$; Min $f(x) = f(-1) = -22$ (B) Max $f(x) = f(1) = -2$; Min $f(x) = f(-1) = -22$
(C) Max $f(x) = f(5) = 14$; Min $f(x) = f(3) = -6$
21. (A) Max $f(x) = f(0) = 126$; Min $f(x) = f(2) = -26$ (B) Max $f(x) = f(7) = 49$; Min $f(x) = f(2) = -26$
(C) Max $f(x) = f(6) = 6$; Min $f(x) = f(3) = -15$
23. Exactly in half   **25.** 15 and -15 **27.** A square of side 25 cm; maximum area $= 625$ cm²
29. If x and y are the dimensions of the rectangle and A is the fixed area, the model is: Minimize $C = 2Bx + 2AB/x$, $x > 0$. This mathematical
problem always has a solution that agrees with our economic intuition that there should be a cheapest way to build the fence.
31. If x and y are the dimensions of the rectangle and C is the fixed amount to be spent, the model is: Maximize $A = x(C - 2Bx)/(2B)$,
$0 \leq x \leq C/(2B)$. This mathematical problem always has a solution that agrees with our economic intuition that there should be a largest area that can
be enclosed with a fixed amount of fencing.
33. (A) Max $R(x) = R(3,000) = \$300,000$ (B) Maximum profit is $\$75,000$ when 2,100 sets are manufactured and sold for $\$130$ each.
(C) Maximum profit is $\$64,687.50$ when 2,025 sets are manufactured and sold for $\$132.50$ each.
35. $\$35$; $\$6,125$ **37.** 40 trees; 1,600 lb   **39.** $(10 - 2\sqrt{7})/3 = 1.57$ in. squares
41. 20 ft by 40 ft (with the expensive side being one of the short sides)   **43.** 10,000 books in 5 printings
45. (A) $x = 5.1$ mi   (B) $x = 10$ mi   **47.** 4 days; 20 bacteria/cm³   **49.** 50 mice per order   **51.** 1 month; 2 ft
53. 4 yr from now

Chapter 9   Review Exercise

1. $(a, c_1), (c_3, c_6)$ *(9-2, 9-3)*   **2.** $(c_1, c_3), (c_6, b)$ *(9-2, 9-3)*   **3.** $(a, c_2), (c_4, c_5), (c_7, b)$ *(9-2, 9-3)*   **4.** c_3 *(9-2)*   **5.** c_6 *(9-5)*
6. c_1, c_3, c_5 *(9-2)*   **7.** c_6 *(9-2)*   **8.** c_2, c_4, c_5, c_7 *(9-3)*   **9.** (A) Does not exist   (B) 3   (C) No *(9-1)*
10. (A) 2   (B) Not defined   (C) No *(9-1)*   **11.** (A) 1   (B) 1   (C) Yes *(9-1)*
12. *(9-4)*   **13.** *(9-4)*

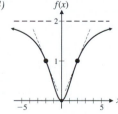

14. $f''(x) = 12x^2 + 30x$ *(9-3)*   **15.** $y'' = 8/x^3$ *(9-3)*   **16.** (A) 4   (B) 6   (C) Does not exist   (D) 6   (E) No *(9-1)*
17. (A) 3   (B) 3   (C) 3   (D) 3   (E) Yes *(9-1)*   **18.** $(-3, 4)$ *(9-1)*   **19.** $(-3, 0) \cup (5, \infty)$ *(9-1)*   **20.** $(-2.34, -0.47) \cup (1.81, \infty)$ *(9-1)*

21. (A) All real numbers   (B) y intercept: 0; x intercepts: 0, 9
   (C) $\lim_{x \to -\infty} f(x) = -\infty$; $\lim_{x \to \infty} f(x) = \infty$ *(9-4)*
22. (A) 3, 9   (B) 3, 9
   (C) Increasing on $(-\infty, 3)$ and $(9, \infty)$; decreasing on $(3, 9)$
   (D) Local maximum at $x = 3$; local minimum at $x = 9$ *(9-4)*
23. (A) Concave downward on $(-\infty, 6)$; concave upward on $(6, \infty)$
   (B) Inflection point at $x = 6$ *(9-4)*

24. *(9-4)*

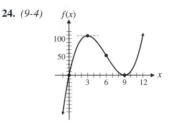

25. (A) All real numbers, except -2 (B) y intercept: 0; x intercept: 0
   (C) $\lim_{x \to -\infty} f(x) = \lim_{x \to \infty} f(x) = 3$; horizontal asymptote: $y = 3$; vertical asymptote: $x = -2$ *(9-4)*
26. (A) None   (B) -2 (C) Increasing on $(-\infty, -2)$ and $(-2, \infty)$ (D) None *(9-4)*
27. (A) Concave upward on $(-\infty, -2)$; concave downward on $(-2, \infty)$
   (B) No inflection points *(9-4)*

28. *(9-4)*

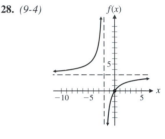

29.

x	$f'(x)$	$f(x)$
$-\infty < x < -2$	Negative and increasing	Decreasing and concave upward
$x = -2$	x intercept	Local minimum
$-2 < x < -1$	Positive and increasing	Increasing and concave upward
$x = -1$	Local maximum	Inflection point
$-1 < x < 1$	Positive and decreasing	Increasing and concave downward
$x = 1$	Local minimum	Inflection point
$1 < x < \infty$	Positive and increasing	Increasing and concave upward

(9-3)

30. (C) *(9-3)*   **31.** Min $f(x) = f(5) = -88$; no absolute maximum *(9-3)*   **32.** Min $f(x) = f(2) = -4$; Max $f(x) = f(5) = 77$ *(9-5)*
33. Min $f(x) = f(2) = 8$ *(9-5)*   **34.** $(-\infty, \infty)$ *(9-1)*   **35.** $(-\infty, -5), (-5, \infty)$ *(9-1)*   **36.** $(-\infty, -2), (-2, 3), (3, \infty)$ *(9-1)*   **37.** $[3, \infty)$ *(9-1)*
38. $(-\infty, \infty)$ *(9-1)*   **39.** ∞ *(9-1)*   **40.** $-\infty$ *(9-1)*   **41.** Does not exist *(9-1)*   **42.** ∞ *(9-1)*   **43.** $-\infty$ *(9-4)*   **44.** ∞ *(9-4)*   **45.** ∞ *(9-4)*
46. 2 *(9-4)*   **47.** ∞ *(9-4)*   **48.** 0 *(9-4)*
49. Horizontal asymptote: $y = 0$; no vertical asymptotes *(9-4)*
50. No horizontal asymptotes; vertical asymptotes: $x = -3$ and $x = 3$ *(9-4)*
51. Yes. Since f is continuous on $[a, b]$, f has an absolute maximum on $[a, b]$. But each end point is a local minimum; hence, the absolute maximum must occur between a and b. *(9-5)*
52. No, increasing/decreasing properties apply to intervals in the domain of f. It is correct to say that $f(x)$ is decreasing on $(-\infty, 0)$ and $(0, \infty)$. *(9-2)*
53. A critical value for $f(x)$ is a partition number for $f'(x)$ that is also in the domain of f. For example, if $f(x) = x^{-1}$, then 0 is a partition number for $f'(x) = -x^{-2}$, but 0 is not a critical value for $f(x)$ since 0 is not in the domain of f. *(9-2)*
54. Max $f'(x) = f'(2) = 12$ *(9-5)*   **55.** Each number is 20; minimum sum is 40 *(9-5)*

56. Domain: All real numbers
   y intercept: -3; x intercepts: -3, 1
   $\lim_{x \to \pm\infty} f(x) = \infty$
   Increasing on $(-2, \infty)$
   Decreasing on $(-\infty, -2)$
   Local minimum at $x = -2$
   Concave upward on $(-\infty, -1)$ and $(1, \infty)$
   Concave downward on $(-1, 1)$
   Inflection points at $x = -1$ and $x = 1$ *(9-4)*

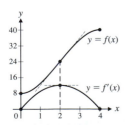

57. Domain: All real numbers
x intercepts: 0.79, 1.64; y intercept: 4
$\lim_{x \to \pm\infty} f(x) = \infty$
Increasing on $(-1.68, -0.35)$ and $(1.28, \infty)$
Decreasing on $(-\infty, -1.68)$ and $(-0.35, 1.28)$
Local minima at $x = -1.68$ and $x = 1.28$
Local maximum at $x = -0.35$
Concave downward on $(-1.10, 0.60)$
Concave upward on $(-\infty, -1.10)$ and $(0.60, \infty)$
Inflection points at $x = -1.10$ and $x = 0.60$ *(9-4)*

58. Domain: All real numbers
x intercepts: 0, 11.10
y intercept: 0
$\lim_{x \to \pm\infty} f(x) = \infty$
Increasing on $(1.87, 4.19)$ and $(8.94, \infty)$
Decreasing on $(-\infty, 1.87)$ and $(4.19, 8.94)$
Local maximum at $x = 4.19$
Local minima at $x = 1.87$ and $x = 8.94$
Concave upward on $(-\infty, 2.92)$ and $(7.08, \infty)$
Concave downward on $(2.92, 7.08)$
Inflection points at $x = 2.92$ and $x = 7.08$ *(9-4)*

59. (A) $p(x)$

(B) p is discontinuous at $x = 12$ and $x = 24$. In each case, the limit from the left is greater than the limit from the right, reflecting the corresponding drop in price at these order quantities.

(C) $C(x)$

(D) C is discontinuous at $x = 12$ and $x = 24$. In each case, the limit from the left is greater than the limit from the right, reflecting savings to the customer due to the corresponding drop in price at these order quantities. *(9-1)*

60. (A) For the first 15 months, the graph of the price is increasing and concave downward, with a local maximum at $t = 15$. For the next 15 months, the graph of the price is decreasing and concave downward, with an inflection point at $t = 30$. For the next 15 months, the graph of the price is decreasing and concave upward, with a local minimum at $t = 45$. For the remaining 15 months, the graph of the price is increasing and concave upward.

(B) $p(t)$

(9-3)

61. (A) Max $R(x) = R(10,000) = \$2,500,000$
(B) Maximum profit is \$175,000 when 3,000 stoves are manufactured and sold for \$425 each.
(C) Maximum profit is \$119,000 when 2,600 stoves are manufactured and sold for \$435 each. *(9-5)*

62. (A) The expensive side is 50 ft; the other side is 100 ft.   (B) The expensive side is 75 ft; the other side is 150 ft. *(9-5)*

63. \$49; \$6,724 *(9-5)*   **64.** 12 orders/yr *(9-5)*

65. (200, 50); the minimal average cost of \$50 occurs at $x = 200$, the point where marginal cost equals average cost. *(9-4)*

66. (A)
```
CubicReg
 y=ax³+bx²+cx+d
 a=-.01
 b=.83
 c=-2.3
 d=221
```

(B) 28 ads to sell 588 refrigerators per month *(9-4)*

67. 3 days *(9-2)*   **68.** 2 yr from now *(9-2)*

■ CHAPTER 10

Exercise 10-1

1. \$1,221.40; \$1,648.72; \$2,225.54   **3.** 11.55   **5.** 10.99   **7.** 0.14

9.

n	$[1 + (1/n)]^n$
10	2.593 74
100	2.704 81
1,000	2.716 92
10,000	2.718 15
100,000	2.718 27
1,000,000	2.718 28
10,000,000	2.718 28
↓	↓
∞	$e = 2.718\ 281\ 828\ 459. . .$

11.

$y_2 = \left(1 + \dfrac{1}{n}\right)^{n+1}$

$y_1 = \left(1 + \dfrac{1}{n}\right)^n$

13. $\lim_{n \to \infty} (1 + n)^{1/n} = 1$
15. \$55,463.90
17. \$9,931.71
19. $r = \frac{1}{4} \ln 1.5 \approx 0.1014$ or 10.14%

21. (A)

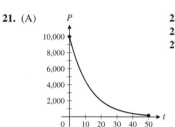

(B) $\lim_{t \to \infty} 10,000e^{-0.08t} = 0$

23. 2.77 yr
25. 13.86%
27. 7.3 yr

29. (A) $A = Pe^{rt}$
$2P = Pe^{rt}$
$2 = e^{rt}$
$rt = \ln 2$
$t = \dfrac{\ln 2}{r}$

(B)

$t = \dfrac{\ln 2}{r}$

Although r could be any positive number, the restrictions on r are reasonable in the sense that most investments would be expected to earn a return of between 2% and 30%.
(C) The doubling times (in years) are 13.86, 6.93, 4.62, 3.47, 2.77, and 2.31, respectively.

31. $t = -(\ln 0.5)/0.000\ 433\ 2 \approx 1,600$ yr   **33.** $r = (\ln 0.5)/30 \approx -0.0231$ **35.** 34.66 yr   **37.** 3.47%   **39.** Approx. 521 yr

Exercise 10-2

1. $6e^x - \dfrac{7}{x}$ **3.** $2exe^{-1} + 3e^x$ **5.** $\dfrac{5}{x}$ **7.** $\dfrac{2 \ln x}{x}$ **9.** $x^3 + 4x^3 \ln x = x^3(1 + 4 \ln x)$ **11.** $x^3e^x + 3x^2e^x = x^2e^x(x + 3)$

13. $\dfrac{(x^2 + 9)e^x - 2xe^x}{(x^2 + 9)^2} = \dfrac{e^x(x^2 - 2x + 9)}{(x^2 + 9)^2}$ **15.** $\dfrac{x^3 - 4x^3 \ln x}{x^8} = \dfrac{1 - 4 \ln x}{x^5}$ **17.** $3(x + 2)^2 \ln x + \dfrac{(x + 2)^3}{x} = (x + 2)^2 \left(3 \ln x + \dfrac{x + 2}{x}\right)$

19. $(x + 1)^3e^x + 3(x + 1)^2e^x = (x + 1)^2e^x(x + 4)$ **21.** $\dfrac{2xe^x - (x^2 + 1)e^x}{(e^x)^2} = \dfrac{2x - x^2 - 1}{e^x}$ **23.** $(\ln x)^3 + 3(\ln x)^2 = (\ln x)^2(\ln x + 3)$

25. $3(4 - 5e^x)^2(-5e^x) = -15e^x(4 - 5e^x)^2$ **27.** $\dfrac{1}{2}(1 + \ln x)^{-1/2}\left(\dfrac{1}{x}\right) = \dfrac{1}{2x(1 + \ln x)^{1/2}}$ **29.** $xe^x + e^x - e^x = xe^x$

31. $2x^2\left(\dfrac{1}{x}\right) + 4x \ln x - 2x = 4x \ln x$ **33.** $y = ex$ **35.** $y = \dfrac{1}{e}x$

37. Yes, she is correct. In fact, for any real number c, the tangent line to $y = e^x$ at the point (c, e^c) has equation $y - e^c = e^c(x - c)$, and thus the tangent line passes through the point $(c - 1, 0)$.
39. Max $f(x) = f(e^3) = e^3 \approx 20.086$ **41.** Min $f(x) = f(1) = e \approx 2.718$ **43.** Max $f(x) = f(e^{1/2}) = 2e^{-1/2} \approx 1.213$
45. $y = 2$ is a horizontal asymptote; no vertical asymptotes; $\lim_{x \to \infty} f(x) = -\infty$; $\lim_{x \to -\infty} f(x) = 2$
47. $y = -5$ is a horizontal asymptote; $x = 1$ is a vertical asymptote; $\lim_{x \to 1^-} f(x) = -\infty$; $\lim_{x \to 1^+} f(x) = \infty$; $\lim_{x \to \infty} f(x) = -5$

49. $y = 3$ is a horizontal asymptote; no vertical asymptotes; $\lim_{x\to\infty} f(x) = 3$; $\lim_{x\to-\infty} f(x) = \infty$

51. $x = 0$ is a vertical asymptote; no horizontal asymptotes; $\lim_{x\to0^+} f(x) = -\infty$; $\lim_{x\to\infty} f(x) = \infty$

53. Domain: All real numbers
y intercept: 0; x intercept: 0
Horizontal asymptote: $y = 1$
Decreasing on $(-\infty, \infty)$
Concave downward on $(-\infty, \infty)$

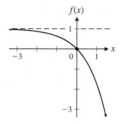

55. Domain: $(0, \infty)$
Vertical asymptote: $x = 0$
Decreasing on $(0, 1)$
Local minimum at $x = 1$
Concave upward on $(0, \infty)$

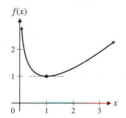

57. Domain: All real numbers
y intercept: 3; x intercept: 3
Horizontal asymptote: $y = 0$
Increasing on $(-\infty, 2)$
Decreasing on $(2, \infty)$
Local maximum at $x = 2$
Concave upward on $(-\infty, 1)$
Concave downward on $(1, \infty)$
Inflection point at $x = 1$

59. Domain: $(0, \infty)$
x intercept: 1
Increasing on $(e^{-1/2}, \infty)$
Decreasing on $(0, e^{-1/2})$
Local minimum at $x = e^{-1/2}$
Concave upward on $(e^{-3/2}, \infty)$
Concave downward on $(0, e^{-3/2})$
Inflection point at $x = e^{-3/2}$

61. Critical values: $x = 0.36$, $x = 2.15$
Increasing on $(-\infty, 0.36)$ and $(2.15, \infty)$
Decreasing on $(0.36, 2.15)$
Local maximum at $x = 0.36$
Local minimum at $x = 2.15$

63. Critical value: $x = 2.21$
Increasing on $(0, 2.21)$
Decreasing on $(2.21, \infty)$
Local maximum at $x = 2.21$

65. $(-0.82, 0.44)$, $(1.43, 4.18)$, $(8.61, 5503.66)$

67. $(4.18, 1.43)$, $(5503.66, 8.61)$

69. Increasing on $(0, 0.62)$ and $(1.51, \infty)$; decreasing on $(0.62, 1.51)$

71. $p = \$2$

73. $p = \$27.57$

75. Min $\overline{C}(x) = \overline{C}(e^7) \approx \$99.91$

77. (A) At $3.68 each, the maximum revenue will be $3,680/wk (in the test city).
(B)

79. At the 40 lb weight level, blood pressure would increase at the rate of 0.44 mm of mercury/lb of weight gain.
At the 90 lb weight level, blood pressure would increase at the rate of 0.19 mm of mercury/lb of weight gain.

81. (A) After 1 hr, the concentration is decreasing at the rate of 1.60 mg/ml/hr; after 4 hr, the concentration is decreasing at the rate of 0.08 mg/ml/hr.
(B)

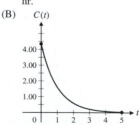

83. $dR/dS = k/S$

Exercise 10-3

1. $y = u^3; u = 2x + 5$ **3.** $y = \ln u; u = 2x^2 + 7$ **5.** $y = e^u; u = x^2 - 2$ **7.** $y = (2 + e^x)^2; dy/dx = 2e^x(2 + e^x)$

9. $y = e^{2-x^4}; dy/dx = -4x^3e^{2-x^4}$ **11.** $y = \ln(4x^5 - 7); \dfrac{dy}{dx} = \dfrac{20x^4}{4x^5 - 7}$ **13.** $\dfrac{1}{x - 3}$ **15.** $\dfrac{-2}{3 - 2t}$ **17.** $6e^{2x}$ **19.** $-8e^{-4t}$

21. $-3e^{-0.03x}$ **23.** $\dfrac{4}{x + 1}$ **25.** $4e^{2x} - 3e^x$ **27.** $(6x - 2)e^{3x^2 - 2x}$ **29.** $\dfrac{2t + 3}{t^2 + 3t}$ **31.** $\dfrac{x}{x^2 + 1}$

33. $\dfrac{4[\ln(t^2 + 1)]^3(2t)}{t^2 + 1} = \dfrac{8t[\ln(t^2 + 1)]^3}{t^2 + 1}$ **35.** $4(e^{2x} - 1)^3(2e^{2x}) = 8e^{2x}(e^{2x} - 1)^3$ **37.** $\dfrac{(x^2 + 1)(2e^{2x}) - e^{2x}(2x)}{(x^2 + 1)^2} = \dfrac{2e^{2x}(x^2 - x + 1)}{(x^2 + 1)^2}$

39. $(x^2 + 1)(-e^{-x}) + e^{-x}(2x) = e^{-x}(2x - x^2 - 1)$ **41.** $\dfrac{e^{-x}}{x} - e^{-x}\ln x = \dfrac{e^{-x}(1 - x\ln x)}{x}$ **43.** $\dfrac{-2x}{(1 + x^2)[\ln(1 + x^2)]^2}$

45. $\dfrac{-2x}{3(1 - x^2)[\ln(1 - x^2)]^{2/3}}$ **47.** f is increasing; g and h are decreasing   **49.** h increases the most rapidly; g the least rapidly

51. Domain: $(-\infty, \infty)$
y intercept: 0; x intercept: 0
Horizontal asymptote: $y = 1$
Increasing on $(-\infty, \infty)$
Concave downward on $(-\infty, \infty)$

53. Domain: $(-\infty, 1)$
y intercept: 0; x intercept: 0
Vertical asymptote: $x = 1$
Decreasing on $(-\infty, 1)$
Concave downward on $(-\infty, 1)$

55. Domain: $(-\infty, \infty)$
y intercept: 1
Horizontal asymptote: $y = 0$
Increasing on $(-\infty, 0)$
Decreasing on $(0, \infty)$
Local maximum at $x = 0$
Concave upward on $(-\infty, -1)$ and $(1, \infty)$
Concave downward on $(-1, 1)$
Inflection points at $x = -1$ and $x = 1$

57. $y = 1 + [\ln(2 + e^x)]^2; \dfrac{dy}{dx} = \dfrac{2e^x\ln(2 + e^x)}{2 + e^x}$

59. $\dfrac{1}{\ln 2}\left(\dfrac{6x}{3x^2 - 1}\right)$

61. $(2x + 1)(10^{x^2 + x})(\ln 10)$

63. $\dfrac{12x^2 + 5}{(4x^3 + 5x + 7)\ln 3}$

65. $2^{x^3 - x^2 + 4x + 1}(3x^2 - 2x + 4)\ln 2$

69. (A) $g(x)$ is not negative when $f(x)$ is decreasing
(B) $f'(x) = \dfrac{2x}{x^2 + 1}$

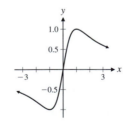

71. $f'(x) = g'(x) = \dfrac{8x}{x^2 + 3}$

73. A maximum revenue of $735.80 is realized at a production level of 20 units at $36.79 each.

75. A maximum profit of $224.61 is realized at a production level of 17 units at $42.74 each.

77. $-$27,145/yr; $-$18,196/yr; $-$11,036/yr

79. (A) 23 days; $26,685; about 50%
(B)

81. 2.27 mm of mercury/yr; 0.81 mm of mercury/yr; 0.41 mm of mercury/yr
83. $A'(t) = 2(\ln 2)5{,}000e^{2t\ln 2} = 10{,}000(\ln 2)2^{2t}$;
  $A'(1) = 27{,}726$ bacteria/hr (rate of change at the end of the first hour);
  $A'(5) = 7{,}097{,}827$ bacteria/hr (rate of change at the end of the fifth hour)

85.

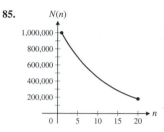

Exercise 10-4

1. $y' = 6x$; 6   **3.** $y' = \dfrac{3x}{y}$; 3   **5.** $y' = \dfrac{1}{2y + 1}$; $\dfrac{1}{3}$ **7.** $y' = -\dfrac{y}{x}$; $-\dfrac{3}{2}$ **9.** $y' = -\dfrac{2y}{2x + 1}$; 4   **11.** $y' = \dfrac{6 - 2y}{x}$; -1

13. $y' = \dfrac{2x}{e^y - 2y}$; 2   **15.** $y' = \dfrac{3x^2 y}{y + 1}$; $\dfrac{3}{2}$ **17.** $y' = \dfrac{6x^2 y - y \ln y}{x + 2y}$; 2   **19.** $x' = \dfrac{2tx - 3t^2}{2x - t^2}$; 8   **21.** $y = -x + 5$

23. $y = \frac{2}{5}x - \frac{12}{5}$; $y = \frac{3}{5}x + \frac{12}{5}$ **25.** $y' = -\dfrac{1}{x}$ **27.** $y' = \dfrac{1}{3(1 + y)^2 + 1}$; $\dfrac{1}{13}$ **29.** $y' = \dfrac{3(x - 2y)^2}{6(x - 2y)^2 + 4y}$; $\dfrac{3}{10}$

31. $y' = \dfrac{3x^2(7 + y^2)^{1/2}}{y}$; 16   **33.** $y' = \dfrac{y}{2xy^2 - x}$; 1   **35.** 2.8; -2.8 **37.** $y = 0.63x + 1.04$ **39.** $p' = \dfrac{1}{2p - 2}$

41. $p' = -\dfrac{\sqrt{10{,}000 - p^2}}{p}$ **43.** $\dfrac{dL}{dV} = \dfrac{-(L + m)}{V + n}$

Exercise 10-5

1. 240   **3.** $\frac{9}{4}$ **5.** $\frac{1}{2}$ **7.** Decreasing at 9 units/sec   **9.** Approx. -3.03 ft/sec   **11.** $dA/dt \approx 126$ ft²/sec   **13.** 3,768 cm³/min

15. 6 lb/in.²/hr   **17.** $-\frac{9}{4}$ ft/sec   **19.** $\dfrac{dy}{dt} = \dfrac{-3x}{\sqrt{100 - x^2}}$; -1.31 ft/sec; -2.25 ft/sec; -4 ft/sec   **21.** $\frac{20}{3}$ ft/sec

23. 0.0214 ft/sec; 0.0135 ft/sec; yes, at $t = 0.000\ 19$ sec   **25.** 3.835 units/sec
27. (A) $dC/dt = \$15{,}000$/wk   (B) $dR/dt = -\$50{,}000$/wk   (C) $dP/dt = -\$65{,}000$/wk   **29.** $ds/dt = \$2{,}207$/wk
31. (A) $dx/dt = -12.73$ units/month   (B) $dp/dt = \$1.53$/month   **33.** Approx. 100 ft³/min

Chapter 10   Review Exercise

1. $\$3{,}136.62$; $\$4{,}919.21$; $\$12{,}099.29$ *(10-1)*   **2.** $\dfrac{2}{x} + 3e^x$ *(10-2)*   **3.** $2e^{2x - 3}$ *(10-2)*   **4.** $\dfrac{2}{2x + 7}$ *(10-2)*

5. (A) $y = \ln(3 + e^x)$ (B) $\dfrac{dy}{dx} = \dfrac{e^x}{3 + e^x}$ *(10-3)*   **6.** $y' = \dfrac{9x^2}{4y}$; $\dfrac{9}{8}$ *(10-4)*   **7.** $dy/dt = 216$ *(10-4)*

8. Domain: All real numbers
  y intercept: 100
  Horizontal asymptote: $y = 0$
  Decreasing on $(-\infty, \infty)$
  Concave upward on $(-\infty, \infty)$

(10-2)

9. $e^2 = 7.389\ldots$ *(10-1)*
10. f is increasing; g and h are decreasing *(10-3)*
11. $\dfrac{7[(\ln z)^6 + 1]}{z}$ *(10-3)*   **12.** $x^5(1 + 6 \ln x)$ *(10-2)*

13. $\dfrac{e^x(x - 6)}{x^7}$ *(10-2)*   **14.** $\dfrac{6x^2 - 3}{2x^3 - 3x}$ *(10-3)*

15. $(3x^2 - 2x)e^{x^3 - x^2}$ *(10-3)*   **16.** $\dfrac{1 - 2x \ln 5x}{xe^{2x}}$ *(10-3)*

17. $y = -x + 2$; $y = -ex + 1$ *(10-2)*   **18.** $y' = \dfrac{3y - 2x}{8y - 3x}$; $\dfrac{8}{19}$ *(10-4)*   **19.** $x' = \dfrac{4tx}{3x^2 - 2t^2}$; -4 *(10-4)*   **20.** $y' = \dfrac{1}{e^y + 2y}$; 1 *(10-4)*

21. $y' = \dfrac{2xy}{1 + 2y^2}$; $\dfrac{2}{3}$ *(10-4)*   **22.** -2; 2 *(10-4)*   **23.** $dy/dt = -2$ units/sec *(10-5)*   **24.** 0.27 ft/sec *(10-5)*

25. $dR/dt = 1/\pi \approx 0.318$ in./min *(10-5)*   **26.** Max $f(x) = f(e^{4.5}) = 2e^{4.5} \approx 180.03$ *(10-2)*   **27.** Max $f(x) = f(0.5) = 5e^{-1} \approx 1.84$ *(10-3)*
28. Max $f(x) = f(1.373) = 2.487$ *(10-2)*   **29.** Max $f(x) = f(1.763) = 0.097$ *(10-2)*

30. Domain: All real numbers
y intercept: 0; x intercept: 0
Horizontal asymptote: $y = 5$
Increasing on $(-\infty, \infty)$
Concave downward on $(-\infty, \infty)$

$(10\text{-}2)$

31. Domain: $(0, \infty)$
x intercept: 1
Increasing on $(e^{-1/3}, \infty)$
Decreasing on $(0, e^{-1/3})$
Local minimum at $x = e^{-1/3}$
Concave upward on $(e^{-5/6}, \infty)$
Concave downward on $(0, e^{-5/6})$
Inflection point at $x = e^{-5/6}$

$(10\text{-}2)$

32. (A) $y = [\ln(4 - e^x)]^3$ (B) $\dfrac{dy}{dx} = \dfrac{-3e^x[\ln(4 - e^x)]^2}{4 - e^x}$ $(10\text{-}3)$ **33.** $2x(5^{x^2-1})(\ln 5)$ $(10\text{-}3)$ **34.** $\left(\dfrac{1}{\ln 5}\right)\dfrac{2x - 1}{x^2 - x}$ $(10\text{-}3)$

35. $\dfrac{2x + 1}{2(x^2 + x)\sqrt{\ln(x^2 + x)}}$ $(10\text{-}3)$ **36.** $y' = \dfrac{2x - e^{xy}y}{xe^{xy} - 1}; 0$ $(10\text{-}4)$

37. The rate of increase of area is proportional to the radius R, so it is smallest when $R = 0$, and has no largest value. $(10\text{-}5)$
38. Yes, for $-\sqrt{3}/3 < x < \sqrt{3}/3$ $(10\text{-}5)$ **39.** (A) 15 yr (B) 13.9 yr $(10\text{-}1)$ **40.** $A'(t) = 10e^{0.1t}$; $A'(1) = \$11.05/\text{yr}$; $A'(10) = \$27.18/\text{yr}$ $(10\text{-}1)$
41. $R'(x) = (1{,}000 - 20x)e^{-0.02x}$ $(10\text{-}3)$
42. A maximum revenue of $18,394 is realized at a production level of 50 units at $367.88 each. $(10\text{-}3)$

43. $(10\text{-}3)$

44. Min $\overline{C}(x) = \overline{C}(e^5) \approx \$49.66$ $(10\text{-}2)$ **45.** $p' = \dfrac{-(5{,}000 - 2p^3)^{1/2}}{3p^2}$ $(10\text{-}4)$

46. $dR/dt = \$110/\text{day}$ $(10\text{-}5)$ **47.** -1.111 mg/ml/hr; -0.335 mg/ml/hr $(10\text{-}3)$

48. $dR/dt = -3/(2\pi)$; approx. 0.477 mm/day $(10\text{-}5)$

49. (A) Increasing at the rate of 2.68 units/day at the end of 1 day of training;   **50.** $dT/dt = -1/27 \approx -0.037$ min/operation hour $(10\text{-}5)$
increasing at the rate of 0.54 unit/day after 5 days of training
(B) N(t)

$(10\text{-}3)$

■ CHAPTER 11

Exercise 11-1

1. $7x + C$ **3.** $(x^7/7) + C$ **5.** $2t^4 + C$ **7.** $u^2 + u + C$ **9.** $x^3 + x^2 - 5x + C$ **11.** $(s^5/5) - \frac{4}{3}s^6 + C$ **13.** $3e^t + C$ **15.** $2 \ln|z| + C$
17.   **19.**   **21.** $y = 40x^5 + C$ **23.** $P = 24x - 3x^2 + C$ **25.** $y = \frac{1}{3}u^6 - u^3 - u + C$
27. $y = e^x + 3x + C$ **29.** $x = 5 \ln|t| + t + C$
31. No, since one graph cannot be obtained from another by a vertical translation.
33. Yes, since one graph can be obtained from another by a vertical translation.
35. $4x^{3/2} + C$ **37.** $-4x^{-2} + C$ **39.** $2\sqrt{u} + C$ **41.** $-(x^{-2}/8) + C$
43. $-(u^{-4}/8) + C$ **45.** $x^3 + 2x^{-1} + C$
47. $2x^5 + 2x^{-4} - 2x + C$ **49.** $2x^{3/2} + 4x^{1/2} + C$ **51.** $\frac{3}{5}x^{5/3} + 2x^{-2} + C$
53. $(e^x/4) - (3x^2/8) + C$ **55.** $-z^{-2} - z^{-1} + \ln|z| + C$

57.

59.

61. $y = x^2 - 3x + 5$ **63.** $C(x) = 2x^3 - 2x^2 + 3,000$ **65.** $x = 40\sqrt{t}$
67. $y = -2x^{-1} + 3\ln|x| - x + 3$ **69.** $x = 4e^t - 2t - 3$
71. $y = 2x^2 - 3x + 1$ **73.** $x^2 + x^{-1} + C$ **75.** $\frac{1}{2}x^2 + x^{-2} + C$
77. $e^x - 2\ln|x| + C$ **79.** $M = t + t^{-1} + \frac{3}{4}$
81. $y = 3x^{5/3} + 3x^{2/3} - 6$ **83.** $p(x) = 10x^{-1} + 10$

85. $\overline{C}(x) = 15 + \dfrac{1,000}{x}$; $C(x) = 15x + 1,000$; $C(0) = \$1,000$

87. (A) The cost function increases from 0 to 8, is concave downward from 0 to 4, and is concave upward from 4 to 8. There is an inflection point at $x = 4$.
(B) $C(x) = x^3 - 12x^2 + 53x + 30$; $C(4) = \$114,000$; $C(8) = \$198,000$
(D) Manufacturing plants are often inefficient at low and high levels of production.

(C) $C(x)$

89. $S(t) = 2,000 - 15t^{5/3}$; $80^{3/5} \approx 14$ mo **91.** $S(t) = 2,000 - 15t^{5/3} - 70t$; $t \approx 8.92$ mo **93.** $L(x) = 4,800x^{1/2}$; $L(25) = 24,000$ labor-hours
95. $W(h) = 0.0005h^3$; $W(70) = 171.5$ lb **97.** 19,400

Exercise 11-2

1. $\frac{1}{6}(x^2 - 4)^6 + C$ **3.** $e^{4x} + C$ **5.** $\ln|2t + 3| + C$ **7.** $\frac{1}{24}(3x - 2)^8 + C$ **9.** $\frac{1}{16}(x^2 + 3)^8 + C$ **11.** $-20e^{-0.5t} + C$
13. $\frac{1}{10}\ln|10x + 7| + C$ **15.** $\frac{1}{4}e^{2x^2} + C$ **17.** $\frac{1}{3}\ln|x^3 + 4| + C$ **19.** $-\frac{1}{18}(3t^2 + 1)^{-3} + C$ **21.** $\frac{1}{3}(4 - x^3)^{-1} + C$
23. $\frac{2}{5}(x + 4)^{5/2} - \frac{8}{3}(x + 4)^{3/2} + C$ **25.** $\frac{2}{3}(x - 3)^{3/2} + 6(x - 3)^{1/2} + C$ **27.** $\frac{1}{11}(x - 4)^{11} + \frac{2}{5}(x - 4)^{10} + C$ **29.** $\frac{1}{8}(1 + e^{2x})^4 + C$
31. $\frac{1}{2}\ln|4 + 2x + x^2| + C$ **33.** $e^{x^2+x+1} + C$ **35.** $\frac{1}{4}(e^x - 2x)^4 + C$ **37.** $-\frac{1}{12}(x^4 + 2x^2 + 1)^{-3} + C$
39. (A) Differentiate the right side to get the integrand on the left side.

(B) Wrong, since $\dfrac{d}{dx}[\ln|2x - 3| + C] = \dfrac{2}{2x - 3} \neq \dfrac{1}{2x - 3}$. If $u = 2x - 3$, then $du = 2\,dx$. The integrand was not adjusted for the missing constant factor 2.

(C) $\displaystyle\int \dfrac{1}{2x - 3}\,dx = \dfrac{1}{2}\int \dfrac{2}{2x - 3}\,dx = \dfrac{1}{2}\ln|2x - 3| + C$ Check: $\dfrac{d}{dx}\left[\dfrac{1}{2}\ln|2x - 3| + C\right] = \dfrac{1}{2x - 3}$

41. (A) Differentiate the right side to get the integrand on the left side.

(B) Wrong, since $\dfrac{d}{dx}[e^{x^4} + C] = 4x^3e^{x^4} \neq x^3e^{x^4}$. If $u = x^4$, then $du = 4x^3$. The integrand was not adjusted for the missing constant factor 4.

(C) $\displaystyle\int x^3e^{x^4}\,dx = \dfrac{1}{4}\int 4x^3e^{x^4}\,dx = \dfrac{1}{4}e^{x^4} + C$ Check: $\dfrac{d}{dx}\left[\dfrac{1}{4}e^{x^4} + C\right] = x^3e^{x^4}$

43. (A) Differentiate the right side to get the integrand on the left side.

(B) Wrong, since $\dfrac{d}{dx}\left[\dfrac{(x^2 - 2)^2}{3x} + C\right] = \dfrac{3x^4 - 4x^2 - 4}{3x^2} \neq 2(x^2 - 2)^2$. If $u = x^2 - 2$, then $du = 2x\,dx$. It appears that the student moved a variable factor across the integral sign as follows (which is *not* valid): $\displaystyle\int 2(x^2 - 2)^2\,dx = \dfrac{1}{x}\int 2x(x^2 - 2)^2\,dx$.

(C) $\displaystyle\int 2(x^2 - 2)^2\,dx = \int (2x^4 - 8x^2 + 8)\,dx = \dfrac{2}{5}x^5 - \dfrac{8}{3}x^3 + 8x + C$ Check: $\dfrac{d}{dx}\left[\dfrac{2}{5}x^5 - \dfrac{8}{3}x^3 + 8x + C\right] = 2x^4 - 8x^2 + 8 = 2(x^2 - 2)^2$

45. $\frac{1}{9}(3x^2 + 7)^{3/2} + C$ **47.** $\frac{1}{8}x^8 + \frac{4}{5}x^5 + 2x^2 + C$ **49.** $\frac{1}{9}(x^3 + 2)^3 + C$ **51.** $\frac{1}{4}(2x^4 + 3)^{1/2} + C$ **53.** $\frac{1}{4}(\ln x)^4 + C$ **55.** $e^{-1/x} + C$
57. $x = \frac{1}{3}(t^3 + 5)^7 + C$ **59.** $y = 3(t^2 - 4)^{1/2} + C$ **61.** $p = -(e^x - e^{-x})^{-1} + C$ **65.** (A) $p(x) = 2,000/(3x + 50)$; (B) 250 bottles
67. $C(x) = 12x + 500\ln(x + 1) + 2,000$; $\overline{C}(1,000) = \$17.45$
69. (A) $S(t) = 10t + 100e^{-0.1t} - 100$, $0 \le t \le 24$ (B) $S(12) \approx \$50$ million (C) 18.41 mo
71. $Q(t) = 100\ln(t + 1) + 5t$, $0 \le t \le 20$; $Q(9) \approx 275$ thousand barrels **73.** $W(t) = 2e^{0.1t}$; $W(8) \approx 4.45$ g
75. $N(t) = 5,000 - 1,000\ln(1 + t^2)$; $N(10) \approx 385$ bacteria/ml **77.** $N(t) = 100 - 60e^{-0.1t}$, $0 \le t \le 15$; $N(15) \approx 87$ words/min
79. $E(t) = 12,000 - 10,000(t + 1)^{-1/2}$; $E(15) = 9,500$ students

Exercise 11-3

1. $y = 2e^{0.5x} + C$ **3.** $y = \dfrac{x^3}{3} - \dfrac{x^2}{2}$ **5.** $y = e^{-x^2} + 3$

7. Figure (B). When $x = 1$, the slope $dy/dx = 1 - 1 = 0$ for any y. When $x = 0$, the slope $dy/dx = 0 - 1 = -1$ for any y. Both are consistent with the slope field shown in Figure (B).

9. $y = \dfrac{x^2}{2} - x + C; y = \dfrac{x^2}{2} - x - 2$ **11.**

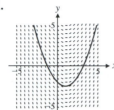

13. $y = Ce^{-0.8x}$ **15.** $y = 1{,}000e^{0.07x}$ **17.** $x = Ce^{-t}$

19. Figure (A). When $y = 1$, the slope $dy/dx = 1 - 1 = 0$ for any x. When $y = 2$, the slope $dy/dx = 1 - 2 = -1$ for any x. Both are consistent with the slope field shown in Figure (A).

21. $y = 1 - e^{-x}$ **23.**

25.

27.

29.

31.

33.

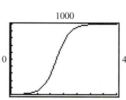

35. $A = 1{,}000e^{0.08t}$

37. $A = 8{,}000e^{0.06t}$ **39.** (A) $p(x) = 100e^{-0.05x}$ (B) \$60.65 per unit   (C) 16 units
41. (A) $N = L(1 - e^{-0.051t})$ (B) 22.5%   (C) 32 days   **43.** $I = I_0e^{-0.00942x}$; $x \approx 74$ ft
45. (A) $Q = 3e^{-0.04t}$ (B) $Q(10) = 2.01$ ml   (C) 27.47 hr   **47.** $-0.023\ 117$ **49.** Approx. 24,200 yr
51. 104 times; 67 times   **53.** (A) 7 people; 353 people   (B) 400

Exercise 11-4

1.

$f(x)$	$g(x)$

3. Figure (A): $L_3 = 13, R_3 = 20, A_3 = 16.5$; Figure (B): $L_3 = 14, R_3 = 7, A_3 = 10.5$

5. $L_3 \leq \displaystyle\int_1^4 f(x)\, dx \leq R_3; R_3 \leq \displaystyle\int_1^4 g(x)\, dx \leq L_3$; since $f(x)$ is increasing, L_3 underestimates the area and R_3 overestimates the area; since $g(x)$ is decreasing, the reverse is true.
7. Both figures: Error bound for L_3 and R_3 is 7; error bound for A_3 is 3.5
9. The exact area under the graph of $y = f(x)$ is within 3.5 units (either way) of the average of the left sum and right sum estimates, $A_3 = 16.5$.

11. (A) $L_4 = 320, R_4 = 192, A_4 = 256$; error bound for L_4 and R_4 is 128; error bound for A_4 is 64
   (B) The height of each rectangle represents an instantaneous rate, and the base of each rectangle
      represents a time interval; rate times time is distance.
   (C) $n > 256$

13. $A_{10} = 311{,}100$ ft^2; error bound is 25,000 ft^2; $n \geqslant 100$ **15.** (A) $A_7 = 298$ ft; error bound $= 55$ ft   (B) $n > 77$

17. (A) $P = 2$
   (B) $f(x)$

(C) To the left of $P = 2$, the left rectangles underestimate the true area and the right rectangles overestimate the true area. To the right of $P = (2, 9)$, the left rectangles overestimate the true area and the right rectangles underestimate the true area.

(D) $26 \leqslant \displaystyle\int_0^5 f(x)\,dx \leqslant 39$

19. $L_6 = -3.53$, $R_6 = -0.91$, $A_6 = -2.22$; error bound for L_6 and R_6 is 2.63; error bound for A_6 is 1.31. Geometrically, the definite integral over the interval [2, 5] is the sum of the areas between the curve and the x axis from $x = 2$ to $x = 5$, with the areas below the x axis counted negatively and those above the x axis counted positively.

21. $n \geqslant 60$ **23.** $n \geqslant 394$

25. If we let I represent the definite integral on the left side, then we know that $|I - L_n| \leqslant |f(b) - f(a)| \dfrac{b-a}{n}$. Since the limit of the right side is 0 as $n \to \infty$, the result follows.

27. $A_2 = \$180{,}000$; error bound $= \$30{,}000$ **29.** $L_4 = \$4{,}251$, $R_4 = \$4{,}605$; $\$4{,}251 \leqslant \displaystyle\int_2^6 800e^{0.08t}\,dt \leqslant \$4{,}605$

31. First 60 days: $A_3 = 3{,}240$ units; error bound $= 660$ units; Second 60 days: $A_3 = 4{,}920$ units; error bound $= 100$ units

33. (A) $\displaystyle\int_{100}^{200} R'(x)\,dx$ represents the area under the marginal revenue curve from $x = 100$ to $x = 200$; it also represents the total change in revenue going from sales of 100 six-packs per day to sales of 200 six-packs per day.

   (B) $A_4 = \$200$; error bound $= \$50$

   (C) $\$200$; both $R(200) - R(100)$ and $\displaystyle\int_{100}^{200} R'(x)\,dx$ represent total change in revenue going from sales of 100 six-packs per day to sales of 200 six-packs per day. This suggests that $\displaystyle\int_{100}^{200} R'(x)\,dx = R(200) - R(100)$.

35. (A) $L_5 = 3.72$ cm^2, $R_5 = 3.37$ cm^2 (B) $3.37 \leqslant \displaystyle\int_0^5 A'(t)\,dt \leqslant 3.72$ **37.** $A_3 = 108$ code symbols; error bound $= 6$ code symbols

Exercise 11-5

1. -2.33 **3.** 8.34   **5.** 1.067   **7.** 5   **9.** 48   **11.** 3.195   **13.** 2.506   **15.** $F(x) = x^2 - 4$ **17.** $g(x) = x^3 - 4x$ **19.** $r(t) = \frac{1}{2}(e^{2t} - 1)$
21. $T_2 = 1{,}730$ m$^2 \leqslant A \leqslant 1{,}820$ m$^2 = M_2$ **23.** -10.67 **25.** -5.04 **27.** 15,625   **29.** 13.550   **31.** 9.333   **33.** 0.466
35. $f(x) = (x^2 - 4)^6$ **37.** $G(x) = 20(e^{0.05x} - e^{0.25})$ **39.** $m(x) = 0.5(\ln|x^2 - 9| - \ln 5)$
41. (A) Real number
   (B) Single function (a particular antiderivative of f)
   (C) Set of functions (the set of all antiderivatives of f)

43. (A) Average $f(x) = 250$
   (B)

45. (A) Average $f(t) = 2$
   (B)

47. (A) Average $f(x) = \frac{45}{28} \approx 1.61$
   (B)

49. (A) Average $f(x) = 2(1 - e^{-2}) \approx 1.73$ **51.** 240 ft   **53.** $L_n = 5.78$, $T_n = 5.89$, $M_n = 6.01$, $R_n = 6.63$

(B)

55. (A)

57. (A)

(B) 1,300 ft

(B) $H = \int_0^8 v(t)\, dt$; $M_4 \leq H \leq T_4$; a smooth curve through the data points is concave up, hence, M_4 underestimates and T_4 overestimates.

(C) $M_4 = 1{,}288$ ft $\leq H = \int_0^8 v(t)\, dt \leq 1{,}326$ ft $= T_4$

59. $\int_2^5 (1 - x^2)\, dx = -36$ **61.** $\int_2^{12} (3x^2 - 2x + 3)\, dx = 1{,}610$ **63.** 2

65. (A) $M_5 = 0.6919 \leq \ln 2 = \int_1^2 \frac{1}{t}\, dt \leq T_5 = 0.6956$

(B) 0.6931   (C) 0.6931

67. (A)

(B) A smooth curve through the data points is concave down; consequently, M_3 overestimates and T_3 underestimates the production.

(C) $T_3 = 7{,}660$ units $\leq \int_0^{120} N(t)\, dt \leq 8{,}440$ units $= M_3$

69. $\int_{300}^{900} \left(500 - \frac{x}{3}\right) dx = \$180{,}000$

71. (A) \$16,000   (B)

$C(x) = 1 + 12x - x^2$

30

0

3

0

Ave $C(x) = 16$

(Cash reserves in thousands of dollars)

73. (A) $F(x) = \int_{400}^x (30 - 0.9\sqrt{t})\, dt = 30x - 0.6x^{3/2} - 7{,}200$

(B)

5,000

400   2,000

Maximum
X=1111.1109 Y=3911.1111

−5,000

By increasing production from 400 hoses per day to 1,111 hoses per day, the company will realize the greatest increase in daily revenue, a gain of \$3,911 per day.

75. $\int_0^5 500(t-12)\,dt = -\$23{,}750;\ \int_5^{10} 500(t-12)\,dt = -\$11{,}250$

77. (A) \$420 (B) \$135,000 (C) Part (A) gives the average cost per unit at a production level of 500 units and part (B) gives the average value of the cost function over the interval [0, 500].

79. (A) 3 yr (B) $P(3) - P(0) = \int_0^3 P'(t)\,dt = \int_0^3 [R'(t) - C'(t)]\,dt = \$7.984$ thousand, or \$7,984 **81.** 2 yr; \$2,272

83. (A)

85. (A)

(B) Midpoint sum underestimates and trapezoidal sum overestimates the marginal cost.

(B) \$101 thousand, or \$101,000

(C) $M_3 = \$98$ thousand $\le \int_2^8 C'(x)\,dx \le \$113$ thousand $= T_3$

87. 290 thousand barrels; 115 thousand barrels **89.** $Q(x) = 100\ln(x+1) + 5x,\ 0 \le x \le 20$ **91.** 0.0485; 0.0563

93. (A)

95. 4.58 cm² **97.** 10,208

(B) $M_3 = 4.58$ cm² $\le \int_0^6 A'(t)\,dt \le 4.61$ cm² $= T_3$

Chapter 11   Review Exercise

1. $t^3 - t^2 + C$ *(11-1)* **2.** 12 *(11-5)* **3.** $-3t^{-1} - 3t + C$ *(11-1)* **4.** $(x^2 - 1)/2$ *(11-5)* **5.** $-2e^{-0.5x} + C$ *(11-2)*
6. $2\ln 5 \approx 3.219$ *(11-5)* **7.** $y = f(x) = x^3 - 2x + 4$ *(11-3)*
8. Increasing on [0, 2]; decreasing on [2, 4]; concave downward on [0, 4]; local maximum at $x = 2$. Antiderivative graphs differ by a vertical translation. *(11-1)*
9. *(11-1)* $f(x)$

10. (A) $2x^4 - 2x^2 - x + C$ (B) $e^t - 4\ln|t| + C$ *(11-1)*

11. $M_2 = 44 \le \int_1^5 (x^2 + 1)\,dx \le 48 = T_2$; the graph is concave up, so M_2 underestimates and T_2 overestimates the value *(11-5)*

12. $I = 45.333;\ (M_2 - I) = -1.333;\ (T_2 - I) = 2.667$ *(11-5)*

13. $R_5 = 20.8 \le \int_7^{17} f(x)\,dx \le 26.6 = L_5$; since $f(x)$ is decreasing, R_5 underestimates and L_5 overestimates the value *(11-4, 11-5)*

14. 7 *(11-5)* **15.** Width $= 2 - (-1) = 3$, Height $=$ Average $f(x) = 7$ *(11-5)* **16.** -10 *(11-4, 11-5)*

17. 0.4 *(11-4, 11-5)* **18.** 1.4 *(11-4, 11-5)* **19.** 0 *(11-4, 11-5)* **20.** 0.4 *(11-4, 11-5)*

21. 2 *(11-4, 11-5)* **22.** -2 *(11-4, 11-5)* **23.** -0.4 *(11-4, 11-5)*

24. Increasing on [0, 1] and [3, 4]; decreasing on [1, 3]; concave downward on [0, 2]; concave upward on [2, 4]; local maximum at $x = 1$; local minimum at $x = 3$; inflection point at $x = 2$; graphs of antiderivatives differ by a vertical translation. *(11-1)*

25. *(11-1)* $f(x)$

26. (A) 1; 1 (B) 4; 4 *(11-3)*

27. $dy/dx = (2y)/x$; the slopes computed in Problem 26A are compatible with the slope field shown. *(11-3)*   **29.** $y = \frac{1}{4}x^2$; $y = -\frac{1}{4}x^2$ *(11-3)*

30. *(11-3)*

31. *(11-3)*

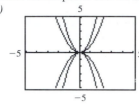

32. $\frac{1}{8}(6x - 5)^{4/3} + C$ *(11-1, 11-2)*

33. $(2x - 1)^5 + 1$ *(11-5)*

34. $-2x^{-1} - e^{x^2} + C$ *(11-2)*

35. $(20^{3/2} - 8)/3 \approx 27.148$ *(11-5)*

36. $-\frac{1}{2}e^{-2x} + \ln|x| + \frac{1}{2}e^{-2}$ *(11-5)*

37. $-500(e^{-0.2} - 1) \approx 90.635$ *(11-5)*

38. $\frac{1}{2}\ln 10 \approx 1.151$ *(11-5)*   **39.** 0.45 *(11-5)*

40. $\frac{1}{48}(2x^4 + 5)^6 + C$ *(11-2)*

41. $-\ln(e^{-x} + 3) + C$ *(11-2)*

42. $-(e^x + 2)^{-1} + C$ *(11-2)*

43. $y = f(x) = 3\ln|x| + x^{-1} + 4$ *(11-2, 11-3)*

44. $y = 3x^2 + x - 4$ *(11-3)*

45. $L_n = 4.79$; $M_n = 4.83$; $T_n = 5.01$; $R_n = 5.11$ *(11-5)*

46. (A) Distance $= \int_0^8 v(t)\, dt$; plot the data set and note that a smooth curve passing through the data points is concave down, so T_n underestimates and M_n overestimates the value.

(B) $T_4 = 67.97$ m $\leq \int_0^8 v(t)\, dt \leq 70.70$ m $= M_4$ *(11-5)*

47. 69.74 m *(11-5)*   **48.** $F'(x) = e^{-0.25x}$ *(11-5)*

49. (A) Average $f(x) = 6.5$ (B) $f(x)$

(11-5)

50. $\frac{1}{3}(\ln x)^3 + C$ *(11-2)*   **51.** $\frac{1}{8}x^8 - \frac{2}{5}x^5 + \frac{1}{2}x^2 + C$ *(11-2)*   **52.** $\frac{2}{3}(6 - x)^{3/2} - 12(6 - x)^{1/2} + C$ *(11-2)*   **53.** $\frac{1,234}{15} \approx 82.267$ *(11-5)*

54. $\frac{1}{6}(x + 1)^6 - \frac{1}{5}(x + 1)^5$ *(11-5)*   **55.** $y = 3e^{x^3} - 1$ *(11-3)*   **56.** $N = 800e^{0.06t}$ *(11-3)*

57.

Since the graph is concave up, M_n underestimates and T_n overestimates I. *(11-5)*

58. $M_5 = 0.134 \leq \int_1^2 e^{-x^2}\, dx \leq 0.137 = T_5$ *(11-5)*

59. $I = 0.135$; $(T_5 - I) = 0.002$; $(M_5 - I) = -0.001$ *(11-5)*

60. $\int_4^9 \frac{1}{x - 2}\, dx = 1.253$ *(11-5)*   **61.** $f(1) = 2$ *(11-5)*

62. Limited growth

(11-3)

63. Exponential decay

(11-3)

64. Unlimited growth

(11-3)

65. Logistic growth

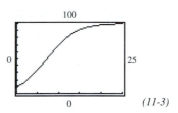

(11-3)

66. $L_2 = \$180,000$; $R_2 = \$140,000$; $\$140,000 \leq \int_{200}^{600} C'(x)\, dx \leq \$180,000$ *(11-4)*

67. The height of the rectangle, $C'(x)$, represents the marginal cost at a production level of x units—that is, the approximate cost per unit at that production level. The width of the rectangle represents the number of units involved in the increase in production over $x = 200$. Thus, the cost per unit times the number of units equals the increase in production costs. The approximation improves as n increases. *(11-4)*

68. $\int_{200}^{600} \left(600 - \dfrac{x}{2}\right) dx = \$160,000$ *(11-5)* **69.** $\int_{10}^{40} \left(150 - \dfrac{x}{10}\right) dx = \$4,425$ *(11-5)* **70.** $P(x) = 100x - 0.01x^2$; $P(10) = \$999$ *(11-3)*

71. $\int_{0}^{15} (60 - 4t)\, dt = 450$ thousand barrels *(11-5)* **72.** 109 items *(11-5)* **73.** $16e^{2.5} - 16e^2 - 8 \approx \$68.70$ *(11-5)*

74. (A) $I = \displaystyle\int_{0}^{30} N(t)\, dt$

(B) Graph the data set and notice that a smooth curve through the data points is concave down, so M_n overestimates and T_n underestimates I.
(C) $T_3 = 485$ components $\leq I \leq 500$ components $= M_3$
(D) 495 components *(11-5)*

75. Useful life $= 10 \ln \frac{20}{3} \approx 19$ yr; total profit $= 143 - 200e^{-1.9} \approx 113.086$ or $\$113,086$ *(11-5)*
76. $S(t) = 50 - 50e^{-0.08t}$; $50 - 50e^{-0.96} \approx \$31$ million; $-(\ln 0.2)/0.8 \approx 20$ mo *(11-3)*
77. (A) $G(x) = 120x - 0.03x^2 - 90,000$

(B) A maximum revenue increase of $\$30,000$ is achieved by increasing production from 1,000 watches per week to 2,000 watches per week. *(11-5)*

78. (A) $\dfrac{1}{12} \displaystyle\int_{0}^{12} C(t)\, dt$

(B) Since the graph is either concave down or a straight line over the interval $[0, 12]$, T_n underestimates I and M_n overestimates I.
(C) $T_6/12 = 6.2 \leq I \leq 6.5 = M_6/12$; during a 12 hour period, the average concentration of particulate matter in the atmosphere is between 6.2 ppm and 6.5 ppm. *(11-5)*

79. 800 gal *(11-5)* **80. (A)** 1,093 million **(B)** About 70 yr *(11-3)* **81.** $\dfrac{-\ln 0.04}{0.000\ 123\ 8} \approx 26,000$ yr *(11-3)*

82. $N(t) = 95 - 70e^{-0.1t}$; $N(15) \approx 79$ words/min *(11-3)*

■ CHAPTER 12

Exercise 12-1

1. $\displaystyle\int_{a}^{b} g(x)\, dx$ **3.** $\displaystyle\int_{a}^{b} [-h(x)]\, dx$

5. Since the shaded region in figure (C) is below the x axis, $h(x) \leq 0$; thus, $\displaystyle\int_{a}^{b} h(x)\, dx$ represents the negative of the area of the region.

7. 7   **9.** 20   **11.** $\frac{7}{3} \approx 2.333$ **13.** 9   **15.** 7.021   **17.** 0.693   **19.** $\displaystyle\int_{a}^{b} [-f(x)]\, dx$ **21.** $\displaystyle\int_{b}^{c} f(x)\, dx + \int_{c}^{d} [-f(x)]\, dx$

23. $\displaystyle\int_{c}^{d} [f(x) - g(x)]\, dx$ **25.** $\displaystyle\int_{a}^{b} [f(x) - g(x)]\, dx + \int_{b}^{c} [g(x) - f(x)]\, dx$

27. Find the intersection points by solving $f(x) = g(x)$ on the interval $[a, d]$ to determine b and c. Then observe that $f(x) \geq g(x)$ over $[a, b]$, $g(x) \geq f(x)$ over $[b, c]$, and $f(x) \geq g(x)$ over $[c, d]$. Thus, Area $= \int_a^b [f(x) - g(x)] \, dx + \int_b^c [g(x) - f(x)] \, dx + \int_c^d [f(x) - g(x)] \, dx$.

29. 2.5   **31.** 7.667   **33.** 23.667   **35.** 15   **37.** 32   **39.** 36   **41.** 9   **43.** 2.832   **45.** 18   **47.** 1.858   **49.** 17   **51.** 8
53. 101.75   **55.** 8   **57.** 17.979   **59.** 5.113   **61.** 8.290
63. Total production from the end of the 5th year to the end of the 10th year is 79 thousand barrels.
65. Total profit over the 5 yr useful life of the game is 13.306 thousand or $13,306.
67. 1935: 0.412; 1947: 0.231; income is more equally distributed in 1947.
69. 1963: 0.818; 1983: 0.846; total assets are less equally distributed in 1983.   **71.** Total weight gain during the first 10 hr is 5.15 g.
73. Average number of words learned during the second 2 hr is 10.

Exercise 12-2

1. (A) .75   (B) .11   (C) $f(x)$

3. 8 yr   **5.** (A) .11   (B) .10   **7.** $P(t \geq 12) = 1 - P(0 \leq t \leq 12) = .89$ **9.** $12,500

11.

If $f(t)$ is the rate of flow of a continuous income stream, then the total income produced from 0 to 5 yr is the area under the graph of $y = f(t)$ from $t = 0$ to $t = 5$.

13. $1,295   **15.**

If $f(t)$ is the rate of flow of a continuous income stream, then the total income produced from 0 to 3 yr is the area under the graph of $y = f(t)$ from $t = 0$ to $t = 3$.

17. $255,562; $175,562   **19.** $7,109   **21.** $1,343
23. Clothing store: $FV = \$77,847$; computer store: $FV = \$72,939$; the clothing store is the better investment.

25. Bond: $FV = \$14,918$; business: $FV = \$12,296$; the bond is the better investment.   **27.** $46,283   **29.** $\dfrac{k}{r}(e^{rT} - 1)$ **31.** $625,000

33.

The shaded area is the consumers' surplus and represents the total savings to consumers who are willing to pay more than $150 for a product but are still able to buy the product for $150.

35. $9,900   **37.**

The area of the region PS is the producers' surplus and represents the total gain to producers who are willing to supply units at a lower price than $67 but are still able to supply the product at $67.

39. $CS = \$3,380$; $PS = \$1,690$ **41.** $CS = \$6,980$; $PS = \$5,041$ **43.** $CS = \$7,810$; $PS = \$8,336$ **45.** $CS = \$8,544$; $PS = \$11,507$

$\bar{x} = 260$

$\bar{x} = 490$

$\bar{x} = 614$

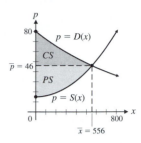
$\bar{x} = 556$

Exercise 12-3

1. $\frac{1}{3}xe^{3x} - \frac{1}{9}e^{3x} + C$ **3.** $\frac{x^3}{3}\ln x - \frac{x^3}{9} + C$ **5.** $-xe^{-x} - e^{-x} + C$ **7.** $\frac{1}{2}e^{x^2} + C$ **9.** $xe^x - 4e^x + 4$ **11.** $x\ln(2x) - x - \ln 2 + 1$

13. $\ln(x^2 + 1) + C$ **15.** $(\ln x)^2/2 + C$ **17.** $\frac{2}{3}x^{3/2}\ln x - \frac{4}{9}x^{3/2} + C$

19.

$y = (t - 3)e^t$

The integral represents the signed area between the graph of $y = (t - 3)e^t$ and the t axis from $t = 0$ to $t = x$, $x \geqslant 0$.

21.
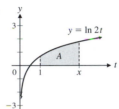
$y = \ln 2t$

The integral represents the area between the graph of $y = \ln 2t$ and the t axis from $t = 1$ to $t = x$, $x \geqslant 1$.

23. $(x^2 - 2x + 2)e^x + C$ **25.** $\dfrac{xe^{ax}}{a} - \dfrac{e^{ax}}{a^2} + C$ **27.** $-\dfrac{\ln x}{x} - \dfrac{1}{x} + 1$ **29.** $(x + 4)\ln(x + 4) - x - 4\ln 4$

31. $xe^{x-2} - e^{x-2} + C$ **33.** $\frac{1}{2}(1 + x^2)\ln(1 + x^2) - \frac{1}{2}(1 + x^2) + C$ **35.** $(1 + e^x)\ln(1 + e^x) - (1 + e^x) + C$

37. $x(\ln x)^2 - 2x\ln x + 2x + C$ **39.** $x(\ln x)^3 - 3x(\ln x)^2 + 6x\ln x - 6x + C$

41. $S(t) = 1,600 + 400e^{0.1t} - 40te^{0.1t}$; 15 mo   **43.** $R(x) = 65x - 6[(x + 1)\ln(x + 1) - x]$; 618 per week; \$29,506

45. $P(x) = \displaystyle\int_0^x (2t - te^{-t})\, dt = x^2 + (x + 1)e^{-x} - 1, x \geqslant 0$

47.

$P'(t) = 2t - te^{-t}$

Total profit

The total profit for the first x yr (in millions of dollars) is the same as the area under the marginal profit function, $P'(t) = 2t - te^{-t}$, from $t = 0$ to $t = x$, $x \geqslant 0$.

49. $C(t) = -5te^{-0.2t} - 25e^{-0.2t} + 25$; 10 min; 22.7 mg   **51.** $N(t) = -4te^{-0.25t} - 40e^{-0.25t} + 80$; 8 wk; 78 words/min

Exercise 12-4

1. $\ln\left|\dfrac{x}{1 + x}\right| + C$ **3.** $\dfrac{1}{3 + x} + 2\ln\left|\dfrac{5 + 2x}{3 + x}\right| + C$ **5.** $\dfrac{2(x - 32)}{3}\sqrt{16 + x} + C$ **7.** $\dfrac{1}{2}\ln\left|\dfrac{2}{2 + \sqrt{x^2 + 4}}\right| + C$

9. $\frac{1}{3}x^3\ln x - \frac{1}{9}x^3 + C$ **11.** $9\ln\frac{3}{2} - 2 = 1.6492$ **13.** $\frac{1}{2}\ln\frac{12}{5} = 0.4377$ **15.** $\ln 3 = 1.0986$

17. $-\dfrac{\sqrt{4x^2 + 1}}{x} + 2\ln|2x + \sqrt{4x^2 + 1}| + C$ **19.** $\frac{1}{2}\ln|x^2 + \sqrt{x^4 - 16}| + C$

21. $\frac{1}{6}(x^3\sqrt{x^6+4}+4\ln|x^3+\sqrt{x^6+4}|)+C$ **23.** $-\dfrac{\sqrt{4-x^4}}{8x^2}+C$ **25.** $\dfrac{1}{5}\ln\left|\dfrac{3+4e^x}{2+e^x}\right|+C$ **27.** $\frac{2}{3}(\ln x-8)\sqrt{4+\ln x}+C$

29. $\frac{1}{5}x^2e^{5x}-\frac{2}{25}xe^{5x}+\frac{2}{125}e^{5x}+C$ **31.** $-x^3e^{-x}-3x^2e^{-x}-6xe^{-x}-6e^{-x}+C$ **33.** $x(\ln x)^3-3x(\ln x)^2+6x\ln x-6x+C$

35. $\frac{1}{3}(x^2-9)^{3/2}$ **37.** $\dfrac{1}{2}\ln\left|\dfrac{x-1}{x+1}\right|-\dfrac{1}{2}\ln\dfrac{1}{3}=\dfrac{1}{2}\ln 3\left|\dfrac{x-1}{x+1}\right|$ **39.** $\frac{1}{2}\ln|x^2+2x|+C$ **41.** $\frac{2}{3}\ln|3+x|+\frac{1}{3}\ln|x|+C$

43. $C(x)=200x+1{,}000\ln(1+0.05x)+25{,}000$; 608 pairs; \$198,773   **45.** $P(x)=\dfrac{2(9x-4)}{135}(2+3x)^{3/2}-2{,}000.83$; 54 cars; $-\$1{,}770$

47. $S(t)=\dfrac{1+2t}{2}-\dfrac{1}{2(1+2t)}-\ln|1+2t|$; \$20.6 million; 15 mo   **49.** $R(t)=100\ln\dfrac{t+\sqrt{t^2+9}}{3}$; 110 ft; 30 min

51. $N(t)=60\ln\dfrac{t+\sqrt{t^2+25}}{5}$; 53 items; 6.5 hr

Chapter 12   Review Exercise

1. $\displaystyle\int_a^b f(x)\,dx$ *(12-1)*   **2.** $\displaystyle\int_b^c [-f(x)]\,dx$ *(12-1)*   **3.** $\displaystyle\int_a^b f(x)\,dx+\int_b^c[-f(x)]\,dx$ *(12-1)*

4. Area = 1.153 *(12-1)*

5. $\frac{1}{4}xe^{4x}-\frac{1}{16}e^{4x}+C$ *(12-3, 12-4)*   **6.** $\frac{1}{2}x^2\ln x-\frac{1}{4}x^2+C$ *(12-3, 12-4)*   **7.** $\dfrac{1}{1+x}+\ln\left|\dfrac{x}{1+x}\right|+C$ *(12-4)*

8. $-\dfrac{\sqrt{1+x}}{x}-\dfrac{1}{2}\ln\left|\dfrac{\sqrt{1+x}-1}{\sqrt{1+x}+1}\right|+C$ *(12-4)*   **9.** $\displaystyle\int_a^b [f(x)-g(x)]\,dx$ *(12-1)*

10. $\displaystyle\int_b^c [g(x)-f(x)]\,dx$ *(12-1)*   **11.** $\displaystyle\int_c^c [g(x)-f(x)]\,dx+\int_c^d[f(x)-g(x)]\,dx$ *(12-1)*

12. $\displaystyle\int_a^b [f(x)-g(x)]\,dx+\int_b^c[g(x)-f(x)]\,dx+\int_c^d[f(x)-g(x)]\,dx$ *(12-1)*

13. Area = 20.833 *(12-1)*   **14.** xe^x-e^x+1 *(12-3, 12-4)*

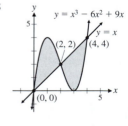

15. $0.5x\sqrt{x^2+16}-8\ln\dfrac{|x+\sqrt{x^2+16}|}{4}$ *(12-4)*

16. $\frac{1}{6}(3x\sqrt{9x^2-49}-49\ln|3x+\sqrt{9x^2-49}|)+C$ *(12-4)*
17. $-2te^{-0.5t}-4e^{-0.5t}+C$ *(12-3, 12-4)*
18. $\frac{1}{3}x^3\ln x-\frac{1}{9}x^3+C$ *(12-3, 12-4)*
19. $x-\ln|1+2e^x|+C$ *(12-4)*

20. (A) Area = 8   (B) Area = 8.38

(12-1)

21. $\frac{1}{3}(\ln x)^3+C$ *(11-2)*   **22.** $\frac{1}{2}x^2(\ln x)^2-\frac{1}{2}x^2\ln x+\frac{1}{4}x^2+C$ *(12-3, 12-4)*   **23.** $\sqrt{x^2-36}+C$ *(11-2)*   **24.** $\frac{1}{2}\ln|x^2+\sqrt{x^4-36}|+C$ *(12-4)*
25. $\frac{1}{2}x^2\ln(2x)-\frac{1}{4}x^2+\frac{1}{4}-\frac{1}{2}\ln 2$ *(12-3, 12-4)*   **26.** $x(\ln x)^2-2x\ln x+2x+C$ *(12-3, 12-4)*   **27.** $-\frac{1}{4}e^{-2x^2}+C$ *(11-2)*
28. $-\frac{1}{2}x^2e^{-2x}-\frac{1}{2}xe^{-2x}-\frac{1}{4}e^{-2x}+C$ *(12-3, 12-4)*   **29.** (A) .189   (B) .154 *(12-2)*

30.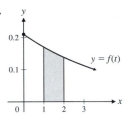

The probability that the product will fail during the second year of warranty is the area under the probability density function $y = f(t)$ from $t = 1$ to $t = 2$. *(12-2)*

31. (A)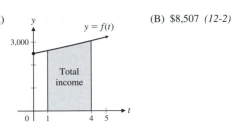

(B) $8,507 *(12-2)*

32. (A) $20,824 **(B)** $6,623 *(12-2)*

33. (A)

(B) More equitably distributed, since the area bounded by the two curves will have decreased.

(C) Current = 0.3; projected = 0.2; income will be more equitably distributed 10 years from now. *(12-1)*

34. (A) $CS = \$2,250$; $PS = \$2,700$ **(B)** $CS = \$2,890$; $PS = \$2,278$

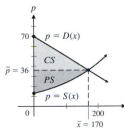

$(12\text{-}2)$

35. $Q(t) = \dfrac{60}{t + 1} - 120 \ln \left| \dfrac{t + 2}{t + 1} \right| + 40 + 120 \ln 2$;
115.73 ml; 2.44 hr *(12-4)*
36. $40 + 120 \ln 2 = 123.18$ ml *(12-4)*
37. .667; .333 *(12-2)*

38.

The probability that the doctor will spend more than an hour with a randomly selected patient is the area under the probability density function $y = f(t)$ from $t = 1$ to $t = 3$. *(12-2)*

39. $N(t) = 1,000(t - 1)\sqrt{1 + 2t} + 351,000$; 364,000; 8 yr *(12-4)* **40.** .368 *(12-2)*

■ CHAPTER 13

Exercise 13-1

1. 10 **3.** 1 **5.** 0 **7.** 1 **9.** 6 **11.** 150 **13.** 16π **15.** 791 **17.** 0.192 **19.** 118 **21.** $100e^{0.8} \approx 222.55$
23. $-1.926, 0.599$ **25.** $2x + h$ **27.** $2y^2$ **29.** $E(0, 0, 3)$; $F(2, 0, 3)$
31. (A) In the plane $y = c$, c any constant, $z = x^2$.
  (B) The y axis; the horizontal line parallel to the y axis and passing through the point $(1, 0, 1)$; the horizontal line parallel to the y axis and passing through the point $(2, 0, 4)$
  (C) A parabolic "trough" lying on top of the y axis

33. (A) Upper semicircles whose centers lie on the y axis
    (B) Upper semicircles whose centers lie on the x axis
    (C) The upper hemisphere of radius 6 with center at the origin
35. $4,400; $6,000; $7,100   **37.** $R(p, q) = -5p^2 + 6pq - 4q^2 + 200p + 300q$; $R(2, 3) = $1,280$; $R(3, 2) = $1,175$ **39.** 30,065 units
41. $272,615.08   **43.** $T(70, 47) \approx 29$ min; $T(60, 27) = 33$ min   **45.** $C(6, 8) = 75$; $C(8.1, 9) = 90$ **47.** $Q(12, 10) = 120$; $Q(10, 12) \approx 83$

Exercise 13-2

1. 3   **3.** 2   **5.** $-4xy$ **7.** -6 **9.** $10xy^3$ **11.** 60   **13.** $2x - 2y + 6$ **15.** 6   **17.** -2 **19.** 2   **21.** $2e^{2x+3y}$
23. $6e^{2x+3y}$ **25.** $6e^2$ **27.** $4e^3$ **29.** $f_x(x, y) = 6x(x^2 - y^3)^2$; $f_y(x, y) = -9y^2(x^2 - y^3)^2$
31. $f_x(x, y) = 24xy(3x^2y - 1)^3$; $f_y(x, y) = 12x^2(3x^2y - 1)^3$ **33.** $f_x(x, y) = 2x/(x^2 + y^2)$; $f_y(x, y) = 2y/(x^2 + y^2)$
35. $f_x(x, y) = y^4e^{xy^2}$; $f_y(x, y) = 2xy^3e^{xy^2} + 2ye^{xy^2}$ **37.** $f_x(x, y) = 4xy^2/(x^2 + y^2)^2$; $f_y(x, y) = -4x^2y/(x^2 + y^2)^2$
41. $f_{xx}(x, y) = 2y^2 + 6x$; $f_{xy}(x, y) = 4xy = f_{yx}(x, y)$; $f_{yy}(x, y) = 2x^2$
43. $f_{xx}(x, y) = -2y/x^3$; $f_{xy}(x, y) = (-1/y^2) + (1/x^2) = f_{yx}(x, y)$; $f_{yy}(x, y) = 2x/y^3$
45. $f_{xx}(x, y) = (2y + xy^2)e^{xy}$; $f_{xy}(x, y) = (2x + x^2y)e^{xy} = f_{yx}(x, y)$; $f_{yy}(x, y) = x^3e^{xy}$ **47.** $x = 2$ and $y = 4$ **49.** $x = 1.200$ and $y = -0.695$
51. (A) 5.25   (B) The function $f(x, 0)$, for example, has values greater than 5.25.
53. $f_{xx}(x, y) + f_{yy}(x, y) = (2y^2 - 2x^2)/(x^2 + y^2)^2 + (2x^2 - 2y^2)/(x^2 + y^2)^2 = 0$ **55.** (A) $2x$ (B) $4y$
57. $P_x(1,200, 1,800) = 24$: profit will increase approx. $24 per unit increase in production of type A calculators at the $(1,200, 1,800)$ output level;
    $P_y(1,200, 1,800) = -48$; profit will decrease approx. $48 per unit increase in production of type B calculators at the $(1,200, 1,800)$ output level
59. $\partial x/\partial p = -5$: a $1 increase in the price of brand A will decrease the demand for brand A by 5 lb at any price level (p, q); $\partial y/\partial p = 2$: a $1 increase
in the price of brand A will increase the demand for brand B by 2 lb at any price level (p, q)
61. (A) $f_x(x, y) = 7.5x^{-0.25}y^{0.25}$; $f_y(x, y) = 2.5x^{0.75}y^{-0.75}$
    (B) Marginal productivity of labor $= f_x(600, 100) \approx 4.79$; Marginal productivity of capital $= f_y(600, 100) \approx 9.58$ (C) Capital
63. Competitive   **65.** Complementary
67. (A) $f_w(w, h) = 6.65w^{-0.575}h^{0.725}$; $f_h(w, h) = 11.34w^{0.425}h^{-0.275}$
    (B) $f_w(65, 57) = 11.31$: for a 65 lb child 57 in. tall, the rate of change in surface area is 11.31 in.^{2} for each pound gained in weight (height is held
    fixed); $f_h(65, 57) = 21.99$: for a 65 lb child 57 in. tall, the rate of change in surface area is 21.99 in.^{2} for each inch gained in height (weight
    is held fixed)
69. $C_w(6, 8) = 12.5$: index increases approx. 12.5 units for 1 in. increase in width of head (length held fixed) when $W = 6$ and $L = 8$;
    $C_L(6, 8) = -9.38$: index decreases approx. 9.38 units for 1 in. increase in length (width held fixed) when $W = 6$ and $L = 8$

Exercise 13-3

1. $f(-2, 0) = 10$ is a local maximum   **3.** $f(-1, 3) = 4$ is a local minimum   **5.** f has a saddle point at $(3, -2)$
7. $f(3, 2) = 33$ is a local maximum   **9.** $f(2, 2) = 8$ is a local minimum   **11.** f has a saddle point at $(0, 0)$
13. f has a saddle point at $(0, 0)$; $f(1, 1) = -1$ is a local minimum
15. f has a saddle point at $(0, 0)$; $f(3, 18) = -162$ and $f(-3, -18) = -162$ are local minima
17. The test fails at $(0, 0)$; f has saddle points at $(2, 2)$ and $(2, -2)$ **19.** f has a saddle point at $(0.614, -1.105)$
21. $f(x, y)$ is nonnegative and equals 0 when $x = 0$, so f has a local minimum at each point of the y axis.
23. (B) Local minimum   **25.** 2,000 type A and 4,000 type B; Max $P = P(2, 4) = 15 million
27. (A)   (B) A maximum weekly profit of $288 is realized for $p = 10 and $q = 12.

p	q	x	y
$10	$12	56	16
$11	$11	6	56

29. $P(x, y) = P(4, 2)$ **31.** 8 by 4 by 2 in.   **33.** 20 by 20 by 40 in.

Exercise 13-4

1. Max $f(x, y) = f(3, 3) = 18$ **3.** Min $f(x, y) = f(3, 4) = 25$
5. Max $f(x, y) = f(3, 3) = f(-3, -3) = 18$; Min $f(x, y) = f(3, -3) = f(-3, 3) = -18$
7. Maximum product is 25 when each number is 5   **9.** Min $f(x, y, z) = f(-4, 2, -6) = 56$
11. Max $f(x, y, z) = f(2, 2, 2) = 6$; Min $f(x, y, z) = f(-2, -2, -2) = -6$ **13.** Max $f(x, y) = f(0.217, 0.885) = 1.055$
15. Maximize $f(x, 5)$, a function of just one independent variable.   **17.** (A) Max $f(x, y) = f(0.707, 0.5) = f(-0.707, 0.5) = .47$
19. 60 of model A and 30 of model B will yield a minimum cost of $32,400 per week

21. (A) 8,000 units of labor and 1,000 units of capital; Max $N(x, y) = N(8,000, 1,000) \approx 263,902$ units
   (B) Marginal productivity of money ≈ 0.6598; increase in production $\approx 32,990$ units
23. 8 by 8 by $\frac{8}{3}$ in.   **25.** $x = 50$ ft and $y = 200$ ft; maximum area is 10,000 ft^2

Exercise 13-5

1.

3.

5.

7. $y = 2.12x + 10.8$; $y = 63.8$ when $x = 25$
9. $y = -1.2x + 12.6$; $y = 10.2$ when $x = 2$
11. $y = -1.53x + 26.67$; $y = 14.4$ when $x = 8$

13.

$y = 0.75x^2 - 3.45x + 4.75$

19. (A) $y = 1.52x - 1.68$; $y = 0.73x^2 - 1.39x + 1.30$ (B) The quadratic function
21. The normal equations form a system of four linear equations in the four variables a, b, c, and d, which can be solved using Gauss–Jordan elimination.
23. (A) $y = 7.15x + 12.62$ (B) 105.57 thousand per month   **25.** (A) $y = -0.48x + 4.38$ (B) \$6.56/bottle
27. (A) $P = -0.66T + 48.8$ (B) 11.18 beats/min   **29.** (A) $D = -3.1A + 54.6$ (B) 45%   **31.** (A) $y = 0.08653x + 10.81$ (B) **20.50 ft**

Exercise 13-6

1. (A) $3x^2y^4 + C(x)$ (B) $3x^2$ **3.** (A) $2x^2 + 6xy + 5x + E(y)$ (B) $35 + 30y$ **5.** (A) $\sqrt{y + x^2} + E(y)$ (B) $\sqrt{y + 4} - \sqrt{y}$ **7.** 9

9. 330   **11.** $(56 - 20\sqrt{5})/3$ **13.** 16   **15.** 49   **17.** $\frac{1}{8}\int_1^5\int_{-1}^1 (x + y)^2\,dy\,dx = \frac{32}{3}$ **19.** $\frac{1}{15}\int_1^4\int_2^7 (x/y)\,dy\,dx = \frac{1}{2}\ln\frac{7}{2} \approx 0.6264$

21. $\frac{4}{3}$ cubic units   **23.** $\frac{32}{3}$ cubic units   **25.** $\int_0^1\int_1^2 xe^{xy}\,dy\,dx = \frac{1}{2} + \frac{1}{2}e^2 - e$ **27.** $\int_0^1\int_{-1}^1 \frac{2y + 3xy^2}{1 + x^2}\,dy\,dx = \ln 2$

31. (A) $\frac{1}{3} + \frac{1}{4}e^{-2} - \frac{1}{4}e^2$ (B)

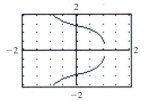

   (C) Points to the right of the graph in part (B) are greater than 0; points to the left of the graph are less than 0.

33. $\frac{1}{0.4}\int_{0.6}^{0.8}\int_5^7 \frac{y}{1 - x}\,dy\,dx = 30\ln 2 \approx \$20.8$ billion   **35.** $\frac{1}{10}\int_{10}^{20}\int_1^2 x^{0.75}y^{0.25}\,dy\,dx = \frac{8}{175}(2^{1.25} - 1)(20^{1.75} - 10^{1.75}) \approx 8.375$ or **8,375 units**

37. $\frac{1}{192}\int_{-8}^8\int_{-6}^6 [10 - \frac{1}{10}(x^2 + y^2)]\,dy\,dx = \frac{20}{3}$ insects/ft^2 **39.** $\frac{1}{8}\int_{-2}^2\int_{-1}^1 [100 - 15(x^2 + y^2)]\,dy\,dx = 75$ ppm

41. $\frac{1}{10,000}\int_{2,000}^{3,000}\int_{50}^{60} 0.000\,013\,3xy^2\,dy\,dx \approx 100.86$ ft   **43.** $\frac{1}{16}\int_8^{16}\int_{10}^{12} 100\frac{x}{y}\,dy\,dx = 600\ln 1.2 \approx 109.4$

Chapter 13 Review Exercise

1. $f(5, 10) = 2,900; f_x(x, y) = 40; f_y(x, y) = 70$ *(13-1, 13-2)*   **2.** $\partial^2 z/\partial x^2 = 6xy^2; \partial^2 z/\partial x \, \partial y = 6x^2 y$ *(13-2)*   **3.** $2xy^3 + 2y^2 + C(x)$ *(13-6)*
4. $3x^2 y^2 + 4xy + E(y)$ *(13-6)*   **5.** 1 *(13-6)*   **6.** $f(2, 3) = 7; f_x(x, y) = -2x + 2y + 3; f_y(2, 3) = 5$ *(13-1, 13-2)*
7. $(-8)(-6) - (4)^2 = 32$ *(13-2)*   **8.** $(1, 3, -\frac{1}{2}), (-1, -3, \frac{1}{2})$ *(13-4)*   **9.** $y = -1.5x + 15.5; y = 0.5$ when $x = 10$ *(13-5)*   **10.** 18 *(13-6)*
11. $f_x(x, y) = 2xe^{x^2+2y}; f_y(x, y) = 2e^{x^2+2y}; f_{xy}(x, y) = 4xe^{x^2+2y}$ *(13-2)*   **12.** $f_x(x, y) = 10x(x^2 + y^2)^4; f_{xy}(x, y) = 80xy(x^2 + y^2)^3$ *(13-2)*
13. $f(2, 3) = -25$ is a local minimum; f has a saddle point at $(-2, 3)$ *(13-3)*   **14.** Max $f(x, y) = f(6, 4) = 24$ *(13-4)*
15. Min $f(x, y, z) = f(2, 1, 2) = 9$ *(13-4)*   **16.** $y = \frac{116}{165}x + \frac{100}{3}$ *(13-5)*   **17.** $\frac{27}{5}$ *(13-6)*   **18.** 4 cubic units *(13-6)*   **19.** 0 *(13-6)*
20. (A) 6.28 (B) No *(13-6)*
21. (A) $P_x(1, 3) = 8$; profit will increase $8,000 for a 100 unit increase in product A if production of product B is held fixed at an output level of $(1, 3)$
    (B) For 200 units of A and 300 units of B, $P(2, 3) = \$100$ thousand is a local maximum. *(13-2, 13-3)*
22. 8 by 6 by 2 in. *(13-3)*   **23.** $y = 0.63x + 1.33$; profit in sixth year is $5.11 million *(13-4)*
24. (A) Marginal productivity of labor ≈ 8.37; marginal productivity of capital ≈ 1.67; management should encourage increased use of labor.
    (B) 80 units of labor and 40 units of capital; Max $N(x, y) = N(80, 40) \approx 696$ units; marginal productivity of money ≈ 0.0696; increase in
    production ≈ 139 units
    (C) $\dfrac{1}{1,000} \displaystyle\int_{50}^{100} \int_{20}^{40} 10x^{0.8} y^{0.2} \, dy \, dx = \dfrac{(40^{1.2} - 20^{1.2})(100^{1.8} - 50^{1.8})}{216} = 621$ items *(13-4)*

25. $T_x(70, 17) = -0.924$ min/ft increase in depth when $V = 70$ ft^3 and $x = 17$ ft *(13-2)*   **26.** $\dfrac{1}{16} \displaystyle\int_{-2}^{2} \int_{-2}^{2} [100 - 24(x^2 + y^2)] \, dy \, dx = 36$ ppm *(13-6)*

27. 50,000 *(13-1)*   **28.** $y = \frac{1}{2}x + 48; y = 68$ when $x = 40$ *(13-5)*
29. (A) $y = 0.4709x + 25.87$ (B) 72.96 persons/mi^2 (C) 74.41 persons/mi^2; 80.13 persons/mi^2 *(13-5)*
30. (A) $y = 1.069x + 0.522$ (B) 64.68 yr (C) 64.78 yr; 64.80 yr *(13-5)*
31. (A) A continuous curve passing through each data point would cross certain horizontal lines 3 times. *(13-5)*

■ CHAPTER 14

Exercise 14-1

5. (C)   **7.** (A)   **9.**   **11.**

13. $y = 7e^x - 5x - 5$
15. $y = e^x - 2e^{2x}$
17. $y = x + \dfrac{2}{x}$
23. $y = \sqrt{9 - x^2}$
25. $y = 2 - e^x$

27. (A) $y = 2 - e^{-x}$ (B) $y = 2$ (C) $y = 2 + e^{-x}$ If $y(0) > 2$, then y decreases and approaches 2 as x approaches ∞; if $y(0) < 2$, then y increases
and approaches 2 as x approaches ∞; if $y(0) = 2$, then $y = 2$ for all x.
29. (A) $y = 2 - e^x$ (B) $y = 2$ (C) $y = 2 + e^x$ If $y(0) > 2$, then y increases and approaches ∞ as x approaches ∞; if $y(0) < 2$, then y decreases
and approaches $-\infty$ as x approaches ∞; if $y(0) = 2$, then $y = 2$ for all x.
31. (A) $y = \dfrac{10}{1 + 9e^{-x}}$ (B) $y = 10$ (C) $y = \dfrac{10}{1 - 0.5e^{-x}}$ If $y(0) > 10$, then y decreases and approaches 10 as x approaches ∞; if $y(0) < 10$, then y
increases and approaches 10 as x approaches ∞; if $y(0) = 10$, then $y = 10$ for all x.
33. (A) $y = Cx^3 + 2$ for any C (B) No particular solution exists. (C) $y = 2 - x^3$
35. (A)   (B) Each graph decreases to a local   (C)   (D) Each graph is
              minimum and then increases,   increasing for all x
              approaching the line $y = x$ as   and approaches the
              x approaches ∞.   line $y = x$ as x
                                                         approaches ∞.

37. (A) $\bar{p} = 5$ (B) $p(t) = 5 - 4e^{-0.1t}$; $p(t) = 5 + 5e^{-0.1t}$ **39.** (B) $A(t) = 2,500e^{0.08t} - 2,500$; $A(t) = 3,500e^{0.08t} - 2,500$
41. (A) $\bar{N} = 200$ (B) $N(t) = 200 - 150e^{-0.5t}$; $N(t) = 200 + 100e^{-0.5t}$ **43.** (A) $\bar{N} = C$ (B) $N(t) = 100e^{2-2e^{-0.5t}}$; $N(t) = 200e^{2-2e^{-0.5t}}$

Exercise 14-2

1. $dy/dt = 100,000$ **3.** $dy/dt = k(10,000 - y)$
5. The annual sales now are \$2 million, and sales are increasing at a rate proportional to the annual sales.
7. In a community of 5,000 people, a single person began the spread of a rumor that is spreading at a rate proportional to the product of the number of people who have heard the rumor and the number who have not.
9. General solution: $y = x + C$; particular solution: $y = x + 2$ **11.** General solution: $y = 2x^{1/2} + C$; particular solution: $y = 2x^{1/2} - 4$
13. General solution: $y = Ce^x$; particular solution: $y = 10e^x$ **15.** General solution: $y = 25 - Ce^{-x}$; particular solution: $y = 25 - 20e^{-x}$
17. General solution: $y = Cx$; particular solution: $y = 5x$ **19.** General solution: $y = (3x + C)^{1/3}$; particular solution: $y = (3x + 24)^{1/3}$
21. General solution: $y = Ce^{e^x}$; particular solution: $y = 3e^{e^x}$ **23.** General solution: $y = \ln(e^x + C)$; particular solution: $y = \ln(e^x + 1)$
25. General solution: $y = Ce^{x^2/2} - 1$; particular solution: $y = 3e^{x^2/2} - 1$
27. General solution: $y = 2 - 1/(e^x + C)$; particular solution: $y = 2 - e^{-x}$ **29.** $y + \frac{1}{3}y^3 = x + \frac{1}{3}x^3 + C$
31. $\frac{1}{2}\ln|1 + y^2| = \ln|x| + (x^2/2) + C$ or $\ln(1 + y^2) = \ln(x^2) + x^2 + C$ **33.** $\ln(1 + e^y) = \frac{1}{2}x^2 + C$ **35.** $y = \sqrt{1 + (\ln x)^2}$
37. $y = \frac{1}{4}[x + \ln(x^2) + 3]^2$ **39.** $y = \sqrt{\ln(x^2 + e)}$ **41.** (A) $M = 3/(1 - e^{-k})$; $M = 4/(1 - e^{-1.5k})$ (B) 7.3 **43.** \$16,600
45. (A) 22 days (B) 29,300 people **47.** 100 days **49.** 29 yr **51.** $k = 0.2$; $M = \$11$ million **53.** 9.6 min **55.** 84.26°F
57. (A) 538 bacteria (B) 6.8 hr **59.** (A) 2,422 people (B) 38.4 days **61.** $I = As^k$ **63.** (A) 400 people (B) 10 days

Exercise 14-3

1. $\int (x^3y' + 3x^2y)\,dx = x^3y$ **3.** $\int (e^{-3x}y' - 3e^{-3x}y)\,dx = e^{-3x}y$ **5.** $\int (x^4y' + 4x^3y)\,dx = x^4y$ **7.** $\int (e^{-0.5x}y' - 0.5e^{-0.5x}y)\,dx = e^{-0.5x}y$

9. $I(x) = e^{2x}$; $y = 2 + Ce^{-2x}$; $y = 2 - e^{-2x}$ **11.** $I(x) = e^x$; $y = -e^{-2x} + Ce^{-x}$; $y = -e^{-2x} + 4e^{-x}$
13. $I(x) = e^{-x}$; $y = 2xe^x + Ce^x$; $y = 2xe^x - 4e^x$ **15.** $I(x) = e^x$; $y = 3x^3e^{-x} + Ce^{-x}$; $y = 3x^3e^{-x} + 2e^{-x}$ **17.** $I(x) = x$; $y = x + (C/x)$; $y = x$
19. $I(x) = x^2$; $y = 2x^3 + (C/x^2)$; $y = 2x^3 - (32/x^2)$ **21.** $I(x) = e^{x^2/2}$; $y = 5 + Ce^{-x^2/2}$ **23.** $I(x) = e^{-2x}$; $y = -2x - 1 + Ce^{2x}$
25. $I(x) = x$; $y = e^x - (e^x/x) + (C/x)$ **27.** $I(x) = x$; $y = \frac{1}{2}x\ln x - \frac{1}{4}x + (C/x)$ **29.** $I(x) = x^{3/2}$; $y = 4x + Cx^{-3/2}$
31. (B) The solution is wrong. The term $1/x$ in front of the integral and the term x inside the integral cannot be canceled.
   (C) $y = \frac{1}{4}x^3 + \frac{2}{3}x^2 + \frac{1}{2}x + (C/x)$
33. (B) The solution is wrong. A constant of integration is omitted.
   (C) $y = \frac{1}{2}e^{-x} + Ce^{-3x}$
35. $y = 1 + (C/x)$ **37.** $y = C(1 + x^2) - 1$ **39.** $y = Ce^{x^2} - 1$ **41.** $y = Ce^{kt}$
43. $A(t) = 100,000 - 80,000e^{0.04t}$; 5.579 yr; \$22,316 **45.** \$11,804 **47.** $A(t) = 32,000e^{0.08t} - 25,000$; \$22,738.39 **49.** 7.13%
51. $p(t) = 20 + 10e^{-3t}$; $\bar{p} = 20$ **53.** 622 lb; 2.5 lb/gal **55.** 479 lb; 2.4 lb/gal **57.** 6.2 hr
59. 154 lb; 58 days. Weight will approach 120 lb if the diet is maintained for a long period of time. **61.** 1,647 cal
63. Student A: 0.8934 or 89.34%; student B: 0.6975 or 69.75%

Chapter 14 Review Exercise

3. (B) *(14-1)* **4.** (A) *(14-1)*
5.

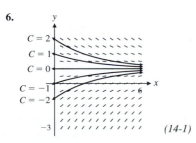

(14-1)

7. $dy/dt = -k(y - 5)$, $k > 0$ *(14-2)* **8.** $dy/dt = ky$, $k > 0$ *(14-2)*
9. A single person began the spread of a disease that is spreading at a rate proportional to the product of the number of people who have contracted the disease and the number who have not. *(14-2)*
10. The amount of radioactive material now is 100 g, and the amount is decreasing at a rate proportional to the amount present. *(14-2)*

11. $y = C/x^4$ *(14-2)*   **12.** $y = \frac{1}{6}x^2 + (C/x^4)$ *(14-3)*   **13.** $y = -1/(x^3 + C)$ *(14-2)*   **14.** $y = e^x + Ce^{2x}$ *(14-3)*   **15.** $y = \frac{1}{2}x^7 + Cx^5$ *(14-3)*
16. $y = C(2 + x) - 3$ *(14-2)*   **17.** $y = 10 - 10e^{-x}$ *(14-2, 14-3)*   **18.** $y = x - 1 + e^{-x}$ *(14-3)*   **19.** $y = e^2 e^{-2e^{-x}}$ *(14-2)*
20. $y = (x^2 + 4)/(x + 4)$ *(14-3)*   **21.** $y = \sqrt{x^2 + 16} - 4$ *(14-2)*   **22.** $y = \frac{1}{3}x \ln x - \frac{1}{9}x + \frac{19}{9}(1/x^2)$ *(14-3)*   **23.** $y = \sqrt{1 + 2x^2}$ *(14-2)*
24. $y = (2x + 1)e^{-x^2}$ *(14-3)*   **25.** $y = -2 + Cx^4$ *(14-2, 14-3)*
28. $y = 2 + Cx^5$ (A) $y = 2 + Cx^5$ for any constant C (B) No particular solution exists. (C) $y = 2 - x^5$ *(14-3)*   **29.** $y = \sqrt{x^2 + 16}$ *(14-2)*
30. (A)

(B) The graphs are increasing and   (C)   (D) Each graph has a
cross the x axis only at $x = 0$.   local maximum, a
                                            local minimum, and
                                            crosses the x axis
                                            three times. *(14-1)*

31. (A) $M = 7/(1 - e^{-5k})$, $M = 4/(1 - e^{-2k})$ (B) 9.2 *(14-2)*   **32.** $236.44 *(14-2)*   **33.** (A) 5 yr (B) $74,000 *(14-2)*
34. (A) $p = 50 + Ae^{-t}$ (B) $\bar{p} = 50$ *(14-1, 14-2)*   **35.** $A(t) = 100{,}000 - 40{,}000e^{0.05t}$; 18.326 yr; $91,630 *(14-3)*
36. 9.12% *(14-3)*   **37.** $y = 100 + te^{-t} - 100e^{-t}$ *(14-3)*   **38.** (A) 211.1 lb (B) 10.3 hr *(14-3)*   **39.** (A) 337 birds (B) 357 birds *(14-2)*
40. (A) 23 people (B) 30 days *(14-2)*

INDEX

APPLICATIONS INDEX

Life Sciences